Algorithms for Communications Systems
and their Applications

Algorithms for Communications Systems and their Applications

Second Edition

Nevio Benvenuto
University of Padua
Italy

Giovanni Cherubini
IBM Research Zurich
Switzerland

Stefano Tomasin
University of Padua
Italy

This edition first published 2021

Registered Offices
John Wiley & Sons, Inc., 111 River Street, Hoboken, NJ 07030, USA
John Wiley & Sons Ltd, The Atrium, Southern Gate, Chichester, West Sussex, PO19 8SQ, UK

Editorial Office
The Atrium, Southern Gate, Chichester, West Sussex, PO19 8SQ, UK

For details of our global editorial offices, customer services, and more information about Wiley products visit us at www.wiley.com.

Wiley also publishes its books in a variety of electronic formats and by print-on-demand. Some content that appears in standard print versions of this book may not be available in other formats.

Library of Congress Cataloging-in-Publication Data

Names: Benvenuto, Nevio, author. | Cherubini, Giovanni, 1957- author. | Tomasin, Stefano, 1975- author.
Title: Algorithms for communications systems and their applications / Nevio Benvenuto, University of Padua, Italy, Giovanni Cherubini, IBM Research Zurich, Switzerland, Stefano Tomasin, University of Padua, Italy.
Description: Second edition. | Hoboken, NJ, USA : Wiley, 2021. | Includes bibliographical references and index.
Identifiers: LCCN 2020004346 (print) | LCCN 2020004347 (ebook) | ISBN 9781119567967 (cloth) | ISBN 9781119567974 (adobe pdf) | ISBN 9781119567981 (epub)
Subjects: LCSH: Signal processing–Mathematics. | Telecommunication systems–Mathematics. | Algorithms.
Classification: LCC TK5102.9 .B46 2020 (print) | LCC TK5102.9 (ebook) | DDC 621.382/2–dc23
LC record available at https://lccn.loc.gov/2020004346
LC ebook record available at https://lccn.loc.gov/2020004347

Cover Design: Wiley
Cover Image: © betibup33/Shutterstock

Set in 9.5/12.5pt STIXGeneral by SPi Global, Chennai, India
Printed and bound by CPI Group (UK) Ltd, Croydon, CR0 4YY

10 9 8 7 6 5 4 3 2 1

To Adriana, to Antonio, Claudia, and Mariuccia, and in memory of Alberto

Contents

Preface

The motivation for writing this book is twofold. On the one hand, we provide a teaching tool for advanced courses in communications systems. On the other hand, we present a collection of fundamental algorithms and structures useful as an in-depth reference for researchers and engineers. The contents reflect our experience in teaching university courses on algorithms for telecommunications, as well as our professional experience acquired in industrial research laboratories.

The text illustrates the steps required for solving problems posed by the design of systems for reliable communications over wired or wireless channels. In particular, we have focused on fundamental developments in the field in order to provide the reader with the necessary insight to design practical systems.

The second edition of this book has been enriched by new solutions in fields of application and standards that have emerged since the first edition of 2002. To name one, the adoption of multiple antennas in wireless communication systems has received a tremendous impulse in recent years, and an entire chapter is now dedicated to this topic. About error correction, *polar* codes have been invented and are considered for future standards. Therefore, they also have been included in this new book edition. On the standards side, cellular networks have evolved significantly, thus we decided to dedicate a large part of a chapter to the new fifth-generation (5G) of cellular networks, which is being finalized at the time of writing. Moreover, a number of transmission techniques that have been designed and studied for application to 5G systems, with special regard to multi-carrier transmission, have been treated in this book. Lastly, many parts have been extensively integrated with new material, rewritten, and improved, with the purpose of illustrating to the reader their connection with current research trends, such as advances in machine learning.

Acknowledgements

We gratefully acknowledge all who have made the realization of this book possible. In particular, the editing of the various chapters would never have been completed without the contributions of numerous students in our courses on Algorithms for Telecommunications. Although space limitations preclude mentioning them all by name, we nevertheless express our sincere gratitude. We also thank Christian Bolis and Chiara Paci for their support in developing the software for the book, Charlotte Bolliger and Lilli M. Pavka for their assistance in administering the project, and Urs Bitterli and Darja Kropaci for their help with the graphics editing. For text processing, also for the Italian version, the contribution of Barbara Sicoli and Edoardo Casarin was indispensable; our thanks also go to Jane Frankenfield Zanin for her help in translating the text into English. We are pleased to thank the following colleagues for their invaluable assistance throughout the revision of the book: Antonio Assalini, Leonardo Bazzaco, Paola Bisaglia, Matthieu Bloch, Alberto Bononi, Alessandro Brighente, Giancarlo Calvagno, Giulio Colavolpe, Roberto Corvaja, Elena Costa, Daniele Forner, Andrea Galtarossa, Antonio Mian, Carlo Monti, Ezio Obetti, Riccardo Rahely, Roberto Rinaldo, Antonio Salloum, Fortunato Santucci, Andrea Scaggiante, Giovanna Sostrato, and Luciano Tomba. We gratefully acknowledge our colleague and mentor Jack Wolf for letting us include his lecture notes in the chapter on channel codes. We also acknowledge the important contribution of Ingmar Land on writing the section on polar codes. An acknowledgement goes also to our colleagues Werner Bux and Evangelos Eleftheriou of the IBM Zurich Research Laboratory, and Silvano Pupolin of the University of Padova, for their continuing support. Finally, special thanks go to Hideki Ochiai of Yokohama National University and Jinhong Yuan of University of New South Wales for hosting Nevio Benvenuto in the Fall 2018 and Spring 2019, respectively: both colleagues provided an ideal setting for developing the new book edition.

To make the reading of the adopted symbols easier, the Greek alphabet is reported below.

The Greek alphabet					
α	A	alpha	ν	N	nu
β	B	beta	ξ	Ξ	xi
γ	Γ	gamma	o	O	omicron
δ	Δ	delta	π	Π	pi
ϵ, ε	E	epsilon	ρ, ϱ	P	rho
ζ	Z	zeta	σ, ς	Σ	sigma
η	H	eta	τ	T	tau
θ, ϑ	Θ	theta	υ	Y	upsilon
ι	I	iota	ϕ, φ	Φ	phi
κ	K	kappa	χ	X	chi
λ	Λ	lambda	ψ	Ψ	psi
μ	M	mu	ω	Ω	omega

Chapter 1

Elements of signal theory

In this chapter, we recall some concepts on signal theory and random processes. For an in-depth study, we recommend the companion book [1]. First, we introduce various forms of the Fourier transform. Next, we provide the complex representation of passband signals and their baseband equivalent. We will conclude with the study of random processes, with emphasis on the statistical estimation of first- and second-order ergodic processes, i.e. periodogram, correlogram, auto-regressive (AR), moving-average (MA), and auto-regressive moving average (ARMA) models.

1.1 Continuous-time linear systems

A time-invariant continuous-time continuous-amplitude linear system, also called analog filter, is represented in Figure 1.1, where x and y are the input and output signals, respectively, and h denotes the filter impulse response.

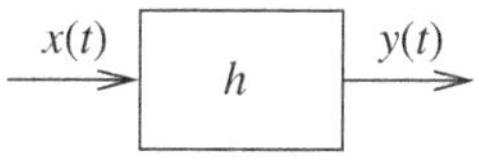

$$x(t) \longrightarrow \boxed{h} \longrightarrow y(t)$$

Figure 1.1 Analog filter as a time-invariant linear system with continuous domain.

The output at a certain instant $t \in \mathbb{R}$, where $\mathbb{R}$ denotes the set of real numbers, is given by the convolution integral

$$y(t) = \int_{-\infty}^{\infty} h(t-\tau)\, x(\tau)\, d\tau = \int_{-\infty}^{\infty} h(\tau)\, x(t-\tau)\, d\tau \tag{1.1}$$

denoted in short

$$y(t) = x * h(t) \tag{1.2}$$

We also introduce the Fourier transform of the signal $x(t)$, $t \in \mathbb{R}$,

$$\mathcal{X}(f) = \mathcal{F}[x(t)] = \int_{-\infty}^{+\infty} x(t)\, e^{-j2\pi ft}\, dt \qquad f \in \mathbb{R} \tag{1.3}$$

where $j = \sqrt{-1}$. The inverse Fourier transform is given by

$$x(t) = \int_{-\infty}^{\infty} \mathcal{X}(f)\, e^{j2\pi ft}\, df \tag{1.4}$$

Algorithms for Communications Systems and their Applications, Second Edition.
Nevio Benvenuto, Giovanni Cherubini, and Stefano Tomasin.

In the frequency domain, (1.2) becomes

$$\mathcal{Y}(f) = \mathcal{X}(f)\,\mathcal{H}(f), \qquad f \in \mathbb{R} \tag{1.5}$$

where $\mathcal{H}$ is the filter frequency response. The magnitude of the frequency response, $|\mathcal{H}(f)|$, is usually called *magnitude response* or *amplitude response*.

General properties of the Fourier transform are given in Table 1.1,[1] where we use two important functions

$$\text{step function:} \qquad 1(t) = \begin{cases} 1 & t > 0 \\ 0 & t < 0 \end{cases} \tag{1.6}$$

$$\text{sign function:} \qquad \text{sgn}(t) = \begin{cases} 1 & t > 0 \\ -1 & t < 0 \end{cases} \tag{1.7}$$

Moreover, we denote by $\delta(t)$ the *Dirac impulse* or delta function,

$$\delta(t) = \frac{d1(t)}{dt} \tag{1.8}$$

where the derivative is taken in the generalized sense.

Definition 1.1

We introduce two functions that will be extensively used:

$$\text{rect}(f) = \begin{cases} 1 & |f| < \frac{1}{2} \\ 0 & \text{elsewhere} \end{cases} \tag{1.9}$$

$$\text{sinc}(t) = \frac{\sin(\pi t)}{\pi t} \tag{1.10}$$

The following relation holds

$$\mathcal{F}[\text{sinc}(Ft)] = \frac{1}{F}\text{rect}\left(\frac{f}{F}\right) \tag{1.11}$$

as illustrated in Figure 1.2. □

Further examples of signals and relative Fourier transforms are given in Table 1.2.

We reserve the notation $H(s)$ to indicate the Laplace transform of $h(t)$, $t \in \mathbb{R}$:

$$H(s) = \int_{-\infty}^{+\infty} h(t)e^{-st}dt \tag{1.12}$$

with s complex variable; $H(s)$ is also called the *transfer function* of the filter. A class of functions $H(s)$ often used in practice is characterized by the ratio of two polynomials in s, each with a finite number of coefficients.

It is easy to observe that if the curve $s = j2\pi f$ in the s-plane belongs to the convergence region of the integral in (1.12), then $\mathcal{H}(f)$ is related to $H(s)$ by

$$\mathcal{H}(f) = H(s)|_{s=j2\pi f} \tag{1.13}$$

1.2 Discrete-time linear systems

A discrete-time time-invariant linear system, with sampling period T_c, is shown in Figure 1.3, where $x(k)$ and $y(k)$ are, respectively, the input and output signals at the time instant kT_c, $k \in \mathbb{Z}$, where $\mathbb{Z}$ denotes

[1] x^* denotes the complex conjugate of x, while $Re(x)$ and $Im(x)$ denote, respectively, the real and imaginary part of x.

Table 1.1: Some general properties of the Fourier transform.

property	*signal*	*Fourier transform*
	$x(t)$	$\mathcal{X}(f)$
linearity	$a\,x(t) + b\,y(t)$	$a\,\mathcal{X}(f) + b\,\mathcal{Y}(f)$
duality	$\mathcal{X}(t)$	$x(-f)$
time inverse	$x(-t)$	$\mathcal{X}(-f)$
complex conjugate	$x^*(t)$	$\mathcal{X}^*(-f)$
real part	$Re[x(t)] = \dfrac{x(t) + x^*(t)}{2}$	$\dfrac{1}{2}\,[\mathcal{X}(f) + \mathcal{X}^*(-f)]$
imaginary part	$Im[x(t)] = \dfrac{x(t) - x^*(t)}{2j}$	$\dfrac{1}{2j}\,[\mathcal{X}(f) - \mathcal{X}^*(-f)]$
time scaling	$x(at),\ a \neq 0$	$\dfrac{1}{\lvert a\rvert}\,\mathcal{X}\left(\dfrac{f}{a}\right)$
time shift	$x(t - t_0)$	$e^{-j2\pi f t_0}\,\mathcal{X}(f)$
frequency shift	$x(t)\,e^{j2\pi f_0 t}$	$\mathcal{X}(f - f_0)$
modulation	$x(t)\cos(2\pi f_0 t + \varphi)$	$\dfrac{1}{2}\,[e^{j\varphi}\mathcal{X}(f - f_0) + e^{-j\varphi}\mathcal{X}(f + f_0)]$
	$x(t)\sin(2\pi f_0 t + \varphi)$	$\dfrac{1}{2j}\,[e^{j\varphi}\mathcal{X}(f - f_0) - e^{-j\varphi}\mathcal{X}(f + f_0)]$
	$Re[x(t)\,e^{j(2\pi f_0 t + \varphi)}]$	$\dfrac{1}{2}[e^{j\varphi}\mathcal{X}(f - f_0) + e^{-j\varphi}\mathcal{X}^*(-f - f_0)]$
differentiation	$\dfrac{d}{dt}\,x(t)$	$j2\pi f\,\mathcal{X}(f)$
integration	$\int_{-\infty}^{t} x(\tau)\,d\tau = 1 * x(t)$	$\dfrac{1}{j2\pi f}\,\mathcal{X}(f) + \dfrac{\mathcal{X}(0)}{2}\,\delta(f)$
convolution	$x * y(t)$	$\mathcal{X}(f)\,\mathcal{Y}(f)$
correlation	$[x(\tau) * y^*(-\tau)](t)$	$\mathcal{X}(f)\,\mathcal{Y}^*(f)$
product	$x(t)\,y(t)$	$\mathcal{X} * \mathcal{Y}(f)$
real signal	$x(t) = x^*(t)$	$\mathcal{X}(f) = \mathcal{X}^*(-f)$, $\mathcal{X}$ Hermitian, $Re[\mathcal{X}(f)]$ even, $Im[\mathcal{X}(f)]$ odd, $\lvert\mathcal{X}(f)\rvert^2$ even
imaginary signal	$x(t) = -x^*(t)$	$\mathcal{X}(f) = -\mathcal{X}^*(-f)$
real and even signal	$x(t) = x^*(t) = x(-t)$	$\mathcal{X}(f) = \mathcal{X}^*(f) = \mathcal{X}(-f)$, $\mathcal{X}$ real and even
real and odd signal	$x(t) = x^*(t) = -x(-t)$	$\mathcal{X}(f) = -\mathcal{X}^*(f) = -\mathcal{X}(-f)$, $\mathcal{X}$ imaginary and odd
Parseval theorem	$E_x = \int_{-\infty}^{+\infty} \lvert x(t)\rvert^2\,dt = \int_{-\infty}^{+\infty} \lvert\mathcal{X}(f)\rvert^2\,df = E_{\mathcal{X}}$	
Poisson sum formula	$\sum_{k=-\infty}^{+\infty} x(kT_c) = \dfrac{1}{T_c}\sum_{\ell=-\infty}^{+\infty} \mathcal{X}\left(\dfrac{\ell}{T_c}\right)$	

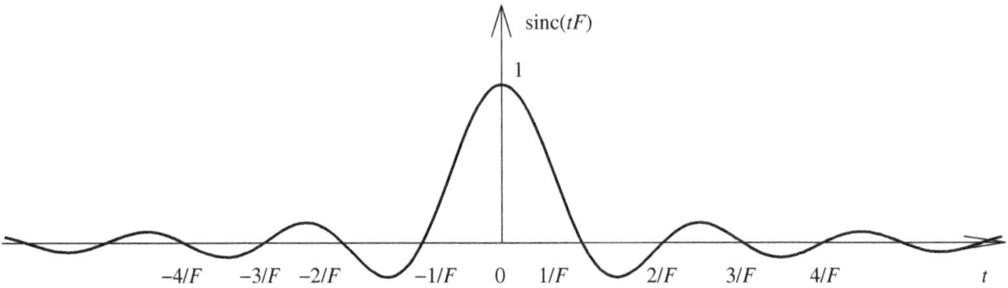

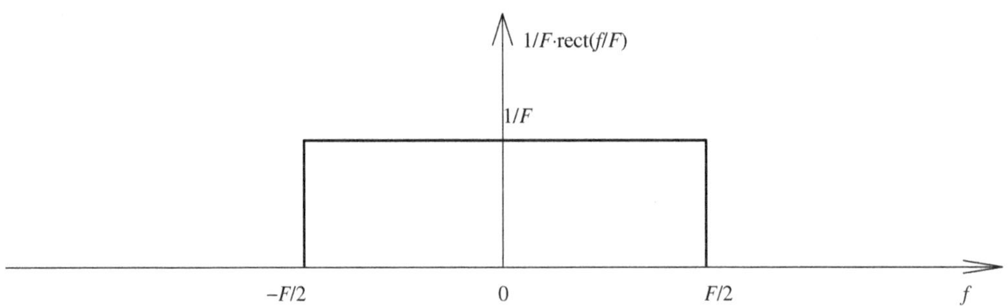

Figure 1.2 Example of signal and Fourier transform pair.

$x(k)$ → h → $y(k)$

T_c T_c

Figure 1.3 Discrete-time linear system (filter).

the set of integers. We denote by $\{x(k)\}$ or $\{x_k\}$ the entire discrete-time signal, also called *sequence*. The impulse response of the system is denoted by $\{h(k)\}$, $k \in \mathbb{Z}$, or more simply by h.

The relation between the input sequence $\{x(k)\}$ and the output sequence $\{y(k)\}$ is given by the convolution operation:

$$y(k) = \sum_{n=-\infty}^{+\infty} h(k-n)x(n) \tag{1.14}$$

denoted as $y(k) = x * h(k)$. In the discrete time, the delta function is simply the *Kronecker impulse*

$$\delta_n = \delta(n) = \begin{cases} 1 & n = 0 \\ 0 & n \neq 0 \end{cases} \tag{1.15}$$

Here are some definitions holding for time-invariant linear systems.

The system is *causal* (anticausal) if $h(k) = 0$, $k < 0$ (if $h(k) = 0$, $k > 0$).

Table 1.2: Examples of Fourier transform signal pairs.

signal	*Fourier transform*
$x(t)$	$\mathcal{X}(f)$
$\delta(t)$	1
1 (constant)	$\delta(f)$
$e^{j2\pi f_0 t}$	$\delta(f-f_0)$
$\cos(2\pi f_0 t)$	$\frac{1}{2}\,[\delta(f-f_0)+\delta(f+f_0)]$
$\sin(2\pi f_0 t)$	$\frac{1}{2j}\,[\delta(f-f_0)-\delta(f+f_0)]$
$1(t)$	$\frac{1}{2}\,\delta(f)+\frac{1}{j2\pi f}$
$\text{sgn}(t)$	$\frac{1}{j\pi f}$
$\text{rect}\left(\frac{t}{T}\right)$	$T\,\text{sinc}(fT)$
$\text{sinc}\left(\frac{t}{T}\right)$	$T\,\text{rect}(fT)$
$\left(1-\frac{\lvert t\rvert}{T}\right)\text{rect}\left(\frac{t}{2T}\right)$	$T\text{sinc}^2(fT)$
$e^{-at}\ 1(t),\ a>0$	$\frac{1}{a+j2\pi f}$
$t\,e^{-at}\ 1(t),\ a>0$	$\frac{1}{(a+j2\pi f)^2}$
$e^{-a\lvert t\rvert},\ a>0$	$\frac{2a}{a^2+(2\pi f)^2}$
$e^{-at^2},\ a>0$	$\sqrt{\frac{\pi}{a}}e^{-\pi\frac{\pi}{a}f^2}$

The *transfer function* of the filter is defined as the z-transform[2] of the impulse response h, given by

$$H(z)=\sum_{k=-\infty}^{+\infty} h(k)z^{-k} \tag{1.16}$$

Let the *frequency response* of the filter be defined as

$$\mathcal{H}(f)=\mathcal{F}[h(k)]=\sum_{k=-\infty}^{+\infty} h(k)e^{-j2\pi fkT_c}=H(z)_{z=e^{j2\pi fT_c}} \tag{1.17}$$

The inverse Fourier transform of the frequency response yields

$$h(k)=T_c\int_{-\frac{1}{2T_c}}^{+\frac{1}{2T_c}} \mathcal{H}(f)e^{j2\pi fkT_c}df \tag{1.18}$$

We note the property that, for $x(k)=b^k$, where b is a complex constant, the output is given by $y(k)=H(b)\ b^k$. In Table 1.3, some further properties of the z-transform are summarized.

For discrete-time linear systems, in the frequency domain (1.14) becomes

$$\mathcal{Y}(f)=\mathcal{X}(f)\mathcal{H}(f) \tag{1.19}$$

where all functions are periodic of period $1/T_c$.

---

[2] Sometimes the D transform is used instead of the z-transform, where $D=z^{-1}$, and $H(z)$ is replaced by $h(D)=\sum_{k=-\infty}^{+\infty} h(k)D^k$.

Table 1.3: Properties of the z-transform.

property	*sequence*	*z transform*
	$x(k)$	$X(z)$
linearity	$ax(k)+by(k)$	$aX(z)+bY(z)$
time shift	$x(k-m)$	$z^{-m}X(z)$
complex conjugate	$x^*(k)$	$X^*(z^*)$
time inverse	$x(-k)$	$X\left(\frac{1}{z}\right)$
	$x^*(-k)$	$X^*\left(\frac{1}{z^*}\right)$
z-domain scaling	$a^{-k}x(k)$	$X(az)$
convolution	$x*y(k)$	$X(z)Y(z)$
correlation	$x*(y^*(-m))(k)$	$X(z)Y^*\left(\frac{1}{z^*}\right)$
real sequence	$x(k)=x^*(k)$	$X(z)=X^*(z^*)$

Example 1.2.1
A fundamental example of z-transform is that of the sequence:

$$h(k)=\begin{cases} a^k & k\geq 0 \\ 0 & k<0 \end{cases}, \qquad |a|<1 \tag{1.20}$$

Applying the transform (1.16), we find

$$H(z)=\frac{1}{1-az^{-1}} \tag{1.21}$$

defined for $|az^{-1}|<1$ or $|z|>|a|$.

Example 1.2.2
Let $q(t)$, $t\in\mathbb{R}$, be a continuous-time signal with Fourier transform $\mathcal{Q}(f)$, $f\in\mathbb{R}$. We now consider the sequence obtained by sampling q, that is

$$h_k=q(kT_c), \qquad k\in\mathbb{Z} \tag{1.22}$$

Using the Poisson formula of Table 1.1, we have that the Fourier transform of the sequence $\{h_k\}$ is related to $\mathcal{Q}(f)$ by

$$\mathcal{H}(f)=\mathcal{F}[h_k]=H\left(e^{j2\pi fT_c}\right)=\frac{1}{T_c}\sum_{\ell=-\infty}^{\infty}\mathcal{Q}\left(f-\ell\frac{1}{T_c}\right) \tag{1.23}$$

Definition 1.2
Let us introduce the useful pulse with parameter N, a positive integer number,

$$\text{sinc}_N(a)=\frac{1}{N}\frac{\sin(\pi a)}{\sin\left(\pi\frac{a}{N}\right)} \tag{1.24}$$

and $\text{sinc}_N(0)=1$. The pulse is periodic with period N ($2N$) if N is odd (even). For N, very large $\text{sinc}_N(a)$ approximates $\text{sinc}(a)$ in the range $|a|\ll N/2$. □

Example 1.2.3
For the signal

$$h_k = \begin{cases} 1 & k = 0, 1, \ldots, N-1 \\ 0 & \text{otherwise} \end{cases} \tag{1.25}$$

with sampling period T_c, it is

$$\mathcal{H}(f) = e^{-j2\pi f \frac{N-1}{2} T_c} N \operatorname{sinc}_N(f N T_c) \tag{1.26}$$

Discrete Fourier transform

For a sequence with a finite number of samples, $\{g_k\}$, $k = 0, 1, \ldots, N-1$, the Fourier transform becomes

$$\mathcal{G}(f) = \sum_{k=0}^{N-1} g_k e^{-j2\pi f k T_c} \tag{1.27}$$

Evaluating $\mathcal{G}(f)$ at the points $f = m/(NT_c)$, $m = 0, 1, \ldots, N-1$, and setting $\mathcal{G}_m = \mathcal{G}(m/(NT_c))$, we obtain:

$$\mathcal{G}_m = \sum_{k=0}^{N-1} g_k W_N^{km}, \qquad W_N = e^{-j\frac{2\pi}{N}} \tag{1.28}$$

The sequence $\{\mathcal{G}_m\}$, $m = 0, 1, \ldots, N-1$, is called the discrete Fourier transform (DFT) of $\{g_k\}$, $k = 0, 1, \ldots, N-1$. The inverse of (1.28) is given by

$$g_k = \frac{1}{N} \sum_{m=0}^{N-1} \mathcal{G}_m W_N^{-km}, \qquad k = 0, 1, \ldots, N-1 \tag{1.29}$$

We note that, besides the factor $1/N$, the expression of the inverse DFT (IDFT) coincides with that of the DFT, provided W_N^{-1} is substituted with W_N.

We also observe that the direct computation of (1.28) requires $N(N-1)$ complex additions and N^2 complex multiplications; however, the algorithm known as fast Fourier transform (FFT) computes the DFT by $N \log_2 N$ complex additions and $\left(\frac{N}{2}\log_2 N - N\right)$ complex multiplications.[3]

A simple implementation is also available when the DFT size is an integer power of some numbers (e.g. 2, 3, and 5). The efficient implementation of a DFT with length power of n (2, 3, and 5) is denoted as radix-n FFT. Moreover, if the DFT size is the product of integer powers of these numbers, the DFT can be implemented as a cascade of FFTs. In particular, by letting $M = 2^{\alpha_2}$, $L = 3^{\alpha_3} \cdot 5^{\alpha_5}$, the DFT of size $N = LM$ can be implemented as the cascade of L M-size DFTs, the multiplication by twiddle factors (operating only on the phase of the signal) and an L-size DFT. Applying again the same approach to the inner M-size DFT, we obtain that the N-size DFT is the cascade of 2^{α_2} FFTs of size $3^{\alpha_3}5^{\alpha_5}$, each implemented by 3^{α_3} FFTs of size 5^{α_5}.

The DFT operator

The DFT operator can be expressed in matrix form as

$$\boldsymbol{F} = \begin{bmatrix} 1 & 1 & 1 & \ldots & 1 \\ 1 & W_N & W_N^2 & \ldots & W_N^{(N-1)} \\ 1 & W_N^2 & W_N^4 & \ldots & W_N^{2(N-1)} \\ \vdots & \vdots & \vdots & \ddots & \vdots \\ 1 & W_N^{(N-1)} & W_N^{(N-1)2} & \ldots & W_N^{(N-1)(N-1)} \end{bmatrix} \tag{1.30}$$

[3] The computational complexity of the FFT is often expressed as $N \log_2 N$.

with elements $[\boldsymbol{F}]_{i,n} = W_N^{in}$, $i, n = 0, 1, \dots, N-1$. The inverse operator (IDFT) is given by[4]

$$\boldsymbol{F}^{-1} = \frac{1}{N}\boldsymbol{F}^* \tag{1.31}$$

We note that $\boldsymbol{F} = \boldsymbol{F}^T$ and $(1/\sqrt{N})\boldsymbol{F}$ is a unitary matrix.[5]

The following property holds: if $\boldsymbol{C}$ is a right circulant square matrix, i.e. its rows are obtained by successive shifts to the right of the first row, then $\boldsymbol{F}\boldsymbol{C}\boldsymbol{F}^{-1}$ is a diagonal matrix whose elements are given by the DFT of the first row of $\boldsymbol{C}$. This property is exploited in the most common modulation scheme (see Chapter 8).

Introducing the vector formed by the samples of the sequence $\{g_k\}$, $k = 0, 1, \dots, N-1$,

$$\boldsymbol{g}^T = [g_0, g_1, \dots, g_{N-1}] \tag{1.32}$$

and the vector of its transform coefficients

$$\boldsymbol{\mathcal{G}}^T = [\mathcal{G}_0, \mathcal{G}_1, \dots, \mathcal{G}_{N-1}] = \text{DFT}[\boldsymbol{g}] \tag{1.33}$$

from (1.28) we have

$$\boldsymbol{\mathcal{G}} = \boldsymbol{F}\boldsymbol{g} \tag{1.34}$$

Moreover, based on (1.31), we obtain

$$\boldsymbol{g} = \frac{1}{N}\boldsymbol{F}^*\boldsymbol{\mathcal{G}} \tag{1.35}$$

Circular and linear convolution via DFT

Let the two sequences x and h have a finite support of L_x and N samples, respectively, (see Figure 1.4) with $L_x > N$:

$$x(k) = 0 \quad k < 0 \quad k > L_x - 1 \tag{1.36}$$

and

$$h(k) = 0 \quad k < 0 \quad k > N - 1 \tag{1.37}$$

We define the periodic signals of period L,

$$x_{rep_L}(k) = \sum_{\ell=-\infty}^{+\infty} x(k - \ell L), \qquad h_{rep_L}(k) = \sum_{\ell=-\infty}^{+\infty} h(k - \ell L) \tag{1.38}$$

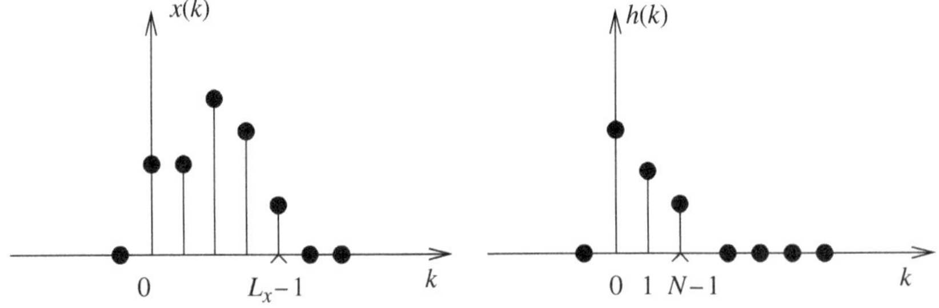

Figure 1.4 Time-limited signals: $\{x(k)\}$, $k = 0, 1, \dots, L_x - 1$, and $\{h(k)\}$, $k = 0, 1, \dots, N-1$.

[4] ^{T} stands for transpose and ^{H} for transpose complex conjugate or Hermitian.

[5] A square matrix $\boldsymbol{A}$ is unitary if $\boldsymbol{A}^H\boldsymbol{A} = \boldsymbol{I}$, where $\boldsymbol{I}$ is the identity matrix, i.e. a matrix for which all elements are zero except the elements on the main diagonal that are all equal to one.

where in order to avoid *time aliasing*, it must be

$$L \geq \max\{L_x, N\} \tag{1.39}$$

Definition 1.3

The *circular convolution* between x and h is a periodic sequence of period L defined as

$$y^{(circ)}(k) = h \overset{L}{\otimes} x(k) = \sum_{i=0}^{L-1} h_{rep_L}(i)\, x_{rep_L}(k-i) \tag{1.40}$$

with *main period* corresponding to $k = 0, 1, \ldots, L-1$. □

Then, if we indicate with $\{\mathcal{X}_m\}$, $\{\mathcal{H}_m\}$, and $\{\mathcal{Y}_m^{(circ)}\}$, $m = 0, 1, \ldots, L-1$, the L-point DFT of sequences x, h, and $y^{(circ)}$, respectively, we obtain

$$\mathcal{Y}_m^{(circ)} = \mathcal{X}_m \mathcal{H}_m, \qquad m = 0, 1, \ldots, L-1 \tag{1.41}$$

In vector notation (1.33), (1.41) becomes[6]

$$\boldsymbol{\mathcal{Y}}^{(circ)} = \left[\mathcal{Y}_0^{(circ)}, \mathcal{Y}_1^{(circ)}, \ldots, \mathcal{Y}_{L-1}^{(circ)}\right]^T = \text{diag}\,\{\text{DFT}[\boldsymbol{x}]\}\,\boldsymbol{\mathcal{H}} \tag{1.42}$$

where $\boldsymbol{\mathcal{H}}$ is the column vector given by the L-point DFT of the sequence h, completed with $L - N$ zeros.

We are often interested in the *linear convolution* between x and h given by (1.14):

$$y(k) = x * h(k) = \sum_{i=0}^{N-1} h(i)x(k-i) \tag{1.43}$$

whose support is $k = 0, 1, \ldots, L_x + N - 2$.

We give below two relations between the circular convolution $y^{(circ)}$ and the linear convolution y.

Relation 1. For

$$L \geq L_x + N - 1 \tag{1.44}$$

by comparing (1.43) with (1.40), the two convolutions $y^{(circ)}$ and y coincide only for the instants $k = 0, 1, \ldots, L-1$, i.e.

$$y(k) = y^{(circ)}(k), \qquad k = 0, 1, \ldots, L-1 \tag{1.45}$$

To compute the convolution between the two finite-length sequences x and h, (1.44) and (1.45) require that both sequences be completed with zeros (zero padding) to get a length of $L = L_x + N - 1$ samples. Then, taking the L-point DFT of the two sequences, performing the product (1.41), and taking the inverse transform of the result, one obtains the desired linear convolution.

Relation 2. For $L = L_x > N$, the two convolutions $y^{(circ)}$ and y coincide only for the instants $k = N-1$, $N, \ldots, L-1$, i.e.

$$y^{(circ)}(k) = y(k) \qquad \text{only for} \qquad k = N-1, N, \ldots, L-1 \tag{1.46}$$

An example of circular convolution is provided in Figure 1.5. Indeed, the result of circular convolution coincides with $\{y(k)\}$, output of the linear convolution, only for a delay k such that it is avoided the product between non-zero samples of the two periodic sequences h_{rep_L} and x_{rep_L}, indicated by • and o, respectively. This is achieved only for $k \geq N-1$ and $k \leq L-1$.

[6] The notation diag$\{\boldsymbol{v}\}$ denotes a diagonal matrix whose elements on the diagonal are equal to the elements of the vector $\boldsymbol{v}$.

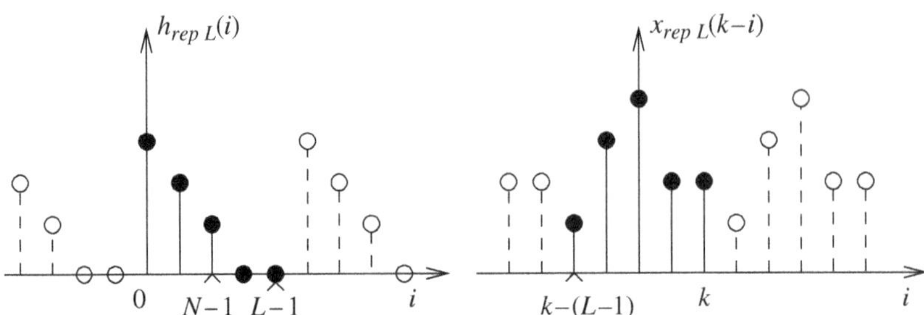

Figure 1.5 Illustration of the *circular convolution* operation between $\{x(k)\}$, $k = 0, 1, \dots, L-1$, and $\{h(k)\}$, $k = 0, 1, \dots, N-1$.

Relation 3. A relevant case wherein the cyclic convolution is equivalent to the linear convolution requires a special structure of the sequence x. Consider $x^{(cp)}$, the extended sequence of x, obtained by partially repeating x with a *cyclic prefix* of N_{cp} samples:

$$x^{(cp)}(k) = \begin{cases} x(k) & k = 0, 1, \dots, L_x - 1 \\ x(L_x + k) & k = -N_{cp}, \dots, -2, -1 \end{cases} \tag{1.47}$$

Let $y^{(cp)}$ be the *linear convolution* between $x^{(cp)}$ and h, with support $\{-N_{cp}, \dots, L_x + N - 2\}$. If $N_{cp} \geq N - 1$, we have

$$y^{(cp)}(k) = y^{(circ)}(k), \qquad k = 0, 1, \dots, L_x - 1 \tag{1.48}$$

Let us define

$$z(k) = \begin{cases} y^{(cp)}(k) & k = 0, 1, \dots, L_x - 1 \\ 0 & \text{elsewhere} \end{cases} \tag{1.49}$$

then from (1.48) and (1.41) the following relation between the corresponding L_x–point DFTs is obtained:

$$\mathcal{Z}_m = \mathcal{X}_m \mathcal{H}_m, \qquad m = 0, 1, \dots, L_x - 1 \tag{1.50}$$

Convolution by the overlap-save method

For a very long sequence x, the application of (1.46) leads to the *overlap-save* method to determine the linear convolution between x and h (with $L = L_x > N$). It is not restrictive to assume that the first $(N-1)$ samples of the sequence $\{x(k)\}$ are zero. If this were not True, it would be sufficient to shift the input by $(N-1)$ samples. A fast procedure to compute the linear convolution $\{y(k)\}$ for instants $k = N-1, N, \dots, L-1$, operates iteratively and processes blocks of L samples, where adjacent blocks are overlapping by $(N-1)$ samples. The procedure operates the following first iteration:[7]

1. *Loading*

$$\boldsymbol{h}'^T = [h(0), h(1), \dots, h(N-1), \overbrace{0, \dots, 0}^{L-N \text{ zeros}}] \tag{1.51}$$

$$\boldsymbol{x}'^T = [x(0), x(1), \dots, x(N-1), x(N), \dots, x(L-1)] \tag{1.52}$$

in which we have assumed $x(k) = 0$, $k = 0, 1, \dots, N-2$.

[7] In this section, the superscript $'$ indicates a vector of L elements.

2. *Transform*

$$\mathcal{H}' = \text{DFT}[\boldsymbol{h}'] \qquad \text{vector} \tag{1.53}$$

$$\mathcal{X}' = \text{diag}\{\text{DFT}[\boldsymbol{x}']\} \quad \text{matrix} \tag{1.54}$$

3. *Matrix product*

$$\mathcal{Y}' = \mathcal{X}'\mathcal{H}' \qquad \text{vector} \tag{1.55}$$

4. *Inverse transform*

$$\boldsymbol{y}'^T = \text{DFT}^{-1}[\mathcal{Y}'^T] = [\overbrace{\sharp, \dots, \sharp}^{N-1 \text{ terms}}, y(N-1), y(N), \dots, y(L-1)] \tag{1.56}$$

where the symbol $\sharp$ denotes a component that is neglected.

The second iteration operates on load

$$\boldsymbol{x}'^T = [x((L-1)-(N-2)), \dots, x(2(L-1)-(N-2))] \tag{1.57}$$

and the desired output samples will be

$$y(k) \qquad k = L, \dots, 2(L-1)-(N-2) \tag{1.58}$$

The third iteration operates on load

$$\boldsymbol{x}'^T = [x(2(L-1)-2(N-2)), \dots, x(3(L-1)-2(N-2))] \tag{1.59}$$

and will yield the desired output samples

$$y(k) \qquad k = 2(L-1)-(N-2)+1, \dots, 3(L-1)-2(N-2) \tag{1.60}$$

The algorithm proceeds iteratively until the entire input sequence is processed.

IIR and FIR filters

An important class of linear systems is identified by the input–output relation

$$\sum_{n=0}^{p} \mathtt{a}_n y(k-n) = \sum_{n=0}^{q} \mathtt{b}_n x(k-n) \tag{1.61}$$

where we will set $\mathtt{a}_0 = 1$ without loss of generality.

If the system is causal, (1.61) becomes

$$y(k) = -\sum_{n=1}^{p} \mathtt{a}_n y(k-n) + \sum_{n=0}^{q} \mathtt{b}_n x(k-n) \qquad k \geq 0 \tag{1.62}$$

and the transfer function for such system is

$$H(z) = \frac{Y(z)}{X(z)} = \frac{\sum_{n=0}^{q} \mathtt{b}_n z^{-n}}{1 + \sum_{n=1}^{p} \mathtt{a}_n z^{-n}} = \frac{\mathtt{b}_0 \prod_{n=1}^{q}(1 - \mathtt{z}_n z^{-1})}{\prod_{n=1}^{p}(1 - \mathtt{p}_n z^{-1})} \tag{1.63}$$

where $\{\mathtt{z}_n\}$ and $\{\mathtt{p}_n\}$ are, respectively, the zeros and poles of $H(z)$. Equation (1.63) generally defines an *infinite impulse response* (IIR) filter. In the case in which $\mathtt{a}_n = 0$, $n = 1, 2, \dots, p$, (1.63) reduces to

$$H(z) = \sum_{n=0}^{q} \mathtt{b}_n z^{-n} \tag{1.64}$$

Table 1.4: Impulse responses of systems having the same magnitude of the frequency response.

	$h(0)$	$h(1)$	$h(2)$	$h(3)$	$h(4)$
h_1 (minimum phase)	$0.9e^{-j1.57}$	0	0	$0.4e^{-j0.31}$	$0.3e^{-j0.63}$
h_2 (maximum phase)	$0.3e^{j0.63}$	$0.4e^{j0.31}$	0	0	$0.9e^{j1.57}$
h_3 (general case)	$0.7e^{-j1.57}$	$0.24e^{j2.34}$	$0.15e^{-j1.66}$	$0.58e^{-j0.51}$	$0.4e^{-j0.63}$

and we obtain a *finite impulse response* (FIR) filter, with $h(n) = \mathtt{b}_n$, $n = 0, 1, \dots, q$. To get the impulse response coefficients, assuming that the z-transform $H(z)$ is known, we can expand $H(z)$ in partial fractions and apply the linear property of the z-transform (see Table 1.3, page 6). If $q < p$, and assuming that all poles are distinct, we obtain

$$H(z) = \sum_{n=1}^{p} \frac{r_n}{1 - \mathtt{p}_n z^{-1}} \implies h(k) = \begin{cases} \sum_{n=1}^{p} r_n \mathtt{p}_n^k & k \geq 0 \\ 0 & k < 0 \end{cases} \tag{1.65}$$

where

$$r_n = H(z)\left[1 - \mathtt{p}_n z^{-1}\right]\Big|_{z=\mathtt{p}_n} \tag{1.66}$$

We give now two definitions.

Definition 1.4
A causal system is *stable* (*bounded input-bounded output* stability) if $|\mathtt{p}_n| < 1$, $\forall n$. □

Definition 1.5
The system is *minimum phase* (*maximum phase*) if $|\mathtt{p}_n| < 1$ and $|\mathtt{z}_n| \leq 1$ ($|\mathtt{p}_n| > 1$ and $|\mathtt{z}_n| > 1$), $\forall n$. □

Among all systems having the same magnitude response $|\mathcal{H}(e^{j2\pi f T_c})|$, the minimum (maximum) phase system presents a phase[8] response, $\arg\ \mathcal{H}(e^{j2\pi f T_c})$, which is below (above) the phase response of all other systems.

Example 1.2.4
It is interesting to determine the phase of a system for a given impulse response. Let us consider the system with transfer function $H_1(z)$ and impulse response $h_1(k)$ shown in Figure 1.6a. After determining the zeros of the transfer function, we factorize $H_1(z)$ as follows:

$$H_1(z) = \mathtt{b}_0 \prod_{n=1}^{4} (1 - \mathtt{z}_n z^{-1}) \tag{1.67}$$

As shown in Figure 1.6a, $H_1(z)$ is minimum phase. We now observe that the magnitude of the frequency response does not change if $1/\mathtt{z}_n^*$ is replaced with $\mathtt{z}_n$ in (1.67). If we move all the zeros outside the unit circle, we get a maximum-phase system $H_2(z)$ whose impulse response is shown in Figure 1.6b. A general case, that is a transfer function with some zeros inside and others outside the unit circle, is given in Figure 1.6c. The coefficients of the impulse responses h_1, h_2, and h_3 are given in Table 1.4. The coefficients are normalized so that the three impulse responses have equal energy.

[8] For a complex number c, $\arg c$ denotes the phase of c.

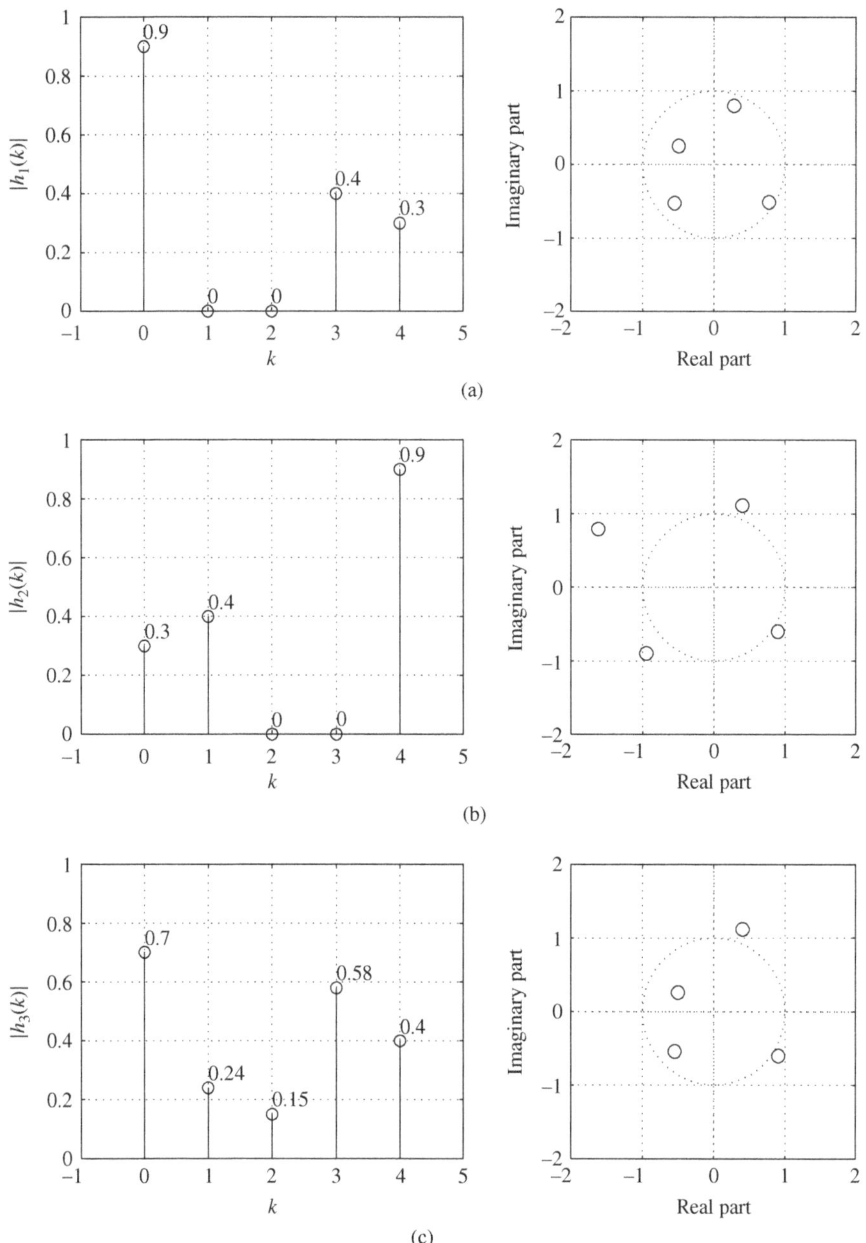

Figure 1.6 Impulse response magnitudes and zero locations for three systems having the same frequency response magnitude. (a) Minimum-phase system, (b) maximum-phase system, and (c) general system.

We define the *partial energy* of a causal impulse response as

$$E(k) = \sum_{i=0}^{k} |h(i)|^2 \tag{1.68}$$

Comparing the partial-energy sequences for the three impulse responses of Figure 1.6, one finds that the minimum (maximum) phase system yields the largest (smallest) $\{E(k)\}$. In other words, the magnitude

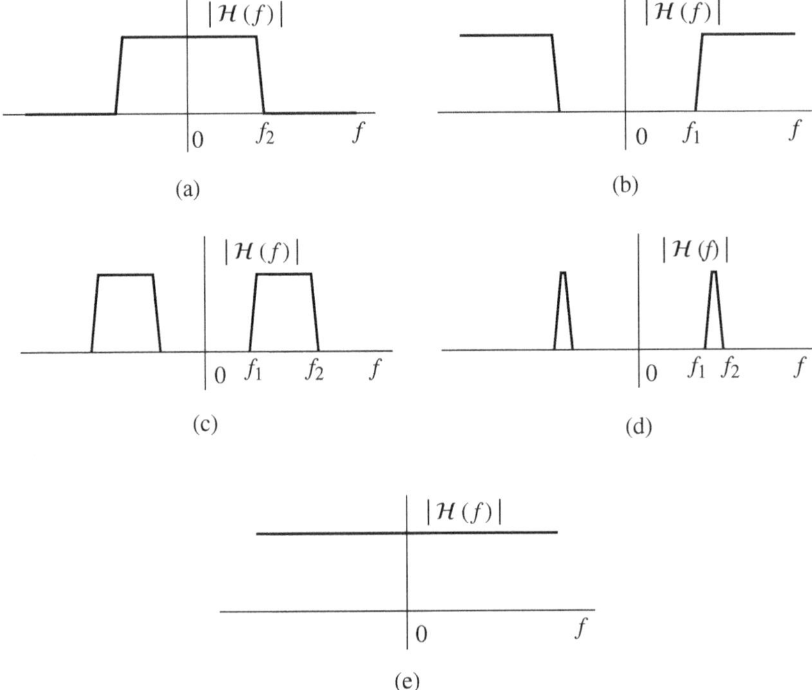

Figure 1.7 Classification of real valued analog filters on the basis of the support of $|\mathcal{H}(f)|$. (a) $f_1 = 0$, $f_2 < \infty$: lowpass filter (LPF). (b) $f_1 > 0$, $f_2 = \infty$: highpass filter (HPF). (c) $f_1 > 0$, $f_2 < \infty$: passband filter (PBF). (d) $B = f_2 - f_1 \ll (f_2 + f_1)/2$: narrowband filter (NBF). (e) $f_1 = 0$, $f_2 = \infty$: allpass filter (APF).

of the frequency responses being equal, a minimum (maximum) phase system concentrates all its energy on the first (last) samples of the impulse response.

Extending our previous considerations also to IIR filters, if h_1 is a causal minimum-phase filter, i.e. $H_1(z) = H_{min}(z)$ is a ratio of polynomials in z^{-1} with poles and zeros inside the unit circle, then $H_{max}(z) = K\, H^*_{min}\left(\frac{1}{z^*}\right)$, where K is a constant, is an anticausal maximum-phase filter, i.e. $H_{max}(z)$ is a ratio of polynomials in z with poles and zeros outside the unit circle.

In the case of a minimum-phase FIR filter with impulse response $h_{min}(n)$, $n = 0, 1, \dots, q$, $H_2(z) = z^{-q}\, H^*_{min}\left(\frac{1}{z^*}\right)$ is a causal maximum-phase filter. Moreover, the relation $\{h_2(n)\} = \{h_1^*(q-n)\}$, $n = 0, 1, \dots, q$, is satisfied. In this text, we use the notation $\{h_2(n)\} = \{h_1^{B*}(n)\}$, where B is the *backward* operator that orders the elements of a sequence from the last to the first.

In Appendix 1.A multirate transformations for systems are described, in which the time domain of the input is different from that of the output. In particular, decimator and interpolator filters are introduced, together with their efficient implementations.

1.3 Signal bandwidth

Definition 1.6
The *support* of a signal $x(\xi)$, $\xi \in \mathbb{R}$, is the set of values $\xi \in \mathbb{R}$ for which $|x(\xi)| \neq 0$. □

Let us consider a filter with impulse response h and frequency response $\mathcal{H}$. If h assumes real values, then $\mathcal{H}$ is Hermitian, $\mathcal{H}(-f) = \mathcal{H}^*(f)$, and $|\mathcal{H}(f)|$ is an even function. Depending on the support

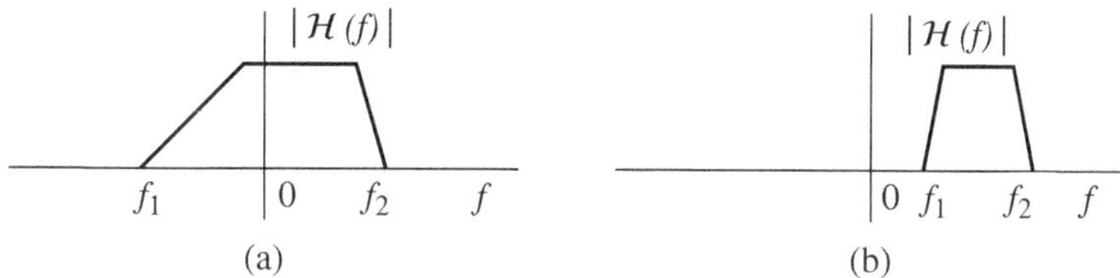

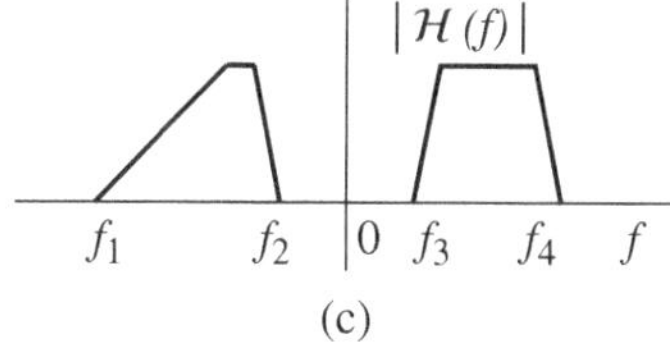

Figure 1.8 Classification of complex-valued analog filters on the basis of support of $|\mathcal{H}(f)|$. (a) $-\infty < f_1 \leq 0$, $0 < f_2 < \infty$: lowpass filter. (b) $f_1 > 0$, $f_2 < \infty$: passband filter. (c) $f_1 > -\infty$, $f_2 < 0$, $f_3 > 0$, $f_4 < \infty$: passband filter.

of $|\mathcal{H}(f)|$, the classification of Figure 1.7 is usually done. If h assumes complex values, the terminology is less standard. We adopt the classification of Figure 1.8, in which the filter is a lowpass filter (LPF) if the support $|\mathcal{H}(f)|$ includes the origin; otherwise, it is a passband filter (PBF).

Analogously, for a signal x, we will use the same denomination and we will say that x is a baseband (BB) or passband (PB) signal depending on whether the support of $|\mathcal{X}(f)|, f \in \mathbb{R}$, includes or not the origin.

Definition 1.7

In general, for a *real-valued signal* x, the set of positive frequencies such that $|\mathcal{X}(f)| \neq 0$ is called *passband* or simply *band* $\mathcal{B}$:

$$\mathcal{B} = \{f \geq 0 : |\mathcal{X}(f)| \neq 0\} \tag{1.69}$$

As $|\mathcal{X}(f)|$ is an even function, we have $|\mathcal{X}(-f)| \neq 0, f \in \mathcal{B}$. We note that $\mathcal{B}$ is equivalent to the support of $\mathcal{X}$ limited to positive frequencies. The bandwidth of x is given by the measure of $\mathcal{B}$:

$$B = \int_{\mathcal{B}} df \tag{1.70}$$

In the case of a *complex-valued signal* x, $\mathcal{B}$ is equivalent to the support of $\mathcal{X}$, and B is thus given by the measure of the entire support. □

Observation 1.1

The signal *bandwidth* may also be given different practical definitions. Let us consider an LPF having frequency response $\mathcal{H}(f)$. The filter gain $\mathcal{H}_0$ is usually defined as $\mathcal{H}_0 = |\mathcal{H}(0)|$; other definitions of gain refer to the average gain of the filter in the passband $\mathcal{B}$, or as $\max_f |\mathcal{H}(f)|$. We give the following four definitions for the bandwidth B of h:

(a) *First zero*:

$$B = \min\{f > 0 : \mathcal{H}(f) = 0\} \tag{1.71}$$

(b) *Based on amplitude*, bandwidth at A dB:

$$B = \max\left\{f > 0 : \frac{|\mathcal{H}(f)|}{\mathcal{H}_0} = 10^{-\frac{A}{20}}\right\} \tag{1.72}$$

Typically, $A = 3, 40$, or 60.

(c) *Based on energy*, bandwidth at $p\%$:

$$\frac{\int_0^B |\mathcal{H}(f)|^2 df}{\int_0^\infty |\mathcal{H}(f)|^2 df} = \frac{p}{100} \tag{1.73}$$

Typically, $p = 90$ or 99.

(d) *Equivalent noise bandwidth*:

$$B = \frac{\int_0^\infty |\mathcal{H}(f)|^2 df}{\mathcal{H}_0^2} \tag{1.74}$$

Figure 1.9 illustrates the various definitions for a particular $|\mathcal{H}(f)|$. For example, with regard to the signals of Figure 1.7, we have that for an LPF $B = f_2$, whereas for a PBF $B = f_2 - f_1$.

For discrete-time filters, for which $\mathcal{H}$ is periodic of period $1/T_c$, the same definitions hold, with the caution of considering the support of $|\mathcal{H}(f)|$ within a period, let us say between $-1/(2T_c)$ and $1/(2T_c)$. In the case of discrete-time highpass filters (HPFs), the passband will extend from a certain frequency f_1 to $1/(2T_c)$.

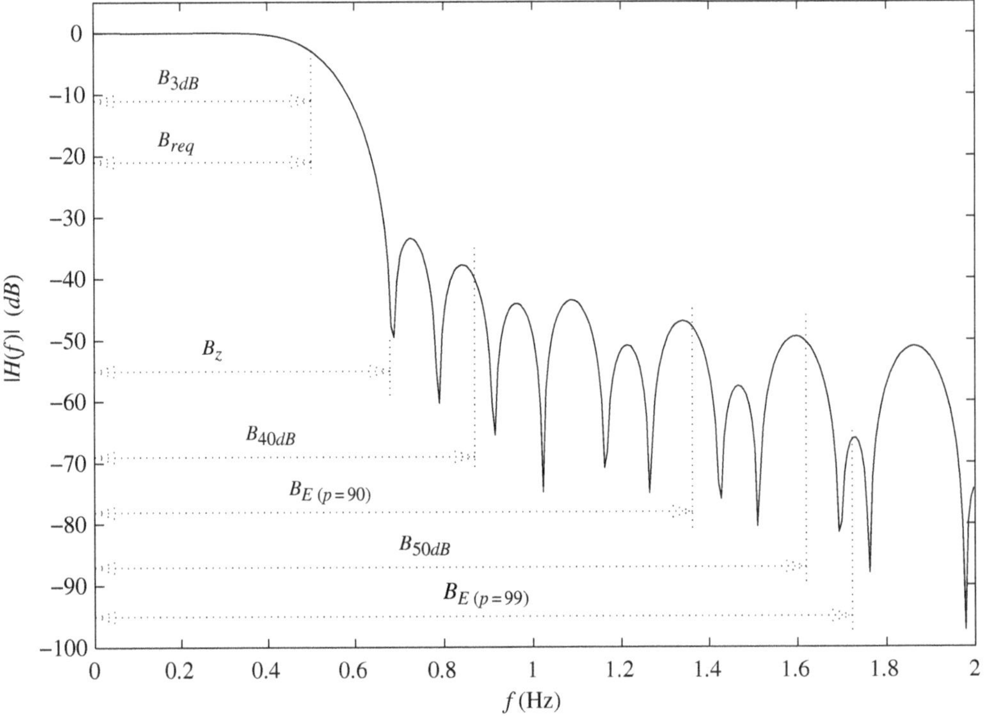

Figure 1.9 The real signal bandwidth following the definitions of (1) bandwidth at first zero: $B_z = 0.652$ Hz; (2) amplitude-based bandwidth: $B_{3\,dB} = 0.5$ Hz, $B_{40\,dB} = 0.87$ Hz, $B_{50\,dB} = 1.62$ Hz; (3) energy-based bandwidth: $B_{E(p=90)} = 1.362$ Hz, $B_{E(p=99)} = 1.723$ Hz; (4) equivalent noise bandwidth: $B_{req} = 0.5$ Hz.

The sampling theorem

As discrete-time signals are often obtained by sampling continuous-time signals, we will state the following fundamental theorem.

Theorem 1.1 (Sampling theorem)
Let $q(t)$, $t \in \mathbb{R}$ be a continuous-time signal, in general complex-valued, whose Fourier transform $\mathcal{Q}(f)$ has support within an interval $\mathcal{B}$ of finite measure B_0. The samples of the signal q, taken with period T_c as represented in Figure 1.10a,

$$h_k = q(kT_c) \tag{1.75}$$

univocally represent the signal $q(t)$, $t \in \mathbb{R}$, under the condition that the *sampling frequency* $1/T_c$ satisfies the relation

$$\frac{1}{T_c} \geq B_0 \tag{1.76}$$

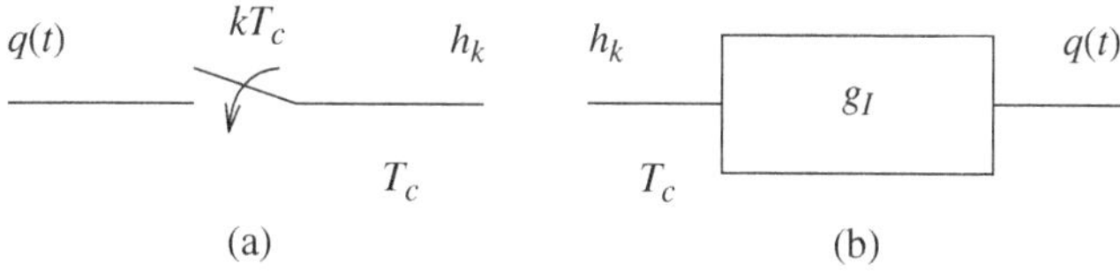

Figure 1.10 Operation of (a) sampling and (b) interpolation.

□

For the proof, which is based on the relation (1.23) between a signal and its samples, we refer the reader to [2].

B_0 is often referred to as the minimum sampling frequency. If $1/T_c < B_0$ the signal cannot be perfectly reconstructed from its samples, originating the so-called *aliasing phenomenon* in the frequency-domain signal representation.

In turn, the signal $q(t)$, $t \in \mathbb{R}$, can be reconstructed from its samples $\{h_k\}$ according to the scheme of Figure 1.10b, where it is employed an interpolation filter having an ideal frequency response given by

$$\mathcal{G}_I(f) = \begin{cases} 1 & f \in \mathcal{B} \\ 0 & \text{elsewhere} \end{cases} \tag{1.77}$$

We note that for *real-valued baseband signals* $B_0 = 2B$. For passband signals, care must be taken in the choice of $B_0 \geq 2B$ to avoid *aliasing* between the positive and negative frequency components of $\mathcal{Q}(f)$.

Heaviside conditions for the absence of signal distortion

Let us consider a filter having frequency response $\mathcal{H}(f)$ (see Figures 1.1 or 1.3) given by

$$\mathcal{H}(f) = \mathcal{H}_0 e^{-j2\pi f t_0}, \quad f \in \mathcal{B} \tag{1.78}$$

where $\mathcal{H}_0$ and t_0 are two non-negative constants, and $\mathcal{B}$ is the *passband* of the filter input signal x. Then the output is given by

$$\mathcal{Y}(f) = \mathcal{H}(f)\mathcal{X}(f) = \mathcal{H}_0\mathcal{X}(f)\, e^{-j2\pi f t_0} \tag{1.79}$$

or, in the time domain,

$$y(t) = \mathcal{H}_0 x(t - t_0) \tag{1.80}$$

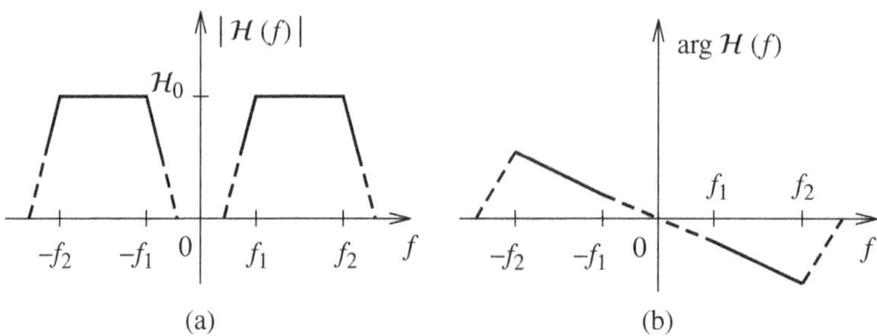

Figure 1.11 Characteristics of a filter satisfying the conditions for the absence of signal distortion in the frequency interval (f_1,f_2). (a) Magnitude and (b) phase.

In other words, for a filter of the type (1.78), the signal at the input is reproduced at the output with a gain factor $\mathcal{H}_0$ and a delay t_0.

A filter of the type (1.78) satisfies the Heaviside conditions for the absence of signal distortion and is characterized by

1. *Constant magnitude*

$$|\mathcal{H}(f)| = \mathcal{H}_0, \qquad f \in \mathcal{B} \tag{1.81}$$

2. *Linear phase*

$$\arg \mathcal{H}(f) = -2\pi f t_0, \qquad f \in \mathcal{B} \tag{1.82}$$

3. *Constant group delay*, also called *envelope delay*

$$\tau(f) = -\frac{1}{2\pi}\frac{d}{df}\arg \mathcal{H}(f) = t_0, \qquad f \in \mathcal{B} \tag{1.83}$$

We underline that it is sufficient that the Heaviside conditions are verified within the support of $\mathcal{X}$; as $|\mathcal{X}(f)| = 0$ outside the support, the filter frequency response may be arbitrary.

We show in Figure 1.11 the frequency response of a PBF, with bandwidth $B = f_2 - f_1$, that satisfies the conditions stated by Heaviside.

1.4 Passband signals and systems

We now provide a compact representation of passband signals and describe their transformation by linear systems.

Complex representation

For a passband signal x, it is convenient to introduce an equivalent representation in terms of a baseband signal $x^{(bb)}$.

Let x be a PB real-valued signal with Fourier transform as illustrated in Figure 1.12. The following two procedures can be adopted to obtain $x^{(bb)}$.

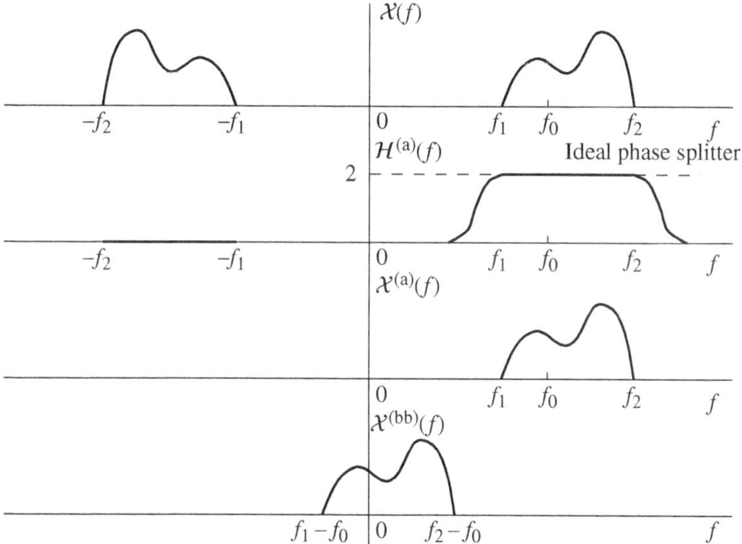

Figure 1.12 Transformations to obtain the baseband equivalent signal $x^{(bb)}$ around the carrier frequency f_0 using a phase splitter.

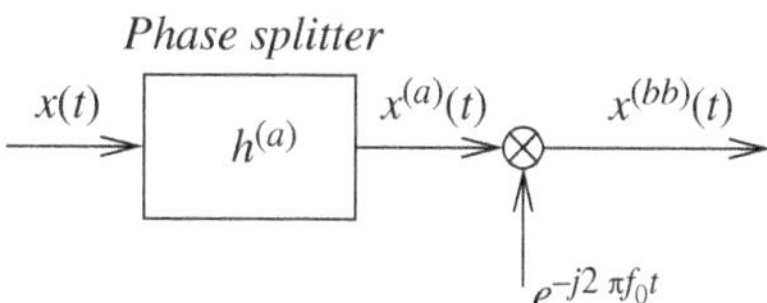

Figure 1.13 Transformations to obtain the baseband equivalent signal $x^{(bb)}$ around the carrier frequency f_0 using a phase splitter.

PB filter. Referring to Figure 1.12 and to the transformations illustrated in Figure 1.13, given x we extract its positive frequency components using an *analytic filter* or *phase splitter*, $h^{(a)}$, having the following ideal frequency response

$$\mathcal{H}^{(a)}(f) = 2 \cdot 1(f) = \begin{cases} 2 & f > 0 \\ 0 & f < 0 \end{cases} \tag{1.84}$$

In practice, it is sufficient that $h^{(a)}$ is a complex PB filter, with $\mathcal{H}^{(a)}(f) \simeq 2$ in the passband that extends from f_1 to f_2, as $\mathcal{X}(f)$, and stopband, in which $|\mathcal{H}^{(a)}(f)| \simeq 0$, that extends from $-f_2$ to $-f_1$. The signal $x^{(a)}$ is called the analytic signal or pre-envelope of x.

It is now convenient to introduce a suitable frequency f_0, called *reference carrier frequency*, which belongs to the passband (f_1, f_2) of x. The filter output, $x^{(a)}$, is frequency shifted by f_0 to obtain a BB signal, $x^{(bb)}$. The signal $x^{(bb)}$ is the *baseband equivalent* of x, also named *complex envelope of x around the carrier frequency f_0*.

Analytically, we have

$$x^{(a)}(t) = x * h^{(a)}(t) \quad \stackrel{\mathcal{F}}{\longleftrightarrow} \quad \mathcal{X}^{(a)}(f) = \mathcal{X}(f)\mathcal{H}^{(a)}(f) \tag{1.85}$$

$$x^{(bb)}(t) = x^{(a)}(t)\, e^{-j2\pi f_0 t} \quad \stackrel{\mathcal{F}}{\longleftrightarrow} \quad \mathcal{X}^{(bb)}(f) = \mathcal{X}^{(a)}(f + f_0) \tag{1.86}$$

and in the frequency domain

$$\mathcal{X}^{(bb)}(f) = \begin{cases} 2\mathcal{X}(f+f_0) & \text{for } f > -f_0 \\ 0 & \text{for } f < -f_0 \end{cases} \tag{1.87}$$

In other words, $x^{(bb)}$ is given by the components of x at positive frequencies, scaled by 2 and frequency shifted by f_0.

BB filter. We obtain the same result using a frequency shift of x followed by a lowpass filter (see Figures 1.14 and 1.15). It is immediate to determine the relation between the frequency responses of the filters of Figures 1.12 and 1.14:

$$\mathcal{H}(f) = \mathcal{H}^{(a)}(f+f_0) \tag{1.88}$$

From (1.88) one can derive the relation between the impulse response of the analytic filter and the impulse response of the lowpass filter:

$$h^{(a)}(t) = h(t)\, e^{j2\pi f_0 t} \tag{1.89}$$

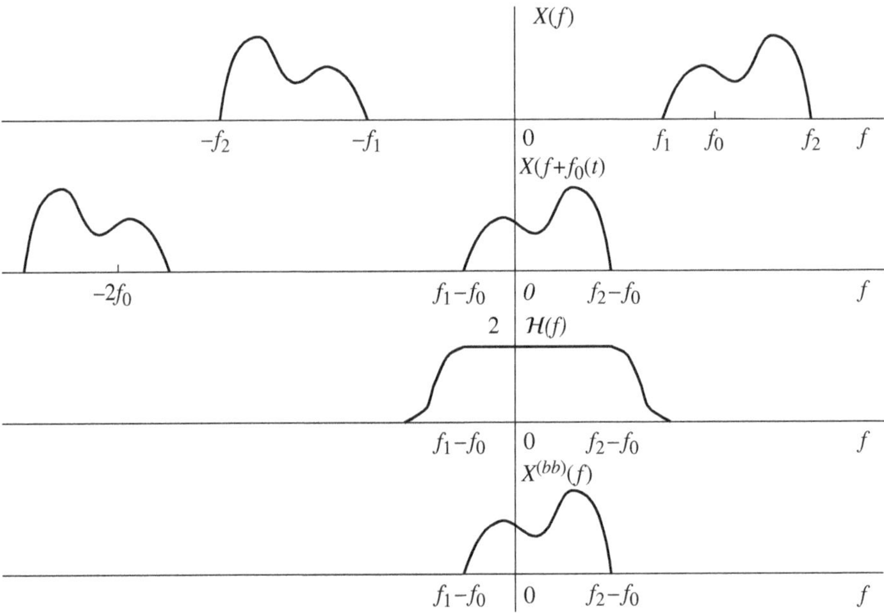

Figure 1.14 Illustration of transformations to obtain the baseband equivalent signal $x^{(bb)}$ around the carrier frequency f_0 using a lowpass filter.

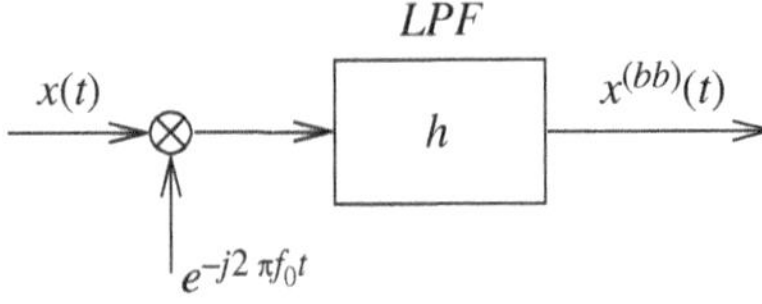

Figure 1.15 Transformations to obtain the baseband equivalent signal $x^{(bb)}$ around the carrier frequency f_0 using a lowpass filter.

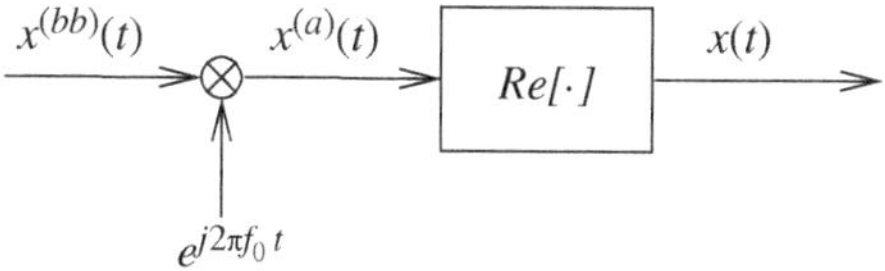

Figure 1.16 Relation between a signal, its complex envelope and the analytic signal.

Relation between a signal and its complex representation

A simple analytical relation exists between a *real signal* x and its complex envelope. In fact, making use of the property $\mathcal{X}(-f) = \mathcal{X}^*(f)$, it follows

$$\mathcal{X}(f) = \mathcal{X}(f)1(f) + \mathcal{X}(f)1(-f) = \mathcal{X}(f)1(f) + \mathcal{X}^*(-f)1(-f) \tag{1.90}$$

or equivalently,

$$x(t) = \frac{x^{(a)}(t) + x^{(a)*}(t)}{2} = Re\,[x^{(a)}(t)] \tag{1.91}$$

Using (1.86) it also follows

$$x(t) = Re\,[x^{(bb)}(t)e^{j2\pi f_0 t}] \tag{1.92}$$

as illustrated in Figure 1.16.

Baseband components of a PB signal. We introduce the notation

$$x^{(bb)}(t) = x_I^{(bb)}(t) + jx_Q^{(bb)}(t) \tag{1.93}$$

where

$$x_I^{(bb)}(t) = Re\,[x^{(bb)}(t)] \tag{1.94}$$

and

$$x_Q^{(bb)}(t) = Im\,[x^{(bb)}(t)] \tag{1.95}$$

are real-valued baseband signals, named *in-phase* and *quadrature components* of x, respectively. Substituting (1.93) in (1.92), we obtain

$$x(t) = x_I^{(bb)}(t)\cos(2\pi f_0 t) - x_Q^{(bb)}(t)\sin(2\pi f_0 t) \tag{1.96}$$

as illustrated in Figure 1.17.

Conversely, given x, one can use the scheme of Figure 1.15 and the relations (1.94) and (1.95) to get the baseband components. If the frequency response $\mathcal{H}(f)$ has Hermitian-symmetric characteristics with respect to the origin, h is real and the scheme of Figure 1.18 holds. The scheme of Figure 1.18 employs instead an ideal Hilbert filter with frequency response given by

$$\mathcal{H}^{(h)}(f) = -j\,\mathrm{sgn}(f) = e^{-j\frac{\pi}{2}\mathrm{sgn}(f)} \tag{1.97}$$

Magnitude and phase of $\mathcal{H}^{(h)}(f)$ are shown in Figure 1.19. We note that $h^{(h)}$ phase-shifts by $-\pi/2$ the positive-frequency components of the input and by $\pi/2$ the negative-frequency components. In practice, these filter specifications are imposed only on the passband of the input signal.[9] To simplify the notation, in block diagrams a Hilbert filter is indicated as $-\pi/2$.

[9] We note that the ideal Hilbert filter in Figure 1.19 has an impulse response given by (see Table 1.2 on page 5):

$$h^{(h)}(t) = \frac{1}{\pi t} \tag{1.98}$$

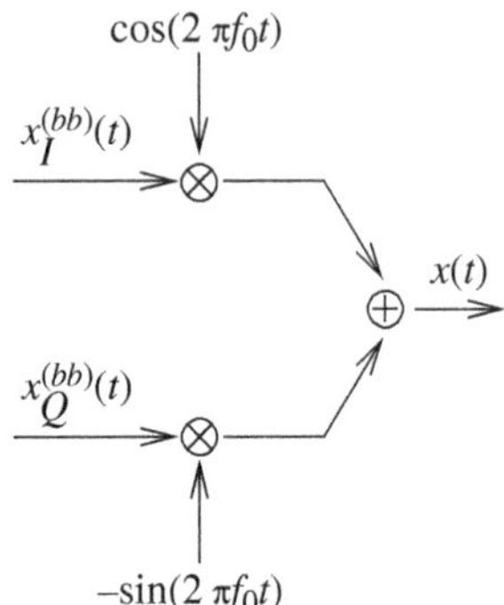

Figure 1.17 Relation between a signal and its baseband components.

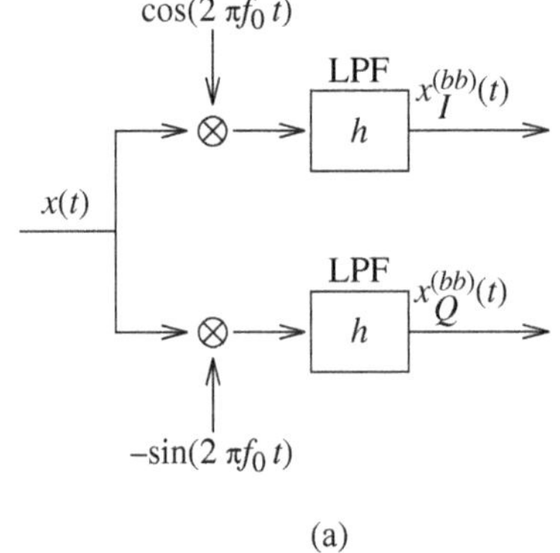

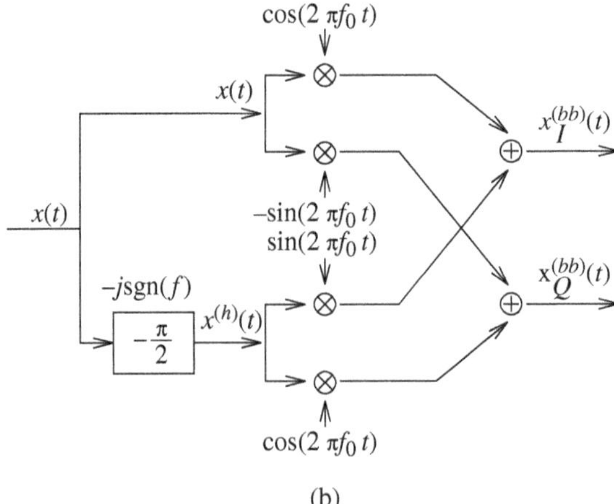

Figure 1.18 Relations to derive the baseband signal components. (a) Implementation using LPF and (b) Implementation using Hilbert filter.

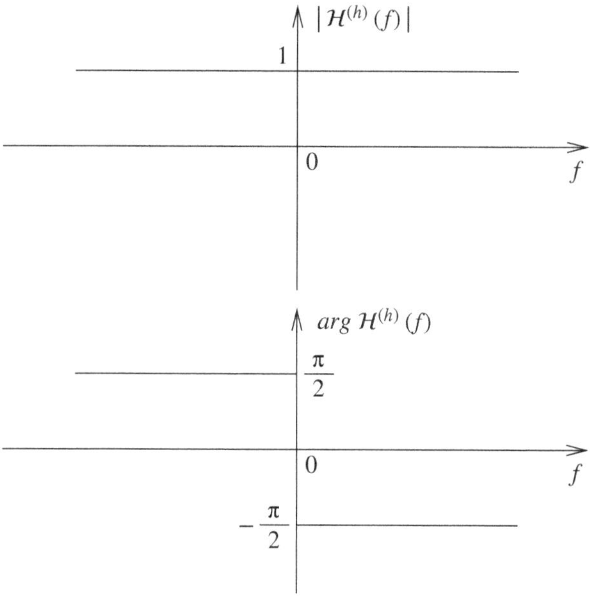

Figure 1.19 Magnitude and phase responses of the ideal Hilbert filter.

Comparing the frequency responses of the analytic filter (1.84) and of the Hilbert filter (1.97), we obtain the relation

$$\mathcal{H}^{(a)}(f) = 1 + j\mathcal{H}^{(h)}(f) \tag{1.101}$$

Then, letting

$$x^{(h)}(t) = x * h^{(h)}(t) \tag{1.102}$$

the analytic signal can be expressed as

$$x^{(a)}(t) = x(t) + jx^{(h)}(t) \tag{1.103}$$

Consequently, from (1.86), (1.94), and (1.95), we have

$$x_I^{(bb)}(t) = x(t)\cos(2\pi f_0 t) + x^{(h)}(t)\sin(2\pi f_0 t) \tag{1.104}$$

$$x_Q^{(bb)}(t) = x^{(h)}(t)\cos(2\pi f_0 t) - x(t)\sin(2\pi f_0 t) \tag{1.105}$$

as illustrated in Figure 1.18.[10]

Consequently, if x is the input signal, the output of the Hilbert filter (also denoted as Hilbert transform of x) is

$$x^{(h)}(t) = \frac{1}{\pi}\int_{-\infty}^{+\infty} \frac{x(\tau)}{t-\tau}\,d\tau \tag{1.99}$$

Moreover, noting that from (1.97) $(-j\,\text{sgn}f)(-j\,\text{sgn}f) = -1$, taking the Hilbert transform of the Hilbert transform of a signal, we get the initial signal with the sign changed. Then it results as

$$x(t) = -\frac{1}{\pi}\int_{-\infty}^{+\infty} \frac{x^{(h)}(\tau)}{t-\tau}\,d\tau \tag{1.100}$$

[10] We recall that the design of a filter, and in particular of a Hilbert filter, requires the introduction of a suitable delay. In other words, we are only able to produce an output with a delay t_D, $x^{(h)}(t-t_D)$. Consequently, in the block diagram of Figure 1.18, also x and the various sinusoidal waveforms must be delayed.

We note that in practical *systems*, transformations to obtain, e.g. the analytic signal, the complex envelope, or the Hilbert transform of a given signal, are implemented by filters. However, it is usually more convenient to perform signal analysis in the frequency domain by the Fourier transform. In the following two examples, we use frequency-domain techniques to obtain the complex envelope of a PB signal.

Example 1.4.1
Consider the sinusoidal signal

$$x(t) = A\cos(2\pi f_0 t + \varphi_0) \tag{1.106}$$

with

$$\mathcal{X}(f) = \frac{A}{2}e^{j\varphi_0}\delta(f - f_0) + \frac{A}{2}e^{-j\varphi_0}\delta(f + f_0) \tag{1.107}$$

The analytic signal is given by

$$\mathcal{X}^{(a)}(f) = Ae^{j\varphi_0}\delta(f - f_0) \quad \overset{\mathcal{F}^{-1}}{\longleftrightarrow} \quad x^{(a)}(t) = Ae^{j\varphi_0}e^{j2\pi f_0 t} \tag{1.108}$$

and

$$\mathcal{X}^{(bb)}(f) = Ae^{j\varphi_0}\delta(f) \quad \overset{\mathcal{F}^{-1}}{\longleftrightarrow} \quad x^{(bb)}(t) = Ae^{j\varphi_0} \tag{1.109}$$

We note that we have chosen as reference carrier frequency of the complex envelope the same carrier frequency as in (1.106).

Example 1.4.2
Let

$$x(t) = A\,\mathrm{sinc}(Bt)\cos(2\pi f_0 t) \tag{1.110}$$

with the Fourier transform given by

$$\mathcal{X}(f) = \frac{A}{2B}\left[\mathrm{rect}\left(\frac{f - f_0}{B}\right) + \mathrm{rect}\left(\frac{f + f_0}{B}\right)\right] \tag{1.111}$$

as illustrated in Figure 1.20. Then, using f_0 as reference carrier frequency,

$$\mathcal{X}^{(bb)}(f) = \frac{A}{B}\mathrm{rect}\left(\frac{f}{B}\right) \tag{1.112}$$

and

$$x^{(bb)}(t) = A\mathrm{sinc}(Bt) \tag{1.113}$$

Another *analytical technique* to get the expression of the signal after the various transformations is obtained by applying the following theorem.

Theorem 1.2
Let the product of two real signals be

$$x(t) = a(t)\;c(t) \tag{1.114}$$

where a is a BB signal with $\mathcal{B}_a = [0, B)$ and c is a PB signal with $\mathcal{B}_c = [f_0, +\infty)$. If $f_0 > B$, then the analytic signal of x is related to that of c by

$$x^{(a)}(t) = a(t)\;c^{(a)}(t) \tag{1.115}$$

□

Proof. We consider the general relation (1.91), valid for every real signal

$$c(t) = \frac{1}{2}\, c^{(a)}(t) + \frac{1}{2}\, c^{(a)*}(t) \tag{1.116}$$

Substituting (1.116) in (1.114) yields

$$x(t) = a(t)\frac{1}{2}c^{(a)}(t) + a(t)\frac{1}{2}c^{(a)*}(t) \tag{1.117}$$

In the frequency domain, the support of the first term in (1.117) is given by the interval $[f_0 - B, +\infty)$, while that of the second is equal to $(-\infty, -f_0 + B]$. Under the hypothesis that $f_0 \geq B$, the two terms in (1.117) have disjoint supports in the frequency domain and (1.115) is immediately obtained. □

Corollary 1.1
From (1.115), we obtain

$$x^{(h)}(t) = a(t)c^{(h)}(t) \tag{1.118}$$

and

$$x^{(bb)}(t) = a(t)c^{(bb)}(t) \tag{1.119}$$

In fact, from (1.103) we get

$$x^{(h)}(t) = Im\,[x^{(a)}(t)] \tag{1.120}$$

which substituted in (1.115) yields (1.118). Finally, (1.119) is obtained by substituting (1.86),

$$x^{(bb)}(t) = x^{(a)}(t)e^{-j2\pi f_0 t} \tag{1.121}$$

in (1.115). □

An interesting application of (1.120) is in the design of a Hilbert filter $h^{(h)}$ starting from a lowpass filter h. In fact, from (1.89) and (1.120), we get

$$h^{(h)}(t) = h(t)\sin(2\pi f_0 t) \tag{1.122}$$

Example 1.4.3
Let a modulated *double sideband* (DSB) signal be expressed as

$$x(t) = a(t)\cos(2\pi f_0 t + \varphi_0) \tag{1.123}$$

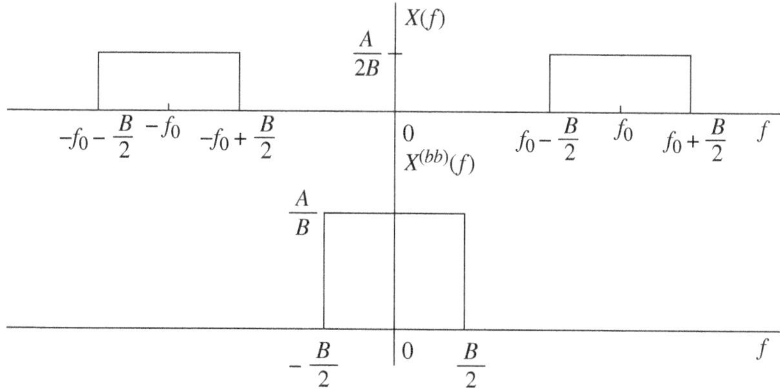

Figure 1.20 Frequency response of a PB signal and corresponding complex envelope.

Table 1.5: Some properties of the Hilbert transform.

property	*(real) signal*	*(real) Hilbert transform*
	$x(t)$	$x^{(h)}(t)$
duality	$x^{(h)}(t)$	$-x(t)$
time inverse	$x(-t)$	$-x^{(h)}(-t)$
even signal	$x(t) = x(-t)$	$x^{(h)}(t) = -x^{(h)}(-t)$, odd
odd signal	$x(t) = -x(-t)$	$x^{(h)}(t) = x^{(h)}(-t)$, even
product (see Theorem 1.2)	$a(t)\; c(t)$	$a(t)\; c^{(h)}(t)$
cosinusoidal signal	$\cos(2\pi f_0 t + \varphi_0)$	$\sin(2\pi f_0 t + \varphi_0)$
energy	$E_x = \int_{-\infty}^{+\infty} \lvert x(t)\rvert^2\, dt =$	$\int_{-\infty}^{+\infty} \lvert x^{(h)}(t)\rvert^2 dt = E_{x^{(h)}}$
orthogonality	$\int_{-\infty}^{+\infty} x(t)\; x^{(h)}(t)\; dt = 0$	

where a is a BB signal with bandwidth B. Then, if $f_0 > B$, from the above theorem we have the following relations:

$$x^{(a)}(t) = a(t) e^{j(2\pi f_0 t + \varphi_0)} \tag{1.124}$$

$$x^{(h)}(t) = a(t) \sin(2\pi f_0 t + \varphi_0) \tag{1.125}$$

$$x^{(bb)}(t) = a(t) e^{j\varphi_0} \tag{1.126}$$

We list in Table 1.5 some properties of the Hilbert transformation (1.102) that are easily obtained by using the Fourier transform and the properties of Table 1.1.

Baseband equivalent of a transformation

Given a transformation involving also passband signals, it is often useful to determine an equivalent relation between baseband complex representations of input and output signals. Three transformations are given in Figure 1.21, together with their baseband equivalent. Note that schemes in Figure 1.21a,b produce very different output signals, although both use a mixer with the same carrier.

We will prove the relation illustrated in Figure 1.21b. Assuming that h is the real-valued impulse response of an LPF and using (1.92),

$$\begin{aligned} y(t) &= \left\{ h * Re\left[x^{(bb)}(\tau) e^{j2\pi f_0 \tau} \left(\cos(2\pi f_0 \tau + \varphi_1) \right) \right] \right\}(t) \\ &= Re\left[\left(h * x^{(bb)} \frac{e^{-j\varphi_1}}{2} + h * x^{(bb)} \frac{e^{+j(2\pi 2 f_0 \tau + \varphi_1)}}{2} \right)(t) \right] \\ &= Re\left[\left(h * x^{(bb)} \frac{e^{-j\varphi_1}}{2} \right)(t) \right] \end{aligned} \tag{1.127}$$

where the last equality follows because the term with frequency components around $2f_0$ is filtered by the LPF.

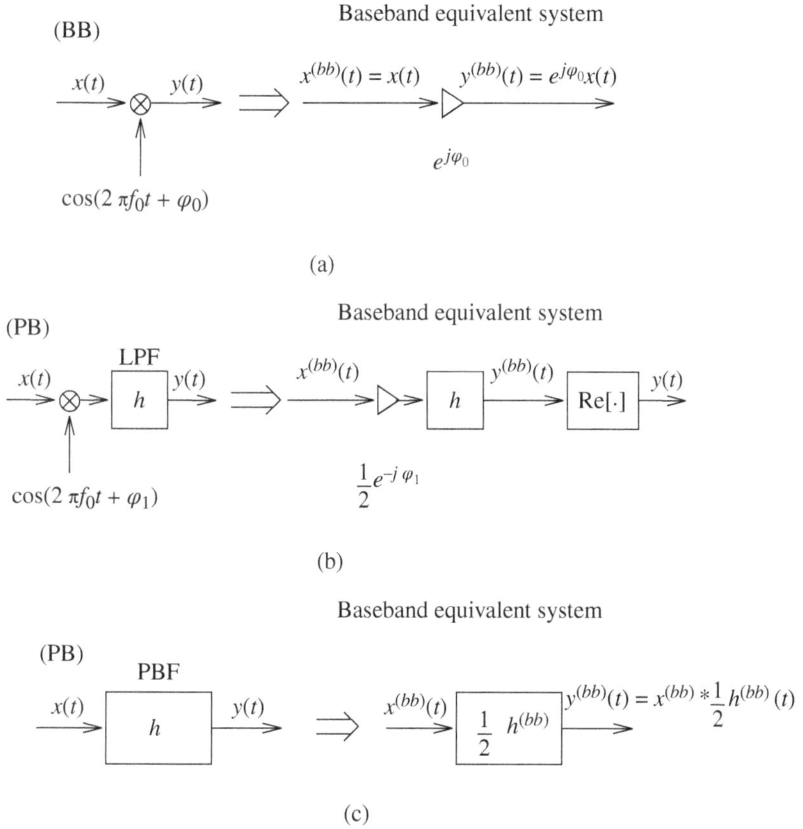

Figure 1.21 Passband transformations and their baseband equivalent. (a) Modulator, (b) demodulator, and (c) passband filtering.

We note, moreover, that the filter $h^{(bb)}$ in Figure 1.21 has in-phase component $h_I^{(bb)}$ and quadrature component $h_Q^{(bb)}$ that are related to $\mathcal{H}^{(a)}$ by (see (1.94) and (1.95))

$$\begin{aligned}\mathcal{H}_I^{(bb)}(f) &= \frac{1}{2}[\mathcal{H}^{(bb)}(f) + \mathcal{H}^{(bb)*}(-f)] \\ &= \frac{1}{2}[\mathcal{H}^{(a)}(f + f_0) + \mathcal{H}^{(a)*}(-f + f_0)]\end{aligned} \tag{1.128}$$

and

$$\begin{aligned}\mathcal{H}_Q^{(bb)}(f) &= \frac{1}{2j}[\mathcal{H}^{(bb)}(f) - \mathcal{H}^{(bb)*}(-f)] \\ &= \frac{1}{2j}[\mathcal{H}^{(a)}(f + f_0) - \mathcal{H}^{(a)*}(-f + f_0)]\end{aligned} \tag{1.129}$$

Consequently, if $\mathcal{H}^{(a)}$ has Hermitian symmetry around f_0, then

$$\mathcal{H}_I^{(bb)}(f) = \mathcal{H}_a^{(a)}(f + f_0)$$

and

$$\mathcal{H}_Q^{(bb)}(f) = 0$$

In other words, $h^{(bb)}(t) = h_I^{(bb)}(t)$ is real and the realization of the filter $\frac{1}{2}h^{(bb)}$ is simplified. In practice, this condition is verified by imposing that the filter $h^{(a)}$ has symmetrical frequency specifications around f_0.

Envelope and instantaneous phase and frequency

We will conclude this section with a few definitions. Given a PB signal x, with reference to the analytic signal we define

1. *Envelope*
$$M_x(t) = |x^{(a)}(t)| \tag{1.130}$$

2. *Instantaneous phase*
$$\varphi_x(t) = \arg x^{(a)}(t) \tag{1.131}$$

3. *Instantaneous frequency*
$$f_x(t) = \frac{1}{2\pi}\frac{d}{dt}\varphi_x(t) \tag{1.132}$$

In terms of the complex envelope signal $x^{(bb)}$, from (1.86) the equivalent relations follow:

$$M_x(t) = |x^{(bb)}(t)| \tag{1.133}$$
$$\varphi_x(t) = \arg x^{(bb)}(t) + 2\pi f_0 t \tag{1.134}$$
$$f_x(t) = \frac{1}{2\pi}\frac{d}{dt}[\arg x^{(bb)}(t)] + f_0 \tag{1.135}$$

Then, from the polar representation, $x^{(a)}(t) = M_x(t)\, e^{j\varphi_x(t)}$ and from (1.91), a PB signal x can be written as

$$x(t) = Re\,[x^{(a)}(t)] = M_x(t)\cos(\varphi_x(t)) \tag{1.136}$$

For example if $x(t) = A\cos(2\pi f_0 t + \varphi_0)$, it follows that

$$M_x(t) = A \tag{1.137}$$
$$\varphi_x(t) = 2\pi f_0 t + \varphi_0 \tag{1.138}$$
$$f_x(t) = f_0 \tag{1.139}$$

With reference to the above relations, three other definitions follow.

1. *Envelope deviation*
$$\Delta M_x(t) = |x^{(a)}(t)| - A = |x^{(bb)}(t)| - A \tag{1.140}$$

2. *Phase deviation*
$$\Delta\varphi_x(t) = \varphi_x(t) - (2\pi f_0 t + \varphi_0) = \arg x^{(bb)}(t) - \varphi_0 \tag{1.141}$$

3. *Frequency deviation*
$$\Delta f_x(t) = f_x(t) - f_0 = \frac{1}{2\pi}\frac{d}{dt}\Delta\varphi_x(t) \tag{1.142}$$

Then (1.136) becomes

$$x(t) = [A + \Delta M_x(t)]\cos(2\pi f_0 t + \varphi_0 + \Delta\varphi_x(t)) \tag{1.143}$$

1.5 Second-order analysis of random processes

We recall the functions related to the statistical description of random processes, especially those functions concerning second-order analysis.

1.5.1 Correlation

Let $x(t)$ and $y(t)$, $t \in \mathbb{R}$, be two continuous-time complex-valued random processes. We indicate the expectation operator with E.

1. *Mean value*

$$\mathtt{m}_x(t) = E[x(t)] \tag{1.144}$$

2. *Statistical power*

$$\mathtt{M}_x(t) = E[|x(t)|^2] \tag{1.145}$$

3. *Autocorrelation*

$$\mathtt{r}_x(t, t-\tau) = E[x(t)x^*(t-\tau)] \tag{1.146}$$

4. *Crosscorrelation*

$$\mathtt{r}_{xy}(t, t-\tau) = E[x(t)y^*(t-\tau)] \tag{1.147}$$

5. *Autocovariance*

$$\begin{aligned} \mathtt{c}_x(t, t-\tau) &= E[(x(t) - \mathtt{m}_x(t))(x(t-\tau) - \mathtt{m}_x(t-\tau))^*] \\ &= \mathtt{r}_x(t, t-\tau) - \mathtt{m}_x(t)\mathtt{m}_x^*(t-\tau) \end{aligned} \tag{1.148}$$

6. *Crosscovariance*

$$\begin{aligned} \mathtt{c}_{xy}(t, t-\tau) &= E[(x(t) - \mathtt{m}_x(t))(y(t-\tau) - \mathtt{m}_y(t-\tau))^*] \\ &= \mathtt{r}_{xy}(t, t-\tau) - \mathtt{m}_x(t)\mathtt{m}_y^*(t-\tau) \end{aligned} \tag{1.149}$$

Observation 1.2

- x and y are *orthogonal* if $\mathtt{r}_{xy}(t, t-\tau) = 0$, $\forall t, \tau$. In this case, we write $x \perp y$.[11]
- x and y are *uncorrelated* if $\mathtt{c}_{xy}(t, t-\tau) = 0$, $\forall t, \tau$.
- if at least one of the two random processes has zero mean, *orthogonality* is equivalent to *uncorrelation.*
- x is *wide-sense stationary* (WSS) if
  1. $\mathtt{m}_x(t) = \mathtt{m}_x$, $\forall t$,
  2. $\mathtt{r}_x(t, t-\tau) = \mathtt{r}_x(\tau)$, $\forall t$.

  In this case, $\mathtt{r}_x(0) = E[|x(t)|^2] = \mathtt{M}_x$ is the *statistical power*, whereas $\mathtt{c}_x(0) = \sigma_x^2 = \mathtt{M}_x - |\mathtt{m}_x|^2$ is the *variance* of x.

[11] We observe that the notion of *orthogonality between two random processes* is quite different from that of *orthogonality between two deterministic signals*. In fact, while in the deterministic case, it is sufficient that $\int_{-\infty}^{\infty} x(t)y^*(t)dt = 0$, in the random case, the crosscorrelation must be zero for all the delays and not only for the zero delay. In particular, we note that the *two random variables* v_1 and v_2 are orthogonal if $E[v_1 v_2^*] = 0$.

- x and y are *jointly wide-sense stationary* if
  1. $\mathtt{m}_x(t) = \mathtt{m}_x$, $\mathtt{m}_y(t) = \mathtt{m}_y$, $\forall t$,
  2. $\mathtt{r}_{xy}(t, t-\tau) = \mathtt{r}_{xy}(\tau)$, $\forall t$.

Properties of the autocorrelation function

1. $\mathtt{r}_x(-\tau) = \mathtt{r}_x^*(\tau)$, $\mathtt{r}_x(\tau)$ is a function with Hermitian symmetry.
2. $\mathtt{r}_x(0) \geq |\mathtt{r}_x(\tau)|$.
3. $\mathtt{r}_x(0)\mathtt{r}_y(0) \geq |\mathtt{r}_{xy}(\tau)|^2$.
4. $\mathtt{r}_{xy}(-\tau) = \mathtt{r}_{yx}^*(\tau)$.
5. $\mathtt{r}_{x^*}(\tau) = \mathtt{r}_x^*(\tau)$.

1.5.2 Power spectral density

Given the WSS random process $x(t)$, $t \in \mathbb{R}$, its *power spectral density* (PSD) is defined as the Fourier transform of its autocorrelation function

$$\mathcal{P}_x(f) = \mathcal{F}[\mathtt{r}_x(\tau)] = \int_{-\infty}^{+\infty} \mathtt{r}_x(\tau) e^{-j2\pi f\tau} d\tau \tag{1.150}$$

The inverse transformation is given by the following formula:

$$\mathtt{r}_x(\tau) = \int_{-\infty}^{+\infty} \mathcal{P}_x(f) e^{j2\pi f\tau} df \tag{1.151}$$

In particular from (1.151), we obtain the statistical power

$$\mathtt{M}_x = \mathtt{r}_x(0) = \int_{-\infty}^{+\infty} \mathcal{P}_x(f) df \tag{1.152}$$

Hence, the name PSD for the function $\mathcal{P}_x(f)$: it represents the distribution of the statistical power in the frequency domain.

The pair of equations (1.150) and (1.151) are obtained from the *Wiener–Khintchine theorem* [3].

Definition 1.8
The passband $\mathcal{B}$ of a random process x is defined with reference to its PSD function. □

Spectral lines in the PSD

In many applications, it is important to detect the presence of sinusoidal components in a random process. With this intent we give the following theorem.

Theorem 1.3
The PSD of a WSS process, $\mathcal{P}_x(f)$, can be uniquely decomposed into a component $\mathcal{P}_x^{(c)}(f)$ without delta functions and a discrete component consisting of delta functions (spectral lines) $\mathcal{P}_x^{(d)}(f)$, so that

$$\mathcal{P}_x(f) = \mathcal{P}_x^{(c)}(f) + \mathcal{P}_x^{(d)}(f) \tag{1.153}$$

where $\mathcal{P}_x^{(c)}(f)$ is an ordinary (piecewise linear) function and

$$\mathcal{P}_x^{(d)}(f) = \sum_{i \in \mathcal{I}} \mathtt{M}_i \delta(f - f_i) \tag{1.154}$$

where $\mathcal{I}$ identifies a discrete set of frequencies $\{f_i\}$, $i \in \mathcal{I}$. □

The inverse Fourier transform of (1.153) yields the relation

$$\mathtt{r}_x(\tau) = \mathtt{r}_x^{(c)}(\tau) + \mathtt{r}_x^{(d)}(\tau) \tag{1.155}$$

with

$$\mathtt{r}_x^{(d)}(\tau) = \sum_{i \in \mathcal{I}} \mathtt{M}_i e^{j2\pi f_i \tau} \tag{1.156}$$

The most interesting consideration is that the following random process decomposition corresponds to the decomposition (1.153) of the PSD:

$$x(t) = x^{(c)}(t) + x^{(d)}(t) \tag{1.157}$$

where $x^{(c)}$ and $x^{(d)}$ are *orthogonal* processes having PSD functions

$$\mathcal{P}_{x^{(c)}}(f) = \mathcal{P}_x^{(c)}(f) \qquad \text{and} \qquad \mathcal{P}_{x^{(d)}}(f) = \mathcal{P}_x^{(d)}(f) \tag{1.158}$$

Moreover, $x^{(d)}$ is given by

$$x^{(d)}(t) = \sum_{i \in \mathcal{I}} x_i e^{j2\pi f_i t} \tag{1.159}$$

where $\{x_i\}$ are orthogonal random variables (r.v.s.) having statistical power

$$E[|x_i|^2] = \mathtt{M}_i, \qquad i \in \mathcal{I} \tag{1.160}$$

where $\mathtt{M}_i$ is defined in (1.154).

Observation 1.3
The spectral lines of the PSD identify the periodic components in the process.

Definition 1.9
A WSS random process is said to be *asymptotically uncorrelated* if the following two properties hold:

$$(1) \quad \lim_{\tau \to \infty} \mathtt{r}_x(\tau) = |\mathtt{m}_x|^2 \tag{1.161}$$

$$(2) \quad \mathtt{c}_x(\tau) = \mathtt{r}_x(\tau) - |\mathtt{m}_x|^2 \quad \text{is absolutely integrable} \tag{1.162}$$

The property (1) shows that $x(t)$ and $x(t - \tau)$ become uncorrelated for $\tau \to \infty$. □

For such processes, one can prove that

$$\mathtt{r}_x^{(c)}(\tau) = \mathtt{c}_x(\tau) \qquad \text{and} \qquad \mathtt{r}_x^{(d)}(\tau) = |\mathtt{m}_x|^2 \tag{1.163}$$

Hence, $\mathcal{P}_x^{(d)}(f) = |\mathtt{m}_x|^2 \delta(f)$, and the process exhibits at most a spectral line at the origin.

Cross power spectral density

One can extend the definition of PSD to two jointly WSS random processes:

$$\mathcal{P}_{xy}(f) = \mathcal{F}[\mathtt{r}_{xy}(\tau)] \tag{1.164}$$

Since $\mathtt{r}_{xy}(-\tau) \neq \mathtt{r}_{xy}^*(\tau)$, $\mathcal{P}_{xy}(f)$ is in general a complex function.

Properties of the PSD

1. $\mathcal{P}_x(f)$ is a *real-valued function*. This follows from property 1 of the autocorrelation.
2. $\mathcal{P}_x(f)$ is generally *not an even function*. However, if the process x is real valued, then both $\mathtt{r}_x(\tau)$ and $\mathcal{P}_x(f)$ are even functions.
3. $\mathcal{P}_x(f)$ is a *non-negative function*.
4. $\mathcal{P}_{yx}(f) = \mathcal{P}_{xy}^*(f)$.
5. $\mathcal{P}_{x^*}(f) = \mathcal{P}_x(-f)$.

Moreover, the following inequality holds:

$$0 \leq |\mathcal{P}_{xy}(f)|^2 \leq \mathcal{P}_x(f)\mathcal{P}_y(f) \tag{1.165}$$

Definition 1.10 (White random process)
The zero-mean random process $x(t)$, $t \in \mathbb{R}$, is called *white* if

$$\mathtt{r}_x(\tau) = K\delta(\tau) \tag{1.166}$$

with K a positive real number. In this case, $\mathcal{P}_x(f)$ is a constant, i.e.

$$\mathcal{P}_x(f) = K \tag{1.167}$$

□

PSD through filtering

With reference to Figure 1.22, by taking the Fourier transform of the various crosscorrelations, the following relations are easily obtained:

$$\mathcal{P}_{yx}(f) = \mathcal{P}_x(f)\mathcal{H}(f) \tag{1.168}$$

$$\mathcal{P}_y(f) = \mathcal{P}_x(f)|\mathcal{H}(f)|^2 \tag{1.169}$$

$$\mathcal{P}_{yz}(f) = \mathcal{P}_x(f)\mathcal{H}(f)\mathcal{G}^*(f) \tag{1.170}$$

The relation (1.169) is of particular interest since it relates the PSDs of the output process of a filter to the PSD of the input process, through the frequency response of the filter. In the particular case in which y and z have disjoint passbands, i.e. $\mathcal{P}_y(f)\mathcal{P}_z(f) = 0$, then from (1.165) $\mathtt{r}_{yz}(\tau) = 0$, and $y \perp z$.

1.5.3 PSD of discrete-time random processes

Let $\{x(k)\}$ and $\{y(k)\}$ be two discrete-time random processes. Definitions and properties of Section 1.5.1 remain valid also for discrete-time processes: the only difference is that the correlation is now defined

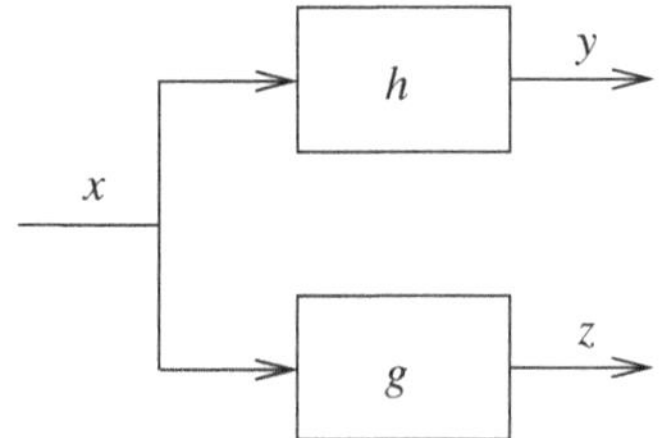

Figure 1.22 Reference scheme of PSD computations.

on discrete time and is called autocorrelation sequence (ACS). It is however interesting to review the properties of PSDs. Given a discrete-time WSS random process x, the PSD is obtained as

$$\mathcal{P}_x(f) = T_c \mathcal{F}[\mathtt{r}_x(n)] = T_c \sum_{n=-\infty}^{+\infty} \mathtt{r}_x(n) e^{-j2\pi f n T_c} \tag{1.171}$$

We note a further property: $\mathcal{P}_x(f)$ is a periodic function of period $1/T_c$. The inverse transformation yields:

$$\mathtt{r}_x(n) = \int_{-\frac{1}{2T_c}}^{\frac{1}{2T_c}} \mathcal{P}_x(f) e^{j2\pi f n T_c} df \tag{1.172}$$

In particular, the statistical power is given by

$$\mathtt{M}_x = \mathtt{r}_x(0) = \int_{-\frac{1}{2T_c}}^{\frac{1}{2T_c}} \mathcal{P}_x(f) df \tag{1.173}$$

Definition 1.11 (White random process)
A discrete-time random process $\{x(k)\}$ is *white* if

$$\mathtt{r}_x(n) = \sigma_x^2 \delta_n \tag{1.174}$$

In this case, the PSD is a constant:

$$\mathcal{P}_x(f) = \sigma_x^2 T_c \tag{1.175}$$

□

Definition 1.12
If the samples of the random process $\{x(k)\}$ are statistically independent and identically Distributed, we say that $\{x(k)\}$ has i.i.d. samples. □

Generating an i.i.d. sequence is not simple; however, it is easily provided by many random number generators [4]. However, generating, storing, and processing a finite length, i.i.d. sequence requires a complex processor and a lot of memory. Furthermore, the deterministic correlation properties of such a subsequence may not be very good. Hence, in Appendix 1.C we introduce a class of pseudonoise (PN) sequences, which are deterministic and periodic, with very good correlation properties. Moreover, the symbol alphabet can be just binary.

Spectral lines in the PSD

Also the PSD of a discrete time random process can be decomposed into ordinary components and spectral lines on a period of the PSD. In particular for a discrete-time WSS *asymptotically uncorrelated* random process, the relation (1.163) and the following are true

$$\mathcal{P}_x^{(c)}(f) = T_c \sum_{n=-\infty}^{+\infty} \mathtt{c}_x(n)\, e^{-j2\pi f n T_c} \tag{1.176}$$

$$\mathcal{P}_x^{(d)}(f) = |\mathtt{m}_x|^2 \sum_{\ell=-\infty}^{+\infty} \delta\left(f - \frac{\ell}{T_c}\right) \tag{1.177}$$

We note that, if the process has non-zero mean value, the PSD exhibits lines at multiples of $1/T_c$.

Example 1.5.1

We calculate the PSD of an i.i.d. sequence $\{x(k)\}$. From

$$\mathtt{r}_x(n) = \begin{cases} \mathtt{M}_x & n = 0 \\ |\mathtt{m}_x|^2 & n \neq 0 \end{cases} \tag{1.178}$$

it follows that

$$\mathtt{c}_x(n) = \begin{cases} \sigma_x^2 & n = 0 \\ 0 & n \neq 0 \end{cases} \tag{1.179}$$

Then

$$\mathtt{r}_x^{(c)}(n) = \sigma_x^2 \delta_n, \quad \mathtt{r}_x^{(d)}(n) = |\mathtt{m}_x|^2 \tag{1.180}$$

$$\mathcal{P}_x^{(c)}(f) = \sigma_x^2 T_c, \quad \mathcal{P}_x^{(d)}(f) = |\mathtt{m}_x|^2 \sum_{\ell=-\infty}^{+\infty} \delta\left(f - \frac{\ell}{T_c}\right) \tag{1.181}$$

PSD through filtering

Given the system illustrated in Figure 1.3, we want to find the relation between the PSDs of the input and output signals, assuming these processes are individually as well as jointly WSS. We introduce the z-transform of the correlation sequence:

$$P_x(z) = \sum_{n=-\infty}^{+\infty} \mathtt{r}_x(n) z^{-n} \tag{1.182}$$

From the comparison of (1.182) with (1.171), the PSD of x is related to $P_x(z)$ by

$$\mathcal{P}_x(f) = T_c P_x(e^{j2\pi f T_c}) \tag{1.183}$$

Using Table 1.3 in page 6, we obtain the relations between ACS and PSD listed in Table 1.6. Let the deterministic autocorrelation of h be defined as[12]

$$\mathtt{r}_h(n) = \sum_{k=-\infty}^{+\infty} h(k)h^*(k-n) = [h(m) * h^*(-m)](n) \tag{1.184}$$

whose z-transform is given by

$$P_h(z) = \sum_{n=-\infty}^{+\infty} \mathtt{r}_h(n)\, z^{-n} = H(z)\, H^*\left(\frac{1}{z^*}\right) \tag{1.185}$$

Table 1.6: Relations between ACS and PSD for discrete-time processes through a linear filter.

ACS	PSD
$\mathtt{r}_{yx}(n) = \mathtt{r}_x * h(n)$	$P_{yx}(z) = P_x(z)H(z)$
$\mathtt{r}_{xy}(n) = [\mathtt{r}_x(m) * h^*(-m)](n)$	$P_{xy}(z) = P_x(z)H^*(1/z^*)$
$\mathtt{r}_y(n) = \mathtt{r}_{xy} * h(n)$	$P_y(z) = P_{xy}(z)H(z)$
$= \mathtt{r}_x * \mathtt{r}_h(n)$	$= P_x(z)H(z)H^*(1/z^*)$

[12] We use the same symbol to indicate the correlation between random processes and the correlation between deterministic signals.

In case $P_h(z)$ is a rational function, from (1.185) one deduces that, if $P_h(z)$ has a pole (zero) of the type $e^{j\varphi}|a|$, it also has a corresponding pole (zero) of the type $e^{j\varphi}/|a|$. Consequently, the poles (and zeros) of $P_h(z)$ come in pairs of the type $e^{j\varphi}|a|$, $e^{j\varphi}/|a|$.

From the last relation in Table 1.6, we obtain the relation between the PSDs of input and output signals, i.e.

$$\mathcal{P}_y(f) = \mathcal{P}_x(f)\left|H(e^{j2\pi fT_c})\right|^2 \tag{1.186}$$

In the case of white noise input

$$P_y(z) = \sigma_x^2 H(z) H^*\left(\frac{1}{z^*}\right) \tag{1.187}$$

and

$$\mathcal{P}_y(f) = T_c \sigma_x^2 \left|H(e^{j2\pi fT_c})\right|^2 \tag{1.188}$$

In other words, $\mathcal{P}_y(f)$ has the same shape as the filter frequency response.

In the case of real filters

$$H^*\left(\frac{1}{z^*}\right) = H(z^{-1}) \tag{1.189}$$

Among the various applications of (1.188), it is worth mentioning the *process synthesis*, which deals with the generation of a random process having a pre-assigned PSD. Two methods are shown in Section 4.1.9.

Minimum-phase spectral factorization

In the previous section, we introduced the relation between an impulse response $\{h(k)\}$ and its ACS $\{\mathtt{r}_h(n)\}$ in terms of the z-transform. In many practical applications, it is interesting to determine the minimum-phase impulse response for a given autocorrelation function: with this intent we state the following theorem [5].

Theorem 1.4 (Spectral factorization for discrete-time processes)
Consider the process y with ACS $\{\mathtt{r}_y(n)\}$ having z-transform $P_y(z)$, which satisfies the Paley–Wiener condition for discrete-time systems, i.e.

$$\int_{1/T_c} \left|\ln P_y(e^{j2\pi fT_c})\right| df < \infty \tag{1.190}$$

where the integration is over an arbitrarily chosen interval $1/T_c$. Then the function $P_y(z)$ can be factorized as follows:

$$P_y(z) = f_0^2\, \tilde{F}(z)\, \tilde{F}^*\left(\frac{1}{z^*}\right) \tag{1.191}$$

where

$$\tilde{F}(z) = 1 + \tilde{f}_1 z^{-1} + \tilde{f}_2 z^{-2} + \cdots \tag{1.192}$$

is monic, minimum phase, and associated with a causal sequence $\{1, \tilde{f}_1, \tilde{f}_2, \dots\}$. The factor f_0 in (1.191) is the geometric mean of $P_y(e^{j2\pi fT_c})$:

$$\ln f_0^2 = T_c \int_{1/T_c} \ln P_y(e^{j2\pi fT_c})\, df \tag{1.193}$$

The logarithms in (1.190) and (1.193) may have any common base.

The Paley–Wiener criterion implies that $P_y(z)$ may have only a discrete set of zeros on the unit circle, and that the spectral factorization (1.191) (with the constraint that $\tilde{F}(z)$ is causal, monic and minimum phase) is unique. For rational $P_y(z)$, the function $f_0\tilde{F}(z)$ is obtained by extracting the poles and zeros of $P_y(z)$ that lie inside the unit circle (see (1.453) and the considerations relative to (1.185)). Moreover, in (1.191) $f_0\tilde{F}^*(1/z^*)$ is the z-transform of an anticausal sequence $f_0\{\dots,\tilde{f}_2^*,\tilde{f}_1^*,1\}$, associated with poles and zeros of $P_y(z)$ that lie outside the unit circle. □

1.5.4 PSD of passband processes

Definition 1.13
A WSS random process x is said to be PB (BB) if its PSD is of PB (BB) type. □

PSD of in-phase and quadrature components

Let x be a real PB WSS process. Our aim is to derive the PSD of the in-phase and quadrature components of the process. We assume that x does not have direct current (DC) components, i.e. a frequency component at $f = 0$, hence, its mean is zero and consequently also $x^{(a)}$ and $x^{(bb)}$ have zero mean.

We introduce the two (ideal) filters with frequency response

$$\mathcal{H}^{(+)}(f) = 1(f) \quad \text{and} \quad \mathcal{H}^{(-)}(f) = 1(-f) \tag{1.194}$$

Note that they have non-overlapping passbands. For the same input x, the output of the two filters is, respectively, $x^{(+)}$ and $x^{(-)}$. We find that

$$x(t) = x^{(+)}(t) + x^{(-)}(t) \tag{1.195}$$

with $x^{(-)}(t) = x^{(+)*}(t)$. The following relations hold

$$\mathcal{P}_{x^{(+)}}(f) = |\mathcal{H}^{(+)}(f)|^2\mathcal{P}_x(f) = \mathcal{P}_x(f)1(f) \tag{1.196}$$

$$\mathcal{P}_{x^{(-)}}(f) = |\mathcal{H}^{(-)}(f)|^2\mathcal{P}_x(f) = \mathcal{P}_x(f)1(-f) \tag{1.197}$$

and

$$\mathcal{P}_{x^{(+)}x^{(-)}}(f) = 0 \tag{1.198}$$

as $x^{(+)}$ and $x^{(-)}$ have non-overlapping passbands. Then $x^{(+)} \perp x^{(-)}$, and (1.195) yields

$$\mathcal{P}_x(f) = \mathcal{P}_{x^{(+)}}(f) + \mathcal{P}_{x^{(-)}}(f) \tag{1.199}$$

where $\mathcal{P}_{x^{(-)}}(f) = \mathcal{P}_{x^{(+)*}}(f) = \mathcal{P}_{x^{(+)}}(-f)$, using Property 5 of the PSD. The analytic signal $x^{(a)}$ is equal to $2x^{(+)}$, hence,

$$\mathtt{r}_{x^{(a)}}(\tau) = 4\mathtt{r}_{x^{(+)}}(\tau) \tag{1.200}$$

and

$$\mathcal{P}_{x^{(a)}}(f) = 4\mathcal{P}_{x^{(+)}}(f) \tag{1.201}$$

Moreover, being $x^{(a)*} = 2x^{(-)}$, it follows that $x^{(a)} \perp x^{(a)*}$ and

$$\mathtt{r}_{x^{(a)}x^{(a)*}}(\tau) = 0 \tag{1.202}$$

The complex envelope $x^{(bb)}$ is related to $x^{(a)}$ by (1.86) and

$$\mathtt{r}_{x^{(bb)}}(\tau) = \mathtt{r}_{x^{(a)}}(\tau)e^{-j2\pi f_0\tau} \tag{1.203}$$

Hence,

$$\mathcal{P}_{x^{(bb)}}(f) = \mathcal{P}_{x^{(a)}}(f + f_0) = 4\mathcal{P}_{x^{(+)}}(f + f_0) \tag{1.204}$$

Moreover, from (1.202), it follows that $x^{(bb)} \perp x^{(bb)*}$.

Using (1.204), (1.199) can be written as

$$\mathcal{P}_x(f) = \frac{1}{4}[\mathcal{P}_{x^{(bb)}}(f - f_0) + \mathcal{P}_{x^{(bb)}}(-f - f_0)] \tag{1.205}$$

Finally, from

$$x_I^{(bb)}(t) = Re\,[x^{(bb)}(t)] = \frac{x^{(bb)}(t) + x^{(bb)*}(t)}{2} \tag{1.206}$$

and

$$x_Q^{(bb)}(t) = Im\,[x^{(bb)}(t)] = \frac{x^{(bb)}(t) - x^{(bb)*}(t)}{2j} \tag{1.207}$$

we obtain the following relations:

$$\mathtt{r}_{x_I^{(bb)}}(\tau) = \frac{1}{2}Re\,[\mathtt{r}_{x^{(bb)}}(\tau)] \tag{1.208}$$

$$\mathcal{P}_{x_I^{(bb)}}(f) = \frac{1}{4}[\mathcal{P}_{x^{(bb)}}(f) + \mathcal{P}_{x^{(bb)}}(-f)] \tag{1.209}$$

$$\mathtt{r}_{x_Q^{(bb)}}(\tau) = \mathtt{r}_{x_I^{(bb)}}(\tau) \tag{1.210}$$

$$\mathtt{r}_{x_Q^{(bb)} x_I^{(bb)}}(\tau) = \frac{1}{2}Im\,[\mathtt{r}_{x^{(bb)}}(\tau)] \tag{1.211}$$

$$\mathcal{P}_{x_Q^{(bb)} x_I^{(bb)}}(f) = \frac{1}{4j}[\mathcal{P}_{x^{(bb)}}(f) - \mathcal{P}_{x^{(bb)}}(-f)] \tag{1.212}$$

$$\mathtt{r}_{x_I^{(bb)} x_Q^{(bb)}}(\tau) = -\mathtt{r}_{x_Q^{(bb)} x_I^{(bb)}}(\tau) = -\mathtt{r}_{x_I^{(bb)} x_Q^{(bb)}}(-\tau) \tag{1.213}$$

The second equality in (1.213) follows from Property 4 of ACS.

From (1.213), we note that $\mathtt{r}_{x_I^{(bb)} x_Q^{(bb)}}(\tau)$ is an odd function. Moreover, from (1.212), we obtain $x_I^{(bb)} \perp x_Q^{(bb)}$ only if $\mathcal{P}_{x^{(bb)}}$ is an even function; in any case, the random variables $x_I^{(bb)}(t)$ and $x_Q^{(bb)}(t)$ are always orthogonal since $\mathtt{r}_{x_I^{(bb)} x_Q^{(bb)}}(0) = 0$. Referring to the block diagram in Figure 1.18b, as

$$\mathcal{P}_{x^{(h)}}(f) = \mathcal{P}_x(f) \quad \text{and} \quad \mathcal{P}_{x^{(h)}x}(f) = -j\,\mathrm{sgn}(f)\,\mathcal{P}_x(f) \tag{1.214}$$

we obtain

$$\mathtt{r}_{x^{(h)}}(\tau) = \mathtt{r}_x(\tau) \quad \text{and} \quad \mathtt{r}_{x^{(h)}x}(\tau) = \mathtt{r}_x^{(h)}(\tau) \tag{1.215}$$

Then

$$\mathtt{r}_{x_I^{(bb)}}(\tau) = \mathtt{r}_{x_Q^{(bb)}}(\tau) = \mathtt{r}_x(\tau)\cos(2\pi f_0\tau) + \mathtt{r}_x^{(h)}(\tau)\sin(2\pi f_0\tau) \tag{1.216}$$

and

$$\mathtt{r}_{x_I^{(bb)} x_Q^{(bb)}}(\tau) = -\mathtt{r}_x^{(h)}(\tau)\cos(2\pi f_0\tau) + \mathtt{r}_x(\tau)\sin(2\pi f_0\tau) \tag{1.217}$$

In terms of statistical power, the following relations hold:

$$\mathtt{r}_{x^{(+)}}(0) = \mathtt{r}_{x^{(-)}}(0) = \frac{1}{2}\mathtt{r}_x(0) \tag{1.218}$$

$$\mathtt{r}_{x^{(bb)}}(0) = \mathtt{r}_{x^{(a)}}(0) = 4\mathtt{r}_{x^{(+)}}(0) = 2\mathtt{r}_x(0) \tag{1.219}$$

$$\mathtt{r}_{x_I^{(bb)}}(0) = \mathtt{r}_{x_Q^{(bb)}}(0) = \mathtt{r}_x(0) \tag{1.220}$$

$$\mathtt{r}_{x^{(h)}}(0) = \mathtt{r}_x(0) \tag{1.221}$$

Example 1.5.2

Let x be a WSS process with PSD

$$\mathcal{P}_x(f) = \frac{N_0}{2}\left[\mathrm{rect}\left(\frac{f - f_0}{B}\right) + \mathrm{rect}\left(\frac{f + f_0}{B}\right)\right] \tag{1.222}$$

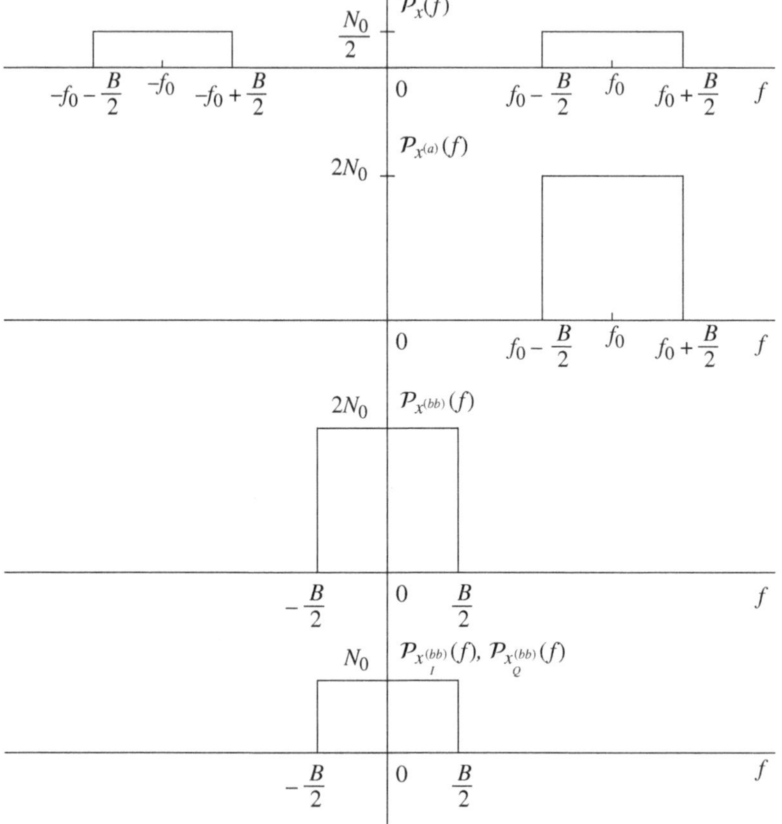

Figure 1.23 Spectral representation of a PB process and its BB components.

depicted in Figure 1.23. It is immediate to get

$$\mathcal{P}_{x^{(a)}}(f) = 2N_0 \text{rect}\left(\frac{f-f_0}{B}\right) \tag{1.223}$$

and

$$\mathcal{P}_{x^{(bb)}}(f) = 2N_0 \text{rect}\left(\frac{f}{B}\right) \tag{1.224}$$

Then

$$\mathcal{P}_{x_I^{(bb)}}(f) = \mathcal{P}_{x_Q^{(bb)}}(f) = \frac{1}{2}\mathcal{P}_{x^{(bb)}}(f) = N_0 \text{rect}\left(\frac{f}{B}\right) \tag{1.225}$$

Moreover, being $\mathcal{P}_{x_I^{(bb)} x_Q^{(bb)}}(f) = 0$, we have that $x_I^{(bb)} \perp x_Q^{(bb)}$.

Cyclostationary processes

We have seen that, if x is a real passband WSS process, then its complex envelope is WSS, and $x^{(bb)} \perp x^{(bb)*}$. The converse is also true: if $x^{(bb)}$ is a WSS process and $x^{(bb)} \perp x^{(bb)*}$, then

$$x(t) = Re\,[x^{(bb)}(t)\; e^{j2\pi f_0 t}] \tag{1.226}$$

is WSS with PSD given by (1.205). If $x^{(bb)}$ is WSS, however, with

$$\mathtt{r}_{x^{(bb)}x^{(bb)*}}(\tau) \neq 0 \tag{1.227}$$

observing (1.226) we find that the autocorrelation of x is a periodic function in t of period $1/f_0$:

$$\mathtt{r}_x(t, t-\tau) = \frac{1}{4}\left[\mathtt{r}_{x^{(bb)}}(\tau)e^{j2\pi f_0\tau} + \mathtt{r}^*_{x^{(bb)}}(\tau)e^{-j2\pi f_0\tau} + \mathtt{r}_{x^{(bb)}x^{(bb)*}}(\tau)e^{-j2\pi f_0\tau}e^{j4\pi f_0 t} + \mathtt{r}^*_{x^{(bb)}x^{(bb)*}}(\tau)e^{j2\pi f_0\tau}e^{-j4\pi f_0 t}\right] \tag{1.228}$$

In other words, x is a *cyclostationary* process of period $T_0 = 1/f_0$.[13]

In this case, it is convenient to introduce the *average correlation*

$$\bar{\mathtt{r}}_x(\tau) = \frac{1}{T_0}\int_0^{T_0} \mathtt{r}_x(t, t-\tau)dt \tag{1.229}$$

whose Fourier transform is the *average PSD*

$$\overline{\mathcal{P}}_x(f) = \mathcal{F}[\bar{\mathtt{r}}_x(\tau)] = \frac{1}{T_0}\int_0^{T_0} \mathcal{P}_x(f, t)dt \tag{1.230}$$

where

$$\mathcal{P}_x(f, t) = \mathcal{F}_\tau[\mathtt{r}_x(t, t-\tau)] \tag{1.231}$$

In (1.231), $\mathcal{F}_\tau$ denotes the Fourier transform with respect to the variable τ. In our case, it is

$$\overline{\mathcal{P}}_x(f) = \frac{1}{4}[\mathcal{P}_{x^{(bb)}}(f - f_0) + \mathcal{P}_{x^{(bb)}}(-f - f_0)] \tag{1.232}$$

as in the stationary case (1.205).

Example 1.5.3

Let x be a modulated DSB signal (see (1.123)), i.e.

$$x(t) = a(t)\cos(2\pi f_0 t + \varphi_0) \tag{1.233}$$

with a real random BB WSS process with bandwidth $B_a < f_0$ and autocorrelation $\mathtt{r}_a(\tau)$. From (1.126) it results $x^{(bb)}(t) = a(t)\, e^{j\varphi_0}$. Hence, we have

$$\mathtt{r}_{x^{(bb)}}(\tau) = \mathtt{r}_a(\tau), \qquad \mathtt{r}_{x^{(bb)}x^{(bb)*}}(\tau) = \mathtt{r}_a(\tau)\, e^{j2\varphi_0} \tag{1.234}$$

Because $\mathtt{r}_a(\tau)$ is not identically zero, observing (1.227) we find that x is cyclostationary with period $1/f_0$. From (1.232), the average PSD of x is given by

$$\overline{\mathcal{P}}_x(f) = \frac{1}{4}[\mathcal{P}_a(f - f_0) + \mathcal{P}_a(f + f_0)] \tag{1.235}$$

Therefore, x has a bandwidth equal to $2B_a$ and an average statistical power

$$\overline{\mathtt{M}}_x = \frac{1}{2}\,\mathtt{M}_a \tag{1.236}$$

We note that one finds the same result (1.235) assuming that φ_0 is a uniform r.v. in $[0, 2\pi)$; in this case x turns out to be WSS.

Example 1.5.4

Let x be a modulated *single sideband* (SSB) with an upper sideband, i.e.

$$\begin{aligned} x(t) &= Re\left[\frac{1}{2}\left(a(t) + ja^{(h)}(t)\right)e^{j(2\pi f_0 t + \varphi_0)}\right] \\ &= \frac{1}{2}\,a(t)\cos(2\pi f_0 t + \varphi_0) - \frac{1}{2}\,a^{(h)}(t)\sin(2\pi f_0 t + \varphi_0) \end{aligned} \tag{1.237}$$

[13] To be precise, x is cyclostationary in mean value with period $T_0 = 1/f_0$, while it is cyclostationary in correlation with period $T_0/2$.

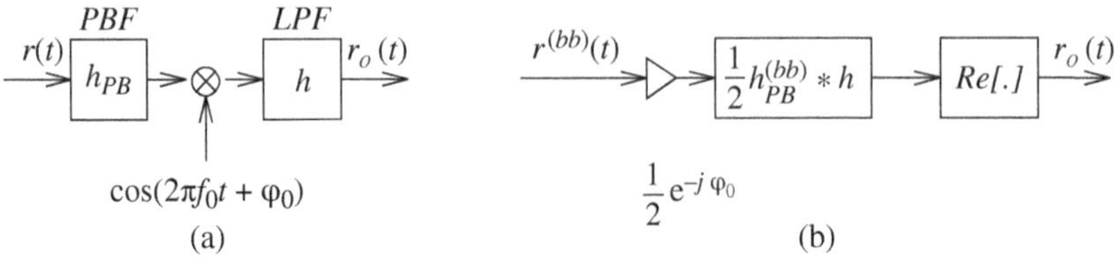

Figure 1.24 Coherent DSB demodulator and baseband-equivalent scheme. (a) Coherent DSB demodulator and (b) baseband-equivalent scheme.

where $a^{(h)}$ is the Hilbert transform of a, a real WSS random process with autocorrelation $\mathtt{r}_a(\tau)$ and bandwidth B_a.

We note that the modulating signal $(a(t) + ja^{(h)}(t))$ coincides with the analytic signal $a^{(a)}$ and its spectral support contains only positive frequencies.

Being

$$x^{(bb)}(t) = \frac{1}{2}(a(t) + ja^{(h)}(t))e^{j\varphi_0}$$

it results that $x^{(bb)}$ and $x^{(bb)*}$ have non-overlapping passbands and

$$\mathtt{r}_{x^{(bb)}x^{(bb)*}}(\tau) = 0 \tag{1.238}$$

The process (1.237) is then stationary with

$$\mathcal{P}_x(f) = \frac{1}{4}[\mathcal{P}_{a^{(+)}}(f - f_0) + \mathcal{P}_{a^{(+)}}(-f - f_0)] \tag{1.239}$$

where $a^{(+)}$ is defined in (1.195). In this case, x has bandwidth equal to B_a and statistical power given by

$$\mathtt{M}_x = \frac{1}{4}\,\mathtt{M}_a \tag{1.240}$$

Example 1.5.5 (DSB and SSB demodulators)

Let the signal r be the sum of a desired part x and additive white noise w with PSD equal to $\mathcal{P}_w(f) = N_0/2$,

$$r(t) = x(t) + w(t) \tag{1.241}$$

where the signal x is modulated DSB (1.233). To obtain the signal a from r, one can use the coherent demodulation scheme illustrated in Figure 1.24 (see Figure 1.21b), where h is an ideal lowpass filter, having a frequency response

$$\mathcal{H}(f) = \mathcal{H}_0 \,\text{rect}\left(\frac{f}{2B_a}\right) \tag{1.242}$$

Let r_o be the output signal of the demodulator, given by the sum of the desired part x_o and noise w_o:

$$r_o(t) = x_o(t) + w_o(t) \tag{1.243}$$

We evaluate now the ratio between the powers of the signals in (1.243),

$$\Lambda_o = \frac{\mathtt{M}_{x_o}}{\mathtt{M}_{w_o}} \tag{1.244}$$

in terms of the reference signal-to-noise ratio

$$\Gamma = \frac{\mathtt{M}_x}{(N_0/2)\,2B_a} \tag{1.245}$$

Using the equivalent block scheme of Figure 1.24 and (1.126), we have

$$r^{(bb)}(t) = a(t)\, e^{j\varphi_0} + w^{(bb)}(t) \tag{1.246}$$

with $\mathcal{P}_{w^{(bb)}}(f) = 2N_0\; 1(f + f_0)$. Being

$$h * a(t) = \mathcal{H}_0\; a(t) \tag{1.247}$$

it results

$$\begin{aligned} x_o(t) &= Re\left[h * \frac{1}{2}\; e^{-j\varphi_1}\; a\, e^{j\varphi_0}\right](t) \\ &= \frac{\mathcal{H}_0}{2}\; a(t)\cos(\varphi_0 - \varphi_1) \end{aligned} \tag{1.248}$$

Hence, we get

$$\mathtt{M}_{x_o} = \frac{\mathcal{H}_0^2}{4}\; \mathtt{M}_a \cos^2(\varphi_0 - \varphi_1) \tag{1.249}$$

In the same baseband equivalent scheme, we consider the noise w_{eq} at the output of filter h; we find

$$\begin{aligned} \mathcal{P}_{w_{eq}}(f) &= \frac{1}{4}\; |\mathcal{H}(f)|^2\; 2N_0\; 1(f + f_0) \\ &= \frac{\mathcal{H}_0^2}{2}\; N_0 \operatorname{rect}\left(\frac{f}{2B_a}\right) \end{aligned} \tag{1.250}$$

Being now w WSS, $w^{(bb)}$ is uncorrelated with $w^{(bb)*}$ and thus w_{eq} with w_{eq}^*. Then, from

$$w_o(t) = w_{eq,I}(t) \tag{1.251}$$

and using (1.209) it follows

$$\mathcal{P}_{w_o}(f) = \frac{\mathcal{H}_0^2}{4}\; N_0 \operatorname{rect}\left(\frac{f}{2B_a}\right) \tag{1.252}$$

and

$$\mathtt{M}_{w_o} = \frac{\mathcal{H}_0^2}{4}\; N_0\; 2B_a \tag{1.253}$$

In conclusion, using (1.236), we have

$$\Lambda_o = \frac{(\mathcal{H}_0^2/4)\; \mathtt{M}_a \cos^2(\varphi_0 - \varphi_1)}{(\mathcal{H}_0^2/4)\; N_0\; 2B_a} = \Gamma \cos^2(\varphi_0 - \varphi_1) \tag{1.254}$$

For $\varphi_1 = \varphi_0$ (1.254) becomes

$$\Lambda_o = \Gamma \tag{1.255}$$

It is interesting to observe that, at the demodulator input, the ratio between the power of the desired signal and the power of the noise in the *passband of* x is given by

$$\Lambda_i = \frac{\mathtt{M}_x}{(N_0/2)\; 4B_a} = \frac{\Gamma}{2} \tag{1.256}$$

For $\varphi_1 = \varphi_0$ then

$$\Lambda_o = 2\Lambda_i \tag{1.257}$$

We will now analyse the case of a SSB signal x (see (1.237)), coherently demodulated, following the scheme of Figure 1.25, where h_{PB} is a filter used to eliminate the noise that otherwise, after the *mixer*, would have fallen within the passband of the desired signal. The ideal frequency response of h_{PB} is given by

$$\mathcal{H}_{PB}(f) = \operatorname{rect}\left(\frac{f - f_0 - B_a/2}{B_a}\right) + \operatorname{rect}\left(\frac{-f - f_0 - B_a/2}{B_a}\right) \tag{1.258}$$

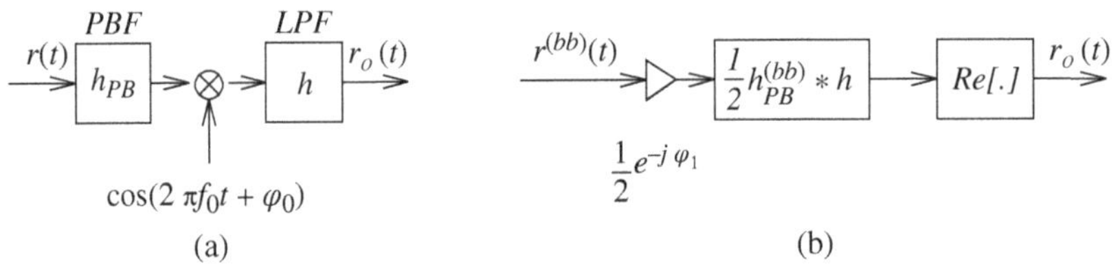

Figure 1.25 (a) Coherent SSB demodulator and (b) baseband-equivalent scheme.

Note that in this scheme, we have assumed the phase of the receiver carrier equal to that of the transmitter, to avoid distortion of the desired signal.

Being

$$\mathcal{H}_{PB}^{(bb)}(f) = 2\,\mathrm{rect}\left(\frac{f - B_a/2}{B_a}\right) \tag{1.259}$$

the filter of the baseband-equivalent scheme is given by

$$h_{eq}(t) = \frac{1}{2}\, h_{PB}^{(bb)} * h(t) \tag{1.260}$$

with frequency response

$$\mathcal{H}_{eq}(f) = \mathcal{H}_0\,\mathrm{rect}\left(\frac{f - B_a/2}{B_a}\right) \tag{1.261}$$

We now evaluate the desired component x_o. Using the fact $x^{(bb)} * h_{eq}(t) = \mathcal{H}_0\, x^{(bb)}(t)$, it results

$$\begin{aligned} x_o(t) &= Re\left[h_{eq} * \frac{1}{2}\, e^{-j\varphi_0}\, \frac{1}{2}(a + j\, a^{(h)})\, e^{j\varphi_0}\right](t) \\ &= \frac{\mathcal{H}_0}{4}\, Re\left[a(t) + j\, a^{(h)}(t)\right] = \frac{\mathcal{H}_0}{4}\, a(t) \end{aligned} \tag{1.262}$$

In the baseband-equivalent scheme, the noise w_{eq} at the output of h_{eq} has a PSD given by

$$\mathcal{P}_{w_{eq}}(f) = \frac{1}{4}\, |\mathcal{H}_{eq}(f)|^2\, 2N_0\, 1(f + f_0) = \frac{N_0}{2}\, \mathcal{H}_0^2\,\mathrm{rect}\left(\frac{f - B_a/2}{B_a}\right) \tag{1.263}$$

From the relation $w_o = w_{eq,I}$ and using (1.209), which is valid because $w_{eq} \perp w_{eq}^*$, we have

$$\mathcal{P}_{w_o}(f) = \frac{1}{4}[\mathcal{P}_{w_{eq}}(f) + \mathcal{P}_{w_{eq}}(-f)] = \frac{\mathcal{H}_0^2}{8}\, N_0\,\mathrm{rect}\left(\frac{f}{2B_a}\right) \tag{1.264}$$

and

$$\mathtt{M}_{w_o} = \frac{\mathcal{H}_0^2}{8}\, N_0\, 2B_a \tag{1.265}$$

Then we obtain

$$\Lambda_o = \frac{(\mathcal{H}_0^2/16)\,\mathtt{M}_a}{(\mathcal{H}_0^2/8)\, N_0\, 2B_a} \tag{1.266}$$

which using (1.240) and (1.245) can be written as

$$\Lambda_o = \Gamma \tag{1.267}$$

We note that the SSB system yields the same performance (for $\varphi_1 = \varphi_0$) of a DSB system, even though half of the bandwidth is required. Finally, it results

$$\Lambda_i = \frac{\mathtt{M}_x}{(N_0/2)\, 2B_a} = \Lambda_o \tag{1.268}$$

Observation 1.4
We note that also for the simple examples considered in this section, the desired signal is analysed via the various transformations, whereas the noise is analysed via the PSD. As a matter of fact, we are typically interested only in the statistical power of the noise at the system output. The demodulated signal x_o, on the other hand, must be expressed as the sum of a desired component proportional to a and an orthogonal component that represents the distortion, which is, typically, small and has the same effects as noise.

In the previous example, the considered systems do not introduce any distortion since x_o is proportional to a.

1.6 The autocorrelation matrix

Definition 1.14
Given the discrete-time wide-sense stationary random process $\{x(k)\}$, we introduce the random vector with N components

$$\boldsymbol{x}^T(k) = [x(k), x(k-1), \dots, x(k-N+1)] \tag{1.269}$$

The $N \times N$ autocorrelation matrix of $\boldsymbol{x}^*(k)$ is given by

$$\boldsymbol{R} = E[\boldsymbol{x}^*(k)\boldsymbol{x}^T(k)] = \begin{bmatrix} \mathtt{r}_x(0) & \mathtt{r}_x(-1) & \dots & \mathtt{r}_x(-N+1) \\ \mathtt{r}_x(1) & \mathtt{r}_x(0) & \dots & \mathtt{r}_x(-N+2) \\ \vdots & \vdots & \ddots & \dots \\ \mathtt{r}_x(N-1) & \mathtt{r}_x(N-2) & \dots & \mathtt{r}_x(0) \end{bmatrix}. \tag{1.270}$$

□

Properties

1. $\boldsymbol{R}$ is *Hermitian*: $\boldsymbol{R}^H = \boldsymbol{R}$. For real random processes $\boldsymbol{R}$ is *symmetric*: $\boldsymbol{R}^T = \boldsymbol{R}$.
2. $\boldsymbol{R}$ is a *Toeplitz matrix*, i.e. all elements along any diagonal are equal.
3. $\boldsymbol{R}$ is *positive semi-definite* and almost always *positive definite*. Indeed, taking an arbitrary vector $\boldsymbol{v}^T = [v_0, \dots, v_{N-1}]$, and letting $y_k = \boldsymbol{x}^T(k)\boldsymbol{v}$, we have

$$E[|y_k|^2] = E[\boldsymbol{v}^H\boldsymbol{x}^*(k)\boldsymbol{x}^T(k)\boldsymbol{v}] = \boldsymbol{v}^H\boldsymbol{R}\boldsymbol{v} = \sum_{i=0}^{N-1}\sum_{j=0}^{N-1} v_i^*\mathtt{r}_x(i-j)v_j \geq 0 \tag{1.271}$$

If $\boldsymbol{v}^H\boldsymbol{R}\boldsymbol{v} > 0$, $\forall \boldsymbol{v} \neq \boldsymbol{0}$, then $\boldsymbol{R}$ is said to be *positive definite* and all its *principal minor* determinants are positive; in particular $\boldsymbol{R}$ is *non-singular*.

Eigenvalues

We indicate with $\det \boldsymbol{R}$ the determinant of a matrix $\boldsymbol{R}$. The eigenvalues of $\boldsymbol{R}$ are the solutions λ_i, $i = 1, \dots, N$, of the *characteristic equation* of order N

$$\det[\boldsymbol{R} - \lambda \boldsymbol{I}] = 0 \tag{1.272}$$

and the corresponding column eigenvectors $\boldsymbol{u}_i$, $i = 1, \dots, N$, satisfy the equation

$$\boldsymbol{R}\boldsymbol{u}_i = \lambda_i \boldsymbol{u}_i \tag{1.273}$$

Table 1.7: Correspondence between eigenvalues and eigenvectors of four matrices.

	$\boldsymbol{R}$	$\boldsymbol{R}^m$	$\boldsymbol{R}^{-1}$	$\boldsymbol{I}-\mu\boldsymbol{R}$
Eigenvalue	λ_i	λ_i^m	λ_i^{-1}	$(1-\mu\lambda_i)$
Eigenvector	$\boldsymbol{u}_i$	$\boldsymbol{u}_i$	$\boldsymbol{u}_i$	$\boldsymbol{u}_i$

Example 1.6.1

Let $\{w(k)\}$ be a white noise process. Its autocorrelation matrix $\boldsymbol{R}$ assumes the form

$$\boldsymbol{R} = \begin{bmatrix} \sigma_w^2 & 0 & \dots & 0 \\ 0 & \sigma_w^2 & \dots & 0 \\ \vdots & \vdots & \ddots & \vdots \\ 0 & 0 & \dots & \sigma_w^2 \end{bmatrix} \tag{1.274}$$

from which it follows that

$$\lambda_1 = \lambda_2 = \cdots = \lambda_N = \sigma_w^2 \tag{1.275}$$

and

$$\boldsymbol{u}_i \text{ can be any arbitrary vector} \quad 1 \le i \le N \tag{1.276}$$

Example 1.6.2

We define a complex-valued sinusoid as

$$x(k) = e^{j(\omega k+\varphi)}, \qquad \omega = 2\pi f T_c \tag{1.277}$$

with φ a uniform r.v. in $[0, 2\pi)$. The autocorrelation matrix $\boldsymbol{R}$ is given by

$$\boldsymbol{R} = \begin{bmatrix} 1 & e^{-j\omega} & \dots & e^{-j(N-1)\omega} \\ e^{j\omega} & 1 & \dots & e^{-j(N-2)\omega} \\ \vdots & \vdots & \ddots & \vdots \\ e^{j(N-1)\omega} & e^{j(N-2)\omega} & \dots & 1 \end{bmatrix} \tag{1.278}$$

One can see that the rank of $\boldsymbol{R}$ is 1 and it will therefore have only one eigenvalue. The solution is given by

$$\lambda_1 = N \tag{1.279}$$

and the relative eigenvector is

$$\boldsymbol{u}_1^T = [1, e^{j\omega}, \dots, e^{j(N-1)\omega}] \tag{1.280}$$

Other properties

1. From $\boldsymbol{R}^m\boldsymbol{u} = \lambda^m\boldsymbol{u}$, we obtain the relations of Table 1.7.
2. If the eigenvalues are distinct, then the eigenvectors are linearly independent:

$$\sum_{i=1}^{N} c_i\boldsymbol{u}_i \neq \boldsymbol{0} \tag{1.281}$$

for all combinations of $\{c_i\}$, $i = 1, 2, \dots, N$, not all equal to zero. Therefore, in this case, the eigenvectors form a basis in $\mathbb{R}^N$.

3. The *trace* of a matrix $\boldsymbol{R}$ is defined as the sum of the elements of the main diagonal, and we indicate it with $\text{tr}\,\boldsymbol{R}$. It holds

$$\text{tr}\,\boldsymbol{R} = \sum_{i=1}^{N} \lambda_i \tag{1.282}$$

Eigenvalue analysis for Hermitian matrices

As previously seen, the autocorrelation matrix $\boldsymbol{R}$ is Hermitian, thus enjoys the following properties:

1. The eigenvalues of a Hermitian matrix are real.
   By left multiplying both sides of (1.273) by $\boldsymbol{u}_i^H$, it follows

   $$\boldsymbol{u}_i^H \boldsymbol{R} \boldsymbol{u}_i = \lambda_i \boldsymbol{u}_i^H \boldsymbol{u}_i \tag{1.283}$$

   from which, by the definition of norm, we obtain

   $$\lambda_i = \frac{\boldsymbol{u}_i^H \boldsymbol{R} \boldsymbol{u}_i}{\boldsymbol{u}_i^H \boldsymbol{u}_i} = \frac{\boldsymbol{u}_i^H \boldsymbol{R} \boldsymbol{u}_i}{\|\boldsymbol{u}_i\|^2} \tag{1.284}$$

   The ratio (1.284) is defined as *Rayleigh quotient*. As $\boldsymbol{R}$ is positive semi-definite, $\boldsymbol{u}_i^H \boldsymbol{R} \boldsymbol{u}_i \geq 0$, from which $\lambda_i \geq 0$.
2. If the eigenvalues of $\boldsymbol{R}$ are distinct, then the eigenvectors are orthogonal. In fact, from (1.273), we obtain:

   $$\boldsymbol{u}_i^H \boldsymbol{R} \boldsymbol{u}_j = \lambda_j \boldsymbol{u}_i^H \boldsymbol{u}_j \tag{1.285}$$

   $$\boldsymbol{u}_i^H \boldsymbol{R} \boldsymbol{u}_j = \lambda_i \boldsymbol{u}_i^H \boldsymbol{u}_j \tag{1.286}$$

   Subtracting the second equation from the first:

   $$0 = (\lambda_j - \lambda_i) \boldsymbol{u}_i^H \boldsymbol{u}_j \tag{1.287}$$

   and since $\lambda_j - \lambda_i \neq 0$ by hypothesis, it follows $\boldsymbol{u}_i^H \boldsymbol{u}_j = 0$.
3. If the eigenvalues of $\boldsymbol{R}$ are distinct and their corresponding eigenvectors are normalized, i.e.

   $$\|\boldsymbol{u}_i\|^2 = \boldsymbol{u}_i^H \boldsymbol{u}_i = \begin{cases} 1 & i = j \\ 0 & i \neq j \end{cases} \tag{1.288}$$

   then the matrix $\boldsymbol{U} = [\boldsymbol{u}_1, \boldsymbol{u}_2, \ldots, \boldsymbol{u}_N]$, whose columns are the eigenvectors of $\boldsymbol{R}$, is a unitary matrix, that is

   $$\boldsymbol{U}^{-1} = \boldsymbol{U}^H \tag{1.289}$$

   This property is an immediate consequence of the orthogonality of the eigenvectors $\{\boldsymbol{u}_i\}$. Moreover, if we define the matrix

   $$\boldsymbol{\Lambda} = \begin{bmatrix} \lambda_1 & 0 & \ldots & 0 \\ 0 & \lambda_2 & \ldots & 0 \\ \vdots & \vdots & \ddots & \vdots \\ 0 & 0 & \ldots & \lambda_N \end{bmatrix} \tag{1.290}$$

   we get

   $$\boldsymbol{U}^H \boldsymbol{R} \boldsymbol{U} = \boldsymbol{\Lambda} \tag{1.291}$$

   From (1.291), we obtain the following important relations:

   $$\boldsymbol{R} = \boldsymbol{U} \boldsymbol{\Lambda} \boldsymbol{U}^H = \sum_{i=1}^{N} \lambda_i \boldsymbol{u}_i \boldsymbol{u}_i^H \tag{1.292}$$

and

$$I - \mu R = U(I - \mu\Lambda)U^H = \sum_{i=1}^{N}(1 - \mu\lambda_i)u_i u_i^H \tag{1.293}$$

4. The eigenvalues of a positive semi-definite autocorrelation matrix R and the PSD of x are related by the inequalities,

$$\min_f\{\mathcal{P}_x(f)\} \leq \lambda_i \leq \max_f\{\mathcal{P}_x(f)\}, \qquad i = 1, \ldots, N \tag{1.294}$$

In fact, let $U_i(f)$ be the Fourier transform of the sequence represented by the elements of u_i, i.e.

$$U_i(f) = \sum_{n=1}^{N} u_{i,n} e^{-j2\pi f n T_c} \tag{1.295}$$

where $u_{i,n}$ is the n-th element of the eigenvector u_i. Observing that

$$u_i^H R u_i = \sum_{n=1}^{N}\sum_{m=1}^{N} u_{i,n}^* \mathtt{r}_x(n-m) u_{i,m} \tag{1.296}$$

and using (1.172) and (1.295), we have

$$\begin{aligned} u_i^H R u_i &= \int_{-\frac{1}{2T_c}}^{\frac{1}{2T_c}} \mathcal{P}_x(f) \sum_{n=1}^{N} u_{i,n}^* e^{j2\pi f n T_c} \sum_{m=1}^{N} u_{i,m} e^{-j2\pi f m T_c} df \\ &= \int_{-\frac{1}{2T_c}}^{\frac{1}{2T_c}} \mathcal{P}_x(f)\, |U_i(f)|^2 df \end{aligned} \tag{1.297}$$

Substituting the latter result in (1.284) one finds

$$\lambda_i = \frac{\displaystyle\int_{-\frac{1}{2T_c}}^{\frac{1}{2T_c}} \mathcal{P}_x(f)\, |U_i(f)|^2 df}{\displaystyle\int_{-\frac{1}{2T_c}}^{\frac{1}{2T_c}} |U_i(f)|^2 df} \tag{1.298}$$

from which (1.294) follows.

If we indicate with λ_{min} and λ_{max}, respectively, the minimum and maximum eigenvalue of R, in view of the latter point, we can define the *eigenvalue spread* as:

$$\chi(R) = \frac{\lambda_{max}}{\lambda_{min}} \leq \frac{\max_f\{\mathcal{P}_x(f)\}}{\min_f\{\mathcal{P}_x(f)\}} \tag{1.299}$$

From (1.299), we observe that $\chi(R)$ may assume large values in the case $\mathcal{P}_x(f)$ exhibits large variations. Moreover, $\chi(R)$ assumes the minimum value of 1 for a white process.

1.7 Examples of random processes

Before reviewing some important random processes, we recall the definition of Gaussian complex-valued random vector.

Example 1.7.1

A complex r.v. with a Gaussian distribution can be generated from two r.v.s. with uniform distribution (see Appendix 1.B for an illustration of the method).

Example 1.7.2

Let $\boldsymbol{x}^T = [x_1, \dots, x_N]$ be a real Gaussian random vector, each component has mean $\mathtt{m}_{x_i}$ and variance $\sigma_{x_i}^2$, denoted as $x_i \sim \mathcal{N}(\mathtt{m}_{x_i}, \sigma_{x_i}^2)$. The joint probability density function (pdf) is

$$p_{\boldsymbol{x}}(\boldsymbol{\xi}) = [(2\pi)^N \det \boldsymbol{C}_N]^{-\frac{1}{2}} e^{-\frac{1}{2}(\boldsymbol{\xi}-\boldsymbol{m}_x)^T \boldsymbol{C}_N^{-1}(\boldsymbol{\xi}-\boldsymbol{m}_x)} \tag{1.300}$$

where $\boldsymbol{\xi}^T = [\xi_1, \dots, \xi_N]$, $\boldsymbol{m}_x = E[\boldsymbol{x}]$ is the vector of its components' mean values and $\boldsymbol{C}_N = E[(\boldsymbol{x} - \boldsymbol{m}_x)(\boldsymbol{x} - \boldsymbol{m}_x)^T]$ is its covariance matrix.

Example 1.7.3

Let $\boldsymbol{x}^T = [x_{1,I} + jx_{1,Q}, \dots, x_{N,I} + jx_{N,Q}]$ be a complex-valued Gaussian random vector. If the in-phase component $x_{i,I}$ and the quadrature component $x_{i,Q}$ are uncorrelated,

$$E[(x_{i,I} - \mathtt{m}_{x_{i,I}})(x_{i,Q} - \mathtt{m}_{x_{i,Q}})] = 0, \qquad i = 1, 2, \dots, N \tag{1.301}$$

Moreover, we have

$$\sigma_{x_{i,I}}^2 = \sigma_{x_{i,Q}}^2 = \frac{1}{2}\sigma_{x_i}^2 \tag{1.302}$$

then the joint pdf is

$$p_{\boldsymbol{x}}(\boldsymbol{\xi}) = [\pi^N \det \boldsymbol{C}_N]^{-1} e^{-(\boldsymbol{\xi}-\boldsymbol{m}_x)^H \boldsymbol{C}_N^{-1}(\boldsymbol{\xi}-\boldsymbol{m}_x)} \tag{1.303}$$

with the vector of mean values and the covariance matrix given by

$$\boldsymbol{m}_x = E[\boldsymbol{x}] = E[\boldsymbol{x}_I] + jE[\boldsymbol{x}_Q] \tag{1.304}$$

$$\boldsymbol{C}_N = E[(\boldsymbol{x} - \boldsymbol{m}_x)(\boldsymbol{x} - \boldsymbol{m}_x)^H] \tag{1.305}$$

Vector $\boldsymbol{x}$ is called *circularly symmetric Gaussian* random vector. For the generic component, we write $x_i \sim \mathcal{CN}(\mathtt{m}_{x_i}, \sigma_{x_i}^2)$ and

$$p_{x_i}(\xi_i) = \frac{1}{\sqrt{2\pi\sigma_{x_{i,I}}^2}} e^{-\frac{|\xi_{i,I} - \mathtt{m}_{x_{i,I}}|^2}{2\sigma_{x_{i,I}}^2}} \frac{1}{\sqrt{2\pi\sigma_{x_{i,Q}}^2}} e^{-\frac{|\xi_{i,Q} - \mathtt{m}_{x_{i,Q}}|^2}{2\sigma_{x_{i,Q}}^2}} \tag{1.306}$$

$$= \frac{1}{\pi\sigma_{x_i}^2} e^{-\frac{|\xi_i - \mathtt{m}_{x_i}|^2}{\sigma_{x_i}^2}} \tag{1.307}$$

with $\xi_i = \xi_{i,I} + j\xi_{i,Q}$ complex valued.

Example 1.7.4

Let $\boldsymbol{x}^T = [x_1, \dots, x_N] = [x_1(t_1), \dots, x_N(t_N)]$ be a complex-valued Gaussian (vector) process, with each element $x_i(t_i)$ having real and imaginary components that are uncorrelated Gaussian r.v.s. whose pdf is with zero mean and equal variance for all values of t_i. The vector $\boldsymbol{x}$ is called *circularly symmetric Gaussian random process*. The joint pdf in this case results

$$p_{\boldsymbol{x}}(\boldsymbol{\xi}) = [\pi^N \det \boldsymbol{C}]^{-1} e^{-\boldsymbol{\xi}^H \boldsymbol{C}^{-1} \boldsymbol{\xi}} \tag{1.308}$$

where $\boldsymbol{C}$ is the covariance matrix of $[x_1(t_1), x_2(t_2), \dots, x_N(t_N)]$.

Example 1.7.5

Let $x(t) = A\sin(2\pi ft + \varphi)$ be a real-valued sinusoidal signal with φ r.v. uniform in $[0, 2\pi)$, for which we will use the notation $\varphi \sim \mathcal{U}[0, 2\pi)$. The mean of x is

$$\begin{aligned} \mathtt{m}_x(t) &= E[x(t)] \\ &= \int_0^{2\pi} \frac{1}{2\pi} A \sin(2\pi ft + a) da \\ &= 0 \end{aligned} \tag{1.309}$$

and the autocorrelation function is given by

$$\begin{aligned} \mathtt{r}_x(\tau) &= \int_0^{2\pi} \frac{1}{2\pi} A \sin(2\pi ft + a) A \sin[2\pi f(t-\tau) + a] da \\ &= \frac{A^2}{2} \cos(2\pi f\tau) \end{aligned} \tag{1.310}$$

Example 1.7.6

Consider the sum of N real-valued sinusoidal signals, i.e.

$$x(t) = \sum_{i=1}^{N} A_i \sin(2\pi f_i t + \varphi_i) \tag{1.311}$$

with $\varphi_i \sim \mathcal{U}[0, 2\pi)$ statistically independent, from Example 1.7.5 it is immediate to obtain the mean

$$\mathtt{m}_x(t) = \sum_{i=1}^{N} \mathtt{m}_{x_i}(t) = 0 \tag{1.312}$$

and the autocorrelation function

$$\mathtt{r}_x(\tau) = \sum_{i=1}^{N} \frac{A_i^2}{2} \cos(2\pi f_i \tau) \tag{1.313}$$

We note that, according to the Definition 1.9, page 31, the process (1.311) is not asymptotically uncorrelated.

Example 1.7.7

Consider the sum of N complex-valued sinusoidal signals, i.e.

$$x(t) = \sum_{i=1}^{N} A_i \, e^{j(2\pi f_i t + \varphi_i)} \tag{1.314}$$

with $\varphi_i \sim \mathcal{U}[0, 2\pi)$ statistically independent. Following a similar procedure to that used in Examples 1.7.5 and 1.7.6, we find

$$\mathtt{r}_x(\tau) = \sum_{i=1}^{N} |A_i|^2 \, e^{j2\pi f_i \tau} \tag{1.315}$$

We note that the process (1.315) is not asymptotically uncorrelated.

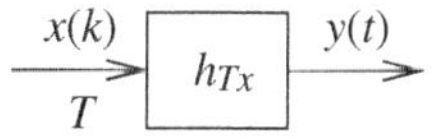

Figure 1.26 Modulator of a PAM system as interpolator filter.

Example 1.7.8

Let the discrete-time random process $y(k) = x(k) + w(k)$ be given by the sum of the random process x of Example 1.7.7 and white noise w with variance σ_w^2. Moreover, we assume x and w uncorrelated. In this case,

$$\mathtt{r}_y(n) = \sum_{i=1}^{N} |A_i|^2 \; e^{j2\pi f_i n T_c} + \sigma_w^2 \delta_n \tag{1.316}$$

Example 1.7.9

We consider a signal obtained by pulse-amplitude modulation (PAM), expressed as

$$y(t) = \sum_{k=-\infty}^{+\infty} x(k) \; h_{Tx}(t - kT) \tag{1.317}$$

The signal y is the output of the system shown in Figure 1.26, where h_{Tx} is a finite-energy pulse and $\{x(k)\}$ is a discrete-time (with T-spaced samples) WSS sequence, having PSD $\mathcal{P}_x(f)$. We note that $\mathcal{P}_x(f)$ is a periodic function of period $1/T$.

Let the deterministic autocorrelation of the signal h_{Tx} be

$$\mathtt{r}_{h_{Tx}}(\tau) = \int_{-\infty}^{+\infty} h_{Tx}(t) h_{Tx}^*(t - \tau) dt = [h_{Tx}(t) * h_{Tx}^*(-t)](\tau) \tag{1.318}$$

with Fourier transform $|\mathcal{H}_{Tx}(f)|^2$. In general, y is a cyclostationary process of period T. In fact, we have

1. *Mean*

$$\mathtt{m}_y(t) = \mathtt{m}_x \sum_{k=-\infty}^{+\infty} h_{Tx}(t - kT) \tag{1.319}$$

2. *Correlation*

$$\mathtt{r}_y(t, t-\tau) = \sum_{i=-\infty}^{+\infty} \mathtt{r}_x(i) \sum_{m=-\infty}^{+\infty} h_{Tx}(t - (i+m)\,T) h_{Tx}^*(t - \tau - mT) \tag{1.320}$$

If we introduce the average spectral analysis

$$\bar{\mathtt{m}}_y = \frac{1}{T} \int_0^T \mathtt{m}_y(t) \; dt = \mathtt{m}_x \mathcal{H}_{Tx}(0) \tag{1.321}$$

$$\bar{\mathtt{r}}_y(\tau) = \frac{1}{T} \int_0^T \mathtt{r}_y(t, t-\tau) dt = \frac{1}{T} \sum_{i=-\infty}^{+\infty} \mathtt{r}_x(i) \mathtt{r}_{h_{Tx}}(\tau - iT) \tag{1.322}$$

and

$$\overline{\mathcal{P}}_y(f) = \mathcal{F}[\bar{\mathtt{r}}_y(\tau)] = \left| \frac{1}{T} \mathcal{H}_{Tx}(f) \right|^2 \mathcal{P}_x(f) \tag{1.323}$$

we observe that the modulator of a PAM system may be regarded as an interpolator filter with frequency response $\mathcal{H}_{Tx}/T$.

3. *Average power for a white noise input* For a white noise input with power $\mathtt{M}_x$, from (1.322), the average statistical power of the output signal is given by

$$\overline{\mathtt{M}}_y = \mathtt{M}_x \frac{E_h}{T} \tag{1.324}$$

where $E_h = \int_{-\infty}^{+\infty} |h_{Tx}(t)|^2 \, dt$ is the energy of h_{Tx}.

4. *Moments of y for a circularly symmetric i.i.d. input*

Let $\{x(k)\}$ be a complex-valued random circularly symmetric sequence with zero mean (see (1.301) and (1.302)), i.e. letting

$$x_I(k) = Re[x(k)], \qquad x_Q(k) = Im[x(k)] \tag{1.325}$$

we have

$$E[x_I^2(k)] = E[x_Q^2(k)] = \frac{E[|x(k)|^2]}{2} \tag{1.326}$$

and

$$E[x_I(k) \, x_Q(k)] = 0 \tag{1.327}$$

These two relations can be merged into the single expression

$$E[x^2(k)] = E[x_I^2(k)] - E[x_Q^2(k)] + 2j \, E[x_I(k) \, x_Q(k)] = 0 \tag{1.328}$$

Filtering the i.i.d. input signal $\{x(k)\}$ by using the system depicted in Figure 1.26, and from the relation

$$\mathtt{r}_{yy^*}(t, t-\tau) = \sum_{i=-\infty}^{+\infty} \mathtt{r}_{xx^*}(i) \sum_{m=-\infty}^{+\infty} h_{Tx}(t-(i+m)T) h_{Tx}(t-\tau-mT) \tag{1.329}$$

we have

$$\mathtt{r}_{xx^*}(i) = E[x^2(k)]\delta(i) = 0 \tag{1.330}$$

and

$$\mathtt{r}_{yy^*}(t, t-\tau) = 0 \tag{1.331}$$

that is $y \perp y^*$. In particular, we have that y is circularly symmetric, i.e.

$$E[y^2(t)] = 0 \tag{1.332}$$

We note that the condition (1.331) can be obtained assuming the less stringent condition that $x \perp x^*$; on the other hand, this requires that the following two conditions are verified

$$\mathtt{r}_{x_I}(i) = \mathtt{r}_{x_Q}(i) \tag{1.333}$$

and

$$\mathtt{r}_{x_I x_Q}(i) = -\mathtt{r}_{x_I x_Q}(-i) \tag{1.334}$$

Observation 1.5

It can be shown that if the filter h_{Tx} has a bandwidth smaller than $1/(2T)$ and $\{x(k)\}$ is a WSS sequence, then $\{y(k)\}$ is WSS with PSD (1.323).

Example 1.7.10

Let us consider a PAM signal sampled with period $T_Q = T/Q_0$, where Q_0 is a positive integer number. Let

$$y_q = y(q \, T_Q), \qquad h_p = h_{Tx}(p \, T_Q) \tag{1.335}$$

from (1.317) it follows

$$y_q = \sum_{k=-\infty}^{+\infty} x(k)\, h_{q-kQ_0} \tag{1.336}$$

If $Q_0 \neq 1$, (1.336) describes the input–output relation of an interpolator filter (see (1.536)). We recall the statistical analysis given in Table 1.6, page 34. We denote with $\mathcal{H}(f)$ the Fourier transform (see (1.17)) and with $\mathtt{r}_h(n)$ the deterministic autocorrelation (see (1.184)) of the sequence $\{h_p\}$. We also assume that $\{x(k)\}$ is a WSS random sequence with mean $\mathtt{m}_x$ and autocorrelation $\mathtt{r}_x(n)$. In general, $\{y_q\}$ is a cyclostationary random sequence of period Q_0 with

1. *Mean*

$$\mathtt{m}_y(q) = \mathtt{m}_x \sum_{k=-\infty}^{+\infty} h_{q-kQ_0} \tag{1.337}$$

2. *Correlation*

$$\mathtt{r}_y(q, q-n) = \sum_{i=-\infty}^{+\infty} \mathtt{r}_x(i) \sum_{m=-\infty}^{+\infty} h_{q-(i+m)Q_0}\, h^*_{q-n-m\,Q_0} \tag{1.338}$$

By the average spectral analysis, we obtain

$$\overline{\mathtt{m}}_y = \frac{1}{Q_0} \sum_{q=0}^{Q_0-1} \mathtt{m}_y(q) = \mathtt{m}_x \frac{\mathcal{H}(0)}{Q_0} \tag{1.339}$$

where

$$\mathcal{H}(0) = \sum_{p=-\infty}^{+\infty} h_p \tag{1.340}$$

and

$$\overline{\mathtt{r}}_y(n) = \frac{1}{Q_0} \sum_{q=0}^{Q_0-1} \mathtt{r}_y(q, q-n) = \frac{1}{Q_0} \sum_{i=-\infty}^{+\infty} \mathtt{r}_x(i)\, \mathtt{r}_h(n - iQ_0) \tag{1.341}$$

Consequently, the average PSD is given by

$$\overline{\mathcal{P}}_y(f) = T_Q\, \mathcal{F}[\overline{\mathtt{r}}_y(n)] = \left| \frac{1}{Q_0} \mathcal{H}(f) \right|^2 \mathcal{P}_x(f) \tag{1.342}$$

If $\{x(k)\}$ is white noise with power $\mathtt{M}_x$, from (1.341) it results

$$\overline{\mathtt{r}}_y(n) = \mathtt{M}_x \frac{\mathtt{r}_h(n)}{Q_0} \tag{1.343}$$

In particular, the average power of the filter output signal is given by

$$\overline{\mathtt{M}}_y = \mathtt{M}_x \frac{E_h}{Q_0} \tag{1.344}$$

where $E_h = \sum_{p=-\infty}^{+\infty} |h_p|^2$ is the energy of $\{h_p\}$. We point out that the condition $\overline{\mathtt{M}}_y = \mathtt{M}_x$ is satisfied if the energy of the filter impulse response is equal to the interpolation factor Q_0.

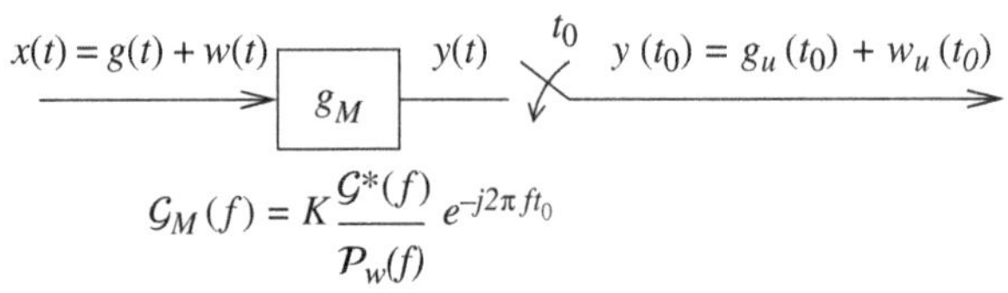

Figure 1.27 Reference scheme for the matched filter.

1.8 Matched filter

Referring to Figure 1.27, we consider a finite-energy signal pulse g in the presence of additive noise w having zero mean and PSD $\mathcal{P}_w$. The signal

$$x(t) = g(t) + w(t) \tag{1.345}$$

is filtered with a filter having impulse response g_M. We indicate with g_u and w_u, respectively, the desired signal and the noise component at the output:

$$g_u(t) = g_M * g(t) \tag{1.346}$$

$$w_u(t) = g_M * w(t) \tag{1.347}$$

The output is

$$y(t) = g_u(t) + w_u(t) \tag{1.348}$$

We now suppose that y is observed at a given instant t_0. The problem is to determine g_M so that the ratio between the square amplitude of $g_u(t_0)$ and the power of the noise component $w_u(t_0)$ is maximum, i.e.

$$g_M : \max_{g_M} \frac{|g_u(t_0)|^2}{E[|w_u(t_0)|^2]} \tag{1.349}$$

The optimum filter has frequency response

$$\mathcal{G}_M(f) = K \frac{\mathcal{G}^*(f)}{\mathcal{P}_w(f)} e^{-j2\pi f t_0} \tag{1.350}$$

where K is a constant. In other words, the best filter selects the frequency components of the desired input signal and weights them with weights that are inversely proportional to the noise level.

Proof. $g_u(t_0)$ coincides with the inverse Fourier transform of $\mathcal{G}_M(f)\mathcal{G}(f)$ evaluated in $t = t_0$, while the power of $w_u(t_0)$ is equal to

$$\mathtt{r}_{w_u}(0) = \int_{-\infty}^{+\infty} \mathcal{P}_w(f) |\mathcal{G}_M(f)|^2 df \tag{1.351}$$

Then we have

$$\begin{aligned} \frac{|g_u(t_0)|^2}{\mathtt{r}_{w_u}(0)} &= \frac{\left| \int_{-\infty}^{+\infty} \mathcal{G}_M(f)\mathcal{G}(f) e^{j2\pi f t_0} df \right|^2}{\int_{-\infty}^{+\infty} \mathcal{P}_w(f) |\mathcal{G}_M(f)|^2 df} \\ &= \frac{\left| \int_{-\infty}^{+\infty} \mathcal{G}_M(f) \sqrt{\mathcal{P}_w(f)} \frac{\mathcal{G}(f)}{\sqrt{\mathcal{P}_w(f)}} e^{j2\pi f t_0} df \right|^2}{\int_{-\infty}^{+\infty} \mathcal{P}_w(f) |\mathcal{G}_M(f)|^2 df} \end{aligned} \tag{1.352}$$

where the integrand at the numerator was divided and multiplied by $\sqrt{\mathcal{P}_w(f)}$. Implicitly, it is assumed that $\mathcal{P}_w(f) \neq 0$. Applying the Schwarz inequality[14] to the functions

$$\mathcal{G}_M(f)\sqrt{\mathcal{P}_w(f)} \tag{1.355}$$

and

$$\frac{\mathcal{G}^*(f)}{\sqrt{\mathcal{P}_w(f)}}\, e^{-j2\pi f t_0} \tag{1.356}$$

it turns out

$$\frac{|g_u(t_0)|^2}{\mathtt{r}_{w_u}(0)} \leq \int_{-\infty}^{+\infty} \left| \frac{\mathcal{G}(f)}{\sqrt{\mathcal{P}_w(f)}} e^{j2\pi f t_0} \right|^2 df = \int_{-\infty}^{+\infty} \left| \frac{\mathcal{G}(f)}{\sqrt{\mathcal{P}_w(f)}} \right|^2 df \tag{1.357}$$

Therefore, the maximum value is equal to the right-hand side of (1.357) and is achieved for

$$\mathcal{G}_M(f)\sqrt{\mathcal{P}_w(f)} = K \frac{\mathcal{G}^*(f)}{\sqrt{\mathcal{P}_w(f)}}\, e^{-j2\pi f t_0} \tag{1.358}$$

where K is a constant. From (1.358), the solution (1.350) follows immediately. □

White noise case

If w is white, then $\mathcal{P}_w(f) = \mathcal{P}_w$ is a constant and the optimum solution (1.350) becomes

$$\mathcal{G}_M(f) = K\mathcal{G}^*(f)e^{-j2\pi f t_0} \tag{1.359}$$

Correspondingly, the filter has impulse response

$$g_M(t) = Kg^*(t_0 - t) \tag{1.360}$$

from which the name of *matched filter* (MF), i.e. matched to the input signal pulse. The desired signal pulse at the filter output has the frequency response

$$\mathcal{G}_u(f) = K|\mathcal{G}(f)|^2 e^{-j2\pi f t_0} \tag{1.361}$$

From the definition of the autocorrelation function of pulse g,

$$\mathtt{r}_g(\tau) = \int_{-\infty}^{+\infty} g(a)g^*(a-\tau)da \tag{1.362}$$

then, as depicted in Figure 1.28,

$$g_u(t) = K\mathtt{r}_g(t - t_0) \tag{1.363}$$

i.e. the pulse at the filter output coincides with the autocorrelation function of the pulse g. If E_g is the energy of g, using the relation $E_g = \mathtt{r}_g(0)$ the maximum of the functional (1.349) becomes

$$\frac{|g_u(t_0)|^2}{\mathtt{r}_{w_u}(0)} = \frac{|K|^2\mathtt{r}_g^2(0)}{\mathcal{P}_w|K|^2\mathtt{r}_g(0)} = \frac{E_g}{\mathcal{P}_w} \tag{1.364}$$

---

[14] Given two signals x and y it holds

$$\left|\int_{-\infty}^{\infty} x(t)y^*(t)dt\right|^2 \leq \int_{-\infty}^{\infty} |x(t)|^2 dt \int_{-\infty}^{\infty} |y(t)|^2 dt \tag{1.353}$$

where equality holds if and only if

$$y(t) = Kx(t) \tag{1.354}$$

with K a complex constant.

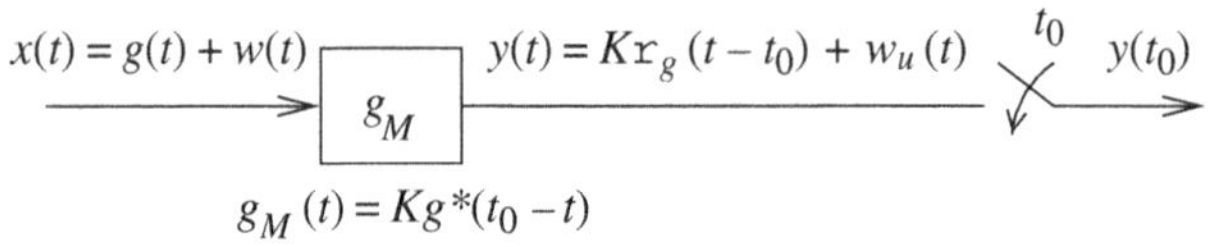

$x(t) = g(t) + w(t)$

g_M

$y(t) = K\mathtt{r}_g(t - t_0) + w_u(t)$

t_0

$y(t_0)$

$g_M(t) = Kg^*(t_0 - t)$

Figure 1.28 Matched filter for an input pulse in the presence of white noise.

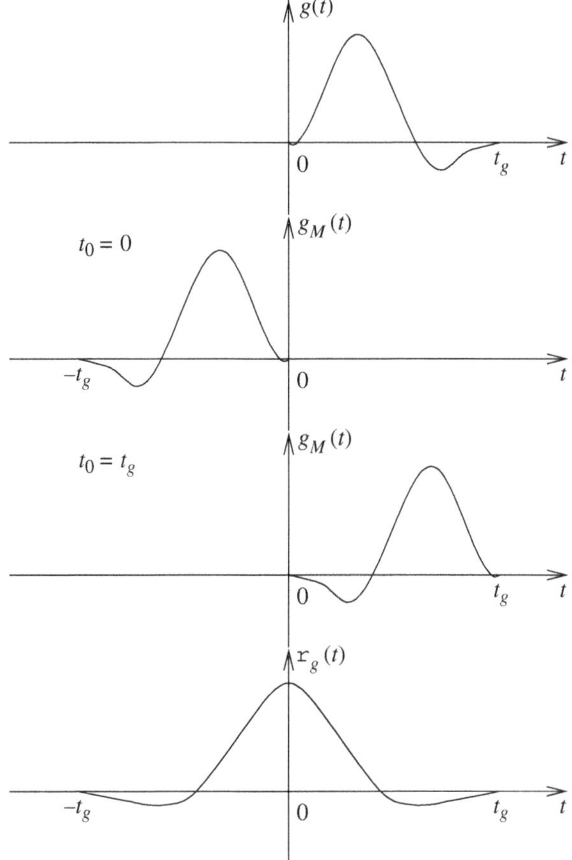

Figure 1.29 Various pulse shapes related to a matched filter.

In Figure 1.29, the different pulse shapes are illustrated for a signal pulse g with limited duration t_g. Note that in this case, the matched filter has also limited duration, and it is causal if $t_0 \geq t_g$.

Example 1.8.1 (MF for a rectangular pulse)

Let

$$g(t) = \mathtt{w}_T(t) = \text{rect}\left(\frac{t - T/2}{T}\right) \tag{1.365}$$

with

$$\mathtt{r}_g(\tau) = T\left(1 - \frac{|\tau|}{T}\right)\text{rect}\left(\frac{\tau}{2T}\right) \tag{1.366}$$

For $t_0 = T$, the matched filter is proportional to g

$$g_M(t) = K\mathtt{w}_T(t) \tag{1.367}$$

and the output pulse in the absence of noise is equal to

$$g_u(t) = KT\left(1 - \left|\frac{t-T}{T}\right|\right) \text{rect}\left(\frac{t-T}{2T}\right) \tag{1.368}$$

1.9 Ergodic random processes

The functions that have been introduced in the previous sections for the analysis of random processes give a valid statistical description of an ensemble of realizations of a random process. We investigate now the possibility of moving from ensemble averaging to time averaging, that is we consider the problem of estimating a statistical descriptor of a random process from the observation of a single realization. Let x be a discrete-time WSS random process having mean $\mathtt{m}_x$. If in the limit it holds[15]

$$\lim_{K\to\infty} \frac{1}{K} \sum_{k=0}^{K-1} x(k) = E[x(k)] = \mathtt{m}_x \tag{1.369}$$

then x is said to be *ergodic in the mean*. In other words, for when the above limit holds, the time-average of samples tends to the statistical mean as the number of samples increases. We note that the existence of the limit (1.369) implies the condition

$$\lim_{K\to\infty} E\left[\left|\frac{1}{K}\sum_{k=0}^{K-1} x(k) - \mathtt{m}_x\right|^2\right] = 0 \tag{1.370}$$

or equivalently

$$\lim_{K\to\infty} \frac{1}{K} \sum_{n=-(K-1)}^{K-1} \left[1 - \frac{|n|}{K}\right] \mathtt{c}_x(n) = 0 \tag{1.371}$$

From (1.371), we see that for a random process to be ergodic in the mean, some conditions on the second-order statistics must be verified. Analogously to definition (1.369), we say that x is *ergodic in correlation* if in the limits it holds:

$$\lim_{K\to\infty} \frac{1}{K} \sum_{k=0}^{K-1} x(k)x^*(k-n) = E[x(k)x^*(k-n)] = \mathtt{r}_x(n) \tag{1.372}$$

Also for processes that are ergodic in correlation, one could get a condition of ergodicity similar to that expressed by the limit (1.371). Let $y(k) = x(k)x^*(k-n)$. Observing (1.372) and (1.369), we find that the ergodicity in correlation of the process x is equivalent to the ergodicity in the mean of the process y. Therefore, it is easy to deduce that the condition (1.371) for y translates into a condition on the statistical moments of the fourth order for x.

In practice, we *will assume all stationary processes to be ergodic*; ergodicity is however difficult to prove for non-Gaussian random processes. We will not consider particular processes that are not ergodic such as $x(k) = A$, where A is a random variable, or $x(k)$ equal to the sum of sinusoidal signals (see (1.311)).

[15] The *limit* is meant in the mean square sense, that is the variance of the r.v. $\left(\frac{1}{K}\sum_{k=0}^{K-1} x(k) - \mathtt{m}_x\right)$ vanishes for $K \to \infty$.

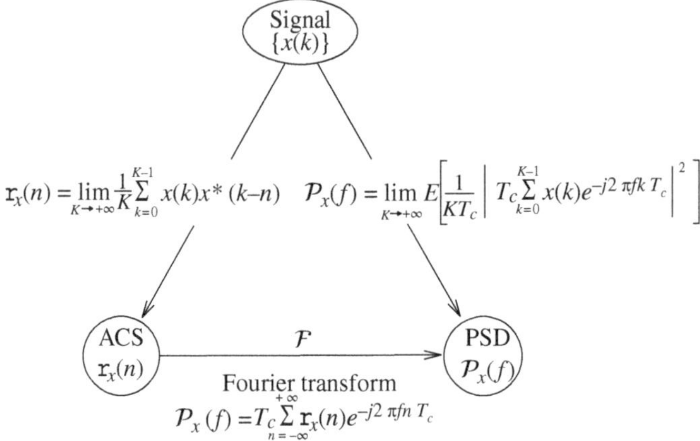

Figure 1.30 Relation between ergodic processes and their statistical description.

The property of ergodicity assumes a fundamental importance if we observe that from a single Realization, it is possible to obtain an estimate of the autocorrelation function, and from this, the PSD. Alternatively, one could prove that under the hypothesis[16]

$$\sum_{n=-\infty}^{+\infty} |n| \; \mathtt{r}_x(n) < \infty \tag{1.373}$$

the following limit holds:

$$\lim_{K\to\infty} E\left[\frac{1}{KT_c}\left|T_c \sum_{k=0}^{K-1} x(k)\; e^{-j2\pi f k T_c}\right|^2\right] = \mathcal{P}_x(f) \tag{1.374}$$

Then, exploiting the ergodicity of a WSS random process, one obtains the relations among the process itself, its autocorrelation function, and PSD shown in Figure 1.30. We note how the direct computation of the PSD, given by (1.374), makes use of a statistical ensemble of the Fourier transform of the process x, while the indirect method via ACS makes use of a single realization.

If we let

$$\tilde{\mathcal{X}}_{KT_c}(f) = T_c\; \mathcal{F}[x(k)\; \mathtt{w}_K(k)] \tag{1.375}$$

where $\mathtt{w}_K$ is the rectangular window of length K (see (1.401)) and $T_d = KT_c$, (1.374) becomes

$$\mathcal{P}_x(f) = \lim_{T_d\to\infty} \frac{E[|\tilde{\mathcal{X}}_{T_d}(f)|^2]}{T_d} \tag{1.376}$$

The relation (1.376) holds also for continuous-time ergodic random processes, where $\tilde{\mathcal{X}}_{T_d}(f)$ denotes the Fourier transform of the windowed realization of the process, with a rectangular window of duration T_d.

[16] We note that for random processes with non-zero mean and/or sinusoidal components this property is not verified. Therefore, it is usually recommended that the deterministic components of the process be removed before the spectral estimation is performed.

1.9.1 Mean value estimators

Given the random process $\{x(k)\}$, we wish to estimate the mean value of a related process $\{y(k)\}$: for example to estimate the statistical power of x we set $y(k) = |x(k)|^2$, while for the estimation of the correlation of x with lag n, we set $y(k) = x(k)x^*(k-n)$. Based on a realization of $\{y(k)\}$, from (1.369) an estimate of the mean value of y is given by

$$\hat{\mathtt{m}}_y = \frac{1}{K}\sum_{k=0}^{K-1} y(k) \tag{1.377}$$

In fact, (1.377) attempts to determine the average component of the signal $\{y(k)\}$. As illustrated in Figure 1.31a, in general, we can think of extracting the average component of $\{y(k)\}$ using an LPF filter h having unit gain, i.e. $\mathcal{H}(0) = 1$, and suitable bandwidth B. Let K be the length of the impulse response with support from $k = 0$ to $k = K - 1$. Note that for a unit step input signal, the transient part of the output signal will last $K - 1$ time instants. Therefore, we assume

$$\hat{\mathtt{m}}_y = z(k) = h * y(k) \qquad \text{for } k \geq K-1 \tag{1.378}$$

We now compute the mean and variance of the estimate. From (1.378), the mean value is given by

$$E[\hat{\mathtt{m}}_y] = \mathtt{m}_y \mathcal{H}(0) = \mathtt{m}_y \tag{1.379}$$

as $\mathcal{H}(0) = 1$. Using the expression in Table 1.6 of the correlation of a filter output signal given the input, the variance of the estimate is given by

$$\text{var}[\hat{\mathtt{m}}_y] = \sigma_y^2 = \sum_{n=-\infty}^{+\infty} \mathtt{r}_h(-n)\mathtt{c}_y(n) \tag{1.380}$$

Assuming

$$S = \sum_{n=-\infty}^{+\infty} |\mathtt{c}_y(n)| = \sigma_y^2 \sum_{n=-\infty}^{+\infty} \frac{|\mathtt{c}_y(n)|}{\sigma_y^2} < \infty \tag{1.381}$$

and being $|\mathtt{r}_h(n)| \leq \mathtt{r}_h(0)$, the variance in (1.380) is *bounded* by

$$\text{var}[\hat{\mathtt{m}}_y] \leq E_h S \tag{1.382}$$

where $E_h = \mathtt{r}_h(0)$.

For an ideal lowpass filter,

$$\mathcal{H}(f) = \text{rect}\left(\frac{f}{2B}\right), \qquad |f| < \frac{1}{2T_c} \tag{1.383}$$

assuming as filter length K that of the principal lobe of $\{h(k)\}$, and neglecting a delay factor, it results as $E_h = 2B$ and $K \simeq 1/B$. Introducing the criterion that for a good estimate, it must be

$$\text{var}\,[\hat{\mathtt{m}}_y] \leq \varepsilon \tag{1.384}$$

with $\varepsilon \ll |\mathtt{m}_y|^2$, from (1.382) it follows

$$B \leq \frac{\varepsilon}{2S} \tag{1.385}$$

and

$$K \geq \frac{2S}{\varepsilon} \tag{1.386}$$

In other words, from (1.381) and (1.386), for a fixed ε, the length K of the filter impulse response must be larger, or equivalently the bandwidth B must be smaller, to obtain estimates for those processes $\{y(k)\}$ that exhibit larger variance and/or larger correlation among samples. Because of their simple implementation, two commonly used filters are the rectangular window and the exponential filter, whose impulse responses are shown in Figure 1.31.

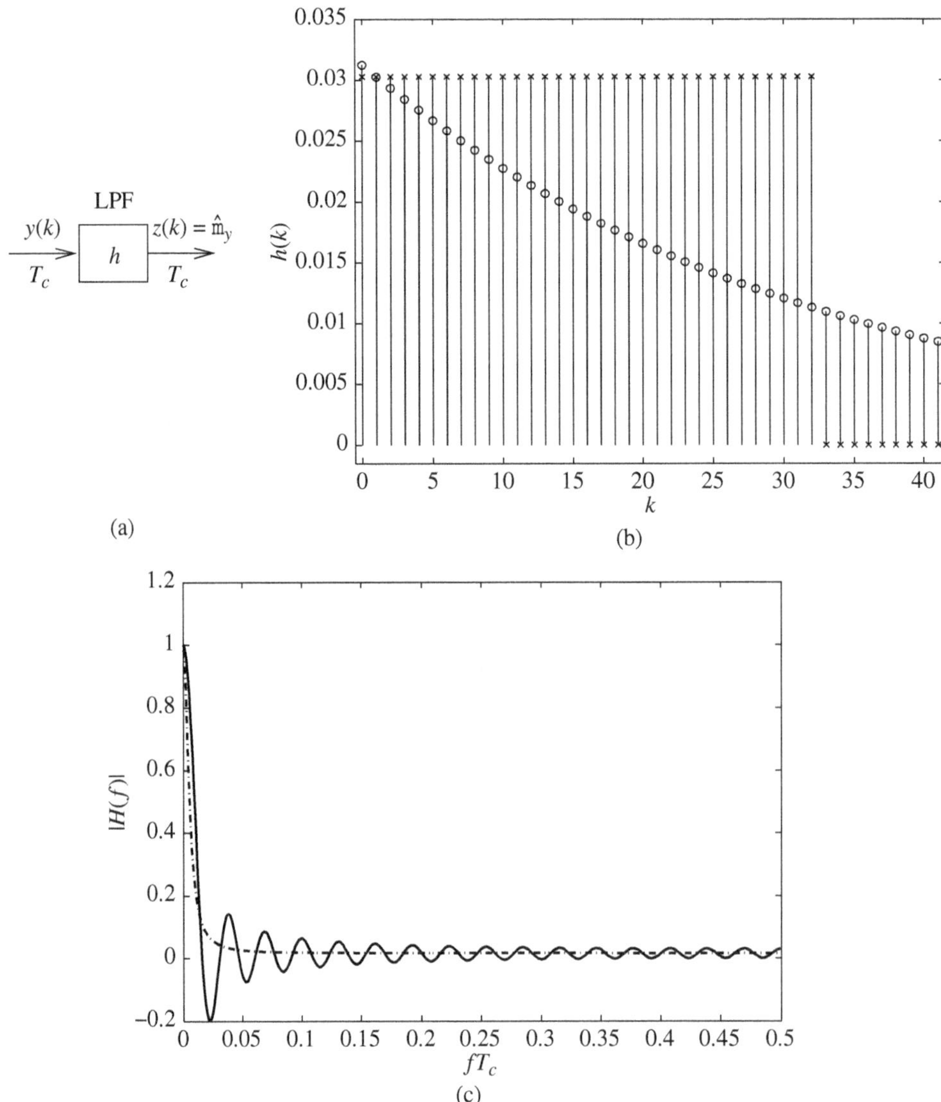

Figure 1.31 (a) Time average as output of a narrow band lowpass filter. (b) Typical impulse responses: exponential filter with parameter $a = 1 - 2^{-5}$ and rectangular window with $K = 33$. (c) Corresponding frequency responses.

Rectangular window

For a rectangular window,

$$h(k) = \begin{cases} \dfrac{1}{K} & k = 0, 1, \ldots, K-1 \\ 0 & \text{elsewhere} \end{cases} \tag{1.387}$$

the frequency response is given by (see (1.24))

$$\mathcal{H}(f) = e^{-j2\pi f\left(\frac{K-1}{2}\right)T_c} \, \text{sinc}_K(fKT_c) \tag{1.388}$$

We have $E_h = 1/K$ and, adopting as bandwidth the frequency of the first zero of $|\mathcal{H}(f)|$, $B = 1/(KT_c)$. The filter output is given by

$$z(k) = \sum_{n=0}^{K-1} \frac{1}{K} y(k-n) \tag{1.389}$$

that can be expressed as

$$z(k) = z(k-1) + \frac{y(k) - y(k-K)}{K} \tag{1.390}$$

Exponential filter

For an exponential filter

$$h(k) = \begin{cases} (1-a)a^k & k \geq 0 \\ 0 & \text{elsewhere} \end{cases} \tag{1.391}$$

with $|a| < 1$, the frequency response is given by

$$\mathcal{H}(f) = \frac{1-a}{1 - ae^{-j2\pi f T_c}} \tag{1.392}$$

Moreover, $E_h = (1-a)/(1+a)$ and, adopting as length of h the time constant of the filter, i.e. the interval it takes for the amplitude of the impulse response to decrease of a factor e,

$$K - 1 = \frac{1}{\ln 1/a} \simeq \frac{1}{1-a} \tag{1.393}$$

where the approximation holds for $a \simeq 1$. The 3 dB filter bandwidth is equal to

$$B = \frac{1-a}{2\pi} \frac{1}{T_c} \qquad \text{for } a > 0.9 \tag{1.394}$$

The filter output has a simple expression given by the recursive equation

$$z(k) = az(k-1) + (1-a)\, y(k) \tag{1.395}$$

We note that choosing a as

$$a = 1 - 2^{-l} \tag{1.396}$$

then (1.395) becomes

$$z(k) = z(k-1) + 2^{-l}(y(k) - z(k-1)) \tag{1.397}$$

whose computation requires only two additions and one *shift* of l bits. Moreover, from (1.393), the filter time constant is given by

$$K - 1 = 2^l \tag{1.398}$$

General window

In addition to the two filters described above, a general window can be defined as

$$h(k) = A\mathtt{w}(k) \tag{1.399}$$

with $\{\mathtt{w}(k)\}$ window[17] of length K. Factor A in (1.399) is introduced to normalize the area of h to 1. We note that, for random processes with slowly time-varying statistics, (1.390) and (1.397) give an expression to update the estimates.

---

[17] We define the *continuous-time rectangular window* with duration T_d as

$$\mathtt{w}_{T_d}(t) = \text{rect}\left(\frac{t - T_d/2}{T_d}\right) = \begin{cases} 1 & 0 < t < T_d \\ 0 & \text{elsewhere} \end{cases} \tag{1.400}$$

Commonly used *discrete-time windows* are:

1.9.2 Correlation estimators

Let $\{x(k)\}$, $k = 0, 1, \dots, K-1$, be a realization of a random process with K samples. We examine two estimates.

Unbiased estimate

The unbiased estimate

$$\hat{\mathtt{r}}_x(n) = \frac{1}{K-n} \sum_{k=n}^{K-1} x(k)x^*(k-n) \qquad n = 0, 1, \dots, K-1 \tag{1.405}$$

has mean

$$E[\hat{\mathtt{r}}_x(n)] = \frac{1}{K-n} \sum_{k=n}^{K-1} E[x(k)x^*(k-n)] = \mathtt{r}_x(n) \tag{1.406}$$

If the process is Gaussian, one can show that the variance of the estimate is approximately given by

$$\mathrm{var}[\hat{\mathtt{r}}_x(n)] \simeq \frac{K}{(K-n)^2} \sum_{m=-\infty}^{+\infty} [\mathtt{r}_x^2(m) + \mathtt{r}_x(m+n)\mathtt{r}_x(m-n)] \tag{1.407}$$

from which it follows

$$\mathrm{var}\,[\hat{\mathtt{r}}_x(n)] \xrightarrow[K\to\infty]{} 0 \tag{1.408}$$

The above limit holds for $n \ll K$. Note that the variance of the estimate increases with the correlation lag n.

Biased estimate

The biased estimate

$$\check{\mathtt{r}}_x(n) = \frac{1}{K} \sum_{k=n}^{K-1} x(k)x^*(k-n) = \left(1 - \frac{|n|}{K}\right) \hat{\mathtt{r}}_x(n) \tag{1.409}$$

---

1. *Rectangular window*

$$\mathtt{w}(k) = \mathtt{w}_D(k) = \begin{cases} 1 & k = 0, 1, \dots, D-1 \\ 0 & \text{elsewhere} \end{cases} \tag{1.401}$$

where D denotes the length of the rectangular window expressed in number of samples.

2. *Raised cosine or Hamming window*

$$\mathtt{w}(k) = \begin{cases} 0.54 + 0.46 \cos\left(2\pi \dfrac{k - \frac{D-1}{2}}{D-1}\right) & k = 0, 1, \dots, D-1 \\ 0 & \text{elsewhere} \end{cases} \tag{1.402}$$

3. *Hann window*

$$\mathtt{w}(k) = \begin{cases} 0.50 + 0.50 \cos\left(2\pi \dfrac{k - \frac{D-1}{2}}{D-1}\right) & k = 0, 1, \dots, D-1 \\ 0 & \text{elsewhere} \end{cases} \tag{1.403}$$

4. *Triangular or Bartlett window*

$$\mathtt{w}(k) = \begin{cases} 1 - 2\left|\dfrac{k - \frac{D-1}{2}}{D-1}\right| & k = 0, 1, \dots, D-1 \\ 0 & \text{elsewhere} \end{cases} \tag{1.404}$$

has mean satisfying the following relations:

$$E[\check{\mathtt{r}}_x(n)] = \left(1 - \frac{|n|}{K}\right) \mathtt{r}_x(n) \xrightarrow[K\to\infty]{} \mathtt{r}_x(n) \tag{1.410}$$

Unlike the unbiased estimate, the mean of the biased estimate is not equal to the autocorrelation function, but approaches it as K increases. Note that the biased estimate differs from the autocorrelation function by one additive constant, denoted as *bias*:

$$\mu_{bias} = E[\check{\mathtt{r}}_x(n)] - \mathtt{r}_x(n) \tag{1.411}$$

For a Gaussian process, the variance of the biased estimate is

$$\text{var}[\check{\mathtt{r}}_x(n)] = \left(\frac{K - |n|}{K}\right)^2 \text{var}[\hat{\mathtt{r}}_x(n)] \simeq \frac{1}{K} \sum_{m=-\infty}^{+\infty} [\mathtt{r}_x^2(m) + \mathtt{r}_x(m+n)\mathtt{r}_x(m-n)] \tag{1.412}$$

In general, the biased estimate of the ACS exhibits a mean-square error[18] larger than the unbiased, especially for large values of n. It should also be noted that the estimate does not necessarily yield sequences that satisfy the properties of autocorrelation functions: for example the following property may not be verified:

$$\hat{\mathtt{r}}_x(0) \geq |\hat{\mathtt{r}}_x(n)|, \qquad n \neq 0 \tag{1.414}$$

1.9.3 Power spectral density estimators

After examining ACS estimators, we review some spectral density estimation methods.

Periodogram or instantaneous spectrum

Let $\tilde{\mathcal{X}}(f) = T_c \mathcal{X}(f)$, where $\mathcal{X}(f)$ is the Fourier transform of $\{x(k)\}$, $k = 0, \dots, K-1$; an estimate of the statistical power of $\{x(k)\}$ is given by

$$\hat{\mathtt{M}}_x = \frac{1}{K} \sum_{k=0}^{K-1} |x(k)|^2 = \frac{1}{KT_c} \int_{-\frac{1}{2T_c}}^{\frac{1}{2T_c}} |\tilde{\mathcal{X}}(f)|^2 \, df \tag{1.415}$$

using the properties of the Fourier transform (Parseval theorem). Based on (1.415), a PSD estimator called *periodogram* is given by

$$\mathcal{P}_{PER}(f) = \frac{1}{KT_c} |\tilde{\mathcal{X}}(f)|^2 \tag{1.416}$$

We can write (1.416) as

$$\mathcal{P}_{PER}(f) = T_c \sum_{n=-(K-1)}^{K-1} \check{\mathtt{r}}_x(n) \, e^{-j2\pi f n T_c} \tag{1.417}$$

---

[18] For example, for the estimator (1.405) the mean-square error is defined as

$$E\left[|\hat{\mathtt{r}}_x(n) - \mathtt{r}_x(n)|^2\right] = \text{var}[\hat{\mathtt{r}}_x(n)] + |\mu_{bias}|^2 \tag{1.413}$$

and, consequently,

$$\begin{aligned} E[\mathcal{P}_{PER}(f)] &= T_c \sum_{n=-(K-1)}^{K-1} E[\check{\mathtt{r}}_x(n)] e^{-j2\pi f n T_c} \\ &= T_c \sum_{n=-(K-1)}^{K-1} \left(1 - \frac{|n|}{K}\right) \mathtt{r}_x(n) e^{-j2\pi f n T_c} \\ &= T_c \mathcal{W}_B * \mathcal{P}_x(f) \end{aligned} \tag{1.418}$$

where $\mathcal{W}_B(f)$ is the Fourier transform of the *symmetric Bartlett window*

$$\mathtt{w}_B(n) = \begin{cases} 1 - \frac{|n|}{K} & |n| \leq K-1 \\ 0 & |n| > K-1 \end{cases} \tag{1.419}$$

and

$$\mathcal{W}_B(f) = K\,[\mathrm{sinc}_K(fKT_c)]^2 \tag{1.420}$$

We note the periodogram estimate is affected by *bias* for finite K. Moreover, it also exhibits a large variance, as $\mathcal{P}_{PER}(f)$ is computed using the samples of $\check{\mathtt{r}}_x(n)$ even for lags up to $K-1$, whose variance is very large.

Welch periodogram

This method is based on applying (1.374) for finite K. Given a sequence of K samples, different subsequences of consecutive D samples are extracted. Subsequences may partially overlap. Let $x^{(s)}$ be the s-th subsequence, characterized by S samples in common with the preceding subsequence $x^{(s-1)}$ and with the following one $x^{(s+1)}$. In general, $0 \leq S \leq D/2$, with the choice $S = 0$ yielding subsequences with no overlap and therefore with less correlation. The number of subsequences N_s is[19]

$$N_s = \left\lfloor \frac{K-D}{D-S} + 1 \right\rfloor \tag{1.421}$$

Let $\mathtt{w}$ be a window (see footnote 17 on page 59) of D samples: then

$$x^{(s)}(k) = \mathtt{w}(k)\, x(k + s(D-S)), \qquad k = 0, 1, \ldots, D-1 \;\; s = 0, 1, \ldots, N_s - 1 \tag{1.422}$$

For each s, compute the Fourier transform

$$\tilde{\mathcal{X}}^{(s)}(f) = T_c \sum_{k=0}^{D-1} x^{(s)}(k) e^{-j2\pi f k T_c} \tag{1.423}$$

and obtain

$$\mathcal{P}_{PER}^{(s)}(f) = \frac{1}{D T_c \mathtt{M}_{\mathtt{w}}} |\tilde{\mathcal{X}}^{(s)}(f)|^2 \tag{1.424}$$

where

$$\mathtt{M}_{\mathtt{w}} = \frac{1}{D} \sum_{k=0}^{D-1} \mathtt{w}^2(k) \tag{1.425}$$

is the normalized energy of the window. As a last step, for each frequency, average the periodograms:

$$\mathcal{P}_{WE}(f) = \frac{1}{N_s} \sum_{s=0}^{N_s-1} \mathcal{P}_{PER}^{(s)}(f) \tag{1.426}$$

[19] The symbol $\lfloor a \rfloor$ denotes the function *floor*, that is the largest integer smaller than or equal to a. The symbol $\lceil a \rceil$ denotes the function *ceiling*, that is the smallest integer larger than or equal to a.

The mean of the estimate is given by

$$E[\mathcal{P}_{WE}(f)] = T_c[|\mathcal{W}|^2 * \mathcal{P}_x](f) \tag{1.427}$$

where

$$\mathcal{W}(f) = \sum_{k=0}^{D-1} \mathtt{w}(k) e^{-j2\pi fkT_c} \tag{1.428}$$

Assuming the process Gaussian and the different subsequences statistically independent, we get[20]

$$\text{var}[\mathcal{P}_{WE}(f)] \propto \frac{1}{N_s}\mathcal{P}_x^2(f) \tag{1.429}$$

Note that the partial overlap introduces correlation between subsequences. From (1.429), we see that the variance of the estimate is reduced by increasing the number of subsequences. In general, D must be large enough so that the *generic subsequence represents the process*[21] and also N_s must be large to obtain a reliable estimate (see (1.429)); therefore, the application of the Welch method requires many samples.

Blackman and Tukey correlogram

For an *unbiased* estimate of the ACS, $\{\hat{\mathtt{r}}_x(n)\}$, $n = -L, \dots, L$, consider the Fourier transform

$$\mathcal{P}_{BT}(f) = T_c \sum_{n=-L}^{L} \mathtt{w}(n)\hat{\mathtt{r}}_x(n)\ e^{-j2\pi fnT_c} \tag{1.430}$$

where $\mathtt{w}$ is a window[22] of length $2L+1$, with $\mathtt{w}(0) = 1$. If K is the number of samples of the realization sequence, we require that $L \le K/5$ to reduce the variance of the estimate. Then if the Bartlett window (1.420) is chosen, one finds that $\mathcal{P}_{BT}(f) \ge 0$.

In terms of the mean value of the estimate, we find

$$E[\mathcal{P}_{BT}(f)] = T_c(\mathcal{W} * \mathcal{P}_x)(f) \tag{1.431}$$

For a Gaussian process, if the Bartlett window is chosen, the variance of the estimate is given by

$$\text{var}[\mathcal{P}_{BT}(f)] = \frac{1}{K}\mathcal{P}_x^2(f)E_{\mathtt{w}} = \frac{2}{3}\frac{L}{K}\mathcal{P}_x^2(f) \tag{1.432}$$

Windowing and window closing

The windowing operation of time sequence in the periodogram, and of the ACS in the correlogram, has a strong effect on the performance of the estimate. In fact, any truncation of a sequence is equivalent to a windowing operation, carried out via the rect function. The choice of the window type in the frequency domain depends on the compromise between a narrow central lobe (to reduce *smearing*) and a fast decay of secondary lobes (to reduce *leakage*). *Smearing* yields a lower spectral resolution, that is the capability to distinguish two spectral lines that are close. On the other hand, *leakage* can mask spectral components that are further apart and have different amplitudes.

The choice of the window length is based on the compromise between spectral resolution and the variance of the estimate. An example has already been seen in the correlogram, where the condition $L \le K/5$ must be satisfied. Another example is the Welch periodogram. For a given observation of K samples, it is

[20] Notation $a \propto b$ means that a is proportional to b.

[21] For example, if x is a sinusoidal process, DT_c must at least be greater than 5 or 10 periods of x.

[22] The windows used in (1.430) are the same introduced in footnote 17: the only difference is that they are now centered around zero instead of $(D-1)/2$. To simplify the notation, we will use the same symbol in both cases.

initially better to choose a small number of samples over which to perform the DFT, and therefore a large number of windows (subsequences) over which to average the estimate. The estimate is then repeated by increasing the number of samples per window, thus decreasing the number of windows. In this way, we get estimates with not only a higher resolution but also characterized by an increasing variance. The procedure is terminated once it is found that the increase in variance is no longer compensated by an increase in the spectral resolution. The aforementioned method is called *window closing*.

Example 1.9.1

Consider a realization of $K = 10\,000$ samples of the signal:

$$y(kT_c) = \frac{1}{A_h} \sum_{n=-16}^{16} h(nT_c)w((k-n)T_c) + A_1 \cos(2\pi f_1 kT_c + \varphi_1) + A_2 \cos(2\pi f_2 kT_c + \varphi_2) \tag{1.433}$$

where $\varphi_1, \varphi_2 \sim \mathcal{U}[0, 2\pi)$, $w(nT_c)$ is a white random process with zero mean and variance $\sigma_w^2 = 5$, $T_c = 0.2$, $A_1 = 1/20$, $f_1 = 1.5$, $A_2 = 1/40$, $f_2 = 1.75$, and

$$A_h = \sum_{-16}^{16} h(kT_c) \tag{1.434}$$

Moreover

$$h(kT_c) = \frac{\sin\left(\pi(1-\rho)\frac{kT_c}{T}\right) + 4\rho\frac{kT_c}{T}\cos\left(\pi(1+\rho)\frac{kT_c}{T}\right)}{\pi\left[1-\left(4\rho\frac{kT_c}{T}\right)^2\right]\frac{kT_c}{T}} \mathrm{rect}\left(\frac{kT_c}{8T+T_c}\right) \tag{1.435}$$

with $T = 4T_c$ and $\rho = 0.32$.

Actually y is the sum of two sinusoidal signals and filtered white noise through h. Consequently, observing (1.188) and (1.313),

$$\begin{aligned} \mathcal{P}_y(f) = {} & \sigma_w^2\, T_c\, \frac{|\mathcal{H}(f)|^2}{A_h^2} + \frac{A_1^2}{4}(\delta(f-f_1) + \delta(f+f_1)) \\ & + \frac{A_2^2}{4}(\delta(f-f_2) + \delta(f+f_2)) \end{aligned} \tag{1.436}$$

where $\mathcal{H}(f)$ is the Fourier transform of $\{h(kT_c)\}$.

The shape of the PSD in (1.436) is shown in Figures 1.32–1.34 as a solid line. A Dirac impulse is represented by an isosceles triangle having a base equal to twice the desired frequency resolution F_q. Consequently, a Dirac impulse, for example of area $A_1^2/4$ will have a height equal to $A_1^2/(4F_q)$, thus, maintaining the equivalence in statistical power between different representations.

We now compare several spectral estimates, obtained using the previously described methods; in particular, we will emphasize the effect on the resolution of the type of window used and the number of samples for each window.

We state beforehand the following important result. Windowing a complex sinusoidal signal $\{e^{j2\pi f_1 kT_c}\}$ with $\{\mathrm{w}(k)\}$ produces a signal having Fourier transform equal to $\mathcal{W}(f-f_1)$, where $\mathcal{W}(f)$ is the Fourier transform of w. Therefore, in the frequency domain, the spectral line of a sinusoidal signal becomes a signal with shape $\mathcal{W}(f)$ centred around f_1.

In general, from (1.424), the periodogram of a real sinusoidal signal with amplitude A_1 and frequency f_1 is

$$\mathcal{P}_{PER}(f) = \frac{T_c}{\mathrm{DM_w}}\left(\frac{A_1}{2}\right)^2 |\mathcal{W}(f-f_1) + \mathcal{W}(f+f_1)|^2 \tag{1.437}$$

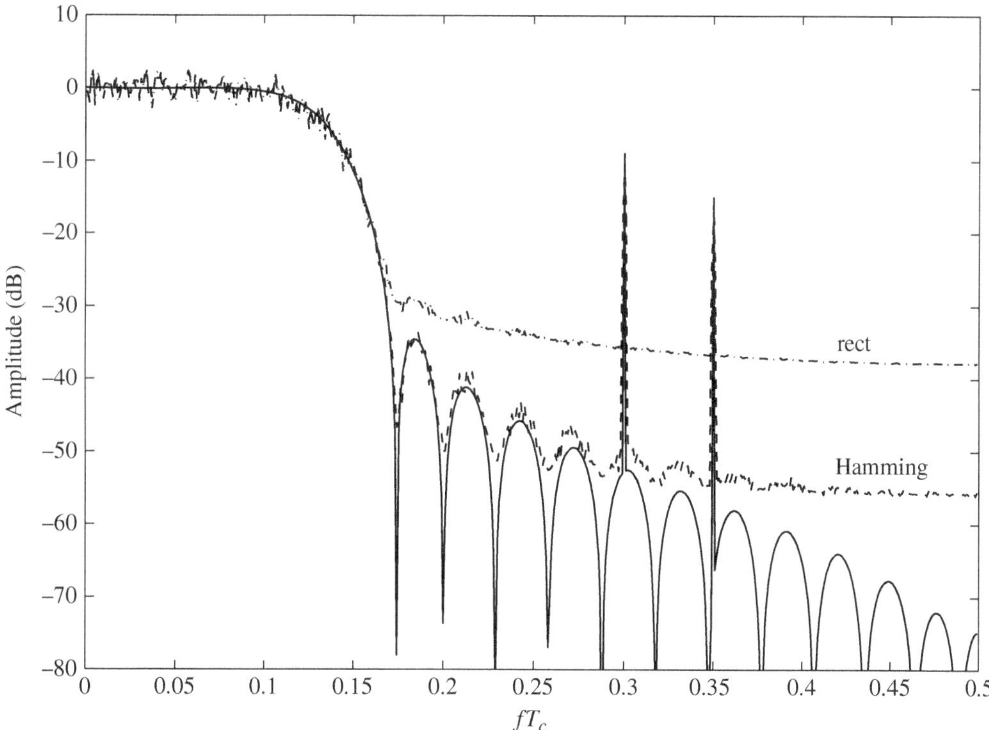

Figure 1.32 Comparison between spectral estimates obtained with Welch periodogram method, using the Hamming or the rectangular window, and the analytical PSD given by (1.436).

Figure 1.32 shows, in addition to the analytical PSD (1.436), the estimate obtained by the Welch periodogram method using the Hamming or the rectangular windows. Parameters used in (1.423) and (1.426) are: $D = 1000$, $N_s = 19$, and 50% overlap between windows. We observe that the use of the Hamming window yields an improvement of the estimate due to less *leakage*. Likewise Figure 1.33 shows how the Hamming window also improves the estimate carried out with the correlogram; in particular, the estimates of Figure 1.33 were obtained using in (1.430) $L = 500$. Lastly, Figure 1.34 shows how the resolution and the variance of the estimate obtained by the Welch periodogram vary with the parameters D and N_s, using the Hamming window. Note that by increasing D, and hence decreasing N_s, both resolution and variance of the estimate increase.

1.10 Parametric models of random processes

ARMA

Let us consider the realization of a random process x according to the *auto-regressive moving average* (ARMA) model illustrated in Figure 1.35. In other words, the process x, also called observed sequence, is the output of an IIR filter having as input white noise with variance σ_w^2, and is given by the recursive

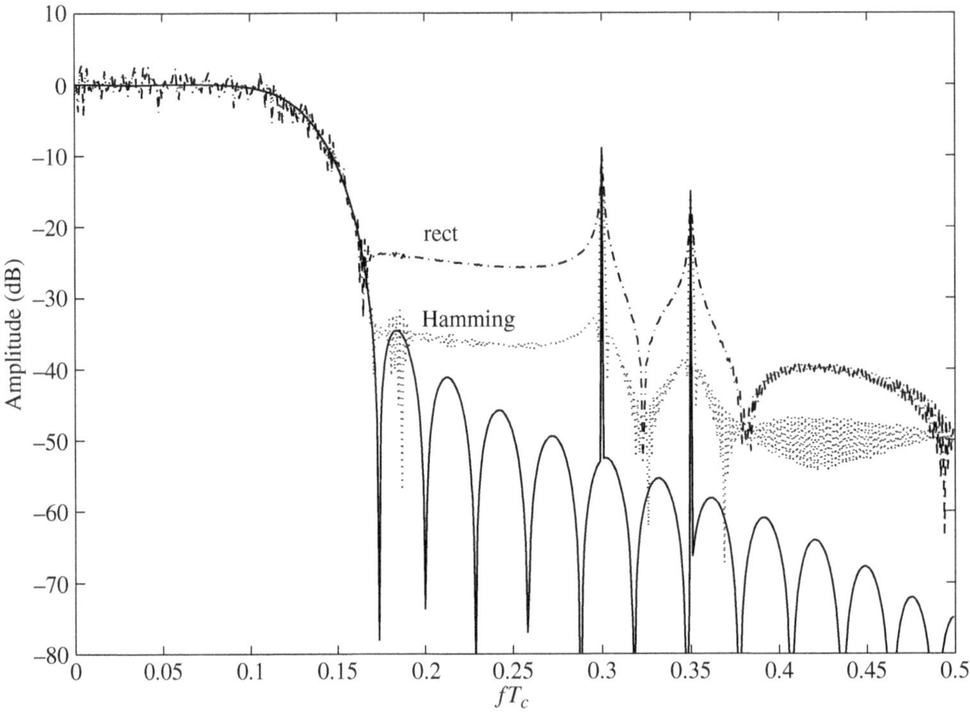

Figure 1.33 Comparison between spectral estimates obtained with the correlogram using the Hamming or the rectangular window, and the analytical PSD given by (1.436).

equation[23]

$$x(k) = -\sum_{n=1}^{p} \mathtt{a}_n x(k-n) + \sum_{n=0}^{q} \mathtt{b}_n w(k-n) \tag{1.438}$$

and the model is denoted as $ARMA(p,q)$.

Rewriting (1.438) in terms of the filter impulse response h_{ARMA}, we find in general

$$x(k) = \sum_{n=0}^{+\infty} h_{ARMA}(n) w(k-n) \tag{1.439}$$

which indicates that the filter used to realize the ARMA model is causal. From (1.63), one finds that the filter transfer function is given by

$$H_{ARMA}(z) = \frac{B(z)}{A(z)} \quad \text{where} \quad \begin{cases} B(z) = \sum_{n=0}^{q} \mathtt{b}_n z^{-n} \\ A(z) = \sum_{n=0}^{p} \mathtt{a}_n z^{-n} \text{ assuming } \mathtt{a}_0 = 1 \end{cases} \tag{1.440}$$

Using (1.188), the PSD of the process x is given by

$$\mathcal{P}_x(f) = T_c \sigma_w^2 \left| \frac{\mathcal{B}(f)}{\mathcal{A}(f)} \right|^2 \quad \text{where} \quad \begin{cases} \mathcal{B}(f) = B(e^{j2\pi f T_c}) \\ \mathcal{A}(f) = A(e^{j2\pi f T_c}) \end{cases} \tag{1.441}$$

[23] In a simulation of the process, the first samples $x(k)$ generated by (1.438) should be neglected because they depend on the initial conditions. Specifically, if N_{ARMA} is the *length* of the filter impulse response h_{ARMA}, the minimum number of samples to be ignored is $N_{ARMA} - 1$, equal to the filter transient.

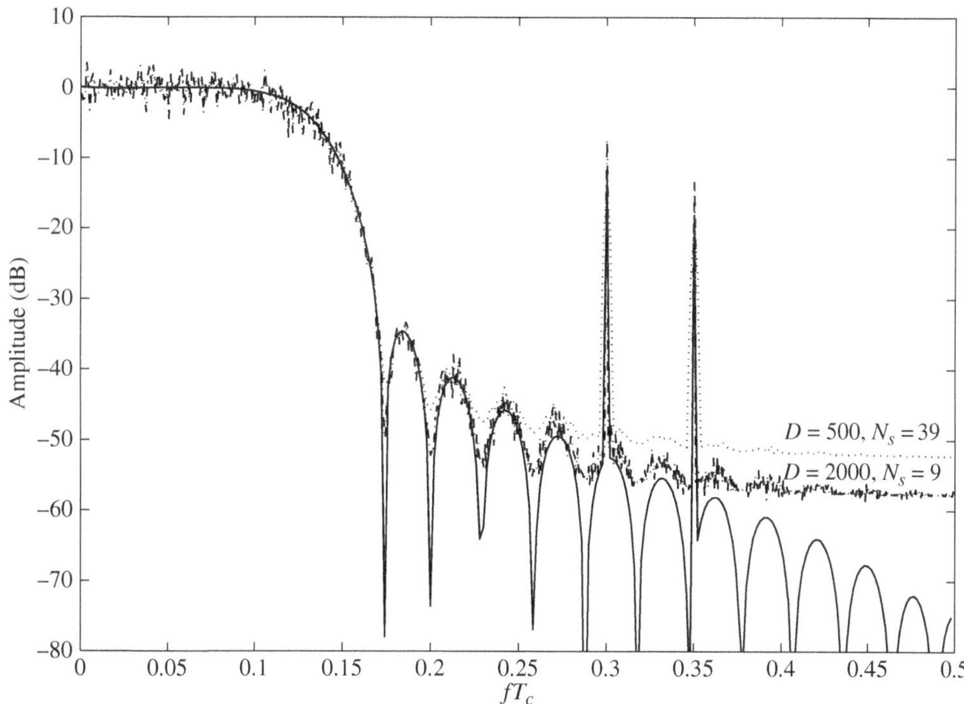

Figure 1.34 Comparison of spectral estimates obtained with the Welch periodogram method, using the Hamming window, by varying parameters D ed N_s.

MA

If we particularize the ARMA model, assuming

$$\mathrm{a}_i = 0, \qquad i = 1, 2, \dots, p \tag{1.442}$$

or $A(z) = 1$, we get the *moving average* (MA) model of order q, also denoted $MA(q)$. The equations of the ARMA model therefore are reduced to

$$H_{MA}(z) = B(z) \tag{1.443}$$

and

$$\mathcal{P}_x(f) = T_c\, \sigma_w^2 |\mathcal{B}(f)|^2 \tag{1.444}$$

If we represent the function $\mathcal{P}_x(f)$ of a process obtained by the MA model, we see that its behaviour is generally characterized by wide *peaks* and narrow *valleys*, as illustrated in Figure 1.36.

AR

The *auto-regressive* (AR) model of order N, also denoted as AR(N) is shown in Figure 1.37.

The output process is described in this case by the recursive equation

$$x(k) = -\sum_{n=1}^{N} \mathrm{a}_n\, x(k-n) + w(k) \tag{1.445}$$

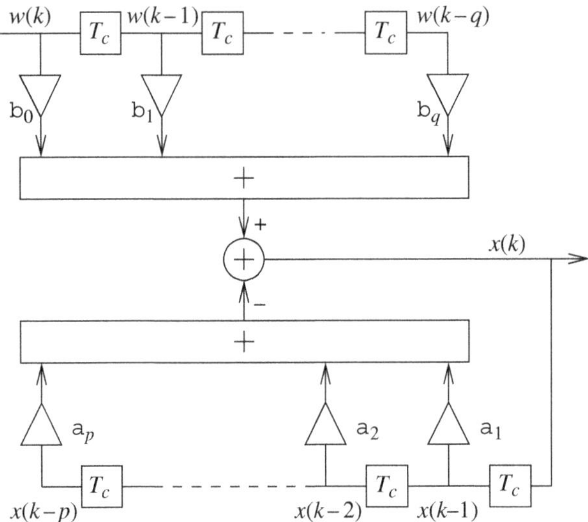

Figure 1.35 ARMA model of a process x.

where w is white noise with variance σ_w^2. The transfer function is given by

$$H_{AR}(z) = \frac{1}{A(z)} \tag{1.446}$$

with

$$A(z) = 1 + \sum_{n=1}^{N} \mathrm{a}_n z^{-n} \tag{1.447}$$

We observe that (1.446) describes a filter having N poles. Therefore, $H_{AR}(z)$ can be expressed as

$$H_{AR}(z) = \frac{1}{(1-\mathrm{p}_1 z^{-1})(1-\mathrm{p}_2 z^{-1}) \dots (1-\mathrm{p}_N z^{-1})} \tag{1.448}$$

For a causal filter, the stability condition is $|\mathrm{p}_i| < 1$, $i = 1, 2, \dots, N$, i.e. all poles must be inside the unit circle of the z plane.

In the case of the AR model, from Table 1.6 the z-transform of the ACS of x is given by

$$P_x(z) = P_w(z) \frac{1}{A(z)A^*\left(\frac{1}{z^*}\right)} = \frac{\sigma_w^2}{A(z)A^*\left(\frac{1}{z^*}\right)} \tag{1.449}$$

Hence, the function $P_x(z)$ has poles of the type

$$|\mathrm{p}_i| e^{j\varphi_i} \quad \text{and} \quad \frac{1}{|\mathrm{p}_i|} e^{j\varphi_i} \tag{1.450}$$

Letting

$$\mathcal{A}(f) = A(e^{j2\pi f T_c}) \tag{1.451}$$

one obtains the PSD of x, given by

$$\mathcal{P}_x(f) = \frac{T_c \sigma_w^2}{|\mathcal{A}(f)|^2} \tag{1.452}$$

Typically, the function $\mathcal{P}_x(f)$ of an AR process will have narrow *peaks* and wide *valleys* (see Figure 1.38), opposite to the MA model.

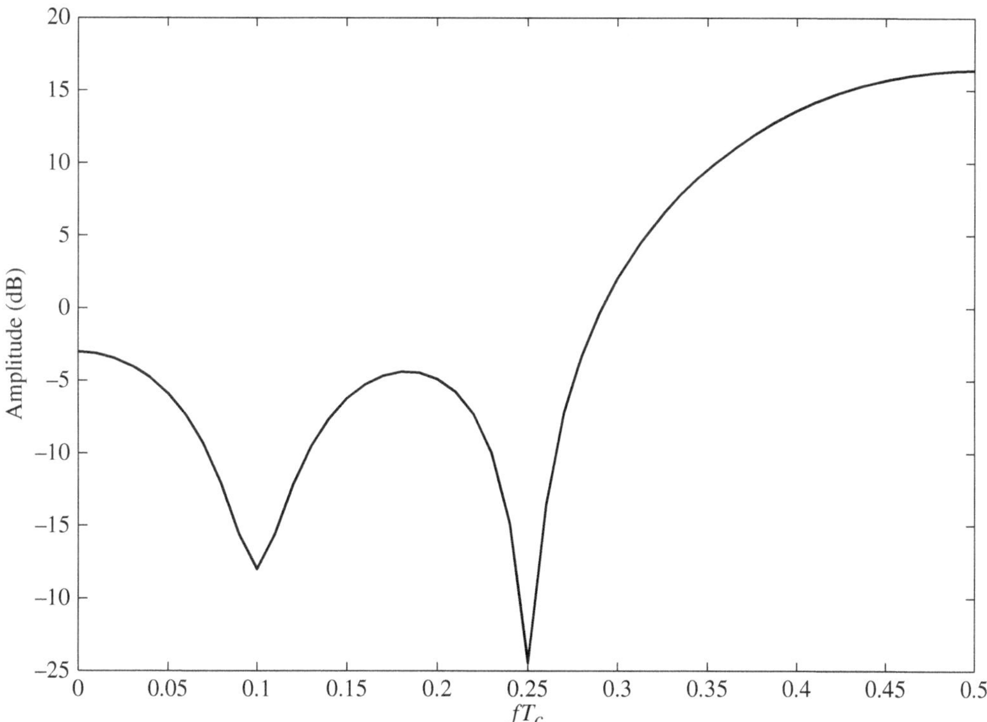

Figure 1.36 PSD of a MA process with $q = 4$.

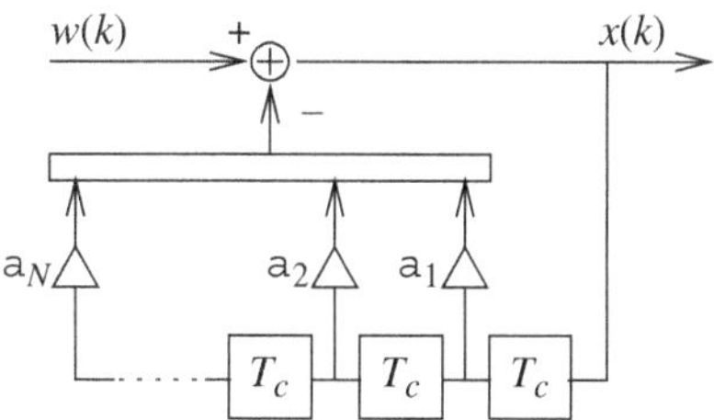

Figure 1.37 AR model of a process x.

Spectral factorization of AR models

Consider the AR process described by (1.449). Observing (1.450), we have the following decomposition:

$$P_x(z) = \frac{\sigma_w^2}{(1 - |p_1|e^{j\varphi_1}z^{-1})...(1 - |p_N|e^{j\varphi_N}z^{-1})\left(1 - \frac{e^{j\varphi_1}z^{-1}}{|p_1|}\right)...\left(1 - \frac{e^{j\varphi_N}z^{-1}}{|p_N|}\right)} \tag{1.453}$$

For a given $P_x(z)$, it is clear that the N zeros of $A(z)$ in (1.449) can be chosen in 2^N different ways. The selection of the zeros of $A(z)$ is called *spectral factorization*. Two examples are illustrated in Figure 1.39.

As stated by the spectral factorization theorem (see page 35), there exists a unique spectral factorization that yields a minimum-phase $A(z)$, which is obtained by associating with $A(z)$ only the poles of $P_x(z)$ that lie inside the unit circle of the z-plane.

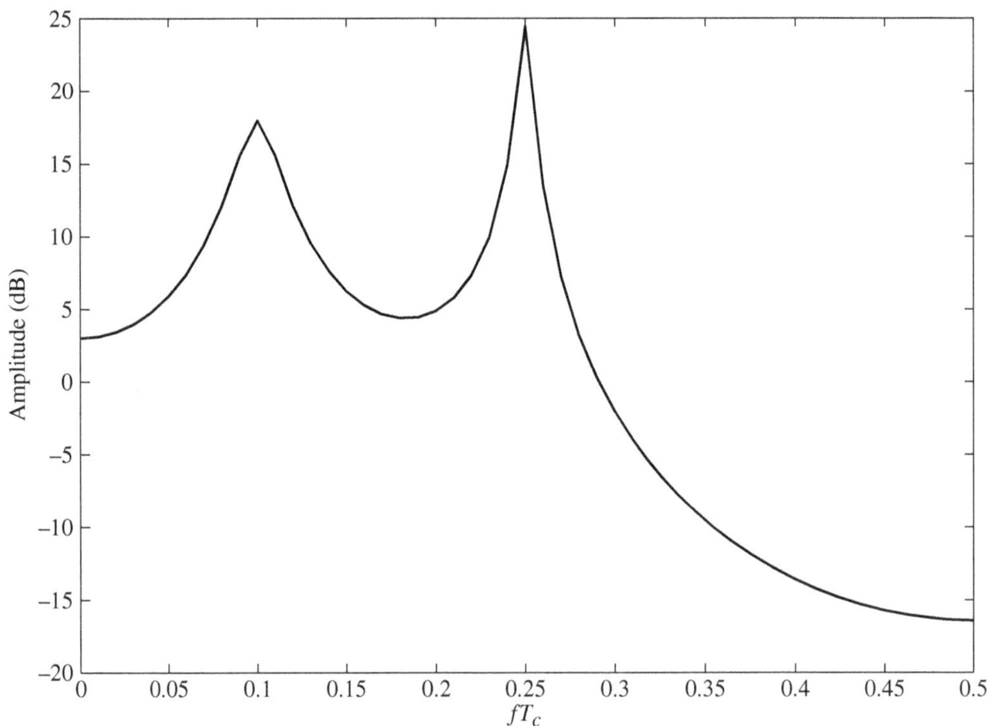

Figure 1.38 PSD of an AR process with $N = 4$.

Whitening filter

We observe an important property illustrated in Figure 1.40. Suppose x is modelled as an AR process of order N and has PSD given by (1.449). If x is input to a filter having transfer function $A(z)$, the output of this latter filter would be white noise. In this case, the filter $A(z)$ is called *whitening filter* (WF).

If $A(z)$ is minimum phase, the white process w is also called *innovation* of the process x, in the sense that the new information associated with the sample $x(k)$ is carried only by $w(k)$.

Relation between ARMA, MA, and AR models

The relations among the three parametric models are expressed through the following propositions.

Wold decomposition. Every WSS random process y can be decomposed into:

$$y(k) = s(k) + x(k) \tag{1.454}$$

where s and x uncorrelated processes. The process s, *completely predictable*, is described by the recursive equation

$$s(k) = -\sum_{n=1}^{+\infty} \alpha_n s(k-n) \tag{1.455}$$

while x is obtained as filtered white noise:

$$x(k) = \sum_{n=0}^{+\infty} h(n) w(k-n) \tag{1.456}$$

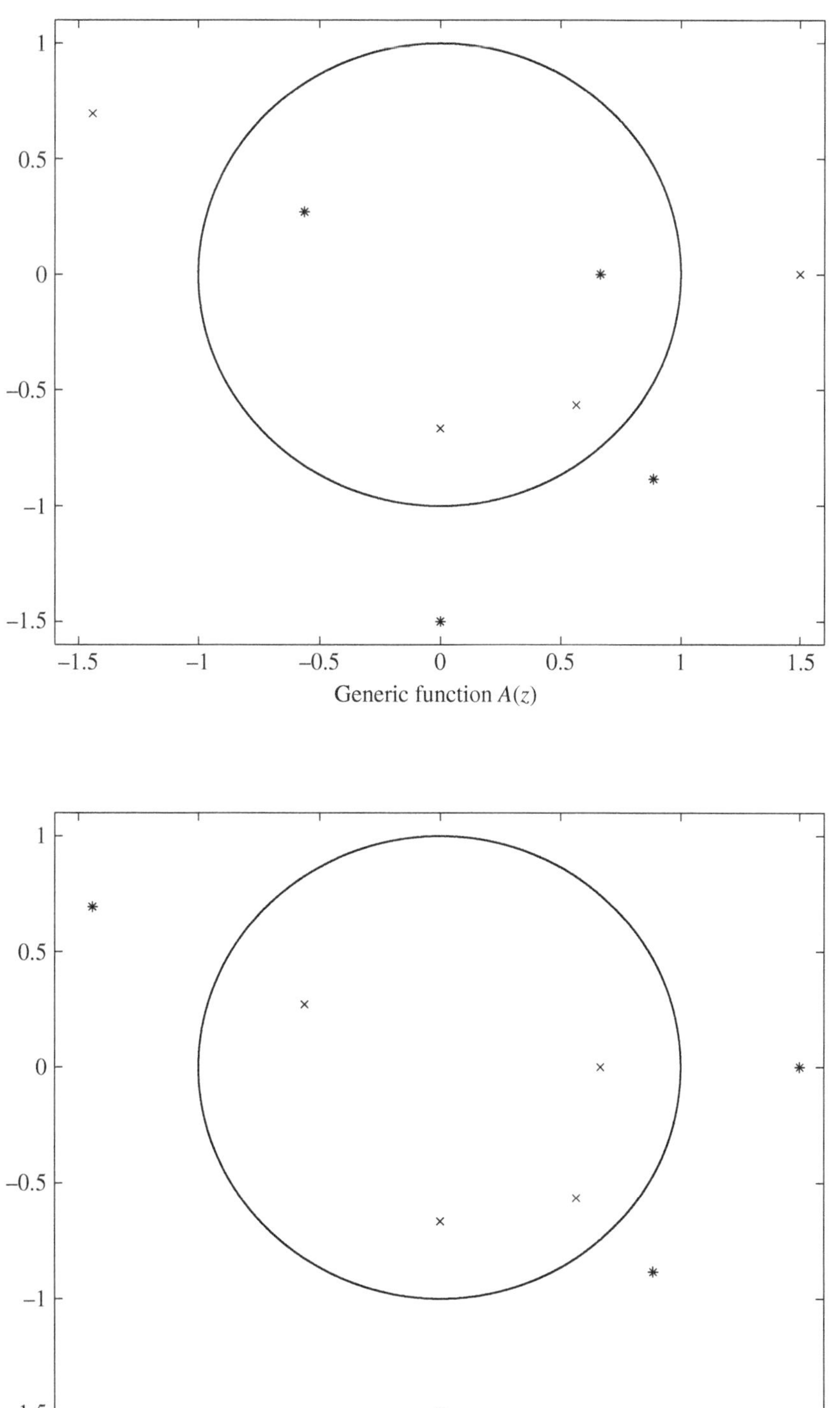

Figure 1.39 Two examples of possible choices of the zeros (×) of $A(z)$, among the poles of $P_x(z)$.

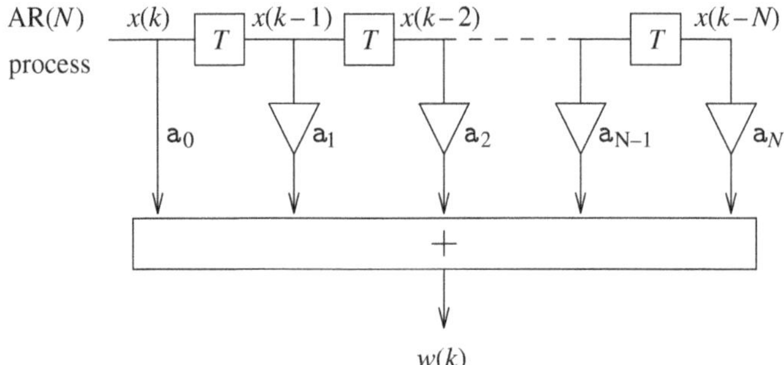

Figure 1.40 Whitening filter for an AR process of order N.

Theorem 1.5 (Kolmogorov theorem)
Any ARMA or MA process can be represented by an AR process of infinite order. □

Therefore, any one of the three descriptions (ARMA, MA, or AR) can be adopted to approximate the spectrum of a process, provided that the order is sufficiently high. However, in practice, determining the coefficients associated with a model when the order is high is not a simple task from a numerical point of view. It is better to start from a quite low order value, evaluating the fitting quality (usually by the mean square error) between the given process and the output model and then increasing the order value up to when the fitting quality does not significantly change, i.e. it reaches approximately its asymptotic value.

1.10.1 Autocorrelation of AR processes

It is interesting to evaluate the autocorrelation function of a process x obtained by the AR model. Multiplying both members of (1.445) by $x^*(k-n)$, we find

$$x(k)x^*(k-n) = -\sum_{m=1}^{N} \mathtt{a}_m x(k-m)x^*(k-n) + w(k)x^*(k-n) \tag{1.457}$$

Taking expectations, and observing that $w(k)$ is uncorrelated with all past values of x, for $n \geq 0$ we obtain

$$E[x(k)x^*(k-n)] = -\sum_{m=1}^{N} \mathtt{a}_m \, E[x(k-m)\, x^*(k-n)] + \sigma_w^2 \delta_n \tag{1.458}$$

From (1.458), it follows

$$\mathtt{r}_x(n) = -\sum_{m=1}^{N} \mathtt{a}_m \mathtt{r}_x(n-m) + \sigma_w^2 \delta_n \tag{1.459}$$

In particular, we have

$$\mathtt{r}_x(n) = \begin{cases} -\sum_{m=1}^{N} \mathtt{a}_m \mathtt{r}_x(n-m) & n > 0 \\ -\sum_{m=1}^{N} \mathtt{a}_m \mathtt{r}_x(-m) + \sigma_w^2 & n = 0 \\ \mathtt{r}_x^*(-n) & n < 0 \end{cases} \tag{1.460}$$

We observe that, for $n > 0$, $\mathtt{r}_x(n)$ satisfies an equation analogous to the (1.445), with the exception of the component $w(k)$. This implies that, if $\{p_i\}$ are zeros of $A(z)$, $\mathtt{r}_x(n)$ can be written as

$$\mathtt{r}_x(n) = \sum_{i=1}^{N} \mathtt{c}_i p_i^n \qquad n > 0 \tag{1.461}$$

Assuming an AR process with $|p_i| < 1$, for $i = 1, 2, \ldots, N$, we get:

$$\mathtt{r}_x(n) \xrightarrow[n\to\infty]{} 0 \tag{1.462}$$

Simplifying the notation $\mathtt{r}_x(n)$ with $\mathtt{r}(n)$, and observing (1.460), for $n = 1, 2, \ldots, N$, we obtain a set of equations that in matrix notation are expressed as

$$\begin{bmatrix} \mathtt{r}(0) & \mathtt{r}(-1) & \ldots & \mathtt{r}(-N+1) \\ \mathtt{r}(1) & \mathtt{r}(0) & \ldots & \mathtt{r}(-N+2) \\ \vdots & \vdots & \ddots & \cdots \\ \mathtt{r}(N-1) & \mathtt{r}(N-2) & \ldots & \mathtt{r}(0) \end{bmatrix} \begin{bmatrix} \mathtt{a}_1 \\ \mathtt{a}_2 \\ \vdots \\ \mathtt{a}_N \end{bmatrix} = - \begin{bmatrix} \mathtt{r}(1) \\ \mathtt{r}(2) \\ \vdots \\ \mathtt{r}(N) \end{bmatrix} \tag{1.463}$$

that is

$$\boldsymbol{R}\boldsymbol{a} = -\boldsymbol{r} \tag{1.464}$$

with obvious definition of the vectors. In the hypothesis that the matrix $\boldsymbol{R}$ has an inverse, the solution for the coefficients $\{\mathtt{a}_i\}$ is given by

$$\boldsymbol{a} = -\boldsymbol{R}^{-1}\boldsymbol{r} \tag{1.465}$$

Equations (1.464) and (1.465), called *Yule-Walker equations*, provide the coefficients of an AR model for a process having autocorrelation function $\mathtt{r}_x$. The variance σ_w^2 of white noise at the input can be obtained from (1.460) for $n = 0$, which yields

$$\begin{aligned} \sigma_w^2 &= \mathtt{r}_x(0) + \sum_{m=1}^{N} \mathtt{a}_m \mathtt{r}_x(-m) \\ &= \mathtt{r}_x(0) + \boldsymbol{r}^H \boldsymbol{a} \end{aligned} \tag{1.466}$$

Observation 1.6

- From (1.464) one finds that $\mathtt{a}$ does not depend on $\mathtt{r}_x(0)$, but only on the correlation coefficients

$$\rho_x(n) = \frac{\mathtt{r}_x(n)}{\mathtt{r}_x(0)}, \qquad n = 1, \ldots, N \tag{1.467}$$

- Exploiting the fact that $\boldsymbol{R}$ is Toeplitz and Hermitian,[24] the set of equations (1.465) and (1.466) can be numerically solved by the Levinson–Durbin or by Delsarte–Genin algorithms, with a computational complexity proportional to N^2 (see Appendix 2.A).
- We note that the knowledge of $\mathtt{r}_x(0), \mathtt{r}_x(1), \ldots, \mathtt{r}_x(N)$ univocally determines the ACS of an AR(N) process; for $n > N$, from (1.460), we get

$$\mathtt{r}_x(n) = -\sum_{m=1}^{N} \mathtt{a}_m \mathtt{r}_x(n-m) \tag{1.468}$$

[24] A matrix is Toeplitz if all elements of any diagonal are equal within the diagonal. Matrix $\boldsymbol{A}$ is Hermitian if $\boldsymbol{A} = \boldsymbol{A}^H$.

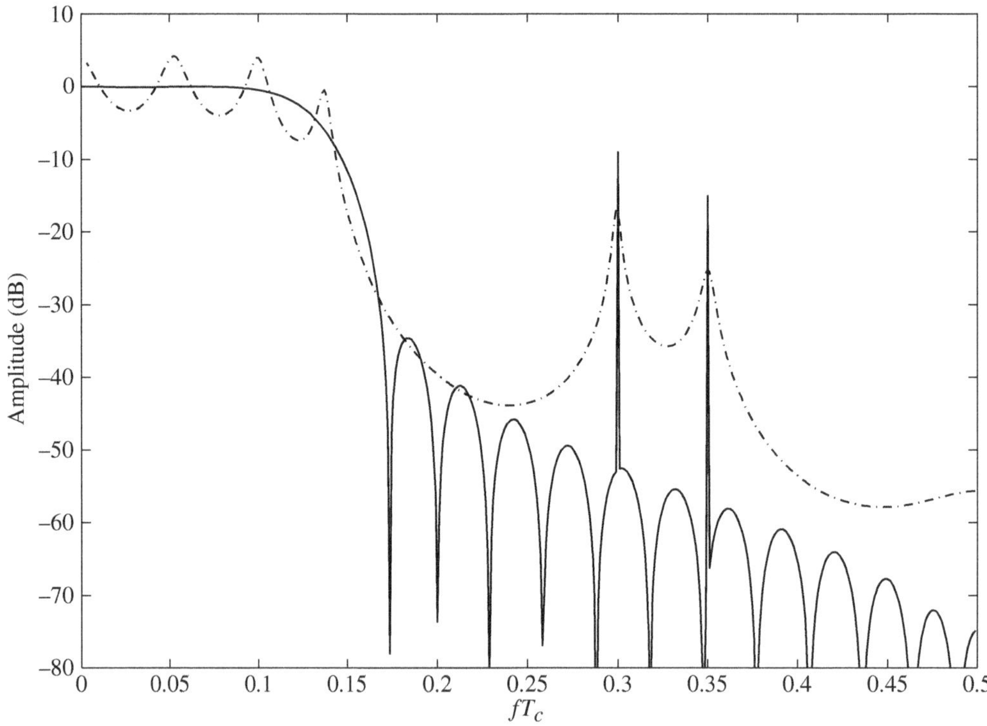

Figure 1.41 Comparison between the spectral estimate obtained by an AR(12) process model and the analytical PSD given by (1.436).

1.10.2 Spectral estimation of an AR process

Assuming an AR(N) model for a process x, (1.465) yields the coefficient vector $\mathbf{a}$, which implies an estimate of the ACS up to lag N is available. From (1.452), we define as spectral estimate

$$\mathcal{P}_{AR}(f) = \frac{T_c \sigma_w^2}{|\mathcal{A}(f)|^2} \tag{1.469}$$

Usually the estimate (1.469) yields a better resolution than estimates obtained by other methods, such as $\mathcal{P}_{BT}(f)$, because it does not show the effects of ACS truncation. In fact, the AR model yields

$$\mathcal{P}_{AR}(f) = T_c \sum_{n=-\infty}^{+\infty} \hat{\mathtt{r}}_x(n)\, e^{-j2\pi f n T_c} \tag{1.470}$$

where $\hat{\mathtt{r}}_x(n)$ is estimated for $|n| \leq N$ with one of the two methods of Section 1.9.2, while for $|n| > N$ the recursive equation (1.468) is used. The AR model accurately estimates processes with a spectrum similar to that given in Figure 1.38. For example, a spectral estimate for the process of Example 1.9.1 on page 64 obtained by an AR(12) model is depicted in Figure 1.41. The correlation coefficients were obtained by a *biased* estimate on 10 000 samples. Note that the continuous part of the spectrum is estimated only approximately; on the other hand, the choice of a larger order N would result in an estimate with larger variance.

Also note that the presence of spectral lines in the original process leads to zeros of the polynomial $A(z)$ near the unit circle (see page 77). In practice, the correlation estimation method and a choice of a large N may result in an ill-conditioned matrix $\boldsymbol{R}$. In this case, the solution may have poles outside the unit circle,

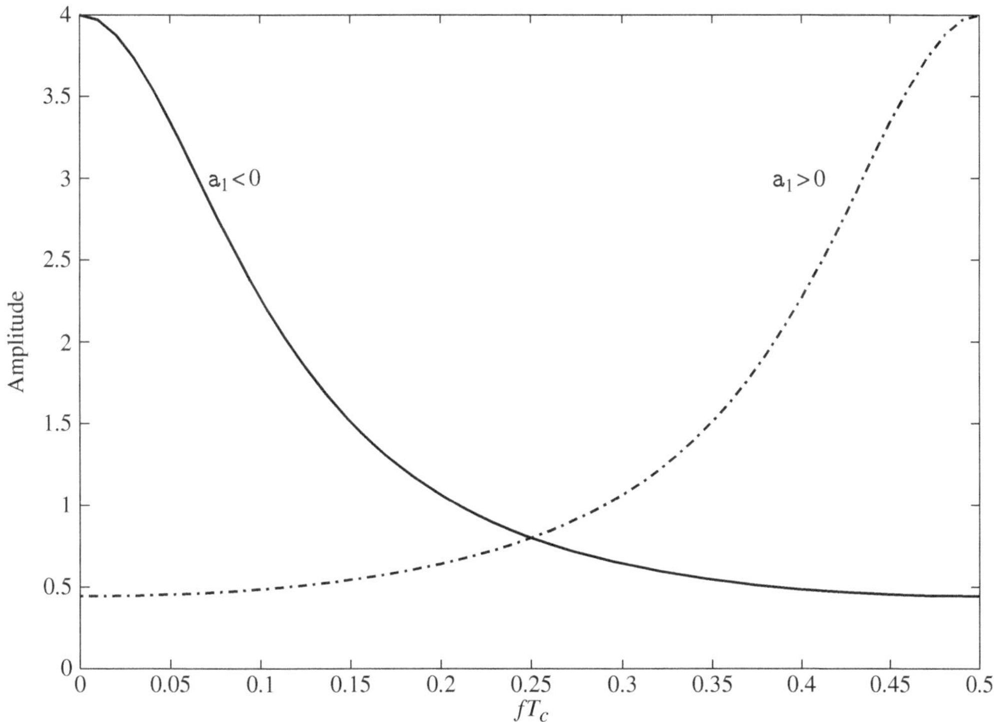

Figure 1.42 PSD of an AR(1) process.

and hence, the system would be unstable. Hence, two observations are in order. If possible, by filtering, we should split the input process into its *continuous* and *discrete* parts and estimate the two components separately. To avoid using an ill-conditioned matrix $\boldsymbol{R}$, yielding unstable numerical solutions, we should use an iterative approach where for increasing values of N, we evaluate σ_w^2 by (1.466), and where the correlation $\hat{\mathtt{r}}_x(n)$ is replaced by its estimate. A suitable value of N is reached when the estimate of σ_w^2 does not significantly change, i.e. it reaches approximately its asymptotic value.

Some useful relations

We will illustrate some examples of AR models. In particular, we will focus on the Yule–Walker equations and the relation (1.466) for $N = 1$ and $N = 2$.

AR(1). From

$$\begin{cases} \mathtt{r}_x(n) = -\mathtt{a}_1 \mathtt{r}_x(n-1) & n > 0 \\ \sigma_w^2 = \mathtt{r}_x(0) + \mathtt{a}_1 \mathtt{r}_x(-1) \end{cases} \tag{1.471}$$

we obtain

$$\mathtt{r}_{AR(1)}(n) = \frac{\sigma_w^2}{1 - |\mathtt{a}_1|^2}(-\mathtt{a}_1)^{|n|} \tag{1.472}$$

from which the PSD is

$$\mathcal{P}_{AR(1)}(f) = \frac{T_c \sigma_w^2}{|1 + \mathtt{a}_1 e^{-j2\pi f T_c}|^2} \tag{1.473}$$

The behaviour of the PSD of an AR(1) process is illustrated in Figure 1.42.

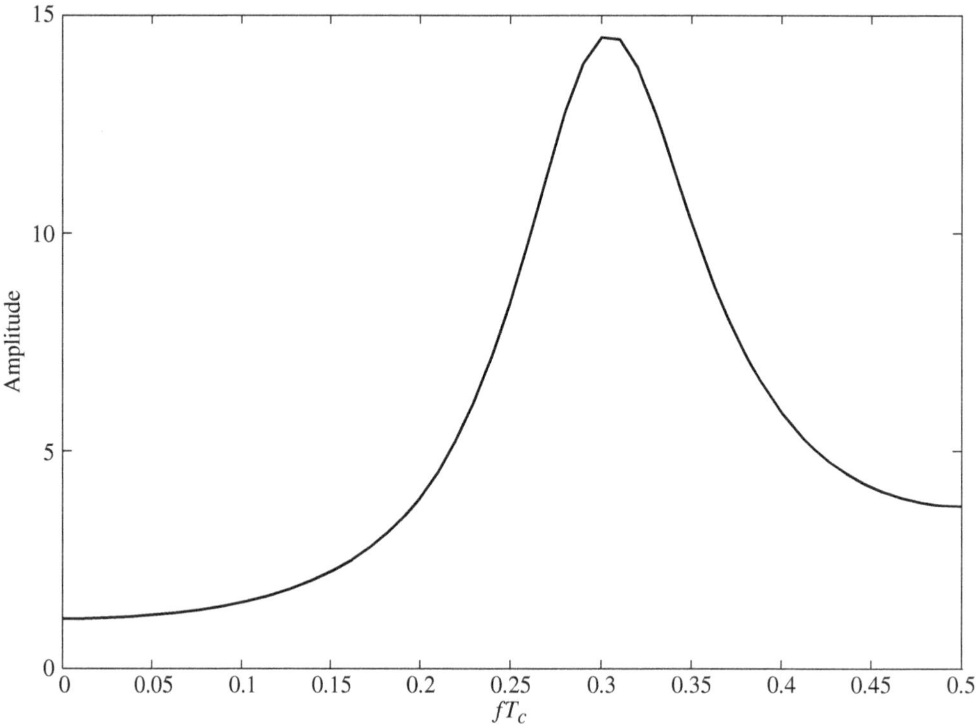

Figure 1.43 PSD of an AR(2) process.

AR(2). Let $p_{1,2} = \varrho e^{\pm j\varphi_0}$, where $\varphi_0 = 2\pi f_0 T_c$, be the two complex roots of $A(z) = 1 + \mathtt{a}_1 z^{-1} + \mathtt{a}_2 z^{-2}$. We consider a *real process*:

$$\begin{cases} \mathtt{a}_1 = -2\varrho\cos(2\pi f_0 T_c) \\ \mathtt{a}_2 = \varrho^2 \end{cases} \tag{1.474}$$

Letting

$$\vartheta = \tan^{-1}\left[\frac{1-\varrho^2}{1+\varrho^2}\tan^{-1}(2\pi f_0 T_c)\right] \tag{1.475}$$

we find

$$\mathtt{r}_{AR(2)}(n) = \sigma_w^2 \frac{\frac{1+\varrho^2}{1-\varrho^2}\sqrt{1+\left(\frac{1-\varrho^2}{1+\varrho^2}\right)^2[\tan^{-1}(2\pi f_0 T_c)]^2}}{1-\varrho^2\cos^2(4\pi f_0 T_c)+\varrho^4}\varrho^{|n|}\cos(2\pi f_0 |n| T_c - \vartheta) \tag{1.476}$$

The PSD is thus given by

$$\mathcal{P}_{AR(2)}(f) = \frac{T_c \sigma_w^2}{|1-\varrho e^{-j2\pi(f-f_0)T_c}|^2 |1-\varrho e^{-j2\pi(f+f_0)T_c}|^2} \tag{1.477}$$

We observe that, as $\varrho \to 1$, $\mathcal{P}_{AR(2)}(f)$ has a peak that becomes more pronounced, as illustrated in Figure 1.43, and $\mathtt{r}_x(k)$ tends to exhibit a sinusoidal behaviour.

Solutions of the Yule–Walker equations are

$$\begin{cases} \mathtt{a}_1 = -\dfrac{\mathtt{r}_x(1)\mathtt{r}_x(0) - \mathtt{r}_x(1)\mathtt{r}_x(2)}{\mathtt{r}_x^2(0) - \mathtt{r}_x^2(1)} \\ \mathtt{a}_2 = -\dfrac{\mathtt{r}_x(0)\mathtt{r}_x(2) - \mathtt{r}_x^2(1)}{\mathtt{r}_x^2(0) - \mathtt{r}_x^2(1)} \\ \sigma_w^2 = \mathtt{r}_x(0) + \mathtt{a}_1\mathtt{r}_x(1) + \mathtt{a}_2\mathtt{r}_x(2) \end{cases} \tag{1.478}$$

Solving the previous set of equations with respect to $\mathtt{r}_x(0)$, $\mathtt{r}_x(1)$, and $\mathtt{r}_x(2)$, one obtains

$$\begin{cases} \mathtt{r}_x(0) = \dfrac{1+\mathtt{a}_2}{1-\mathtt{a}_2} \dfrac{\sigma_w^2}{(1+\mathtt{a}_2)^2 - \mathtt{a}_1^2} \\ \mathtt{r}_x(1) = -\dfrac{\mathtt{a}_1}{1+\mathtt{a}_2}\mathtt{r}_x(0) \\ \mathtt{r}_x(2) = \left(-\mathtt{a}_2 + \dfrac{\mathtt{a}_1^2}{1+\mathtt{a}_2}\right)\mathtt{r}_x(0) \end{cases} \tag{1.479}$$

In general, for $n > 0$, we have

$$\mathtt{r}_x(n) = \mathtt{r}_x(0)\left[\frac{p_1(p_2^2-1)}{(p_2-p_1)(p_1p_2+1)}p_1^n - \frac{p_2(p_1^2-1)}{(p_2-p_1)(p_1p_2+1)}p_2^n\right] \tag{1.480}$$

AR model of sinusoidal processes

The general formulation of a sinusoidal process is

$$x(k) = A\cos(2\pi f_0 kT_c + \varphi) \tag{1.481}$$

with $\varphi \sim \mathcal{U}[0, 2\pi)$. We observe that the process described by (1.481) satisfies the following difference equation for $k \geq 0$:

$$x(k) = 2\cos(2\pi f_0 T_c)x(k-1) - x(k-2) + \delta_k A\cos\varphi - \delta_{k-1}A\cos(2\pi f_0 T_c - \varphi) \tag{1.482}$$

with $x(-2) = x(-1) = 0$. We note that the Kronecker impulses determine only the amplitude and phase of x.

In the z-domain, we get the homogeneous equation

$$A(z) = 1 - 2\cos(2\pi f_0 T_c)z^{-1} + z^{-2} \tag{1.483}$$

The zeros of $A(z)$ are

$$p_{1,2} = e^{\pm j2\pi f_0 T_c} \tag{1.484}$$

It is immediate to verify that these zeros belong to the unit circle of the z plane. Consequently, the representation of a sinusoidal process via the AR model is not possible, as the stability condition, $|p_i| < 1$, is not satisfied. Moreover, the input (1.482) is not white noise. In any case, we can try to find an approximation. In the hypothesis of uniform φ, from Example 1.7.5,

$$\mathtt{r}_x(n) = \frac{A^2}{2}\cos(2\pi f_0 nT_c) \tag{1.485}$$

This autocorrelation function can be approximated by the autocorrelation of an AR(2) process for $\varrho \to 1$ and $\sigma_w^2 \to 0$. Using (1.476), for $\varrho \simeq 1$ we find

$$\mathtt{r}_{AR(2)}(n) \simeq \left[\frac{\sigma_w^2 \frac{2}{1-\varrho^2}}{2 - \varrho^2\cos^2(4\pi f_0 T_c)}\right]\cos(2\pi f_0 nT_c) \tag{1.486}$$

and impose the condition

$$\lim_{\rho\to 1,\sigma_w^2\to 0} \frac{\sigma_w^2 \frac{2}{1-\rho^2}}{2-\rho^2\cos^2(4\pi f_0 T_c)} = \frac{A^2}{2} \tag{1.487}$$

Observation 1.7

We can observe the following facts about the order of an AR model approximating a sinusoidal process.

- From (1.315), we find that an AR(N) process is required to model N *complex sinusoids*; on the other hand, from (1.313), we see that an AR process of order $2N$ is required to model N real sinusoids.
- An ARMA($2N, 2N$) process is required to model N real sinusoids plus white noise having variance σ_b^2. Observing (1.440), it results $\sigma_w^2 \to \sigma_b^2$ and $B(z) \to A(z)$.

1.11 Guide to the bibliography

The basics of signal theory and random processes can be found in the companion book [1]. Many of the topics surveyed in this chapter are treated in general in several texts on digital communications, in particular [6–8].

In-depth studies on deterministic systems and signals are found in [5, 9–13]. For a statistical analysis of random processes, we refer to [2, 14, 15]. Finally, the subject of spectral estimation is discussed in detail in [3, 16–18].

Bibliography

[1] Benvenuto, N. and Zorzi, M. (2011). *Principles of Communications Networks and Systems*. Wiley.

[2] Papoulis, A. (1991). *Probability, Random Variables and Stochastic Processes*, 3e. New York, NY: McGraw-Hill.

[3] Priestley, M.B. (1981). *Spectral Analysis and Time Series*. New York, NY: Academic Press.

[4] Gentle, J. (2006). *Random Number Generation and Monte Carlo Methods, Statistics and Computing*. New York, NY: Springer.

[5] Papoulis, A. (1984). *Signal Analysis*. New York, NY: McGraw-Hill.

[6] Benedetto, S. and Biglieri, E. (1999). *Principles of Digital Transmission with Wireless Applications*. New York, NY: Kluwer Academic Publishers.

[7] Messerschmitt, D.G. and Lee, E.A. (1994). *Digital Communication*, 2e. Boston, MA: Kluwer Academic Publishers.

[8] Proakis, J.G. (1995). *Digital Communications*, 3e. New York, NY: McGraw-Hill.

[9] Cariolaro, G. (1996). *La teoria unificata dei segnali*. Torino: UTET.

[10] Papoulis, A. (1962). *The Fourier Integral and Its Applications*. New York, NY: McGraw-Hill.

[11] Oppenheim, A.V. and Schafer, R.W. (1989). *Discrete-Time Signal Processing*. Englewood Cliffs, NJ: Prentice-Hall.

[12] Vaidyanathan, P.P. (1993). *Multirate Systems and Filter Banks*. Englewood Cliffs, NJ: Prentice-Hall.

[13] Crochiere, R.E. and Rabiner, L.R. (1983). *Multirate Digital Signal Processing*. Englewood Cliffs, NJ: Prentice-Hall.

[14] Davenport, J.W.B. and Root, W.L. (1987). *An Introduction to the Theory of Random Signals and Noise*. New York, NY: IEEE Press.

[15] Shiryayev, A.N. (1984). *Probability*. New York, NY: Springer-Verlang.

[16] Kay, S.M. (1988). *Modern Spectral Estimation-Theory and Applications*. Englewood Cliffs, NJ: Prentice-Hall.

[17] Marple, L.S. Jr. (1987). *Digital Spectral Analysis with Applications*. Englewood Cliffs, NJ: Prentice-Hall.

[18] Stoica, P. and Moses, R. (1997). *Introduction to Spectral Analysis*. Englewood Cliffs, NJ: Prentice-Hall.

[19] Erup, L., Gardner, F.M., and Harris, R.A. (1993). Interpolation in digital modems - Part II: implementation and performance. *IEEE Transactions on Communications* 41: 998–1008.

[20] Meyr, H., Moeneclaey, M., and Fechtel, S.A. (1998). *Digital Communication Receivers*. New York, NY: Wiley.

[21] Golomb, S.W. (1967). *Shift Register Sequences*. San Francisco, CA: Holden-Day.

[22] Fan, P. and Darnell, M. (1996). *Sequence Design for Communications Applications*. Taunton: Research Studies Press.

[23] Chu, D.C. (1972). Polyphase codes with good periodic correlation properties. *IEEE Transactions on Information Theory* 18: 531–532.

[24] Frank, R.L. and Zadoff, S.A. (1962). Phase shift pulse codes with good periodic correlation properties. *IRE Transactions on Information Theory* 8: 381–382.

[25] Milewsky, A. (1983). Periodic sequences with optimal properties for channel estimation and fast start-up equalization. *IBM Journal of Research and Development* 27: 426–431.

[26] Peterson, R.L., Ziemer, R.E., and Borth, D.E. (1995). *Introduction to Spread Spectrum Communications*. Englewood Cliffs, NJ: Prentice-Hall.

[27] Gold, R. (1967). Optimal binary sequences for spread spectrum multiplexing. *IEEE Transactions on Information Theory* 13: 619–621.

[28] Gold, R. (1968). Maximal recursive sequences with 3-valued recursive cross correlation functions. *IEEE Transactions on Information Theory* 14: 154–155.

Appendix 1.A Multirate systems

The first part of this appendix is a synthesis from [12, 13].

1.A.1 Fundamentals

We consider the discrete-time linear transformation of Figure 1.44, with impulse response $h(t)$, $t \in \mathbb{R}$; the sampling period of the input signal is T_c, whereas that of the output signal is T'_c.

The input–output relation is given by the equation

$$y(kT'_c) = \sum_{n=-\infty}^{+\infty} h(kT'_c - nT_c)x(nT_c) \tag{1.488}$$

We will use the following simplified notation:

$$x_n = x(nT_c) \tag{1.489}$$

$$y_k = y(kT'_c) \tag{1.490}$$

If we assume that h has a finite support, say between t_1 and t_2, that is $h(kT'_c - nT_c) \neq 0$ for

$$kT'_c - nT_c < t_2, \qquad kT'_c - nT_c > t_1 \tag{1.491}$$

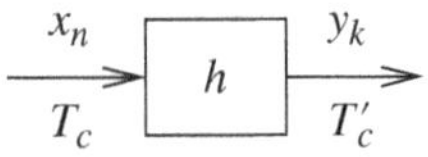

Figure 1.44 Discrete-time linear transformation.

or equivalently for

$$n > \frac{kT'_c - t_2}{T_c}, \qquad n < \frac{kT'_c - t_1}{T_c} \tag{1.492}$$

then, letting

$$n_1 = \left\lceil \frac{kT'_c - t_2}{T_c} \right\rceil \tag{1.493}$$

$$n_2 = \left\lfloor \frac{kT'_c - t_1}{T_c} \right\rfloor \tag{1.494}$$

(1.488) can be written as

$$y_k = \sum_{n=n_1}^{n_2} h(kT'_c - nT_c)x_n = x_{n_1} h(kT'_c - n_1 T_c) + \cdots + x_{n_2} h(kT'_c - n_2 T_c) \tag{1.495}$$

One observes from (1.488) that

- the values of h that contribute to y_k are equally spaced by T_c;
- the limits of the summation (1.495) are a complicated function of T_c, T'_c, t_1, and t_2.

Introducing the change of variable

$$i = \left\lfloor \frac{kT'_c}{T_c} \right\rfloor - n \tag{1.496}$$

and setting

$$\Delta_k = \frac{kT'_c}{T_c} - \left\lfloor \frac{kT'_c}{T_c} \right\rfloor \tag{1.497}$$

$$I_1 = \left\lceil \frac{t_1}{T_c} - \Delta_k \right\rceil = \left\lceil \frac{t_1}{T_c} - \frac{kT'_c}{T_c} \right\rceil + \left\lfloor \frac{kT'_c}{T_c} \right\rfloor \tag{1.498}$$

$$I_2 = \left\lfloor \frac{t_2}{T_c} - \Delta_k \right\rfloor = \left\lfloor \frac{t_2}{T_c} - \frac{kT'_c}{T_c} \right\rfloor + \left\lfloor \frac{kT'_c}{T_c} \right\rfloor \tag{1.499}$$

(1.495) becomes

$$y_k = \sum_{i=I_1}^{I_2} h((i + \Delta_k)T_c) x_{\lfloor kT'_c/T_c \rfloor - i} \tag{1.500}$$

From the definition (1.497), it is clear that Δ_k represents the truncation error of kT'_c/T_c and that $0 \le \Delta_k < 1$. In the special case,

$$\frac{T'_c}{T_c} = \frac{M}{L} \tag{1.501}$$

with M and L integers, we get[25]

$$\begin{aligned} \Delta_k &= k\frac{M}{L} - \left\lfloor k\frac{M}{L} \right\rfloor \\ &= \frac{1}{L}\left(kM - \left\lfloor k\frac{M}{L} \right\rfloor L\right) \\ &= \frac{1}{L}(kM)_{mod\ L} \end{aligned} \tag{1.502}$$

[25] We denote by $(k)_{mod\ n}$ or $k\ mod\ n$ the reminder of the division of integer k by integer n.

We observe that Δ_k can assume L values $\{0, 1/L, 2/L, \dots, (L-1)/L\}$ for any value of k. Hence, there are only L univocally determined sets of values of h that are used in the computation of $\{y_k\}$; in particular, if $L = 1$ only one set of coefficients exists, while if $M = 1$ the sets are L. Summarizing, the output of a filter with impulse response h and with different input and output time domains can be expressed as

$$y_k = \sum_{i=-\infty}^{+\infty} g_{k,i} x_{\lfloor \frac{kM}{L} \rfloor - i} \tag{1.503}$$

where

$$g_{k,i} = h((i + \Delta_k)T_c) \tag{1.504}$$

We note that the system is *linear* and *periodically time varying*. For $T'_c = T_c$, that is for $L = M = 1$, we get $\Delta_k = 0$, and the input–output relation is the usual convolution

$$y_k = \sum_{i=-\infty}^{+\infty} g_{0,i} x_{k-i} \tag{1.505}$$

We will now analyse a few elementary multirate transformations.

1.A.2 Decimation

Figure 1.45 represents a decimator or downsampler, with the output sequence related to the input sequence $\{x_n\}$ by

$$y_k = x_{kM} \tag{1.506}$$

where M, the *decimation factor*, is an integer number.

We now obtain an expression for the z-transform of the output $Y(z)$ in terms of $X(z)$. We will show that

$$Y(z) = \frac{1}{M} \sum_{m=0}^{M-1} X\left(z^{\frac{1}{M}} W_M^m\right) \tag{1.507}$$

where $W_M = e^{-j\frac{2\pi}{M}}$ is defined in (1.28). Equivalently, in terms of the radian frequency normalized by the sampling frequency, $\omega' = 2\pi f / F'_c$, (1.507) can be written as

$$Y(e^{j\omega'}) = \frac{1}{M} \sum_{m=0}^{M-1} X\left(e^{j\frac{\omega' - 2\pi m}{M}}\right) \tag{1.508}$$

A graphical interpretation of (1.508) is shown in Figure 1.46:

- expand $X(e^{j\omega})$ by a factor M, obtaining $X(e^{j\omega'/M})$;
- create $M - 1$ replicas of the expanded version, and frequency-shift them uniformly with increments of 2π for each replica;
- sum all the replicas and divide the result by M.

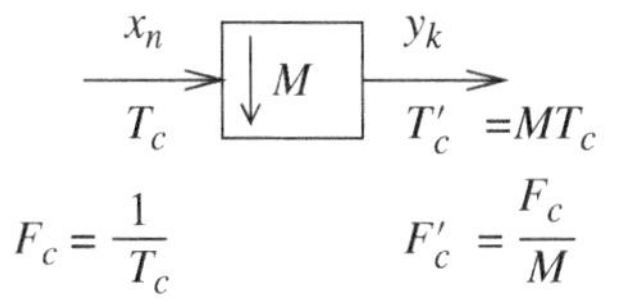

Figure 1.45 Decimation or downsampling transformation by a factor M.

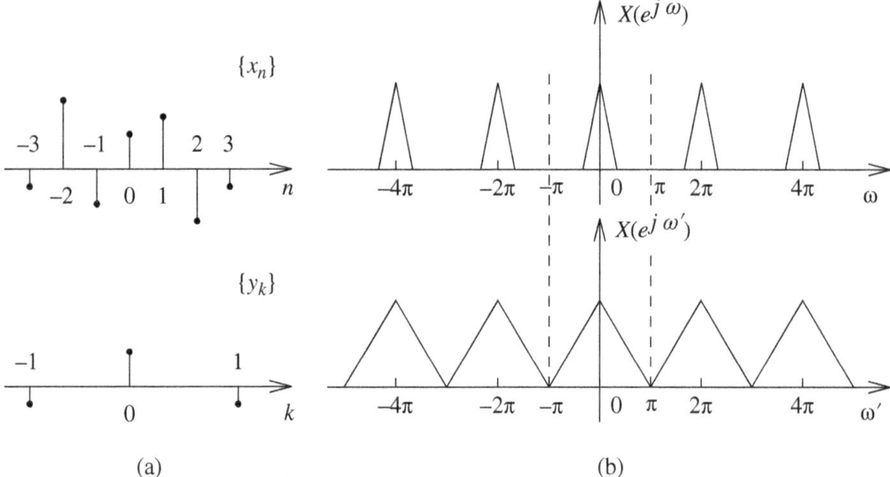

Figure 1.46 Decimation by a factor $M = 3$: (a) in the time domain, and (b) in the normalized radian frequency domain.

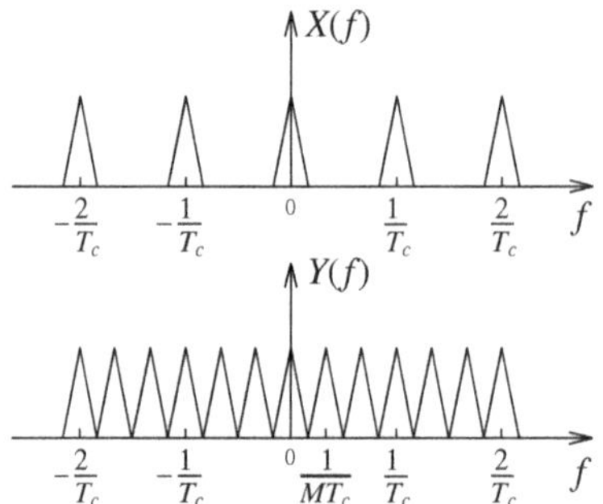

Figure 1.47 Effect of decimation in the frequency domain.

We observe that, after summation, the result is periodic in ω' with period 2π, as we would expect from a discrete-time Fourier transform.

It is also useful to give the expression of the output sequence in the frequency domain; we get

$$\mathcal{Y}(f) = \frac{1}{M} \sum_{m=0}^{M-1} \mathcal{X}\left(f - \frac{m}{MT_c}\right) \tag{1.509}$$

where

$$\mathcal{X}(f) = X(e^{j2\pi f T_c}) \tag{1.510}$$

$$\mathcal{Y}(f) = Y(e^{j2\pi f M T_c}) \tag{1.511}$$

The relation (1.509) for the signal of Figure 1.46 is represented in Figure 1.47. Note that the only difference with respect to the previous representation is that all frequency responses are now functions of the frequency f.

Proof of (1.507). The z-transform of $\{y_k\}$ can be written as

$$Y(z) = \sum_{k=-\infty}^{+\infty} y_k z^{-k} = \sum_{k=-\infty}^{+\infty} x_{Mk} z^{-k} \tag{1.512}$$

We define the intermediate sequence

$$x'_k = \begin{cases} x_k & k = 0, \pm M, \pm 2M, \dots \\ 0 & \text{otherwise} \end{cases} \tag{1.513}$$

so that $y_k = x_{Mk} = x'_{Mk}$. With this position, we get

$$Y(z) = \sum_{k'=-\infty}^{+\infty} x'_{k'M} z^{-k'} = \sum_{k=-\infty}^{+\infty} x'_k z^{-k/M} = X'(z^{1/M}) \tag{1.514}$$

This relation is valid, because x' is non-zero only at multiples of M. It only remains to express $X'(z)$ in terms of $X(z)$; to do this, we note that (1.513) can be expressed as

$$x'_k = c_k x_k \tag{1.515}$$

where c_k is defined as

$$c_k = \begin{cases} 1 & k = 0, \pm M, \pm 2M, \dots \\ 0 & \text{otherwise} \end{cases} \tag{1.516}$$

Note that the (1.516) can be written as

$$c_k = \frac{1}{M} \sum_{m=0}^{M-1} W_M^{-km} \tag{1.517}$$

Hence, we obtain

$$X'(z) = \frac{1}{M} \sum_{m=0}^{M-1} \sum_{k=-\infty}^{+\infty} x_k W_M^{-km} z^{-k} = \frac{1}{M} \sum_{m=0}^{M-1} \sum_{k=-\infty}^{+\infty} x_k (z W_M^m)^{-k} \tag{1.518}$$

The inner summation yields $X(zW_M^m)$: hence, using (1.514) we get (1.507).

1.A.3 Interpolation

Figure 1.48 represents an interpolator or upsampler, with the input sequence $\{x_n\}$ related to the output sequence by

$$y_k = \begin{cases} x\left(\frac{k}{L}\right) & k = 0, \pm L, \pm 2L, \dots \\ 0 & \text{otherwise} \end{cases} \tag{1.519}$$

where L, the *interpolation factor*, is an integer number.

We will show that the input–output relation in terms of the z-transforms $Y(z)$ and $X(z)$ is given by

$$Y(z) = X(z^L) \tag{1.520}$$

Equivalently, in terms of radian frequency normalized by the sampling frequency, $\omega' = 2\pi f / F'_c$, the (1.520) can be expressed as

$$Y(e^{j\omega'}) = X(e^{j\omega' L}) \tag{1.521}$$

The graphical interpretation of (1.521) is illustrated in Figure 1.49: $Y(e^{j\omega})$ is the compressed version by a factor L of $X(e^{j\omega})$; moreover, there are $L-1$ replicas of the compressed spectrum, called *images*. The creation of images implies that a lowpass signal does not remain lowpass after interpolation.

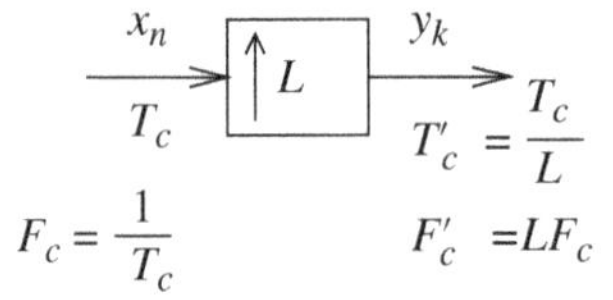

Figure 1.48 Interpolation or upsampling transformation by a factor L.

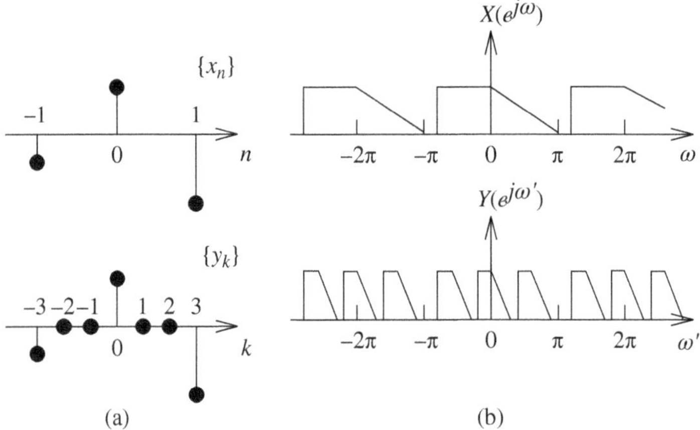

Figure 1.49 Interpolation by a factor $L = 3$: (a) in the time domain, (b) in the normalized radian frequency domain.

It is also useful to give the expression of the output sequence in the frequency domain; we get

$$\mathcal{Y}(f) = \mathcal{X}(f) \tag{1.522}$$

where

$$\mathcal{X}(f) = X(e^{j2\pi f T_c}) \tag{1.523}$$

$$\mathcal{Y}(f) = Y(e^{j2\pi f \frac{T_c}{L}}) \tag{1.524}$$

Relation (1.522) for the signal of Figure 1.49 is illustrated in Figure 1.50. We note that the only effect of the interpolation is that the signal $\mathcal{X}$ must be regarded as periodic with period F'_c rather than F_c.

Proof of (1.520). Observing (1.519), we get

$$Y(z) = \sum_{k=-\infty}^{+\infty} y_k z^{-k} = \sum_{n=-\infty}^{+\infty} y_{nL} z^{-nL} = \sum_{n=-\infty}^{+\infty} x_n z^{-nL} = X(z^L) \tag{1.525}$$

1.A.4 Decimator filter

In most applications, a downsampler is preceded by a lowpass digital filter, to form a *decimator filter* as illustrated in Figure 1.51. The filter ensures that the signal v_n is bandlimited, to avoid *aliasing* in the downsampling process.

Let $h_n = h(nT_c)$. Then we have

$$y_k = v_{kM} \tag{1.526}$$

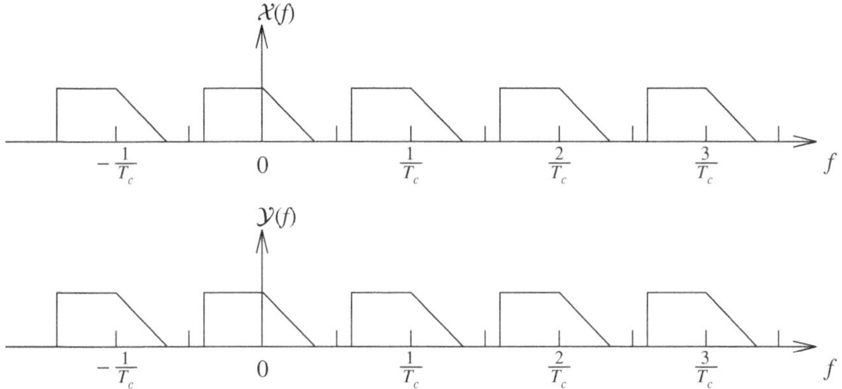

Figure 1.50 Effect of interpolation in the frequency domain.

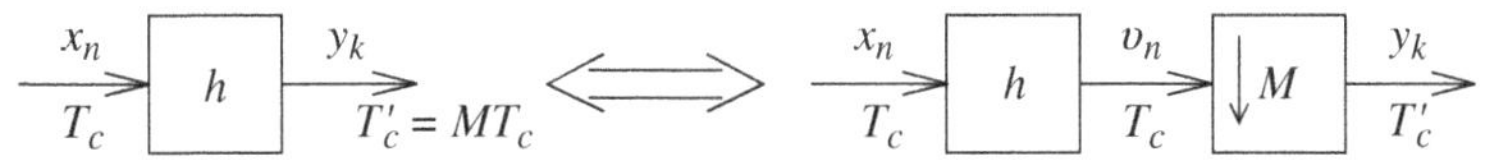

x_n T_c → h → y_k $T'_c = MT_c$ ⟺ x_n T_c → h → v_n T_c → $\downarrow M$ → y_k T'_c

Figure 1.51 Decimator filter.

and

$$v_n = \sum_{i=-\infty}^{+\infty} h_i x_{n-i} \tag{1.527}$$

The output can be expressed as

$$y_k = \sum_{i=-\infty}^{+\infty} h_i x_{kM-i} = \sum_{n=-\infty}^{+\infty} h_{kM-n} x_n \tag{1.528}$$

Using definition (1.504), we get

$$g_{k,i} = h_i \qquad \forall k, i \tag{1.529}$$

Note that the overall system is not time invariant, unless the delay applied to the input is constrained to be a multiple of M.

From $V(z) = X(z)H(z)$, it follows that

$$Y(z) = \frac{1}{M} \sum_{m=0}^{M-1} H(z^{1/M} W_M^m) X(z^{1/M} W_M^m) \tag{1.530}$$

or, equivalently, recalling that $\omega' = 2\pi f M T_c$,

$$Y(e^{j\omega'}) = \frac{1}{M} \sum_{m=0}^{M-1} H\left(e^{j\frac{\omega'-2\pi m}{M}}\right) X\left(e^{j\frac{\omega'-2\pi m}{M}}\right) \tag{1.531}$$

If

$$H(e^{j\omega}) = \begin{cases} 1 & |\omega| \le \dfrac{\pi}{M} \\ 0 & \text{otherwise} \end{cases} \tag{1.532}$$

we obtain

$$Y(e^{j\omega'}) = \frac{1}{M} X\left(e^{\frac{j\omega'}{M}}\right) \qquad |\omega'| \le \pi \tag{1.533}$$

In this case, h is a lowpass filter that avoids *aliasing* caused by sampling; if x is bandlimited, the specifications of h can be made less stringent.

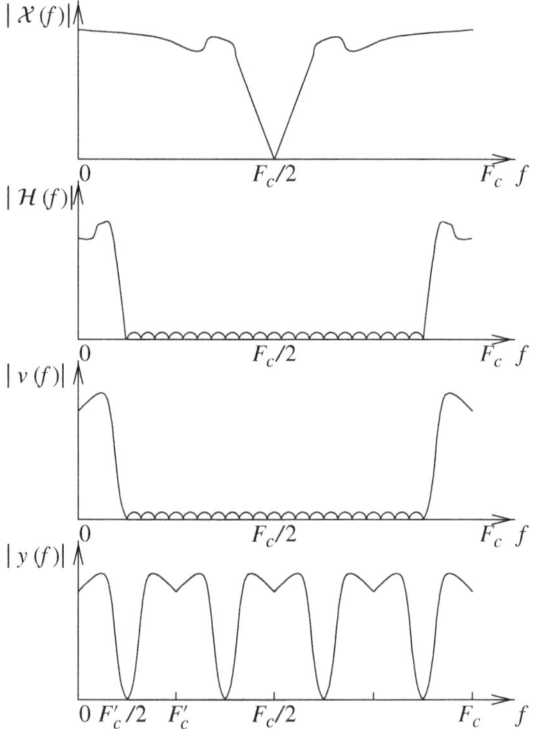

Figure 1.52 Frequency responses related to the transformations in a decimator filter for $M = 4$.

The decimator filter transformations are illustrated in Figure 1.52 for $M = 4$.

1.A.5 Interpolator filter

An interpolator filter is given by the cascade of an upsampler and a digital filter, as illustrated in Figure 1.53; the task of the digital filter is to suppress images created by upsampling [19].

Let $h_n = h(nT'_c)$. Then we have the following input–output relations:

$$y_k = \sum_{j=-\infty}^{+\infty} h_{k-j} w_j \tag{1.534}$$

$$w_k = \begin{cases} x\left(\dfrac{k}{L}\right) & k = 0, \pm L, \dots \\ 0 & \text{otherwise} \end{cases} \tag{1.535}$$

Therefore,

$$y_k = \sum_{r=-\infty}^{+\infty} h_{k-rL} x_r \tag{1.536}$$

Let $i = \lfloor k/L \rfloor - r$ and $g_{k,i} = h_{iL+(k)_{mod\ L}}$. From (1.536), we get

$$y_k = \sum_{i=-\infty}^{+\infty} g_{k,i} x_{\lfloor \frac{k}{L} \rfloor - i} \tag{1.537}$$

We note that $g_{k,i}$ is periodic in k of period L.

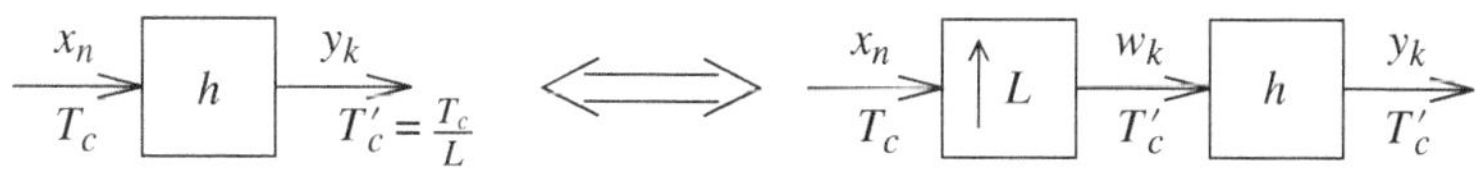

Figure 1.53 Interpolator filter.

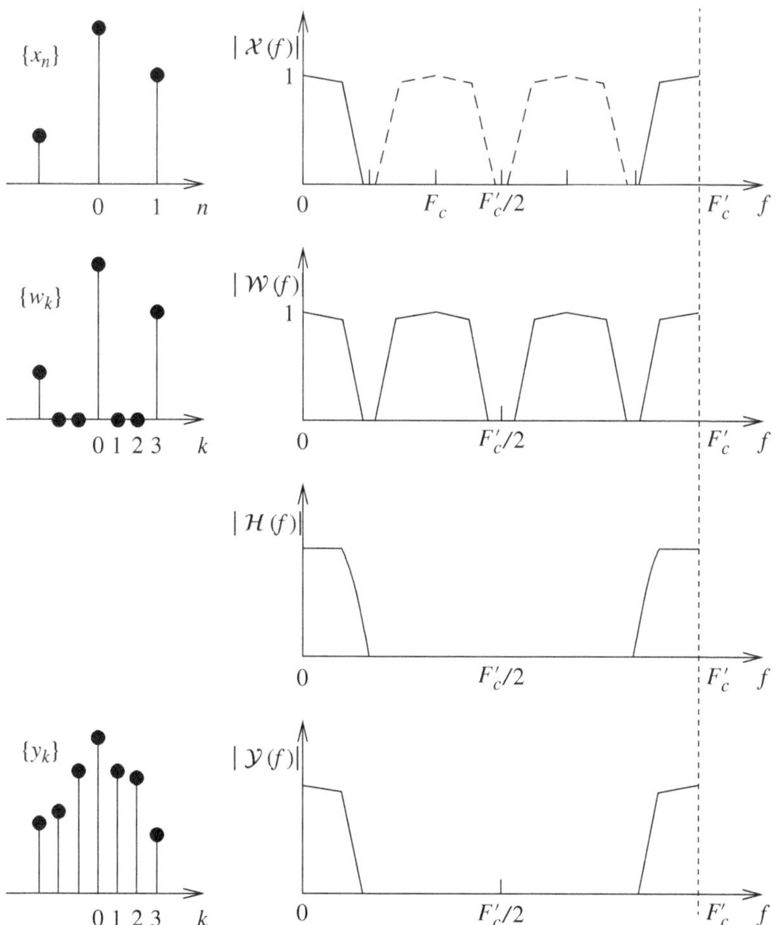

Figure 1.54 Time and frequency responses related to the transformations in an interpolator filter for $L = 3$.

In the z-transform domain, we find

$$W(z) = X(z^L) \tag{1.538}$$

$$Y(z) = H(z)W(z) = H(z)X(z^L) \tag{1.539}$$

or, equivalently,

$$Y(e^{j\omega'}) = H(e^{j\omega'})X(e^{j\omega' L}) \tag{1.540}$$

where $\omega' = 2\pi fT/L = \omega/L$.

The interpolator filter transformations in the time and frequency domains are illustrated in Figure 1.54 for $L = 3$.

If

$$H(e^{j\omega'}) = \begin{cases} 1 & |\omega'| \le \frac{\pi}{L} \\ 0 & \text{elsewhere} \end{cases} \tag{1.541}$$

we find

$$Y(e^{j\omega'}) = \begin{cases} X(e^{j\omega'}) & |\omega'| \le \frac{\pi}{L} \\ 0 & \text{elsewhere} \end{cases} \tag{1.542}$$

The relation between the input and output signal power for an interpolator filter is expressed by (1.344).

1.A.6 Rate conversion

Decimator and interpolator filters can be employed to vary the sampling frequency of a signal by an integer factor; in some applications, however, it is necessary to change the sampling frequency by a rational factor L/M. A possible procedure consists of first converting a discrete-time signal into a continuous-time signal by a *digital-to-analog converter* (DAC), then re-sampling it at the new frequency. It is however easier and more convenient to change the sampling frequency by discrete-time transformations, for example using the structure of Figure 1.55.

This system can be thought of as the cascade of an interpolator and decimator filter, as illustrated in Figure 1.56, where $h(t) = h_1 * h_2(t)$. We obtain that

$$H(e^{j\omega'}) = \begin{cases} 1 & |\omega'| \le \min\left(\frac{\pi}{L}, \frac{\pi}{M}\right) \\ 0 & \text{elsewhere} \end{cases} \tag{1.543}$$

In the time domain, the following relation holds:

$$y_k = \sum_{i=-\infty}^{+\infty} g_{k,i} x_{\lfloor \frac{kM}{L} \rfloor - i} \tag{1.544}$$

where $g_{k,i} = h((iL + (kM)_{mod\ L})T'_c)$ is the time-varying impulse response.

In the frequency domain, we get

$$Y(e^{j\omega''}) = \frac{1}{M} \sum_{l=0}^{M-1} V\left(e^{j\frac{\omega''-2\pi l}{M}}\right) \tag{1.545}$$

As

$$V(e^{j\omega'}) = H(e^{j\omega'})X(e^{j\omega' L}) \tag{1.546}$$

we obtain

$$Y(e^{j\omega''}) = \frac{1}{M} \sum_{l=0}^{M-1} H\left(e^{j\frac{\omega''-2\pi l}{M}}\right) X\left(e^{j\frac{\omega'' L-2\pi l}{M}}\right) \tag{1.547}$$

From (1.543), we have

$$Y(e^{j\omega''}) = \begin{cases} \frac{1}{M} X\left(e^{\frac{j\omega'' L}{M}}\right) & |\omega''| \le \min\left(\pi, \frac{\pi M}{L}\right) \\ 0 & \text{elsewhere} \end{cases} \tag{1.548}$$

or

$$\mathcal{Y}(f) = \frac{1}{M}\mathcal{X}(f) \quad \text{for} \quad |f| \le \min\left(\frac{1}{2T_c}, \frac{L}{2MT_c}\right) \tag{1.549}$$

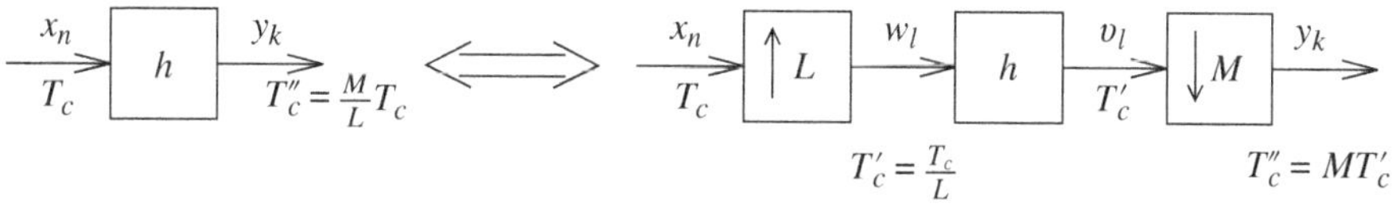

Figure 1.55 Sampling frequency conversion by a rational factor.

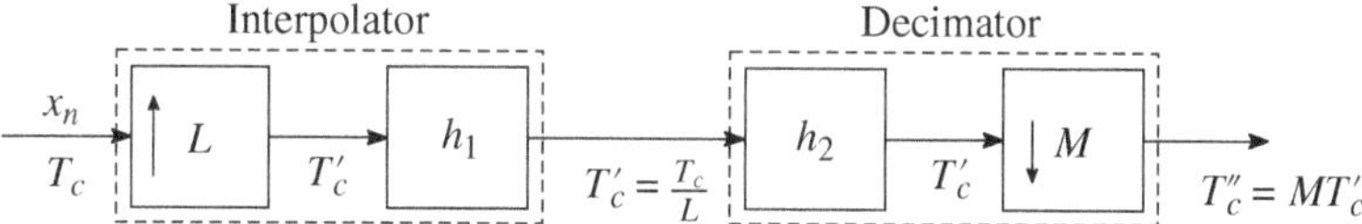

Figure 1.56 Decomposition of the system of Figure 1.55.

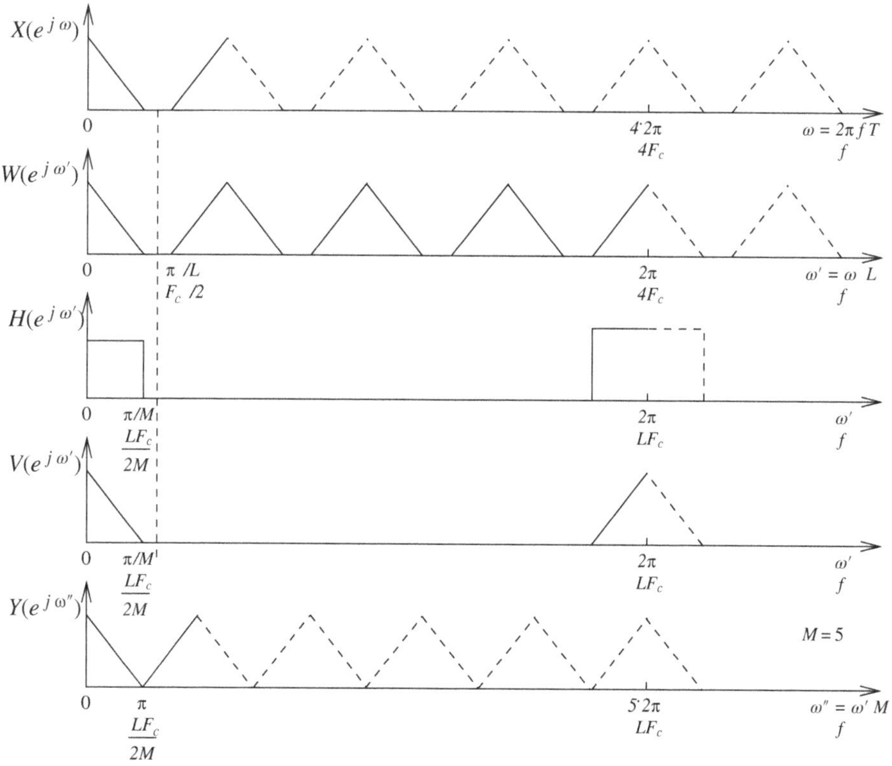

Figure 1.57 Rate conversion by a rational factor L/M, where $M > L$.

Example 1.A.1 ($M > L$: $M = 5$, $L = 4$)

Transformations for $M = 5$ and $L = 4$ are illustrated in Figure 1.57. Observing the fact that $W(e^{j\omega'})$ is zero for $\frac{\pi}{M} \leq \omega' \leq 2\frac{\pi}{L} - \frac{\pi}{M}$, the desired result is obtained by a response $H(e^{j\omega'})$ that has the stopband cutoff frequency within this interval.

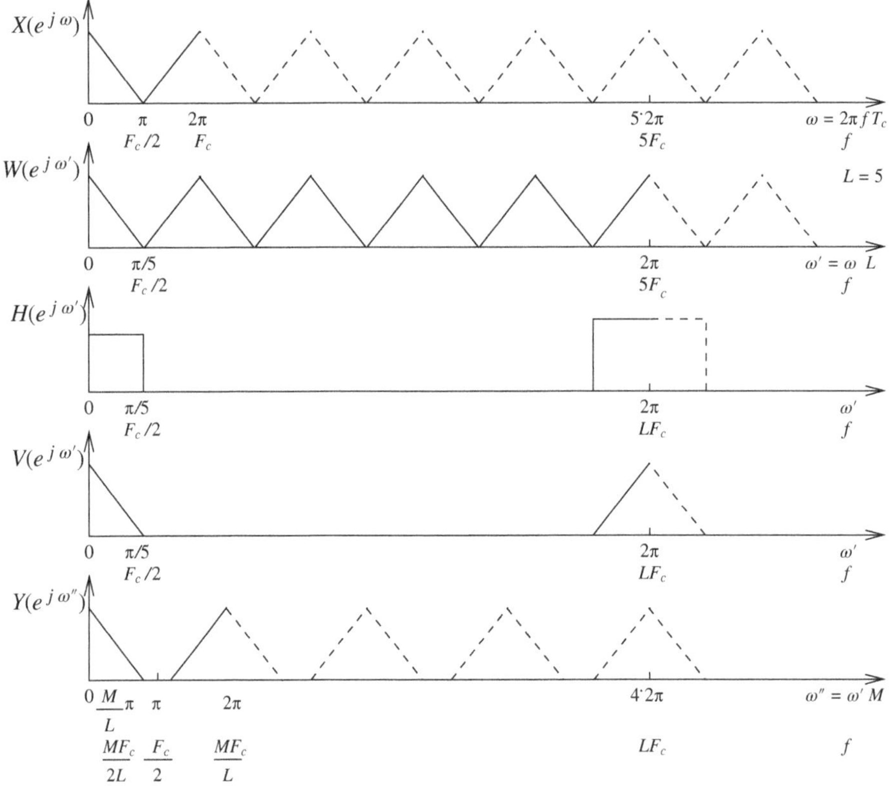

Figure 1.58 Rate conversion by a rational factor L/M, where $M < L$.

Example 1.A.2 ($M < L$: $M = 4$, $L = 5$)
The inverse transformation of the above example is obtained by a transformation with $M = 4$ and $L = 5$, as depicted in Figure 1.58.

1.A.7 Time interpolation

Referring to the interpolator filter h of Figure 1.53, one finds that if L is large the filter implementation may require non-negligible complexity; in fact, the number of coefficients required for a FIR filter implementation can be very large. Consequently, in the case of a very large interpolation factor L, after a first interpolator filter with a moderate value of the interpolation factor, the samples $\{y_k = y(kT_c')\}$ may be further time interpolated until the desired sampling accuracy is reached [19].

As shown in Figure 1.59, let $\{y_k\}$ be the sequence that we need to interpolate to produce the signal $z(t)$, $t \in \mathbb{R}$; we describe two time interpolation methods: linear and quadratic.

Linear interpolation

Given two samples y_{k-1} and y_k, the continuous signal z defined on the interval $[(k-1)\,T_c', kT_c']$, is obtained by the linear interpolation

$$z(t) = y_{k-1} + \frac{t-(k-1)T_c'}{T_c'}\,(y_k - y_{k-1}) \tag{1.550}$$

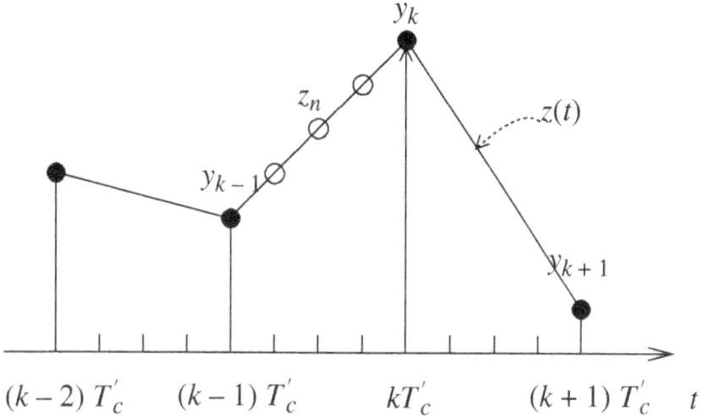

Figure 1.59 Linear interpolation in time by a factor $P = 4$.

For an interpolation factor P of y_k, we need to consider the sampling instants

$$nT_c'' = n\frac{T_c'}{P} \tag{1.551}$$

and the values of $z_n = z(nT_c'')$ are given by

$$z_n = y_{k-1} + \frac{n-(k-1)P}{P}(y_k - y_{k-1}) \tag{1.552}$$

where $n = (k-1)P, (k-1)P+1, \dots, kP-1$. The case $k = 1$ is of particular interest:

$$z_n = y_0 + \frac{n}{P}(y_1 - y_0), \qquad n = 0, 1, \dots, P-1 \tag{1.553}$$

In fact, regarding y_0 and y_1 as the two most recent input samples, their linear interpolation originates the sequence of P values given by (1.553).

Quadratic interpolation

In many applications linear interpolation does not always yield satisfactory results. Therefore, instead of connecting two points with a straight line, one resorts to a polynomial of degree $Q-1$ passing through Q points that are determined by the samples of the input sequence. For this purpose, the Lagrange interpolation is widely used. As an example, we report here the case of quadratic interpolation. In this case, we consider a polynomial of degree 2 that passes through three points that are determined by the input samples. Let y_{k-1}, y_k, and y_{k+1} be the samples to interpolate by a factor P in the interval $[(k-1)T_c', (k+1)T_c']$. The quadratic interpolation yields the values

$$z_n = \frac{n'}{2P}\left(\frac{n'}{P}-1\right)y_{k-1} + \left(1-\frac{n'}{P}\right)\left(1+\frac{n'}{P}\right)y_k + \frac{n'}{2P}\left(\frac{n'}{P}+1\right)y_{k+1} \tag{1.554}$$

with $n' = 0, 1, \dots, P-1$ and $n = (k-1)\,P + n'$.

1.A.8 The noble identities

We recall some important properties of decimator and interpolator filters, known as *noble identities*; they will be used extensively in the Appendix 1.A.9 on polyphase decomposition.

Let $G(z)$ be a rational transfer function, i.e. a function expressed as the ratio of two polynomials in z or in z^{-1}; it is possible to exchange the order of downsampling and filtering, or the order of upsampling

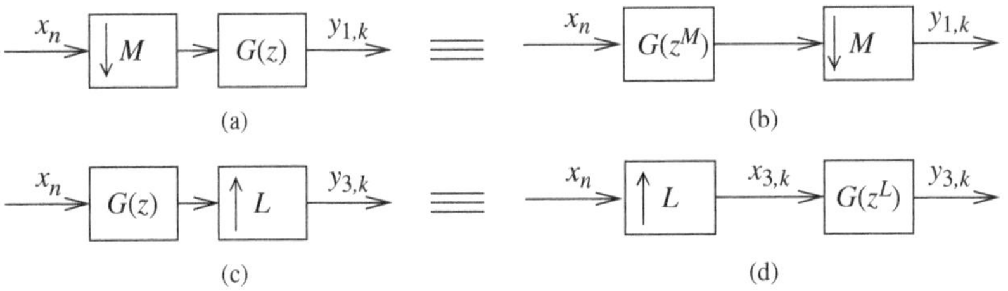

Figure 1.60 Noble identities.

and filtering as illustrated in Figure 1.60; in other words, the system of Figure 1.60a is equivalent to that of Figure 1.60b, and the system of Figure 1.60c is equivalent to that of Figure 1.60d.

The proof of the noble identities is simple. For the first identity, it is sufficient to note that $W_M^{-mM} = 1$, hence,

$$\begin{aligned} Y_2(z) &= \frac{1}{M}\sum_{m=0}^{M-1} X(z^{1/M} W_M^{-m}) G((z^{1/M} W_M^{-m})^M) \\ &= \frac{1}{M}\sum_{m=0}^{M-1} X(z^{1/M} W_M^{-m}) G(z) = Y_1(z) \end{aligned} \tag{1.555}$$

For the second identity, it is sufficient to observe that

$$Y_4(z) = G(z^L)X_4(z) = G(z^L)X(z^L) = Y_3(z) \tag{1.556}$$

1.A.9 The polyphase representation

The polyphase representation allows considerable simplifications in the analysis of transformations via interpolator and decimator filters, as well as the efficient implementation of such filters. To explain the basic concept, let us consider a filter having transfer function $H(z) = \sum_{n=0}^{\infty} h_n z^{-n}$. Separating the coefficients with even and odd time indices, we get

$$H(z) = \sum_{m=0}^{\infty} h_{2m} z^{-2m} + z^{-1} \sum_{m=0}^{\infty} h_{2m+1} z^{-2m} \tag{1.557}$$

Defining

$$E^{(0)}(z) = \sum_{m=0}^{\infty} h_{2m} z^{-m} \qquad E^{(1)}(z) = \sum_{m=0}^{\infty} h_{2m+1} z^{-m} \tag{1.558}$$

we can write $H(z)$ as

$$H(z) = E^{(0)}(z^2) + z^{-1} E^{(1)}(z^2) \tag{1.559}$$

To expand this idea, let M be an integer; we can always decompose $H(z)$ as

$$\begin{aligned} H(z) &= \sum_{m=0}^{\infty} h_{mM} z^{-mM} \\ &\quad + z^{-1} \sum_{m=0}^{\infty} h_{mM+1} z^{-mM} + \cdots \\ &\quad + z^{-(M-1)} \sum_{m=0}^{\infty} h_{mM+M-1} z^{-mM} \end{aligned} \tag{1.560}$$

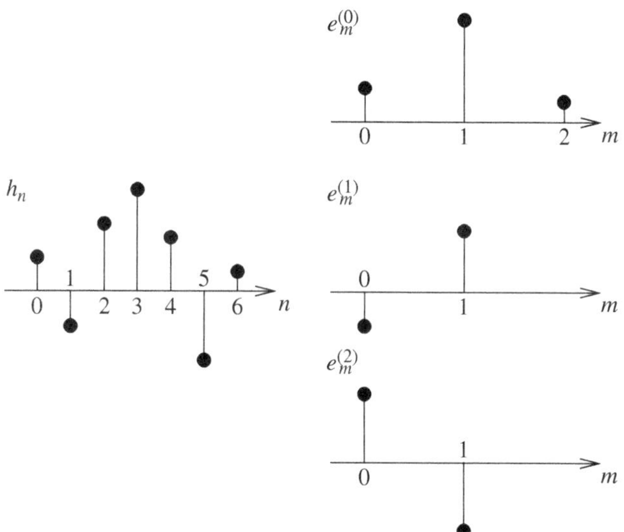

Figure 1.61 Polyphase representation of the impulse response $\{h_n\}$, $n = 0, \ldots, 6$, for $M = 3$.

Letting

$$e_m^{(\ell)} = h_{mM+\ell} \qquad 0 \le \ell \le M-1 \tag{1.561}$$

we can express compactly the previous equation as

$$H(z) = \sum_{\ell=0}^{M-1} z^{-\ell} E^{(\ell)}(z^M) \tag{1.562}$$

where

$$E^{(\ell)}(z) = \sum_{i=0}^{\infty} e_i^{(\ell)} z^{-i} \tag{1.563}$$

The expression (1.562) is called the *type-1 polyphase representation* (with respect to M), and $E^{(\ell)}(z)$, where $\ell = 0, 1, \ldots, M-1$, the *polyphase components* of $H(z)$.

The polyphase representation of an impulse response $\{h_n\}$ with seven coefficients is illustrated in Figure 1.61 for $M = 3$.

A variation of (1.562), named *type-2 polyphase representation*, is given by

$$H(z) = \sum_{\ell=0}^{M-1} z^{-(M-1-\ell)} R^{(\ell)}(z^M) \tag{1.564}$$

where the components $R^{(\ell)}(z)$ are permutations of $E^{(\ell)}(z)$, that is $R^{(\ell)}(z) = E^{(M-1-\ell)}(z)$.

Efficient implementations

The polyphase representation is key to obtain efficient implementation of decimator and interpolator filters. In the following, we will first consider the efficient implementations for $M = 2$ and $L = 2$, then we will extend the results to the general case.

Decimator filter. Referring to Figure 1.51, we consider a decimator filter with $M = 2$. By the (1.562), we can represent $H(z)$ as illustrated in Figure 1.62; by the noble identities, the filter representation can be drawn as in Figure 1.63a. The structure can be also drawn as in Figure 1.63b, where input samples $\{x_n\}$ are alternately presented at the input to the two filters $e^{(0)}$ and $e^{(1)}$; this latter operation is generally called *serial-to-parallel* (S/P) conversion. Note that the system output is now given by the sum of the outputs of two filters, each operating at half the input frequency and having half the number of coefficients as the original filter.

To formalize the above ideas, let N be the number of coefficients of h, and $N^{(0)}$ and $N^{(1)}$ be the number of coefficients of $e^{(0)}$ and $e^{(1)}$, respectively, so that $N = N^{(0)} + N^{(1)}$. In this implementation, $e^{(\ell)}$ requires $N^{(\ell)}$ multiplications and $N^{(\ell)} - 1$ additions; the total cost is still N multiplications and $N - 1$ additions, but, as $e^{(\ell)}$ operates at half the input rate, the computational complexity in terms of multiplications per second (MPS) is

$$\text{MPS} = \frac{NF_c}{2} \tag{1.565}$$

while the number of additions per second (APS) is

$$\text{APS} = \frac{(N-1)F_c}{2} \tag{1.566}$$

Therefore, the complexity is about one-half the complexity of the original filter. The efficient implementation for the general case is obtained as an extension of the case for $M = 2$ and is shown in Figure 1.64.

Interpolator filter. With reference to Figure 1.53, we consider an interpolator filter with $L = 2$. By the (1.562), we can represent $H(z)$ as illustrated in Figure 1.65; by the noble identities, the filter representation can be drawn as in Figure 1.66a. The structure can be also drawn as in Figure 1.66b, where output samples are alternately taken from the output of the two filters $e^{(0)}$ and $e^{(1)}$; this latter operation is generally called *parallel-to-serial* (P/S) conversion. Remarks on the computational complexity are analogous to those of the decimator filter case.

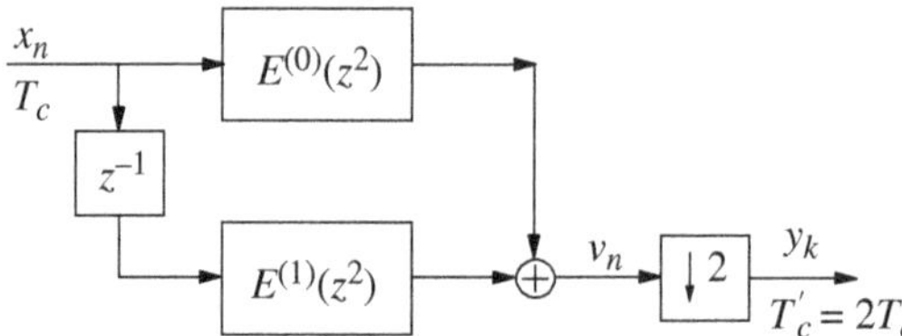

Figure 1.62 Implementation of a decimator filter using the type-1 polyphase representation for $M = 2$.

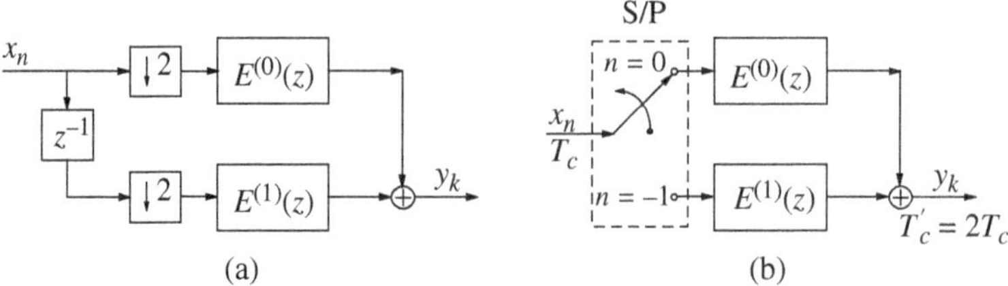

Figure 1.63 Optimized implementation of a decimator filter using the type-1 polyphase representation for $M = 2$.

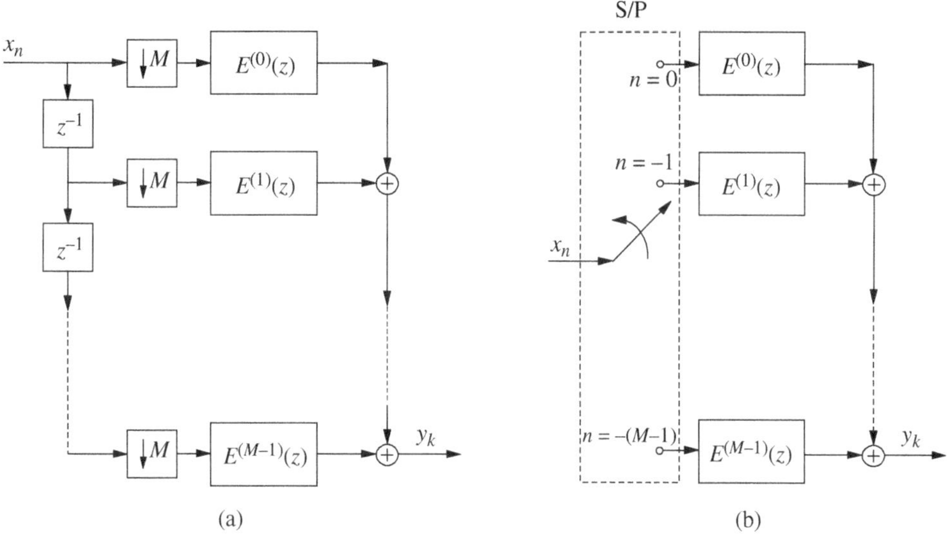

Figure 1.64 Implementation of a decimator filter using the type-1 polyphase representation.

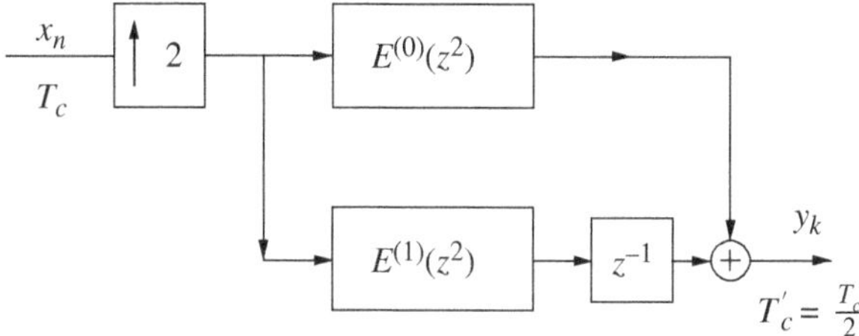

Figure 1.65 Implementation of an interpolator filter using the type-1 polyphase representation for $L = 2$.

In the general case, efficient implementations are easily obtainable as extensions of the case for $L = 2$ and are shown in Figure 1.67. The type-2 polyphase implementations of interpolator filters are depicted in Figure 1.68.

Interpolator-decimator filter. As illustrated in Figure 1.69, at the receiver of a transmission system, it is often useful to interpolate the signal $\{r(nT_Q)\}$ from T_Q to T'_Q to get the signal $\{x(qT'_Q)\}$.

Let

$$r_n = r(nT_Q), \qquad x_q = x(qT'_Q) \tag{1.567}$$

The sequence $\{x(qT'_Q)\}$ is then downsampled with *timing phase* t_0. Let y_k be the output with sampling period T_c,

$$y_k = x(kT_c + t_0) \tag{1.568}$$

To simplify the notation, we assume the following relations:

$$L = \frac{T_Q}{T'_Q}, \qquad M = \frac{T_c}{T_Q} \tag{1.569}$$

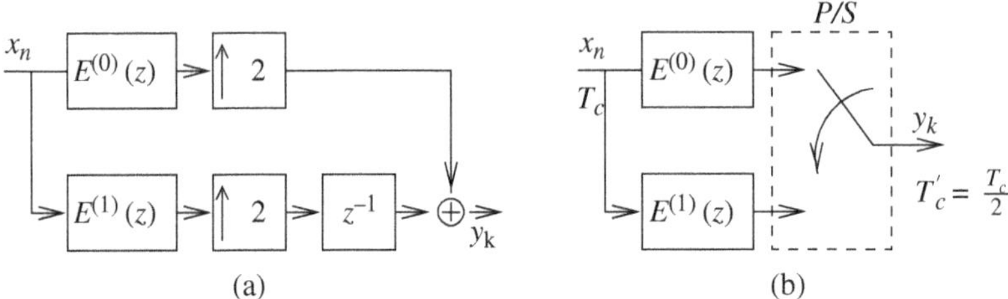

Figure 1.66 Optimized implementation of an interpolator filter using the type-1 polyphase representation for $L = 2$.

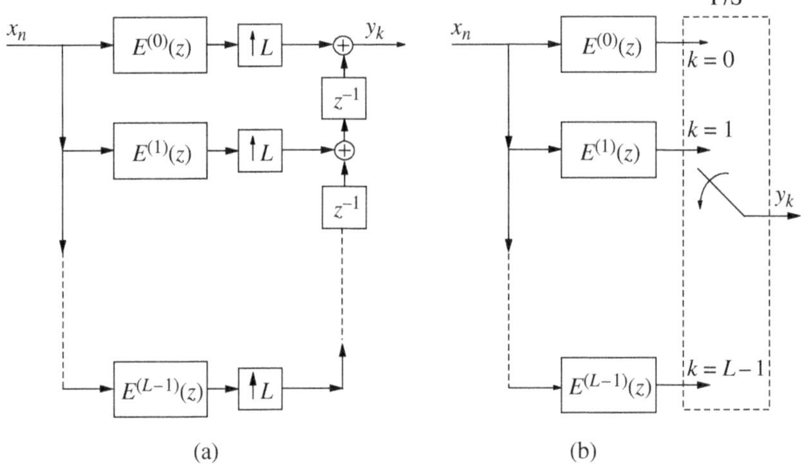

Figure 1.67 Implementation of an interpolator filter using the type-1 polyphase representation.

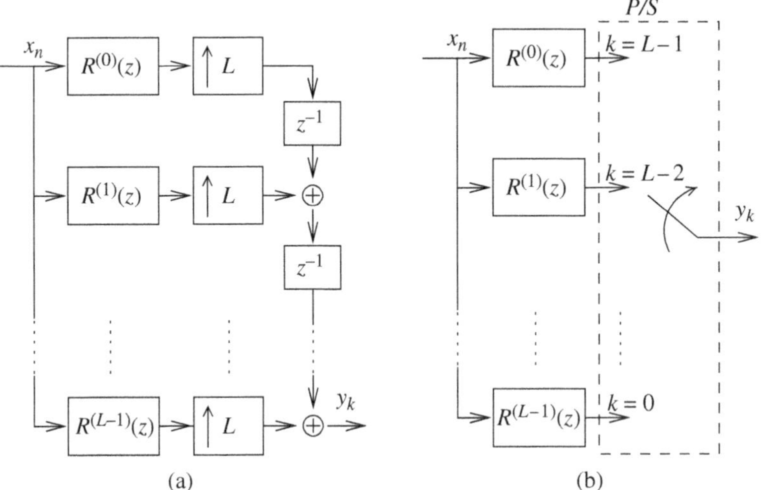

Figure 1.68 Implementation of an interpolator filter using the type-2 polyphase representation.

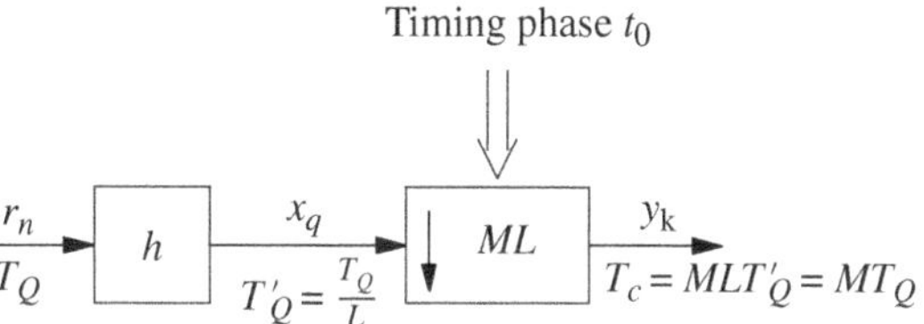

Figure 1.69 Interpolator-decimator filter.

with L and M positive integer numbers. Moreover, we assume that t_0 is a multiple of T'_Q, i.e.

$$\frac{t_0}{T'_Q} = \ell_0 + \mathcal{L}_0 L \tag{1.570}$$

where $\ell_0 \in \{0, 1, \dots, L-1\}$, and $\mathcal{L}_0$ is a non-negative integer number. For the general case of an interpolator-decimator filter, where t_0 and the ratio T_c/T'_Q are not constrained, we refer to [20] (see also Chapter 14).

Based on the above equations we have

$$y_k = x_{kML+\ell_0+\mathcal{L}_0 L} \tag{1.571}$$

We now recall the polyphase representation of $\{h(nT'_Q)\}$ with L phases

$$\{E^{(\ell)}(z)\}, \qquad \ell = 0, 1, \dots, L-1 \tag{1.572}$$

The interpolator filter structure from T_Q to T'_Q is illustrated in Figure 1.67. For the special case $M = 1$, that is for $T_c = T_Q$, the implementation of the interpolator-decimator filter is given in Figure 1.70, where

$$y_k = v_{k+\mathcal{L}_0} \tag{1.573}$$

In other words, $\{y_k\}$ coincides with the signal $\{v_n\}$ at the output of branch ℓ_0 of the polyphase structure. In practice, we need to ignore the first $\mathcal{L}_0$ samples of $\{v_n\}$, as the relation between $\{v_n\}$ and $\{y_k\}$ must take into account a lead, $z^{\mathcal{L}_0}$, of $\mathcal{L}_0$ samples. With reference to Figure 1.70, the output $\{x_q\}$ at instants that are multiples of T'_Q is given by the outputs of the various polyphase branches in sequence.

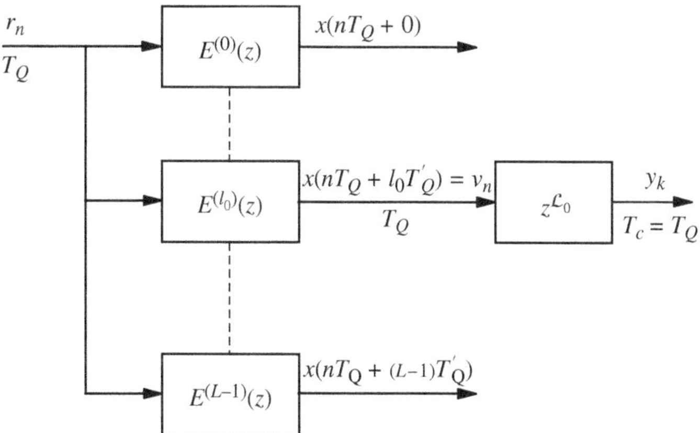

Figure 1.70 Polyphase implementation of an interpolator-decimator filter with timing phase $t_0 = (\ell_0 + \mathcal{L}_0 L)T'_Q$.

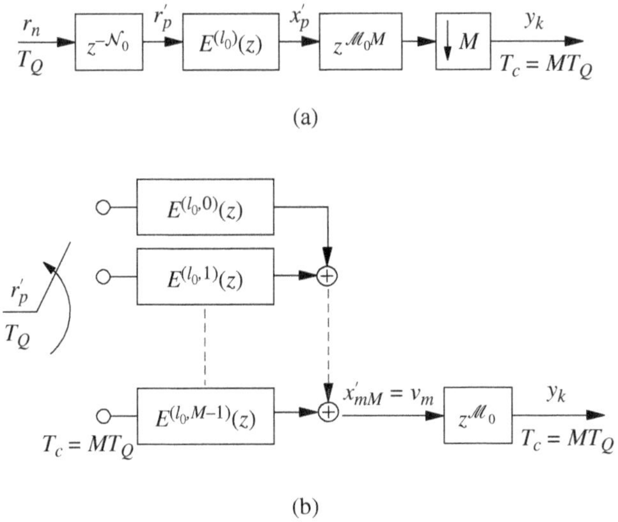

(a)

(b)

Figure 1.71 Implementation of an interpolator-decimator filter with timing phase $t_0 = (\ell_0 + \mathcal{L}_0 L)T'_Q$. (a) Basic scheme and (b) efficient implementation.

In fact, let $q = \ell + nL$, $\ell = 0, 1, \dots, L-1$, and n integer, we have

$$x_{\ell+nL} = x(nLT'_Q + \ell T'_Q) = x(nT_Q + \ell T'_Q) \tag{1.574}$$

We now consider the general case $M \neq 1$. First, to downsample the signal interpolated at T'_Q one can still use the polyphase structure of Figure 1.70. In any case, once t_0 is chosen, the branch is identified (say ℓ_0) and its output must be downsampled by a factor ML. Notice that there is the timing lead $\mathcal{L}_0 L$ in (1.570) to be considered. Given $\mathcal{L}_0$, we determine a positive integer $\mathcal{N}_0$ so that $\mathcal{L}_0 + \mathcal{N}_0$ is a multiple of M, that is

$$\mathcal{L}_0 + \mathcal{N}_0 = \mathcal{M}_0 M \tag{1.575}$$

The structure of Figure 1.70, considering only branch ℓ_0, is equivalent to that given in Figure 1.71a, in which we have introduced a lag of $\mathcal{N}_0$ samples on the sequence $\{r_n\}$ and a further lead of $\mathcal{N}_0$ samples before the downsampler. In particular, we have

$$r'_p = r_{p-\mathcal{N}_0} \quad \text{and} \quad x'_p = x_{p-\mathcal{N}_0} \tag{1.576}$$

As a result, the signal is not modified before the downsampler.

Using now the representation of $E^{(\ell_0)}(z^L)$ in M phases:

$$E^{(\ell_0,m)}(z^{LM}), \qquad m = 0, 1, \dots, M-1 \tag{1.577}$$

an efficient implementation of the interpolator-decimator filter is given in Figure 1.71b.

Appendix 1.B Generation of a complex Gaussian noise

Let $\overline{w} = \overline{w}_I + j\overline{w}_Q$ be a complex Gaussian r.v. with zero mean and unit variance; note that $\overline{w}_I = Re\,[\overline{w}]$ and $\overline{w}_Q = Im\,[\overline{w}]$. In polar notation,

$$\overline{w} = A\, e^{j\varphi} \tag{1.578}$$

It can be shown that φ is a uniform r.v. in $[0, 2\pi)$, and A is a Rayleigh r.v. with probability distribution

$$P[A \leq a] = \begin{cases} 1 - e^{-a^2} & a > 0 \\ 0 & a < 0 \end{cases} \tag{1.579}$$

Observing (1.579) and (1.578), if $u_1, u_2 \sim \mathcal{U}(0, 1)$ independent, then

$$A = \sqrt{-\ln(1 - u_1)} \tag{1.580}$$

and

$$\varphi = 2\pi\, u_2 \tag{1.581}$$

In terms of real components, it results that

$$\overline{w}_I = A \cos \varphi \qquad \text{and} \qquad \overline{w}_Q = A \sin \varphi \tag{1.582}$$

are two statistically independent Gaussian r.v.s., each with zero mean and variance equal to 0.5.

The r.v. $\overline{w}$ is also called circularly symmetric Gaussian r.v., as the real and imaginary components, being statistically independent with equal variance, have a circularly symmetric Gaussian joint probability density function.

Appendix 1.C Pseudo-noise sequences

In this Appendix, we introduce three classes of *deterministic periodic sequences* having spectral characteristics similar to those of a white noise, hence, the name *pseudo-noise* (PN) sequences.

Maximal-length

A maximal-length sequence (MLS), also denoted m-sequence or r-sequence, is a binary PN sequence generated recursively, e.g. using a shift-register, and has period $L = 2^r - 1$. Let $\{p(\ell)\}$, $\ell = 0, 1, \ldots, L-1$, $p(\ell) \in \{0, 1\}$, be the values assumed by the sequence in a period. It enjoys the following properties [7, 21].

- Every non-zero sequence of r bits appears exactly once in each period; Therefore, all binary sequences of r bits are generated, except the all zero sequence.
- The number of bits equal to *1* in a period is 2^{r-1}, and the number of bits equal to *0* is $2^{r-1} - 1$.
- A *subsequence* is intended here as a set of consecutive bits of the MLS. The relative frequency of any *non-zero* subsequence of length $i \leq r$ is

$$\frac{2^{r-i}}{2^r - 1} \simeq 2^{-i} \tag{1.583}$$

  and the relative frequency of a subsequence of length $i < r$ with *all bits equal to zero* is

$$\frac{2^{r-i} - 1}{2^r - 1} \simeq 2^{-i} \tag{1.584}$$

  In both formulas the approximation is valid for a sufficiently large r.
- The sum of two MLSs, which are generated by the same shift-register, but with different initial conditions, is still an MLS.
- The *linear span*, that determines the predictability of a sequence, is equal to r [22]. In other words, the elements of a sequence can be determined by any $2r$ consecutive elements of the sequence itself, while the remaining elements can be produced by a recursive algorithm.

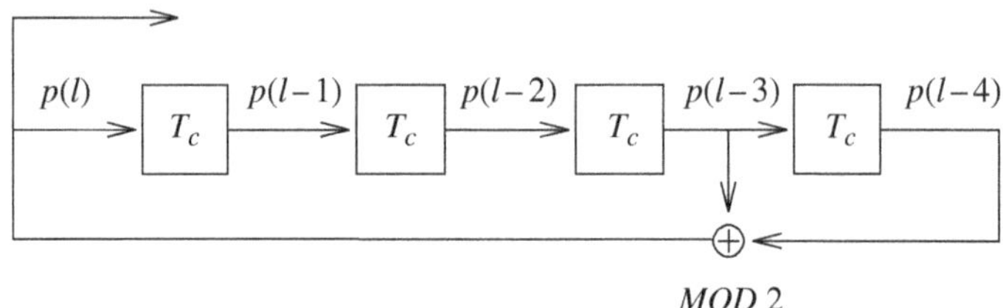

Figure 1.72 Generation of an MLS with period $L = 15$.

Table 1.8: Recursive equations to generate PN sequences of length $L = 2^r - 1$, for different values of r. $p(\ell) \in \{0, 1\}$ and $\oplus$ denotes the modulus-2 sum.

r	period $L = 2^r - 1$
1	$p(\ell) = p(\ell - 1)$
2	$p(\ell) = p(\ell - 1) \oplus p(\ell - 2)$
3	$p(\ell) = p(\ell - 2) \oplus p(\ell - 3)$
4	$p(\ell) = p(\ell - 3) \oplus p(\ell - 4)$
5	$p(\ell) = p(\ell - 3) \oplus p(\ell - 5)$
6	$p(\ell) = p(\ell - 5) \oplus p(\ell - 6)$
7	$p(\ell) = p(\ell - 6) \oplus p(\ell - 7)$
8	$p(\ell) = p(\ell - 2) \oplus p(\ell - 3) \oplus p(\ell - 4) \oplus p(\ell - 8)$
9	$p(\ell) = p(\ell - 5) \oplus p(\ell - 9)$
10	$p(\ell) = p(\ell - 7) \oplus p(\ell - 10)$
11	$p(\ell) = p(\ell - 9) \oplus p(\ell - 11)$
12	$p(\ell) = p(\ell - 2) \oplus p(\ell - 10) \oplus p(\ell - 11) \oplus p(\ell - 12)$
13	$p(\ell) = p(\ell - 1) \oplus p(\ell - 11) \oplus p(\ell - 12) \oplus p(\ell - 13)$
14	$p(\ell) = p(\ell - 2) \oplus p(\ell - 12) \oplus p(\ell - 13) \oplus p(\ell - 14)$
15	$p(\ell) = p(\ell - 14) \oplus p(\ell - 15)$
16	$p(\ell) = p(\ell - 11) \oplus p(\ell - 13) \oplus p(\ell - 14) \oplus p(\ell - 16)$
17	$p(\ell) = p(\ell - 14) \oplus p(\ell - 17)$
18	$p(\ell) = p(\ell - 11) \oplus p(\ell - 18)$
19	$p(\ell) = p(\ell - 14) \oplus p(\ell - 17) \oplus p(\ell - 18) \oplus p(\ell - 19)$
20	$p(\ell) = p(\ell - 17) \oplus p(\ell - 20)$

A practical example is given in Figure 1.72 for a sequence with $L = 15$ ($r = 4$), which is generated by the recursive equation:

$$p(\ell) = p(\ell - 3) \oplus p(\ell - 4) \tag{1.585}$$

where $\oplus$ denotes *modulo 2 sum*. Assuming initial conditions $p(-1) = p(-2) = p(-3) = p(-4) = 1$, applying (1.585) we obtain the sequence

$$\underbrace{0}_{p(0)} \; \underbrace{0}_{p(1)} \; 0 \; 1 \; 0 \; 0 \; 1 \; 1 \; 0 \; 1 \; 0 \; 1 \; 1 \; \underbrace{1}_{p(L-1)} \; 1 \; \ldots \tag{1.586}$$

Obviously, the all zero initial condition must be avoided. To generate MLSs with a larger period L, we refer to Table 1.8. The above properties make an MLS, even if deterministic and periodic, appear as a random i.i.d. sequence from the point of view of the relative frequency of subsequences of bits. It turns out that an MLS appears as random i.i.d. also from the point of view of the autocorrelation function. In fact, mapping '0' to '−1' and '1' to '+1', we get the following correlation properties (see also Theorem 11.1).[26]

1. *Mean*
$$\frac{1}{L}\sum_{\ell=0}^{L-1} p(\ell) = \frac{1}{L} \tag{1.587}$$

2. *Correlation* (periodic of period L, see (11.159))
$$\mathtt{r}_p(n) = \frac{1}{L}\sum_{\ell=0}^{L-1} p(\ell)p^*(\ell - n)_{mod\ L} = \begin{cases} 1 & \text{for } (n)_{mod\ L} = 0 \\ -\dfrac{1}{L} & \text{otherwise} \end{cases} \tag{1.588}$$

3. *PSD* (periodic of period L)
$$\mathcal{P}_p\left(m\frac{1}{LT_c}\right) = T_c \sum_{n=0}^{L-1} \mathtt{r}_p(n) e^{-j2\pi m \frac{1}{LT_c} nT_c} = \begin{cases} T_c\dfrac{1}{L} & \text{for } (m)_{mod\ L} = 0 \\ T_c\left(1 + \dfrac{1}{L}\right) & \text{otherwise} \end{cases} \tag{1.589}$$

We note that, with the exception of the values assumed for $(m)_{mod\ L} = 0$, the PSD of an MLS is constant.

CAZAC

The *constant amplitude zero autocorrelation* (CAZAC) sequences, also denoted as Zadoff–Chu sequences, are complex-valued PN sequences with constant amplitude (assuming values on the unit circle) and autocorrelation function $\mathtt{r}_p(n)$, with $\mathtt{r}_p(n) = 0$ for $(n)_{mod\ L} \neq 0$. Because of these characteristics they are also called *polyphase sequences* [23–25]. Let L and M be two integer numbers that are relatively prime. The CAZAC sequences are defined as

$$\text{for L even } p(\ell) = e^{j\frac{M\pi\ell^2}{L}}, \quad \ell = 0, 1, \ldots, L-1 \tag{1.590}$$

$$\text{for L odd } p(\ell) = e^{j\frac{M\pi\ell(\ell+1)}{L}}, \quad \ell = 0, 1, \ldots, L-1 \tag{1.591}$$

It can be shown that these sequences have the following properties:

1. *Mean* The mean of a CAZAC sequence is usually zero, although this property does not hold in general.
2. *Correlation*
$$\mathtt{r}_p(n) = \begin{cases} 1 & \text{for } (n)_{mod\ L}=0 \\ 0 & \text{otherwise} \end{cases} \tag{1.592}$$
3. *PSD*
$$\mathcal{P}_p\left(m\frac{1}{LT_c}\right) = T_c \tag{1.593}$$

We note the complexity of generating, storing and processing a CAZAC sequence w.r.t. an MLS.

[26] Note that $p(\ell) \in \{0, 1\}$ when generated, while $p(\ell) \in \{-1, 1\}$ when used as a signal. The context will clarify the alphabet of $p(\ell)$.

Gold

In a large number of applications, as for example in *spread-spectrum* systems with code-division multiple access, *sets of sequences* having one or both of the following properties [26] are required.

- Each sequence of the set must be easily distinguishable from *its own time shifted versions*.
- Each sequence of the set must be easily distinguishable from *any other sequence* of the set and from its time-shifted versions.

An important class of periodic binary sequences that satisfy these properties, or, in other words, that have good autocorrelation and crosscorrelation characteristics, is the set of Gold sequences [27, 28].

Construction of pairs of preferred MLSs. In general, the crosscorrelation sequence (CCS) between two MLSs may assume three, four or a larger number of values. We show now the construction of a pair of MLSs, called *preferred MLSs* [22], whose CCS assumes only three values. Let $a = \{a(\ell)\}$ be an MLS with period $L = 2^r - 1$. We define now another MLS of length $L = 2^r - 1$ obtained from the sequence a by decimation by a factor M, that is

$$b = \{b(\ell)\} = \{a(M\ell)_{mod\ L}\} \tag{1.594}$$

We make the following assumptions:

- $r_{mod\ 4} \neq 0$, that is r must be odd or equal to odd multiples of 2, i.e. $r_{mod\ 4} = 2$.
- The factor M satisfies one of the following properties:

$$M = 2^k + 1 \quad \text{or} \quad M = 2^{2k} - 2^k + 1 \quad k \text{ integer} \tag{1.595}$$

- For k determined as in the (1.595), defining g.c.d.(r, k) as the greatest common divisor of r and k, let

$$e = \text{g.c.d.}(r, k) = \begin{cases} 1 & r \text{ odd} \\ 2 & r_{mod\ 4} = 2 \end{cases} \tag{1.596}$$

Then the CCS between the two MLS a and b assumes only three values [21, 22]:

$$\begin{aligned} \mathtt{r}_{ab}(n) &= \frac{1}{L}\sum_{\ell=0}^{L-1} a(\ell)b^*(\ell - n)_{mod\ L} \\ &= \frac{1}{L}\begin{cases} -1 + 2^{\frac{r+e}{2}} & \text{(value assumed} \quad 2^{r-e-1} + 2^{\frac{r+e-2}{2}} \quad \text{times} \\ -1 & \text{(value assumed} \quad 2^r - 2^{r-e} - 1 \quad \text{times} \\ -1 - 2^{\frac{r+e}{2}} & \text{(value assumed} \quad 2^{r-e-1} + 2^{\frac{r+e-2}{2}} \quad \text{times} \end{cases} \end{aligned} \tag{1.597}$$

Example 1.C.1 (Construction of a pair of preferred MLSs*)*
Let the following MLS of period $L = 2^5 - 1 = 31$ be given:

$$\{a(\ell)\} = (0000100101100111110001101110101) \tag{1.598}$$

As $r = 5$ and $r_{mod\ 4} = 1$, we take $k = 1$. Therefore, $e = \text{g.c.d.}(r, k) = \text{g.c.d.}(5, 1) = 1$ and $M = 2^k + 1 = 2^1 + 1 = 3$. The sequence $\{b(\ell)\}$ obtained by decimation of the sequence $\{a(\ell)\}$ is then given by

$$\{b(\ell)\} = \{a(3\ell)_{mod\ L}\} = (0001010110100001100100111110111) \tag{1.599}$$

The CCS between the two sequences, assuming 0 is mapped to -1, is

$$\{\mathtt{r}_{ab}(n)\} = \frac{1}{31}(7,7,-1,-1,-1,-9,7,-9,7,7,-1,-1,7,7,-1,7,-1,-1,-9,-1,-1,-1,-1,-9,\\ -1,7,-1,-9,-9,7,-1) \tag{1.600}$$

We note that, if we had chosen $k = 2$, then $e = \text{g.c.d.}(5,2) = 1$ and $M = 2^2 + 1 = 5$, or else $M = 2^{2\cdot 2} - 2^2 + 1 = 13$.

Construction of a set of Gold sequences. A set of Gold sequences can be constructed from any pair $\{a(\ell)\}$ and $\{b(\ell)\}$ of preferred MLSs of period $L = 2^r - 1$. We define the set of sequences:

$$G(a,b) = \{a, b, a \oplus b, a \oplus Zb, a \oplus Z^2 b, \dots, a \oplus Z^{L-1} b\} \tag{1.601}$$

where Z is the *shift* operator that cyclically shifts a sequence to the left by a position. The set (1.601) contains $L + 2 = 2^r + 1$ sequences of length $L = 2^r - 1$ and is called the *set of Gold sequences*. It can be proven [27, 28] that, for the two sequences $\{a'(\ell)\}$ and $\{b'(\ell)\}$ belonging to the set $G(a,b)$, the CCS as well as the ACS, with the exception of zero lag, assume only three values:

$$\mathtt{r}_{a'b'}(n) = \frac{1}{L}\begin{cases} -1, -1-2^{\frac{r+1}{2}}, -1+2^{\frac{r+1}{2}} & r \text{ odd} \\ -1, -1-2^{\frac{r+2}{2}}, -1+2^{\frac{r+2}{2}} & r_{mod\ 4} = 2 \end{cases} \tag{1.602}$$

Clearly, the ACS of a Gold sequence no longer has the characteristics of an MLS, as is seen in the next example.

Example 1.C.2 (Gold sequence properties)

Let $r = 5$, hence, $L = 2^5 - 1 = 31$. From Example 1.C.1, the two sequences (1.598) and (1.599) are a pair of preferred MLSs, from which it is possible to generate the whole set of Gold sequences. For example we calculate the ACS of $\{a(\ell)\}$and $\{b'(\ell)\} = \{a(\ell) \oplus b(\ell - 2)\} = a \oplus Z^2 b$, and the CCS between $\{a(\ell)\}$ and $\{b'(\ell)\}$:

$$\{a(\ell)\} = (-1,-1,-1,-1,1,-1,-1,1,-1,1,1,-1,-1,1,1,1,1,1,-1,\\ -1,-1,1,1,-1,1,1,1,-1,1,-1,1) \tag{1.603}$$

$$\{b'(\ell)\} = (-1,-1,1,-1,-1,-1,1,-1,-1,-1,1,-1,-1,1,-1,-1,1,\\ 1,1,-1,-1,-1,-1,1,-1,-1,1,1,-1,1,1) \tag{1.604}$$

$$\{\mathtt{r}_a(n)\} = \frac{1}{31}(31,-1,-1,-1,-1,-1,-1,-1,-1,-1,-1,-1,-1,-1,-1,\\ -1,-1,-1,-1,-1,-1,-1,-1,-1,-1,-1,-1,-1,-1,-1,-1) \tag{1.605}$$

$$\{\mathtt{r}_{b'}(n)\} = \frac{1}{31}(31,-1,-9,7,7,-9,-1,7,-1,-9,7,7,-1-1,7,-1,-1,7,\\ -1,-1,7,7,-9,-1,7,-1,9,7,7,-9,-1) \tag{1.606}$$

$$\{\mathtt{r}_{ab'}(n)\} = \frac{1}{31}(-1,7,7,7,-1,-1,-1,-1,-1,7,-9,-1,-1,7,-1,-9,7,\\ -1-9,7,7,-9,-1,7,-1,-9,-1,-1,-1,-9,-1) \tag{1.607}$$

Chapter 2

The Wiener filter

The theory of the Wiener filter [1, 2] that will be presented in this chapter is fundamental for the comprehension of several important applications. The development of this theory assumes the knowledge of the correlation functions of the relevant processes. An approximation of the Wiener filter can be obtained by the least squares (LS) method, through realizations of the processes involved.

The section on estimation extends the Wiener filter to a structure with multiple inputs and outputs. Next, some examples of application of the developed theory are illustrated.

2.1 The Wiener filter

With reference to Figure 2.1, let x and d be two individually and jointly wide sense stationary (WSS) random processes with zero mean; the problem is to determine the finite impulse system (FIR) filter so that, if the filter input is $x(k)$, the output $y(k)$ replicates as closely as possible $d(k)$, for each k. The Wiener theory provides the means to design the required filter.

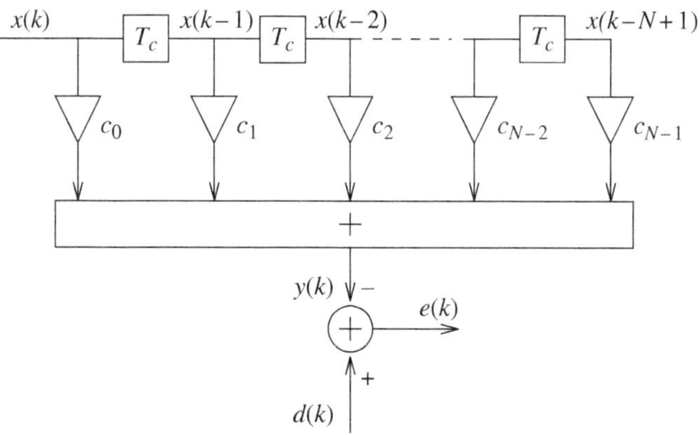

Figure 2.1 The Wiener filter with N coefficients.

The FIR filter in Figure 2.1 is called *transversal filter*, as the output is formed by summing the products of the delayed input samples by suitable coefficients. If we indicate with $\{c_n\}$, $n = 0, 1, \dots, N-1$, the

Algorithms for Communications Systems and their Applications, Second Edition.
Nevio Benvenuto, Giovanni Cherubini, and Stefano Tomasin.

N coefficients of the filter, we have

$$y(k) = \sum_{n=0}^{N-1} c_n x(k-n) \tag{2.1}$$

If $d(k)$ is the *desired sample* at the filter output at instant k, we define the *estimation error* as

$$e(k) = d(k) - y(k) \tag{2.2}$$

In the Wiener theory, in order that the filter output $y(k)$ replicates as closely as possible $d(k)$, the coefficients of the filter are determined using the *minimum mean-square error* (MMSE) criterion. Therefore, the cost function is defined as the *mean square error* (MSE)

$$J = E[|e(k)|^2] \tag{2.3}$$

and the coefficients of the optimum filter minimize J, i.e.[1]

$$\boldsymbol{c}_{opt} = \arg \min_{\{c_n\}, n=0,1,\ldots,N-1} J \tag{2.4}$$

The Wiener filter problem can be formulated as the problem of estimating $d(k)$ by a linear combination of $x(k), \ldots, x(k-N+1)$. A brief introduction to estimation theory is given in Section 2.4, where the formulation of the Wiener theory is further extended to the case of vector signals.

Matrix formulation

The problem introduced in the previous section is now formulated using matrix notation. We define[2]:

1. *Coefficient vector*

$$\boldsymbol{c}^T = [c_0, c_1, \ldots, c_{N-1}] \tag{2.5}$$

2. *Filter input vector at instant k*[3]

$$\boldsymbol{x}^T(k) = [x(k), x(k-1), \ldots, x(k-N+1)] \tag{2.6}$$

The filter output at instant k is expressed as

$$y(k) = \boldsymbol{c}^T \boldsymbol{x}(k) = \boldsymbol{x}^T(k)\, \boldsymbol{c} \tag{2.7}$$

and the estimation error as

$$e(k) = d(k) - \boldsymbol{c}^T \boldsymbol{x}(k) \tag{2.8}$$

Moreover,

$$y^*(k) = \boldsymbol{c}^H \boldsymbol{x}^*(k) = \boldsymbol{x}^H(k)\boldsymbol{c}^*, \qquad e^*(k) = d^*(k) - \boldsymbol{c}^H \boldsymbol{x}^*(k) \tag{2.9}$$

We express now the cost function J as a function of the vector $\boldsymbol{c}$. We will then seek vector $\boldsymbol{c}_{opt}$ that minimizes J. Recalling the definition

$$J = E[e(k)e^*(k)] = E[(d(k) - \boldsymbol{x}^T(k)\boldsymbol{c})(d^*(k) - \boldsymbol{c}^H \boldsymbol{x}^*(k))] \tag{2.10}$$

---

[1] Here $\arg\min_a f(a)$ provides the value of a that minimizes $f(a)$. A similar definition holds for $\arg\max_a f(a)$.

[2] The components of an N-dimensional vector are usually identified by an index varying either from 1 to N or from 0 to $N-1$.

[3] Indeed, the theory is more general than the one developed here, and the various inputs to the transversal filter taps need not to be shifted versions of a given process $x(k)$. In fact, we can replace $\boldsymbol{x}^T(k)$ in (2.6) by $\boldsymbol{x}^T(k) = [x_1(k), x_2(k), \ldots, x_N(k)]$, where $\{x_n(k)\}$, $n = 1, \ldots, N$, are random processes, and carry out a more general derivation than the one developed here. Although this general approach is seldom useful, two examples of application are given in (7.193) and Section 2.5.4.

and computing the products, it follows

$$J = E[d^*(k)d(k)] - c^H E[x^*(k)d(k)] + \\ - E[d^*(k)x^T(k)]c + c^H E[x^*(k)x^T(k)]c \tag{2.11}$$

Assuming that x and d are individually and jointly WSS, we introduce the following quantities:

1. *Variance of the desired signal*

$$\sigma_d^2 = E[d(k)d^*(k)] \tag{2.12}$$

2. *Correlation between the desired output at instant k and the filter input vector at the same instant*

$$r_{dx} = E\left[d(k)x^*(k)\right] = \begin{bmatrix} E\left[d(k)x^*(k)\right] \\ E\left[d(k)x^*(k-1)\right] \\ \vdots \\ E\left[d(k)x^*(k-N+1)\right] \end{bmatrix} = p \tag{2.13}$$

The components of p are given by

$$[p]_n = E\left[d(k)x^*(k-n)\right] = \mathtt{r}_{dx}(n), \qquad n = 0, 1, \dots, N-1 \tag{2.14}$$

Moreover, it holds

$$p^H = E\left[d^*(k)x^T(k)\right] \tag{2.15}$$

3. $N \times N$ *autocorrelation matrix* of the complex-conjugate filter input vector $x^*(k)$, as defined in (1.270),

$$R = E\left[x^*(k)x^T(k)\right] \tag{2.16}$$

Then

$$J = \sigma_d^2 - c^H p - p^H c + c^H R c \tag{2.17}$$

The cost function J, considered as a function of c, is a quadratic function. Then, if R is positive definite, J admits one and only one minimum value. J as a function of c is shown in Figure 2.2, for the particular cases $N = 1$ and $N = 2$.

Optimum filter design

It is now a matter of finding the minimum of (2.17) with respect to c. Recognizing that, as $c = c_I + jc_Q$, the real independent variables are $2N$, to accomplish this task, we define the derivative of J with respect to c as the gradient vector

$$\nabla_c J = \frac{\partial J}{\partial c} = \begin{bmatrix} \frac{\partial J}{\partial c_{0,I}} + j\frac{\partial J}{\partial c_{0,Q}} \\ \frac{\partial J}{\partial c_{1,I}} + j\frac{\partial J}{\partial c_{1,Q}} \\ \vdots \\ \frac{\partial J}{\partial c_{N-1,I}} + j\frac{\partial J}{\partial c_{N-1,Q}} \end{bmatrix} \tag{2.18}$$

and also

$$\nabla_c J = \nabla_{c_I} J + j\nabla_{c_Q} J \tag{2.19}$$

In general, because the vector p and the autocorrelation matrix R are complex, we also write

$$p = p_I + jp_Q \tag{2.20}$$

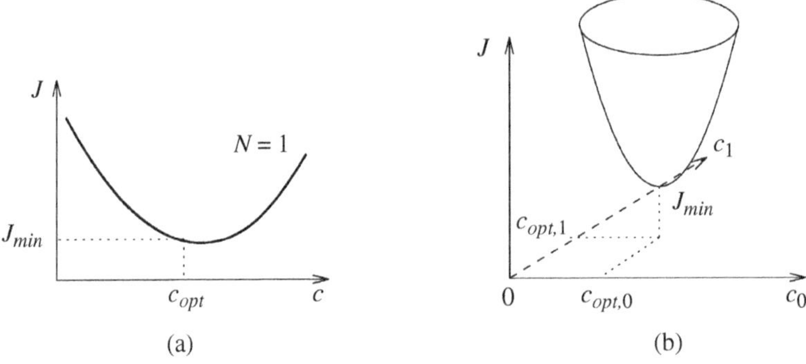

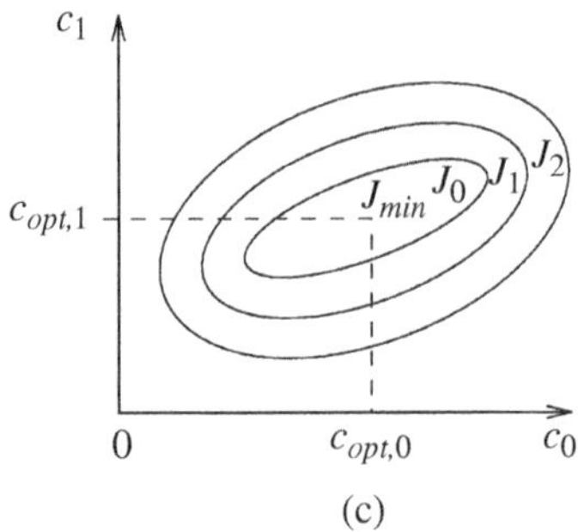

Figure 2.2 J as a function of $\boldsymbol{c}$ for the cases $N = 1$ and $N = 2$. (a) J versus $\boldsymbol{c}$ for $N = 1$, (b) J versus $\boldsymbol{c}$ for $N = 2$, and (c) equal-J contours for $N = 2$.

and

$$\boldsymbol{R} = \boldsymbol{R}_I + j\boldsymbol{R}_Q \tag{2.21}$$

If now we take the derivative of the terms of (2.17) by the real and imaginary part of $\boldsymbol{c}$, we find

$$\nabla_{c_I}\boldsymbol{p}^H\boldsymbol{c} = \nabla_{c_I}\boldsymbol{p}^H(\boldsymbol{c}_I + j\boldsymbol{c}_Q) = \boldsymbol{p}^H \tag{2.22a}$$

$$\nabla_{c_Q}\boldsymbol{p}^H\boldsymbol{c} = \nabla_{c_Q}\boldsymbol{p}^H(\boldsymbol{c}_I + j\boldsymbol{c}_Q) = j\boldsymbol{p}^H \tag{2.22b}$$

$$\nabla_{c_I}\boldsymbol{c}^H\boldsymbol{p} = \nabla_{c_I}(\boldsymbol{c}_I^T - j\boldsymbol{c}_Q{}^T)\boldsymbol{p} = \boldsymbol{p} \tag{2.22c}$$

$$\nabla_{c_Q}\boldsymbol{c}^H\boldsymbol{p} = \nabla_{c_Q}(\boldsymbol{c}_I^T - j\boldsymbol{c}_Q^T)\boldsymbol{p} = -j\boldsymbol{p} \tag{2.22d}$$

$$\nabla_{c_I}(\boldsymbol{c}^H\boldsymbol{R}\boldsymbol{c}) = 2\boldsymbol{R}_I\boldsymbol{c}_I - 2\boldsymbol{R}_Q\boldsymbol{c}_Q \tag{2.22e}$$

$$\nabla_{c_Q}(\boldsymbol{c}^H\boldsymbol{R}\boldsymbol{c}) = 2\boldsymbol{R}_I\boldsymbol{c}_Q + 2\boldsymbol{R}_Q\boldsymbol{c}_I \tag{2.22f}$$

From (2.22) and (2.19), we obtain

$$\nabla_c\boldsymbol{p}^H\boldsymbol{c} = \boldsymbol{0} \tag{2.23}$$

$$\nabla_c\boldsymbol{c}^H\boldsymbol{p} = 2\boldsymbol{p} \tag{2.24}$$

$$\nabla_c\boldsymbol{c}^H\boldsymbol{R}\boldsymbol{c} = 2\boldsymbol{R}\boldsymbol{c} \tag{2.25}$$

Substituting the above results into (2.17), it turns out

$$\nabla_c J = -2\boldsymbol{p} + 2\boldsymbol{R}\boldsymbol{c} = 2(\boldsymbol{R}\boldsymbol{c} - \boldsymbol{p}) \tag{2.26}$$

For the optimum coefficient vector $\boldsymbol{c}_{opt}$, it must be $\nabla_c J = \boldsymbol{0}$, where $\boldsymbol{0}$ is a vector of all zeros, hence, we get

$$\boldsymbol{R}\boldsymbol{c}_{opt} = \boldsymbol{p} \tag{2.27}$$

Equation (2.27) is called the *Wiener–Hopf* (W–H) *equation.*

Observation 2.1

The computation of the optimum coefficients $\boldsymbol{c}_{opt}$ requires the knowledge only of the input correlation matrix $\boldsymbol{R}$ and of the crosscorrelation vector $\boldsymbol{p}$ between the desired output and the input vector.

In scalar form, the W–H equation is a linear system of N equations in N unknowns:

$$\sum_{i=0}^{N-1} c_{opt,i}\mathtt{r}_x(n-i) = \mathtt{r}_{dx}(n), \qquad n = 0, 1, \dots, N-1 \tag{2.28}$$

If $\boldsymbol{R}^{-1}$ exists, the solution of (2.27) is

$$\boldsymbol{c}_{opt} = \boldsymbol{R}^{-1}\boldsymbol{p} \tag{2.29}$$

The principle of orthogonality

It is interesting to observe the relation

$$E\left[e(k)\boldsymbol{x}^*(k)\right] = E\left[\boldsymbol{x}^*(k)(d(k) - \boldsymbol{x}^T(k)\boldsymbol{c})\right] = \boldsymbol{p} - \boldsymbol{R}\boldsymbol{c} \tag{2.30}$$

which for $\boldsymbol{c} = \boldsymbol{c}_{opt}$ yields

$$E\left[e(k)\boldsymbol{x}^*(k)\right] = \boldsymbol{0} \tag{2.31}$$

We now state the following important result.

Theorem 2.1 (Principle of orthogonality)

The condition of optimality for $\boldsymbol{c}$ is satisfied if $e(k)$ and $\boldsymbol{x}(k)$ are orthogonal.[4] In scalar form, the filter is optimum if $e(k)$ is orthogonal to $\{x(k-n)\}$, $n = 0, 1, \dots, N-1$, that is

$$E\left[e(k)x^*(k-n)\right] = 0, \qquad n = 0, 1, \dots, N-1 \tag{2.32}$$

□

This is equivalent to say that the optimum-filter output $y(k)$ reproduces in some statistical sense all components of $d(k)$ proportional to $x(k), x(k-1), \dots, x(k-N+1)$.

Corollary 2.1

For $\boldsymbol{c} = \boldsymbol{c}_{opt}$, $e(k)$ and $y(k)$ are orthogonal, i.e.

$$E\left[e(k)y^*(k)\right] = 0 \qquad \text{for } \boldsymbol{c} = \boldsymbol{c}_{opt} \tag{2.33}$$

□

In fact, using the orthogonality principle,

$$\begin{aligned} E\left[e(k)y^*(k)\right] &= E\left[e(k)\boldsymbol{c}^H\boldsymbol{x}^*(k)\right] \\ &= \boldsymbol{c}^H E\left[e(k)\boldsymbol{x}^*(k)\right] \\ &= \boldsymbol{c}^H\boldsymbol{0} \\ &= 0 \end{aligned} \tag{2.34}$$

For an optimum filter, Figure 2.3 depicts the relation between the three signals d, e, and y.

---

[4] Note that orthogonality holds only if e and $\boldsymbol{x}$ are considered at the same instant. Indeed, there could be statistical correlation between $e(k)$ and $x(k-M)$ with $M \geq N$.

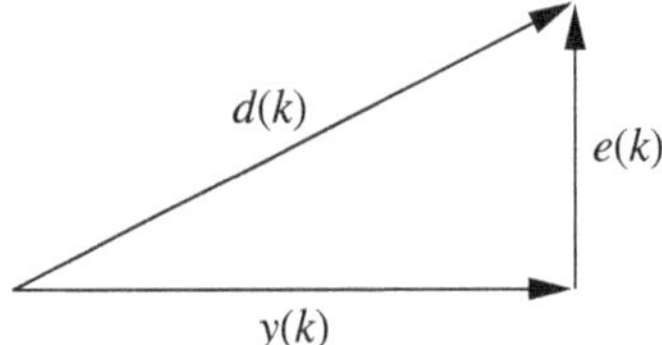

Figure 2.3 Orthogonality of signals for an optimum filter.

Expression of the minimum mean-square error

We now determine the value of the cost function J in correspondence of $\boldsymbol{c}_{opt}$. Substituting the expression (2.27) of $\boldsymbol{c}_{opt}$ in (2.17), we get

$$\begin{aligned} J_{min} &= \sigma_d^2 - \boldsymbol{c}_{opt}^H \boldsymbol{p} - \boldsymbol{p}^H \boldsymbol{c}_{opt} + \boldsymbol{c}_{opt}^H \boldsymbol{p} \\ &= \sigma_d^2 - \boldsymbol{p}^H \boldsymbol{c}_{opt} \end{aligned} \tag{2.35}$$

We note that $\boldsymbol{R}$ in (2.27) can be ill-conditioned if N is very large. Hence, as for the order of the auto-regressive moving-average (ARMA) model (see Section 1.10), also here N should be determined in an iterative way by evaluating J_{min} (2.35) for increasing values of N. We stop when J_{min} does not significantly change, i.e. it reaches approximately its asymptotic value.

Another useful expression of J_{min} is obtained from (2.2):

$$d(k) = e(k) + y(k) \tag{2.36}$$

As $e(k)$ and $y(k)$ are orthogonal for $\boldsymbol{c} = \boldsymbol{c}_{opt}$, then

$$\sigma_d^2 = J_{min} + \sigma_y^2 \tag{2.37}$$

whereby it follows

$$J_{min} = \sigma_d^2 - \sigma_y^2 \tag{2.38}$$

Using (2.35), we can find an alternative expression to (2.17) for the cost function J:

$$J = J_{min} + (\boldsymbol{c} - \boldsymbol{c}_{opt})^H \boldsymbol{R} (\boldsymbol{c} - \boldsymbol{c}_{opt}) \tag{2.39}$$

Recalling that the autocorrelation matrix is positive semi-definite, it follows that the quantity $(\boldsymbol{c} - \boldsymbol{c}_{opt})^H \boldsymbol{R} (\boldsymbol{c} - \boldsymbol{c}_{opt})$ is non-negative and in particular it vanishes for $\boldsymbol{c} = \boldsymbol{c}_{opt}$.

Characterization of the cost function surface

The result (2.39) calls for further observations on J. In fact, using the decomposition (1.292), we get

$$J = J_{min} + (\boldsymbol{c} - \boldsymbol{c}_{opt})^H \boldsymbol{U} \boldsymbol{\Lambda} \boldsymbol{U}^H (\boldsymbol{c} - \boldsymbol{c}_{opt}) \tag{2.40}$$

Let us now define

$$\boldsymbol{\nu} = \begin{bmatrix} \nu_1 \\ \vdots \\ \nu_N \end{bmatrix} = \boldsymbol{U}^H (\boldsymbol{c} - \boldsymbol{c}_{opt}) \tag{2.41}$$

where $v_i = \boldsymbol{u}_i^H(\boldsymbol{c} - \boldsymbol{c}_{opt})$. Vector $\boldsymbol{v}$ may be interpreted as a translation and a rotation of vector $\boldsymbol{c}$. Then J assumes the form:

$$\begin{aligned} J &= J_{min} + \boldsymbol{v}^H \boldsymbol{\Lambda} \boldsymbol{v} \\ &= J_{min} + \sum_{i=1}^{N} \lambda_i |v_i|^2 \\ &= J_{min} + \sum_{i=1}^{N} \lambda_i \left| \boldsymbol{u}_i^H(\boldsymbol{c} - \boldsymbol{c}_{opt}) \right|^2 \end{aligned} \tag{2.42}$$

The result (2.42) expresses the excess mean-square error $J - J_{min}$ as the sum of N components in the direction of each eigenvector of $\boldsymbol{R}$. Note that each component is proportional to the corresponding eigenvalue.

From the above observation, we conclude that J increases more rapidly in the direction of the eigenvector corresponding to the maximum eigenvalue λ_{max}. Likewise, the increase is slower in the direction of the eigenvector corresponding to the minimum eigenvalue λ_{min}. Let $\boldsymbol{u}_{\lambda_{max}}$ and $\boldsymbol{u}_{\lambda_{min}}$ denote the eigenvectors of $\boldsymbol{R}$ in correspondence of λ_{max} and λ_{min}, respectively; it follows that $\nabla_c J$ is largest along $\boldsymbol{u}_{\lambda_{max}}$. This is also observed in Figure 2.4, where sets (loci) of points $\boldsymbol{c}$ for which a constant value of J is obtained are graphically represented. In the two-dimensional case, they trace ellipses with axes parallel to the direction of the eigenvectors and the ratio of axes length is related to the eigenvalues.

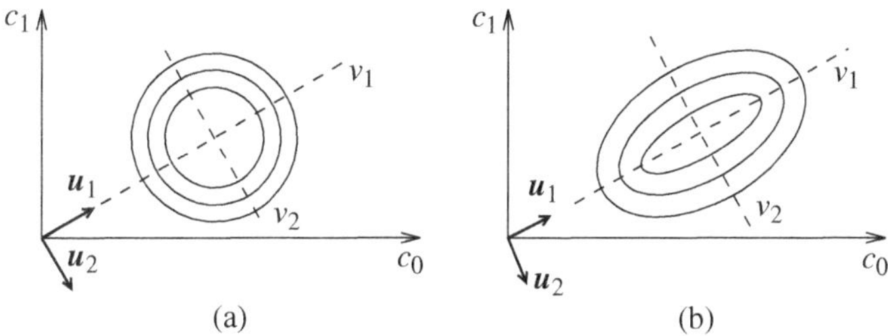

Figure 2.4 Loci of points with constant J (contour plots). (a) Case $\lambda_1 \simeq \lambda_2$ and (b) case $\lambda_1 < \lambda_2$.

The Wiener filter in the z-domain

For a filter with an infinite number of coefficients, not necessarily causal, (2.28) of the optimum filter becomes

$$\sum_{i=-\infty}^{+\infty} c_{opt,i} \mathtt{r}_x(n-i) = \mathtt{r}_{dx}(n) \qquad \forall n \tag{2.43}$$

Taking the z-transform of both sides of the equation yields

$$C_{opt}(z) P_x(z) = P_{dx}(z) \tag{2.44}$$

Then the transfer function of the optimum filter is

$$C_{opt}(z) = \frac{P_{dx}(z)}{P_x(z)} \tag{2.45}$$

We note that while (2.29) is useful to evaluate the coefficients of the optimum FIR filter, (2.45) is employed to analyse the system in the general case of an infinite impulse response (IIR) filter.

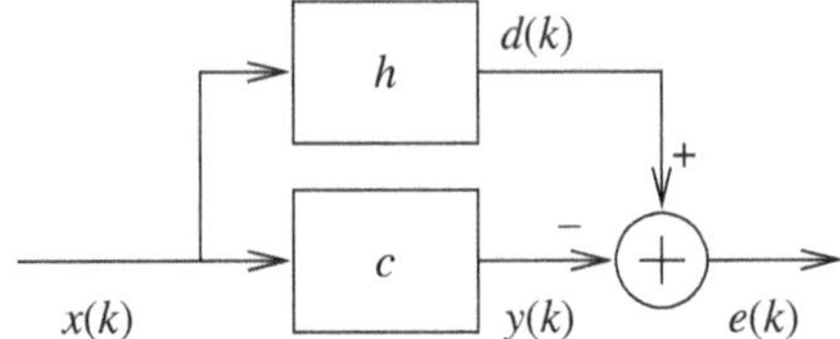

Figure 2.5 An application of the Wiener filter theory.

Example 2.1.1
Let $d(k) = h * x(k)$, as shown in Figure 2.5. In this case, from Table 1.3,

$$P_{dx}(z) = P_x(z)H(z) \tag{2.46}$$

and the optimum filter is given by

$$C_{opt}(z) = H(z) \tag{2.47}$$

From (2.35) in scalar notation, applying Fourier transform properties, we get

$$\begin{aligned} J_{min} &= \sigma_d^2 - \sum_{i=0}^{N-1} c_{opt,i} \mathtt{r}_{dx}^*(i) \\ &= \sigma_d^2 - \int_{-\frac{1}{2T_c}}^{\frac{1}{2T_c}} \mathcal{P}_{dx}^*(f) C_{opt}(e^{j2\pi f T_c})\, df \end{aligned} \tag{2.48}$$

where $\mathcal{P}_{dx}^*(f)$ is the Fourier transform of $\mathtt{r}_{dx}^*(i)$.

Using (2.45), we have

$$\begin{aligned} J_{min} &= \sigma_d^2 - \int_{-\frac{1}{2T_c}}^{\frac{1}{2T_c}} \mathcal{P}_{dx}^*(f) \frac{\mathcal{P}_{dx}(f)}{\mathcal{P}_x(f)}\, df \\ &= \sigma_d^2 - \int_{-\frac{1}{2T_c}}^{\frac{1}{2T_c}} \frac{|\mathcal{P}_{dx}(f)|^2}{\mathcal{P}_x(f)}\, df \\ &= \sigma_d^2 - T_c \int_{-\frac{1}{2T_c}}^{\frac{1}{2T_c}} \frac{|P_{dx}(e^{j2\pi f T_c})|^2}{P_x(e^{j2\pi f T_c})}\, df \end{aligned} \tag{2.49}$$

where $\mathcal{P}_x(f)$ is the Fourier transform of $\mathtt{r}_x(n)$.

Example 2.1.2
We desire to filter the noise from a signal given by one complex sinusoid (tone) plus noise, i.e.

$$x(k) = Ae^{j(\omega_0 k + \varphi)} + w(k) \tag{2.50}$$

In (2.50) $\omega_0 = 2\pi f_0 T_c$ is the tone radian frequency normalized to the sampling period, in radians. We assume the desired signal is given by

$$d(k) = B\ e^{j[\omega_0(k-D)+\varphi]} \tag{2.51}$$

where D is a known delay. We also assume that $\varphi \sim \mathcal{U}(0, 2\pi)$, and w is white noise with zero mean and variance σ_w^2, uncorrelated with φ. The autocorrelation function of x and the crosscorrelation between d and x are given by

$$\mathtt{r}_x(n) = A^2\ e^{j\omega_0 n} + \sigma_w^2 \delta_n \tag{2.52}$$

$$\mathtt{r}_{dx}(n) = AB\ e^{j\omega_0 (n-D)} \tag{2.53}$$

For a Wiener filter with N coefficients, the autocorrelation matrix $\boldsymbol{R}$ and the vector $\boldsymbol{p}$ have the following structure:

$$\boldsymbol{R} = \begin{bmatrix} A^2+\sigma_w^2 & A^2e^{-j\omega_0} & \dots & A^2e^{-j\omega_0(N-1)} \\ A^2e^{j\omega_0} & A^2+\sigma_w^2 & \dots & A^2e^{-j\omega_0(N-2)} \\ \vdots & \vdots & \ddots & \dots \\ A^2e^{j\omega_0(N-1)} & A^2e^{j\omega_0(N-2)} & \dots & A^2+\sigma_w^2 \end{bmatrix} \tag{2.54}$$

$$\boldsymbol{p} = \begin{bmatrix} 1 \\ e^{j\omega_0} \\ \vdots \\ e^{j\omega_0(N-1)} \end{bmatrix} AB\ e^{-j\omega_0 D} \tag{2.55}$$

Defining

$$\boldsymbol{E}^T(\omega) = \left[1, e^{j\omega}, \dots, e^{j\omega(N-1)}\right] \tag{2.56}$$

we can express $\boldsymbol{R}$ and $\boldsymbol{p}$ as

$$\boldsymbol{R} = \sigma_w^2\boldsymbol{I} + A^2\boldsymbol{E}(\omega_0)\boldsymbol{E}^H(\omega_0) \tag{2.57}$$

$$\boldsymbol{p} = ABe^{-j\omega_0 D}\boldsymbol{E}(\omega_0) \tag{2.58}$$

Observing that $\boldsymbol{E}^H(\omega)\boldsymbol{E}(\omega) = N$, the inverse of $\boldsymbol{R}$ is given by

$$\boldsymbol{R}^{-1} = \frac{1}{\sigma_w^2}\left[\boldsymbol{I} - \frac{A^2}{\sigma_w^2 + NA^2}\boldsymbol{E}(\omega_0)\boldsymbol{E}^H(\omega_0)\right] \tag{2.59}$$

Hence, using (2.29):

$$\boldsymbol{c}_{opt} = \frac{ABe^{-j\omega_0 D}}{\sigma_w^2 + NA^2}\boldsymbol{E}(\omega_0) = \frac{B}{A}\frac{\Lambda e^{-j\omega_0 D}}{1+N\Lambda}\boldsymbol{E}(\omega_0) \tag{2.60}$$

where $\Lambda = A^2/\sigma_w^2$ is the signal-to-noise ratio. From (2.35) the minimum value of the cost function J is given by

$$J_{min} = B^2 - ABe^{j\omega_0 D}\boldsymbol{E}^H(\omega_0)\boldsymbol{E}(\omega_0)\frac{ABe^{-j\omega_0 D}}{\sigma_w^2 + NA^2} = \frac{B^2}{1+N\Lambda} \tag{2.61}$$

Defining $\omega = 2\pi f T_c$, the optimum filter frequency response is given by

$$\begin{aligned} C_{opt}(e^{j\omega}) &= \sum_{i=0}^{N-1} c_{opt,i}e^{-j\omega i} \\ &= \boldsymbol{E}^H(\omega)\boldsymbol{c}_{opt} \\ &= \frac{B}{A}\frac{\Lambda e^{-j\omega_0 D}}{1+N\Lambda}\sum_{i=0}^{N-1} e^{-j(\omega-\omega_0)i} \end{aligned} \tag{2.62}$$

that is,

$$C_{opt}(e^{j\omega}) = \begin{cases} \dfrac{B}{A}\dfrac{N\Lambda e^{-j\omega_0 D}}{1+N\Lambda} & \omega = \omega_0 \\ \dfrac{B}{A}\dfrac{\Lambda e^{-j\omega_0 D}}{1+N\Lambda}\dfrac{1-e^{-j(\omega-\omega_0)N}}{1-e^{-j(\omega-\omega_0)}} & \omega \neq \omega_0 \end{cases} \tag{2.63}$$

We observe that, for $\Lambda \gg 1$,

1. J_{min} becomes negligible;

2. $\boldsymbol{c}_{opt} = \frac{B}{AN} e^{-j\omega_0 D} \boldsymbol{E}(\omega_0)$;
3. $|C_{opt}(e^{j\omega_0})| = \frac{B}{A}$.

Conversely, for $\Lambda \to 0$, i.e. when the power of the useful signal is negligible with respect to the power of the additive noise, it results in

1. $J_{min} = B^2$;
2. $\boldsymbol{c}_{opt} = \boldsymbol{0}$;
3. $|C_{opt}(e^{j\omega_0})| = 0$.

Indeed, as the signal-to-noise ratio vanishes, the best choice is to set the output y to zero. The plot of $|C_{opt}(e^{j2\pi f T_c})|$ is given in Figure 2.6 for some values of the parameters.

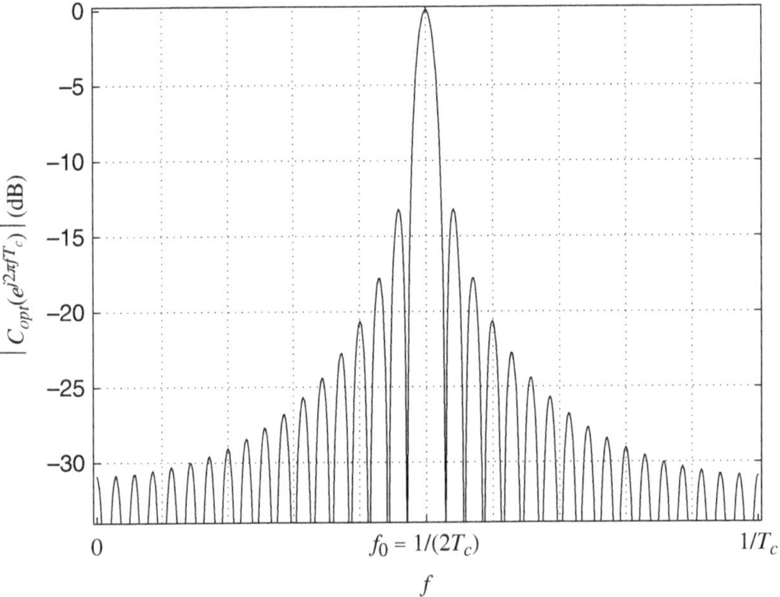

Figure 2.6 Magnitude of $C_{opt}(e^{j2\pi fT_c})$ given by (2.63) for $f_0 T_c = 1/2$, $B = A$, $\Lambda = 30$ dB, and $N = 35$.

2.2 Linear prediction

The Wiener theory considered in the Section 2.1 has an important application to the solution of the following problem. Let x be a discrete-time WSS random process with zero mean; prediction (or *regression*) consists in estimating a *future* value of the process starting from a set of known *past* values. In particular, let us define the vector

$$\boldsymbol{x}^T(k-1) = [x(k-1), x(k-2), \dots, x(k-N)] \tag{2.64}$$

The *one-step forward predictor of order* N, given $\boldsymbol{x}^T(k-1)$, attempts to estimate the value of $x(k)$. There exists also the problem of predicting $x(k-N)$, given the values of $x(k-N+1), \dots, x(k)$. In this case, the system is called the *one step backward predictor of order* N.

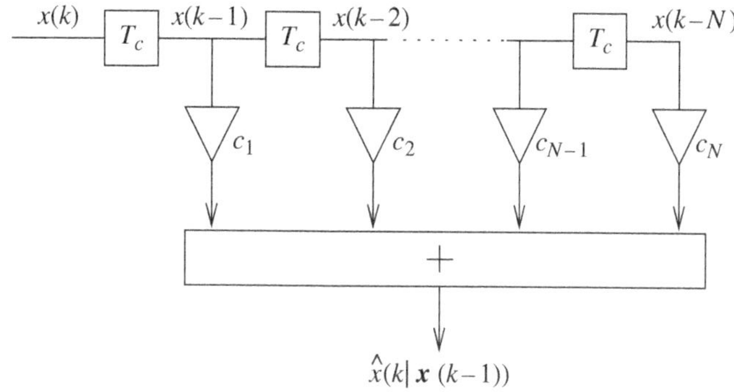

Figure 2.7 Linear predictor of order N.

Forward linear predictor

The estimate of $x(k)$ is expressed as a linear combination of the preceding N samples:

$$\hat{x}(k|\boldsymbol{x}(k-1)) = \sum_{i=1}^{N} c_i x(k-i) \tag{2.65}$$

The block diagram of the linear predictor is represented in Figure 2.7.

This estimate will be subject to a *forward prediction error* given by

$$\begin{aligned} f_N(k) &= x(k) - \hat{x}(k|\boldsymbol{x}(k-1)) \\ &= x(k) - \sum_{i=1}^{N} c_i x(k-i) \end{aligned} \tag{2.66}$$

Optimum predictor coefficients

If we adopt the criterion of minimizing the mean-square prediction error,

$$J = E\left[|f_N(k)|^2\right] \tag{2.67}$$

to determine the predictor coefficients, we can use the optimization results according to Wiener. We recall the following definitions.

1. *Desired signal*

$$d(k) = x(k) \tag{2.68}$$

2. *Filter input vector* (defined by (2.64))

$$\boldsymbol{x}^T(k-1) \tag{2.69}$$

3. *Cost function* J given by (2.67).

Then it turns out:

$$\sigma_d^2 = E\left[x(k)x^*(k)\right] = \sigma_x^2 = \mathtt{r}_x(0) \tag{2.70}$$

$$E\left[\boldsymbol{x}^*(k-1)\boldsymbol{x}^T(k-1)\right] = \boldsymbol{R}_N \tag{2.71}$$

with $\boldsymbol{R}_N$ the $N \times N$ correlation matrix of $\boldsymbol{x}^*$, and

$$\boldsymbol{p} = E\left[d(k)\boldsymbol{x}^*(k-1)\right] = E\left[x(k)\boldsymbol{x}^*(k-1)\right] = \begin{bmatrix} \mathtt{r}_x(1) \\ \mathtt{r}_x(2) \\ \vdots \\ \mathtt{r}_x(N) \end{bmatrix} = \boldsymbol{r}_N \tag{2.72}$$

Applying (2.27) the optimum coefficients satisfy

$$\boldsymbol{R}_N \boldsymbol{c}_{opt} = \boldsymbol{r}_N \tag{2.73}$$

Moreover, from (2.35), we get the minimum value of the cost function

$$J_{min} = J_N = \mathtt{r}_x(0) - \boldsymbol{r}_N^H \boldsymbol{c}_{opt} \tag{2.74}$$

We can combine the latter two equations to get an augmented form of the W–H equation (2.27) for the linear predictor:

$$\begin{bmatrix} \mathtt{r}_x(0) & \boldsymbol{r}_N^H \\ \boldsymbol{r}_N & \boldsymbol{R}_N \end{bmatrix} \begin{bmatrix} 1 \\ -\boldsymbol{c}_{opt} \end{bmatrix} = \begin{bmatrix} J_N \\ \boldsymbol{0}_N \end{bmatrix} \tag{2.75}$$

where $\boldsymbol{0}_N$ is the column vector of N zeros.

Forward prediction error filter

We determine the filter that gives the forward linear prediction error f_N. For an optimum predictor,

$$f_N(k) = x(k) - \sum_{i=1}^{N} c_{opt,i} x(k-i) \tag{2.76}$$

We introduce the vector

$$\mathtt{a}'_{i,N} = \begin{cases} 1 & i = 0 \\ -c_{opt,i} & i = 1, 2, \dots, N \end{cases} \tag{2.77}$$

which can be rewritten as

$$\boldsymbol{a}'_N = \begin{bmatrix} 1 \\ \boldsymbol{a} \end{bmatrix} \tag{2.78}$$

where $\boldsymbol{a} = -\boldsymbol{c}_{opt}$. Substituting (2.77) in (2.76) and taking care of extending the summation also to $i = 0$, we obtain

$$f_N(k) = \sum_{i=0}^{N} \mathtt{a}'_{i,N} x(k-i) \tag{2.79}$$

as shown in Figure 2.8.

The coefficients $\boldsymbol{a}'^T_N = [\mathtt{a}'_{0,N}, \mathtt{a}'_{1,N}, \dots, \mathtt{a}'_{N,N}]$ are directly obtained by substituting (2.78) in (2.75):

$$\boldsymbol{R}_{N+1} \boldsymbol{a}'_N = \begin{bmatrix} J_N \\ \boldsymbol{0}_N \end{bmatrix} \tag{2.80}$$

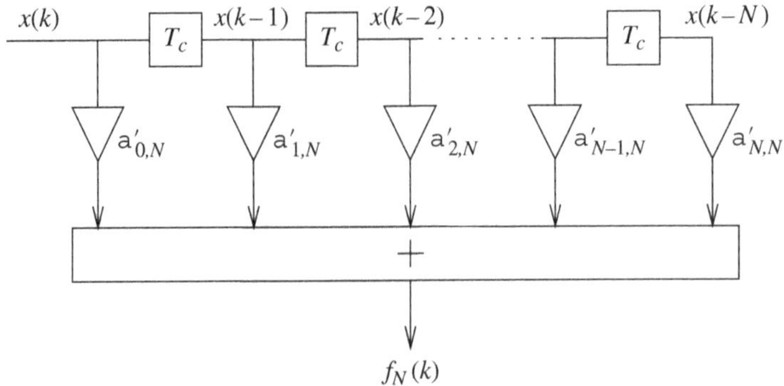

Figure 2.8 Forward prediction error filter.

With a similar procedure, we can derive the filter that gives the *backward linear prediction error*,

$$b_N(k) = x(k-N) - \sum_{i=1}^{N} g_i x(k-i+1) \tag{2.81}$$

It can be shown that the optimum coefficients are given by

$$\boldsymbol{g}_{opt} = \boldsymbol{c}_{opt}^{B*} \tag{2.82}$$

where ^{B} is the *backward* operator that orders the elements of a vector backward, from the last to the first (see page 14).

Efficient algorithms to solve (2.80) are given in Appendix 2.A.

Relation between linear prediction and AR models

The similarity of (2.73) with the Yule–Walker equation (1.464) allows us to state that, given an auto-regressive (AR) process x of order N, the optimum prediction coefficients $\boldsymbol{c}_{opt}$ coincide with the parameters $-\boldsymbol{a}$ of the process and $J_N = \sigma_w^2$. Actually, for $\boldsymbol{c}_{opt} = -\boldsymbol{a}$, comparing (2.76) with (1.445) we have $f_N(k) = w(k)$, that is, the *prediction error* f_N *coincides with the white noise* $w(k)$ at the input of the AR system. In general, if the prediction error filter has a large enough order, it has *whitening* properties, as it is capable of removing the correlated signal component at the input, producing at the output only the uncorrelated or *white* component. Moreover, while *prediction* can be interpreted as the *analysis* of an AR process, the *AR model* may be regarded as the *synthesis* of the process. As illustrated in Figure 2.9, given a realization of the process $\{x(k)\}$, by estimating the autocorrelation sequence over a suitable observation window, the parameters $\boldsymbol{c}_{opt}$ and J_N can be determined. Using the predictor then we determine the prediction error $\{f_N(k)\}$. To reproduce $\{x(k)\}$, an all-pole filter with coefficients $\boldsymbol{a}_N'^T = [1, -\boldsymbol{c}_{opt}]$, having white noise $\{w(k) = f_N(k)\}$ of power $\sigma_w^2 = J_N$ as input, can be used.

An interesting application of this scheme is in the voice-coding [3–5], where the generic subsequence of input voice samples $\{x(k)\}$, say of length L_x, yields a given $\{f_N(k)\}$ which is compared against a set of 2^b predetermined vectors of length L_x and the *closest one* is selected. This information of b bits together with quantized values of $\boldsymbol{c}_{opt}$ and J_N is sent to the receiver where the synthesis scheme is applied to reproduce an approximation of the input voice signal.

First- and second-order solutions

We give now formulas to compute the predictor filter coefficients and prediction error filter coefficients for orders $N = 1$ and $N = 2$. In the next equations, $\rho(n)$ is the correlation coefficient of x, introduced in (1.467).

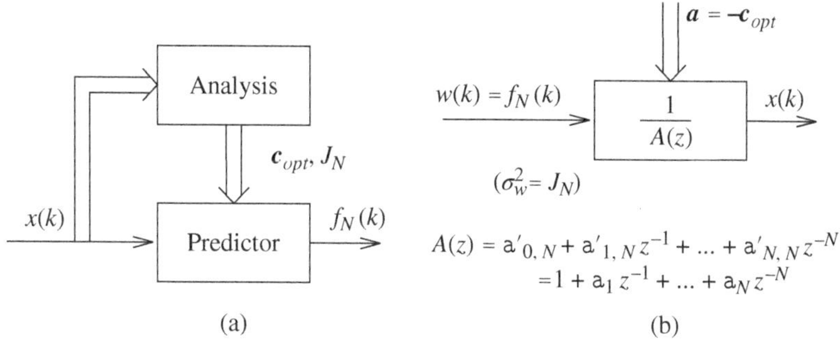

Figure 2.9 (a) Analysis and (b) synthesis of AR(N) processes.

These results extend to the complex case the formulas obtained in Section 1.10.2.

- $N = 1$. From

$$\begin{bmatrix} \mathtt{r}_x(0) & \mathtt{r}_x^*(1) \\ \mathtt{r}_x(1) & \mathtt{r}_x(0) \end{bmatrix} \begin{bmatrix} \mathtt{a}'_{0,1} \\ \mathtt{a}'_{1,1} \end{bmatrix} = \begin{bmatrix} J_1 \\ 0 \end{bmatrix} \tag{2.83}$$

it results

$$\begin{cases} \mathtt{a}'_{0,1} = \dfrac{J_1}{\Delta\mathtt{r}} \mathtt{r}_x(0) \\ \mathtt{a}'_{1,1} = -\dfrac{J_1}{\Delta\mathtt{r}} \mathtt{r}_x(1) \end{cases} \tag{2.84}$$

where

$$\Delta\mathtt{r} = \begin{vmatrix} \mathtt{r}_x(0) & \mathtt{r}_x^*(1) \\ \mathtt{r}_x(1) & \mathtt{r}_x(0) \end{vmatrix} = \mathtt{r}_x^2(0) - |\mathtt{r}_x(1)|^2 \tag{2.85}$$

As $\mathtt{a}'_{0,1} = 1$, it turns out

$$\begin{cases} J_1 = \dfrac{\Delta\mathtt{r}}{\mathtt{r}_x(0)} \\ \mathtt{a}'_{1,1} = -\dfrac{\mathtt{r}_x(1)}{\mathtt{r}_x(0)} \end{cases} \Rightarrow \begin{cases} c_{opt,1} = -\mathtt{a}'_{1,1} = \rho(1) \\ \dfrac{J_1}{\mathtt{r}_x(0)} = 1 - |\rho(1)|^2 \end{cases} \tag{2.86}$$

- $N = 2$.

$$\begin{cases} \mathtt{a}'_{1,2} = -\dfrac{\mathtt{r}_x(1)\mathtt{r}_x(0) - \mathtt{r}_x^*(1)\mathtt{r}_x(2)}{\mathtt{r}_x^2(0) - |\mathtt{r}_x(1)|^2} \\ \mathtt{a}'_{2,2} = -\dfrac{\mathtt{r}_x(0)\mathtt{r}_x(2) - \mathtt{r}_x^2(1)}{\mathtt{r}_x^2(0) - |\mathtt{r}_x(1)|^2} \end{cases} \Rightarrow \begin{cases} c_{opt,1} = \dfrac{\rho(1) - \rho^*(1)\rho(2)}{1 - |\rho(1)|^2} \\ c_{opt,2} = \dfrac{\rho(2) - \rho^2(1)}{1 - |\rho(1)|^2} \end{cases} \tag{2.87}$$

and

$$\frac{J_2}{\mathtt{r}_x(0)} = \frac{1 - 2|\rho(1)|^2 + \rho^{*2}(1)\rho(2) + |\rho(1)|^2\rho^*(2) - |\rho(2)|^2}{1 - |\rho(1)|^2} \tag{2.88}$$

2.3 The least squares method

The Wiener filter will prove to be a powerful *analytical* tool in various applications, one of which is indeed prediction. However, from a practical point of view, often only realizations of the processes $\{x(k)\}$ and $\{d(k)\}$ are available. Therefore, to get the solution, it is necessary to determine estimates of $\mathtt{r}_x$ and $\mathtt{r}_{dx}$, and various alternatives emerge. Two possible methods are (1) the *autocorrelation method*, in which from the estimate of $\mathtt{r}_x$, we construct $\boldsymbol{R}$ as a Toeplitz correlation matrix, and (2) the *covariance method*, in which we estimate each element of $\boldsymbol{R}$ by (2.94). In this case, the matrix is only Hermitian and the solution that we are going to illustrate is of the *LS* type [1, 2].

We reconsider the problem of Section 2.1, introducing a new cost function. Based on the observation of the sequences

$$\{x(k)\} \quad \text{and} \quad \{d(k)\}, \quad k = 0, \dots, K-1 \tag{2.89}$$

and of the error

$$e(k) = d(k) - y(k) \tag{2.90}$$

where y is given by (2.1), according to the LS method the optimum filter coefficients yield the *minimum of the sum of the square errors*:

$$\boldsymbol{c}_{LS} = \arg \min_{\{c_n\}, n=0,1,\ldots,N-1} \mathcal{E} \tag{2.91}$$

where

$$\mathcal{E} = \sum_{k=N-1}^{K-1} |e(k)|^2 \tag{2.92}$$

Note that in the LS method, a time average is substituted for the expectation (2.3), which gives the MSE.

Data windowing

In matrix notation, the output $\{y(k)\}$, $k = N-1, \ldots, K-1$, given by (2.1), can be expressed as

$$\begin{bmatrix} y(N-1) \\ y(N) \\ \vdots \\ y(K-1) \end{bmatrix} = \underbrace{\begin{bmatrix} x(N-1) & x(N-2) & \ldots & x(0) \\ x(N) & x(N-1) & \ldots & x(1) \\ \vdots & \vdots & \ddots & \vdots \\ x(K-1) & x(K-2) & \ldots & x(K-N) \end{bmatrix}}_{\text{input data matrix } \mathcal{I}} \begin{bmatrix} c_0 \\ c_1 \\ \vdots \\ c_{N-1} \end{bmatrix} \tag{2.93}$$

In (2.93), we note that the input data sequence goes from $x(0)$ to $x(K-1)$. Other choices are possible for the input data window. The examined case is called the *covariance method*, and the input data matrix $\mathcal{I}$, defined by (2.93), is $L \times N$, where $L = K - N + 1$.

Matrix formulation

We define

$$\Phi(i, n) = \sum_{k=N-1}^{K-1} x^*(k-i)\, x(k-n), \qquad i, n = 0, 1, \ldots, N-1 \tag{2.94}$$

$$\vartheta(n) = \sum_{k=N-1}^{K-1} d(k)\, x^*(k-n), \qquad n = 0, 1, \ldots, N-1 \tag{2.95}$$

Using (1.405) for an *unbiased* estimate of the correlation, the following identities hold:

$$\Phi(i, n) = (K - N + 1)\hat{\mathtt{r}}_x(i-n) \tag{2.96}$$

$$\vartheta(n) = (K - N + 1)\hat{\mathtt{r}}_{dx}(n) \tag{2.97}$$

in which the values of $\Phi(i, n)$ depend on both indices (i, n) and not only upon their difference, especially if K is not very large. We give some definitions:

1. *Energy of* $\{d(k)\}$

$$\mathcal{E}_d = \sum_{k=N-1}^{K-1} |d(k)|^2 \tag{2.98}$$

2. *Crosscorrelation vector between* d *and* $\boldsymbol{x}$

$$\boldsymbol{\vartheta}^T = [\vartheta(0), \vartheta(1), \ldots, \vartheta(N-1)] \tag{2.99}$$

3. *Input autocorrelation matrix*

$$\mathbf{\Phi} = \begin{bmatrix} \Phi(0,0) & \Phi(0,1) & \dots & \Phi(0,N-1) \\ \Phi(1,0) & \Phi(1,1) & \dots & \Phi(1,N-1) \\ \vdots & \vdots & \ddots & \vdots \\ \Phi(N-1,0) & \Phi(N-1,1) & \dots & \Phi(N-1,N-1) \end{bmatrix} \tag{2.100}$$

Then the cost function can be written as

$$\mathcal{E} = \mathcal{E}_d - \boldsymbol{c}^H \boldsymbol{\vartheta} - \boldsymbol{\vartheta}^H \boldsymbol{c} + \boldsymbol{c}^H \mathbf{\Phi} \boldsymbol{c} \tag{2.101}$$

Correlation matrix

$\mathbf{\Phi}$ is the time average of $\boldsymbol{x}^*(k)\boldsymbol{x}^T(k)$, i.e.

$$\mathbf{\Phi} = \sum_{k=N-1}^{K-1} \boldsymbol{x}^*(k)\boldsymbol{x}^T(k) \tag{2.102}$$

Properties of $\mathbf{\Phi}$

1. $\mathbf{\Phi}$ is Hermitian.
2. $\mathbf{\Phi}$ is positive semi-definite.
3. Eigenvalues of $\mathbf{\Phi}$ are real and non-negative.
4. $\mathbf{\Phi}$ can be written as

$$\mathbf{\Phi} = \mathcal{I}^H \mathcal{I} \tag{2.103}$$

with $\mathcal{I}$ the input data matrix defined by (2.93). We note that matrix $\mathcal{I}$ is Toeplitz.

Determination of the optimum filter coefficients

By analogy of (2.101) with (2.17), the gradient of (2.101) is given by

$$\nabla_c \mathcal{E} = 2(\mathbf{\Phi}\boldsymbol{c} - \boldsymbol{\vartheta}) \tag{2.104}$$

Then the vector of optimum coefficients based on the LS method, $\boldsymbol{c}_{LS}$, satisfies the *normal equation*:

$$\mathbf{\Phi}\boldsymbol{c}_{LS} = \boldsymbol{\vartheta} \tag{2.105}$$

If $\mathbf{\Phi}^{-1}$ exists, then the solution to (2.105) is given by

$$\boldsymbol{c}_{LS} = \mathbf{\Phi}^{-1}\boldsymbol{\vartheta} \tag{2.106}$$

In the solution of the LS problem, (2.105) corresponds to the W–H equation (2.27). For an ergodic process, (2.96) yields:

$$\frac{1}{K-N+1}\mathbf{\Phi} \xrightarrow[k\to\infty]{} \boldsymbol{R} \tag{2.107}$$

and

$$\frac{1}{K-N+1}\boldsymbol{\vartheta} \xrightarrow[k\to\infty]{} \boldsymbol{p} \tag{2.108}$$

Hence, we have that the LS solution converges to the Wiener solution as K goes to infinity, that is

$$\boldsymbol{c}_{LS} \xrightarrow[k\to\infty]{} \boldsymbol{c}_{opt} \tag{2.109}$$

In other words, for $K \to \infty$, the *covariance method* gives the same solution as the *autocorrelation method*.

In scalar notation, (2.105) becomes a system of N equations in N unknowns:

$$\sum_{i=0}^{N-1} \Phi(n,i)c_{LS,i} = \vartheta(n), \qquad n = 0, 1, \ldots, N-1 \tag{2.110}$$

2.3.1 The principle of orthogonality

From (2.92), taking the gradient with respect to c_n, we have

$$\begin{aligned}
\nabla_{c_n}\mathcal{E} &= \nabla_{c_n,I}\mathcal{E} + j\nabla_{c_n,Q}\mathcal{E} \\
&= \sum_{k=N-1}^{K-1} \big[-x^*(k-n)e(k) - x(k-n)e^*(k) \\
&\quad + j\,(jx^*(k-n)e(k) - jx(k-n)e^*(k))\big] \\
&= -2\sum_{k=N-1}^{K-1} x^*(k-n)e(k)
\end{aligned} \tag{2.111}$$

If we denote with $\{e_{min}(k)\}$, the estimation error found with the optimum coefficient $\boldsymbol{c}_{LS}$, they must satisfy the conditions:

$$\sum_{k=N-1}^{K-1} e_{min}(k)x^*(k-n) = 0, \qquad n = 0, 1, \ldots, N-1 \tag{2.112}$$

which represent the time-average version of the *statistical orthogonality principle* (2.31). Moreover, being $y(k)$ a linear combination of $\{x(k-n)\}$, $n = 0, 1, \ldots, N-1$, we have

$$\sum_{k=N-1}^{K-1} e_{min}(k)y^*(k) = 0 \tag{2.113}$$

which expresses the fundamental result: the *optimum filter output* sequence is orthogonal to the *minimum estimation error* sequence.

Minimum cost function

Substituting (2.105) in (2.101), the minimum cost function can be written as

$$\mathcal{E}_{min} = \mathcal{E}_d - \boldsymbol{\vartheta}^H \boldsymbol{c}_{LS} \tag{2.114}$$

An alternative expression to $\mathcal{E}_{min}$ uses the energy of the output sequence:

$$\mathcal{E}_y = \sum_{k=N-1}^{K-1} |y(k)|^2 = \boldsymbol{c}^H \boldsymbol{\Phi} \boldsymbol{c} \tag{2.115}$$

Note that for $\boldsymbol{c} = \boldsymbol{c}_{LS}$, we have

$$d(k) = y(k) + e_{min}(k) \tag{2.116}$$

then, because of the orthogonality (2.113) between y and e_{min}, it follows that

$$\mathcal{E}_d = \mathcal{E}_y + \mathcal{E}_{min} \tag{2.117}$$

from which, substituting (2.105) in (2.115), we get

$$\mathcal{E}_{min} = \mathcal{E}_d - \mathcal{E}_y \tag{2.118}$$

where $\mathcal{E}_y = \boldsymbol{c}_{LS}^H \boldsymbol{\vartheta}$.

The normal equation using the data matrix

Defining the *vector of desired samples*

$$d^T = [d(N-1), d(N), \dots, d(K-1)] \tag{2.119}$$

from the definition (2.95) of $\vartheta(n)$ we get

$$\begin{bmatrix} \vartheta(0) \\ \vartheta(1) \\ \vdots \\ \vartheta(N-1) \end{bmatrix} = \begin{bmatrix} x^*(N-1) & x^*(N) & \dots & x^*(K-1) \\ x^*(N-2) & x^*(N-1) & \dots & x^*(K-2) \\ \vdots & \vdots & \ddots & \vdots \\ x^*(0) & x^*(2) & \dots & x^*(K-N) \end{bmatrix} \begin{bmatrix} d(N-1) \\ d(N) \\ \vdots \\ d(K-1) \end{bmatrix} \tag{2.120}$$

that is

$$\boldsymbol{\vartheta} = \mathcal{I}^H \boldsymbol{d} \tag{2.121}$$

Thus, using (2.103) and (2.121), the normal equation (2.105) becomes

$$\mathcal{I}^H \mathcal{I} \boldsymbol{c}_{LS} = \mathcal{I}^H \boldsymbol{d} \tag{2.122}$$

Associated with system (2.122), it is useful to introduce the system of equations for the minimization of $\mathcal{E}$,

$$\mathcal{I} \boldsymbol{c} = \boldsymbol{d} \tag{2.123}$$

From (2.122), if $(\mathcal{I}^H \mathcal{I})^{-1}$ exists, the solution is

$$\boldsymbol{c}_{LS} = (\mathcal{I}^H \mathcal{I})^{-1}\, \mathcal{I}^H \boldsymbol{d} \tag{2.124}$$

and correspondingly (2.114) becomes

$$\mathcal{E}_{min} = \boldsymbol{d}^H \boldsymbol{d} - \boldsymbol{d}^H \mathcal{I} (\mathcal{I}^H \mathcal{I})^{-1} \mathcal{I}^H \boldsymbol{d} \tag{2.125}$$

We note how both formulas (2.124) and (2.125) depend only on the desired signal samples and input samples. Moreover, the solution $\boldsymbol{c}$ is unique only if the columns of $\mathcal{I}$ are linearly independent, that is the case of non-singular $\mathcal{I}^H \mathcal{I}$. This requires at least $K - N + 1 > N$, that is the system of equations (2.123) must be overdetermined (more equations than unknowns).

Geometric interpretation: the projection operator

In general, from (2.93) the vector of filter output samples

$$\boldsymbol{y}^T = \left[y(N-1), y(N), \dots, y(K-1)\right] \tag{2.126}$$

can be related to the input data matrix $\mathcal{I}$ as

$$\boldsymbol{y} = \mathcal{I} \boldsymbol{c} \tag{2.127}$$

This relation still holds for $\boldsymbol{c} = \boldsymbol{c}_{LS}$, and from (2.124), we get

$$\boldsymbol{y} = \mathcal{I} \boldsymbol{c}_{LS} = \mathcal{I} (\mathcal{I}^H \mathcal{I})^{-1} \mathcal{I}^H \boldsymbol{d} \tag{2.128}$$

Correspondingly, the estimation vector error is given by

$$\boldsymbol{e}_{min} = \boldsymbol{d} - \boldsymbol{y} \tag{2.129}$$

The matrix $\mathcal{O} = \mathcal{I}(\mathcal{I}^H \mathcal{I})^{-1} \mathcal{I}^H$ can be thought of as a projection operator defined on the space generated by the columns of $\mathcal{I}$. Let $\boldsymbol{I}$ be the identity matrix. The difference

$$\mathcal{O}_\perp = \boldsymbol{I} - \mathcal{O} = \boldsymbol{I} - \mathcal{I} (\mathcal{I}^H \mathcal{I})^{-1} \mathcal{I}^H \tag{2.130}$$

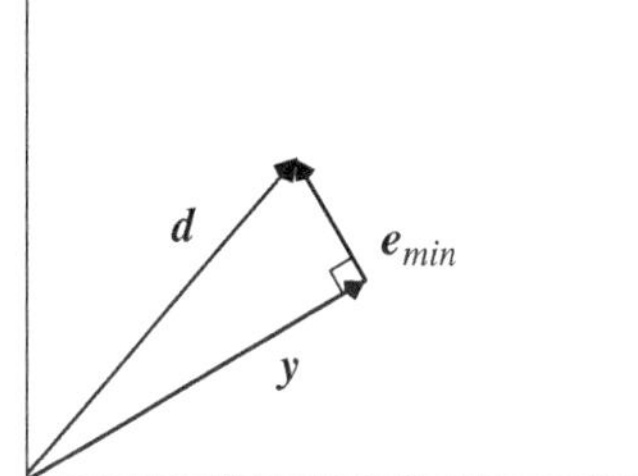

Figure 2.10 Relations among vectors in the LS minimization.

is the complementary projection operator, orthogonal to $\mathcal{O}$. In fact, from (2.128)

$$y = \mathcal{O}d \tag{2.131}$$

and from (2.129)

$$e_{min} = d - y = \mathcal{O}_{\perp}d \tag{2.132}$$

where $e_{min} \perp y$ (see (2.113)). Moreover, (2.125) can be written as

$$\mathcal{E}_{min} = e^H_{min} e_{min} = d^H e_{min} = d^H \mathcal{O}_{\perp} d \tag{2.133}$$

Figure 2.10 shows an example illustrating the relation among d, y, and e_{min}.

2.3.2 Solutions to the LS problem

If the inverse of $(\mathcal{I}^H \mathcal{I})$ does not exist, the solution of the LS problem (2.124) must be re-examined. This is what we will do in this section after taking a closer look at the associated system of equations (2.123). In general, let us consider the solutions to a linear system of equations

$$\mathcal{I}c = d \tag{2.134}$$

with $\mathcal{I}$ on $N \times N$ *square matrix*. If $\mathcal{I}^{-1}$ exists, the solution $c = \mathcal{I}^{-1}d$ is unique and can be obtained in various ways [6]:

1. If $\mathcal{I}$ is *triangular* and non-singular, a solution to the system (2.134) can be found by the successive substitutions method with $O(N^2)$ operations.
2. In general, if $\mathcal{I}$ is non-singular, one can use the Gauss method, which involves three steps:
   a. Factorization of $\mathcal{I}$
   $$\mathcal{I} = LU \tag{2.135}$$
   with L lower triangular having all ones along the diagonal and U upper triangular;
   b. Solution of the system in z
   $$Lz = d \tag{2.136}$$
   through the successive substitutions method;
   c. Solution of the system in c
   $$Uc = z \tag{2.137}$$
   through the successive substitutions method.

   This method requires $O(N^3)$ operations and $O(N^2)$ memory locations.

3. If $\mathcal{I}$ is *Hermitian* and non-singular, the factorization (2.135) becomes the *Cholesky decomposition*:

$$\mathcal{I} = \boldsymbol{L}\boldsymbol{L}^H \tag{2.138}$$

with $\boldsymbol{L}$ lower triangular having non-zero elements on the diagonal. This method requires $O(N^3)$ operations, about half as many as the Gauss method.

4. If $\mathcal{I}$ is *Toeplitz* and non-singular, the generalized Schur algorithm can be used, with a complexity of $O(N^2)$: generally, it is applicable to all $\mathcal{I}$ *structured matrices* [7]. We also recall the Kumar fast algorithm [8].

However, if $\mathcal{I}^{-1}$ does not exist, e.g. because $\mathcal{I}$ is not a square matrix, it is necessary to use alternative methods to solve the system (2.134) [6]: in particular, we will consider the method of the pseudoinverse. First, we will state the following result:

Singular value decomposition

We have seen in (1.292) how the $N \times N$ Hermitian matrix $\boldsymbol{R}$ can be decomposed in terms of a matrix $\boldsymbol{U}$ of eigenvectors and a diagonal matrix $\boldsymbol{\Lambda}$ of eigenvalues. Now, we extend this concept to an arbitrary complex matrix $\mathcal{I}$. Given an $L \times N$ matrix $\mathcal{I}$ of rank R, two unitary matrices $\boldsymbol{V}$ and $\boldsymbol{U}$ exist, the singular value decomposition (SVD) of $\mathcal{I}$ is

$$\begin{aligned} \mathcal{I} &= \boldsymbol{U}\boldsymbol{\Sigma}\boldsymbol{V}^H \\ &= \sum_{i=1}^{R} \sigma_i \boldsymbol{u}_i \boldsymbol{v}_i^H \end{aligned} \tag{2.139}$$

with

$$\boldsymbol{\Sigma} = \left[\begin{array}{c|c} \mathbf{D} & \mathbf{0} \\ \hline \mathbf{0} & \mathbf{0} \end{array}\right]_{L\times N} \tag{2.140}$$

$$\boldsymbol{D} = \text{diag}(\sigma_1, \sigma_2, \dots, \sigma_R), \qquad \sigma_1 \geq \sigma_2 \geq \cdots \geq \sigma_R > 0 \tag{2.141}$$

$$\boldsymbol{U} = [\boldsymbol{u}_1, \boldsymbol{u}_2, \dots, \boldsymbol{u}_L]_{L\times L}, \qquad \boldsymbol{U}^H\boldsymbol{U} = \boldsymbol{I}_{L\times L} \tag{2.142}$$

$$\boldsymbol{V} = [\boldsymbol{v}_1, \boldsymbol{v}_2, \dots, \boldsymbol{v}_N]_{N\times N}, \qquad \boldsymbol{V}^H\boldsymbol{V} = \boldsymbol{I}_{N\times N} \tag{2.143}$$

In (2.141) $\{\sigma_i\}$, $i = 1, \dots, R$, are *singular values* of $\mathcal{I}$. Being $\boldsymbol{U}$ and $\boldsymbol{V}$ unitary, it follows

$$\boldsymbol{U}^H \mathcal{I} \boldsymbol{V} = \boldsymbol{\Sigma} \tag{2.144}$$

as illustrated in Figure 2.11.

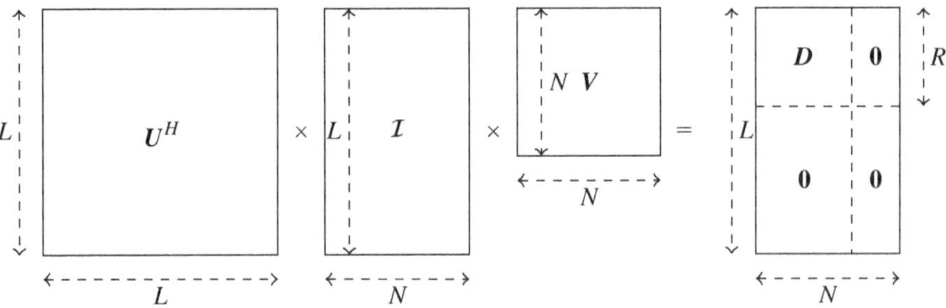

Figure 2.11 Singular value decomposition of matrix $\mathcal{I}$.

The following facts hold

- Matrices $\mathcal{I}^H\mathcal{I}$ and $\mathcal{I}\mathcal{I}^H$ have the same non-zero eigenvalues, λ_i, $i = 1, \dots, R$, and
$$\sigma_i = \sqrt{\lambda_i} \quad i = 1, \dots, R \tag{2.145}$$
- $\boldsymbol{u}_i$, $i = 1, \dots, R$, can be normalized (to unit norm) eigenvectors of $\mathcal{I}\mathcal{I}^H$. Moreover, $\boldsymbol{u}_j$, $j = R + 1, \dots, L$, are obtained by including successive vectors of length L orthonormal to previous selected vectors, i.e. $\boldsymbol{u}_i^H\boldsymbol{u}_j = 0$, $i = 1, \dots, j-1$, $j = R+1, \dots, L$. $\{\boldsymbol{u}_i\}$ are denoted *left singular vectors* of $\mathcal{I}$.
- $\boldsymbol{v}_i$, $i = 1, \dots, R$, can be normalized eigenvectors of $\mathcal{I}^H\mathcal{I}$. Moreover, $\boldsymbol{v}_j$, $j = R + 1, \dots, N$, are obtained by including successive vectors of length N orthonormal to previous selected vectors, i.e. $\boldsymbol{v}_i^H\boldsymbol{v}_j = 0$, $i = 1, \dots, j-1$, $j = R+1, \dots, N$. $\{\boldsymbol{v}_i\}$ are denoted *right singular vectors* of $\mathcal{I}$. Moreover, from (2.139)
$$\boldsymbol{v}_i = \mathcal{I}^H\frac{\boldsymbol{u}_i}{\sigma_i} \tag{2.146}$$

Definition 2.1
The *pseudoinverse* of $\mathcal{I}$, $L \times N$, of rank R, is given by the matrix

$$\mathcal{I}^\# = \boldsymbol{V}\boldsymbol{\Sigma}^\#\boldsymbol{U}^H = \sum_{i=1}^{R} \sigma_i^{-1}\boldsymbol{v}_i\boldsymbol{u}_i^H \tag{2.147}$$

where

$$\boldsymbol{\Sigma}^\# = \begin{bmatrix} \boldsymbol{D}^{-1} & \boldsymbol{0} \\ \boldsymbol{0} & \boldsymbol{0} \end{bmatrix}, \quad \boldsymbol{D}^{-1} = \text{diag}\left(\sigma_1^{-1}, \sigma_2^{-1}, \dots, \sigma_R^{-1}\right) \tag{2.148}$$

□

We find an expression of $\mathcal{I}^\#$ for the two cases in which $\mathcal{I}$ has full rank,[5] that is $R = \min(L, N)$.

Case of an overdetermined system ($L > N$) and $R = N$. Note that in this case, the system (2.134) has more equations than unknowns. Using the above relations, it can be shown that

$$\mathcal{I}^\# = (\mathcal{I}^H\mathcal{I})^{-1}\mathcal{I}^H \tag{2.149}$$

In this case, $\mathcal{I}^\#\boldsymbol{d}$ coincides with the solution of system (2.105).

Case of an underdetermined system ($L < N$) and $R = L$. Note that in this case, there are fewer equations than unknowns, hence, there are infinite solutions to the system (2.134). Again, it can be shown that

$$\mathcal{I}^\# = \mathcal{I}^H(\mathcal{I}\mathcal{I}^H)^{-1} \tag{2.150}$$

Minimum norm solution

Definition 2.2
The *solution of a least squares problem* is given by the vector

$$\boldsymbol{c}_{LS} = \mathcal{I}^\#\boldsymbol{d} \tag{2.151}$$

where $\mathcal{I}^\#$ is the pseudoinverse of $\mathcal{I}$. □

By applying (2.151), the *pseudoinverse matrix $\mathcal{I}^\#$ gives the LS solution of minimum norm*; in other words, it provides the vector $\boldsymbol{c}$ that minimizes the square error (2.92), $\mathcal{E} = \|\boldsymbol{e}\|^2 = \|\boldsymbol{y} - \boldsymbol{d}\|^2 = \|\mathcal{I}\boldsymbol{c} - \boldsymbol{d}\|^2$, and simultaneously minimizes the norm of the solution, $\|\boldsymbol{c}\|^2$. The constraint on $\|\boldsymbol{c}\|^2$ is needed in those cases wherein there is more than one vector that minimizes $\|\mathcal{I}\boldsymbol{c} - \boldsymbol{d}\|^2$.

[5] We will denote the rank of $\mathcal{I}$ by rank($\mathcal{I}$).

We summarize the different cases:

1. If $L = N$ and $\text{rank}(\mathcal{I}) = N$, i.e. $\mathcal{I}$ is non-singular,
$$\mathcal{I}^{\#} = \mathcal{I}^{-1} \tag{2.152}$$
2. If $L > N$ and
   a. $\text{rank}(\mathcal{I}) = N$, then
   $$\mathcal{I}^{\#} = (\mathcal{I}^H\mathcal{I})^{-1}\mathcal{I}^H \tag{2.153}$$
   and $\boldsymbol{c}_{LS}$ is the LS solution of an overdetermined system of equations (2.134).
   b. $\text{rank}(\mathcal{I}) = R$ (also $< N$), from (2.151)
   $$\boldsymbol{c}_{LS} = \sum_{i=1}^{R} \frac{\boldsymbol{v}_i^H\mathcal{I}^H\boldsymbol{d}}{\sigma_i^2}\boldsymbol{v}_i \tag{2.154}$$
3. If $L < N$ and
   a. $\text{rank}(\mathcal{I}) = L$, then
   $$\mathcal{I}^{\#} = \mathcal{I}^H(\mathcal{I}\mathcal{I}^H)^{-1} \tag{2.155}$$
   and $\boldsymbol{c}_{LS}$ is the minimum norm solution of an underdetermined system of equations.
   b. $\text{rank}(\mathcal{I}) = R$ (also $< L$),
   $$\boldsymbol{c}_{LS} = \sum_{i=1}^{R} \frac{\boldsymbol{u}_i^H\boldsymbol{d}}{\sigma_i^2}\mathcal{I}^H\boldsymbol{u}_i \tag{2.156}$$

Only solutions (2.151) in the cases (2.152) and (2.153) coincide with the solution (2.106).

The expansion of $\boldsymbol{c}$ in terms of $\{\boldsymbol{u}_i\}$, $\{\boldsymbol{v}_i\}$, and $\{\sigma_i^2\}$ have two advantages with respect to applying (2.151) by the computation of $\mathcal{I}^{\#}$ (in the form (2.153), for $L > N$ and $\text{rank}(\mathcal{I}) = N$, or in the form (2.155), for $L < N$ and $\text{rank}(\mathcal{I}) = L$).

1. The SVD also gives the rank of $\mathcal{I}$ through the number of non-zero singular values.
2. The required accuracy in computing $\mathcal{I}^{\#}$ via SVD is almost halved with respect to the computation of $(\mathcal{I}^H\mathcal{I})^{-1}$ or $(\mathcal{I}\mathcal{I}^H)^{-1}$.

There are two algorithms to determine the SVD of $\mathcal{I}$: the Jacobi algorithm and the Householder transformation [9].

We conclude this section citing two texts [10, 11], which report examples of realizations of the algorithms described in this section.

2.4 The estimation problem

This section extends the formulation of the Wiener problem to a more general multiple-input multiple-output framework.

Estimation of a random variable

Let d and x be two r.v., somehow related. Two examples are $x = f(d)$ with $f(\cdot)$ an unknown function, or the relation between d and x is explicit, as in $x = d + w$, with w another random variable. On the basis of an observation, given $x = \beta$, the estimation problem is to determine what is the corresponding value of d, given that we know only the joint probability density function (pdf) of the two r.v.s., $p_{dx}(\alpha, \beta)$. In any case, using as *estimate of* d the function
$$\hat{d} = h(x) \tag{2.157}$$

the estimation error is given by

$$e = d - \hat{d} \tag{2.158}$$

The estimator is called *unbiased* if

$$E[\hat{d}] = E[d] \tag{2.159}$$

Otherwise, the *bias* is

$$\mu_{bias} = E[\hat{d}] - E[d] \tag{2.160}$$

MMSE estimation

Let $p_d(\alpha)$ and $p_x(\beta)$ be the pdfs of d and x, respectively, and $p_{d|x}(\alpha \mid \beta)$ the conditional probability density function of d given $x = \beta$; moreover, let $p_x(\beta) \neq 0$, $\forall \beta$. We wish to determine the function h that minimizes the mean-square error, that is

$$\begin{aligned} J = E[e^2] &= \int_{-\infty}^{+\infty} \int_{-\infty}^{+\infty} [\alpha - h(\beta)]^2 \, p_{dx}(\alpha, \beta) \, d\alpha \, d\beta \\ &= \int_{-\infty}^{+\infty} p_x(\beta) \int_{-\infty}^{+\infty} [\alpha - h(\beta)]^2 \, p_{d|x}(\alpha \mid \beta) \, d\alpha \, d\beta \end{aligned} \tag{2.161}$$

where the relation $p_{dx}(\alpha, \beta) = p_{d|x}(\alpha \mid \beta) p_x(\beta)$ is used.

Theorem 2.2

The estimator $h(\beta)$ that minimizes J is given by the expected value of d given $x = \beta$,

$$h(\beta) = E[d \mid x = \beta] \tag{2.162}$$

□

Proof. The integral (2.161) is minimum when the function

$$\int_{-\infty}^{+\infty} [\alpha - h(\beta)]^2 \, p_{d|x}(\alpha \mid \beta) \, d\alpha \tag{2.163}$$

is minimized for every value of β. Using the variational method (see Ref. [12]), we find that this occurs if

$$2 \int_{-\infty}^{+\infty} [\alpha - h(\beta)] \, p_{d|x}(\alpha \mid \beta) \, d\alpha = 0 \qquad \forall \beta \tag{2.164}$$

that is for

$$h(\beta) = \int_{-\infty}^{+\infty} \alpha \, p_{d|x}(\alpha \mid \beta) \, d\alpha = \int_{-\infty}^{+\infty} \alpha \, \frac{p_{dx}(\alpha, \beta)}{p_x(\beta)} \, d\alpha \tag{2.165}$$

from which (2.162) follows. □

An alternative to the MMSE criterion for determining $\hat{d}$ is given by the *maximum a posteriori probability* (MAP) criterion, which yields

$$\hat{d} = \arg\max_{\alpha} p_{d|x}(\alpha \mid \beta) \tag{2.166}$$

If the distribution of d is uniform, the MAP criterion becomes the *maximum-likelihood* (ML) criterion, where

$$\hat{d} = \arg\max_{\alpha} p_{x|d}(\beta \mid \alpha) \tag{2.167}$$

Examples of both MAP and ML criteria are given in Chapters 6 and 14.

Example 2.4.1
Let d and x be two jointly Gaussian r.v.s. with mean values $\mathtt{m}_d$ and $\mathtt{m}_x$, respectively, and covariance $\mathtt{c} = E[(d - \mathtt{m}_d)(x - \mathtt{m}_x)]$. After several steps, it can be shown that [13]

$$h(\beta) = \mathtt{m}_d + \frac{\mathtt{c}}{\sigma_x^2}\,(\beta - \mathtt{m}_x) \tag{2.168}$$

The corresponding mean-square error is equal to

$$J_{min} = \sigma_d^2 - \left(\frac{\mathtt{c}}{\sigma_x}\right)^2 \tag{2.169}$$

Example 2.4.2
Let $x = d + w$, where d and w are two statistically independent r.v.s. For $w \sim \mathcal{N}(0,1)$ and $d \in \{-1, 1\}$ with $P[d = -1] = P[d = 1] = 1/2$, it can be shown that

$$h(\beta) = \tanh \beta = \frac{e^{\beta} - e^{-\beta}}{e^{\beta} + e^{-\beta}} \tag{2.170}$$

Extension to multiple observations

Let $\boldsymbol{x} = [x_1, \dots, x_N]^T$ and $\boldsymbol{\beta} = [\beta_1, \dots, \beta_N]^T$. In case of multiple observations,

$$\boldsymbol{x} = \boldsymbol{\beta} \tag{2.171}$$

the estimation of d is obtained by applying the following theorem, whose proof is similar to that given in the case of a single observation.

Theorem 2.3
The estimator of d, $\hat{d} = h(\boldsymbol{x})$, that minimizes $J = E[(d - \hat{d})^2]$ is given by

$$\begin{aligned} h(\boldsymbol{\beta}) &= E[d \mid \boldsymbol{x} = \boldsymbol{\beta}] \\ &= \int_{-\infty}^{+\infty} \alpha\, p_{d|\boldsymbol{x}}(\alpha \mid \boldsymbol{\beta})\, d\alpha \\ &= \int \alpha\, \frac{p_{d,\boldsymbol{x}}(\alpha, \boldsymbol{\beta})}{p_{\boldsymbol{x}}(\boldsymbol{\beta})}\, d\alpha \end{aligned} \tag{2.172}$$

□

In the following, to simplify the formulation, we will refer to r.v.s. with zero mean.

Example 2.4.3
Let $d, \boldsymbol{x} = [x_1, \dots, x_N]^T$, be real-valued jointly Gaussian r.v.s. with *zero mean* and the following second-order description:

- *Correlation matrix of observations*

$$\boldsymbol{R} = E[\boldsymbol{x}\,\boldsymbol{x}^T] \tag{2.173}$$

- *Crosscorrelation vector*

$$\boldsymbol{p} = E[d\boldsymbol{x}] \tag{2.174}$$

For $\boldsymbol{x} = \boldsymbol{\beta}$, it can be shown that

$$h(\boldsymbol{\beta}) = \boldsymbol{p}^T\,\boldsymbol{R}^{-1}\,\boldsymbol{\beta} \tag{2.175}$$

and

$$J_{min} = \sigma_d^2 - \boldsymbol{p}^T\,\boldsymbol{R}^{-1}\,\boldsymbol{p} \tag{2.176}$$

Linear MMSE estimation of a random variable

For a low-complexity implementation, it is often convenient to consider a linear function h. Letting $\boldsymbol{c} = [c_1, \dots, c_N]^T$, in the case of multiple observations, the estimate is a linear combination of observations, and

$$\hat{d} = \boldsymbol{c}^T \boldsymbol{x} + b \tag{2.177}$$

where b is a constant.

In the case of real-valued r.v.s., using the definitions (2.173) and (2.174), it is easy to prove the following theorem:

Theorem 2.4

Given the vector of observations $\boldsymbol{x}$, the linear minimum mean-square error (LMMSE) estimator of d has the following expression:

$$\hat{d} = \boldsymbol{p}^T \boldsymbol{R}^{-1} \boldsymbol{x} \tag{2.178}$$

In other words, $b = 0$ and

$$\boldsymbol{c}_{opt} = \boldsymbol{R}^{-1} \boldsymbol{p} \tag{2.179}$$

The corresponding mean-square error is

$$J_{min} = \sigma_d^2 - \boldsymbol{p}^T \boldsymbol{R}^{-1} \boldsymbol{p} \tag{2.180}$$

Note that the r.v.s. are assumed to have zero mean. □

Observation 2.2

Comparing (2.178) and (2.180) with (2.175) and (2.176), respectively, we note that, in the case of jointly Gaussian r.v.s., the linear estimator coincides with the optimum MMSE estimator.

Linear MMSE estimation of a random vector

We extend the results of the previous section to the case of complex-valued r.v.s., and for a desired vector signal. Let $\boldsymbol{x}$ be an observation, modelled as a vector of N r.v.s.,

$$\boldsymbol{x}^T = [x_1, x_2, \dots, x_N] \tag{2.181}$$

Moreover, let $\boldsymbol{d}$ be the desired vector, modelled as a vector of M r.v.s.,

$$\boldsymbol{d}^T = [d_1, d_2, \dots, d_M] \tag{2.182}$$

We introduce the following correlation matrices:

$$\boldsymbol{r}_{d_i x} = E[d_i\, \boldsymbol{x}^*] \tag{2.183}$$

$$\boldsymbol{R}_{xd} = E[\boldsymbol{x}^* \boldsymbol{d}^T] = [\boldsymbol{r}_{d_1 x}, \boldsymbol{r}_{d_2 x}, \dots, \boldsymbol{r}_{d_M x}] \tag{2.184}$$

$$\boldsymbol{R}_{dx} = E[\boldsymbol{d}^* \boldsymbol{x}^T] = \boldsymbol{R}_{xd}^H \tag{2.185}$$

$$\boldsymbol{R}_x = E[\boldsymbol{x}^* \boldsymbol{x}^T] \tag{2.186}$$

$$\boldsymbol{R}_d = E[\boldsymbol{d}^* \boldsymbol{d}^T] \tag{2.187}$$

The problem is to determine a linear transformation of $\boldsymbol{x}$, given by

$$\hat{\boldsymbol{d}} = \boldsymbol{C}^T \boldsymbol{x} + \boldsymbol{b} \tag{2.188}$$

such that $\hat{\boldsymbol{d}}$ is a close replica of $\boldsymbol{d}$ in the mean-square error sense.

Definition 2.3

The LMMSE estimator of a random vector, consisting of the $N \times M$ matrix $\boldsymbol{C}$, and of the $M \times 1$ vector $\boldsymbol{b}$, coincides with the linear function of the observations (2.188) that minimizes the cost function:

$$J = E\left[\| \boldsymbol{d} - \hat{\boldsymbol{d}}\|^2\right] = \sum_{m=1}^{M} E\left[\left\|d_m - \hat{d}_m\right\|^2\right] \tag{2.189}$$

In other words, the optimum coefficients $\boldsymbol{C}$ and $\boldsymbol{b}$ are the solution of the following problem:

$$\min_{\boldsymbol{C},\boldsymbol{b}} J \tag{2.190}$$

□

We note that in the formulation of Section 2.1, we have

$$\boldsymbol{x}^T = [x(k), \dots, x(k-N+1)] \tag{2.191}$$

$$\boldsymbol{d}^T = [d(k)] \tag{2.192}$$

that is $M = 1$, and the matrix $\boldsymbol{C}$ becomes a column vector.

We determine now the expression of $\boldsymbol{C}$ and $\boldsymbol{b}$ in terms of the correlation matrices introduced above. First of all, we observe that if $\boldsymbol{d}$ and $\boldsymbol{x}$ have zero mean, then $\boldsymbol{b} = \boldsymbol{0}$, since the choice of $\boldsymbol{b} = \tilde{\boldsymbol{b}} \neq \boldsymbol{0}$ implies an estimator $\tilde{\boldsymbol{C}}^T \boldsymbol{x} + \tilde{\boldsymbol{b}}$ with a suboptimal value of the cost function. In fact,

$$\begin{aligned} J &= E\left[\|\boldsymbol{d} - \tilde{\boldsymbol{C}}^T\boldsymbol{x} - \tilde{\boldsymbol{b}}\|^2\right] \\ &= E\left[\|\boldsymbol{d} - \tilde{\boldsymbol{C}}^T\boldsymbol{x}\|^2\right] - 2Re\left\{E\left[(\boldsymbol{d} - \tilde{\boldsymbol{C}}^T\boldsymbol{x})^H\, \tilde{\boldsymbol{b}}\right]\right\} + \|\tilde{\boldsymbol{b}}\|^2 \\ &= E\left[\|\boldsymbol{d} - \tilde{\boldsymbol{C}}^T\boldsymbol{x}\|^2\right] + \|\tilde{\boldsymbol{b}}\|^2 \end{aligned} \tag{2.193}$$

being $E[\boldsymbol{d}] = \boldsymbol{0}$ and $E[\boldsymbol{x}] = \boldsymbol{0}$. Now, (2.193) implies that the choice $\tilde{\boldsymbol{b}} = \boldsymbol{0}$ yields the minimum value of J. Without loss of generality, we will assume that both $\boldsymbol{x}$ and $\boldsymbol{d}$ are zero mean random vectors.

Scalar case For $M = 1$, $d = d_1$, we have

$$\hat{d} = \hat{d}_1 = \boldsymbol{c}_1^T\boldsymbol{x} = \boldsymbol{x}^T\boldsymbol{c}_1 \tag{2.194}$$

with $\boldsymbol{c}_1$ column vector with N coefficients. In this case, the problem (2.190) leads again to the Wiener filter; the solution is given by

$$\boldsymbol{R}_x\boldsymbol{c}_1 = \boldsymbol{r}_{d_1x} \tag{2.195}$$

where $\boldsymbol{r}_{d_1x}$ is defined by (2.183).

Vector case For $M > 1$, $\boldsymbol{d}$, and $\hat{\boldsymbol{d}}$ are M-dimensional vectors. Nevertheless, since function (2.189) operates on single components, the vector problem (2.190) leads to M scalar problems, each with input $\boldsymbol{x}$ and output $\hat{d}_1, \hat{d}_2, \dots, \hat{d}_M$, respectively. Therefore, the columns of matrix $\boldsymbol{C}$, $\{\boldsymbol{c}_m\}$, $m = 1, \dots, M$, satisfy equations of the type (2.195), namely

$$\boldsymbol{R}_x\boldsymbol{c}_m = \boldsymbol{r}_{d_mx}, \qquad m = 1, \dots, M \tag{2.196}$$

hence, based on the definition (2.184), we have

$$\boldsymbol{C} = \boldsymbol{R}_x^{-1}\,\boldsymbol{R}_{xd} \tag{2.197}$$

Thus, *the optimum estimator in the LMMSE sense* is given by

$$\hat{\boldsymbol{d}} = \left(\boldsymbol{R}_x^{-1}\,\boldsymbol{R}_{xd}\right)^T\boldsymbol{x} \tag{2.198}$$

Value of the cost function On the basis of the estimation error

$$\boldsymbol{e} = \boldsymbol{d} - \hat{\boldsymbol{d}} \tag{2.199}$$

with correlation matrix

$$\boldsymbol{R}_e = E[\boldsymbol{e}^* \boldsymbol{e}^T] = \boldsymbol{R}_d - \boldsymbol{R}_{dx}\boldsymbol{C} - \boldsymbol{C}^H \boldsymbol{R}_{xd} + \boldsymbol{C}^H \boldsymbol{R}_x \boldsymbol{C} \tag{2.200}$$

substituting (2.197) in (2.200), yields

$$\boldsymbol{R}_{e,\min} = \boldsymbol{R}_d - \boldsymbol{R}_{dx}\, \boldsymbol{R}_x^{-1}\, \boldsymbol{R}_{xd} \tag{2.201}$$

The cost function (2.189) is given by the trace of $\boldsymbol{R}_e$, and

$$J_{min} = \operatorname{tr} \boldsymbol{R}_{e,\min} \tag{2.202}$$

2.4.1 The Cramér–Rao lower bound

Until now we have considered the estimation of a random variable or vector, under the assumption of knowing the joint distribution of $\boldsymbol{d}$ and $\boldsymbol{x}$. However, in some cases, we only know the distribution of $\boldsymbol{x}$, and its relation to $\boldsymbol{d}$, without specific knowledge of the distribution of $\boldsymbol{d}$. Hence, $\boldsymbol{d}$ is modelled as *an unknown parameter vector*. For example, we may have multiple noisy observations of a scalar d by

$$x_n = d + w_n, \quad n = 1, \dots, N \tag{2.203}$$

where we know the distribution of w_n, e.g. $w_n \sim \mathcal{N}(0, 1)$. An example of estimator of d is (see (1.377))

$$\hat{d} = h(\boldsymbol{x}) = \frac{1}{N} \sum_{n=1}^{N} x_n \tag{2.204}$$

In this section, we focus on the estimation of a scalar parameter d by $\boldsymbol{x}$, as $\hat{d} = h(\boldsymbol{x})$. In particular, given $p_{\boldsymbol{x}|d}(\boldsymbol{\beta}|\alpha)$, we focus on the best performance of any estimator, when the metric is the variance of the estimator, i.e. the design objective is

$$\min_h \mathtt{var}(\hat{d}) \tag{2.205}$$

Note that (2.205) coincides with the MMSE only if

$$E[\hat{d}] = d \tag{2.206}$$

which, from (2.159), is equivalent to consider an unbiased estimator of parameter d. Note that in order to find the estimator, we must design the function $h(\boldsymbol{x})$, which is in general a challenging task.

The *Cramér–Rao lower bound* (CRLB) theorem provides a limit on the minimum variance that *any* estimator can achieve, given the number of observations N and the relation between observations and the parameter. Moreover, when the unbiased estimator exists, under mild conditions, the theorem provides the expression of the minimum-variance estimator.

Theorem 2.5 (Cramér–Rao lower bound)
Consider a conditional pdf $p_{\boldsymbol{x}|d}(\boldsymbol{\beta}|\alpha)$ satisfying

$$E\left[\frac{\partial \ln p_{\boldsymbol{x}|d}(\boldsymbol{x}|\alpha)}{\partial \alpha}\right] = 0 \quad \forall \alpha \tag{2.207}$$

where the expectation is taken with respect to $\boldsymbol{x}$, given $d = \alpha$. Let

$$I(\alpha) = -E\left[\frac{\partial^2 \ln p_{\boldsymbol{x}|d}(\boldsymbol{x}|\alpha)}{\partial \alpha^2}\right] \tag{2.208}$$

Then, any unbiased estimator satisfies

$$\mathrm{var}(\hat{d}) \geq \frac{1}{I(\alpha)} \quad \forall \alpha \tag{2.209}$$

Moreover, if and only if there exists a function $h_{opt}(\boldsymbol{x})$ such that

$$\frac{\partial \ln p_{\boldsymbol{x}|d}(\boldsymbol{\beta}|\alpha)}{\partial \alpha} = I(\alpha)[h_{opt}(\boldsymbol{\beta}) - \alpha] \quad \forall \alpha \tag{2.210}$$

the estimator $\hat{d} = h_{opt}(\boldsymbol{x})$ is unbiased and reaches the minimum variance (2.209). □

Example 2.4.4
Consider $x_n = d + w_n$, $n = 1, \ldots, N$, with $w_n \sim \mathcal{N}\left(0, \sigma_w^2\right)$, statistically independent, for a known σ_w^2. First, we observe that

$$p_{\boldsymbol{x}|d}(\boldsymbol{\beta}|\alpha) = \frac{1}{\left(\sqrt{2\pi\sigma_w^2}\right)^N} e^{-\frac{1}{2\sigma_w^2}(\boldsymbol{\beta}-\alpha\mathbf{1})^T(\boldsymbol{\beta}-\alpha\mathbf{1})} \tag{2.211}$$

where $\mathbf{1}$ is the N-size column vector with all entries equal to 1. We have

$$\frac{\partial \ln p_{\boldsymbol{x}|d}(\boldsymbol{\beta}|\alpha)}{\partial \alpha} = \frac{1}{\sigma_w^2}(\boldsymbol{\beta} - d\mathbf{1})^T\mathbf{1} \tag{2.212}$$

and clearly (2.207) is satisfied. Then we can apply Theorem 2.5 and

$$I(\alpha) = \frac{N}{\sigma_w^2}, \qquad \hat{d} = h(\boldsymbol{x}) = \frac{1}{N}\boldsymbol{x}^T\mathbf{1} \tag{2.213}$$

Note that $h(\boldsymbol{\beta})$ corresponds to (2.204), which is therefore the minimum-variance estimator, and the error variance is $1/I(\alpha) = \sigma_w^2/N$ for all values of α.

Extension to vector parameter

We now focus on the estimate of M real parameters, collected into the *vector parameter* $\boldsymbol{d}$ (2.182) from a vector of observations $\boldsymbol{x}$, i.e. we look for the estimator

$$\hat{\boldsymbol{d}} = \boldsymbol{h}(\boldsymbol{x}) \tag{2.214}$$

A simple example is the estimation of mean and variance ($\boldsymbol{d} = [\mathtt{m}_x, \sigma_x^2]^T$) of a Gaussian random variable from N observations collected into vector $\boldsymbol{x}$, as in (2.181). Theorem 2.5 can be extended to the case of the estimate of a vector parameter, as follows:

Theorem 2.6 (Cramér–Rao lower bound for vector parameter)
Consider a conditional pdf $p_{\boldsymbol{x}|\boldsymbol{d}}(\boldsymbol{\beta}|\boldsymbol{\alpha})$ satisfying

$$E\left[\frac{\partial \ln p_{\boldsymbol{x}|\boldsymbol{d}}(\boldsymbol{x}|\boldsymbol{\alpha})}{\partial \alpha_m}\right] = 0 \quad \forall \alpha_m,\ m = 1, \ldots, M \tag{2.215}$$

where the expectation is taken with respect to $\boldsymbol{x}$, given $\boldsymbol{d} = \boldsymbol{\alpha}$. Let us define the $M \times M$ *Fisher information matrix*, with entry $m_1, m_2 \in \{1, \ldots, M\}$, given by

$$[\boldsymbol{I}(\boldsymbol{\alpha})]_{m_1,m_2} = -E\left[\frac{\partial^2 \ln p_{\boldsymbol{x}|\boldsymbol{d}}(\boldsymbol{x}|\boldsymbol{\alpha})}{d\alpha_{m_1}\, d\alpha_{m_2}}\right] \tag{2.216}$$

Then for any unbiased estimator $\hat{\boldsymbol{d}}$ with covariance matrix $\boldsymbol{C}_{\hat{d}} = E[\hat{\boldsymbol{d}}\hat{\boldsymbol{d}}^T]$, we have that matrix $\boldsymbol{C}_{\hat{d}} - [\boldsymbol{I}(\boldsymbol{\alpha})]^{-1}$ is *positive semidefinite*.

Moreover, if and only if (here $[\cdot]_m$ denotes entry m of the vector argument)

$$\frac{\partial \ln p_{x|d}(\beta|\alpha)}{\partial \alpha_m} = [I(\alpha)(h(\beta) - \alpha)]_m \quad \forall \alpha_m, \quad m = 1, \dots, M \tag{2.217}$$

for some function h, the estimator $\hat{d} = h(x)$ is unbiased and its covariance matrix is $[I(\alpha)]^{-1}$. □

Before proceeding, we observe that in some cases, we can directly compute the Fisher information matrix. Then, if (2.217) and the semi-definitive condition on $C_{\hat{d}}$ are verified, and (2.215) holds, then $h(x)$ in (2.217) is the desired estimator. Otherwise, the theorem does not apply.

Example 2.4.5

Consider x_n, $n = 1, \dots, N$, uncorrelated with $x_n \sim \mathcal{N}\left(\mathtt{m}_x, \sigma_x^2\right)$, where both $\mathtt{m}_x$ and σ_x^2 are unknown parameters; hence, $d = [\mathtt{m}_x, \sigma_x^2]^T$. We have

$$p_{x|d}(\beta|\alpha) = \frac{1}{\left(\sqrt{2\pi\sigma_x^2}\right)^N} e^{-\frac{1}{2\sigma_x^2}(\beta - \mathtt{m}_x \mathbf{1})^T(\beta - \mathtt{m}_x \mathbf{1})} \tag{2.218}$$

First, we observe that

$$\frac{\partial \ln p_{x|d}(\beta|\alpha)}{\partial \alpha_1} = \frac{1}{\sigma_x^2} \sum_{n=1}^{N} [\beta_n - \mathtt{m}_x] \tag{2.219}$$

$$\frac{\partial \ln p_{x|d}(\beta|\alpha)}{\partial \alpha_2} = \frac{N}{2\sigma_x^4} - \frac{1}{\sigma_x^6} \sum_{n=1}^{N} [\beta_n - \mathtt{m}_x]^2 \tag{2.220}$$

and it can easily concluded that (2.215) holds. The Fisher information matrix in this case is

$$I(\alpha) = -\begin{bmatrix} E\left[\frac{\partial^2 \ln p_{x|d}(x|\alpha)}{d\alpha_1 d\alpha_1}\right] & E\left[\frac{\partial^2 \ln p_{x|d}(x|\alpha)}{d\alpha_1 d\alpha_2}\right] \\ E\left[\frac{\partial^2 \ln p_{x|d}(x|\alpha)}{d\alpha_2 d\alpha_1}\right] & E\left[\frac{\partial^2 \ln p_{x|d}(x|\alpha)}{d\alpha_2 d\alpha_2}\right] \end{bmatrix} = \begin{bmatrix} \frac{N}{\sigma_x^2} & 0 \\ 0 & \frac{N}{2\sigma_x^4} \end{bmatrix} \tag{2.221}$$

with inverse

$$[I(\alpha)]^{-1} = \begin{bmatrix} \frac{\sigma_x^2}{N} & 0 \\ 0 & \frac{2\sigma_x^4}{N} \end{bmatrix} \tag{2.222}$$

which provides on the diagonal the minimum variances of any unbiased estimator.

Inserting (2.221), (2.219), and (2.220) into (2.217), we conclude that (2.217) holds for $m = 1$, while for $m = 2$ for any function $h_2(x)$, we have

$$\frac{N}{2\sigma_x^4} - \frac{1}{\sigma_x^6} \sum_{n=1}^{N} [x_n - \mathtt{m}_x]^2 \neq \frac{2\sigma^4}{N}(h_2(x) - \sigma_x^2) \tag{2.223}$$

Therefore, we can conclude that in this case, no unbiased estimator can have covariance matrix $[I(\alpha)]^{-1}$, and any estimator of σ_x^2 will have a variance strictly larger than $\frac{2\sigma_x^4}{N}$.

Example 2.4.6

Consider a known complex sequence s_n, with $|s_n| = 1$, $n = 1, \dots, N$, and two observation vectors $x_1 = [x_{1,1}, \dots, x_{1,N}]^T$ and $x_2 = [x_{2,1}, \dots, x_{2,N}]^T$ with

$$x_{1,n} = As_n + w_{1,n}, \qquad x_{2,n} = As_n e^{j\epsilon} + w_{2,n} \tag{2.224}$$

where A it a complex number, ϵ is a real number, and $w_{1,n}$ and $w_{2,n} \sim \mathcal{CN}\left(0,\sigma^2\right)$, with $w_{1,n}$ and $w_{2,n}$ statistically independent. Assume that σ^2 is known, while both parameters A and ϵ are to be estimated, thus $\boldsymbol{d}^T = [A, \epsilon]$. Let $\boldsymbol{s}^T = [s_1, \dots, s_N]$, $\boldsymbol{m}_1 = A\boldsymbol{s}$, and $\boldsymbol{m}_2 = Ae^{j\epsilon}\boldsymbol{s}$, we have

$$\ln p_{\boldsymbol{x}_1,\boldsymbol{x}_2|\boldsymbol{d}}(\boldsymbol{x}_1,\boldsymbol{x}_2|\boldsymbol{\alpha}) = -N\ln(\pi\sigma^2) - \frac{1}{2\sigma^2}(\boldsymbol{x}_1 - \boldsymbol{m}_1)^H(\boldsymbol{x}_1 - \boldsymbol{m}_1) - \frac{1}{2\sigma^2}(\boldsymbol{x}_2 - \boldsymbol{m}_2)^H(\boldsymbol{x}_2 - \boldsymbol{m}_2) \quad (2.225)$$

First observe that

$$\frac{\partial \ln p_{\boldsymbol{x}_1,\boldsymbol{x}_2|\boldsymbol{d}}(\boldsymbol{x}_1,\boldsymbol{x}_2|\boldsymbol{\alpha})}{\partial A} = \frac{1}{\sigma^2}(\boldsymbol{x}_1 - \boldsymbol{m}_1)^H\boldsymbol{s} + \frac{e^{j\epsilon}}{\sigma^2}(\boldsymbol{x}_2 - \boldsymbol{m}_2)^H\boldsymbol{s} \quad (2.226)$$

$$\frac{\partial \ln p_{\boldsymbol{x}_1,\boldsymbol{x}_2|\boldsymbol{d}}(\boldsymbol{x}_1,\boldsymbol{x}_2|\boldsymbol{\alpha})}{\partial \epsilon} = \frac{jAe^{j\epsilon}}{\sigma^2}(\boldsymbol{x}_2 - \boldsymbol{m}_2)^H\boldsymbol{s} \quad (2.227)$$

and it can be easily verified that (2.215) is satisfied. Moreover, using also $|s_n| = 1$, we have

$$\boldsymbol{I}(\boldsymbol{\alpha}) = \begin{bmatrix} \frac{2N}{\sigma^2} & \frac{AN}{\sigma^2} \\ \frac{AN}{\sigma^2} & \frac{A^2N}{\sigma^2} \end{bmatrix} \quad \text{and} \quad [\boldsymbol{I}(\boldsymbol{\alpha})]^{-1} = \frac{\sigma^2}{NA^2}\begin{bmatrix} A^2 & -\frac{A^3N}{\sigma^2} \\ -\frac{A^3N}{\sigma^2} & 2 \end{bmatrix} \quad (2.228)$$

2.5 Examples of application

We give now some examples of application of the Wiener filter theory [14–17].

2.5.1 Identification of a linear discrete-time system

We want to determine the relation between the input x and the output z of the system illustrated in Figure 2.12. We note that the observation d is affected by additive noise w, having zero mean and variance σ_w^2, assumed statistically independent of x.

Assuming the system between z and x can be modelled as a FIR filter, the configuration experiment illustrated in Figure 2.12 can be adopted to estimate the filter impulse response.

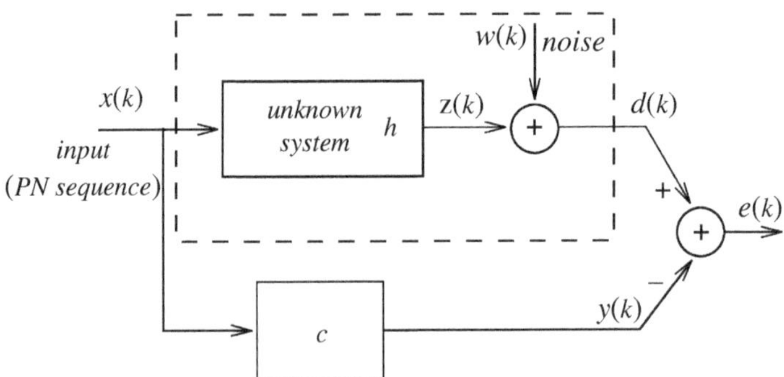

Figure 2.12 Scheme to estimate the impulse response of an unknown system.

Using an input x, known to both systems, we determine the output of the transversal filter c with N coefficients

$$y(k) = \sum_{i=0}^{N-1} c_i x(k-i) = \boldsymbol{c}^T\boldsymbol{x}(k) \quad (2.229)$$

and the estimation error

$$e(k) = d(k) - y(k) \tag{2.230}$$

We analyse the specific case of an unknown linear FIR system whose impulse response has N_h coefficients. Assuming $N \geq N_h$, we introduce the vector $\boldsymbol{h}$ with N components,

$$\boldsymbol{h}^T = [h_0, h_1, \ldots, h_{N_h-1}, 0, \ldots, 0] \tag{2.231}$$

In this case,

$$\begin{aligned} d(k) &= h_0 x(k) + h_1 x(k-1) + \cdots + h_{N_h-1} x\left(k - (N_h - 1)\right) + w(k) \\ &= \boldsymbol{h}^T \boldsymbol{x}(k) + w(k) \end{aligned} \tag{2.232}$$

For $N \geq N_h$, and assuming the input $\boldsymbol{x}$ is white noise with statistical power $\mathtt{r}_x(0)$, we get

$$\boldsymbol{R} = E\left[\boldsymbol{x}^*(k)\boldsymbol{x}^T(k)\right] = \mathtt{r}_x(0)\boldsymbol{I} \tag{2.233}$$

and

$$\boldsymbol{p} = E\left[d(k)\boldsymbol{x}^*(k)\right] = \mathtt{r}_x(0)\boldsymbol{h} \tag{2.234}$$

Then the W–H equation (2.29) yields the desired solution as

$$\boldsymbol{c}_{opt} = \boldsymbol{R}^{-1}\boldsymbol{p} = \boldsymbol{h} \tag{2.235}$$

and

$$J_{min} = \sigma_w^2 \tag{2.236}$$

On the other hand, if $N < N_h$, then $\boldsymbol{c}_{opt}$ in (2.235) coincides with the first N coefficients of h, and

$$J_{min} = \sigma_w^2 + \mathtt{r}_x(0)\ \|\boldsymbol{\Delta h}(\infty)\|^2, \tag{2.237}$$

where $\boldsymbol{\Delta h}(\infty)$ represents the residual error vector,

$$\boldsymbol{\Delta h}(\infty) = [0, \ldots, 0, -h_N, -h_{N+1}, \ldots, -h_{N_h-1}]^T \tag{2.238}$$

2.5.2 Identification of a continuous-time system

We now consider the identification of a continuous-time system.

With reference to Figure 2.13 [18], a pseudo-noise (PN) sequence p of period L, repeated several times, is used to modulate in amplitude the pulse

$$g(t) = \mathtt{w}_{T_c}(t) = \text{rect}\left(\frac{t - T_c/2}{T_c}\right) \tag{2.239}$$

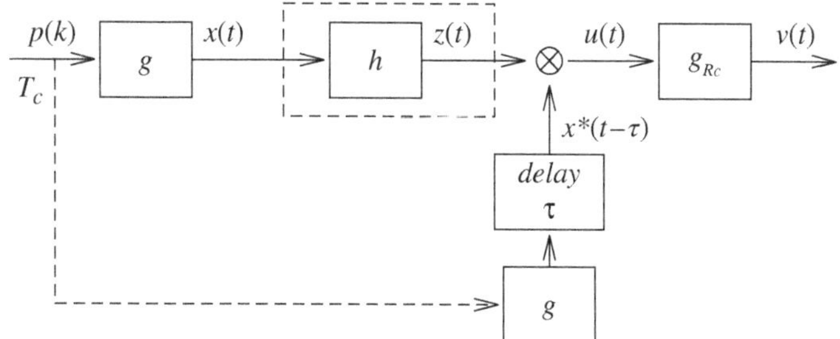

Figure 2.13 Basic scheme to measure the impulse response of an unknown system.

The modulated output signal is therefore given by

$$x(t) = \sum_{i=0}^{+\infty} p(i)_{mod\ L}\, g(t - iT_c) \tag{2.240}$$

The autocorrelation of x, periodic function of period LT_c, is

$$\mathtt{r}_x(t) = \frac{1}{LT_c} \int_{-\frac{LT_c}{2}}^{+\frac{LT_c}{2}} x(\eta)x^*(\eta - t)\, d\eta \tag{2.241}$$

As g has finite support of length T_c, we have

$$\mathtt{r}_x(t) = \frac{1}{T_c} \sum_{\ell=0}^{L-1} \mathtt{r}_p(\ell)\mathtt{r}_g(t - \ell T_c) \qquad 0 \le t \le LT_c \tag{2.242}$$

where, in the case of g given by (2.239), we have

$$\mathtt{r}_g(t) = \int_0^{T_c} g(\eta)g(\eta - t)\, d\eta = T_c \left(1 - \frac{|t|}{T_c}\right) \mathrm{rect}\left(\frac{t}{2T_c}\right) \tag{2.243}$$

Substituting (2.243) in (2.242) and assuming that p is a maximal-length sequence (MLS) (see Appendix 1.C), with $\mathtt{r}_p(0) = 1$ and $\mathtt{r}_p(\ell) = -1/L$, for $\ell = 1, \dots, L-1$, we obtain

$$\mathtt{r}_x(t) = -\frac{1}{L} + \left(1 + \frac{1}{L}\right)\left(1 - \frac{|t|}{T_c}\right) \mathrm{rect}\left(\frac{t}{2T_c}\right), \qquad |t| \le \frac{LT_c}{2} \tag{2.244}$$

as shown in Figure 2.14 for $L = 8$. If the output z of the unknown system to be identified is multiplied by a delayed version of the input, $x^*(t - \tau)$, and the result is filtered by an ideal integrator between 0 and LT_c with impulse response

$$g_{Rc}(t) = \frac{1}{LT_c} \mathrm{w}_{LT_c}(t) = \frac{1}{LT_c} \mathrm{rect}\left(\frac{t - LT_c/2}{LT_c}\right) \tag{2.245}$$

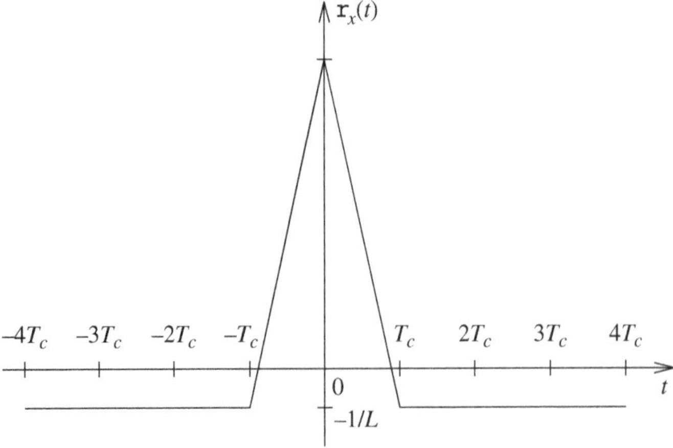

Figure 2.14 Autocorrelation function of x.

we obtain

$$\begin{aligned} v(t) &= \frac{1}{LT_c} \int_{t-LT_c}^{t} u(\eta)\, d\eta \\ &= \frac{1}{LT_c} \int_{t-LT_c}^{t} \int_0^{+\infty} [h(\xi)x(\eta-\xi)\, d\xi]\, x^*(\eta-\tau)\, d\eta \\ &= \int_0^{+\infty} h(\xi)\mathtt{r}_x(\tau-\xi)\, d\xi = h * \mathtt{r}_x(\tau) = v_\tau \end{aligned} \tag{2.246}$$

Therefore, the output assumes a constant value v_τ equal to the convolution between the unknown system h and the autocorrelation of x evaluated in τ. Assuming $1/T_c$ is larger than the maximum frequency of the spectral components of h, and L is sufficiently large, the output v_τ is approximately proportional to $h(\tau)$.

The scheme represented in Figure 2.15 is alternative to that of Figure 2.13, of simpler implementation because it does not require synchronization of the two PN sequences at transmitter and receiver. In this latter scheme, the output z of the unknown system is multiplied by a PN sequence having the same characteristics of the transmitted sequence, but a different clock frequency $f_0' = 1/T_c'$, related to the clock frequency $f_0 = 1/T_c$ of the transmitter by the relation

$$f_0' = f_0 \left(1 - \frac{1}{K}\right) \tag{2.247}$$

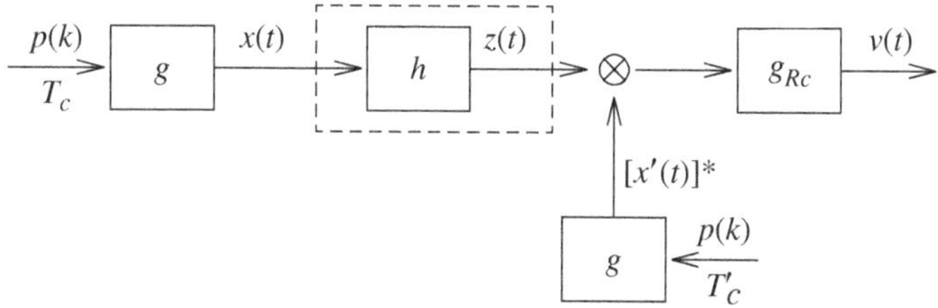

Figure 2.15 Sliding window method to measure the impulse response of an unknown system.

where K is a parameter of the system. We consider the function

$$\mathtt{r}_{x'x}(\tau) = \frac{1}{LT_c} \int_0^{LT_c} [x'(\eta)]^* x(\eta-\tau)\, d\eta \tag{2.248}$$

where τ is the delay at time $t = 0$ between the two sequences. As time elapses, the delay between the two sequence diminishes by $(t/T_c')(T_c' - T_c) = t/K$, so that

$$\frac{1}{LT_c} \int_{t-LT_c}^{t} [x'(\eta)]^* x(\eta-\tau)\, d\eta \simeq \mathtt{r}_{x'x}\left(\tau - \frac{t-LT_c}{K}\right) \qquad \text{for } t \geq LT_c \tag{2.249}$$

If K is sufficiently large, we can assume that

$$\mathtt{r}_{x'x}(\tau) \simeq \mathtt{r}_x(\tau) \tag{2.250}$$

At the output of the filter g_{Rc}, given by (2.245), therefore, we have

$$\begin{aligned} v(t) &= \frac{1}{LT_c} \int_{t-LT_c}^{t} \int_0^{+\infty} [h(\xi)x(\eta-\xi)\, d\xi]\, [x'(\eta)]^*\, d\eta \\ &= \int_0^{+\infty} h(\xi)\mathtt{r}_{x'x}\left(\xi - \frac{t-LT_c}{K}\right) d\xi \\ &\simeq \int_0^{+\infty} h(\xi)\mathtt{r}_x\left(\frac{t-LT_c}{K} - \xi\right) d\xi \end{aligned} \tag{2.251}$$

or with the substitution $t' = (t - LT_c)/K$,

$$y(Kt' + LT_c) \simeq \int_0^{+\infty} h(\xi) \mathrm{r}_x(t' - \xi)\, d\xi \tag{2.252}$$

where the integral in (2.252) coincides with the integral in (2.246). If K is sufficiently large (an increase of K clearly requires a greater precision and hence a greater cost of the frequency synthesizer to generate f_0'), it can be shown that the approximations in (2.249) and (2.250) are valid. Therefore, the systems in Figures 2.15 and 2.13 are equivalent.

2.5.3 Cancellation of an interfering signal

With reference to Figure 2.16, we consider two sensors:

1. *Primary input*, consisting of the desired signal s corrupted by additive noise w_0,

$$d(k) = s(k) + w_0(k) \qquad \text{with } s \perp w_0 \tag{2.253}$$

2. *Reference input*, consisting of the noise signal w_1, with $s \perp w_1$. We assume that w_0 and w_1 are in general correlated.

w_1 is filtered by filter $\boldsymbol{c}$ with N coefficients, so that the filter output y, given by

$$y(k) = \sum_{i=0}^{N-1} c_i w_1(k-i) \tag{2.254}$$

is the most accurate replica of w_0. Defining the error

$$e(k) = d(k) - y(k) = s(k) + w_0(k) - y(k) \tag{2.255}$$

the cost function, assuming real-valued signals and recalling that s is orthogonal to the noise signals, is given by

$$J = E[e^2(k)] = E[s^2(k)] + E\left[(w_0(k) - y(k))^2\right] \tag{2.256}$$

We have two cases.

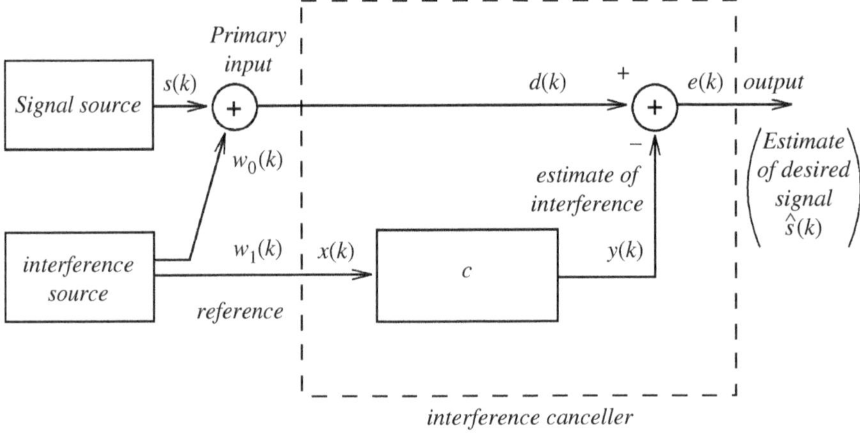

Figure 2.16 General configuration of an interference canceller.

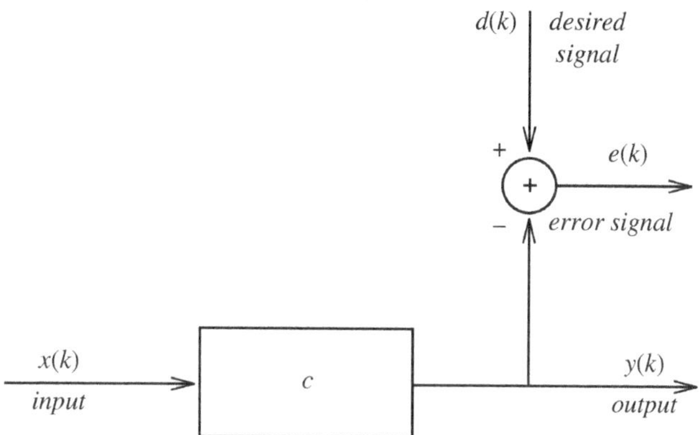

Figure 2.17 Block diagram of an interference canceller.

1. w_1 and w_0 are correlated:

$$\min_{c} J = \mathtt{r}_s(0) + \min_{c} E\left[(w_0(k) - y(k))^2\right] = \mathtt{r}_s(0) \tag{2.257}$$

for $y(k) = w_0(k)$. In this case, $e(k) = s(k)$.

2. w_1 and w_0 are uncorrelated:

$$\begin{aligned}\min_{c} J &= E\left[(s(k) + w_0(k))^2\right] + \min_{c} E\left[y^2(k)\right] \\ &= E\left[(s(k) + w_0(k))^2\right]\end{aligned} \tag{2.258}$$

for $y(k) = 0$. In this case, $e(k) = d(k)$, and the noise w_0 is not cancelled.

With reference to Figure 2.17, for a general input x to the filter $\boldsymbol{c}$, the W–H solution in the z-transform domain is given by (see (2.45))

$$C_{opt}(z) = \frac{P_{dx}(z)}{P_x(z)} \tag{2.259}$$

Adopting for d and x, the model of Figure 2.18, in which w'_0 and w'_1 are additive noise signals uncorrelated with w and s, and using Table 1.3, (2.259) becomes

$$C_{opt}(z) = \frac{P_w(z)H^*(1/z^*)}{P_{w'_1}(z) + P_w(z)H(z)H^*(1/z^*)} \tag{2.260}$$

If $w'_1(k) = 0$, $\forall k$, (2.260) becomes

$$C_{opt}(z) = \frac{1}{H(z)} \tag{2.261}$$

2.5.4 Cancellation of a sinusoidal interferer with known frequency

Let

$$d(k) = s(k) + A\cos(2\pi f_0 k T_c + \varphi_0) \tag{2.262}$$

where s is the desired signal, and the sinusoidal term is the interferer. We take as reference signals

$$x_1(k) = B\cos(2\pi f_0 k T_c + \varphi) \tag{2.263}$$

and

$$x_2(k) = B\sin(2\pi f_0 k T_c + \varphi) \tag{2.264}$$

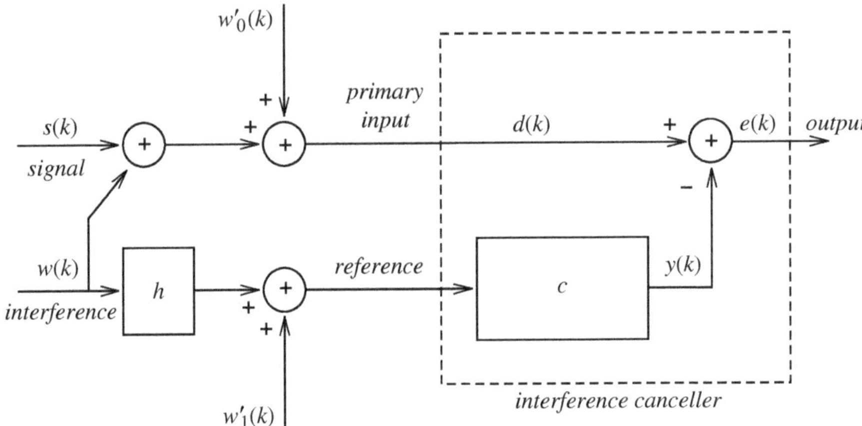

Figure 2.18 Specific configuration of an interference canceller.

The two coefficients c_1 and c_2, which respectively multiply x_1 and x_2, change the amplitude and phase of the reference signal to cancel the interfering tone.

It is easy to see that x_2 is obtained from x_1 via a Hilbert filter (see Figure 1.19). We note that in this case, x_2 can be obtained as a delayed version of x_1.

2.5.5 Echo cancellation in digital subscriber loops

With reference to the simplified scheme of Figure 2.19, the signal of user A is transmitted over a transmission line consisting of a pair of wires (*local loop*) [19] to the central office A, where the signals in the two directions of transmission, i.e. the signal transmitted by user A and the signal received from user B, are separated by a device called *hybrid*. A similar situation takes place at the central office B, with the roles of the signals A and B reversed. Because of impedance mismatch, the hybrids give origin to echo signals that are added to the desired signals, a problem that will be addressed in details in Chapter 17. A method to remove echo signals is illustrated in Figure 2.20, where y is a replica of the echo. At convergence, e will consist of the signal B only.

We note that this problem is a case of *system identification*, however, in echo cancellation, we are interested in the filter output y rather than the impulse response.

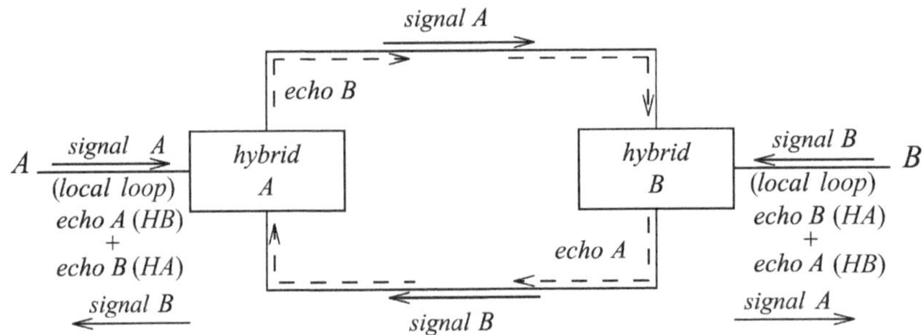

Figure 2.19 Transmission between two users in the public network.

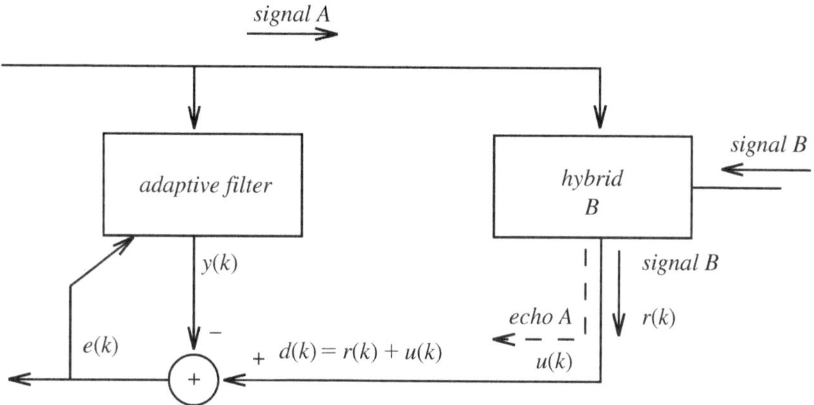

Figure 2.20 Configuration to remove the echo of signal A caused by the hybrid B.

2.5.6 Cancellation of a periodic interferer

For the cancellation of a periodic interfering signal, here a sinusoidal signal, we can use the scheme of Figure 2.21, where

- we note the absence of an external reference signal; the reference signal is generated by delaying the primary input;
- a delay $\Delta = DT_c$, where D is an integer, is needed to decorrelate the desired component of the *primary signal* from that of the *reference signal*, otherwise, part of the desired signal would also be cancelled.

On the other hand, to cancel a wideband interferer from a periodic signal it is sufficient to take the output of the filter.

Note that in both schemes, the filter acts as predictor of the periodic signal.

Exploiting the general concept described above, an alternative approach to that of Section 2.5.4 is illustrated in Figure 2.21, where the knowledge of the frequency of the interfering signal is not required. In general, the scheme of Figure 2.21 requires many more than two coefficients, therefore, it has a higher implementation complexity than that of Section 2.5.4. If the wideband signal can be modelled as white noise, then $D = 1$.

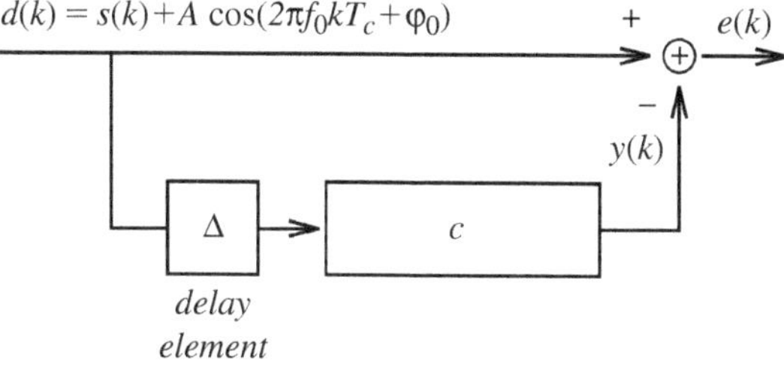

Figure 2.21 Scheme to remove a sinusoidal interferer from a wideband signal.

Bibliography

[1] Haykin, S. (1996). *Adaptive Filter ThEory*, 3e. Englewood Cliffs, NJ: Prentice-Hall.

[2] Honig, M.L. and Messerschmitt, D.G. (1984). *Adaptive Filters: Structures, Algorithms and Applications*. Boston, MA: Kluwer Academic Publishers.

[3] Atal, B.S. and Remde, J.R. (1982). A new model of LPC excitation for producing natural-sounding speech at low bit rates. *Proceedings of ICASSP*, pp. 614–617.

[4] Atal, B.S., Cuperman, V., and Gersho, A. (eds.) (1991). *Advances in Speech Coding*. Boston, MA: Kluwer Academic Publishers.

[5] Jayant, N.S. and Noll, P. (1984). *Digital Coding of Waveforms*. Englewood Cliffs, NJ: Prentice-Hall.

[6] Golub, G.H. and van Loan, C.F. (1989). *Matrix Computations*, 2e. Baltimore, MD and London: The Johns Hopkins University Press.

[7] Al-Dhahir, N. and Cioffi, J.M. (1995). Fast computation of channel-estimate based equalizers in packet data transmission. *IEEE Transactions on Signal Processing* 43: 2462–2473.

[8] Kumar, R. (1985). A fast algorithm for solving a Toeplitz system of equations. *IEEE Transactions on Acoustics, Speech, and Signal Processing* 33: 254–267.

[9] Press, S.A.T.W.H., Flannery, B.P., and Vetterling, W.T. (1988). *Numerical Recipes*, 3e. New York, NY: Cambridge University Press.

[10] Marple, L.S. Jr. (1987). *Digital Spectral Analysis with Applications*. Englewood Cliffs, NJ: Prentice-Hall.

[11] Kay, S.M. (1988). *Modern Spectral Estimation-Theory and Applications*. Englewood Cliffs, NJ: Prentice-Hall.

[12] Franks, L.E. (1981). *Signal Theory*, revised ed. Stroudsburg, PA: Dowden and Culver.

[13] Kay, S.M. (1993). *Fundamentals of Statistical Signal Processing: Estimation Theory*. Henceforth, NJ: Prentice-Hall.

[14] Treichler, J.R., Johnson, C.R. Jr., and Larimore, M.G. (1987). *Theory and Design of Adaptive Filters*. New York, NY: Wiley.

[15] Cowan, C.F.N. and Grant, P.M. (1985). *Adaptive Filters*. Englewood Cliffs, NJ: Prentice-Hall.

[16] Widrow, B. and Stearns, S.D. (1985). *Adaptive Signal Processing*. Englewood Cliffs, NJ: Prentice-Hall.

[17] Macchi, O. (1995). *Adaptive Processing: The LMS Approach with Applications in Transmission*. New York, NY: Wiley.

[18] Benvenuto, N. (1984). Distortion analysis on measuring the impulse response of a system using a cross correlation method. *AT&T Bell Laboratories Technical Journal* 63: 2171–2192.

[19] Messerschmitt, D.G. (1984). Echo cancellation in speech and data transmission. *IEEE Journal on Selected Areas in Communications* 2: 283–297.

[20] Delsarte, P. and Genin, Y.V. (1986). The split Levinson algorithm. *IEEE Transactions on Acoustics, Speech, and Signal Processing* 34: 470–478.

Appendix 2.A The Levinson–Durbin algorithm

The Levinson–Durbin algorithm (LDA) yields the solution of matrix equations like (2.80), in which $\boldsymbol{R}_{N+1}$ is positive definite, Hermitian, and Toeplitz, with a computational complexity proportional to N^2, instead of N^3 as it happens with algorithms that make use of the inverse matrix. In the case of real signals, $\boldsymbol{R}_{N+1}$ is symmetric and the computational complexity of the Delsarte–Genin algorithm (DGA) is halved with respect to that of LDA.

Here, we start by a step-by-step description of the LDA:

1. *Initialization*: We set

$$J_0 = r_x(0) \tag{2.265}$$
$$\Delta_0 = r_x(1) \tag{2.266}$$

2. *n-th iteration*, $n = 1, 2, \ldots, N$: We calculate

$$C_n = -\frac{\Delta_{n-1}}{J_{n-1}} \tag{2.267}$$
$$\boldsymbol{a}'_n = \begin{bmatrix} \boldsymbol{a}'_{n-1} \\ 0 \end{bmatrix} + C_n \begin{bmatrix} 0 \\ \boldsymbol{a}'^{B*}_{n-1} \end{bmatrix} \tag{2.268}$$

The (2.268) corresponds to the scalar equations:

$$\mathrm{a}'_{k,n} = \mathrm{a}'_{k,n-1} + C_n \mathrm{a}'^{*}_{n-k,\, n-1} k = 0, 1, \ldots, n \tag{2.269}$$

with $\mathrm{a}'_{0,n-1} = 1$ and $\mathrm{a}'_{n,n-1} = 0$. Moreover,

$$\Delta_n = (\boldsymbol{r}^B_{n+1})^T \boldsymbol{a}'_n \tag{2.270}$$
$$J_n = J_{n-1}\left(1 - |C_n|^2\right) \tag{2.271}$$

We now interpret the physical meaning of the parameters in the algorithm. J_n represents the *statistical power of the forward prediction error* at the n-th iteration:

$$J_n = E\left[|f_n(k)|^2\right] \tag{2.272}$$

It results

$$0 \le J_n \le J_{n-1} \qquad n \ge 1 \tag{2.273}$$

with

$$J_0 = r_x(0) \tag{2.274}$$

and

$$J_N = J_0 \prod_{n=1}^{N} \left(1 - |C_n|^2\right) \tag{2.275}$$

The following relation holds for Δ_n:

$$\Delta_{n-1} = E\left[f_{n-1}(k) b^*_{n-1}(k-1)\right] \tag{2.276}$$

In other words, Δ_n can be interpreted as the crosscorrelation between the forward linear prediction error and the backward linear prediction error delayed by one sample. C_n satisfies the following property:

$$C_n = \mathrm{a}'_{n,n} \tag{2.277}$$

Last, by substitution, from (2.267), along with (2.272) and (2.276), we get

$$C_n = -\frac{E[f_{n-1}(k) b^*_{n-1}(k-1)]}{E\left[|f_{n-1}(k)|^2\right]} \tag{2.278}$$

and noting that $E[|f_{n-1}(k)|^2] = E[|b_{n-1}(k)|^2] = E[|b_{n-1}(k-1)|^2]$, from (2.278), we have

$$|C_n| \le 1 \tag{2.279}$$

The coefficients $\{C_n\}$ are named *reflection coefficients* or *partial correlation coefficients* (PARCOR).

Lattice filters

We have just described the LDA. Its analysis permits to implement the prediction error filter via a modular structure. Defining

$$\boldsymbol{x}_{n+1}(k) = [x(k), \dots, x(k-n)]^T \tag{2.280}$$

we can write

$$\boldsymbol{x}_{n+1}(k) = \begin{bmatrix} \boldsymbol{x}_n(k) \\ x(k-n) \end{bmatrix} = \begin{bmatrix} x(k) \\ \boldsymbol{x}_n(k-1) \end{bmatrix} \tag{2.281}$$

We recall the relation for forward and backward linear prediction error filters of order n:

$$\begin{cases} f_n(k) = \mathrm{a}'_{0,n} x(k) + \cdots + \mathrm{a}'_{n,n} x(k-n) = \boldsymbol{a}'^T_n \, \boldsymbol{x}_{n+1}(k) \\ b_n(k) = \mathrm{a}'^*_{n,n} x(k) + \cdots + \mathrm{a}'^*_{0,n} x(k-n) = \boldsymbol{a}'^{BH}_n \, \boldsymbol{x}_{n+1}(k) \end{cases} \tag{2.282}$$

From (2.268), we obtain

$$\begin{aligned} f_n(k) &= \begin{bmatrix} \boldsymbol{a}'_{n-1} \\ 0 \end{bmatrix}^T \boldsymbol{x}_{n+1}(k) + C_n \begin{bmatrix} 0 \\ \boldsymbol{a}'^{B*}_{n-1} \end{bmatrix}^T \boldsymbol{x}_{n+1}(k) \\ &= f_{n-1}(k) + C_n b_{n-1}(k-1) \end{aligned} \tag{2.283}$$

By a similar procedure, we also find

$$b_n(k) = b_{n-1}(k-1) + C_n^* f_{n-1}(k) \tag{2.284}$$

Finally, taking into account the initial conditions,

$$f_0(k) = b_0(k) = x(k) \quad \text{and} \quad \mathrm{a}'_{0,0} = 1 \tag{2.285}$$

the block diagram of Figure 2.22 is obtained, in which the output is given by f_N.

We list the following fundamental properties:

1. The optimum coefficients C_n, $n = 1, \dots, N$, are independent of the order of the filter; therefore, one can change N without having to re-calculate all the coefficients. This property is useful if the filter length is unknown and must be estimated.
2. If conditions $|C_n| \le 1$, $n = 1, \dots, N$, are verified, the filter is minimum phase.
3. The lattice filters are quite insensitive to coefficient quantization.

Observation 2.3

From the above property 2 and (2.279), we have that all predictor error filters are *minimum phase*.

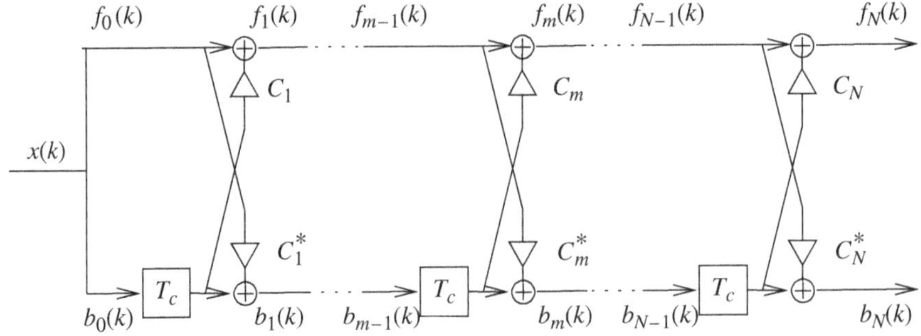

Figure 2.22 Lattice filter.

The Delsarte–Genin algorithm

In the case of real signals, the DGA, also known as the *split Levinson algorithm* [20], further reduces the number of operations with respect to the LDA, at least for $N \geq 10$.[6] Here is the step-by-step description.

1. *Initialization*: We set

$$\boldsymbol{v}_0 = 1 \quad \beta_0 = r_x(0) \quad \gamma_0 = r_x(1) \tag{2.286}$$

$$\boldsymbol{v}_1 = [1, 1]^T \quad \beta_1 = r_x(0) + r_x(1) \quad \gamma_1 = r_x(1) + r_x(2) \tag{2.287}$$

2. *n-th iteration*, $n = 2, \dots, N$: We compute

$$\alpha_n = \frac{(\beta_{n-1} - \gamma_{n-1})}{(\beta_{n-2} - \gamma_{n-2})} \tag{2.288}$$

$$\beta_n = 2\beta_{n-1} - \alpha_n \beta_{n-2} \tag{2.289}$$

$$\boldsymbol{v}_n = \begin{bmatrix} \boldsymbol{v}_{n-1} \\ 0 \end{bmatrix} + \begin{bmatrix} 0 \\ \boldsymbol{v}_{n-1} \end{bmatrix} - \alpha_n \begin{bmatrix} 0 \\ \boldsymbol{v}_{n-2} \\ 0 \end{bmatrix} \tag{2.290}$$

$$\gamma_n = \boldsymbol{r}_{n+1}^T \boldsymbol{v}_n = \left(r_x(1) + r_x(n+1)\right) + [\boldsymbol{v}_n]_2 \left(r_x(2) + r_x(n)\right) + \cdots \tag{2.291}$$

$$\lambda_n = \frac{\beta_n}{\beta_{n-1}} \tag{2.292}$$

$$\boldsymbol{a}'_n = \boldsymbol{v}_n - \lambda_n \begin{bmatrix} 0 \\ \boldsymbol{v}_{n-1} \end{bmatrix} \tag{2.293}$$

$$J_n = \beta_n - \lambda_n \gamma_{n-1} \tag{2.294}$$

$$C_n = 1 - \lambda_n \tag{2.295}$$

We note that (2.291) exploits the symmetry of the vector $\boldsymbol{v}_n$; in particular it is $[\boldsymbol{v}_n]_1 = [\boldsymbol{v}_n]_{n+1} = 1$.

6 Faster algorithms, with a complexity proportional to $N(\log N)^2$, have been proposed by Kumar [8].

Chapter 3

Adaptive transversal filters

We reconsider the Wiener filter introduced in Section 2.1. Given two random processes x and d, we want to determine the coefficients of a finite impulse response (FIR) filter having input x, so that its output y is a replica as accurate as possible of the process d. Adopting, for example, the mean-square error (MSE) criterion, the knowledge of the autocorrelation matrix $\boldsymbol{R}$ of the filter input vector and the crosscorrelation $\boldsymbol{p}$ between the desired output and the input vector are required. Next, the filter coefficients are determined by solving the system of equations (2.27). When the autocorrelation and crosscorrelation are not known a priori, but only (long) realizations of x and d are available, two approaches are possible. The *direct approach* provides that both autocorrelation and crosscorrelation are estimated,[1] and then the system of equations is solved. Clearly, estimation errors are propagated to the designed filter coefficients. An alternative approach discussed in this chapter provides that the filter coefficients are directly obtained from x and d by a *learning* and *adaptive* procedure. In this case, neither autocorrelation nor crosscorrelation are explicitly estimated, and no linear system of equations is solved. It turns out that in most cases adaptive methods are more robust to estimate errors than the direct approach. Note that the solutions presented in this chapter can be seen as *machine learning* approaches, where we have a machine (i.e. a parametric function, in this case an FIR filter) that learns by given input and desired output (x and d).

We will consider transversal FIR filters[2] with N coefficients. In general, the coefficients may vary with time. The filter structure at instant k is illustrated in Figure 3.1. We define:

1. Coefficient vector at instant k:

$$\boldsymbol{c}^T(k) = [c_0(k), c_1(k), \dots, c_{N-1}(k)] \tag{3.1}$$

2. Input vector at instant k:

$$\boldsymbol{x}^T(k) = [x(k), x(k-1), \dots, x(k-N+1)] \tag{3.2}$$

The output signal is given by

$$y(k) = \sum_{i=0}^{N-1} c_i(k)x(k-i) = \boldsymbol{x}^T(k)\boldsymbol{c}(k) \tag{3.3}$$

[1] Two estimation methods are presented in Section 1.9.2.
[2] For the analysis of infinite impulse response (IIR) adaptive filters, we refer the reader to [1, 2].

Algorithms for Communications Systems and their Applications, Second Edition.
Nevio Benvenuto, Giovanni Cherubini, and Stefano Tomasin.

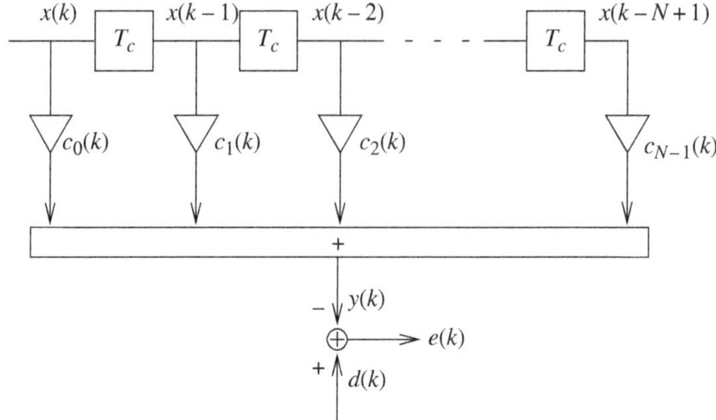

Figure 3.1 Structure of an adaptive transversal filter at instant k.

Comparing $y(k)$ with the desired response $d(k)$, we obtain the estimation error[3]

$$e(k) = d(k) - y(k) \tag{3.4}$$

Depending on the cost function associated with $\{e(k)\}$, in Chapter 2 two classes of algorithms have been developed:

1. mean-square error (MSE),
2. least squares (LS).

In the following sections, we will present iterative algorithms for each of the two classes.

3.1 The MSE design criterion

The cost function, or functional, to minimize is

$$J(k) = E[|e(k)|^2] \tag{3.5}$$

Assuming that x and d are individually and jointly wide-sense stationary (WSS), analogously to (2.17), $J(k)$ can be written as

$$J(k) = \sigma_d^2 - \boldsymbol{c}^H(k)\boldsymbol{p} - \boldsymbol{p}^H\boldsymbol{c}(k) + \boldsymbol{c}^H(k)\boldsymbol{R}\boldsymbol{c}(k) \tag{3.6}$$

where $\boldsymbol{R}$ and $\boldsymbol{p}$ are defined respectively in (2.16) and (2.13). The optimum Wiener–Hopf solution is $\boldsymbol{c}(k) = \boldsymbol{c}_{opt}$, where $\boldsymbol{c}_{opt}$ is given by (2.29). The corresponding minimum value of $J(k)$ is J_{min}, given by (2.35).

3.1.1 The steepest descent or gradient algorithm

Our first step is to realize a *deterministic iterative* procedure to compute $\boldsymbol{c}_{opt}$. We will see that this method avoids the computation of the inverse $\boldsymbol{R}^{-1}$; however, it requires that $\boldsymbol{R}$ and $\boldsymbol{p}$ be known.

[3] In this chapter, the estimation error is defined as as the difference between the desired signal and the filter output. Depending on the application, the estimation error may be defined using the opposite sign. Some caution is therefore necessary in using the equations of an adaptive filter.

The steepest descent or gradient algorithm is defined as:

$$\boldsymbol{c}(k+1) = \boldsymbol{c}(k) - \frac{1}{2}\mu\nabla_{\boldsymbol{c}(k)}J(k) \tag{3.7}$$

where $\nabla_{\boldsymbol{c}(k)}J(k)$ denotes the gradient of $J(k)$ with respect to $\boldsymbol{c}$ (see (2.18)), μ is the adaptation gain, a real-valued positive constant, and k is the iteration index, in general not necessarily coinciding with time instants.

As $J(k)$ is a quadratic function of the vector of coefficients, from (2.26) we get

$$\nabla_{\boldsymbol{c}(k)}J(k) = 2[\boldsymbol{R}\boldsymbol{c}(k) - \boldsymbol{p}] \tag{3.8}$$

hence

$$\boldsymbol{c}(k+1) = \boldsymbol{c}(k) - \mu[\boldsymbol{R}\boldsymbol{c}(k) - \boldsymbol{p}] \tag{3.9}$$

In the scalar case ($N = 1$), for real-valued signals the above relations become:

$$J(k) = J_{min} + \mathtt{r}_x(0)[c_0(k) - c_{opt,0}]^2 \tag{3.10}$$

and

$$\nabla_{c_0}J(k) = \frac{\partial}{\partial c_0}J(k) = 2\mathtt{r}_x(0)[c_0(k) - c_{opt,0}] \tag{3.11}$$

The iterative algorithm is given by:

$$c_0(k+1) = c_0(k) - \mu\mathtt{r}_x(0)[c_0(k) - c_{opt,0}] \tag{3.12}$$

The behaviour of J and the sign of $\nabla_{c_0}J(k)$ as a function of c_0 is illustrated in Figure 3.2. In the two dimensional case ($N = 2$), the trajectory of $\nabla_{\boldsymbol{c}}J(k)$ is illustrated in Figure 3.3. We recall that in general the gradient vector for $\boldsymbol{c} = \boldsymbol{c}(k)$ is orthogonal to the locus of points with constant J that includes $\boldsymbol{c}(k)$.

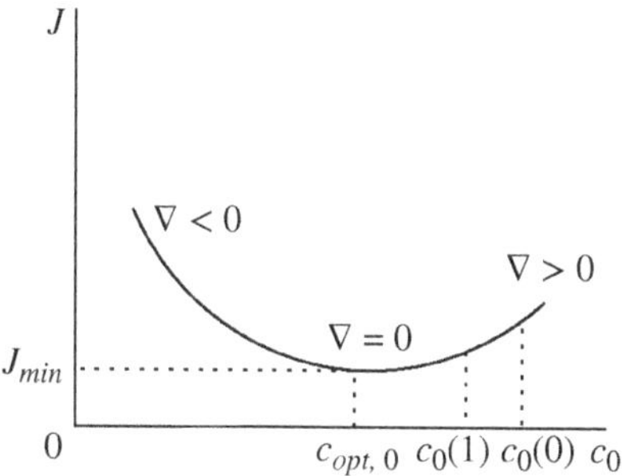

Figure 3.2 Behaviour of J and sign of the gradient vector ∇_c in the scalar case ($N = 1$).

Note that the steepest descent algorithm (3.7) is commonly used to update the parameters of other parametric functions (such as *neural networks*) in machine learning solutions.

Stability

Substituting for $\boldsymbol{p}$ the expression given by (2.27), the iterative algorithm (3.9) can be written as

$$\begin{aligned}\boldsymbol{c}(k+1) &= \boldsymbol{c}(k) - \mu[\boldsymbol{R}\boldsymbol{c}(k) - \boldsymbol{R}\boldsymbol{c}_{opt}] \\ &= [\boldsymbol{I} - \mu\boldsymbol{R}]\boldsymbol{c}(k) + \mu\boldsymbol{R}\boldsymbol{c}_{opt}\end{aligned} \tag{3.13}$$

Defining the *coefficient error* vector as

$$\Delta\boldsymbol{c}(k) = \boldsymbol{c}(k) - \boldsymbol{c}_{opt} \tag{3.14}$$

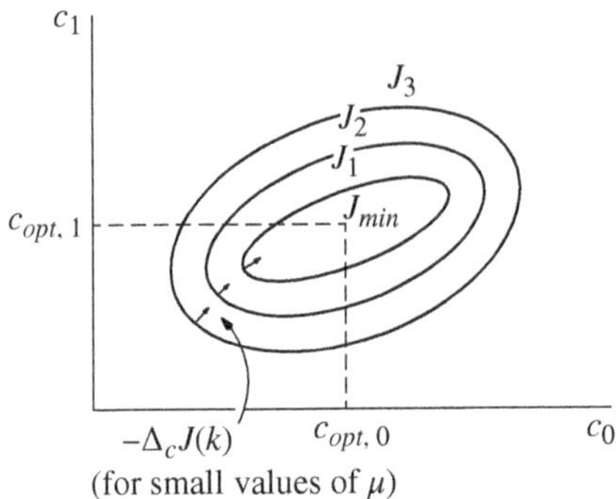

Figure 3.3 Loci of points with constant J and trajectory of $\nabla_c J$ in the two dimensional case ($N = 2$).

from (3.13) we obtain

$$\begin{aligned} \Delta c(k+1) &= [I - \mu R]c(k) + [\mu R - I]c_{opt} \\ &= [I - \mu R]\Delta c(k) \end{aligned} \tag{3.15}$$

Starting at $k = 0$ with arbitrary $c(0)$, that is with $\Delta c(0) = c(0) - c_{opt}$, we determine now the conditions for the convergence of $c(k)$ to c_{opt} or, equivalently, for the convergence of $\Delta c(k+1)$ to $\mathbf{0}$.

Using the decomposition (1.292), $R = U\Lambda U^H$, where U is the unitary matrix formed of eigenvectors of R, and Λ is the diagonal matrix of eigenvalues $\{\lambda_i\}$, $i = 1, \dots, N$, and setting (see (2.41))

$$v(k) = U^H \Delta c(k) \tag{3.16}$$

Eq. (3.15) becomes

$$v(k+1) = [I - \mu\Lambda]v(k) \tag{3.17}$$

Conditions for convergence

As Λ is diagonal, the i-th component of the vector $v(k)$ in (3.17) satisfies the difference equation:

$$v_i(k+1) = (1 - \mu\lambda_i)v_i(k), \qquad i = 1, 2, \dots, N \tag{3.18}$$

Hence, $v_i(k)$ as a function of k is given by (see Figure 3.4):

$$v_i(k) = (1 - \mu\lambda_i)^k v_i(0) \tag{3.19}$$

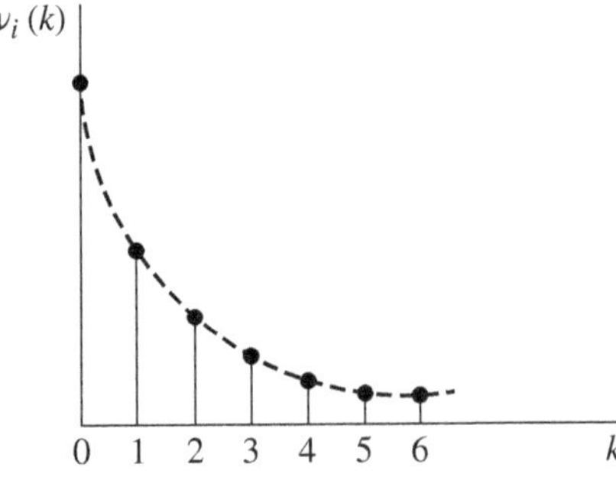

Figure 3.4 $v_i(k)$ as a function of k for $\mu\lambda_i < 1$ and $|1 - \mu\lambda_i| < 1$. In the case $\mu\lambda_i > 1$ and $|1 - \mu\lambda_i| < 1$, $v_i(k)$ is still decreasing in magnitude, but it assumes alternating positive and negative values.

The i-th component of the vector $\boldsymbol{v}(k)$ converges, that is

$$v_i(k) \xrightarrow[k\to\infty]{} 0 \qquad \forall v_i(0) \tag{3.20}$$

if and only if

$$|1 - \mu\lambda_i| < 1 \tag{3.21}$$

or, equivalently,

$$-1 < 1 - \mu\lambda_i < 1 \tag{3.22}$$

As λ_i is positive (see (1.284)), we have that the algorithm converges if and only if

$$0 < \mu < \frac{2}{\lambda_i}, \qquad i = 1, 2, \dots, N \tag{3.23}$$

If λ_{max} (λ_{min}) is the largest (smallest) eigenvalue of $\boldsymbol{R}$, observing (3.23) the convergence condition can be expressed as

$$0 < \mu < \frac{2}{\lambda_{max}} \tag{3.24}$$

Correspondingly, observing (3.16) and (3.19) we obtain the expression of the vector of coefficients, given by

$$\begin{aligned} \boldsymbol{c}(k) &= \boldsymbol{c}_{opt} + \boldsymbol{U}\boldsymbol{v}(k) = \boldsymbol{c}_{opt} + [\boldsymbol{u}_1, \boldsymbol{u}_2, \dots, \boldsymbol{u}_N]\boldsymbol{v}(k) \\ &= \boldsymbol{c}_{opt} + \sum_{i=1}^{N} \boldsymbol{u}_i v_i(k) = \boldsymbol{c}_{opt} + \sum_{i=1}^{N} \boldsymbol{u}_i (1 - \mu\lambda_i)^k v_i(0) \end{aligned} \tag{3.25}$$

Therefore, for each coefficient it results

$$c_n(k) = c_{opt,n} + \sum_{i=1}^{N} u_{i,n}(1 - \mu\lambda_i)^k v_i(0), \qquad n = 0, \dots, N-1 \tag{3.26}$$

where $u_{i,n}$ is the n-th component of $\boldsymbol{u}_i$. In (3.26), the term $u_{i,n}(1 - \mu\lambda_i)^k v_i(0)$ characterizes the *i-th mode of convergence*. Note that, if the convergence condition (3.23) is satisfied, each coefficient c_n, $n = 0, \dots, N-1$, converges to the optimum solution as a weighted sum of N exponentials, each with the time constant[4]

$$\tau_i = -\frac{1}{\ln|1 - \mu\lambda_i|} \simeq \frac{1}{\mu\lambda_i}, \qquad i = 1, 2, \dots, N \tag{3.27}$$

where the approximation is valid if $\mu\lambda_i \ll 1$.

Adaptation gain

The *speed of convergence*, which is related to the inverse of the convergence time, depends on the choice of μ. We define as μ_{opt} the value of μ that minimizes the time constant of the slowest mode. If we let

$$\xi(\mu) = \max_i |1 - \mu\lambda_i| \tag{3.28}$$

then we need to determine

$$\min_\mu \xi(\mu) \tag{3.29}$$

As illustrated in Figure 3.5, the solution is obtained for

$$1 - \mu\lambda_{min} = -(1 - \mu\lambda_{max}) \tag{3.30}$$

from which we get

$$\mu_{opt} = \frac{2}{\lambda_{max} + \lambda_{min}} \tag{3.31}$$

[4] The time constant is the number of iterations needed to reduce the signal associated with the i-th mode by a factor e.

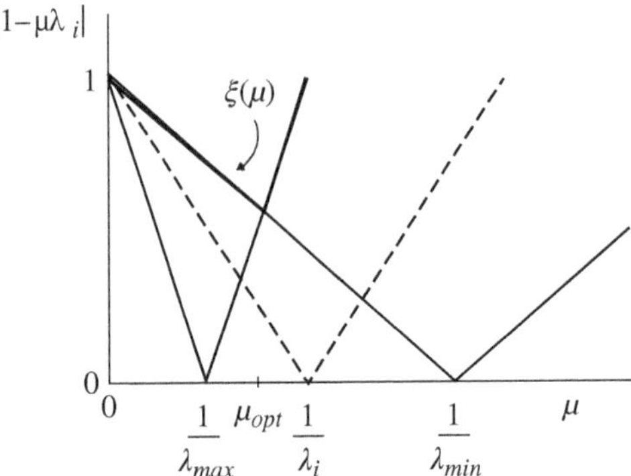

Figure 3.5 ξ and $|1-\mu\lambda_i|$ as a function of μ for different values of λ_i: $\lambda_{min} < \lambda_i < \lambda_{max}$.

and

$$\xi(\mu_{opt}) = 1 - \frac{2}{\lambda_{max} + \lambda_{min}}\lambda_{min} = \frac{\lambda_{max} - \lambda_{min}}{\lambda_{max} + \lambda_{min}} = \frac{\chi(\boldsymbol{R}) - 1}{\chi(\boldsymbol{R}) + 1} \tag{3.32}$$

where $\chi(\boldsymbol{R}) = \lambda_{max}/\lambda_{min}$ is the *eigenvalue spread* (1.299). We note that other values of μ (associated with λ_{max} or λ_{min}) cause a slower mode.

We emphasize that $\xi(\mu_{opt})$ is a monotonic function of the eigenvalue spread (see Figure 3.6), and consequently, the larger the eigenvalue spread, the slower the convergence.

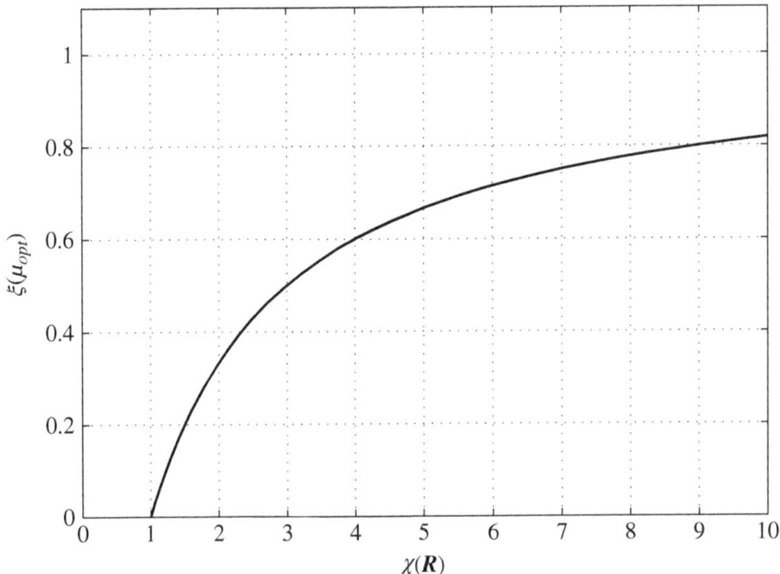

Figure 3.6 $\xi(\mu_{opt})$ as a function of the eigenvalue spread $\chi(\boldsymbol{R}) = \lambda_{max}/\lambda_{min}$.

Transient behaviour of the MSE

From (2.42) the general relation holds

$$J(k) = J_{min} + \sum_{i=1}^{N} \lambda_i |\nu_i(k)|^2 \tag{3.33}$$

Now using (3.19) we have

$$J(k) = J_{min} + \sum_{i=1}^{N} \lambda_i (1 - \mu\lambda_i)^{2k} v_i^2(0) \tag{3.34}$$

Consequently, if the condition for convergence is verified, for $k \to \infty$, $J(k) \to J_{min}$ as a weighted sum of exponentials. The i-th mode will have a time constant given by

$$\tau_{MSE,i} = -\frac{1}{2 \ln |1 - \mu\lambda_i|} \simeq \frac{1}{2\mu\lambda_i} \tag{3.35}$$

assuming $\mu\lambda_i \ll 1$. We note that (3.34) is different from (3.26) because of the presence of λ_i as weight of the i-th mode: consequently, modes associated with small eigenvalues tend to weigh less in the convergence of $J(k)$. In particular, let us examine the two dimensional case ($N = 2$). Recalling the observation that $J(k)$ increases more rapidly (slowly) in the direction of the eigenvector corresponding to $\lambda = \lambda_{max}$ ($\lambda = \lambda_{min}$) (see Figure 2.4), we have the following two cases.

Case 1 for $\lambda_1 \ll \lambda_2$. Choosing $\boldsymbol{c}(0)$ on the v_2 axis (in correspondence of λ_{max}), the iterative algorithm has following behaviour

$$\text{if } \mu \begin{cases} < \dfrac{1}{\lambda_{max}} & \text{non-oscillatory} \\ = \dfrac{1}{\lambda_{max}} & \text{convergent in one iteration} \\ > \dfrac{1}{\lambda_{max}} & \text{oscillatory around the minimum} \end{cases} \tag{3.36}$$

Let $\boldsymbol{u}_{\lambda_{min}}$ and $\boldsymbol{u}_{\lambda_{max}}$ be the eigenvectors corresponding to λ_{min} and λ_{max}, respectively. If no further information is given regarding the initial condition $\boldsymbol{c}(0)$, choosing $\mu = \mu_{opt}$ the algorithm exhibits monotonic convergence along $\boldsymbol{u}_{\lambda_{min}}$, and an oscillatory behaviour around the minimum along $\boldsymbol{u}_{\lambda_{max}}$.

Case 2 for $\lambda_2 = \lambda_1$. Choosing $\mu = 1/\lambda_{max}$, the algorithm converges in one iteration, independently of the initial condition $\boldsymbol{c}(0)$.

3.1.2 The least mean square algorithm

The least mean square (LMS) or stochastic gradient algorithm is an algorithm with low computational complexity that provides an approximation to the optimum Wiener–Hopf solution without requiring the knowledge of $\boldsymbol{R}$ and $\boldsymbol{p}$. Actually, the following instantaneous estimates are used:

$$\hat{\boldsymbol{R}}(k) = \boldsymbol{x}^*(k)\boldsymbol{x}^T(k) \tag{3.37}$$

and

$$\hat{\boldsymbol{p}}(k) = d(k)\boldsymbol{x}^*(k) \tag{3.38}$$

The gradient vector (3.8) is thus estimated as[5]

$$\begin{aligned} \widehat{\nabla_{\boldsymbol{c}(k)} J(k)} &= -2d(k)\boldsymbol{x}^*(k) + 2\boldsymbol{x}^*(k)\boldsymbol{x}^T(k)\boldsymbol{c}(k) \\ &= -2\boldsymbol{x}^*(k)[d(k) - \boldsymbol{x}^T(k)\boldsymbol{c}(k)] \\ &= -2\boldsymbol{x}^*(k)e(k) \end{aligned} \tag{3.39}$$

[5] We note that (3.39) represents an *unbiased* estimate of the gradient (3.8), but in general it also exhibits a large variance.

The equation for the adaptation of the filter coefficients, where k now denotes a given instant, becomes

$$\boldsymbol{c}(k+1) = \boldsymbol{c}(k) - \frac{1}{2}\mu\widehat{\nabla_{\boldsymbol{c}(k)}J(k)} \tag{3.40}$$

that is

$$\boldsymbol{c}(k+1) = \boldsymbol{c}(k) + \mu e(k)\boldsymbol{x}^*(k) \tag{3.41}$$

Observation 3.1

The same equation is obtained for a cost function equal to $|e(k)|^2$, whose gradient is given by

$$\widehat{\nabla_{\boldsymbol{c}(k)}J(k)} = -2e(k)\boldsymbol{x}^*(k) \tag{3.42}$$

Implementation

The block diagram for the implementation of the LMS algorithm is shown in Figure 3.7, with reference to the following parameters and equations.

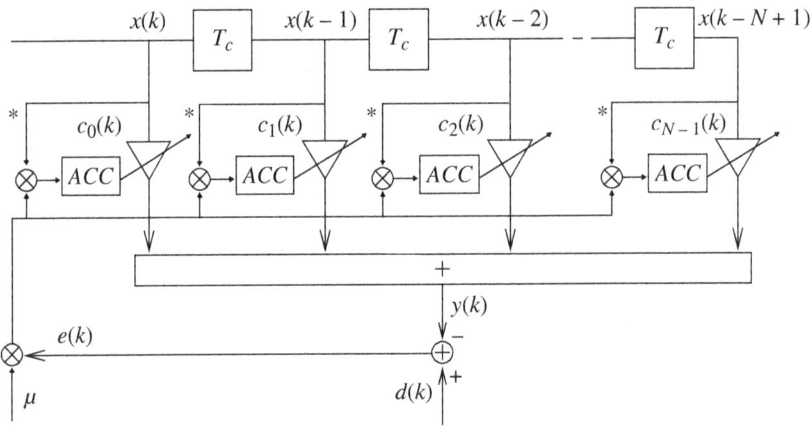

Figure 3.7 Block diagram of an adaptive transversal filter adapted according to the LMS algorithm.

Parameters. Required parameters are:

1. N, number of coefficients of the filter.
2. $0 < \mu < \dfrac{2}{\text{statistical power of input vector}}$.

Filter. The filter output is given by

$$y(k) = \boldsymbol{x}^T(k)\boldsymbol{c}(k) \tag{3.43}$$

Adaptation.

1. Estimation error

$$e(k) = d(k) - y(k) \tag{3.44}$$

2. Coefficient vector adaptation

$$\boldsymbol{c}(k+1) = \boldsymbol{c}(k) + \mu e(k)\boldsymbol{x}^*(k) \tag{3.45}$$

Initialization. If no a priori information is available, we set

$$c(0) = \mathbf{0} \tag{3.46}$$

The accumulators (ACCs) in Figure 3.7 are used to memorize the coefficients, which are updated by the current value of $\mu e(k)\boldsymbol{x}^*(k)$.

Computational complexity

For every iteration, we have $2N + 1$ complex multiplication (N due to filtering and $N + 1$ to adaptation) and $2N$ complex additions. Therefore, the LMS algorithm has a complexity of $O(N)$.

Conditions for convergence

Recalling that the objective of the LMS algorithm is to approximate the Wiener–Hopf solution, we introduce two criteria for convergence.

Convergence of the mean.

$$E[\boldsymbol{c}(k)] \xrightarrow[k\to\infty]{} \boldsymbol{c}_{opt} \tag{3.47}$$

$$E[e(k)] \xrightarrow[k\to\infty]{} 0 \tag{3.48}$$

In other words, it is required that the mean of the iterative solution converges to the Wiener–Hopf solution and the mean of the estimation error approaches zero. To show the weakness of this criterion, in Figure 3.8 we illustrate the results of a simple experiment for an input x given by a real-valued AR(1) random process:

$$x(k) = -\mathsf{a}\ x(k-1) + w(k), \qquad w(k) \sim \mathcal{N}(0, \sigma_w^2) \tag{3.49}$$

where $\mathsf{a} = 0.95$, and $\sigma_x^2 = 1$ (i.e. $\sigma_w^2 = 0.097$).

For a first-order predictor, we adapt the coefficient, $c(k)$, according to the LMS algorithm with $\mu = 0.1$. For $c(0) = 0$ and $k \geq 0$, we compute

1. Predictor output

$$y(k) = c(k)x(k-1) \tag{3.50}$$

2. Prediction error

$$e(k) = d(k) - y(k) = x(k) - y(k) \tag{3.51}$$

3. Coefficient update

$$c(k+1) = c(k) + \mu e(k)x(k-1) \tag{3.52}$$

In Figure 3.8, realizations of the processes $\{x(k)\}$, $\{c(k)\}$, and $\{|e(k)|^2\}$ are illustrated, as well as mean values $E[c(k)]$ and $J(k) = E[|e(k)|^2]$, estimated by averaging over 500 realizations; c_{opt} and J_{min} represent the Wiener–Hopf solution. From the plots in Figure 3.8, we observe two facts:

1. The random processes x and c exhibit a completely different behaviour, for which they may be considered uncorrelated. It is interesting to observe that this hypothesis corresponds to assuming the filter input vectors, $\{\boldsymbol{x}(k)\}$, statistically independent. Actually, for small values of μ, $\boldsymbol{c}$ tracks *mean statistical parameters* associated with the process x and not the process itself.
2. Convergence of the mean is an easily reachable objective. By itself, however, it does not yield the desired results, because the iterative solution $\boldsymbol{c}$ may exhibit very large oscillations around the optimum solution. A constraint on the amplitude of the oscillations must be introduced.

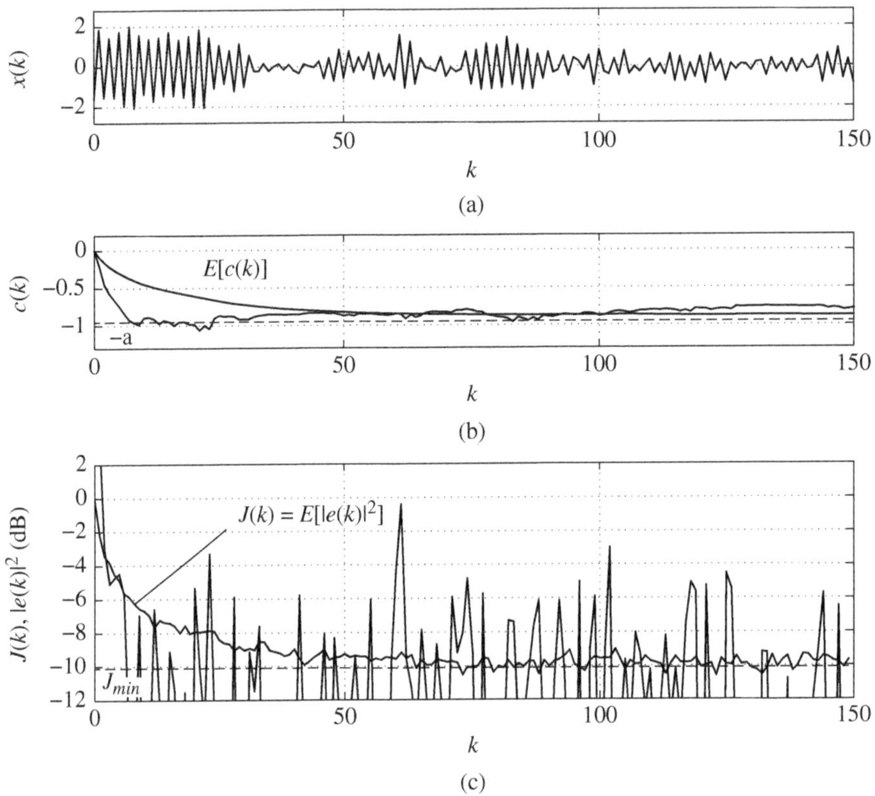

Figure 3.8 Realizations of (a) input $\{x(k)\}$, (b) coefficient $\{c(k)\}$, and (c) square error $\{|e(k)|^2\}$ for a one-coefficient predictor ($N = 1$), adapted according to the LMS algorithm.

Convergence in the mean-square sense.

$$E[\|\boldsymbol{c}(k) - \boldsymbol{c}_{opt}\|^2] \xrightarrow[k\to\infty]{} \text{constant} \tag{3.53}$$

$$J(k) = E[|e(k)|^2] \xrightarrow[k\to\infty]{} J(\infty) \qquad \text{constant} \tag{3.54}$$

In other words, at convergence, both the mean of the coefficient error vector norm and the output MSE must be finite. The quantity $J(\infty) - J_{min} = J_{ex}(\infty)$ is the MSE in excess and represents the price paid for using a random adaptation algorithm for the coefficients rather than a deterministic one, such as the steepest-descent algorithm. In any case, we will see that the ratio $J_{ex}(\infty)/J_{min}$ can be made small by choosing a small adaptation gain μ. We note that the coefficients are obtained by averaging in time the quantity $\mu e(k)\boldsymbol{x}^*(k)$. Choosing a small μ the adaptation will be slow and the effect of the gradient noise on the coefficients will be strongly attenuated.

3.1.3 Convergence analysis of the LMS algorithm

We recall the following definitions.

1. Coefficient error vector

$$\Delta\boldsymbol{c}(k) = \boldsymbol{c}(k) - \boldsymbol{c}_{opt} \tag{3.55}$$

2. Optimum filter output error

$$e_{min}(k) = d(k) - \boldsymbol{x}^T(k)\boldsymbol{c}_{opt} \tag{3.56}$$

We also make the following assumptions.

1. $\boldsymbol{c}(k)$ is statistically independent of $\boldsymbol{x}(k)$.
2. The components of the coefficient error vector, transformed according to $\boldsymbol{U}^H$, $\boldsymbol{v}(k) = [v_1(k), \dots, v_N(k)] = \boldsymbol{U}^H \Delta\boldsymbol{c}(k)$ (see (3.16)), are orthogonal:

$$E[v_i(k)v_n^*(k)] = 0, \qquad i \neq n \; i,n = 1,\dots,N \tag{3.57}$$

   This assumption is justified by the observation that the linear transformation that orthogonalizes both $\boldsymbol{x}(k)$ (see (1.291)) and $\Delta\boldsymbol{c}(k)$ (see (3.16)) in the gradient algorithm is given by $\boldsymbol{U}^H$.
3. Fourth-order moments can be expressed as products of second-order moments (see (3.86)).

The adaptation equation of the LMS algorithm (3.41) can thus be written as

$$\Delta\boldsymbol{c}(k+1) = \Delta\boldsymbol{c}(k) + \mu\boldsymbol{x}^*(k)[d(k) - \boldsymbol{x}^T(k)\boldsymbol{c}(k)] \tag{3.58}$$

Adding and subtracting $\boldsymbol{x}^T(k)\boldsymbol{c}_{opt}$ to the terms within parentheses, we obtain

$$\begin{aligned} \Delta\boldsymbol{c}(k+1) &= \Delta\boldsymbol{c}(k) + \mu\boldsymbol{x}^*(k)[e_{min}(k) - \boldsymbol{x}^T(k)\Delta\boldsymbol{c}(k)] \\ &= [\boldsymbol{I} - \mu\boldsymbol{x}^*(k)\boldsymbol{x}^T(k)]\Delta\boldsymbol{c}(k) + \mu e_{min}(k)\boldsymbol{x}^*(k) \end{aligned} \tag{3.59}$$

We note that $\Delta\boldsymbol{c}(k)$ depends only on the terms $\boldsymbol{x}(k-1), \boldsymbol{x}(k-2), \dots$

Moreover, with the change of variables (3.16), observing (3.33), the cost function[6]

$$J(k) = \underset{\boldsymbol{x},\boldsymbol{c}}{E}[|e(k)|^2] \tag{3.60}$$

can be written as

$$J(k) = J_{min} + \sum_{i=1}^{N} \lambda_i E[|v_i(k)|^2] \tag{3.61}$$

Convergence of the mean

Taking the expectation of (3.59) and exploiting the statistical independence between $\boldsymbol{x}(k)$ and $\Delta\boldsymbol{c}(k)$, we get

$$E[\Delta\boldsymbol{c}(k+1)] = [\boldsymbol{I} - \mu E[\boldsymbol{x}^*(k)\boldsymbol{x}^T(k)]]E[\Delta\boldsymbol{c}(k)] + \mu E[e_{min}(k)\boldsymbol{x}^*(k)] \tag{3.62}$$

As $E[\boldsymbol{x}^*(k)\boldsymbol{x}^T(k)] = \boldsymbol{R}$ and the second term on the right-hand side of (3.62) vanishes for the orthogonality property (2.31) of the optimum filter, we obtain the same equation as in the case of the *steepest descent* algorithm:

$$E[\Delta\boldsymbol{c}(k+1)] = [\boldsymbol{I} - \mu\boldsymbol{R}]E[\Delta\boldsymbol{c}(k)] \tag{3.63}$$

Consequently, for the LMS algorithm, the convergence of the mean is obtained if

$$0 < \mu < \frac{2}{\lambda_{max}} \tag{3.64}$$

Observing (3.25) and (3.32), and choosing the value of $\mu = 2/(\lambda_{max} + \lambda_{min})$, the vector $E[\Delta\boldsymbol{c}(k)]$ is reduced at each iteration at least by the factor $(\lambda_{max} - \lambda_{min})/(\lambda_{max} + \lambda_{min})$. We can therefore assume that $E[\Delta\boldsymbol{c}(k)]$ becomes rapidly negligible with respect to the MSE during the process of convergence.

Convergence in the mean-square sense: real scalar case

The assumption of a filter with real-valued input and only one coefficient $c(k)$ provides by a simple analysis important properties of the convergence in the mean-square sense. From

$$\Delta c(k+1) = (1 - \mu x^2(k))\Delta c(k) + \mu e_{min}(k)x(k) \tag{3.65}$$

[6] $\underset{\boldsymbol{x},\boldsymbol{c}}{E}$ denotes the expectation with respect to both $\boldsymbol{x}$ and $\boldsymbol{c}$.

because $x(k)$ and $\Delta c(k)$ are assumed to be statistically independent and $x(k)$ is orthogonal to $e_{min}(k)$, and *assuming* furthermore that $x(k)$ and $e_{min}(k)$ are statistically independent, we get

$$\begin{aligned} E[\Delta c^2(k+1)] &= E[1 + \mu^2 x^4(k) - 2\mu x^2(k)]E[\Delta c^2(k)] + \mu^2 E[x^2(k)]J_{min} \\ &\quad + 2E[(1 - \mu x^2(k))\Delta c(k)\mu x(k)e_{min}(k)] \end{aligned} \tag{3.66}$$

where the last term vanishes, as $\Delta c(k)$ has zero mean and is statistically independent of all other terms. Assuming[7] moreover $E[x^4(k)] = E[x^2(k)x^2(k)] = \mathtt{r}_x^2(0)$, (3.66) becomes

$$E[\Delta c^2(k+1)] = (1 + \mu^2 \mathtt{r}_x^2(0) - 2\mu \mathtt{r}_x(0))E[\Delta c^2(k)] + \mu^2 \mathtt{r}_x(0) J_{min} \tag{3.67}$$

Let

$$\gamma = 1 + \mu^2 \mathtt{r}_x^2(0) - 2\mu \mathtt{r}_x(0) \tag{3.68}$$

whose behaviour as a function of μ is given in Figure 3.9. Then for the convergence of the difference equation (3.67), it must be $|\gamma| < 1$. Consequently, μ must satisfy the condition

$$0 < \mu < \frac{2}{\mathtt{r}_x(0)} \tag{3.69}$$

Moreover, assuming $\mu \mathtt{r}_x(0) \ll 1$, we get

$$\begin{aligned} E[\Delta c^2(\infty)] &= \frac{\mu^2 \mathtt{r}_x(0)}{\mu \mathtt{r}_x(0)(2 - \mu \mathtt{r}_x(0))} J_{min} \\ &= \frac{\mu}{(2 - \mu \mathtt{r}_x(0))} J_{min} \\ &\simeq \frac{\mu}{2} J_{min} \end{aligned} \tag{3.70}$$

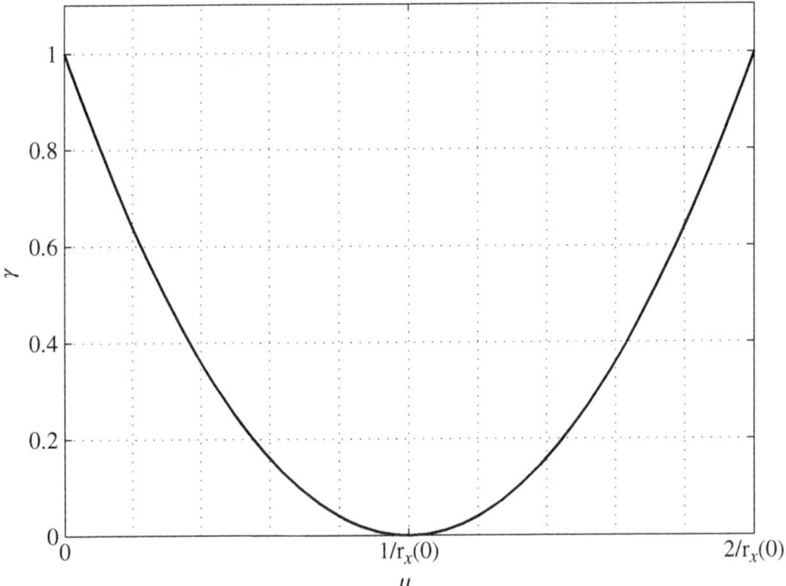

Figure 3.9 γ as a function of μ.

[7] In other books the Gaussian assumption is made, whereby $E[x^4(k)] = 3\mathtt{r}_x^2(0)$. The conclusions of the analysis are similar.

Likewise, from

$$e(k) = d(k) - x(k)c(k) = e_{min}(k) - \Delta c(k)x(k) \tag{3.71}$$

it turns out

$$E[e^2(k)] = E[e^2_{min}(k)] + E[x^2(k)]E[\Delta c^2(k)] \tag{3.72}$$

that is

$$J(k) = J_{min} + \mathtt{r}_x(0)E[\Delta c^2(k)] \tag{3.73}$$

In particular, for $k \to \infty$, we have

$$J(\infty) \simeq J_{min} + \mathtt{r}_x(0)\frac{\mu}{2}J_{min} \tag{3.74}$$

The relative MSE deviation, or *misadjustment*, is:

$$\text{MSD} = \frac{J(\infty) - J_{min}}{J_{min}} = \frac{J_{ex}(\infty)}{J_{min}} = \mathtt{r}_x(0)\frac{\mu}{2} \tag{3.75}$$

Convergence in the mean-square sense: general case

The convergence theory given here follows the method developed in [3]. With the change of variables (3.16), (3.59) becomes

$$\boldsymbol{v}(k+1) = [\boldsymbol{I} - \mu \boldsymbol{U}^H \boldsymbol{x}^*(k)\boldsymbol{x}^T(k)\boldsymbol{U}]\boldsymbol{v}(k) + \mu e_{min}(k)\boldsymbol{U}^H\boldsymbol{x}^*(k) \tag{3.76}$$

Let us define

$$\tilde{\boldsymbol{x}}(k) = [\tilde{x}_1(k), \dots, \tilde{x}_N(k)]^T = \boldsymbol{U}^T\boldsymbol{x}(k) \tag{3.77}$$

and

$$\boldsymbol{\Omega} = \tilde{\boldsymbol{x}}^*(k)\tilde{\boldsymbol{x}}^T(k) = \boldsymbol{U}^H\boldsymbol{x}^*(k)\boldsymbol{x}^T(k)\boldsymbol{U} \tag{3.78}$$

$N \times N$ matrix with elements

$$\boldsymbol{\Omega}(i,n) = \tilde{x}_i^*(k)\tilde{x}_n(k), \qquad i,n = 1,\dots,N \tag{3.79}$$

From (1.291) we get

$$E[\boldsymbol{\Omega}] = \boldsymbol{\Lambda} \tag{3.80}$$

hence the components $\{\tilde{x}_i(k)\}$ are mutually orthogonal. Then (3.76) becomes

$$\boldsymbol{v}(k+1) = [\boldsymbol{I} - \mu\boldsymbol{\Omega}]\boldsymbol{v}(k) + \mu e_{min}(k)\tilde{\boldsymbol{x}}^*(k) \tag{3.81}$$

Recalling Assumption 1 of the convergence analysis, and assuming $e_{min}(k)$ and $\boldsymbol{x}(k)$ are not only orthogonal but also statistically independent, and consequently $e_{min}(k)$ independent of $\tilde{\boldsymbol{x}}(k)$, the correlation matrix of $\boldsymbol{v}^*$ at the instant $k+1$ can be expressed as

$$E[\boldsymbol{v}^*(k+1)\boldsymbol{v}^T(k+1)] = E[(\boldsymbol{I} - \mu\boldsymbol{\Omega}^*)\boldsymbol{v}^*(k)\boldsymbol{v}^T(k)(\boldsymbol{I} - \mu\boldsymbol{\Omega}^T)] + \mu^2 J_{min}E[\boldsymbol{\Omega}^*] \tag{3.82}$$

Observing (3.80), the second term at the right-hand side of (3.82) is equal to $\mu^2 J_{min}\boldsymbol{\Lambda}$. Moreover, considering that $\boldsymbol{v}(k)$ is statistically independent of $\boldsymbol{x}(k)$ and $\tilde{\boldsymbol{x}}(k)$, the first term can be written as

$$\begin{aligned} \underset{\boldsymbol{x},\boldsymbol{v}}{E}[(\boldsymbol{I} - \mu\boldsymbol{\Omega}^*)\boldsymbol{v}^*(k)\boldsymbol{v}^T(k)(\boldsymbol{I} - \mu\boldsymbol{\Omega}^T)] &= \underset{\boldsymbol{x}}{E}\{(\boldsymbol{I} - \mu\boldsymbol{\Omega}^*)\underset{\boldsymbol{v}}{E}[\boldsymbol{v}^*(k)\boldsymbol{v}^T(k)](\boldsymbol{I} - \mu\boldsymbol{\Omega}^T)\} \\ &= E[(\boldsymbol{I} - \mu\boldsymbol{\Omega}^*)(\boldsymbol{I} - \mu\boldsymbol{\Omega}^T)]E[\boldsymbol{v}^*(k)\boldsymbol{v}^T(k)] \end{aligned} \tag{3.83}$$

Recalling Assumption 2 of the convergence analysis, we get that the matrix $E[\boldsymbol{v}^*(k)\ \boldsymbol{v}^T(k)]$ is diagonal, with elements on the main diagonal given by the vector

$$\boldsymbol{\eta}^T(k) = [\eta_1(k), \dots, \eta_N(k)] = [E[|v_1(k)|^2], \dots, E[|v_N(k)|^2]] \tag{3.84}$$

Observing (3.79), the elements with indices (i, i) of the matrix expressed by (3.83) are given by

$$E\left[\eta_i(k) + \mu^2 \sum_{n=1}^{N} \mathbf{\Omega}^*(i,n)\mathbf{\Omega}(n,i)\eta_n(k) - 2\mu\mathbf{\Omega}(i,i)\eta_i(k)\right]$$
$$= \eta_i(k) + \mu^2 \sum_{n=1}^{N} E[|\tilde{x}_i(k)|^2|\tilde{x}_n(k)|^2]\eta_n(k) - 2\mu\lambda_i\eta_i(k) \tag{3.85}$$
$$= \eta_i(k) + \mu^2 \sum_{n=1}^{N} \lambda_i\lambda_n\eta_n(k) - 2\mu\lambda_i\eta_i(k)$$

where, recalling Assumption 3 of the convergence analysis,

$$E[|\tilde{x}_i(k)|^4] = E[|\tilde{x}_i(k)|^2]E[|\tilde{x}_i(k)|^2] = \lambda_i^2 \tag{3.86}$$

Let

$$\boldsymbol{\lambda}^T = [\lambda_1, \dots, \lambda_N] \tag{3.87}$$

be the vector of eigenvalues of $\boldsymbol{R}$, and

$$\boldsymbol{B} = \begin{bmatrix} (1-\mu\lambda_1)^2 & \mu^2\lambda_1\lambda_2 & \dots & \mu^2\lambda_1\lambda_N \\ \mu^2\lambda_2\lambda_1 & (1-\mu\lambda_2)^2 & \dots & \mu^2\lambda_2\lambda_N \\ \vdots & \vdots & \ddots & \vdots \\ \mu^2\lambda_N\lambda_1 & \mu^2\lambda_N\lambda_2 & \dots & (1-\mu\lambda_N)^2 \end{bmatrix} \tag{3.88}$$

$N \times N$ symmetric positive definite matrix with positive elements. From (3.82) and (3.85), we obtain the relation

$$\boldsymbol{\eta}(k+1) = \boldsymbol{B}\boldsymbol{\eta}(k) + \mu^2 J_{min}\boldsymbol{\lambda} \tag{3.89}$$

Using the properties of $\boldsymbol{B}$, the general decomposition (2.139) becomes

$$\boldsymbol{B} = \boldsymbol{V}\,\mathrm{diag}(\sigma_1, \dots, \sigma_N)\boldsymbol{V}^H \tag{3.90}$$

where $\{\sigma_i\}$ denote the eigenvalues of $\boldsymbol{B}$, and $\boldsymbol{V}$ is the unitary matrix formed by the eigenvectors $\{\boldsymbol{v}_i\}$ of $\boldsymbol{B}$. After simple steps, similar to those applied to get (3.25) from (3.13), and using the relation

$$N\mathtt{r}_x(0) = \mathrm{tr}\,\boldsymbol{R} = \sum_{i=1}^{N} \lambda_i \tag{3.91}$$

the solution of the vector difference equation (3.89) is given by:

$$\boldsymbol{\eta}(k) = \sum_{i=1}^{N} \mathcal{K}_i\sigma_i^k\boldsymbol{v}_i + \frac{\mu J_{min}}{2-\mu N\mathtt{r}_x(0)}\mathbf{1}, \qquad k \geq 0 \tag{3.92}$$

where $\mathbf{1} = [1, 1, \dots, 1]^T$. In (3.92), the constants $\{\mathcal{K}_i\}$ are determined by the initial conditions

$$\mathcal{K}_i = \boldsymbol{v}_i^H\left(\boldsymbol{\eta}(0) - \frac{\mu J_{min}}{2-\mu N\mathtt{r}_x(0)}\mathbf{1}\right), \qquad i = 1, \dots, N \tag{3.93}$$

where the components of $\boldsymbol{\eta}(0)$ depend on the choice of $\boldsymbol{c}(0)$ according to (3.84) and (3.16):

$$\eta_n(0) = E[|\nu_n(0)|^2] = E[|\boldsymbol{u}_n^H\Delta\boldsymbol{c}(0)|^2], \qquad n = 1, \dots, N \tag{3.94}$$

Using (3.84) and (3.87), the cost function $J(k)$ given by (3.61) becomes

$$J(k) = J_{min} + \boldsymbol{\lambda}^T\boldsymbol{\eta}(k) \tag{3.95}$$

Substituting the result (3.92) in (3.95), we get

$$J(k) = \sum_{i=1}^{N} C_i \sigma_i^k + \frac{2}{2 - \mu N \mathtt{r}_x(0)} J_{min} \tag{3.96}$$

where

$$C_i = \mathcal{K}_i \boldsymbol{\lambda}^T \boldsymbol{v}_i \tag{3.97}$$

The first term on the right-hand side of (3.96) describes the convergence behaviour of the mean-square error, whereas the second term gives the steady-state value. Therefore, the further investigation of the properties of the matrix $\boldsymbol{B}$ will allow us to characterize the transient behaviour of J.

Fundamental results

From the above convergence analysis, we will obtain some fundamental properties of the LMS algorithm.

1. The transient behaviour of J does not exhibit oscillations; this result is obtained by observing the properties of the eigenvalues of $\boldsymbol{B}$.
2. The LMS algorithm converges if the adaptation gain μ satisfies the condition

$$0 < \mu < \frac{2}{\text{statistical power of input vector}} \tag{3.98}$$

In fact the adaptive system is stable and J converges to a constant steady-state value under the conditions $|\sigma_i| < 1$, $i = 1, \dots, N$. This happens if

$$0 < \mu < \frac{2}{N \mathtt{r}_x(0)} \tag{3.99}$$

Conversely, if μ satisfies (3.99), from (3.88) the sum of the elements of the i-th row of $\boldsymbol{B}$ satisfies

$$\sum_{n=1}^{N} [\boldsymbol{B}]_{i,n} = 1 - \mu \lambda_i (2 - \mu N \mathtt{r}_x(0)) < 1 \tag{3.100}$$

A matrix with these properties and whose elements are all positive has eigenvalues with absolute value less than one. In particular, being

$$\begin{aligned} \sum_{i=1}^{N} \lambda_i &= \mathrm{tr}[\boldsymbol{R}] \\ &= N \mathtt{r}_x(0) \\ &= \sum_{i=0}^{N-1} E[|x(k-i)|^2] \end{aligned} \tag{3.101}$$

the statistical power of input vector, Eq. (3.99) becomes (3.98).

We recall that, for convergence of the mean, it must be

$$0 < \mu < \frac{2}{\lambda_{max}} \tag{3.102}$$

but since

$$\sum_{i=1}^{N} \lambda_i > \lambda_{max} \tag{3.103}$$

the condition for convergence in the mean square implies convergence of the mean. In other words, convergence in the mean square imposes a tighter bound to allowable values of the adaptation gain

μ than that imposed by convergence of the mean (3.102). The new bound depends on the number of coefficients, rather than on the eigenvalue distribution of the matrix $\boldsymbol{R}$. The relation (3.99) can be intuitively explained noting that, for a given value of μ, an increase in the number of coefficients causes an increase in the excess MSE due to fluctuations of the coefficients around the mean value. Increasing the number of coefficients without reducing the value of μ eventually leads to instability of the adaptive system.

3. Equation (3.96) reveals a simple relation between the adaptation gain μ and the value $J(k)$ in the steady state ($k \to \infty$):

$$J(\infty) = \frac{2}{2 - \mu N \mathtt{r}_x(0)} J_{min} \tag{3.104}$$

from which the excess MSE is given by

$$J_{ex}(\infty) = J(\infty) - J_{min} = \frac{\mu N \mathtt{r}_x(0)}{2 - \mu N \mathtt{r}_x(0)} J_{min} \tag{3.105}$$

and the misadjustment is

$$\text{MSD} = \frac{J_{ex}(\infty)}{J_{min}} = \frac{\mu N \mathtt{r}_x(0)}{2 - \mu N \mathtt{r}_x(0)} \simeq \frac{\mu}{2} N \mathtt{r}_x(0) \tag{3.106}$$

for $\mu \ll 2/(N\mathtt{r}_x(0))$.

Observations

1. For $\mu \to 0$ all eigenvalues of $\boldsymbol{B}$ tend towards 1.
2. As shown below, a small eigenvalue of the matrix $\boldsymbol{R}$ ($\lambda_i \to 0$) determines a large time constant for one of the convergence modes of J, as $\sigma_i \to 1$. However, a large time constant of one of the modes implies a low probability that the corresponding term contributes significantly to the MSE.

Proof. If $\lambda_i = 0$, the i-th row of $\boldsymbol{B}$ becomes $(0, \dots, 0, [\boldsymbol{B}]_{i,i} = 1, 0, \dots, 0)$. Consequently, $\sigma_i = 1$ and $\boldsymbol{v}_i^T = (0, \dots, 0, v_{i,i} = 1, 0, \dots, 0)$. As $\boldsymbol{\lambda}^T \boldsymbol{v}_i = 0$, from (3.97) we get $C_i = 0$. □

It is generally correct to state that a large *eigenvalue spread* of $\boldsymbol{R}$ determines a slow convergence of J to the steady state. However, the fact that modes with a large time constant usually contribute to J less than the modes that converge more rapidly, mitigates this effect. Therefore, the convergence of J is less influenced by the eigenvalue spread of $\boldsymbol{R}$ than would be the convergence of $\Delta \boldsymbol{c}(k)$.

3. If all eigenvalues of the matrix $\boldsymbol{R}$ are equal,[8] $\lambda_i = \mathtt{r}_x(0)$, $i = 1, \dots, N$, the maximum eigenvalue of the matrix $\boldsymbol{B}$ is given by

$$\sigma_{i_{max}} = 1 - \mu \mathtt{r}_x(0)(2 - \mu N \mathtt{r}_x(0)) \tag{3.107}$$

The remaining eigenvalues of $\boldsymbol{B}$ do not influence the transient behaviour of J, since $C_i = 0$, $i \neq i_{max}$.

Proof. It is easily verified that $\sigma_{i_{max}}$ is an eigenvalue of $\boldsymbol{B}$ and $\boldsymbol{v}_{i_{max}}^T = N^{-1/2}[1, 1, \dots, 1]$ is the corresponding eigenvector. Moreover, the Perron–Frobenius theorem affirms that the maximum eigenvalue of a positive matrix $\boldsymbol{B}$ is a positive real number and that the elements of the corresponding eigenvector are positive real numbers [4]. Since all elements of $\boldsymbol{v}_{i_{max}}$ are positive, it follows that $\sigma_{i_{max}}$ is the maximum eigenvalue of $\boldsymbol{B}$. Moreover, because $\boldsymbol{v}_{i_{max}}$ is parallel to $\boldsymbol{\lambda}^T$, the other eigenvectors of $\boldsymbol{B}$ are orthogonal to $\boldsymbol{\lambda}$. Hence, $C_i = 0$, $i \neq i_{max}$. □

[8] This occurs, for example, if the input x is white noise.

4. If all eigenvalues of the matrix $\boldsymbol{R}$ are equal, $\lambda_i = \mathtt{r}_x(0)$, $i = 1, \dots, N$, combining (3.96) with the (3.107) and considering a time varying adaptation gain $\mu(k)$, we obtain

$$J(k+1) \simeq [1 - \mu(k)\mathtt{r}_x(0)(2 - \mu(k)N\mathtt{r}_x(0))]J(k) + 2\mu(k)\mathtt{r}_x(0)J_{min} \tag{3.108}$$

The maximum convergence rate of J is obtained for the adaptation gain

$$\mu_{opt}(k) = \frac{1}{N\mathtt{r}_x(0)} \frac{J(k) - J_{min}}{J(k)} \tag{3.109}$$

As the condition $J(k) \gg J_{min}$ is normally verified at the beginning of the iteration process, it results

$$\mu_{opt}(k) \simeq \frac{1}{N\mathtt{r}_x(0)} \tag{3.110}$$

and

$$J(k+1) \simeq \left(1 - \frac{1}{N}\right) J(k) \tag{3.111}$$

We note that the number of iterations required to reduce the value of $J(k)$ by one order of magnitude is approximately $2.3N$.

5. Equation (3.92) indicates that at steady state, all elements of $\boldsymbol{\eta}$ become equal. Consequently, recalling Assumption 2 of the convergence analysis, in steady state the filter coefficients are uncorrelated random variables with equal variance. The mean corresponds to the optimum vector $\boldsymbol{c}_{opt}$.
6. In case the LMS algorithm is used to estimate the coefficients of a system that slowly changes in time, the adaptation gain μ must be strictly larger than 0. In this case, the value of J *in steady state* varies with time and is given by the sum of three terms:

$$J_{tot}(k) = J_{min}(k) + J_{ex}(\infty) + J_{\ell} \tag{3.112}$$

where $J_{min}(k)$ corresponds to the Wiener–Hopf solution, $J_{ex}(\infty)$ depends instead on the LMS algorithm and is directly proportional to μ, and J_{ℓ} depends on the ability of the LMS algorithm to track the system variations and expresses the lag error in the estimate of the coefficients. It turns out that J_{ℓ} is inversely proportional to μ. Therefore, for time-varying systems μ must be chosen as a compromise between J_{ex} and J_{ℓ} and cannot be arbitrarily small [5–7].

Final remarks

1. The LMS algorithm is easy to implement.
2. The relatively slow convergence is influenced by μ, the number of coefficients and the eigenvalues of $\boldsymbol{R}$. In particular it must be

$$0 < \mu < \frac{2}{N\mathtt{r}_x(0)} = \frac{2}{\text{statistical power of input vector}} \tag{3.113}$$

3. Choosing a small μ results in a slow adaptation, and in a small excess MSE at convergence $J_{ex}(\infty)$. For a large μ, instead, the adaptation is fast at the expense of a large $J_{ex}(\infty)$.
4. $J_{ex}(\infty)$ is determined by the large eigenvalues of $\boldsymbol{R}$, whereas the speed of convergence of $E[\boldsymbol{c}(k)]$ is imposed by λ_{min}. If the eigenvalue spread of $\boldsymbol{R}$ increases, the convergence of $E[\boldsymbol{c}(k)]$ becomes slower; on the other hand, the convergence of $J(k)$ is less sensitive to this parameter. Note, however, that the convergence behaviour depends on the initial condition $\boldsymbol{c}(0)$ [8].

3.1.4 Other versions of the LMS algorithm

In the Section 3.1.3, the basic version of the LMS algorithm has been presented. We now give a brief introduction to other versions that can be used for various applications.

Leaky LMS

The leaky LMS algorithm is a variant of the LMS algorithm that uses the following adaptation equation:

$$\boldsymbol{c}(k+1) = (1-\mu\alpha)\boldsymbol{c}(k) + \mu e(k)\boldsymbol{x}^*(k) \tag{3.114}$$

with $0 < \alpha \ll \mathtt{r}_x(0)$. This equation corresponds to the following cost function:

$$J(k) = E[|e(k)|^2] + \alpha E[\|\boldsymbol{c}(k)\|^2] \tag{3.115}$$

where, as usual,

$$e(k) = d(k) - \boldsymbol{c}^T(k)\boldsymbol{x}(k) \tag{3.116}$$

In other words, the cost function includes an additional term proportional to the norm of the vector of coefficients. In steady state, we get

$$\lim_{k\to\infty} E[\boldsymbol{c}(k)] = (\boldsymbol{R} + \alpha\boldsymbol{I})^{-1}\boldsymbol{p} \tag{3.117}$$

It is interesting to give another interpretation to what has been stated. Observing (3.117), the application of the leaky LMS algorithm results in the addition of a small constant α to the terms on the main diagonal of the correlation matrix of the input process; one obtains the same result by summing white noise with statistical power α to the input process. Both approaches are useful to make invertible an ill-conditioned matrix $\boldsymbol{R}$, or to accelerate the convergence of the LMS algorithm. It is usually sufficient to choose α two or three orders of magnitude smaller than $\mathtt{r}_x(0)$, in order not to modify substantially the original Wiener–Hopf solution. Therefore, the leaky LMS algorithm is used when the Wiener problem is ill-conditioned, and multiple solutions exist.

Sign algorithm

There are adaptation equations that are simpler to implement, at the expense of a lower speed of convergence, for the same $J(\infty)$. Three versions are[9]

$$\boldsymbol{c}(k+1) = \boldsymbol{c}(k) + \mu \begin{cases} \operatorname{sgn}(e(k))\boldsymbol{x}^*(k) \\ e(k)\operatorname{sgn}(\boldsymbol{x}^*(k)) \\ \operatorname{sgn}(e(k))\operatorname{sgn}(\boldsymbol{x}^*(k)) \end{cases} \tag{3.119}$$

Note that the function $\operatorname{sgn}(e(k))$ never goes to zero, hence μ must be very small to have good convergence properties. The first version has as objective the minimization of the cost function

$$J(k) = E[|e(k)|] \tag{3.120}$$

Normalized LMS

In the LMS algorithm, if some $\boldsymbol{x}(k)$ assume large values, the adaptation algorithm is affected by strong noise in the gradient. This problem can be overcome by choosing an adaptation gain μ of the type:

$$\mu = \frac{\tilde{\mu}}{p + \hat{\mathtt{M}}_x(k)} \tag{3.121}$$

where $0 < \tilde{\mu} < 2$, and

$$\hat{\mathtt{M}}_x(k) = \|\boldsymbol{x}\|^2 = \sum_{i=0}^{N-1} |x(k-i)|^2 \tag{3.122}$$

---

[9] The sign of a complex-valued vector is defined as

$$\operatorname{sgn}(\boldsymbol{x}(k)) = [\operatorname{sgn}(x_I(k)) + j\operatorname{sgn}(x_Q(k)), \dots, \operatorname{sgn}(x_I(k-N+1)) + j\operatorname{sgn}(x_Q(k-N+1))] \tag{3.118}$$

or, alternatively,

$$\hat{\mathtt{M}}_{\mathbf{x}}(k) = N\hat{\mathtt{M}}_x(k) \tag{3.123}$$

where $\hat{\mathtt{M}}_x(k)$ is the estimate of the statistical power of $x(k)$. A simple estimate is obtained by the iterative equation (see (1.395)):

$$\hat{\mathtt{M}}_x(k) = a\hat{\mathtt{M}}_x(k-1) + (1-a)|x(k)|^2 \tag{3.124}$$

where $0 < a < 1$, with time constant given by

$$\tau = \frac{-1}{\ln a} \simeq \frac{1}{1-a} \tag{3.125}$$

for $a \simeq 1$. In (3.121), p is a positive parameter that is introduced to avoid that the denominator becomes too small; typically

$$p \simeq \frac{1}{10}\mathtt{M}_x \tag{3.126}$$

The normalized LMS algorithm has a speed of convergence that is potentially higher than the standard algorithm, for uncorrelated as well as correlated input signals [9]. To be able to apply the normalized algorithm, however, some knowledge of the input process is necessary, in order to assign the values of $\mathtt{M}_x$ and p so that the adaptation process does not become unstable.

Variable adaptation gain

In the following variants of the LMS algorithms, the coefficient μ varies with time.

1. *Two values of* μ:
   a. Initially, a large value of μ is chosen for fast convergence, for example $\mu = 1/(N\mathtt{r}_x(0))$.
   b. Subsequently, μ is reduced to achieve a smaller $J(\infty)$.

   For a choice of μ of the type

$$\mu = \begin{cases} \mu_1 & 0 \le k \le K_1 \\ \mu_2 & k \ge K_1 \end{cases} \tag{3.127}$$

   the behaviour of J is illustrated in Figure 3.10.

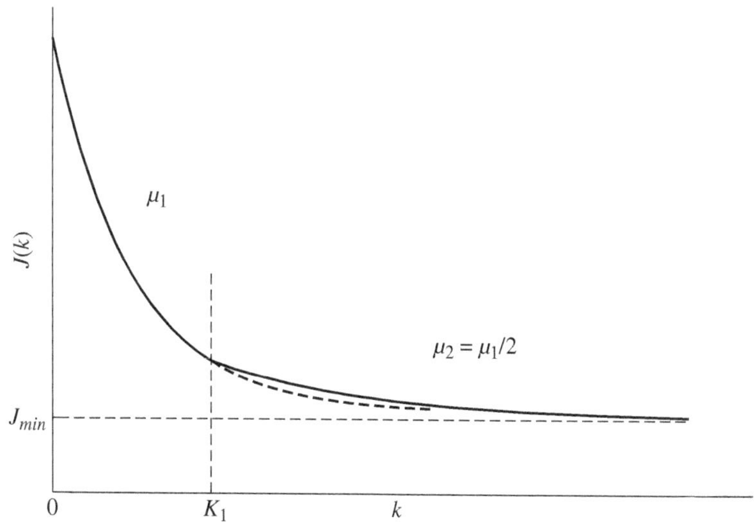

Figure 3.10 Behaviour of $J(k)$ obtained by using two values of μ.

2. *Decreasing* μ: For a time-invariant system, the adaptation gain usually selected for application with the sign algorithm (3.119) is given by

$$\mu(k) = \frac{\mu_1}{\mu_2 + k}, \qquad k \geq 0 \tag{3.128}$$

3. μ *proportional to* $e(k)$: The following expression of μ is used:

$$\mu(k+1) = \alpha_1 \mu(k) + \alpha_2 |e(k)|^2 \tag{3.129}$$

with μ limited to the range $[\mu_{min}, \mu_{max}]$. Typical values are $\alpha_1 \simeq 1$ and $\alpha_2 \ll 1$.

4. *Vector of values of* μ: Let $\boldsymbol{\mu}^T = [\mu_0, \dots, \mu_{N-1}]$; two approaches are possible.
   a. Initially, larger values μ_i are chosen in correspondence of those coefficients c_i that have larger amplitude.
   b. μ_i changes with time following the rule

$$\mu_i(k+1) = \begin{cases} \mu_i(k)\dfrac{1}{\alpha} & \text{if the } i\text{-th component of the gradient } \textit{has always} \text{ changed sign in the} \\ & \text{last } m_0 \text{ iterations,} \\ \mu_i(k)\alpha & \text{if the } i\text{-th component of the gradient } \textit{has never} \text{ changed sign in the} \\ & \text{last } m_1 \text{ iterations,} \end{cases} \tag{3.130}$$

with μ limited to the range $[\mu_{min}, \mu_{max}]$. Typical values are $m_0, m_1 \in \{1, 3\}$ and $\alpha = 2$.

3.1.5 Example of application: the predictor

We consider a real AR(2) process of unit power, described by the equation

$$x(k) = -\mathtt{a}_1 x(k-1) - \mathtt{a}_2 x(k-2) + w(k) \tag{3.131}$$

with w additive white Gaussian noise (AWGN), and

$$\mathtt{a}_1 = -1.3, \qquad \mathtt{a}_2 = 0.995 \tag{3.132}$$

From (1.474), the roots of $A(z)$ are given by $\varrho e^{\pm\varphi_0}$, where

$$\varrho = \sqrt{\mathtt{a}_2} = 0.997 \tag{3.133}$$

and

$$\varphi_0 = \cos^{-1}\left(-\frac{\mathtt{a}_1}{2\varrho}\right) = 2.28 \text{ rad} \tag{3.134}$$

Being $\mathtt{r}_x(0) = \sigma_x^2 = 1$, from the (1.479) we have that the statistical power of w is given by

$$\sigma_w^2 = \frac{1-\mathtt{a}_2}{1+\mathtt{a}_2}[(1+\mathtt{a}_2)^2 - \mathtt{a}_1^2] = 0.0057 = -22.4 \text{ dB} \tag{3.135}$$

We construct a predictor for x of order $N = 2$ with coefficients $\boldsymbol{c}^T = [c_1, c_2]$, as illustrated in Figure 3.11, using the LMS algorithm and some of its variants [10]. From (2.78), we expect to find in steady state

$$\boldsymbol{c} \simeq -\boldsymbol{a} \tag{3.136}$$

that is $c_1 \simeq -\mathtt{a}_1$, $c_2 \simeq -\mathtt{a}_2$, and $\sigma_e^2 \simeq \sigma_w^2$. In any case, the predictor output is given by

$$\begin{aligned} y(k) &= \boldsymbol{c}^T(k)\boldsymbol{x}(k-1) \\ &= c_1(k)x(k-1) + c_2(k)\,x(k-2) \end{aligned} \tag{3.137}$$

with prediction error

$$e(k) = x(k) - y(k) \tag{3.138}$$

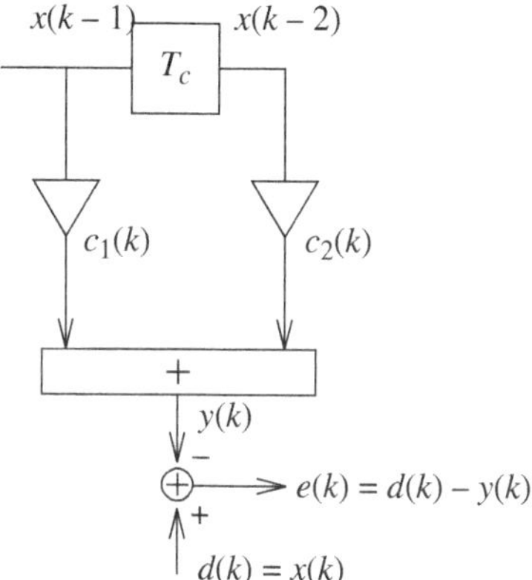

Figure 3.11 Predictor of order $N = 2$.

For the predictor of Figure 3.11, we consider now various versions of the adaptive LMS algorithm and their relative performance.

Example 3.1.1 (Standard LMS)
The equation for updating the coefficient vector is

$$\boldsymbol{c}(k+1) = \boldsymbol{c}(k) + \mu e(k)\boldsymbol{x}(k-1) \tag{3.139}$$

Convergence curves are plotted in Figure 3.12 for a single realization and for the mean (estimated over 500 realizations) of the coefficients and of the square prediction error, for $\mu = 0.04$.

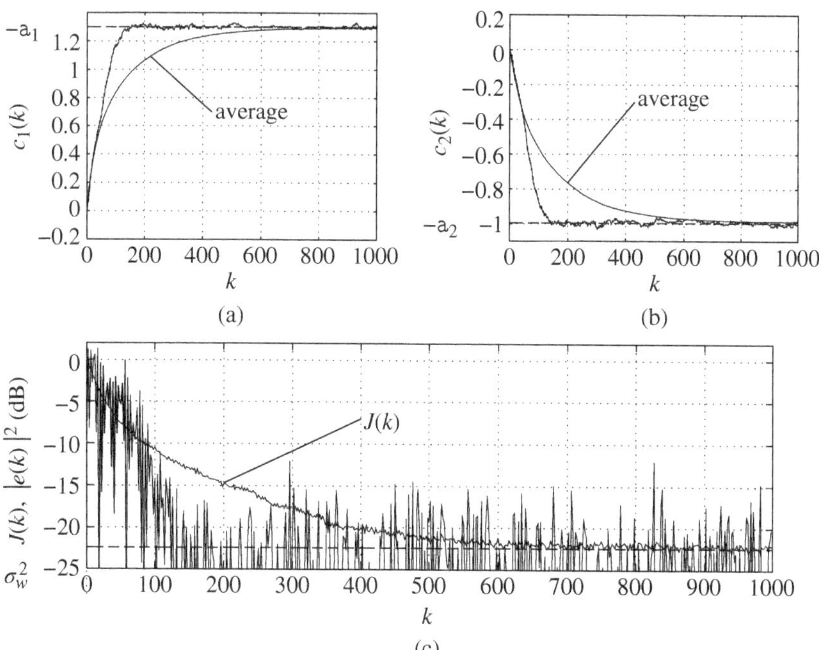

Figure 3.12 Convergence curves for the predictor of order $N = 2$, obtained by the standard LMS algorithm. (a) $c_1(k)$, (b) $c_2(k)$, (c) $J(k)$ and $|e(k)|^2$, all versus k.

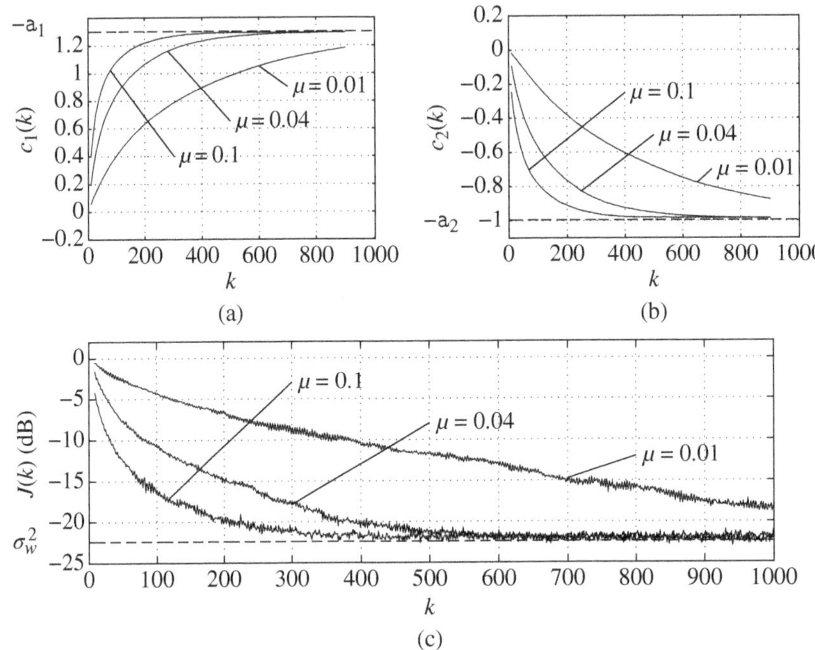

Figure 3.13 Comparison among curves of convergence of the mean obtained by the standard LMS algorithm for three values of μ. (a) $c_1(k)$, (b) $c_2(k)$, (c) $J(k)$, all versus k.

In Figure 3.13, a comparison is made among the curves of convergence of the mean for three values of μ. We observe that, by decreasing μ, the excess error decreases, thus giving a more accurate solution, but the convergence time increases.

Example 3.1.2 (Leaky LMS)
The equation for updating the coefficient vector is

$$\boldsymbol{c}(k+1) = (1 - \mu\alpha)\boldsymbol{c}(k) + \mu e(k)\boldsymbol{x}(k-1) \tag{3.140}$$

Convergence curves are plotted in Figure 3.14 for a single realization and for the mean (estimated over 500 realizations) of the coefficients and of the square prediction error, for $\mu = 0.04$ and $\alpha = 0.01$. We note that the steady-state values are worse than in the previous case.

Example 3.1.3 (Normalized LMS)
The equation for updating the coefficient vector is (3.139) with adaptation gain μ is of the type

$$\mu(k) = \frac{\tilde{\mu}}{p + N\hat{\sigma}_x^2(k)} \tag{3.141}$$

where

$$\hat{\sigma}_x^2(k) = a\hat{\sigma}_x^2(k-1) + (1-a)|x(k)|^2, s \qquad k \geq 0 \tag{3.142}$$

$$\hat{\sigma}_x^2(-1) = \frac{1}{2}[|x(-1)|^2 + |x(-2)|^2] \tag{3.143}$$

with $a = 1 - 2^{-5} = 0.97$ and

$$p = \frac{1}{10}E[\|\boldsymbol{x}\|^2] = 0.2 \tag{3.144}$$

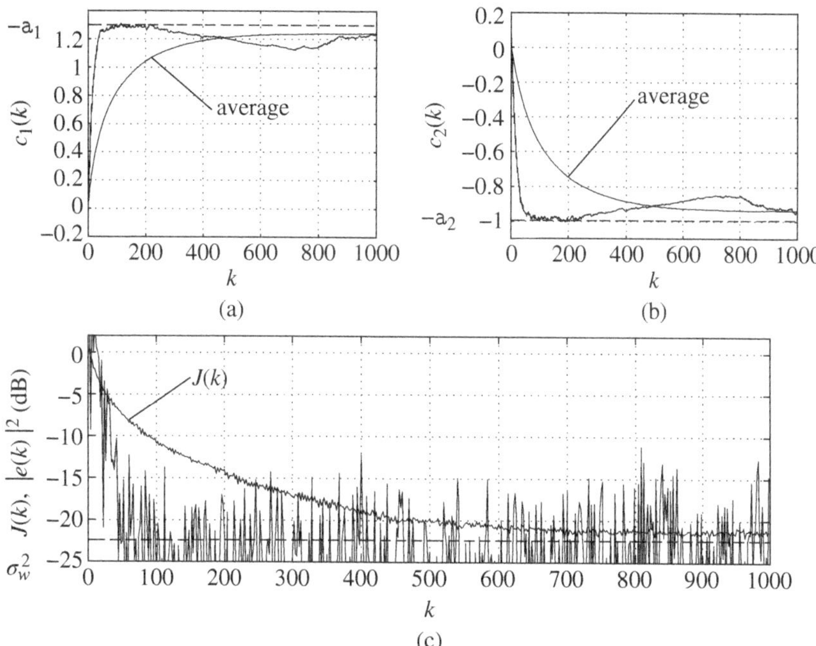

Figure 3.14 Convergence curves for the predictor of order $N = 2$, obtained by the leaky LMS. (a) $c_1(k)$, (b) $c_2(k)$, (c) $J(k)$ and $|e(k)|^2$, all versus k.

Convergence curves are plotted in Figure 3.15 for a single realization and for the mean (estimated over 500 realizations) of the coefficients and of the square prediction error, for $\tilde{\mu} = 0.08$. We note that, with respect to the standard LMS algorithm, the convergence is considerably faster.

A direct comparison of the convergence curves obtained in the previous examples is given in Figure 3.16.

Example 3.1.4 (Sign LMS algorithm)
We consider the three versions of the sign LMS algorithm:

(1) $\boldsymbol{c}(k+1) = \boldsymbol{c}(k) + \mu\,\text{sgn}(e(k))\boldsymbol{x}(k-1)$,
(2) $\boldsymbol{c}(k+1) = \boldsymbol{c}(k) + \mu e(k)\,\text{sgn}(\boldsymbol{x}(k-1))$,
(3) $\boldsymbol{c}(k+1) = \boldsymbol{c}(k) + \mu\,\text{sgn}(e(k))\,\text{sgn}(\boldsymbol{x}(k-1))$.

A comparison of convergence curves is given in Figure 3.17 for the three versions of the sign LMS algorithm, for $\mu = 0.04$.

It turns out that version (2), where the estimation error in the adaptation equation is not quantized, yields the best performance in steady state. Version (3), however, yields fastest convergence. To decrease the prediction error in steady state for versions (1) and (3), the value of μ could be further lowered, at the expense of reducing the speed of convergence.

Observation 3.2
As observed on page 74, for an AR process x, if the order of the predictor is greater than the required minimum, the correlation matrix results ill-conditioned with a large eigenvalue spread. Thus, the convergence of the LMS prediction algorithm may turn out to be extremely slow and can lead to a solution

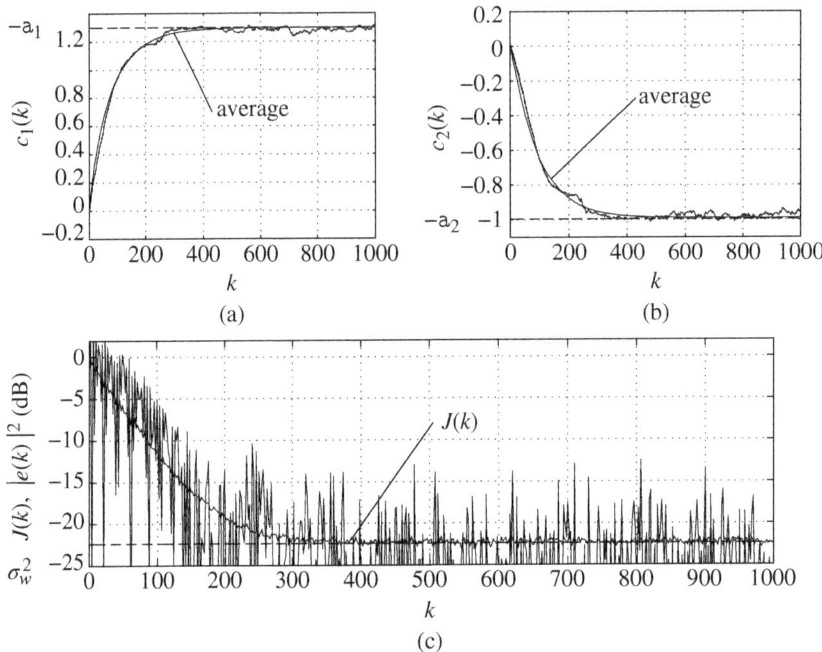

Figure 3.15 Convergence curves for the predictor of order $N = 2$, obtained by the normalized LMS algorithm. (a) $c_1(k)$, (b) $c_2(k)$, (c) $J(k)$ and $|e(k)|^2$, all versus k.

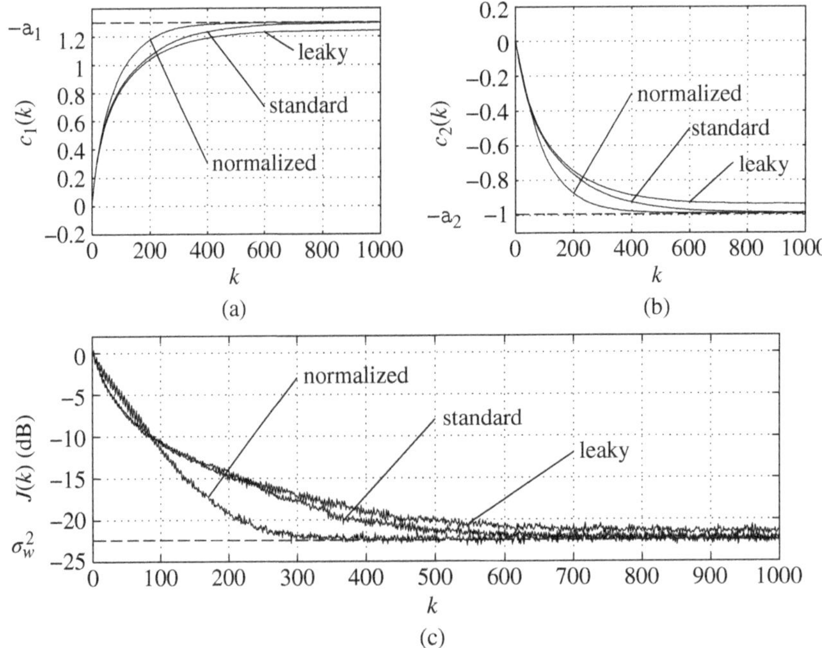

Figure 3.16 Comparison of convergence curves for the predictor of order $N = 2$, obtained by three versions of the LMS algorithm. (a) $c_1(k)$, (b) $c_2(k)$, (c) $J(k)$, all versus k.

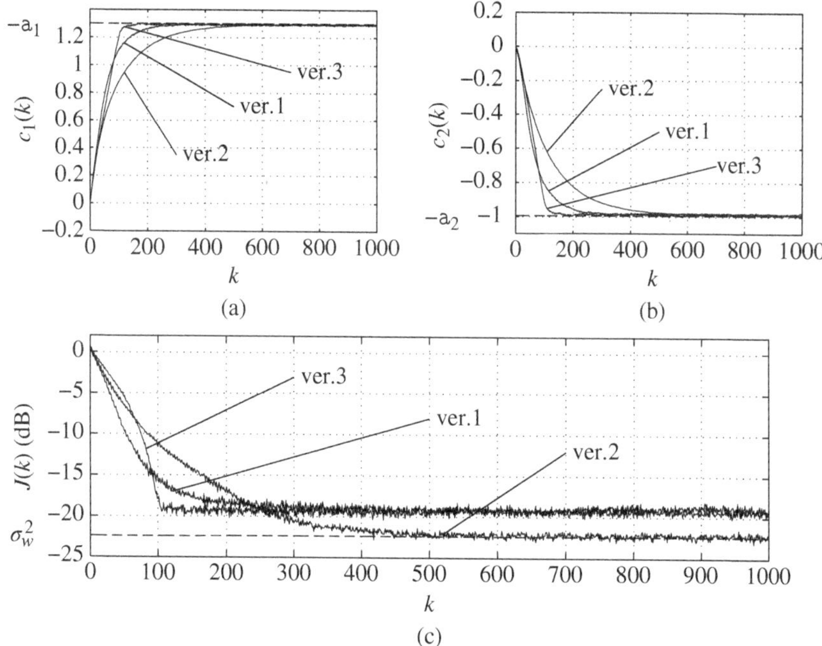

Figure 3.17 Comparison of convergence curves obtained by three versions of the sign LMS algorithm. (a) $c_1(k)$, (b) $c_2(k)$, (c) $J(k)$, all versus k.

quite different from the Yule–Walker solution. In this case, it is necessary to adopt a method that ensures the stability of the error prediction filter, such as the leaky LMS.

3.2 The recursive least squares algorithm

We now consider a recursive algorithm to estimate the vector of coefficients $\boldsymbol{c}$ by an LS method, named *recursive least squares* (RLS) algorithm. Note that this algorithm is an implementation of the Kalman filter algorithm, for a non-evolving state. The RLS algorithm is characterized by a speed of convergence that can be one order of magnitude faster than the LMS algorithm, obtained at the expense of a larger computational complexity.

With reference to the system illustrated in Figure 3.18, we introduce the following quantities:

1. Input vector at instant i

$$\boldsymbol{x}^T(i) = [x(i), x(i-1), \ldots, x(i-N+1)] \tag{3.145}$$

2. Coefficient vector at instant k

$$\boldsymbol{c}^T(k) = [c_0(k), c_1(k), \ldots, c_{N-1}(k)] \tag{3.146}$$

3. Filter output signal at instant i, obtained for the vector of coefficients $\boldsymbol{c}(k)$

$$y(i) = \boldsymbol{c}^T(k)\boldsymbol{x}(i) = \boldsymbol{x}^T(i)\boldsymbol{c}(k) \tag{3.147}$$

4. Desired output at instant i

$$d(i) \tag{3.148}$$

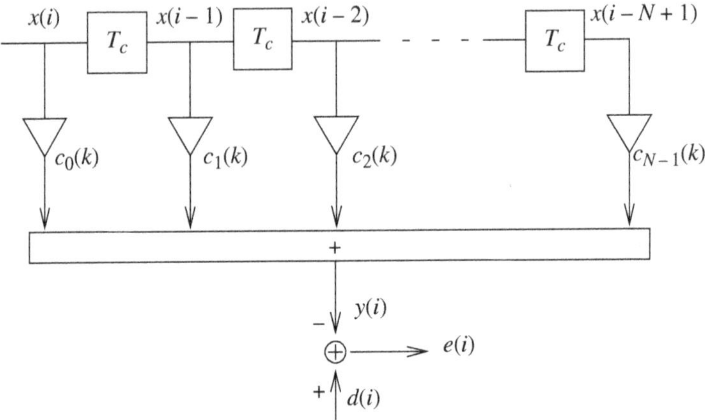

Figure 3.18 Reference system for a RLS adaptive algorithm.

At instant k, based on the observation of the sequences

$$\{x(i)\}\{d(i)\}, \qquad i = 1, 2, \dots, k \tag{3.149}$$

the criterion for the optimization of the vector of coefficients $\boldsymbol{c}(k)$ is the *minimum sum of square errors up to instant* k. Defining

$$\mathcal{E}(k) = \sum_{i=1}^{k} \lambda^{k-i} |e(i)|^2 \tag{3.150}$$

we want to find

$$\min_{\boldsymbol{c}(k)} \mathcal{E}(k) \tag{3.151}$$

where the error signal is $e(i) = d(i) - \boldsymbol{x}^T(i)\boldsymbol{c}(k)$.

Two observations arise:

- λ is a forgetting factor, that enables proper filtering operations even with non-stationary signals or slowly time-varying systems. The memory of the algorithm is approximately $1/(1 - \lambda)$.
- This problem is the classical LS problem (2.92), applied to a sequence of pre-windowed samples with the exponential weighting factor λ^k.

Normal equation

Using the gradient method, the optimum value of $\boldsymbol{c}(k)$ satisfies the *normal equation*

$$\boldsymbol{\Phi}(k)\boldsymbol{c}(k) = \boldsymbol{\vartheta}(k) \tag{3.152}$$

where

$$\boldsymbol{\Phi}(k) = \sum_{i=1}^{k} \lambda^{k-i} \boldsymbol{x}^*(i)\boldsymbol{x}^T(i) \tag{3.153}$$

$$\boldsymbol{\vartheta}(k) = \sum_{i=1}^{k} \lambda^{k-i} d(i)\boldsymbol{x}^*(i) \tag{3.154}$$

From (3.152), if $\boldsymbol{\Phi}^{-1}(k)$ exists, the solution is given by

$$\boldsymbol{c}(k) = \boldsymbol{\Phi}^{-1}(k)\boldsymbol{\vartheta}(k) \tag{3.155}$$

Derivation

To solve the normal equation by the inversion of $\mathbf{\Phi}(k)$ may be too hard, especially if N is large. Therefore, we seek a recursive algorithm for $k = 1, 2, \ldots$.

Both expressions of $\mathbf{\Phi}(k)$ and $\boldsymbol{\vartheta}(k)$ can be written recursively. From

$$\mathbf{\Phi}(k) = \sum_{i=1}^{k-1} \lambda^{k-i} \mathbf{x}^*(i)\mathbf{x}^T(i) + \mathbf{x}^*(k)\mathbf{x}^T(k) \tag{3.156}$$

it follows that

$$\mathbf{\Phi}(k) = \lambda\mathbf{\Phi}(k-1) + \mathbf{x}^*(k)\mathbf{x}^T(k) \tag{3.157}$$

and similarly

$$\boldsymbol{\vartheta}(k) = \lambda\boldsymbol{\vartheta}(k-1) + d(k)\mathbf{x}^*(k) \tag{3.158}$$

We now recall the following identity known as *matrix inversion lemma* [11]. Let

$$\mathbf{A} = \mathbf{B}^{-1} + \mathbf{C}\mathbf{D}^{-1}\mathbf{C}^H \tag{3.159}$$

where $\mathbf{A}$, $\mathbf{B}$, and $\mathbf{D}$ are positive definite matrices. Then

$$\mathbf{A}^{-1} = \mathbf{B} - \mathbf{B}\mathbf{C}(\mathbf{D} + \mathbf{C}^H\mathbf{B}\mathbf{C})^{-1}\mathbf{C}^H\mathbf{B} \tag{3.160}$$

For

$$\mathbf{A} = \mathbf{\Phi}(k), \qquad \mathbf{B}^{-1} = \lambda\mathbf{\Phi}(k-1), \qquad \mathbf{C} = \mathbf{x}^*(k), \qquad \mathbf{D} = [1] \tag{3.161}$$

Eq. (3.160) becomes

$$\mathbf{\Phi}^{-1}(k) = \lambda^{-1}\mathbf{\Phi}^{-1}(k-1) - \frac{\lambda^{-1}\mathbf{\Phi}^{-1}(k-1)\mathbf{x}^*(k)\mathbf{x}^T(k)\lambda^{-1}\mathbf{\Phi}^{-1}(k-1)}{1 + \mathbf{x}^T(k)\lambda^{-1}\mathbf{\Phi}^{-1}(k-1)\mathbf{x}^*(k)} \tag{3.162}$$

We introduce two quantities:

$$\mathbf{P}(k) = \mathbf{\Phi}^{-1}(k) \tag{3.163}$$

and

$$\mathbf{k}^*(k) = \frac{\lambda^{-1}\mathbf{\Phi}^{-1}(k-1)\mathbf{x}^*(k)}{1 + \lambda^{-1}\mathbf{x}^T(k)\mathbf{\Phi}^{-1}(k-1)\mathbf{x}^*(k)} \tag{3.164}$$

also called the *Kalman vector gain*. From (3.162), we have the recursive relation

$$\mathbf{P}(k) = \lambda^{-1}\mathbf{P}(k-1) - \lambda^{-1}\mathbf{k}^*(k)\mathbf{x}^T(k)\mathbf{P}(k-1) \tag{3.165}$$

We derive now a simpler expression for $\mathbf{k}^*(k)$. From (3.164), we obtain

$$\mathbf{k}^*(k)[1 + \lambda^{-1}\mathbf{x}^T(k)\mathbf{\Phi}^{-1}(k-1)\mathbf{x}^*(k)] = \lambda^{-1}\mathbf{\Phi}^{-1}(k-1)\mathbf{x}^*(k) \tag{3.166}$$

from which we get

$$\begin{aligned} \mathbf{k}^*(k) &= \lambda^{-1}\mathbf{P}(k-1)\mathbf{x}^*(k) - \lambda^{-1}\mathbf{k}^*(k)\mathbf{x}^T(k)\mathbf{P}(k-1)\mathbf{x}^*(k) \\ &= [\lambda^{-1}\mathbf{P}(k-1) - \lambda^{-1}\mathbf{k}^*(k)\mathbf{x}^T(k)\mathbf{P}(k-1)]\mathbf{x}^*(k) \end{aligned} \tag{3.167}$$

Using (3.165), it follows

$$\mathbf{k}^*(k) = \mathbf{P}(k)\mathbf{x}^*(k) \tag{3.168}$$

Using (3.158), the recursive equation to update the estimate of $\mathbf{c}$ is given by

$$\begin{aligned} \mathbf{c}(k) &= \mathbf{\Phi}^{-1}(k)\boldsymbol{\vartheta}(k) \\ &= \mathbf{P}(k)\boldsymbol{\vartheta}(k) \\ &= \lambda\mathbf{P}(k)\boldsymbol{\vartheta}(k-1) + \mathbf{P}(k)\mathbf{x}^*(k)d(k) \end{aligned} \tag{3.169}$$

Substituting the recursive expression for $\boldsymbol{P}(k)$ in the first term, we get

$$\begin{aligned}\boldsymbol{c}(k) &= \lambda[\lambda^{-1}\boldsymbol{P}(k-1) - \lambda^{-1}\boldsymbol{k}^*(k)\boldsymbol{x}^T(k)\boldsymbol{P}(k-1)]\boldsymbol{\vartheta}(k-1) + \boldsymbol{P}(k)\boldsymbol{x}^*(k)d(k) \\ &= \boldsymbol{P}(k-1)\boldsymbol{\vartheta}(k-1) - \boldsymbol{k}^*(k)\boldsymbol{x}^T(k)\boldsymbol{P}(k-1)\boldsymbol{\vartheta}(k-1) + \boldsymbol{P}(k)\boldsymbol{x}^*(k)d(k) \\ &= \boldsymbol{c}(k-1) + \boldsymbol{k}^*(k)[d(k) - \boldsymbol{x}^T(k)\boldsymbol{c}(k-1)]\end{aligned} \tag{3.170}$$

where in the last step (3.168) has been used. Defining the *a priori estimation error*,

$$\epsilon(k) = d(k) - \boldsymbol{x}^T(k)\boldsymbol{c}(k-1) \tag{3.171}$$

we note that $\boldsymbol{x}^T(k)\boldsymbol{c}(k-1)$ is the filter output at instant k obtained by using the old coefficient estimate. In other words, from the *a posteriori estimation error*

$$e(k) = d(k) - \boldsymbol{x}^T(k)\boldsymbol{c}(k) \tag{3.172}$$

we could say that $\epsilon(k)$ is an approximated value of $e(k)$, that is computed before updating $\boldsymbol{c}$. In any case, the following relation holds

$$\boldsymbol{c}(k) = \boldsymbol{c}(k-1) + \boldsymbol{k}^*(k)\epsilon(k) \tag{3.173}$$

In summary, the RLS algorithm consists of four equations:

$$\boldsymbol{k}^*(k) = \frac{\boldsymbol{P}(k-1)\boldsymbol{x}^*(k)}{\lambda + \boldsymbol{x}^T(k)\boldsymbol{P}(k-1)\boldsymbol{x}^*(k)} \tag{3.174}$$

$$\epsilon(k) = d(k) - \boldsymbol{x}^T(k)\boldsymbol{c}(k-1) \tag{3.175}$$

$$\boldsymbol{c}(k) = \boldsymbol{c}(k-1) + \epsilon(k)\boldsymbol{k}^*(k) \tag{3.176}$$

$$\boldsymbol{P}(k) = \lambda^{-1}\boldsymbol{P}(k-1) - \lambda^{-1}\boldsymbol{k}^*(k)\boldsymbol{x}^T(k)\boldsymbol{P}(k-1) \tag{3.177}$$

In (3.174), $\boldsymbol{k}(k)$ is the input vector filtered by $\boldsymbol{P}(k-1)$ and normalized by the $\lambda + \boldsymbol{x}^T(k)\boldsymbol{P}(k-1)\boldsymbol{x}^*(k)$. The term $\boldsymbol{x}^T(k)\boldsymbol{P}(k-1)\boldsymbol{x}^*(k)$ may be interpreted as the energy of the filtered input.

Initialization

We need to assign a value to $\boldsymbol{P}(0)$. We modify the definition of $\boldsymbol{\Phi}(k)$ as

$$\boldsymbol{\Phi}(k) = \sum_{i=1}^{k} \lambda^{k-i}\boldsymbol{x}^*(i)\boldsymbol{x}^T(i) + \delta\lambda^k\boldsymbol{I} \qquad \text{with } \delta \ll 1 \tag{3.178}$$

so that

$$\boldsymbol{\Phi}(0) = \delta\boldsymbol{I} \tag{3.179}$$

This is equivalent to having for $k \leq 0$ an all-zero input with the exception of $x(-N+1) = (\lambda^{-N+1}\delta)^{1/2}$, therefore

$$\boldsymbol{P}(0) = \delta^{-1}\boldsymbol{I} \qquad \delta \ll \mathtt{r}_x(0) \tag{3.180}$$

Typically, we have

$$\delta^{-1} = \frac{100}{\mathtt{r}_x(0)} \tag{3.181}$$

where $\mathtt{r}_x(0)$ is the statistical power of the input signal. Table 3.1 shows a version of the RLS algorithm that exploits the fact that $\boldsymbol{P}(k)$ (inverse of the Hermitian matrix $\boldsymbol{\Phi}(k)$) is Hermitian, hence

$$\boldsymbol{x}^T(k)\boldsymbol{P}(k-1) = [\boldsymbol{P}(k-1)\boldsymbol{x}^*(k)]^H = \boldsymbol{\pi}^T(k) \tag{3.182}$$

Table 3.1: The RLS algorithm.

Initialization	$\boldsymbol{c}(0) = \boldsymbol{0}$ $\boldsymbol{P}(0) = \delta^{-1}\boldsymbol{I}$
For $k = 1, 2, \ldots$	$\boldsymbol{\pi}^*(k) = \boldsymbol{P}(k-1)\boldsymbol{x}^*(k)$ $r(k) = \dfrac{1}{\lambda + \boldsymbol{x}^T(k)\boldsymbol{\pi}^*(k)}$ $\boldsymbol{k}^*(k) = r(k)\boldsymbol{\pi}^*(k)$ $\epsilon(k) = d(k) - \boldsymbol{x}^T(k)\boldsymbol{c}(k-1)$ $\boldsymbol{c}(k) = \boldsymbol{c}(k-1) + \epsilon(k)\boldsymbol{k}^*(k)$ $\boldsymbol{P}(k) = \lambda^{-1}(\boldsymbol{P}(k-1) - \boldsymbol{k}^*(k)\boldsymbol{\pi}^T(k))$

Recursive form of the minimum cost function

We set

$$\mathcal{E}_d(k) = \sum_{i=1}^{k} \lambda^{k-i}|d(i)|^2 = \lambda\mathcal{E}_d(k-1) + |d(k)|^2 \tag{3.183}$$

From the general LS expression (2.114),

$$\mathcal{E}_{min}(k) = \mathcal{E}_d(k) - \boldsymbol{\vartheta}^H(k)\boldsymbol{c}(k) \tag{3.184}$$

and from (3.158) and (3.176) we get

$$\begin{aligned}\mathcal{E}_{min}(k) &= \lambda\mathcal{E}_d(k-1) + |d(k)|^2 \\ &\quad - [\lambda\boldsymbol{\vartheta}^H(k-1) + \boldsymbol{x}^T(k)d^*(k)][\boldsymbol{c}(k-1) + \epsilon(k)\boldsymbol{k}^*(k)] \\ &= \lambda\mathcal{E}_d(k-1) - \lambda\boldsymbol{\vartheta}^H(k-1)\boldsymbol{c}(k-1) \\ &\quad + d(k)d^*(k) - d^*(k)\boldsymbol{x}^T(k)\boldsymbol{c}(k-1) - \boldsymbol{\vartheta}^H(k)\boldsymbol{k}^*(k)\epsilon(k) \\ &= \lambda\mathcal{E}_{min}(k-1) + d^*(k)\epsilon(k) - \boldsymbol{\vartheta}^H(k)\boldsymbol{k}^*(k)\epsilon(k)\end{aligned} \tag{3.185}$$

Using (3.163), and recalling that $\boldsymbol{\Phi}(k)$ is Hermitian, from (3.168) we obtain

$$\begin{aligned}\boldsymbol{\vartheta}^H(k)\boldsymbol{k}^*(k) &= \boldsymbol{\vartheta}^H(k)\boldsymbol{\Phi}^{-1}(k)\boldsymbol{x}^*(k) \\ &= [\boldsymbol{\Phi}^{-1}(k)\boldsymbol{\vartheta}(k)]^H\boldsymbol{x}^*(k)\end{aligned} \tag{3.186}$$

Moreover from (3.168) and (3.155), it follows that

$$\boldsymbol{\vartheta}^H(k)\boldsymbol{k}^*(k) = \boldsymbol{c}^H(k)\boldsymbol{x}^*(k) = \boldsymbol{x}^H(k)\boldsymbol{c}^*(k) \tag{3.187}$$

Then, (3.185) becomes

$$\begin{aligned}\mathcal{E}_{min}(k) &= \lambda\mathcal{E}_{min}(k-1) + d^*(k)\epsilon(k) - \boldsymbol{c}^H(k)\boldsymbol{x}^*(k)\epsilon(k) \\ &= \lambda\mathcal{E}_{min}(k-1) + \epsilon(k)[d^*(k) - (\boldsymbol{x}^T(k)\boldsymbol{c}(k))^*]\end{aligned} \tag{3.188}$$

Finally, the recursive relation is given by

$$\mathcal{E}_{min}(k) = \lambda\mathcal{E}_{min}(k-1) + \epsilon(k)e^*(k) \tag{3.189}$$

We note that, as $\mathcal{E}_{min}(k)$ is real, we get

$$\epsilon(k)e^*(k) = \epsilon^*(k)e(k) \tag{3.190}$$

that is $\epsilon(k)e^*(k)$ is a real scalar value.

Convergence

We make some remarks on the convergence of the RLS algorithm.

- The RLS algorithm converges in the mean-square sense in about $2N$ iterations, independently of the eigenvalue spread of $\boldsymbol{R}$.
- For $k \to \infty$, there is no excess error and the misadjustment MSD is zero. This is true for $\lambda = 1$.
- In any case, when $\lambda < 1$ the *memory* of the algorithm is approximately $1/(1-\lambda)$ and

$$\text{MSD} = \frac{1-\lambda}{1+\lambda} N \tag{3.191}$$

- From the above observation, it follows that the RLS algorithm for $\lambda < 1$ gives origin to noisy estimates.
- On the other hand, the RLS algorithm for $\lambda < 1$ can be used for tracking slowly time-varying systems.

Computational complexity

Exploiting the symmetry of $\boldsymbol{P}(k)$, the computational complexity of the RLS algorithm, expressed as the number of complex multiplications per output sample, is given by

$$\text{CC}_{RLS} = 2N^2 + 4N \tag{3.192}$$

For a number of $(K - N + 1)$ output samples, the direct method (3.155) requires instead

$$\text{CC}_{DIR} = N^2 + N + \frac{N^3}{K-N+1} \tag{3.193}$$

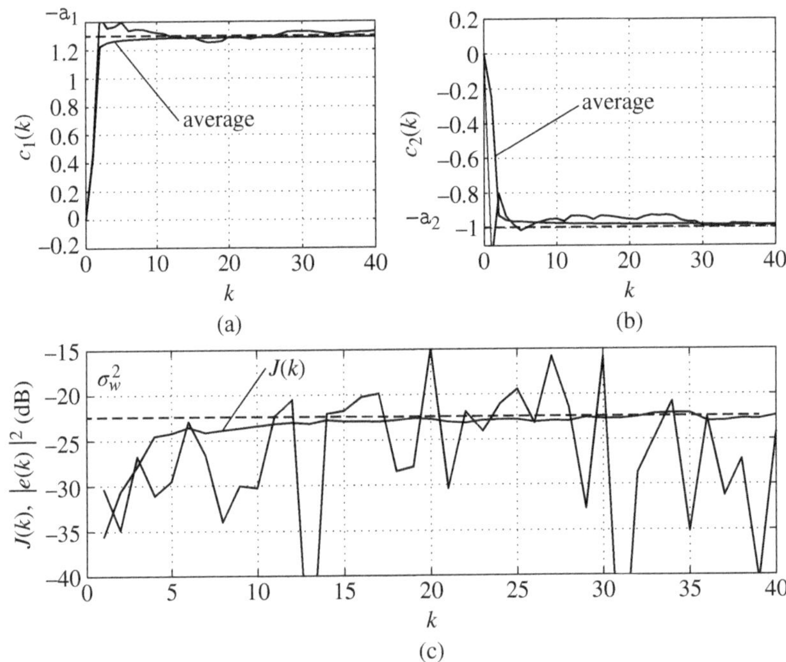

Figure 3.19 Convergence curves for the predictor of order $N = 2$, obtained by the RLS algorithm. (a) $c_1(k)$, (b) $c_2(k)$, (c) $J(k)$ and $|e(k)|^2$, all versus k.

We note that, if $K \gg N$, the direct method is more convenient. In any case, the RLS solution has other advantages:

1. it can be numerically more stable than the direct method;
2. it provides an estimate of the coefficients at each step and not only at the end of the data sequence;
3. for $\lambda < 1$ and $1/(1-\lambda)$ much smaller than the time interval it takes for the input samples to change statistics, the algorithm is capable of *tracking* the changes.

Example of application: the predictor

With reference to the AR(2) process considered in Section 3.1.5, convergence curves for the RLS algorithm are plotted in Figure 3.19 for a single realization and for the mean (estimated over 500 realizations) of the coefficients and of the square estimation error, for $\lambda = 1$. We note that a different scale is used for the abscissa as compared to the LMS method; in fact the RLS algorithm converges in a number of iterations of the order of N.

3.3 Fast recursive algorithms

As observed in the Section 3.2, the RLS algorithm has the disadvantage of requiring $(2N^2 + 4N)$ multiplications per iteration. Therefore, we will list a few *fast* algorithms, whose computational complexity increases linearly with N, the length of the coefficient vector $\boldsymbol{c}$.

1. *Algorithms for transversal filters*: The *fast Kalman algorithm* has the same speed of convergence as the RLS, but with a computational complexity comparable to that of the LMS algorithm. Exploiting some properties of the correlation matrix $\boldsymbol{\Phi}(k)$, Falconer and Ljung [12] have shown that the recursive equation (3.177) requires only $10(2N+1)$ multiplications. Cioffi and Kailath [13], with their *fast transversal filter* (FTF), have further reduced the number of multiplications to $7(2N+1)$. The implementation of these algorithms still remains relatively simple; their weak point resides in the sensitivity of the operations to round off errors in the various coefficients and signals. As a consequence, the fast algorithms may become numerically unstable.
2. *Algorithms for transversal filters based on systolic structures*: A particular structure is the *QR-decomposition-based LSL*. The name comes from the use of an orthogonal triangularization process, usually known as QR decomposition, that leads to a systolic-type structure with the following characteristics:
   - high speed of convergence;
   - numerical stability, owing to the QR decomposition and lattice structure;
   - a very efficient and modular structure, which does not require the a priori knowledge of the filter order and is suitable for implementation in *very large-scale integration* (VLSI) technology.

   For further study on the subject, we refer the reader to [14–19].

3.3.1 Comparison of the various algorithms

In practice, the choice of an algorithm must be made bearing in mind some fundamental aspects:

- computational complexity;
- performance in terms of speed of convergence, error in steady state, and tracking capabilities under non-stationary conditions;
- robustness, that is good performance achieved in the presence of a large eigenvalue spread and finite-precision arithmetic [5, 20].

Table 3.2: Comparison of three adaptive algorithms in terms of computational complexity.

cost function	*algorithm*	*multiplications*	*divisions*	*additions subtractions*
MSE	LMS	$2N+1$	0	$2N$
LS	RLS	$2N^2+7N+5$	N^2+4N+3	$2N^2+6N+4$
	FTF	$7(2N+1)$	4	$6(2N+1)$

Regarding the computational complexity per output sample, a brief comparison among LMS, RLS, and FTF is given in Table 3.2. Although the FTF method exhibits a lower computational complexity than the RLS method, its implementation is rather laborious, therefore it is rarely used.

3.4 Examples of application

This section re-examines two examples of Section 2.5 from an adaptive (LMS) point of view.

3.4.1 Identification of a linear discrete-time system

From Figure 2.12 and definition of $d(k)$, $e(k)$, and $\boldsymbol{x}(k)$, the LMS adaptation equation follows,

$$\boldsymbol{c}(k+1) = \boldsymbol{c}(k) + \mu e(k)\boldsymbol{x}^*(k) \tag{3.194}$$

From (2.235), we see that the noise w does not affect the solution $\boldsymbol{c}_{opt}$, consequently the expectation of (3.194) for $k \to \infty$ (equal to $\boldsymbol{c}_{opt}$) is also not affected by w. Anyway, as seen in Section 3.1.3, the noise influences the convergence process and the solution obtained by the adaptive LMS algorithm. The larger the power of the noise, the smaller μ must be so that $\boldsymbol{c}(k)$ approaches $E[\boldsymbol{c}(k)]$. In any case, $J(\infty) \neq 0$.

As the input x is white, the convergence behaviour of the LMS algorithm (3.194) is easily determined. Let γ be defined as in (3.68):

$$\gamma = 1 + \mathtt{r}_x(0)(\mu^2 N \mathtt{r}_x(0) - 2\mu)$$

Let $\boldsymbol{\Delta c}(k) = \boldsymbol{c}(k) - \boldsymbol{c}_{opt}$; then we get

$$J(k) = E[|e(k)|^2] = J_{min} + \mathtt{r}_x(0) E[\|\boldsymbol{\Delta c}(k)\|^2] \tag{3.195}$$

where

$$E[\|\boldsymbol{\Delta c}(k)\|^2] = \gamma^k\, E[\|\boldsymbol{\Delta c}(0)\|^2] + \mu^2 N\, \mathtt{r}_x(0)\, J_{min}\, \frac{1-\gamma^k}{1-\gamma} \qquad k \geq 0 \tag{3.196}$$

The result (3.196) is obtained by (3.59) and the following assumptions:

1. $\boldsymbol{\Delta c}(k)$ is statistically independent of $\boldsymbol{x}(k)$;
2. $e_{min}(k)$ is orthogonal to $\boldsymbol{x}(k)$;
3. the approximation $\boldsymbol{x}^T(k)\, \boldsymbol{x}^*(k) \simeq N\, \mathtt{r}_x(0)$ holds.

Indeed, (3.196) is an extension of (3.67). At convergence, for $\mu\, \mathtt{r}_x(0) \ll 1$, it results

$$E[\|\boldsymbol{\Delta c}(\infty)\|^2] = \mu\, \frac{N}{2}\, J_{min} \tag{3.197}$$

and

$$J(\infty) = J_{min}\left(1 + \mu\, \mathtt{r}_x(0)\frac{N}{2}\right) \tag{3.198}$$

A faster convergence and a more accurate estimate, for fixed μ, is obtained by choosing a smaller value of N; this, however, may increase the residual estimation error (2.237).

Example 3.4.1

Consider an unknown system whose impulse response, given in Table 1.4 on page 12 as h_1, has energy equal to 1.06. AWGN is present with statistical power $\sigma_w^2 = 0.01$. Identification via standard LMS and RLS adaptive algorithms is obtained using as input a maximal-length pseudo-noise (PN) sequence of length $L = 31$ and unit power, $\mathtt{M}_x = 1$. For a filter with $N = 5$ coefficients, the convergence curves of the MSE (estimated over 500 realizations) are shown in Figure 3.20. For the LMS algorithm, $\mu = 0.1$ is chosen, which leads to a misadjustment equal to MSD = 0.26.

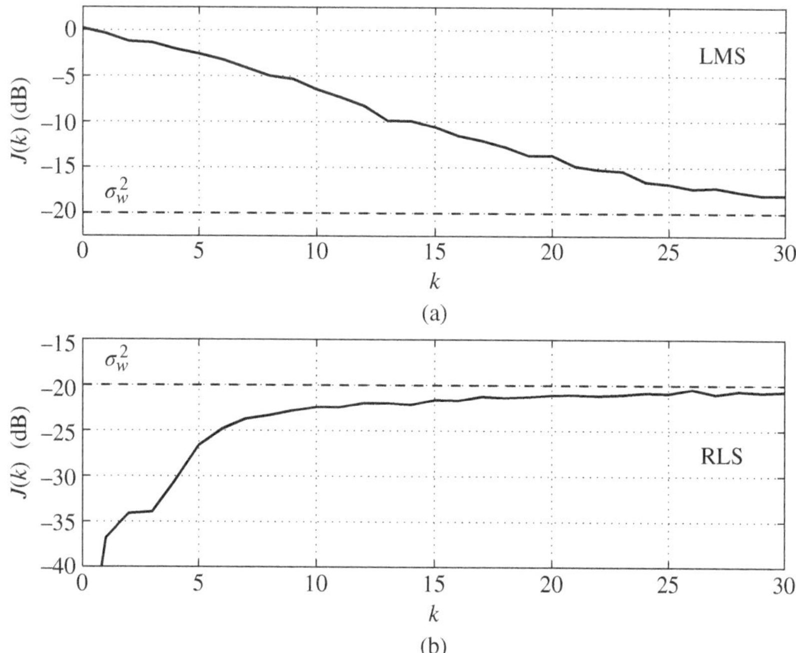

Figure 3.20 Convergence curves of the MSE for system identification using (a) LMS and (b) RLS.

As discussed in Section 7.8.3, as metric of the estimate quality we adopt the ratio:

$$\Lambda_n = \frac{\sigma_w^2}{E[\|\Delta h\|^2]} \tag{3.199}$$

where $\Delta h = c - h$ is the estimate error vector. At convergence, that is for $k = 30$ in our example, it results:

$$\Lambda_n = \begin{cases} 3.9 & \text{for LMS} \\ 7.8 & \text{for RLS} \end{cases} \tag{3.200}$$

We note that, even if the input signal is white, the RLS algorithm usually yields a better estimate than LMS. However, for systems with a large noise power and/or slow time-varying impulse responses, the two methods tend to give the same performance in terms of speed of convergence and error in steady state. As a result, it is usually preferable to adopt the LMS algorithm, as it leads to an easier implementation.

Finite alphabet case

Assume a more general, non-linear relation between the unknown system output z and input x, given by

$$z(k) = g[x(k), x(k-1), x(k-2)] = g(\boldsymbol{x}(k)) \tag{3.201}$$

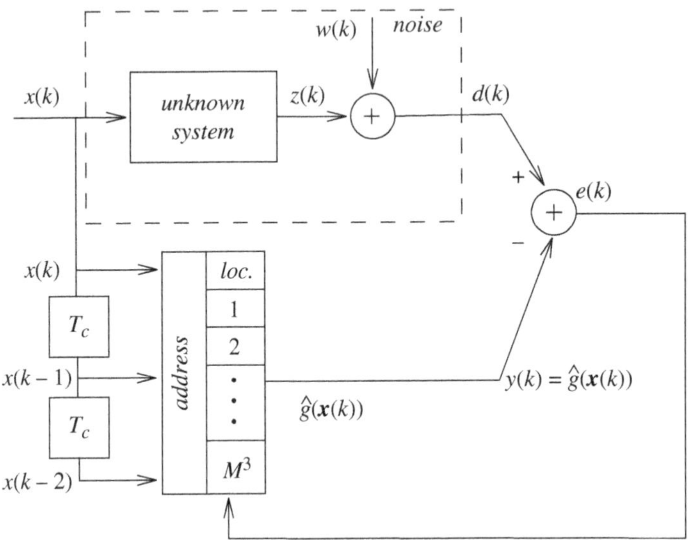

Figure 3.21 Adaptive scheme to estimate the input–output relation of a system.

where $x(i) \in \mathcal{A}$, finite alphabet with M elements. Then, z assumes values in an alphabet with at most M^3 values, which can be identified by a table or *random-access memory* (RAM) method, as illustrated in Figure 3.21. The cost function to be minimized is expressed as

$$E[|e(k)|^2] = E[|d(k) - \hat{g}(\boldsymbol{x}(k))|^2] \tag{3.202}$$

and the gradient estimate is given by

$$\nabla_{\hat{g}} |e(k)|^2 = -2e(k) \tag{3.203}$$

Therefore, the LMS adaptation equation becomes

$$\hat{g}(\boldsymbol{x}(k)) = \hat{g}(\boldsymbol{x}(k)) + \mu e(k) \tag{3.204}$$

In other words, the input vector $\boldsymbol{x}$ identifies a particular RAM location whose content is updated by adding a term proportional to the error. In the absence of noise, if the RAM is initialized to zero, the content of a memory location can be immediately identified by looking at the output. In practice, however, it is necessary to access each memory location several times to average out the noise. We note that, if the sequence $\{x(k)\}$ is i.i.d., $\boldsymbol{x}$ selects in the average each RAM location the same number of times.

An alternative method consists in setting $y(k) = 0$ during the entire time interval devoted to system identification, and to update the RAM as

$$\hat{g}(\boldsymbol{x}(k)) = \hat{g}(\boldsymbol{x}(k)) + d(k) \qquad k = 0, 1, \ldots \tag{3.205}$$

To complete the identification process, the value at each RAM location is scaled by the number of updates that have taken place for that location. This is equivalent to considering

$$\hat{g}(\boldsymbol{x}) = E[g(\boldsymbol{x}) + w] \tag{3.206}$$

We note that this method is a block version of the LMS algorithm with block length equal to the input sequence, where the RAM is initialized to zero, so that $e(k) = d(k)$, and μ is given by the relative frequency of each address.

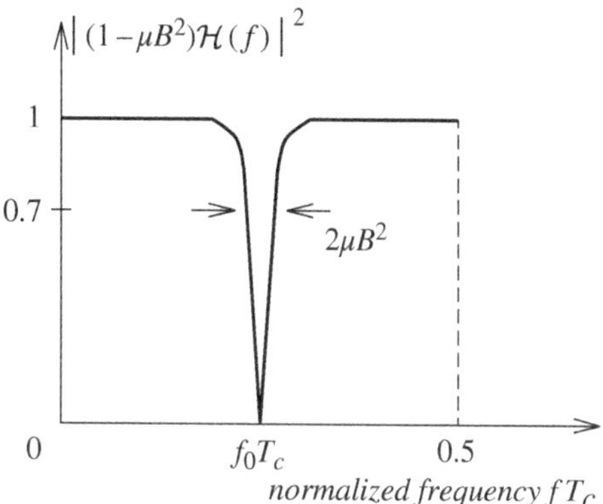

Figure 3.22 Frequency response of a notch filter.

3.4.2 Cancellation of a sinusoidal interferer with known frequency

With reference to Figure 3.23, the adaptation equations of the LMS algorithm are

$$c_1(k+1) = c_1(k) + \mu e(k) x_1(k) \tag{3.207}$$

$$c_2(k+1) = c_2(k) + \mu e(k) x_2(k) \tag{3.208}$$

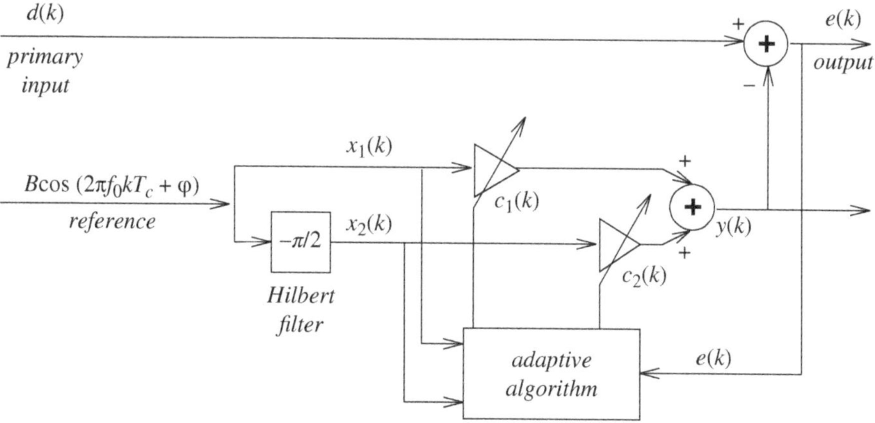

Figure 3.23 Configuration to cancel a sinusoidal interferer of known frequency.

The relation between d and output e corresponds to a *notch* filter as illustrated in Figure 3.22.

Bibliography

[1] Treichler, J.R., Johnson, C.R. Jr., and Larimore, M.G. (1987). *Theory and Design of Adaptive Filters*. New York, NY: Wiley.

[2] Shynk, J.J. (1989). Adaptive IIR filtering. *IEEE ASSP Magazine* 6: 4–21.

[3] Ungerboeck, G. (1972). Theory on the speed of convergence in adaptive equalizers for digital communication. *IBM Journal of Research and Development* 16: 546–555.

[4] Golub, G.H. and van Loan, C.F. (1989). *Matrix Computations*, 2e. Baltimore, MD and London: The Johns Hopkins University Press.

[5] Ardalan, S.H. and Alexander, S.T. (1987). Fixed-point round-off error analysis of the exponentially windowed RLS algorithm for time varying systems. *IEEE Transactions on Acoustics, Speech, and Signal Processing* 35: 770–783.

[6] Eweda, E. (1994). Comparison of RLS, LMS and sign algorithms for tracking randomly time varying channels. *IEEE Transactions on Signal Processing* 42: 2937–2944.

[7] Gardner, W.A. (1987). Nonstationary learning characteristics of the LMS algorithm. *IEEE Transactions on Circuits and Systems* 34: 1199–1207.

[8] Solo, V. (1989). The limiting behavior of LMS. *IEEE Transactions on Acoustics, Speech, and Signal Processing* 37: 1909–1922.

[9] Douglas, S.C. (1994). A family of normalized LMS algorithms. *IEEE Signal Processing Letters* 1: 49–51.

[10] Zeidler, J.R. (1990). Performance analysis of LMS adaptive prediction filters. *IEEE Proceedings* 78: 1781–1806.

[11] Haykin, S. (1996). *Adaptive Filter Theory*, 3e. Englewood Cliffs, NJ: Prentice-Hall.

[12] Falconer, D. and Ljung, L. (1978). Application of fast Kalman estimation to adaptive equalization. *IEEE Transactions on Communications* 26: 1439–1446.

[13] Cioffi, J.M. and Kailath, T. (1984). Fast, recursive-least-squares transversal filter for adaptive filtering. *IEEE Transactions on Acoustics, Speech, and Signal Processing* 32: 304–337.

[14] Cioffi, J.M. (1988). High speed systolic implementation of fast QR adaptive filters. *Proceedings of ICASSP*, pp. 1584–1588.

[15] Ling, F., Manolakis, D., and Proakis, J.G. (1986). Numerically robust least-squares lattice-ladder algorithm with direct updating of the reflection coefficients. *IEEE Transactions on Acoustics, Speech, and Signal Processing* 34: 837–845.

[16] Regalia, P.A. (1993). Numerical stability properties of a QR-based fast least squares algorithm. *IEEE Transactions on Signal Processing* 41: 2096–2109.

[17] Alexander, S.T. and Ghirnikar, A.L. (1993). A method for recursive least-squares filtering based upon an inverse QR decomposition. *IEEE Transactions on Signal Processing* 41: 20–30.

[18] Cioffi, J.M. (1990). The fast adaptive rotor's RLS algorithm. *IEEE Transactions on Acoustics, Speech, and Signal Processing* 38: 631–653.

[19] Liu, Z.-S. (1995). QR methods of *O*(*N*) complexity in adaptive parameter estimation. *IEEE Transactions on Signal Processing* 43: 720–729.

[20] Bucklew, J.A., Kurtz, T.G., and Sethares, W.A. (1993). Weak convergence and local stability properties of fixed step size recursive algorithms. *IEEE Transactions on Information Theory* 39: 966–978.

Chapter 4

Transmission channels

This chapter introduces the radio channel with its statistical description and associated numerous parameters, including the use of multiple antennas at the transmitter and/or receiver. Briefly, we also mention the telephone channel, which however is gradually replaced by optical fibres.

4.1 Radio channel

The term *radio* is used to indicate the transmission of an electromagnetic field that propagates in free space. Some examples of radio transmission systems are as follows:

- point-to-point terrestrial links [1];
- mobile terrestrial communication systems [2–6];
- earth–satellite links (with satellites employed as signal repeaters) [7];
- deep-space communication systems (with space probes at a large distance from earth).

A radio channel model is illustrated in Figure 4.1, where we assume that the transmit antenna input impedance and the receive antenna output impedance are matched for maximum transfer of power.

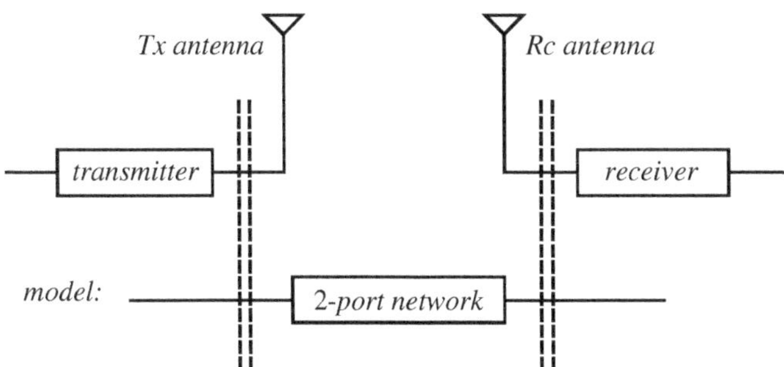

Figure 4.1 Radio channel model.

4.1.1 Propagation and used frequencies in radio transmission

Propagation over the various layers of the atmosphere is denoted respectively as *ionospheric* and *tropospheric propagation*, while propagation occurring very close to the earth's surface is denoted as *ground*

Algorithms for Communications Systems and their Applications, Second Edition.
Nevio Benvenuto, Giovanni Cherubini, and Stefano Tomasin.

propagation. The importance of each of propagation methods depends on the length of the path and the carrier frequency [8].

Basic propagation mechanisms

The mechanisms behind electromagnetic wave propagation are diverse, but can generally be attributed to *reflection*, *diffraction*, and *scattering* [3].

Reflection. This phenomenon occurs when an electromagnetic wave impinges on an interface between two media having different properties and the incident wave is partially reflected and partially transmitted. In general, the dimensions of the object where the wave impinges must be large in comparison to the wavelength for reflection to occur. In radio propagation, reflections typically occur from the surface of the earth and from buildings and walls.

Diffraction. It occurs when the path between transmitter and receiver is obstructed by a surface that has sharp irregularities. The waves originating from the obstructing surface are present throughout the space and behind the obstacle, even when a line of sight (LOS) does not exist between transmitter and receiver.

Scattering. It occurs when the medium through which the wave travels consists of objects whose dimensions are small compared to the wavelength. When a wave is scattered, it is spread out in many different directions. In mobile communication systems, scattering is due for example to foliage.

Frequency ranges

Frequencies used for radio transmission are in the range from about 100 kHz to hundreds of GHz. To a frequency f_0 corresponds a wavelength $\lambda = c/f_0$, where c is the speed of light in free space (that is, in a vacuum). The choice of the carrier frequency depends on various factors, among which the size of the transmit antenna plays an important role. Indeed, for a dipole antenna both the efficiency and its resistance grows with the dipole length up to the wavelength λ, and typically a length of $\lambda/2$ is chosen, as it provides both high efficiency and a good resistance (to be matched by the amplifier). For example, smartphones operating at a carrier frequency $f_0 = 900$ MHz and wavelength $\lambda = 0.3$ m, usually have dipole antennas of 0.15 m.

Recall that, if the atmosphere is not homogeneous (in terms of temperature, pressure, humidity, …), the radio propagation depends on the changes of its refraction index: this gives origin to reflections. We speak of diffusion or scattering phenomena if molecules that are present in the atmosphere absorb part of the power of the incident wave and then re-emit it in all directions. Obstacles such as mountains, buildings, etc., give also origin to signal reflection and/or scattering. In any case, these phenomena permit transmission between two points that are not in LOS.

We will now consider the types of propagation associated with major frequency bands, where f_0 is the signal carrier frequency.

Low frequency (LF), $0.03 < f_0 < 0.3$ *MHz*: The earth and the ionosphere form a waveguide for the electromagnetic waves. At these frequencies, the signals propagate around the earth.

Medium frequency (MF), $0.3 < f_0 < 3$ *MHz*: The signal propagates as a ground wave up to a distance of 160 km. Low and medium frequency bands are utilized for long-distance communications in navigation, aircraft beacons, and weather monitoring systems.

High frequency (HF), $3 < f_0 < 30$ MHz: The signal is reflected by the ionosphere at an altitude that may vary between 50 and 400 km. However, propagation depends on the time of the day/sunlight/darkness, season of the year, and geographical location. They are used for shortwave radio broadcasts.

Very high frequency (VHF), $30 < f_0 < 300$ MHz: For $f_0 > 30$ MHz, the signal propagates through the ionosphere with small attenuation. Therefore, these frequencies are adopted for frequency modulation (FM) radio transmissions. They are also employed for LOS transmissions, using high towers where the antennas are positioned to *cover* a wide area. The limit to the coverage is set by the earth curvature. If h is the height of the tower in meters, the range covered expressed in km is $d = 4.12\sqrt{h}$: for example, if $h = 100$ m, coverage is up to about $d = 41.2$ km. However, ionospheric and tropospheric scattering (at an altitude of 16 km or less) are present at frequencies in the range 30–60 MHz and 40–300 MHz, respectively, which cause the signal to propagate over long distances, however with a large attenuation.

Ultra high frequency (UHF), 300 MHz $< f_0 < 3$ GHz: These frequencies are used for video broadcasting and unlicensed transmissions in the industrial, scientific, and medical (ISM) band with $2.4 < f_0 < 2.483$ GHz, such as cordless phones, Bluetooth, wireless local area network (LAN) or WiFi, door openers, key-less vehicle locks, baby monitors, and many other applications.

Super high frequency (SHF), $3 < f_0 < 30$ GHz: At frequencies of about 10 GHz, atmospheric conditions play an important role in signal propagation. We note the following *absorption phenomena*, which cause additional signal attenuation:

1. *Due to water vapour*: For $f_0 > 20$ GHz, with peak attenuation at around 20 GHz;
2. *Due to rain*: For $f_0 > 10$ GHz, assuming the diameter of the rain drops is of the order of the signal wavelength.

We note that, if the antennas are not positioned high enough above the ground, the electromagnetic field propagates not only into the free space but also through ground waves. These frequencies are used for satellite communications, point-to-point terrestrial links, radars, and short-range communications.

Extremely high frequency (EHF), $30 < f_0 < 300$ GHz: These frequencies are used for radio astronomy, wireless LAN, and millimetre waves (mmWave) cellular transmissions. They are affected by absorption due to oxygen for $f_0 > 30$ GHz, with peak attenuation at 60 GHz.

4.1.2 Analog front-end architectures

In this section, an overview of a general front-end architecture for radio receivers is given.

Radiation masks

A radio channel by itself does not set constraints on the frequency band that can be used for transmission. In any case, to prevent interference among radio transmissions, regulatory bodies specify power radiation masks: a typical example is given in Figure 4.2, where the plot represents the limit on the power spectrum of the transmitted signal with reference to the power of a unmodulated carrier for the global system for mobile communications (GSM). The shown frequency offset is with respect to the carrier frequency. To comply with these limits, a filter is usually employed at the transmitter front-end.

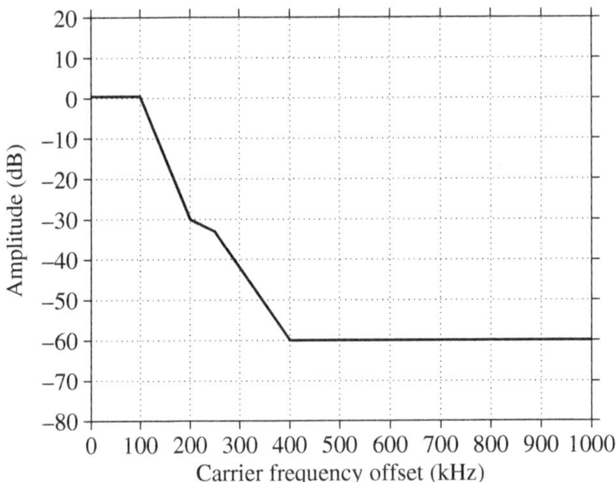

Figure 4.2 Radiation mask of the GSM system with a bandwidth of 200 kHz around the carrier frequency.

Conventional superheterodyne receiver

Many receivers for radio frequency (RF) communications use the conventional superheterodyne scheme illustrated in Figure 4.3 [9]. A first *mixer* employs a *local oscillator* (LO), whose frequency f_1 is variable, to shift the signal to around a fixed *intermediate frequency* (IF) f_{IF}. The output signal of the first mixer is filtered by a passband filter centred around the frequency f_{IF} to eliminate out of band components. The cascade of RF filter, *linear amplifier* (LNA), and *image rejection* (IR) filter performs the tasks of amplifying the desired signal as well as eliminating the noise outside of the desired band and rejecting the spectral image components introduced by the LO. The IF filter output signal is shifted into baseband through a second mixer, which makes use of an oscillator with a fixed frequency f_2, whose output is filtered by a lowpass filter to eliminate high frequency components. The signal is then sent to the analog-to-digital converter (ADC).

A drawback of the superheterodyne receiver is the high selectivity in frequency (high Q-factor) that the various elements must exhibit, which makes its integrated implementation at high frequency problematic.

Issues related to a full integration can be divided into two categories.

1. Integration of the acquisition structure and signal processing requires the elimination of both IR and IF filters.
2. For channel selection, a synthesizer can be integrated using an on-chip voltage-controlled oscillator (VCO) with low Q-factor, which, however, yields poor performance in terms of phase noise (see Section 4.1.3).

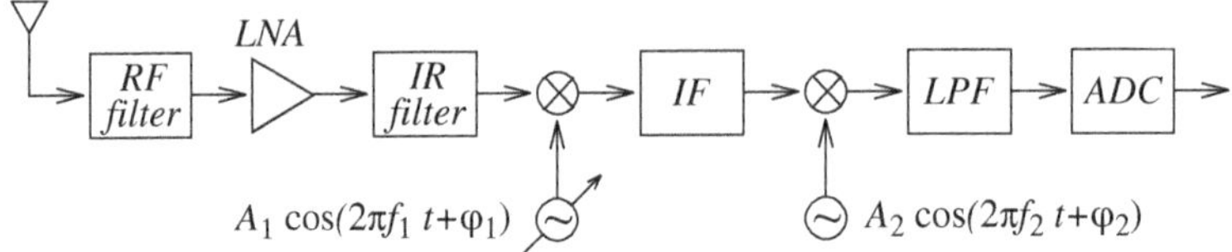

Figure 4.3 Conventional superheterodyne receiver.

Alternative architectures

In this section, three receiver architectures are considered that attempt to integrate the largest number of receiver elements. To reach this objective, the analog-to-digital conversion of the scheme of Figure 4.3 must be shifted from baseband to IF or RF. In this case, implementation of the channel selection by an analog filter bank is not efficient; indeed, it is more convenient to use a wideband analog filter followed by a *digital filter bank*. Reference is also made to multi-standard *software-defined radio* (SDR) receivers, or to the possibility of receiving signals that have different bandwidths and carriers, that are defined according to different standard systems.

We can identify two approaches to the analog-to-digital conversion.

1. *Full bandwidth digital conversion*: By using a very high sampling frequency, the whole bandwidth of the SDR system is available in the digital domain, i.e. all desired channels are considered. Considering that this bandwidth can easily achieve 100 MHz, and taking into account the characteristics of interference signals, the dynamic of the ADC should exceed 100 dB. This solution, even though is the most elegant, cannot be easily implemented as it leads to high complexity and high power consumption.
2. *Partial bandwidth digital conversion*: In this approach, the sampling frequency is determined by the radio channel with the most extended bandwidth within the different systems we want to implement.

The second approach will be considered, as its implementation has a moderate complexity, and it is used in SDR systems.

Direct conversion receiver

This architecture eliminates many off-chip components. As illustrated in Figure 4.4, all desired channels are shifted to baseband through a mixer using an oscillator with a varying frequency. Undesired spectral components are removed by an on-chip baseband filter. Clearly, this structure provides a higher integration level with respect to the superheterodyne scheme with two important advantages:

1. the problem of image components is bypassed and the IR filter is not needed;
2. the IF filter and following operations are replaced by a lowpass filter and a baseband amplifier, which are suitable for monolithic integration.

This structure has still numerous issues; in fact, the oscillator operates at the RF carrier and it may leak signal towards both the mixer input as well as towards the antenna, with consequent radiation. The interference signal generated by the oscillator could be reflected on surrounding objects and *re-received*: this spurious signal would produce a time variant direct current (DC) offset [9] at the mixer output.

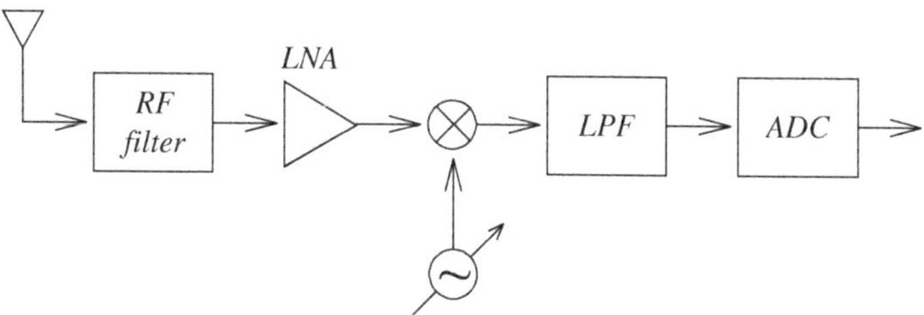

Figure 4.4 Direct conversion receiver.

To understand the origin and consequences of this offset, we can make the following two observations.

1. The isolation between the oscillator input and the inputs of the mixer and linear amplifier is not ideal; in fact an LO leakage is determined by both capacitive coupling and device substrate. The spurious signal appears at the linear amplifier and mixer inputs and is then mixed with the signal generated by the oscillator, creating a DC component; this phenomenon is called *self-mixing*. A similar effect is present when there is a strong signal coming from the linear amplifier or from the mixer that couples with the LO input: this signal would then be multiplied with itself.
2. In order to amplify the input signal of Figure 4.4 of the order of microvolts, to a level such that it can be digitized by a low cost and low power ADC, the total gain, from the antenna to the LPF output, is about 100 dB. 30 dB of this gain are usually provided by the linear amplifier/mixer combination. With these data, we can make a first computation of the offset due to self-mixing. We assume that the oscillator generates a signal with a peak-to-peak value of 0.63 V and undergoes an attenuation of 60 dB when it couples with the LNA input. If the gain of the linear amplifier/mixer combination is 30 dB, then the offset produced at the mixer output is of the order of 10 mV; if directly amplified with a gain of 70 dB, the voltage offset would saturate the following circuits.

The problem is even worse if self-mixing is time variant. This event, as previously mentioned, occurs if the oscillator signal leaks to the antenna, thus being irradiated and reflected back to the receiver from moving objects.

Finally, the direct conversion receiver needs a tuner for the channel frequency selection that works at high frequency and with low phase noise; this is hardly obtainable with an integrated VCO with low Q.

Single conversion to low-IF

The low-IF architecture reduces the problem of DC offset of the direct conversion receiver. This system has the structure of Figure 4.4 and, similarly to direct conversion, utilizes a single mixer stage; the main difference is that the frequency shift is not made to DC but to a small IF. The main advantage of the low-IF system is that the desired channel has no DC components; therefore, the usual problems due to the DC offset in the direct conversion are avoided.

Double conversion and wideband IF

This architecture, illustrated in Figure 4.5, takes all desired channels and shifts them from RF to IF using the first mixer by a fixed frequency oscillator; a simple lowpass filter is used to eliminate high frequency image components. All channels are then shifted down to baseband by a second mixer, this time with a variable frequency oscillator. The following baseband filter has a variable gain and eliminates spectral components outside the desired signal band. Channel tuning is obtained by the second low frequency LO; in this case, the first RF oscillator can be implemented by a quartz oscillator with a fixed frequency,

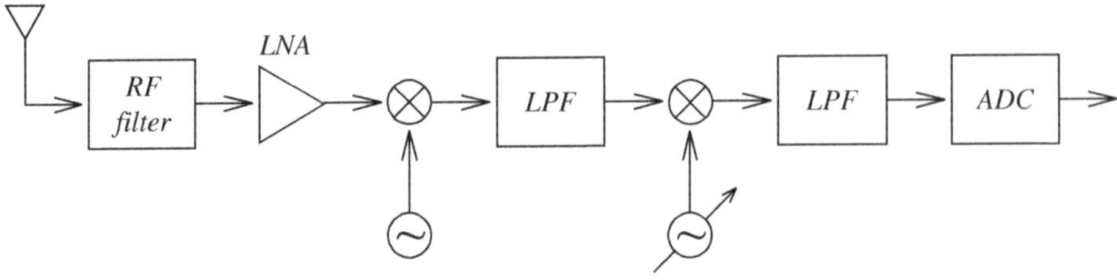

Figure 4.5 Double conversion with wideband IF.

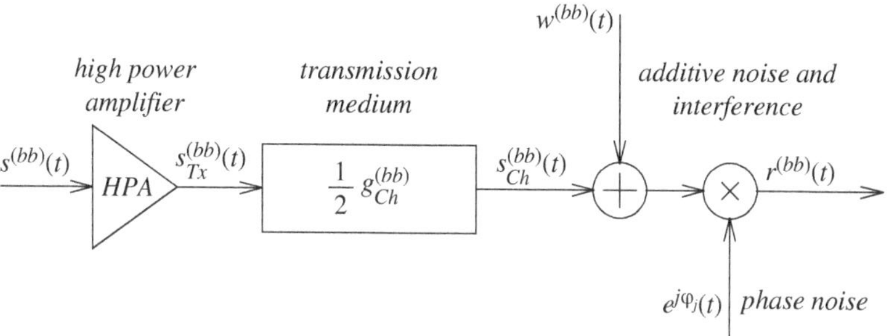

Figure 4.6 Baseband equivalent model of a transmission channel including a non-linear device.

while the second can be implemented by on-chip techniques that provide low-frequency oscillators with low phase noise and low Q.

It is important to emphasize the absence of oscillators that operate at the same frequency of the RF carrier; this eliminates the potential problem of *re-radiation* of the oscillator. We can conclude that the double conversion architecture is the most attractive for the analog front-end.

4.1.3 General channel model

In this section, we will describe a transmission channel model that takes into account the non-linear effects due to the transmitter and the disturbance introduced by the receiver and by the radio channel. We will now analyse the various blocks of the baseband equivalent model illustrated in Figure 4.6 (see Section 1.4).

High power amplifier

The final transmitter stage in a communication system usually consists of a *high power amplifier* (HPA). The HPA is a non-linear device with saturation i.e. in addition to not amplifying the input signal above a certain value, it introduces non-linear distortion of the signal itself. The non-linearity of a HPA can be described by a *memoryless envelope model*. Let the input signal of the HPA be

$$s(t) = A(t)\cos[2\pi f_0 t + \varphi(t)] \tag{4.1}$$

where $A(t) \geq 0$ is the signal envelope and φ is the instantaneous phase deviation.

The envelope and the phase of the output signal, s_{Tx}, depend on the instantaneous, i.e. without memory, transformation of the input, i.e.

$$s_{Tx}(t) = G[A(t)]\cos(2\pi f_0 t + \varphi(t) + \Phi[A(t)]) \tag{4.2}$$

It is usually more convenient to refer to baseband equivalent signals

$$s^{(bb)}(t) = A(t)e^{j\varphi(t)} \tag{4.3}$$

and

$$s_{Tx}^{(bb)}(t) = G[A(t)]e^{j(\varphi(t)+\Phi[A(t)])} \tag{4.4}$$

The functions $G[A]$ and $\Phi[A]$, called *envelope transfer functions*, represent respectively the amplitude/amplitude (AM/AM) conversion and the amplitude/phase (AM/PM) conversion of the amplifier. In practice, the HPAs are of two types. For each type, we give the AM/AM and AM/PM functions commonly adopted for the analysis. First, however, we need to introduce some normalizations. As a rule, the point at which the amplifier operates is identified by the *back-off*. We adopt here the following definitions for the *input back-off* (IBO) and the *output back-off* (OBO):

$$\text{IBO} = 20 \log_{10} \left(\frac{S}{\sqrt{\mathtt{M}_s}} \right) \quad \text{(dB)} \tag{4.5}$$

$$\text{OBO} = 20 \log_{10} \left(\frac{S_{Tx}}{\sqrt{\mathtt{M}_{s_{Tx}}}} \right) \quad \text{(dB)} \tag{4.6}$$

where $\mathtt{M}_s$ is the statistical power of the input signal s, $\mathtt{M}_{s_{Tx}}$ is the statistical power of the output signal s_{Tx}, and S and S_{Tx} are the amplitudes of the input and output signals, respectively, that lead to saturation of the amplifier. Here, we assume $S = 1$ and $S_{Tx} = G[1]$.

A posteriori, these definitions measure the trade-off between transmitting at the highest possible power, for a longer coverage of the system, and avoiding distortion (due to clamping) of the transmit signal. Hence, we neglect distortion due to the fact that in practice $G[A(t)]$ is not a straight line for $0 \le A \le 1$, and consider only the problem of saturation of $G[A(t)]$ for $A(t) > 1$.

Indeed, the above trade-off depends on the amplitude distribution of s. If for example s has a constant envelope, say $A(t) = A$, we can transmit s_{Tx} at the highest power without clamping of the signal and OBO = 0 dB. If, however, s has a Gaussian amplitude distribution, we have to transmit s_{Tx} at a much lower average power than the maximum value in order to control distortion due to clamping and OBO could be even 3 or 4 dB. In general, we want to select a working point of the HPA average output power level, where OBO is minimized and at the same time keep distortion under control. However, this needs a measure of the effect of distortion of the system, e.g. the level of out-of-band leakage of power spectral density (PSD) of s_{Tx}.

Travelling wave tube The *travelling wave tube* (TWT) is a device characterized by a strong AM/PM conversion. The conversion functions are

$$G[A] = \frac{\alpha_A A}{1 + \beta_A A^2} \tag{4.7}$$

$$\Phi[A] = \frac{\alpha_\Phi A^2}{1 + \beta_\Phi A^2} \tag{4.8}$$

where α_A, β_A, α_Φ, and β_Φ are suitable parameters.

Functions (4.7) and (4.8) are illustrated in Figure 4.7, for $\alpha_A = 1$, $\beta_A = 0.25$, $\alpha_\Phi = 0.26$, and $\beta_\Phi = 0.25$.

Solid state power amplifier A *solid state power amplifier* (SSPA) has a more linear behaviour in the region of small signals as compared to the TWT. The AM/PM conversion is usually negligible. Therefore, the conversion functions are

$$G[A] = \frac{A}{(1 + A^{2p})^{1/2p}} \tag{4.9}$$

$$\Phi[A] = 0 \tag{4.10}$$

where p is a suitable parameter.

In Figure 4.8, the function $G[A]$ is plotted for three values of p; the superimposed dashed line is an ideal curve given by

$$G[A] = \begin{cases} A, & 0 < A < 1 \\ 1, & A \ge 1 \end{cases} \tag{4.11}$$

It is interesting to compare the above analytical models with the behaviour of a practical HPA. Figure 4.9 illustrates the AM/AM characteristics of two waveguide HPA operating at frequencies of 38 and 40 GHz.

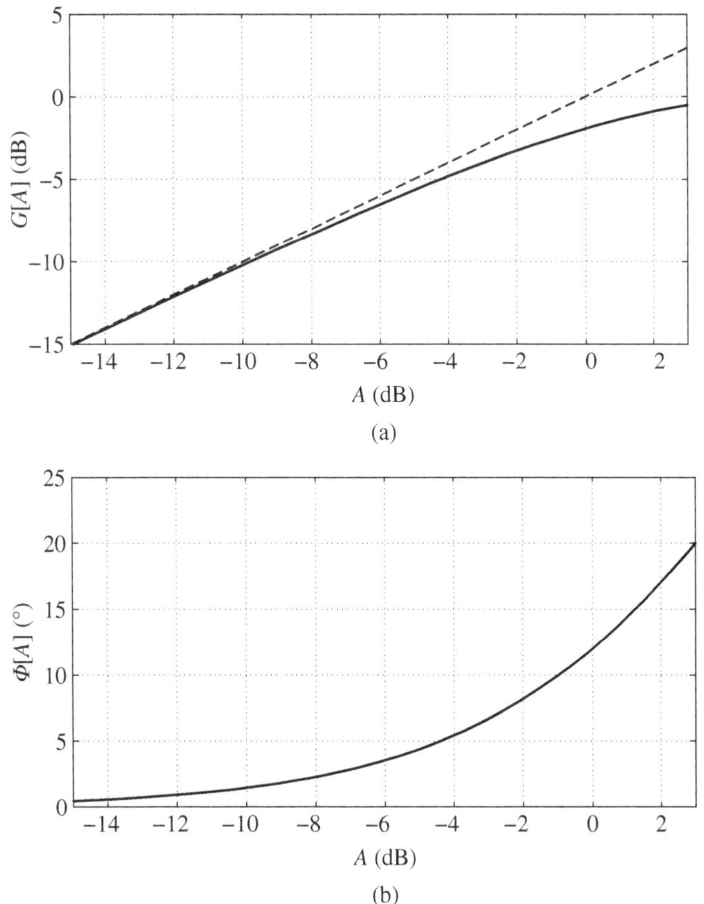

Figure 4.7 AM/AM and AM/PM characteristics of a TWT for $\alpha_A = 1$, $\beta_A = 0.25$, $\alpha_\Phi = 0.26$, and $\beta_\Phi = 0.25$.

Transmission medium

The transmission medium is typically modelled as a filter, as detailed in Section 4.1.

Additive noise

Several noise sources that cause a degradation of the received signal may be present in a transmission system. Consider for example the noise introduced by a receive antenna or the thermal noise and shot noise generated by the pre-amplifier stage of a receiver. At the receiver input, all these noise signals are modelled as an effective additive white Gaussian noise (AWGN), statistically independent of the desired signal. The power spectral density of the AWGN noise can be obtained by the analysis of the system devices, or by experimental measurements.

Phase noise

The demodulators used at the receivers are classified as *coherent* or *non-coherent*, depending on whether they use or not a carrier signal, which ideally should have the same phase and frequency as the carrier

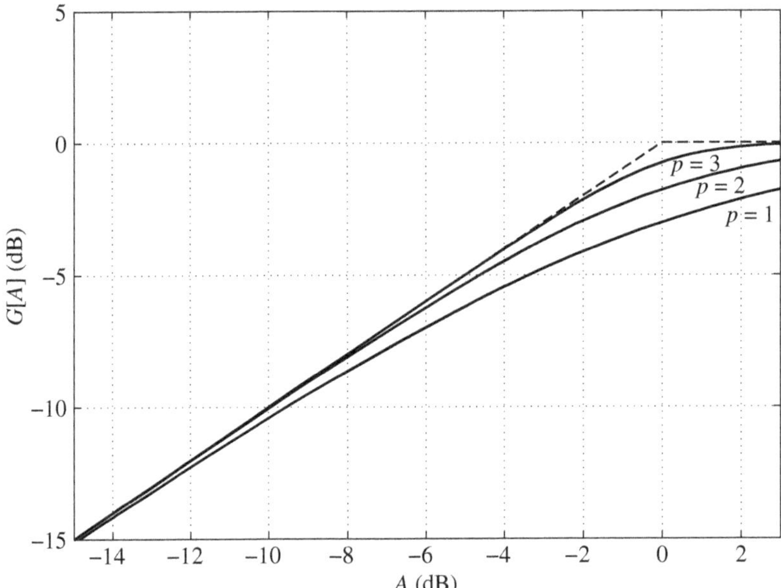

Figure 4.8 AM/AM characteristic of an SSPA.

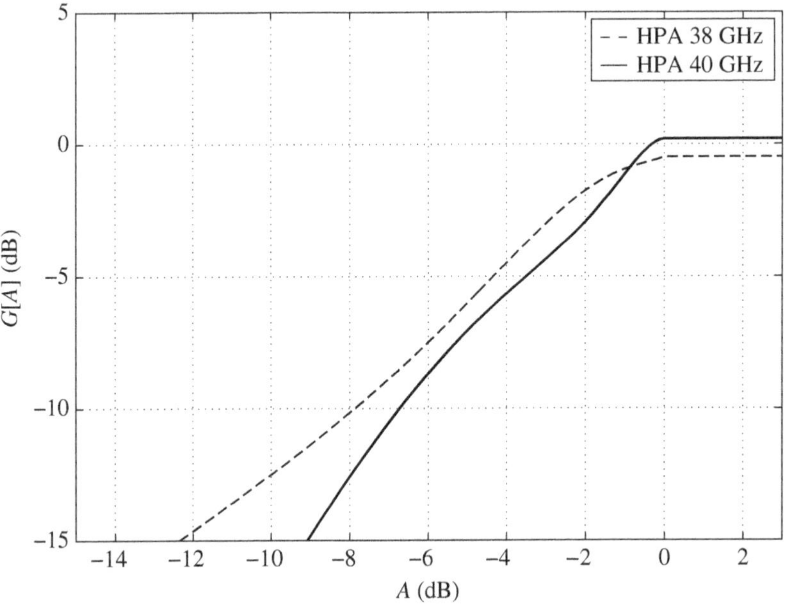

Figure 4.9 AM/AM experimental characteristic of two amplifiers operating at 38 and 40 GHz.

at the transmitter, to demodulate the received signal. Typically, both phase and frequency are recovered from the received signal by a *phase locked loop* (PLL) system, which employs a local oscillator. The recovered carrier may differ from the transmitted carrier frequency f_0 because of the *phase noise*,[1] due

[1] Sometimes also called *phase jitter*.

to short-term stability, i.e. *frequency drift*, of the oscillator, and because of the dynamics and transient behaviour of the PLL.

The recovered carrier is

$$v(t) = V_o[1 + a(t)] \cos \left(2\pi f_0 t + \varphi_j(t) + \frac{\ell_D t^2}{2} \right) \tag{4.12}$$

where ℓ_D (long term drift) represents the effect of the oscillator ageing, a is the amplitude noise, and φ_j denotes the phase noise. Often both the amplitude noise and the ageing effect can be neglected.

The *phase noise* φ_j, represented in Figure 4.6, consists of deterministic components and random noise. For example, among deterministic components, we have effects of temperature, supply voltage, and output impedance of the oscillator.

Ignoring the deterministic effects, with the exception of the *frequency drift*, the PSD of φ_j comprises five terms:

$$\mathcal{P}_{\varphi_j}(f) = \underbrace{k_{-4}\frac{f_0^2}{f^4}}_{\substack{\text{random}\\\text{frequency}\\\text{walk}}} + \underbrace{k_{-3}\frac{f_0^2}{f^3}}_{\substack{\text{flicker}\\\text{frequency}\\\text{noise}}} + \underbrace{k_{-2}\frac{f_0^2}{f^2}}_{\substack{\text{random phase}\\\text{walk or white}\\\text{frequency noise}}} + \underbrace{k_{-1}\frac{f_0^2}{f}}_{\substack{\text{flicker}\\\text{phase noise}}} + \underbrace{k_0 f_0^2}_{\substack{\text{white}\\\text{phase noise}}} \tag{4.13}$$

for $f_\ell \leq f \leq f_h$.

A simplified model, often used, is given by

$$\mathcal{P}_{\varphi_j}(f) = c + \begin{cases} a, & |f| \leq f_1 \\ b\dfrac{1}{f^2}, & f_1 \leq |f| < f_2 \end{cases} \tag{4.14}$$

where parameters a and c are typically of the order of -65 and -125 dBc/Hz, respectively, and b is a scaling factor that depends on f_1 and f_2 and ensures continuity of the PSD. dBc means dB *carrier*, i.e. the statistical power of the *phase noise*, expressed in dB with respect to the statistical power of the desired signal received in the passband.

Typical values of the statistical power of φ_j are in the range from 10^{-2} to 10^{-4}. Figure 4.10 shows the PSD (4.14) for $f_1 = 0.1$ MHz, $f_2 = 2$ MHz, $a = -65$ dBc/Hz, and $c = -125$ dBc/Hz.

4.1.4 Narrowband radio channel model

The propagation of electromagnetic waves should be studied using the Maxwell equations with appropriate boundary conditions. Nevertheless, for our purposes a very simple model, approximating the electromagnetic wave as a ray (in the optical sense), is often adequate.

The deterministic model is used to evaluate the power of the received signal in the absence of obstacles between the transmitter and receiver, that is in LOS: in this case we can think of only one wave that propagates from the transmitter to the receiver. This situation is typical of transmissions between satellites and terrestrial radio stations in the SHF band.

Let P_{Tx} be the power of the signal transmitted by an ideal *isotropic* antenna, which uniformly radiates in all directions in the free space. At a distance d from the antenna, the power density (per area) is

$$\Phi_0 = \frac{\mathrm{P}_{Tx}}{4\pi d^2} \tag{4.15}$$

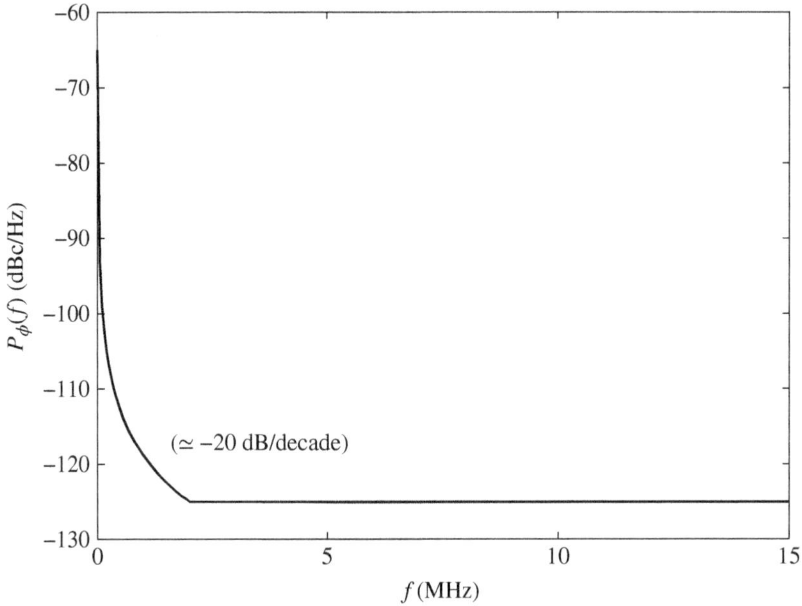

Figure 4.10 Simplified model of the phase-noise power spectral density.

where $4\pi d^2$ is the surface of a sphere of radius d that is uniformly illuminated by the antenna. We observe that the power density decreases with the square of the distance. On a logarithmic scale (dB), this is equivalent to a decrease of 20 dB-per-decade with the distance.

In the case of a *directional* antenna, the power density is concentrated within a cone and is given by

$$\Phi = \mathtt{G}_{Tx}\Phi_0 = \frac{\mathtt{G}_{Tx}\mathtt{P}_{Tx}}{4\pi d^2} \tag{4.16}$$

where $\mathtt{G}_{Tx}$ is the transmit antenna *gain*. Obviously, $\mathtt{G}_{Tx} = 1$ for an isotropic antenna; usually, $\mathtt{G}_{Tx} \gg 1$ for a directional antenna.

At the receive antenna, the *available power* in conditions of matched impedance is given by

$$\mathtt{P}_{Rc} = \Phi \mathtt{A}_{Rc}\eta_{Rc} \tag{4.17}$$

where $\mathtt{P}_{Rc}$ is the received power, $\mathtt{A}_{Rc}$ is the *effective area* of the receive antenna, and η_{Rc} is the *efficiency* of the receive antenna. The factor $\eta_{Rc} < 1$ takes into account the fact that the antenna does not capture all the incident radiation, because a part is reflected or lost. To conclude, the power of the received signal is given by

$$\mathtt{P}_{Rc} = \mathtt{P}_{Tx}\frac{\mathtt{A}_{Rc}}{4\pi d^2}\mathtt{G}_{Tx}\eta_{Rc} \tag{4.18}$$

The antenna gain can be expressed as [10]

$$\mathtt{G} = \frac{4\pi \mathtt{A}}{\lambda^2}\eta \tag{4.19}$$

where $\mathtt{A}$ is the effective area of the antenna, $\lambda = c/f_0$ is the wavelength of the transmitted signal, f_0 is the carrier frequency, and η is the efficiency factor. Equation (4.19) holds for both transmit and receive antennas.

We note that, because of the factor $\mathtt{A}/\lambda^2$, at higher frequencies we can use smaller antennas, for a given $\mathtt{G}$. Usually, $\eta \in [0.5, 0.6]$ for *parabolic* antennas, while $\eta \simeq 0.8$ for *horn* antennas.

Observing (4.19), we get

$$\mathtt{P}_{Rc} = \mathtt{P}_{Tx}\mathtt{G}_{Tx}\mathtt{G}_{Rc}\left(\frac{\lambda}{4\pi d}\right)^2 \tag{4.20}$$

known as the *Friis transmission equation* and valid in conditions of maximum transfer of power. The term $(\lambda/4\pi d)^2$ is called *free-space path loss*. Later, we will use the following definition:

$$\mathtt{P}_0 = \mathtt{P}_{Tx}\mathtt{G}_{Tx}\mathtt{G}_{Rc}\left(\frac{\lambda}{4\pi}\right)^2 \tag{4.21}$$

which represents the power of a signal received at the distance $d = 1$ from the transmitter. In any case, (4.20) does not take into account the attenuation due to rain or other environmental factors, nor the possibility of wrong positioning of the antennas.

The *(available) attenuation* of the medium, expressed in dB, is

$$(\mathtt{a})_{dB} = 10\log_{10}\frac{\mathtt{P}_{Tx}}{\mathtt{P}_{Rc}} = 32.4 + 20\log_{10}d|_{km} + 20\log_{10}f_0|_{MHz} - (\mathtt{G}_{Tx})_{dB} - (\mathtt{G}_{Rc})_{dB} \tag{4.22}$$

where $32.4 = 10\log_{10}(4\pi/c)^2$, d is expressed in km, f_0 in MHz, and $\mathtt{G}_{Tx}$ and $\mathtt{G}_{Rc}$ in dB.

It is worthwhile to make the following observations on the attenuation as from (4.22): (i) it increases with distance as $\log_{10}d$ (whereas for metallic transmission lines the dependency is linear); (ii) it increases with frequency as $\log_{10}f_0$.

For $\mathtt{G}_{Tx} = \mathtt{G}_{Rc} = 1$, $(\mathtt{a})_{dB}$ is the *free space path loss* due to distance, i.e.

$$(\bar{\mathtt{a}}_{PL,fs})_{dB} = 32.4 + 20\log_{10}d|_{km} + 20\log_{10}f_0|_{MHz} \tag{4.23}$$

In many cases, this is also the attenuation in the presence of LOS. In practice, there is a minimum value of the path loss due to distance, defined as the minimum coupling loss (MCL). For example, $(\mathtt{a}_{MCL})_{dB} = 70$ dB (80 dB) for a macro cell urban (rural) area [11]. In general, if we denote by $(\mathtt{a}_{PL})_{dB}$ the *path loss of the effective link* (in dB), whose expression will be given in (4.53), the global attenuation is

$$(\mathtt{a})_{dB} = \max\{(\mathtt{a}_{PL})_{dB} - (\mathtt{G}_{Tx})_{dB} - (\mathtt{G}_{Rc})_{dB}, (\mathtt{a}_{MCL})_{dB}\} \tag{4.24}$$

and

$$(\mathtt{P}_{Rc})_{dBm} = (\mathtt{P}_{Tx})_{dBm} - (\mathtt{a})_{dB} \tag{4.25}$$

where, if $(\mathtt{P})_{mW}$ is the power in mW, $(\mathtt{P})_{dBm} = 10\log_{10}(\mathtt{P})_{mW}$.

Equivalent circuit at the receiver

Figure 4.11 shows the electrical equivalent circuit at the receiver. The antenna produces the desired signal s, and w represents the total noise due to both the antenna and the amplifier. The amplifier has a nominal bandwidth B around the carrier frequency f_0. The spectral density of the open circuit noise voltage is $\mathcal{P}_w(f) = 2\mathtt{k}\,\mathtt{T}_w R_i$, and the available noise power per unit of frequency is $\mathtt{p}_w(f) = (\mathtt{k}/2)\mathtt{T}_w$, where $\mathtt{k} = 1.3806\ 10^{-23}$ J/K is the Boltzmann constant. The effective noise temperature at the input is $\mathtt{T}_w = \mathtt{T}_S + (\mathtt{F} - 1)\mathtt{T}_0$, where $\mathtt{T}_S$ is the effective noise temperature of the antenna, and $\mathtt{T}_A = (\mathtt{F} - 1)\mathtt{T}_0$ is the noise temperature of the amplifier; $\mathtt{T}_0$ is the room temperature and $\mathtt{F}$ is the *noise figure* of the amplifier.

For matched input and output circuits, the signal-to-noise ratio at the amplifier output is

$$\Gamma = \frac{\mathtt{P}_{Rc}}{\mathtt{k}\mathtt{T}_w B} = \frac{\text{available power of received desired signal}}{\text{nominal available power of effective input noise}} \tag{4.26}$$

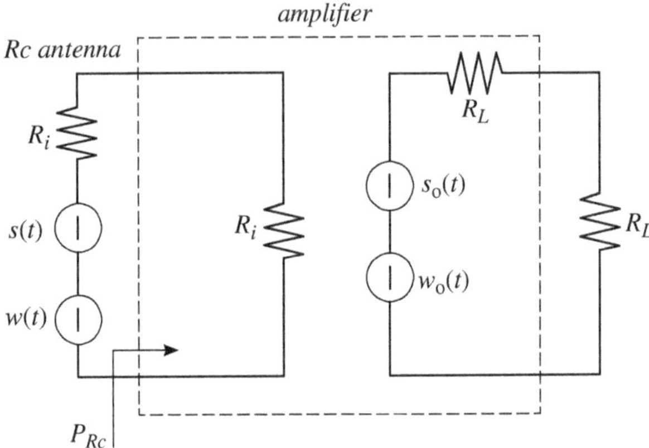

Figure 4.11 Electrical equivalent circuit at the receiver.

Multipath

It is useful to study the propagation of a sinusoidal signal over the one-ray channel model, which implies using a directional antenna. Let the narrowband (NB) transmitted signal be

$$s_{Tx}(t) = Re\,[A_{Tx} e^{j2\pi f_0 t}] \tag{4.27}$$

The received signal at a distance d from the transmitter is

$$s_{Rc}(t) = Re[A_{Rc} e^{j2\pi f_0 (t-\tau_0)}] = Re[A_{Rc} e^{j\varphi_{Rc}} e^{j2\pi f_0 t}] \tag{4.28}$$

where $\tau_0 = d/c$ denotes the propagation delay, A_{Rc} is the amplitude of the received signal, and

$$\varphi_{Rc} = -2\pi f_0 \tau_0 = -2\pi \frac{d}{\lambda} \tag{4.29}$$

is the phase of the received signal.

Using the definition (1.84) of $h^{(a)}$, the radio channel associated with (4.28) has impulse response

$$g_{Ch,overall}(\tau) = Re\left[\frac{A_{Rc}}{A_{Tx}}\, h^{(a)}(\tau - \tau_0)\right] \quad \text{ANY INPUT} \tag{4.30}$$

that is the channel attenuates the signal and introduces a delay equal to τ_0. Indeed, in the above *ideal* case, it is simply $g_{Ch,overall}(\tau) = \frac{A_{Rc}}{A_{Tx}} \delta(\tau - \tau_0)$. However, (4.30) yields a more general mathematical model, better analysed in the frequency domain, where $h^{(a)}$ ($h^{(a)*}$) accounts for the channel behaviour for positive (negative) frequencies.

Choosing f_0 as the carrier frequency, the baseband equivalent of $g_{Ch,overall}$ is given by[2]

$$g^{(bb)}_{Ch,overall}(\tau) = \frac{2A_{Rc}}{A_{Tx}}\, e^{-j2\pi f_0 \tau}\, \delta(\tau - \tau_0) = \frac{2A_{Rc}}{A_{Tx}}\, e^{-j2\pi f_0 \tau_0}\, \delta(\tau - \tau_0) \tag{4.31}$$

Limited to *signals* s_{Tx} *of the type* (4.27), (4.30) can be rewritten as

$$g_{Ch,overall}(\tau) = Re\left[\frac{A_{Rc}}{A_{Tx}}\, e^{j\varphi_{Rc}}\, h^{(a)}(\tau)\right] \quad \text{NB INPUT} \tag{4.32}$$

[2] The constraint $\mathcal{G}^{(bb)}_{Ch,overall}(f) = 0$ for $f < -f_0$ was removed because the input already satisfies the condition $\mathcal{S}^{(bb)}_{Tx}(f) = 0$ for $f < -f_0$.

Equation (4.32) indicates that the received signal exhibits a phase shift of $\varphi_{Rc} = -2\pi f_0 \tau_0$ with respect to the transmitted signal, because of the propagation delay. In general, as the propagation delay is given by d/c, the delay per unit of distance is 3.3 ns/m.

Moreover, from (4.20) in free space, as the power decreases with the square of the distance between transmitter and receiver, the amplitude of the received signal decreases with the distance, hence $A_{Rc} \propto A_{Tx}/d$; in particular, if A_0 is the amplitude of the received signal at the distance $d = 1$ from the transmitter, then

$$A_{Rc} = A_0/d \tag{4.33}$$

and the power of the received signal is given by

$$\mathrm{P}_{Rc} = \frac{1}{2}A_{Rc}^2 = \frac{1}{2}\frac{A_0^2}{d^2} = \frac{\mathrm{P}_0}{d^2} \tag{4.34}$$

where P_0 is given by (4.21).

Reflection and scattering phenomena imply that the one-ray model is applicable only to propagation in free space, and is not adequate to characterize radio channels, such as for example the channel between a fixed radio station and a mobile receiver.

We will now consider the propagation of a signal in the presence of reflections. In particular, a signal propagates from the transmitter to the receiver through N_c paths (hence the *multipath effect*). Path i, $i = 0, \ldots, N_c - 1$, of length d_i, is characterized by K_i reflections caused by surfaces, where at each reflection part of its power is absorbed by the surface, while the rest is re-transmitted in another direction. If a_{ij} is a complex number denoting the *reflection coefficient* of the j-th reflection of the i-th ray, the total reflection factor is defined as

$$a_i = \prod_{j=1}^{K_i} a_{ij} \tag{4.35}$$

Path i attenuates the signal proportionally to $|a_i| < 1$ besides $1/d_i$ (see (4.33)). Moreover, the total phase shift associated with path i is the sum of the phase shift due to the path length d_i (see (4.29)) and an additional phase shift $\arg a_i$, due to reflections. Since signals over the N_c paths add at the receiver, the resulting channel is modelled as

$$g_{Ch,overall}(\tau) = Re\left[\frac{A_0}{A_{Tx}} \sum_{i=0}^{N_c-1} \frac{a_i}{d_i} h^{(a)}(\tau - \tau_i)\right] \quad \text{ANY INPUT} \tag{4.36}$$

where $\tau_i = d_i/c$ is the delay of the i-th ray. The complex envelope of the channel impulse response (4.36) around f_0 is equal to

$$g_{Ch,overall}^{(bb)}(\tau) = \frac{2A_0}{A_{Tx}} \sum_{i=0}^{N_c-1} \frac{a_i}{d_i} e^{-j2\pi f_0 \tau_i} \delta(\tau - \tau_i) \quad \text{ANY INPUT} \tag{4.37}$$

We note that the only difference between the passband model and its baseband equivalent is constituted by the additional phase term $e^{-j2\pi f_0 \tau_i}$ for the i-th ray.

Path loss as a function of distance

Limited to *NB signals*, extending the channel model (4.32) to the case of multipath, the received signal can still be written as

$$s_{Rc}(t) = Re\,[A_{Rc}\, e^{j\varphi_{Rc}}\, e^{j2\pi f_0 t}] \tag{4.38}$$

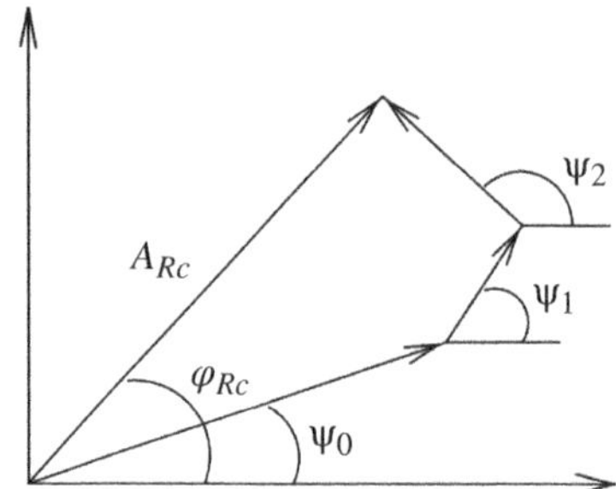

Figure 4.12 Representation of $A_{Rc}\, e^{j\varphi_{Rc}}$ in the complex plane.

where from (4.36) amplitude and phase are given by

$$A_{Rc}\, e^{j\varphi_{Rc}} = A_0 \sum_{i=0}^{N_c-1} \frac{a_i}{d_i}\, e^{j\varphi_i} = \sum_{i=0}^{N_c-1} A_i e^{j\psi_i} \tag{4.39}$$

with $\varphi_i = -2\pi f_0 \tau_i$, and A_i and ψ_i are the amplitude and phase, respectively, of the term $A_0(a_i/d_i)e^{j\varphi_i}$. Equation (4.39) is represented in Figure 4.12. As $\mathtt{P}_0 = A_0^2/2$, the received power is

$$\mathtt{P}_{Rc} = \mathtt{P}_0 \left| \sum_{i=0}^{N_c-1} \frac{a_i}{d_i}\, e^{j\varphi_i} \right|^2 \tag{4.40}$$

and is independent of the total phase of the first ray.

The simplest relation between the transmitted power and *average* (within a short time/distance) received power due to distance is

$$\overline{\mathtt{P}}_{Rc} = \frac{\mathtt{P}_0}{d^\alpha} \tag{4.41}$$

or equivalently by

$$(\overline{\mathtt{a}}_{PL})_{dB} = 32.4 + \alpha 10 \log_{10} d|_{km} + 20 \log_{10} f_0|_{\mathrm{MHz}} \tag{4.42}$$

In the free space model, $\alpha = 2$.

Ground reflection model A slightly more complex model is the ground reflection model, a *two-ray model* shown in Figure 4.13. It is based on geometric optics and considers both the direct path and a ground reflected propagation path. This model yields the average received power over distances of several kilometres for mobile radio systems that use tall towers with heights exceeding 50 m, as well as for LOS micro-cell channels in urban environments [3]. h_1 and h_2 denote the heights of transmit and receive antennas, respectively, that are placed at distance d, with

$$d \gg h_1, \qquad d \gg h_2 \tag{4.43}$$

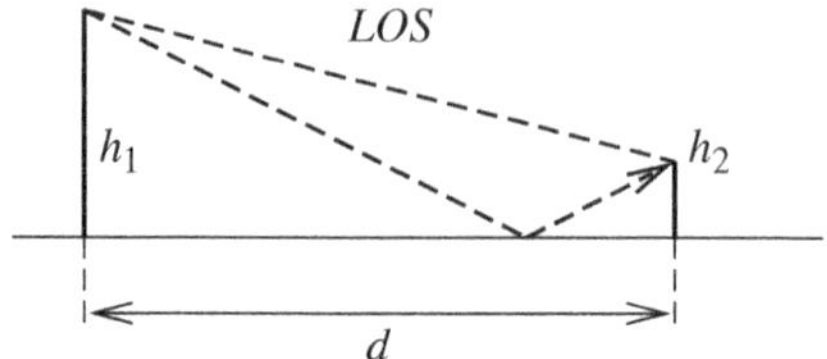

Figure 4.13 Ground reflection model.

At the receiver, the signal is given by the sum of the direct LOS and the ground reflected component.

We suppose that the earth acts as an ideal reflecting surface and does not absorb power, so reflection coefficient in this case is $a_1 = -1$. Observing (4.40), and considering that for the above assumptions the lengths of both paths are approximately equal to d, the received power is given by

$$\mathrm{P}_{Rc} \simeq \frac{\mathrm{P}_0}{d^2}|1 - e^{j\Delta\varphi}|^2 \tag{4.44}$$

where $\Delta\varphi = 2\pi f_0 \Delta d/c = 2\pi\Delta d/\lambda$ is the phase shift between the two paths, and $\Delta d = 2h_1 h_2/d$ is the difference between the lengths of the two paths. For small values of $\Delta\varphi$, we obtain

$$|1 - e^{j\Delta\varphi}|^2 \simeq |\Delta\varphi|^2 = 16\pi^2 \frac{h_1^2 h_2^2}{\lambda^2 d^2} \tag{4.45}$$

from which, by substituting (4.21) in (4.44), we get

$$\mathrm{P}_{Rc} = \frac{\mathrm{P}_0}{d^2}|\Delta\varphi|^2 = \mathrm{P}_{Tx}\mathrm{G}_{Tx}\mathrm{G}_{Rc}\frac{h_1^2 h_2^2}{d^4} \tag{4.46}$$

We note that the received power decreases as the fourth-power of the distance d, that is 40 dB/decade instead of 20 dB/decade as in the case of free space. Therefore, the law of power attenuation as a function of distance changes in the presence of multipath with respect to the case of propagation in free space.

Example of radio propagation in indoor environments We now consider an indoor environment, where both transmitter and receiver are positioned in a room, so that (4.43) does not hold any more. It is assumed, moreover, that the rays that reach the receive antenna are due, respectively, to LOS, reflection from the floor, and reflection from the ceiling, as shown in Figure 4.14.

As a result, the instantaneous received power is

$$\mathrm{P}_{Rc} = \mathrm{P}_0 \left| \sum_{i=0}^{2} \frac{a_i e^{j\varphi_i}}{d_i} \right|^2 \tag{4.47}$$

where the reflection coefficients are $a_0 = 1$ for the LOS path, and $a_1 = a_2 = -0.7$. With these assumptions, the power decreases with the distance in an erratic way, in the sense that by varying the position of the antennas the received power presents fluctuations of about 20÷30 dB. In fact, depending on the

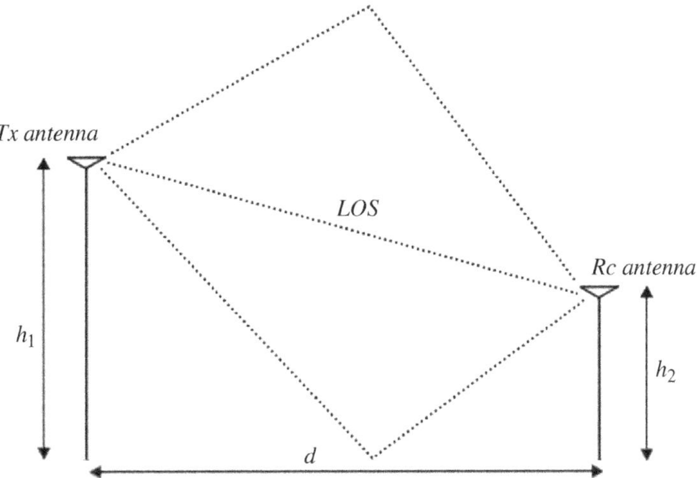

Figure 4.14 Example of propagation in indoor environment.

position, the phases of the various rays change and the sum in (4.39) also varies: in some positions all rays are aligned in phase and the received power is high, whereas in others the rays cancel each other and the received power is low. In the previous example, this phenomenon is not observed because the distance d is much larger than the antenna heights, and the phase difference between the two rays remains always small.

To summarize, relation (4.41), that describes the average received power as a function of the path loss exponent α, yields $\alpha = 2$ in the free space model and $\alpha = 4$ in the ground reflection model, so the path loss is, respectively, 20 and 40 dB/decade. In general, we can also see the path loss as the attenuation at a given distance and frequency, averaged over all other phenomena (such as fading described in Section 4.1.5). For example, in real cellular environments, where more complicated multipath phenomena occur, (4.41) still holds with exponent $\alpha \in [2.5, 6]$.

Moreover, according to the 3GPP channel model for 5G cellular systems, in a urban area if h_{BS} is the base station antenna elevation measured from the average rooftop level [11], the path loss, as the average attenuation at a given distance, is given by

$$\begin{aligned}(\bar{\mathtt{a}}_{PL})_{dB} = {} & 40(1 - 4 \times 10^{-3} h_{BS}|_m) \log_{10} d|_{km} \\ & -18 \log_{10} h_{BS}|_m + 21 \log_{10} f_0|_{MHz} + 80\end{aligned} \tag{4.48}$$

For a carrier frequency $f_0 = 900$ MHz and $h_{BS} = 15$ m, $(\bar{\mathtt{a}}_{PL})_{dB} = 120.9 + 37.6 \log_{10} d|_{km}$. This expression should be compared with $(\bar{\mathtt{a}}_{PL})_{dB}$ in LOS

$$(\bar{\mathtt{a}}_{PL})_{dB} = 31.95 + 20 \log_{10} d|_{km} \tag{4.49}$$

In rural areas, the Hata model is used and

$$\begin{aligned}(\bar{\mathtt{a}}_{PL})_{dB} = {} & 69.55 + 26.16 \log_{10} f_0|_{MHz} - 13.82 \log_{10} h_{BS}|_m \\ & + (44.9 - 6.55 \log_{10} h_{BS}|_m) \log_{10} d|_{km} - 4.78 (\log_{10} f_0|_{MHz})^2 \\ & + 18.33 \log_{10} f_0|_{MHz} - 40.94s\end{aligned} \tag{4.50}$$

For $f_0 = 900$ MHz and $h_{BS} = 95$ m above ground, $(\bar{\mathtt{a}}_{PL})_{dB} = 95.4 + 31.94 \log_{10} d|_{km}$. In any case, it must be $(\bar{\mathtt{a}}_{PL})_{dB} \geq (\bar{\mathtt{a}}_{PL,fs})_{dB}$.

4.1.5 Fading effects in propagation models

In addition to path loss due to distance, the short-term received power exhibits fluctuations called *fading*; these fluctuations are typically composed of two multiplicative components, *macroscopic* and *microscopic fading*.

Macroscopic fading represents the long-term variation (within a distance of many wavelengths) of the received power level, while microscopic fading represents short-term variation (within a distance of a fraction of wavelength), as described next.

Macroscopic fading or shadowing

Macroscopic fading is generally attributed to *shadowing* effects of buildings or natural features, and an example is shown in Figure 4.15. Indeed, it is determined by the local mean of fast fading and its statistical distribution has been derived experimentally. Its distribution depends on the antenna height, the carrier frequency, and the specific environment. Experimentally, it is seen that the attenuation component in dB due to macroscopic fading follows a normal distribution, hence it is also called *log-normal fading*. In terms of attenuation, we have

$$(\mathtt{a}_{PL})_{dB} = (\bar{\mathtt{a}}_{PL})_{dB} + (\xi)_{dB} \tag{4.51}$$

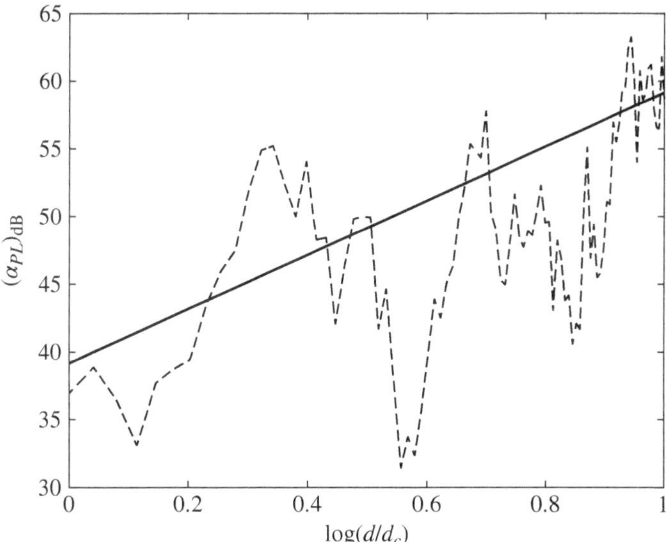

Figure 4.15 Attenuation fluctuation as function of distance in wireless channels. Mean propagation loss increases monotonically with distance (solid line), while fluctuations are due to macroscopic fading (dashed line).

where $(\xi)_{dB} \sim \mathcal{N}(0, \sigma^2_{(\xi)_{dB}})$ models the shadowing component, while $(\overline{\mathtt{a}}_{PL})_{dB}$ is the path loss attenuation due to distance. The shadowing components for two different receiver locations at distance Δ are correlated with correlation given by the Gudmundson model [12]

$$\mathtt{r}_{(\xi)_{dB}}(\Delta) = \sigma^2_{(\xi)_{dB}} e^{-|\Delta|/D_{coh}} \tag{4.52}$$

where D_{coh} is the coherence distance.[3] The standard deviation of $(\xi)_{dB}$ depends on the particular environment, usually being $\sigma_{(\xi)_{dB}} \in [4, 12]$.

In practice, shadowing provides a measure of the adequacy of the adopted deterministic model. A propagation model that completely ignores any information on land configuration, and therefore is based only on the distance between transmitter and receiver, has *shadowing* with $\sigma_{(\xi)_{dB}} = 12$. Improving the accuracy of the propagation model, for example by using more details regarding the environmental configuration, *shadowing* can be reduced; in case we had an enormous amount of topographic data and the means to elaborate them, we would have a model with $\sigma_{(\xi)_{dB}} = 0$. Hence, *shadowing* should be considered in the performance evaluation of mobile radio systems, whereas for the correct design of a network it is good practice to make use of the largest possible quantity of topographic data.

Microscopic fading

Microscopic fading refers to rapid fluctuations of the received signal power in space, time, and frequency domain, and is caused by the signal scattering off objects along the propagation path and by changes in the relative position of the transmitter and/or receiver and environment (e.g. foliage).

[3] For a receiver moving along a straight line departing from the transmitter, if we denote the distance as $d = id_0$, with d_0 a suitable step size, we can model $(\xi)_{dB}$ by an AR(1) process (see Section 1.10) and

$$(\xi)_{dB}[id_0] = -\mathtt{a}_1(\xi)_{dB}[(i-1)d_0] + w[id_0] \tag{4.53}$$

with $w(id_0)$ white noise with variance $\sigma_w^2 = (1 - |a_1|^2)\sigma^2_{(\xi)_{dB}}$ and $\mathtt{a}_1 = -e^{-d_0/D_{coh}}$. For a suitable representation of $(\xi)_{dB}$ versus d, it must be $d_0 < D_{coh}/5$.

For a NB signal, assuming a large number of statistically independent components in the propagation paths, the central limit theorem can be used to model $A_{Rc}\, e^{j\varphi_{Rc}}$ in (4.39) as complex Gaussian. Hence, phase $\varphi_{Rc} \sim \mathcal{U}((0, 2\pi])$, and amplitude A_{Rc} has a Rayleigh probability density function (pdf)

$$p_{A_{Rc}}(a) = \frac{2a}{\overline{\mathtt{P}}_{Rc}} e^{-\frac{a^2}{\overline{\mathtt{P}}_{Rc}}} 1(a) \tag{4.54}$$

where $\overline{\mathtt{P}}_{Rc} = \mathtt{P}_{Tx}/\mathtt{a}$ is the short-term average received power.

If in addition to the scattering components, there is an LOS component, the signal envelope A_{Rc} will have a Ricean distribution. Defining the *Rice factor* K as to the ratio between the power of the direct component and the power of the reflected and/or scattered components, the pdf of A_{Rc} is

$$p_{A_{Rc}}(a) = \frac{2a(K+1)}{\overline{\mathtt{P}}_{Rc}} e^{-K-\frac{(K+1)a^2}{\overline{\mathtt{P}}_{Rc}}} I_0\left(2a\sqrt{\frac{K(K+1)}{\overline{\mathtt{P}}_{Rc}}}\right) 1(a) \tag{4.55}$$

where $I_0(a)$ is the modified Bessel function of the first type and order zero defined as

$$I_0(x) = \frac{1}{\pi} \int_0^{\pi} e^{x \cos\theta}\, d\theta \tag{4.56}$$

Note that in the absence of a direct path ($K = 0$), the Ricean pdf in (4.55) reduces to the Rayleigh pdf in (4.54), since $I_0(0) = 1$.

We note that both (4.54) and (4.55) refer to the NB channel model for a sinusoidal transmitted signal, and a more general wideband channel model will be given in Section 4.1.7.

4.1.6 Doppler shift

In the presence of relative motion between transmitter and receiver, the frequency of the received signal undergoes a shift with respect to the frequency of the transmitted signal, known as *Doppler shift*.

We now analyze in detail the Doppler shift. With reference to Figure 4.16, we consider a transmitter T_x and a receiver that moves with speed v_p from a point P to a point Q. The variation in distance between the transmitter and the receiver is $\Delta\ell = v_p \Delta t \cos\theta$, where v_p is the speed of the receiver relative to the transmitter, Δt is the time required for the receiver to go from P to Q, and θ is the angle of incidence of the signal with respect to the direction of motion. θ is assumed to be approximately the same in P and in Q. The phase variation of the received signal because of the different path length in P and Q is

$$\Delta\varphi = \frac{2\pi\Delta\ell}{\lambda} = \frac{2\pi v_p \Delta t}{\lambda} \cos\theta \tag{4.57}$$

and hence the apparent change in frequency or Doppler shift is

$$f_s = \pm\frac{1}{2\pi}\frac{\Delta\varphi}{\Delta t} = \pm\frac{v_p}{\lambda}\cos\theta \tag{4.58}$$

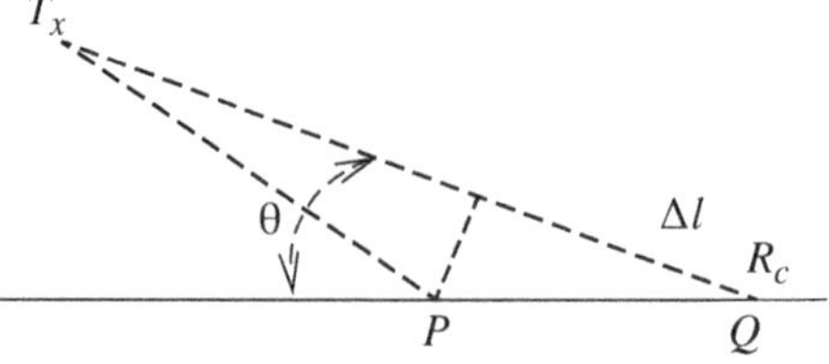

Figure 4.16 Illustration of the Doppler shift.

where the sign is positive if Q is a further away from the transmitter than P, and negative otherwise. This implies that if the NB signal (4.27) is transmitted, the received signal is

$$s_{Rc}(t) = Re[A_{Rc} e^{j2\pi(f_0 - f_s)t}] \tag{4.59}$$

With typically used unit measures, (4.58) becomes

$$f_s = \pm 9.259 \ 10^{-4} \ v_p|_{\text{km/h}} \ f_0|_{MHz} \cos\theta \qquad (Hz) \tag{4.60}$$

where $v_p|_{km/h}$ is the speed of the mobile in km/h, and $f_0|_{MHz}$ is the carrier frequency in MHz. For example, if $v_p = 100$ km/h, $f_0 = 900$ MHz, and $\theta = 0$, with the receiver moving away from the transmitter, we have $f_s = 83$ Hz.

Indeed, from (4.59), the Doppler can also be interpreted as transmitting the NB signal at frequency f_0, through a *time varying channel* with

$$g^{(bb)}_{Ch,overall}(t, \tau) = \frac{2A_{Rc}}{A_{Tx}} \ e^{-j2\pi f_s t} \delta(\tau) \tag{4.61}$$

We now consider a NB signal transmitted in an *indoor* environment,[4] where the signal received by the antenna is given by the contribution of many rays, each with a different length. In the presence of a single ray, the received signal would undergo only one Doppler shift. However, according to (4.58) the frequency shift f_s depends on the angle θ. Therefore, because of the different paths, the received signal is no longer monochromatic, and we speak of a *Doppler spectrum* to indicate the spectrum of the received signal around f_0. This phenomenon occurs even when if both transmitter and receiver are static, but a person or an object is moving around them, modifying the signal propagation.

The Doppler spectrum is characterized by the *Doppler spread*, which measures the dispersion in the frequency domain that is experienced by a transmitted sinusoidal signal. For example, from (4.28), for one ray channel model affected by Doppler, the PSD of the received amplitude is $\delta(f - f_s)$ up to a constant factor. It is intuitive that the more the characteristics of the radio channel are varying with time, the larger the Doppler spread will be. An important consequence of this observation is that the convergence time of channel-dependent algorithms used in receivers (e.g. adaptive equalization) must be much smaller than the inverse of the Doppler spread of the channel, in order to follow the channel variations.

Example 4.1.1 (Doppler shift)

Consider a transmitter that radiates a sinusoidal carrier at frequency $f_0 = 1850$ MHz. For a vehicle travelling at 96.55 km/h (26.82 m/s), we want to evaluate the frequency of the received carrier if the vehicle is moving: (a) approaching the transmitter, (b) going away from the transmitter, and (c) perpendicular to the direction of arrival of the transmitted signal.

(a) The Doppler shift is positive and $\theta = 0$; the received frequency is

$$f_{Rc} = f_0 + f_s = 1850 \cdot 10^6 + 9.259 \cdot 10^{-4} \cdot 96.55 \cdot 1850 = 1850.000\ 166 \text{ MHz} \tag{4.62}$$

(b) The Doppler shift is negative and $\theta = 0$; the rceived frequency is

$$f_{Rc} = f_0 - f_s = 1850 \times 10^6 - 9.259 \cdot 10^{-4} \cdot 96.55 \cdot 1850 = 1849.999\ 834 \text{ MHz} \tag{4.63}$$

(c) In this case, $\cos\theta = 0$; therefore, there is no Doppler shift.

[4] The term *indoor* is usually referred to areas inside buildings, possibly separated by walls of various thickness, material, and height. The term *outdoor*, instead, is usually referred to areas outside buildings: these environments can be of various types, for example, urban, suburban, rural, etc.

4.1.7 Wideband channel model

For a wideband signal with spectrum centred around the carrier frequency f_0, the channel model (4.37) still holds. If $\mathtt{a}$ is the attenuation due to both path loss and shadowing, we now write the overall baseband-equivalent channel impulse response as

$$g_{Ch,overall}^{(bb)}(t,\tau) = \frac{1}{\sqrt{\mathtt{a}}} g_{Ch}^{(bb)}(t,\tau) \tag{4.64}$$

where $g_{Ch}^{(bb)}(t,\tau)$ is normalized to a unit statistical power. Moreover,

$$g_{Ch}^{(bb)}(t,\tau) = \sum_{i=0}^{N_c-1} g_i(t)\delta(\tau - \tau_i(t)) \tag{4.65}$$

and g_i represents the normalized complex-valued gain of the i-th ray that arrives with delay τ_i. Power normalization yields

$$\sum_{i=0}^{N_c-1} E[|g_i|^2] = 1 \tag{4.66}$$

For a given receiver location, (4.65) models the channel as a linear filter having time-varying impulse response, where the channel variability is due to the motion of transmitter and/or receiver, or to changes in the surrounding environment, or to both factors.

If the channel is time-invariant, or at least it is time-invariant within a short time interval, in this time interval, the impulse response is only a function of τ.

The transmitted signal undergoes three phenomena: (i) *fading* of some gains g_i due to multipath, which implies rapid changes of the received signal power over short distances (of the order of the carrier wavelength) and brief time intervals, (ii) *time dispersion* of the impulse response caused by diverse propagation delays of multipath rays, and (iii) *Doppler shift*, which introduces a random frequency modulation that is in general different for each ray.

In a digital transmission system, the effect of multipath depends on both the relative duration of the symbol period and the channel impulse response. If the duration of the channel impulse response is very small with respect to the duration of the symbol period, i.e. the transmitted signal is NB with respect to the channel, then the one-ray model is appropriate, and if the gain of the single ray varies in time we have a *flat fading* channel. Otherwise, an adequate model must include several rays, and if the gains vary in time, we have of a *frequency selective fading* channel.

Neglecting the group delay τ_0, and letting $\tilde{\tau}_1 = \tau_1 - \tau_0$, a simple two-ray radio channel model has impulse response

$$g_{Ch}^{(bb)}(t,\tau) = g_0(t)\delta(\tau) + g_1(t)\delta(\tau - \tilde{\tau}_1(t)) \tag{4.67}$$

At a given instant t, the channel is equivalent to a filter with impulse response shown in Figure 4.17 and frequency response given by[5]:

$$\mathcal{G}_C(t,f) = g_0(t) + g_1(t)e^{-j2\pi f\tilde{\tau}_1(t)} \tag{4.69}$$

It is evident that the channel has a selective frequency behaviour, as the attenuation depends on frequency. For g_0 and g_1 real-valued and time-invariant, from (4.69) the following frequency response is obtained

$$|\mathcal{G}_C(f)|^2 = g_0^2 + g_1^2 + 2g_0g_1\cos(2\pi f\tilde{\tau}_1) \tag{4.70}$$

[5] If we normalize the coefficients with respect to g_0, (4.69) becomes

$$\mathcal{G}_C(f) = 1 + b\, e^{-j2\pi f\tilde{\tau}_1} \tag{4.68}$$

where b is a complex number. In the literature (4.68) is called *Rummler model* of the radio channel.

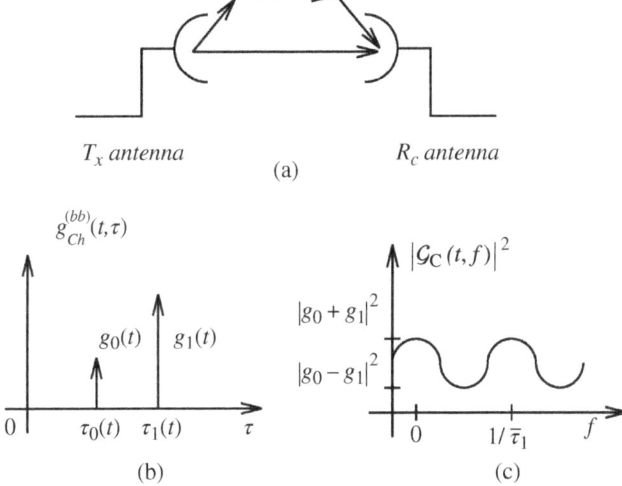

Figure 4.17 Physical representation (a) and model in time (b) and frequency (c) of a two-ray radio channel, where g_0 and g_1 are assumed to be positive.

shown in Figure 4.17c. In any case, the signal distortion depends on the transmitted signal bandwidth. In particular, a non-negligible signal distortion is observed if the signal bandwidth is of the order of $1/\tilde{\tau}_1$.

Going back to the general case, for wideband communications, rays with different delays are assumed to be independent, i.e. they do not interact with each other. In this case, by using a suitable receiver, from (4.65) the received power is

$$\mathtt{P}_{Rc} = \frac{\mathtt{P}_{Tx}}{\mathtt{a}} \sum_{i=0}^{N_c-1} |g_i|^2 \tag{4.71}$$

From (4.71), we note that the received power is given by the sum of the square amplitude of all the rays. Conversely, in the transmission of NB signals, the received power is the square of the vector amplitude resulting from the vector sum of all the received rays. Therefore, for a given transmitted power, the received power will be lower for a NB signal as compared to a wideband signal.

Multipath channel parameters

To study the performance of mobile radio systems, it is convenient to introduce a measure of the channel dispersion in the time domain known as *multipath delay spread* (MDS). In order to simplify notation, we consider here time-invariant channels, dropping dependency on time t. The MDS is the measure of the time interval that elapses between the arrival of the first and the last ray. Another measure is the delay time that it takes for the amplitude of the ray to decrease by x dB below the maximum value; this time is also called *excess delay spread* (EDS).

However, the most common measure is the -mean square (rms) *delay spread*

$$\tau_{rms} = \sqrt{\overline{\tau^2} - \overline{\tau^1}^2} \tag{4.72}$$

Table 4.1: Values of $E[|g_i|^2]$ (in dB) and τ_i (in ns) for three typical channels. In order to satisfy constraint (4.75) it is necessary to scale all entries by a suitable constant

Standard GSM	
τ_i	$E[\lvert g_i\rvert^2]$
0	−3.0
200	0
500	−2.0
1600	−6.0
2300	−8.0
5000	−10.0

Indoor offices	
τ_i	$E[\lvert g_i\rvert^2]$
0	0.0
50	−1.6
150	−4.7
325	−10.1
550	−17.1
700	−21.7

Indoor business	
τ_i	$E[\lvert g_i\rvert^2]$
0	−4.6
50	0
150	−4.3
225	−6.5
400	−3.0
525	−15.2
750	−21.7

where

$$\overline{\tau^n} = \frac{\sum_{i=0}^{N_c-1} |g_i|^2 \tau_i^n}{\sum_{i=0}^{N_c-1} |g_i|^2}, \qquad n = 1, 2 \tag{4.73}$$

Typical values of (average) rms delay spread are of the order of microseconds in *outdoor* mobile radio channels (precisely on the order of 0.05/0.2/2.5 μs in rural/urban/hilly areas), and of the order of some tenths of nanoseconds in *indoor* channels.

We define as *power delay profile* (PDP) $\mathtt{M}(\tau)$, also called *delay power spectrum* or *multipath intensity profile*, the expectation of the square amplitude of the channel gain as a function of given delays τ_i,

$$\mathtt{M}(\tau_i) = E(|g_i|^2), \qquad i = 0, \ldots, N_c - 1 \tag{4.74}$$

Associated to the function $\mathtt{M}(\tau_i)$, we define the rms delay spread $\overline{\tau}_{rms}$ obtained replacing $|g_i|^2$ by $\mathtt{M}(\tau_i)$ in (4.73). From (4.66), it is

$$\sum_{i=0}^{N_c-1} \mathtt{M}(\tau_i) = 1 \tag{4.75}$$

Table 4.1 shows PDPs for some typical channels.

Statistical description of fading channels

Typically, gains $\{g_i\}$ are given by the sum of a large number of random components, that can be approximated by independent Gaussian random variables. The most widely used statistical description is

$$\begin{aligned} g_0 &= C_0 + \tilde{g}_0 & i &= 0 \\ g_i &= \tilde{g}_i & i &= 1, \ldots, N_c - 1 \end{aligned} \tag{4.76}$$

where C_0 is a real-valued constant and $\tilde{g}_i$, $i = 0, \ldots, N_c - 1$, are independent complex circular Gaussian-distributed random variables, with zero mean and variance σ_i^2 (see Example 1.7.3 on page 47). Moreover, in order to ensure the power normalization (4.77), we must have

$$C_0^2 + \sum_{i=0}^{N_c-1} \sigma_i^2 = 1 \tag{4.77}$$

Therefore, the phase of $\tilde{g}_i$ is uniformly distributed in $[0, 2\pi)$, the distribution of $|g_i|$ varies for each index i. Since the first ray contains a direct (deterministic) component in addition to a random component as other rays, $|g_i|$ will have a Rice distribution for $i = 0$, and a Rayleigh distribution for $i \neq 0$. In particular, introducing the normalized random variables

$$\overline{g}_i = \frac{g_i}{\sqrt{\mathtt{M}(\tau_i)}} \tag{4.78}$$

we have (see also (4.54) and (4.55))

$$\begin{aligned} p_{|\overline{g}_0|}(a) &= 2(1+K)\, a e^{-K-(1+K)a^2} I_0(2a\sqrt{K(1+K)}) 1(a) \\ p_{|\overline{g}_i|}(a) &= 2a\, e^{-a^2} 1(a) \end{aligned} \tag{4.79}$$

where K is the *Rice factor*, with

$$K = \frac{C_0^2}{\mathtt{M}_d} \qquad \mathtt{M}_d = \sum_{i=0}^{N_c-1} \sigma_i^2 \tag{4.80}$$

In turn, from (4.80) and (4.77), given K it is

$$C_0 = \sqrt{\frac{K}{K+1}} \tag{4.81}$$

and

$$\mathtt{M}_d = 1 - C_0^2 \tag{4.82}$$

The pdf (4.79) is shown in Figure 4.18 for various values of K (where $K = 0$ provides the Rayleigh distribution).

It may be useful to recall that if $|g_i|$ is Rayleigh, then

$$E[|g_i|] = \sqrt{\frac{\pi}{4}} \sqrt{\mathtt{M}(\tau_i)} = 0.8862\ \sqrt{\mathtt{M}(\tau_i)} \tag{4.83}$$

$$E[(|g_i| - E[|g_i|])^2] = \left(1 - \frac{\pi}{4}\right) \mathtt{M}(\tau_i) = 0.2146\ \mathtt{M}(\tau_i) \tag{4.84}$$

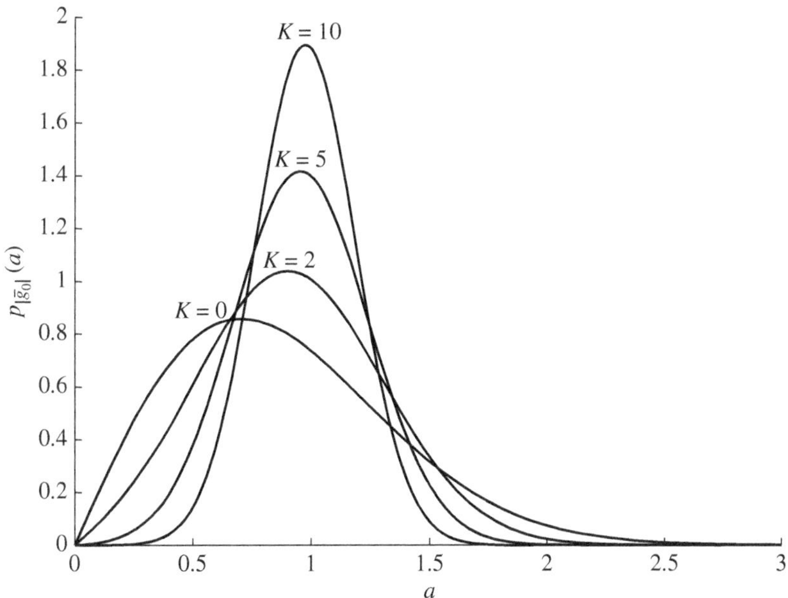

Figure 4.18 The Rice probability density function for various values of K. The Rayleigh density function is obtained for $K = 0$.

Typical reference values for K are 2 and 10. If $C_0 = 0$, i.e. no direct component exists, it is $K = 0$, and the Rayleigh distribution is obtained for all the gains $\{g_i\}$. For $K \to \infty$, i.e. with no reflected and/or scattered components and, hence, $C_0 = 1$, we find the model having only the deterministic component. In general, K falls off exponentially with distance, with $K = 20$ near the transmitter and $K = 0$ at large distances.

4.1.8 Channel statistics

We evaluate the autocorrelation function of g_i in (4.65) at two different instants and delays

$$\mathtt{r}_g(t, t-\Delta t, \tau_i, \tau_j) = E[g_i(t) g_j^*(t-\Delta t)] \tag{4.85}$$

According to the model known as the *wide-sense stationary uncorrelated scattering* (WSSUS), the values of g_i for rays that arrive with different delays are uncorrelated, and g_i is stationary in t. Therefore, we have

$$\mathtt{r}_g(t, t-\Delta t, \tau_i, \tau_j) = \begin{cases} \mathtt{r}_g(\Delta t; \tau_i), & \tau_i = \tau_j \\ 0, & \tau_i \neq \tau_j \end{cases} \tag{4.86}$$

In other words, the correlation is not zero only for gains at the same delay time and autocorrelation of g_i only depends on the difference of the times at which the gain is considered.

Power delay profile

In (4.86) for $\Delta t = 0$, we find the PDP (4.74). A typical channel model is the *two rays model* (see (4.68)) with two equal-power rays, i.e.

$$\mathtt{M}(\tau) = \begin{cases} \frac{1}{2}, & \tau = 0 \\ \frac{1}{2}, & \tau = 2\overline{\tau}_{rms} \end{cases} \tag{4.87}$$

Sometimes the discrete-time function $\mathtt{M}(\tau_i)$, rather than experimentally, is assigned analytically by sampling a continuous-time function $\mathtt{M}_{cnt}(\tau) = E[|g_\tau(t)|^2]$ defined as in (4.74), where g_i, with some abuse of notation, is replaced by g_τ, the channel gain at lag τ, with τ now a continuous variable.

As in the case of the previously studied channel model, we define the *rms delay spread* associated to $\mathtt{M}_{cnt}(\tau)$ as

$$\overline{\tau}_{rms,cnt} = \sqrt{\frac{\int_{-\infty}^{\infty} (\tau - \overline{\tau}_{cnt})^2 \mathtt{M}_{cnt}(\tau) d\tau}{\int_{-\infty}^{\infty} \mathtt{M}_{cnt}(\tau) d\tau}} \tag{4.88}$$

where

$$\overline{\tau}_{cnt} = \frac{\int_{-\infty}^{\infty} \tau \mathtt{M}_{cnt}(\tau) d\tau}{\int_{-\infty}^{\infty} \mathtt{M}_{cnt}(\tau) d\tau} \tag{4.89}$$

For a Rayleigh channel model, two typical curves are now given for $\mathtt{M}_{cnt}(\tau)$.

1. *Gaussian, unilateral*

$$\mathtt{M}_{cnt}(\tau) = \begin{cases} \sqrt{\frac{2}{\pi}} \frac{1}{\overline{\tau}_{rms,cnt}} e^{-\tau^2/(2\overline{\tau}^2_{rms,cnt})}, & \tau \geq 0 \\ 0, & \text{otherwise} \end{cases} \tag{4.90}$$

2. *Exponential, unilateral*

$$\mathtt{M}_{cnt}(\tau) = \begin{cases} \frac{1}{\overline{\tau}_{rms,cnt}} e^{-\tau/\overline{\tau}_{rms,cnt}}, & \tau \geq 0 \\ 0, & \text{otherwise} \end{cases} \tag{4.91}$$

As $\mathtt{M}_{cnt}(\tau)$ does not model LOS, once we select a suitable sampling period T_Q and a corresponding number of gains N_c, for $\tau_i = iT_Q$, we set

$$\mathtt{M}(iT_Q) = \begin{cases} s\mathtt{M}_{cnt}(0) + C_0^2 & i = 0 \\ s\mathtt{M}_{cnt}(iT_Q) & i = 1, \ldots, N_c - 1 \end{cases} \tag{4.92}$$

where s is a scaling factor to satisfy constraint (4.82).

Coherence bandwidth

The dispersion of $\mathtt{M}(\tau)$ is often measured by an equivalent description of the channel in the frequency domain using a parameter called *coherence bandwidth*, B_{coh}, which is inversely proportional to $\overline{\tau}_{rms}$. B_{coh} gives the range of frequencies over which the channel can be assumed to be flat, with approximately constant gain and linear phase. If B_{coh} is defined as the bandwidth over which the frequency correlation function is above 0.5, then approximately it is [3]

$$B_{coh} = \frac{1}{5\overline{\tau}_{rms}} \tag{4.93}$$

For example, in rural and hilly areas, B_{coh} is on the order of MHz and hundreds of kHz, respectively. For digital transmissions over such channels, we observe that if $\overline{\tau}_{rms}$ is about 20% of the symbol period or larger, then signal distortion is not negligible. Equivalently, if B_{coh} defined as (4.93) is lower than the modulation rate, then we speak of a *frequency selective fading* channel. Otherwise the channel is denoted as *flat fading*. Note that in the presence of flat fading, the received signal may vanish completely, whereas frequency selective fading produces several replicas of the transmitted signal at the receiver, so that a suitably designed receiver can recover the transmitted information.

Example 4.1.2 (Power delay profile)

We compute the average rms delay spread for the multipath delay profile $\mathtt{M}(0) = -20$ dB, $\mathtt{M}(T_Q) = -10$ dB, $\mathtt{M}(2T_Q) = -10$ dB, $\mathtt{M}(5T_Q) = 0$ dB, with $T_Q = 1$ μs, and determine the *coherence bandwidth*, defined as $B_{coh} = \frac{1}{5\overline{\tau}_{rms}}$.

From (4.73), we have

$$\overline{\tau} = \frac{(1)(5) + (0.1)(1) + (0.1)(2) + (0.01)(0)}{0.01 + 0.1 + 0.1 + 1} = 4.38 \text{ μs} \tag{4.94}$$

and

$$\overline{\tau^2} = \frac{(1)(5)^2 + (0.1)(1)^2 + (0.1)(2)^2 + (0.01)(0)}{0.01 + 0.1 + 0.1 + 1} = 21.07 \text{ (μs)}^2 \tag{4.95}$$

Therefore, we get

$$\overline{\tau}_{rms} = \sqrt{21.07 - (4.38)^2} = 1.37 \text{ μs} \tag{4.96}$$

Consequently, the coherence bandwidth of the channel is equal to $B_{coh} = 146$ kHz.

Doppler spectrum

We now analyse the WSSUS channel model with reference to time variations due to the motion of terminals and/or surrounding objects. First, we introduce the correlation function of the channel frequency response taken at instants t and $t - \Delta t$, and, respectively, at frequencies f and $f - \Delta f$ as

$$\mathtt{r}_{\mathcal{G}_C}(t, t - \Delta t; f, f - \Delta f) = E[\mathcal{G}_C(t,f)\mathcal{G}_C^*(t - \Delta t, f - \Delta f)] \tag{4.97}$$

with

$$\mathcal{G}_C(t,f) = \int_{-\infty}^{+\infty} g_{Ch}^{(bb)}(t, \tau)\, e^{-j2\pi f \tau}\, d\tau \tag{4.98}$$

By inserting (4.98) into (4.97) we find that $\mathtt{r}_{\mathcal{G}_C}$ depends only on Δt and Δf. Moreover, it holds

$$\mathtt{r}_{\mathcal{G}_C}(\Delta t; \Delta f) = \sum_{i=0}^{Nc-1} \mathtt{r}_g(\Delta t; \tau_i) e^{-j2\pi \Delta f \tau_i} \tag{4.99}$$

i.e. $\mathtt{r}_{\mathcal{G}_C}(\Delta t; \Delta f)$ is the Fourier transform of $\mathtt{r}_g(\Delta t; \tau_i)$ as a function of τ_i. The Fourier transform of $\mathtt{r}_{\mathcal{G}_C}$ as a function of Δt, for a given Δf, is given by

$$\mathcal{P}_{\mathcal{G}_C}(\alpha; \Delta f) = \int_{-\infty}^{\infty} \mathtt{r}_{\mathcal{G}_C}(\Delta t; \Delta f) e^{-j2\pi\alpha(\Delta t)} d(\Delta t) \tag{4.100}$$

The time variation of the frequency response is measured by $\mathcal{P}_{\mathcal{G}_C}(\alpha; 0)$.

Now, we define the *Doppler spectrum* $\mathcal{D}(\alpha)$, which represents the power of the Doppler shift for different values of the frequency α as the Fourier transform of the autocorrelation function of the impulse response in correspondence of a given delay τ_i, evaluated at two different instants, separated by Δt, i.e.

$$\mathcal{D}(\alpha) = \int_{-\infty}^{+\infty} \frac{\mathtt{r}_g(\Delta t; \tau_i)}{\mathtt{r}_g(0; \tau_i)}\, e^{-j2\pi\alpha\Delta t}\, d(\Delta t) \tag{4.101}$$

Note that we omitted the delay τ_i in $\mathcal{D}(\alpha)$, as we assume the same Doppler spectrum for all delays.[6]

The term $\mathtt{r}_g(0; \tau_i)$ in (4.101) represents a normalization factor such that

$$\int_{-\infty}^{+\infty} \mathcal{D}(\alpha)\, d\alpha = 1 \tag{4.102}$$

By letting $\mathtt{r}_d(\Delta t) = \mathcal{F}^{-1}[\mathcal{D}(\alpha)]$, from (4.101) we note that $\mathtt{r}_g(\Delta t; \tau_i)$ is a separable function,

$$\mathtt{r}_g(\Delta t; \tau_i) = \mathtt{r}_d(\Delta t) \cdot \mathtt{r}_g(0; \tau_i) = \mathtt{r}_d(\Delta t)\, \mathtt{M}(\tau_i) \tag{4.103}$$

with (see (4.102))

$$\mathtt{r}_d(0) = 1 \tag{4.104}$$

and $\mathtt{M}(\tau)$ such that (4.75) holds. With the above assumptions, the following equality holds

$$\mathcal{D}(\alpha) = \mathcal{P}_{\mathcal{G}_C}(\alpha; 0) \tag{4.105}$$

Therefore, from (4.100) $\mathcal{D}(\alpha)$ can also be obtained as the Fourier transform of $\mathtt{r}_{\mathcal{G}_C}(\Delta t; 0)$, which in turn can be determined by transmitting a sinusoidal signal (hence $\Delta f = 0$) and estimating the autocorrelation function of the amplitude of the received signal.

---

6 In very general terms, we could have a different Doppler spectrum for each path, or gain g_i, of the channel.

Coherence time

The maximum frequency of a Doppler spectrum support is called the Doppler spread of a given channel as from (4.58) for $\theta = 0$, i.e.

$$f_d = \frac{v_p}{\lambda} = \frac{v_p f_0}{c} \tag{4.106}$$

The Doppler spread gives a measure of the channel fading rate.

Variations in time of the channel are often measured by the coherence time T_{coh}, as the time interval within which a channel can be assumed to be time invariant or static. Hence, T_{coh} is inversely proportional to f_d.

For a digital modulation system with symbol period T, we usually say that the channel is *slow fading* if $f_d T < 10^{-3}$ and *fast fading* if $f_d T > 10^{-2}$.

Doppler spectrum models

A widely used model, known as the *Jakes model* or *classical Doppler spectrum*, to represent the Doppler spectrum is due to Clarke. It is given in Figure 4.19 with

$$\mathcal{D}(f) = \begin{cases} \dfrac{1}{\pi f_d} \dfrac{1}{\sqrt{1-(f/f_d)^2}}, & |f| \le f_d \\ 0, & \text{otherwise} \end{cases} \tag{4.107}$$

with inverse Fourier transform (see also (4.103))

$$\mathtt{r}_d(\Delta t) = J_0(2\pi f_d \Delta t) \tag{4.108}$$

where

$$J_0(x) = \frac{1}{\pi} \int_0^{\pi} e^{jx\cos\theta} d\theta \tag{4.109}$$

is the Bessel function of the first type and order zero. The model of the Doppler spectrum described above agrees with the experimental results obtained for mobile radio channels.

For *indoor* radio channels, the Doppler spectrum can be modelled as

$$\mathcal{D}(f) = \begin{cases} \dfrac{1}{2f_d}, & |f| \le f_d \\ 0, & \text{elsewhere} \end{cases} \tag{4.110}$$

with a corresponding

$$\mathtt{r}_d(\Delta t) = \text{sinc}(2f_d \Delta t) \tag{4.111}$$

A further model assumes that the Doppler spectrum is described by a second- or third-order Butterworth filter with the 3 dB cutoff frequency equal to f_d.

Power angular spectrum

A receive antenna sees waves impinging at different angles; so for each angle, there is a certain amount of the overall power that the antenna receives.

If we indicate with ϑ the horizontal angle of aperture at the receive antenna (or array antenna), we introduce the density function $\mathcal{A}(\vartheta)$, called *angular spectrum*, which indicates the percentage of power received at angle ϑ. As in previous analogous definitions, we also define the rms angle spread associated to $\mathcal{A}(\vartheta)$. The rms angular spread depends on the specific environment, and is high in urban centres and low in rural scenarios.

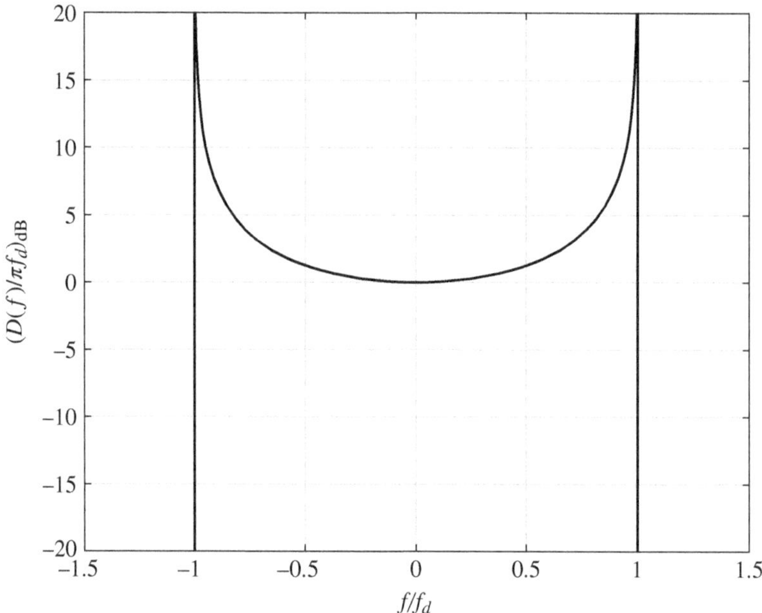

Figure 4.19 Classical Doppler spectrum.

Coherence distance

The phenomena behind the angular spread have also an impact on space-selective fading, characterized by the *coherence distance* D_{coh} defined as the distance in space within which received signals are strongly correlated, i.e. fading effects are similar. D_{coh} is inversely proportional to the rms angle spread. Hence, antenna array elements that are spaced above D_{coh} tend to experience uncorrelated fading. Note that the coherence distance is used for example to model space correlation of shadowing (see (4.52)).

For example, in Section 4.1.11, we will see that in a cellular scenario with uniformly distributed scatterers around a terminal, $D_{coh} = \lambda/2$; hence, for $f_0 = 1.8$ GHz, we have $D_{coh} = 0.08$ m. In general, typical values of D_{coh} at the base station are from 3λ to 20λ, respectively, in hilly/dense urban areas and flat rural areas, while at terminals $D_{coh} \in [\lambda/4,\ 5\lambda]$. Overall, D_{coh} strongly depends on both the terrain and the antenna height, and in any case the angle spread does not exceed the antenna beamwidth.

On fading

A signal that propagates in a radio channel for mobile communications undergoes a fading that depends on the channel characteristics. In particular, whereas the delay spread due to multipath leads to dispersion in the time domain and therefore frequency-selective fading, the Doppler spread causes time-selective fading.

Fading can be divided into *flat fading* and *frequency selective fading*. In the first case, the channel has a constant gain; in other words, the inverse of the transmitted signal bandwidth is much larger than the delay spread of the channel and $g(t, \tau)$ can be approximated by a delta function, with random amplitude and phase, centred at $\tau = 0$. In the second case, instead the channel has a varying frequency response within the passband of the transmitted signal and consequently the signal undergoes frequency selective fading; these conditions occur when the inverse of the transmitted signal bandwidth is of the same order or larger than the coherence bandwidth of the channel. The received signal consists of several attenuated and delayed versions of the transmitted signal.

A channel can be *fast fading* or *slow fading*. In a *fast fading* channel, the impulse response of the channel changes very fast, with coherence time of the channel on the order of the symbol period; this condition leads to signal distortion, which increases with increasing Doppler spread. Usually, there are no remedies to compensate for such distortion unless the symbol period is decreased; on the other hand, this choice leads to larger intersymbol interference. In a *slow fading* channel, the impulse response changes much more slowly with respect to the symbol period. In general, the channel can be assumed as time invariant for a time interval that is proportional to the inverse of the rms Doppler spread.

4.1.9 Discrete-time model for fading channels

Our aim is to model a transmission channel defined in the continuous-time domain by a channel in the discrete-time domain characterized by sampling period T_Q. We immediately notice that the various delays in (4.65) must be a multiple of T_Q and consequently we need to approximate the delays of the power delay profile (see, e.g. Table 4.1). The discrete-time model of the radio channel is represented, as illustrated in Figure 4.20, by a time-varying linear filter whose coefficient g_i corresponds to the complex gain of the ray i with delay iT_Q, $i = 0, 1, \dots, N_c - 1$. In the case of flat fading, $N_c = 1$.

If the channel is time invariant ($f_d = 0$), all coefficients $\{g_i\}$, $i = 0, \dots, N_c - 1$, are constant, and are obtained as realizations of N_c random variables. In general, however, the coefficients $\{g_i\}$ are random processes. To generate each process $g_i(kT_Q)$, the scheme of Figure 4.21 is used, where $\overline{w}_i(\ell T_P)$ is a complex-valued Gaussian white noise with zero mean and unit variance, h_{ds} is a narrowband filter that produces a signal g'_i with the desired Doppler spectrum, and h_{int} is an interpolator filter. We choose T_P such that and $1/10 \leq f_d T_P \leq 1/5$. Note that, as $f_d T_Q \ll 1$, a direct realization of $g'_i(kT_Q)$ by filtering $\overline{w}_i(lT_p)$ is not feasible.

The interpolator output signal is then multiplied by σ_i, which imposes the desired power delay profile.

If the channel model includes a deterministic component for ray i, a constant C_i must be added to the random component $\tilde{g}_i$. Here we assume that only C_0 may be different from zero. Furthermore, if the channel model includes a Doppler shift f_{s_i} for the i-th branch, then we need to multiply the term $C_i + \tilde{g}_i$ by the exponential function $\exp(j2\pi f_{s_i} kT_Q)$. In order to ensure average unit power of the channel (see (4.66)), g'_i has unit statistical power[7] and $\{\sigma_i\}$ satisfy (4.77). If we start from a continuous-time model

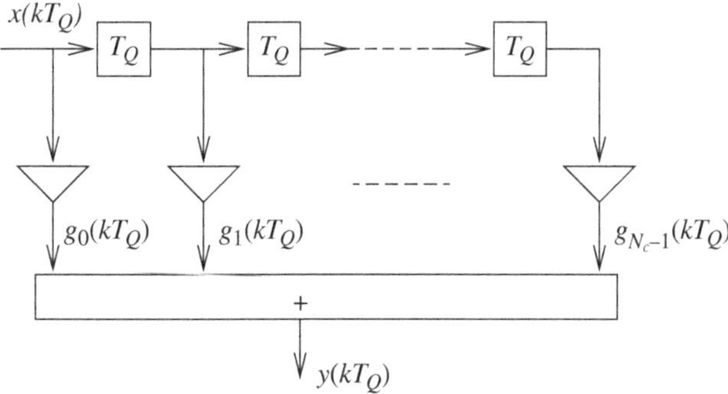

Figure 4.20 Discrete time model of a radio channel.

[7] It is $\mathtt{M}_{g'_i} = 1$ if $\mathtt{M}_{\overline{w}_i} = 1$, and filter has unit energy ($\sum_{i=0}^{\infty} |h_{ds}(iT_P)|^2 = 1$). If T_P/T_Q is very large, interpolation by h_{int} does not change the short-time signal power.

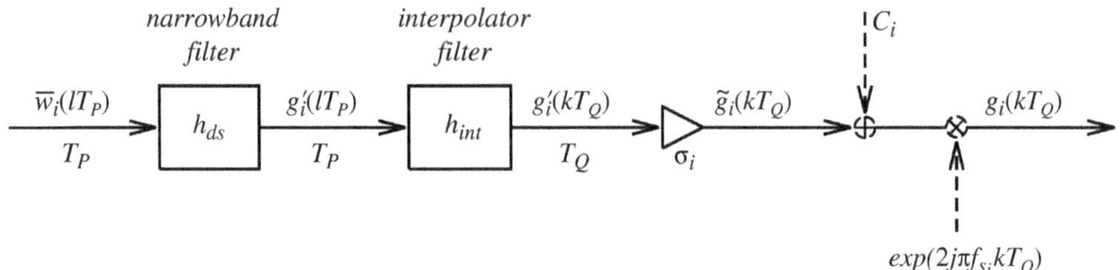

Figure 4.21 Model to generate the i-th coefficient of a time-varying channel.

of the power delay profile, we need to obtain a sampled version as shown in (4.92). Hence, from the assigned value of the Rice factor K, we determine C_0 by (4.81). Next, by imposing the power delay profile $\mathtt{M}(\tau_i)$, we have

$$\sigma_i^2 = \frac{\mathtt{M}_{cnt}(iT_Q)}{\sum_{j=0}^{N_c-1} \mathtt{M}_{cnt}(jT_Q)}[1 - C_0^2], \qquad i = 0, 1 \dots, N_c - 1 \tag{4.112}$$

The same equation (4.112) can be used if $\mathtt{M}_{cnt}(iT_Q)$ is replaced by a table of values as in Table 4.1.

Generation of a process with a pre-assigned spectrum

The signal g'_i with the desired Doppler spectrum can be obtained in two ways: (1) use the scheme of Figure 4.21 with $|\mathcal{H}_{ds}(f)|^2 = \mathcal{D}(f)$ and (2) generate a set of N_f (at least 10) complex sinusoids with frequencies $\{\pm f_m\}$, $m = 1, \dots, N_f$, in the range from $-f_d$ to f_d.

We analyse the two methods.

(1) *Using a filter:* We give the description of h_{ds} for two cases:

(1.a) *IIR filter with classical Doppler spectrum*: Now h_{ds} is implemented as the cascade of two filters. The first, $H_{ds_1}(z)$, is a finite-impulse-response (FIR) shaping filter with amplitude characteristic of the frequency response given by the square root of (4.107). The second, $H_{ds_2}(z)$, is an infinite-impulse-response (IIR) Chebychev lowpass filter, with cutoff frequency f_d. Table 4.2 reports values of the overall filter parameters for $f_d T_P = 0.1$ [13].

(1.b) *Doppler spectrum defined by a second-order Butterworth filter*: Given $\omega_d = 2\pi f_d$, where f_d is the Doppler spread, the transfer function of the discrete-time filter is

$$H_{ds}(z) = \frac{c_0(1 + z^{-1})^2}{\left(1 + \sum_{n=1}^{2} \mathtt{a}_n z^{-n}\right)} \tag{4.113}$$

where, defining

$$\omega_0 = \tan(\omega_d T_P/2)$$

we have [14]

$$\mathtt{a}_1 = -\frac{2(1 - \omega_0^2)}{1 + \omega_0^2 + \sqrt{2}\,\omega_0} \tag{4.114}$$

$$\mathtt{a}_2 = \frac{1 + \omega_0^4}{(1 + \omega_0^2 + \sqrt{2}\,\omega_0)^2} \tag{4.115}$$

$$c_0 = \frac{1}{4}\,(1 + \mathtt{a}_1 + \mathtt{a}_2) \tag{4.116}$$

Table 4.2: Parameters of an IIR filter which implements a classical Doppler spectrum

$H_{ds}(z) = B(z)/A(z)$ $f_d T_P = 0.1$			
$\{a_n\}$, $n = 0, \ldots, 11$:			
1.0000 e + 0	−4.4153 e + 0	8.6283 e + 0	−9.4592 e + 0
6.1051 e + 0	−1.3542 e + 0	−3.3622 e + 0	7.2390 e + 0
−7.9361 e + 0	5.1221 e + 0	−1.8401 e + 0	2.8706 e − 1
$\{b_n\}$, $n = 0, \ldots, 21$:			
1.3651 e − 4	8.1905 e − 4	2.0476 e − 3	2.7302 e − 3
2.0476 e − 3	9.0939 e − 4	6.7852 e − 4	1.3550 e − 3
1.8067 e − 3	1.3550 e − 3	5.3726 e − 4	6.1818 e − 5
−7.1294 e − 5	−9.5058 e − 5	−7.1294 e − 5	−2.5505 e − 5
1.3321 e − 5	4.5186 e − 5	6.0248 e − 5	4.5186 e − 5
1.8074 e − 5	3.0124 e − 6		

The filter output gives

$$g_i'(\ell T_P) = -a_1\, g_i'((\ell-1)T_P) - a_2\, g_i'((\ell-2)T_P) + c_0(\tilde{w}_i(\ell T_P) + 2\tilde{w}_i((\ell-1)T_P) + \tilde{w}_i((\ell-2)T_P)) \tag{4.117}$$

(2) *Using sinusoidal signals:* Let

$$g_i'(\ell T_P) = \sum_{m=1}^{N_f} A_{i,m}[e^{j(2\pi f_m \ell T_P + \alpha_{i,m})} + e^{-j(2\pi f_m \ell T_P + \beta_{i,m})}] \tag{4.118}$$

The spacing between the different frequencies is Δf_m; letting $f_1 = \Delta f_1/2$, for $m > 1$ we have $f_m = f_{m-1} + \Delta f_m$. Each Δf_m can be chosen as a constant,

$$\Delta f_m = \frac{f_d}{N_f} \tag{4.119}$$

or, defining $K_d = \int_0^{f_d} \mathcal{D}^{1/3}(f)df$, as

$$\Delta f_m = \frac{K_d}{N_f \mathcal{D}^{1/3}(f_m)}, \qquad m = 1, \ldots, N_f \tag{4.120}$$

Suppose $\bar{f}_0 = 0$ and $\bar{f}_m = \bar{f}_{m-1} + \Delta f_m$, $m = 1, \ldots, N_f$, the choice (4.120) corresponds to minimizing the error

$$\sum_{m=1}^{N_f} \int_{\bar{f}_{m-1}}^{\bar{f}_m} (f_m - f)^2 \mathcal{D}(f)df \tag{4.121}$$

The phases $\alpha_{i,m}$, $\beta_{i,m}$ are uniformly distributed in $[0, 2\pi)$ and statistically independent to ensure that the real and imaginary parts of g_i' are *statistically independent*. The amplitude is given by

$$A_{i,m} = \sqrt{\mathcal{D}(f_m)\Delta f_m} \tag{4.122}$$

If $\mathcal{D}(f)$ is flat, by the central limit theorem (for $N_f \to \infty$), we can claim that g_i' is a Gaussian process; if instead $\mathcal{D}(f)$ presents some frequencies with large amplitude, $A_{i,m}$ must be generated as a Gaussian random variable with zero mean and variance $\mathcal{D}(f_m)\Delta f_m$.

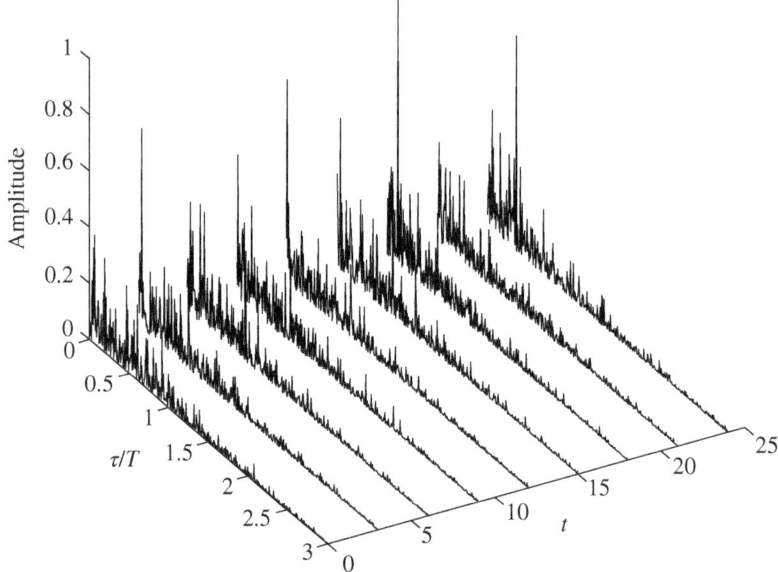

Figure 4.22 Nine realizations of $|g_{Ch}^{(bb)}(t, \tau)|$ for a Rayleigh channel with an exponential power delay profile having $\overline{\tau}_{rms,cnt} = 0.5\ T$.

Figure 4.22 shows nine realizations of the amplitude of the impulse response of a Rayleigh channel obtained by the simulation model of Figure 4.20, for an exponential power delay profile with $\overline{\tau}_{rms,cnt} = 0.5\ T$. The Doppler frequency f_d was assumed zero. We point out that the parameter $\overline{\tau}_{rms}$ provides scarce information on the actual behaviour of $g_{Ch}^{(bb)}$, which can scatter for a duration equal to 4–5 times $\overline{\tau}_{rms}$.

4.1.10 Discrete-space model of shadowing

The aim is to simulate $(\xi)_{dB}[\boldsymbol{n}]$ in a two-dimensional (2D) space with discrete coordinates

$$\boldsymbol{c} = (c_1, c_2) = (n_1 A, n_2 A) = (n_1, n_2)A, \qquad (n_1, n_2) \in \mathcal{S} \subset \mathbb{Z}^2 \tag{4.123}$$

with A a suitable (space) step size, given the autocorrelation function $\mathtt{r}_{(\xi)_{dB}}(\Delta)$, a circularly symmetric function, where

$$\Delta = \sqrt{c_1^2 + c_2^2} = A\sqrt{n_1^2 + n_2^2} \tag{4.124}$$

We need to introduce the 2D discrete Fourier transform (DFT). Given a pulse $h_{sh}(\boldsymbol{n})$, its 2D DFT on the domain

$$\mathcal{S} = \left\{-\frac{N_1}{2}, \dots, 0, \dots, \frac{N_1}{2} - 1\right\} \times \left\{-\frac{N_2}{2}, \dots, 0, \dots, \frac{N_2}{2} - 1\right\} \tag{4.125}$$

is denoted as $\mathcal{H}_{sh}(\boldsymbol{k})$, $\boldsymbol{k} = (k_1, k_2) \in \mathcal{S}$, and defined as [15]

$$\mathcal{H}_{sh}(\boldsymbol{k}) = \mathrm{DFT}_2[h_{sh}(\boldsymbol{n})] = \sum_{n_1=-\frac{N_1}{2}}^{\frac{N_2}{2}-1} \sum_{n_2=-\frac{N_2}{2}}^{\frac{N_2}{2}-1} h_{sh}(n_1, n_2) e^{-k2\pi \frac{k_1 n_1}{N_1}} e^{-j2\pi \frac{k_2 n_2}{N_2}} \tag{4.126}$$

while in turn, we have

$$h_{sh}(\boldsymbol{n}) = \mathrm{DFT}_2^{-1}[\mathcal{H}_{sh}(\boldsymbol{k})] = \frac{1}{N_1 N_2} \sum_{k_1=-\frac{N_1}{2}}^{\frac{N_1}{2}-1} \sum_{k_2=-\frac{N_2}{2}}^{\frac{N_2}{2}-1} \mathcal{H}_{sh}(k_1, k_2) e^{j2\pi \frac{k_1 n_1}{N_1}} e^{j2\pi \frac{k_2 n_2}{N_2}} \tag{4.127}$$

We note that $\mathcal{H}_{sh}(\boldsymbol{k})$ is periodic in k_1 and k_2, of period N_1 and N_2, respectively. We note that a similar property holds for $h_{sh}(\boldsymbol{n})$ in (4.127). In fact, (4.126) leads to

$$\sum_{\ell_1=-\infty}^{+\infty} \sum_{\ell_2=-\infty}^{+\infty} h_{sh}(n_1 - \ell_1 N_1, n_1 - \ell_2 N_2) = h_{sh}(n_1, n_2) \tag{4.128}$$

Given $\mathtt{r}_{(\xi)_{dB}}\left(A\sqrt{n_1^2+n_2^2}\right)$, we evaluate its PSD

$$\mathcal{P}(\boldsymbol{k}) = \mathrm{DFT}_2[\mathtt{r}_{(\xi)_{dB}}(A\sqrt{n_1^2+n_2^2})] \qquad \boldsymbol{k} \in \mathcal{S} \tag{4.129}$$

It is interesting to recall that the Fourier transform of a circularly symmetric function is still circularly symmetric [15].

Now, we consider

$$\mathcal{H}_{sh}(\boldsymbol{k}) = \mathcal{K}\sqrt{\mathcal{P}(\boldsymbol{k})} \quad \text{and} \quad h_{sh}(\boldsymbol{n}) = \mathrm{DFT}_2^{-1}[\mathcal{H}_{sh}(\boldsymbol{k})] \tag{4.130}$$

where $\mathcal{K}$ is such that h_{sh} has unit energy, i.e.

$$\sum_{\boldsymbol{n}\in\mathcal{S}} |h_{sh}(\boldsymbol{n})|^2 = \frac{1}{N_1 N_2} \sum_{\boldsymbol{k}\in\mathcal{S}} |\mathcal{H}_{sh}(\boldsymbol{k})|^2 = 1 \tag{4.131}$$

Coming back to the problem of simulating $(\xi)_{dB}[\boldsymbol{n}]$, we consider two solutions.

1. *Filtering approach*: If by setting to zero small values of h_{sh} it turns out that its support, $\mathcal{S}_h$, is much smaller that $\mathcal{S}$, the following derivation of $(\xi)_{dB}[\boldsymbol{n}]$ is very efficient from a computational point of view.
   We generate an i.i.d. Gaussian noise across the space $\mathcal{S}$,

   $$w(\boldsymbol{n}) \sim \mathcal{CN}(0, \sigma^2_{(\xi)_{dB}}), \quad \boldsymbol{n} \in \mathcal{S} \tag{4.132}$$

   Then, filtering $w(\boldsymbol{n})$ by $h_{sh}(\boldsymbol{n})$ in (4.130) yields the desired result, i.e.

   $$(\xi)_{dB}[\boldsymbol{n}] = \sum_{\boldsymbol{p}\in\mathcal{S}_h} h_{sh}(\boldsymbol{p}) w(\boldsymbol{n}-\boldsymbol{p}), \quad \boldsymbol{n} \in \mathcal{S} \tag{4.133}$$

   A word of caution: on the boundaries of $\mathcal{S}$ the convolution (4.133) is affected by the transient of h_{sh}, hence the corresponding values should be dropped.
   The filtering approach to generate $(\xi)_{dB}[\boldsymbol{n}]$ is also very efficient if we are interested in evaluating it only on a small subset of $\mathcal{S}$ rather than over the whole region, e.g. along a given path inside $\mathcal{S}$.
2. *Frequency domain approach*: If the above conditions are not verified, the general approach is preferable [16].
   Let $\mathcal{W}$ be an i.i.d. Gaussian noise in the frequency domain,

   $$\mathcal{W}(\boldsymbol{k}) \sim \mathcal{CN}(0, N_1 N_2 \sigma^2_{(\xi)_{dB}}), \quad \boldsymbol{k} \in \mathcal{S} \tag{4.134}$$

   Then, we multiply $\mathcal{W}(\boldsymbol{k})$ by $\mathcal{H}_{sh}(\boldsymbol{k})$ in (4.130). By taking the inverse DFT, we obtain the desired results, i.e.

   $$(\xi)_{dB}[\boldsymbol{n}] = \mathrm{DFT}_2^{-1}[\mathcal{H}_{sh}(\boldsymbol{k})\mathcal{W}(\boldsymbol{k})], \quad \boldsymbol{n} \in \mathcal{S} \tag{4.135}$$

   As (4.135) is implementing a circular convolution, now (4.135) is not affected by the boundary effect.

4.1.11 Multiantenna systems

A way to overcome fading consists in using more antennas (namely an array of omnidirectional antennas) at the either the transmitter or the receiver, or both devices. Due to its importance, the entire Chapter 9 is devoted to this topic. Indeed, the model described until now holds for any couple of transmit-receive antennas of the arrays. Here we just provide the geometrical or statistical relations between the rays for various antenna couples.

Line of sight

For a single source (that can be either a transmit antenna or a scatterer), we consider the scenario shown in Figure 4.23, with an array of receive antennas, and where α represents the angle of arrival (AoA), i.e. the angle between the direction of arrival (DoA) of the ray and the line of the antenna array. Note that we consider a direct transmission between the source and the receive antenna, thus this can be considered as an LOS scenario. Let $D^{(r_1,r_2)}$ be the distance between receive antennas r_1 and r_2, and $d^{(r)}$ the distance between the source and receive antenna r. Neglecting attenuation, signals at the antennas are affected by phase shifts due to their distance from the source. For the NB model of (4.28), the signal received by antenna r is affected by the phase offset (see (4.29))

$$\mathcal{G}_C^{(r)} = e^{-j2\pi\frac{d^{(r)}}{\lambda}} \tag{4.136}$$

Hence, the correlation function of the channel at antenna r_1 and r_2, for a given source is

$$\mathtt{r}_{\mathcal{G}_C}^{(r_1,r_2)} = \mathcal{G}_C^{(r_1)}\mathcal{G}_C^{(r_2)*} = e^{-j2\pi\frac{d^{(r_1)}-d^{(r_2)}}{\lambda}} = \begin{cases} e^{-j2\pi\frac{D^{(r_1,r_2)}\sin\alpha}{\lambda}}, & r_1 \neq r_2 \\ 1, & r_1 = r_2 \end{cases} \tag{4.137}$$

where we used the geometric approximation (see Figure 4.23)

$$d^{(r_1)} - d^{(r_2)} \simeq D^{(r_1,r_2)}\sin\alpha \tag{4.138}$$

Let us now compute the correlation of the channel at antennas r_1 and r_2, by averaging also on the AoA α, over its angular spectrum $\mathcal{A}(\alpha)$, with support $(-\vartheta_{Rc}/2, \vartheta_{Rc}/2)$, thus obtaining

$$\mathtt{r}_{\mathcal{G}_C}^{(r_1,r_2)} = \begin{cases} \int_{-\vartheta_{Rc}/2}^{+\vartheta_{Rc}/2} e^{-j2\pi\frac{D^{(r_1,r_2)}\sin\alpha}{\lambda}}\mathcal{A}(\alpha)d\alpha, & r_1 \neq r_2 \\ 1, & r_1 = r_2 \end{cases} \tag{4.139}$$

When the probability density function of direction of arrival is uniform and $\vartheta_{Rc} = 2\pi$, we have $\mathtt{r}^{(r_1,r_2)} = J_0(2\pi D^{(r_1,r_2)}/\lambda)$, where $J_0(\cdot)$ is defined in (4.109). As $J_0(\pi) \simeq 0$, to achieve zero correlation, the antenna

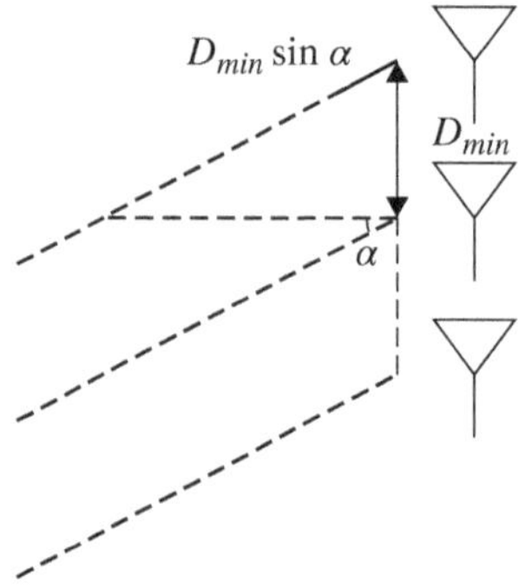

Figure 4.23 Propagation model for an LOS scenario.

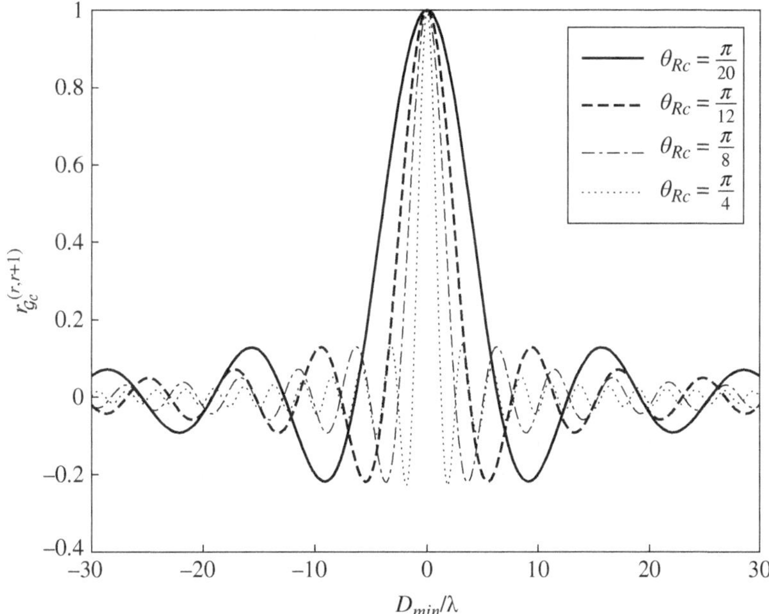

Figure 4.24 Channel correlation (4.139) at adjacent antennas in a fading MIMO channel with a uniformly distributed direction of arrival.

elements should be spaced by $\lambda/2$. Figure 4.24 shows the correlation between adjacent antennas, where $D^{(r,r+1)} = D_{\min}$, for small values of ϑ_{Rc}.

As it has already been observed, antennas at a large enough distance will receive essentially uncorrelated signals. However, in handsets or terminals, large separations among the antennas may not be feasible, whereas, when either the transmitter or receiver are not surrounded by scatterers, the antenna signals are correlated. Overall, small antenna spacing and lack of scattering cause the antenna signals to be correlated [4].

Similar considerations hold in the presence of multiple antennas transmitting signals that hit a source, before been reflected to a receive antenna. The geometric observations now hold for the phase shifts introduced by the path from the various transmit antennas to the source, where α now becomes the *angle of departure* (AoD) between the transmit antenna array and the source.

Discrete-time model

We aim at modelling a transmission channel between N_{Tx} transmit antennas and N_{Rc} receive antennas. For the sake of simplicity, we assume that all antennas are on the same plane, and scatterers are at a far distance from both the transmit and the receive antennas, thus a single AoD (AoA) is enough to characterize the path between the transmit (receiver) antennas and each scatterer. In general, a discrete-time model for multiple-input-multiple-output (MIMO) channel includes all the effects of channels between two antennas, i.e. path loss, fading, shadowing, and Doppler. However, it must also take into account the correlation of channels among different antennas, as described above.

For each couple of transmit-receive antennas (r, t), we have a channel with taps $g_i^{(r,t)}(kT_Q)$, with $r = 1, \dots, N_{Rc}$, and $t = 1, \dots, N_{Tx}$. Two approaches are possible for the generation of the channel, according to the number of scatterers of the channel.

Small number of scatterers

When the number of scatterers is small, they can be grouped into N_c *clusters*, each including the same number of scatterers N_{sct}. Rays of scatterers of cluster i yield the same delay τ_i and the same power σ_i^2. Ray $n = 0, \dots, N_{sct} - 1$, within cluster i has AoD $\alpha_{i,n}^{(Tx)}$ and AoA $\alpha_{i,n}^{(Rc)}$. In general, they are correlated variables: see [17] for a detailed description.

Let us define

$$\boldsymbol{c}_{Tx,i,n} = [\cos \alpha_{i,n}^{(Tx)}, \sin \alpha_{i,n}^{(Tx)}]^T, \qquad \boldsymbol{c}_{Rc,i,n} = [\cos \alpha_{i,n}^{(Rc)}, \sin \alpha_{i,n}^{(Rc)}]^T \tag{4.140}$$

and the position of antennas t and r on the plane at time kT_Q as

$$\boldsymbol{p}_{Tx}^{(t)}(kT_Q) = [p_{Tx,1}^{(t)}(kT_Q), p_{Tx,2}^{(t)}(kT_Q)]^T, \qquad \boldsymbol{p}_{Rc}^{(r)}(kT_Q) = [p_{Rc,1}^{(r)}(kT_Q), p_{Rc,2}^{(r)}(kT_Q)]^T \tag{4.141}$$

where in any case we assume that antennas are close to each other (within a few λ). We indicate the velocities of the transmitter and receiver (we assume that scatterers are static) as

$$\boldsymbol{v}_{Tx}^{(t)}(kT_Q) = [v_{Tx,1}^{(t)}(kT_Q), v_{Tx,2}^{(t)}(kT_Q)]^T, \qquad \boldsymbol{v}_{Rc}^{(r)}(kT_Q) = [v_{Rc,1}^{(r)}(kT_Q), v_{Rc,2}^{(r)}(kT_Q)]^T \tag{4.142}$$

Up to an irrelevant phase shift and neglecting scaling by the path loss due to distance and shadowing components (see (4.64)), the model for channel tap i at time kT_Q between antennas t and r is (see also (4.136))

$$\begin{aligned} g_i^{(r,t)}(kT_Q) = {} & \frac{1}{\sqrt{N_{sct}}} \sum_{n=0}^{N_{sct}-1} e^{-j2\pi \frac{\boldsymbol{c}_{Tx,i,n}^T \boldsymbol{p}_{Tx}^{(t_A)}(kT_Q)}{\lambda}} e^{-j2\pi \frac{\boldsymbol{c}_{Rc,i,n}^T \boldsymbol{p}_{Rc}^{(r_A)}(kT_Q)}{\lambda}} \\ & \times e^{-j2\pi \frac{\boldsymbol{c}_{Tx,i,n}^T \boldsymbol{v}_{Tx}^{(t_A)}(kT_Q)}{\lambda} kT_Q} e^{-j2\pi \frac{\boldsymbol{c}_{Rc,i,n}^T \boldsymbol{v}_{Rc}^{(r_A)}(kT_Q)}{\lambda} kT_Q} \end{aligned} \tag{4.143}$$

which includes both the antenna correlation and the Doppler effect. This model still assumes a planar wave hitting the receive antennas (see Figure 4.23), i.e. the signal source (cluster i) is at far distance from the antennas. Therefore, (4.143) describes the resulting geometrical model, including the movement of the antennas through their velocity. A similar observation holds from the transmit antennas.

The model with a small number of scatters is particularly suitable for transmissions at mmWave, i.e. with $\lambda \in [1, 10]$ mm, or $f_0 = c/\lambda \in [30{,}300]$ GHz. At these frequencies, the free-space attenuation per unit length is very high, therefore multiple reflections give negligible contributions and only very few ($N_c = 1$ and $N_{sct} = 2$ or 3) scatterers characterize the channel. Hence at these frequencies, multiple antennas are required, as will be seen in more details in Chapter 9.

In this case, (4.143) yields the channel at time kT_Q, which in Chapter 9 will be denoted as

$$\mathcal{G}_C^{(r,t)}(kT_Q) = g_0^{(r,t)}(kT_Q) \tag{4.144}$$

This notation reflects the fact that $\mathcal{G}_C^{(r,t)}(kT_Q)$ is the channel frequency response between the transmit-receive antenna couple. The corresponding flat fading MIMO channel matrix at time kT_Q is $N_{Rc} \times N_{Tx}$ with entries

$$[\mathcal{G}_C(kT_Q)]_{r,t} = \mathcal{G}_C^{(r,t)}(kT_Q), \qquad r = 1, \dots, N_{Rc}, \qquad t = 1, \dots, N_{Tx} \tag{4.145}$$

Large number of scatterers

Here to simplify notation, we drop the Doppler effect, which in any case can be introduced for each channel coefficient next, as described in Section 4.1.9.

For a large number scatterers, we extend the general model of Section 4.1.9; however, we only give the model of the MIMO channel matrix in correspondence of cluster $i = 0$ with lag τ_0. The channel model for other clusters (at different lags) will be the same, with a proper power assignment among various clusters. Let us define the following matrices.

Spatially white noise matrix $\boldsymbol{w}$, $N_{Rc} \times N_{Tx}$, with entries

$$w^{(r,t)} \sim \mathcal{CN}(0,1) \qquad \text{i.i.d.} \tag{4.146}$$

Receive antennas correlation matrix $\boldsymbol{R}^{(Rc)}$, $N_{Rc} \times N_{Rc}$, with entries given by (4.139).

Transmit antennas correlation matrix $\boldsymbol{R}^{(T_x)}$, $N_{Tx} \times N_{Tx}$, with entries given by (4.139) and receive antennas replaced by transmit antennas.

Spatial channel matrix $\boldsymbol{g}_0$, $N_{Rc} \times N_{Tx}$, with entries $g_0^{(r,t)}$.

We have

$$\boldsymbol{g}_0 = (\boldsymbol{R}^{(Rc)})^{1/2}\ \boldsymbol{w}\ (\boldsymbol{R}^{(Tx)T})^{1/2} \tag{4.147}$$

We analyse the correlation matrix of entries of $\boldsymbol{g}_0$. Introducing the notation

$$\text{vec}(\boldsymbol{g}_0) = [g_0^{(1,1)}, \ldots, g_0^{(1,N_{Tx})}, g_0^{(2,1)}, \ldots, g_0^{(N_{Rc},N_{Tx})}]^T \tag{4.148}$$

the correlation among entries of $\boldsymbol{g}_0$ is simply the $N_{Rc}N_{Tx} \times N_{Rc}N_{Tx}$ matrix

$$\boldsymbol{R}_{\text{vec}(\boldsymbol{g}_0)} = E[\text{vec}(\boldsymbol{g}_0)(\text{vec}(\boldsymbol{g}_0))^H] \tag{4.149}$$

If $\otimes$ denotes the Kronecker matrix product[8]

$$\boldsymbol{R}_{\text{vec}(\boldsymbol{g}_0)} = \boldsymbol{R}^{(Rc)} \otimes \boldsymbol{R}^{(Tx)} \tag{4.152}$$

In short notation, it is common to write

$$\boldsymbol{g}_0 \sim \mathcal{CN}(\boldsymbol{0}, \boldsymbol{R}^{(Rc)} \otimes \boldsymbol{R}^{(Tx)}) \tag{4.153}$$

This model approximately holds at frequencies up to 1 GHz.

It is interesting to note that in (4.143) for $N_{sct} \to \infty$, $g_i^{(r_A,t_A)}$ tends to become Gaussian with correlation matrix (4.152). In this case, the two approaches (4.143) and (4.147) yield the same statistical model for the channel matrix.

When the antennas at both the transmitter and the receiver are spaced enough far apart and in the presence of numerous scatters, $\boldsymbol{R}^{(Rc)} = \boldsymbol{I}$ (of size $N_{Rc} \times N_{Rc}$), $\boldsymbol{R}^{(Tx)} = \boldsymbol{I}$ (of size $N_{Tx} \times N_{Tx}$), and

$$\boldsymbol{g}_0 \sim \mathcal{CN}(\boldsymbol{0}, \boldsymbol{I}) \tag{4.154}$$

This is a *spatially rich* MIMO channel model, often used as mathematical model, but seldom observed in practice. Lastly, $g_i^{(r,t)}$ obtained as in (4.153) can be used as input $\overline{w}_i^{(r,t)}(lT_P)$ of $N_{Tx}N_{Rc}$ signal generators in correspondence of the i-th tap of the MIMO dispersive channel, for the extended model including power delay profile and Doppler as detailed in Section 4.1.9.

---

[8] The symbol $\otimes$ denotes the Kronecker matrix product. If

$$\boldsymbol{A} = \begin{bmatrix} a_{1,1} & \ldots & a_{1,M} \\ \vdots & \ddots & \vdots \\ a_{N,1} & \ldots & a_{N,M} \end{bmatrix} \tag{4.150}$$

and $\boldsymbol{B}$ is any matrix, it is

$$\boldsymbol{A} \otimes \boldsymbol{B} = \begin{bmatrix} a_{1,1}\boldsymbol{B} & \ldots & a_{1,M}\boldsymbol{B} \\ \vdots & \ddots & \vdots \\ a_{N,1}\boldsymbol{B} & \ldots & a_{N,M}\boldsymbol{B} \end{bmatrix} \tag{4.151}$$

Blockage effect

An important effect typically observed when operating at mmWave, is the *blockage effect*, i.e. some paths are obstructed by objects. In the discrete-time model, this is taken into account by removing some clusters having AoD and/or AoA within ranges that are generated randomly [17]. An alternative simpler blockage model [18] provides that the cluster is in blockage with probability $e^{-\beta_{bl}d}$ (and not in blockage with probability $1 - e^{-\beta_{bl}d}$), where β_{bl} is a suitable parameter (e.g. $\beta_{bl} = 0.006$), and d is the transmitter–receiver distance. Then, clusters in blockage have different path loss from those not in blockage.

4.2 Telephone channel

A telephone channel is the channel between two end users of the telephone service. It was originally conceived for the transmission of voice, and today is extensively used also for data transmission. The telephone channel utilizes several transmission media, such as symmetrical transmission lines, coaxial cables, optical fibres, radio, and satellite links. Therefore, channel characteristics depend on the particular connection established. As a statistical analysis made in 1983 indicated [19], a telephone channel is characterized by the following disturbances and distortions.

4.2.1 Distortion

The telephone channel has a linear distortion, where the frequency response $\mathcal{G}_{Ch}(f)$ of a telephone channel can be approximated by a passband filter with band in the range of frequencies from 300 to 3400 Hz. The attenuation

$$(\mathrm{a}(f))_{dB} = -20 \log_{10} |\mathcal{G}_{Ch}(f)| \tag{4.155}$$

and of the group delay or envelope delay (see (1.83))

$$\tau(f) = -\frac{1}{2\pi} \frac{d}{df} \arg \mathcal{G}_{Ch}(f) \tag{4.156}$$

as a function of f is shown in Figure 4.25 for two typical channels. The attenuation and envelope delay distortion are normalized by the values obtained for $f = 1004$ Hz and $f = 1704$ Hz, respectively.

Moreover, the channel is also affected by a non-linear distortion caused by amplifiers and by non-linear A-law and μ-law converters (see Ref. [20]).

The channel is also affected by frequency offset caused by the use of carriers for frequency up and downconversion. The relation between the channel input x and output y signals is given in the frequency domain by

$$Y(f) = \begin{cases} X(f - f_{off}), & f > 0 \\ X(f + f_{off}), & f < 0 \end{cases} \tag{4.157}$$

Usually, $f_{off} \leq 5$ Hz.

Lastly, phase jitter is present, which is a generalization of the frequency *offset* (see (4.12)).

4.2.2 Noise sources

Quantization noise:

It is introduced by the digital representation of voice signals and is the dominant noise in telephone channels (see Ref. [20]). For a single quantizer, the signal-to-quantization noise ratio Λ_q has the behaviour illustrated in Figure 4.26.

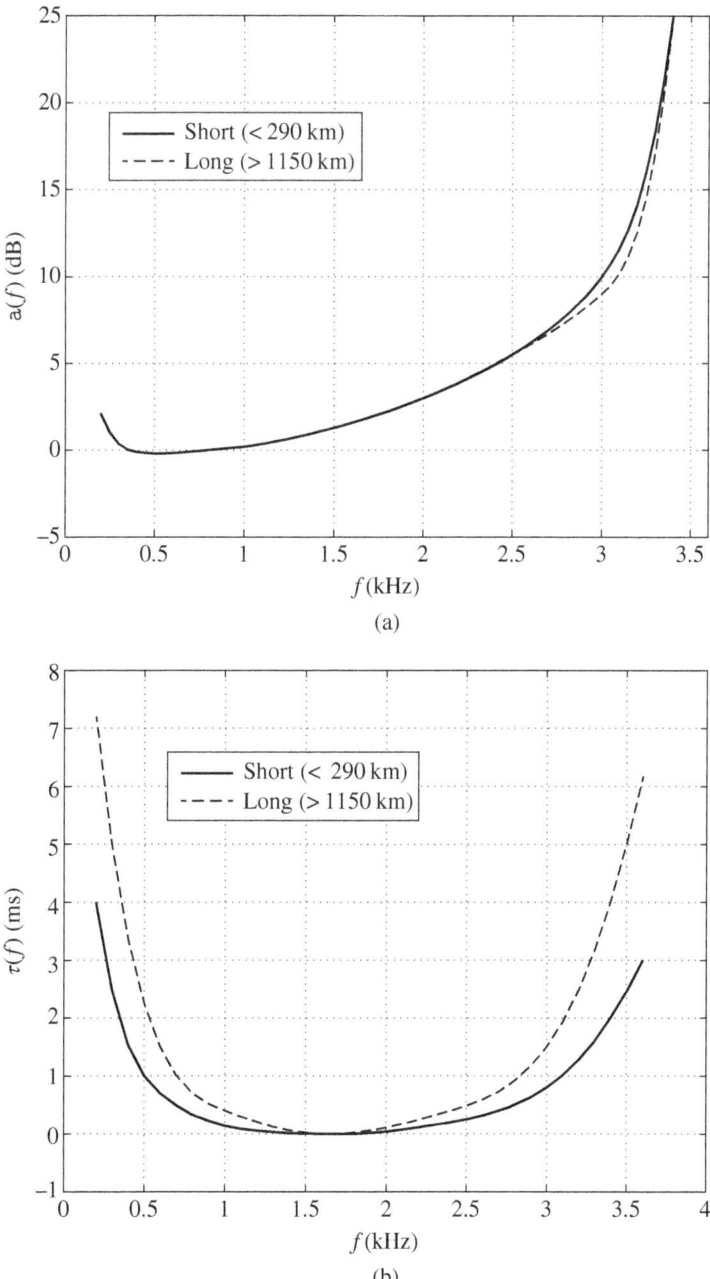

Figure 4.25 (a) Attenuation and (b) envelope delay distortion for two typical telephone channels.

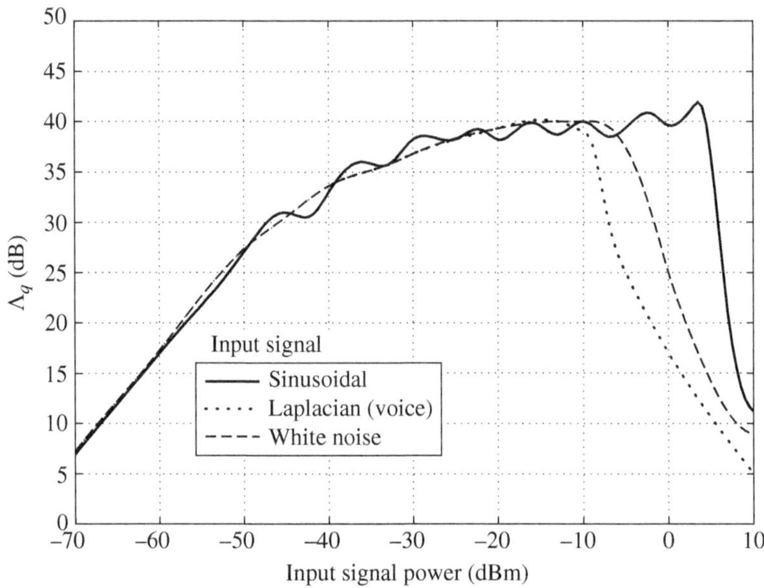

Figure 4.26 Signal to quantization noise ratio as a function of the input signal power for three different inputs.

Thermal noise:

It is present at a level of $20 \div 30$ dB below the desired signal.

4.2.3 Echo

It is caused by the mismatched impedances of the *hybrid*. As illustrated in Figure 4.27, there are two types of echoes:

1. *Talker echo*: Part of the signal is reflected and is received at the transmit side. If the echo is not significantly delayed, then it is practically indistinguishable from the original voice signal, however, it cannot be neglected for data transmission;
2. *Listener echo*: If the echo is reflected a second time, it returns to the listener and disturbs the original signal.

On terrestrial channels, the round-trip delay of echoes is of the order of 10–60 ms, whereas on satellite links, it may be as large as 600 ms. We note that the effect of echo is similar to multipath fading in radio

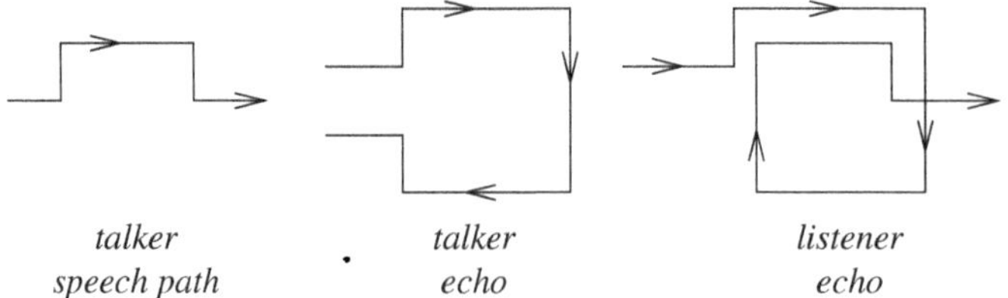

Figure 4.27 Three of the many signal paths in a simplified telephone channel with a single two-to-four wire conversion at each end.

systems. However in the telephone channel, the echo signal can be derived directly from the transmitted signal, while in the radio channel, it can be obtained at the receiver only after detection. Therefore, while in radio system, equalization solutions must be used (with noise propagation effect), specific solutions using the noise-less signal available at the transmitter can be used, in the telephone channel namely:

- echo suppressors that attenuate the unused connection of a four-wire transmission line;
- echo cancellers that cancel the echo at the source (see Section 2.5.5).

Bibliography

[1] Messerschmitt, D.G. and Lee, E.A. (1994). *Digital Communication*, 2e. Boston, MA: Kluwer Academic Publishers.

[2] Feher, K. (1995). *Wireless Digital Communications*. Upper Saddle River, NJ: Prentice-Hall.

[3] Rappaport, T.S. (1996). *Wireless Communications: Principles and Practice*. Englewood Cliffs, NJ: Prentice-Hall.

[4] Pahlavan, K. and Levesque, A.H. (1995). *Wireless Information Networks*. New York, NY: Wiley.

[5] Jakes, W.C. (1993). *Microwave Mobile Communications*. New York, NY: IEEE Press.

[6] Stuber, G.L. (1996). *Principles of Mobile Communication*. Norwell, MA: Kluwer Academic Publishers.

[7] Spilker, J.J. (1977). *Digital Communications by Satellite*. Englewood Cliffs, NJ: Prentice-Hall.

[8] Parsons, J.D. (2000). *The Mobile Radio Propagation Channel*. Chichester: Wiley.

[9] Razavi, B. (1997). *RF Microelectronics*. Englewood Cliffs, NJ: Prentice-Hall.

[10] Someda, C.G. (1998). *Electromagnetic Waves*. London: Chapman & Hall.

[11] 3GPP (2017). E-UTRA; radio frequency (RF) system scenarios.

[12] Gudmundson, M. (1991). Correlation model for shadow fading in mobile radio systems. *Electronics Letters* 27 (1): 2145–2146.

[13] Anastasopoulos, A. and Chugg, K. (1997). An efficient method for simulation of frequency selective isotropic Rayleigh fading. *Proceedings of 1997 IEEE Vehicular Technology Conference*, pp. 2084–2088.

[14] Oppenheim, A.V. and Schafer, R.W. (1989). *Discrete-Time Signal Processing*. Englewood Cliffs, NJ: Prentice-Hall.

[15] Cariolaro, G. (2011). *Unified Signal Theory*. Springer.

[16] Zordan, D., Quer, G., Zorzi, M., and Rossi, M. (2011). Modeling and generation of space-time correlated signals for sensor network fields. *Proceedings of IEEE Global Telecommunications Conference (GLOBECOM)*, pp. 1–6.

[17] 3GPP. (2018) Study on channel model for frequencies from 0.5 to 100 GHz. 3GPP TR 38.901 version 15.0.0 Release 15.

[18] Bai, T., Desai, V., and Heath, R.W. (2014). Millimeter wave cellular channel models for system evaluation. *Proceedings International Conference on Computing, Networking and Communications (ICNC)*, pp. 178–182.

[19] Laboratories, B.T. (1982). *Transmission Systems for Communications*, 5e. Winston, NC: Bell Telephone Laboratories.

[20] Benvenuto, N. and Zorzi, M. (2011). *Principles of Communications Networks and Systems*. Wiley.

Appendix 4.A Discrete-time NB model for mmWave channels

For a transmitter and receiver, both equipped with linear antenna arrays, as in Figure 4.23, let $D_{min}^{(Tx)}$ ($D_{min}^{(Rc)}$) be the distance between adjacent antennas at the transmitter (receiver).

Let us define the suitable phase shift between adjacent transmit (receiver) antennas for ray i with AoD (AoA) $\alpha_i^{(Tx)}$ ($\alpha_i^{(Rc)}$) as

$$\eta_i^{(Tx)} = \frac{D_{min}^{(Tx)}}{\lambda} \sin \alpha_i^{(Tx)}, \qquad \eta_i^{(Rc)} = \frac{D_{min}^{(Rc)}}{\lambda} \sin \alpha_i^{(Rc)} \tag{4.158}$$

Entry (r, t) of matrix $\mathcal{G}_C$ is

$$\mathcal{G}_C^{(r,t)} = \sum_{i=0}^{N_c-1} \breve{g}_i e^{-j2\pi(r-1)\eta_i^{(Rc)}} e^{-j2\pi(t-1)\eta_i^{(Tx)}} \tag{4.159}$$

$r = 1, \dots, N_{Rc}$, $t = 1, \dots, N_{Tx}$, where $\breve{g}_i$ is a complex gain. When comparing (4.143) and (4.159), we observe that (4.143) describes fading due to Doppler using a geometric model (through velocities), while (4.159) describes it using the gain $\breve{g}_i$. We remark once more that, while (4.159) holds for radio transmission at any frequency (with N_c very large), it becomes of practical use at mmWave frequencies, where $N_c = 2$ or 3.

Other antenna geometries, such as rectangular planar arrays require multiple angles to describe the departure (arrival) directions of the signals; still, $\mathcal{G}_C^{(r,t)}$ is the sum of N_c exponential products, one for each angle.

4.A.1 Angular domain representation

The mmWave channel model finds a simple representation in the *angular domain* (also denoted *virtual channel*), i.e. by taking the 2D DFT of the channel matrix. For example, with reference to the case of

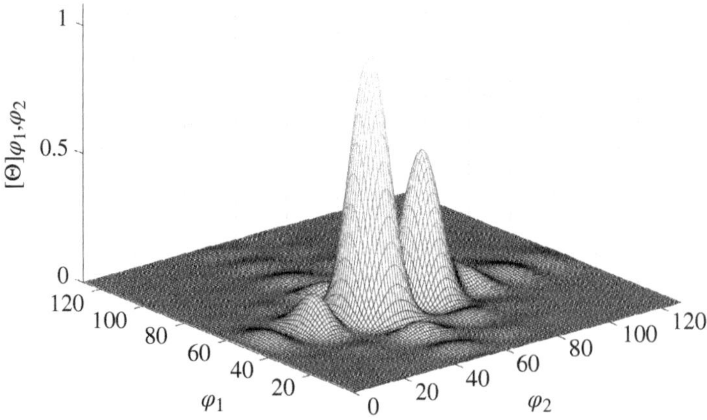

Figure 4.28 Example of mmWave channel in the angular domain.

transmit and receive antennas implemented by linear arrays, let us introduce matrix $\boldsymbol{W}(\boldsymbol{\Omega})$, $\boldsymbol{\Omega} = (\Omega_1, \Omega_2)$, with entries

$$[\boldsymbol{W}(\boldsymbol{\Omega})]_{\varphi_1,\varphi_2} = \left[N_{Rc}\,\mathrm{sinc}_{N_{Rc}}\left(\frac{N_{Rc}(\varphi_1 - \Omega_1)}{M_1}\right) e^{-j\pi(\varphi_1-\Omega_1)\frac{N_{Rc}-1}{M_1}}\right] \times \left[N_{Tx}\,\mathrm{sinc}_{N_{Tx}}\left(\frac{N_{Tx}(\varphi_2 - \Omega_2)}{M_2}\right) e^{-j\pi(\varphi_2-\Omega_2)\frac{N_{Tx}-1}{M_2}}\right] \tag{4.160}$$

where the periodic function $\mathrm{sinc}_N(x)$ is defined in (1.24). By taking the $M_1 \times M_2$ DFT_2 of (4.159), the $M_1 \times M_2$ matrix $\boldsymbol{\Theta}$ is obtained, with entries[9]

$$[\boldsymbol{\Theta}]_{\varphi_1,\varphi_2} = \sum_{r=1}^{N_{Rc}} \sum_{t=1}^{N_{Tx}} \mathcal{G}_C^{(r,t)} e^{-\frac{j2\pi\varphi_1(r-1)}{M_1}} e^{-\frac{j2\pi\varphi_2(t-1)}{M_2}} = \sum_{i=0}^{N_c-1} \breve{g}_i [\boldsymbol{W}(\boldsymbol{\Omega}^{(i)})]_{\varphi_1,\varphi_2} \tag{4.161}$$

where $\varphi_1 = 0, 1, \dots, M_1 - 1$ and $\varphi_2 = 0, 1, \dots, M_2 - 1$, define the angular domain, and $\boldsymbol{\Omega}^{(i)} = (-M_1\eta_i^{(Rc)}, -M_2\eta_i^{(Tx)})$. For a large number of antennas, the periodic sinc_N functions that compose $\boldsymbol{\Theta}$ become simple sinc functions. In this asymptotic regime, $\boldsymbol{\Theta}$ presents N_c peaks at $\boldsymbol{\Omega}^{(i)}$, $i = 0, \dots, N_c - 1$, i.e. it is a *sparse matrix*. This representation can be particularly useful to design efficient channel estimation and resource allocation techniques. An example of mmWave channel in the angular domain is shown in Figure 4.28, wherein we can clearly identify a small number of peaks.

[9] Rather than using the definition of DFT, it is simpler to first evaluate the discrete-time Fourier transform of (4.159), and then sampling it at *angular frequencies* $(\varphi_1/M_1, \varphi_2/M_2)$.

Chapter 5

Vector quantization

Vector quantization (VQ) is introduced as a natural extension of the scalar quantization (SQ) [1] concept. However, using multidimensional signals opens the way to many techniques and applications that are not found in the scalar case [2, 3].

5.1 Basic concept

The basic concept is that of associating with an input vector $\boldsymbol{s} = [s_1, \dots, s_N]^T$, generic sample of a vector random process $\boldsymbol{s}(k)$, a reproduction vector $\boldsymbol{s}_q = \boldsymbol{Q}[\boldsymbol{s}]$ chosen from a finite set of L elements (*code vectors*), $\mathcal{A} = \{\boldsymbol{Q}_1, \dots, \boldsymbol{Q}_L\}$, called codebook, so that a given distortion measure $d(\boldsymbol{s}, \boldsymbol{Q}[\boldsymbol{s}])$ is minimized.

Figure 5.1 exemplifies the encoder and decoder functions of a VQ scheme. The encoder computes the distortion associated with the representation of the input vector $\boldsymbol{s}$ by each reproduction vector of $\mathcal{A}$ and decides for the vector $\boldsymbol{Q}_i$ of the codebook $\mathcal{A}$ that minimizes it; the decoder associates the vector $\boldsymbol{Q}_i$ to the index i received. We note that the information transmitted over the digital channel identifies the code vector $\boldsymbol{Q}_i$; therefore, it depends only on the codebook size L and not on N, dimension of the code vectors.

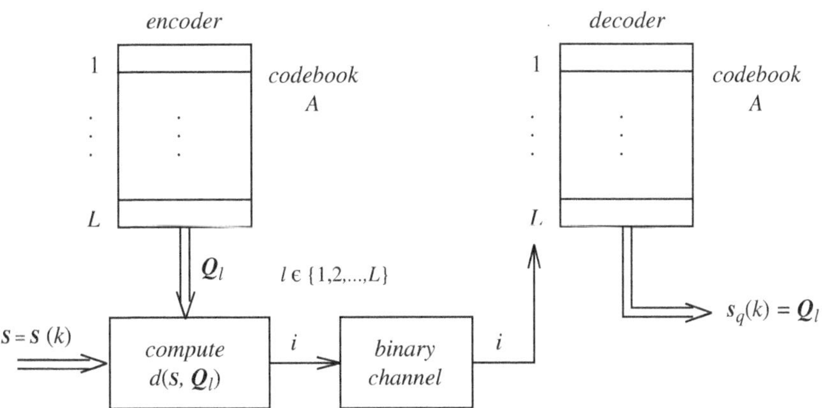

Figure 5.1 Block diagram of a vector quantizer.

Algorithms for Communications Systems and their Applications, Second Edition.
Nevio Benvenuto, Giovanni Cherubini, and Stefano Tomasin.

An example of input vector s is obtained by considering N samples at a time of a signal, $s(k) = [s(kNT_c), \dots, s((kN - N + 1)T_c)]^T$, or the N linear predictive coding (LPC) coefficients (see Section 2.2), $s(k) = [c_1(k), \dots, c_N(k)]^T$, associated with an observation window of a signal.

5.2 Characterization of VQ

Considering the general case of complex-valued signals, a vector quantizer is characterized by

- *Source or input vector* $s = [s_1, s_2, \dots, s_N]^T \in \mathbb{C}^N$.
- *Codebook* $\mathcal{A} = \{Q_i\}$, $i = 1, \dots, L$, where $Q_i \in \mathbb{C}^N$ is a code vector.
- *Distortion measure* $d(s, Q_i)$.
- *Quantization rule* (minimum distortion)

$$Q : \mathbb{C}^N \to \mathcal{A} \qquad \text{with } Q_i = Q[s] \text{ if } i = \arg\min_{\ell} d(s, Q_\ell) \tag{5.1}$$

Definition 5.1 (Partition of the source space)
The equivalence relation

$$\mathcal{Q} = \{(s_1, s_2) : Q[s_1] = Q[s_2]\} \tag{5.2}$$

which associates input vector pairs having the same reproduction vector, identifies a partition $\mathcal{R} = \{\mathcal{R}_1, \dots, \mathcal{R}_L\}$ of the source space $\mathbb{C}^N$, whose elements are the sets

$$\mathcal{R}_\ell = \{s \in \mathbb{C}^N : Q[s] = Q_\ell\} \qquad \ell = 1, \dots, L \tag{5.3}$$

The sets $\{\mathcal{R}_\ell\}$, $\ell = 1, \dots, L$, are called *Voronoi regions*.

It can be easily demonstrated that the sets $\{\mathcal{R}_\ell\}$ are non-overlapping and cover the entire space $\mathbb{C}^N$:

$$\bigcup_{\ell=1}^{L} \mathcal{R}_\ell = \mathbb{C}^N \qquad \mathcal{R}_i \cap \mathcal{R}_j = \emptyset \quad i \neq j \tag{5.4}$$

In other words, as indicated by (5.3) every subset $\mathcal{R}_\ell$ contains all input vectors associated by the quantization rule with the code vector Q_ℓ. An example of partition for $N = 2$ and $L = 4$ is illustrated in Figure 5.2. □

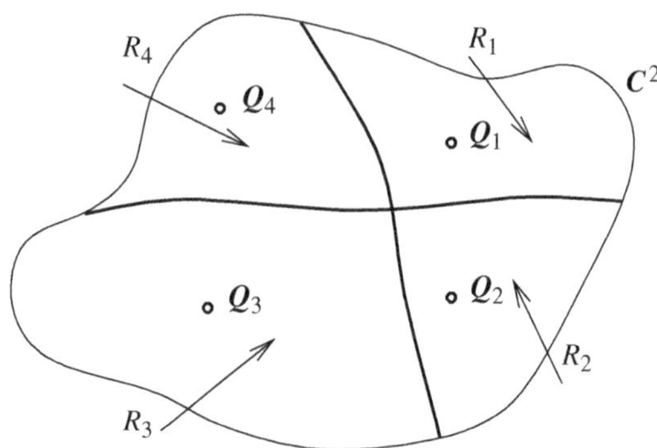

Figure 5.2 Partition of the source space $\mathbb{C}^2$ in four subsets or Voronoi regions.

Parameters determining VQ performance

We define the following parameters.

- Quantizer rate

$$R_q = \log_2 L \qquad \text{(bit/vector) or (bit/symbol)} \tag{5.5}$$

- Rate per dimension

$$R_I = \frac{R_q}{N} = \frac{\log_2 L}{N} \qquad \text{(bit/sample)} \tag{5.6}$$

- Rate in bit/s

$$R_b = \frac{R_I}{T_c} = \frac{\log_2 L}{NT_c} \qquad \text{(bit/s)} \tag{5.7}$$

  where T_c denotes the time interval between two consecutive samples of a vector. In other words, in (5.7) NT_c is the sampling period of the vector sequence $\{\boldsymbol{s}(k)\}$.
- Distortion

$$d(\boldsymbol{s}, \boldsymbol{Q}_i) \tag{5.8}$$

  The distortion is a non-negative scalar function of a vector variable,

$$d : \mathbb{C}^N \times \mathcal{A} \rightarrow \mathfrak{R}_+ \tag{5.9}$$

If the input process $\boldsymbol{s}(k)$ is stationary and the probability density function $p_{\boldsymbol{s}}(\boldsymbol{a})$ is known, we can compute the mean distortion as

$$D(\mathcal{R}, \mathcal{A}) = E[d(\boldsymbol{s}, \boldsymbol{Q}[\boldsymbol{s}])] \tag{5.10}$$

$$= \sum_{i=1}^{L} E[d(\boldsymbol{s}, \boldsymbol{Q}_i), \boldsymbol{s} \in \mathcal{R}_i] \tag{5.11}$$

$$= \sum_{i=1}^{L} \overline{d}_i \tag{5.12}$$

where

$$\overline{d}_i = \int_{\mathcal{R}_i} d(\boldsymbol{a}, \boldsymbol{Q}_i) p_{\boldsymbol{s}}(\boldsymbol{a}) \, d\boldsymbol{a} \tag{5.13}$$

If the source is also ergodic, we obtain

$$D(\mathcal{R}, \mathcal{A}) = \lim_{K \to \infty} \frac{1}{K} \sum_{k=1}^{K} d(\boldsymbol{s}(k), \boldsymbol{Q}[\boldsymbol{s}(k)]) \tag{5.14}$$

In practice, we always assume that the process $\{\boldsymbol{s}\}$ is stationary and ergodic, and we use the average distortion (5.14) as an estimate of the expectation (5.10).

Defining

$$\boldsymbol{Q}_i = [Q_{i,1}, Q_{i,2}, \dots, Q_{i,N}]^T \tag{5.15}$$

we give below two measures of distortion of particular interest.

1. Distortion as the ℓ_ν norm to the μ-th power:

$$d(\boldsymbol{s}, \boldsymbol{Q}_i) = \|\boldsymbol{s} - \boldsymbol{Q}_i\|_\nu^\mu = \left[\sum_{n=1}^{N} |s_n - Q_{i,n}|^\nu\right]^{\mu/\nu} \tag{5.16}$$

The most common version is the *square distortion*[1]:

$$d(\boldsymbol{s}, \boldsymbol{Q}_i) = \|\boldsymbol{s} - \boldsymbol{Q}_i\|_2^2 = \sum_{n=1}^{N} |s_n - Q_{i,n}|^2 \tag{5.17}$$

2. Itakura–Saito distortion:

$$d(\boldsymbol{s}, \boldsymbol{Q}_i) = (\boldsymbol{s} - \boldsymbol{Q}_i)^H \boldsymbol{R}_s (\boldsymbol{s} - \boldsymbol{Q}_i) = \sum_{n=1}^{N} \sum_{m=1}^{N} (s_n - Q_{i,n})^* [\boldsymbol{R}_s]_{n,m} (s_m - Q_{i,m}) \tag{5.18}$$

where $\boldsymbol{R}_s$ is the autocorrelation matrix of the vector $\boldsymbol{s}^*(k)$, defined in (1.270), with entries $[\boldsymbol{R}_s]_{n,m}$, $n, m = 1, 2, \dots, N$.

Comparison between VQ and scalar quantization

Defining the mean distortion per dimension as $\tilde{D} = D/N$, for a given rate R_I we find (see [4] and references therein)

$$\frac{\tilde{D}_{SQ}}{\tilde{D}_{VQ}} = F(N)\ S(N)\ M(N) \tag{5.19}$$

where

- $F(N)$ is the space filling gain. In the scalar case, the partition regions must necessarily be intervals. In an N-dimensional space, R_i can be *shaped* very closely to a sphere. The asymptotic value for $N \to \infty$ equals $F(\infty) = 2\pi e/12 = 1.4233 = 1.53$ dB.
- $S(N)$ is the gain related to the shape of $p_s(\boldsymbol{a})$, defined as[2]

$$S(N) = \frac{\|\tilde{p}_s(\boldsymbol{a})\|_{1/3}}{\|\tilde{p}_s(\boldsymbol{a})\|_{N/(N+2)}} \tag{5.21}$$

  where $\tilde{p}_s(\boldsymbol{a})$ is the probability density function of the input $\boldsymbol{s}$ considered with uncorrelated components. $S(N)$ does not depend on the variance of the random variables of $\boldsymbol{s}$, but only on the norm order $N/(N+2)$ and shape $\tilde{p}_s(\boldsymbol{a})$. For $N \to \infty$, we obtain

$$S(\infty) = \frac{\|\tilde{p}_s(\boldsymbol{a})\|_{1/3}}{\|\tilde{p}_s(\boldsymbol{a})\|_1} \tag{5.22}$$

- $M(N)$ is the memory gain, defined as

$$M(N) = \frac{\|\tilde{p}_s(\boldsymbol{a})\|_{N/(N+2)}}{\|p_s(\boldsymbol{a})\|_{N/(N+2)}} \tag{5.23}$$

  where $p_s(\boldsymbol{a})$ is the probability density function of the input $\boldsymbol{s}$. The expression of $M(N)$ depends on the two functions $\tilde{p}_s(\boldsymbol{a})$ and $p_s(\boldsymbol{a})$, which differ for the correlation among the various vector components; obviously if the components of $\boldsymbol{s}$ are statistically independent, we have $M(N) = 1$; otherwise $M(N)$ increases as the correlation increases.

---

[1] Although the same symbol is used, the metric defined by (5.17) is the square of the Euclidean distance between the two vectors.

[2] Extending (5.16) to the continuous case, we obtain

$$\|\tilde{p}_s(\boldsymbol{a})\|^{\mu/\nu} = \left[\int \cdots \int \tilde{p}_s^{\nu}(a_1, \dots, a_N)\, da_1 \cdots da_N\right]^{\mu/\nu} \tag{5.20}$$

5.3 Optimum quantization

Our objective is to design a vector quantizer, choosing the codebook $\mathcal{A}$ and the partition $\mathcal{R}$ so that the mean distortion given by (5.10) is minimized.

Two necessary conditions arise.

Rule A (Optimum partition): Assuming the codebook, $\mathcal{A} = \{\boldsymbol{Q}_1, \dots, \boldsymbol{Q}_L\}$ is given, we want to find the optimum partition $\mathcal{R}$ that minimizes $D(\mathcal{R}, \mathcal{A})$. Observing (5.10), the solution is given by

$$\mathcal{R}_i = \{\boldsymbol{s} : d(\boldsymbol{s}, \boldsymbol{Q}_i) = \min_{\boldsymbol{Q}_\ell \in \mathcal{A}} d(\boldsymbol{s}, \boldsymbol{Q}_\ell)\}, \qquad i = 1, \dots, L \tag{5.24}$$

As illustrated in Figure 5.3, $\mathcal{R}_i$ contains all the points $\boldsymbol{s}$ *nearest* to $\boldsymbol{Q}_i$.

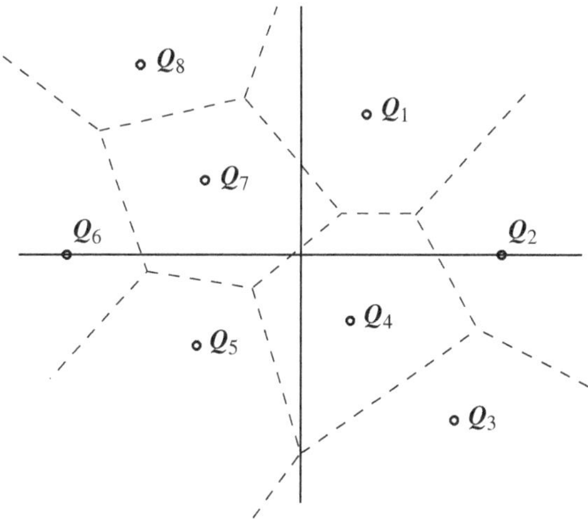

Figure 5.3 Example of partition for $K = 2$ and $N = 8$.

Rule B (Optimum codebook): Assuming the partition $\mathcal{R}$ is given, we want to find the optimum codebook $\mathcal{A}$. By minimizing (5.13), we obtain the solution

$$\boldsymbol{Q}_i = \arg \min_{\boldsymbol{Q} \in \mathbb{C}^N} E[d(\boldsymbol{s}, \boldsymbol{Q}), \boldsymbol{s} \in \mathcal{R}_i] \tag{5.25}$$

In other words, $\boldsymbol{Q}_i$ coincides with the centroid of the region $\mathcal{R}_i$.

As a particular case, choosing the square distortion (5.17), (5.13) becomes

$$\overline{d}_i = \int_{\mathcal{R}_i} \|\boldsymbol{s} - \boldsymbol{Q}_i\|_2^2 \, p_{\boldsymbol{s}}(\boldsymbol{a}) \, d\boldsymbol{a} \tag{5.26}$$

and (5.25) yields

$$\boldsymbol{Q}_i = \frac{\displaystyle\int_{\mathcal{R}_i} \boldsymbol{a} \, p_{\boldsymbol{s}}(\boldsymbol{a}) \, d\boldsymbol{a}}{\displaystyle\int_{\mathcal{R}_i} p_{\boldsymbol{s}}(\boldsymbol{a}) \, d\boldsymbol{a}} \tag{5.27}$$

Generalized Lloyd algorithm

The generalized Lloyd algorithm, given in Figure 5.4, generates a sequence of suboptimum quantizers specified by $\{\mathcal{R}_i\}$ and $\{\boldsymbol{Q}_i\}$ using the previous two rules.

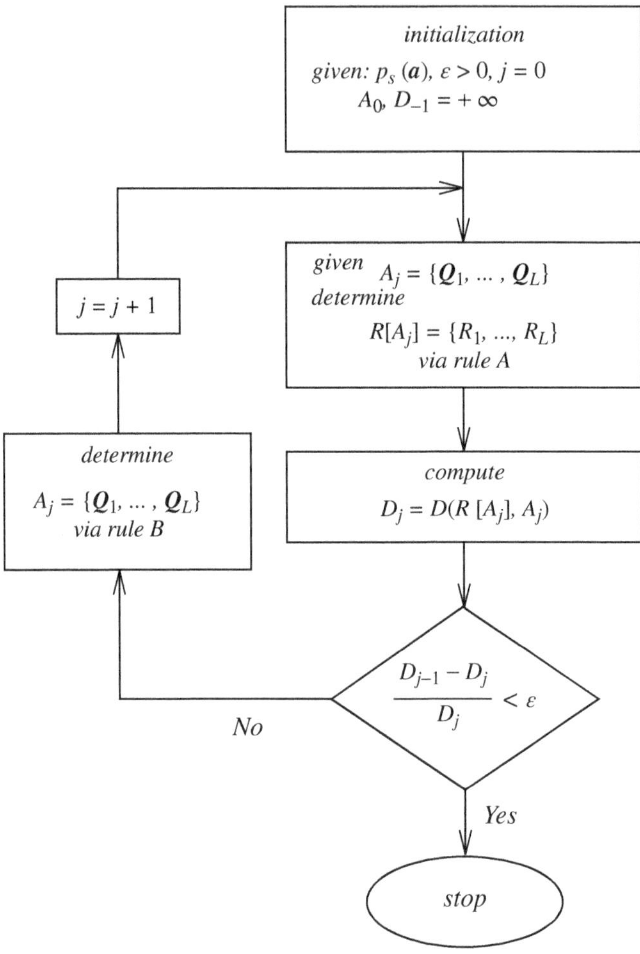

Figure 5.4 Generalized Lloyd algorithm for designing a vector quantizer.

1. Initialization. We choose an initial codebook $\mathcal{A}_0$ and a termination criterion based on a relative error ϵ between two successive iterations. The iteration index is denoted by j.
2. Using the *rule A*, we determine the optimum partition $\mathcal{R}[\mathcal{A}_j]$ using the codebook $\mathcal{A}_j$.
3. We evaluate the distortion associated with the choice of $\mathcal{A}_j$ and $\mathcal{R}[\mathcal{A}_j]$ using (5.12).
4. If

$$\frac{D_{j-1} - D_j}{D_j} < \epsilon \tag{5.28}$$

   we stop the procedure, otherwise we update the value of j.
5. Using *rule B*, we evaluate the optimum codebook associated to the partition $\mathcal{R}[\mathcal{A}_{j-1}]$.
6. We go back to step 2.

The solution found is at least locally optimum; nevertheless, given that the number of locally optimum codes can be rather large, and some of the locally optimum codes may give rather poor performance, it is often advantageous to provide a good codebook to the algorithm to start with, as well as trying different initial codebooks.

The algorithm is clearly a generalization of the Lloyd algorithm: the only difference is that the vector version begins with a codebook (alphabet) rather than with an initial partition of the input space. However, the implementation of this algorithm is difficult for the following reasons.

- The algorithm assumes that $p_s(\boldsymbol{a})$ is known. In the scalar quantization, it is possible, in many applications, to develop an appropriate model of $p_s(a)$, but this becomes a more difficult problem with the increase of the number of dimensions N: in fact the identification of the distribution type, for example Gaussian or Laplacian, is no longer sufficient, as we also need to characterize the statistical dependence among the elements of the source vector.
- The computation of the input space partition is much harder for the VQ. In fact, whereas in the scalar quantization the partition of the real axis is completely specified by a set of $(L-1)$ points, in the two-dimensional case the partition is specified by a set of straight lines, and for the multi-dimensional case to find the optimum solution becomes very hard. For VQ with a large number of dimensions, the partition becomes also harder to describe geometrically.
- Also in the particular case (5.27), the calculation of the centroid is difficult for the VQ, because it requires evaluating a multiple integral on the region $\mathcal{R}_i$.

5.4 The Linde, Buzo, and Gray algorithm

An alternative approach led Linde, Buzo, and Gray (LBG) [5] to consider some very long realizations of the input signal and to substitute (5.10) with (5.14) for K sufficiently large. The sequence used to design the VQ is called *training sequence* (TS) and is composed of K vectors

$$\{\boldsymbol{s}(m)\} \qquad m = 1, \dots, K \tag{5.29}$$

The average distortion is now given by

$$D = \frac{1}{K} \sum_{k=1}^{K} d(\boldsymbol{s}(k), \boldsymbol{Q}[\boldsymbol{s}(k)]) \tag{5.30}$$

and the two rules to minimize D become:

Rule A

$$\mathcal{R}_i = \{\boldsymbol{s}(k) : d(\boldsymbol{s}(k), \boldsymbol{Q}_i) = \min_{\boldsymbol{Q}_\ell \in \mathcal{A}} d(\boldsymbol{s}(k), \boldsymbol{Q}_\ell)\}, \qquad i = 1, \dots, L \tag{5.31}$$

that is $\mathcal{R}_i$ is given by all the elements of the TS nearest to $\boldsymbol{Q}_i$.

Rule B

$$\boldsymbol{Q}_i = \arg \min_{\boldsymbol{Q} \in \mathbb{C}^N} \frac{1}{m_i} \sum_{\boldsymbol{s}(k) \in \mathcal{R}_i} d(\boldsymbol{s}(k), \boldsymbol{Q}) \tag{5.32}$$

where m_i is the number of elements of the TS that are inside $\mathcal{R}_i$

Using the structure of the Lloyd algorithm with the new cost function (5.30) and the two new rules (5.31) and (5.32), we arrive at the LBG algorithm. Before discussing the details, it is worthwhile pointing out some aspects of this new algorithm.

- It converges to a minimum, which is not guaranteed to be a global minimum, and generally depends on the choice of the TS.
- It does not require any stationarity assumption.
- The partition is determined without requiring the computation of expectations over $\mathbb{C}^N$.
- The computation of $\boldsymbol{Q}_i$ in (5.32) is still burdensome.

However, for the square distortion (5.17), we have

$$d(s(k), \boldsymbol{Q}_i) = \sum_{n=1}^{N} |s_n(k) - Q_{i,n}|^2 \tag{5.33}$$

and (5.32) simply becomes

Rule B

$$\boldsymbol{Q}_i = \frac{1}{m_i} \sum_{s(k)\in\mathcal{R}_i} s(k) \tag{5.34}$$

that is $\boldsymbol{Q}_i$ coincides with the arithmetic mean of the TS vectors that are inside $\mathcal{R}_i$.

Choice of the initial codebook

With respect to the choice of the initial codebook, the first L vectors of the TS can be used; however, if the data are highly correlated, it is necessary to use L vectors that are sufficiently spaced in time from each other.

A more effective alternative is that of taking as initial value the centroid of the TS and start with a codebook with a number of elements $L = 1$. Slightly changing the components of this code vector (*splitting procedure*), we derive two code vectors and an initial alphabet with $L = 2$; at this point, using the LBG algorithm, we determine the optimum VQ for $L = 2$. At convergence, each optimum code vector is changed to obtain two code vectors and the LBG algorithm is used for $L = 4$. Iteratively the splitting procedure and optimization is repeated until the desired number of elements for the codebook is obtained.

Let $\mathcal{A}_j = \{\boldsymbol{Q}_1, \dots, \boldsymbol{Q}_L\}$ be the codebook at iteration j-th. The splitting procedure generates $2L$ N-dimensional vectors yielding the new codebook

$$\mathcal{A}_{j+1} = \{\mathcal{A}_j^-\} \cup \{\mathcal{A}_j^+\} \tag{5.35}$$

where

$$\mathcal{A}_j^- = \{\boldsymbol{Q}_i - \boldsymbol{\varepsilon}_-\}, \qquad i = 1, \dots, L, \tag{5.36}$$

$$\mathcal{A}_j^+ = \{\boldsymbol{Q}_i - \boldsymbol{\varepsilon}_+\}, \qquad i = 1, \dots, L \tag{5.37}$$

Typically, $\boldsymbol{\varepsilon}_-$ is the zero vector,

$$\boldsymbol{\varepsilon}_- = \boldsymbol{0} \tag{5.38}$$

and

$$\boldsymbol{\varepsilon}_+ = \frac{1}{10}\sqrt{\frac{\mathtt{M}_s}{N}} \cdot \boldsymbol{1} \tag{5.39}$$

where $\boldsymbol{1}$ is an all-one vector, so that

$$\| \boldsymbol{\varepsilon}_+ \|_2^2 \leq 0.01\ \mathtt{M}_s \tag{5.40}$$

where $\mathtt{M}_s$ is the power of the TS.

Splitting procedure

Choosing $\epsilon > 0$ (typically $\epsilon = 10^{-3}$) and an initial alphabet given by the splitting procedure applied to the average of the TS, we obtain the LBG algorithm, whose block diagram is shown in Figure 5.5, whereas its operations are depicted in Figure 5.6.

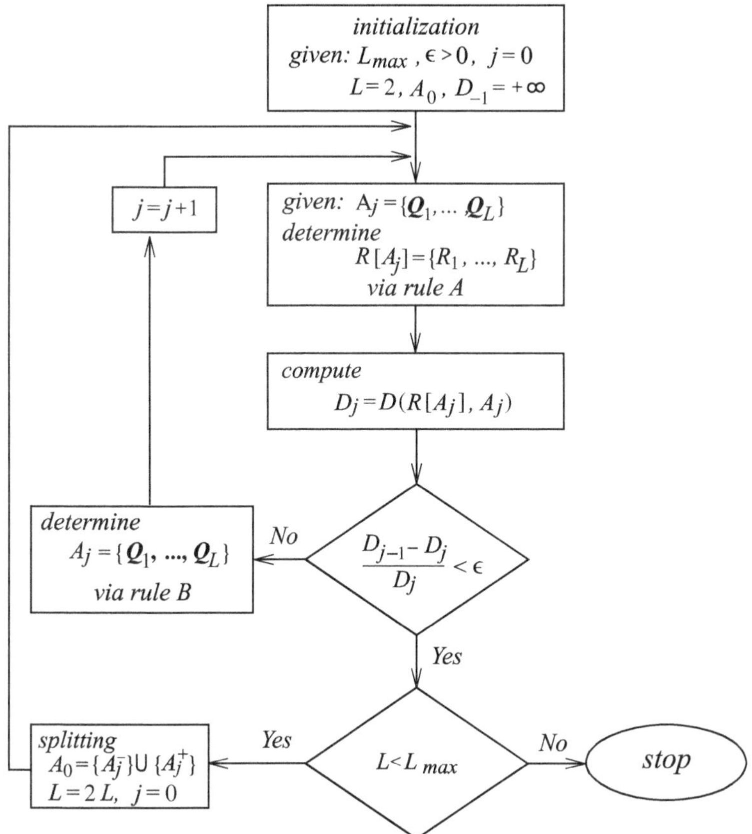

Figure 5.5 LBG algorithm with splitting procedure.

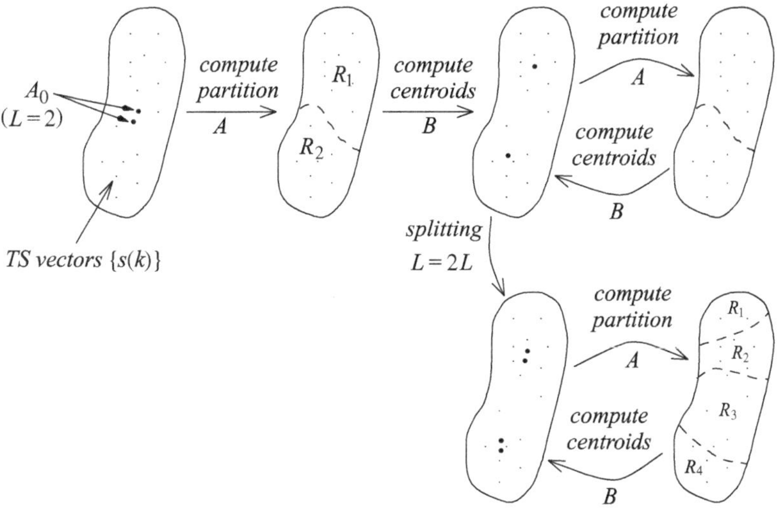

Figure 5.6 Operations of the LBG algorithm with splitting procedure.

Selection of the training sequence

A rather important problem associated with the use of a TS is that of empty cells. It is in fact possible that some regions $\mathcal{R}_i$ contain few or no elements of the TS: in this case, the code vectors associated with these regions contribute little or nothing at all to the reduction of the total distortion.

Possible causes of this phenomenon are:

- *TS too short*: the training sequence must be sufficiently long, so that every region $\mathcal{R}_i$ contains at least 30–40 vectors;
- *Poor choice of the initial alphabet*: in this case, in addition to the obvious solution of modifying this choice, we can limit the problem through the following splitting procedure.

Let $\mathcal{R}_i$ be a region that contains $m_i < m_{min}$ elements. We eliminate the code vector $\boldsymbol{Q}_i$ from the codebook and apply the splitting procedure limited only to the region that gives the largest contribution to the distortion; then we compute the new partition and proceed in the usual way.

We give some practical rules, taken from [6] for LPC applications,[3] that can be useful in the design of a vector quantizer.

- If $K/L \leq 30$, in practice there is a possibility of empty regions, where we recall K is the number of vectors of the TS, and L is the number of code vectors.
- If $K/L \leq 600$, an appreciable difference between the distortion calculated with the TS and that calculated with a new sequence may exist.

In the latter situation, it may in fact happen that, for a very short TS, the distortion computed for vectors of the TS is very small; the extreme case is obtained by setting $K = L$, hence $D = 0$. In this situation, for a sequence different from the TS (*outside* TS), the distortion is in general very high. As illustrated in Figure 5.7, only if K is large enough the TS adequately represents the input process and no substantial difference appears between the distortion measured with vectors inside or outside TS [5].[4]

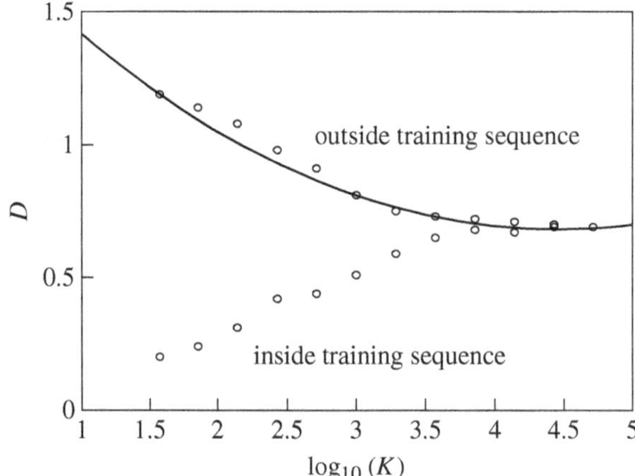

Figure 5.7 Values of the distortion as a function of the number of vectors K in the *inside* and *outside* training sequences.

[3] These rules were derived in the VQ of LPC vectors. They can be considered valid in the case of strongly correlated vectors.

[4] This situation is similar to that obtained by the least squares (LS) method (see Section 3.2).

Finally, we find that the LBG algorithm, even though very simple, requires numerous computations. We consider for example as vector source the LPC coefficients with $N = 10$, computed over windows of duration equal to 20 ms of a speech signal sampled at 8 kHz. Taking $L = 256$, we have a rate $R_b = 8$ bit/20 ms equal to 400 bit/s. As a matter of fact, the LBG algorithm requires a minimum $K = 600 \cdot 256 \simeq 155\,000$ vectors for the TS, which roughly corresponds to three minutes of speech.

5.4.1 k-means clustering

k-means clustering is used in pattern recognition [7] and aggregates a vector sequence into L (in the literature L is often denoted as k) clusters. This is equivalent to the LBG algorithm of Section 5.4.

5.5 Variants of VQ

Tree search VQ

A random VQ, determined according to the LBG algorithm, requires:

- a large memory to store the codebook;
- a large computational complexity to evaluate the L distances for encoding.

A variant of VQ that requires a lower computational complexity, at the expense of a larger memory, is the tree search VQ. As illustrated in Figure 5.8, whereas in the memoryless VQ case the comparison of the input vector $\boldsymbol{s}$ must occur with all the elements of the codebook, thus determining a full search, in the tree search VQ we proceed by levels: first we compare $\boldsymbol{s}$ with $\boldsymbol{Q}_{A1}$ and $\boldsymbol{Q}_{A2}$, then we proceed along the branch whose node has a representative vector *closest* to $\boldsymbol{s}$.

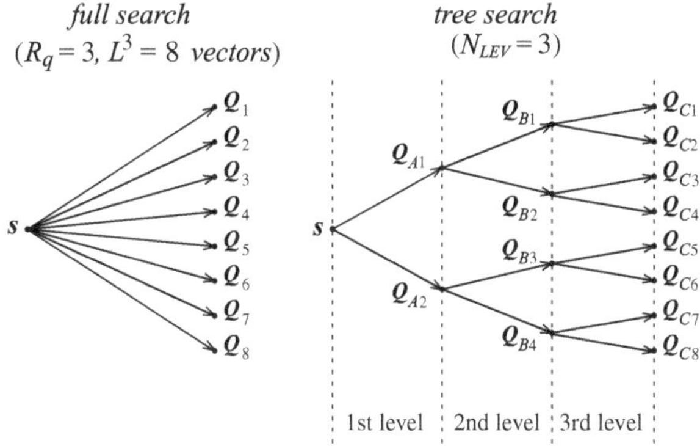

Figure 5.8 Comparison between full search and tree search.

To determine the code vectors at different nodes, for a binary tree the procedure consists of the following steps.

1. Calculate the optimum quantizer for the first level by the LBG algorithm; the codebook contains two code vectors.
2. Divide the training sequence into subsequences relative to every node of level n ($n = 2, 3, \ldots, N_{LEV}$, $N_{LEV} = \log_2 L$); in other words, collect all vectors that are associated with the same code vector.
3. Apply the LGB algorithm to every subsequence to calculate the codebook of level n.

Table 5.1: Comparison between full search and tree search.

	Computation of $d(\cdot,\cdot)$	No. of vectors to memorize
Full search	2^{R_q}	2^{R_q}
Tree search	$2R_q$	$\sum_{i=1}^{R_q} 2^i \simeq 2^{R_q+1}$
For $R_q = 10$ (bit/vector)	Computation of $d(\cdot,\cdot)$	No. of vectors to memorize
Full search	1024	1024
Tree search	20	2046

As an example, the memory requirements and the number of computations of $d(\cdot,\cdot)$ are shown in Table 5.1 for a given value of R_q (bit/vector) in the cases of full search and tree search. Although the performance is slightly lower, the computational complexity of the encoding scheme for a tree search is considerably reduced.

Multistage VQ

The multistage VQ technique presents the advantage of reducing both the encoder computational complexity and the memory required. The idea consists in dividing the encoding procedure into successive stages, where the first stage performs quantization with a codebook with a reduced number of elements. Successively, the second stage performs quantization of the error vector $\boldsymbol{e} = \boldsymbol{s} - \boldsymbol{Q}[\boldsymbol{s}]$: the quantized error gives a more accurate representation of the input vector. A third stage could be used to quantize the error of the second stage and so on.

We compare the complexity of a one-stage scheme with that of a two-stage scheme, illustrated in Figure 5.9. Let $R_q = \log_2 L$ be the rate in bit/vector for both systems and assume that all the code vectors have the same dimension $N = N_1 = N_2$.

- Two-stage:
  $R_q = \log_2 L_1 + \log_2 L_2$, hence $L_1 L_2 = L$.
  Computations of $d(\cdot,\cdot)$ for encoding: $L_1 + L_2$.
  Memory: $L_1 + L_2$ locations.
- One-stage:
  $R_q = \log_2 L$.
  Computations of $d(\cdot,\cdot)$ for encoding: L.
  Memory: L locations.

The advantage of a multistage approach in terms of cost of implementation is evident; however, it has lower performance than a one-stage VQ.

Product code VQ

The input vector is split into subvectors that are quantized independently, as illustrated in Figure 5.10. This technique is useful if (i) there are input vector components that can be encoded separately because of their different effects, e.g. prediction gain and LPC coefficients, or (ii) the input vector has too large a dimension to be encoded directly. It presents the disadvantage that it does not consider the correlation that may exist between the various subvectors, that could bring about a greater coding efficiency.

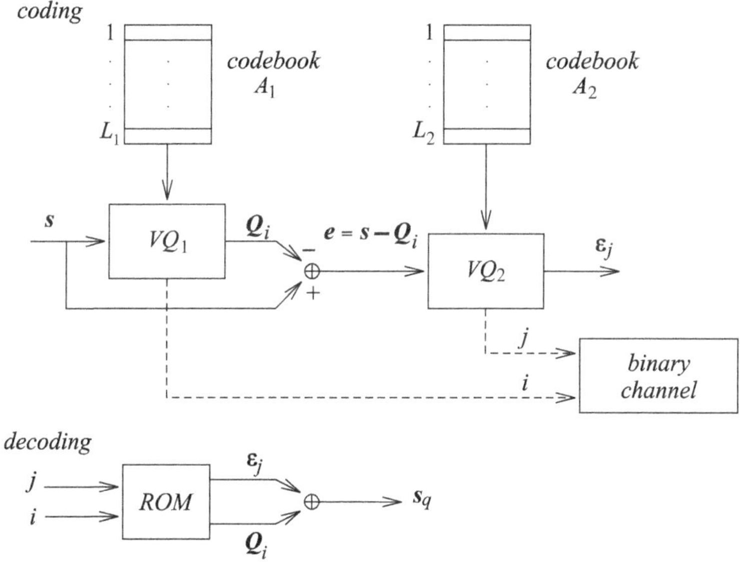

Figure 5.9 Multistage (two-stage) VQ.

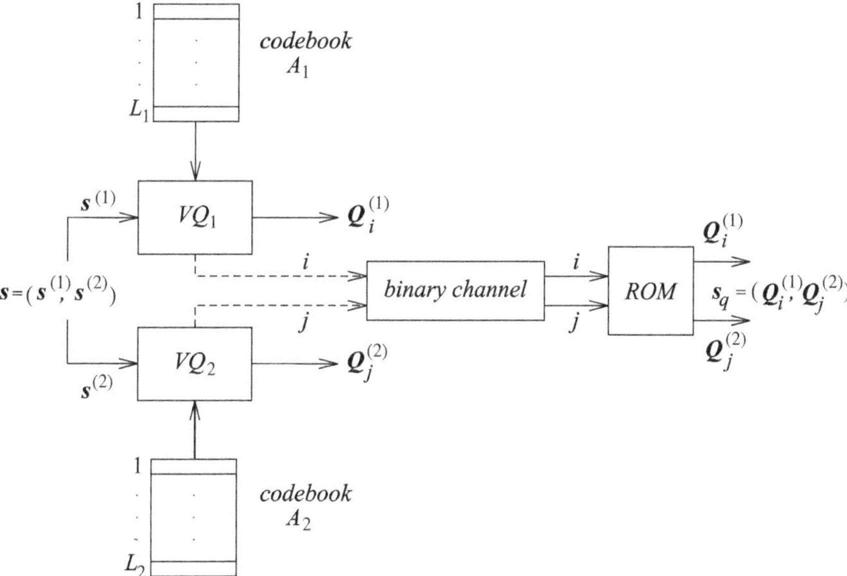

Figure 5.10 Product code VQ.

A more general approach is the *sequential search product code* [2], where the quantization of the subvector n depends also on the quantization of previous subvectors. With reference to Figure 5.10, assuming $L = L_1 L_2$ and $N = N_1 + N_2$, we note that the rate per dimension for the VQ is given by

$$R_q = \frac{\log_2 L}{N} = \frac{\log_2 L_1}{N} + \frac{\log_2 L_2}{N} \tag{5.41}$$

whereas for the product code VQ, it is given by

$$R_q = \frac{\log_2 L_1}{N_1} + \frac{\log_2 N_2}{N_2} \tag{5.42}$$

5.6 VQ of channel state information

We present now two examples of use of VQ, where the vectors to be quantized represent sampled frequency responses of the channel between two devices, also denoted as *channel state information* (CSI). In particular, we consider a cellular system, with a base station (BS) aiming at transmitting data to user equipments (UEs) using multicarrier (MC) transmission (see Chapter 8), where the discrete frequency response of the channel is a vector of $\mathcal{M}$ complex numbers, one for each subcarrier.

As described in details in Section 13.1.2, power and bit adaptation on a per-subcarrier basis greatly increases the achievable data rate, but it requires the knowledge of CSI before transmission. In time division multiplexing (TDD) systems (see Appendix 18.A), where uplink and downlink transmissions occur in the same band, although at different times, the BS estimates the uplink channel, and the downlink channel is the same (by reciprocity), within a short time interval. However, when frequency division duplexing (FDD) (see Appendix 18.A) is used, uplink and downlink transmissions occur in two separate bands, thus the BS cannot exploit the channel reciprocity. Indeed, the UE must (i) estimate the downlink CSI, (ii) quantize the CSI, and (iii) *feedback* the quantization index to the BS. Therefore, VQ is needed.

When the BS is equipped with $N_{Tx} > 1$ antennas (obtaining a multiple-input single-output, MISO system, see also Chapter 9), the UE must estimate the N_{Tx} channel vectors and stack them into a single vector, to be quantized. Clearly, this problem is even more challenging, as the vector size grows with N_{Tx}, and a larger codebook is needed. In this case, CSI availability at the transmitter is relevant also to exploit space diversity by precoding (see Section 9.3.2), or to implement spatial multiuser transmission (see Section 9.6).

We will first consider the problem of MISO channel quantization, investigating the more appropriate VQ distortion measure. Then, we consider a more elaborate CSI reporting scheme, where we exploit the correlation between uplink and downlink channels (although being still in two separate bands).

MISO channel quantization

First consider the quantization of channel estimates in MISO orthogonal frequency division multiplexing (OFDM) systems with $\mathcal{M}$ subcarriers, modelling the downlink of a cellular system with a BS equipped with $N_{Tx} > 1$ antennas and single-antenna UEs. Let $\mathcal{G}_C[m]$ be the $1 \times N_{Tx}$ (space) channel vector at subcarrier m, $m = 0, \dots, \mathcal{M} - 1$, and let

$$\mathcal{G}_C = [\mathcal{G}_C[0], \dots, \mathcal{G}_C[\mathcal{M} - 1]]^T \tag{5.43}$$

be the overall $\mathcal{M}N_{Tx}$ channel vector.

For CSI feedback, the solution adopted for example in the long-term evolution (LTE) of the 3GPP cellular systems [8] provides that the UE feeds back: the *channel quality indicator* (CQI), and the *channel direction indicator* (CDI). The CQI is a quantized version of $||\mathcal{G}_C[m]||$, $m = 0, \dots, \mathcal{M} - 1$, i.e. the norm of the space channel vector at subcarrier m. Next, we define the normalized channel vectors and their collection as

$$\overline{\mathcal{G}}_C[m] = \frac{\mathcal{G}_C[m]}{||\mathcal{G}_C[m]||}, \qquad s = [\overline{\mathcal{G}}_C[0], \dots, \overline{\mathcal{G}}_C[\mathcal{M} - 1]]^T \tag{5.44}$$

The CDI is a quantized version $\boldsymbol{Q}_i = [\boldsymbol{Q}_i[0], \dots, \boldsymbol{Q}_i[\mathcal{M}-1]]$ of $\boldsymbol{s}$, where $\boldsymbol{Q}_i[m]$, $m = 0, \dots, \mathcal{M}-1$, are $1 \times N_{Tx}$-size vectors. Using the MISO linear precoding $\boldsymbol{B}[m] = \boldsymbol{Q}_i^H[m]$ the signal-to-noise-plus-interference ratio (SNIR) at the detector input is [9]

$$\gamma[m] = \frac{\gamma_{MF} ||\mathcal{G}_C[m]||^2 |\overline{\mathcal{G}}_C[m]\boldsymbol{Q}_i^H[m]|^2}{1 + \gamma_{MF} ||\mathcal{G}_C[m]||^2 (1 - |\overline{\mathcal{G}}_C[m]\boldsymbol{Q}_i^H[m]|^2)} \tag{5.45}$$

where γ_{MF} is the signal-to-noise-ratio (SNR) at the detection point for an ideal additive white Gaussian noise (AWGN) channel. Differently from (9.109), note that the denominator of $\gamma[m]$ accounts also for the contribution of multiuser interference.

The novelty here is that on quantizing the channel we aim at maximizing the sum-rate of data transmission, i.e.

$$R(\boldsymbol{s}, \boldsymbol{Q}_i) = \sum_{m=0}^{\mathcal{M}-1} \log_2(1 + \gamma[m]) \tag{5.46}$$

Thus, for VQ we use the *rate-based* distortion measure

$$d(\boldsymbol{s}, \boldsymbol{Q}_i) = -R(\boldsymbol{s}, \boldsymbol{Q}_i) \tag{5.47}$$

However, using optimum quantization with this distortion measure is rather complicated, therefore an approximate simplified expression is derived [9].

Now, we can exploit the Jensen inequality[5] on $\log_2(1+x)$ to obtain the following upper bound on $R(\boldsymbol{s}, \boldsymbol{Q}_i)$

$$R(\boldsymbol{s}, \boldsymbol{Q}_i) \leq \mathcal{M} \log_2 \left(1 + \frac{1}{\mathcal{M}} \sum_{m=0}^{\mathcal{M}-1} \gamma[m] \right) \tag{5.48}$$

Then, by exploiting the fact that $\log_2(1+x)$ is strictly monotonically increasing, minimizing $d(\boldsymbol{s}, \boldsymbol{Q}_i)$ is equivalent to minimizing

$$d'(\boldsymbol{s}, \boldsymbol{Q}_i) = -\sum_{m=0}^{\mathcal{M}-1} \gamma[m] \tag{5.49}$$

When the multiuser interference is negligible with respect to the channel noise, the expectation of (5.49) yields

$$E[d'(\boldsymbol{s}, \boldsymbol{Q}_i)] = -E\left[\sum_{m=0}^{\mathcal{M}-1} \gamma[m]\right] = -\gamma_{MF} \sum_{m=0}^{\mathcal{M}-1} E\,[||\mathcal{G}_C[m]||^2]\; E[|\overline{\mathcal{G}}_C[m]\boldsymbol{Q}_i^H[m]|^2] \tag{5.50}$$

where we have exploited the independence between the norm and the direction of the MISO channel. Therefore, the maximization of the expectation in (5.50) is achieved by the new distortion measure

$$d''(\boldsymbol{s}, \boldsymbol{Q}_i) = -\sum_{m=0}^{\mathcal{M}-1} |\overline{\mathcal{G}}_C[m]\boldsymbol{Q}_i^H[m]|^2 \tag{5.51}$$

When multiuser interference becomes predominant, it can be shown [9] that again using (5.51) as distortion measure is reasonable.

A further simplified approach is obtained by ignoring the correlation among the channels at different frequencies and applying a vector quantizer on each $\overline{\mathcal{G}}_C[m]$, separately, using the following *inner product* distortion measure

$$d(\overline{\mathcal{G}}_C[m], \boldsymbol{Q}_i[m]) = -|\overline{\mathcal{G}}_C[m]\boldsymbol{Q}_i^H[m]|^2 \tag{5.52}$$

Note that since $\overline{\mathcal{G}}_C[m]$ has unit norm, $\boldsymbol{Q}_i[m]$ will have unit norm, and minimizing (5.52) is equivalent to minimizing the square distortion (5.17) between $\overline{\mathcal{G}}_C[m]$ and $\boldsymbol{Q}_i[m]$.

[5] For a strictly concave function h, $E[h(x)] < h(E[x])$.

Channel feedback with feedforward information

We consider another scenario, wherein both the BS and the UE are equipped with a single antenna ($N_{Tx} = N_{Rc} = 1$), but in order to reduce the feedback rate (thus, the number of bits used to describe the downlink channel), we exploit the correlation between the uplink and downlink channels. Consider two MC systems[6] for both uplink and downlink, with $\mathcal{M}$ subcarriers each, and let $\mathcal{G}_C^{(U)} = [\mathcal{G}_C^{(U)}[0], \dots, \mathcal{G}_C^{(U)}[\mathcal{M} - 1]]^T$ and $\mathcal{G}_C^{(D)} = [\mathcal{G}_C^{(D)}[0], \dots, \mathcal{G}_C^{(D)}[\mathcal{M} - 1]]^T$, be the uplink and downlink channel vectors, respectively.

The standard procedure provides the feedback of CQI and CDI, as described in the previous part of this section (for $N_{Tx} = 1$ in this case). We now consider a modified procedure [10], wherein both BS and UE obtain a quantized version of the channel, and then share their knowledge. The procedure works as follows:

1. The BS transmits b_F *feedforward* bits to the UE representing the index $c^{(F)}$ of vector-quantized uplink channel.
2. The UE, making use of the quantized uplink channel, transmits b_B *feedback* bits to the BS representing the index $c^{(B)}$ of vector-quantized downlink channel.
3. The BS uses both the quantized uplink channel and the feedback bits of step 2 to reconstruct the downlink channel as $\hat{\mathcal{G}}_C^{(D)}$.

Note that when $b_F = 0$, i.e. no feedforward bits are transmitted, we find the CDI feedback described in the previous part of this section.

For step 1, the codebook of size 2^{b_F} for the uplink channel is constructed over the statistics of $\mathcal{G}_C^{(U)}$, using as distortion measure, for example (5.52). Let $\mathcal{R}^{(U)} = \{\mathcal{R}_\ell^{(U)}\}$ be the partition of the uplink channel vector quantizer.

For step 2 instead we need 2^{b_F} different *codebooks* $\mathcal{C}_\ell^{(D)}$, $\ell = 1, \dots, 2^{b_F}$, each designed on the statistics of the downlink channel, conditioned to the fact that the uplink channel belongs to region $\mathcal{R}_\ell^{(U)}$. Note that we are exploiting the correlation of uplink and downlink channels, and also in this case we can use the inner-product distortion. Let $\mathcal{R}_{c^{(F)}}^{(D)} = \{\mathcal{R}_{c^{(F)},i}^{(D)}\}$, $i = 1, \dots, 2^{b_B}$, be the partition of the downlink channel vector quantizer conditioned to the fact that the uplink channel belongs to the region $\mathcal{R}_{c^{(F)}}^{(U)}$.

Finally, in step 3, the BS will put together the dependent and independent components to obtain a complete estimate of the downlink channel, i.e.

$$\hat{\mathcal{G}}_C^{(D)} = E[\mathcal{G}_C^{(D)} | \mathcal{G}_C^{(D)} \in \mathcal{R}_{c^{(F)},c^{(B)}}^{(D)}, \mathcal{G}_C^{(U)} \in \mathcal{R}_{c^{(F)}}^{(U)}] \tag{5.53}$$

An enhancement of the procure is obtained by observing the BS perfectly knows the uplink channel $\mathcal{G}_C^{(U)}$, therefore (5.53) can be replaced by

$$\hat{\mathcal{G}}_C^{(D)} = E[\mathcal{G}_C^{(D)} | \mathcal{G}_C^{(D)} \in \mathcal{R}_{c^{(F)},c^{(B)}}^{(D)}, \mathcal{G}_C^{(U)}] \tag{5.54}$$

5.7 Principal component analysis

Given a set of K vectors, $\{\mathbf{s}(k)\}$, $k = 1, \dots, K$, with $\mathbf{s}(k) \in \mathcal{C}^N$, we *encode* $\mathbf{s}(k)$ into the corresponding codevector $\mathbf{r}(k)$, i.e.

$$\mathbf{r}(k) = \text{enc}(\mathbf{s}(k)) \tag{5.55}$$

[6] In LTE systems, in the downlink orthogonal frequency division multiple access (OFDMA) is used, while in uplink single-carrier frequency division multiple access (SC-FDMA) is used.

If $\boldsymbol{r}(k) \in \mathcal{C}^M$ with $M < N$, a sort of compression of the sequence $\{\boldsymbol{s}(k)\}$ is established, in the sense that we can represent $\{\boldsymbol{r}(k)\}$ by fewer elements than the original sequence $\{\boldsymbol{s}(k)\}$. Decoding implies a reconstruction

$$\hat{\boldsymbol{s}}(k) = \text{dec}(\boldsymbol{r}(k)) \tag{5.56}$$

with $\hat{\boldsymbol{s}}(k) \simeq \boldsymbol{s}(k)$. The simplest decoding function is given by matrix $\boldsymbol{B}$, such that $\boldsymbol{B}^H$ has orthonormal columns, and

$$\hat{\boldsymbol{s}}(k) = \boldsymbol{B}^H \boldsymbol{r}(k) \quad \text{with} \quad \boldsymbol{B}^H \;\; N \times M \tag{5.57}$$

under constraint

$$\boldsymbol{B}\boldsymbol{B}^H = \boldsymbol{I} \tag{5.58}$$

where $\boldsymbol{I}$ is the $M \times M$ identity matrix. The encoder is

$$\boldsymbol{r}(k) = \boldsymbol{B}\boldsymbol{s}(k) \tag{5.59}$$

Hence,

$$\hat{\boldsymbol{s}}(k) = \boldsymbol{B}^H \boldsymbol{B}\boldsymbol{s}(k)) \tag{5.60}$$

Selection of columns of $\boldsymbol{B}^H$ is by the square of the Euclidean distance (5.17)

$$\boldsymbol{B}_{opt} = \arg\min_{\boldsymbol{B}} \sum_{k=1}^{K} ||\boldsymbol{s}(k) - \boldsymbol{B}^H \boldsymbol{B}\boldsymbol{s}(k)||^2 \tag{5.61}$$

under constraint (5.58).

Before giving the solution, we introduce some matrices. We define the data matrix

$$\boldsymbol{S} = [\boldsymbol{s}(1), \dots, \boldsymbol{s}(K)], \quad N \times K \tag{5.62}$$

and its covariance matrix (see also (2.102))

$$\boldsymbol{\Phi}_s = \sum_{k=1}^{K} \boldsymbol{s}(k)\boldsymbol{s}^H(k) = \boldsymbol{S}\boldsymbol{S}^H, \quad N \times N \tag{5.63}$$

In (5.63), we assume $\{\boldsymbol{s}(k)\}$ has zero mean, otherwise we subtract from $\boldsymbol{s}(k)$ the estimated mean $\frac{1}{K}\sum_{k=1}^{K} \boldsymbol{s}(k)$. Next, we perform the decomposition of $\boldsymbol{\Phi}_s$ by determining eigenvalues $\{\lambda_1, \lambda_2, \dots\}$, and corresponding eigenvectors $\{\boldsymbol{u}_1, \boldsymbol{u}_2, \dots\}$, ordered by decreasing values. It is (see (1.292))

$$\boldsymbol{\Phi}_s = \boldsymbol{U}\boldsymbol{\Lambda}\boldsymbol{U}^H \tag{5.64}$$

with

$$\boldsymbol{U}^H \boldsymbol{U} = \boldsymbol{I} \tag{5.65}$$

where $\boldsymbol{I}$ is the $N \times N$ identity matrix and

$$\boldsymbol{\Lambda} = \text{diag}\,[[\lambda_1, \dots, \lambda_N]^T], \quad \boldsymbol{U} = [\boldsymbol{u}_1, \dots, \boldsymbol{u}_N], \quad N \times N \tag{5.66}$$

Finally, the optimum solution of $\boldsymbol{B}^H$ is given by the first M eigenvectors $\{\boldsymbol{u}_i\}$, $i = 1, \dots, M$,[7]

$$\boldsymbol{B}^H_{opt} = [\boldsymbol{u}_1, \dots, \boldsymbol{u}_M], \quad N \times M \tag{5.67}$$

or

$$\boldsymbol{B}_{opt} = \begin{bmatrix} \boldsymbol{u}_1^H \\ \vdots \\ \boldsymbol{u}_M^H \end{bmatrix} \tag{5.68}$$

[7] The solution is determined iteratively for increasing values of M, starting with $M = 1$.

An important property of the principal component analysis (PCA) is that the covariance matrix of $\boldsymbol{r}(k)$ is diagonal, the components of $\boldsymbol{r}(k)$ for $\boldsymbol{B} = \boldsymbol{B}_{opt}$ are uncorrelated. In fact, let

$$\boldsymbol{r}^{(ext)}(k) = \boldsymbol{U}^H \boldsymbol{s}(k), \quad N \times N \tag{5.69}$$

where $\boldsymbol{r}^{(ext)}(k) = [\boldsymbol{r}^T(k) \;\; \boldsymbol{0}^T]^T$, i.e. we are interested only on the first M components of $\boldsymbol{r}^{(ext)}(k)$, and set to zero the remaining $(N - M)$ components. The covariance matrix of

$$\boldsymbol{R}^{(ext)} = [\boldsymbol{r}^{(ext)}(1), \ldots, \boldsymbol{r}^{(ext)}(K)] = \boldsymbol{U}^H \boldsymbol{S} \tag{5.70}$$

is

$$\boldsymbol{\Phi}_{\boldsymbol{r}^{(ext)}} = \boldsymbol{R}^{(ext)} \boldsymbol{R}^{(ext)H} = \boldsymbol{U}^H \boldsymbol{S}\boldsymbol{S}^H \boldsymbol{U} \tag{5.71}$$

Using (5.64) and (5.65), it is

$$\boldsymbol{\Phi}_{\boldsymbol{r}^{(ext)}} = \boldsymbol{U}^H \boldsymbol{U} \boldsymbol{\Lambda} \boldsymbol{U}^H \boldsymbol{U} \tag{5.72}$$

which from constraint (5.65) yields

$$\boldsymbol{\Phi}_{\boldsymbol{r}^{(ext)}} = \boldsymbol{\Lambda} \tag{5.73}$$

a diagonal matrix. Entries of $\boldsymbol{r}^{(ext)}(k)$ are called principal components and are obtained by projecting the data vector $\boldsymbol{s}(k)$ onto matrix $\boldsymbol{U}^H$. Using the singular value decomposition (SVD) of $\boldsymbol{S}$ (see (2.139))

$$\boldsymbol{S} = \boldsymbol{U} \boldsymbol{\Lambda}^{1/2} \boldsymbol{V}^H = \sum_{i=1}^{N} \sqrt{\lambda_i} \boldsymbol{u}_i \boldsymbol{v}_i^H \tag{5.74}$$

where $\boldsymbol{U}$ is given in from (5.63). From (5.70), it is also

$$\boldsymbol{R}^{(ext)} = \begin{bmatrix} \sqrt{\lambda_1} \boldsymbol{v}_1^H \\ \vdots \\ \sqrt{\lambda_M} \boldsymbol{v}_M^H \\ \boldsymbol{0} \end{bmatrix} \tag{5.75}$$

5.7.1 PCA and k-means clustering

The PCA unsupervised dimension reduction is used in very broad areas of science such as meteorology, image processing, genomic analysis, and information retrieval. It is also common that PCA is used to project data to a lower-dimensional subspace and k-means is then applied in the subspace. However, this space reduction property alone is inadequate to explain the effectiveness of PCA. It turns out that principal components are actually the continuous solution of the cluster membership indicators in the k-means clustering method, i.e. the PCA dimension reduction automatically performs data clustering according to the k-means objective function [11]. This provides an important justification of PCA-based data reduction.

The notation here is that of Section 5.4. We recall the average distortion (5.30) for a number of clusters L that we write as

$$D_L = \sum_{i=1}^{L} \frac{1}{m_i} \sum_{s \in \mathcal{R}_i} ||s - \boldsymbol{Q}_i||^2 \tag{5.76}$$

If $\boldsymbol{Q}_i$ is the centroid of cluster $\mathcal{R}_i$ (see (5.32))

$$\boldsymbol{Q}_i = \frac{1}{m_i} \sum_{s \in \mathcal{R}_i} s \tag{5.77}$$

it minimizes the general term in (5.76)

$$\frac{1}{m_i}\sum_{s\in\mathcal{R}_i}||s-Q_i||^2 = \frac{1}{m_i}\sum_{s\in\mathcal{R}_i}||s||^2 + ||Q_i||^2 - s^H Q_i - sQ_i^H \tag{5.78}$$

which can be written as

$$\frac{1}{m_i^2}\sum_{s_i\in\mathcal{R}_i}\sum_{s_j\in\mathcal{R}_j}||s_i||^2 - s_i^H s_j = \frac{1}{2m_i^2}\sum_{s_i\in\mathcal{R}_i}\sum_{s_j\in\mathcal{R}_j}||s_i - s_j||^2 \tag{5.79}$$

We now consider the case for $L = 2$ and introduce the average distortion (square distance) between elements of clusters $\mathcal{R}_1$ and $\mathcal{R}_2$

$$D(\mathcal{R}_1, \mathcal{R}_2) = \frac{1}{m_1 m_2}\sum_{s_1\in\mathcal{R}_1}\sum_{s_2\in\mathcal{R}_2}||s_1 - s_2||^2 \tag{5.80}$$

Moreover, let

$$J_D = \frac{m_1 m_2}{2K}[2D(\mathcal{R}_1, \mathcal{R}_2) - D(\mathcal{R}_1, \mathcal{R}_1) - D(\mathcal{R}_2, \mathcal{R}_2)] \tag{5.81}$$

and

$$\mathtt{M}_s = \frac{1}{K}\sum_{k=1}^{K}||s(k)||^2 \tag{5.82}$$

it turns out that by substituting (5.79) into (5.76), we can write

$$D_L = \mathtt{M}_s - J_D \tag{5.83}$$

Therefore, for $L = 2$, finding the minimum of the k-means cluster objective function D_L is equivalent to finding the maximum of the distance objective J_D, which can be shown to be always positive. In J_D, the first term represents average distances between clusters that are maximized, forcing the resulting clusters as separated as possible. Furthermore, the second and third terms represent the average distances within clusters that will be minimized, forcing the resulting clusters as compact as possible. Finally, the presence of the factor $m_1 m_2$ favours a balanced number of elements in the clusters. These results lead to a solution of k-means clustering via PCA, as stated by the following theorem, which we give without proof.

Theorem 5.1
We consider real-values signals and $M = L - 1$. For k-means clustering with $L = 2$, the continuous solution of the *cluster indicator vector* is the principal component $\boldsymbol{v}_1$, i.e. the clusters are given by

$$\mathcal{R}_1 = \{s(k)|(\boldsymbol{v}_1)_k \leq 0\} \tag{5.84}$$

$$\mathcal{R}_2 = \{s(k)|(\boldsymbol{v}_1)_k > 0\} \tag{5.85}$$

Further, the optimal value of the k-means objective satisfies the bounds

$$\mathtt{M}_s - \lambda_1 < D_{L=2} < \mathtt{M}_s \tag{5.86}$$

□

The result for $L = 2$ can be generalized to the case $L > 2$, using $L - 1$ cluster indicator vectors by a suitable representation [11].

Theorem 5.2
When optimizing the k-means objective function, the continuous solutions for the transformed discrete cluster membership indicator vectors are the $L - 1$ principal components $\{\boldsymbol{v}_1, \ldots, \boldsymbol{v}_{L-1}\}$. Furthermore,

the optimal value of the k-means objective satisfies the bounds

$$\mathtt{M}_s - \sum_{i=1}^{L-1} \lambda_i < D_L < \mathtt{M}_s \tag{5.87}$$

□

Principal components can be viewed as approximate cluster membership indicators in k-means clustering. The two views of PCA are consistent, as data clustering is also a form of data reduction. Standard data reduction is obtained in Euclidean space by applying SVD, whereas clustering is obtained as data reduction in classification space. In fact, in Section 5.4, it is shown how the high-dimensional space of signal feature vectors is divided into Voronoi cells by vector quantization. Signal feature vectors are approximated by the cluster centroids represented by the code-vectors. PCA plays a crucial role in both types of data reduction.

Bibliography

[1] Benvenuto, N. and Zorzi, M. (2011). *Principles of Communications Networks and Systems*. Wiley.

[2] Gersho, A. and Gray, R.M. (1992). *Vector Quantization and Signal Compression*. Boston, MA: Kluwer Academic Publishers.

[3] Gray, R.M. (1984). Vector quantization. *IEEE ASSP Magazine* 1: 4–29.

[4] Lookabaugh, T.D. and Gray, R.M. (1989). High–resolution quantization theory and the vector quantizer advantage. *IEEE Transactions on Information Theory* 35: 1020–1033.

[5] Linde, Y., Buzo, A., and Gray, R.M. (1980). An algorithm for vector quantizer design. *IEEE Transactions on Communications* 28: 84–95.

[6] Sereno, D. and Valocchi, P. (1996). *Codifica numerica del segnale audio*. Scuola Superiore G. Reiss Romoli: L'Aquila.

[7] MacQueen, J. (1967). Some methods for classification and analysis of multivariate observations. In: *Proceedings of the 5th Berkeley Symposium on Mathematical Statistics and Probability: Statistics*, vol. 1, 281–297. Berkeley, CA: University of California Press.

[8] 3GPP Technical Specification Group Radio Access Network (2010). Physical layer procedures (FDD) (release 8).

[9] Trivellato, M., Tomasin, S., and Benvenuto, N. (2009). On channel quantization and feedback strategies for multiuser MIMO-OFDM downlink systems. *IEEE Transactions on Communications* 57: 2645–2654.

[10] Zhang, X., Centenaro, M., Tomasin, S. et al. (2019). A study on CSI feedback schemes exploiting feedforward information in FDD cellular systems. *Transactions on Emerging Telecommunications Technologies*. e3628. https://doi.org/10.1002/ett.3628.

[11] Ding, C. and He, X. (2004). K-means clustering via principal component analysis. *Proceedings 21st International Conference on Machine Learning*, ICML '04, 29-2-. New York, NY: ACM.

Chapter 6

Digital transmission model and channel capacity

In this chapter, we define a general model of a digital transmission system, where data must be transmitted reliably to a receiver using an analog transmission medium. We provide a description of the main system parameters. We then focus on detection, a key operation performed at the receiver to properly recover the data. Relevant performance metrics (error probability, capacity, and achievable rates) are introduced and discussed for various channel models and modulation systems.

6.1 Digital transmission model

With reference to Figure 6.1, we discuss some common aspects of a communication systems.

Source. We assume that a source generates a message $\{b_\ell\}$ composed of a sequence of binary symbols $b_\ell \in \{0, 1\}$, that are emitted every T_b seconds:

$$\{b_\ell\} = \{\dots, b_{-1}, b_0, b_1, b_2, \dots\} \tag{6.1}$$

Usually, $\{b_\ell\}$ is a sequence of i.i.d. symbols.

The system bit rate, which is associated with the message $\{b_\ell\}$, is equal to

$$R_b = \frac{1}{T_b} \quad \text{[bit/s]} \tag{6.2}$$

Channel encoder. In order to detect and/or correct errors introduced by the transmission over the channel, a mapping of information message $\{b_\ell\}$ into a coded message[1] $\{c_m\}$ is introduced, according to rules investigated in Chapter 11. Here the code bit is $c_m \in \{0, 1\}$, and the period between two consecutive code bits is T_{cod}, with $T_{cod} < T_b$.

Bit interleaver (or scrambler). In order to randomize the positions of the errors introduced by the channel, sometimes a permutation of bits $\{c_m\}$ is performed. The simplest scheme provides that, given an $M_1 \times M_2$ matrix and writing in code bits $\{c_m\}$ *row-wise*, they are read out *column-wise*. We denote

[1] We distinguish three types of coding: (i) source or entropy coding, (ii) channel coding, and (iii) line coding. Their objectives are respectively: (i) *compress* the digital message by lowering the bit rate without losing the original signal information (see Ref. [1]), (ii) increase the *reliability* of the transmission by inserting redundancy in the transmitted message, so that errors can be detected and/or corrected at the receiver (see Chapter 11), and (iii) *shape* the spectrum of the transmitted signal (see Appendix 7.C).

Algorithms for Communications Systems and their Applications, Second Edition.
Nevio Benvenuto, Giovanni Cherubini, and Stefano Tomasin.

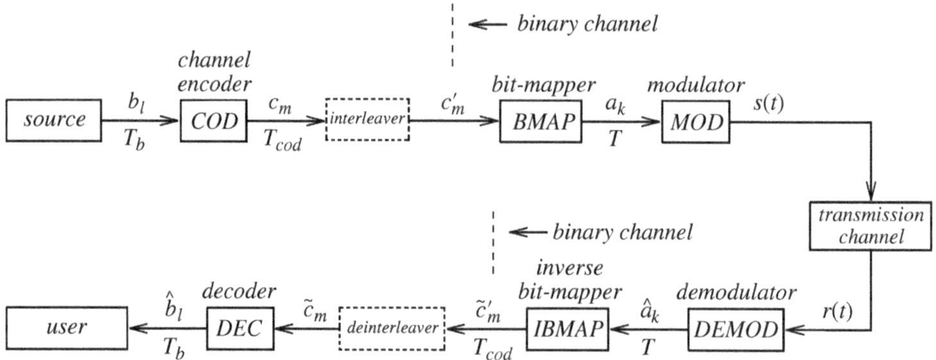

Figure 6.1 Simplified model of a transmission system.

the new sequence $\{c'_m\}$. At the receiver, a deinteleaver writes in the detected bits $\{\tilde{c}'_m\}$ column-wise and reads them out row-wise, to yield $\tilde{c}_m$. As a result, an error burst of length iM_1 is broken into shorten bursts of length i. In the following analysis, to simplify notation, the interleaver is omitted. Note that originally the term scrambler indicated an interleaver designed to avoid long sequences of bits with the same value.

Bit mapper. The bit mapper (BMAP) uses a one-to-one map to associate a multilevel symbol a_k, emitted every T seconds (symbol period or modulation interval), to an input bit pattern, i.e. a collection of bits $\{c_m\}$. $1/T$ is the modulation rate or symbol rate of the system and is measured in Baud: it indicates the number of symbols per second that are transmitted. In the following, we describe the BMAP for baseband (PAM) and passband (QAM and PSK) modulations, while further examples will be provided in Chapter 16.

Bit mapper (PAM). Let us consider for example symbols $\{a_k\}$ from a quaternary pulse amplitude modulation (PAM) alphabet, $a_k \in \mathcal{A} = \{-3, -1, 1, 3\}$. To select the values of a_k, we consider pairs of input bits and map them into quaternary symbols as indicated in Table 6.1. Note that this correspondence is one-to-one. Usually, Gray labelling or mapping is used (see Appendix 6.A), where binary labelling associated to adjacent symbol values differ only in one bit.

For quaternary transmission, the symbol period T is given by $T = 2T_{cod}$. In general, if the values of a_k belong to an alphabet $\mathcal{A}$ with M elements, then

$$T = T_{cod} \log_2 M \tag{6.3}$$

Table 6.1: Example of quaternary bit map (Gray labelling).

c_{2k}	c_{2k-1}	a_k
0	0	−3
0	1	−1
1	1	1
1	0	3

Moreover, the alphabet is

$$\mathcal{A} = \{-(M-1), \ldots, -1, 1, \ldots, (M-1)\} \tag{6.4}$$

i.e. symbol values are equally spaced around zero, with distance $d_{min} = 2$ between adjacent values.

Bit mapper (QAM). For quadrature amplitude modulation (QAM) systems, values of a_k are complex, and the bit mapper associates a complex-valued symbol to an input bit pattern. Figure 6.2 shows two examples of constellations and corresponding binary labelling, where

$$a_k = a_{k,I} + j a_{k,Q} \tag{6.5}$$

and $a_{k,I} = Re\,[a_k]$ and $a_{k,Q} = Im\,[a_k]$. Values of $a_{k,I}$ and $a_{k,Q}$ are from an alphabet similar to (6.4).

For the 4-phase shift keying (4-PSK) (or quadrature phase shift keying, QPSK), symbols are taken from an alphabet with four elements, each identified by 2 bits. Similarly each element in a 16-QAM constellation is uniquely identified by 4 bits. Gray labelling of values of a_k is obtained by Gray labelling values of $a_{k,I}$ and $a_{k,Q}$, independently. In Figure 6.2, the first (second) 2 bits refer to values of $a_{k,I}$ ($a_{k,Q}$).

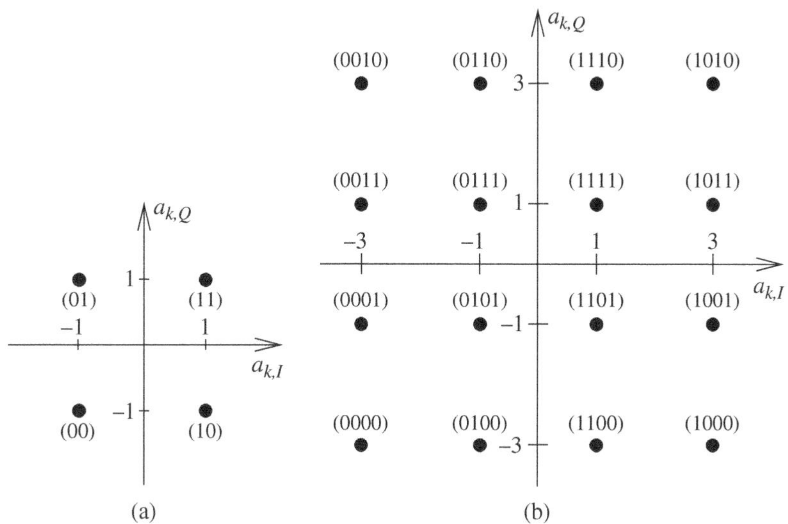

Figure 6.2 Two rectangular QAM constellations and corresponding bit map (Gray labelling). (a) QPSK, (b) 16-QAM.

In general, for a rectangular constellation, each value of a_k is identified by $2\mathtt{m} = \log_2 M$ bits by

$$a_k \overset{\text{BMAP}}{\longleftrightarrow} (c_{k2\mathtt{m}}, \ldots, c_{k2\mathtt{m}-2\mathtt{m}+1}) \tag{6.6}$$

or by

$$(a_{k,I}; a_{k,Q}) \overset{\text{BMAP}}{\longleftrightarrow} (c_{k\mathtt{m},I}, \ldots, c_{k\mathtt{m}-\mathtt{m}+1,I}; c_{k\mathtt{m},Q}, \ldots, c_{k\mathtt{m}-\mathtt{m}+1,Q}) \tag{6.7}$$

Bit mapper (PSK). Phase-shift keying is a particular example of QAM with

$$a_k = e^{j\vartheta_k} \tag{6.8}$$

where $\vartheta_k \in \{\pi/M, 3\pi/M, \ldots, (2M-1)\pi/M\}$ or equivalently if α_n, $n = 1, \ldots M$, are the values of a_k, and φ_n are the values of ϑ_k, it is [2]

$$\alpha_n = e^{j\varphi_n} \tag{6.9}$$

[2] A more general definition of φ_n is given by $\varphi_n = \frac{\pi}{M}(2n-1) + \varphi_0$, where φ_0 is a constant phase.

and

$$\varphi_n = \frac{\pi}{M}(2n-1), \qquad n = 1, \dots, M \tag{6.10}$$

i.e. symbol values lie equispaced on a circle.

The minimum distance between symbol values is

$$d_{min} = 2 \sin \frac{\pi}{M} \tag{6.11}$$

PSK is rarely used for $M > 8$, as d_{min} gets quite small.

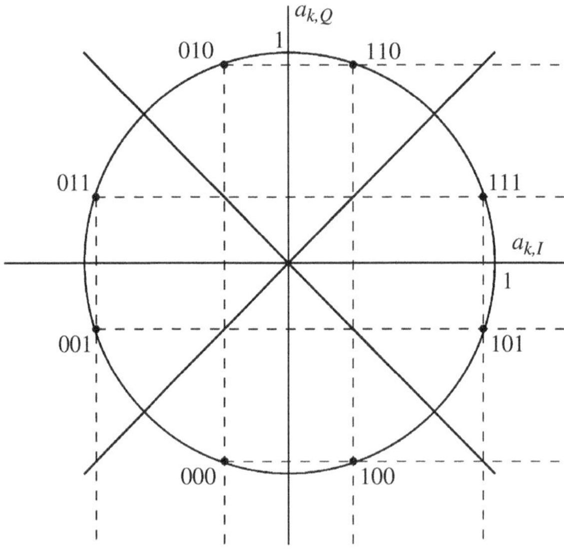

Figure 6.3 8-PSK constellation and decision regions.

An example of M-PSK constellation with $M = 8$ is illustrated in Figure 6.3, where Gray labelling is used.

Modulator. The basic principle of a digital modulator is to map each symbol a_k, generated every T seconds, to a property of a given pulse, e.g. its amplitude. The digital modulated signal requires an up-frequency conversion for PSK and QAM, where the signal is allocated in the pass band around a certain carrier frequency. The corresponding analog modulated signal s is transmitted over the channel. Two relevant examples of modulation systems are discussed in Chapters 7 and 8.

Transmission channel. The channel may (i) distort the transmitted signal s to yield the received *useful signal* s_{Ch} and (ii) introduce noise w modelled as (real value) additive white Gaussian noise (AWGN) with power spectral density (PSD) $N_0/2$. Hence, the received signal is

$$r(t) = s_{Ch}(t) + w(t) \tag{6.12}$$

For a more detailed description of the transmission channel, see Chapter 4. In particular, for an *ideal AWGN channel* $s_{Ch}(t) = s(t)$.

Receiver. Given an observation of r, the final task of the receiver is to recover the information bits $\{b_\ell\}$ with as few errors as possible. In order to accomplish this task with a feasible complexity, the receiver performs demodulation (DEMOD) (including down-frequency conversion for PSK and QAM, filtering to remove noise and/or shape the received signal, equalization to shape the received signal and remove inter-symbol interference (ISI), mainly in the digital domain) to recover the transmitted data, which will be denoted as $\{\hat{a}_k\}$. Inverse bit-mapping (IBMAP) yields bits $\{\tilde{c}_m\}$ (or $\{\tilde{c}'_m\}$ if a de-interleaver is used). Next, the decoder (DEC) for an input $\{\tilde{c}_m\}$ will generate the *code sequence* $\{\hat{c}_m\}$ with corresponding decoded (or detected) information message $\{\hat{b}_\ell\}$ (see Chapter 11). Sometimes, some of these operations are performed jointly, and sometimes they are iterated.

6.2 Detection

At the receiver, the demodulator block includes operations in the analog and digital domain that strongly depend on the specific channel and modulation technique. In general, after analog filtering, a sampler working at T or $T/2$ is usually employed. Next, an equalizer tries to compensate the distortion introduced by the channel, taking into account also the noise. The output of these operations is signal z_k, with sampling period T, whose simplest model is given by a_k plus noise and/or interference, i.e.

$$z_k = a_k + w_k, \quad a_k \in \mathcal{A} \tag{6.13}$$

If σ_I^2 is the noise variance per dimension, it is

$$w_k \sim \begin{cases} \mathcal{N}(0, \sigma_I^2) & \text{PAM} \\ \mathcal{CN}(0, 2\sigma_I^2) & \text{QAM, PSK} \end{cases} \tag{6.14}$$

Note that in (6.13) the *equivalent channel gain* between the modulator input and the equalizer output is equal to 1. Furthermore, we assume $\{w_k\}$ to be a sequence of independent random variables. The following derivation will be for complex-valued (i.e. QAM and PSK) symbols, where also z_k is complex. Let the received observed sample corresponding to z_k be ρ_k, i.e.

$$z_k = \rho_k, \quad \rho_k \in \mathbb{C} \tag{6.15}$$

For PAM, we just consider the real part of (6.15).

By reflecting the operations done at the transmitter, the detector should output the *hard symbols* $\hat{a}_k \in \mathcal{A}$, corresponding to a_k. Next, IBMAP yields bits $\tilde{c}_m$, and the decoder aims at removing errors introduced by the channel, to yield $\hat{b}_\ell$. This approach is denoted as *hard detection*, where the decoder is fed with hard detected bits $\tilde{c}_m$.

Another detection/decoding strategy provides a first elaboration of z_k to obtain a statistical *a posteriori* description of the encoded bits, avoiding the two steps of hard detection and IBMAP. It turns out that suitably designed decoders yield better performance (i.e. typically correct more errors) by exploiting this statistical description. This approach is denoted as *soft detection*, where the decoder is fed with *soft* detected bits (i.e. the statistics).

6.2.1 Optimum detection

Let M be the cardinality of alphabet $\mathcal{A}$. We divide the space $\mathbb{C}$ of the received sample z_k, into M non-overlapping regions

$$\mathcal{R}_\alpha \quad \alpha \in \mathcal{A} \quad \text{with} \quad \bigcup_{\alpha \in \mathcal{A}} \mathcal{R}_\alpha = \mathbb{C} \tag{6.16}$$

such that, if ρ_k belongs to $\mathcal{R}_\alpha$, then the symbol α is detected:

$$\text{if } \rho_k \in \mathcal{R}_\alpha \quad \Longrightarrow \quad \hat{a}_k = \alpha \tag{6.17}$$

The probability of correct decision is computed as follows:

$$\begin{aligned} P[\mathrm{C}] &= P[\hat{a}_k = a_k] \\ &= \sum_{\alpha\in\mathcal{A}} P[\hat{a}_k = \alpha \mid a_k = \alpha]P[a_k = \alpha] \\ &= \sum_{\alpha\in\mathcal{A}} P[z_k \in \mathcal{R}_\alpha \mid a_k = \alpha]P[a_k = \alpha] \\ &= \sum_{\alpha\in\mathcal{A}} \int_{\mathcal{R}_\alpha} p_{z|a}(\rho_k \mid \alpha)d\rho_k P[a_k = \alpha] \end{aligned} \tag{6.18}$$

Two criteria are typically adopted to maximize (6.18) with respect to $\{\mathcal{R}_\alpha\}$:

1. *Maximum a posteriori probability* (MAP) *detection*:

$$\rho_k \in \mathcal{R}_{\alpha_{opt}} \qquad (\text{and } \hat{a}_k = \alpha_{opt}) \qquad \text{if} \qquad \alpha_{opt} = \arg\max_{\alpha\in\mathcal{A}} p_{z|a}(\rho_k \mid \alpha)P[a_k = \alpha] \tag{6.19}$$

   where argmax is defined in page 106. Using the identity

$$p_{z|a}(\rho_k \mid \alpha) = p_z(\rho_k)\frac{P[a_k = \alpha \mid z_k = \rho_k]}{P[a_k = \alpha]} \tag{6.20}$$

   the decision criterion becomes:

$$\hat{a}_k = \arg\max_{\alpha\in\mathcal{A}} P[a_k = \alpha \mid z_k = \rho_k] \tag{6.21}$$

2. *Maximum-likelihood (ML) detection*: If all symbol values are equally likely, the MAP criterion becomes the ML criterion

$$\hat{a}_k = \arg\max_{\alpha\in\mathcal{A}} p_{z|a}(\rho_k \mid \alpha) \tag{6.22}$$

   In other words, the symbol α is chosen, for which the probability to observe $z_k = \rho_k$ is maximum.

We now present computationally efficient methods for both ML and MAP criteria.

ML

As w_k is a complex-valued Gaussian r.v. with zero mean and variance $2\sigma_I^2$, it follows (see (1.307))

$$p_{z|a}(\rho_k \mid \alpha) = \frac{1}{\pi 2\sigma_I^2} e^{-\frac{1}{2\sigma_I^2}|\rho_k-\alpha|^2} \tag{6.23}$$

Taking the logarithm, which is a monotonic increasing function, of both members we get

$$-\ln p_{z|a}(\rho_k \mid \alpha) \propto |\rho_k - \alpha|^2 \tag{6.24}$$

where non-essential constant terms have been neglected.

Then the ML criterion (6.22) is formulated as

$$\hat{a}_k = \arg\min_{\alpha\in\mathcal{A}} |\rho_k - \alpha|^2 \tag{6.25}$$

which corresponds to the *minimum-distance detector*, as it suggests to detect symbol α of the constellation that is closest to the observed sample ρ_k. If M is very large, it is simpler to use the decision regions, where to each region a symbol value corresponds. Two examples are reported in Figure 6.4. Detection (6.25) is also denoted *hard detection*.

MAP

We introduce the *likelihood* function, associated to a_k as

$$\mathrm{L}_{a_k}(\alpha) = P\,[a_k = \alpha \mid z_k = \rho_k], \quad \alpha \in \mathcal{A} \tag{6.26}$$

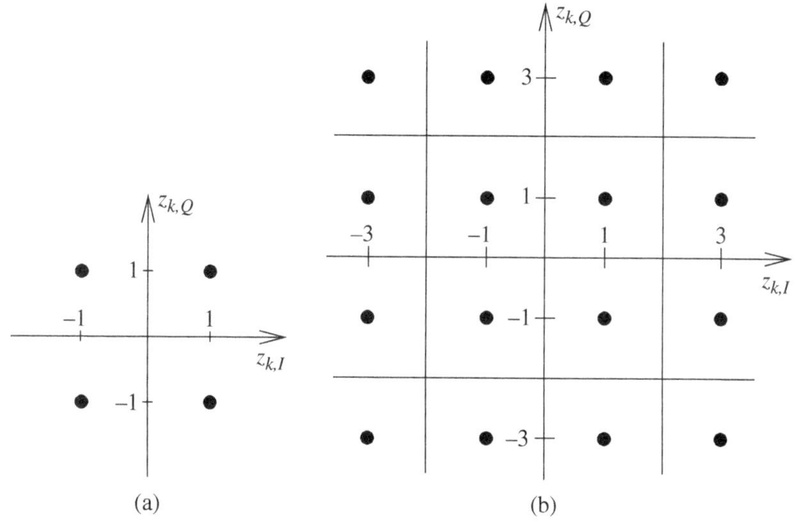

Figure 6.4 Decision regions for (a) QPSK and (b) 16-QAM constellations.

Then, (6.21) becomes

$$\hat{a}_k = \arg\max_{\alpha \in \mathcal{A}} \mathrm{L}_{a_k}(\alpha) \tag{6.27}$$

Here for $\alpha \in \mathcal{A}$,

$$\begin{aligned} \mathrm{L}_{a_k}(\alpha) &= P[a_k = \alpha, z_k = \rho_k]/P[z_k = \rho_k] \\ &= K_k e^{-\frac{|\rho_k - \alpha|^2}{2\sigma_I^2}} \end{aligned} \tag{6.28}$$

where K_k does not depend on α. Note that (6.28) holds for both real-valued and complex-valued signals if σ_I^2 is the noise variance per dimension. As $\mathrm{L}_{a_k}(\alpha)$ is a probability, (6.28) can be written as

$$\mathrm{L}_{a_k}(\alpha) = \frac{e^{-\frac{1}{2\sigma_I^2}|\rho_k - \alpha|^2}}{\sum_{\eta \in \mathcal{A}} e^{-\frac{1}{2\sigma_I^2}|\rho_k - \eta|^2}}, \quad \alpha \in \mathcal{A} \tag{6.29}$$

It is easier working with the log-likelihood (LL) function by taking the logarithm of $\mathrm{L}_{a_k}(\alpha)$

$$\ell_{a_k}(\alpha) = \ln \mathrm{L}_{a_k}(\alpha) = -\frac{|\rho_k - \alpha|^2}{2\sigma_I^2} \tag{6.30}$$

and (6.21) becomes

$$\hat{a}_k = \arg\min_{\alpha \in \mathcal{A}} |\rho_k - \alpha|^2 \tag{6.31}$$

which corresponds again to the minimum-distance detector.

Hard detection (decoder with hard input). In summary, by using symbol decisions, for example based on the minimum-distance detector (6.25), once $\hat{a}_k$ has been chosen, the corresponding inverse bit-map yields the bits $\tilde{c}_m \in \{0, 1\}$, as from (6.6). The sequence $\{\tilde{c}_m\}$ will form the input to the decoder, which in turn provides the decoded (or detected) information bits $\{\hat{b}_\ell\}$.

6.2.2 Soft detection

In the binary case, where $\alpha \in \{-1, 1\}$, it is convenient to introduce the *likelihood ratio* by (see (6.26))

$$\mathrm{L}_{a_k} = \frac{P\,[a_k = 1 \mid z_k = \rho_k]}{P\,[a_k = -1 \mid z_k = \rho_k]} = \frac{\mathrm{L}_{a_k}(1)}{\mathrm{L}_{a_k}(-1)} \tag{6.32}$$

Then, the decision rule (6.27) becomes

$$\hat{a}_k = \begin{cases} 1 & \text{if } \mathrm{L}_{a_k} \geq 1 \\ -1 & \text{if } \mathrm{L}_{a_k} < 1 \end{cases} \tag{6.33}$$

The analysis is simplified by the introduction of the log-likelihood ratio (LLR),

$$\ell_{a_k} = \ln \mathrm{L}_{a_k} = \ell_{a_k}(1) - \ell_{a_k}(-1) \tag{6.34}$$

where $\ell_{a_k}(\alpha)$ is given by (6.30). Then (6.33) becomes

$$\hat{a}_k = \mathrm{sgn}(\ell_{a_k}) \tag{6.35}$$

Observing (6.35), we can write

$$\ell_{a_k} = \hat{a}_k |\ell_{a_k}| \tag{6.36}$$

In other words, the sign of the LLR yields the hard-decision, while its magnitude indicates the decision reliability.

Indeed, most of the time we are not interested on $\{\hat{a}_k\}$ but on extracting from $\{z_k\}$ statistics of c_m to be used by a decoder with *soft input*. This procedure is called *soft detection*. In particular, the considered statistics are summarized by the LLR

$$\ell_{c_m} = \ln \frac{P[c_m = 1 \mid z_k = \rho_k]}{P[c_m = 0 \mid z_k = \rho_k]} \tag{6.37}$$

Next, we derive techniques to compute (6.37).

One word of caution: on labelling each code bit assumes $\{0, 1\}$ values. However, LLRs refer to $\{-1, 1\}$, where the original 0 has become -1.

LLRs associated to bits of BMAP

We give correct and simplified LLR expressions of bits of BMAP (6.7) for square constellations. We start with the simplest case.

1. *Binary*: For the mapping $c_k = 1 \to a_k = 1$ and $c_k = 0 \to a_k = -1$, and ρ_k real, evaluation of (6.30) for $\alpha = -1, 1$ yields

$$\ell_{c_k}(1) = -\frac{(\rho_k - 1)^2}{2\sigma_I^2}, \quad \ell_{c_k}(-1) = -\frac{(\rho_k + 1)^2}{2\sigma_I^2} \tag{6.38}$$

   Hence from (6.34)

$$\ell_{c_k} = \frac{2\rho_k}{\sigma_I^2} \tag{6.39}$$

2. *QPSK*: Using the BMAP (6.7) with $\mathtt{m} = 1$, it is like having a binary constellation in each dimension. By letting $\rho_k = \rho_{k,I} + j\rho_{k,Q}$, it is

$$\ell_{c_{k,I}} = \frac{2\rho_{k,I}}{\sigma_I^2}, \quad \ell_{c_{k,Q}} = \frac{2\rho_{k,Q}}{\sigma_I^2} \tag{6.40}$$

3. *16-QAM and above*: For simplicity, $c_{k\mathbb{m}-m,I}$ and $c_{k\mathbb{m}-m,Q}$ in (6.7) will be denoted here as $c_{m,I}$ and $c_{m,Q}$, respectively. Moreover, ρ will replace ρ_k. For each bit $c_{m,I}$ in (6.7), we split the constellation symbols into two sets $\mathcal{R}^{(0)}_{m,I}$ and $\mathcal{R}^{(1)}_{m,I}$, which, respectively, collect symbols whose BMAP labelling has $c_{m,I} = 0$ and $c_{m,I} = 1$. Similarly for $c_{m,Q}$. For a 16-QAM constellation, where the BMAP is

$$(a_{k,I}; a_{k,Q}) \stackrel{BMAP}{\longleftrightarrow} (c_{0,I}, c_{1,I}; c_{0,Q}, c_{1,Q}) \tag{6.41}$$

as reported in Figure 6.2, the four partitions $\{\mathcal{R}^{(0)}_{0,I}, \mathcal{R}^{(1)}_{0,I}\}$, $\{\mathcal{R}^{(0)}_{1,I}, \mathcal{R}^{(1)}_{1,I}\}$, $\{\mathcal{R}^{(0)}_{0,Q}, \mathcal{R}^{(1)}_{0,Q}\}$, and $\{\mathcal{R}^{(0)}_{1,Q}, \mathcal{R}^{(1)}_{1,Q}\}$ are shown in Figure 6.5.
We introduce the following approximation: for the generic bit $c_{m,I}$, the LLs are

$$\ell_{c_{m,I}}(1) = \ln\left[\sum_{\alpha\in\mathcal{R}^{(1)}_{m,I}} \mathrm{L}_{a_k}(\alpha)\right], \qquad \ell_{c_{m,I}}(-1) = \ln\left[\sum_{\alpha\in\mathcal{R}^{(0)}_{m,I}} \mathrm{L}_{a_k}(\alpha)\right] \tag{6.42}$$

By the Max-Log-MAP approximation, we replace the sums in (6.42) by their largest terms, and

$$\tilde{\ell}_{c_{m,I}}(1) = \max_{\alpha\in\mathcal{R}^{(1)}_{m,I}} \ell_{a_k}(\alpha), \qquad \tilde{\ell}_{c_{m,I}}(-1) = \max_{\alpha\in\mathcal{R}^{(0)}_{m,I}} \ell_{a_k}(\alpha) \tag{6.43}$$

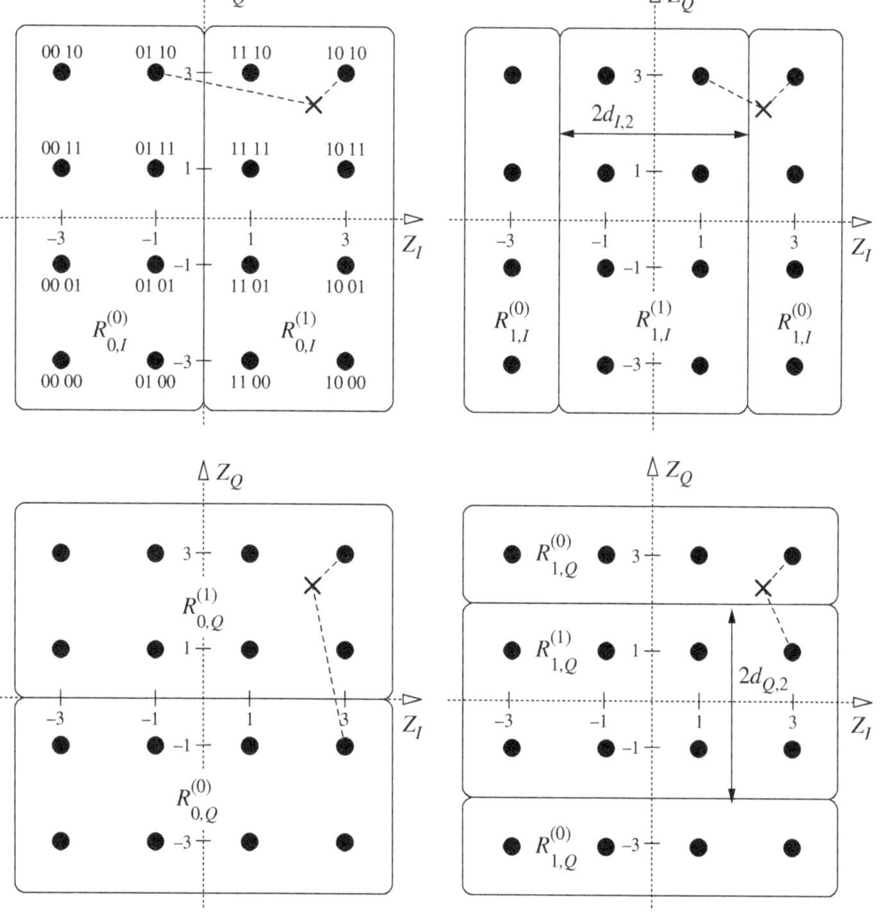

Figure 6.5 Partitions of the 16-QAM constellation.

Finally, the approximated LLR is

$$\tilde{\ell}_{c_{m,I}} = \tilde{\ell}_{c_{m,I}}(1) - \tilde{\ell}_{c_{m,I}}(-1) \tag{6.44}$$

To simplify notation, we remove the symbol time index k. Substituting (6.30) in (6.44) yields, for $m = 0, 1, \dots, \mathtt{m} - 1$,

$$\tilde{\ell}_{c_{m,I}} = \frac{1}{2\sigma_I^2}\left\{ \min_{\alpha \in \mathcal{R}^{(1)}_{m,I}} |\rho - \alpha|^2 - \min_{\alpha \in \mathcal{R}^{(0)}_{m,I}} |\rho - \alpha|^2 \right\} \tag{6.45}$$

and

$$\tilde{\ell}_{c_{m,Q}} = \frac{1}{2\sigma_I^2}\left\{ \min_{\alpha \in \mathcal{R}^{(1)}_{m,Q}} |\rho - \alpha|^2 - \min_{\alpha \in \mathcal{R}^{(0)}_{m,Q}} |\rho - \alpha|^2 \right\} \tag{6.46}$$

For square constellations, boundaries of partitions are vertical lines for bits $c_{m,I}$, while are horizontal lines for bits $c_{m,Q}$. Let $\alpha = \alpha_I + j\alpha_Q$, $\rho = \rho_I + j\rho_Q$, and $Re\,[\mathcal{R}^{(i)}_{m,I}]$ ($Im\,[\mathcal{R}^{(i)}_{m,I}]$) containing the real (imaginary) parts of symbols in $\mathcal{R}^{(i)}_{m,I}$. Then, (6.45) and (6.46) can be written as

$$\tilde{\ell}_{c_{m,I}} = \frac{1}{2\sigma_I^2}\left\{ \min_{\alpha_I \in Re[\mathcal{R}^{(1)}_{m,I}]} |\rho_I - \alpha_I|^2 - \min_{\alpha_I \in Re[\mathcal{R}^{(0)}_{m,I}]} |\rho_I - \alpha_I|^2 \right\} \tag{6.47}$$

and

$$\tilde{\ell}_{c_{m,Q}} = \frac{1}{2\sigma_I^2}\left\{ \min_{\alpha_Q \in Im[\mathcal{R}^{(1)}_{m,Q}]} |\rho_Q - \alpha_Q|^2 - \min_{\alpha_Q \in Im[\mathcal{R}^{(0)}_{m,Q}]} |\rho_Q - \alpha_Q|^2 \right\} \tag{6.48}$$

Simplified expressions

Now, we concentrate on simplifying (6.47), keeping in mind that problem (6.48) is similar with index I replaced by Q. Introduced the metric

$$d_{m,I} = \frac{1}{4}\left\{ \min_{\alpha_I \in Re[\mathcal{R}^{(1)}_{m,I}]} |\rho_I - \alpha_I|^2 - \min_{\alpha_I \in Re[\mathcal{R}^{(0)}_{m,I}]} |\rho_I - \alpha_I|^2 \right\} \tag{6.49}$$

it is

$$\tilde{\ell}_{c_{m,I}} = \frac{2d_{m,I}}{\sigma_I^2} \tag{6.50}$$

For 16-QAM, evaluation of (6.49) yields [2]

$$d_{0,I} = \begin{cases} \rho_I & |\rho_I| \le 2 \\ 2(\rho_I - 1) & \rho_I > 2 \\ 2(\rho_I + 1) & \rho_I < -2 \end{cases}, \qquad d_{1,I} = -|\rho_I| + 2 \tag{6.51}$$

Similarly, for 64-QAM

$$d_{0,I} = \begin{cases} \rho_I & |\rho_I| \le 2 \\ 2(\rho_I - 1) & 2 < \rho_I \le 4 \\ 3(\rho_I - 2) & 4 < \rho_I \le 6 \\ 4(\rho_I - 3) & \rho_I > 6 \\ 2(\rho_I + 1) & -4 < \rho_I \le -2 \\ 3(\rho_I + 2) & -6 < \rho_I \le -4 \\ 4(\rho_I + 3) & \rho_I < -6 \end{cases} \tag{6.52}$$

$$d_{1,I} = \begin{cases} 2(-|\rho_I| + 3) & |\rho_I| \le 2 \\ 4 - |\rho_I| & 2 < |\rho_I| \le 4 \\ 2(-|\rho_I| + 5) & |\rho_I| > 6 \end{cases}, \quad d_{2,I} = \begin{cases} |\rho_I| - 2 & |\rho_I| \le 4 \\ -|\rho_I| + 6 & |\rho_I| > 4 \end{cases} \tag{6.53}$$

Above expressions are still cumbersome to evaluate, so we introduce a further simplification [3]. First, for 16-QAM we have the following approximation

$$d_{0,I} \simeq \rho_I, \qquad d_{1,I} = -|\rho_I| + 2 \tag{6.54}$$

while for 64-QAM it is

$$d_{0,I} \simeq \rho_I, \quad d_{1,I} \simeq -|\rho_I| + 4, \quad d_{2,I} = -||\rho_I| - 4| + 2 \tag{6.55}$$

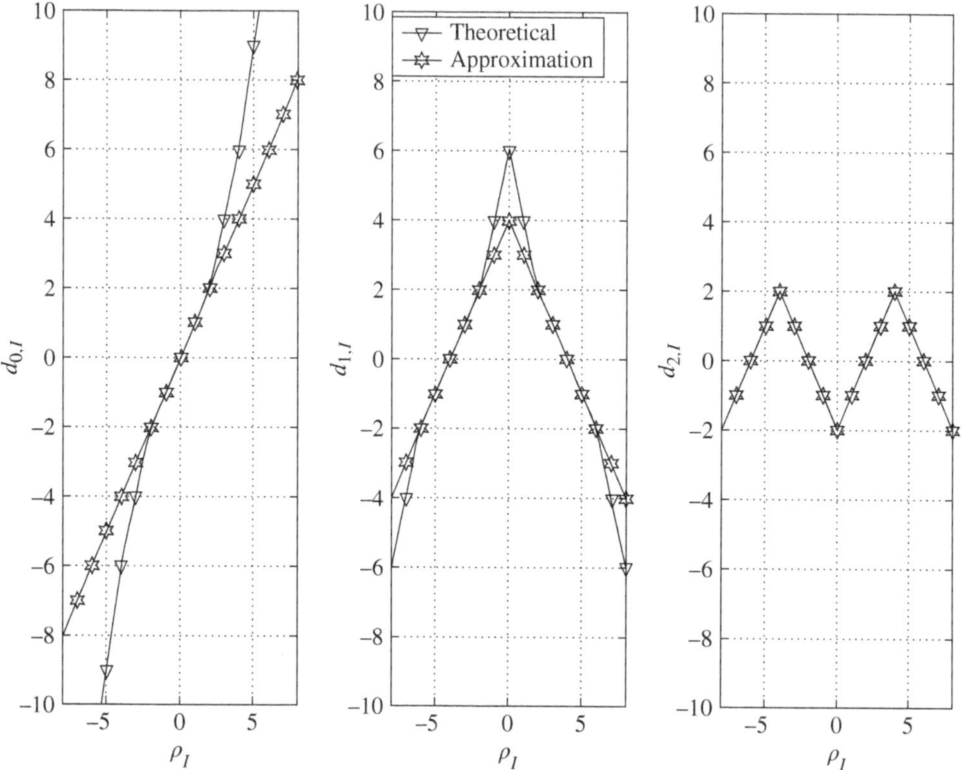

Figure 6.6 Approximate versus exact metrics.

These approximations correspond to evaluating $|d_{m,I}|$ as the distance of ρ_I from the nearest partition boundary and assigning to $d_{m,I}$ the sign according to the partition wherein ρ_I falls. For 16-QAM, Figure 6.6 shows the exact and approximated functions. These simplifications can be generalized to any square QAM constellation. For $m > 0$, let $D_{m,I}$ denote half the distance between the partition boundary relative to bit $c_{m,I}$. For example, for 16-QAM $D_{1,I} = 2$, see Figure 6.5, while for 64-QAM $D_{1,I} = 4$ and $D_{2,I} = 2$. Then, $d_{m,I}$, $m = 0, 1, \dots, \mathtt{m} - 1$, can be written in a recursive form as

$$d_{m,I} \simeq \begin{cases} \rho_I & m = 0 \\ -|d_{m-1,I}| + D_{m,I} & m > 1 \end{cases} \tag{6.56}$$

Comparing (6.56) with the Max-LogMAP approach (6.45), we note that the amount of complexity of the proposed scheme, in terms of real multiplications, is independent of the number of bits per symbol, as well as the dimension of the constellation. From simulation results [3], it is seen that for both 16- and 64-QAM there is almost no performance degradation using the approximate metric (6.56).

We also mention alternative approximation methods for evaluating LLRs. Some make use of a different bit mapper [4]. Moreover, LLRs for PSK and differential PSK (DPSK) can be found in [5–7].

6.2.3 Receiver strategies

The receiver may adopt one of the following strategies to obtain $\hat{b}_\ell$:

1. Equalization of the received signal, hard detection (to yield $\hat{a}_k$), IBMAP (to yield $\hat{c}_m$), and hard-input decoding. In general, this approach, although rather simple, yields poor performance.
2. Equalization of the digital signal, soft detection of code bits (to yield their LLRs), and soft-input decoding. The two operations of equalization and detection/decoding are, sometimes, iterated (see Section 11.7).
3. Soft sequence detection (to yield the LLRs starting from the received signal, without equalization) and soft-input decoding. This strategy significantly improves performance and is feasible for a small constellation size and simple transmission channel (see Section 7.5).
4. Combined detection and decoding. This yields best performance (see Section 11.3.2); however, this approach is not *modular* in the constellation, and is usually quite complex to implement.

6.3 Relevant parameters of the digital transmission model

We report below several general definitions widely used to describe a modulation system.

- T_b: bit period (s): time interval between two consecutive information bits. We assume that the message $\{b_\ell\}$ is composed of binary i.i.d. symbols.
- $R_b = 1/T_b$: bit rate of the system (bit/s).
- T: modulation interval or symbol period (s).
- $1/T$: modulation rate or symbol rate (Baud).
- L_b: number of *information message bits* b_ℓ per modulation interval T.
- M: cardinality of the alphabet of symbols at the transmitter.
- I: number of dimensions of the symbol constellation $\mathcal{A}$, or, equivalently, of the modulator space (e.g. for PAM $I = 1$, for QAM $I = 2$).
- R_I: rate of the encoder-modulator (bit/dim).
- $\mathtt{M}_{s_{Ch}}$: statistical power of the desired signal at the receiver input (V^2).
- $E_{s_{Ch}}$: average pulse energy (V^2 s).
- E_I: average energy per dimension (V^2 s/dim).
- E_b: average energy per information bit (V^2 s/bit).
- $N_0/2$: PSD of additive white noise introduced by the channel (V^2/Hz).
- B_{min}: *conventional* minimum bandwidth of the modulated signal (Hz).
- ν: spectral efficiency of the system (bit/s/Hz).
- Γ: *conventional* signal-to-noise ratio (SNR) at the receiver input.
- Γ_I: SNR per dimension.
- $\mathtt{P}_{s_{Ch}}$: available power of the desired signal at the receiver input (W).
- $\mathtt{T}_{w_i}$: effective receiver noise temperature (K).
- $\mathcal{S}$: *sensitivity* (W): the minimum value of $\mathtt{P}_{s_{Ch}}$ (see (4.26)) such that the system achieves a given performance in terms of bit error probability.

Relations among parameters

1. *Rate of the encoder-modulator*:
$$R_I = \frac{L_b}{I} \tag{6.57}$$
2. *Number of information bits per modulation interval*: via the channel encoder and the bit-mapper, L_b information bits of the message $\{b_\ell\}$ are mapped in an M-ary symbol, a_k. In general we have
$$L_b \leq \log_2 M \tag{6.58}$$
where the equality holds for a system without coding, or, with abuse of language, for an uncoded system. In this case, we also have
$$R_I = \frac{\log_2 M}{I} \quad \text{(uncoded case)} \tag{6.59}$$
3. *Symbol period*:
$$T = T_b L_b \tag{6.60}$$
4. *Statistical power of the desired signal at the receiver input*:
$$\mathtt{M}_{s_{Ch}} = \frac{E_{s_{Ch}}}{T} \tag{6.61}$$
For a *continuous* transmission, $\mathtt{M}_{s_{Ch}}$ is finite and we define $E_{s_{Ch}} = \mathtt{M}_{s_{Ch}} T$.
5. *Average energy per dimension*:
$$E_I = \frac{E_{s_{Ch}}}{I} \tag{6.62}$$
6. *Average energy per information bit*:
$$E_b = \frac{E_I}{R_I} = \frac{E_{s_{Ch}}}{L_b} \tag{6.63}$$
For an uncoded system, (6.63) becomes
$$E_b = \frac{E_{s_{Ch}}}{\log_2 M} \quad \text{(uncoded case)} \tag{6.64}$$
7. *Conventional minimum bandwidth of the modulated signal*:
$$B_{min} = \frac{1}{2T} \text{ for baseband signals (PAM),} \tag{6.65}$$
$$B_{min} = \frac{1}{T} \text{ for passband signals (QAM, PSK)} \tag{6.66}$$
8. *Spectral efficiency*:
$$\nu = \frac{R_b}{B_{min}} = \frac{L_b}{B_{min} T} \tag{6.67}$$
In practice, ν measures how many information bits per unit of time are sent over a channel with the conventional bandwidth B_{min}. In terms of R_I, from (6.57), we have
$$\nu = \frac{R_I I}{B_{min} T} \tag{6.68}$$
For most modulation systems, $R_I = \nu/2$.
9. *Conventional SNR at the receiver input*:
$$\Gamma = \frac{\mathtt{M}_{s_{Ch}}}{(N_0/2) 2 B_{min}} = \frac{E_{s_{Ch}}}{N_0 B_{min} T} \tag{6.69}$$
In general, Γ expresses the ratio between the statistical power of the desired signal at the receiver input and the statistical power of the noise measured with respect to the conventional bandwidth B_{min}. We stress that the noise power is infinite, from a theoretical point of view.

Note that, for the same value of $N_0/2$, if B_{min} doubles the statistical power must also double to maintain a given ratio Γ.

10. *SNR per dimension*:

$$\Gamma_I = \frac{E_I}{N_0/2} = \frac{2E_{s_{Ch}}}{N_0 I} \tag{6.70}$$

is the ratio between the energy per dimension E_I and the channel noise PSD (per dimension) $N_0/2$. Using (6.63), the general relation becomes

$$\Gamma_I = 2R_I \frac{E_b}{N_0} \tag{6.71}$$

It is interesting to observe that in most modulation systems it turns out $\Gamma_I = \Gamma$.

6.4 Error probability

We now consider a metric to assess the performance of uncoded communication systems with hard detection.

Definition 6.1

The transformation that maps c_m into $\tilde{c}_m$ is called *binary channel*. It is characterized by the bit rate $1/T_{cod}$, which is the transmission rate of the bits of the sequence $\{c_m\}$, and by the bit error probability

$$P_{bit} = P_{BC} = P[\tilde{c}_m \neq c_m], \qquad \tilde{c}_m, c_m \in \{0, 1\} \tag{6.72}$$

In the case of a *binary symmetric channel* (BSC), it is assumed that $P[\tilde{c}_m \neq c_m \mid c_m = 0] = P[\tilde{c}_m \neq c_m \mid c_m = 1]$. We say that the BSC is *memoryless* if, for every choice of N distinct instants $m_1, m_2, \ldots, m_N$, the following relation holds:

$$\begin{aligned} P[\tilde{c}_{m_1} &\neq c_{m_1}, \tilde{c}_{m_2} \neq c_{m_2}, \ldots, \tilde{c}_{m_N} \neq c_{m_N}] \\ &= P[\tilde{c}_{m_1} \neq c_{m_1}]\, P[\tilde{c}_{m_2} \neq c_{m_2}] \ldots P[\tilde{c}_{m_N} \neq c_{m_N}] \end{aligned} \tag{6.73}$$

□

In a *memoryless binary symmetric channel*, the probability distribution of $\{\tilde{c}_m\}$ is obtained from that of $\{c_m\}$ and P_{BC} according to the statistical model shown in Figure 6.7.

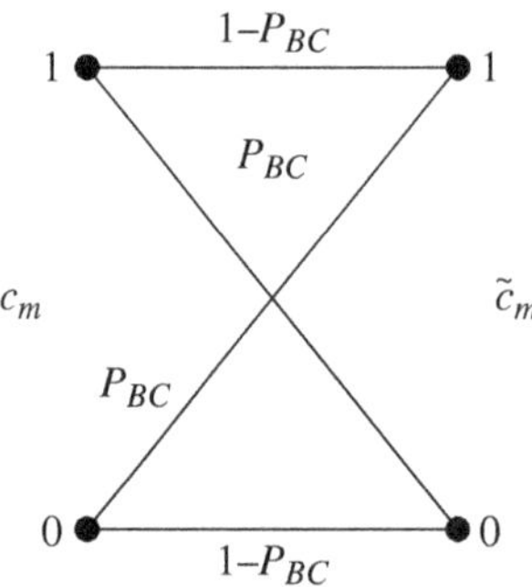

Figure 6.7 Memoryless binary symmetric channel.

We note that the aim of the channel encoder is to introduce redundancy in the sequence $\{c_m\}$, which is exploited by the decoder to detect and/or correct errors introduced by the binary channel.

The overall objective of the transmission system is to reproduce the sequence of information bits $\{b_\ell\}$ with a high degree of reliability, measured by the bit error probability

$$P_{bit}^{(dec)} = P[\hat{b}_\ell \neq b_\ell] \tag{6.74}$$

An approximated expression of P_{bit} is derived, while the relation between $P_{bit}^{(dec)}$ and P_{bit} is detailed in Chapter 11 for some error correcting code classes. We assume the signal at the detection point is given by (6.13). If d_{min} is the distance between adjacent constellation points, we define the SNR at the detection point as

$$\gamma = \left(\frac{d_{min}/2}{\sigma_I}\right)^2 \tag{6.75}$$

We recall that here $d_{min} = 2$ for both PAM and QAM. The symbol error probability

$$P_e = 1 - P[C] = P[\hat{a}_k \neq a_k] \tag{6.76}$$

has the following expressions for PAM and for rectangular QAM constellations (see Ref. [1])

$$P_e = 2\left(1 - \frac{1}{M}\right) Q(\sqrt{\gamma}), \qquad M - \text{PAM} \tag{6.77}$$

$$P_e \simeq 4\left(1 - \frac{1}{\sqrt{M}}\right) Q(\sqrt{\gamma}), \qquad \text{QAM } (M = 4, 16, 64, \ldots) \tag{6.78}$$

where $Q(\cdot)$ is the complementary Gaussian cumulative distribution function (see Appendix 6.B).

In narrowband transmission systems, it is (see the observation after (7.107))

$$\gamma = \frac{\Gamma}{\mathtt{M}_a}, \qquad \text{PAM} \tag{6.79}$$

$$\gamma = \frac{\Gamma}{\mathtt{M}_a/2}, \qquad \text{QAM, PSK} \tag{6.80}$$

where $\mathtt{M}_a/2$ is the statistical power per-dimension of the symbol sequence $\{a_k\}$. In this case,

$$P_e = 2\left(1 - \frac{1}{M}\right) Q\left(\sqrt{\frac{3}{(M^2-1)}\Gamma}\right), \qquad \text{PAM} \tag{6.81}$$

$$P_e \simeq 4\left(1 - \frac{1}{\sqrt{M}}\right) Q\left(\sqrt{\frac{3}{(M-1)}\Gamma}\right), \qquad \text{QAM} \tag{6.82}$$

while for PSK (see Ref. [1])

$$P_e \simeq 2Q\left[\sqrt{2\Gamma}\sin\left(\frac{\pi}{M}\right)\right] \tag{6.83}$$

is a good approximation of P_e for $M \geq 4$ and $\Gamma >> 1$. For $M = 2$ (binary PSK, BPSK), we have

$$P_e = P_{bit} = Q(\sqrt{2\Gamma}) \tag{6.84}$$

If an error occurs on detecting a_k, $\hat{a}_k$ is most probably one of the adjacent symbols of a_k, and for Gray labelling only one bit will be in error, therefore

$$P_{bit} \simeq \frac{P_e}{\log_2 M} \tag{6.85}$$

Curves of P_{bit} for PAM, QAM, and PSK, are reported, respectively, in Figures 6.8–6.10.

In order to achieve a given P_{bit} with QAM, if M is increased by a factor 4, we need to increase Γ by 6 dB: in other words, if we increment by one the number of bits per symbol, Γ should be increased on average by 3 dB.

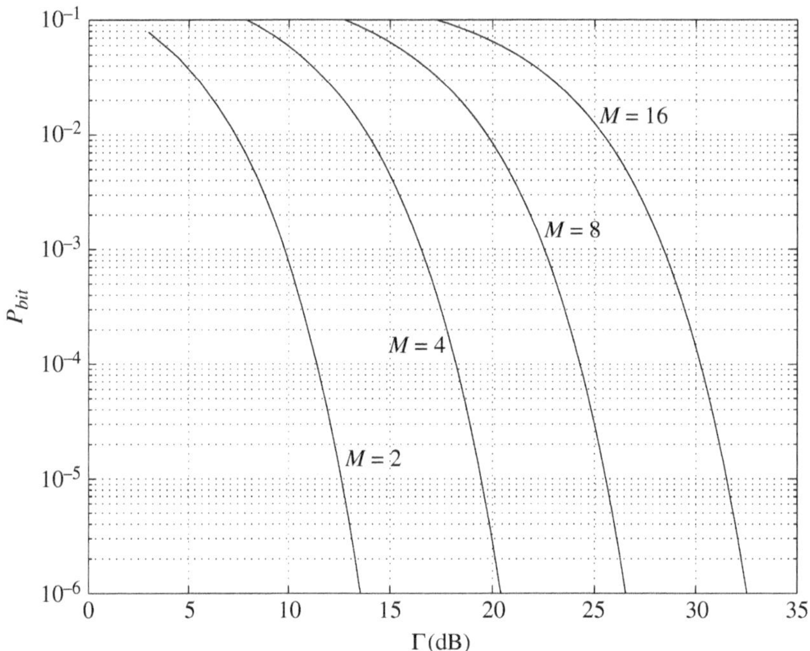

Figure 6.8 Bit error probability as a function of Γ for M-PAM.

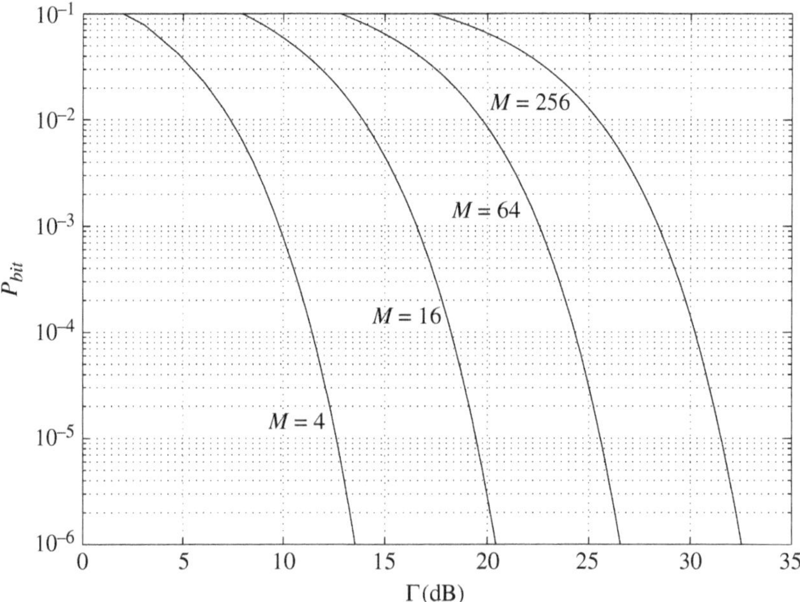

Figure 6.9 Bit error probability as a function of Γ for M-QAM with a rectangular constellation.

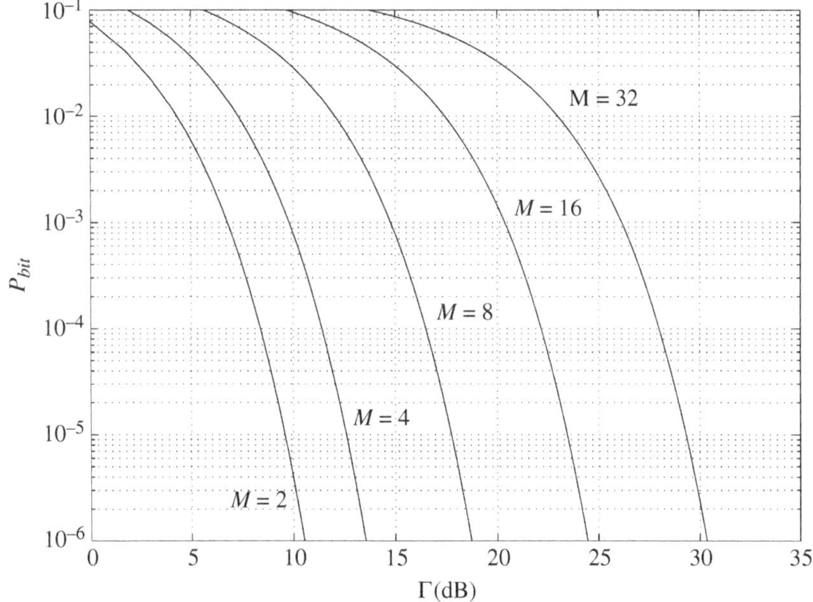

Figure 6.10 Bit error probability as a function of Γ for M-PSK.

6.5 Capacity

The *channel capacity* is defined as the maximum of the *mutual information* between the input and output signals of the channel [8, 9], where the maximum is taken with respect to the statistical distribution of the channel input. The channel capacity will be denoted as C when measured in bit/symbol/dim or simply in bit/dim, or as $C_{[bit/s]}$ when measured in bit/s.

Let R_I be the rate of the encoder–modulator (in bit/dim) (see (6.57)) of a communication system over a channel with capacity C. We give, without proof, the following fundamental theorem [8, 10].

Theorem 6.1 (Shannon's theorem)
For any rate $R_I < C$, with transmitted *symbols having a Gaussian distribution*, there exists a channel coding that provides an arbitrarily small probability of error; such coding does not exist if $R_I > C$. □

We note that Shannon's theorem indicates the bounds, in terms of encoder–modulator rate or, equivalently, in terms of *transmission bit rate* (see (6.96)), within which we can develop systems for the reliable transmission of information, but it does not give any indication about the practical realization of channel coding.

The bound $R_I < C$ for a reliable transmission corresponds to $R_b < C_{[bit/s]}$.

We now report expressions on the channel capacity for AWGN channels. In particular, we consider (i) the discrete-time AWGN channel, (ii) the single-input-single-output (SISO) narrowband (NB) AWGN channel, (iii) the SISO dispersive AGN channel, and (iv) the discrete-time MIMO NB AWGN channel, and its extension to continuous-time channels.

6.5.1 Discrete-time AWGN channel

With reference to a message $\{a_k\}$ composed of a sequence of i.i.d. symbols, with zero mean and power $\mathtt{M}_a$, which belongs to a I-dimensional constellation, let the received signal be as in (6.13), where the noise sequence w_k is i.i.d. and each w_k is Gaussian distributed with power per dimension σ_I^2, i.e. $\sigma_w^2 = I\sigma_I^2$. It is

$$C = \frac{1}{2}\log_2\left(1 + \frac{\mathtt{M}_a}{\sigma_w^2}\right) \quad \text{[bit/dim]} \tag{6.86}$$

The capacity is reached if a_k is Gaussian distributed. From (6.69) and (6.70), assuming $\mathtt{M}_{s_{Ch}} = \mathtt{M}_a$ and $\sigma_w^2 = \frac{N_0}{2}2B_{min}$, it is also

$$\Gamma = \frac{\mathtt{M}_a}{\sigma_w^2} = \frac{\mathtt{M}_a/I}{\sigma_I^2} = \Gamma_I \tag{6.87}$$

and

$$C = \frac{1}{2}\log_2(1 + \Gamma_I) \tag{6.88}$$

In particular, for PAM and QAM constellations, we have

$$C = \begin{cases} \frac{1}{2}\log_2\left(1 + \frac{\mathtt{M}_a}{\sigma_I^2}\right) & \text{PAM} \\ \frac{1}{2}\log_2\left(1 + \frac{\mathtt{M}_a/2}{\sigma_I^2}\right) & \text{QAM} \end{cases} \tag{6.89}$$

Moreover, if the symbol rate is $1/T$

$$C_{[bit/s]} = \begin{cases} \frac{1}{T}C = \frac{1}{2T}\log_2(1 + \Gamma) & \text{PAM } (I = 1) \\ \frac{1}{T}2C = \frac{1}{T}\log_2(1 + \Gamma) & \text{QAM } (I = 2) \end{cases} \tag{6.90}$$

From (6.88), C can be upper bounded and approximated for small values of Γ_I by a linear function and also lower bounded and approximated for large values of Γ_I by a logarithmic function as follows:

$$\Gamma_I \ll 1 : C \leq \frac{1}{2} \log_2(e)\, \Gamma_I \tag{6.91}$$

$$\Gamma_I \gg 1 : C \geq \frac{1}{2} \log_2 \Gamma_I \tag{6.92}$$

6.5.2 SISO narrowband AWGN channel

Let us consider an ideal SISO NB AWGN channel for $s_{Ch}(t) = s(t)$ with, see (6.12),

$$r(t) = s(t) + w(t) \tag{6.93}$$

where s is the real-valued transmitted signal, and w is AWGN with PSD $N_0/2$. We further assume that the transmitted signal has PSD $\mathcal{P}_s(f)$ and passband $\mathcal{B}$, collecting only positive sequences (see (1.69)), with bandwidth given by (1.70), i.e. $B = \int_{\mathcal{B}} df$. We further assume a power constraint on the signal s, i.e.

$$\frac{1}{2}E[s^2(t)] = \int_{\mathcal{B}} \mathcal{P}_s(f)\, df \leq V_P \tag{6.94}$$

The capacity of this channel is achieved when s is Gaussian distributed with zero-mean and constant PSD over the band. Thus, recalling the power constraint (6.94), we have for $f > 0$,

$$\mathcal{P}_{s,opt}(f) = \begin{cases} \frac{V_P}{B} & f \in \mathcal{B} \\ 0 & \text{otherwise} \end{cases} \tag{6.95}$$

and for $f < 0$ we have $\mathcal{P}_{s,opt}(f) = \mathcal{P}_{s,opt}(-f)$.

The resulting capacity of the SISO NB AWGN channel (6.93), under constraint (6.94) is

$$C_{[bit/s]} = B \log_2(1 + \Gamma) \quad \text{[bit/s]} \tag{6.96}$$

where Γ is the SNR at the receiver input (6.69) for $B_{min} = B$, i.e.

$$\Gamma = \frac{V_P}{\frac{N_0}{2} B} \tag{6.97}$$

Channel gain

A first generalization of the ideal channel (6.93) is obtained by considering a channel with gain $\mathcal{G}_{Ch}$, i.e.

$$r(t) = \mathcal{G}_{Ch} s(t) + w(t) \tag{6.98}$$

Under power constraint (6.94), the capacity of channel (6.98) is given by (6.96), still achieved for a Gaussian signal s with constant PSD over $\mathcal{B}$, where

$$\Gamma = \frac{\mathcal{G}_{Ch}^2 V_P}{\frac{N_0}{2} B} \tag{6.99}$$

6.5.3 SISO dispersive AGN channel

In general, a linear dispersive channel with additive Gaussian noise (AGN) not necessarily white, is characterized by a filter with impulse response g_{Ch} and input–output relation

$$r(t) = s * g_{Ch}(t) + w(t) \tag{6.100}$$

where the noise w has a PSD $\mathcal{P}_w(f)$. Let $\mathcal{B}_{Tx}$ be the transmit signal band, and consider the transmit power constraint

$$\int_{\mathcal{B}_{Tx}} \mathcal{P}_s(f) \, df \le V_P \tag{6.101}$$

The channel-gain-to-noise-PSD ratio as a function of the frequency is defined as

$$\Theta_{Ch}(f) = \frac{|\mathcal{G}_{Ch}(f)|^2}{\mathcal{P}_w(f)} \tag{6.102}$$

where $\mathcal{G}_{Ch}(f)$ is the Fourier transform of g_{Ch}.

In order to find the capacity of a dispersive AGN channel, we divide the passband $\mathcal{B}_{Tx}$, having measure B, into N subbands $\mathcal{B}_i$, $i = 1, \ldots, N$, of width $\Delta f = B/N$, where Δf is chosen sufficiently small so that $\mathcal{P}_s(f)$ and $\Theta_{Ch}(f)$ are, to a first approximation, constant within $\mathcal{B}_i$, i.e. we assume $\exists f_i \in \mathcal{B}_i$ such that $\mathcal{P}_s(f) \simeq \mathcal{P}_s(f_i)$ and $\Theta_{Ch}(f) \simeq \Theta_{Ch}(f_i)$, $\forall f \in \mathcal{B}_i$.

Since on each subchannel we have a SISO NB AGN channel, from (6.96) and (6.98) subchannel i has an (approximated) capacity

$$C_{[bit/s]}[i] = \Delta f \log_2 \left[1 + \frac{\Delta f\, \mathcal{P}_{s,opt}(f_i)\, |\mathcal{G}_{Ch}(f_i)|^2}{\Delta f\, \mathcal{P}_w(f_i)}\right] \tag{6.103}$$

with s Gaussian on each subchannel and the $\mathcal{P}_{s,opt}(f)$ is optimized to satisfy the power constraint (as detailed in the following). Since the noise components are independent on each subchannel, we have a *parallel AGN channel* and the capacity of the overall dispersive AGN channel is obtained by adding terms $C_{[bit/s]}[i]$, $i = 1, 2, \dots, N$, that is

$$\sum_{i=1}^{N} C_{[bit/s]}[i] = \Delta f \sum_{i=1}^{N} \log_2 \left[1 + \frac{\Delta f\, \mathcal{P}_{s,opt}(f_i)\, |\mathcal{G}_{Ch}(f_i)|^2}{\Delta f\, \mathcal{P}_w(f_i)}\right] \tag{6.104}$$

By letting Δf tend to zero, we obtain

$$C_{[bit/s]} = \int_{\mathcal{B}_{Tx}} \log_2 \left[1 + \mathcal{P}_{s,opt}(f) \frac{|\mathcal{G}_{Ch}(f)|^2}{\mathcal{P}_w(f)}\right] df = \int_{\mathcal{B}_{Tx}} \log_2 \left[1 + \mathcal{P}_{s,opt}(f)\, \Theta_{Ch}(f)\right] df \tag{6.105}$$

We now find the PSD $\mathcal{P}_s(f)$ that maximizes (6.105), under the constraint (6.101). By applying the method of Lagrange multipliers, the optimum $\mathcal{P}_{s,opt}(f)$ maximizes the integral

$$\int_{\mathcal{B}_{Tx}} \{\log_2 \left[1 + \mathcal{P}_{s,opt}(f)\, \Theta_{Ch}(f)\right] + \lambda\, \mathcal{P}_{s,opt}(f)\} df \tag{6.106}$$

where λ is a Lagrange multiplier. Using the calculus of variations [11], we find the following condition:

$$\frac{\Theta_{Ch}(f)}{1 + \mathcal{P}_{s,opt}(f)\, \Theta_{Ch}(f)} + \lambda \ln 2 = 0 \tag{6.107}$$

Therefore, the PSD providing the capacity is given for $f \geq 0$ by[3]

$$\mathcal{P}_{s,opt}(f) = \left[L - \frac{1}{\Theta_{Ch}(f)}\right]^+ \tag{6.109}$$

where L is a constant such that (6.101) is satisfied with the equal sign. As the channel impulse response is assumed real valued, for $f < 0$ we get $\mathcal{P}_{s,opt}(f) = \mathcal{P}_{s,opt}(-f)$. The function $\mathcal{P}_{s,opt}(f)$ is illustrated in Figure 6.11 for a typical behaviour of the function $\Theta_{Ch}(f)$.

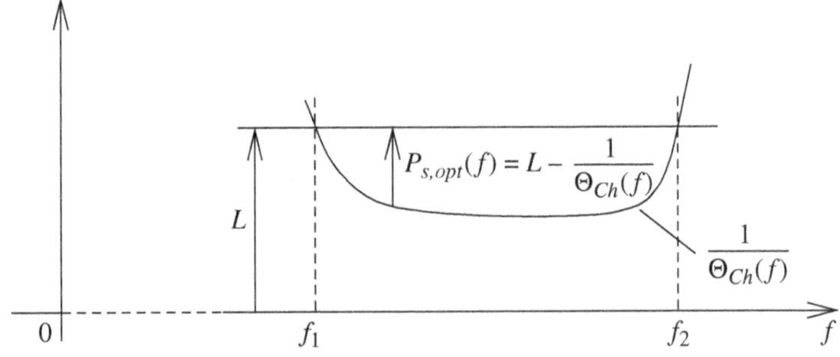

Figure 6.11 Illustration of $\mathcal{P}_{s,opt}(f)$ for a typical behaviour of the function $\Theta_{Ch}(f)$.

[3] We use the notation

$$(x)^+ = \begin{cases} x & \text{if } x > 0 \\ 0 & \text{otherwise} \end{cases} \tag{6.108}$$

It results that $\mathcal{P}_{s,opt}(f)$ should assume large values (small values) at frequencies for which $\Theta_{Ch}(f)$ assumes large values (small values). From the figure we note that if $1/\Theta_{Ch}(f)$ is the profile of a cup in which if we pour a quantity of water equivalent to V_P, the distribution of the water in the cup takes place according to the behaviour predicted by (6.109); this observation leads to the *water pouring interpretation* of the optimum distribution of $\mathcal{P}_{s,opt}(f)$ as a function of frequency.

Note that from (6.109) $\mathcal{P}_{s,opt}(f) > 0$ only for a subset of frequencies $f \in \mathcal{B} \subset \mathcal{B}_{Tx}$; therefore, the effective transmission passband of s will be further restricted to $\mathcal{B}$.

In analogy with the case of an ideal (non-dispersive) AWGN channel with limited bandwidth, a linear dispersive channel can be roughly characterized by two parameters: the bandwidth $B = \int_{\mathcal{B}} df$ and the *effective SNR* Γ_{eff}, implicitly defined so that

$$C_{[bit/s]} = B \log_2(1 + \Gamma_{eff}) \tag{6.110}$$

The comparison with (6.105) yields

$$\Gamma_{eff} = \exp\left\{ \frac{1}{B} \int_{\mathcal{B}} \ln\left[1 + \mathcal{P}_{s,opt}(f)\Theta_{Ch}(f)\right] df \right\} - 1 \tag{6.111}$$

Note that $1 + \Gamma_{eff}$ corresponds to the geometric mean of the function $1 + \mathcal{P}_{s,opt}(f)\Theta_{Ch}(f)$ in the band $\mathcal{B}$.

The passband $\mathcal{B}$ represents the most important parameter of the spectrum obtained by water pouring. For many channels utilized in practice, the passband $\mathcal{B}$ is composed of only one frequency interval $[f_1, f_2]$, as illustrated in Figure 6.11; in other cases, for example in the presence of high-power narrowband interference signals, $\mathcal{B}$ may be formed by the union of disjoint frequency intervals. In practice, we find that the dependence of the capacity on $\mathcal{P}_{s,opt}(f)$ is not so critical as the dependence on $\mathcal{B}$; a constant power spectral density in the band $\mathcal{B}$ usually allows a system to closely approach capacity. Therefore, the application of *the water pouring criterion may be limited to the computation of the passband.*

Expression (6.105) for any $\mathcal{P}_s(f) \neq \mathcal{P}_{s,opt}(f)$ under constraint (6.101) is called *achievable bit rate*. More in general, the achievable bit rate refers to the performance obtained with suboptimal coding and/or modulation schemes.

6.5.4 MIMO discrete-time NB AWGN channel

We now refer to the discrete-time (with symbol period T) multiple-input-multiple-output (MIMO) NB (i.e. non-dispersive) AWGN channel, between N_{Tx} inputs and N_{Rc} outputs, that can be modelled as (see (9.17))

$$\boldsymbol{x}_k = \mathcal{G}_C \boldsymbol{s}_k + \boldsymbol{w}_k \tag{6.112}$$

where $\boldsymbol{x}_k$ is the $N_{Rc} \times 1$ vector of received signals, $\mathcal{G}_C$ is the $N_{Rc} \times N_{Tx}$ channel matrix, $\boldsymbol{s}_k = [s_k^{(1)}, \ldots, s_k^{(N_{Tx})}]^T$ is the $N_{Tx} \times 1$ transmitted signal vector, and $\boldsymbol{w}_k$ is the $N_{Rc} \times 1$ noise vector. We consider the transmit power constraint

$$\frac{1}{2} E\left[||\boldsymbol{s}_k||^2\right] = \frac{1}{2} \sum_{i=1}^{N_{Tx}} E\left[|s_k^{(i)}|^2\right] \leq V_P \tag{6.113}$$

Let $N_{S,max} = \text{rank}\, \mathcal{G}_C$, hence $N_{S,max} \leq \min(N_{Rc}, N_{Tx})$. The singular value decomposition (SVD) of the channel matrix yields (see Section 2.3.2)

$$\mathcal{G}_C = \boldsymbol{U}_{ext} \boldsymbol{\Sigma}_{ext} \boldsymbol{V}_{ext}^H \tag{6.114}$$

with $\boldsymbol{U}_{ext}$ $N_{Rc} \times N_{Rc}$, and $\boldsymbol{V}_{ext}$ $N_{Tx} \times N_{Tx}$ unitary matrices, and $\boldsymbol{\Sigma}_{ext}$ $N_{Rc} \times N_{Tx}$ diagonal matrix such that

$$\boldsymbol{\Sigma}_{ext} = \begin{bmatrix} \boldsymbol{\Sigma} & \boldsymbol{0} \\ \boldsymbol{0} & \boldsymbol{0} \end{bmatrix} \tag{6.115}$$

with $\boldsymbol{\Sigma} = \text{diag}([\varsigma_1, \dots, \varsigma_{N_{S,max}}]^T)$ and $\boldsymbol{0}$ an all-zero matrix of suitable size. Let us also define $\boldsymbol{V}$ as the sub-matrix of $\boldsymbol{V}_{ext}$ defined by its first $N_{S,max}$ columns. If $\boldsymbol{a}_k = [a_{k,1}, \dots, a_{k,N_{S,max}}]^T$ is the vector of data streams, we set $\boldsymbol{s}_k = \boldsymbol{V}\boldsymbol{a}_k$. Then, the capacity of the MIMO NB AWGN channel is achieved when entries of $\boldsymbol{s}_k$ are zero-mean Gaussian distributed with constant PSD and (spatial) correlation matrix

$$\boldsymbol{R}_s = E[\boldsymbol{s}_k \boldsymbol{s}_k^H] = \boldsymbol{V}\text{diag}([\mathtt{M}_1, \dots, \mathtt{M}_{N_{S,max}}]^T)\boldsymbol{V}^H \tag{6.116}$$

with $\mathtt{M}_n = E[|a_{k,n}|^2] \geq 0$. Note that $E[||\boldsymbol{s}_k||^2] = \text{tr}\,\boldsymbol{R}_s = \sum_{n=1}^{N_{S,max}} \mathtt{M}_n$. Define the SNR on stream n as

$$\Gamma_n = \mathtt{M}_n \Theta_n, \qquad \Theta_n = \frac{|\varsigma_n|^2}{\frac{N_0}{2} 2B_{min}} \tag{6.117}$$

with $B_{min} = 1/T$, the capacity is [12, 13]

$$C_{[bit/s]} = \frac{1}{T} \sum_{n=1}^{N_{S,max}} \log_2(1 + \Gamma_n) \tag{6.118}$$

where values of $\mathtt{M}_n = \mathtt{M}_{opt,n}$, $n = 1, \dots, N_{S,max}$, maximize (6.118) under the power constraint (6.113), which can be written as

$$\frac{1}{2} \sum_{n=1}^{N_{S,max}} \mathtt{M}_n \leq V_P \tag{6.119}$$

From (6.118), we observe that the achievable bit rate of the MIMO NB AWGN channel is that of a *parallel AWGN channel* with $N_{S,max}$ subchannels.

By maximizing (6.118) under constraint (6.119), we have the following solution

$$\mathtt{M}_{opt,n} = \left[L - \frac{1}{\Theta_n}\right]^+ \tag{6.120}$$

where the constant L is such that (6.119) is satisfied with the equal sign.

With N_{Tx} inputs and N_{Rc} outputs we can transmit up to $N_{S,max}$ independent data streams, thus exploiting the *multiplexing* capabilities of the MIMO channel. Since $N_{S,max} \leq \min(N_{Rc}, N_{Tx})$, we observe that a larger number of streams can be transmitted only when both N_{Tx} and N_{Rc} increase. In Section 9.4, it will be seen that the MIMO capacity provides also a specific transmitter and receiver architecture.

Continuous-time model

For the continuous-time NB AWGN MIMO channel model, the derivation of the capacity follows the same line of the discrete-time case.

MIMO dispersive channel

For a MIMO dispersive channel, we can still consider its frequency-domain representation, and extend above capacity formulas. In particular, capacity is achieved by transmitting a Gaussian signal, with correlation given by (6.116), where $\boldsymbol{V}$ is replaced by $\boldsymbol{V}(f)$, the right singular vector matrix of $\mathcal{G}_C(f)$, the matrix frequency response of the channel.

6.6 Achievable rates of modulations in AWGN channels

We now provide some derivations on the achievable rate (in bit/dim) of communication systems using specific modulation techniques over the NB AWGN channel.

6.6.1 Rate as a function of the SNR per dimension

For a given value of the symbol error probability, we now derive Γ_I as a function of R_I for some multilevel modulations in an ideal AWGN channel. The result will be compared with the Shannon limit (6.88)

$$\Gamma_{I,bound} = 2^{2R_I} - 1 \tag{6.121}$$

that represents the *minimum theoretical value* of Γ_I, in correspondence of a given R_I, for which P_{bit} can be made arbitrarily small by using channel coding without constraints in complexity and latency.

A first comparison is made by assuming the same symbol error probability, $P_e = 10^{-6}$, for all systems. Since $P_e = Q(\sqrt{\zeta})$, where in turn ζ is a function of Γ_I, according to the considered modulation, imposing $Q(\sqrt{\zeta}) = 10^{-6}$ provides $\zeta \simeq 22$. We now consider M-PAM, M-QAM, and M-PSK modulations.

1. *M-PAM*: From (6.81), neglecting terms in from of the Q function, we have

$$\frac{3}{M^2 - 1}\Gamma = \zeta \tag{6.122}$$

and

$$\Gamma_I = \Gamma, \qquad R_I = \log_2 M \tag{6.123}$$

we obtain the following relation

$$\Gamma_I = \frac{\zeta}{3}(2^{2R_I} - 1) \tag{6.124}$$

2. *M-QAM*: From (6.82) we have

$$\frac{3}{M - 1}\Gamma = \zeta \tag{6.125}$$

and

$$\Gamma_I = \Gamma, \qquad R_I = \frac{1}{2}\log_2 M \tag{6.126}$$

we obtain

$$\Gamma_I = \frac{\zeta}{3}(2^{2R_I} - 1) \tag{6.127}$$

We note that for QAM, a certain R_I is obtained with a number of symbols equal to $M_{QAM} = 2^{2R_I}$, whereas for PAM the same efficiency is reached for $M_{PAM} = 2^{R_I} = \sqrt{M_{QAM}}$.

3. *M-PSK*: From (6.83) it turns out

$$\Gamma_I = \frac{\zeta}{20}2^{4R_I} \tag{6.128}$$

Equation (6.128) is obtained by approximating $\sin(\pi/M)$ with π/M, and π^2 with 10.

We note that, for a given value of R_I, PAM, and QAM require the same value of Γ_I, whereas PSK requires a much larger value of Γ_I.

An exact comparison is now made for a given bit error probability. Using the P_{bit} curves previously obtained, the behaviour of R_I as a function of Γ_I for $P_{bit} = 10^{-6}$ is illustrated in Figure 6.12. We observe that the required Γ_I is much larger than the minimum value obtained by the Shannon limit. As will be discussed next, the gap can be reduced by channel coding.

We also note that, for large R_I, PAM and QAM allow a lower Γ_I with respect to PSK.

PAM, QAM, and PSK are bandwidth-efficient modulation methods as they cover the region for $R_I > 1$, as illustrated in Figure 6.12. The bandwidth is traded off with the power, that is Γ, by increasing the number of levels: we note that, in this region, higher values of Γ are required to increase R_I.

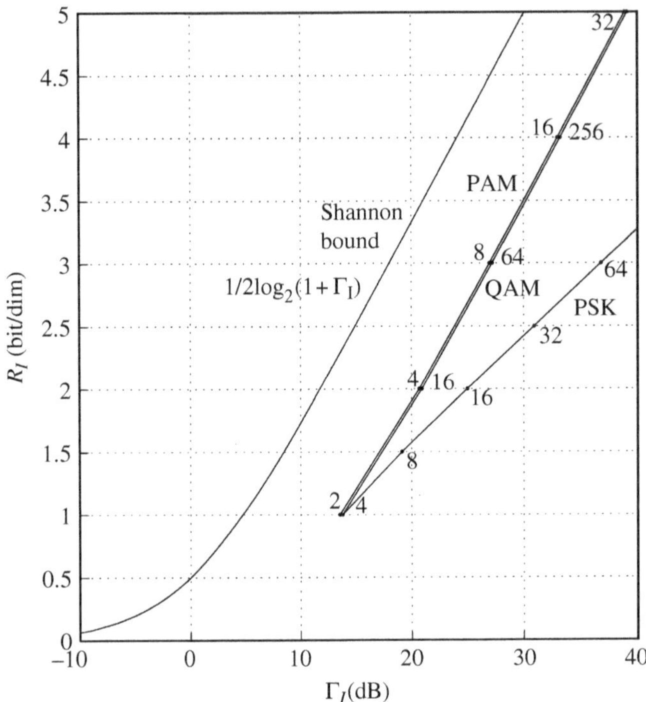

Figure 6.12 Γ_I required for a given rate R_I, for different uncoded modulation methods and bit error probability $P_{bit} = 10^{-6}$. The parameter in the figure denotes the number of symbols M of the constellation.

6.6.2 Coding strategies depending on the signal-to-noise ratio

The formula of the NB AWGN channel capacity (6.88) can be expressed as $\Gamma_I/(2^{2C} - 1) = 1$. This relation suggests the definition of the normalized SNR

$$\overline{\Gamma}_I = \frac{\Gamma_I}{2^{2R_I} - 1} \tag{6.129}$$

for a given R_I given by (6.57). For a scheme that achieves the capacity, R_I is equal to the capacity of the channel C and $\overline{\Gamma}_I = 1$ (0 dB); if $R_I < C$, as it must be in practice, then $\overline{\Gamma}_I > 1$. Therefore, the value of $\overline{\Gamma}_I$ indicates how far from the Shannon limit a system operates, or, in other words, the gap that separates the system from capacity. We now consider two cases.

High signal-to-noise ratios. We note from Figure 6.12 that for high values of Γ_I, it is possible to find coding methods for the reliable transmission of several bits per dimension. For an *uncoded* M-PAM system,

$$R_I = \log_2 M \quad \text{or} \quad 2^{R_I} = M \tag{6.130}$$

bits of information are mapped into each transmitted symbol.

We note that P_e is function only of M and Γ_I. Moreover, using (6.130) and (6.129) we obtain

$$\overline{\Gamma}_I = \frac{\Gamma_I}{M^2 - 1} \quad \text{(uncoded case)} \tag{6.131}$$

For large M, P_e can therefore be expressed as

$$P_e = 2\left(1 - \frac{1}{M}\right) Q\left(\sqrt{3\overline{\Gamma}_I}\right) \simeq 2Q\left(\sqrt{3\overline{\Gamma}_I}\right) \quad \text{(uncoded case)} \tag{6.132}$$

We note that the relation between P_e and $\overline{\Gamma}_I$ is almost independent of M, if M is large. This relation is used in the comparison illustrated in Figure 6.13 between uncoded systems and the Shannon limit given by $\overline{\Gamma}_I = 1$.

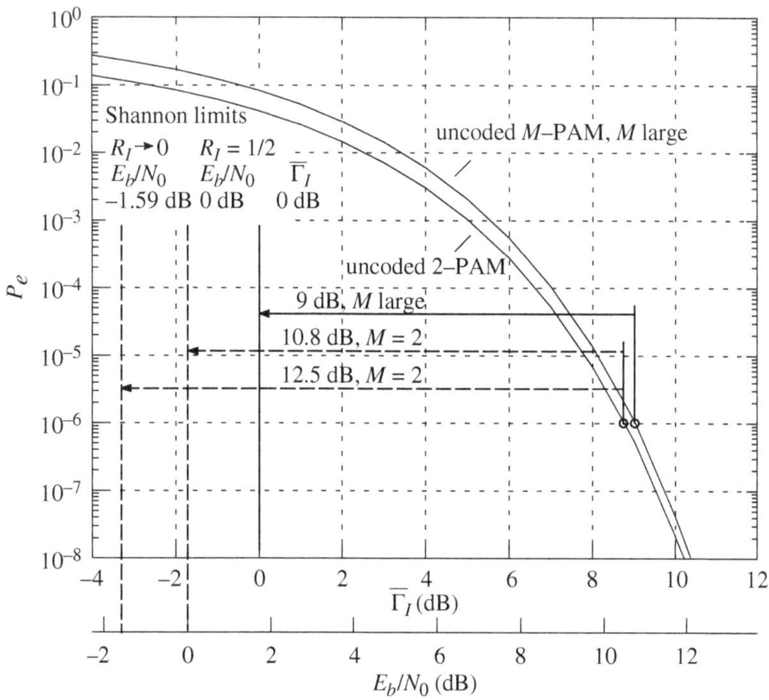

Figure 6.13 Bit error probability as a function of E_b/N_0 for an uncoded 2-PAM system, and symbol error probability as a function of $\overline{\Gamma}_I$ for an uncoded M-PAM system. Source: Reproduced with permission from Forney and Ungerboeck [14]. 1998, IEEE.

Low SNRs. For low values of Γ_I, the capacity is less than 1 and can be almost reached by binary transmission systems: consequently we refer to coding methods that employ more binary symbols to obtain the reliable transmission of 1 bit. For low values of Γ_I, it is customary to introduce the following ratio (see (6.71)):

$$\frac{E_b}{N_0} = \frac{2^{2R_I} - 1}{2R_I} \overline{\Gamma}_I \tag{6.133}$$

We note the following particular cases:

- if $R_I \ll 1$, then $E_b/N_0 \simeq (\ln 2)\, \overline{\Gamma}_I$;
- if $R_I = 1/2$, then $E_b/N_0 = \overline{\Gamma}_I$;
- if $R_I = 1$, then $E_b/N_0 = (3/2)\, \overline{\Gamma}_I$.

For low Γ_I, if the bandwidth can be extended without limit for a given power, for example by using an orthogonal modulation (e.g. see Chapter 10) with $T \to 0$, then by increasing the bandwidth, or equivalently the number of dimensions I of input signals, both Γ_I and R_I tend to zero. For systems with limited power and unlimited bandwidth, usually E_b/N_0 is adopted as a figure of merit.

From (6.133) and the Shannon limit $\overline{\Gamma}_I > 1$, we obtain the Shannon limit in terms of E_b/N_0 for a given rate R_I as

$$\frac{E_b}{N_0} > \frac{2^{2R_I} - 1}{2R_I} \tag{6.134}$$

This lower bound monotonically decreases with R_I.

In particular, we examine again the three cases:

- if R_I tends to zero, the *ultimate Shannon limit* is given by

$$\frac{E_b}{N_0} > \ln 2 \quad (-1.59 \text{ dB}) \tag{6.135}$$

  in other words, (6.135) states that even though an infinitely large bandwidth is used, reliable transmission can be achieved only if $E_b/N_0 > -1.59$ dB;
- if the bandwidth is limited, from (6.134) we find that the Shannon limit in terms of E_b/N_0 is higher; for example, if $R_I = 1/2$, the limit becomes $E_b/N_0 > 1$ (0 dB);
- if $R_I = 1$, as $E_b/N_0 = (3/2)\,\overline{\Gamma}_I$, the symbol error probability or bit error probability for an *uncoded* 2-PAM system can be expressed in two equivalent ways:

$$P_{bit} \simeq Q\left(\sqrt{3\overline{\Gamma}_I}\right) = Q\left(\sqrt{\frac{2E_b}{N_0}}\right) \tag{6.136}$$

Coding gain

Definition 6.2

The *coding gain* of a coded modulation scheme is equal to the reduction in the value of E_b/N_0, or in the value of Γ or $\overline{\Gamma}_I$ (see (11.84)), that is required to obtain a given probability of error relative to a reference uncoded system. If the modulation rate of the coded system remains unchanged, we typically refer to Γ or $\overline{\Gamma}_I$. □

Let us consider as reference systems a 2-PAM system and an M-PAM system with $M \gg 1$, for small and large values of $\overline{\Gamma}_I$, respectively. Figure 6.13 illustrates the bit error probability for an uncoded 2-PAM system as a function of both E_b/N_0 and $\overline{\Gamma}_I$. For $P_{bit} = 10^{-6}$, the reference uncoded 2-PAM system operates at about 12.5 dB from the ultimate Shannon limit. Thus, a coding gain up to 12.5 dB is possible, in principle, at this probability of error, if the bandwidth can be sufficiently extended to allow the use of binary codes with $R_I \ll 1$; if, instead, the bandwidth can be extended only by a factor 2 with respect to an uncoded system, then a binary code with rate $R_I = 1/2$ can yield a coding gain up to about 10.8 dB.

Figure 6.13 also shows the symbol error probability for an uncoded M-PAM system as a function of $\overline{\Gamma}_I$ for large M. For $P_e = 10^{-6}$, a reference uncoded M-PAM system operates at about 9 dB from the Shannon limit: in other words, assuming a limited bandwidth system, the Shannon limit can be achieved by a code having a gain of about 9 dB.

For a dispersive channel, in analogy with the case of an ideal AWGN channel, it is useful to define a *normalized* SNR as (see (6.111))

$$\overline{\Gamma}_{eff} = \frac{\Gamma_{eff}}{2^{2R_I} - 1} \tag{6.137}$$

where $\overline{\Gamma}_{eff} > 1$ measures the gap that separates the considered system from capacity.

6.6.3 Achievable rate of an AWGN channel using PAM

We now consider the achievable rate of a AWGN channel, when the input is forced to be a PAM signal, thus not being Gaussian as required by a capacity-achieving solution. The maximum achievable rate of a real-valued ideal AWGN channel having as input an M-PAM signal is given in bit/dim by [15]

$$C = \max_{p_1,\ldots,p_M} \sum_{n=1}^{M} p_n \int_{-\infty}^{+\infty} p_{z|a_0}(\eta \mid \alpha_n) \log_2 \left[\frac{p_{z|a_0}(\eta \mid \alpha_n)}{\sum_{i=1}^{M} p_i \, p_{z|a_0}(\eta \mid \alpha_i)} \right] d\eta \tag{6.138}$$

where p_n indicates the probability of transmission of the symbol $a_0 = \alpha_n$. By the hypothesis of white Gaussian noise, we have

$$p_{z|a_0}(\eta \mid \alpha_n) \propto \exp\left\{ -\frac{(\eta - \alpha_n)^2}{2\sigma_I^2} \right\} \tag{6.139}$$

With the further hypothesis that only codes with equally likely symbols (els) are of practical interest, the computation of the maximum of C with respect to the probability distribution of the input signal can be omitted. The channel achievable rate is therefore given by

$$C_{els} = \log_2 M - \frac{1}{M} \sum_{n=1}^{M} \int_{-\infty}^{+\infty} \frac{1}{\sqrt{2\pi}\sigma_I} e^{-\frac{\xi^2}{2\sigma_I^2}} \log_2 \left[\sum_{i=1}^{M} \exp\left(-\frac{(\alpha_n + \xi - \alpha_i)^2 - \xi^2}{2\sigma_I^2} \right) \right] d\xi \tag{6.140}$$

The capacity C_{els} is illustrated in Figure 6.14, where the Shannon limit given by (6.88), as well as the SNR given by (6.124) for which a symbol error probability equal to 10^{-6} is obtained for uncoded transmission, are also shown [14]. We note that the curves saturate as information cannot be transmitted with a rate larger than $R_I = \log_2 M$.

Let us consider for example the uncoded transmission of 1 bit of information per modulation interval by a 2-PAM system, where we have a symbol error probability equal to 10^{-6} for $\Gamma_I = 13.5$ dB. If

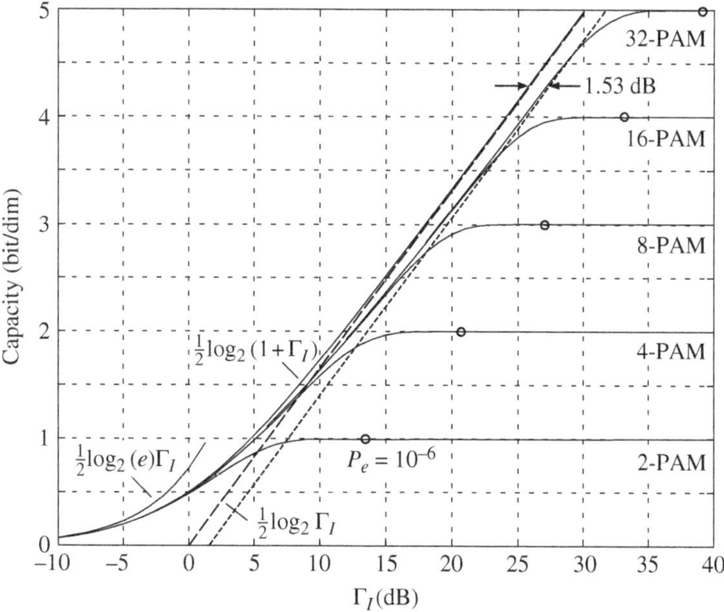

Figure 6.14 Capacity C_{els} of an ideal AWGN channel for Gaussian for M-PAM input signals. Source: Reproduced with permission from Forney and Ungerboeck [14]. © 1998, IEEE.

the number of symbols in the alphabet $\mathcal{A}$ doubles, choosing 4-PAM modulation, we see that the *coded transmission* of 1 bit of information per modulation interval with rate $R_I = 1$ is possible, and an arbitrarily small error probability can be obtained for $\Gamma_I = 5$ dB. This indicates that a coded 4-PAM system may achieve a gain of about 8.5 dB in SNR over an uncoded 2-PAM system, at an error probability of 10^{-6}. If the number of symbols is further increased, the additional achievable gain is negligible. Therefore we conclude that, by doubling the number of symbols with respect to an uncoded system, we obtain in practice the entire gain that would be expected from the expansion of the input alphabet.

We see from Figure 6.14 that for small values of Γ_I the choice of a binary alphabet is almost optimum: in fact for $\Gamma_I < 1$ (0 dB), the capacity given by (6.88) is essentially equivalent to the capacity given by (6.140) with a binary alphabet of input symbols.

For large values of Γ_I, the capacity of multilevel systems asymptotically approximates a straight line that is parallel to the capacity of the AWGN channel. The asymptotic loss of $\pi e/6$ (1.53 dB) is due to the choice of a uniform distribution rather than Gaussian for the set of input symbols. To achieve the Shannon limit, it is not sufficient to use coding techniques with equally likely input symbols, no matter how sophisticated they are: to bridge the gap of 1.53 dB, *shaping* techniques are required [16] that produce a distribution of the input symbols similar to a Gaussian distribution.

Coding techniques for small Γ_I and large Γ_I are therefore quite different: for low Γ_I, the binary codes are almost optimum and the shaping of the constellation is not necessary; for high Γ_I instead constellations with more than two elements must be used. To reach capacity, coding must be extended with shaping techniques; moreover, to reach the capacity in channels with limited bandwidth, techniques are required that combine coding, shaping, and equalization.

Bibliography

[1] Benvenuto, N. and Zorzi, M. (2011). *Principles of Communications Networks and Systems*. Wiley.

[2] Pyndiah, R., Glavieux, A., Picart, A., and Jacq, S. (1994). Near optimum decoding of product codes. *1994 IEEE GLOBECOM. Communications: The Global Bridge*, volume 1, pp. 339–343.

[3] Tosato, F. and Bisaglia, P. (2002). Simplified soft-output demapper for binary interleaved COFDM with application to HIPERLAN/2. *2002 IEEE International Conference on Communications. Conference Proceedings. ICC 2002 (Cat. No. 02CH37333)*, Volume 2, pp. 664–668.

[4] Alhabsi, A.H. (2005). Spectrally efficient modulation and turbo coding for communication systems.

[5] Pursley, M.B. and Shea, J.M. (1997). Bit-by-bit soft-decision decoding of trellis-coded M-DPSK modulation. *IEEE Communications Letters* 1: 133–135.

[6] Blasiak, D. (1995). A soft decision metric for coherently detected, differentially encoded MPSK systems. *Proceedings 1995 IEEE 45th Vehicular Technology Conference. Countdown to the Wireless 21st Century*, Volume 2, pp. 664–668.

[7] Stuber, G.L. (1988). Soft decision direct-sequence DPSK receivers. *IEEE Transactions on Vehicular Technology* 37: 151–157.

[8] Gallager, R.G. (1968). *Information Theory and Reliable Communication*. New York, NY: Wiley.

[9] Cover, T.M. and Thomas, J. (1991). *Elements of Information Theory*. New York, NY: Wiley.

[10] Shannon, C.E. (1948). A mathematical theory of communication. *Bell System Technical Journal* 27 (Part I): 379–427 and (Part II) 623–656.

[11] Franks, L.E. (1981). *Signal Theory*, revised ed. Stroudsburg, PA: Dowden and Culver.

[12] Foschini, G.J. and Gans, M.J. (1998). On limits of wireless communications in a fading environment when using multiple antennas. *Wireless Personal Communications* 6: 311–335.

[13] Telatar, E. (1999). Capacity of multi-antenna Gaussian channels. *European Transactions on Telecommunications* 10: 585–595.

[14] Forney, G.D. Jr. and Ungerboeck, G. (1998). Modulation and coding for linear Gaussian channels. *IEEE Transactions on Information Theory* 44: 2384–2415.
[15] Ungerboeck, G. (1982). Channel coding with multilevel/phase signals. *IEEE Transactions on Information Theory* 28: 55–67.
[16] Forney, G.D. Jr. (1992). Trellis shaping. *IEEE Transactions on Information Theory* 38: 281–300.
[17] Simon, M.K. and Alouini, M.-S. (2000). Exponential-type bounds on the generalized Marcum Q-function with application to error probability analysis over fading channels. *IEEE Transactions on Communications* 48: 359–366.
[18] McGee, W.F. (1970). Another recursive method of computing the Q–function. *IEEE Transactions on Information Theory* 16: 500–501.

Appendix 6.A Gray labelling

In this appendix, we give the procedure to construct a list of 2^n binary words of n bits, where adjacent words differ in only one bit.

The case for $n = 1$ is immediate. We have two words with two possible values

$$\begin{array}{c} 0 \\ 1 \end{array} \tag{6.141}$$

The list for $n = 2$ is constructed by considering first the list of $(1/2)2^2 = 2$ words that are obtained by appending a 0 in front of the words of the list (6.141):

$$\begin{array}{cc} 0 & 0 \\ 0 & 1 \end{array} \tag{6.142}$$

The remaining two words are obtained by inverting the order of the words in (6.141) and appending a 1 in front:

$$\begin{array}{cc} 1 & 1 \\ 1 & 0 \end{array} \tag{6.143}$$

The final result is the following list of words:

$$\begin{array}{cc} 0 & 0 \\ 0 & 1 \\ 1 & 1 \\ 1 & 0 \end{array} \tag{6.144}$$

Iterating the procedure for $n = 3$, the first four words are obtained by repeating the list (6.144) and appending a 0 in front of the words of the list.

Inverting then the order of the list (6.144) and appending a 1 in front, the final result is the list of eight words

$$\begin{array}{ccc} 0 & 0 & 0 \\ 0 & 0 & 1 \\ 0 & 1 & 1 \\ 0 & 1 & 0 \\ 1 & 1 & 0 \\ 1 & 1 & 1 \\ 1 & 0 & 1 \\ 1 & 0 & 0 \end{array} \tag{6.145}$$

It is easy to extend this procedure to any value of n. By induction, it is just as easy to prove that two adjacent words in each list differ by 1 bit at most.

Appendix 6.B The Gaussian distribution and Marcum functions

6.B.1 The Q function

The probability density function of a Gaussian variable w with mean $\mathtt{m}$ and variance σ^2 is given by

$$p_w(b) = \frac{1}{\sqrt{2\pi}\sigma} e^{-\frac{(b-\mathtt{m})^2}{2\sigma^2}} \tag{6.146}$$

We define *normalized Gaussian distribution* ($\mathtt{m} = 0$ and $\sigma^2 = 1$) the function

$$\Phi(a) = \int_{-\infty}^{a} p_w(b)db = \int_{-\infty}^{a} \frac{1}{\sqrt{2\pi}} e^{-b^2/2} db \tag{6.147}$$

It is often convenient to use the *complementary Gaussian distribution* function, defined as

$$Q(a) = 1 - \Phi(a) = \int_{a}^{+\infty} \frac{1}{\sqrt{2\pi}} e^{-b^2/2} db \tag{6.148}$$

Two other functions that are widely used are the *error function*

$$\begin{aligned} \text{erf}(a) &= -1 + 2\int_{-\infty}^{a\sqrt{2}} p_w(b)db \\ &= 1 - 2\int_{a\sqrt{2}}^{+\infty} \frac{1}{\sqrt{2\pi}} e^{-b^2} db \end{aligned} \tag{6.149}$$

and the *complementary error function*

$$\text{erfc}(a) = 1 - \text{erf}(a) \tag{6.150}$$

which are related to Φ and Q by the following equations

$$\Phi(a) = \frac{1}{2}\left[1 + \text{erf}\left(\frac{a}{\sqrt{2}}\right)\right] \tag{6.151}$$

$$Q(a) = \frac{1}{2}\,\text{erfc}\left(\frac{a}{\sqrt{2}}\right) \tag{6.152}$$

In Table 6.2, the values assumed by the complementary Gaussian distribution are given for values of the argument between 0 and 8. We present below some bounds of the Q function.

$$bound_1: \quad Q_1(a) = \frac{1}{\sqrt{2\pi}a}\left(1 - \frac{1}{a^2}\right)\exp\left(-\frac{a^2}{2}\right) \tag{6.153}$$

$$bound_2: \quad Q_2(a) = \frac{1}{\sqrt{2\pi}a}\exp\left(-\frac{a^2}{2}\right) \tag{6.154}$$

$$bound_3: \quad Q_3(a) = \frac{1}{2}\exp\left(-\frac{a^2}{2}\right) \tag{6.155}$$

The Q function and the above bounds are illustrated in Figure 6.15.

Table 6.2: Complementary Gaussian distribution.

a	$Q(a)$	a	$Q(a)$	a	$Q(a)$
0.0	5.0000(−01)[a]	2.7	3.4670(−03)	5.4	3.3320(−08)
0.1	4.6017(−01)	2.8	2.5551(−03)	5.5	1.8990(−08)
0.2	4.2074(−01)	2.9	1.8658(−03)	5.6	1.0718(−08)
0.3	3.8209(−01)	3.0	1.3499(−03)	5.7	5.9904(−09)
0.4	3.4458(−01)	3.1	9.6760(−04)	5.8	3.3157(−09)
0.5	3.0854(−01)	3.2	6.8714(−04)	5.9	1.8175(−09)
0.6	2.7425(−01)	3.3	4.8342(−04)	6.0	9.8659(−10)
0.7	2.4196(−01)	3.4	3.3693(−04)	6.1	5.3034(−10)
0.8	2.1186(−01)	3.5	2.3263(−04)	6.2	2.8232(−10)
0.9	1.8406(−01)	3.6	1.5911(−04)	6.3	1.4882(−10)
1.0	1.5866(−01)	3.7	1.0780(−04)	6.4	7.7688(−11)
1.1	1.3567(−01)	3.8	7.2348(−05)	6.5	4.0160(−11)
1.2	1.1507(−01)	3.9	4.8096(−05)	6.6	2.0558(−11)
1.3	9.6800(−02)	4.0	3.1671(−05)	6.7	1.0421(−11)
1.4	8.0757(−02)	4.1	2.0658(−05)	6.8	5.2310(−12)
1.5	6.6807(−02)	4.2	1.3346(−05)	6.9	2.6001(−12)
1.6	5.4799(−02)	4.3	8.5399(−06)	7.0	1.2798(−12)
1.7	4.4565(−02)	4.4	5.4125(−06)	7.1	6.2378(−13)
1.8	3.5930(−02)	4.5	3.3977(−06)	7.2	3.0106(−13)
1.9	2.8717(−02)	4.6	2.1125(−06)	7.3	1.4388(−13)
2.0	2.2750(−02)	4.7	1.3008(−06)	7.4	6.8092(−14)
2.1	1.7864(−02)	4.8	7.9333(−07)	7.5	3.1909(−14)
2.2	1.3903(−02)	4.9	4.7918(−07)	7.6	1.4807(−14)
2.3	1.0724(−02)	5.0	2.8665(−07)	7.7	6.8033(−15)
2.4	8.1975(−03)	5.1	1.6983(−07)	7.8	3.0954(−15)
2.5	6.2097(−03)	5.2	9.9644(−08)	7.9	1.3945(−15)
2.6	4.6612(−03)	5.3	5.7901(−08)	8.0	6.2210(−16)

[a]Writing 5.0000(−01) means 5.0000×10^{-1}.

6.B.2 Marcum function

We define the first order *Marcum function* as

$$Q_1(a,b) = \int_b^{+\infty} x\, e^{-\frac{x^2+a^2}{2}} I_0(ax)dx \tag{6.156}$$

where I_0 is the *modified Bessel function* of the first type and order zero, defined in (4.56).

From (6.156), two particular cases follow:

$$Q_1(0,b) = e^{-\frac{b^2}{2}} \tag{6.157}$$

$$Q_1(a,0) = 1 \tag{6.158}$$

Moreover, for $b \gg 1$ and $b \gg b - a$ the following approximation holds

$$Q_1(a,b) \simeq Q(b-a) \tag{6.159}$$

where the Q function is given by (6.148).

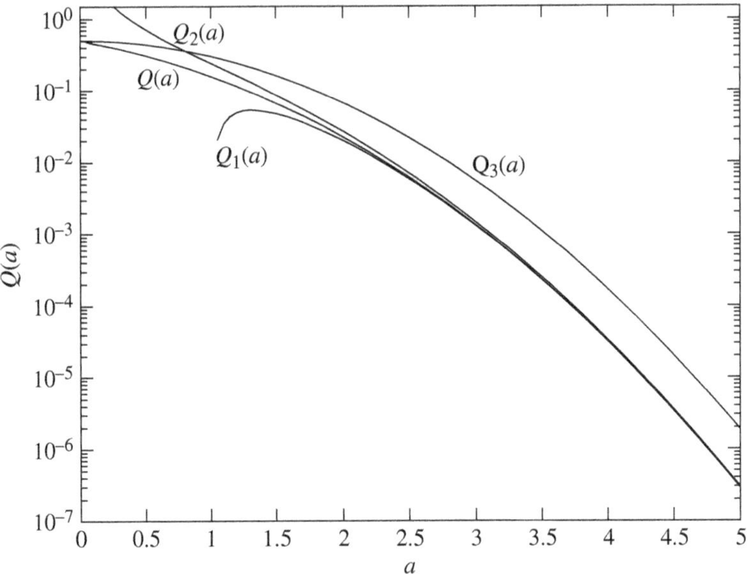

Figure 6.15 The Q function and relative bounds.

A useful approximation valid for $b \gg 1$, $a \gg 1$, $b \gg b - a > 0$, is given by

$$1 + Q_1(a, b) - Q_1(b, a) \simeq 2Q(b - a) \tag{6.160}$$

We also give the Simon bound [17]

$$e^{-\frac{(b+a)^2}{2}} \leq Q_1(a, b) \leq e^{-\frac{(b-a)^2}{2}}, \qquad b > a > 0 \tag{6.161}$$

and

$$1 - \frac{1}{2}\left[e^{-\frac{(a-b)^2}{2}} - e^{-\frac{(a+b)^2}{2}}\right] \leq Q_1(a, b), \qquad a > b \geq 0 \tag{6.162}$$

We observe that in (6.161), the upper bound is very tight, and the lower bound for a given value of b becomes looser as a increases. In (6.162), the lower bound is very tight. A recursive method for computing the Marcum function is given in [18].

Chapter 7

Single-carrier modulation

This chapter is split into two parts. The first part considers digital pulse amplitude modulation (PAM) and quadrature amplitude modulation (QAM) for continuous transmission by using a single carrier, taking into account the possibility that the channel may distort the transmitted signal [1–3]. The Nyquist criterion will be given, which states conditions on the pulse at the receiver for the absence of intersymbol interference. The second part instead first presents several equalization methods to compensate for linear distortion introduced by the channel. Then, we analyse detection methods that operate on sequences of samples. Last, we present techniques to estimate the discrete-time channel impulse response, a starting point of receiver design.

7.1 Signals and systems

With reference to the general scheme of Figure 6.1, here we omit the channel encoder to simplify the analysis, hence, the information message $\{b_\ell\}$ is input to the bit mapper to yield the data sequence $\{a_k\}$.

7.1.1 Baseband digital transmission (PAM)

The fundamental blocks of a baseband digital transmission system are illustrated in Figure 7.1, while signals at the various points of a quaternary PAM system are shown in Figure 7.2.

Modulator

For a PAM system, the modulator associates the symbol a_k, having alphabet (6.4), with the amplitude of a given pulse h_{Tx}:

$$a_k \rightarrow a_k h_{Tx}(t - kT) \tag{7.1}$$

Therefore, the modulated signal s that is input to the transmission channel is given by

$$s(t) = \sum_{k=-\infty}^{+\infty} a_k h_{Tx}(t - kT) \tag{7.2}$$

Algorithms for Communications Systems and their Applications, Second Edition.
Nevio Benvenuto, Giovanni Cherubini, and Stefano Tomasin.

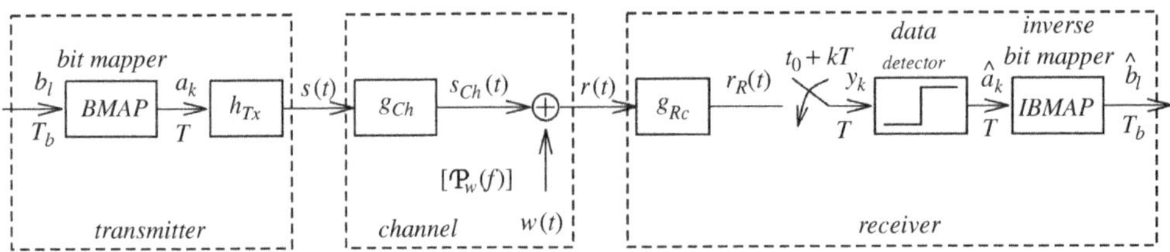

Figure 7.1 Block diagram of a baseband digital transmission system.

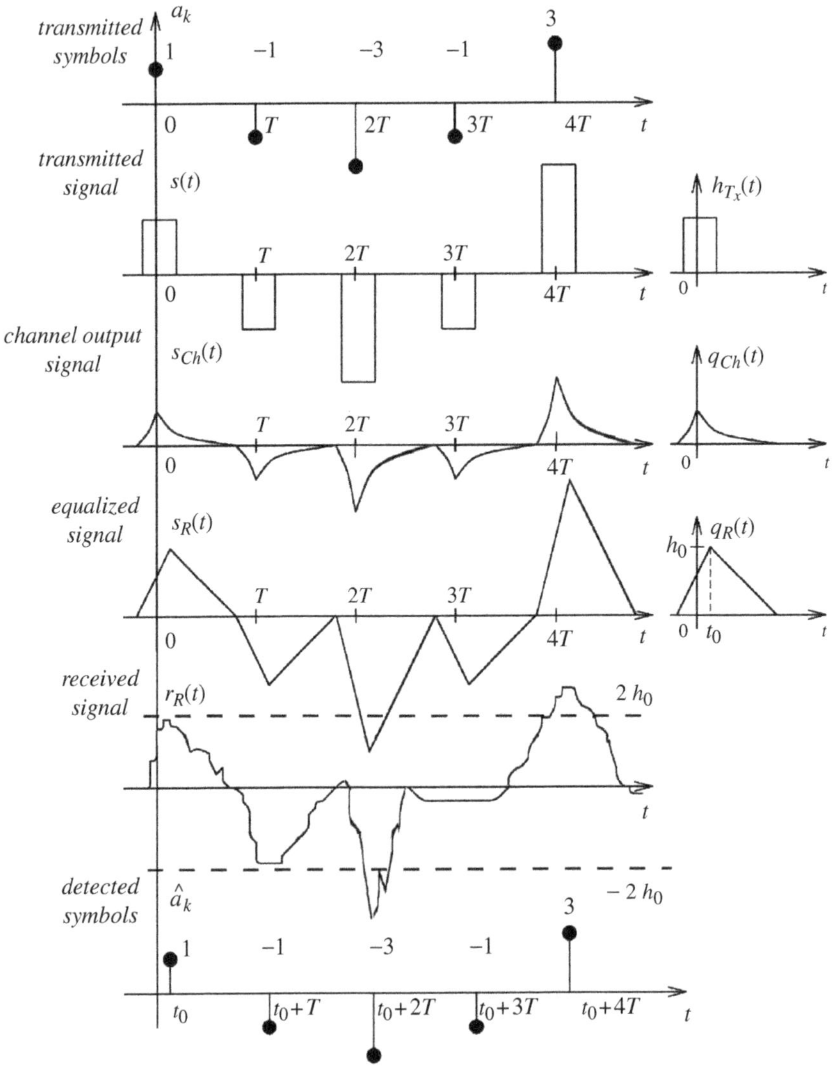

Figure 7.2 Signals at various points of a quaternary PAM transmission system with alphabet $\mathcal{A} = \{-3, -1, 1, 3\}$.

Transmission channel

The transmission channel is assumed to be *linear* and *time invariant*, with impulse response g_{Ch}. Therefore, the desired signal at the output of the transmission channel still has a PAM structure. From the relation,

$$s_{Ch}(t) = g_{Ch} * s(t) \tag{7.3}$$

we define

$$q_{Ch}(t) = h_{Tx} * g_{Ch}(t) \tag{7.4}$$

then, we have

$$s_{Ch}(t) = \sum_{k=-\infty}^{+\infty} a_k q_{Ch}(t - kT) \tag{7.5}$$

The transmission channel introduces an effective noise w. Therefore, the signal at the input of the receive filter is given by

$$r(t) = s_{Ch}(t) + w(t) \tag{7.6}$$

Receiver

The receiver consists of three functional blocks:

1. *Amplifier-equalizer filter*: This block is assumed *linear* and *time invariant* with impulse response g_{Rc}. Then, the desired signal is given by

$$s_R(t) = g_{Rc} * s_{Ch}(t) \tag{7.7}$$

Let the *overall impulse response* of the system be

$$q_R(t) = q_{Ch} * g_{Rc}(t) = h_{Tx} * g_{Ch} * g_{Rc}(t) \tag{7.8}$$

then

$$s_R(t) = \sum_{k=-\infty}^{+\infty} a_k q_R(t - kT) \tag{7.9}$$

In the presence of noise,

$$r_R(t) = s_R(t) + w_R(t) \tag{7.10}$$

where $w_R(t) = w * g_{Rc}(t)$.

2. *Sampler*: Sampling r_R at instants $t_0 + kT$ yields[1]:

$$y_k = r_R(t_0 + kT) = \sum_{i=-\infty}^{\infty} a_i q_R(t_0 + (k - i)T) + w_R(t_0 + kT) \tag{7.11}$$

Sometimes, y_k can be simplified as

$$y_k = h_0 a_k + w_{R,k} \tag{7.12}$$

where $h_0 = q_R(t_0)$ is the amplitude of the overall impulse response at the sampling instant t_0 and $w_{R,k} = w_R(t_0 + kT)$. The parameter t_0 is called *timing phase*, and its choice is fundamental for system performance. We scale now y_k by h_0 to yield

$$\bar{y}_k = \frac{y_k}{h_0} = a_k + \frac{w_{R,k}}{h_0} \tag{7.13}$$

---

[1] To simplify the notation, the sample index k, associated with the instant $t_0 + kT$, here appears as a subscript.

3. *Threshold detector*: From the sequence $\{\overline{y}_k\}$, we detect the transmitted sequence $\{a_k\}$. The simplest structure is the instantaneous non-linear threshold detector:

$$\hat{a}_k = \mathcal{T}[\overline{y}_k] \tag{7.14}$$

where $\mathcal{T}[\overline{y}_k]$ is the quantizer characteristic with $\overline{y}_k \in \mathbb{R}$ and $\hat{a}_k \in \mathcal{A}$, alphabet of a_k. An example of quantizer characteristic for $\mathcal{A} = \{-3, -1, 1, 3\}$ is given in Figure 7.3.

Indeed, under some assumptions examined in Section 6.2.1, the threshold detector, implementing the minimum distance detection criterion, is optimum.

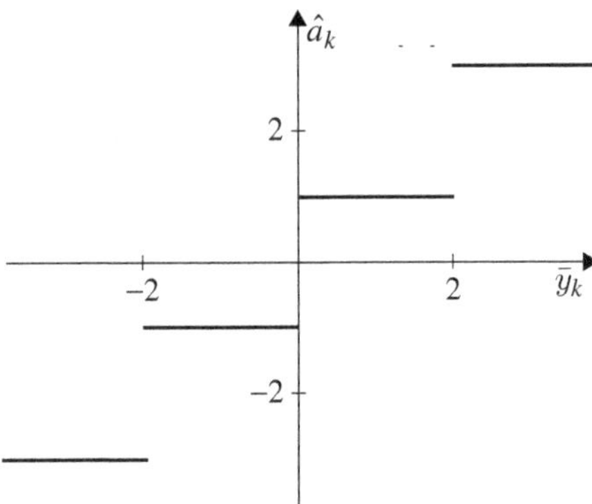

Figure 7.3 Characteristic of a threshold detector for quaternary symbols with alphabet $\mathcal{A} = \{-3, -1, 1, 3\}$.

From the sequence $\{\hat{a}_k\}$, using an inverse bit mapper (IBMAP), the detected binary information message $\{\hat{b}_\ell\}$ is obtained. Incidentally, as y_k is in general affected by interference from other symbols than a_k and noise, it may be $\hat{a}_k \neq a_k$.

Power spectral density

The PAM signals found in various points of the system, s_R, s_{Ch}, and s, have the following structure:

$$s(t) = \sum_{k=-\infty}^{+\infty} a_k q(t - kT) \tag{7.15}$$

where q is the impulse response of a suitable filter. In other words, a PAM signal s may be regarded as a signal generated by an interpolator filter with impulse response $q(t)$, $t \in \mathbb{R}$, as shown in Figure 7.4.

From the spectral analysis (see Example 1.7.9 on page 49), we know that s is a *cyclostationary* process with *average power spectral density* (PSD) given by (see (1.323)):

$$\overline{\mathcal{P}}_s(f) = \left|\frac{1}{T}\mathcal{Q}(f)\right|^2 \mathcal{P}_a(f) \tag{7.16}$$

where $\mathcal{P}_a$ is the PSD of the message and $\mathcal{Q}$ is the Fourier transform of q.

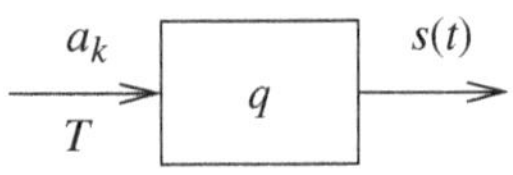

a_k → [q] → $s(t)$; T

Figure 7.4 The PAM signal as output of an interpolator filter.

From Figure 7.4, it is immediate to verify that by filtering s we obtain a signal that is still PAM, with a pulse given by the convolution of the filter impulse responses. The PSD of a filtered PAM signal is obtained by multiplying $\mathcal{P}_a(f)$ in (7.16) by the square magnitude of the filter frequency response. As $\mathcal{P}_a(f)$ is periodic of period $1/T$, then the *bandwidth B of the transmitted signal* is equal to that of h_{Tx}.

Example 7.1.1 (PSD of an i.i.d. symbol sequence)
Let $\{a_k\}$ be a sequence of i.i.d. symbols with values from the alphabet

$$\mathcal{A} = \{\alpha_1, \alpha_2, \dots, \alpha_M\} \tag{7.17}$$

and $p(\alpha)$, $\alpha \in \mathcal{A}$, be the probability distribution of each symbol.

The mean value and the statistical power of the sequence are given by

$$\mathtt{m}_a = \sum_{\alpha\in\mathcal{A}} \alpha p(\alpha), \qquad \mathtt{M}_a = \sum_{\alpha\in\mathcal{A}} |\alpha|^2 p(\alpha) \tag{7.18}$$

Consequently, $\sigma_a^2 = \mathtt{M}_a - |\mathtt{m}_a|^2$.

Following Example 1.5.1 on page 34, that describes the decomposition of the PSD of a message into ordinary and impulse functions, the decomposition of the PSD of s is given by

$$\overline{\mathcal{P}}_s^{(c)}(f) = \left|\frac{1}{T}\mathcal{Q}(f)\right|^2 \sigma_a^2 T = |\mathcal{Q}(f)|^2 \frac{\sigma_a^2}{T} \tag{7.19}$$

and

$$\begin{aligned} \overline{\mathcal{P}}_s^{(d)}(f) &= \left|\frac{1}{T}\mathcal{Q}(f)\right|^2 |\mathtt{m}_a|^2 \sum_{\ell=-\infty}^{+\infty} \delta\left(f - \frac{\ell}{T}\right) \\ &= \left|\frac{\mathtt{m}_a}{T}\right|^2 \sum_{\ell=-\infty}^{+\infty} \left|\mathcal{Q}\left(\frac{\ell}{T}\right)\right|^2 \delta\left(f - \frac{\ell}{T}\right) \end{aligned} \tag{7.20}$$

We note that spectral lines occur in the PSD of s if $\mathtt{m}_a \neq 0$ and $\mathcal{Q}(f)$ is non-zero at frequencies multiple of $1/T$.

Typically, the presence of spectral lines is not desirable for a transmission scheme as it implies a higher transmit power without improving system performance. We obtain $\mathtt{m}_a = 0$ by choosing an alphabet with symmetric values with respect to zero, and by assigning equal probabilities to antipodal values.

In some applications, the PSD of the transmitted signal is shaped by introducing correlation among transmitted symbols by a *line encoder*, see (7.16): in this case, the sequence has memory. We refer the reader to Appendix 7.12 for a description of the more common line codes.

7.1.2 Passband digital transmission (QAM)

We consider the QAM transmission system illustrated in Figure 7.5. Now, values of a_k are complex, and two examples of constellations with corresponding bit map are shown in Figure 6.2.

Modulator

Typically, the pulse h_{Tx} is real-valued, however, the baseband modulated signal is complex-valued:

$$\begin{aligned} s^{(bb)}(t) &= \sum_{k=-\infty}^{+\infty} a_k h_{Tx}(t-kT) \\ &= \sum_{k=-\infty}^{+\infty} a_{k,I} h_{Tx}(t-kT) + j \sum_{k=-\infty}^{+\infty} a_{k,Q} h_{Tx}(t-kT) \end{aligned} \tag{7.21}$$

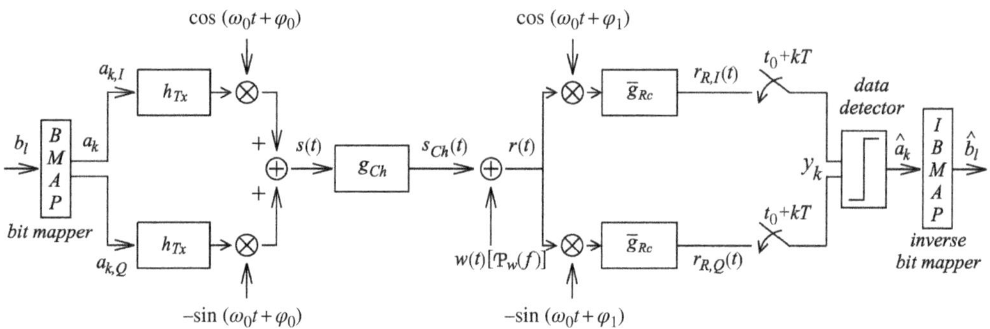

Figure 7.5 Block diagram of a passband digital transmission system.

Let f_0 ($\omega_0 = 2\pi f_0$) and φ_0 be, respectively, the carrier frequency (radian frequency) and phase. We define the following signal and its corresponding Fourier transform

$$s^{(+)}(t) = \frac{1}{2}s^{(bb)}(t)e^{j(\omega_0 t+\varphi_0)} \quad \stackrel{\mathcal{F}}{\longleftrightarrow} \quad S^{(+)}(f) = \frac{1}{2}S^{(bb)}(f - f_0)e^{j\varphi_0} \tag{7.22}$$

then the real-valued transmitted signal is given by

$$s(t) = 2Re\{s^{(+)}(t)\} \quad \stackrel{\mathcal{F}}{\longleftrightarrow} \quad S(f) = S^{(+)}(f) + S^{(+)*}(-f) \tag{7.23}$$

The transformation in the frequency domain from $s^{(bb)}$ to s is illustrated in Figure 7.6.

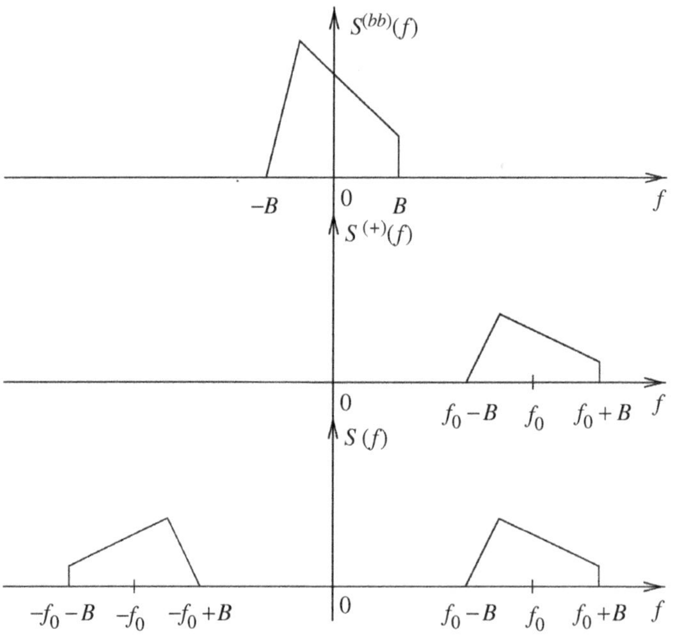

Figure 7.6 Fourier transforms of baseband signal and modulated signal.

Power spectral density

From the analysis leading to (1.320), $s^{(bb)}$ is a cyclostationary process of period T with an average PSD (see (1.323)):

$$\overline{\mathcal{P}}_{s^{(bb)}}(f) = \left|\frac{1}{T}\mathcal{H}_{Tx}(f)\right|^2 \mathcal{P}_a(f) \tag{7.24}$$

Moreover, starting from a relation similar to (1.228), we get that s is a cyclostationary random process with an average PSD given by[2]

$$\overline{\mathcal{P}}_s(f) = \frac{1}{4}\left[\overline{\mathcal{P}}_{s^{(bb)}}(f-f_0) + \overline{\mathcal{P}}_{s^{(bb)}}(-f-f_0)\right] \tag{7.25}$$

We note that the bandwidth of the transmitted signal is equal to *twice* the bandwidth of h_{Tx}.

Three equivalent representations of the modulator

1. From (7.22) and (7.23), using (7.21), it turns out

$$\begin{aligned} s(t) &= Re\left[s^{(bb)}(t)e^{j(\omega_0 t+\varphi_0)}\right] \\ &= Re\left[e^{j(\omega_0 t+\varphi_0)} \sum_{k=-\infty}^{+\infty} a_k h_{Tx}(t-kT)\right] \end{aligned} \tag{7.26}$$

   The block-diagram representation of (7.26) is shown in Figure 7.7. As $s^{(bb)}$ is in general a complex-valued signal, an implementation based on this representation requires a processor capable of complex arithmetic.

2. As

$$e^{j(\omega_0 t+\varphi_0)} = \cos(\omega_0 t+\varphi_0) + j\sin(\omega_0 t+\varphi_0) \tag{7.27}$$

   (7.26) becomes:

$$s(t) = \cos(\omega_0 t+\varphi_0) \sum_{k=-\infty}^{+\infty} a_{k,I} h_{Tx}(t-kT) - \sin(\omega_0 t+\varphi_0) \sum_{k=-\infty}^{+\infty} a_{k,Q} h_{Tx}(t-kT) \tag{7.28}$$

   The block-diagram representation of (7.28) is shown in Figure 7.5. The implementation of a QAM transmitter based on (7.28) is discussed in Appendix 7.D.

3. Using the polar notation $a_k = |a_k|e^{j\theta_k}$, (7.26) becomes

$$\begin{aligned} s(t) &= Re\left[e^{j(\omega_0 t+\varphi_0)} \sum_{k=-\infty}^{+\infty} |a_k|e^{j\theta_k} h_{Tx}(t-kT)\right] \\ &= Re\left[\sum_{k=-\infty}^{+\infty} |a_k|e^{j(\omega_0 t+\varphi_0+\theta_k)} h_{Tx}(t-kT)\right] \\ &= \sum_{k=-\infty}^{+\infty} |a_k|\cos(\omega_0 t+\varphi_0+\theta_k) h_{Tx}(t-kT) \end{aligned} \tag{7.29}$$

   If $|a_k|$ is a constant, we obtain the phase-shift keying (PSK) signal, where the information bits select only the value of the carrier phase.

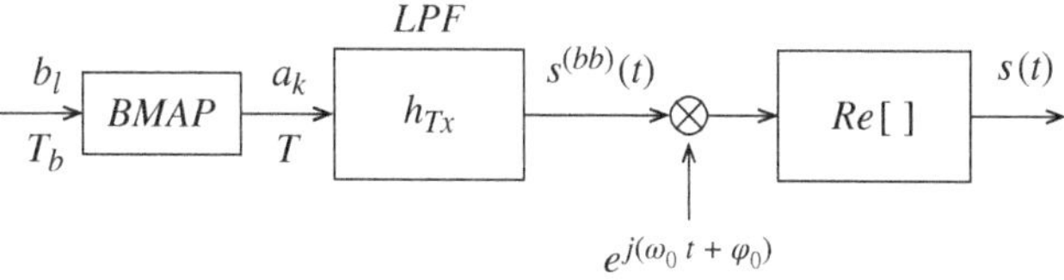

Figure 7.7 QAM transmitter: complex-valued representation.

[2] The result (7.25) needs a clarification. We assume that $\mathtt{r}_{s^{(bb)}s^{(bb)*}}(t, t-\tau) = 0$ is satisfied, as for example in the case of QAM with i.i.d. circularly symmetric symbols (see (1.332)). From the equation (similar to (1.228)) that relates $\mathtt{r}_s$ to $\mathtt{r}_{s^{(bb)}}$ and $\mathtt{r}_{s^{(bb)}s^{(bb)*}}$, as the crosscorrelations are zero, we find that the process s is cyclostationary in t of period T. Taking the average correlation in a period T, the results (7.24) and (7.25) follow.

Coherent receiver

In the *absence of noise*, the general scheme of a coherent receiver is shown in Figure 7.8.

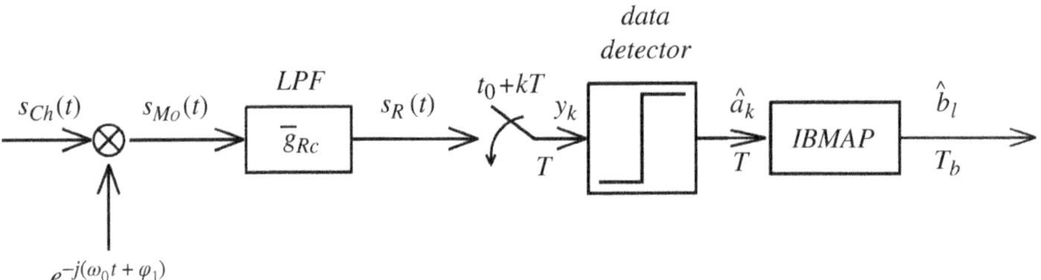

Figure 7.8 QAM receiver: complex-valued representation.

The received signal is given by

$$s_{Ch}(t) = s * g_{Ch}(t) \quad \stackrel{\mathcal{F}}{\longleftrightarrow} \quad \mathcal{S}_{Ch}(f) = \mathcal{S}(f)\mathcal{G}_{Ch}(f) \tag{7.30}$$

First the received signal is translated to baseband by a frequency shift,

$$s_{M_0}(t) = s_{Ch}(t)e^{-j(\omega_0 t+\varphi_1)} \quad \stackrel{\mathcal{F}}{\longleftrightarrow} \quad \mathcal{S}_{M_0}(f) = \mathcal{S}_{Ch}(f+f_0)e^{-j\varphi_1} \tag{7.31}$$

then it is filtered by a lowpass filter (LPF), $\overline{g}_{Rc}$,

$$s_R(t) = s_{M_0} * \overline{g}_{Rc}(t) \quad \stackrel{\mathcal{F}}{\longleftrightarrow} \quad \mathcal{S}_R(f) = \mathcal{S}_{M_0}(f)\overline{\mathcal{G}}_{Rc}(f) \tag{7.32}$$

We note that, if $\overline{g}_{Rc}$ is a non-distorting ideal filter with unit gain, then $s_R(t) = (1/2)s^{(bb)}_{Ch}(t)$. In the particular case where $\overline{g}_{Rc}$ is a real-valued filter, then the receiver in Figure 7.8 is simplified into that of Figure 7.5. Figure 7.9 illustrates these transformations.

We note that in the above analysis, as the channel may introduce a phase offset, the receive carrier phase φ_1 may be different from the transmit carrier phase φ_0.

7.1.3 Baseband equivalent model of a QAM system

Recalling the relations of Figure 1.21, we illustrate the baseband equivalent scheme with reference to Figure 7.10:[3] by assuming that the transmit and receive carriers have the same frequency, we can study QAM systems by the same method that we have developed for PAM systems.

Signal analysis

We refer to Section 1.5.4 for an analysis of passband signals; we recall here that if for $f > 0$ the PSD of w, $\mathcal{P}_w(f)$, is an even function around the frequency f_0, then the real and imaginary parts of $w^{(bb)}(t) = w_I(t) + jw_Q(t)$ the PSD of w_I and w_Q are

$$\mathcal{P}_{w_I}(f) = \mathcal{P}_{w_Q}(f) = \frac{1}{2}\mathcal{P}_{w^{(bb)}}(f) = \begin{cases} 2\mathcal{P}_w(f+f_0), & f \geq -f_0 \\ 0, & \text{elsewhere} \end{cases} \tag{7.33}$$

[3] We note that the term $e^{j\varphi_0}$ has been moved to the receiver; Therefore, the signals $s^{(bb)}$ and $r^{(bb)}$ in Figure 7.10 are defined apart from the term $e^{j\varphi_0}$. This is the same as assuming as reference carrier $e^{j(2\pi f_0 t+\varphi_0)}$.

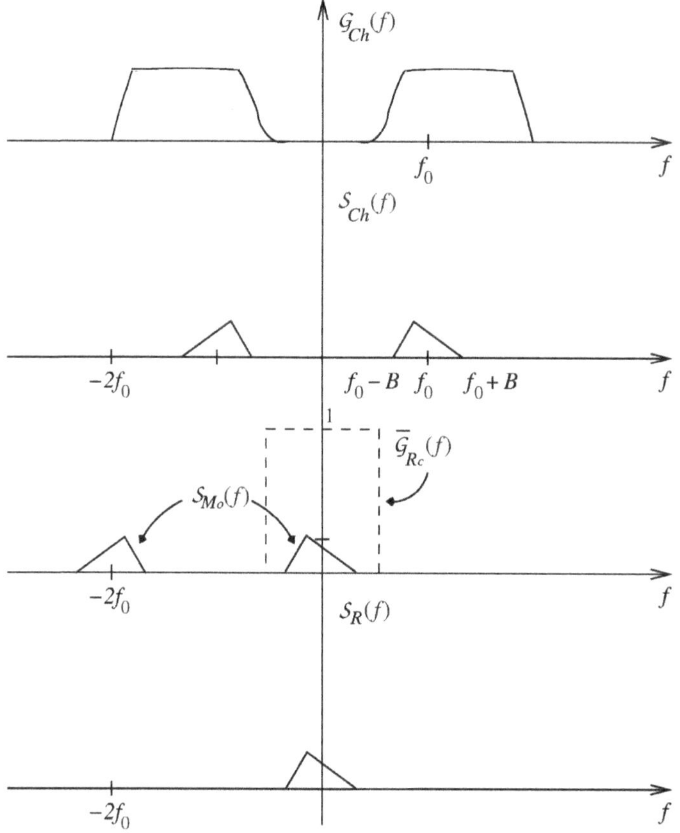

Figure 7.9 Frequency responses of channel and signals at various points of the receiver.

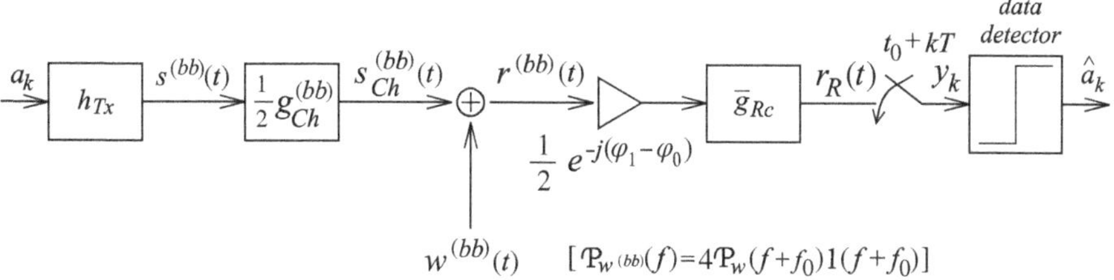

Figure 7.10 Baseband equivalent model of a QAM transmission system.

Moreover, $\mathcal{P}_{w_I w_Q}(f) = 0$, that is w_I and w_Q are uncorrelated, hence,

$$\sigma_{w_I}^2 = \sigma_{w_Q}^2 = \frac{1}{2}\sigma_{w^{(bb)}}^2 = \sigma_w^2 \tag{7.34}$$

To simplify the analysis, for the study of a QAM system, we will adopt the PAM model of Figure 7.1, assuming that all signals and filters are in general complex. We note that the factor $(1/2)e^{-j(\varphi_1-\varphi_0)}$ appears in Figure 7.10. We will include the factor $e^{-j(\varphi_1-\varphi_0)}/\sqrt{2}$ in the impulse response of the transmission channel g_{Ch}, and the factor $1/\sqrt{2}$ in the impulse response g_{Rc}. Consequently, the additive noise has a PSD equal to $(1/2)\mathcal{P}_{w^{(bb)}}(f) = 2\mathcal{P}_w(f + f_0)$ for $f \geq -f_0$. Therefore, the scheme of Figure 7.1 holds also for QAM in the presence of additive noise: the only difference is that in the case of a QAM system the noise is complex-valued with uncorrelated in-phase and quadrature components, each having PSD $\mathcal{P}_w(f+f_0)$ for $f \geq -f_0$.

Hence, the scheme of Figure 7.11 is a reference scheme for both PAM and QAM, where

$$\mathcal{G}_C(f) = \begin{cases} \mathcal{G}_{Ch}(f) & \text{for PAM} \\ \dfrac{e^{-j(\varphi_1-\varphi_0)}}{\sqrt{2}}\mathcal{G}_{Ch}(f+f_0)1(f+f_0) & \text{for QAM} \end{cases} \tag{7.35}$$

We note that for QAM, we have

$$g_C(t) = \frac{e^{-j(\varphi_1-\varphi_0)}}{2\sqrt{2}} g_{Ch}^{(bb)}(t) \tag{7.36}$$

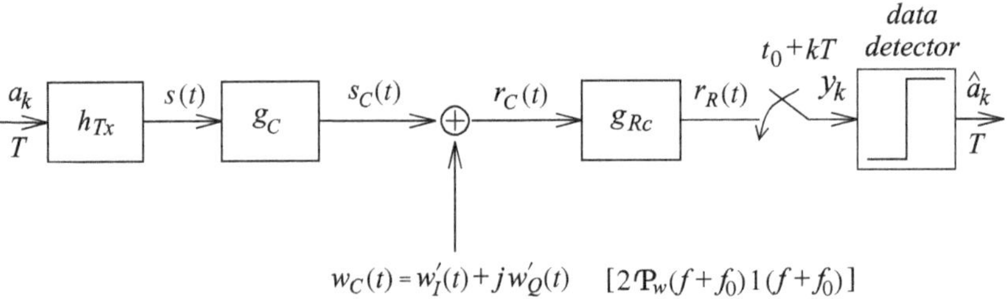

Figure 7.11 Baseband equivalent model of PAM and QAM transmission systems.

With reference to the scheme of Figure 7.8, the relation between the impulse responses of the receive filters is

$$g_{Rc}(t) = \frac{1}{\sqrt{2}}\bar{g}_{Rc}(t) \tag{7.37}$$

We summarize the definitions of the various signals in QAM systems.

1. *Sequence of input symbols*, $\{a_k\}$, sequence of symbols with values from a complex-valued alphabet $\mathcal{A}$. In PAM systems, the symbols of the sequence $\{a_k\}$ assume real values.
2. *Modulated signal*,[4]

$$s(t) = \sum_{k=-\infty}^{+\infty} a_k h_{Tx}(t-kT) \tag{7.38}$$

3. *Signal at the channel output*,

$$s_C(t) = \sum_{k=-\infty}^{+\infty} a_k q_C(t-kT) \qquad q_C(t) = h_{Tx} * g_C(t) \tag{7.39}$$

4. *Circularly-symmetric, complex-valued, additive Gaussian noise*, $w_C(t) = w_I'(t) + jw_Q'(t)$, with PSD $\mathcal{P}_{w_C}$. In the case of white noise, it is[5]:

$$\mathcal{P}_{w_I'}(f) = \mathcal{P}_{w_Q'}(f) = \frac{N_0}{2} \qquad (\text{V}^2/\text{Hz}) \tag{7.40}$$

and

$$\mathcal{P}_{w_C}(f) = N_0 \qquad (\text{V}^2/\text{Hz}) \tag{7.41}$$

In the model of PAM systems, only the component w_I' is considered.

---

4 We point out that for QAM s is in fact $s^{(bb)}$.

5 In fact (7.40) should include the condition $f > -f_0$. Because the bandwidth of g_{Rc} is smaller than f_0, this condition can be omitted.

5. *Received or observed signal*,

$$r_C(t) = s_C(t) + w_C(t) \tag{7.42}$$

6. *Signal at the output of the complex-valued amplifier-equalizer filter* g_{Rc},

$$r_R(t) = s_R(t) + w_R(t) \tag{7.43}$$

where

$$s_R(t) = \sum_{k=-\infty}^{+\infty} a_k q_R(t - kT) \tag{7.44}$$

with

$$q_R(t) = q_C * g_{Rc}(t) \quad \text{and} \quad w_R(t) = w_C * g_{Rc}(t) \tag{7.45}$$

In PAM systems, g_{Rc} is a real-valued filter.

7. *Signal at the decision point at instant* $t_0 + kT$,

$$y_k = r_R(t_0 + kT) \tag{7.46}$$

8. *Data detector*: Based on the sequence y_k, we examine various methods for data detection. Here we recall the simplest one, which is based on the fact that y_k can be approximated as (see Section 7.3)

$$y_k = h_0 a_k + w_{R,k} \tag{7.47}$$

where

$$w_{R,k} = w_R(t_0 + kT) \tag{7.48}$$

Scaling y_k by h_0 yields

$$\overline{y}_k = a_k + \frac{w_{R,k}}{h_0} \tag{7.49}$$

If $w_{R,k}$ is i.i.d., decision is based on reporting $\overline{y}_k$ on the data constellation and deciding for the closest symbol value. Two examples of decision region are reported in Figure 6.4.

9. *Sequence of detected symbols*,

$$\{\hat{a}_k\} \tag{7.50}$$

7.1.4 Characterization of system elements

We consider some characteristics of the signals in the scheme of Figure 7.11.

Transmitter

The choice of the transmit pulse is quite important because it determines the bandwidth of the system (see (7.16) and (7.25)). Two choices are shown in Figure 7.12, where

1. $h_{Tx}(t) = \mathtt{w}_T(t) = \text{rect}\left(\frac{t-\frac{T}{2}}{T}\right)$, with wide spectrum;
2. h_{Tx} with longer duration and smaller bandwidth.

Transmission channel

The transmission channel is modelled as a time invariant linear system. Therefore, it is represented by a filter having impulse response g_{Ch}. As described the majority of channels are characterized by a

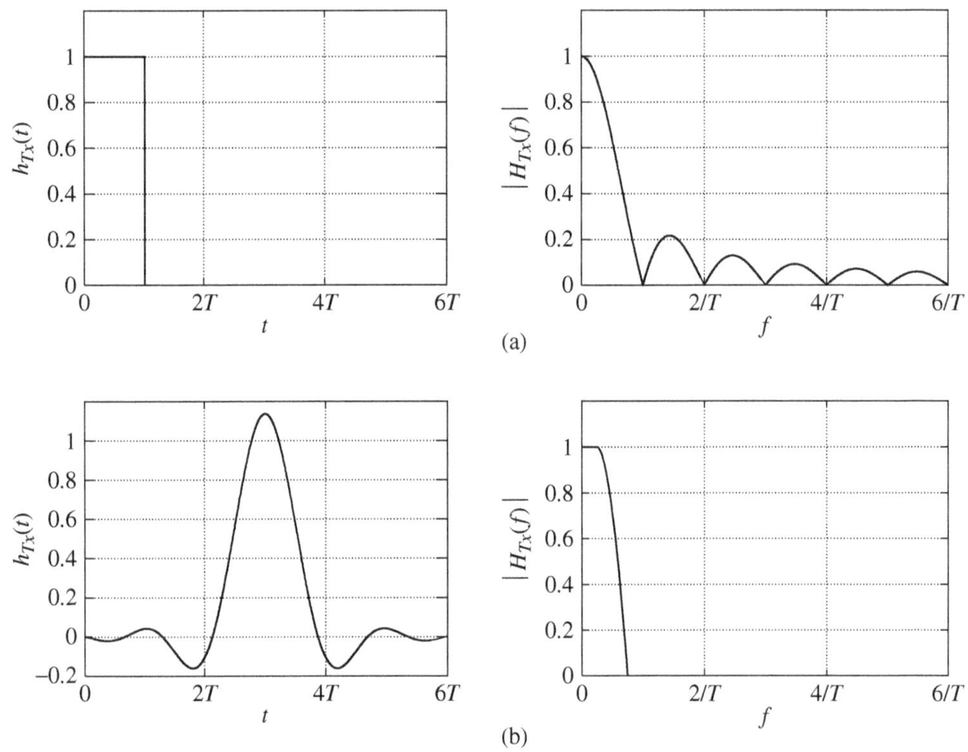

Figure 7.12 Two examples of transmit pulse h_{Tx}. (a) Wideband. (b) Narrowband.

frequency response having a null at DC, the shape of $\mathcal{G}_{Ch}(f)$ is as represented in Figure 7.13, where the passband goes from f_1 to f_2. For transmission over cables f_1 may be of the order of a few hundred Hz, whereas for radio links f_1 may be in the range of MHz or GHz. Consequently, PAM (possibly using a line code) as well as QAM transmission systems may be considered over cables; for transmission over radio, instead a QAM system may be used, assuming as carrier frequency f_0 the centre frequency of the passband (f_1, f_2). In any case, the channel is bandlimited with a finite bandwidth $f_2 - f_1$.

With reference to the general model of Figure 7.11, we adopt the polar notation for $\mathcal{G}_C$:

$$\mathcal{G}_C(f) = |\mathcal{G}_C(f)| e^{j \arg \mathcal{G}_C(f)} \tag{7.51}$$

Let B be the bandwidth of s. According to (1.78), a channel presents *ideal characteristics*, known as *Heaviside conditions for the absence of distortion*, if the following two properties are satisfied:

1. the magnitude response is a constant for $|f| < B$,

$$|\mathcal{G}_C(f)| = \mathcal{G}_0 \qquad \text{for } |f| < B \tag{7.52}$$

2. the phase response is proportional to f for $|f| < B$,

$$\arg \mathcal{G}_C(f) = -2\pi f t_0 \qquad \text{for } |f| < B \tag{7.53}$$

Under these conditions, s is reproduced at the output of the channel without distortion, that is:

$$s_C(t) = \mathcal{G}_0 s(t - t_0) \tag{7.54}$$

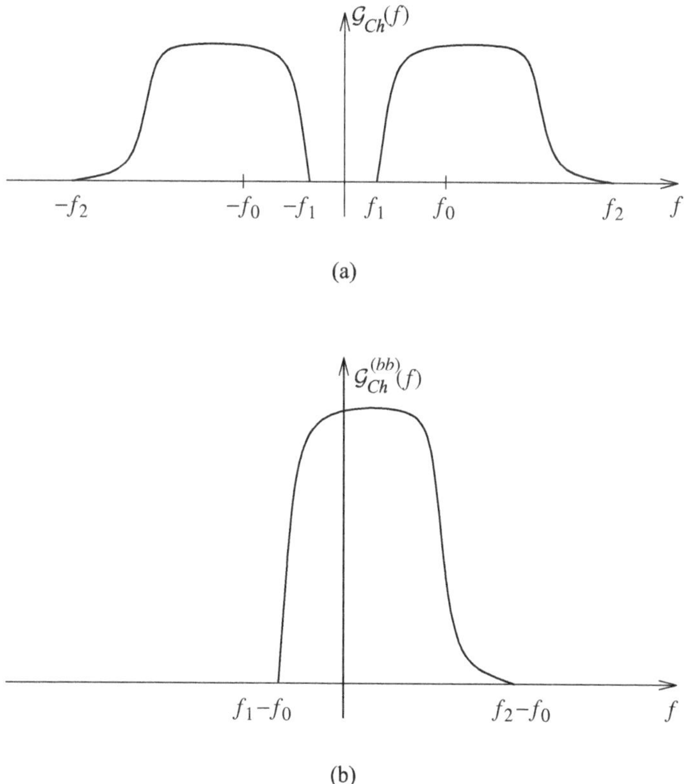

Figure 7.13 Frequency response of the transmission channel. (a) Channel. (b) Baseband equivalent channel.

In practice, channels introduce both *amplitude distortion* and *phase distortion*. An example of frequency response of a radio channel is given in Figure 4.17: the overall effect is that the signal s_C may be very different from s.

In short, for channels encountered in practice conditions (7.52) and (7.53) are too stringent; for PAM and QAM transmission systems, we will refer instead to the Nyquist criterion (7.72).

Receiver

We return to the receiver structure of Figure 7.11, consisting of a filter g_{Rc} followed by a sampler with sampling rate $1/T$, and a data detector.

In general, if the frequency response of the receive filter $\mathcal{G}_{R_c}(f)$ contains a factor $C(e^{j2\pi fT})$, periodic of period $1/T$, such that the following factorization holds:

$$\mathcal{G}_{Rc}(f) = \mathcal{G}_M(f)C(e^{j2\pi fT}) \tag{7.55}$$

where $\mathcal{G}_M(f)$ is a generic function, then the filter-sampler block before the data detector of Figure 7.11 can be represented as in Figure 7.14, where the sampler is followed by a discrete-time filter and the impulse response c is the inverse discrete-time Fourier transform of $C(e^{j2\pi fT})$. It is easy to prove that in the two systems, the relation between r_C and y_k is the same.

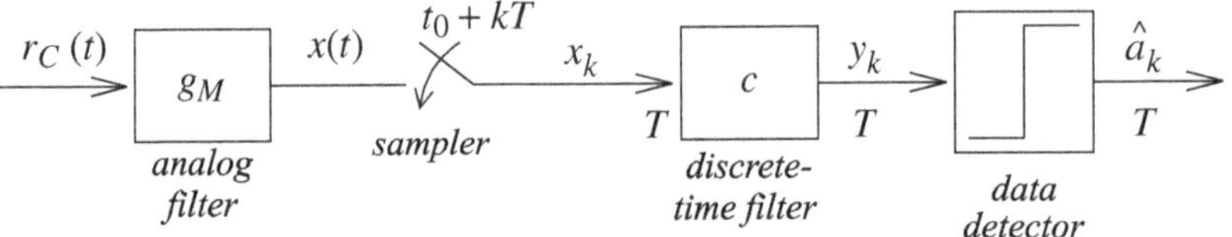

Figure 7.14 Receiver structure with analog and discrete-time filters.

Ideally, in the system of Figure 7.14, y_k should be equal to a_k. In practice, linear distortion and additive noise, the only disturbances considered here, may determine a significant deviation of y_k from the desired symbol a_k.

The last element in the receiver is the data detector. The simplest scheme is the *threshold detector*, as described earlier.

7.2 Intersymbol interference

Discrete-time equivalent system

From (7.43), we define

$$s_{R,k} = s_R(t_0 + kT) = \sum_{i=-\infty}^{+\infty} a_i q_R(t_0 + (k-i)T) \tag{7.56}$$

and

$$w_{R,k} = w_R(t_0 + kT) \tag{7.57}$$

Then, from (7.56), at the decision point, the generic sample is expressed as

$$y_k = s_{R,k} + w_{R,k} \tag{7.58}$$

Introducing the version of q_R, time-shifted by t_0, as

$$h(t) = q_R(t_0 + t) \tag{7.59}$$

and defining

$$h_i = h(iT) = q_R(t_0 + iT) \tag{7.60}$$

it follows that

$$s_{R,k} = \sum_{i=-\infty}^{+\infty} a_i h_{k-i} = a_k h_0 + \mathtt{i}_k \tag{7.61}$$

where

$$\mathtt{i}_k = \sum_{i=-\infty,\ i\neq k}^{+\infty} a_i h_{k-i} = \cdots + h_{-1}a_{k+1} + h_1 a_{k-1} + h_2 a_{k-2} + \cdots \tag{7.62}$$

represents the intersymbol interference (ISI). The coefficients $\{h_i\}_{i\neq 0}$ are called *interferers*. Moreover, (7.58) becomes

$$y_k = a_k h_0 + \mathtt{i}_k + w_{R,k} \tag{7.63}$$

We observe that, even in the absence of noise, the detection of a_k from y_k by a threshold detector takes place in the presence of the term $\mathtt{i}_k$, which behaves as a disturbance with respect to the desired term $a_k h_0$.

For the analysis, it is often convenient to approximate $\mathtt{i}_k$ as noise with a Gaussian distribution: the more numerous and similar in amplitude are the interferers, the more valid is this approximation. In the case of i.i.d. symbols, the first two moments of $\mathtt{i}_k$ are easily determined.

$$\text{Mean value of } \mathtt{i}_k: \qquad \mathtt{m}_{\mathtt{i}} = \mathtt{m}_a \sum_{i=-\infty,\ i\neq 0}^{+\infty} h_i \tag{7.64}$$

$$\text{Variance of } \mathtt{i}_k: \qquad \sigma_{\mathtt{i}}^2 = \sigma_a^2 \sum_{i=-\infty,\ i\neq 0}^{+\infty} |h_i|^2 \tag{7.65}$$

From (7.58), with $\{s_{R,k}\}$ given by (7.61), we derive the discrete-time equivalent scheme, with period T (see Figure 7.15), that relates the signal at the decision point to the data transmitted over a discrete-time channel with impulse response given by the sequence $\{h_i\}$, called *overall discrete-time equivalent impulse response* of the system.

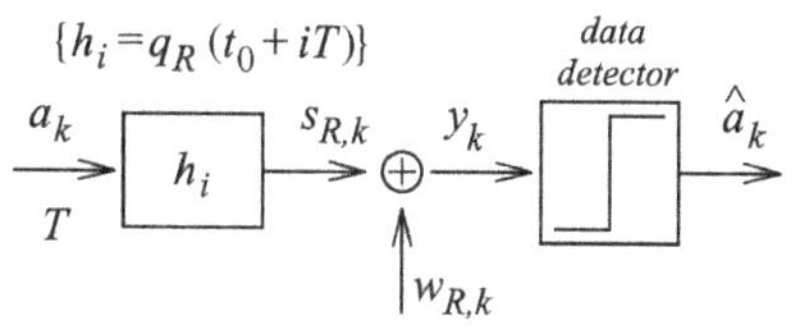

Figure 7.15 Discrete-time equivalent scheme, with period T, of a QAM system.

Concerning the additive noise $\{w_{R,k}\}$,[6] being

$$\mathcal{P}_{w_R}(f) = \mathcal{P}_{w_C}(f)|\mathcal{G}_{Rc}(f)|^2 \tag{7.66}$$

the PSD of $\{w_{R,k}\}$ is given by

$$\mathcal{P}_{w_{R,k}}(f) = \sum_{\ell=-\infty}^{+\infty} \mathcal{P}_{w_R}\left(f - \ell\frac{1}{T}\right) \tag{7.67}$$

In any case, the variance of $w_{R,k}$ is equal to that of w_R and is given by

$$\sigma_{w_{R,k}}^2 = \sigma_{w_R}^2 = \int_{-\infty}^{+\infty} \mathcal{P}_{w_C}(f)|\mathcal{G}_{Rc}(f)|^2 df \tag{7.68}$$

In particular, the *variance per dimension* of the noise is given by

$$\text{PAM} \qquad \sigma_I^2 = E[w_{R,k}^2] = \sigma_{w_R}^2 \tag{7.69}$$

$$\text{QAM} \qquad \sigma_I^2 = E\left[(Re[w_{R,k}])^2\right] = E\left[(Im[w_{R,k}])^2\right] = \frac{1}{2}\sigma_{w_R}^2 \tag{7.70}$$

In the case of PAM (QAM) transmission over a channel with white noise, where $\mathcal{P}_{w_C}(f) = N_0/2$ (N_0), (7.68) yields a variance per dimension equal to

$$\sigma_I^2 = \frac{N_0}{2} E_{g_{Rc}} \tag{7.71}$$

where $E_{g_{Rc}}$ is the energy of the receive filter. We observe that (7.71) holds for PAM as well as for QAM.

Nyquist pulses

The problem we wish to address consists in finding the conditions on the various filters of the system, so that, *in the absence of noise*, y_k is a replica of a_k. The solution is the *Nyquist criterion* for the absence of distortion in digital transmission.

---

[6] See Observation 1.4 on page 43.

From (7.61), to obtain $y_k = a_k$, it must be

Nyquist criterion in the time domain

$$\begin{cases} h_0 = 1 \\ h_i = 0, & i \neq 0 \end{cases} \tag{7.72}$$

and ISI vanishes. A pulse h that satisfies the conditions (7.72) is said to be a *Nyquist pulse with modulation interval T*

The conditions (7.72) have their equivalent in the frequency domain. They can be derived using the Fourier transform of the sequence $\{h_i\}$ (1.23),

$$\sum_{i=-\infty}^{+\infty} h_i e^{-j2\pi fiT} = \frac{1}{T}\sum_{\ell=-\infty}^{+\infty} \mathcal{H}\left(f-\frac{\ell}{T}\right) \tag{7.73}$$

where $\mathcal{H}(f)$ is the Fourier transform of h. From the conditions, (7.72) the left-hand side of (7.73) is equal to 1, hence, the condition for the absence of ISI is formulated in the frequency domain for the generic pulse h as

Nyquist criterion in the frequency domain

$$\sum_{\ell=-\infty}^{+\infty} \mathcal{H}\left(f-\frac{\ell}{T}\right) = T \tag{7.74}$$

From (7.74), we deduce an important fact: the Nyquist pulse with minimum bandwidth is given by

$$h(t) = h_0 \operatorname{sinc} \frac{t}{T} \quad \stackrel{\mathcal{F}}{\longleftrightarrow} \quad \mathcal{H}(f) = Th_0 \operatorname{rect} \frac{f}{1/T} \tag{7.75}$$

Definition 7.1
The frequency $1/(2T)$, which coincides with half of the modulation frequency, is called *Nyquist frequency*. □

A family of Nyquist pulses widely used in telecommunications is composed of the *raised cosine* pulses whose time and frequency plots, for three values of the parameter ρ, are illustrated in Figure 7.16a.

We define

$$\operatorname{rcos}(x,\rho) = \begin{cases} 1, & 0 \le |x| \le \dfrac{1-\rho}{2} \\ \cos^2\left(\dfrac{\pi}{2}\dfrac{|x|-\frac{1-\rho}{2}}{\rho}\right), & \dfrac{1-\rho}{2} < |x| \le \dfrac{1+\rho}{2} \\ 0, & |x| > \dfrac{1+\rho}{2} \end{cases} \tag{7.76}$$

then

$$\mathcal{H}(f) = T \operatorname{rcos}\left(\frac{f}{1/T},\rho\right) \tag{7.77}$$

with inverse Fourier transform

$$\begin{aligned} h(t) &= \operatorname{sinc}\left(\frac{t}{T}\right)\frac{\pi}{4}\left[\operatorname{sinc}\left(\rho\frac{t}{T}+\frac{1}{2}\right)+\operatorname{sinc}\left(\rho\frac{t}{T}-\frac{1}{2}\right)\right] \\ &= \operatorname{sinc}\left(\frac{t}{T}\right)\cos\left(\pi\rho\frac{t}{T}\right)\frac{1}{1-\left(2\rho\frac{t}{T}\right)^2} \end{aligned} \tag{7.78}$$

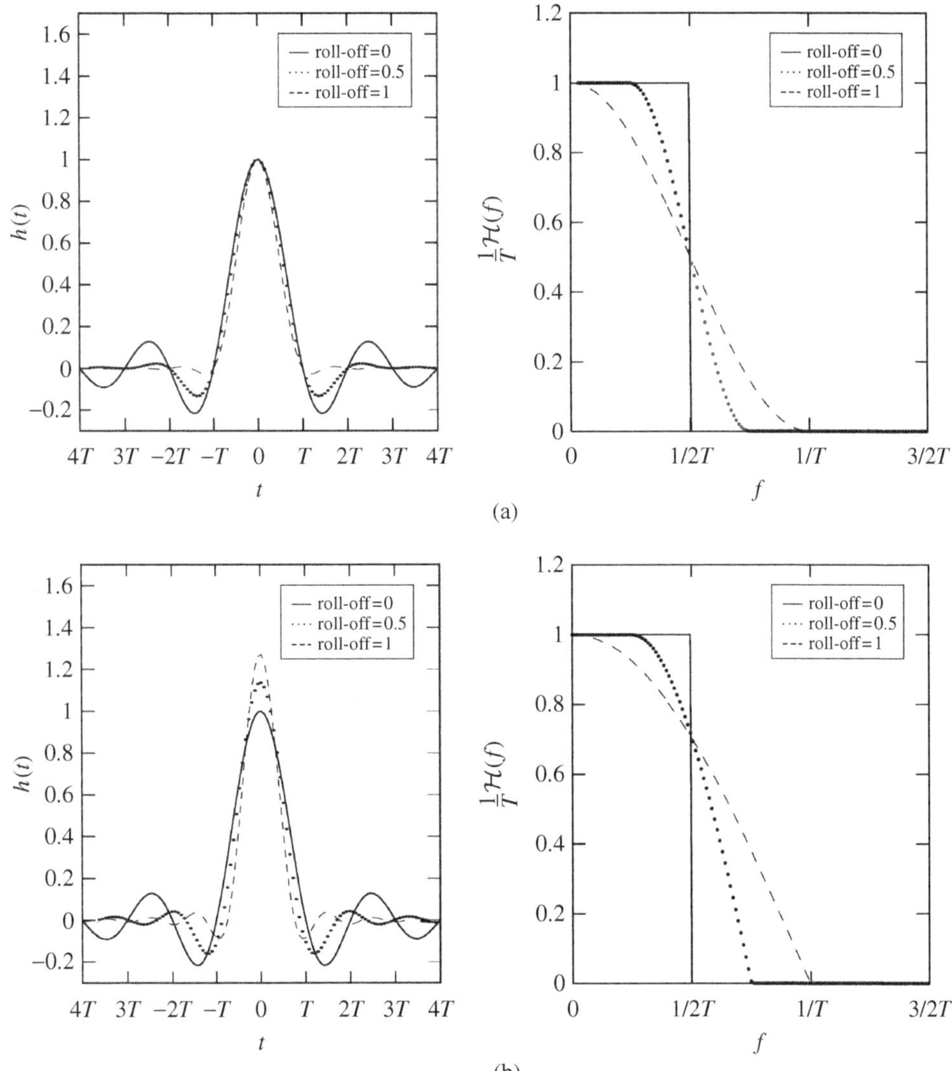

Figure 7.16 Time and frequency plots of (a) *raised cosine* and (b) *square root raised cosine* pulses for three values of the roll-off factor ρ.

It is easily proven that, from (7.76), the area of $\mathcal{H}(f)$ in (7.77) is equal to 1, that is

$$h(0) = \int_{-\infty}^{+\infty} T \,\mathrm{rcos}\left(\frac{f}{1/T}, \rho\right) df = 1 \tag{7.79}$$

and the energy is

$$E_h = \int_{-\infty}^{+\infty} T^2 \,\mathrm{rcos}^2\left(\frac{f}{1/T}, \rho\right) df = T\left(1 - \frac{\rho}{4}\right) = \mathcal{H}(0)\,h(0)\left(1 - \frac{\rho}{4}\right) \tag{7.80}$$

We note that, from (7.77), the bandwidth of the baseband equivalent system is equal to $(1 + \rho)/(2T)$. Consequently, for a QAM system, the required bandwidth is $(1 + \rho)/T$.

Later, we will also refer to *square root raised cosine* pulses, with frequency response given by

$$\mathcal{H}(f) = T\sqrt{\text{rcos}\left(\frac{f}{1/T}, \rho\right)} \tag{7.81}$$

and inverse Fourier transform

$$\begin{aligned} h(t) &= (1-\rho)\text{sinc}\left[(1-\rho)\frac{t}{T}\right] + \rho\cos\left[\pi\left(\frac{t}{T}+\frac{1}{4}\right)\right]\text{sinc}\left(\rho\frac{t}{T}+\frac{1}{4}\right) \\ &\quad + \rho\cos\left[\pi\left(\frac{t}{T}-\frac{1}{4}\right)\right]\text{sinc}\left(\rho\frac{t}{T}-\frac{1}{4}\right) \\ &= \frac{\sin\left[\pi(1-\rho)\frac{t}{T}\right] + 4\rho\frac{t}{T}\cos\left[\pi(1+\rho)\frac{t}{T}\right]}{\pi\left[1-\left(4\rho\frac{t}{T}\right)^2\right]\frac{t}{T}} \end{aligned} \tag{7.82}$$

In this case,

$$h(0) = \int_{-\infty}^{+\infty} T\sqrt{\text{rcos}\left(\frac{f}{1/T}, \rho\right)}df = 1 - \rho\left(1 - \frac{4}{\pi}\right) \tag{7.83}$$

and the pulse energy is given by

$$E_h = \int_{-\infty}^{+\infty} T^2 \,\text{r}\cos\left(\frac{f}{1/T}, \rho\right) df = T \tag{7.84}$$

We note that $\mathcal{H}(f)$ in (7.81) is not the frequency response of a Nyquist pulse. Plots of h and $\mathcal{H}(f)$, given respectively by (7.82) and (7.81), for various values of ρ are shown in Figure 7.16b.

The parameter ρ, called *excess bandwidth parameter* or *roll-off* factor, is in the range between 0 and 1. We note that ρ determines how fast the pulse decays in time.

Observation 7.1
From the Nyquist conditions, we deduce that

1. a data sequence can be transmitted with modulation rate $1/T$ without errors if $\mathcal{H}(f)$ satisfies the Nyquist criterion, and there is no noise;
2. the channel, with frequency response $\mathcal{G}_C$, must have a bandwidth equal to at least $1/(2T)$, otherwise, intersymbol interference cannot be avoided.

Eye diagram

From (7.61), we observe that if the samples $\{h_i\}$, for $i \neq 0$, are not sufficiently small with respect to h_0, the ISI may result a dominant disturbance with respect to noise and impair the system performance. On the other hand, from (7.59) and (7.60), the discrete-time impulse response $\{h_i\}$ depends on the choice of the *timing phase* t_0 (see Chapter 14) and on the pulse shape q_R.

In the absence of noise, at the decision point, the sample y_0, as a function of t_0, is given by

$$\begin{aligned} y_0 = y(t_0) &= \sum_{i=-\infty}^{+\infty} a_i q_R(t_0 - iT) \\ &= a_0 q_R(t_0) + \mathtt{i}_0(t_0) \end{aligned} \tag{7.85}$$

where

$$\begin{aligned} \mathtt{i}_0(t_0) &= \sum_{i=-\infty,\ i\neq 0}^{+\infty} a_i q_R(t_0 - iT) \\ &= \cdots + a_{-1}q_R(t_0+T) + a_1 q_R(t_0 - T) + a_2 q_R(t_0 - 2T) + \cdots \end{aligned} \tag{7.86}$$

is the ISI.

We now illustrate, through an example, a graphic method to represent the effect of the choice of t_0 for a given pulse q_R. We consider a PAM transmission system, where $y(t_0)$ is real: for a QAM system, both $Re\ [y(t_0)]$ and $Im\ [y(t_0)]$ need to be represented. We consider a quaternary transmission with

$$a_k = \alpha_n \in \mathcal{A} = \{-3, -1, 1, 3\} \tag{7.87}$$

and pulse q_R as shown in Figure 7.17.

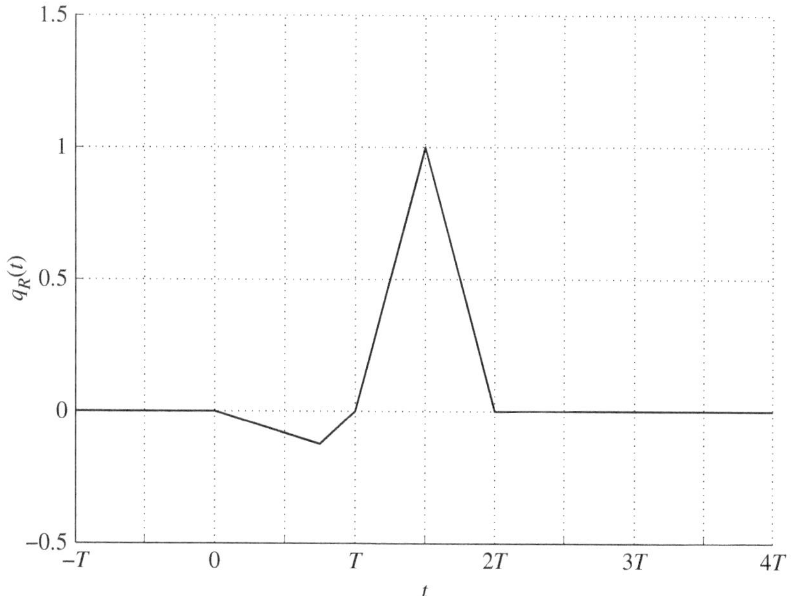

Figure 7.17 Pulse shape for the computation of the eye diagram.

In the absence of ISI, $\mathtt{i}_0(t_0) = 0$ and $y_0 = a_0 q_R(t_0)$. In correspondence to each possible value α_n of a_0, the pattern of y_0 as a function of t_0 is shown in Figure 7.18: it is seen that the possible values of y_0, for $\alpha_n \in \mathcal{A}$, are farther apart, therefore, they offer a greater margin against noise in correspondence of the peak of q_R, which in this example occurs at instant $t_0 = 1.5T$. In fact, for a given t_0 and for a given message $\ldots, a_{-1}, a_1, a_2, \ldots$, it may result $\mathtt{i}_0(t_0) \neq 0$, and this value is added to the desired sample $a_0 q_R(t_0)$.

The range of variations of $y_0(t_0)$ around the desired sample $\alpha_n q_R(t_0)$ is determined by the values

$$\mathtt{i}_{max}(t_0; \alpha_n) = \max_{\{a_k\},\ a_0=\alpha_n} \mathtt{i}_0(t_0) \tag{7.88}$$

$$\mathtt{i}_{min}(t_0; \alpha_n) = \min_{\{a_k\},\ a_0=\alpha_n} \mathtt{i}_0(t_0) \tag{7.89}$$

The *eye diagram* is characterized by the $2M$ profiles

$$\alpha_n q_R(t_0) + \begin{cases} \mathtt{i}_{max}(t_0; \alpha_n) \\ \mathtt{i}_{min}(t_0; \alpha_n) \end{cases}, \qquad \alpha_n \in \mathcal{A} \tag{7.90}$$

If the symbols $\{a_k\}$ are statistically independent with balanced values, that is both α_n and $-\alpha_n$ belong to $\mathcal{A}$, defining

$$\alpha_{max} = \max_n \alpha_n \tag{7.91}$$

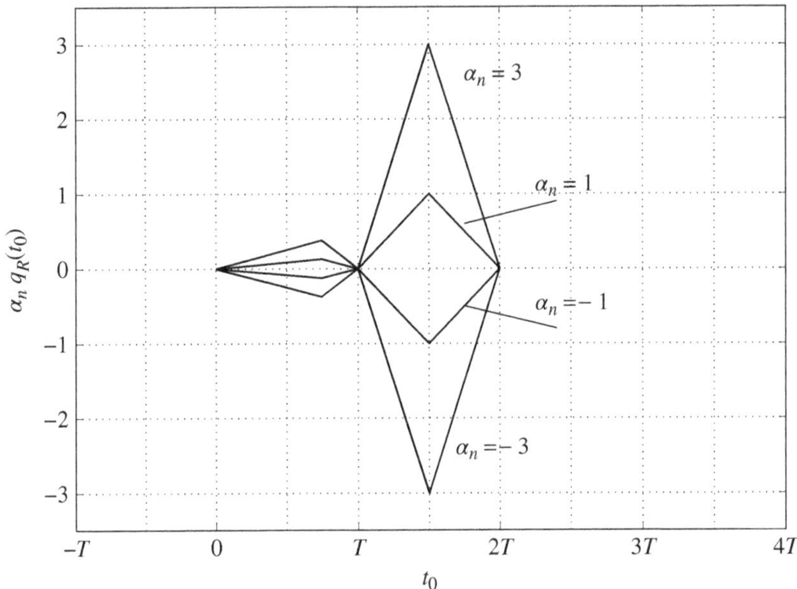

Figure 7.18 Desired component $\alpha_n q_R(t_0)$ as a function of t_0, $\alpha_n \in \{-3, -1, 1, 3\}$.

and

$$\mathtt{i}_{abs}(t_0) = \alpha_{max} \sum_{i=-\infty,\ i\neq 0}^{+\infty} |q_R(t_0 - iT)| \tag{7.92}$$

we have that

$$\mathtt{i}_{max}(t_0) = \mathtt{i}_{abs}(t_0) \tag{7.93}$$

$$\mathtt{i}_{min}(t_0) = -\mathtt{i}_{abs}(t_0) \tag{7.94}$$

We note that both functions do not depend on $a_0 = \alpha_n$.

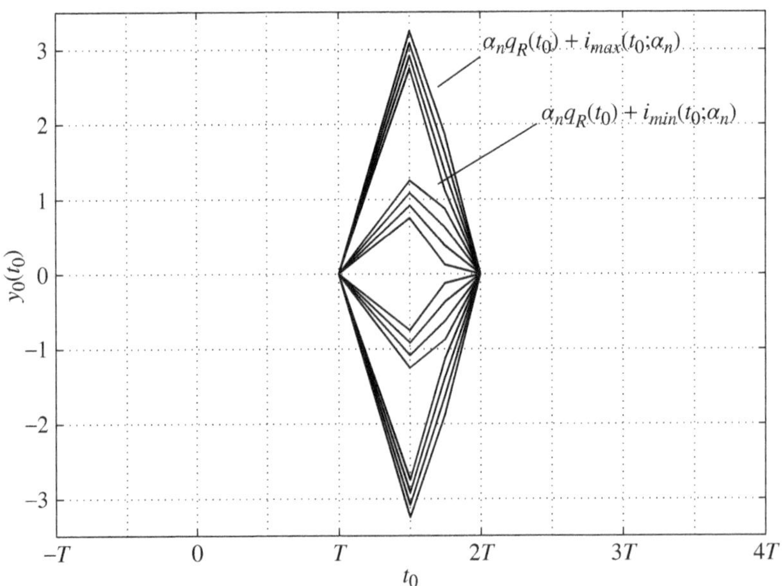

Figure 7.19 Eye diagram for quaternary transmission and pulse q_R of Figure 7.17.

For the considered pulse, the eye diagram is given in Figure 7.19. We observe that as a result of the presence of ISI, the values of y_0 may be very close to each other, and therefore, reduce considerably the margin against noise. We also note that, in general, the timing phase that offers the largest margin against noise is not necessarily found in correspondence of the peak of q_R. In this example, however, the choice $t_0 = 1.5T$ guarantees the largest margin against noise.

In the general case, where there exists correlation between the symbols of the sequence $\{a_k\}$, it is easy to show that $\mathtt{i}_{max}(t_0; \alpha_n) \leq \mathtt{i}_{abs}(t_0)$ and $\mathtt{i}_{min}(t_0; \alpha_n) \geq -\mathtt{i}_{abs}(t_0)$. Consequently, the eye may be wider as compared to the case of i.i.d. symbols.

For quaternary transmission, we show in Figure 7.20 the eye diagram obtained with a raised cosine pulse q_R, for two values of the roll-off factor.

In general, the $M-1$ *pupils* of the eye diagram have a shape as illustrated in Figure 7.21, where two parameters are identified: the height a and the width b. The height a is an indicator of the *noise immunity* of the system. The width b indicates the *immunity* with respect to *deviations from the optimum timing phase*. For example a raised cosine pulse with $\rho = 1$ offers greater immunity against errors in the choice of t_0 as compared to $\rho = 0.125$. The price we pay is a larger bandwidth of the transmission channel.

We now illustrate an alternative method to obtain the eye diagram. A long random sequence of symbols $\{a_k\}$ is transmitted over the channel, and the portions of the curve $y(t) = s_R(t)$ relative to the various intervals $[t_1, t_1 + T), [t_1 + T, t_1 + 2T), [t_1 + 2T, t_1 + 3T), \ldots]$, are mapped on the same interval, for example

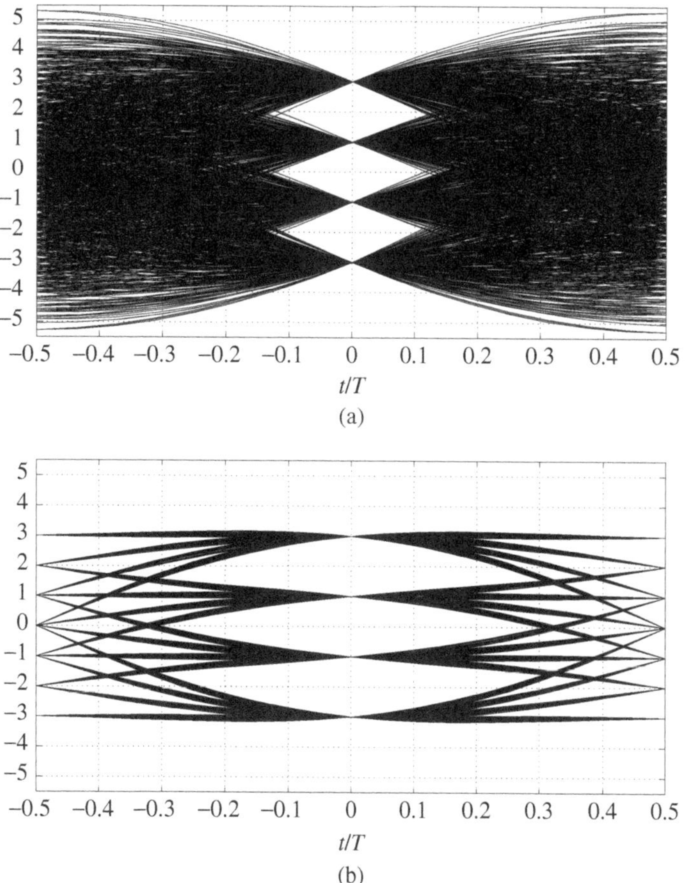

Figure 7.20 Eye diagram for quaternary transmission and raised cosine pulse q_R with roll-off factor: (a) $\rho = 0.125$ and (b) $\rho = 1$.

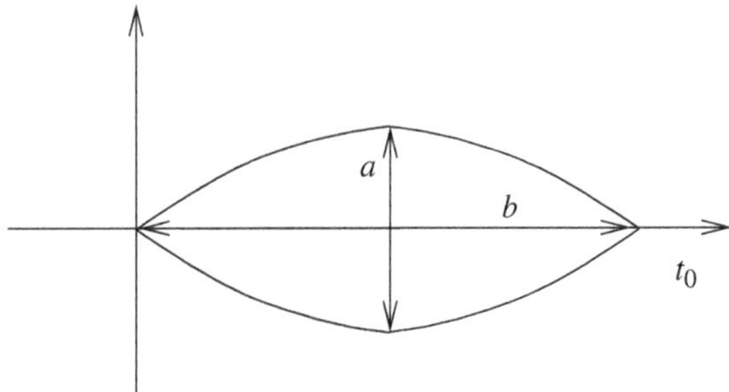

Figure 7.21 Height a and width b of the *pupil* of an eye diagram.

on $[t_1, t_1 + T)$. Typically, we select t_1 so that the centre of the eye falls in the centre of the interval $[t_1, t_1 + T)$. Then the contours of the obtained eye diagram correspond to the different profiles (7.90). If the contours of the eye do not appear, it means that for all values of t_0 the worst-case ISI is larger than the desired component and the *eye* is *shut*. We note that, if the pulse $q_R(t)$, $t \in \mathbb{R}$, has a duration equal to t_h, and define $N_h = \lceil t_h/T \rceil$, we must omit plotting the values of y for the first and last $N_h - 1$ modulation intervals, as they would be affected by the transient behaviour of the system. Moreover, for transmission with i.i.d. symbols, at every instant $t \in \mathbb{R}$ the number of symbols $\{a_k\}$ that contribute to $y(t)$ is at most equal to N_h. To plot the eye diagram, we need in principle to generate all the M-ary symbol sequences of length N_h: in this manner, we will reproduce the values of $y(t)$ in correspondence of the various profiles.

7.3 Performance analysis

Signal-to-noise ratio

The performance of a system is expressed by P_e or P_{bit} (see definitions (6.76) and (6.85), respectively) as a function of the signal-to-noise ratio Γ (6.69). We express now $E_{s_{Ch}} = T\mathtt{M}_{s_{Ch}}$ in PAM and QAM systems.

PAM systems. For an i.i.d. input symbol sequence, using (1.324) we have

$$E_{s_{Ch}} = \mathtt{M}_a E_{q_{Ch}} \tag{7.95}$$

and, for baseband signals $B_{min} = 1/(2T)$, we obtain

$$\Gamma = \frac{\mathtt{M}_a E_{q_{Ch}}}{N_0/2} \tag{7.96}$$

where $\mathtt{M}_a$ is the statistical power of the data and $E_{q_{Ch}}$ is the energy of the pulse $q_{Ch}(t) = h_{Tx} * g_{Ch}(t)$. Because for PAM, observing (7.35), we get $q_{Ch}(t) = q_C(t)$, then (7.96) can be expressed as

$$\Gamma = \frac{\mathtt{M}_a E_{q_C}}{N_0/2} \tag{7.97}$$

QAM systems. From (1.219), using (7.35), we obtain

$$E\left[s_{Ch}^2(t)\right] = \frac{1}{2}E\left[|s_{Ch}^{(bb)}(t)|^2\right] = E\left[|s_C(t)|^2\right] \tag{7.98}$$

and

$$\Gamma = \frac{E_{s_C}}{N_0(B_{min}T)} \tag{7.99}$$

As for passband signals $B_{min} = 1/T$, (7.99) becomes

$$\Gamma = \frac{\mathtt{M}_a E_{q_C}}{N_0} \tag{7.100}$$

We note that (7.100), expressed as[7]

$$\Gamma = \frac{\mathtt{M}_a E_{q_C}/2}{N_0/2} \tag{7.101}$$

represents the ratio between the energy per component of s_C, given by $TE[(Re[s_C(t)])^2]$ and $TE[(Im[s_C(t)])^2]$, and the PSD of the noise components, equal to $N_0/2$. Then (7.100), observing also (7.35), coincides with (7.96) of PAM systems.

Symbol error probability in the absence of ISI

If the Nyquist conditions (7.72) are verified, from (7.58), the samples of the received signal at the decision point are given by (see also (6.13))

$$y_k = a_k + w_{R,k}, \qquad a_k \in \mathcal{A} \tag{7.102}$$

We consider a *memoryless decision rule* on y_k, i.e. regarding y_k as an isolated sample. If $w_{R,k}$ has a circularly symmetric Gaussian probability density function, as is the case if equation (7.40) holds, and the values assumed by a_k are equally likely, then given the observation y_k, the minimum-distance detector leads to choosing the value of $\alpha_n \in \mathcal{A}$ that is closest to y_k.[8] Moreover, the error probability depends on the distance d_{min} between adjacent values $\alpha_n \in \mathcal{A}$, and on the *variance per dimension* of the noise $w_{R,k}$, σ_I^2. Hence, from the definition (6.75) of the signal-to-noise ratio (SNR) at detection point γ, the bit error probability for M-PAM and M-QAM are given in (6.77) and (6.78), respectively. We note that, for the purpose of computing P_e, only the variance of the noise $w_{R,k}$ is needed and not its PSD.

With reference to the constellations of Table 6.1 and Figure 6.2, we consider $d_{min} = 2h_0 = 2$. Now, for a channel with white noise, σ_I^2 is given by (7.71), hence, from (6.75) it follows that

$$\gamma = \frac{2}{N_0 E_{g_{Rc}}} \tag{7.103}$$

Apparently, the above equation could lead to choosing a filter g_{Rc} with very low energy so that $\gamma \gg 1$. However, here g_{Rc} is not arbitrary, but it must be chosen such that the condition (7.72) for the absence of ISI is satisfied. We will see later a criterion to design the filter g_{Rc}.

Matched filter receiver

Assuming absence of ISI, (6.77) and (6.78) imply that the best performance, that is the minimum value of P_e, is obtained when the ratio γ is maximum.

Assuming that the pulse q_C that determines the signal at the channel output is *given*, the solution (see Section 1.8) is provided by the receive filter g_{Rc} matched to q_C: hence, the name matched filter (MF). In particular, with reference to the scheme of Figure 7.11 and for white noise w_C, we have

$$\mathcal{G}_{Rc}(f) = K Q_C^*(f)\, e^{-j2\pi f t_0} \tag{7.104}$$

---

[7] The term $\mathtt{M}_a E_{q_C}/2$ represents the energy of both $Re\,[s_C(t)]$ and $Im\,[s_C(t)]$, assuming that s_C is circularly symmetric (see (1.332)).

[8] We observe that this memoryless decision criterion is optimum only if the noise samples $\{w_{R,k}\}$ are statistically independent.

where K is a constant. In this case, from the condition

$$h_0 = \mathcal{F}^{-1}[\mathcal{G}_{Rc}(f)\mathcal{Q}_C(f)]|_{t=t_0} = K\, \mathtt{r}_{q_C}(0) = 1 \tag{7.105}$$

we obtain

$$K = \frac{1}{E_{q_C}} \tag{7.106}$$

Substitution of (7.106) in (7.104) yields $E_{g_{Rc}} = 1/E_{q_C}$. Therefore, (7.103) assumes the form

$$\gamma = \gamma_{MF} = \frac{2E_{q_C}}{N_0} \tag{7.107}$$

The matched filter receiver is of interest also for another reason. Using (7.97) and (7.100), it is possible to determine the relations (6.79) and (6.80) between the SNRs γ at the decision point and Γ at the receiver input.

We stress that, for a given modulation with pulse q_C and Γ, it is not possible by varying the filter g_{Rc} to obtain a higher γ at the decision point than (6.80), and consequently a lower P_e. The equation (6.80) is often used as an upper bound of the system performance. However, we note that we have ignored the possible presence of ISI at the decision point that the choice of (7.104) might imply.

7.4 Channel equalization

We analyse methods which approximate the ISI-free condition (7.72) at detection point.

7.4.1 Zero-forcing equalizer

From (7.59), assuming that $\mathcal{H}_{Tx}(f)$ and $\mathcal{G}_C(f)$ are known, and imposing $\mathcal{H}(f)$ as in (7.77), then the equation

$$\mathcal{H}(f) = \mathcal{Q}_R(f)e^{j2\pi f t_0} = \mathcal{H}_{Tx}(f)\mathcal{G}_C(f)\mathcal{G}_{Rc}(f)e^{j2\pi f t_0} \tag{7.108}$$

can be solved with respect to the receive filter, yielding

$$\mathcal{G}_{Rc}(f) = \frac{T\,\mathrm{rcos}\left(\frac{f}{1/T}, \rho\right)}{\mathcal{H}_{Tx}(f)\mathcal{G}_C(f)} e^{-j2\pi f t_0} \tag{7.109}$$

From (7.109), the magnitude and phase responses of $\mathcal{G}_{Rc}$ can be obtained. In practice, however, although condition (7.109) leads to the suppression of the ISI, hence, the filter g_{Rc} is called *linear equalizer zero-forcing* (LE-ZF)), it may also lead to the enhancement of the noise power at the decision point, as expressed by (7.68).

In fact, if the frequency response $\mathcal{G}_C(f)$ exhibits strong attenuation at certain frequencies in the range $[-(1+\rho)/(2T), (1+\rho)/(2T)]$, then $\mathcal{G}_{Rc}(f)$ presents peaks that determine a large value of $\sigma^2_{w_R}$. In any event, the choice (7.109) guarantees the absence of ISI at the decision point, and from (7.103), we get

$$\gamma_{LE-ZF} = \frac{2}{N_0 E_{g_{Rc}}} \tag{7.110}$$

Obviously, based on the considerations of Section 7.1.3, it is

$$\gamma_{LE-ZF} \le \gamma_{MF} \tag{7.111}$$

where γ_{MF} is defined in (7.107).

In the particular case of an ideal NB channel, with $\mathcal{G}_{Ch}(f) = \mathcal{G}_0$ in the passband of the system, and assuming h_{Tx}

$$\mathcal{H}_{Tx}(f) = T\sqrt{\text{rcos}\left(\frac{f}{1/T}, \rho\right)} \tag{7.112}$$

from (7.39)

$$\mathcal{Q}_C(f) = \mathcal{H}_{Tx}(f)\ \mathcal{G}_C(f) = k_1\mathcal{H}_{Tx}(f) \tag{7.113}$$

where from (7.35), $k_1 = \mathcal{G}_0$ for a PAM system, whereas $k_1 = (\mathcal{G}_0/\sqrt{2})e^{-j(\varphi_1-\varphi_0)}$ for a QAM system. Moreover, from (7.109), neglecting a constant delay, i.e. for $t_0 = 0$, it results that

$$\mathcal{G}_{Rc}(f) = \frac{1}{k_1}\sqrt{\text{rcos}\left(\frac{f}{1/T}, \rho\right)} \tag{7.114}$$

In other words, g_{Rc} is matched to $q_C(t) = k_1 h_{Tx}(t)$, and

$$\gamma_{LE-ZF} = \gamma_{MF} \tag{7.115}$$

Methods for the design of a LE-ZF filter with a finite number of coefficients are given in Section 7.4.6.

7.4.2 Linear equalizer

We introduce an optimization criterion for $\mathcal{G}_{Rc}$ that takes into account the ISI as well as the statistical power of the noise at the decision point.

Optimum receiver in the presence of noise and ISI

With reference to the scheme of Figure 7.11 for a QAM system, the criterion of choosing the receive filter such that the signal y_k is *as close as possible to a_k in the mean-square sense* is widely used.[9]

Let h_{Tx} and g_C be known. Defining the error

$$e_k = a_k - y_k \tag{7.116}$$

the receive filter g_{Rc} is chosen such that the mean-square error

$$J = E[|e_k|^2] = E[|a_k - y_k|^2] \tag{7.117}$$

is minimized.

The following assumptions are made:

1. the sequence $\{a_k\}$ is wide sense stationary (WSS) with spectral density $\mathcal{P}_a(f)$;
2. the noise w_C is complex-valued and WSS. In particular, we assume it is white with PSD $\mathcal{P}_{w_C}(f) = N_0$;
3. the sequence $\{a_k\}$ and the noise w_C are statistically independent.

The minimization of J in this situation differs from the classical problem of the optimum Wiener filter because h_{Tx} and g_C are continuous-time pulses. By resorting to the calculus of variations [4], we obtain the general solution

$$\mathcal{G}_{Rc}(f) = \frac{\mathcal{Q}_C^*(f)}{N_0}e^{-j2\pi f t_0}\frac{\mathcal{P}_a(f)}{T + \mathcal{P}_a(f)\frac{1}{T}\sum_{\ell=-\infty}^{+\infty}\frac{1}{N_0}\left|\mathcal{Q}_C\left(f - \frac{\ell}{T}\right)\right|^2} \tag{7.118}$$

where $\mathcal{Q}_C(f) = \mathcal{H}_{Tx}(f)\mathcal{G}_C(f)$.

9 It would be desirable to find the filter such that $P[\hat{a}_k \neq a_k]$ is minimum. This problem, however, is usually very difficult to solve. Therefore, we resort to the criterion of minimizing $E[|y_k - a_k|^2]$ instead.

If the symbols are *statistically independent* and have *zero mean*, then $\mathcal{P}_a(f) = T\sigma_a^2$, and (7.118) becomes:

$$\mathcal{G}_{Rc}(f) = \mathcal{Q}_C^*(f)e^{-j2\pi f t_0} \frac{\sigma_a^2}{N_0 + \sigma_a^2 \frac{1}{T} \sum_{\ell=-\infty}^{+\infty} \left|\mathcal{Q}_C\left(f - \frac{\ell}{T}\right)\right|^2} \tag{7.119}$$

The expression of the cost function J in correspondence of the optimum filter (7.119) is given in (7.147).

From the decomposition (7.55) of $\mathcal{G}_{Rc}(f)$, in (7.119), we have the following correspondences:

$$\mathcal{G}_M(f) = \mathcal{Q}_C^*(f)e^{-j2\pi f t_0} \tag{7.120}$$

and

$$C(e^{j2\pi fT}) = \frac{\sigma_a^2}{N_0 + \sigma_a^2 \frac{1}{T} \sum_{\ell=-\infty}^{+\infty} \left|\mathcal{Q}_C\left(f - \frac{\ell}{T}\right)\right|^2} \tag{7.121}$$

The optimum receiver thus assumes the structure of Figure 7.22. We note that g_M is the matched filter to the impulse response of the QAM system at the receiver input.[10]

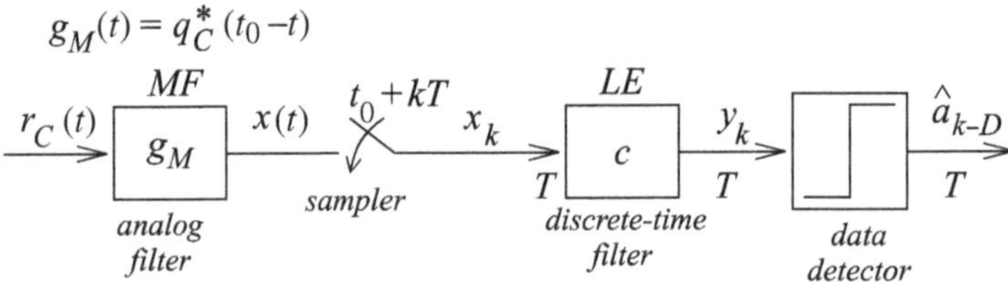

Figure 7.22 Optimum receiver structure for a channel with additive white noise.

The filter c is called *linear equalizer* (LE). It attempts to find the optimum trade-off between removing the ISI and enhancing the noise at the decision point.

We analyse two particular cases of the solution (7.119).

1. In the absence of noise, $w_C(t) \simeq 0$, and

$$C(e^{j2\pi fT}) = \frac{1}{\frac{1}{T} \sum_{\ell=-\infty}^{+\infty} \left|\mathcal{Q}_C\left(f - \frac{\ell}{T}\right)\right|^2} \tag{7.122}$$

   Note that the system is perfectly equalized, i.e. there is no ISI. In this case, the filter (7.122) is the *LE-ZF*, as it completely eliminates the ISI.
2. In the absence of ISI at the ouput of g_M, that is if $|\mathcal{Q}_C(f)|^2$ is a Nyquist pulse, then $C(e^{j2\pi fT})$ is constant and the equalizer can be removed.

Alternative derivation of the IIR equalizer

Starting from the receiver of Figure 7.22 and for *any type of filter* g_M, not necessarily matched, it is possible to determine the coefficients of the finite-impulse-response (FIR) equalizer filter c using the Wiener formulation, with the following definitions:

- filter input signal, x_k;
- filter output signal, y_k;

[10] As derived later in the text (see Observation 7.14 on page 357), the output signal of the matched filter, sampled at the modulation rate $1/T$, forms a *sufficient statistic* if all the channel parameters are known.

- desired output signal, $d_k = a_{k-D}$;
- estimation error, $e_k = d_k - y_k$.

Parameter D that denotes the lag of the desired signal, i.e. the delay introduced by the equalizer in numbers of symbol intervals. The overall delay from the emission of a_k to the generation of the detected symbol $\hat{a}_k$ is equal to $t_0 + DT$.

However, the particular case of a matched filter, for which $g_M(t) = q_C^*(-(t - t_0))$, is very interesting from a theoretical point of view. We assume that the filter c may have an infinite number of coefficients, i.e. it may be with infinite impulse response (IIR). With reference to the scheme of Figure 7.23a, q is the overall impulse response of the system at the sampler input:

$$q(t) = h_{Tx} * g_C * g_M(t) = q_C * g_M(t) = \mathtt{r}_{q_C}(t - t_0) \tag{7.123}$$

where the autocorrelation of the deterministic pulse q_C is

$$\mathtt{r}_{q_C}(t) = [q_C(t') * q_C^*(-t')](t) \tag{7.124}$$

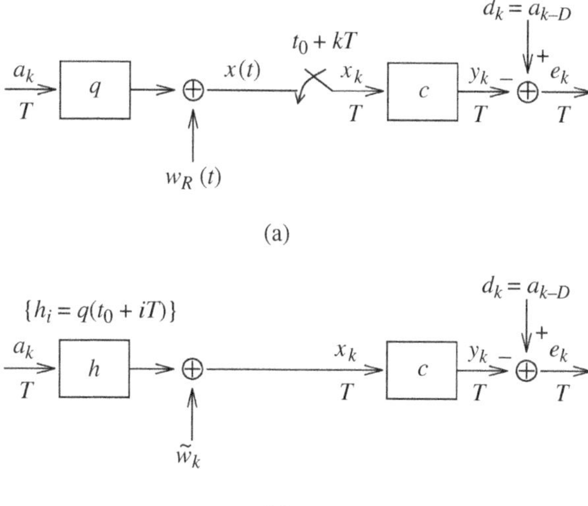

Figure 7.23 Linear equalizer as a Wiener filter. (a) Continuous-time model. (b) Discrete-time equivalent.

The Fourier transform of $\mathtt{r}_{q_C}(t)$ is given by

$$\mathcal{P}_{q_C}(f) = |\mathcal{Q}_C(f)|^2 \tag{7.125}$$

We note that if q_C has a finite support $(0, t_{q_C})$, then $g_M(t) = q_C^*(t_0 - t)$ has support $(t_0 - t_{q_C}, t_0)$. Hence, to obtain a causal filter g_M the minimum value of t_0 is t_{q_C}. In any case, from (7.123), as $\mathtt{r}_{q_C}$ is a correlation function, the desired sample $q(t_0)$ is taken in correspondence of the maximum value of $|q(t)|$.

Assuming w_C is white, we have (see Figure 7.23a)

$$w_R(t) = w_C * g_M(t) \tag{7.126}$$

with autocorrelation function given by

$$\begin{aligned} \mathtt{r}_{w_R}(\tau) &= \mathtt{r}_{w_C} * \mathtt{r}_{q_C}(\tau) \\ &= N_0 \mathtt{r}_{q_C}(\tau) \end{aligned} \tag{7.127}$$

and PSD

$$\mathcal{P}_{w_R}(f) = N_0 \mathcal{P}_{q_C}(f) = N_0 |\mathcal{Q}_C(f)|^2 \tag{7.128}$$

In Figure 7.23a, sampling at instants $t_k = t_0 + kT$ yields the sampled QAM signal

$$x_k = \sum_{i=-\infty}^{+\infty} a_i h_{k-i} + \tilde{w}_k \tag{7.129}$$

The discrete-time equivalent model is illustrated in Figure 7.23b. The discrete-time overall impulse response is given by

$$h_n = q(t_0 + nT) = \mathtt{r}_{q_C}(nT) \tag{7.130}$$

In particular, it results

$$h_0 = \mathtt{r}_{q_C}(0) = E_{q_C} \tag{7.131}$$

The sequence $\{h_n\}$ has z-transform

$$\Phi(z) = \mathcal{Z}[h_n] = P_{q_C}(z) \tag{7.132}$$

which, by the Hermitian symmetry of an autocorrelation sequence, $\mathtt{r}_{q_C}(nT) = \mathtt{r}^*_{q_C}(-nT)$, satisfies the relation:

$$\Phi(z) = \Phi^*\left(\frac{1}{z^*}\right) \tag{7.133}$$

On the other hand, from (1.23), the Fourier transform of (7.130) is given by

$$\Phi(e^{j2\pi fT}) = \mathcal{F}[h_n] = \frac{1}{T} \sum_{\ell=-\infty}^{+\infty} \left| \mathcal{Q}_C\left(f - \frac{\ell}{T}\right)\right|^2 \tag{7.134}$$

Moreover, the correlation sequence of $\{h_n\}$ has z-transform equal to

$$\mathcal{Z}[\mathtt{r}_h(m)] = \Phi(z)\Phi^*\left(\frac{1}{z^*}\right) \tag{7.135}$$

Also, from (7.127), the z-transform of the autocorrelation of the noise samples $\tilde{w}_k = w_R(t_0 + kT)$ is given by

$$\mathcal{Z}[\mathtt{r}_{\tilde{w}}(n)] = \mathcal{Z}[\mathtt{r}_{w_R}(nT)] = N_0 \Phi(z) \tag{7.136}$$

The Wiener solution that gives the optimum coefficients is given in the z-transform domain by (see (2.259))

$$C_{opt}(z) = \mathcal{Z}[c_n] = \frac{P_{dx}(z)}{P_x(z)} \tag{7.137}$$

where

$$P_{dx}(z) = \mathcal{Z}[\mathtt{r}_{dx}(n)] \quad \text{and} \quad P_x(z) = \mathcal{Z}[\mathtt{r}_x(n)] \tag{7.138}$$

We assume

1. The sequence $\{a_k\}$ is WSS, with symbols that are statistically independent and with zero mean,

$$\mathtt{r}_a(n) = \sigma_a^2 \delta_n \quad \text{and} \quad \mathcal{P}_a(f) = T\sigma_a^2 \tag{7.139}$$

2. $\{a_k\}$ and $\{\tilde{w}_k\}$ are statistically independent and hence uncorrelated.

Using assumption 2, the crosscorrelation between $\{d_k\}$ and $\{x_k\}$ is given by

$$\begin{aligned}
\mathtt{r}_{dx}(n) &= E[d_k x_{k-n}^*] \\
&= E\left[a_{k-D}\left(\sum_{i=-\infty}^{+\infty} a_i h_{k-n-i} + \tilde{w}_{k-n}\right)^*\right] \\
&= \sum_{i=-\infty}^{+\infty} h_{k-n-i}^* E[a_{k-D} a_i^*]
\end{aligned} \tag{7.140}$$

Lastly, from assumption 1,

$$\mathtt{r}_{dx}(n) = \sigma_a^2 h_{D-n}^* \tag{7.141}$$

Under the same assumptions 1 and 2, the computation of the autocorrelation of the process $\{x_k\}$ yields

$$\mathtt{r}_x(n) = E[x_k x_{k-n}^*] = \sigma_a^2 \mathtt{r}_h(n) + \mathtt{r}_{\tilde{w}}(n) \tag{7.142}$$

Thus, using (7.135), we obtain

$$\begin{aligned}
P_{dx}(z) &= \sigma_a^2 \Phi^*\left(\frac{1}{z^*}\right) z^{-D} \\
P_x(z) &= \sigma_a^2 \Phi(z) \Phi^*\left(\frac{1}{z^*}\right) + N_0 \Phi(z)
\end{aligned} \tag{7.143}$$

Therefore, from (7.137),

$$C_{opt}(z) = \frac{\sigma_a^2 \Phi^*\left(\frac{1}{z^*}\right) z^{-D}}{\sigma_a^2 \Phi(z) \Phi^*\left(\frac{1}{z^*}\right) + N_0 \Phi(z)} \tag{7.144}$$

Taking into account the property (7.133), (7.144) is simplified as

$$C_{opt}(z) = \frac{\sigma_a^2 z^{-D}}{N_0 + \sigma_a^2 \Phi(z)} \tag{7.145}$$

It can be observed that, for $z = e^{j2\pi fT}$, (7.145) corresponds to (7.121), apart from the term z^{-D}, which accounts for a possible delay introduced by the equalizer.

In correspondence of the *optimum filter* $C_{opt}(z)$, we determine the minimum value of the cost function. We recall the general expression for the Wiener filter (2.48)

$$\begin{aligned}
J_{min} &= \sigma_d^2 - \sum_{i=0}^{N-1} c_{opt,i}\, \mathtt{r}_{dx}^*(i) \\
&= \sigma_d^2 - \int_{-\frac{1}{2T}}^{\frac{1}{2T}} \mathcal{P}_{dx}^*(f)\, \mathcal{C}_{opt}(e^{j2\pi fT})\, df
\end{aligned} \tag{7.146}$$

Finally, substitution of the relations (7.143) in (7.146) yields

$$\begin{aligned}
J_{min} &= \sigma_d^2 - T \int_{-\frac{1}{2T}}^{\frac{1}{2T}} P_{dx}^*(e^{j2\pi fT}) C_{opt}(e^{j2\pi fT}) df \\
&= \sigma_a^2 T \int_{-\frac{1}{2T}}^{\frac{1}{2T}} \frac{N_0}{N_0 + \sigma_a^2 \Phi(e^{j2\pi fT})} df
\end{aligned} \tag{7.147}$$

If $\Phi(z)$ is a rational function of z, the integral (7.147) may be computed by evaluating the coefficient of the term z^0 of the function $\sigma_a^2 N_0 / (N_0 + \sigma_a^2 \Phi(z))$, which can be obtained by series expansion of the integrand, or by using the partial fraction expansion method (see (1.65)).

We note that in the absence of ISI, at the output of the MF, we get $\Phi(z) = h_0 = E_{q_C}$, and

$$J_{min} = \frac{\sigma_a^2 N_0}{N_0 + \sigma_a^2 E_{q_C}} \tag{7.148}$$

Signal-to-noise ratio at detector

We define the overall impulse response at the equalizer output, sampled with a sampling rate equal to the modulation rate $1/T$, as

$$\psi_i = (h_n * c_{opt,n})_i \tag{7.149}$$

where $\{h_n\}$ is given by (7.130) and $c_{opt,n}$ is the impulse response of the optimum filter (7.145).

At the decision point, we have

$$y_k = \psi_D a_{k-D} + \sum_{\substack{i=-\infty \\ i \neq D}}^{+\infty} \psi_i a_{k-i} + (\tilde{w}_n * c_{opt,n})_k \tag{7.150}$$

We assume that in (7.150), the total disturbance $w_{eq,k}$, given by ISI plus noise, is modelled as Gaussian noise with variance $2\sigma_I^2$. As $e_k = a_{k-D} - y_k = a_{k-D} - \psi_D a_{k-D} - w_{eq,k}$, with $w_{eq,k}$ uncorrelated to a_{k-D}, it follows

$$2\sigma_I^2 = J_{min} - \sigma_a^2 |1 - \psi_D|^2 \tag{7.151}$$

where J_{min} is given by (7.147) or in general (7.196). Hence, for a minimum distance among symbol values of the constellation equal to 2, (6.75) yields the SNR at the detector

$$\gamma_{LE} = \left(\frac{|\psi_D|}{\sigma_I}\right)^2 = \frac{2|\psi_D|^2}{J_{min} - \sigma_a^2 |1 - \psi_D|^2} \tag{7.152}$$

In case the approximation $\psi_D \simeq 1$ holds, the total disturbance in (7.150) coincides with e_k, hence, $2\sigma_I^2 \simeq J_{min}$, and (7.152) becomes

$$\gamma_{LE} \simeq \frac{2}{J_{min}} \tag{7.153}$$

7.4.3 LE with a finite number of coefficients

In practice, if the channel is either unknown a priori or it is time variant, it is necessary to design a receiver that tries to identify the channel characteristics and at the same time equalize it through suitable adaptive algorithms.

Two alternative approaches are usually considered.

First solution. The classical block diagram of an adaptive receiver is shown in Figure 7.24.

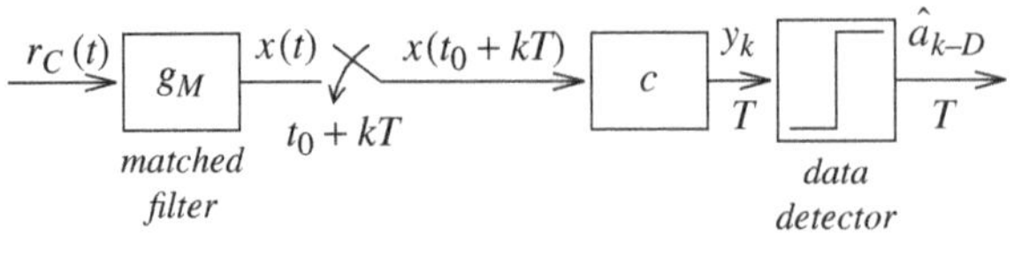

Figure 7.24 Receiver implementation by an analog matched filter followed by a sampler and a discrete-time linear equalizer.

The *matched filter* g_M is designed assuming an ideal channel. Therefore, the equalization task is left to the filter c; otherwise, if it is possible to rely on some a priori knowledge of the channel, the filter g_M

may be designed according to the average characteristics of the channel. The filter c is then an adaptive transversal filter that attempts, in real time, to equalize the channel by adapting its coefficients to the channel variations.

Second solution. The receiver is represented in Figure 7.25.

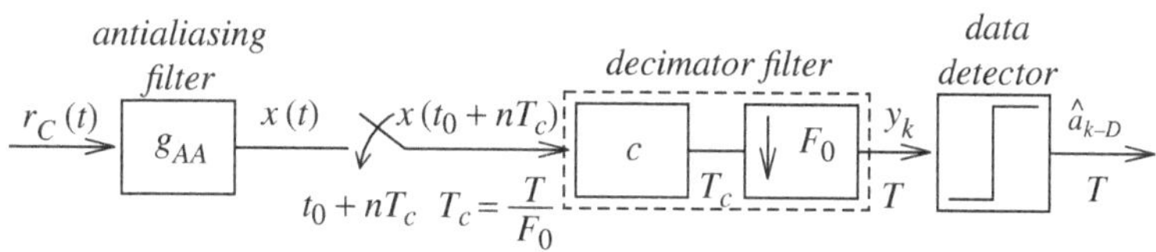

Figure 7.25 Receiver implementation by discrete-time filters.

The *anti-aliasing* filter g_{AA} is designed according to specifications imposed by the sampling theorem. In particular if the desired signal s_C has a bandwidth B and x is sampled with period $T_c = T/F_0$, where F_0 is the *oversampling index*, with $F_0 \geq 2$, then the passband of g_{AA} should extend at least up to frequency B. Moreover, because the noise w_C is considered as a wideband signal, g_{AA} should also attenuate the noise components outside the passband of the desired signal s_C, hence the cut-off frequency of g_{AA} is between B and $F_0/(2T)$. In practice, to simplify the implementation of the filter g_{AA}, it is convenient to consider a wide transition band.

Thus, the discrete-time filter c needs to accomplish the following tasks:

1. filter the residual noise outside the passband of the desired signal s_C;
2. act as a matched filter;
3. equalize the channel.

Note that the filter c of Figure 7.25 is implemented as a decimator filter (see Appendix 1.A), where the input signal $x_n = x(t_0 + nT_c)$ is defined over a discrete-time domain with period $T_c = T/F_0$, and the output signal y_k is defined over a discrete-time domain with period T.

In turn two strategies may be used to determine an equalizer filter c with N coefficients:

1. *direct method*, which employs the Wiener formulation and requires the computation of the matrix $\boldsymbol{R}$ and the vector $\boldsymbol{p}$. The description of the direct method is postponed to Section 7.4.4 (see Observation 7.3 on page 319);
2. *adaptive method*, which we will describe next (see Chapter 3).

Adaptive LE

We analyse now the solution illustrated in Figure 7.24: the discrete-time equivalent scheme is illustrated in Figure 7.26, where $\{h_n\}$ is the discrete-time impulse response of the overall system, given by

$$h_n = q(t_0 + nT) = h_{Tx} * g_C * g_M(t)|_{t=t_0+nT} \tag{7.154}$$

and

$$\tilde{w}_k = w_R(t_0 + kT) \tag{7.155}$$

with $w_R(t) = w_C * g_M(t)$.

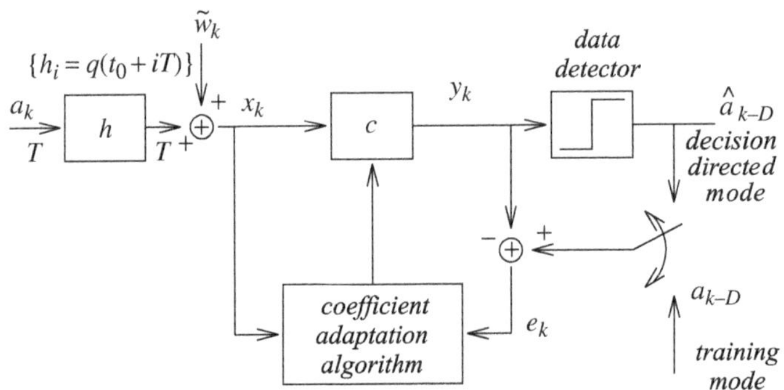

Figure 7.26 Discrete-time equivalent channel model and linear equalizer.

The design strategy consists of the following steps.

1. Define the performance measure of the system. The mean square error (MSE) criterion is typically adopted:
$$J(k) = E\,[|e_k|^2] \tag{7.156}$$
2. Select the law of coefficient update. For example, for an FIR filter c with N coefficients using the least mean square (LMS) algorithm (see Section 3.1.2), we have
$$\boldsymbol{c}_{k+1} = \boldsymbol{c}_k + \mu e_k \boldsymbol{x}_k^* \tag{7.157}$$
where
   (a) *input vector*
$$\boldsymbol{x}_k^T = [x_k, x_{k-1}, \ldots, x_{k-N+1}] \tag{7.158}$$
   (b) *coefficient vector*
$$\boldsymbol{c}_k^T = [c_{0,k}, c_{1,k}, \ldots, c_{N-1,k}] \tag{7.159}$$
   (c) *adaptation gain*
$$0 < \mu < \frac{2}{N\mathtt{r}_x(0)} \tag{7.160}$$
3. To evaluate the signal error e_k to be used in the adaptive algorithm, we distinguish two modes.
   (a) *Training mode*:
$$e_k = a_{k-D} - y_k, \qquad k = D, \ldots, L_{TS} + D - 1 \tag{7.161}$$
Evaluation of the error in training mode is possible if a sufficiently long sequence of L_{TS} symbols known at the receiver, called *training sequence* (TS), $\{a_k\}$, $k = 0, 1, \ldots, L_{TS} - 1$,
is transmitted. The duration of the transmission of TS is equal to $L_{TS}T$. During this time interval, the automatic identification of the channel characteristics takes place, allowing the computation of the optimum coefficients of the equalizer filter c, and consequently channel equalization. As the PSD of the training sequence must be wide, typically a pseudo-noise (PN) sequence is used (see Appendix 1.C). We note that even the direct method requires a training sequence to determine the vector $\boldsymbol{p}$ and the matrix $\boldsymbol{R}$ (see Observation 7.4 on page 319).
   (b) *Decision directed mode*:
$$e_k = \hat{a}_{k-D} - y_k, \qquad k \geq L_{TS} + D \tag{7.162}$$

Once the transmission of the TS is completed, we assume that the equalizer has reached convergence. Therefore, $\hat{a}_k \simeq a_k$, and the transmission of information symbols may start. In (7.161), we then substitute the known transmitted symbol with the detected symbol to obtain (7.162).

Unfortunately, in the presence of a very dispersive channel, the signal x_k is *far from being white* and the LMS algorithm will have a very slow convergence. Hence, it is seldom used.

Fractionally spaced equalizer

We consider the receiver structure with oversampling illustrated in Figure 7.25, denoted fractionally spaced equalizer (FSE). The discrete-time overall system, shown in Figure 7.27, has impulse response given by

$$h_i = q(t_0 + iT_c) \tag{7.163}$$

where

$$q(t) = h_{Tx} * g_C * g_{AA}(t) \tag{7.164}$$

The noise is given by

$$\tilde{w}_n = w_R(t_0 + nT_c) = w_C * g_{AA}(t)|_{t=t_0+nT_c} \tag{7.165}$$

Only for analysis purposes, filter c is decomposed into a discrete-time filter with sampling period T_c that is cascaded with a downsampler.

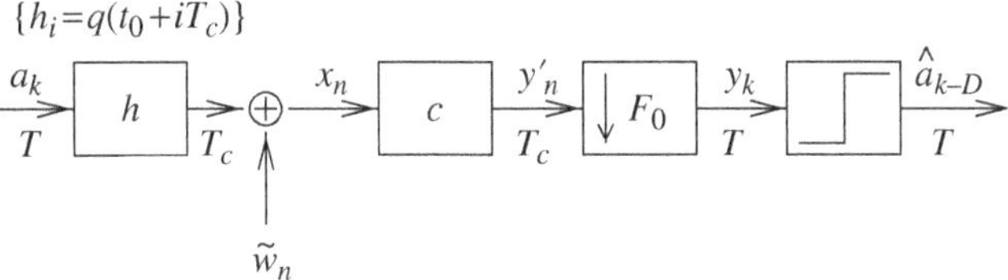

Figure 7.27 Fractionally spaced linear equalizer.

The input signal to the filter c is the sequence $\{x_n\}$ with sampling period $T_c = T/F_0$; the n-th sample of the sequence is given by

$$x_n = \sum_{k=-\infty}^{+\infty} h_{n-kF_0} a_k + \tilde{w}_n \tag{7.166}$$

The output of the filter c is given by

$$y'_n = \sum_{i=0}^{N-1} c_i x_{n-i} \tag{7.167}$$

The overall impulse response at the filter output, defined on the discrete-time domain with sampling period T_c, is given by

$$\psi_i = h * c_i \tag{7.168}$$

If we denote by $\{y'_n\}$, the sequence of samples at the filter output, and by $\{y_k\}$ the downsampled sequence, we have

$$y_k = y'_{kF_0} \tag{7.169}$$

As mentioned earlier, in a practical implementation of the filter the sequence $\{y'_n\}$ is not explicitly generated; only the sequence $\{y_k\}$ is produced (see Appendix 1.A). However, introducing the downsampler helps to illustrate the advantages of operating with an oversampling index $F_0 > 1$.

Before analysing this system, we recall the Nyquist problem. Let us consider a QAM system with pulse h:

$$s_R(t) = \sum_{n=-\infty}^{+\infty} a_n h(t - nT), \qquad t \in \mathbb{R} \tag{7.170}$$

In Section 7.1.3, we considered continuous-time Nyquist pulses $h(t)$, $t \in \mathbb{R}$. Let h be defined now on a discrete-time domain $\{nT_c\}$, n integer, where $T_c = T/F_0$. If F_0 is an integer, the discrete-time pulse satisfies the Nyquist criterion if $h(0) \neq 0$, and $h(\ell F_0 T_c) = 0$, for all integers $\ell \neq 0$. In the particular case $F_0 = 1$, we have $T_c = T$, and the Nyquist conditions impose that $h(0) \neq 0$, and $h(nT) = 0$, for $n \neq 0$. Recalling the input–output downsampler relations in the frequency domain, it is easy to deduce the behaviour of a discrete-time Nyquist pulse in the frequency domain: two examples are given in Figure 7.28, for $F_0 = 2$ and $F_0 = 1$. We note that, for $F_0 = 1$, in the frequency domain a discrete-time Nyquist pulse is equal to a constant.

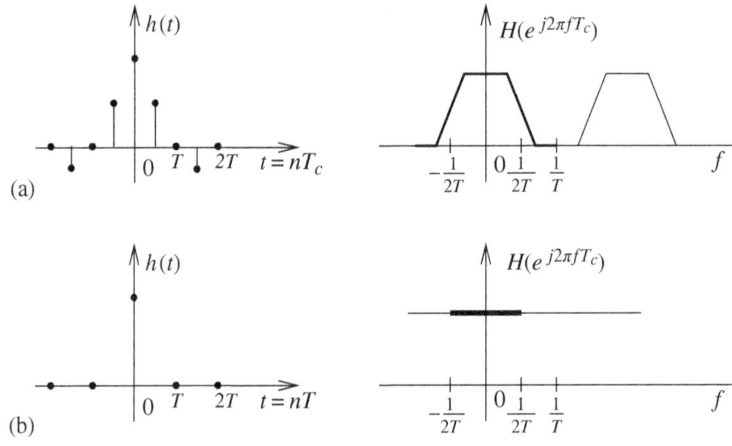

Figure 7.28 Discrete-time Nyquist pulses and relative Fourier transforms. (a) $F_0 = 2$. (b) $F_0 = 1$.

With reference to the scheme of Figure 7.27, the QAM pulse defined on the discrete-time domain with period T_c is given by (7.163), where q is defined in (7.164). From (7.168), using (7.163), the pulse $\{\psi_i\}$ at the equalizer output before the downsampler has the following Fourier transform:

$$\begin{aligned}\Psi(e^{j2\pi fT_c}) &= C(e^{j2\pi fT_c})H(e^{j2\pi fT_c}) \\ &= C(e^{j2\pi fT_c})\frac{1}{T}\sum_{\ell=-\infty}^{+\infty} \mathcal{Q}\left(f - \ell\frac{F_0}{T}\right) e^{j2\pi\left(f-\ell\frac{F_0}{T}\right)t_0}\end{aligned} \tag{7.171}$$

The task of the equalizer is to yield a pulse $\{\psi_i\}$ that approximates a Nyquist pulse, i.e. a pulse of the type shown in Figure 7.28. We see that choosing $F_0 = 1$, i.e. sampling the equalizer input signal with period equal to T, it may happen that $H(e^{j2\pi fT_c})$ assumes very small values at frequencies near $f = 1/(2T)$, because of an *incorrect choice of the timing phase* t_0. In fact, let us assume q real with a bandwidth smaller than $1/T$. Using the polar notation for $\mathcal{Q}\left(\frac{1}{2T}\right)$, we have

$$\mathcal{Q}\left(\frac{1}{2T}\right) = Ae^{j\varphi} \quad \text{and} \quad \mathcal{Q}\left(-\frac{1}{2T}\right) = Ae^{-j\varphi} \tag{7.172}$$

as $\mathcal{Q}(f)$ is Hermitian. Therefore, from (7.171),

$$\begin{aligned}H(e^{j2\pi fT})\Big|_{f=\frac{1}{2T}} &= \mathcal{Q}\left(\frac{1}{2T}\right)e^{j2\pi\frac{t_0}{2T}} + \mathcal{Q}\left(\frac{1}{2T} - \frac{1}{T}\right)e^{j2\pi\left(\frac{1}{2T}-\frac{1}{T}\right)t_0} \\ &= 2A\cos\left(\varphi + \pi\frac{t_0}{T}\right)\end{aligned} \tag{7.173}$$

If t_0 is such that

$$\varphi + \pi\frac{t_0}{T} = \frac{2i+1}{2}\pi, \qquad i \text{ integer} \tag{7.174}$$

then

$$H\left(e^{j2\pi\frac{1}{2T}T}\right) = 0 \tag{7.175}$$

In this situation, the equalizer will enhance the noise around $f = 1/(2T)$, or converge with difficulty in the adaptive case. If $F_0 \geq 2$ is chosen, this problem is avoided because *aliasing* between replicas of $Q(f)$ does not occur. Therefore, the choice of t_0 may be less accurate. In fact, as we will see in Chapter 14, if c has an input signal sampled with sampling period $T/2$ it also acts as an interpolator filter, whose output can be used to determine the optimum timing phase.

In conclusion, the FSE receiver presents two advantages over T-spaced equalizers:

1. it is an optimum structure according to the MSE criterion, in the sense that it carries out the task of both matched filter (better rejection of the noise) and equalizer (reduction of ISI);
2. it is less sensitive to the choice of t_0. In fact, the correlation method (7.602) with accuracy $T'_Q = T/2$ is usually sufficient to determine the timing phase.

The direct method to compute the coefficients of FSE is described in Section 7.4.4. Here we just mention that the design of FSE may incur a difficulty in the presence of noise with variance that is small with respect to the level of the desired signal: in this case some eigenvalues of the autocorrelation matrix of $\boldsymbol{x}^*_{2k}$ may assume a value that is almost zero and consequently, the problem of finding the optimum coefficients become ill-conditioned, with numerous solutions that present the same minimum value of the cost function. This effect can be illustrated also in the frequency domain: outside the passband of the input signal the filter c may assume arbitrary values, in the limit case of absence of noise. In general, simulations show that FSE outperform equalizers operating at the symbol rate [5].

7.4.4 Decision feedback equalizer

We consider the sampled signal at the output of the analog receive filter (see Figure 7.26 or Figure 7.27). For example, in the scheme of Figure 7.26, the desired signal is

$$s_k = s_R(t_0 + kT) = \sum_{i=-\infty}^{+\infty} a_i h_{k-i} \tag{7.176}$$

where the sampled pulse $\{h_n\}$ is defined in (7.154). In the presence of noise, we have

$$x_k = s_k + \tilde{w}_k \tag{7.177}$$

where $\tilde{w}_k$ is the noise, given by (7.155).

We assume, as illustrated in Figure 7.29, that $\{h_n\}$ has finite duration and support $[-N_1, -N_1 + 1, \dots, N_2 - 1, N_2]$. The samples with positive time indices are called *postcursors*, and those with negative time indices *precursors*. Explicitly writing terms that include precursors and postcursors, (7.177) becomes

$$x_k = h_0 a_k + (h_{-N_1} a_{k+N_1} + \cdots + h_{-1} a_{k+1}) + (h_1 a_{k-1} + \cdots + h_{N_2} a_{k-N_2}) + \tilde{w}_k \tag{7.178}$$

In addition to the *actual symbol* a_k that we desire to detect from the observation of x_k, in (7.178) two terms are identified in parentheses: one that depends only on *past symbols* $a_{k-1}, \dots, a_{k-N_2}$, and another that depends only on *future symbols*, $a_{k+1}, \dots, a_{k+N_1}$.

If the *past symbols* and the impulse response $\{h_n\}$ were perfectly known, we could use an ISI cancellation scheme limited only to postcursors. Substituting the *past symbols* with their detected versions

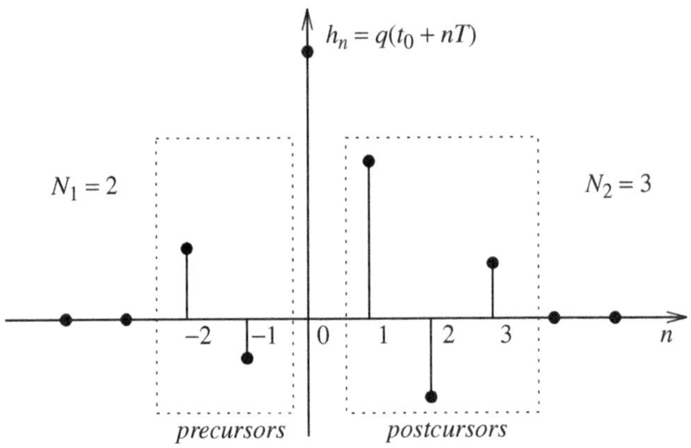

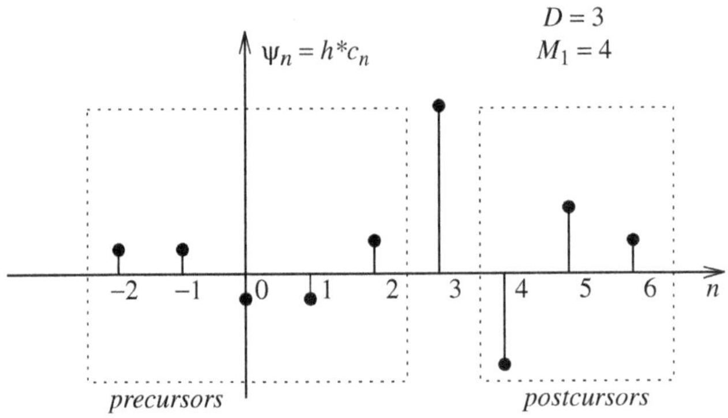

Figure 7.29 Discrete-time pulses in a DFE. (a) Before the FF filter. (b) After the FF filter.

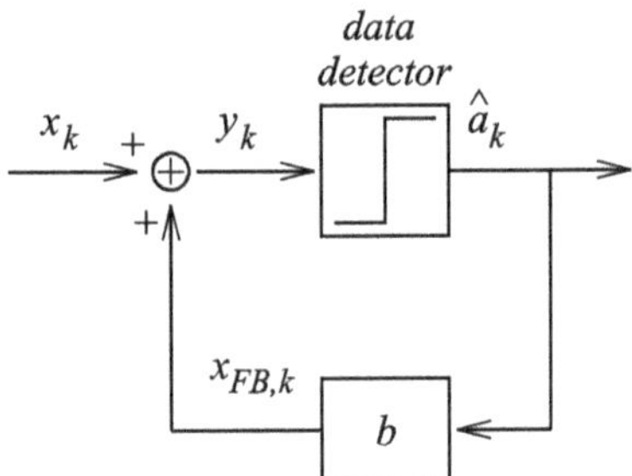

Figure 7.30 Simplified scheme of a DFE, where only the feedback filter is included.

$\{\hat{a}_{k-1}, \dots, \hat{a}_{k-N_2}\}$, we obtain a scheme to cancel in part ISI, as illustrated in Figure 7.30, where, in general, the *feedback filter* has impulse response $\{b_n\}$, $n = 1, \cdots, M_2$, and output given by

$$x_{FB,k} = b_1\hat{a}_{k-1} + \dots + b_{M_2}\hat{a}_{k-M_2} \tag{7.179}$$

If $M_2 \geq N_2$, $b_n = -h_n$, for $n = 1, \dots, N_2$, $b_n = 0$, for $n = N_2 + 1, \dots, M_2$, and $\hat{a}_{k-i} = a_{k-i}$, for $i = 1, \dots, N_2$, then the decision feedback equalizer (DFE) cancels the ISI due to postcursors. We note that this is done without changing the noise $\tilde{w}_k$ that is present in x_k.

The general structure of a DFE is shown in Figure 7.31, where two filters and the detection delay are outlined:

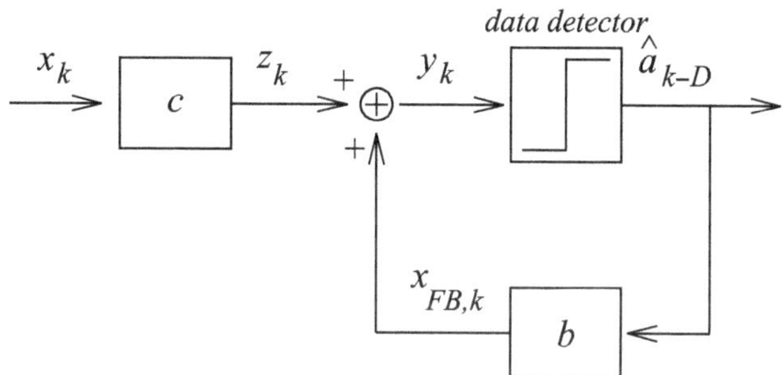

Figure 7.31 General structure of the DFE.

1. *Feedforward (FF) filter c*, with M_1 coefficients,

$$z_k = x_{\mathrm{FF},k} = \sum_{i=0}^{M_1-1} c_i x_{k-i} \tag{7.180}$$

2. *Feedback (FB) filter b*, with M_2 coefficients,

$$x_{FB,k} = \sum_{i=1}^{M_2} b_i a_{k-i-D} \tag{7.181}$$

Moreover,

$$y_k = x_{FF,k} + x_{FB,k} \tag{7.182}$$

which from (7.180) and (7.181) can also be written as

$$y_k = \sum_{i=0}^{M_1-1} c_i x_{k-i} + \sum_{j=1}^{M_2} b_j a_{k-D-j} \tag{7.183}$$

We recall that for a LE, the goal is to obtain a pulse $\{\psi_n\}$ free of ISI, with respect to the desired sample ψ_D. Now (see Figure 7.29) ideally the task of the feedforward filter is to obtain an overall impulse response $\{\psi_n = h * c_n\}$ with very small precursors and a transfer function $\mathcal{Z}[\psi_n]$ that is minimum phase (see Example 1.2.4). In this manner, almost all the ISI is cancelled by the FB filter. We note that the FF filter may be implemented as an FSE, whereas the feedback filter operates with sampling period equal to T.

The choice of the various parameters depends on $\{h_n\}$. The following guidelines, however, are usually observed.

1. $M_1 T/F_0$ (time-span of the FF filter) at least equal to $(N_1 + N_2 + 1)T/F_0$ (time-span of h), so the FF filter can effectively equalize;
2. $M_2 T$ (time-span of the FB filter) equal to or less than $(M_1 - 1)T/F_0$ (time-span of the FF filter minus one); M_2 depends also on the delay D, which determines the number of postcursors.
3. For very dispersive channels, for which we have $N_1 + N_2 \gg M_1$, it results $M_2 T \gg \frac{M_1 T}{F_0}$.
4. The choice of DT, equal to the detection delay, is obtained by initially choosing a large delay, $D \geq (M_1 - 1)/F_0$, to simplify the FB filter. If the precursors are not negligible, to reduce the constraints on the coefficients of the FF filter the value of D is lowered and the system is iteratively designed. In practice, DT is equal to or smaller than $(M_1 - 1)T/F_0$.
   For a LE, instead, DT is approximately equal to $\frac{N-1}{2}\frac{T}{F_0}$; the criterion is that the centre of gravity of the coefficients of the filter c is approximately equal to $(N - 1)/2$.

The detection delays discussed above are referred to a pulse $\{h_n\}$ *centered* in the origin, that does not introduce any delay.

Design of a DFE with a finite number of coefficients

If the channel impulse response $\{h_i\}$ and the autocorrelation function of the noise $\mathtt{r}_{\tilde{w}}(n)$ are known, for the MSE criterion with

$$J = E[|a_{k-D} - y_k|^2] \tag{7.184}$$

the Wiener filter theory may be applied to determine the optimum coefficients of the DFE filter in the case $\hat{a}_k = a_k$, with the usual assumptions of symbols that are i.d.d. and statistically independent of the noise.

For a generic sequence $\{h_i\}$ in (7.176), we recall the following results:

1. *cross correlation* between a_k and x_k

$$\mathtt{r}_{ax}(n) = \sigma_a^2 h_{-n}^* \tag{7.185}$$

2. *auto correlation* of x_k

$$\mathtt{r}_x(n) = \sigma_a^2 \mathtt{r}_h(n) + \mathtt{r}_{\tilde{w}}(n) \tag{7.186}$$

where

$$\mathtt{r}_h(n) = \sum_{j=-N_1}^{N_2} h_j h_{j-n}^*, \qquad \mathtt{r}_{\tilde{w}}(n) = N_0 \, \mathtt{r}_{g_M}(nT) \tag{7.187}$$

Defining

$$\psi_p = h * c_p = \sum_{\ell=0}^{M_1-1} c_\ell h_{p-\ell} \tag{7.188}$$

Equation (7.183) becomes

$$y_k = \sum_{p=-N_1}^{N_2+M_1-1} \psi_p a_{k-p} + \sum_{i=0}^{M_1-1} c_i \tilde{w}_{k-i} + \sum_{j=1}^{M_2} b_j a_{k-D-j} \tag{7.189}$$

From (7.189), the optimum choice of the feedback filter coefficients is given by

$$b_i = -\psi_{i+D}, \qquad i = 1, \dots, M_2 \tag{7.190}$$

Substitution of (7.190) in (7.183) yields

$$y_k = \sum_{i=0}^{M_1-1} c_i \left(x_{k-i} - \sum_{j=1}^{M_2} h_{j+D-i} a_{k-j-D} \right) \tag{7.191}$$

To obtain the Wiener–Hopf solution, the following correlations are needed:

$$\begin{aligned}
[\boldsymbol{p}]_p &= E\left[a_{k-D} \left(x_{k-p} - \sum_{j=1}^{M_2} h_{j+D-p} a_{k-D-j} \right)^* \right] = \sigma_a^2 h_{D-p}^*, \\
&\qquad p = 0, 1, \dots, M_1 - 1 \\
[\boldsymbol{R}]_{p,q} &= E\left[\left(x_{k-q} - \sum_{j_1=1}^{M_2} h_{j_1+D-q} a_{k-D-j_1} \right) \right. \\
&\qquad \left. \left(x_{k-p} - \sum_{j_2=1}^{M_2} h_{j_2+D-p} a_{k-D-j_2} \right)^* \right]
\end{aligned} \tag{7.192}$$

$$= \sigma_a^2 \left(\sum_{j=-N_1}^{N_2} h_j h_{j-(p-q)}^* - \sum_{j=1}^{M_2} h_{j+D-q} h_{j+D-p}^* \right) + \mathtt{r}_{\tilde{w}}(p-q),$$
$$p, q = 0, 1, \dots, M_1 - 1 \tag{7.193}$$

Therefore, the optimum FF filter coefficients are given by

$$\boldsymbol{c}_{opt} = \boldsymbol{R}^{-1}\boldsymbol{p} \tag{7.194}$$

and, from (7.190), the optimum feedback filter coefficients are given by

$$b_i = -\sum_{\ell=0}^{M_1-1} c_{opt,\ell} h_{i+D-\ell} \qquad i = 1, 2, \dots, M_2 \tag{7.195}$$

Moreover, using (7.192), we get

$$\begin{aligned} J_{min} &= \sigma_a^2 - \sum_{\ell=0}^{M_1-1} c_{opt,\ell} [\boldsymbol{p}]_\ell^* \\ &= \sigma_a^2 \left(1 - \sum_{\ell=0}^{M_1-1} c_{opt,\ell} h_{D-\ell} \right) \end{aligned} \tag{7.196}$$

Observation 7.2

In the particular case, wherein all the postcursors are cancelled by the feedback filter, that is for

$$M_2 + D = N_2 + M_1 - 1 \tag{7.197}$$

(7.193) is simplified into

$$[\boldsymbol{R}]_{p,q} = \sigma_a^2 \sum_{j=-N_1}^{D} h_{j-q} h_{j-p}^* + \mathtt{r}_{\tilde{w}}(p-q) \tag{7.198}$$

For determining optimum values of M_1, M_2, and D, rather than following the generic guidelines of Section 7.4.4, a better approach is repeating the design method for increasing values of M_1 and M_2 (and determining the corresponding optimum value of D) until J_{min} does not significantly change, i.e. it reaches approximately its asymptotic value.

Observation 7.3

The equations to determine $\boldsymbol{c}_{opt}$ for a LE are identical to (7.192)–(7.196), with $M_2 = 0$. In particular, the vector $\boldsymbol{p}$ in (7.192) is not modified, while the expression of the elements of the matrix $\boldsymbol{R}$ in (7.193) are modified by the terms including the detected symbols.

Observation 7.4

For white noise w_C, the autocorrelation of $\tilde{w}_k$ is proportional to the autocorrelation of the receive filter impulse response: consequently, if the statistical power of $\tilde{w}_k$ is known, the autocorrelation $\mathtt{r}_{\tilde{w}}(n)$ is easily determined. Lastly, the coefficients of the channel impulse response $\{h_n\}$ and the statistical power of $\tilde{w}_k$ used in $\boldsymbol{R}$ and $\boldsymbol{p}$ can be determined by the methods given in Section 7.8.

Observation 7.5

For a LE, the matrix $\boldsymbol{R}$ is Hermitian and Toeplitz, while for a DFE, it is only Hermitian; in any case it is (semi)definite positive. Efficient methods to determine the inverse of the matrix are described in Section 2.3.2.

Observation 7.6

The definition of $\{h_n\}$ depends on the value of t_0, which is determined by methods described in Chapter 14. A particularly useful method to determine the impulse response $\{h_n\}$ in wireless systems resorts to a short training sequence to achieve fast synchronization. We recall that a fine estimate of $t_{0,MF}$ is needed if the sampling period of the signal at the MF output is equal to T. The overall discrete-time system impulse response obtained by sampling the output signal of the anti-aliasing filter g_{AA} (see Figure 7.25) is assumed to be known, e.g. by estimation. The sampling period, for example $T/8$, is in principle determined by the accuracy with which we desire to estimate the timing phase $t_{0,MF}$ at the MF output. To reduce implementation complexity, however, a larger sampling period of the signal at the MF input is considered, for example $T/2$. We then implement the MF g_M by choosing, among the four polyphase components (see Section 1.A.9 on page 92) of the impulse response, the component with the largest energy, thus realizing the MF criterion (see also (7.123)). This is equivalent to selecting the component with largest statistical power among the four possible components with sampling period $T/2$ of the sampled output signal of the filter g_{AA}. This method is similar to the timing estimator (14.117).

The timing phase $t_{0,AA}$ for the signal at the input of g_M is determined during the estimation of the channel impulse response. It is usually chosen either as the time at which the first useful sample of the overall impulse response occurs, or the time at which the peak of the impulse response occurs, shifted by a number of modulation intervals corresponding to a given number of precursors. Note that, if t_{MF} denotes the duration of g_M, then the timing phase at the output of g_M is given by $t_{0,MF} = t_{0,AA} + t_{MF}$. The criterion (7.602), according to which t_0 is chosen in correspondence of the correlation peak, is a particular case of this procedure.

Observation 7.7

In systems where the training sequence is placed at the end of a block of data (see the Global System for Mobile Communications, GSM, frame in Section 18.2), it is convenient to process the observed signal $\{x_k\}$ starting from the end of the block, let us say from $k = K - 1$ to 0, thus exploiting the knowledge of the training sequence. Now, if $\{\psi_n = h * c_n\}$ and $\{b_n\}$ are the optimum impulse responses if the signal is processed in the forward mode, i.e. for $k = 0, 1, \dots, K - 1$, it is easy to verify that $\{\psi_n^{B*}\}$ and $\{b_n^{B*}\}$, where B is the backward operator defined on page 14, are the optimum impulse responses in the backward mode for $k = K - 1, \dots, 1, 0$, apart from a constant delay. In fact, if $\{\psi_n\}$ is ideally minimum phase and causal with respect to the timing phase, now $\{\psi_n^{B*}\}$ is maximum phase and anticausal with respect to the new instant of optimum sampling. Also the FB filter will result anticausal. In the particular case $\{h_n\}$ is a correlation sequence, then $\{\psi_n^{B*}\}$ can be obtained using as FF filter the filter having impulse response $\{c_n^{B*}\}$.

Design of a fractionally spaced DFE

We briefly describe the equations to determine the coefficients of a DFE comprising a fractionally spaced FF filter with M_1 coefficients and sampling period of the input signal equal to $T/2$, and an FB filter with M_2 coefficients; the extension to an FSE with an oversampling factor $F_0 > 2$ is straightforward. The resulting structure is denoted as fractionally spaced DFE (FS-DFE).

We consider the scheme of Figure 7.31, where the FF filter c is now a fractionally spaced filter, as illustrated in Figure 7.27. The overall receiver structure is shown in Figure 7.32, where the filter g_{AA} may be more sophisticated than a simple anti-aliasing filter and partly perform the function of the MF. Otherwise, the function of the MF may be performed by a discrete-time filter placed in front of the filter c. We recall that the MF, besides reducing the complexity of c (see task 2 on page 311), facilitates the optimum choice of t_0 (see Chapter 14).

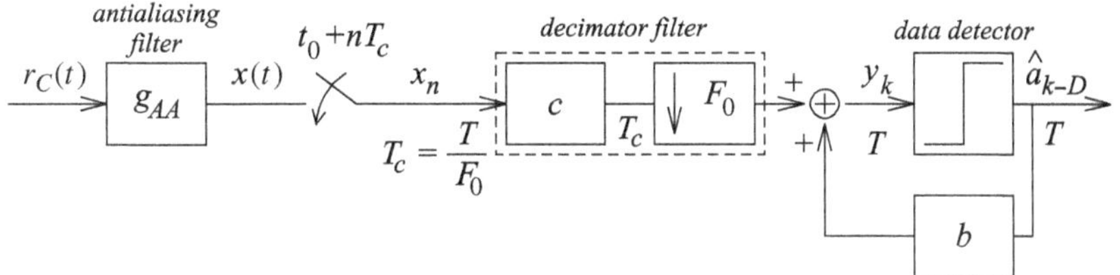

Figure 7.32 FS-DFE structure.

As usual, the symbols are assumed i.i.d. and statistically independent of the noise signal. If $\{x_n\}$ is the input signal of the FF filter, we have

$$x_n = \sum_{k=-\infty}^{+\infty} h_{n-2k} a_k + \tilde{w}_n \tag{7.199}$$

The signal $\{y_k\}$ at the DFE output is given by

$$y_k = \sum_{i=0}^{M_1-1} c_i x_{2k-i} + \sum_{j=1}^{M_2} b_j a_{k-D-j} \tag{7.200}$$

Let

$$\psi_p = h * c_p = \sum_{\ell=0}^{M_1-1} c_\ell h_{p-\ell} \tag{7.201}$$

then the optimum choice of the coefficients $\{b_i\}$ is given by

$$b_i = -\psi_{2(i+D)} = -\sum_{\ell=0}^{M_1-1} c_\ell h_{2(i+D)-\ell}, \qquad i = 1, \dots, M_2 \tag{7.202}$$

With this choice (7.200) becomes

$$y_k = \sum_{i=0}^{M_1-1} c_i \left(x_{2k-i} - \sum_{j=1}^{M_2} h_{2(j+D)-i} a_{k-(j+D)} \right) \tag{7.203}$$

Using the following relations (see also Example 1.7.10 on page 51)

$$E\,[a_{k-K}\, x_{2k-i}^*] = \sigma_a^2\, h_{2K-i}^* \tag{7.204}$$

$$E\,[x_{2k-q}\, x_{2k-p}^*] = \sigma_a^2 \sum_{n=-\infty}^{+\infty} h_{2n-q}\, h_{2n-p}^* + \mathtt{r}_{\tilde{w}}(p-q) \tag{7.205}$$

where, from (7.165),

$$\mathtt{r}_{\tilde{w}}(m) = N_0\, \mathtt{r}_{g_{AA}}\left(m\, \frac{T}{2}\right) \tag{7.206}$$

the components of the vector $\boldsymbol{p}$ and the matrix $\boldsymbol{R}$ of the Wiener problem associated with (7.203) are given by

$$[\boldsymbol{p}]_p = \sigma_a^2 h_{2D-p}^*, \qquad p = 0, 1, \dots, M_1 - 1 \tag{7.207}$$

$$[\boldsymbol{R}]_{p,q} = \sigma_a^2 \left(\sum_{n=-\infty}^{+\infty} h_{2n-q} h_{2n-p}^* - \sum_{j=1}^{M_2} h_{2(j+D)-q} h_{2(j+D)-p}^* \right) + \mathtt{r}_{\tilde{w}}(p-q),$$
$$p, q = 0, 1, \dots, M_1 - 1 \tag{7.208}$$

The feedforward filter is obtained by solving the system of equations

$$\boldsymbol{R}\,\boldsymbol{c}_{opt} = \boldsymbol{p} \tag{7.209}$$

and the feedback filter is determined from (7.202). The minimum value of the cost function is given by

$$\begin{aligned} J_{min} &= \sigma_a^2 - \sum_{\ell=0}^{M_1-1} c_{opt,\ell}[\boldsymbol{p}]_\ell^* \\ &= \sigma_a^2 \left(1 - \sum_{\ell=0}^{M_1-1} c_{opt,\ell} h_{2D-\ell}\right) \end{aligned} \tag{7.210}$$

A problem encountered with this method is the inversion of the matrix $\boldsymbol{R}$ in (7.209), because it may be ill-conditioned. Similarly to the procedure outlined on page 164, a solution consists in adding a positive constant to the elements on the diagonal of R, so that $\boldsymbol{R}$ becomes invertible; obviously, the value of this constant must be rather small, so that the performance of the optimum solution does not change significantly.

Observation 7.8

Observations 7.2–7.4 on page 319 hold also for a FS-DFE, with appropriate changes. In this case, the timing phase t_0 after the filter g_{AA} can be determined with accuracy $T/2$, for example by the correlation method (7.603). For an FSE, or FS-LE, the equations to determine $\boldsymbol{c}_{opt}$ are given by (7.207)–(7.209) with $M_2 = 0$. Note that the matrix $\boldsymbol{R}$ is Hermitian, but in general, it is no longer Toeplitz.

Signal-to-noise ratio at the decision point

Using FF and FB filters with an infinite number of coefficients, it is possible to achieve the minimum value of J_{min}. Salz derived the expression of J_{min} for this case, given by [6]

$$J_{min} = \sigma_a^2 \exp\left(T \int_{-\frac{1}{2T}}^{\frac{1}{2T}} \ln \frac{N_0}{N_0 + \sigma_a^2 \Phi(e^{j2\pi fT})} df\right) \tag{7.211}$$

where Φ is defined in (7.134).

Applying the Jensen's inequality,

$$e^{\int f(a) da} \leq \int e^{f(a)} da \tag{7.212}$$

to (7.211), we can compare the performance of a linear equalizer given by (7.147) with that of a DFE given by (7.211): the result is that, assuming $\Phi(e^{j2\pi fT})$ not constant and the absence of detection errors in the DFE, for infinite-order filters the value J_{min} of a DFE is always smaller than J_{min} of a LE.

If FF and FB filters with a finite number of coefficients are employed, in analogy with (7.152), also for a DFE, we have

$$\gamma_{DFE} = \frac{2|\psi_D|^2}{J_{min} - \sigma_a^2 |1 - \psi_D|^2} \tag{7.213}$$

where J_{min} is given by (7.196). An analogous relation holds for an FS-DFE, with J_{min} given by (7.210).

Remarks

1. In the absence of errors of the data detector, the DFE has better asymptotic ($M_1, M_2 \to \infty$) performance than the linear equalizer. However, for a given finite number of coefficients $M_1 + M_2$, the performance is a function of the channel and of the choice of M_1.

2. The DFE is definitely superior to the linear equalizer for channels that exhibit large variations of the attenuation in the passband, as in such cases the linear equalizer tends to enhance the noise.
3. Detection errors tend to propagate, because they produce *incorrect cancellations*. Error propagation leads to an increase of the error probability. However, simulations indicate that for typical channels and symbol error probability smaller than 5×10^{-2}, error propagation is not catastrophic.
4. For channels with impulse response $\{h_i\}$, such that detection errors may propagate catastrophically, instead of the DFE structure, it is better to implement the linear FF equalizer at the receiver, and the FB filter at the transmitter as a precoder, using the *precoding* method discussed in Appendix 7.C and Chapter 13.

7.4.5 Frequency domain equalization

We consider now an equalization method, which is attractive for its low complexity for both determining the filter coefficients and performing the filtering operations.

DFE with data frame using a unique word

In this section, a DFE with the FF filter operating in the frequency domain (FD) is given [7, 8]. The first step concerns the format and size of signals that must be mapped in the FD by the discrete Fourier transform (DFT) (1.28). With reference to Figure 7.23, we take as new timing phase at the receiver the instant t_0^0, corresponding to the first *significant* sample of the equivalent channel impulse response, and define the new discrete time channel impulse response as h_i^0. It is

$$t_0^0 = t_0 - N_1 \quad \text{and} \quad h_i^0 = h_{i-N_1}, \quad i = 0, 1, \dots, N_h - 1 \tag{7.214}$$

with $N_h = N_1 + N_2 + 1$. Consequently,

$$x_k = x(t_0^0 + kT), \quad k \geq 0 \tag{7.215}$$

Next, we select a suitable information block size $\mathcal{M}$. Then, we collect $P = \mathcal{M} + N_{uw}$, with

$$N_{uw} \geq N_h - 1 \tag{7.216}$$

samples of x_k. In the absence of noise, from (7.176)

$$x_k = s_k = \sum_{i=0}^{N_h-1} h_i^0 a_{k-i}^{(aug)}, \quad k = N_{uw}, N_{uw} + 1, \dots, N_{uw} + P - 1 \tag{7.217}$$

where $\{a_k^{(aug)}\}$ is the augmented transmitted data sequence. Convolution (7.217) is *circular* on a block size P if (see (1.46))

$$a_k^{(aug)} = a_{k+P}^{(aug)} \quad \text{for} \quad k = 0, 1, \dots, N_{uw} - 1 \tag{7.218}$$

Defined the following DFTs, all of size P,

$$\mathcal{H}_p = \text{DFT}\{h_i^0\} = \sum_{i=0}^{P-1} h_i^0 e^{-j2\pi \frac{ip}{P}} \tag{7.219}$$

$$\mathcal{A}_p^{(aug)} = \text{DFT}\{a_i^{(aug)}, i = N_{uw}, N_{uw} + 1, \dots, N_{uw} + P - 1\} \tag{7.220}$$

$$\mathcal{X}_p = \text{DFT}\{x_k, k = N_{uw}, N_{uw} + 1, \dots, N_{uw} + P - 1\} \tag{7.221}$$

it is

$$\mathcal{X}_p = \mathcal{H}_p \mathcal{A}_p^{(aug)}, \quad p = 0, 1, \dots, P - 1 \tag{7.222}$$

and

$$x_k = \text{DFT}^{-1}\{\mathcal{X}_p\} \quad k = N_{uw}, N_{uw}+1, \dots, N_{uw}+P-1 \tag{7.223}$$

To satisfy property (7.218), we augment each data block of size $\mathcal{M}$ with a unique word given by a PN sequence of length N_{uw}, $\{p_k\}$, $k = 0, 1, \dots, N_{uw} - 1$. The augmented transmit block of $P = \mathcal{M} + N_{uw}$ data samples is

$$\boldsymbol{a}^{(aug)}(\ell) = [a_{\ell\mathcal{M}}, a_{\ell\mathcal{M}+1}, \dots, a_{\ell\mathcal{M}+\mathcal{M}-1}, p_0, p_1, \dots, p_{N_{uw}-1}]^T \tag{7.224}$$

Moreover, the PN sequence is transmitted before the first data block. In other words, block $\boldsymbol{a}^{(aug)}(0)$ is to be sent at time $k = N_{uw}$. Consequently, samples x_k start to be collected at $k = N_{uw}$ up to $k = N_{uw} + P - 1$. In fact, the first PN sequence absorbs the transient due to the channel impulse response $\{h_i^0\}$.

Let

$$\boldsymbol{x}(\ell) = [x_{N_{uw}+\ell P}, \dots, x_{N_{uw}+\ell P+P-1}]^T \tag{7.225}$$

with

$$\boldsymbol{\mathcal{X}}(\ell) = \text{DFT}[\boldsymbol{x}(\ell)] \tag{7.226}$$

and

$$\boldsymbol{C} = [\mathcal{C}_0, \mathcal{C}_1, \dots, \mathcal{C}_{P-1}]^T \tag{7.227}$$

the FF filter in the FD. We define the DFT of the FF output

$$\boldsymbol{\mathcal{Z}}(\ell) = \text{diag}[\boldsymbol{C}]\boldsymbol{\mathcal{X}}(\ell) \tag{7.228}$$

whose inverse yields

$$\boldsymbol{z}(\ell) = [z_{N_{uw}+\ell P}, \dots, z_{N_{uw}+\ell P+P-1}]^T = \text{DFT}^{-1}[\boldsymbol{\mathcal{Z}}(\ell)] \tag{7.229}$$

Out of the P samples of $\boldsymbol{z}(\ell)$, we only retain the first $\mathcal{M}$, i.e. z_k for $k = N_{uw} + \ell P, \dots, N_{uw} + \mathcal{M} - 1 + \ell P$, as the remaining N_{uw} are not relevant for detection. In fact, they would correspond to the PN sequence. A block diagram of the whole receiver is shown in Figure 7.33. Note how the use of the PN sequence is very useful at the beginning of each detection block where interference is due to the PN sequence transmitted in the previous block.

Surely transmitting the PN sequence reduces the spectral efficiency by $\mathcal{M}/(\mathcal{M} + N_{uw})$. There is also a power inefficiency as the transmitted data corresponding to the PN sequence do not carry information. The choice of the PN sequence may be done according to different criteria, including the reduction of the peak to average power ration of the transmitted signal.

In Figure 7.33 $\hat{a}_{k-N_{uw}}$, for $k = N_{uw}, \dots, N_{uw} + \mathcal{M} - 1$, is the detected value of the information sequence $a_0, a_1, \dots, a_{\mathcal{M}-1}$. Moreover, for $k = 0, 1, \dots, N_{uw} - 1$, the PN sequence is filling up the shift register of the FB filter and $z_{FB,k}$ can be evaluated starting at $k = N_{uw}$. Next values of $z_{FB,k}$ use also detected data.

Concerning the design derivation of $\{\mathcal{C}_p\}$, $p = 0, 1, \dots, P - 1$, and $\{b_i\}$, $i = 1, \dots, N_{uw}$, the interested reader is referenced to [7]. Here we report the final equations. We introduce the DFT of correlation of noise $\tilde{w}_k$ (see (7.187) and (1.23)), for $p = 0, 1, \dots, P - 1$, as

$$\mathcal{P}_{\tilde{w},p} = \text{DFT}[\mathtt{r}_{\tilde{w}}(n)] = \frac{1}{T}\sum_{\ell=-\infty}^{\infty} N_0 \left|\mathcal{G}_M\left(f - \frac{\ell}{T}\right)\right|^2 \Bigg|_{f=\frac{p}{PT}} \tag{7.230}$$

For an MSE criterion,

$$J = E[|y_k - a_k|^2] \tag{7.231}$$

defined the matrix $\boldsymbol{R}$ $N_{uw} \times N_{uw}$ and the vector $\boldsymbol{p}$ $N_{uw} \times 1$ by

$$[\boldsymbol{R}]_{m,\ell} = \sum_{p=0}^{P-1} \frac{e^{-j2\pi\frac{p(\ell-m)}{P}}}{|\mathcal{H}_p|^2 + \mathcal{P}_{\tilde{w},p}/\sigma_a^2}, \quad m, \ell = 1, 2, \dots, N_{uw} \tag{7.232}$$

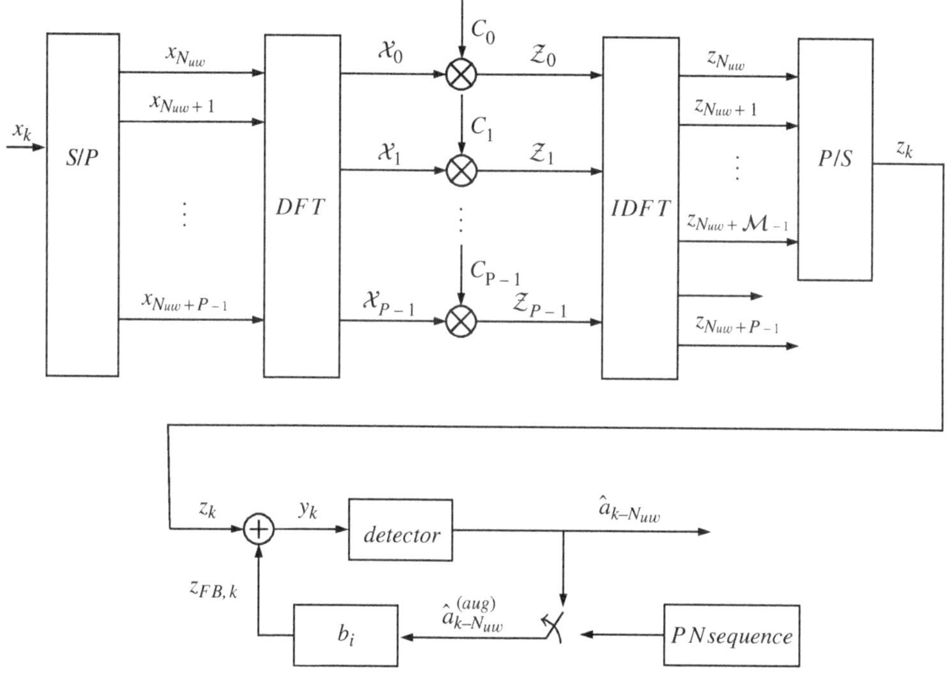

Figure 7.33 Block diagram of a DFE with FF in the FD. To simplify notation the block index ℓ is assumed 0.

$$[\boldsymbol{p}]_m = \sum_{p=0}^{P-1} \frac{e^{j2\pi\frac{pm}{P}}}{|\mathcal{H}_p|^2 + \mathcal{P}_{\tilde{w},p}/\sigma_a^2}, \quad m = 1, 2, \ldots, N_{uw} \tag{7.233}$$

the vector with FB coefficients

$$\boldsymbol{b} = [b_1, \ldots, b_{N_{uw}}]^T \tag{7.234}$$

with DFT $\mathcal{B} = [\mathcal{B}_0, \mathcal{B}_1 \ldots, \mathcal{B}_{P-1}]$, satisfies the following system of N_{uw} equations with N_{uw} unknowns

$$\boldsymbol{R}\boldsymbol{b} = \boldsymbol{p} \tag{7.235}$$

Moreover,

$$\mathcal{C}_p = \frac{\mathcal{H}_p^*(1 - \mathcal{B}_p)}{|\mathcal{H}_p|^2 + \mathcal{P}_{\tilde{w},p}/\sigma_a^2}, \quad p = 0, 1, \ldots, P-1 \tag{7.236}$$

and

$$J_{min} = \frac{1}{P} \sum_{p=0}^{P-1} \frac{|1 - \mathcal{B}_p|^2 \mathcal{P}_{\tilde{w},p}}{|\mathcal{H}_p|^2 + \mathcal{P}_{\tilde{w},p}/\sigma_a^2} \tag{7.237}$$

We conclude this section by mentioning a DFE where also the FB operates in the FD [9]. Now, filters design and detection are iterated (for a few times) up to reaching convergence. Computational complexity is further reduced w.r.t. previous DFE structures, and more importantly, there are no systems of equations to be solved. Moreover, this new iterative DFE yields better performance than the above DFEs. A version with the FF working at $T/2$ was also proposed [10].

Observation 7.9

For a DFE with the FF operating in the FD, the channel estimate of $\mathcal{H}_p$ to be used in filter design and processing can be FD based. Starting from (7.222), methods to estimate $\mathcal{H}_p$, $p = 0, 1, \ldots, P-1$, are given in Section 8.9. A simple and effective approach determines a first rough estimate by sending a block of known data symbols and estimating the channel by $\hat{\mathcal{H}}_p = \mathcal{X}_p / \mathcal{A}_p^{(aug)}$. Let the inverse discrete Fourier transform (IDFT) of $\{\hat{\mathcal{H}}_p\}$ be $\{\hat{h}_i^0\}$, $i = 0, 1, \dots, P-1$. Next, we truncate $\hat{h}_i^0$ to a length $N_{uw} + 1$, and set to zero the remaining taps. The DFT of size P of the truncated $\hat{h}_i^0$ usually yields a very good estimate of $\{\mathcal{H}_p\}$.

Observation 7.10

In [11, 12], a LE working in the FD was proposed whose structure is similar to the above DFE, obviously, for $b_i = 0$, $i = 1, \dots, N_{uw}$. However, as for all LEs, its performance may be much worse than that of DFE [7]. We recall this LE structure, as there is no cancellation of ISI using known/detected data, the augmented data structure may use a cyclic prefix of length N_{cp}, with $N_{cp} \geq N_h - 1$, where the information data block of size $\mathcal{M}$ is partially repeated up front,

$$\boldsymbol{a}^{(aug)}(\ell) = [a_{\ell\mathcal{M}+\mathcal{M}-1-N_{cp}}, \dots, a_{\ell\mathcal{M}+\mathcal{M}-1}, a_{\ell\mathcal{M}}, \dots, a_{\ell\mathcal{M}+\mathcal{M}-1}]^T \tag{7.238}$$

As data is circular of period $\mathcal{M}$, the advantages is that all DFTs are of size $\mathcal{M}$. In fact, (7.222) holds for $P = \mathcal{M}$. If (7.238) is transmitted starting at time $\ell(\mathcal{M} + N_{cp})T$, the receiver drops the first N_{cp} samples, due to the channel transient, and stores the remaining $\mathcal{M}$ samples for the LE in the FD. This approach is actually used also in orthogonal frequency division multiplexing (OFDM) (see Section 8.7.1).

7.4.6 LE-ZF

Neglecting the noise, the signal at the output of a LE with N coefficients (see (7.189)) is given by

$$y_k = \sum_{i=0}^{N-1} c_i \, x_{k-i} = \sum_{p=-N_1}^{N_2+N-1} \psi_p \, a_{k-p} \tag{7.239}$$

where from (7.188),

$$\psi_p = \sum_{\ell=0}^{N-1} c_\ell \, h_{p-\ell} = c_0 \, h_p + c_1 \, h_{p-1} + \dots + c_{N-1} \, h_{p-(N-1)} \qquad p = -N_1, \dots, 0, \dots, N_2 + N - 1 \tag{7.240}$$

For a LE-ZF, it must be

$$\psi_p = \delta_{p-D} \tag{7.241}$$

where D is a suitable delay.

If the overall impulse response $\{h_n\}$, $n = -N_1, \dots, N_2$, is known, a method to determine the coefficients of the LE-ZF consists in considering the system (7.240) with $N_t = N_1 + N_2 + N$ equations and N unknowns, that can be solved by the method of the pseudoinverse (see (2.151)); alternatively, the solution can be found in the frequency domain by taking the N_t-point DFT of the various signals (see (1.43)), and the result windowed in the time domain so that the filter coefficients are given by the N consecutive coefficients that maximize the energy of the filter impulse response.

An approximate solution is obtained by forcing the condition (7.241) only for N values of p in (7.240) centred around D; then the matrix of the system (7.240) is square and, if the determinant is different from zero, it can be inverted.

Note that all these methods require an accurate estimate of the overall impulse response, otherwise, the equalizer coefficients may deviate considerably from the desired values.

7.4.7 DFE-ZF with IIR filters

We determine the expressions of FF and FB filters of a DFE in the case of IIR filter structure.

We consider a receiver with a matched filter g_M (see Figure 7.22) followed by the DFE illustrated in Figure 7.34, where, to simplify the notation, we assume $t_0 = 0$ and $D = 0$. The z-transform of the QAM system impulse response is given by $\Phi(z)$, as defined in (7.132).

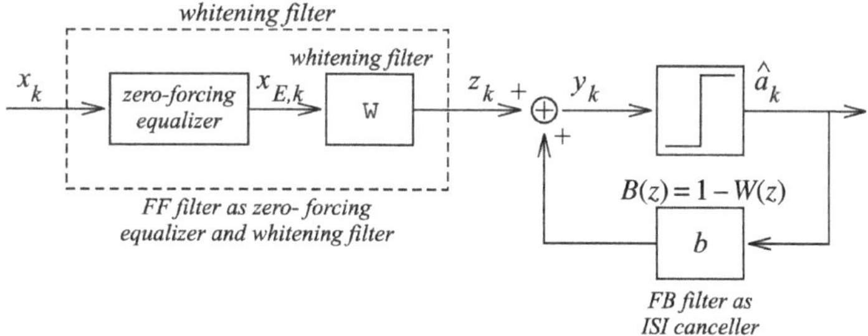

Figure 7.34 DFE zero-forcing.

With reference to Figure 7.34, the matched filter output x_k is input to a linear equalizer *zero forcing* (LE-ZF) with transfer function $1/\Phi(z)$ to remove ISI: therefore, the LE-ZF output is given by

$$x_{E,k} = a_k + w_{E,k} \tag{7.242}$$

From (7.136), using the property (7.133), we obtain that the PSD $P_{w_E}(z)$ of $w_{E,k}$ is given by

$$\begin{aligned} P_{w_E}(z) &= N_0 \Phi(z) \frac{1}{\Phi(z)\Phi^*\left(\frac{1}{z^*}\right)} \\ &= N_0 \frac{1}{\Phi(z)} \end{aligned} \tag{7.243}$$

As $\Phi(z)$ is the z-transform of a correlation sequence, it can be factorized as (see page 35)

$$\Phi(z) = F(z)\, F^*\left(\frac{1}{z^*}\right) \tag{7.244}$$

where

$$F(z) = \sum_{n=0}^{+\infty} \mathtt{f}_n\, z^{-n} \tag{7.245}$$

is a minimum-phase function, that is with poles and zeros inside the unit circle, associated with a causal sequence $\{\mathtt{f}_n\}$.

Observation 7.11

A useful method to determine the filter $F(z)$ in (7.244), with a computational complexity that is proportional to the square of the number of filter coefficients, is obtained by considering a minimum-phase prediction error filter $A(z) = 1 + \mathtt{a}'_{1,N}\, z^{-1} + \cdots + \mathtt{a}'_{N,N}\, z^{-N}$ (see page 116), designed using the auto-correlation sequence $\{\mathtt{r}_{q_C}(nT)\}$ defined by (7.130). The equation to determine the coefficients of $A(z)$ is given by (2.80), where $\{\mathtt{r}_x(n)\}$ is now substituted by $\{\mathtt{r}_{q_C}(nT)\}$. The final result is $F(z) = \mathtt{f}_0/A(z)$.

On the other hand, $F^*(1/z^*)$ is a function with zeros and poles outside the unit circle, associated with an anticausal sequence $\{\mathtt{f}^*_{-n}\}$.

We choose as transfer function of the filter $\mathtt{w}$ in Figure 7.34 the function

$$W(z) = F(z)\frac{1}{\mathtt{f}_0} \tag{7.246}$$

The ISI term in z_k is determined by $W(z) - 1$, hence *there are no precursors*; the noise is white with statistical power$(N_0/\mathtt{f}_0^2)$. Therefore, the filter $\mathtt{w}$ is called *whitening filter* (WF). In any case, the filter composed of the cascade of LE-ZF and $\mathtt{w}$ is also a WF.

If $\hat{a}_k = a_k$ then, for

$$B(z) = 1 - W(z) \tag{7.247}$$

the FB filter removes the ISI present in z_k and leaves the white noise unchanged. As y_k is not affected by ISI, this structure is called DFE-ZF, for which we obtain

$$\gamma_{DFE-ZF} = \frac{2|\mathtt{f}_0|^2}{N_0} \tag{7.248}$$

Summarizing, the relation between x_k and z_k is given by

$$\frac{1}{\Phi(z)}F(z)\frac{1}{\mathtt{f}_0} = \frac{1}{\mathtt{f}_0 F^*\left(\frac{1}{z^*}\right)} \tag{7.249}$$

With this filter, the noise in z_k is white. The relation between a_k and the desired signal in z_k is instead governed by $\Psi(z) = F(z)/\mathtt{f}_0$ and $B(z) = 1 - \Psi(z)$. In other words, the overall discrete-time system is causal and minimum phase, that is the energy of the impulse response is mostly concentrated at the beginning of the pulse. The overall receiver structure is illustrated in Figure 7.35, where the block including the *matched filter*, *sampler*, and *whitening filter*, is called *whitened matched Filter* (WMF). Note that the impulse response at the WF output has no precursors.

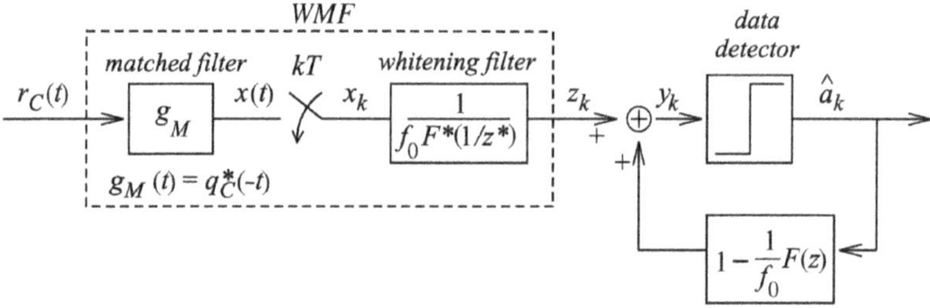

Figure 7.35 DFE-ZF as whitened matched filter followed by a canceller of ISI.

In principle, the WF of Figure 7.35 is non-realizable, because it is anticausal. In practice, we can implement it in two ways:

(a) introducing an appropriate delay in the impulse response of an FIR WF, and processing the output samples in the forward mode for $k = 0, 1, \ldots$;
(b) processing the output samples of the IIR WF in the backward mode, for $k = K - 1, K - 2, \ldots, 0$, starting from the end of the block of samples.

We observe that the choice $F(z) = \mathtt{f}_0/A(z)$, where $A(z)$ is discussed in Observation 7.11 of page 327, leads to an FIR WF with transfer function $\frac{1}{\mathtt{f}_0^2}A^*(1/z^*)$.

Observation 7.12

With reference to the scheme of Figure 7.35, using a LE-ZF instead of a DFE structure means that a filter with transfer function $\mathtt{f}_0/F(z)$ is placed after the WF to produce the signal $x_{E,k}$ given by (7.242). For a data detector based on $x_{E,k}$, the SNR at the detector input γ is

$$\gamma_{LE-ZF} = \frac{2}{\mathtt{r}_{w_E}(0)} \tag{7.250}$$

where $\mathtt{r}_{w_E}(0)$ is determined as the coefficient of z^0 in $N_0/\Phi(z)$. This expression is alternative to (7.110).

Example 7.4.1 (WF for a channel with exponential impulse response)

A method to determine the WF in the scheme of Figure 7.35 is illustrated by an example. Let

$$q_C(t) = \sqrt{E_{q_C} 2\beta} e^{-\beta t} 1(t) \tag{7.251}$$

be the overall system impulse response at the MF input; in (7.251) E_{q_C} is the energy of q_C.

The autocorrelation of q_C, sampled at instant nT, is given by

$$\mathtt{r}_{q_C}(nT) = E_{q_C} a^{|n|} \qquad a = e^{-\beta T} < 1 \tag{7.252}$$

Then

$$\begin{aligned} \Phi(z) &= \mathcal{Z}[\mathtt{r}_{q_C}(nT)] \\ &= E_{q_C} \frac{(1-a^2)}{-az^{-1} + (1+a^2) - az} \\ &= E_{q_C} \frac{(1-a^2)}{(1-az^{-1})(1-az)} \end{aligned} \tag{7.253}$$

We note that the frequency response of (7.252) is

$$\Phi(e^{j2\pi fT}) = E_{q_C} \frac{(1-a^2)}{1+a^2 - 2a\cos(2\pi fT)} \tag{7.254}$$

and presents a minimum for $f = 1/(2T)$.

With reference to the factorization (7.244), it is easy to identify the poles and zeros of $\Phi(z)$ inside the unit circle, hence

$$\begin{aligned} F(z) &= \sqrt{E_{q_C}(1-a^2)} \frac{1}{1-az^{-1}} \\ &= \sqrt{E_{q_C}(1-a^2)} \sum_{n=0}^{+\infty} a^n z^{-n} \end{aligned} \tag{7.255}$$

In particular, the coefficient of z^0 is given by

$$\mathtt{f}_0 = \sqrt{E_{q_C}(1-a^2)} \tag{7.256}$$

The WF of Figure 7.35 is expressed as

$$\begin{aligned} \frac{1}{\mathtt{f}_0 F^*\left(\frac{1}{z^*}\right)} &= \frac{1}{E_{q_C}(1-a^2)}(1-az) \\ &= \frac{1}{E_{q_C}(1-a^2)} z(-a+z^{-1}) \end{aligned} \tag{7.257}$$

In this case the WF, apart from a delay of one sample ($D = 1$), can be implemented by a simple FIR with two coefficients, whose values are equal to $-a/(E_{q_C}(1-a^2))$ and $1/(E_{q_C}(1-a^2))$.

The FB filter is a first-order IIR filter with transfer function

$$\begin{aligned} 1 - \frac{1}{\mathtt{f}_0} F(z) &= -\sum_{n=1}^{+\infty} a^n z^{-n} \\ &= 1 - \frac{1}{1 - a z^{-1}} \\ &= \frac{-a z^{-1}}{1 - a z^{-1}} \end{aligned} \tag{7.258}$$

Example 7.4.2 (WF for a two-ray channel)

In this case, we directly specify the autocorrelation sequence at the matched filter output:

$$\Phi(z) = Q_{CC}(z) Q_{CC}^*\left(\frac{1}{z^*}\right) \tag{7.259}$$

where

$$Q_{CC}(z) = \sqrt{E_{q_C}}(q_0 + q_1 z^{-1}) \tag{7.260}$$

with q_0 and q_1 such that

$$|q_0|^2 + |q_1|^2 = 1 \tag{7.261}$$

$$|q_0| > |q_1| \tag{7.262}$$

In this way, E_{q_C} is the energy of $\{q_{CC}(nT)\}$ and $Q_{CC}(z)$ is minimum phase.

Equation (7.260) represents the discrete-time model of a wireless system with a two-ray channel. The impulse response is given by

$$q_{CC}(nT) = \sqrt{E_{q_C}}(q_0 \delta_n + q_1 \delta_{n-1}) \tag{7.263}$$

The frequency response is given by

$$\mathcal{Q}_{CC}(f) = \sqrt{E_{q_C}}(q_0 + q_1 e^{-j2\pi f T}) \tag{7.264}$$

We note that if $q_0 = q_1$, the frequency response has a zero for $f = 1/(2T)$.

From (7.259) and (7.260), we get

$$\Phi(z) = E_{q_C}(q_0 + q_1 z^{-1})(q_0^* + q_1^* z) \tag{7.265}$$

hence, recalling assumption (7.262),

$$F(z) = \sqrt{E_{q_C}}(q_0 + q_1 z^{-1}) = Q_{CC}(z) \tag{7.266}$$

and

$$\mathtt{f}_0 = \sqrt{E_{q_C}}\, q_0 \tag{7.267}$$

The WF is given by

$$\frac{1}{\mathtt{f}_0 F^*\left(\frac{1}{z^*}\right)} = \frac{1}{E_{q_C}|q_0|^2(1 - bz)} \tag{7.268}$$

where

$$b = \left(-\frac{q_1}{q_0}\right)^*$$

We note that the WF has a pole for $z = b^{-1}$, which, recalling (7.262), lies outside the unit circle. In this case, in order to have a stable filter, it is convenient to associate the z-transform $1/(1 - bz)$ with an anticausal sequence,

$$\frac{1}{1 - bz} = \sum_{i=-\infty}^{0} (bz)^{-i} = \sum_{n=0}^{\infty} (bz)^n \tag{7.269}$$

On the other hand, as $|b| < 1$, we can approximate the series by considering only the first $(N-1)$ terms, obtaining

$$\begin{aligned}\frac{1}{\mathtt{f}_0 F^*\left(\frac{1}{z^*}\right)} &\simeq \frac{1}{E_{q_C}|q_0|^2}\sum_{n=0}^{N}(bz)^n \\ &= \frac{1}{E_{q_C}|q_0|^2} z^N [b^N + \cdots + bz^{-(N-1)} + z^{-N}]\end{aligned} \tag{7.270}$$

Consequently, the WF, apart from a delay $D = N$, can be implemented by an FIR filter with $N+1$ coefficients.

The FB filter in this case is a simple FIR filter with one coefficient

$$1 - \frac{1}{\mathtt{f}_0}F(z) = -\frac{q_1}{q_0}z^{-1} \tag{7.271}$$

DFE-ZF as noise predictor

Let $A(z) = \mathcal{Z}[a_k]$. From the identity

$$\begin{aligned}Y(z) &= X_E(z)W(z) + (1 - W(z))A(z) \\ &= X_E(z) + (1 - W(z))(A(z) - X_E(z))\end{aligned} \tag{7.272}$$

the scheme of Figure 7.34 is redrawn as in Figure 7.36, where the FB filter acts as a noise predictor. In fact, for $\hat{a}_k = a_k$, the FB filter input is coloured noise. By removing the correlated noise from $x_{E,k}$, we obtain y_k that is composed of white noise, with minimum variance, plus the desired symbol a_k.

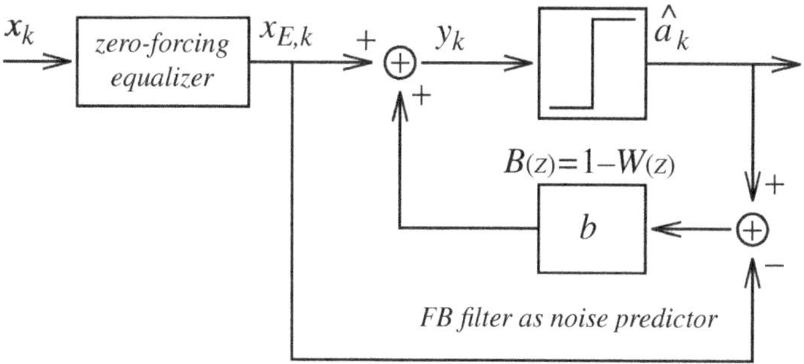

Figure 7.36 Predictive DFE: the FF filter is a linear equalizer zero forcing.

DFE as ISI and noise predictor

A variation of the scheme of Figure 7.36 consists in using as FF filter, a minimum-MSE linear equalizer. We refer to the scheme of Figure 7.37, where the filter c is given by (7.145). The z-transform of the overall impulse response at the FF filter output is given by

$$\Phi(z)C(z) = \frac{\sigma_a^2\Phi(z)}{N_0 + \sigma_a^2\Phi(z)} \tag{7.273}$$

As $t_0 = 0$, the ISI in z_k is given by

$$\Phi(z)C(z) - 1 = -\frac{N_0}{N_0 + \sigma_a^2\Phi(z)} \tag{7.274}$$

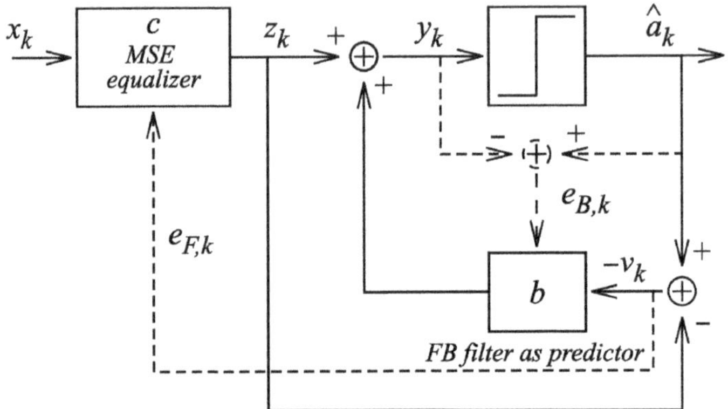

Figure 7.37 Predictive DFE with the FF filter as a minimum-MSE linear equalizer.

Hence, the PSD of the ISI has the following expression:

$$\begin{aligned} P_{ISI}(z) &= P_a(z)\frac{N_0}{N_0+\sigma_a^2\Phi(z)}\frac{N_0}{N_0+\sigma_a^2\Phi^*\left(\frac{1}{z^*}\right)} \\ &= \frac{\sigma_a^2(N_0)^2}{(N_0+\sigma_a^2\Phi(z))^2} \end{aligned} \tag{7.275}$$

using (7.133) and the fact that the symbols are uncorrelated with $P_a(z) = \sigma_a^2$.

The PSD of the noise in z_k is given by

$$\begin{aligned} P_{noise}(z) &= N_0\Phi(z)C(z)C^*\left(\frac{1}{z^*}\right) \\ &= N_0\frac{\Phi(z)\sigma_a^4}{(N_0+\sigma_a^2\Phi(z))^2} \end{aligned} \tag{7.276}$$

Therefore, the PSD of the disturbance v_k in z_k, composed of ISI and noise, is given by

$$P_v(z) = \frac{N_0\sigma_a^2}{N_0+\sigma_a^2\Phi(z)} \tag{7.277}$$

We note that the FF filter could be an FSE and the result (7.277) would not change.

To minimize the power of the disturbance in y_k, the FB filter, with input $\hat{a}_k - z_k = a_k - z_k = -v_k$ (assuming $\hat{a}_k = a_k$), needs to remove the predictable components of z_k. For a predictor of infinite length, we set

$$B(z) = \sum_{n=1}^{+\infty} \mathrm{b}_n z^{-n} \tag{7.278}$$

An alternative form is

$$B(z) = 1 - A(z) \tag{7.279}$$

where

$$A(z) = \sum_{n=0}^{+\infty} \mathrm{a}'_n z^{-n}, \qquad \mathrm{a}'_0 = 1 \tag{7.280}$$

is the forward prediction error filter defined in (2.78). To determine $B(z)$, we use the spectral factorization in (1.453):

$$\begin{aligned} P_v(z) &= \frac{\sigma_y^2}{A(z)A^*\left(\frac{1}{z^*}\right)} \\ &= \frac{\sigma_y^2}{[1-B(z)]\left[1-B^*\left(\frac{1}{z^*}\right)\right]} \end{aligned} \tag{7.281}$$

with $P_v(z)$ given by (7.277).

In conclusion, it results that the *prediction error* signal y_k is a white noise process with statistical power σ_y^2.

An adaptive version of the basic scheme of Figure 7.37 suggests that the two filters, c and b, are separately adapted through the error signals $\{e_{F,k}\}$ and $\{e_{B,k}\}$, respectively. This configuration, although sub-optimum with respect to the DFE, is used in conjunction with trellis-coded modulation [13].

7.4.8 Benchmark performance of LE-ZF and DFE-ZF

We compare bounds on the performance of LE-ZF and DFE-ZF, in terms of the SNR at the decision point, γ.

Comparison

From (7.243), the noise sequence $\{w_{E,k}\}$ can be modelled as the output of a filter having transfer function

$$C_F(z) = \frac{1}{F(z)} \tag{7.282}$$

and input given by white noise with PSD N_0. Because $F(z)$ is causal, also $C_F(z)$ is causal:

$$C_F(z) = \sum_{n=0}^{\infty} c_{F,n} z^{-n} \tag{7.283}$$

where $\{c_{F,n}\}$, $n \geq 0$ is the filter impulse response.

Then, we can express the statistical power of $w_{E,k}$ as

$$\mathtt{r}_{w_E}(0) = N_0 \sum_{n=0}^{\infty} |c_{F,n}|^2 \tag{7.284}$$

where, from (7.282),

$$c_{F,0} = C(\infty) = \frac{1}{F(\infty)} = \frac{1}{\mathtt{f}_0} \tag{7.285}$$

Using the inequality

$$\sum_{n=0}^{\infty} |c_{F,n}|^2 \geq |c_{F,0}|^2 = \frac{1}{|\mathtt{f}_0|^2} \tag{7.286}$$

the comparison between (7.248) and (7.250) yields

$$\gamma_{LE-ZF} \leq \gamma_{DFE-ZF} \tag{7.287}$$

Performance for two channel models

We analyse now the value of γ for the two simple systems introduced in Examples 7.4.1 and 7.4.2.

LE-ZF

Channel with exponential power delay profile. From (7.253), the coefficient of z^0 in $N_0/\Phi(z)$ is equal to

$$\mathtt{r}_{w_E}(0) = \frac{N_0}{E_{q_C}} \frac{1+a^2}{1-a^2} \tag{7.288}$$

and consequently, from (7.250)

$$\gamma_{LE-ZF} = 2\frac{E_{q_C}}{N_0} \frac{1-a^2}{1+a^2} \tag{7.289}$$

Using the expression of γ_{MF} (7.107), obtained for a MF receiver in the absence of ISI, we get

$$\gamma_{LE-ZF} = \gamma_{MF} \frac{1-a^2}{1+a^2} \tag{7.290}$$

Therefore, the loss due to the ISI, given by the factor $(1-a^2)/(1+a^2)$, can be very large if a is close to 1. In this case, the minimum value of the frequency response in (7.254) is close to 0.

Two-ray channel. From (7.265), we have

$$\frac{1}{\Phi(z)} = \frac{1}{E_{q_C}(q_0 + q_1 z^{-1})(q_0^* + q_1^* z)} \tag{7.291}$$

By a partial fraction expansion, we find only the pole for $z = -q_1/q_0$ lying inside the unit circle, hence,

$$\mathtt{r}_{w_E}(0) = \frac{N_0}{E_{q_C} q_0 \left(q_0^* - q_1^* \frac{q_1}{q_0}\right)} = \frac{N_0}{E_{q_C}(|q_0|^2 - |q_1|^2)} \tag{7.292}$$

Then

$$\gamma_{LE-ZF} = \gamma_{MF}(|q_0|^2 - |q_1|^2) \tag{7.293}$$

Also in this case, we find that the LE is unable to equalize channels with a spectral zero.

DFE-ZF

Channel with exponential power delay profile. Substituting the expression of $\mathtt{f}_0$ given by (7.256) in (7.248), we get

$$\begin{aligned}\gamma_{DFE-ZF} &= \frac{2}{N_0} E_{q_C}(1-a^2) \\ &= \gamma_{MF}(1-a^2)\end{aligned} \tag{7.294}$$

We note that γ_{DFE-ZF} is better than γ_{LE-ZF} by the factor $(1+a^2)$.

Two-ray channel. Substitution of (7.267) in (7.248) yields

$$\gamma_{DFE-ZF} = \frac{2}{N_0} E_{q_C} |q_0|^2 = \gamma_{MF} |q_0|^2 \tag{7.295}$$

In this case, the advantage with respect to LE-ZF is given by the factor $|q_0|^2/(|q_0|^2 - |q_1|^2)$, which may be substantial if $|q_1| \simeq |q_0|$.

We recall that in case $E_{q_C} \gg N_0$, that is for low noise levels, LE and DFE yields similar performance to LE-ZF and DFE-ZF, respectively. Anyway, for the two systems of Examples 7.4.1 and 7.4.2 the values of γ in terms of J_{min} are given in [13], for both LE and DFE.

7.4.9 Passband equalizers

For QAM signals, we analyse alternatives to the baseband equalizers considered at the beginning of this chapter. We refer to the QAM transmitter scheme of Figure 7.5 and we consider the model of Figure 7.38, where the transmission channel has impulse response g_{Ch} and introduces additive white noise w. The received passband signal is given by

$$r(t) = Re\left[\sum_{k=-\infty}^{+\infty} a_k q_{Ch}(t-kT) e^{j(2\pi f_0 t+\varphi)}\right] + w(t) \tag{7.296}$$

where, from (7.39) or equivalently from Figure 7.10, q_{Ch} is a *baseband equivalent pulse* with frequency response

$$\mathcal{Q}_{Ch}(f) = \mathcal{H}_{Tx}(f)\,\frac{1}{2}\,\mathcal{G}^{(bb)}_{Ch}(f) = \mathcal{H}_{Tx}(f)\,\mathcal{G}_{Ch}(f+f_0)\,1(f+f_0) \tag{7.297}$$

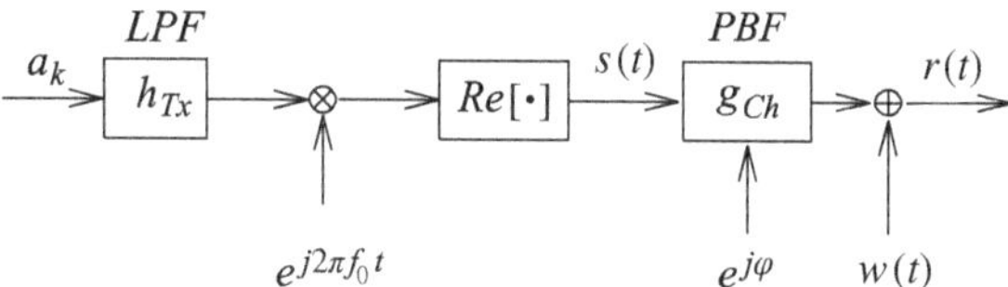

Figure 7.38 Passband modulation scheme with phase offset introduced by the channel.

In this model, the phase φ in (7.296) implies that arg $\mathcal{Q}_{Ch}(0) = 0$, Hence, it includes also the phase offset introduced by the channel frequency response at $f = f_0$, in addition to the carrier phase offset between the transmit and receive carriers. Because of this impairment, which as a first approximation is equivalent to a rotation of the symbol constellation, a suitable receiver structure needs to be developed [14].

As φ may be time varying, it can be decomposed as the sum of three terms

$$\varphi(t) = \Delta\varphi + 2\pi\Delta f t + \psi(t) \tag{7.298}$$

where $\Delta\varphi$ is a fixed phase offset, Δf is a fixed frequency offset, and ψ is a random or quasi-periodic term (see the definition of phase noise in (4.13)).

For example, over telephone channels typically $|\psi(t)| \leq \pi/20$. Moreover, the highest frequency of the spectral components of ψ is usually lower than $0.1/T$: in other words, if ψ were a sinusoidal signal, it would have a period larger than $10T$.

Therefore, φ may be regarded as a constant, or at least as *slowly time varying*, at least for a time interval equal to the duration of the overall system impulse response.

Passband receiver structure

Filtering and equalization of the signal r are performed in the passband. First, a filter extracts the positive frequency components of r, acting also as a matched filter. It is a complex-valued passband filter, as illustrated in Figure 7.39.

In particular, from the theory of the optimum receiver, we have

$$g_M^{(pb)}(t) = g_M(t) e^{j2\pi f_0 t} \tag{7.299}$$

where

$$g_M(t) = q^*_{Ch}(t_0 - t) \tag{7.300}$$

with q_{Ch} defined in (7.297).

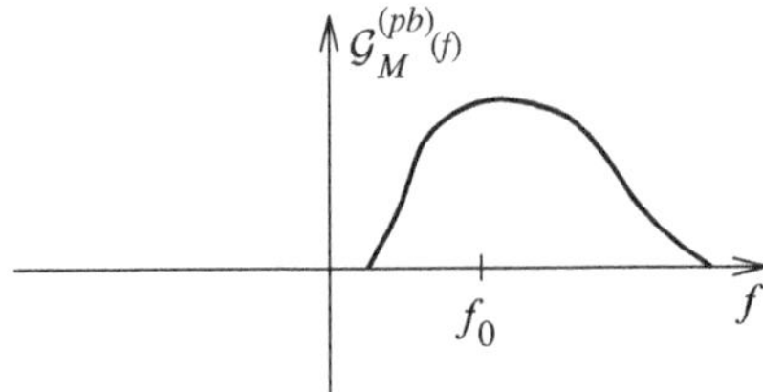

Figure 7.39 Frequency response of a passband matched filter.

Then, with reference to Figure 7.40, we have

$$g_{M,I}^{(pb)}(t) = Re\{g_M^{(pb)}(t)\} \quad \text{and} \quad g_{M,Q}^{(pb)}(t) = Im\{g_M^{(pb)}(t)\} \tag{7.301}$$

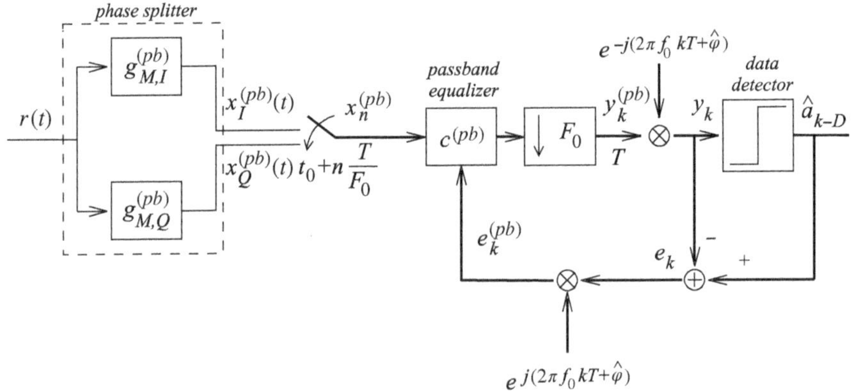

Figure 7.40 QAM passband receiver.

If g_M in (7.300) is real-valued, then $g_{M,I}^{(pb)}$ and $g_{M,Q}^{(pb)}$ are related by the Hilbert transform (1.97), that is

$$g_{M,Q}^{(pb)}(t) = \mathcal{H}^{(h)}[g_{M,I}^{(pb)}(t)] \tag{7.302}$$

After the passband matched filter, the signal is oversampled with sampling period T/F_0; oversampling is suggested by the following two reasons:

1. if q_{Ch} is unknown and $g_M^{(pb)}$ is a simple passband filter, matched filtering is carried out by the filter $c^{(pb)}$, hence the need for oversampling;
2. if the timing phase t_0 is not accurate, it is convenient to use an FSE.

Let the overall baseband equivalent impulse response of the system at the sampler input be

$$q(t) = \mathcal{F}^{-1}[\mathcal{Q}(f)] \tag{7.303}$$

where

$$\mathcal{Q}(f) = \mathcal{Q}_{Ch}(f)\ \mathcal{G}_M(f) = \mathcal{H}_{Tx}(f)\ \mathcal{G}_{Ch}(f+f_0)\ \mathcal{G}_M(f)\ 1(f+f_0) \tag{7.304}$$

The sampled passband signal is given by

$$\begin{aligned} x_n^{(pb)} &= x_I^{(pb)}\left(t_0 + n\frac{T}{F_0}\right) + jx_Q^{(pb)}\left(t_0 + n\frac{T}{F_0}\right) \\ &= \sum_{k=-\infty}^{+\infty} a_k h_{n-kF_0} e^{j\left(2\pi f_0 n\frac{T}{F_0}+\varphi\right)} + \tilde{w}_n \end{aligned} \tag{7.305}$$

where[11]

$$h_n = q\left(t_0 + n\frac{T}{F_0}\right) \tag{7.306}$$

In (7.305), $\tilde{w}_n$ denotes the noise component

$$\tilde{w}_n = w_R\left(t_0 + n\frac{T}{F_0}\right) \tag{7.307}$$

where $w_R(t) = w * g_M^{(pb)}(t)$.

For a passband equalizer with N coefficients, the output signal with sampling period equal to T is given by

$$y_k^{(pb)} = \sum_{i=0}^{N-1} c_i^{(pb)} x_{kF_0-i}^{(pb)} = \boldsymbol{x}_{kF_0}^{(pb)T} \boldsymbol{c}^{(pb)} \tag{7.308}$$

with the usual meaning of the two vectors $\boldsymbol{x}_n^{(pb)}$ and $\boldsymbol{c}^{(pb)}$.

Ideally, it should result

$$y_k^{(pb)} = a_{k-D} e^{j(2\pi f_0 kT + \varphi)} \tag{7.309}$$

where the phase offset φ needs to be estimated as $\hat{\varphi}$. In Figure 7.40, the signal $y_k^{(pb)}$ is shifted to baseband by multiplication with the function $e^{-j(2\pi f_0 kT + \hat{\varphi})}$. Then the data detector follows.

At this point, some observations can be made: as seen in Figure 7.41, by demodulating the received signal, that is by multiplying it by $e^{-j2\pi f_0 t}$, before the equalizer or the receive filter we obtain a scheme equivalent to that of Figure 7.24, with a baseband equalizer. As we will see at the end of this section, the only advantage of a passband equalizer is that the computational complexity of the receiver is reduced; in any case, it is desirable to compensate for the presence of the phase offset, that is to multiply the received signal by $e^{-j\hat{\varphi}}$, as near as possible to the decision point, so that the delay in the loop for the update of the phase offset estimate is small.

Optimization of equalizer coefficients and carrier phase offset

To simplify the notation, the analysis is carried out for $F_0 = 1$. As usual, for an error signal defined as

$$e_k = a_{k-D} - y_k \tag{7.310}$$

where y_k is given by (7.308), we desire to minimize the following cost function:

$$\begin{aligned} J &= E[|e_k|^2] \\ &= E\left[\left|a_{k-D} - \boldsymbol{x}_k^{(pb)T} \boldsymbol{c}^{(pb)} e^{-j(2\pi f_0 kT + \hat{\varphi})}\right|^2\right] \\ &= E\left[\left|a_{k-D} e^{j(2\pi f_0 kT + \hat{\varphi})} - \boldsymbol{x}_k^{(pb)T} \boldsymbol{c}^{(pb)}\right|^2\right] \end{aligned} \tag{7.311}$$

Equation (7.311) expresses the classical Wiener problem for a desired signal expressed as

$$a_{k-D} e^{j(2\pi f_0 kT + \hat{\varphi})} \tag{7.312}$$

and input $\boldsymbol{x}_k^{(pb)}$. Assuming φ known, the Wiener–Hopf solution is given by

$$\boldsymbol{c}_{opt}^{(pb)} = \boldsymbol{R}^{-1} \boldsymbol{p} \tag{7.313}$$

where

$$\boldsymbol{R} = E\left[\boldsymbol{x}_k^{(pb)*} \boldsymbol{x}_k^{(pb)T}\right] \tag{7.314}$$

[11] In (7.305), φ takes also into account the phase $2\pi f_0 t_0$.

has elements for $\ell, m = 0, 1, \dots, N-1$, given by

$$[\boldsymbol{R}]_{\ell,m} = E\left[x_{k-m}^{(pb)} x_{k-\ell}^{(pb)*}\right] = \mathtt{r}_{x^{(pb)}}((\ell - m)T)$$
$$= \sigma_a^2 \sum_{i=-\infty}^{+\infty} h_i h_{i-(\ell-m)}^* e^{j2\pi f_0(\ell-m)T} + \mathtt{r}_{w_R}((\ell - m)T) \tag{7.315}$$

and

$$\boldsymbol{p} = E\left[a_{k-D} e^{j(2\pi f_0 kT + \hat{\varphi})} \boldsymbol{x}_k^{(pb)*}\right] \tag{7.316}$$

has elements for $\ell = 0, 1, \dots, N-1$, given by

$$[\boldsymbol{p}]_\ell = [\boldsymbol{p}']_\ell e^{-j(\varphi - \hat{\varphi})} \tag{7.317}$$

where

$$[\boldsymbol{p}']_\ell = \sigma_a^2 h_{D-\ell}^* e^{j2\pi f_0 \ell T} \tag{7.318}$$

From (7.317), the optimum solution (7.313) is expressed as

$$\boldsymbol{c}_{opt}^{(pb)} = \boldsymbol{R}^{-1} \boldsymbol{p}' e^{-j(\varphi - \hat{\varphi})} \tag{7.319}$$

where $\boldsymbol{R}^{-1}$ and $\boldsymbol{p}'$ do not depend on the phase offset φ.

From (7.319), it can be verified that if φ is a constant the equalizer automatically compensates for the phase offset introduced by the channel, and the output signal $y_k^{(pb)}$ remains unchanged.

A difficulty appears if φ varies (slowly) in time and the equalizer attempts to track it. In fact, to avoid that the output signal is affected by convergence errors, typically an equalizer in the steady state must not vary its coefficients by more than 1% within a symbol interval. Therefore, another algorithm is needed to estimate φ.

Adaptive method

The adaptive LMS algorithm is used for an instantaneous square error defined as

$$|e_k|^2 = |a_{k-D} - \boldsymbol{x}_k^{(pb)T} \boldsymbol{c}^{(pb)} e^{-j(2\pi f_0 kT + \hat{\varphi})}|^2 \tag{7.320}$$

The gradient of the function in (7.320) with respect to $\boldsymbol{c}^{(pb)}$ is equal to

$$\nabla_{\boldsymbol{c}^{(pb)}} |e_k|^2 = -2 e_k (\boldsymbol{x}_k^{(pb)} e^{-j(2\pi f_0 kT + \hat{\varphi}_k)})^*$$
$$= -2 e_k e^{j(2\pi f_0 kT + \hat{\varphi}_k)} \boldsymbol{x}_k^{(pb)*}$$
$$= -2 e_k^{(pb)} \boldsymbol{x}_k^{(pb)*} \tag{7.321}$$

where

$$e_k^{(pb)} = e_k e^{j(2\pi f_0 kT + \hat{\varphi}_k)} \tag{7.322}$$

The law for coefficient adaptation is given by

$$\boldsymbol{c}_{k+1}^{(pb)} = \boldsymbol{c}_k^{(pb)} + \mu e_k^{(pb)} \boldsymbol{x}_k^{(pb)*} \tag{7.323}$$

We now compute the gradient with respect to $\hat{\varphi}_k$. Let

$$\theta = 2\pi f_0 kT + \hat{\varphi}_k \tag{7.324}$$

then

$$|e_k|^2 = (a_{k-D} - y_k)(a_{k-D} - y_k)^*$$
$$= (a_{k-D} - y_k^{(pb)} e^{-j\theta})(a_{k-D}^* - y_k^{(pb)*} e^{j\theta}) \tag{7.325}$$

Therefore, we obtain

$$\begin{aligned}\nabla_{\hat{\varphi}}|e_k|^2 &= \frac{\partial}{\partial\theta}\,|e_k|^2 = jy_k^{(pb)}e^{-j\theta}e_k^* - e_k jy_k^{(pb)*}e^{j\theta}\\ &= 2Im\left[e_k\left(y_k^{(pb)}e^{-j\theta}\right)^*\right]\\ &= 2Im[e_k y_k^{\,*}]\end{aligned} \tag{7.326}$$

As $e_k = a_{k-D} - y_k$, (7.326) may be rewritten as

$$\nabla_{\hat{\varphi}}|e_k|^2 = 2Im[a_{k-D}y_k^*] \tag{7.327}$$

We note that $Im[a_{k-D}y_k^*]$ is related to the sine of the phase difference between a_{k-D} and y_k, therefore, the algorithm has reached convergence only if the phase of y_k coincides (on average) with that of a_{k-D}.

The law for updating the phase offset estimate is given by

$$\hat{\varphi}_{k+1} = \hat{\varphi}_k - \mu_\varphi Im[e_k y_k^*] \tag{7.328}$$

or

$$\hat{\varphi}_{k+1} = \hat{\varphi}_k - \mu_\varphi Im[a_{k-D}y_k^*] \tag{7.329}$$

The adaptation gain is typically normalized as

$$\mu_\varphi = \frac{\tilde{\mu}_\varphi}{|a_{k-D}|\;|y_k|} \tag{7.330}$$

In general, μ_φ is chosen larger than μ, so that the variations of φ are tracked by the carrier phase offset estimator, and not by the equalizer.

In the ideal case, we have

$$y_k^{(pb)} = a_{k-D}e^{j(2\pi f_0 kT+\varphi)} \tag{7.331}$$

and

$$y_k = a_{k-D}e^{j(\varphi-\hat{\varphi}_k)} \tag{7.332}$$

Therefore, the adaptive algorithm becomes:

$$\begin{aligned}\hat{\varphi}_{k+1} &= \hat{\varphi}_k - \tilde{\mu}_\varphi Im[e^{-j(\varphi-\hat{\varphi}_k)}]\\ &= \hat{\varphi}_k + \tilde{\mu}_\varphi \sin(\varphi-\hat{\varphi}_k)\end{aligned} \tag{7.333}$$

which is the equation of a first order *phase-locked-loop* (PLL) (see Section 14.7). For a constant phase offset, at convergence we get $\hat{\varphi} = \varphi$.

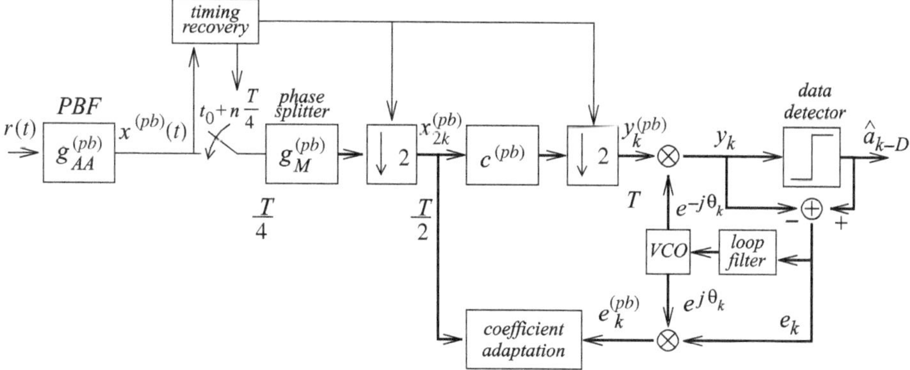

Figure 7.41 QAM passband receiver for transmission over telephone channels.

7.5 Optimum methods for data detection

Adopting an *MSE criterion* at the decision point, we have derived the configuration of Figure 7.23 for an LE, and that of Figure 7.31 for a DFE. In both cases, the decision on a transmitted symbol a_{k-D} is based only on y_k through a memoryless threshold detector.

Actually, the decision criterion that minimizes the probability that an error occurs in the detection of a symbol of the sequence $\{a_k\}$ requires in general that the entire sequence of received samples is considered for symbol detection.

We assume a *sampled signal* having the following structure:

$$z_k = u_k + w_k \quad k = 0, 1, \ldots, K-1 \tag{7.334}$$

where

- u_k is the desired signal that carries the information,

$$u_k = \sum_{n=-L_1}^{L_2} \eta_n \, a_{k-n} \tag{7.335}$$

  where η_0 is the sample of the overall system impulse response, obtained in correspondence of the optimum timing phase; $\{\eta_{-L_1}, \ldots, \eta_{-1}\}$ are the precursors, which are typically negligible with respect to η_0. Recalling the expression of the pulse $\{\psi_p\}$ given by (7.188), we have $\eta_n = \psi_{n+D}$. We assume the coefficients $\{\eta_n\}$ are known; in practice, however, they are estimated by the methods discussed in Section 7.8;
- w_k is a circularly symmetric *white Gaussian* noise, with equal statistical power in the two I and Q components given by $\sigma_I^2 = \sigma_w^2/2$; hence, the samples $\{w_k\}$ are uncorrelated and therefore, *statistically independent*.

In this section, a general derivation of optimum detection methods is considered; possible applications span, e.g. decoding of convolutional codes (see Chapter 11), coherent demodulation of constant-phase modulation (CPM) signals (see Appendix 16.A), and obviously detection of sequences transmitted over channels with ISI.

We introduce the following vectors with K components (K may be very large):

1. Sequence of transmitted symbols, or information message, modelled as a sequence of r.v.s. from a finite alphabet

$$\boldsymbol{a} = [a_{L_1}, a_{L_1+1}, \ldots, a_{L_1+K-1}]^T, \qquad a_i \in \mathcal{A} \tag{7.336}$$

2. Sequence of detected symbols, modelled as a sequence of r.v.s. from a finite alphabet

$$\hat{\boldsymbol{a}} = [\hat{a}_{L_1}, \hat{a}_{L_1+1}, \ldots, \hat{a}_{L_1+K-1}]^T, \qquad \hat{a}_i \in \mathcal{A} \tag{7.337}$$

3. Sequence of detected symbol values

$$\boldsymbol{\alpha} = [\alpha_{L_1}, \alpha_{L_1+1}, \ldots, \alpha_{L_1+K-1}]^T, \qquad \alpha_i \in \mathcal{A} \tag{7.338}$$

4. Sequence of received samples, modelled as a sequence of complex r.v.s.

$$\boldsymbol{z} = [z_0, z_1, \ldots, z_{K-1}]^T \tag{7.339}$$

5. Sequence of received sample values, or observed sequence

$$\boldsymbol{\rho} = [\rho_0, \rho_1, \ldots, \rho_{K-1}]^T, \qquad \rho_i \in C, \quad \boldsymbol{\rho} \in C^K \tag{7.340}$$

Let M be the cardinality of the alphabet $\mathcal{A}$. We divide the vector space of the received samples, $\mathcal{C}^K$, into M^K non-overlapping regions

$$\mathcal{R}_{\alpha}, \qquad \boldsymbol{\alpha} \in \mathcal{A}^K \tag{7.341}$$

such that, if $\boldsymbol{\rho}$ belongs to $\mathcal{R}_{\alpha}$, then the sequence $\boldsymbol{\alpha}$ is detected:

$$\text{if } \boldsymbol{\rho} \in \mathcal{R}_{\alpha} \qquad \Longrightarrow \qquad \hat{\boldsymbol{a}} = \boldsymbol{\alpha} \tag{7.342}$$

The probability of a correct decision is computed as follows:

$$\begin{aligned} P[\mathrm{C}] &= P[\hat{\boldsymbol{a}} = \boldsymbol{a}] \\ &= \sum_{\boldsymbol{\alpha} \in \mathcal{A}^K} P[\hat{\boldsymbol{a}} = \boldsymbol{\alpha} \mid \boldsymbol{a} = \boldsymbol{\alpha}] P[\boldsymbol{a} = \boldsymbol{\alpha}] \\ &= \sum_{\boldsymbol{\alpha} \in \mathcal{A}^K} P[\boldsymbol{z} \in \mathcal{R}_{\alpha} \mid \boldsymbol{a} = \boldsymbol{\alpha}] P[\boldsymbol{a} = \boldsymbol{\alpha}] \\ &= \sum_{\boldsymbol{\alpha} \in \mathcal{A}^K} \int_{\mathcal{R}_{\alpha}} p_{\boldsymbol{z}|\boldsymbol{a}}(\boldsymbol{\rho} \mid \boldsymbol{\alpha}) d\boldsymbol{\rho} P[\boldsymbol{a} = \boldsymbol{\alpha}] \end{aligned} \tag{7.343}$$

The following criteria may be adopted to maximize (7.343).

Maximum a posteriori probability (MAP) criterion

$$\boldsymbol{\rho} \in \mathcal{R}_{\alpha_{opt}} \qquad (\text{and } \hat{\boldsymbol{a}} = \boldsymbol{\alpha}_{opt}) \qquad \text{if} \qquad \boldsymbol{\alpha}_{opt} = \arg\max_{\boldsymbol{\alpha} \in \mathcal{A}^K} p_{\boldsymbol{z}|\boldsymbol{a}}(\boldsymbol{\rho} \mid \boldsymbol{\alpha}) P[\boldsymbol{a} = \boldsymbol{\alpha}] \tag{7.344}$$

Using the identity

$$p_{\boldsymbol{z}|\boldsymbol{a}}(\boldsymbol{\rho} \mid \boldsymbol{\alpha}) = p_{\boldsymbol{z}}(\boldsymbol{\rho}) \frac{P[\boldsymbol{a} = \boldsymbol{\alpha} \mid \boldsymbol{z} = \boldsymbol{\rho}]}{P[\boldsymbol{a} = \boldsymbol{\alpha}]} \tag{7.345}$$

the decision criterion becomes

$$\hat{\boldsymbol{a}} = \arg\max_{\boldsymbol{\alpha} \in \mathcal{A}^K} P[\boldsymbol{a} = \boldsymbol{\alpha} \mid \boldsymbol{z} = \boldsymbol{\rho}] \tag{7.346}$$

An efficient realization of the MAP criterion will be developed in Section 7.5.2.

If all data sequences are equally likely, the MAP criterion coincides with the maximum-likelihood sequence detection (MLSD) criterion

$$\hat{\boldsymbol{a}} = \arg\max_{\boldsymbol{\alpha} \in \mathcal{A}^K} p_{\boldsymbol{z}|\boldsymbol{a}}(\boldsymbol{\rho} \mid \boldsymbol{\alpha}) \tag{7.347}$$

In other words, *the sequence $\boldsymbol{\alpha}$ is chosen, for which the probability to observe $\boldsymbol{z} = \boldsymbol{\rho}$ is maximum.*

Note that the above derivation is simply a vector extension of Section 6.2.1.

7.5.1 Maximum-likelihood sequence detection

We now discuss a computationally efficient method to find the solution indicated by (7.347). As the vector $\boldsymbol{a}$ of transmitted symbols is hypothesized to assume the value $\boldsymbol{\alpha}$, both $\boldsymbol{a}$ and $\boldsymbol{u} = [u_0, \dots, u_{K-1}]^T$ are fixed. Recalling that the noise samples are statistically independent, from (7.334), we get[12]

$$p_{\boldsymbol{z}|\boldsymbol{a}}(\boldsymbol{\rho} \mid \boldsymbol{\alpha}) = \prod_{k=0}^{K-1} p_{z_k|\boldsymbol{a}}(\rho_k \mid \boldsymbol{\alpha}) \tag{7.348}$$

[12] We note that (7.348) formally requires that the vector $\boldsymbol{a}$ is extended to include the symbols $a_{-L_2}, \dots, a_{L_1-1}$. In fact, the symbols may be known (see (7.395) or unknown (see (7.391)).

As w_k is a complex-valued Gaussian r.v. with zero mean and variance σ_w^2, it follows

$$p_{z|a}(\rho \mid \alpha) = \prod_{k=0}^{K-1} \frac{1}{\pi\sigma_w^2} e^{-\frac{1}{\sigma_w^2}|\rho_k - u_k|^2} \tag{7.349}$$

Taking the logarithm, which is a monotonic increasing function, of both members, we get

$$-\ln p_{z|a}(\rho \mid \alpha) \propto \sum_{k=0}^{K-1} |\rho_k - u_k|^2 \tag{7.350}$$

where non-essential constant terms have been neglected.

Then the MLSD criterion is formulated as

$$\hat{a} = \arg\min_{\alpha} \sum_{k=0}^{K-1} |\rho_k - u_k|^2 \tag{7.351}$$

where u_k, defined by (7.335), is a function of the transmitted symbols expressed by the general relation

$$u_k = \tilde{f}(a_{k+L_1}, \ldots, a_k, \ldots, a_{k-L_2}) \tag{7.352}$$

We note that (7.351) is a particular case of the minimum distance criterion, and it suggests to detect the vector $\boldsymbol{u}$ that is closest to the observed vector $\boldsymbol{\rho}$. However, we are interested in detecting the symbols $\{a_k\}$ and not the components $\{u_k\}$.

A *direct computation method* requires that, given the sequence of observed samples, for each possible data sequence $\boldsymbol{\alpha}$ of length K, the corresponding K output samples, elements of the vector $\boldsymbol{u}$, should be determined, and the relative distance, or metric, should be computed as

$$\Gamma_{K-1} = \sum_{k=0}^{K-1} |\rho_k - u_k|^2 \tag{7.353}$$

The detected sequence is the sequence that yields the smallest value of $\Gamma(K-1)$; as in the case of i.i.d. symbols, there are M^K possible sequences, this method has a complexity $O(M^K)$.

Lower bound to error probability using MLSD

We interpret the vector $\boldsymbol{u}$ as a function of the sequence $\boldsymbol{a}$, that is $\boldsymbol{u} = \boldsymbol{u}(\boldsymbol{a})$. In the signal space spanned by $\boldsymbol{u}$, we compute the distances

$$d^2(\boldsymbol{u}(\boldsymbol{\alpha}), \boldsymbol{u}(\boldsymbol{\beta})) = \|\boldsymbol{u}(\boldsymbol{\alpha}) - \boldsymbol{u}(\boldsymbol{\beta})\|^2 \tag{7.354}$$

for each possible pair of distinct $\boldsymbol{\alpha}$ and $\boldsymbol{\beta}$ in $\mathcal{A}^K$. As the noise is white, we define

$$\Delta_{min}^2 = \min_{\boldsymbol{\alpha},\boldsymbol{\beta}} d^2(\boldsymbol{u}(\boldsymbol{\alpha}), \boldsymbol{u}(\boldsymbol{\beta})) \tag{7.355}$$

Then the minimum-distance lower bound on the error probability [3, (5.116)] can be used, and we get

$$P_e \geq \frac{N_{min}}{M^K} Q\left(\frac{\Delta_{min}}{2\sigma_I}\right) \tag{7.356}$$

where M^K is the number of vectors $\boldsymbol{u}(\boldsymbol{a})$, and N_{min} is the number of vectors $\boldsymbol{u}(\boldsymbol{a})$ whose distance from another vector is Δ_{min}.

In practice, the exhaustive method for the computation of the expressions in (7.353) and (7.355) is not used; the Viterbi algorithm, that will be discussed in the next section, is utilized instead.

The Viterbi algorithm

The Viterbi algorithm (VA) efficiently implements the maximum likelihood (ML) criterion. With reference to (7.352), it is convenient to describe $\{u_k\}$ as the output sequence of a *finite-state machine* (FSM), as discussed in Appendix 7.B. In this case the input is a_{k+L_1}, the state is

$$s_k = (a_{k+L_1}, a_{k+L_1-1}, \ldots, a_k, \ldots, a_{k-L_2+1}) \tag{7.357}$$

and the output is given in general by (7.352).

We denote by $\mathcal{S}$ the set of the states, that is the set of possible values of s_k:

$$s_k \in \mathcal{S} = \{\sigma_1, \sigma_2, \ldots, \sigma_{N_s}\} \tag{7.358}$$

With the assumption of i.i.d. symbols, the number of states is equal to $N_s = M^{L_1+L_2}$.

Observation 7.13

We denote by s'_{k-1} the vector that is obtained by removing from s_{k-1} the oldest symbol, a_{k-L_2}. Then

$$s_k = (a_{k+L_1}, s'_{k-1}) \tag{7.359}$$

From (7.352) and (7.357), we may define u_k as a function of s_k and s_{k-1} as

$$u_k = f(s_k, s_{k-1}) \tag{7.360}$$

Defining the metric

$$\Gamma_k = \sum_{i=0}^{k} |\rho_i - u_i|^2 \tag{7.361}$$

the following recursive equation holds:

$$\Gamma_k = \Gamma_{k-1} + |\rho_k - u_k|^2 \tag{7.362}$$

or, using (7.360),

$$\Gamma_k = \Gamma_{k-1} + |\rho_k - f(s_k, s_{k-1})|^2 \tag{7.363}$$

Thus, we have interpreted $\{u_k\}$ as the output sequence of a finite state machine, and we have expressed recursively the metric (7.353). Note that the metric is a function of the sequence of states $s_0, s_1, \ldots, s_k$, associated with the sequence of output samples $u_0, u_1, \ldots, u_k$. The following example illustrates how to describe the transitions between states of the FSM.

Example 7.5.1

Let us consider a transmission system with symbols taken from a binary alphabet, that is $M = 2$, $a_k \in \{-1, 1\}$, and overall impulse response characterized by $L_1 = L_2 = 1$.

In this case, we have

$$s_k = (a_{k+1}, a_k) \tag{7.364}$$

and the set of states contains $N_s = 2^2 = 4$ elements:

$$\mathcal{S} = \{\sigma_1 = (-1, -1), \sigma_2 = (-1, 1), \sigma_3 = (1, -1), \sigma_4 = (1, 1)\} \tag{7.365}$$

The possible transitions

$$s_{k-1} = \sigma_i \quad \longrightarrow \quad s_k = \sigma_j \tag{7.366}$$

are represented in Figure 7.42, where a dot indicates a possible value of the state at a certain instant, and a branch indicates a possible transition between two states at consecutive instants. According to (7.359), the variable that determines a transition is a_{k+L_1}. The diagram of Figure 7.42, extended for all instants k, is called *trellis diagram*. We note that in this case there are exactly M transitions that leave each state s_{k-1}; likewise, there are M transitions that arrive to each state s_k.

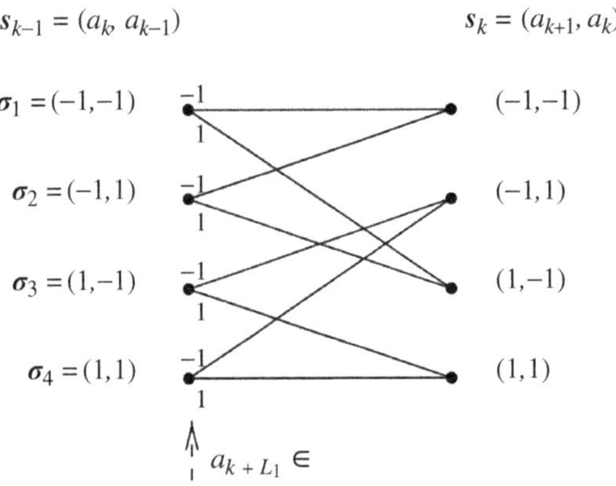

Figure 7.42 Portion of the trellis diagram showing the possible transitions from state s_{k-1} to state s_k, as a function of the symbol $a_{k+L_1} \in \mathcal{A}$.

With each state σ_j, $j = 1, \dots, N_s$, at instant k we associate two quantities:

1. the *path metric*, or *cost* function, defined as:

$$\Gamma(s_k = \sigma_j) = \min_{s_0, s_1, \dots, s_k = \sigma_j} \Gamma_k \; ; \tag{7.367}$$

2. the *survivor sequence*, defined as the sequence of symbols that ends in that state and determines $\Gamma(s_k = \sigma_j)$:

$$\mathcal{L}(s_k = \sigma_j) = (s_0, s_1, \dots, s_k = \sigma_j) = (a_{L_1}, \dots, a_{k+L_1}) \tag{7.368}$$

Note that the notion of survivor sequence can be equivalently applied to a sequence of symbols or to a sequence of states.

These two quantities are determined recursively. In fact, it is easy to verify that if, at instant k, a survivor sequence of states includes $s_k = \sigma_j$ then, at instant $k-1$, the same sequence includes $s_{k-1} = \sigma_{i_{opt}}$, which is determined as follows:

$$\sigma_{i_{opt}} = \arg \min_{s_{k-1} = \sigma_i \in \mathcal{S} \to s_k = \sigma_j} \Gamma(s_{k-1} = \sigma_i) + |\rho_k - f(\sigma_j, \sigma_i)|^2 \tag{7.369}$$

The term $|\rho_k - f(\sigma_j, \sigma_i)|^2$ is called *branch metric*. Therefore, we obtain

$$\Gamma(s_k = \sigma_j) = \Gamma(s_{k-1} = \sigma_{i_{opt}}) + |\rho_k - f(\sigma_j, \sigma_{i_{opt}})|^2 \tag{7.370}$$

and the survivor sequence is augmented as follows:

$$\mathcal{L}(s_k = \sigma_j) = (\mathcal{L}(s_{k-1} = \sigma_{i_{opt}}), \sigma_j) \tag{7.371}$$

Starting from $k = 0$, with initial state s_{-1}, which may be known or arbitrary, the procedure is repeated until $k = K - 1$. The optimum sequence of states is given by the survivor sequence $\mathcal{L}(s_{K-1} = \sigma_{j_{opt}})$ associated with $s_{K-1} = \sigma_{j_{opt}}$ having minimum cost.

If the state s_{-1} is known and equal to σ_{i_0}, it is convenient to assign to the states s_{-1} the following costs:

$$\Gamma(s_{-1} = \sigma_i) = \begin{cases} 0, & \text{for } \sigma_i = \sigma_{i_0} \\ \infty, & \text{otherwise} \end{cases} \tag{7.372}$$

Analogously, if the final state s_K is equal to σ_{f_0}, the optimum sequence of states coincides with the survivor sequence associated with the state $s_{K-1} = \sigma_j$ having minimum cost *among* those that admit a transition into $s_K = \sigma_{f_0}$.

Example 7.5.2

Let us consider a system with the following characteristics: $a_k \in \{-1, 1\}$, $L_1 = 0$, $L_2 = 2$, $K = 4$, and $s_{-1} = (-1, -1)$. The development of the survivor sequences on the *trellis diagram* from $k = 0$ to $k = K - 1 = 3$ is represented in Figure 7.43a. The branch metric, $|\rho_k - f(\sigma_j, \sigma_i)|^2$, associated with each transition is given in this example and is written above each branch. The survivor paths associated with each state are represented in bold; we note that some paths are abruptly interrupted and not extended at the following instant: for example the path ending at state σ_3 at instant $k = 1$.

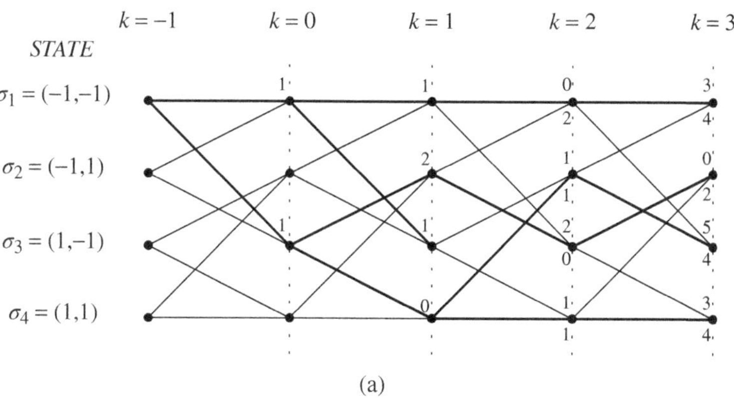

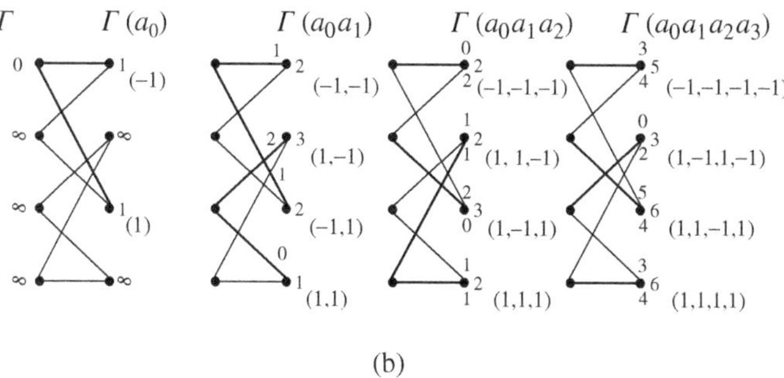

Figure 7.43 Trellis diagram and determination of the survivor sequences. (a) Trellis diagram from $k = -1$ to $k = K - 1 = 3$. (b) Determination of the survivors.

Figure 7.43b illustrates how the survivor sequences of Figure 7.43a are determined. Starting with $s_{-1} = (-1, -1)$, we have

$$\begin{aligned} \Gamma(s_{-1} = \sigma_1) &= 0 \\ \Gamma(s_{-1} = \sigma_2) &= \infty \\ \Gamma(s_{-1} = \sigma_3) &= \infty \\ \Gamma(s_{-1} = \sigma_4) &= \infty \end{aligned} \tag{7.373}$$

We apply (7.369) for $k = 0$; starting with $s_0 = \sigma_1$, we obtain

$$\begin{aligned} \Gamma(s_0 = \sigma_1) &= \min\{\Gamma(s_{-1} = \sigma_1) + 1, \Gamma(s_{-1} = \sigma_2) + 3\} \\ &= \min\{1, \infty\} \\ &= 1 \end{aligned} \tag{7.374}$$

We observe that the result (7.374) is obtained for $s_{-1} = \sigma_1$. Then the survivor sequence associated with $s_0 = \sigma_1$ is $\mathcal{L}(s_0 = \sigma_1) = (-1)$, expressed as a sequence of symbols rather than states.

Considering now $s_0 = \sigma_2$, σ_3, and σ_4, in sequence, and applying (7.369), the first iteration is completed. Obviously, there is no interest in determining the survivor sequence for states with metric equal to ∞, because the corresponding path will not be extended.

Next, for $k = 1$, the final metrics and the survivor sequences are shown in the second diagram of Figure 7.43b, where we have

$$\Gamma(s_1 = \sigma_1) = \min\{1 + 1, \infty\} = 2 \tag{7.375}$$

$$\Gamma(s_1 = \sigma_2) = \min\{1 + 2, \infty\} = 3 \tag{7.376}$$

$$\Gamma(s_1 = \sigma_3) = \min\{1 + 1, \infty\} = 2 \tag{7.377}$$

$$\Gamma(s_1 = \sigma_4) = \min\{1 + 0, \infty\} = 1 \tag{7.378}$$

The same procedure is repeated for $k = 2, 3$, and the trellis diagram is completed. The minimum among the values assumed by $\Gamma(s_3)$ is

$$\min_{i \in \{1,\dots,N_s\}} \Gamma(s_3 = \sigma_i) = \Gamma(s_3 = \sigma_2) = 3 \tag{7.379}$$

The associated optimum survivor sequence is

$$a_0 = 1, \qquad a_1 = -1, \qquad a_2 = 1, \qquad a_3 = -1 \tag{7.380}$$

In practice, if the parameter K is large, the length of the survivor sequences is limited to a value K_d, called *transient length* or *trellis depth* or *path memory depth*, that typically is between 3 and 10 times the length of the channel impulse response: this means that at every instant k, we decide on $a_{k-K_d+L_1}$, a value that is then removed from the diagram. The decision is based on the survivor sequence associated with the minimum among the values of $\Gamma(s_k)$.

Indeed, in most cases, we should find that survivor sequences associated with *all* states at time k, contain the same symbol value in correspondence of $a_{k-K_d+L_1}$, hence, this symbol can be detected.

In practice, k and $\Gamma(s_k)$ may become very large; then the latter value is usually normalized by subtracting the same amount from all the metrics, for example the smallest of the metrics $\Gamma(s_k = \sigma_i)$, $i = 1, \dots, N_s$.

Computational complexity of the VA

Memory. The memory to store the metrics $\Gamma(s_k)$, $\Gamma(s_{k-1})$ and the survivor sequences is proportional to the number of states N_s.

Computational complexity. The number of additions and comparisons is proportional to the number of transitions, $MN_s = M^{(L_1+L_2+1)}$. In any case, the complexity is linear in K.

We note that the values of $f(\sigma_j, \sigma_i)$, that determine u_k in (7.360) can be memorized in a table, limited to the possible transitions $s_{k-1} = \sigma_i \to s_k = \sigma_j$, for $\sigma_i, \sigma_j \in \mathcal{S}$. However, at every instant k and for every transition $\sigma_i \to \sigma_j$, the branch metric $|\rho_k - f(\sigma_j, \sigma_i)|^2$ needs to be evaluated: this computation, however, can be done outside the recursive algorithm.

7.5.2 Maximum a posteriori probability detector

We now discuss an efficient method to compute the a posteriori probability (APP) in (7.346) for an isolated symbol. Given a vector $z = [z_0, z_1, \dots, z_{K-1}]^T$, the notation z_ℓ^m indicates the vector formed only by the components $[z_\ell, z_{\ell+1}, \dots, z_m]^T$.

We introduce the *likelihood* function,

$$\mathrm{L}_{a_{k+L_1}}(\alpha) = P[a_{k+L_1} = \alpha \mid z_0^{K-1} = \rho_0^{K-1}], \qquad \alpha \in \mathcal{A} \tag{7.381}$$

Then, we have

$$\hat{a}_{k+L_1} = \arg\max_{\alpha \in \mathcal{A}} \mathrm{L}_{a_{k+L_1}}(\alpha), \qquad k = 0, 1, \dots, K-1 \tag{7.382}$$

As for the VA, it is convenient to define the state

$$s_k = (a_{k+L_1}, a_{k+L_1-1}, \dots, a_{k-L_2+1}) \tag{7.383}$$

so that the desired signal at instant k, u_k, given by (7.335), turns out to be only a function of s_k and s_{k-1} (see (7.360)).

The formulation that we give for the solution of the problem (7.382) is called *forward–backward algorithm* (FBA) and it follows the work by Rabiner [15]: it is seen that it coincides with the Bahl-Cocke–Jelinek–Raviv (BCJR) algorithm [16] for the decoding of convolutional codes.

We observe that, unlike in (7.383), in the two formulations [15, 16] the definition of state also includes the symbol a_{k-L_2} and, consequently, u_k is only a function of the state at the instant k.

Statistical description of a sequential machine

Briefly, we give a statistical description of the sequential machine associated with the state s_k.

1. Let N_s be the number of values that s_k can assume. For a sequence of i.i.d. symbols $\{a_k\}$, we have $N_s = M^{L_1+L_2}$; as in (7.359), the values assumed by the state are denoted by σ_j, $j = 1, \dots, N_s$.
2. The sequence $\{s_k\}$ is obtained by a time invariant sequential machine with transition probabilities

$$\Pi(j \mid i) = P[s_k = \sigma_j \mid s_{k-1} = \sigma_i] \tag{7.384}$$

Now, if there is a transition from $s_{k-1} = \sigma_i$ to $s_k = \sigma_j$, determined by the symbol $a_{k+L_1} = \alpha$, $\alpha \in \mathcal{A}$, then

$$\Pi(j \mid i) = P[a_{k+L_1} = \alpha] \tag{7.385}$$

which represents the a priori probability of the generic symbol. There are algorithms to iteratively estimate this probability from the output of another decoder or equalizer (see Section 11.6). Here, for the time being, we assume there is no a priori knowledge on the symbols. Consequently, for i.i.d. symbols, we have that every state $s_k = \sigma_j$ can be reached through M states, and

$$P[a_{k+L_1} = \alpha] = \frac{1}{M}, \qquad \alpha \in \mathcal{A} \tag{7.386}$$

If there is no transition from $s_{k-1} = \sigma_i$ to $s_k = \sigma_j$, we set

$$\Pi(j \mid i) = 0 \tag{7.387}$$

3. The channel transition probabilities are given by

$$p_{z_k}(\rho_k \mid j, i) = P[z_k = \rho_k \mid s_k = \sigma_j, s_{k-1} = \sigma_i] \tag{7.388}$$

assuming that there is a transition from $s_{k-1} = \sigma_i$ to $s_k = \sigma_j$. For a channel with complex-valued additive Gaussian noise, (6.23) holds, and

$$p_{z_k}(\rho_k \mid j, i) = \frac{1}{\pi\sigma_w^2} e^{-\frac{1}{\sigma_w^2} |\rho_k - u_k|^2} \tag{7.389}$$

where $u_k = f(\sigma_j, \sigma_i)$.

4. We merge (7.384) and (7.388) by defining the variable

$$\begin{aligned} C_k(j \mid i) &= P[z_k = \rho_k, s_k = \sigma_j \mid s_{k-1} = \sigma_i] \\ &= P[z_k = \rho_k \mid s_k = \sigma_j, s_{k-1} = \sigma_i] \; P[s_k = \sigma_j \mid s_{k-1} = \sigma_i] \\ &= p_{z_k}(\rho_k \mid j, i) \; \Pi(j \mid i) \end{aligned} \tag{7.390}$$

5. Initial and final conditions are given by

$$\overline{p}_j = P[s_{-1} = \sigma_j] = \frac{1}{N_s}, \quad j = 1, \dots, N_s \tag{7.391}$$

$$\overline{q}_j = P[s_K = \sigma_j] = \frac{1}{N_s}, \quad j = 1, \dots, N_s \tag{7.392}$$

$$C_K(j \mid i) = \Pi(j \mid i), \quad i,j = 1, \dots, N_s \tag{7.393}$$

If the initial and/or final state are known, for example

$$s_{-1} = \sigma_{i_0}, \quad s_K = \sigma_{f_0} \tag{7.394}$$

we set

$$\overline{p}_j = \begin{cases} 1 & \text{for } \sigma_j = \sigma_{i_0} \\ 0 & \text{otherwise} \end{cases} \tag{7.395}$$

and

$$\overline{q}_j = \begin{cases} 1 & \text{for } \sigma_j = \sigma_{f_0} \\ 0 & \text{otherwise} \end{cases} \tag{7.396}$$

The forward–backward algorithm

We consider the following four metrics.

(A) Forward metric:

$$F_k(j) = P[z_0^k = \rho_0^k, s_k = \sigma_j] \tag{7.397}$$

Equation (7.397) gives the probability of observing the sequence $\rho_0, \rho_1, \dots, \rho_k$, up to instant k, and the state σ_j at instant k.

Theorem 7.1

$F_k(j)$ can be defined recursively as follows:

1. Initialization

$$F_{-1}(j) = \overline{p}_j, \quad j = 1, \dots, N_s \tag{7.398}$$

2. Updating for $k = 0, 1, \dots, K-1$,

$$F_k(j) = \sum_{\ell=1}^{N_s} C_k(j \mid \ell)\, F_{k-1}(\ell), \qquad j = 1, \dots, N_s \tag{7.399}$$

□

Proof. Using the total probability theorem, and conditioning the event on the possible values of $\boldsymbol{s}_{k-1}$, we express the probability in (7.397) as

$$\begin{aligned} F_k(j) &= \sum_{\ell=1}^{N_s} P[\boldsymbol{z}_0^{k-1} = \boldsymbol{\rho}_0^{k-1}, z_k = \rho_k, \boldsymbol{s}_k = \boldsymbol{\sigma}_j, \boldsymbol{s}_{k-1} = \boldsymbol{\sigma}_\ell] \\ &= \sum_{\ell=1}^{N_s} P[\boldsymbol{z}_0^{k-1} = \boldsymbol{\rho}_0^{k-1}, z_k = \rho_k \mid \boldsymbol{s}_k = \boldsymbol{\sigma}_j, \boldsymbol{s}_{k-1} = \boldsymbol{\sigma}_\ell]\, P[\boldsymbol{s}_k = \boldsymbol{\sigma}_j, \boldsymbol{s}_{k-1} = \boldsymbol{\sigma}_\ell] \end{aligned} \tag{7.400}$$

Because the noise samples are i.i.d., once the values of $\boldsymbol{s}_k$ and $\boldsymbol{s}_{k-1}$ are assigned, the event $[\boldsymbol{z}_0^{k-1} = \boldsymbol{\rho}_0^{k-1}]$ is independent of the event $[z_k = \rho_k]$, and it results

$$\begin{aligned} F_k(j) = \sum_{\ell=1}^{N_s} & P[\boldsymbol{z}_0^{k-1} = \boldsymbol{\rho}_0^{k-1} \mid \boldsymbol{s}_k = \boldsymbol{\sigma}_j, \boldsymbol{s}_{k-1} = \boldsymbol{\sigma}_\ell] \\ & P[z_k = \rho_k \mid \boldsymbol{s}_k = \boldsymbol{\sigma}_j, \boldsymbol{s}_{k-1} = \boldsymbol{\sigma}_\ell]\, P[\boldsymbol{s}_k = \boldsymbol{\sigma}_j, \boldsymbol{s}_{k-1} = \boldsymbol{\sigma}_\ell] \end{aligned} \tag{7.401}$$

Moreover, given $\boldsymbol{s}_{k-1}$, the event $[\boldsymbol{z}_0^{k-1} = \boldsymbol{\rho}_0^{k-1}]$ is independent of $\boldsymbol{s}_k$, and we have

$$\begin{aligned} F_k(j) = \sum_{\ell=1}^{N_s} & P[\boldsymbol{z}_0^{k-1} = \boldsymbol{\rho}_0^{k-1} \mid \boldsymbol{s}_{k-1} = \boldsymbol{\sigma}_\ell] \\ & P[z_k = \rho_k \mid \boldsymbol{s}_k = \boldsymbol{\sigma}_j, \boldsymbol{s}_{k-1} = \boldsymbol{\sigma}_\ell]\, P[\boldsymbol{s}_k = \boldsymbol{\sigma}_j, \boldsymbol{s}_{k-1} = \boldsymbol{\sigma}_\ell] \end{aligned} \tag{7.402}$$

By applying Bayes' rule, (7.402) becomes

$$\begin{aligned} F_k(j) &= \sum_{\ell=1}^{N_s} P[\boldsymbol{z}_0^{k-1} = \boldsymbol{\rho}_0^{k-1}, \boldsymbol{s}_{k-1} = \boldsymbol{\sigma}_\ell] \frac{1}{P[\boldsymbol{s}_{k-1} = \boldsymbol{\sigma}_\ell]} P[z_k = \rho_k, \boldsymbol{s}_k = \boldsymbol{\sigma}_j, \boldsymbol{s}_{k-1} = \boldsymbol{\sigma}_\ell] \\ &= \sum_{\ell=1}^{N_s} P[\boldsymbol{z}_0^{k-1} = \boldsymbol{\rho}_0^{k-1}, \boldsymbol{s}_{k-1} = \boldsymbol{\sigma}_\ell] P[z_k = \rho_k, \boldsymbol{s}_k = \boldsymbol{\sigma}_j \mid \boldsymbol{s}_{k-1} = \boldsymbol{\sigma}_\ell] \end{aligned} \tag{7.403}$$

Replacing (7.390) in (7.403) yields (7.399). □

(B) Backward metric:

$$B_k(i) = P[\boldsymbol{z}_{k+1}^{K-1} = \boldsymbol{\rho}_{k+1}^{K-1} \mid \boldsymbol{s}_k = \boldsymbol{\sigma}_i] \tag{7.404}$$

Equation (7.404) is the probability of observing the sequence $\rho_{k+1}, \dots, \rho_{K-1}$, from instant $k+1$ onwards, given the state $\boldsymbol{\sigma}_i$ at instant k.

Theorem 7.2

A recursive expression also exists for $B_k(i)$.

1. Initialization

$$B_K(i) = \overline{q}_i, \qquad i = 1, \dots, N_s \tag{7.405}$$

2. Updating for $k = K-1, K-2, \dots, 0$,

$$B_k(i) = \sum_{m=1}^{N_s} B_{k+1}(m)\, C_{k+1}(m \mid i), \qquad i = 1, \dots, N_s \tag{7.406}$$

□

Proof. Using the total probability theorem, and conditioning the event on the possible values of s_{k+1}, we express the probability in (7.404) as

$$B_k(i) = \sum_{m=1}^{N_s} P[z_{k+1}^{K-1} = \rho_{k+1}^{K-1}, s_{k+1} = \sigma_m \mid s_k = \sigma_i]$$
$$= \sum_{m=1}^{N_s} P[z_{k+1}^{K-1} = \rho_{k+1}^{K-1} \mid s_{k+1} = \sigma_m, s_k = \sigma_i] \; P[s_{k+1} = \sigma_m \mid s_k = \sigma_i] \tag{7.407}$$

Now, given the values of s_{k+1} and s_k, the event $[z_{k+2}^{K-1} = \rho_{k+2}^{K-1}]$ is independent of $[z_{k+1} = \rho_{k+1}]$. In turn, assigned the value of s_{k+1}, the event $[z_{k+2}^{K-1} = \rho_{k+2}^{K-1}]$ is independent of s_k.

Then (7.407) becomes

$$B_k(i) = \sum_{m=1}^{N_s} P[z_{k+2}^{K-1} = \rho_{k+2}^{K-1} \mid s_{k+1} = \sigma_m, s_k = \sigma_i]$$
$$P[z_{k+1} = \rho_{k+1} \mid s_{k+1} = \sigma_m, s_k = \sigma_i] \; P[s_{k+1} = \sigma_m \mid s_k = \sigma_i]$$
$$= \sum_{m=1}^{N_s} P[z_{k+2}^{K-1} = \rho_{k+2}^{K-1} \mid s_{k+1} = \sigma_m]$$
$$P[z_{k+1} = \rho_{k+1} \mid s_{k+1} = \sigma_m, s_k = \sigma_i] \; P[s_{k+1} = \sigma_m \mid s_k = \sigma_i] \tag{7.408}$$

Observing (7.404) and (7.390), (7.406) follows. □

(C) State metric:

$$V_k(i) = P[s_k = \sigma_i \mid z_0^{K-1} = \rho_0^{K-1}] \tag{7.409}$$

Equation (7.409) expresses the probability of being in the state σ_i at instant k, given the whole observation ρ_0^{K-1}.

Theorem 7.3

Equation (7.409) can be expressed as a function of the forward and backward metrics,

$$V_k(i) = \frac{F_k(i)\, B_k(i)}{\sum_{n=1}^{N_s} F_k(n)\, B_k(n)}, \qquad i = 1, \dots, N_s \tag{7.410}$$

□

Proof. Using the fact that, given the value of s_k, the r.v.s. $\{z_t\}$ with $t > k$ are statistically independent of $\{z_t\}$ with $t \le k$, from (7.409) it follows

$$V_k(i) = P[z_0^k = \rho_0^k, z_{k+1}^{K-1} = \rho_{k+1}^{K-1}, s_k = \sigma_i] \frac{1}{P[z_0^{K-1} = \rho_0^{K-1}]}$$
$$= P[z_0^k = \rho_0^k, s_k = \sigma_i] \; P[z_{k+1}^{K-1} = \rho_{k+1}^{K-1} \mid s_k = \sigma_i] \frac{1}{P[z_0^{K-1} = \rho_0^{K-1}]} \tag{7.411}$$

Observing the definitions of forward and backward metrics, (7.410) follows. □

We note that the normalization factor

$$P[z_0^{K-1} = \rho_0^{K-1}] = \sum_{n=1}^{N_s} F_k(n)\, B_k(n) \tag{7.412}$$

makes $V_k(i)$ a probability, so that

$$\sum_{i=1}^{N_s} V_k(i) = 1 \tag{7.413}$$

(D) Likelihood function of the generic symbol: Applying the total probability theorem to (7.381), we obtain the relation

$$\mathrm{L}_{a_{k+L_1}}(\alpha) = \sum_{i=1}^{N_s} P[a_{k+L_1} = \alpha, s_k = \boldsymbol{\sigma}_i \mid \mathbf{z}_0^{K-1} = \boldsymbol{\rho}_0^{K-1}] \tag{7.414}$$

From the comparison of (7.414) with (7.409), indicating with $[\boldsymbol{\sigma}_i]_m$, $m = 0, \dots, L_1 + L_2 - 1$, the mth component of the state $\boldsymbol{\sigma}_i$ (see (7.483)), we have

$$\mathrm{L}_{a_{k+L_1}}(\alpha) = \sum_{\substack{i=1 \\ \text{condition} \\ [\boldsymbol{\sigma}_i]_0 = \alpha}}^{N_s} V_k(i), \qquad \alpha \in \mathcal{A} \tag{7.415}$$

In other words, at instant k, the likelihood function coincides with the sum of the metrics $V_k(i)$ associated with the states whose first component is equal to the symbol of value α. Note that $\mathrm{L}_{a_{k+L_1}}(\alpha)$ can also be obtained using the state metrics evaluated at different instants, that is

$$\mathrm{L}_{a_{k+L_1}}(\alpha) = \sum_{\substack{i=1 \\ [\boldsymbol{\sigma}_i]_m = \alpha}}^{N_s} V_{k+m}(i), \qquad \alpha \in \mathcal{A} \tag{7.416}$$

for $m \in \{0, \dots, L_1 + L_2 - 1\}$.

Scaling

We see that, due to the exponential form of $p_{\mathbf{z}_k}(\boldsymbol{\rho}_k \mid j, i)$ in $C_k(j \mid i)$, in a few iterations, the forward and backward metrics may assume very small values; this leads to numerical problems in the computation of the metrics: therefore, we need to substitute equations (7.399) and (7.406) with analogous expressions that are scaled by a suitable coefficient.

We note that the state metric (7.410) does not change if we multiply $F_k(i)$ and $B_k(i)$, $i = 1, \dots, N_s$, by the same coefficient $\mathcal{K}_k$. The idea [15] is to choose

$$\mathcal{K}_k = \frac{1}{\sum_{n=1}^{N_s} F_k(n)} \tag{7.417}$$

Indicating with $\overline{F}_k(i)$ and $\overline{B}_k(i)$ the normalized metrics, for

$$F_k(j) = \sum_{\ell=1}^{N_s} C_k(j \mid \ell)\, \overline{F}_{k-1}(\ell) \tag{7.418}$$

Equation (7.399) becomes

$$\overline{F}_k(j) = \frac{\displaystyle\sum_{\ell=1}^{N_s} C_k(j \mid \ell)\, \overline{F}_{k-1}(\ell)}{\displaystyle\sum_{n=1}^{N_s} \sum_{\ell=1}^{N_s} C_k(n \mid \ell)\, \overline{F}_{k-1}(\ell)}, \qquad \begin{array}{l} j = 1, \dots, N_s \\ k = 0, 1, \dots, K-1 \end{array} \tag{7.419}$$

Correspondingly, (7.406) becomes

$$\overline{B}_k(i) = \frac{\sum_{m=1}^{N_s} \overline{B}_{k+1}(m)\, C_{k+1}(m \mid i)}{\sum_{n=1}^{N_s} \sum_{\ell=1}^{N_s} C_k(n \mid \ell)\, \overline{F}_{k-1}(\ell)}, \qquad \begin{matrix} i = 1, \dots, N_s \\ k = K-1, K-2, \dots, 0 \end{matrix} \tag{7.420}$$

Hence,

$$V_k(i) = \frac{\overline{F}_k(i)\, \overline{B}_k(i)}{\sum_{n=1}^{N_s} \overline{F}_k(n)\, \overline{B}_k(n)}, \qquad \begin{matrix} i = 1, \dots, N_s \\ k = 0, 1, \dots, K-1 \end{matrix} \tag{7.421}$$

The log likelihood function and the Max-Log-MAP criterion

We introduce the Log-MAP criterion, which employs the logarithm of the variables $F_k(i)$, $B_k(i)$, $C_k(j \mid i)$, $V_k(i)$, and $\mathrm{L}_{a_{k+L_1}}(\alpha)$. The logarithmic variables are indicated with the corresponding lower-case letters; in particular, from (7.410), we have

$$v_k(i) = \ln V_k(i) \tag{7.422}$$

and (7.415) becomes

$$\ell_{a_{k+L_1}}(\alpha) = \ln \mathrm{L}_{a_{k+L_1}}(\alpha) = \ln \left(\sum_{\substack{i=1 \\ [\sigma_i]_0 = \alpha}}^{N_s} e^{v_k(i)} \right), \qquad \alpha \in \mathcal{A} \tag{7.423}$$

The function $\ell_k(\alpha)$ is called *log-likelihood* (LL).

The exponential emphasize the difference between the metrics $v_k(i)$: typically a term dominates within each sum; this suggests the approximation

$$\ln \sum_{i=1}^{N_s} e^{v_k(i)} \simeq \max_{i \in \{1,\dots,N_s\}} v_k(i) \tag{7.424}$$

Consequently, (7.423) is *approximated* as

$$\tilde{\ell}_{a_{k+L_1}}(\alpha) = \max_{\substack{i \in \{1, \dots, N_s\} \\ [\sigma_i]_0 = \alpha}} v_k(i) \tag{7.425}$$

and the Log-MAP criterion is replaced by the Max-Log–MAP criterion

$$\hat{a}_{k+L_1} = \arg\max_{\alpha \in \mathcal{A}} \tilde{\ell}_{a_{k+L_1}}(\alpha) \tag{7.426}$$

Apart from non-essential constants, the Max-Log–MAP algorithm in the case of transmission of i.i.d. symbols over a dispersive channel with additive white Gaussian noise is formulated as follows:

1. *Computation of channel transition metrics*: For $k = 0, 1, \dots, K-1$,

$$c_k(j \mid i) = -|\rho_k - u_k|^2, \qquad i,j = 1, \dots, N_s \tag{7.427}$$

where $u_k = f(\sigma_j, \sigma_i)$, assuming there is a transition between $s_{k-1} = \sigma_i$ and $s_k = \sigma_j$. For $k = K$, we let

$$c_K(j \mid i) = 0 \qquad i,j = 1, \dots, N_s \tag{7.428}$$

again, assuming there is a transition between σ_i and σ_j.

2. *Backward procedure*: For $k = K-1, K-2, \dots, 0$,

$$\tilde{b}_k(i) = \max_{m\in\{1,\dots,N_s\}} [\tilde{b}_{k+1}(m) + c_{k+1}(m \mid i)], \qquad i = 1, \dots, N_s \tag{7.429}$$

If the final state is known, then

$$\tilde{b}_K(i) = \begin{cases} 0, & \boldsymbol{\sigma}_i = \boldsymbol{\sigma}_{f_0} \\ -\infty, & \text{otherwise} \end{cases} \tag{7.430}$$

If the final state is unknown, we set $\tilde{b}_K(i) = 0$, $i = 1, \dots, N_s$.

3. *Forward procedure*: For $k = 0, 1, \dots, K-1$,

$$\tilde{f}_k(j) = \max_{\ell\in\{1,\dots,N_s\}} [\tilde{f}_{k-1}(\ell) + c_k(j \mid \ell)], \qquad j = 1, \dots, N_s \tag{7.431}$$

If the initial state is known, then

$$\tilde{f}_{-1}(j) = \begin{cases} 0, & \boldsymbol{\sigma}_j = \boldsymbol{\sigma}_{i_0} \\ -\infty, & \text{otherwise} \end{cases} \tag{7.432}$$

If the initial state is unknown, we set $\tilde{f}_{-1}(j) = 0$, $j = 1, \dots, N_s$.

4. *State metric*: For $k = 0, 1, \dots, K-1$,

$$\tilde{v}_k(i) = \tilde{f}_k(i) + \tilde{b}_k(i), \qquad i = 1, \dots, N_s \tag{7.433}$$

5. *LL function of an isolated symbol*: For $k = 0, 1, \dots, K-1$, the LL function is given by (7.425), with $\tilde{v}_k(i)$ in substitution of $v_k(i)$; the decision rule is given by (7.427).

In practical implementations of the algorithm, steps 3–5 can be carried out in sequence for each value of k: this saves memory locations. To avoid overflow, for each value of k, a common value can be added to all variables $\tilde{f}_k(i)$ and $\tilde{b}_k(i)$, $i = 1, \dots, N_s$.

We observe that the two procedures, backward and forward, can be efficiently implemented by a VA trellis diagram run both in backward and forward directions. The simplified MAP algorithm requires about twice the complexity of the VA implementing the MLSD criterion. Memory requirements are considerably increased with respect to the VA, because the backward metrics must be stored before evaluating the state metrics. However, methods for an efficient use of the memory are proposed in [17]. The most common approach is splitting the very long receiver sequence into shorter subsequences over which the algorithm is run. If K_d is the transient length, now at the beginning and at the end of each detected subsequence, the various receiver subsequences must overlap by $2K_d$ samples.

LLRs associated to bits of BMAP

In the multilevel case, we recall from (6.6) and (6.7) that each value of a_k is identified by $2\mathtt{m} = \log_2 M$ bits by

$$a_k \overset{\text{BMAP}}{\longleftrightarrow} (b_{k2\mathtt{m}}, \dots, b_{k2\mathtt{m}-2\mathtt{m}+1}) \tag{7.434}$$

or by

$$(a_{k,I}; a_{k,Q}) \overset{\text{BMAP}}{\longleftrightarrow} (b_{k\mathtt{m},I}, \dots, b_{k\mathtt{m}-\mathtt{m}+1,I}; b_{k\mathtt{m},Q}, \dots, b_{k\mathtt{m}-\mathtt{m}+1,Q}) \tag{7.435}$$

If we denote the generic state value by a bit pattern $\boldsymbol{\sigma}_i^{bit}$ rather than a symbol pattern $\boldsymbol{\sigma}_i$, with regards to (7.434) the log-likelihood-ratio (LLR) associated with bits of bit mapper (BMAP) $b_{(k+L_1)2\mathtt{m}-m}$ is given by

$$\tilde{\ell}_{b_{(k+L_1)2\mathtt{m}-m}} = \max_{\substack{i\in\{1,\dots,N_s\}\\ [\boldsymbol{\sigma}_i^{bit}]_m = 1}} v_k(i) - \max_{\substack{i\in\{1,\dots,N_s\}\\ [\boldsymbol{\sigma}_i^{bit}]_m = 0}} v_k(i), \qquad m = 0, 1, \dots, 2\mathtt{m}-1 \tag{7.436}$$

One word of caution: on labelling a symbol value each bit assumes $\{0,1\}$ values. However, LLRs refer to $\{-1,1\}$, where the original 0 has become -1. Moreover, $v_k(i)$ in (7.436), as given by (7.422), could be replaced by its approximation $\tilde{v}_k(i)$ as given by (7.433). Note that (7.436), when $\{b_\ell\}$ is replaced by $\{c_m\}$, i.e. in the presence of coding, extends expressions of LLRs of Section 6.2.2 to the more general case of a channel with ISI. Indeed, their computational complexity is much higher, due to the computation of $\{v_k(i)\}$.

Relation between Max-Log–MAP and Log–MAP

We define the following function of two variables [17]

$$\max{}^*(x,y) = \ln(e^x + e^y) \tag{7.437}$$

it can be verified that the following relation holds:

$$\max{}^*(x,y) = \max(x,y) + \ln(1 + e^{-|x-y|}) \tag{7.438}$$

We now extend the above definition to the case of three variables,

$$\max{}^*(x,y,z) = \ln(e^x + e^y + e^z)\ ; \tag{7.439}$$

then we have

$$\max{}^*(x,y,z) = \max{}^*(\max{}^*(x,y),z) \tag{7.440}$$

The extension to more variables is readily obtained by induction. So, if in the backward and forward procedures of page (7.429), we substitute the max function with the max* function, we obtain the exact Log–MAP formulation that relates $v_k(i) = \ln V_k(i)$ to $b_k(i) = \ln B_k(i)$ and $f_k(i) = \ln F_k(i)$, using the branch metric $c_k(j \mid i)$.

7.5.3 Optimum receivers

Definition 7.2

Given a desired signal corrupted by noise, in general the notion of *sufficient statistics* applies to any signal, or sequence of samples, that allows the optimum detection of the desired signal. In other words, no information is lost in considering a set of sufficient statistics instead of the received signal. □

It is possible to identify two different receiver structures that supply the signal z_k given by (7.336).

1. The first, illustrated in Figure 7.44a, is considered for the low implementation complexity; it refers to the receiver of Figure 7.11, where

$$\mathcal{G}_{Rc}(f) = \sqrt{\text{rcos}\left(\frac{f}{1/T}, \rho\right)} \tag{7.441}$$

and w_C is white Gaussian noise with PSD N_0. Recalling that $r_R(t) = s_R(t) + w_R$, with

$$\mathcal{P}_{w_R}(f) = \mathcal{P}_w(f)\ |\mathcal{G}_{Rc}(f)|^2 = N_0\ \text{r}\cos\left(\frac{f}{1/T}, \rho\right) \tag{7.442}$$

it is immediate to verify that the noise sequence $\{w_k = w_R(t_0 + kT)\}$ has a constant PSD equal to N_0, and the variance of the noise samples is $\sigma_w^2 = N_0/T$.
Although the filter defined by (7.441) does not necessarily yield a sufficient statistic (see Observation 7.14 on page 357), it considerably reduces the noise and this may be useful in estimating the channel impulse response. Another problem concerns the optimum timing phase, which may be difficult to determine for non-minimum phase channels.

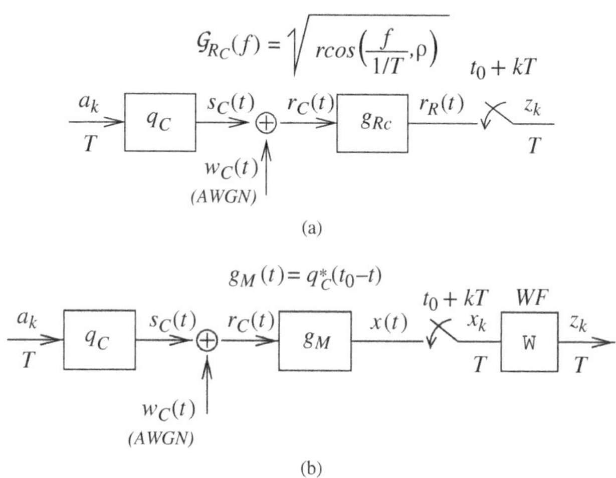

Figure 7.44 Two receiver structures with i.i.d. noise samples at the decision point.

2. An alternative, also known as the *Forney receiver*, is represented in Figure 7.35 and repeated in Figure 7.44b (see also Section 13.2). To construct the WF, however, it is necessary to determine poles and zeros of the function $\Phi(z)$; this can be rather complicated in real-time applications. A practical method is based on Observation 7.11 of page 327. From the knowledge of the autocorrelation sequence of the channel impulse response, the prediction error filter $A(z)$ is determined. The WF of Figure 7.44 is given by $W(z) = \frac{1}{\mathtt{f}_0^2} A^*(1/z^*)$; therefore, it is an FIR filter. The impulse response $\{\eta_n\}$ is given by the inverse z-transform of $\frac{1}{\mathtt{f}_0} F(z) = 1/A(z)$. For the realization of symbol detection algorithms, a windowed version of the impulse response is considered.
A further method, which usually requires a filter $\mathtt{w}$ with a smaller number of coefficients than the previous method, is based on the observation that, for channels with low noise level, the DFE solution determined by the MSE method coincides with the DFE-ZF solution; in this case, the FF filter plays the role of the filter $\mathtt{w}$ in Figure 7.44b. We consider two cases.
   a. At the output of the MF g_M, let $\{h_n\}$ be the system impulse response with sampling period T, determined, for example through the method described in Observation 7.6 on page 320. Using (7.194), a DFE is designed with filter parameters (M_1, M_2), and consequently $\mathtt{w}(t) = c_{opt}(t)$ in Figure 7.44b. At the output of the filter $\mathtt{w}$, the ideally minimum phase impulse response $\{\eta_n\}$ corresponds to the translated, by D sampling intervals, and windowed, with $L_1 = 0$ and $L_2 = M_2$, version of $\psi(t) = h * c_{opt}(t)$.
   b. If the impulse response of the system is unknown, we can resort to the FS-DFE structure of Figure 7.32 with

$$\eta_n = \psi_{n+D} \simeq \begin{cases} 1, & n = 0 \\ -b_n, & n = 1, \dots, M_2 \end{cases} \tag{7.443}$$

Actually, unless the length of $\{\eta_n\}$ is shorter than that of the impulse response q_C, it is convenient to use Ungerboeck's formulation of the MLSD [18] that utilizes only samples $\{x_k\}$ at the MF output; now, however, the metric is no longer Euclidean. The derivation of the non-Euclidean metric is the subject of the next section.

We note that, as it is not immediate to obtain the likelihood of an isolated symbol from the non-Euclidean metric, there are cases in which this method is not adequate. We refer in particular to the case in which decoding with soft input is performed separately from symbol detection in the presence of ISI (see Section 11.3.2). However, joint decoding and detection is always possible using a suitable trellis (see Section 11.3.2).

7.5.4 The Ungerboeck's formulation of MLSD

We refer to the transmission of K symbols and to an observation interval $T_K = KT$ sufficiently large, so that the transient of filters at the beginning and at the end of the transmission has a negligible effect.

The derivation of the likelihood is based on the received signal (7.42),

$$r_C(t) = s_C(t) + w_C(t) \tag{7.444}$$

where w_C is white noise with PSD N_0, and

$$s_C(t) = \sum_{k=0}^{K-1} a_k\, q_C(t-kT) \tag{7.445}$$

For a suitable basis, we consider for (7.444) the following vector representation:

$$\boldsymbol{r} = \boldsymbol{s} + \boldsymbol{w}$$

Assuming that the transmitted symbol sequence $\boldsymbol{a} = [a_0, a_1, \dots, a_{K-1}]$ is equal to $\boldsymbol{\alpha} = [\alpha_0, \alpha_1, \dots, \alpha_{K-1}]$, the probability density function of $\boldsymbol{r}$ is given by

$$p_{\boldsymbol{r}|\boldsymbol{a}}(\boldsymbol{\rho} \mid \boldsymbol{\alpha}) = K \exp\left(-\frac{1}{N_0}\, \|\boldsymbol{\rho} - \boldsymbol{s}\|^2\right) \tag{7.446}$$

Using the equivalence between a signal x and its vector representation $\boldsymbol{x}$,

$$E_x = \int_{-\infty}^{+\infty} |x(t)|^2 dt = \|\boldsymbol{x}\|^2 = E_{\boldsymbol{x}} \tag{7.447}$$

and observing $r_C(t) = \rho(t)$, we get

$$p_{\boldsymbol{r}|\boldsymbol{a}}(\boldsymbol{\rho} \mid \boldsymbol{\alpha}) = K \exp\left(-\frac{1}{N_0} \int_{T_K} |\rho(t) - s_C(t)|^2\, dt\right) \tag{7.448}$$

Taking the logarithm in (7.448), the log-likelihood (to be maximized) is

$$\ell(\boldsymbol{\alpha}) = -\int_{T_K} \left|\rho(t) - \sum_{k=0}^{K-1} \alpha_k\, q_C(t-kT)\right|^2 dt \tag{7.449}$$

Correspondingly, the detected sequence is given by

$$\hat{\boldsymbol{a}} = \arg\max_{\boldsymbol{\alpha}}\ \ell(\boldsymbol{\alpha}) \tag{7.450}$$

Expanding the square term in (7.449), we obtain

$$\begin{aligned}\ell(\boldsymbol{\alpha}) = -\Bigg\{ &\int_{T_K} |\rho(t)|^2\, dt - 2Re\left[\int_{T_K} \rho(t) \sum_{k=0}^{K-1} \alpha_k^*\, q_C^*(t-kT)\, dt\right] \\ &+ \int_{T_K} \sum_{k_1=0}^{K-1} \sum_{k_2=0}^{K-1} \alpha_{k_1} \alpha_{k_2}^*\, q_C(t-k_1T)\, q_C^*(t-k_2T)\, dt \Bigg\}\end{aligned} \tag{7.451}$$

We now introduce the MF

$$g_M(t) = q_C^*(t_0 - t) \tag{7.452}$$

and the overall impulse response at the MF output

$$q(t) = (q_C * g_M)(t) = \mathtt{r}_{q_C}(t - t_0) \tag{7.453}$$

where $\mathtt{r}_{q_C}$ is the autocorrelation of q_C, whose samples are given by (see (7.130))

$$h_n = q(t_0 + nT) = \mathtt{r}_{q_C}(nT) \tag{7.454}$$

The MF output signal is expressed as

$$x(t) = (r_C * g_M)(t) \tag{7.455}$$

with samples given by

$$x_k = x(t_0 + kT) \tag{7.456}$$

In (7.451), the first term can be neglected since it does not depend on $\boldsymbol{\alpha}$, while the other two terms are rewritten in the following form:

$$\ell(\boldsymbol{\alpha}) = -\left\{-2Re\left[\sum_{k=0}^{K-1} \alpha_k^* \, x_k\right] + \sum_{k_1=0}^{K-1}\sum_{k_2=0}^{K-1} \alpha_{k_1}\alpha_{k_2}^* \, h_{k_2-k_1}\right\} \tag{7.457}$$

Observing (7.457), we obtain the following important result.

Observation 7.14
The sequence of samples $\{x_k\}$, taken by sampling the MF output signal with sampling period equal to the symbol period T, forms a *sufficient statistic* to detect the message $\{a_k\}$ associated with the signal r_C defined in (7.444).

We express the double summation in (7.457) as the sum of three terms, the first for $k_1 = k_2$, the second for $k_1 < k_2$, and the third for $k_1 > k_2$:

$$\begin{aligned} A &= \sum_{k_1=0}^{K-1}\sum_{k_2=0}^{K-1} \alpha_{k_1}\alpha_{k_2}^* h_{k_2-k_1} \\ &= \sum_{k_1=0}^{K-1} \alpha_{k_1}\alpha_{k_1}^* h_0 + \sum_{k_1=1}^{K-1}\sum_{k_2=0}^{k_1-1} \alpha_{k_1}\alpha_{k_2}^* h_{k_2-k_1} + \sum_{k_2=1}^{K-1}\sum_{k_1=0}^{k_2-1} \alpha_{k_1}\alpha_{k_2}^* h_{k_2-k_1} \end{aligned} \tag{7.458}$$

Because the sequence $\{h_n\}$ is an autocorrelation, it enjoys the Hermitian property, i.e. $h_{-n} = h_n^*$; consequently, the third term in (7.458) is the complex conjugate of the second, and

$$A = \sum_{k=0}^{K-1} \alpha_k^* \alpha_k h_0 + 2Re\left[\sum_{k=1}^{K-1}\sum_{k_2=0}^{k-1} \alpha_k^* \alpha_{k_2} h_{k-k_2}\right] \tag{7.459}$$

By the change of indices $n = k - k_2$, assuming $\alpha_k = 0$ for $k < 0$, we get

$$\begin{aligned} A &= Re\left\{\sum_{k=0}^{K-1} \alpha_k^* \alpha_k h_0 + 2\left[\sum_{k=1}^{K-1}\sum_{n=1}^{k} \alpha_k^* \alpha_{k-n} h_n\right]\right\} \\ &= Re\left\{\sum_{k=0}^{K-1} \alpha_k^* \left[h_0 \alpha_k + 2\sum_{n=1}^{k} h_n \alpha_{k-n}\right]\right\} \end{aligned} \tag{7.460}$$

In particular, if

$$|h_n| \simeq 0 \qquad \text{for } |n| > N_h \tag{7.461}$$

(7.460) is simplified in the following expression

$$A = Re\left\{\sum_{k=0}^{K-1} \alpha_k^* \left[h_0\alpha_k + 2\sum_{n=1}^{N_h} h_n\alpha_{k-n}\right]\right\} \tag{7.462}$$

Then the log-likelihood (7.457) becomes

$$\ell(\boldsymbol{\alpha}) = -\sum_{k=0}^{K-1} Re\left\{\alpha_k^* \left[-2x_k + h_0\alpha_k + 2\sum_{n=1}^{N_h} h_n\alpha_{k-n}\right]\right\} \tag{7.463}$$

To maximize $\ell(\boldsymbol{\alpha})$ or, equivalently, to minimize $-\ell(\boldsymbol{\alpha})$ with respect to $\boldsymbol{\alpha}$ we apply the VA (see page 343) with the state vector defined as

$$s_k = (a_k, a_{k-1}, \dots, a_{k-N_h+1}) \tag{7.464}$$

and branch metric given by

$$Re\left\{a_k^* \left[-2x_k + h_0 a_k + 2\sum_{n=1}^{N_h} h_n a_{k-n}\right]\right\} \tag{7.465}$$

Rather than detecting the data sequences by (7.450), the extraction of LLRs associated to bits of BMAP follows the same lines of (7.436).

7.5.5 Error probability achieved by MLSD

In the VA, we have an error if a wrong sequence of states is chosen in the trellis diagram; it is of interest the probability that one or more states of the detected ML sequence are in error. The error probability is dominated by the probability that a sequence at the minimum Euclidean distance from the correct sequence is chosen as ML sequence. We note, however, that by increasing the sequence length K the number of different paths in the trellis diagram associated with sequences that are at the minimum distance also increases. Therefore, by increasing K, the probability that the chosen sequence is in error usually tends to 1.

The probability that the whole sequence of states is not received correctly is rarely of interest; instead, we consider the probability that the detection of a generic symbol is in error. For the purpose of determining the symbol error probability, the concept of *error event* is introduced. Let $\{\boldsymbol{\sigma}\} = (\boldsymbol{\sigma}_{i_0}, \dots, \boldsymbol{\sigma}_{i_{K-1}})$ be the realization of the state sequence associated with the information sequence, and let $\{\hat{\boldsymbol{\sigma}}\}$ be the sequence chosen by the VA. In a sufficiently long time interval, the paths in the trellis diagram associated with $\{\boldsymbol{\sigma}\}$ and $\{\hat{\boldsymbol{\sigma}}\}$ diverge and converge several times: every distinct separation from the correct path is called an error event.

Definition 7.3

An *error event* $\boldsymbol{e}$ is defined as a path in the trellis diagram that has only the initial and final states in common with the correct path; the length of an error event is equal to the number of nodes visited in the trellis before rejoining with the correct path. □

Error events of length two and three are illustrated in a trellis diagram with two states, where the correct path is represented by a continuous line, in Figure 7.45a and Figure 7.45b, respectively.

Let E be the set of all error events beginning at instant i. Each element $\boldsymbol{e}$ of E is characterized by a correct path $\{\boldsymbol{\sigma}\}$ and a wrong path $\{\hat{\boldsymbol{\sigma}}\}$, which diverges from $\{\boldsymbol{\sigma}\}$ at instant i and converges at $\{\boldsymbol{\sigma}\}$ after

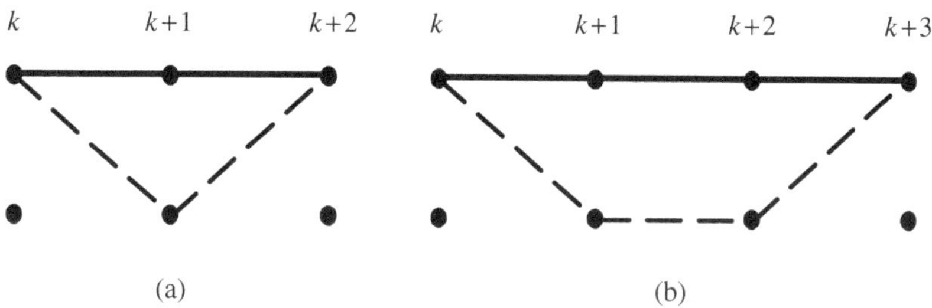

Figure 7.45 Error events of length (a) two and (b) three in a trellis diagram with two states.

a certain number of steps in the trellis diagram. We assume that the probability $P[\boldsymbol{e}]$ is independent of instant i: this hypothesis is verified with good approximation if the trellis diagram is much longer than the length of the significant error events. An error event produces one or more errors in the detection of symbols of the input sequence. We have a *detection error* at instant k if the detection of the input at the k-th stage of the trellis diagram is not correct. We define the function [14]

$$c_m(\boldsymbol{e}) = \begin{cases} 1, & \text{if } \boldsymbol{e} \text{ causes a detection error at the instant } i+m\,, \text{ with } m \geq 0 \\ 0, & \text{otherwise} \end{cases} \tag{7.466}$$

The probability of a particular error event that starts at instant i and causes a detection error at instant k is given by $c_{k-i}(\boldsymbol{e})P[\boldsymbol{e}]$. Because the error events in E are disjoint, we have

$$P_e = P[\hat{a}_k \neq a_k] = \sum_{i=-\infty}^{k} \sum_{\boldsymbol{e}\in \mathrm{E}} c_{k-i}(\boldsymbol{e})\, P[\boldsymbol{e}] \tag{7.467}$$

Assuming that the two summations can be exchanged, we obtain

$$P_e = \sum_{\boldsymbol{e}\in \mathrm{E}} P[\boldsymbol{e}] \sum_{i=-\infty}^{k} c_{k-i}(\boldsymbol{e}) \tag{7.468}$$

With a change of variables, it turns out

$$\sum_{i=-\infty}^{k} c_{k-i}(\boldsymbol{e}) = \sum_{m=0}^{\infty} c_m(\boldsymbol{e}) = N(\boldsymbol{e}) \tag{7.469}$$

which indicates the total number of detection errors caused by the error event $\boldsymbol{e}$. Therefore,

$$P_e = \sum_{\boldsymbol{e}\in \mathrm{E}} N(\boldsymbol{e})P[\boldsymbol{e}] \tag{7.470}$$

where the dependence on the time index k vanishes. We therefore have that the detection error probability is equal to the average number of errors caused by all the possible error events initiating at a given instant i; this result is expected, because the detection error probability at a particular instant k must take into consideration all error events that initiate at previous instants and are not yet terminated.

If $\{s\} = (s_0, \dots, s_{K-1})$ denotes the random variable sequence of states at the transmitter and $\{\hat{s}\} = (\hat{s}_0, \dots, \hat{s}_{K-1})$ denotes the random variable sequence of states selected by the ML receiver, the probability of an error event $\boldsymbol{e}$ beginning at a given instant i depends on the joint probability of the correct and incorrect path, and it can be written as

$$P[\boldsymbol{e}] = P[\{\hat{s}\} = \{\hat{\sigma}\} \mid \{s\} = \{\sigma\}]P[\{s\} = \{\sigma\}] \tag{7.471}$$

Because it is usually difficult to find the exact expression for $P[\{\hat{s}\} = \{\hat{\sigma}\} \mid \{s\} = \{\sigma\}]$, we resort to upper and lower bounds.

Upper bound. Because detection of the sequence of states $\{s\}$ is obtained by observing the sequence $\{\mathbf{u}\}$, for the signal in (7.334) with zero mean additive white Gaussian noise having variance σ_I^2 per dimension, we have the upper bound

$$P[\{\hat{s}\} = \{\hat{\sigma}\} \mid \{s\} = \{\sigma\}] \leq Q\left(\frac{d[\boldsymbol{u}(\{\sigma\}), \boldsymbol{u}(\{\hat{\sigma}\})]}{2\sigma_I}\right) \tag{7.472}$$

where $d[\boldsymbol{u}(\{\sigma\}), \boldsymbol{u}(\{\hat{\sigma}\})]$ is the Euclidean distance between signals $\boldsymbol{u}(\{\sigma\})$ and $\boldsymbol{u}(\{\hat{\sigma}\})$, given by (7.354). Substitution of the upper bound in (7.470) yields

$$P_e \leq \sum_{e \in \mathrm{E}} N(\boldsymbol{e})\, P[\{s\} = \{\sigma\}] Q\left(\frac{d[\boldsymbol{u}(\{\sigma\}), \boldsymbol{u}(\{\hat{\sigma}\})]}{2\sigma_I}\right) \tag{7.473}$$

which can be rewritten as follows, by giving prominence to the more significant terms,

$$P_e \leq \sum_{e \in \mathrm{E}_{min}} N(\boldsymbol{e})\, P[\{s\} = \{\sigma\}] Q\left(\frac{\Delta_{min}}{2\sigma_I}\right) + \text{other terms} \tag{7.474}$$

where E_{min} is the set of error events at minimum distance Δ_{min} defined in (7.355), and the remaining terms are characterized by arguments of the Q function larger than $\Delta_{min}/(2\sigma_I)$. For higher values of the signal-to-noise ratio, these terms are negligible and the following approximation holds

$$P_e \leq \mathcal{K}_1\, Q\left(\frac{\Delta_{min}}{2\sigma_I}\right) \tag{7.475}$$

where

$$\mathcal{K}_1 = \sum_{e \in \mathrm{E}_{min}} N(\boldsymbol{e})\, P[\{s\} = \{\sigma\}] \tag{7.476}$$

Lower bound. A lower bound to the error probability is obtained by considering the probability that any error event may occur rather than the probability of a particular error event. Since $N(\boldsymbol{e}) \geq 1$ for all the error events $\boldsymbol{e}$, from (7.470) we have

$$P_e \geq \sum_{e \in \mathrm{E}} P[\boldsymbol{e}] \tag{7.477}$$

Let us consider a particular path in the trellis diagram determined by the sequence of states $\{\sigma\}$. We set

$$\Delta_{min}(\{\sigma\}) = \min_{\{\tilde{\sigma}\}} d[\boldsymbol{u}(\{\sigma\}), \boldsymbol{u}(\{\tilde{\sigma}\})] \tag{7.478}$$

i.e. for this path, $\Delta_{min}(\{\sigma\})$ is the Euclidean distance of the minimum distance error event. We have $\Delta_{min}(\{\sigma\}) \geq \Delta_{min}$, where Δ_{min} is the minimum distance obtained considering all the possible state sequences. If $\{\sigma\}$ is the correct state sequence, the probability of an error event is lower bounded by

$$P[\boldsymbol{e} \mid \{s\} = \{\sigma\}] \geq Q\left(\frac{\Delta_{min}(\{\sigma\})}{2\sigma_I}\right) \tag{7.479}$$

Consequently,

$$P_e \geq \sum_{\{\sigma\}} P[\{s\} = \{\sigma\}] Q\left(\frac{\Delta_{min}(\{\sigma\})}{2\sigma_I}\right) \tag{7.480}$$

If some terms are omitted in the summation, the lower bound is still valid, because the terms are non-negative. Therefore, taking into consideration only those state sequences $\{\sigma\}$ for which $\Delta_{min}(\{\sigma\}) = \Delta_{min}$, we obtain

$$P_e \geq \sum_{\{\sigma\} \in A} P[\{s\} = \{\sigma\}] Q\left(\frac{\Delta_{min}}{2\sigma_I}\right) \tag{7.481}$$

where A is the set of state sequences that admit an error event with minimum distance Δ_{min}, for an arbitrarily chosen initial instant of the given error event. Defining

$$\mathcal{K}_2 = \sum_{\{\sigma\} \in A} P[\{s\} = \{\sigma\}] \tag{7.482}$$

as the probability that a path $\{\sigma\}$ admits an error event with minimum distance, it is

$$P_e \geq \mathcal{K}_2 \, Q\left(\frac{\Delta_{min}}{2\sigma_I}\right) \tag{7.483}$$

Combining upper and lower bounds, we obtain

$$\mathcal{K}_2 \, Q\left(\frac{\Delta_{min}}{2\sigma_I}\right) \leq P_e \leq \mathcal{K}_1 \, Q\left(\frac{\Delta_{min}}{2\sigma_I}\right) \tag{7.484}$$

For large values of the signal-to-noise ratio, therefore, we have

$$P_e \simeq \mathcal{K} \, Q\left(\frac{\Delta_{min}}{2\sigma_I}\right) \tag{7.485}$$

for some value of the constant $\mathcal{K}$ between $\mathcal{K}_1$ and $\mathcal{K}_2$.

We stress that the error probability, expressed by (7.485) and (7.356), is determined by the ratio between the minimum distance Δ_{min} and the standard deviation of the noise σ_I.

Here the expressions of the constants $\mathcal{K}_1$ and $\mathcal{K}_2$ are obtained by resorting to various approximations. An accurate method to calculate upper and lower bounds of the error probability is proposed in [19].

Computation of the minimum distance

The application of the VA to MLSD in transmission systems with ISI requires that the overall impulse response is FIR, otherwise the number of states, and hence also the complexity of the detector, becomes infinite. From (7.334), the samples at the detector input, conditioned on the event that the sequence of symbols $\{a_k\}$ is transmitted, are statistically independent Gaussian random variables with mean

$$\sum_{n=-L_1}^{L_2} \eta_n \, a_{k-n} \tag{7.486}$$

and variance σ_I^2 per dimension. The metric that the VA attributes to the sequence of states corresponding to the sequence of input symbols $\{a_k\}$ is given by the square Euclidean distance between the sequence of samples $\{z_k\}$ at the detector input and its mean value, which is known given the sequence of symbols (see (6.24)),

$$\sum_{k=0}^{\infty} \left| z_k - \sum_{n=-L_1}^{L_2} \eta_n \, a_{k-n} \right|^2 \tag{7.487}$$

In the previous section, it was demonstrated that the symbol error probability is given by (7.485). In particularly simple cases, the minimum distance can be determined by direct inspection of the trellis diagram; in practice, however, this situation is rarely verified in channels with ISI. To evaluate the minimum distance, it is necessary to resort to simulations. To find the minimum distance error event with initial instant $k = 0$, we consider the desired signal u_k under the condition that the sequence $\{a_k\}$ is transmitted, and we compute the square Euclidean distance between this signal and the signal obtained for another sequence $\{\tilde{a}_k\}$,

$$d^2[\boldsymbol{u}(\{a_k\}), \boldsymbol{u}(\{\tilde{a}_k\})] = \sum_{k=0}^{\infty} \left| \sum_{n=-L_1}^{L_2} \eta_n \, a_{k-n} - \sum_{n=-L_1}^{L_2} \eta_n \, \tilde{a}_{k-n} \right|^2 \tag{7.488}$$

where it is assumed that the two paths identifying the state sequences are identical for $k < 0$.

It is possible to avoid computing the minimum distance for each sequence $\{a_k\}$ if we exploit the linearity of the ISI. Defining

$$\epsilon_k = a_k - \tilde{a}_k \tag{7.489}$$

we have

$$d^2(\{\epsilon_k\}) = d^2[\boldsymbol{u}(\{a_k\}), \boldsymbol{u}(\{\tilde{a}_k\})] = \sum_{k=0}^{\infty} \left| \sum_{n=-L_1}^{L_2} \eta_n \, \epsilon_{k-n} \right|^2 \tag{7.490}$$

The minimum among the square Euclidean distances relative to all error events that initiate at $k = 0$ is

$$\Delta^2_{min} = \min_{\{\epsilon_k\}:\ \epsilon_k=0,\ k<L_1, \epsilon_{L_1} \neq 0} d^2(\{\epsilon_k\}) \tag{7.491}$$

It is convenient to solve this minimization without referring to the symbol sequences. In particular, we define the state $\boldsymbol{s}_k = (\epsilon_{k+L_1}, \epsilon_{k+L_1-1}, \dots, \epsilon_{k-L_2+1})$, and a trellis diagram that describes the development of this state. Adopting the branch metric

$$\left| \sum_{n=-L_1}^{L_2} \eta_n \, \epsilon_{k-n} \right|^2 \tag{7.492}$$

the minimization problem is equivalent to determining the path in a trellis diagram that has minimum metric (7.490) and differs from the path that joins states corresponding to correct decisions: the resulting metric is Δ^2_{min}. We note, however, that the cardinality of ϵ_k is larger than M, and this implies that the complexity of this trellis diagram can be much larger than that of the original trellis diagram. In the PAM case, the cardinality of ϵ_k is equal to $2M - 1$.

In practice, as the terms of the series in (7.490) are non-negative, if we truncate the series after a finite number of terms, we obtain a result that is smaller than or equal to the effective value. Therefore, a lower bound to the minimum distance is given by

$$\Delta^2_{min} \geq \min_{\{\epsilon_k\}:\ \epsilon_k=0,\ k<L_1, \epsilon_{L_1} \neq 0} \sum_{k=0}^{K-1} \left| \sum_{n=-L_1}^{L_2} \eta_n \, \epsilon_{k-n} \right|^2 \tag{7.493}$$

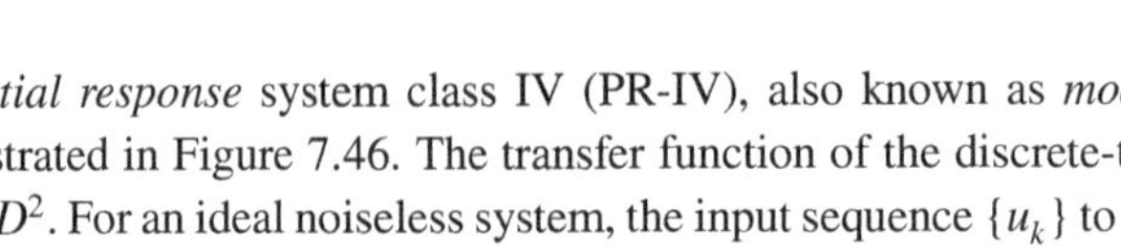

Figure 7.46 PR-IV (modified duobinary) transmission system.

Example 7.5.3

We consider the *partial response* system class IV (PR-IV), also known as *modified duobinary* (see Appendix 7.12), illustrated in Figure 7.46. The transfer function of the discrete-time overall system is given by $\eta(D) = 1 - D^2$. For an ideal noiseless system, the input sequence $\{u_k\}$ to the detector is formed by random variables taking values in the set $\{-2, 0, +2\}$, as shown in Figure 7.47.

Assuming that the sequence of noise samples $\{w_k\}$ is composed of real-valued, statistically independent, Gaussian random variables with mean zero and variance σ_I^2, and observing that u_k for k even (odd) depends only on symbols with even (odd) indices, the MLSD receiver for a PR-IV system is usually implemented by considering two interlaced dicode independent channels, each having a transfer function given by $1 - D$. As seen in Figure 7.48, for detection of the two interlaced input symbol sequences,

b_k	... 0 1 1 0 1 0 0 ...
a_k	... −1 +1 +1 −1 +1 −1 −1 ...
$u_k = a_k - a_{k-2}$	... +2 −2 0 0 −2 ...

Figure 7.47 Input and output sequences for an ideal PR-IV system.

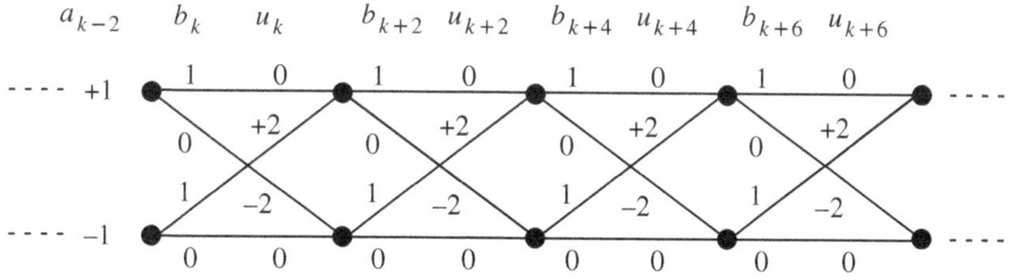

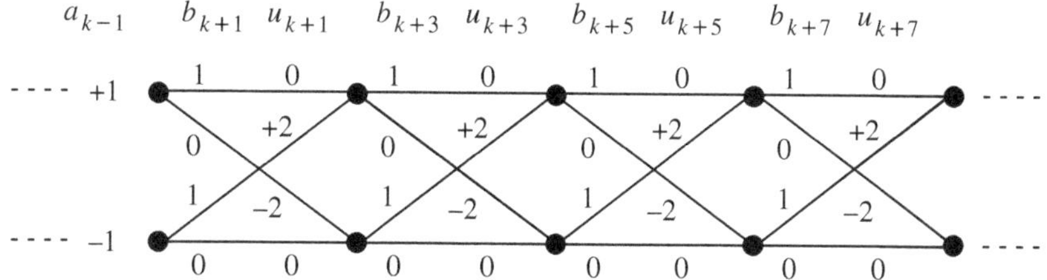

Figure 7.48 Trellis diagrams for detection of interlaced sequences.

two trellis diagrams are used. The state at instant k is given by the symbol $s_k = a_k$, where k is even-valued in one of the two diagrams and odd-valued in the other. Each branch of the diagram is marked with a label that represents the binary input symbol b_k or, equivalently, the value of the dicode signal u_k. For a particular realization of the output signal of a dicode channel, the two survivor sequences at successive iterations of the VA are represented in Figure 7.49.

It is seen that the minimum square Euclidean distance between two separate paths in the trellis diagram is given by $\Delta^2_{min} = 2^2 + 2^2 = 8$. However, we note that for the same initial instant, there are an infinite number of error events with minimum distance from the effective path: this fact is evident in the trellis diagram of Figure 7.50a, where the state $s_k = (\epsilon_k) \in \{-2, 0, +2\}$ characterizes the development of the error event and the labels indicate the branch metrics associated with an error event. It is seen that at every instant an error event may be extended along a path, for which the metric is equal to zero, parallel to the path corresponding to the zero sequence. Paths of this type correspond to a sequence of errors having the same polarity. Four error events with minimum distance are shown in Figure 7.50b.

The error probability is given by $\mathcal{K} Q\left(\frac{\sqrt{8}}{2\sigma_I}\right)$, where $\mathcal{K}_2 \leq \mathcal{K} \leq \mathcal{K}_1$. The constant $\mathcal{K}_2$ can be immediately determined by noting that every effective path admits at every instant at least one error event with minimum distance: consequently, $\mathcal{K}_2 = 1$. To find $\mathcal{K}_1$, we consider the contribution of an error event with m consecutive errors. For this event to occur, it is required that m consecutive input symbols have the same polarity, which happens with probability 2^{-m}. Since such an error event determines m symbol errors and two error events with identical characteristics can be identified in the trellis diagram, we have

$$\mathcal{K}_1 = 2 \sum_{m=1}^{\infty} m\, 2^{-m} = 4 \tag{7.494}$$

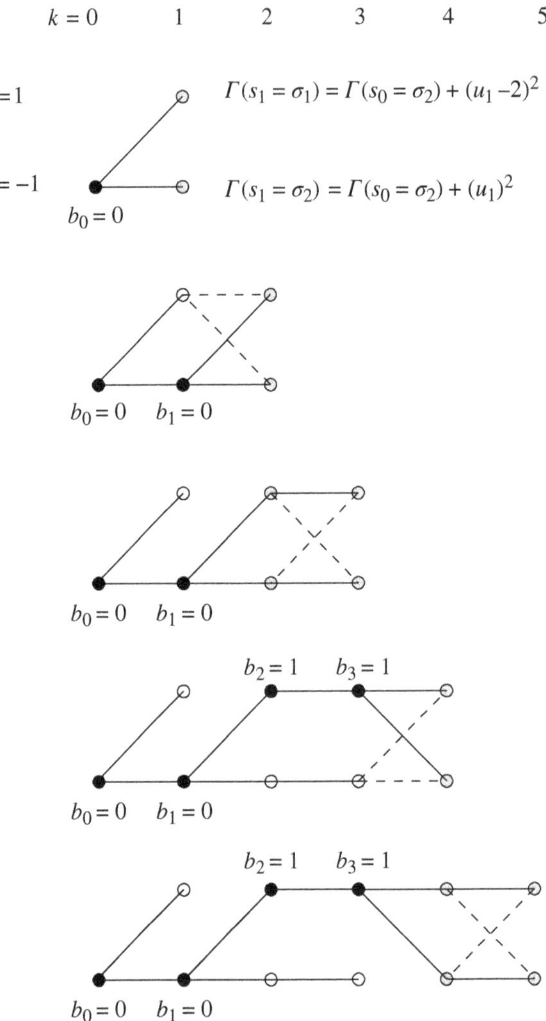

Figure 7.49 Survivor sequences at successive iterations of the VA for a dicode channel.

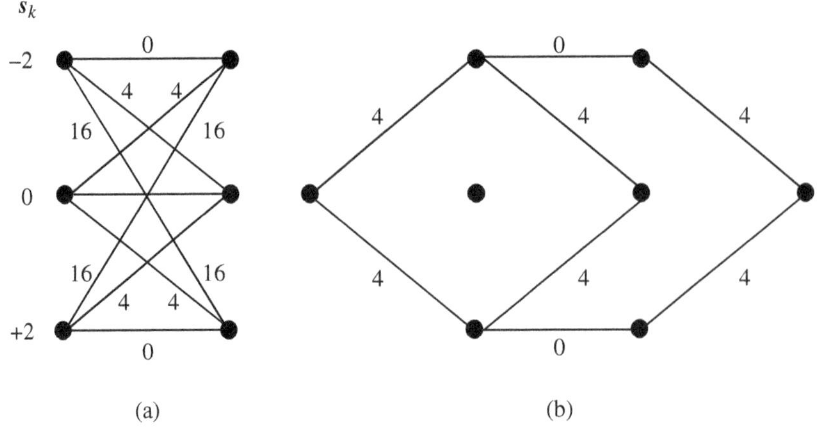

Figure 7.50 Examples of (a) trellis diagram to compute the minimum distance for a dicode channel and (b) four error events with minimum distance.

Besides the error events associated with the minimum distance $\Delta_{min} = \sqrt{8}$, the occurrence of long sequences of identical output symbols from a dicode channel raises two problems.

- The system becomes catastrophic in the sense that in the trellis diagram used by the detector, valid sequences of arbitrary length are found having square Euclidean distance from the effective path equal to 4; therefore, a MLSD receiver with finite memory will make additional errors with probability proportional to $Q(\sqrt{4}/(2\sigma_I))$ if the channel produces sequences $\{u_k = 0\}$ of length larger than the memory of the detector.
- The occurrence of these sequences is detrimental for the control of receive filter gain and sampling instants: the problem is solved by suitable coding of the input binary sequence, that sets a limit on the number of consecutive identical symbols that are allowed at the channel input.

7.5.6 The reduced-state sequence detection

For transmissions over channels characterized by strong ISI, the maximum-likelihood detection implemented by the VA typically gives better performance than the decision-feedback equalizer. On the other hand, if we indicate with $N = L_1 + L_2$ the length of the channel memory and with M the cardinality of the alphabet of transmitted symbols, the implementation complexity of the VA is of the order of M^N, which can be too large for practical applications if N and/or M assume large values.

To find receiver structures that are characterized by performance similar to that of a MLSD receiver, but lower complexity, two directions may be taken:

- decrease N;
- decrease the number of paths considered in the trellis.

The first direction leads to the application of pre-processing techniques, e.g. LE or DFE [20], to reduce the length of the channel impulse response. However, they only partially solve the problem for bandwidth-efficient modulation systems that utilize a large set of signals, the minimization of N typically does not provide a solution with manageable complexity.

The second direction leads to a further study of the MLSD method. An interesting algorithm is the $\mathcal{M}$-algorithm [21, 22]: it operates in the same way as the MLSD, i.e. using the full trellis diagram, but at every step it takes into account only $\mathcal{M} \leq M^N$ states, that is, those associated with paths with smaller metrics. The performance of the $\mathcal{M}$-algorithm is close to that of the MLSD even for small values of $\mathcal{M}$, however, it works only if the channel impulse response is minimum phase: otherwise, it is likely that among the paths that are eliminated in the trellis diagram, the optimum path is also included.

The *reduced-state sequence estimator* (RSSE)[13] [23, 24] yields performance very close to that of MLSD, with significantly reduced complexity even though it retains the fundamental structure of MLSD. It can be used for modulations with a large symbol alphabet $\mathcal{A}$ and/or for channels with a long impulse response. The basic idea consists in using a trellis diagram with a reduced number of states, obtained by combining the states of the ML trellis diagram in a manner suggested by the principle of *partitioning* the set $\mathcal{A}$, and possibly including the decision-feedback method in the computation of the branch metrics. In this way, the RSSE guarantees a performance/complexity trade-off that can vary from that characterizing a DFE zero-forcing to that characterizing MLSD.

Trellis diagram

We consider the transmission system depicted in Figure 7.44b. Contrary to an MLSD receiver, the performance of an RSSE receiver may be poor if the overall channel impulse response is not minimum

[13] We will maintain the name RSSE, although the algorithm is applied to perform a *detection* rather than an *estimation*.

phase: to underline this fact, we slightly change the notation adopted in Section 7.5. We indicate with $f^T = [f_1, f_2, \dots, f_N]$ the coefficients of the impulse response that determine the ISI and assume, without loss of generality, the desired sample is $f_0 = 1$. Hence, the observed signal is

$$z_k = \sum_{n=0}^{N} f_n a_{k-n} + w_k = a_k + s_{k-1}^T f + w_k \tag{7.495}$$

where the state at instant $k - 1$ is

$$s_{k-1}^T = [a_{k-1}, a_{k-2}, \dots, a_{k-N}] \tag{7.496}$$

Observation 7.15

In the transmission of sequences of blocks of data, the RSSE yields its best performance by imposing the final state, for example using the knowledge of a training sequence. Therefore, the formulation of this section is suited for the case of a training sequence placed at the end of the data block. In the case the training sequence is placed at the beginning of a data block, it is better to process the signals in backward mode as described in Observation 7.7 on page 320.

The RSSE maintains the fundamental structure of MLSD unaltered, corresponding to the search in the trellis diagram of the path with minimum cost.

To reduce the number of states to be considered, we introduce for every component a_{k-n}, $n = 1, \dots, N$, of the vector s_{k-1} defined in (7.496) a suitable partition $\Omega(n)$ of the two-dimensional set $\mathcal{A}$ of possible values of a_{k-n}: a partition is composed of J_n subsets, with J_n an integer between 1 and M. The index c_n indicates of the subset of the partition $\Omega(n)$ to which the symbol a_{k-n} belongs, an integer value between 0

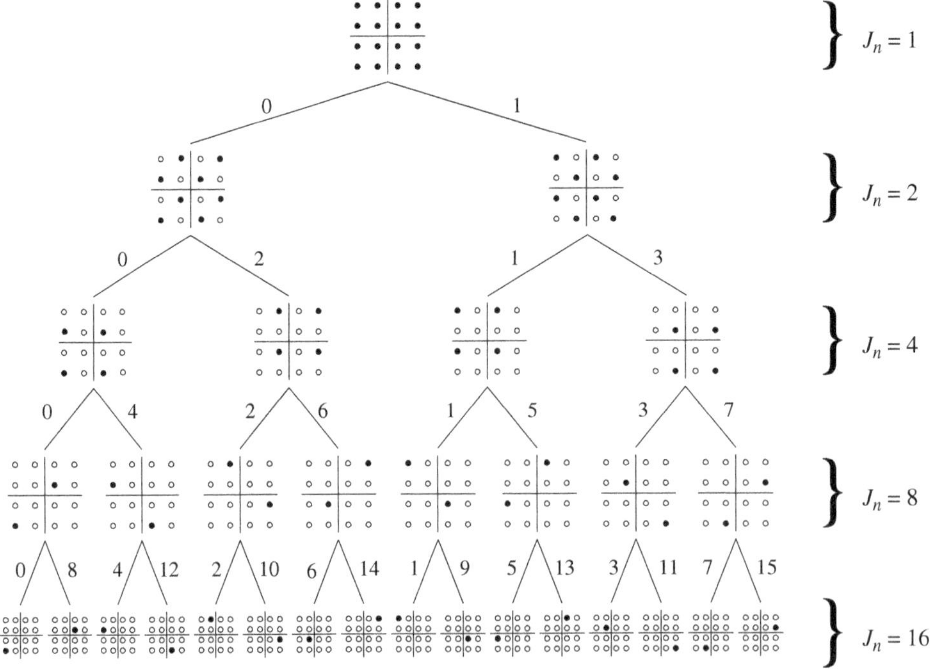

Figure 7.51 Ungerboeck's partitioning of the symbol set associated with a 16-QAM system. The various subsets are identified by the value of $c_n \in \{0, 1, \dots, J_n - 1\}$.

and $J_n - 1$. Ungerboeck's partitioning of the symbol set associated with a 16-QAM system is illustrated in Figure 7.51. The partitions must satisfy the following two conditions [25]:

1. the numbers J_n are non-increasing, that is $J_1 \geq J_2 \geq \cdots \geq J_N$;
2. the partition $\Omega(n)$ is obtained by subdividing the subsets that make up the partition $\Omega(n+1)$, for every n between 1 and $N-1$.

Therefore, we define as *reduced-state at instant* $k-1$ the vector $\boldsymbol{t}_{k-1}$ that has as n-th element the index c_n of the subset of the partition $\Omega(n)$ to which the n-th element of $\boldsymbol{s}_{k-1}$ belongs, for $n = 1, 2, \ldots, N$, that is

$$\boldsymbol{t}_{k-1}^T = [c_1, c_2, \ldots, c_N], \qquad c_n \in \{0, 1, \ldots, J_n - 1\} \tag{7.497}$$

and we write

$$\boldsymbol{s}_{k-1} = \boldsymbol{s}(\boldsymbol{t}_{k-1}) \tag{7.498}$$

It is useful to stress that the reduced state $\boldsymbol{t}_{k-1}$ does not uniquely identify a state $\boldsymbol{s}_{k-1}$, but all the states $\boldsymbol{s}_{k-1}$ that include as n-th element one of the symbols belonging to the subset c_n of partition $\Omega(n)$.

The conditions imposed on the partitions guarantee that, given a reduced state at instant $k-1$, $\boldsymbol{t}_{k-1}$, and the subset j of partition $\Omega(1)$ to which the symbol a_k belongs, the reduced state at instant k, $\boldsymbol{t}_k$, can be uniquely determined. In fact, observing (7.497), we have

$$\boldsymbol{t}_k^T = [c'_1, c'_2, \ldots, c'_N] \tag{7.499}$$

where $c'_1 = j$, c'_2 is the index of the subset of the partition $\Omega(2)$ to which belongs the subset with index c_1 of the partition $\Omega(1)$, c'_3 is the index of the subset of the partition $\Omega(3)$ to which belongs the subset with index c_2 of the partition $\Omega(2)$, and so forth. In this way, the reduced states $\boldsymbol{t}_{k-1}$ define a proper *reduced-state trellis diagram*, that represents all the possible sequences $\{a_k\}$.

As the symbol c_n can only assume one of the integer values between 0 and $J_n - 1$, the total number of possible reduced states of the trellis diagram of the RSSE is given by the product $N_s = J_1 J_2 \ldots J_N$, with $J_n \leq M$, for $n = 1, 2, \ldots, N$.

We know that in the VA, for uncoded transmission of i.i.d. symbols, there are M possible transitions from a state, one for each of the values that a_k can assume. In the reduced-state trellis diagram, M transitions are still possible from a state, however, to only J_1 distinct states, thus giving origin to parallel transitions.[14] In fact, if $J_1 < M$, J_1 sets of branches depart from every state $\boldsymbol{t}_{k-1}$, each set consisting of as many parallel transitions as there are symbols belonging to the subset of $\Omega(1)$ is associated with the reduced state.

Therefore, partitions must be obtained such that two effects are guaranteed: (i) minimum performance degradation with respect to MLSD, and (ii) easy search of the optimum path among the various parallel transitions. Usually, the Ungerboeck's set partitioning method is adopted, which, for every partition Ω, maximizes the minimum distance Δ among the symbols belonging to the same subset. For QAM systems and J_n a power of 2, the maximum distance Δ_n relative to partition $\Omega(n)$ is obtained through a tree diagram with binary partitioning. An example of partitioning of the symbol set associated with a 16-QAM system is illustrated in Figure 7.51.

In Figure 7.52, two examples of reduced-state trellis diagram are shown, both referring to the partition of Figure 7.51.

The RSSE algorithm

As in MLSD, for each transition that originates from a state $\boldsymbol{t}_{k-1}$, the RSSE computes the branch metric according to the expression

$$|z_k - a_k - \boldsymbol{s}_{k-1}^T \boldsymbol{f}|^2 = |z_k - a_k - \boldsymbol{s}^T(\boldsymbol{t}_{k-1})\boldsymbol{f}|^2 \tag{7.500}$$

[14] Parallel transitions are present when two or more branches connect the same pair of states in the trellis diagram.

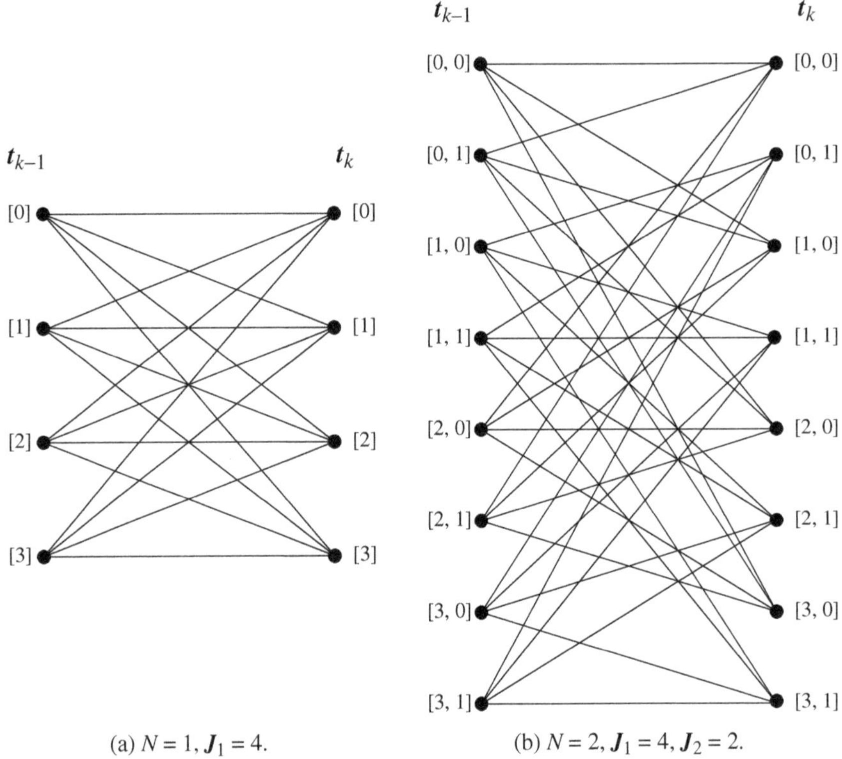

Figure 7.52 Reduced-state trellis diagrams.

However, whereas in MLSD, a survivor sequence may be described in terms of the *sequence of states* that led to a certain state, in the RSSE, if $J_1 < M$, it is convenient to memorize a *survivor sequence* as a *sequence of symbols*. As a matter of fact, there is no one-to-one correspondence between state sequences and symbol sequences. Therefore, backward tracing the optimum path in the trellis diagram in terms of the sequence of states does not univocally establish the optimum sequence of symbols.

Moreover, if $J_1 < M$, and therefore, there are parallel transitions in the trellis diagram, for every branch, the RSSE selects the symbol a_k in the subset of the partition $\Omega(1)$ that yields the minimum metric.[15] Thus, the RSSE already makes a decision in selecting one of the parallel transitions, using also the past decisions memorized in the survivor sequence associated with the considered state.

At every iteration, these decisions reduce the number of possible extensions of the N_s states from N_sM to N_sJ_1. This number is further reduced to N_s by selecting, for each state t_k, the path with the minimum metric among the J_1 entering paths, operation which requires $N_s(J_1 - 1)$ comparisons. As in the case of the VA, *final* decisions are taken with a certain delay by tracing the history of the path with lowest metric.

We note that if $J_n = 1, n = 1, \dots, N$, then the RSSE becomes a DFE-ZF, and if $J_n = M, n = 1, \dots, N$, the RSSE performs full MLSD. Therefore, the choice of $\{J_n\}$ determines a trade-off between performance and computational complexity. The error probability for an RSSE is rather difficult to evaluate

[15] We note that if the points of the subsets of the partition $\Omega(1)$ are on a rectangular grid, as in the example of Figure 7.51, the value of the symbol a_k that minimizes the metric is determined through simple *quantization rules*, without explicitly evaluating the branch metric for every symbol of the subset. Hence, for every state, only J_1 explicit computations of the branch metric are needed.

because of the presence of the decision-feedback mechanism. For the analysis of RSSE performance, we refer the reader to [26].

Further simplification: DFSE

There are many applications in which the complexity of MLSD is mainly due to the length of the channel impulse response. In these cases, a method to reduce the number of states consists of cancelling part of the ISI by a DFE; better results are obtained by incorporating the decision-feedback mechanism in the VA, that is using for each state a different feedback symbol sequence given by the survivor sequence.

The idea of cancelling the residual ISI based on the survivor sequence is a particular application of a general principle that is applied when the branch metric is affected by some uncertainty that can be eliminated or reduced by (possibly adaptive) *data-aided* estimation techniques. Typical examples are the non-perfect knowledge of some channel characteristics, as the carrier phase, the timing phase, or the impulse response: all these cases can be solved in part by a general approach called *per-survivor processing* (PSP) [27]. It represents an effective alternative to classical methods to estimate channel parameters, as the effects of error propagation are significantly reduced. Other interesting aspects characterizing this class of algorithms are the following:

a. The estimator associated with the survivor sequence uses symbols that can be considered decisions with no delay and high reliability, making the PSP a suitable approach for channels that are fast time-varying.
b. *Blind* techniques, i.e. without knowledge of the training sequence, may be adopted as part of the PSP to estimate the various parameters.

A variant of the RSSE, belonging to the class of PSP algorithms, is the *decision feedback sequence estimator* (DFSE) [23] which considers as reduced state the vector s'_{k-1} formed simply by truncating the ML state vector at a length $N_1 \leq N$

$$s'^T_{k-1} = [a_{k-1}, a_{k-2}, \dots, a_{k-N_1}] \tag{7.501}$$

This is the same as considering $J_n = M$, for $n = 1, \dots, N_1$, and $J_n = 1$, for $n = N_1 + 1, \dots, N$.

We discuss the main points of this algorithm. We express the received sequence $\{z_k\}$, always under the hypothesis of a minimum-phase overall impulse response with $f_0 = 1$, as

$$z_k = a_k + s'^T_{k-1} f' + s''^T_{k-1} f'' + w_k \tag{7.502}$$

where f', f'', s'_{k-1}, and s''_{k-1} are defined as follows:

$$f^T = [f'^T \mid f''^T] = [f_1, \dots, f_{N_1} \mid f_{N_1+1}, \dots, f_N] \tag{7.503}$$

$$s^T_{k-1} = [s'^T_{k-1} \mid s''^T_{k-1}] = [a_{k-1}, \dots, a_{k-N_1} \mid a_{k-(N_1+1)}, \dots, a_{k-N}] \tag{7.504}$$

The trellis diagram is built by assuming the reduced state s'_{k-1}. The term $s''^T_{k-1} f''$ represents the residual ISI that is estimated by considering as s''_{k-1} the symbols that are memorized in the survivor sequence associated with each state. We write

$$\hat{s}''_{k-1} = s''(s'_{k-1}) \tag{7.505}$$

In fact, with respect to z_k, it is as if we would have cancelled the term $s''^T_{k-1} f''$ by a FB filter associated with each state s'_{k-1}. With respect to the optimum path, the feedback sequence s''_{k-1} is expected to be very reliable.

The branch metric of the DFSE is computed as follows:

$$|z_k - a_k - s'^T_{k-1} f' - s''^T(s'_{k-1}) f''|^2 \tag{7.506}$$

We note that the reduced state s'_{k-1} may be further reduced by adopting the RSSE technique (7.498).

The primary difference between an MLSD receiver and the DFSE is that in the trellis diagram used by the DFSE two paths may merge earlier, as it is sufficient that they share the more recent N_1 symbols, rather than N as in MLSD. This increases the error probability; however, in general DFSE outperforms a classical DFE.

7.6 Numerical results obtained by simulations

Using Monte Carlo simulations (see Appendix 7.A) we give a comparison, in terms of P_{bit} as a function of the signal-to-noise Γ, of the various equalization and data detection methods described in the previous sections. We refer to the system model of Figure 7.26 with an overall impulse response $\{h_n\}$ having five coefficients, as given in Table 1.4 on page 12, and additive white Gaussian noise (AWGN) $\tilde{w}$. Recalling the definition of the signal-to-noise $\Gamma = \sigma_a^2 \, E_h / \sigma_{\tilde{w}}^2$, we examine four cases.

QPSK over a minimum-phase channel

We examine the following receivers, where the delay D is optimized by applying the MSE criterion.

We anticipate that the ZF equalizer is designed by the DFT method (see Section 7.4.6). We also introduce the abbreviation DFE-VA to indicate the method consisting of the FF filter of a DFE, followed by MLSD implemented by the VA, with M^{M_2} states determined by the impulse response $\eta_n = \psi_{n+D}$ for $n = 0, \ldots, M_2$: this technique is commonly used to shorten the overall channel impulse response and thus simplify the VA (see case (2a) on page 355).

1. ZF with $N = 7$ and $D = 0$;
2. LE with $N = 7$ and $D = 0$;
3. DFE with $M_1 = 7$, $M_2 = 4$, and $D = 6$;
4. VA with $4^4 = 256$ states and path memory depth equal to 15, i.e. approximately three times the length of $\{h_n\}$;
5. DFSE with $J_1 = J_2 = 4$, $J_3 = J_4 = 1$ (16 states);
6. DFE-VA with $M_1 = 7$, $M_2 = 2$, and $D = 0$ (VA with $4^{M_2} = 16$ states).

The performance of the various receivers is illustrated in Figure 7.53; for comparison, the performance achieved by transmission over an ideal AWGN channel is also given.

QPSK over a non-minimum phase channel

We examine the following receivers.

1. ZF with $N = 7$ and $D = 4$;
2. LE with $N = 7$ and $D = 4$;
3. DFE with $M_1 = 7$, $M_2 = 4$, and $D = 6$;
4. VA with $4^4 = 256$ states and path memory depth equal to 15;
5. DFSE with $J_1 = J_2 = 4$, $J_3 = J_4 = 1$ (16 states);
6. DFE-VA with $M_1 = 7$, $M_2 = 2$, and $D = 4$ (VA with 16 states).

The performance of the various receivers is illustrated in Figure 7.54. Comparing the error probability curves shown in Figures 7.53 and 7.54, we note that VA outperforms the other methods; moreover, performance of the VA is almost independent of the phase of the overall channel impulse response; however, if the channel is minimum phase even a simple DFE or DFSE can give performance close to the optimum. We also note that in these simulations, the DFE-VA gives poor performance because the value of M_2 is too small, hence, the DFE is unable to equalize the channel.

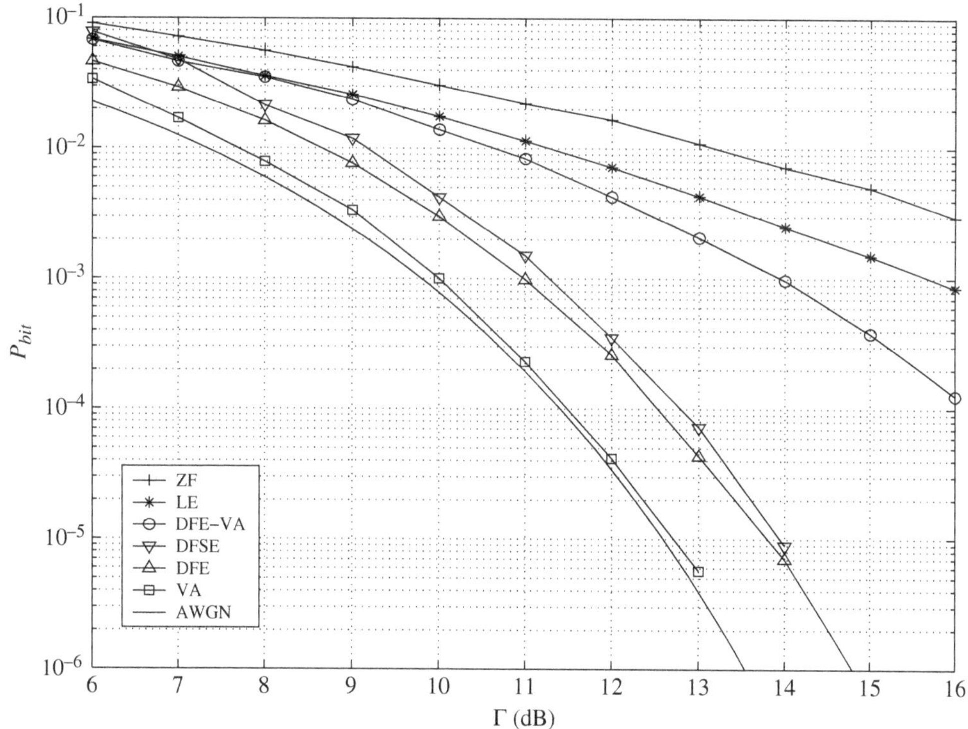

Figure 7.53 Bit error probability, P_{bit}, as a function of Γ for quadrature phase-shift-keying (QPSK) transmission over a minimum phase channel, using various equalization and data detection methods.

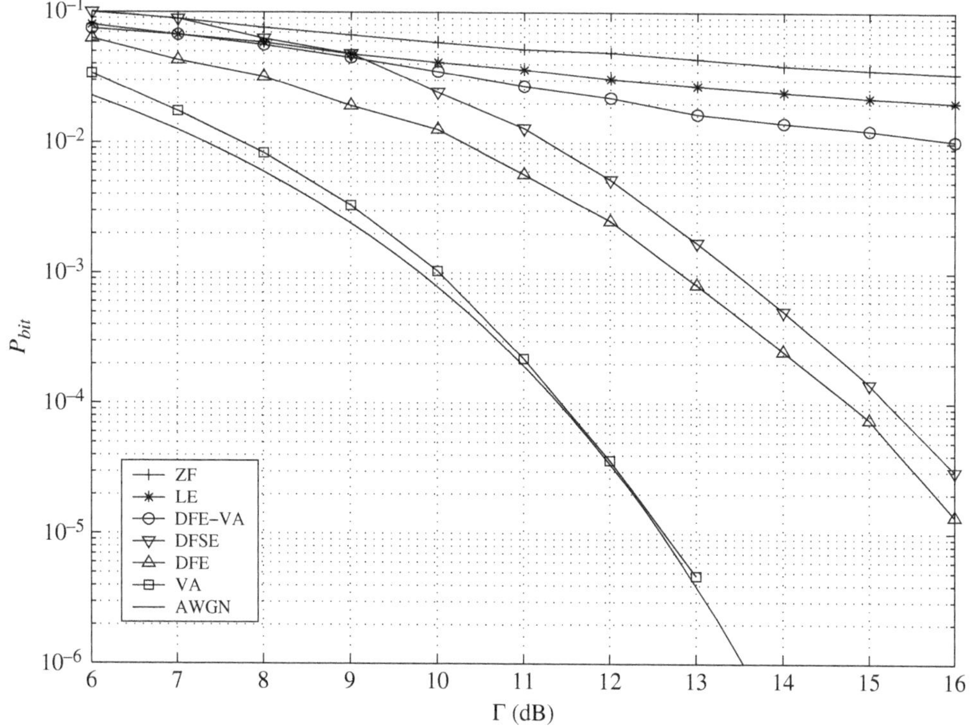

Figure 7.54 Bit error probability, P_{bit}, as a function of Γ for QPSK transmission over a non-minimum phase channel, using various equalization and data detection methods.

8-PSK over a minimum phase channel

We examine the following receivers.

1. DFE with $M_1 = 6$, $M_2 = 4$, and $D = 0$;
2. DFSE with $J_1 = J_2 = 8$, $J_3 = J_4 = 1$ (64 states);
3. DFE-VA with $M_1 = 7$, $M_2 = 2$, and $D = 0$ (VA with $8^{M_2} = 64$ states);
4. RSSE with $J_1 = 8$, $J_2 = 4$, $J_3 = 2$, $J_4 = 1$ (64 states).

The performance of the various receivers is illustrated in Figure 7.55.

8-PSK over a non-minimum phase channel

We examine the following receivers:

1. DFSE with $J_1 = J_2 = 8$, $J_3 = J_4 = 1$ (64 states);
2. DFE with $M_1 = 12$, $M_2 = 4$, and $D = 11$;
3. RSSE with $J_1 = 8$, $J_2 = 4$, $J_3 = 2$, $J_4 = 1$ (64 states).

In these simulations, the error probability for a DFE-VA is in the range between 0.08 and 0.2, and it is not shown; the performance of the various receivers is illustrated in Figure 7.56.

By comparison of the results of Figures 7.55 and 7.56, we observe that the RSSE and DFSE may be regarded as valid approximations of the VA as long as the overall channel impulse response is minimum phase.

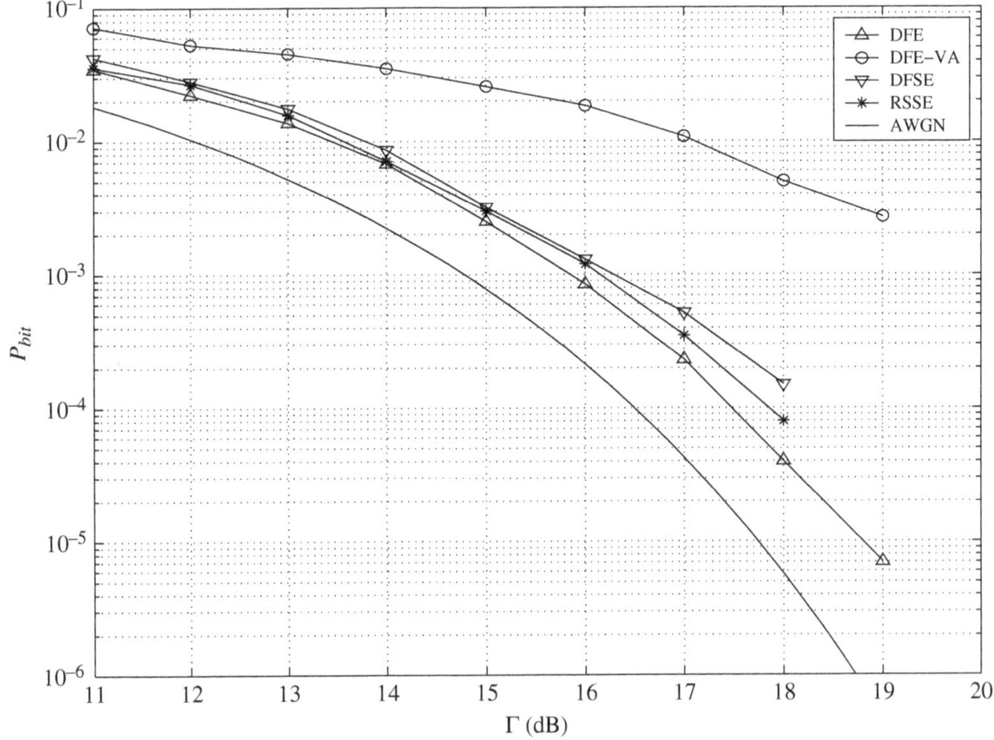

Figure 7.55 Bit probability error, P_{bit}, as a function of Γ for 8-PSK transmission over a minimum phase channel, using various equalization and data detection methods.

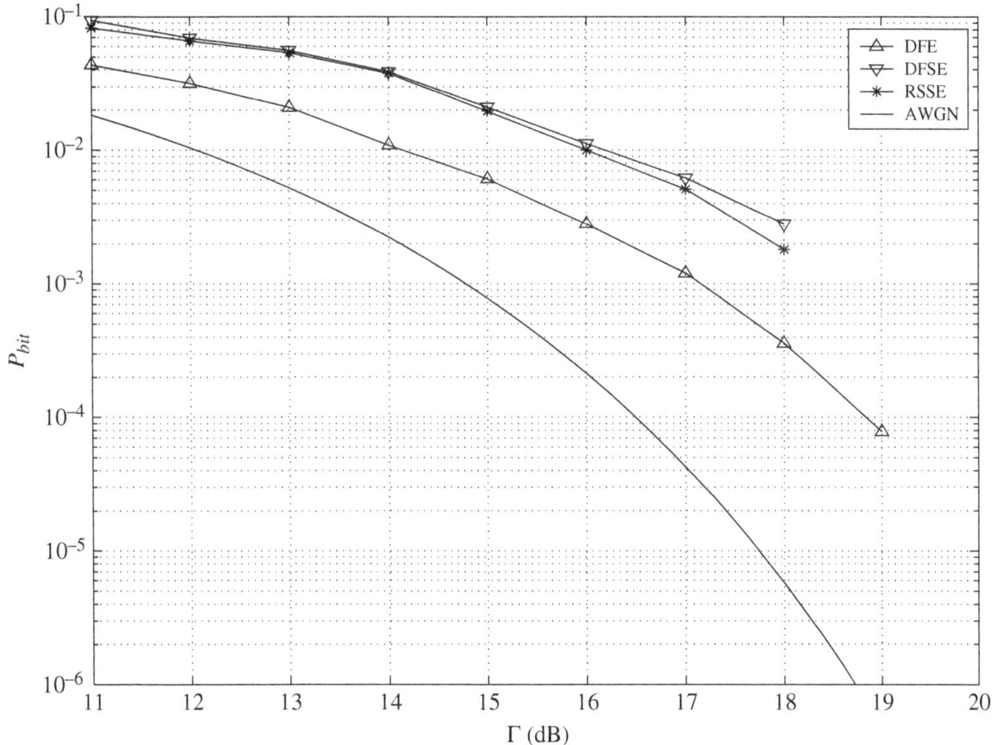

Figure 7.56 Bit probability error rate, P_{bit}, as a function of Γ for 8-PSK transmission over a non-minimum phase channel, using various equalization and data detection methods.

7.7 Precoding for dispersive channels

From Section 7.4.7 (see also Section 13.2), the discrete-time equivalent channel model at the receiver after a whitened matched filter can be written in the D-transform domain as

$$z(D) = a(D)\tilde{\mathtt{f}}(D) + w(D) \tag{7.507}$$

where $\{a_i\}$ is the data sequence, $\{\tilde{\mathtt{f}}_i\}$ is causal and minimum-phase, with $\tilde{\mathtt{f}}_0 = 1$, and $\{w_i\}$ is a white Gaussian noise.

The convolution $u(D) = a(D)\ \tilde{\mathtt{f}}(D)$ that determines the output sequence in the absence of noise can be viewed as the transform of a vector by the channel matrix, that is $\boldsymbol{u} = \tilde{\boldsymbol{F}}\boldsymbol{a}$, where $\boldsymbol{a}$ and $\boldsymbol{u}$ are vectors whose components are given by the transmitted symbols and channel output samples, respectively. As the canonical response $\tilde{\mathtt{f}}(D)$ is causal and monic, the matrix $\tilde{\boldsymbol{F}}$ is triangular with all elements equal to 1 on the diagonal and therefore, has determinant equal to 1; thus, it follows that the matrix $\tilde{\boldsymbol{F}}$ identifies a linear transformation that preserves the volume between the input and output spaces. In other words, the channel matrix $\tilde{\boldsymbol{F}}$ transforms a hypersphere into a hyperellipsoid having the same volume and containing the same number of constellation points.

Coding methods for linear dispersive channels [28–34] that yield high values of coding gain can be obtained by requiring that the channel output vectors in the absence of noise $\boldsymbol{u}$ are points of a set Λ' with good properties in terms of Euclidean distance, for example a lattice identified by integers. From the model of Figure 13.6c, an intuitive explanation of the objectives of coding for linear dispersive channels is obtained by considering the signal sequences $\boldsymbol{a}$, $\boldsymbol{u}$, $\boldsymbol{w}$, and z as vectors with a finite number of components. If the matrix $\tilde{F}$ is known at the transmitter, the input vector $\boldsymbol{a}$ can be predistorted so that the points of the vector $\boldsymbol{a}$ correspond to points of a signal set $\tilde{F}^{-1}\Lambda$; the volumes $V(\Lambda)$ in the output signal space, and $V(\tilde{F}^{-1}\Lambda)$ in the input signal space are equal. In any case, the output channel vectors are observed in the presence of additive white Gaussian noise vectors $\boldsymbol{w}$. If a detector chooses the point of the lattice Λ' with minimum distance from the output vector we obtain a coding gain, relative to Λ, as in the case of an ideal AWGN channel (see Section 6.5).

Recall that to achieve capacity, it is necessary that the distribution of the transmitted signal approximates a Gaussian distribution. We mention that commonly used methods of shaping, which also minimize the transmitted power, require that the points of the input constellation are uniformly distributed within a hypersphere [31]. Coding and shaping thus occur in two Euclidean spaces related by a known linear transformation that preserves the volume. Coding and shaping can be separately optimized, by choosing a method for predistorting the signal set in conjunction with a coding scheme that leads to a large coding gain for an ideal AWGN channel in the signal space where coding takes place, and a method that leads to a large shaping gain in the signal space where shaping takes place. In the remaining part of this chapter, we focus on precoding and coding methods to achieve large coding gains for transmission over channels with ISI, assuming the channel impulse response is known.

7.7.1 Tomlinson–Harashima precoding

Precoding is a method of pre-equalization that may achieve also for transmission over linear dispersive channels the objectives of coding illustrated in Section 6.5 for transmission over ideal AWGN channels. A special case of precoding is illustrated in Appendix 7.C.

We now consider the Tomlinson–Harashima precoding (THP) scheme. Let $\tilde{\mathtt{f}}(D)$ be the response of a discrete-time equivalent channel, with ISI and additive noise, and assume that the canonical response of the channel $\tilde{\mathtt{f}}(D)$ is known at the transmitter. Let $\tilde{\mathtt{f}}(D) = 1 + D\tilde{\mathtt{f}}'(D)$, and furthermore, assume that for every pair of symbols of the input constellation $\alpha_i, \alpha_j \in \mathcal{A}$ the following relation holds:[16] $\alpha_i = \alpha_j \bmod \Lambda_0$, where $\mathcal{A} \subset \Lambda_0 + \lambda$ is a finite set of symbols (constellation), Λ_0 represents the lattice associated with $\mathcal{A}$, and λ is a given offset value, possibly non-zero.

The objective of all precoding methods, with and without channel coding, consists in transmitting a pre-equalized sequence

$$a^{(p)}(D) = \frac{u^{(p)}(D)}{\tilde{\mathtt{f}}(D)} \tag{7.508}$$

so that, in the absence of noise, the channel output sequence $u(D) = a^{(p)}(D)\,\tilde{\mathtt{f}}(D) = u^{(p)}(D)$ represents the output of an ideal channel apparently ISI-free, with input sequence $u^{(p)}(D)$; $u^{(p)}(D)$ is a sequence of symbols belonging to a set $\mathcal{A}^{(p)} \subset \Lambda_0 + \lambda$. To achieve this objective with channel input signal samples that belong to a given finite region, the cardinality of the set $\mathcal{A}^{(p)}$ must be larger than that of the set $\mathcal{A}$. The redundancy in $u^{(p)}(D)$ can therefore be used to minimize the average power of the sequence $a^{(p)}(D)$ given by (7.508), or to obtain other desirable characteristics of $a^{(p)}(D)$, for example a low peak-to-average ratio. In the case of systems that adopt trellis coding (see Chapter 12), the channel output sequence in the absence of noise $u(D)$ must be a valid code sequence and can then be decoded by a decoder designed for an ideal channel. Note that, in a system with precoding, the elements of the transmitted sequence $a^{(p)}(D)$ are not in general symbols with discrete values.

The first precoding method was independently proposed for uncoded systems by Tomlinson [35] and Harashima and Miyakawa [36] (TH precoding). Initially, TH precoding was not used in practice because in an uncoded transmission system, the preferred method to cancel ISI employs a DFE, as it does not

[16] The expression $\alpha_i = \alpha_j \bmod \Lambda_0$ denotes that the two symbols α_i and α_j differ by a quantity that belongs to Λ_0.

require to send information on the channel impulse response to the transmitter. However, if trellis coding is adopted, decision-feedback equalization is no longer a very attractive solution, as reliable decisions are made available by the Viterbi decoder only with a certain delay.

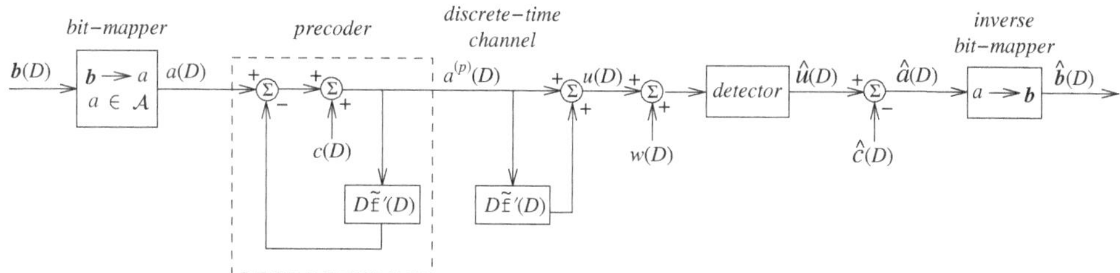

Figure 7.57 Block diagram of a system with TH precoding.

TH precoding, illustrated in Figure 7.57, uses memoryless operations at the transmitter and at the receiver to obtain samples of both the transmitted sequence $a^{(p)}(D)$ and the detected sequence $\hat{a}(D)$ within a finite region that contains $\mathcal{A}$. In principle, TH precoding can be applied to arbitrary symbol sets $\mathcal{A}$; however, unless it is possible to define an efficient extension of the region containing $\mathcal{A}$, the advantages of TH precoding are reduced by the increase of the transmit signal power (*transmit power penalty*). An efficient extension exists only if the signal space of $a^{(p)}(D)$ can be *tiled*, that is completely covered without overlapping with translated versions of a finite region containing $\mathcal{A}$, given by the union of the Voronoi regions of symbols of $\mathcal{A}$, and defined as $R(\mathcal{A})$. Figure 7.58 illustrates the efficient extension of a two-dimensional 16-QAM constellation, where Λ_T denotes the sublattice of Λ_0 that identifies the efficient extension.

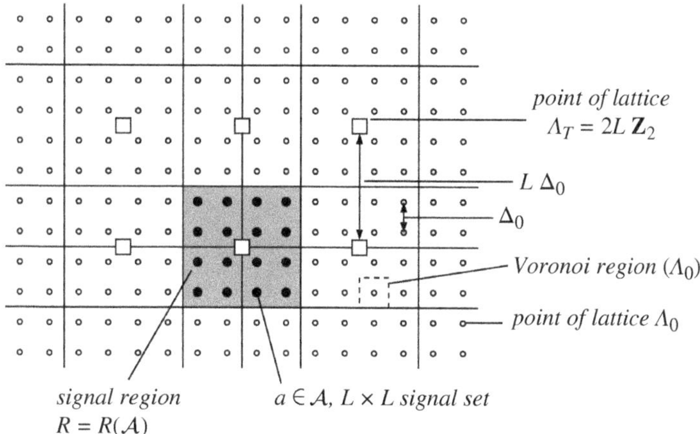

Figure 7.58 Illustration of the efficient extension of a two-dimensional 16-QAM constellation ($L = \sqrt{M} = 4$).

With reference to Figure 7.57, the precoder computes the sequence of channel input signal samples $a^{(p)}(D)$ as

$$a^{(p)}(D) = a(D) - p(D) + c(D) \tag{7.509}$$

where the sequence

$$p(D) = [\tilde{f}(D) - 1]\, a^{(p)}(D) = D\tilde{f}'(D)\, a^{(p)}(D) \tag{7.510}$$

represents the ISI at the channel output that must be compensated at the transmitter. The elements of the sequence $c(D)$ are points of the sublattice Λ_T used for the efficient extension of the region $R(\mathcal{A})$ that contains $\mathcal{A}$.[17] The k-th element $c_k \in \Lambda_T$ of the sequence $c(D)$ is chosen so that the statistical power of the channel input sample $a_k^{(p)}$ is minimum; in other words, the element c_k is chosen so that $a_k^{(p)}$ belongs to the region $R = R(\mathcal{A})$, as illustrated in Figure 7.58. From (7.508), (7.509), and (7.510), the channel output sequence in the absence of noise is given by

$$\begin{aligned} u(D) &= a^{(p)}(D)\ \tilde{\mathtt{f}}(D) = a^{(p)}[1 + D\ \tilde{\mathtt{f}}'(D)] \\ &= a^{(p)}(D) + p(D) = a(D) + c(D) \end{aligned} \tag{7.511}$$

Note that from (7.511), we get the relation $u_k = a_k \bmod \Lambda_T$, which is equivalent to (7.651).

The samples of the sequence $a^{(p)}(D)$ can be considered, with a good approximation, uniformly distributed in the region R. Assuming a constellation with $M = L \times L$ points for a QAM system, the power of the transmitted sequence is equal to that of a complex-valued signal with both real and imaginary parts that are uniformly distributed in $[-(L/2)\ \Delta_0, (L/2)\ \Delta_0]$, where Δ_0 denotes the minimum distance between points of the lattice Λ_0. Therefore, using the fact the quantization error power can be approximated as 1/12 of the square of the quantization step, it follows that

$$E[|a_k^{(p)}|^2] \simeq 2\ \frac{L^2}{12}\ \Delta_0^2 \tag{7.512}$$

Recalling that the statistical power of a transmitted symbol in a QAM system is given by $2\frac{L^2-1}{12}\Delta_0^2$ (see [[3], (5.197)] for $\Delta_0 = 2$), we find that the transmit power penalty in a system that applies TH precoding is equal to $\Delta_0^2/12$ per dimension.

From (7.507), the channel output signal is given by

$$z(D) = u(D) + w(D) \tag{7.513}$$

In the case of TH precoding for an uncoded system, the detector yields a sequence $\hat{u}(D)$ of symbols belonging to the constellation $\mathcal{A}^{(p)}$; from (7.511) the detected sequence $\hat{a}(D)$ of transmitted symbols is therefore, given by the memoryless operation

$$\hat{a}(D) = \hat{u}(D) - \hat{c}(D) \tag{7.514}$$

The k-th element $\hat{c}_k \in \Lambda_T$ of the sequence $\hat{c}(D)$ is chosen so that the symbol $\hat{a}_k = \hat{u}_k - \hat{c}_k$ belongs to the constellation $\mathcal{A}$. As the inverse operation of precoding is memoryless, error propagation at the receiver is completely avoided. Moreover, as the inversion of $\tilde{\mathtt{f}}(D)$ is not required at the receiver, $\tilde{\mathtt{f}}(D)$ may exhibit spectral nulls, that is it can contain factors of the form $(1 \pm D)$.

7.7.2 Flexible precoding

Two versions are introduced [32, 33].

First version [32]. In the first version, illustrated in Figure 7.59, the transmitted signal $a^{(p)}(D)$ can be expressed as

$$a^{(p)}(D) = a(D) + d(D) \tag{7.515}$$

where $d(D)$ is called the dither signal and is given by

$$d(D) = c(D) - p(D) \tag{7.516}$$

[17] Equation (7.509) represents the extension of (7.648) to the general case, in which the operation $\bmod M$ is substituted by the addition of the sequence $c(D)$.

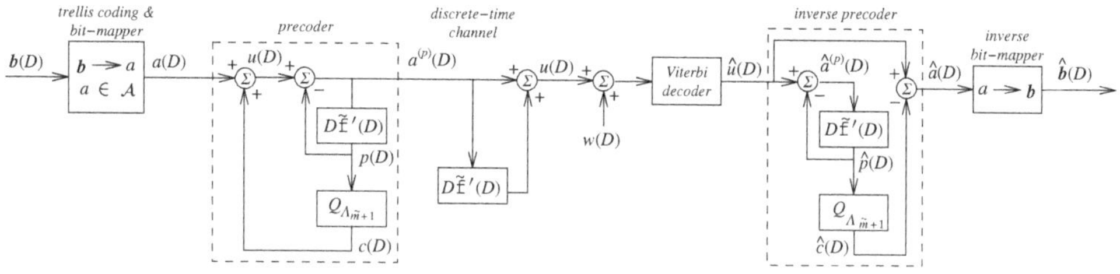

Figure 7.59 Block diagram of a system with flexible precoding.

where $p(D)$ is given by (7.510) and $c(D)$ is obtained from the quantization of $p(D)$ with quantizer $Q_{\Lambda_{\tilde{m}+1}}$. The quantizer $Q_{\Lambda_{\tilde{m}+1}}$ yields the k-th element of the sequence $c(D)$ by quantizing the sample p_k to the closest point of the lattice $\Lambda_{\tilde{m}+1}$, which corresponds to the $(\tilde{m}+1)$-th level of partitioning of the signal set; in the case of an uncoded sequence, it is $\Lambda_{\tilde{m}+1} = \Lambda_0$. Note that the dither signal can be interpreted as the signal with minimum amplitude that must be added to the sequence $a(D)$ to obtain a valid code sequence at the channel output in the absence of noise. In fact, at the channel output, we get the sequence $z(D) = u(D) + w(D)$, where $u(D)$ is obtained by adding a sequence of points taken from the lattice $\Lambda_{\tilde{m}+1}$ to the code sequence $a(D)$, and therefore, it represents a valid code sequence.

The sequence $z(D)$ is input to a Viterbi decoder, which yields the detected sequence $\hat{u}(D)$. To obtain a detection of the sequence $a(D)$, it is first necessary to detect the sequence $\hat{c}(D)$ of the lattice points $\Lambda_{\tilde{m}+1}$ added to the sequence $a(D)$. Observing

$$\hat{a}^{(p)}(D) = \frac{\hat{u}(D)}{\tilde{\mathtt{f}}(D)} \tag{7.517}$$

and

$$\hat{p}(D) = D\,\tilde{\mathtt{f}}'(D)\,\hat{a}^{(p)}(D) \tag{7.518}$$

the sequence $\hat{c}(D)$ is obtained by quantizing with a quantizer $Q_{\Lambda_{\tilde{m}+1}}$ the sequence $\hat{p}(D)$, as illustrated in Figure 7.59. Then subtracting $\hat{c}(D)$ from $\hat{u}(D)$, we obtain the sequence $\hat{a}(D)$, that is used to detect the sequence of information bits.

At this point, we can make the following observations with respect to the first version of flexible precoding:

1. an efficient extension of the region $R(\mathcal{A})$ is not necessary;
2. it is indispensable that the implementation of the blocks that perform similar functions in the precoder and in the inverse precoder (see Figure 7.59) is identical with regard to the binary representation of input and output signals;
3. as the dither signal can be assumed uniformly distributed in the Voronoi region $V(\Lambda_{\tilde{m}+1})$ of the point of the lattice $\Lambda_{\tilde{m}+1}$ corresponding to the origin, the transmit power penalty is equal to $\Delta^2_{\tilde{m}+1}/12$ per dimension; this can significantly reduce the coding gain if the cardinality of the constellation $\mathcal{A}$ is small;
4. to perform the inverse of the precoding operation, the inversion of the channel transfer function is required (see (7.517)); if $\tilde{\mathtt{f}}(D)$ is minimum phase, then $1/\tilde{\mathtt{f}}(D)$ is stable and the effect of an error event at the Viterbi decoder output vanishes after a certain number of iterations; on the other hand, if $\tilde{\mathtt{f}}(D)$ has zeros on the unit circle (spectral nulls), error events at the Viterbi decoder output can result in an unlimited propagation of errors in the detection of the sequence $a(D)$.

Second version [33]. The second version of flexible precoding includes trellis coding with feedback.

7.8 Channel estimation

Equalization, sequence detection, and precoding techniques for systems with ISI require the knowledge of the channel impulse response. Here techniques to obtain an estimate of the channel are described.

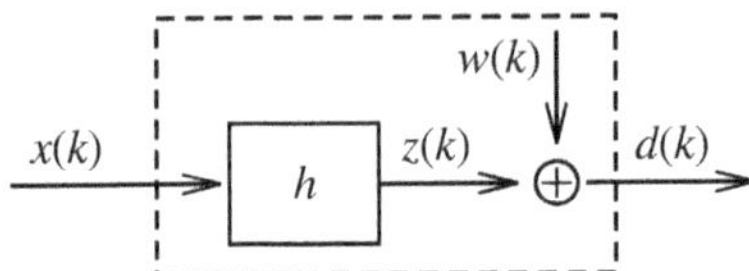

Figure 7.60 System model (see also the front-end of Figure 7.26).

7.8.1 The correlation method

With reference to Figure 7.60, where nomenclature is from Chapter 2, and (2.232), which describes the relation between input and output of an unknown linear system with impulse response $\{h_i\}$, $i = 0, 1, \dots, N-1$, we take as an input signal *white noise* with statistical power $\mathtt{r}_x(0)$. As anticipated in (2.235), to estimate the impulse response of such a system, we evaluate the crosscorrelation between d and x. In fact, as w and x are uncorrelated, we have

$$\mathtt{r}_{dx}(n) = \mathtt{r}_{zx}(n) = \mathtt{r}_x * h(n) = \mathtt{r}_x(0)h_n \tag{7.519}$$

In practice, instead of *noise* a PN sequence $\{p(i)\}$, with period L, is used as input. Moreover, the statistical correlation is replaced by the time average. We recall that the autocorrelation of a PN sequence is given by (see Appendix 1.C):

$$\mathtt{r}_p(n) \begin{cases} = 1, & n = 0, L, 2L, \dots \\ \simeq 0, & n = 1, \dots, L-1, L+1, \dots \end{cases} \tag{7.520}$$

As input we assume

$$x(k) = p(k), \qquad k = 0, 1, \dots, L-1, \dots, L-1+N-1 \tag{7.521}$$

i.e. just a full period plus $(N-1)$ samples of $\{p(i)\}$ are used. At the output, we collect L samples

$$d(k), \qquad k = N-1, \dots, N-1+L-1 \tag{7.522}$$

The first $(N-1)$ samples of x are used to fill up the buffer of the unknown FIR filter. In fact, at time $k = N-1$, the filter transient is over. The correlation method exploits of the following estimate

$$\hat{h}_{COR,n} = \hat{\mathtt{r}}_{dx}(n) = \frac{1}{L} \sum_{k=(N-1)}^{(N-1)+(L-1)} d(k)\, p^*(k-n)_{mod\ L} \simeq h_n, \qquad n = 0, 1, \dots, N-1 \tag{7.523}$$

We denote the output of the estimated system as

$$\hat{d}(k) = \sum_{i=0}^{N-1} \hat{h}_i x(k-i) \tag{7.524}$$

which is an estimate of the unknown system output $z(k)$.

Mean and variance of the estimate (7.523) are obtained as follows. First, we replace $d(k)$ with (2.232) to yield

$$\begin{aligned} \hat{h}_{COR,n} &= \frac{1}{L} \sum_{k=N-1}^{N-1+L-1} \sum_{i=0}^{N-1} h_i p(k-i)_{mod\ L} p^*(k-n)_{mod\ L} + \frac{1}{L} \sum_{k=N-1}^{N-1+L-1} w(k) p^*(k-n)_{mod\ L} \\ &= \sum_{i=0}^{N-1} h_i \mathtt{r}_p(n-i)_{mod\ L} + \frac{1}{L} \sum_{k=N-1}^{N-1+L-1} w(k) p^*(k-n)_{mod\ L} \end{aligned} \tag{7.525}$$

If $L \gg 1$, the second term on the right-hand side of (7.525) can be neglected, hence, observing (7.520) we get

$$\hat{h}_{COR,n} \simeq h_n \tag{7.526}$$

Then, we obtain:

1. *Mean*:
$$E[\hat{h}_{COR,n}] = \sum_{i=0}^{N-1} h_i \ \mathtt{r}_p(n-i)_{mod\ L} \tag{7.527}$$
assuming w has zero mean.
2. *Variance*:
$$\text{var}[\hat{h}_{COR,n}] = \text{var}\left[\frac{1}{L}\sum_{k=N-1}^{N-1+L-1} w(k)p^*(k-n)_{mod\ L}\right] = \frac{\sigma_w^2}{L} \tag{7.528}$$
assuming w white and $|p(i)| = 1$.

If $\hat{h}_{COR,n} \simeq h_n$, it is $d(k) - \hat{d}(k) \simeq w(k)$, the noise term. This approximation can be used to estimate the variance of $w(k)$, e.g. as

$$\hat{\sigma}_w^2 = \frac{1}{L}\sum_{k=N-1}^{N-1+L-1} |d(k) - \hat{d}(k)|^2 \tag{7.529}$$

7.8.2 The LS method

With reference to the system of Figure 7.60, letting $\boldsymbol{x}^T(k) = [x(k), x(k-1), \dots, x(k-(N-1))]$, the noisy output of the unknown system can be written as

$$d(k) = \boldsymbol{h}^T\boldsymbol{x}(k) + w(k), \qquad k = (N-1), \dots, N-1+L-1 \tag{7.530}$$

The unknown system can be identified using the least squares (LS) method of Section 2.3 [37–39]. For a certain estimate $\hat{\boldsymbol{h}}$ of the unknown system, the sum of square output errors is given by

$$\mathcal{E} = \sum_{k=N-1}^{N-1+L-1} |d(k) - \hat{d}(k)|^2 \tag{7.531}$$

where, from (7.524)

$$\hat{d}(k) = \hat{\boldsymbol{h}}^T\boldsymbol{x}(k) \tag{7.532}$$

As for the analysis of Section 2.3, we introduce the following quantities.

1. *Energy of the desired signal*:
$$\mathcal{E}_d = \sum_{k=N-1}^{N-1+L-1} |d(k)|^2 \tag{7.533}$$
2. *Correlation matrix of the input signal*:
$$\boldsymbol{\Phi} = [\Phi(i,n)], \qquad i,n = 0, \dots, N-1 \tag{7.534}$$
where
$$\Phi(i,n) = \sum_{k=N-1}^{N-1+L-1} x^*(k-i)\ x(k-n) \tag{7.535}$$
3. *Crosscorrelation vector*:
$$\boldsymbol{\vartheta}^T = [\vartheta(0), \dots, \vartheta(N-1)] \tag{7.536}$$

where

$$\vartheta(n) = \sum_{k=N-1}^{N-1+L-1} d(k)\, x^*(k-n) \tag{7.537}$$

A comparison with (7.523) shows that $\frac{1}{L}\vartheta(n)$ coincides with $\hat{h}_{COR,n}$. Then the cost function (7.531) becomes

$$\mathcal{E} = \mathcal{E}_d - \hat{\boldsymbol{h}}^H \boldsymbol{\vartheta} - \boldsymbol{\vartheta}^H \hat{\boldsymbol{h}} + \hat{\boldsymbol{h}}^H \boldsymbol{\Phi} \hat{\boldsymbol{h}} \tag{7.538}$$

As the matrix $\boldsymbol{\Phi}$ is determined by a suitably chosen training sequence, we can assume that $\boldsymbol{\Phi}$ is positive definite and therefore, the inverse exists. The solution to the LS problem yields

$$\hat{\boldsymbol{h}}_{LS} = \boldsymbol{\Phi}^{-1} \boldsymbol{\vartheta} \tag{7.539}$$

with a corresponding error energy equal to

$$\mathcal{E}_{min} = \mathcal{E}_d - \boldsymbol{\vartheta}^H \hat{\boldsymbol{h}}_{LS} \tag{7.540}$$

We observe that the matrix $\boldsymbol{\Phi}^{-1}$ in the (7.540) can be pre-computed and memorized, because it depends only on the known input sequence. For an estimate of the variance of the system noise w, observing (7.531), for $\hat{\boldsymbol{h}}_{LS} \simeq \boldsymbol{h}$ we can assume

$$\hat{\sigma}_w^2 = \frac{1}{L}\, \mathcal{E}_{min} \tag{7.541}$$

which coincides with (7.529).

Formulation using the data matrix

From the general analysis given on page 122, we recall the following definitions:

1. *$L \times N$ input data matrix*:

$$\mathcal{I} = \begin{bmatrix} x(N-1) & \dots & x(0) \\ \vdots & \ddots & \vdots \\ x(N-1+L-1) & \dots & x(L-1) \end{bmatrix} \tag{7.542}$$

2. *Desired sample vector*:

$$\boldsymbol{o}^T = [d(N-1), \dots, d((N-1)+(L-1))] \tag{7.543}$$

where $d(k)$ is given by (7.530).
Observing (2.103), (2.95), and (2.124), we have

$$\boldsymbol{\Phi} = \mathcal{I}^H \mathcal{I} \quad \boldsymbol{\vartheta} = \mathcal{I}^H \boldsymbol{o} \tag{7.544}$$

and

$$\hat{\boldsymbol{h}}_{LS} = (\mathcal{I}^H \mathcal{I})^{-1} \mathcal{I}^H \boldsymbol{o} \tag{7.545}$$

which coincides with (7.539).

We note the introduction of the new symbol $\boldsymbol{o}$, in relation to an alternative linear minimium mean square error (LMMSE) estimation method, which will be given in Section 7.8.5.

7.8.3 Signal-to-estimation error ratio

Let $\boldsymbol{h}^T = [h_0, h_1, \dots, h_{N-1}]$ be the filter coefficients and $\hat{\boldsymbol{h}}^T = [\hat{h}_0, \hat{h}_1, \dots, \hat{h}_{N-1}]$ their estimates. Let $\boldsymbol{\Delta h}$ be the estimation error vector

$$\boldsymbol{\Delta h} = \hat{\boldsymbol{h}} - \boldsymbol{h}$$

The quality of the estimate is measured by the signal-to-estimation error ratio

$$\Lambda_e = \frac{\|\boldsymbol{h}\|^2}{E[\|\boldsymbol{\Delta h}\|^2]} \tag{7.546}$$

The noise present in the observed system is measured by (see Figure 7.60)

$$\Lambda = \frac{\mathtt{M}_x \|\boldsymbol{h}\|^2}{\sigma_w^2} \tag{7.547}$$

where $\mathtt{M}_x$ is the statistical power of the input signal. In our case, $\mathtt{M}_x = 1$. Lastly, we refer to the normalized ratio

$$\Lambda_n = \frac{\Lambda_e}{\Lambda} = \frac{\sigma_w^2}{\mathtt{M}_x \, E[\|\boldsymbol{\Delta h}\|^2]} \tag{7.548}$$

We note that if we indicate with $\hat{d}(k)$ the output of the identified system,

$$\hat{d}(k) = \sum_{i=0}^{N-1} \hat{h}_i \, x(k-i) \tag{7.549}$$

the fact that $\hat{\boldsymbol{h}} \neq \boldsymbol{h}$ causes $\hat{d}(k) \neq z(k)$, with an error given by

$$z(k) - \hat{d}(k) = \sum_{i=0}^{N-1} (h_i - \hat{h}_i) \, x(k-i) \tag{7.550}$$

having variance $\mathtt{M}_x \, E[\|\boldsymbol{\Delta h}\|^2]$ for a white noise input. As a consequence, (7.548) measures the ratio between the variance of the additive noise of the observed system and the variance of the error at the output of the identified system. From (7.550), we note that the difference

$$d(k) - \hat{d}(k) = (z(k) - \hat{d}(k)) + w(k) \tag{7.551}$$

consists of two terms, one due to the estimation error and one due to the system noise.

Computation of the signal-to-estimation error ratio

We now evaluate the LS method performance for the estimate of $\boldsymbol{h}$. From (7.537), (7.536) can be rewritten as

$$\boldsymbol{\vartheta} = \sum_{k=N-1}^{N-1+L-1} d(k) \, \boldsymbol{x}^*(k) \tag{7.552}$$

Substituting (7.530) in (7.552), and letting

$$\boldsymbol{\xi} = \sum_{k=N-1}^{N-1+L-1} w(k) \, \boldsymbol{x}^*(k) \tag{7.553}$$

observing (7.534), we obtain the relation

$$\boldsymbol{\vartheta} = \boldsymbol{\Phi h} + \boldsymbol{\xi} \tag{7.554}$$

Consequently, substituting (7.554) in (7.539), the estimation error vector can be expressed as

$$\boldsymbol{\Delta h} = \boldsymbol{\Phi}^{-1} \boldsymbol{\xi} \tag{7.555}$$

If w is zero-mean white noise with variance σ_w^2, $\boldsymbol{\xi}^*$ is a zero-mean random vector with correlation matrix

$$\boldsymbol{R}_{\xi} = E[\boldsymbol{\xi}^* \boldsymbol{\xi}^T] = \sigma_w^2 \boldsymbol{\Phi}^* \tag{7.556}$$

Therefore, $\boldsymbol{\Delta h}$ has zero mean and correlation matrix

$$R_{\boldsymbol{\Delta h}} = \sigma_w^2(\boldsymbol{\Phi}^*)^{-1} \tag{7.557}$$

In particular,

$$E[\|\boldsymbol{\Delta h}\|^2] = \sigma_w^2 \text{tr}[(\boldsymbol{\Phi}^*)^{-1}] \tag{7.558}$$

and, from (7.548), we get

$$\Lambda_n = (\text{tr}[\boldsymbol{\Phi}^{-1}])^{-1} \quad \text{LS} \tag{7.559}$$

Using as input sequence a *CAZAC sequence*, the matrix $\boldsymbol{\Phi}$ is diagonal,

$$\boldsymbol{\Phi} = L\,\boldsymbol{I} \tag{7.560}$$

where $\boldsymbol{I}$ is the $N \times N$ identity matrix. The elements on the diagonal of $\boldsymbol{\Phi}^{-1}$ are equal to $1/L$, and (7.559) yields

$$\Lambda_n = \frac{L}{N} \quad \text{LS–CAZAC} \tag{7.561}$$

The (7.561) gives a good indication of the relation between the number of observations L, the number of system coefficients N, and Λ_n. For example, doubling the length of the PN sequence, Λ_n also doubles.

Now, using as PN sequence a maximum-length sequence (MLS) of periodicity L, and indicating with $\mathbf{1}_{N\times N}$ the matrix with all elements equal to 1, the correlation matrix $\boldsymbol{\Phi}$ can be written as

$$\boldsymbol{\Phi} = (L+1)\boldsymbol{I} - \mathbf{1}_{N\times N} \tag{7.562}$$

whose inverse is given by

$$\boldsymbol{\Phi}^{-1} = \frac{1}{L+1}\left(\boldsymbol{I} + \frac{\mathbf{1}_{N\times N}}{L+1-N}\right) \tag{7.563}$$

which, substituted in (7.563), yields

$$\Lambda_n = \frac{(L+1)(L+1-N)}{N(L+2-N)} \quad \text{LS-MLS} \tag{7.564}$$

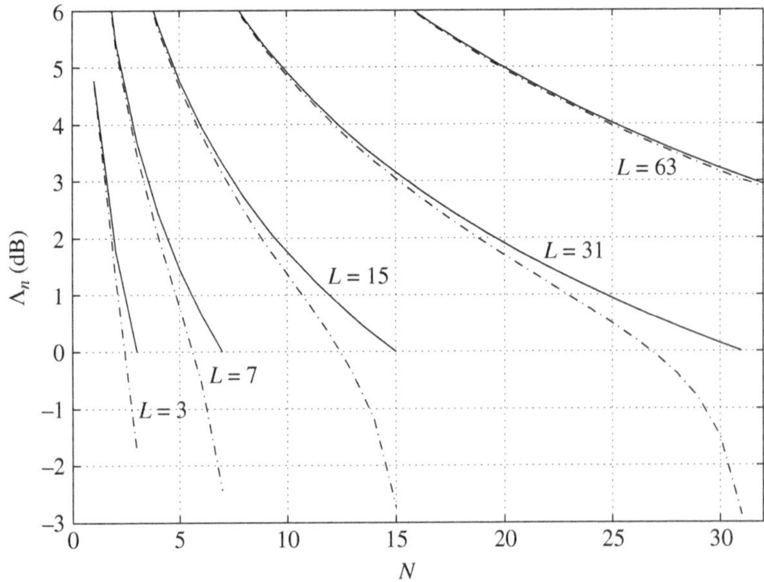

Figure 7.61 LS channel estimate. Λ_n vs. N for CAZAC sequences (solid line) and MLSs (dotted-dashed line), for various values of L.

In Figure 7.61, the behaviour of Λ_n is represented as a function of N, for CAZAC sequences (solid line) and for MLSs (dotted-dashed line), with parameter L. We make the following observations.

1. For a given N, choosing $L \gg N$, the two sequences yield approximately the same Λ_n. The worst case is obtained for $L = N$; for example, for $L = 15$ the MLS yields a value of Λ_n that is about 3 dB lower than the upper bound (7.561). However, in order to get a good estimation performance, usually L is from 5 to 10 times N.
2. For a given value of L, because of the presence of the noise w, the estimate of the coefficients becomes worse if the number of coefficients N is larger than the number of coefficients of the system N_h. On the other hand, if N is smaller than N_h, the estimation error may assume large values (see (2.238)).
3. For sparse systems, where the number of coefficients may be large, but only a few of them are non-zero, the estimate is usually very noisy. Therefore, after obtaining the estimate, it is necessary to set to zero all coefficients whose amplitude is below a certain threshold.

The correlation method (7.523) is now analysed. From

$$\hat{\boldsymbol{h}}_{COR} = \frac{1}{L}\,\boldsymbol{\vartheta} \tag{7.565}$$

where $\boldsymbol{\vartheta}$ is given by (7.536) and observing (7.554), we get

$$\boldsymbol{\Delta h} = \left(\frac{1}{L}\,\boldsymbol{\Phi} - \boldsymbol{I}\right)\boldsymbol{h} + \frac{1}{L}\,\boldsymbol{\xi} \tag{7.566}$$

Consequently, the estimate is affected by a bias term equal to $((1/L)\,\boldsymbol{\Phi} - \boldsymbol{I})\boldsymbol{h}$, and has a covariance matrix equal to $(1/L^2)\,\boldsymbol{R}_{\boldsymbol{\xi}}$. In particular, using (7.556), it turns out

$$E[\|\boldsymbol{\Delta h}\|^2] = \left\|\left(\frac{1}{L}\,\boldsymbol{\Phi} - \boldsymbol{I}\right)\boldsymbol{h}\right\|^2 + \frac{\sigma_w^2}{L^2}\mathrm{tr}[\boldsymbol{\Phi}] \tag{7.567}$$

and

$$\Lambda_n = \frac{1}{\dfrac{1}{L^2}\mathrm{tr}[\boldsymbol{\Phi}] + \left\|\left(\dfrac{1}{L}\,\boldsymbol{\Phi} - \boldsymbol{I}\right)\boldsymbol{h}\right\|^2 \dfrac{1}{\sigma_w^2}} \qquad \text{COR} \tag{7.568}$$

Using a *CAZAC sequence*, from (7.560) the second term of the denominator in (7.568) vanishes, and

$$\Lambda_n = \frac{L}{N} \qquad \text{COR–CAZAC} \tag{7.569}$$

as in (7.561). In fact, for a CAZAC sequence, as (7.520) is strictly true and $\boldsymbol{\Phi}^{-1}$ is diagonal, the LS method coincides with the correlation method.

Using instead an MLS, from (7.562), we get

$$\begin{aligned}\left\|\left(\frac{1}{L}\,\boldsymbol{\Phi} - \boldsymbol{I}\right)\boldsymbol{h}\right\|^2 &= \frac{1}{L^2}\,\|(\boldsymbol{1} - \boldsymbol{I})\,\boldsymbol{h}\|^2 = \frac{1}{L^2}\sum_{i=0}^{N-1}|h_i - \mathcal{H}(0)|^2 \\ &= \frac{1}{L^2}(\|\boldsymbol{h}\|^2 + (N-2)\,|\mathcal{H}(0)|^2)\end{aligned} \tag{7.570}$$

where $\mathcal{H}(0) = \sum_{i=0}^{N-1} h_i$. Moreover, we have

$$\mathrm{tr}[\boldsymbol{\Phi}] = NL \tag{7.571}$$

hence

$$\Lambda_n = \frac{L}{N + \frac{1}{L}\left[\Lambda + (N-2)\,\dfrac{|\mathcal{H}(0)|^2}{\sigma_w^2}\right]} \qquad \text{COR–MLS} \tag{7.572}$$

where Λ is defined in (7.547). We observe that using the correlation method, we obtain the same values Λ_n (7.564) as the LS method, if L is large enough to satisfy the condition

$$\Lambda + (N-2)\,\frac{|\mathcal{H}(0)|^2}{\sigma_w^2} < L$$

On the selection of the channel length

Indeed, both COR and LS methods provide an estimate of the channel coefficients once the channel length has been selected. For a suitable choice of the channel length N_h, we can start from a small value of N, increase it gradually and stop when $\mathcal{E}/L$ does not significantly change, i.e. it reaches approximately its asymptotic value. As from Observation 2 at page 383, we assume that for $N < N_h$ $\mathcal{E}/L$ is affected by a bias due to the truncation of the channel impulse response (see (2.238)), while for $N > N_h$ some coefficients of the estimate are just noise.

Note that by a conservative choice of $N = N_{max} > N_h$, with corresponding PN input sequence length L, we can evaluate the channel estimate for all values of N up to N_{max} and select N_h by the above procedure.

7.8.4 Channel estimation for multirate systems

Let us assume the input to the channel is at T while its output is at $T/2$ as illustrated in Figure 7.62 with

$$d(n) = z(n) + w(n) \tag{7.573}$$

$$z(n) = \sum_{k=-\infty}^{\infty} x(k)h(n-2k) \tag{7.574}$$

Extension to outputs at T/M, with M integer, is straightforward.

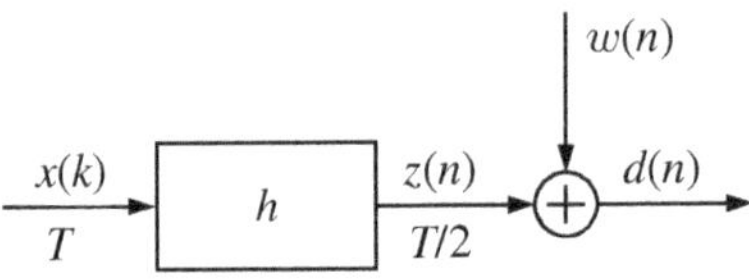

Figure 7.62 Multirate system.

By the polyphase representation (see Section 1.A.9) of d, w, and h, with $i = 0, 1$,

$$d^{(i)}(k) = d(2k+i) \tag{7.575}$$

$$w^{(i)}(k) = w(2k+i) \tag{7.576}$$

$$z^{(i)}(k) = z(2k+i) \tag{7.577}$$

$$h^{(i)}(m) = h(2m+i) \tag{7.578}$$

the system of Figure 7.62 can be redrawn as in Figure 7.63.

For an input PN-sequence $x(k)$, the output $d(n)$ at $T/2$ is split into its even ($d^{(0)}$) and odd ($d^{(1)}$) number samples at T. Next, for each subsequence $d^{(i)}$ by using one of the above methods (LS or COR) we can obtain an estimate of the corresponding polyphase channel component $h^{(i)}$. Note that in Figure 7.62 for a WSS noise w it is

$$\sigma^2_{w^{(0)}} = \sigma^2_{w^{(1)}} = \sigma^2_w \tag{7.579}$$

Moreover, as

$$||\boldsymbol{h}||^2 = ||\boldsymbol{h}^{(0)}||^2 + ||\boldsymbol{h}^{(1)}||^2 \tag{7.580}$$

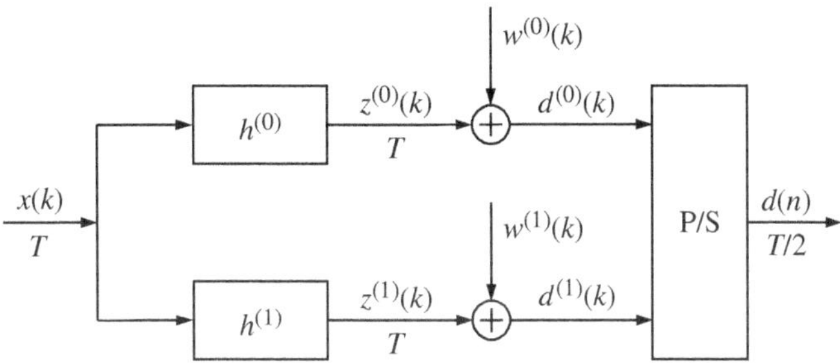

Figure 7.63 Polyphase decomposition of a multirate system.

it is easy to show that

$$\Lambda_n = \frac{\sigma_w^2}{\mathtt{M}_x ||\boldsymbol{h}||^2} = \frac{1}{\dfrac{1}{\Lambda_n^{(0)}} + \dfrac{1}{\Lambda_n^{(1)}}} \tag{7.581}$$

where

$$\Lambda_n^{(i)} = \frac{\sigma_{w^{(i)}}^2}{\mathtt{M}_x ||\boldsymbol{h}^{(i)}||^2} \tag{7.582}$$

7.8.5 The LMMSE method

We refer to the system model of Figure 7.60. Let us assume that w and h are statistically independent random processes, whose *second-order statistic is known.*

For a known input sequence x, we estimate h using the LMMSE method given in Section 2.4, from the *observation of the noisy output sequence* d, which now will be denoted as o. Let $\boldsymbol{C}$ be a $L \times N$ matrix and $\hat{\boldsymbol{h}} = \boldsymbol{C}^T \boldsymbol{o}$ with w and h both random. The problem is the following:

$$\hat{\boldsymbol{h}}_{LMMSE} = \arg\min_{\boldsymbol{C}} E[||\boldsymbol{h} - \hat{\boldsymbol{h}}||^2] \tag{7.583}$$

We note that in the LS method, the observation was the transmitted signal x, while the desired signal was given by the system noisy output d. The observation is now given by d and the desired signal is the system impulse response h. Consequently, some caution is needed to apply (2.198) to the problem under investigation. Recalling the definition (7.543) of the observation vector $\boldsymbol{o}$, and (2.198), the LMMSE estimator is given by

$$\hat{\boldsymbol{h}}_{LMMSE} = (\boldsymbol{R}_o^{-1} \boldsymbol{R}_{oh})^T \boldsymbol{o} \tag{7.584}$$

where we have assumed $E[\boldsymbol{o}] = \boldsymbol{0}$ and $E[\boldsymbol{h}] = \boldsymbol{0}$. Consequently, (7.584) provides an estimate only of the random (i.e. non-deterministic) component of the channel impulse response.

Now, from (7.530), defining the random vector with noise components

$$\boldsymbol{w}^T = [w(N-1), \ldots, w((N-1) + (L-1))] \tag{7.585}$$

we can write

$$\boldsymbol{o} = \boldsymbol{\mathcal{I}} \boldsymbol{h} + \boldsymbol{w} \tag{7.586}$$

Assuming that the sequence $\{x(k)\}$ is known, we have

$$\boldsymbol{R}_o = E[\boldsymbol{o}^* \boldsymbol{o}^T] = \boldsymbol{\mathcal{I}}^* \boldsymbol{R}_h \boldsymbol{\mathcal{I}}^T + \boldsymbol{R}_w \tag{7.587}$$

and

$$\boldsymbol{R}_{oh} = E[\boldsymbol{o}^* \boldsymbol{h}^T] = \mathcal{I}^* \boldsymbol{R}_h \tag{7.588}$$

Then (7.584) becomes

$$\hat{\boldsymbol{h}}_{LMMSE} = [(\mathcal{I}^* \boldsymbol{R}_h \mathcal{I}^T + \boldsymbol{R}_w)^{-1} \mathcal{I}^* \boldsymbol{R}_h]^T \boldsymbol{o} \tag{7.589}$$

Using the matrix inversion lemma (3.160), (7.589) can be rewritten as

$$\hat{\boldsymbol{h}}_{LMMSE} = [(\boldsymbol{R}_h^*)^{-1} + \mathcal{I}^H (\boldsymbol{R}_w^*)^{-1} \mathcal{I}]^{-1} \mathcal{I}^H (\boldsymbol{R}_w^*)^{-1} \boldsymbol{o} \tag{7.590}$$

If $\boldsymbol{R}_w = \sigma_w^2 \boldsymbol{I}$, we have

$$\hat{\boldsymbol{h}}_{LMMSE} = [\sigma_w^2 (\boldsymbol{R}_h^*)^{-1} + \mathcal{I}^H \mathcal{I}]^{-1} \mathcal{I}^H \boldsymbol{o} \tag{7.591}$$

We note that with respect to the LS method (7.545), the LMMSE method (7.591) introduces a weighting of the components given by $\boldsymbol{\vartheta} = \mathcal{I}^H \boldsymbol{o}$, which depends on the ratio between the noise variance and the variance of $\boldsymbol{h}$. If the variance of the components of $\boldsymbol{h}$ is large, then $\boldsymbol{R}_h$ is also large and likely $\boldsymbol{R}_h^{-1}$ can be neglected in (7.591).

We conclude recalling that $\boldsymbol{R}_h$ is diagonal for a wide-sense stationary uncorrelated scattering (WSSUS) radio channel model (see (4.86)), whose components are derived by the *power delay profile*.

For an analysis of the estimation error, we can refer to (2.201), which uses the error vector $\boldsymbol{\Delta h} = \hat{\boldsymbol{h}}_{LMMSE} - \boldsymbol{h}$ having a correlation matrix

$$\boldsymbol{R}_{\Delta h} = \{[(\boldsymbol{R}_h^*)^{-1} + \mathcal{I}^H (\boldsymbol{R}_w^*)^{-1} \mathcal{I}]^*\}^{-1} \tag{7.592}$$

If $\boldsymbol{R}_w = \sigma_w^2 \boldsymbol{I}$, we get

$$\boldsymbol{R}_{\Delta h} = \sigma_w^2 \{[\sigma_w^2 (\boldsymbol{R}_h^*)^{-1} + \mathcal{I}^H \mathcal{I}]^*\}^{-1} \tag{7.593}$$

Moreover, in general

$$E[\|\boldsymbol{\Delta h}\|^2] = \text{tr}[\boldsymbol{R}_{\Delta h}] \tag{7.594}$$

For a comparison with other estimators, from (7.548),

$$\Lambda_n = \frac{\sigma_w^2}{\mathtt{M}_x \text{tr}[\boldsymbol{R}_{\Delta h}]} \tag{7.595}$$

Given the required information about the statistical description of the channel, this method is not extensively used in the single carrier system of Figure 7.60.

7.9 Faster-than-Nyquist Signalling

Introduced in 1975 by Mazo [40], faster-than-Nyquist (FTN) signalling, for an ideal non-dispersive channel, utilizes as transmit and receive filter a square-root Nyquist pulse with Nyquist frequency $1/(2T')$, while the transmit symbol rate is $1/T$, with $T = \tau T'$, where τ is a system parameter such that $0 < \tau \leq 1$. Note that for $\tau = 1$ we obtain the traditional ISI free system. The system transmission bandwidth is $(1 + \rho)1/T'$, with ρ being the roll-off factor of the transmit and receive filters. At the output of the receiver matched filter the sampled signal at rate $1/(\tau T')$ is affected by substantial ISI, especially for a small τ. However, the system spectral efficiency is increased by a factor $1/\tau$. Similarly to other proposed modulation schemes (e.g. Section 8.8), a trade-off is obtained between increased performance (here spectral efficiency) and receiver complexity [41, 42]. We note that the choice of coding and of codeword length play a fundamental role in these systems.

Bibliography

[1] Couch, L.W. (1997). *Digital and Analog Communication Systems*. Upper Saddle River, NJ: Prentice-Hall.

[2] Proakis, J.G. and Salehi, M. (1994). *Communication System Engineering*. Englewood Cliffs, NJ: Prentice-Hall.

[3] Benvenuto, N. and Zorzi, M. (2011). *Principles of Communications Networks and Systems*. Wiley.

[4] Franks, L.E. (1981). *Signal Theory*, revised ed. Stroudsburg, PA: Dowden and Culver.

[5] Gitlin, R.D. and Weinstein, S.B. (1981). Fractionally-spaced equalization: an improved digital transversal equalizer. *Bell System Technical Journal* 60. 275–296.

[6] Mazo, J.E. and Salz, J. (1965). Probability of error quadratic detector. *Bell System Technical Journal* 44. 2165–2186.

[7] Benvenuto, N. and Tomasin, S. (2002). On the comparison between OFDM and single carrier modulation with a DFE using a frequency domain feedforward filter. *IEEE Transactions on Communications*. 50: (6) 947–955.

[8] Benvenuto, N., Dinis, R., Falconer, D., and Tomasin, S. (2010). Single carrier modulation with nonlinear frequency domain equalization: an idea whose time has come–again. *Proceedings of the IEEE* 98: 69–96.

[9] Benvenuto, N. and Tomasin, S. (2002). Block iterative DFE for single carrier modulation. *Electronics Letters* 38: 1144–1145.

[10] Benvenuto, N., Ciccotosto, S., and Tomasin, S. (2015). Iterative block fractionally spaced nonlinear equalization for wideband channels. *IEEE Wireless Communications Letters* 4: 489–492.

[11] Sari, H., Karam, G., and Jeanclaude, I. (1995). Transmission techniques for digital terrestrial TV broadcasting. *IEEE Communications Magazine* 33: 100–109.

[12] Huemer, M., Reindl, L., Springer, A., and Weigel, R. (2000). Frequency domain equalization of linear polyphase channels. *Proceedings of 2000 IEEE Vehicular Technology Conference*, Tokyo, Japan.

[13] Proakis, J.G. (1995). *Digital Communications*, 3e. New York, NY: McGraw-Hill.

[14] Messerschmitt, D.G. and Lee, E.A. (1994). *Digital Communication*, 2e. Boston, MA: Kluwer Academic Publishers.

[15] Rabiner, L.R. (1989). A tutorial on hidden Markov models and selected applications in speech recognition. *IEEE Proceedings* 77: 257–285.

[16] Bahl, L.R., Cocke, J., Jelinek, F., and Raviv, J. (1974). Optimal decoding of linear codes for minimizing symbol error rate. *IEEE Transactions on Information Theory* 20: 284–287.

[17] Viterbi, A.J. (1998). An intuitive justification and simplified implementation of the MAP decoder for convolutional codes. *IEEE Journal on Selected Areas in Communications* 16: 260–264.

[18] Ungerboeck, G. (1974). Adaptive maximum likelihood receiver for carrier modulated data transmission systems. *IEEE Transactions on Communications* 22: 624–635.

[19] Chugg, K.M. and Anastasopoulos, A. (2001). On symbol error probability bounds for ISI-like channels. *IEEE Transactions on Communications* 49: 1704–1709.

[20] Lee, W.U. and Hill, F.S. (1977). A maximum-likelihood sequence estimator with decision feedback equalization. *IEEE Transactions on Communications* 25: 971–979.

[21] Foschini, G.J. (1977). A reduced-state variant of maximum-likelihood sequence detection attaining optimum performance for high signal-to-noise ratio performance. *IEEE Transactions on Information Theory* 24: 505–509.

[22] Anderson, J.B. and Mohan, S. (1984). Sequential coding algorithms: a survey and cost analysis. *IEEE Transactions on Communications* 32: 169–176.

[23] Eyuboglu, M.V. and Qureshi, S.U.H. (1988). Reduced-state sequence estimator with set partitioning and decision feedback. *IEEE Transactions on Communications* 36: 13–20.

[24] Eyuboglu, M.V. (1989). Reduced-state sequence estimator for coded modulation on intersymbol interference channels. *IEEE Transactions on Communications* 7: 989–995.

[25] Kamel, R.E. and Bar-Ness, Y. (1996). Reduced-complexity sequence estimation using state partitioning. *IEEE Transactions on Communications* 44: 1057–1063.

[26] Sheen, W. and Stuber, G.L. (1992). Error probability for reduced-state sequence estimation. *IEEE Journal on Selected Areas in Communications* 10: 571–578.

[27] Raheli, R., Polydoros, A., and Tzou, C.-K. (1995). Per-survivor processing: a general approach to MLSE in uncertain environments. *IEEE Transactions on Communications* 43: 354–364.

[28] Cioffi, J.M., Dudevoir, G.P., Eyuboglu, M.V., and Forney, G.D. Jr. (1995). MMSE decision-feedback equalizers and coding. *IEEE Transactions on Communications* 43: 2582–2604.

[29] Ungerboeck, G. (1982). Channel coding with multilevel/phase signals. *IEEE Transactions on Information Theory* 28: 55–67.

[30] Forney, G.D. Jr. and Ungerboeck, G. (1998). Modulation and coding for linear Gaussian channels. *IEEE Transactions on Information Theory* 44: 2384–2415.

[31] Eyuboglu, M.V. and Forney, G.D. Jr. (1992). Trellis precoding: combined coding, precoding and shaping for intersymbol interference channels. *IEEE Transactions on Information Theory* 38: 301–314.

[32] Laroia, R., Tretter, S.A., and Farvardin, N. (1993). A simple and effective precoding scheme for noise whitening on intersymbol interference channels. *IEEE Transactions on Communications* 41: 1460–1463.

[33] Laroia, R. (1996). Coding for intersymbol interference channels - combined coding and precoding. *IEEE Transactions on Information Theory* 42: 1053–1061.

[34] Cherubini, G., Ölcer, S., and Ungerboeck, G. (1997). Trellis precoding for channels with spectral nulls. *Proceedings 1997 IEEE Internaitonal Symposium on Information Theory*, Ulm, Germany, p. 464.

[35] Tomlinson, M. (1971). New automatic equalizer employing modulo arithmetic. *Electronics Letters* 7: 138–139.

[36] Harashima, H. and Miyakawa, H. (1972). Matched transmission technique for channels with intersymbol interference. *IEEE Transactions on Communications* 20: 774–780.

[37] Marple, S.L. Jr. (1981). Efficient least squares FIR system identification. *IEEE Transactions on Acoustics, Speech, and Signal Processing* 29: 62–73.

[38] Nagumo, J.I. and Noda, A. (1967). A learning method for system identification. *IEEE Transactions on Automatic Control* 12: 282–287.

[39] Crozier, S.N., Falconer, D.D., and Mahmoud, S.A. (1991). Least sum of squared errors (LSSE) channel estimation. *IEE Proceedings-F* 138: 371–378.

[40] Mazo, J.E. (1975). Faster-than-Nyquist signaling. *Bell System Technical Journal* 54: 1451–1462.

[41] Anderson, J.B. (2016). Faster-than-Nyquist signaling for 5G communications. In: *Signal Processing for 5G: Algorithms and Implementations*, (ed. Fa-Long Luo, Charlie (Jianzhong) Zhang) 24–26. Wiley-IEEE Press.

[42] Li, S., Bai, B., Zhou, J. et al. (2018). Reduced-complexity equalization for faster-than Nyquist signaling: new methods based on Ungerboeck observation model. *IEEE Transactions on Communications* 66: 1190–1204.

[43] Jeruchim, M.C., Balaban, P., and Shanmugan, K.S. (1992). *Simulation of Communication Systems*. New York, NY: Plenum Press.

[44] Benedetto, S. and Biglieri, E. (1999). *Principles of Digital Transmission with Wireless Applications*. New York, NY: Kluwer Academic Publishers.

[45] Kabal, P. and Pasupathy, P. (1975). Partial-response signaling. *IEEE Transactions on Communications* 23: 921–934.

[46] Duttweiler, D.L., Mazo, J.E., and Messerschmitt, D.G. (1974). An upper bound on the error probability in decision-feedback equalization. *IEEE Transactions on Information Theory* 20: 490–497.

[47] Birkoff, G. and MacLane, S. (1965). *A Survey of Modern Algebra*, 3e. New York, NY: Macmillan Publishing Company.

Appendix 7.A Simulation of a QAM system

In Figure 7.11, we consider the baseband equivalent scheme of a QAM system. The aim is to simulate the various transformations in the discrete-time domain and to estimate the bit error probability.

This simulation method, also called Monte Carlo, is simple and general because it does not require any special assumption on the processes involved; however, it is intensive from the computational point of view. For alternative methods, for example semi-analytical, to estimate the error probability, we refer to specific texts on the subject [43].

We describe the various transformations in the overall discrete-time system depicted in Figure 7.64, where the only difference with respect to the scheme of Figure 7.11 is that the filters are discrete-time with quantum $T_Q = T/Q_0$, which is chosen to accurately represent the various signals.

(a)

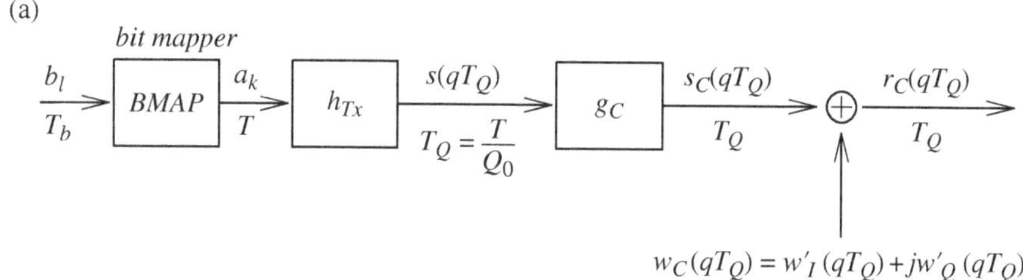

(b)

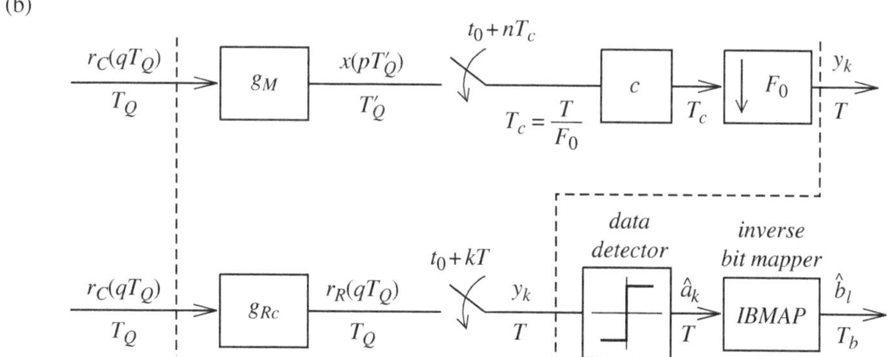

Figure 7.64 Baseband equivalent model of a QAM system with discrete-time filters and sampling period $T_Q = T/Q_0$. At the receiver, in addition to the general scheme, a multirate structure to obtain samples of the received signal at the timing phase t_0 is also shown. (a) Transmitter and channel block diagram. (b) Receiver block diagram.

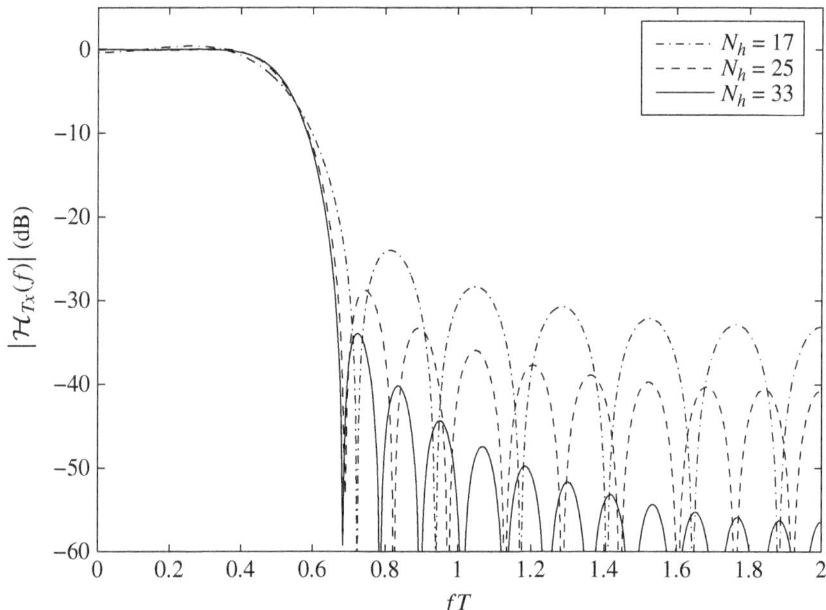

Figure 7.65 Magnitude of the transmit filter frequency response, for a windowed *square root raised cosine* pulse with roll-off factor $\rho = 0.3$, for three values of N_h ($T_Q = T/4$).

Binary sequence $\{b_\ell\}$*:* The sequence $\{b_\ell\}$ is generated as a random sequence or as a PN sequence (see Appendix 1.C), and has length K.

Bit mapper: The bit mapper maps patterns of information bits to symbols; the symbol constellation depends on the modulator (see Figure 6.2 for two constellations).

Interpolator filter h_{Tx} *from period* T *to* T_Q*:* The interpolator filter is efficiently implemented by using the polyphase representation (see Appendix 1.A). For a bandlimited pulse of the *raised cosine* or *square root raised cosine* type the maximum value of T_Q, submultiple of T, is $T/2$. In any case, the implementation of filters, for example the filter representing the channel, and non-linear transformations, for example the transformation due to a power amplifier operating near saturation (not considered in Figure 7.64), typically require a larger bandwidth, leading, for example to the choice $T_Q = T/4$ or $T/8$. In the following examples we choose $T_Q = T/4$.

For the design of h_{Tx}, the window method can be used (N_h odd):

$$h_{Tx}(qT_Q) = h_{id}\left[\left(q - \frac{N_h - 1}{2}\right)T_Q\right] \mathtt{w}_{N_h}(q), \qquad q = 0, 1, \dots, N_h - 1 \tag{7.596}$$

where typically $\mathtt{w}_{N_h}$ is the discrete-time rectangular window or the Hamming window, and h_{id} is the ideal impulse response.

Frequency responses of h_{Tx} are illustrated in Figure 7.65 for h_{id} *square root raised cosine* pulse with roll-off factor $\rho = 0.3$, and $\mathtt{w}_{N_h}$ rectangular window of length N_h, for various values of N_h ($T_Q = T/4$). The corresponding impulse responses are shown in Figure 7.66.

Transmission channel. For a radio channel, the discrete-time model of Figure 4.20 can be used, where in the case of channel affected by fading, the coefficients of the FIR filter that model the channel impulse response are random variables with a given power delay profile.

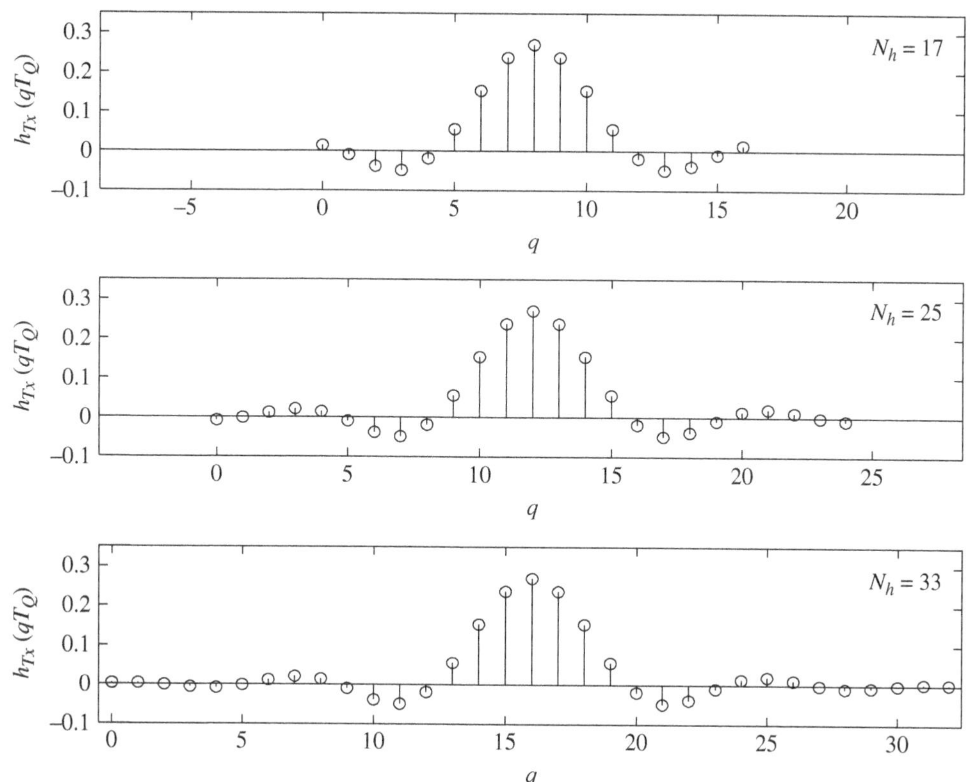

Figure 7.66 Transmit filter impulse response, $\{h_{Tx}(qT_Q)\}$, $q = 0, \dots, N_h - 1$, for a windowed *square root raised cosine* pulse with roll-off factor $\rho = 0.3$, for three values of N_h ($T_Q = T/4$).

We assume the statistical power of the signal at output of the transmission channel is given by $\mathtt{M}_{s_{Ch}} = \mathtt{M}_{s_C}$.

Additive white Gaussian noise. Let $\overline{w}_I(qT_Q)$ and $\overline{w}_Q(qT_Q)$ be two Gaussian statistically independent r.v.s., each with zero mean and variance $1/2$, generated according to (1.582). To generate the complex-valued noise signal $\{w_C(qT_Q)\}$ with spectrum N_0, it is sufficient to use the relation

$$w_C(qT_Q) = \sigma_{w_C}[\overline{w}_I(qT_Q) + j\overline{w}_Q(qT_Q)] \tag{7.597}$$

where

$$\sigma^2_{w_C} = N_0 \frac{1}{T_Q} \tag{7.598}$$

Usually, the signal-to-noise ratio Γ of (6.69) is given. For a QAM system, from (7.98), we have

$$\Gamma = \frac{\mathtt{M}_{s_C}}{N_0(1/T)} = \frac{\mathtt{M}_{s_C}}{\sigma^2_{w_C}(T_Q/T)} \tag{7.599}$$

The standard deviation of the noise to be inserted in (7.597) is given by

$$\sigma_{w_C} = \sqrt{\frac{\mathtt{M}_{s_C} Q_0}{\Gamma}} \tag{7.600}$$

We note that σ_{w_C} is a function of $\mathtt{M}_{s_C}$, of the oversampling ratio $Q_0 = T/T_Q$, and of the given ratio Γ. In place of Γ, the ratio E_b/N_0 (equal to $\Gamma/\log_2 M$ for a uncoded system, see also (6.63)) may be assigned.

Receive filter. As discussed in this chapter, there are several possible solutions for the receive filter. The most common choice is a matched filter g_M, matched to h_{Tx}, of the *square root raised cosine* type. Alternatively, the receive filter may be a simple *anti-aliasing FIR filter* g_{AA}, with passband at least equal to that of the desired signal. The filter attenuation in the stopband must be such that the statistical power of the noise evaluated in the passband is larger by a factor of 5–10 with respect to the power of the noise evaluated in the stopband, so that we can neglect the contribution of the noise in the stopband at the output of the filter g_{AA}.

If we adopt as bandwidth of g_{AA} the Nyquist frequency $1/(2T)$, the stopband of an ideal filter with unit gain goes from $1/(2T)$ to $1/(2T_Q)$: therefore, the ripple δ_s in the stopband must satisfy the constraint

$$\frac{N_0 \frac{1}{2T}}{\delta_s^2 N_0 \left(\frac{1}{2T_Q} - \frac{1}{2T}\right)} > 10 \tag{7.601}$$

from which we get the condition

$$\delta_s^2 < \frac{10^{-1}}{Q_0 - 1} \tag{7.602}$$

Usually, the presence of other interfering signals forces the selection of a value of δ_s that is smaller than that obtained in (7.602).

Interpolator filter. The interpolator filter is used to increase the sampling rate from $1/T_Q$ to $1/T'_Q$: this is useful when T_Q is insufficient to obtain the accuracy needed to represent the timing phase t_0. This filter can be part of g_M or g_{AA}. From Appendix 1.A, the efficient implementation of $\{g_M(pT'_Q)\}$ is obtained by the polyphase representation with T_Q/T'_Q branches.

To improve the accuracy of the desired timing phase, further interpolation, for example linear, may be employed.

Timing phase. Assuming a training sequence is available, for example of the PN type $\{a_0 = p(0), a_1 = p(1), \dots, a_{L-1} = p(L-1), a_L = p(0), a_{L+N_h-1} = p(N_h - 1)\}$, with $N_h T$ duration of the overall system impulse response, a simple method to determine t_0 is to choose the timing phase in correspondence of the peak of the overall impulse response. Let $\{x(pT'_Q)\}$ be the signal before downsampling. If we evaluate

$$\begin{aligned} m_{opt} &= \arg\max_m |\mathtt{r}_{xa}(mT'_Q)| \\ &= \arg\max_m \left| \frac{1}{L} \sum_{\ell=0}^{L-1} x(\ell T + mT'_Q) p^*(\ell) \right|, \quad m_{min}T'_Q < mT'_Q < m_{max}T'_Q \end{aligned} \tag{7.603}$$

then

$$t_0 = m_{opt} T'_Q \tag{7.604}$$

In (7.603) $m_{min}T'_Q$ and $m_{max}T'_Q$ are estimates of minimum and maximum system delay, respectively. Moreover, we note that the accuracy of t_0 is equal to T'_Q and that the amplitude of the desired signal is $h_0 = \mathtt{r}_{xa}(m_{opt}T'_Q)/\mathtt{r}_a(0)$. Note that (7.603) is similar to the channel estimate by the correlation method in Section 7.8, where now the received signal rather than the PN sequence is shifted in time.

Downsampler. The sampling period after downsampling is usually T or $T_c = T/2$, with timing phase t_0. The interpolator filter and the downsampler can be jointly implemented, according to the scheme of Figure 1.71. For example, for $T_Q = T/4$, $T'_Q = T/8$, and $T_c = T/2$ the polyphase representation of the interpolator filter with output $\{x(pT'_Q)\}$ requires two branches. Also the polyphase representation of the interpolator–decimator requires two branches.

Equalizer. After downsampling, the signal is usually input to an equalizer (LE, FSE, or DFE). The output signal of the equalizer has always sampling period equal to T. As observed several times, to decimate simply means to evaluate the output at the desired instants.

Data detection. The simplest method resorts to a threshold detector, which operates on a sample by sample basis.

Viterbi algorithm. An alternative to the threshold detector is represented by maximum likelihood sequence detection by the Viterbi algorithm.

Inverse bit mapper. The inverse bit mapper performs the inverse function of the bit mapper. It translates the detected symbols into bits that represent the recovered information bits.

Simulations are typically used to estimate the bit error probability of the system, for a certain set of values of Γ. We recall that caution must be taken at the beginning and at the end of a simulation to consider transients of the system. Let $\overline{K}$ be the number of recovered bits. The estimate of the bit error probability P_{bit} is given by

$$\hat{P}_{bit} = \frac{\text{number of bits received with errors}}{\text{number of received bits, } \overline{K}} \tag{7.605}$$

It is known that as $\overline{K} \to \infty$ the estimate $\hat{P}_{bit}$ has a Gaussian probability distribution with mean P_{bit} and variance $P_{bit}(1 - P_{bit})/\overline{K}$. From this we can deduce, by varying $\overline{K}$, the confidence interval $[P_-, P_+]$ within which the estimate $\hat{P}_{bit}$ approximates P_{bit} with an assigned probability, that is

$$P[P_- \leq \hat{P}_{bit} \leq P_+] = P_{conf} \tag{7.606}$$

For example, we have that with $P_{bit} = 10^{-\ell}$ and $\overline{K} = 10^{\ell+1}$, we get a confidence interval of about a factor 2 with a probability of 95%, that is $P[1/2P_{bit} \leq \hat{P}_{bit} \leq 2P_{bit}] \simeq 0.95$. This is in good agreement with the experimental rule of selecting

$$\overline{K} = 3 \cdot 10^{\ell+1} \tag{7.607}$$

For a channel affected by fading, the average P_{bit} is not very significant: in this case, it is meaningful to compute the distribution of P_{bit} for various channel realizations. In practice, we assume the transmission of a sequence of N_p packets, each one with $\overline{K}_p$ information bits to be recovered: typically $\overline{K}_p$ = 1000-10 000 bits and N_p = 100–1000 packets. Moreover, the channel realization changes at every packet. For a given average signal-to-noise ratio $\overline{\Gamma}$, the probability $\hat{P}_{bit}(n_p)$, $n_p = 1, \dots, N_p$, is computed for each packet. As a performance measure, we use the percentage of packets with $\hat{P}_{bit}(n_p) < P_{bit}$, also called bit error probability cumulative distribution function (cdf), where P_{bit} assumes values in a certain set.

This performance measure is more significant than the average P_{bit} evaluated for a very long, continuous transmission of $N_p\overline{K}_p$ information bits. In fact, the average P_{bit} does not show that, in the presence of fading, the system may occasionally have a very large P_{bit}, and consequently an outage.

Appendix 7.B Description of a finite-state machine

We consider a discrete-time system with input sequence $\{i_k\}$ and output sequence $\{o_k\}$ with sampling time T. We say that the output sequence is generated by a finite-state machine (FSM) if there are a sequence of states $\{s_k\}$ and two functions f_o and f_s, such that

$$o_k = f_o(i_k, s_{k-1}) \tag{7.608}$$

$$s_k = f_s(i_k, s_{k-1}) \tag{7.609}$$

as illustrated in Figure 7.67. The first equation describes the fact that the output sample depends on the current input and the state of the system. The second equation represents the memory part of the FSM and describes the state evolution.

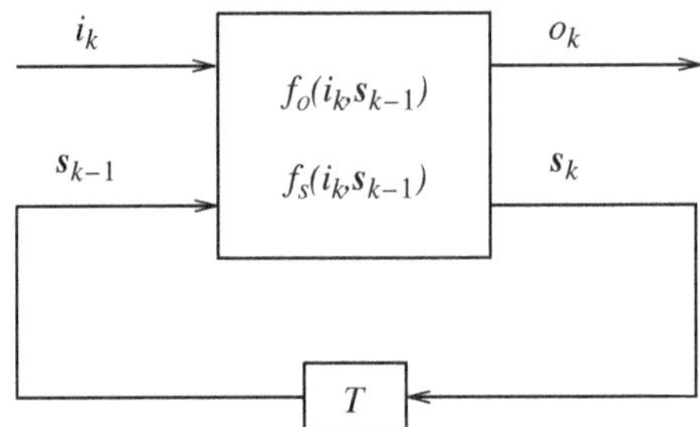

Figure 7.67 General block diagram of a finite state sequential machine.

We note that if f_s is a one-to-one function of i_k, that is if every transition between two states is determined by a unique input value, then (7.608) can be written as

$$o_k = f(s_k, s_{k-1}) \tag{7.610}$$

Appendix 7.C Line codes for PAM systems

The objectives of line codes are

1. to shape the spectrum of the transmitted signals, and match it to the characteristics of the channel (see (7.16)); this task may be performed also by the transmit filter;
2. to facilitate synchronization at the receiver, especially in case the information message contains long sequences of ones or zeros;
3. to improve system performance in terms of P_e.

This appendix is divided in two parts: in the first several representations of binary symbols are listed; in the second, partial response systems are introduced. For in-depth study and analysis of spectral properties of line codes we refer to the bibliography, in particular [1, 44].

7.C.1 Line codes

With reference to Figure 7.68, the binary sequence $\{b_\ell\}$, $b_\ell \in \{0, 1\}$, could be directly generated by a source, or be the output of a channel encoder. The sequence $\{a_k\}$ is produced by a line encoder. The channel input is the PAM signal s, obtained by modulating a pulse h_{Tx}, here a rectangular pulse.

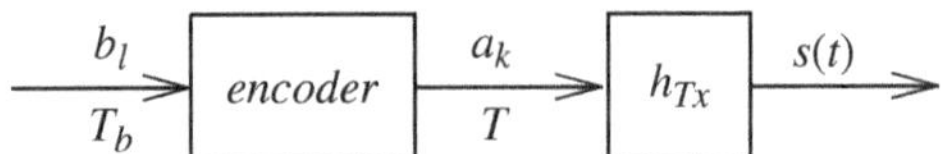

Figure 7.68 PAM transmitter with line encoder.

Non-return-to-zero format

The non-return-to-zero (NRZ) signals are antipodal signals: therefore, NRZ line codes are characterized by the lowest error probability, for transmission over AWGN channels in the absence of ISI. Four formats are illustrated in Figure 7.69.

1. NRZ level (NRZ-L) or, simply, NRZ:
   1 and 0 are represented by two different levels.
2. NRZ mark (NRZ-M):
   1 is represented by a level transition, 0 by no level transition.
3. NRZ space (NRZ-S):
   1 is represented by no level transition, 0 by a level transition.
4. Dicode NRZ:
   A change of polarity in the sequence $\{b_\ell\}$, 1–0 or 0–1, is represented by a level transition; every other case is represented by the zero level.

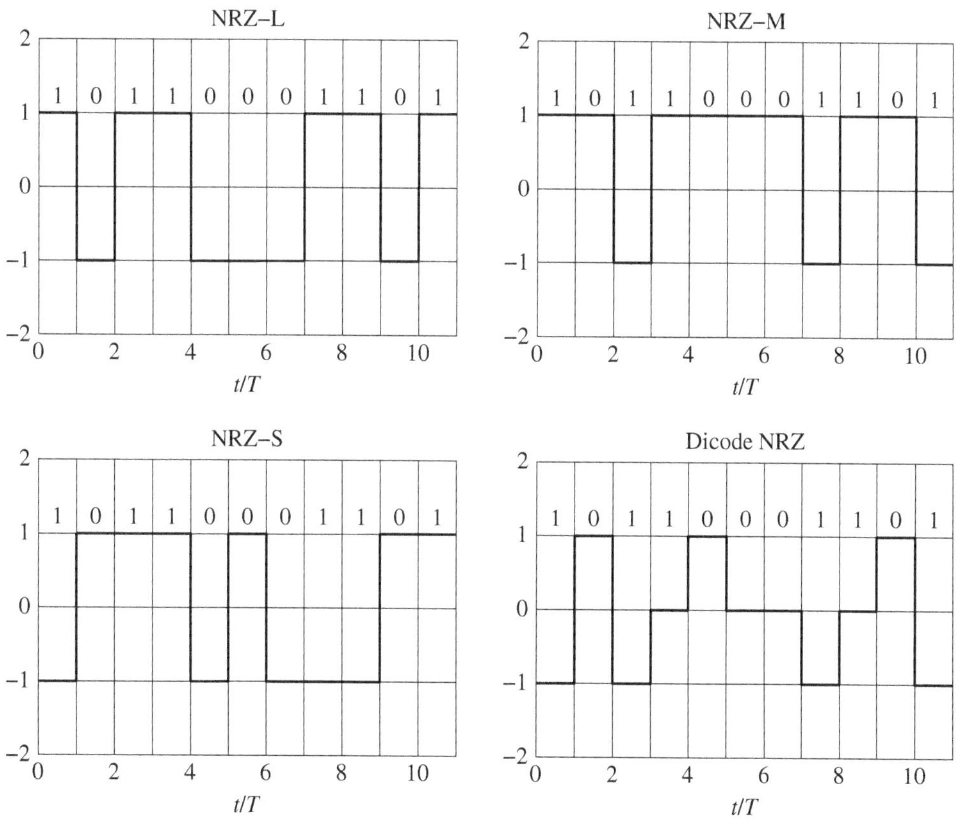

Figure 7.69 NRZ line codes.

Return-to-zero format

1. *Unipolar return-to-zero (RZ)*:
   1 is represented by a pulse having duration equal to half a bit interval, 0 by a zero pulse; we observe that the signal does not have zero mean. This property is usually not desirable, as, for example for transmission over coaxial cables.
2. *Polar RZ*:
   1 and 0 are represented by opposite pulses with duration equal to half a bit interval.
3. *Bipolar RZ or alternate mark inversion (AMI)*:
   Bits equal to 1 are represented by rectangular pulses having duration equal to half a bit interval, sequentially alternating in sign, bits equal to 0 by the zero level.
4. *Dicode RZ*:
   A change of polarity in the sequence $\{b_\ell\}$, 1–0 or 0–1, is represented by a level transition, using a pulse having duration equal to half a bit interval; every other case is represented by the zero level.

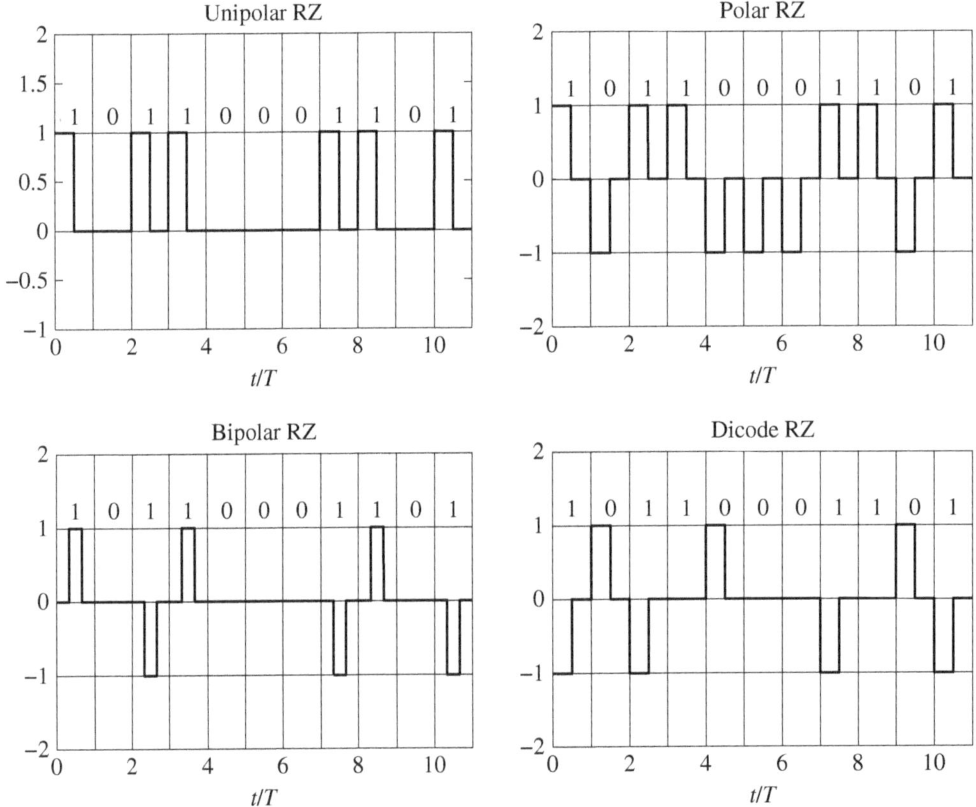

Figure 7.70 RZ line codes.

RZ line codes are illustrated in Figure 7.70.

Biphase format

1. *Biphase level (B–ϕ–L) or Manchester NRZ*:
   1 is represented by a transition from high level to low level, 0 by a transition from low level to high level. Long sequences of ones or zeros in the sequence $\{b_\ell\}$ do not create synchronization problems. It is easy to see, however, that this line code leads to a doubling of the transmission bandwidth.
2. *Biphase mark (B–ϕ–M) or Manchester 1*:
   A transition occurs at the beginning of every bit interval; 1 is represented by a second transition within the bit interval, 0 is represented by a constant level.
3. *Biphase space (B–ϕ–S)*:
   A transition occurs at the beginning of every bit interval; 0 is represented by a second transition within the bit interval, 1 is represented by a constant level.

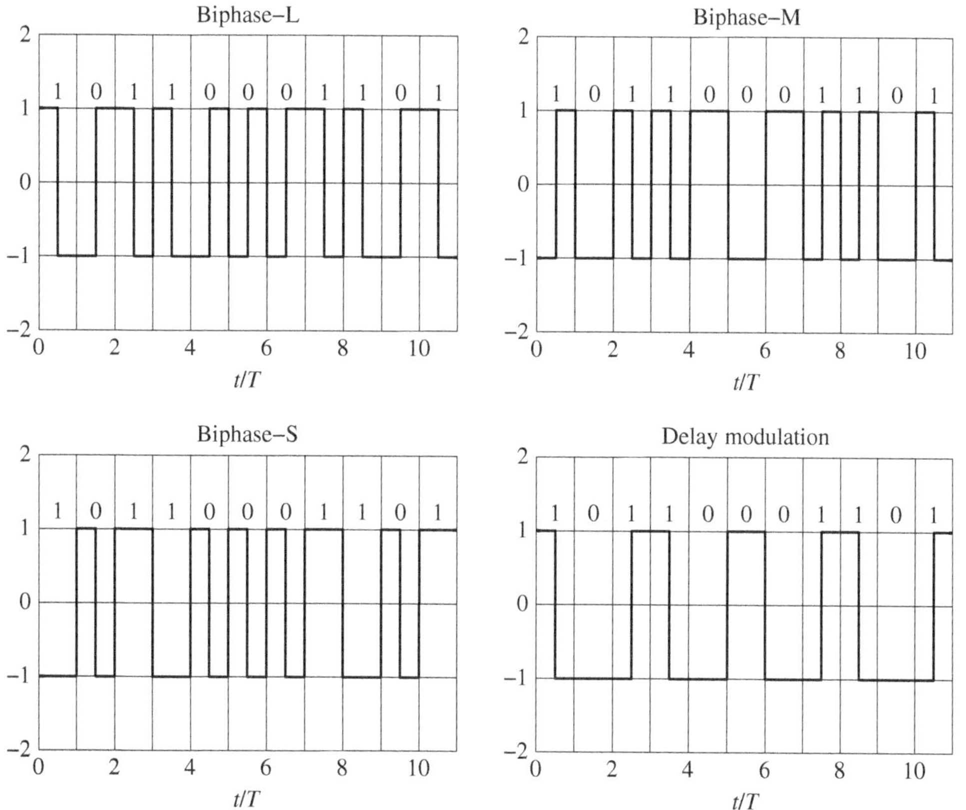

Figure 7.71 B–ϕ and delay modulation line codes.

Biphase line codes are illustrated in Figure 7.71.

Delay modulation or Miller code

1 is represented by a transition at midpoint of the bit interval, 0 is represented by a constant level; if 0 is followed by another 0, a transition occurs at the end of the bit interval. This code shapes the spectrum similar to the Manchester code, but requires a lower bandwidth. The delay modulation line code is illustrated in Figure 7.71.

Block line codes

The input sequence $\{b_\ell\}$ is divided into blocks of K bits. Each block of K bits is then mapped into a block of N symbols belonging to an alphabet of cardinality M, with the constraint

$$2^K \leq M^N \tag{7.611}$$

The KBNT codes are an example of block line codes where the output symbol alphabet is ternary $\{-1, 0, 1\}$.

Alternate mark inversion

We consider the alternate mark inversion (AMI) *differential binary* encoder, that is

$$a_k = b_k - b_{k-1} \qquad \text{with } b_k \in \{0, 1\} \tag{7.612}$$

At the decoder, the bits of the information sequence may be recovered by

$$\hat{b}_k = \hat{a}_k + \hat{b}_{k-1} \tag{7.613}$$

Note that $a_k \in \{-1, 0, 1\}$; in particular

$$a_k = \begin{cases} \pm 1, & \text{if } b_k \neq b_{k-1} \\ 0, & \text{if } b_k = b_{k-1} \end{cases} \tag{7.614}$$

From (7.612), the relation between the PSDs of the sequences $\{a_k\}$ and $\{b_k\}$ is given by

$$\begin{aligned} \mathcal{P}_a(f) &= \mathcal{P}_b(f)\, |1 - e^{-j2\pi fT}|^2 \\ &= \mathcal{P}_b(f)\, 4 \sin^2(\pi f T) \end{aligned}$$

Therefore, $\mathcal{P}_a(f)$ exhibits zeros at frequencies that are integer multiples of $1/T$, in particular at $f = 0$. Moreover, from (7.612), we have $\mathtt{m}_a = 0$, independently of the distribution of $\{b_k\}$.

If the power of the transmitted signals is constrained, a disadvantage of the encoding method (7.612) is a reduced noise immunity with respect to antipodal transmission, that is for $a_k \in \{-1, 1\}$, because a detector at the receiver must now decide among three levels. Moreover, long sequences of information bits $\{\hat{b}_k\}$ that are all equal to 1 or 0 generate sequences of symbols $\{a_k\}$ that are all equal: this is not desirable for synchronization.

In any case, the biggest problem is the error propagation at the decoder, which, observing (7.613), given that an error occurs in $\{\hat{a}_k\}$, generates a sequence of bits $\{\hat{b}_k\}$ that are in error until another error occurs in $\{\hat{a}_k\}$. This problem can be solved by precoding: from the sequence of bits $\{b_k\}$, we first generate the sequence of bits $\{c_k\}$, with $c_k \in \{0, 1\}$, by

$$c_k = b_k \oplus c_{k-1} \tag{7.615}$$

where $\oplus$ denotes the modulo 2 sum. Next,

$$a_k = c_k - c_{k-1} \tag{7.616}$$

with $a_k \in \{-1, 0, 1\}$. Hence, it results

$$a_k = \begin{cases} \pm 1, & \text{if } b_k = 1 \\ 0, & \text{if } b_k = 0 \end{cases} \tag{7.617}$$

In other words, a bit $b_k = 0$ is mapped into the symbol $a_k = 0$, and a bit $b_k = 1$ is mapped alternately in $a_k = +1$ or $a_k = -1$. Consequently, from (7.617) decoding may be performed simply by taking the magnitude of the detected symbol:

$$\hat{b}_k = |\hat{a}_k| \tag{7.618}$$

It is easy to prove that for a message $\{b_k\}$ with statistically independent symbols, and $p = P[b_k = 1]$, we have

$$\mathcal{P}_a(f) = 2p(1-p)\frac{\sin^2(\pi f T)}{p^2 + (1-2p)\sin^2(\pi f T)} \tag{7.619}$$

The plot of $\mathcal{P}_a(f)$ is shown in Figure 7.72 for different values of p. Note that the PSD presents a zero at $f = 0$. Also in this case, $\mathtt{m}_a = 0$.

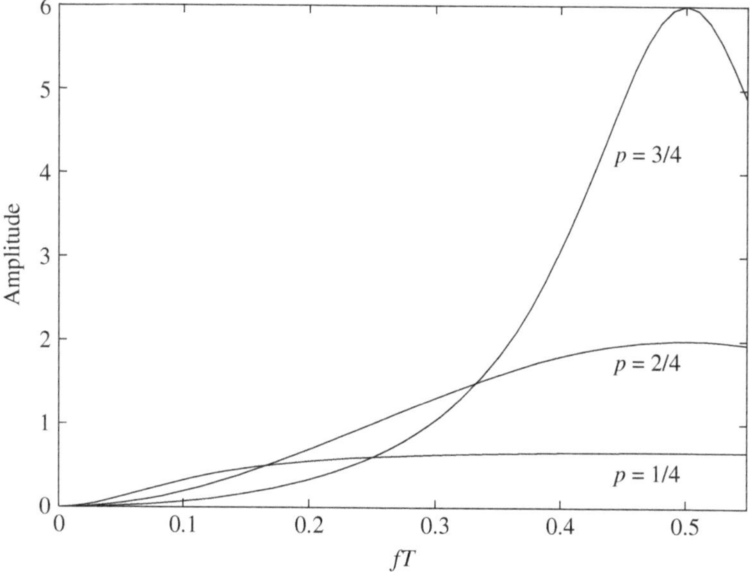

Figure 7.72 Power spectral density of an AMI encoded message.

We observe that the AMI line code is a particular case of the partial response system named *dicode* [45].

7.C.2 Partial response systems

From Figure 7.1, we recall the block diagram of a baseband transmission system, where the symbols $\{a_k\}$ belong to the alphabet (6.4)[18] of cardinality M and w is an additive white Gaussian noise (Figure 7.73).

[18] In the present analysis, only M-PAM systems are considered; for M-QAM systems, the results can be extended to the signals on the I and Q branches.

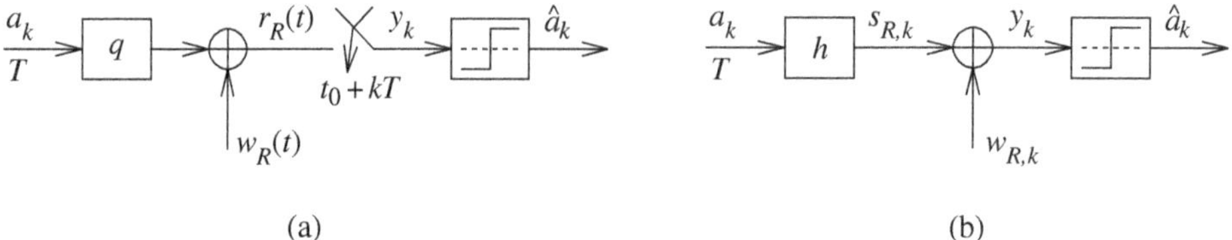

Figure 7.73 Schemes equivalent to Figure 7.1. (a) Analog equivalent. (b) Digital equivalent.

We assume that the transmission channel is ideal: the overall system can then be represented as an interpolator filter having impulse response

$$q(t) = h_{Tx} * g_{Rc}(t) \tag{7.620}$$

A noise w_R, obtained by filtering w by the receive filter, is added to the desired signal. Sampling the received signal at instants $t_0 + kT$ yields the sequence $\{y_k\}$, as illustrated in Figure 7.73a. The discrete-time equivalent of the system is shown in Figure 7.73b, where $\{h_i = q(t_0 + iT)\}$, and $w_{R,k} = w_R(t_0 + kT)$. We assume that $\{h_i\}$ is equal to zero for $i < 0$ and $i \geq N$.

The *partial response (PR) polynomial* of the system is defined as

$$l(D) = \sum_{i=0}^{N-1} \mathtt{l}_i \, D^i \tag{7.621}$$

where the coefficients $\{\mathtt{l}_i\}$ are equal to the samples $\{h_i\}$, and D is the unit delay operator.

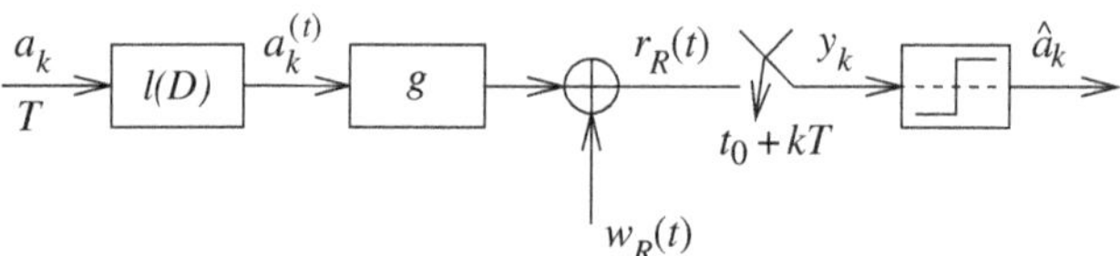

Figure 7.74 Schemes equivalent to Figure 7.1. (a) Analog equivalent. (b) Digital equivalent.

A PR system is illustrated in Figure 7.74, where $l(D)$ is defined in (7.621), and g is an analog filter satisfying the Nyquist criterion for the absence of ISI (see (7.74))

$$\sum_{m=-\infty}^{+\infty} \mathcal{G}\left(f - \frac{m}{T}\right) = T \tag{7.622}$$

The symbols at the output of the filter $l(D)$ in Figure 7.74 are given by

$$a_k^{(t)} = \sum_{i=0}^{N-1} \mathtt{l}_i \, a_{k-i} \tag{7.623}$$

Note that the overall scheme of Figure 7.74 is equivalent to that of Figure 7.73a with

$$q(t) = \sum_{i=0}^{N-1} \mathtt{l}_i \, g(t - iT) \tag{7.624}$$

Also, observing (7.74), the equivalent discrete-time model is obtained for $h_i = \mathtt{l}_i$.

In other words, from (7.624), the system is decomposed into two parts:

- a filter with frequency response $l(e^{-j2\pi fT})$, periodic of period $1/T$, that forces the system to have an overall discrete-time impulse response equal to $\{h_i\}$;

• an analog filter g that does not modify the overall filter $h(D)$ and limits the system bandwidth.

As it will be clear from the analysis, the decomposition of Figure 7.74 on the one hand simplifies the study of the properties of the filter $h(D)$, and on the other hand allows us to to design an efficient receiver.

The scheme of Figure 7.74 suggests two possible ways to implement the system of Figure 7.1:

1. *Analog*: the system is implemented in analog form; therefore, the transmit filter h_{Tx} and the receive filter g_{Rc} must satisfy the relation

$$\mathcal{H}_{Tx}(f)\ \mathcal{G}_{Rc}(f) = \mathcal{Q}(f) = l(e^{-j2\pi fT})\ \mathcal{G}(f) \tag{7.625}$$

2. *Digital*: the filter $l(D)$ is implemented as a component of the transmitter by a digital filter; then the transmit filter $h_{Tx}^{(PR)}$ and receive filter $g_{Rc}^{(PR)}$ must satisfy the relation

$$\mathcal{H}_{Tx}^{(PR)}(f)\ \mathcal{G}_{Rc}^{(PR)}(f) = \mathcal{G}(f) \tag{7.626}$$

The implementation of a PR system using a digital filter is shown in Figure 7.75.

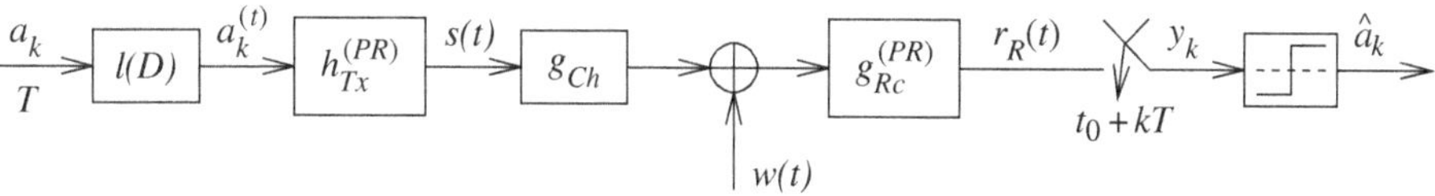

Figure 7.75 Implementation of a PR system using a digital filter.

Note from (7.622) that in both relations (7.625) and (7.626) g is a Nyquist filter.

The choice of the PR polynomial

Several considerations lead to the selection of the polynomial $l(D)$.

(A) System bandwidth: With the aim of maximizing the transmission bit rate, many PR systems are designed for *minimum bandwidth*, i.e. from (7.625) it must be

$$l(e^{-j2\pi fT})\ \mathcal{G}(f) = 0, \qquad |f| > \frac{1}{2T} \tag{7.627}$$

Substitution of (7.627) into (7.622) yields the following conditions on the filter g

$$\mathcal{G}(f) = \begin{cases} T, & |f| \le \dfrac{1}{2T} \\ 0, & \text{elsewhere} \end{cases} \quad \overset{\mathcal{F}^{-1}}{\longleftrightarrow} \quad g(t) = \text{sinc}\left(\frac{t}{T}\right) \tag{7.628}$$

Correspondingly, observing (7.624) the filter q assumes the expression

$$q(t) = \sum_{i=0}^{N-1} \mathtt{l}_i \ \text{sinc}\left(\frac{t-iT}{T}\right) \tag{7.629}$$

(B) Spectral zeros at $f = 1/(2T)$: From signal theory, it is known that if $\mathcal{Q}(f)$ and its first $(n-1)$ derivatives are continuous and the n-th derivative is discontinuous, then $|q(t)|$ asymptotically decays as $1/|t|^{n+1}$. The continuity of $\mathcal{Q}(f)$ and of its derivatives helps to reduce the portion of energy contained in the tails of q.

It is easily proven that in a *minimum bandwidth system*, the $(n-1)$-th derivative of $\mathcal{Q}(f)$ is continuous if and only if $l(D)$ has $(1+D)^n$ as a factor. On the other hand, if $l(D)$ has a 0 of multiplicity greater than 1 in $D=-1$, then the transition band of $\mathcal{G}(f)$ around $f = 1/(2T)$ can be widened, thus simplifying the design of the analog filters.

(C) Spectral zeros at $f = 0$: A transmitted signal with attenuated spectral components at low frequencies is desirable in many cases, e.g. for transmission over channels with frequency responses that exhibit a spectral null at the frequency $f = 0$. Note that a zero of $l(D)$ in $D = 1$ corresponds to a zero of $l(e^{-j2\pi fT})$ at $f = 0$.

(D) Number of output levels: From (7.623), the symbols at the output of the filter $l(D)$ have an alphabet $\mathcal{A}^{(t)}$ of cardinality $M^{(t)}$. If we indicate with n_1 the number of coefficients of $l(D)$ different from zero, then the following inequality for $M^{(t)}$ holds

$$n_1(M-1)+1 \leq M^{(t)} \leq M^{n_1} \tag{7.630}$$

In particular, if the coefficients $\{\mathtt{l}_i\}$ are all equal, then $M^{(t)} = n_1(M-1)+1$.

We note that, if $l(D)$ contains more than one factor $(1 \pm D)$, then n_1 increases and, observing (7.630), also the number of output levels increases. If the power of the transmitted signal is constrained, detection of the sequence $\{a_k^{(t)}\}$ by a threshold detector will cause a loss in system performance.

(E) Some examples of minimum bandwidth systems: In the case of *minimum bandwidth* systems, it is possible to evaluate the expression of $Q(f)$ and q once the polynomial $l(D)$ has been selected.

As the coefficients $\{\mathtt{l}_i\}$ are generally symmetric or antisymmetric around $i = (N-1)/2$, it is convenient to consider the time-shifted pulse

$$\begin{aligned} \tilde{q}(t) &= q\left(t + \tfrac{(N-1)T}{2}\right) \\ \tilde{Q}(f) &= e^{j\pi f(N-1)T}\, Q(f) \end{aligned} \tag{7.631}$$

In Table 7.1, the more common polynomials $l(D)$ are described, as well as the corresponding expressions of $\tilde{Q}(f)$ and $\tilde{q}$, and the cardinality $M^{(t)}$ of the output alphabet $\mathcal{A}^{(t)}$. In the next three examples, polynomials $l(D)$ that are often found in practical applications of PR systems are considered.

Table 7.1: Properties of several minimum bandwidth systems

$l(D)$	$\tilde{Q}(f)$ for $\lvert f\rvert \leq 1/(2T)$	$\tilde{q}(t)$	$M^{(t)}$
$1+D$	$2T\cos(\pi fT)$	$\frac{4T^2}{\pi}\frac{\cos(\pi t/T)}{T^2-4t^2}$	$2M-1$
$1-D$	$j2T\sin(\pi fT)$	$\frac{8Tt}{\pi}\frac{\cos(\pi t/T)}{4t^2-T^2}$	$2M-1$
$1-D^2$	$j2T\sin(2\pi fT)$	$\frac{2T^2}{\pi}\frac{\sin(\pi t/T)}{t^2-T^2}$	$2M-1$
$1+2D+D^2$	$4T\cos^2(\pi fT)$	$\frac{2T^3}{\pi t}\frac{\sin(\pi t/T)}{T^2-t^2}$	$4M-3$
$1+D-D^2-D^3$	$j4T\cos(\pi fT)\sin(2\pi fT)$	$-\frac{64T^3 t}{\pi}\frac{\cos(\pi t/T)}{(4t^2-9T^2)(4t^2-T^2)}$	$4M-3$
$1-D-D^2+D^3$	$-4T\sin(\pi fT)\sin(2\pi fT)$	$\frac{16T^2}{\pi}\frac{\cos(\pi t/T)(4t^2-3T^2)}{(4t^2-9T^2)(4t^2-T^2)}$	$4M-3$
$1-2D^2+D^4$	$-4T\sin^2(2\pi fT)$	$\frac{8T^3}{\pi t}\frac{\sin(\pi t/T)}{t^2-4T^2}$	$4M-3$
$2+D-D^2$	$T+T\cos(2\pi fT)+j3T\sin(2\pi fT)$	$\frac{T^2}{\pi t}\sin(\pi t/T)\left(\frac{3t-T}{t^2-T^2}\right)$	$4M-3$
$2-D^2-D^4$	$-T+T\cos(4\pi fT)+j3T\sin(4\pi fT)$	$\frac{2T^2}{\pi t}\sin(\pi t/T)\left(\frac{2T-3t}{t^2-4T^2}\right)$	$4M-3$

Example 7.C.1 (Dicode filter)
The dicode filter introduces a zero at frequency $f = 0$ and has the following expression:

$$l(D) = 1 - D \tag{7.632}$$

The frequency response, obtained by setting $D = e^{-j2\pi fT}$, is given by

$$l(e^{-j2\pi fT}) = 2j\, e^{-j\pi fT} \sin(\pi fT) \tag{7.633}$$

Example 7.C.2 (Duobinary filter)
The duobinary filter introduces a zero at frequency $f = 1/(2T)$ and has the following expression:

$$l(D) = 1 + D \tag{7.634}$$

The frequency response is given by

$$l(e^{-j2\pi fT}) = 2e^{-j\pi fT} \cos(\pi fT) \tag{7.635}$$

Observing (7.629), we have

$$q(t) = \mathrm{sinc}\left(\frac{t}{T}\right) + \mathrm{sinc}\left(\frac{t-T}{T}\right) \tag{7.636}$$

The plot of the impulse response of a duobinary filter is shown in Figure 7.76 with a continuous line. We notice that the tails of the two sinc functions cancel each other, in line with what was stated at point B) regarding the aymptotical decay of the pulse of a PR system with a zero in $D = -1$.

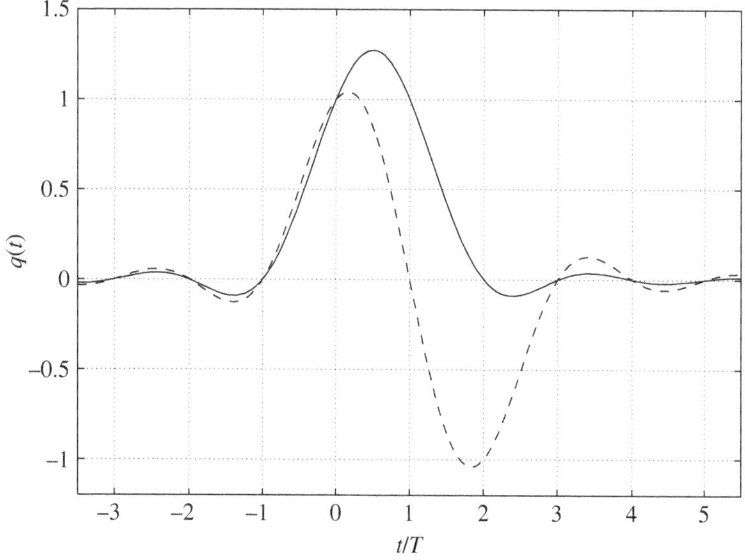

Figure 7.76 Plot of q for duobinary (–) and modified duobinary (- -) filters.

Example 7.C.3 (Modified duobinary filter)
The modified duobinary filter combines the characteristics of duobinary and dicode filters and has the following expression

$$l(D) = (1 - D)\,(1 + D) = 1 - D^2 \tag{7.637}$$

The frequency response becomes

$$l(e^{-j2\pi fT}) = 1 - e^{-j4\pi fT} = 2j\, e^{-j2\pi fT} \sin(2\pi fT) \tag{7.638}$$

Using (7.629) it results

$$q(t) = \text{sinc}\left(\frac{t}{T}\right) - \text{sinc}\left(\frac{t-2T}{T}\right) \tag{7.639}$$

The plot of the impulse response of a modified duobinary filter is shown in Figure 7.76 with a dashed line.
(F) Transmitted signal spectrum: With reference to the PR system of Figure 7.75, the spectrum of the transmitted signal is given by (see (7.16))

$$\mathcal{P}_s(f) = \left|\frac{1}{T}\, l(e^{-j2\pi fT})\, \mathcal{H}_{Tx}^{(PR)}(f)\right|^2 \mathcal{P}_a(f) \tag{7.640}$$

For a minimum bandwidth system, with $\mathcal{H}_{Tx}^{(PR)}(f)$ given by (7.628), (7.640) simplifies into

$$\mathcal{P}_s(f) = \begin{cases} |l(e^{-j2\pi fT})|^2\, \mathcal{P}_a(f), & |f| \le \dfrac{1}{2T} \\ 0, & |f| > \dfrac{1}{2T} \end{cases} \tag{7.641}$$

In Figure 7.77, the PSD of a minimum bandwidth PR system is compared with that of a PAM system. The spectrum of the sequence of symbols $\{a_k\}$ is assumed white. For the PR system, a modified duobinary filter is considered, so that the spectrum is obtained as the product of the functions $|l(-e^{j2\pi fT})|^2 = |2\sin(2\pi fT)|^2$ and $|\mathcal{H}_{Tx}^{(PR)}(f)|^2 = T^2 \text{rect}(fT)$, plotted with continuous lines. For the PAM system, the transmit filter h_{Tx} is a square root raised cosine with roll-off factor $\rho = 0.5$, and the spectrum is plotted with a dashed line.

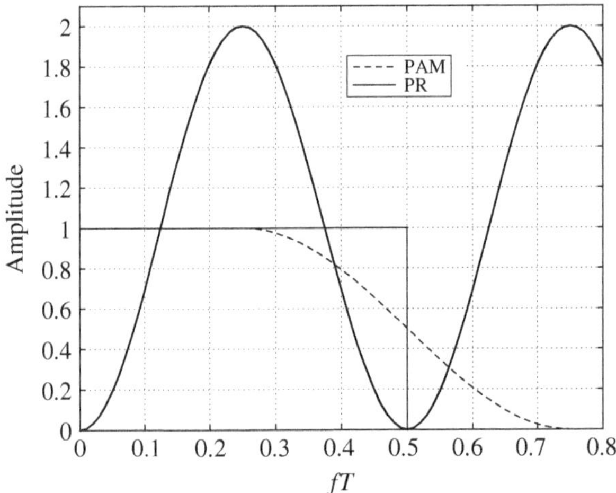

Figure 7.77 PSD of a modified duobinary PR system and of a PAM system.

Symbol detection and error probability

We consider the discrete-time equivalent scheme of Figure 7.73b; the signal $s_{R,k}$ can be expressed as a function of symbols $\{a_k\}$ and coefficients $\{\mathtt{l}_i\}$ of the filter $l(D)$ in the following form:

$$s_{R,k} = a_k^{(t)} = \mathtt{l}_0\, a_k + \sum_{i=1}^{N-1} \mathtt{l}_i\, a_{k-i} \tag{7.642}$$

The term $\mathtt{l}_0\, a_k$ is the desired part of the signal $s_{R,k}$, whereas the summation represents the ISI term that is often designated as *controlled ISI*, as it is deliberately introduced.

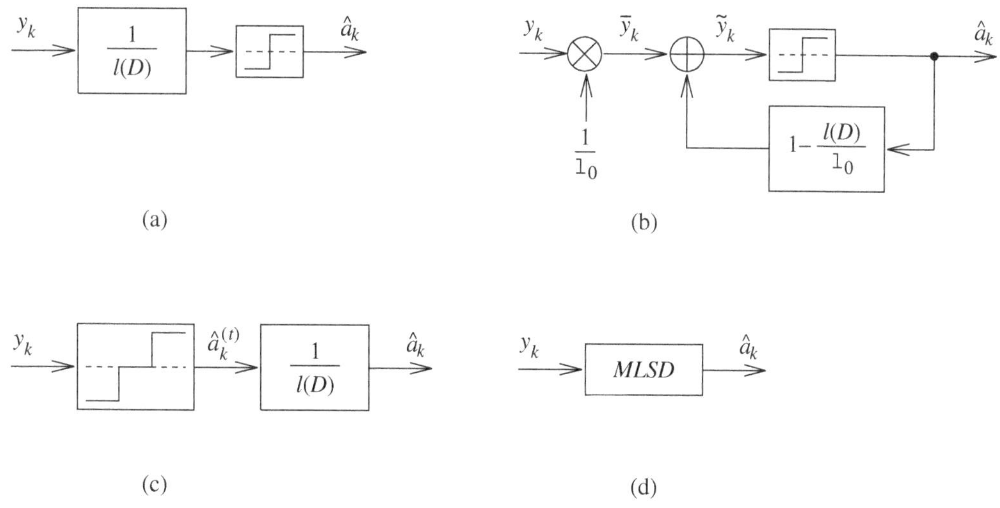

Figure 7.78 Four possible solutions to the detection problem in the presence of *controlled* ISI. (a) LE-ZF followed by M-level threshold detector. (b) DFE. (c) $M^{(t)}$-level threshold detector followed by LE-ZF. (d) Viterbi algorithm.

The receiver detects the symbols $\{a_k\}$ using the sequence of samples $\{y_k = a_k^{(t)} + w_{R,k}\}$. We discuss four possible solutions.[19]

1. *LE-ZF*: A zero-forcing linear equalizer (LE-ZF) having D transform equal to $1/l(D)$ is used. At the equalizer output, at instant k the symbol a_k plus a noise term is obtained; the detected symbols $\{\hat{a}_k\}$ are obtained by an M-level threshold detector, as illustrated in Figure 7.78a. We note however that the amplification of noise by the filter $1/l(D)$ is infinite at frequencies f such that $l(e^{-j2\pi fT}) = 0$.
2. *DFE*: A second solution resorts to a decision-feedback equalizer (DFE), as shown in Figure 7.78b. An M-level threshold detector is also employed by the DFE, but there is no noise amplification as the ISI is removed by the feedback filter, having D transform equal to $1 - l(D)/\mathtt{l}_0$. We observe that at the decision point, the signal $\bar{y}_k$ has the expression

$$\tilde{y}_k = \frac{1}{\mathtt{l}_0}\left(a_k^{(t)} + w_{R,k} - \sum_{i=1}^{N-1} \mathtt{l}_i\, \hat{a}_{k-i}\right) \tag{7.643}$$

If we indicate with $e_k = a_k - \hat{a}_k$ a detection error, then substituting (7.642) in (7.643), we obtain

$$\tilde{y}_k = a_k + \frac{1}{\mathtt{l}_0}\left(w_{R,k} + \sum_{i=1}^{N-1} \mathtt{l}_i\, e_{k-i}\right) \tag{7.644}$$

Equation (7.644) shows that a wrong decision negatively influence successive decisions: this phenomenon is known as *error propagation*.

[19] For a first reading, it is suggested that only solution 3 is considered. The study of the other solutions falls within the equalization methods.

3. *Threshold detector with $M^{(t)}$ levels*: This solution, shown in Figure 7.78c, exploits the $M^{(t)}$-ary nature of the symbols $a_k^{(t)}$, and makes use of a threshold detector with $M^{(t)}$ levels followed by a LE-ZF. This structure does not lead to noise amplification as solution 1, because the noise is eliminated by the threshold detector; however, there is still the problem of error propagation.
4. *Viterbi algorithm*: This solution, shown in Figure 7.78d, corresponds to maximum-likelihood sequence detection (MLSD) of $\{a_k\}$. It yields the best performance.

Solution 2 using the DFE is often adopted in practice: in fact it avoids noise amplification and is simpler to implement than the Viterbi algorithm. However, the problem of *error propagation* remains.

In this case, using (7.644), the error probability can be written as

$$P_e = \left(1 - \frac{1}{M}\right) P \left[\left| w_{R,k} + \sum_{i=1}^{N-1} \mathtt{l}_i \, e_{k-i} \right| > \mathtt{l}_0 \right] \tag{7.645}$$

A *lower bound* $P_{e,L}$ can be computed for P_e by assuming the error propagation is absent, or setting $\{e_k\} = 0$, $\forall k$, in (7.645). If we denote by σ_{w_R}, the standard deviation of the noise $w_{R,k}$, we obtain

$$P_{e,L} = 2\left(1 - \frac{1}{M}\right) Q\left(\frac{\mathtt{l}_0}{\sigma_{w_R}}\right) \tag{7.646}$$

Assuming $w_{R,k}$ white noise, an *upper bound* $P_{e,U}$ is given in [46] in terms of $P_{e,L}$:

$$P_{e,U} = \frac{M^{N-1}\, P_{e,L}}{(M/(M-1))\, P_{e,L}(M^{N-1} - 1) + 1} \tag{7.647}$$

From (7.647), we observe that the effect of the error propagation is that of increasing the error probability by a factor M^{N-1} with respect to $P_{e,L}$.

A solution to the problem of error propagation is represented by *precoding*, which will be investigated in depth in Section 7.7.

Precoding

We make use here of the following two simplifications:

1. the coefficients $\{\mathtt{l}_i\}$ are integer numbers;
2. the symbols $\{a_k\}$ belong to the alphabet $\mathcal{A} = \{0, 1, \dots, M-1\}$; this choice is made because arithmetic modulo M is employed.

We define the sequence of precoded symbols $\{\overline{a}_k^{(p)}\}$ as

$$\overline{a}_k^{(p)}\, \mathtt{l}_0 = \left(a_k - \sum_{i=1}^{N-1} \mathtt{l}_i\, \overline{a}_{k-i}^{(p)} \right) \bmod M \tag{7.648}$$

We note that (7.648) has only one solution if and only if $\mathtt{l}_0$ and M are relatively prime [47]. In case $\mathtt{l}_0 = \cdots = \mathtt{l}_{j-1} = 0 \bmod M$, and $\mathtt{l}_j$ and M are relatively prime, (7.648) becomes

$$\overline{a}_{k-j}^{(p)}\, \mathtt{l}_j = \left(a_k - \sum_{i=j+1}^{N-1} \mathtt{l}_i\, \overline{a}_{k-i}^{(p)} \right) \bmod M \tag{7.649}$$

For example, if $l(D) = 2 + D - D^2$ and $M = 2$, (7.648) is not applicable as $\mathtt{l}_0 \bmod M = 0$. Therefore, (7.649) is used.

Applying the PR filter to $\{\overline{a}_k^{(p)}\}$, we obtain the sequence

$$a_k^{(t)} = \sum_{i=0}^{N-1} \mathtt{l}_i \, \overline{a}_{k-i}^{(p)} \tag{7.650}$$

From the comparison between (7.648) and (7.650), or in general (7.649), we have the fundamental relation

$$a_k^{(t)} \bmod M = a_k \tag{7.651}$$

Equation (7.651) shows that, as in the absence of noise we have $y_k = a_k^{(t)}$, the symbol a_k can be detected by considering the received signal y_k modulo M; this operation is *memoryless*, therefore, the detection of $\hat{a}_k$ is independent of the previous detections $\{\hat{a}_{k-i}\}$, $i = 1, \dots, N-1$. Therefore, the problem of error propagation is solved. Moreover, the desired signal is not affected by ISI.

If the instantaneous transformation

$$a_k^{(p)} = 2\overline{a}_k^{(p)} - (M-1) \tag{7.652}$$

is applied to the symbols $\{\overline{a}_k^{(p)}\}$, then we obtain a sequence of symbols that belong to the alphabet $\mathcal{A}^{(p)}$ in (6.4). The sequence $\{a_k^{(p)}\}$ is then input to the filter $l(D)$. *Precoding* consists of the operation (7.648) followed by the transformation (7.652).

However, we note that (7.651) is no longer valid. From (7.652), (7.650), and (7.648), we obtain the new *decoding* operation, given by

$$a_k = \left(\frac{a_k^{(t)}}{2} + \mathcal{K} \right) \bmod M \tag{7.653}$$

where

$$\mathcal{K} = (M-1) \sum_{i=0}^{N-1} \mathtt{l}_i \tag{7.654}$$

A PR system with precoding is illustrated in Figure 7.79. The receiver is constituted by a threshold detector with $M^{(t)}$ levels that provides the symbols $\{\hat{a}_k^{(t)}\}$, followed by a block that realizes (7.653) and yields the detected data $\{\hat{a}_k\}$.

Figure 7.79 PR system with *precoding*.

Error probability with precoding

To evaluate the error probability of a system with precoding, the statistics of the symbols $\{a_k^{(t)}\}$ must be known; it is easy to prove that if the symbols $\{a_k\}$ are i.i.d., the symbols $\{a_k^{(t)}\}$ are also i.i.d.

If we assume that the cardinality of the set $\mathcal{A}^{(t)}$ is maximum, i.e. $M^{(t)} = M^{n_1}$, then the output levels are equally spaced and the symbols $a_k^{(t)}$ result equally likely with probability

$$P[a_k^{(t)} = \alpha] = \frac{1}{M^{n_1}}, \qquad \alpha \in \mathcal{A}^{(t)} \tag{7.655}$$

In general, however, the symbols $\{a_k^{(t)}\}$ are not equiprobable, because several output levels are redundant, as can be deduced from the following example:

Example 7.C.4 (Dicode filter)
We assume $M = 2$, therefore, $a_k = \{0, 1\}$; the precoding law (7.648) is simply an *exclusive or* and

$$\overline{a}_k^{(p)} = a_k \oplus \overline{a}_{k-1}^{(p)} \tag{7.656}$$

The symbols $\{a_k^{(p)}\}$ are obtained from (7.652),

$$a_k^{(p)} = 2\overline{a}_k^{(p)} - 1 \tag{7.657}$$

they are antipodal as $a_k^{(p)} = \{-1, +1\}$. Finally, the symbols at the output of the filter $l(D)$ are given by

$$a_k^{(t)} = a_k^{(p)} - a_{k-1}^{(p)} = 2(\overline{a}_k^{(p)} - \overline{a}_{k-1}^{(p)}) \tag{7.658}$$

Table 7.2: Precoding for the dicode filter

$\overline{a}_{k-1}^{(p)}$	a_k	$\overline{a}_k^{(p)}$	$a_k^{(t)}$
0	0	0	0
0	1	1	+2
1	0	1	0
1	1	0	−2

The values of $\overline{a}_{k-1}^{(p)}$, a_k, $\overline{a}_k^{(p)}$ and $a_k^{(t)}$ are given in Table 7.2. We observe that both output levels ± 2 correspond to the symbol $a_k = 1$ and therefore, are redundant; the three levels are not equally likely. The symbol probabilities are given by

$$P[a_k^{(t)} = \pm 2] = \frac{1}{4}$$
$$P[a_k^{(t)} = 0] = \frac{1}{2} \tag{7.659}$$

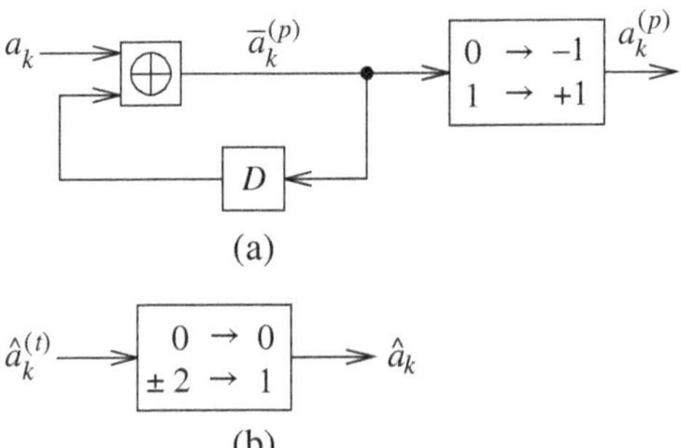

Figure 7.80 Precoder and decoder for a dicode filter $l(D)$ with $M = 2$. (a) Precoder. (b) Decoder.

Figure 7.80a shows the precoder that realizes equations (7.656) and (7.657). The *decoder*, realized as a map that associates the symbol $\hat{a}_k = 1$ to ± 2, and the symbol $\hat{a}_k = 0$ to 0, is illustrated in Figure 7.80b.

Alternative interpretation of PR systems

Up to now we have considered a general transmission system, and looked for an efficient design method. We now assume that the system is given, i.e. that the transmit filter as well as the receive filter are assigned. The scheme of Figure 7.74 can be regarded as a tool for the optimization of a given system, where $l(D)$ includes the characteristics of the transmit and receive filters: as a result, the symbols $\{a_k^{(t)}\}$

are no longer the *transmitted* symbols, but must be interpreted as the symbols that are *ideally received*. In the light of these considerations, the assumption of an ideal channel can also be removed. In this case, the filter $l(D)$ will also include the ISI introduced by the channel.

We observe that the *precoding/decoding* technique is an alternative equalization method to the DFE that presents the advantage of eliminating error propagation, which can considerably deteriorate system performance.

In the following two examples [14], additive white Gaussian noise $w_{R,k} = \tilde{w}_k$ is assumed, and various systems are studied for the same signal-to-noise ratio at the receiver.

Example 7.C.5 (Ideal channel g)
(A) *Antipodal signals*: We transmit a sequence of symbols from a binary alphabet, $a_k \in \{-1, 1\}$. The received signal is

$$y_k = a_k + \tilde{w}_{A,k} \tag{7.660}$$

where the variance of the noise is given by $\sigma^2_{\tilde{w}_A} = \sigma^2_I$.

At the receiver, using a threshold detector with threshold set to zero, we obtain

$$P_{bit} = Q\left(\frac{1}{\sigma_I}\right) \tag{7.661}$$

(B) *Duobinary signal with precoding*: The transmitted signal is now given by $a_k^{(t)} = a_k^{(p)} + a_{k-1}^{(p)} \in \{-2, 0, 2\}$, where $a_k^{(p)} \in \{-1, 1\}$ is given by (7.652) and (7.648).

The received signal is given by

$$y_k = a_k^{(t)} + \tilde{w}_{B,k} \tag{7.662}$$

where the variance of the noise is $\sigma^2_{\tilde{w}_B} = 2\sigma^2_I$, as $\sigma^2_{a_k^{(t)}} = 2$.

At the receiver, using a threshold detector with thresholds set at ± 1, we have the following conditional error probabilities:

$$P[\mathrm{E} \mid a_k^{(t)} = 0] = 2Q\left(\frac{1}{\sigma_{\tilde{w}_B}}\right)$$

$$P[\mathrm{E} \mid a_k^{(t)} = -2] = P[\mathrm{E} \mid a_k^{(t)} = 2] = Q\left(\frac{1}{\sigma_{\tilde{w}_B}}\right)$$

Consequently, at the detector output, we have

$$\begin{aligned} P_{bit} &= P[\hat{a}_k \neq a_k] \\ &= P\,[\mathrm{E} \mid a_k^{(t)} = 0]\frac{1}{2} + P\,[\mathrm{E} \mid a_k^{(t)} = \pm 2]\frac{1}{2} \\ &= 2Q\left(\frac{1}{\sqrt{2}\,\sigma_I}\right) \end{aligned}$$

We observe a worsening of about 3 dB in terms of the signal-to-noise ratio with respect to case A).
(C) *Duobinary signal*. The transmitted signal is $a_k^{(t)} = a_k + a_{k-1}$. The received signal is given by

$$y_k = a_k + a_{k-1} + \tilde{w}_{C,k} \tag{7.663}$$

where $\sigma^2_{\tilde{w}_C} = 2\sigma^2_I$. We consider using a receiver that applies MLSD to recover the data; from Example 7.5.3 on page 365 it results

$$P_{bit} = K\,Q\left(\frac{\sqrt{8}}{2\sigma_{\tilde{w}_C}}\right) = K\,Q\left(\frac{1}{\sigma_I}\right) \tag{7.664}$$

where K is a constant.

We note that the PR system employing MLSD at the receiver achieves a performance similar to that of a system transmitting antipodal signals, as MLSD exploits the correlation between symbols of the sequence $\{a_k^{(t)}\}$.

Example 7.C.6 (Equivalent channel g of the type 1 + D)
In this example, it is the channel itself that forms a *duobinary* signal.

(D) *Antipodal signals*: Transmitting $a_k \in \{-1, 1\}$, the received signal is given by

$$y_k = a_k + a_{k-1} + \tilde{w}_{D,k} \tag{7.665}$$

where $\sigma^2_{\tilde{w}_D} = 2\sigma_I^2$.

An attempt of pre-equalizing the signal at the transmitter by inserting a filter $l(D) = 1/(1 + D) = 1 - D + D^2 + \cdots$ would yield symbols $a_k^{(t)}$ with unbounded amplitude; therefore, such configuration cannot be used. Equalization at the receiver using the scheme of Figure 7.78a would require a filter of the type $1/(1 + D)$, which would lead to unbounded noise enhancement.

Therefore, we resort to the scheme of Figure 7.78c, whose detector thresholds is set at ± 1. To avoid error propagation, we precode the message and transmit the sequence $\{a_k^{(p)}\}$ instead of $\{a_k\}$. At the receiver, we have

$$y_k = a_k^{(p)} + a_{k-1}^{(p)} + \tilde{w}_{D,k} \tag{7.666}$$

We are therefore in the same conditions as in case (B), and

$$P_{bit} = 2Q\left(\frac{1}{\sqrt{2}\,\sigma_I}\right) \tag{7.667}$$

(E) *MLSD receiver*: To detect the sequence of information bits from the received signal (7.665), MLSD can be adopted. P_{bit} is in this case given by (7.664).

Appendix 7.D Implementation of a QAM transmitter

Three structures, which differ by the position of the digital-to-analog converter (DAC), may be considered for the implementation of a QAM transmitter. In Figure 7.81, the modulator employs for both in-phase and quadrature signals a DAC after the interpolator filter h_{Tx}, followed by an analog mixer that shifts the signal to passband. This scheme works if the sampling frequency $1/T_c$ is much greater than twice the bandwidth B of h_{Tx}.

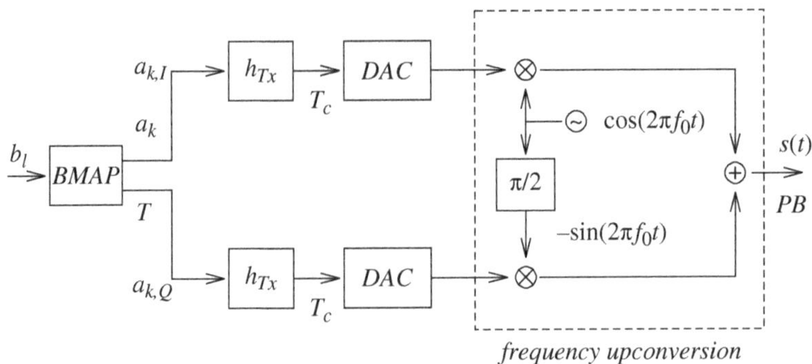

Figure 7.81 QAM with analog mixer.

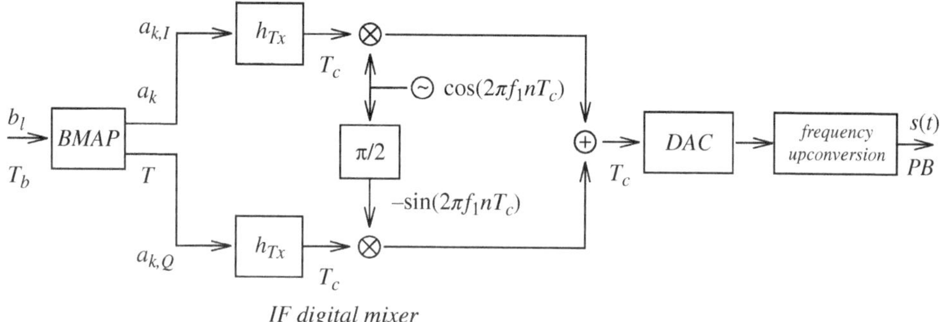

Figure 7.82 QAM with digital and analog mixers.

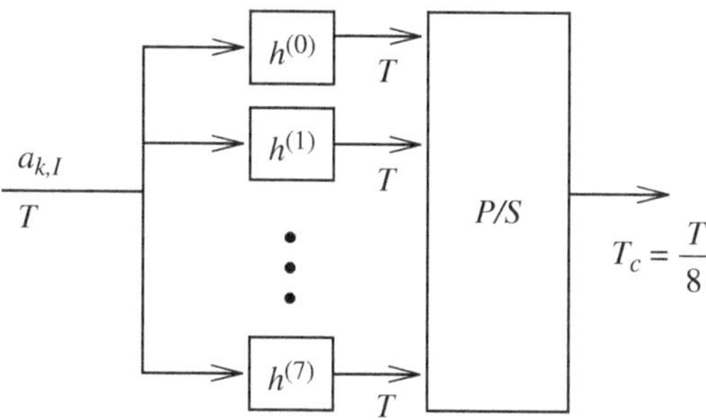

Figure 7.83 Polyphase implementation of the filter h_{Tx} for $T_c = T/8$.

For applications where the symbol rate is very high, the DAC is placed right after the bit mapper and filters are implemented in the analog domain (see Chapter 19).

In the implementation illustrated in Figure 7.82, the DAC is placed instead at an intermediate stage with respect to the case of Figure 7.81. Samples are premodulated by a digital mixer to an intermediate frequency f_1, interpolated by the DAC and subsequently remodulated by a second analog mixer that shifts the signal to the desired band. The intermediate frequency f_1 must be greater than the bandwidth B and smaller than $1/(2T_c) - B$, thus avoiding overlap among spectral components. We observe that this scheme requires only one DAC, but the sampling frequency must be at least double as compared to the previous scheme.

For the first implementation, as the system is typically oversampled with a sampling interval $T_c = T/4$ or $T_c = T/8$, the frequency response of the DAC, $\mathcal{G}_I(f)$, may be considered as a constant in the passband of both the in-phase and quadrature signals. For the second implementation, unless $f_1 \ll 1/T_c$, the distortion introduced by the DAC should be considered and equalized by one of these methods:

- including the compensation for $\mathcal{G}_I(f)$ in the frequency response of the filter h_{Tx},
- inserting a digital filter before the DAC,
- inserting an analog filter after the DAC.

We recall that an efficient implementation of interpolator filters h_{Tx} is obtained by the polyphase representation, as shown in Figure 7.83 for $T_c = T/8$, where

$$h^{(\ell)}(m) = h_{Tx}\left(mT + \ell\frac{T}{8}\right), \quad \ell = 0, 1, \dots, 7, \qquad m = -\infty, \dots, +\infty \tag{7.668}$$

To implement the scheme of Figure 7.83, once the impulse response is known, it may be convenient to pre-compute the possible values of the filter output and store them in a table or RAM. The symbols $\{a_{k,I}\}$ are then used as pointers for the table itself. The same approach may be followed to generate the values of the signals $\cos(2\pi f_1 nT_c)$ and $\sin(2\pi f_1 nT_c)$ in Figure 7.82, using an additional table and the index n as a cyclic pointer.

Chapter 8

Multicarrier modulation

For channels that exhibit high signal attenuation at frequencies within the passband, a valid alternative to single carrier (SC) is represented by a modulation technique based on filter banks, known as *multicarrier* (MC) *modulation*. As the term implies, MC modulation is obtained in principle by modulating several carriers in parallel using blocks of symbols, therefore using a symbol period that is typically much longer than the symbol period of an SC system transmitting at the same bit rate. The resulting narrowband signals around the frequencies of the carriers are then added and transmitted over the channel. The narrowband signals are usually referred to as subchannel (or subcarrier) signals.

An advantage of MC modulation with respect to SC is represented by the lower complexity required for equalization, that under certain conditions can be performed by a single coefficient per subchannel. A long symbol period also yields a greater immunity of an MC system to impulse noise; however the symbol duration, and hence the number of subchannels, is limited for transmissions over time-variant channels. As we will see in this chapter, another important advantage is the efficient implementation of modulator and demodulator, obtained by sophisticated signal processing algorithms.

8.1 MC systems

In MC systems, blocks of $\mathcal{M}$ symbols are transmitted in parallel over $\mathcal{M}$ *subchannels*, using $\mathcal{M}$ modulation filters [1] with frequency responses $\mathcal{H}_i(f)$, $i = 0, \dots, \mathcal{M}-1$. We consider the discrete-time baseband equivalent MC system illustrated in Figure 8.1, where for now the channel has been omitted and the received signal r_n coincides with the transmitted signal s_n. The $\mathcal{M}$ input symbols at the k-th modulation interval are represented by the vector

$$\boldsymbol{a}_k = \left[a_k[0], \dots, a_k[\mathcal{M}-1]\right]^T \tag{8.1}$$

where $a_k[i] \in \mathcal{A}[i]$, $i = 0, \dots, \mathcal{M}-1$. The alphabets $\mathcal{A}[i]$, $i = 0, \dots, \mathcal{M}-1$, correspond to two dimensional constellations, which are not necessarily identical.

The symbol rate of each subchannel is equal to $1/T$ and corresponds to the rate of the input symbol vectors $\{\boldsymbol{a}_k\}$: $1/T$ is also called MC *modulation rate*. The sampling period of each input sequence is changed from T to $T/\mathcal{M}$ by upsampling, that is by inserting $\mathcal{M}-1$ zeros between consecutive symbols (see Appendix 1.A); each sequence thus obtained is then filtered by a filter properly allocated in

Algorithms for Communications Systems and their Applications, Second Edition.
Nevio Benvenuto, Giovanni Cherubini, and Stefano Tomasin.

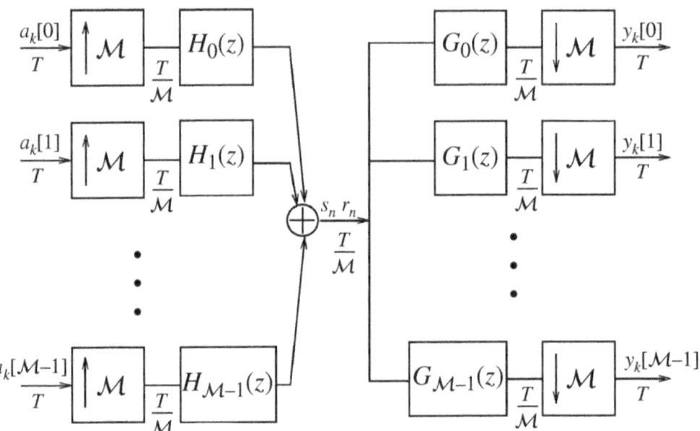

Figure 8.1 Block diagram of an MC system, where the channel has been omitted.

frequency. The filter on the i-th subchannel is characterized by the impulse response $\{h_n[i]\}$, with transfer function

$$H_i(z) = \sum_{n=-\infty}^{+\infty} h_n[i]\, z^{-n} \tag{8.2}$$

and frequency response $\mathcal{H}_i(f) = H_i(z)|_{z=e^{j2\pi fT/\mathcal{M}}}$, $i = 0, \ldots, \mathcal{M}-1$. The transmit filters, also called interpolator filters, work in parallel at the rate $\mathcal{M}/T$, called *transmission rate*. The transmitted signal s_n is given by the sum of the filter output signals, that is

$$s_n = \sum_{i=0}^{\mathcal{M}-1} \sum_{k=-\infty}^{+\infty} a_k[i]\, h_{n-k\mathcal{M}}[i] \tag{8.3}$$

The received signal $r_n = s_n$ is filtered by $\mathcal{M}$ filters in parallel at the transmission rate. The receive filters have impulse responses $\{g_n[i]\}$, $i = 0, \ldots, \mathcal{M}-1$, and transfer functions given by $G_i(z) = \sum_n g_n[i] z^{-n}$, $i = 0, \ldots, \mathcal{M}-1$. By downsampling the $\mathcal{M}$ output signals of the receive filters at the modulation rate $1/T$, we obtain the vector sequence $\{\boldsymbol{y}_k = [y_k[0], \ldots, y_k[\mathcal{M}-1]]^T\}$, which is used to detect the vector sequence $\{\boldsymbol{a}_k\}$; receive filters are also known as decimator filters.

We consider as transmit filters finite-impulse-response (FIR) causal filters having length $\gamma\mathcal{M}$, and support $\{0, 1, \ldots, \gamma\mathcal{M}-1\}$, that is $h_n[i]$ may be non-zero for $n = 0, \ldots, \gamma\mathcal{M}-1$. We assume matched receive filters; as we are dealing with causal filters, we consider the expression of the matched filters $\{g_n[i]\}$, for $i = 0, \ldots, \mathcal{M}-1$, given by

$$g_n[i] = h^*_{\gamma\mathcal{M}-n}[i], \qquad \forall n \tag{8.4}$$

We observe that the support of $\{g_n[i]\}$ is $\{1, 2, \ldots, \gamma\mathcal{M}\}$.

8.2 Orthogonality conditions

If the transmit and receive filter banks are designed such that certain orthogonality conditions are satisfied, the subchannel output signals are delayed versions of the transmitted symbol sequences at the corresponding subchannel inputs. The orthogonality conditions, also called perfect reconstruction conditions, can be interpreted as a more general form of the Nyquist criterion.

Time domain

With reference to the general scheme of Figure 8.1, at the output of the j-th receive subchannel, before downsampling, the impulse response relative to the i-th input is given by

$$\sum_{p=0}^{\gamma\mathcal{M}-1} h_p[i]\, g_{n-p}[j] = \sum_{p=0}^{\gamma\mathcal{M}-1} h_p[i]\, h^*_{\gamma\mathcal{M}+p-n}[j], \qquad \forall n \tag{8.5}$$

We note that, for $j = i$, the peak of the sequence given by (8.5) is obtained for $n = \gamma\mathcal{M}$. Observing (8.5), transmission in the absence of intersymbol interference (ISI) over a subchannel, as well as absence of interchannel interference (ICI) between subchannels, is achieved if orthogonality conditions are satisfied, that in the time domain are expressed as

$$\sum_{p=0}^{\gamma\mathcal{M}-1} h_p[i]\, h^*_{p+\mathcal{M}(\gamma-k)}[j] = \delta_{i-j}\,\delta_{k-\gamma}, \qquad i,j = 0,\dots,\mathcal{M}-1 \tag{8.6}$$

Hence, in the ideal channel case considered here, the vector sequence at the output of the decimator filter bank is a replica of the transmitted vector sequence with a delay of γ modulation intervals, that is $\{\boldsymbol{y}_k\} = \{\boldsymbol{a}_{k-\gamma}\}$. Sometimes the elements of a set of orthogonal impulse responses that satisfy (8.6) are called *wavelets*.

Frequency domain

In the frequency domain, the conditions (8.6) are expressed as

$$\sum_{\ell=0}^{\mathcal{M}-1} \mathcal{H}_i\left(f - \frac{\ell}{T}\right) \mathcal{H}_j^*\left(f - \frac{\ell}{T}\right) = \delta_{i-j}, \qquad i,j = 0,\dots,\mathcal{M}-1 \tag{8.7}$$

z-Transform domain

Setting

$$\boldsymbol{a}(z) = \left[\sum_k a_k[0]z^{-k},\dots,\sum_k a_k[\mathcal{M}-1]z^{-k}\right]^T \tag{8.8}$$

and

$$\boldsymbol{y}(z) = \left[\sum_k y_k[0]z^{-k},\dots,\sum_k y_k[\mathcal{M}-1]z^{-k}\right]^T \tag{8.9}$$

the input–output relation is given by (see Figure 8.2),

$$\boldsymbol{y}(z) = \boldsymbol{S}(z)\,\boldsymbol{a}(z) \tag{8.10}$$

where the element $[\boldsymbol{S}(z)]_{p,q}$ of the matrix $\boldsymbol{S}(z)$ is the transfer function between the q-th input and the p-th output. We note that the orthogonality conditions, expressed in the time domain by (8.6), are satisfied if

$$\boldsymbol{S}(z) = z^{-\gamma}\,\boldsymbol{I} \tag{8.11}$$

where $\boldsymbol{I}$ is the $\mathcal{M} \times \mathcal{M}$ identity matrix.

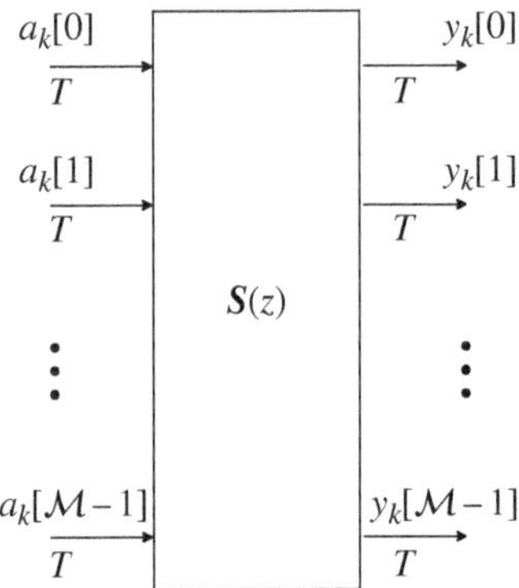

Figure 8.2 Equivalent MC system with input–output relation expressed in terms of the matrix $S(z)$.

8.3 Efficient implementation of MC systems

For large values of $\mathcal{M}$, a direct implementation of an MC system as shown in Figure 8.1 would require an exceedingly large computational complexity, as all filtering operations are performed at the high $\mathcal{M}/T$ rate. A reduction in the number of operations per unit of time is obtained by resorting to the polyphase representation of the various filters [2].

MC implementation employing matched filters

Using the polyphase representation of both transmit and receive filter banks introduced in Appendix 1.A, we have (see (1.562))

$$H_i(z) = \sum_{\ell=0}^{\mathcal{M}-1} z^{-\ell}\, E_i^{(\ell)}(z^{\mathcal{M}}), \qquad i = 0, \dots, \mathcal{M}-1 \tag{8.12}$$

where $E_i^{(\ell)}(z)$ is the transfer function of the ℓ-th polyphase component of $H_i(z)$. Defined the vector of the transfer functions of transmit filters as

$$\boldsymbol{h}(z) = [H_0(z), \dots, H_{\mathcal{M}-1}(z)]^T \tag{8.13}$$

and $\boldsymbol{E}(z)$ the matrix of the transfer functions of the polyphase components of the transmit filters, given by

$$\boldsymbol{E}(z) = \begin{bmatrix} E_0^{(0)}(z) & E_1^{(0)}(z) & \cdots & E_{\mathcal{M}-1}^{(0)}(z) \\ E_0^{(1)}(z) & E_1^{(1)}(z) & \cdots & E_{\mathcal{M}-1}^{(1)}(z) \\ \vdots & \vdots & & \vdots \\ E_0^{(\mathcal{M}-1)}(z) & E_1^{(\mathcal{M}-1)}(z) & \cdots & E_{\mathcal{M}-1}^{(\mathcal{M}-1)}(z) \end{bmatrix} \tag{8.14}$$

$\boldsymbol{h}^T(z)$ can be expressed as

$$\boldsymbol{h}^T(z) = [1, z^{-1}, \dots, z^{-(\mathcal{M}-1)}]\boldsymbol{E}(z^{\mathcal{M}}) \tag{8.15}$$

Observing (8.15), the $\mathcal{M}$ interpolated input sequences are filtered by filters whose transfer functions are represented in terms of the polyphase components, expressed as columns of the matrix $\boldsymbol{E}(z^{\mathcal{M}})$. Because the transmitted signal $\{s_n\}$ is given by the sum of the filter outputs, the operations of the transmit filter bank are illustrated in Figure 8.3, where the signal $\{s_n\}$ is obtained from a delay line that collects the outputs of the $\mathcal{M}$ vectors of filters with transfer functions given by the rows of $\boldsymbol{E}(z^{\mathcal{M}})$, and input given by the vector of interpolated input symbols.

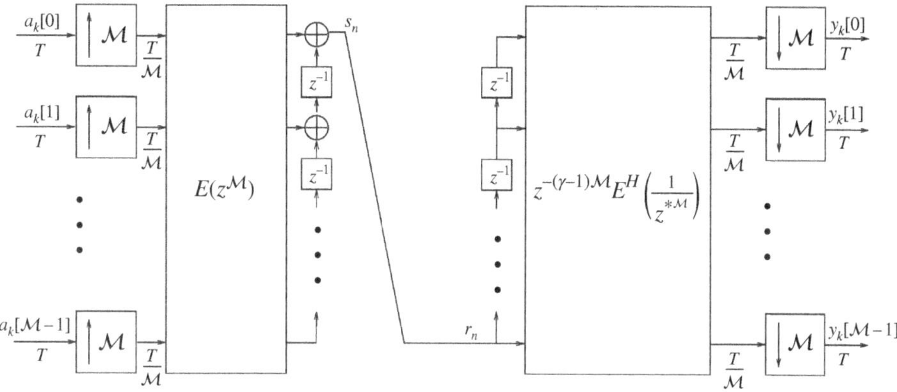

Figure 8.3 Implementation of an MC system using the polyphase representation.

At the receiver, using a representation equivalent to (1.564), that is obtained by a permutation of the polyphase components, we get

$$G_i(z) = \sum_{\ell=0}^{\mathcal{M}-1} z^{-(\mathcal{M}-1-\ell)} R_i^{(\ell)}(z^{\mathcal{M}}), \qquad i = 0, \dots, \mathcal{M}-1 \tag{8.16}$$

where $R_i^{(\ell)}(z)$ is the transfer function of the $(\mathcal{M}-1-\ell)$-th polyphase component of $G_i(z)$. Then, the vector of the transfer functions of the receive filters $\boldsymbol{g}(z) = [G_0(z), \dots, G_{\mathcal{M}-1}(z)]^T$ can be expressed as

$$\boldsymbol{g}(z) = \boldsymbol{R}(z^{\mathcal{M}})[z^{-(\mathcal{M}-1)}, \dots, z^{-1}, 1]^T \tag{8.17}$$

where $\boldsymbol{R}(z)$ is the matrix of the transfer functions of the polyphase components of the receive filters, and is given by

$$\boldsymbol{R}(z) = \begin{bmatrix} R_0^{(0)}(z) & R_0^{(1)}(z) & \cdots & R_0^{(\mathcal{M}-1)}(z) \\ R_1^{(0)}(z) & R_1^{(1)}(z) & \cdots & R_1^{(\mathcal{M}-1)}(z) \\ \vdots & \vdots & & \vdots \\ R_{\mathcal{M}-1}^{(0)}(z) & R_{\mathcal{M}-1}^{(1)}(z) & \cdots & R_{\mathcal{M}-1}^{(\mathcal{M}-1)}(z) \end{bmatrix} \tag{8.18}$$

Observing (8.17), the $\mathcal{M}$ signals at the output of the receive filter bank before downsampling are obtained by filtering in parallel the received signal by filters whose transfer functions are represented in terms of the polyphase components, expressed as rows of the matrix $\boldsymbol{R}(z^{\mathcal{M}})$. Therefore, the $\mathcal{M}$ output signals are equivalently obtained by filtering the vector of received signal samples $[r_{n-(\mathcal{M}-1)}, r_{n-\mathcal{M}+2}, \dots, r_n]^T$ by the $\mathcal{M}$ vectors of filters with transfer functions given by the rows of the matrix $\boldsymbol{R}(z^{\mathcal{M}})$.

In particular, recalling that the receive filters have impulse responses given by (8.4), we obtain

$$G_i(z) = z^{-\gamma\mathcal{M}} H_i^*\left(\frac{1}{z^*}\right), \qquad i = 0, 1, \dots, \mathcal{M}-1 \tag{8.19}$$

Substituting (8.19) in (8.17), and using (8.15), the expression of the vector $\boldsymbol{g}(z)$ of the transfer functions of the receive filters becomes

$$\begin{aligned} \boldsymbol{g}(z) &= z^{-\gamma\mathcal{M}} \boldsymbol{h}^H\left(\frac{1}{z^*}\right) \\ &= z^{-\gamma\mathcal{M}+\mathcal{M}-1} \boldsymbol{E}^H\left(\frac{1}{z^{*\mathcal{M}}}\right)[z^{-(\mathcal{M}-1)}, \dots, z^{-1}, 1]^T \end{aligned} \tag{8.20}$$

Apart from a delay term z^{-1}, the operations of the receive filter bank are illustrated by the scheme of Figure 8.3.

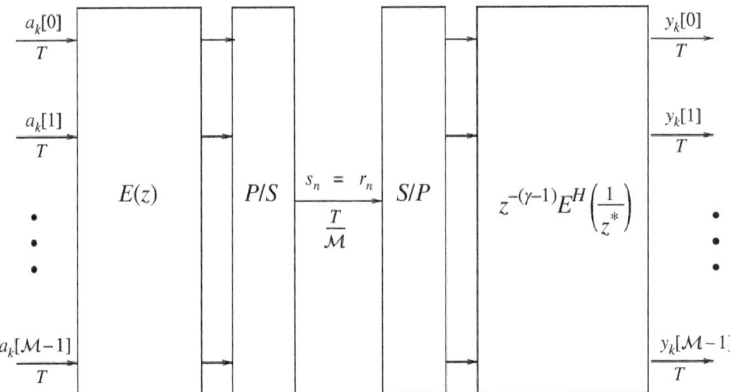

Figure 8.4 Equivalent MC system implementation.

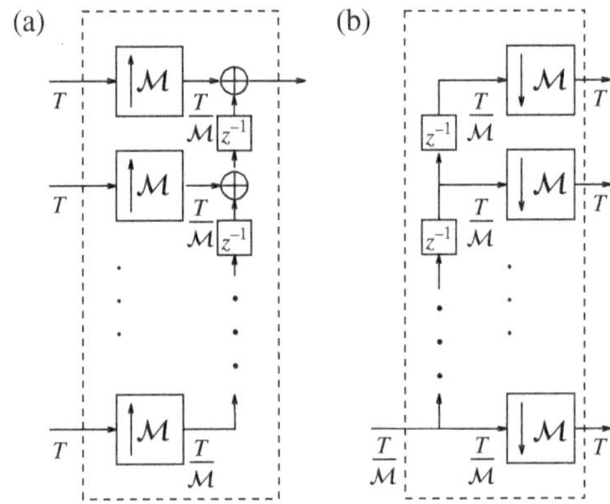

Figure 8.5 Block diagrams of: (a) parallel to serial converter, (b) serial to parallel converter.

From Figure 8.3 using (8.20), and applying the noble identities given in Figure 1.60, we obtain the system represented in Figure 8.4, where parallel to serial (P/S) and serial to parallel (S/P) converters are illustrated in Figure 8.5a,b, respectively.

Orthogonality conditions in terms of the polyphase components

We observe that the input–output relation of the cascade of the P/S converter and the S/P converter is expressed by the matrix $z^{-1}\boldsymbol{I}$, that means that any input is obtained at the output on the same branch with a delay equal to T: therefore, the system of Figure 8.4 is equivalent to that of Figure 8.2, with matrix transfer function given by

$$\boldsymbol{S}(z) = z^{-(\gamma-1)}\, \boldsymbol{E}^H\left(\frac{1}{z^*}\right) z^{-1}\, \boldsymbol{I}\, \boldsymbol{E}(z) = z^{-\gamma}\, \boldsymbol{E}^H\left(\frac{1}{z^*}\right) \boldsymbol{E}(z) \tag{8.21}$$

From (8.11), in terms of the polyphase components of the transmit filters in (8.14), the orthogonality conditions in the z-transform domain are therefore expressed as

$$\boldsymbol{E}^H\left(\frac{1}{z^*}\right) \boldsymbol{E}(z) = \boldsymbol{I} \tag{8.22}$$

Equivalently, using the polyphase representation of the receive filters in (8.18), we find the condition

$$\boldsymbol{R}(z)\, \boldsymbol{R}^H\left(\frac{1}{z^*}\right) = \boldsymbol{I} \tag{8.23}$$

MC implementation employing a prototype filter

The complexity of an MC system can be further reduced by resorting to uniform filter banks [1, 2]. In this case, the frequency responses of the various filters are obtained by shifting the frequency response of a *prototype filter* around the carrier frequencies, given by

$$f_i = \frac{i}{T}, \qquad i = 0, \dots, \mathcal{M} - 1 \tag{8.24}$$

In other words, the spacing in frequency between the subchannels is $1/T$.

To derive an efficient implementation of uniform filter banks we consider the scheme represented in Figure 8.6a, where $H(z)$ and $G(z)$ are the transfer functions of the transmit and receive prototype filters, respectively. We consider as transmit prototype filter a causal FIR filter with impulse response $\{h_n\}$, having length $\gamma\mathcal{M}$ and support $\{0, 1, \dots, \gamma\mathcal{M} - 1\}$; the receive prototype filter is a causal FIR filter matched to the transmit prototype filter, with impulse response given by

$$g_n = h^*_{\gamma\mathcal{M}-n}, \qquad \forall n \tag{8.25}$$

With reference to Figure 8.6a, the i-th subchannel signal at the channel input is given by

$$s_n[i] = e^{j2\pi \frac{in}{\mathcal{M}}} \sum_{k=-\infty}^{\infty} a_k[i]\, h_{n-k\mathcal{M}} \tag{8.26}$$

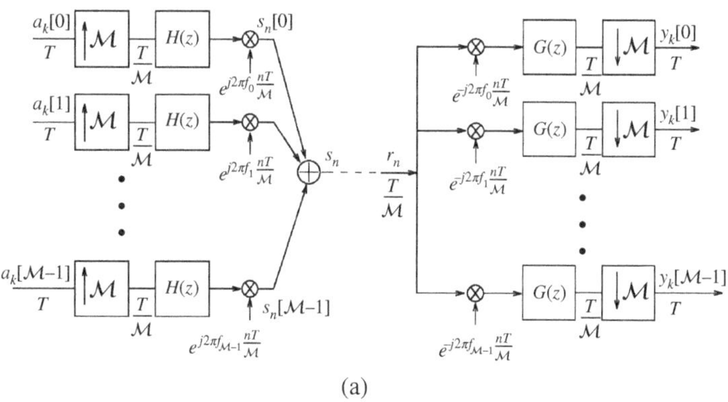

(a)

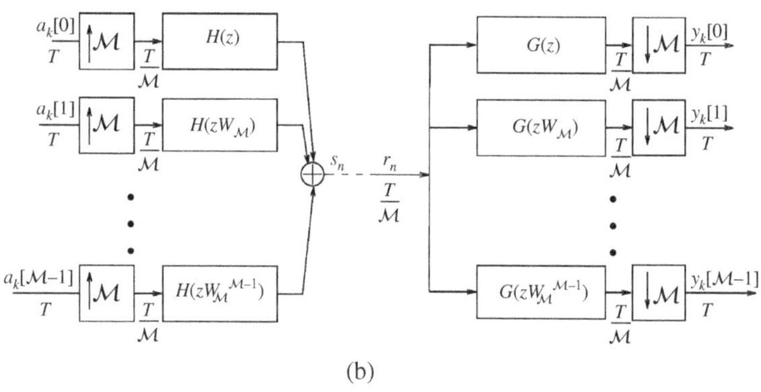

(b)

Figure 8.6 Block diagram of an MC system with uniform filter banks: (a) general scheme, (b) equivalent scheme for $f_i = i/T$, $i = 0, \dots, \mathcal{M} - 1$.

As

$$e^{j2\pi \frac{in}{\mathcal{M}}} = e^{j2\pi \frac{i(n-k\mathcal{M})}{\mathcal{M}}} \tag{8.27}$$

we obtain

$$s_n[i] = \sum_{k=-\infty}^{\infty} a_k[i]\, h_{n-k\mathcal{M}}\, e^{j2\pi \frac{i(n-k\mathcal{M})}{\mathcal{M}}} = \sum_{k=-\infty}^{\infty} a_k[i]\, h_{n-k\mathcal{M}}[i] \tag{8.28}$$

where

$$h_n[i] = h_n\, e^{j2\pi \frac{in}{\mathcal{M}}} \tag{8.29}$$

Recalling the definition $W_{\mathcal{M}} = e^{-j\frac{2\pi}{\mathcal{M}}}$, the z-transform of $\{h_n[i]\}$ is expressed as

$$H_i(z) = H(zW_{\mathcal{M}}^i), \qquad i = 0, \ldots, \mathcal{M}-1 \tag{8.30}$$

Observing (8.28) and (8.30), we obtain the equivalent scheme of Figure 8.6b.

The scheme of Figure 8.6b may be considered as a particular case of the general MC scheme represented in Figure 8.1; in particular, the transfer functions of the filters can be expressed using the polyphase representations, which are however simplified with respect to the general case expressed by (8.14) and (8.18), as we show in the following. Observing (8.26), we express the overall signal s_n as

$$s_n = \sum_{i=0}^{\mathcal{M}-1} e^{j2\pi \frac{in}{\mathcal{M}}} \sum_{k=-\infty}^{+\infty} h_{n-k\mathcal{M}}\, a_k[i] \tag{8.31}$$

With the change of variables $n = m\mathcal{M} + \ell$, for $m = -\infty, \ldots, +\infty$, and $\ell = 0, 1, \ldots, \mathcal{M}-1$, we get

$$s_{m\mathcal{M}+\ell} = \sum_{i=0}^{\mathcal{M}-1} e^{j2\pi \frac{i}{\mathcal{M}}(m\mathcal{M}+\ell)} \sum_{k=-\infty}^{+\infty} h_{(m-k)\mathcal{M}+\ell}\, a_k[i] \tag{8.32}$$

Observing that $e^{j2\pi im} = 1$, setting $s_m^{(\ell)} = s_{m\mathcal{M}+\ell}$ and $h_m^{(\ell)} = h_{m\mathcal{M}+\ell}$, and interchanging the order of summations, we express the ℓ-th polyphase component of $\{s_n\}$ as

$$s_m^{(\ell)} = \sum_{k=-\infty}^{+\infty} h_{m-k}^{(\ell)} \sum_{i=0}^{\mathcal{M}-1} W_{\mathcal{M}}^{-i\ell}\, a_k[i] \tag{8.33}$$

The sequences $\{h_m^{(\ell)}\}$, $\ell = 0, \ldots, \mathcal{M}-1$, denote the polyphase components of the prototype filter impulse response, with transfer functions given by

$$H^{(\ell)}(z) = \sum_{m=0}^{\gamma-1} h_m^{(\ell)}\, z^{-m}, \qquad \ell = 0, \ldots, \mathcal{M}-1 \tag{8.34}$$

Recalling the definition (1.30) of the discrete Fourier transform (DFT) operator as $\mathcal{M} \times \mathcal{M}$ matrix, the inverse discrete Fourier transform (IDFT) of the vector $\boldsymbol{a}_k$ is expressed as

$$\boldsymbol{F}_{\mathcal{M}}^{-1}\, \boldsymbol{a}_k = \boldsymbol{A}_k = [A_k[0], \ldots, A_k[\mathcal{M}-1]]^T \tag{8.35}$$

We find that the inner summation in (8.33) yields

$$\frac{1}{\mathcal{M}} \sum_{i=0}^{\mathcal{M}-1} W_{\mathcal{M}}^{-i\ell}\, a_k[i] = A_k[\ell], \qquad \ell = 0, 1, \ldots, \mathcal{M}-1 \tag{8.36}$$

and

$$s_m^{(\ell)} = \sum_{k=-\infty}^{\infty} \mathcal{M}\, h_{m-k}^{(\ell)}\, A_k[\ell] = \sum_{p=-\infty}^{\infty} \mathcal{M}\, h_p^{(\ell)}\, A_{m-p}[\ell] \tag{8.37}$$

Including the factor $\mathcal{M}$ in the definition of the prototype filter impulse response, or in $\mathcal{M}$ gain factors that establish the statistical power levels to be assigned to each subchannel signal, an efficient implementation

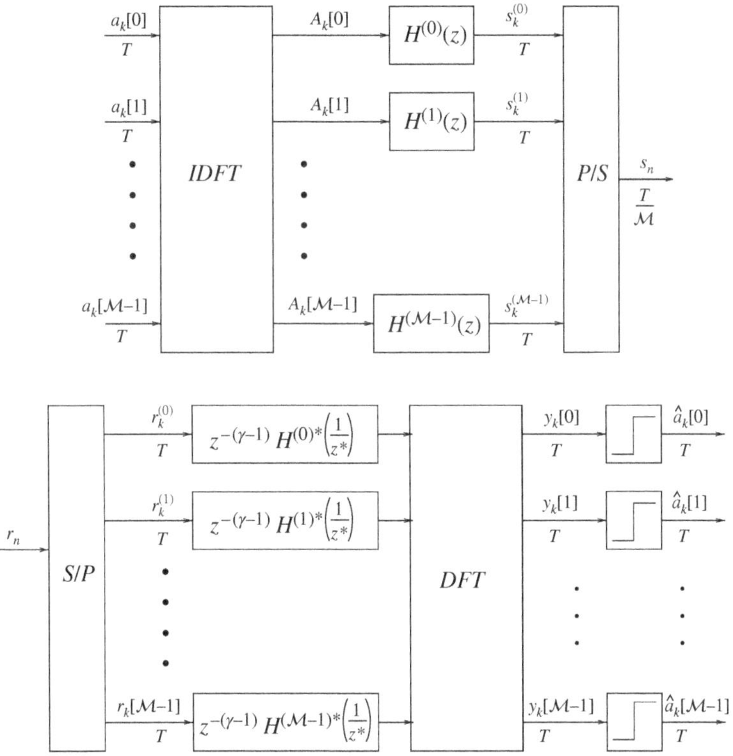

Figure 8.7 Block diagram of an MC system with efficient implementation.

of a uniform transmit filter bank is given by an IDFT, a polyphase network with $\mathcal{M}$ branches, and a P/S converter, as illustrated in Figure 8.7.

Observing (8.33), the transfer functions of the polyphase components of the transmit filters are given by

$$\boldsymbol{E}(z) = \text{diag}\,\{H^{(0)}(z), \dots, H^{(\mathcal{M}-1)}(z)\}\boldsymbol{F}_{\mathcal{M}}^{-1} \tag{8.38}$$

Therefore, the vector of the transfer functions of the transmit filters is expressed as

$$\boldsymbol{h}^T(z) = [1, z^{-1}, \dots, z^{-(\mathcal{M}-1)}]\,\text{diag}\,\{H^{(0)}(z^{\mathcal{M}}), \dots, H^{(\mathcal{M}-1)}(z^{\mathcal{M}})\}\boldsymbol{F}_{\mathcal{M}}^{-1} \tag{8.39}$$

We note that we would arrive at the same result by applying the notion of prototype filter with the condition (8.24) to (8.14).

The vector of the transfer functions of a receive filter bank, which employs a prototype filter with impulse response given by (8.25), is immediately obtained by applying (8.20), with the matrix of the transfer functions of the polyphase components given by (8.38). Therefore, we get

$$\begin{aligned}\boldsymbol{g}(z) = z^{-\gamma\mathcal{M}+\mathcal{M}-1}\,\boldsymbol{F}_{\mathcal{M}}\,\text{diag}\left\{H^{(0)*}\left(\frac{1}{z^{*\mathcal{M}}}\right), \dots\right.\\ \left.\dots, H^{(\mathcal{M}-1)*}\left(\frac{1}{z^{*\mathcal{M}}}\right)\right\}[z^{-(\mathcal{M}-1)}, \dots, z^{-1}, 1]^T\end{aligned} \tag{8.40}$$

Hence, an efficient implementation of a uniform receive filter bank, also illustrated in Figure 8.7, is given by a S/P converter, a polyphase network with $\mathcal{M}$ branches, and a DFT; we note that the filter of the i-th branch at the receiver is matched to the filter of the corresponding branch at the transmitter.

With reference to Figure 8.6a, it is interesting to derive (8.40) by observing the relation between the received sequence r_n and the output of the i-th subchannel $y_k[i]$, given by

$$y_k[i] = \sum_{n=-\infty}^{+\infty} g_{k\mathcal{M}-n}\, e^{-j2\pi \frac{i}{T} n \frac{T}{\mathcal{M}}}\, r_n \tag{8.41}$$

With the change of variables $n = m\mathcal{M} + \ell$, for $m = -\infty, \dots, +\infty$, and $\ell = 0, \dots, \mathcal{M} - 1$, and recalling the expression of the prototype filter impulse response given by (8.25), we obtain

$$y_k[i] = \sum_{\ell=0}^{\mathcal{M}-1} \sum_{m=-\infty}^{+\infty} h^*_{(\gamma-k+m)\mathcal{M}+\ell}\, e^{-j2\pi \frac{i}{\mathcal{M}}(m\mathcal{M}+\ell)}\, r_{m\mathcal{M}+\ell} \tag{8.42}$$

Observing that $e^{-j2\pi \frac{i}{\mathcal{M}} m\mathcal{M}} = 1$, setting $r_m^{(\ell)} = r_{m\mathcal{M}+\ell}$, and $h_m^{(\ell)*} = h^*_{m\mathcal{M}+\ell}$, and interchanging the order of summations, we get

$$y_k[i] = \sum_{\ell=0}^{\mathcal{M}-1} e^{-j\frac{2\pi}{\mathcal{M}} i\ell} \sum_{m=-\infty}^{\infty} h^{(\ell)*}_{\gamma+m-k}\, r_m^{(\ell)} \tag{8.43}$$

Using the relation $e^{-j\frac{2\pi}{\mathcal{M}} i\ell} = W_{\mathcal{M}}^{i\ell}$, we finally find the expression

$$y_k[i] = \sum_{\ell=0}^{\mathcal{M}-1} W_{\mathcal{M}}^{i\ell} \sum_{m=-\infty}^{\infty} h^{(\ell)*}_{\gamma+m-k}\, r_m^{(\ell)} \tag{8.44}$$

Provided the orthogonality conditions are satisfied, from the output samples $y_k[i]$, $i = 0, \dots, \mathcal{M} - 1$, threshold detectors may be employed to yield the detected symbols $\hat{a}_k[i]$, $i = 0, \dots, \mathcal{M} - 1$, with a suitable delay.

As illustrated in Figure 8.7, all filtering operations are carried out at the low rate $1/T$. Also note that in practice the fast Fourier transform (FFT) and the inverse FFT (IFFT) are used in place of the DFT and the IDFT, respectively, thus further reducing the computational complexity.

8.4 Non-critically sampled filter banks

For the above discussion on the efficient realization of MC systems, we referred to Figure 8.7, where the number of subchannels, $\mathcal{M}$, coincides with the interpolation factor of the modulation filters in the transmit filter bank. These systems are called filter-bank systems with critical sampling or in short *critically sampled filter banks*.

We now examine the general case of $\mathcal{M}$ modulators where each uses an interpolation filter by a factor $\mathcal{K} > \mathcal{M}$: this system is called *non-critically sampled*. In principle, the schemes of transmit and receive non-critically sampled filter banks are illustrated in Figure 8.8. As in critically sampled systems, also in non-critically sampled systems it is advantageous to choose each transmit filter as the frequency-shifted version of a prototype filter with impulse response $\{h_n\}$, defined over a discrete-time domain with sampling period $T_c = T/\mathcal{K}$. At the receiver, each filter is the frequency-shifted version of a prototype filter with impulse response $\{g_n\}$, also defined over a discrete-time domain with sampling period T_c.

As depicted in Figure 8.9, each subchannel filter has a bandwidth equal to $\mathcal{K}/(\mathcal{M}T)$, larger than $1/T$. Maintaining a spacing between subchannels of $\mathcal{K}/(\mathcal{M}T)$, it is easier to avoid spectral overlapping between subchannels and consequently to avoid ICI. It is also possible to choose $\{h_n\}$, e.g. as the impulse response of a square root raised cosine filter, such that, at least for an ideal channel, the orthogonality conditions are satisfied and ISI is also avoided. We note that this advantage is obtained at the expense of a larger bandwidth required for the transmission channel, that changes from $\mathcal{M}/T$ for critically sampled

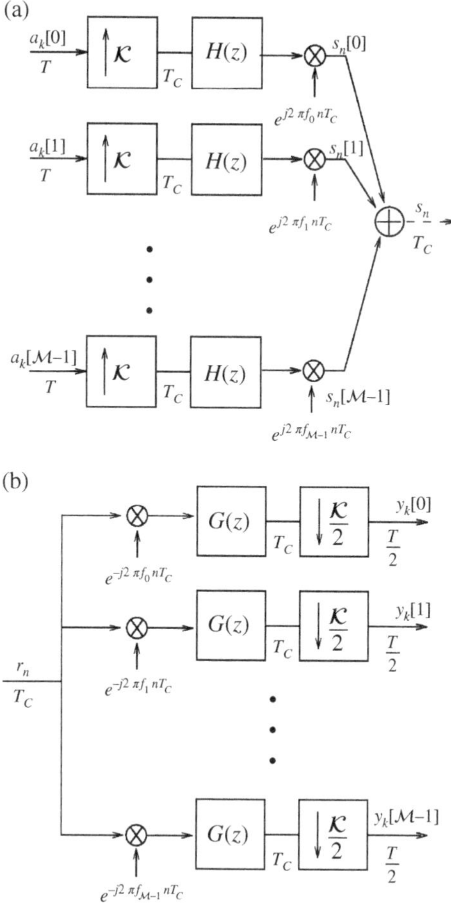

Figure 8.8 Block diagram of (a) transmitter and (b) receiver in a transmission system employing non-critically sampled filter banks, with $\mathcal{K} > \mathcal{M}$ and $f_i = (i\mathcal{K})/(\mathcal{M}T) = i/(\mathcal{M}T_c)$.

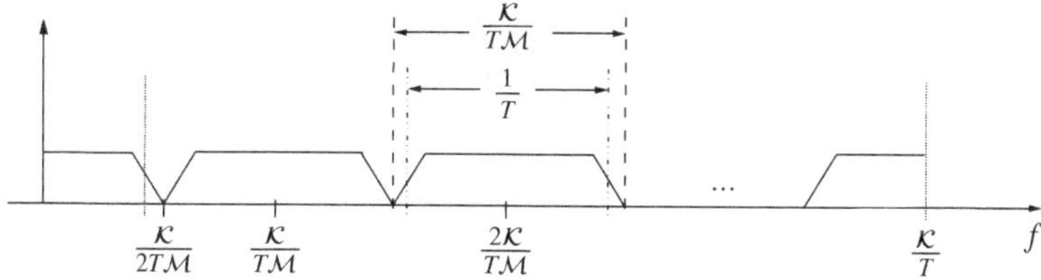

Figure 8.9 Filter frequency responses in a non-critically sampled system.

systems to $\mathcal{K}/T$ for non-critically sampled systems. Therefore, the system requires an excess bandwidth given by $(\mathcal{K} - \mathcal{M})/\mathcal{M}$.

Also for non-critically sampled filter banks, it is possible to obtain an efficient implementation using the discrete Fourier transform [3, 4]. The transmitted signal is expressed as a function of the input symbol sequences as

$$s_n = \sum_{i=0}^{\mathcal{M}-1} e^{j2\pi \frac{i\mathcal{K}}{\mathcal{M}T} n \frac{T}{\mathcal{K}}} \sum_{k=-\infty}^{\infty} a_k[i]\, h_{n-k\mathcal{K}} \tag{8.45}$$

or, equivalently,

$$s_n = \sum_{k=-\infty}^{\infty} h_{n-k\mathcal{K}} \sum_{i=0}^{\mathcal{M}-1} a_k[i]\, W_{\mathcal{M}}^{-in} \tag{8.46}$$

With the change of indices

$$n = m\mathcal{M} + \ell, \quad m \in \mathcal{Z}, \quad \ell = 0, 1, \dots, \mathcal{M}-1 \tag{8.47}$$

(8.46) becomes

$$s_{m\mathcal{M}+\ell} = \sum_{k=-\infty}^{\infty} h_{m\mathcal{M}-k\mathcal{K}+\ell} \sum_{i=0}^{\mathcal{M}-1} W_{\mathcal{M}}^{-i\ell}\, a_k[i] \tag{8.48}$$

Using the definition of the IDFT (8.36), apart from a factor $\mathcal{M}$ that can be included in the impulse response of the filter, and introducing the following polyphase representation of the transmitted signal

$$s_m^{(\ell)} = s_{m\mathcal{M}+\ell} \tag{8.49}$$

we obtain

$$s_m^{(\ell)} = \sum_{k=-\infty}^{\infty} h_{m\mathcal{M}-k\mathcal{K}+\ell}\, A_k[\ell] \tag{8.50}$$

In analogy with (1.488), (8.50) is obtained by interpolation of the sequence $\{A_k[\ell]\}$ by a factor $\mathcal{K}$, followed by decimation by a factor $\mathcal{M}$. From (1.496) and (1.497), we introduce the change of indices

$$p = \left\lfloor \frac{m\mathcal{M}}{\mathcal{K}} \right\rfloor - k \tag{8.51}$$

and

$$\Delta_m = \frac{m\mathcal{M}}{\mathcal{K}} - \left\lfloor \frac{m\mathcal{M}}{\mathcal{K}} \right\rfloor = \frac{(m\mathcal{M})_{mod\ \mathcal{K}}}{\mathcal{K}} \tag{8.52}$$

Using (1.503), it results

$$\begin{aligned} s_m^{(\ell)} &= \sum_{p=-\infty}^{+\infty} h_{(p+\Delta_m)\mathcal{K}+\ell}\, A_{\left\lfloor \frac{m\mathcal{M}}{\mathcal{K}} \right\rfloor - p}[\ell] \\ &= \sum_{p=-\infty}^{+\infty} h_{p\mathcal{K}+\ell+(m\mathcal{M})_{mod\ \mathcal{K}}}\, A_{\left\lfloor \frac{m\mathcal{M}}{\mathcal{K}} \right\rfloor - p}[\ell] \end{aligned}$$

Letting

$$h_{p,m}^{(\ell)} = h_{p\mathcal{K}+\ell+(m\mathcal{M})_{mod\ \mathcal{K}}}, \quad p, m \in \mathcal{Z}, \quad \ell = 0, 1, \dots, \mathcal{M}-1 \tag{8.53}$$

we get

$$s_m^{(\ell)} = \sum_{p=0}^{\infty} h_{p,m}^{(\ell)}\, A_{\left\lfloor \frac{m\mathcal{M}}{\mathcal{K}} \right\rfloor - p}[\ell] \tag{8.54}$$

The efficient implementation of the transmit filter bank is illustrated in Figure 8.10. We note that the system is now periodically time-varying, i.e. the impulse response of the filter components cyclically changes. The $\mathcal{M}$ elements of an IDFT output vector are input to $\mathcal{M}$ delay lines. Also note that within a modulation interval of duration T, the samples stored in some of the delay lines are used to produce more than one sample of the transmitted signal. Therefore, the P/S element used for the realization of critically sampled filter banks needs to be replaced by a commutator. At instant nT_c, the commutator is linked to the $\ell = n_{mod\ \mathcal{M}}$-th filtering element. The transmit signal s_n is then computed by convolving the signal samples stored in the ℓ-th delay line with the $n_{mod\ \mathcal{K}}$-th polyphase component of the T_c-spaced-coefficients prototype filter. In other terms, each element of the IDFT output frame is filtered by a periodically time-varying filter with period equal to $[l.c.m.(\mathcal{M}, \mathcal{K})]T/\mathcal{K}$, where $l.c.m.(\mathcal{M}, \mathcal{K})$ denotes the least common multiple of $\mathcal{M}$ and $\mathcal{K}$.

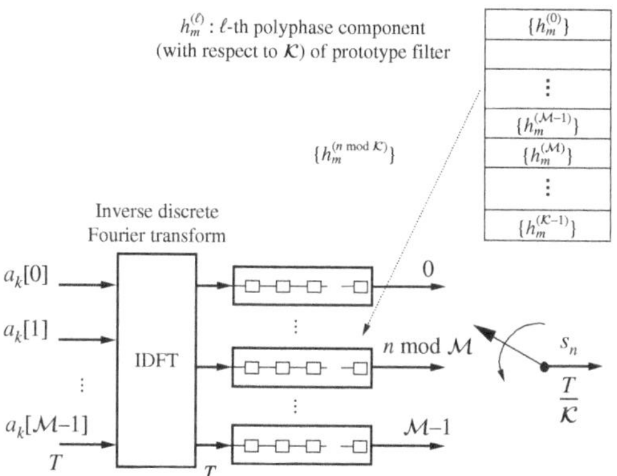

Figure 8.10 Efficient implementation of the transmitter of a system employing non-critically sampled filter banks; the filter components are periodically time-varying.

Likewise, the non-critically sampled filter bank at the receiver can also be efficiently implemented using the DFT. In particular, we consider the case of downsampling of the subchannel output signals by a factor $\mathcal{K}/2$, which yields samples at each subchannel output at an (over)sampling rate equal to $2/T$. With reference to Figure 8.8b, we observe that the output sequence of the i-th subchannel is given by

$$y_{n'}[i] = \sum_{n=-\infty}^{\infty} g_{n'\frac{\mathcal{K}}{2}-n} \, e^{-j2\pi\frac{in}{\mathcal{M}}} \, r_n \tag{8.55}$$

where $g_n = h^*_{\gamma\mathcal{M}-n}$.

With the change of indices

$$n = m\mathcal{M} + \ell, \qquad m \in \mathcal{Z}, \quad \ell = 0, 1, \ldots, \mathcal{M}-1 \tag{8.56}$$

and letting $r_m^{(\ell)} = r_{m\mathcal{M}+\ell}$, from (8.55) we get

$$y_{n'}[i] = \sum_{\ell=0}^{\mathcal{M}-1} \left(\sum_{m=-\infty}^{\infty} g_{n'\frac{\mathcal{K}}{2}-m\mathcal{M}-\ell} \, r_m^{(\ell)} \right) W_{\mathcal{M}}^{i\ell} \tag{8.57}$$

We note that in (8.57) the term within parenthesis may be viewed as an interpolation by a factor $\mathcal{M}$ followed by a decimation by a factor $\mathcal{K}/2$.

Letting

$$q = \left\lfloor \frac{n'\mathcal{K}}{2\mathcal{M}} \right\rfloor - m \tag{8.58}$$

and

$$\Delta_{n'} = \frac{n'\mathcal{K}}{2\mathcal{M}} - \left\lfloor \frac{n'\mathcal{K}}{2\mathcal{M}} \right\rfloor = \frac{(n'\mathcal{K}/2)_{mod\ \mathcal{M}}}{\mathcal{M}} \tag{8.59}$$

the terms within parenthesis in (8.57) can be written as

$$\sum_{q=-\infty}^{\infty} g_{q\mathcal{M}+(n'\frac{\mathcal{K}}{2})_{mod\ \mathcal{M}}-\ell} \; r^{(\ell)}_{\left\lfloor \frac{n'\mathcal{K}/2}{\mathcal{M}} \right\rfloor - q} \tag{8.60}$$

Introducing the $\mathcal{M}$ periodically time-varying filters,

$$g_{q,n'}^{(\ell)} = g_{q\mathcal{M}+(n'\frac{\mathcal{K}}{2})_{mod\ \mathcal{M}}-\ell}, \qquad q, n' \in \mathbb{Z}, \quad \ell = 0, 1, \ldots, \mathcal{M}-1 \tag{8.61}$$

and defining the DFT input samples

$$u_{n'}^{(\ell)} = \sum_{q=-\infty}^{\infty} g_{q,n'}^{(\ell)} \, r_{\left\lfloor \frac{n'\mathcal{K}}{2\mathcal{M}} \right\rfloor - q}^{(\ell)} \tag{8.62}$$

(8.57) becomes

$$y_{n'}[i] = \sum_{\ell=0}^{\mathcal{M}-1} u_{n'}^{(\ell)} \, W_{\mathcal{M}}^{i\ell} \tag{8.63}$$

The efficient implementation of the receive filter bank is illustrated in Figure 8.11, where we assume for the received signal the same sampling rate of $\mathcal{K}/T$ as for the transmitted signal, and a downsampling factor $\mathcal{K}/2$, so that the samples at each subchannel output are obtained at a sampling rate equal to $2/T$. Note that the optimum timing phase for each subchannel can be recovered by using per-subchannel fractionally spaced equalization, as discussed in Section 7.4.3 for SC modulation. Also note that within a modulation interval of duration T, more than one sample is stored in some of the delay lines to produce the DFT input vectors. Therefore, the S/P element used for the realization of critically sampled filter banks needs to be replaced by a commutator. After the $\mathcal{M}$ elements of a DFT input vector are produced, the commutator is circularly rotated $\mathcal{K}/2$ steps clockwise from its current position, allowing a set of $\mathcal{K}/2$ consecutive received samples to be input into the delay lines. The content of each delay line is then convolved with one of the $\mathcal{M}$ polyphase components of the $T/\mathcal{K}$-spaced coefficients receive prototype filter. A similar structure is obtained if in general a downsampling factor $\mathcal{K}' \leq \mathcal{K}$ is considered.

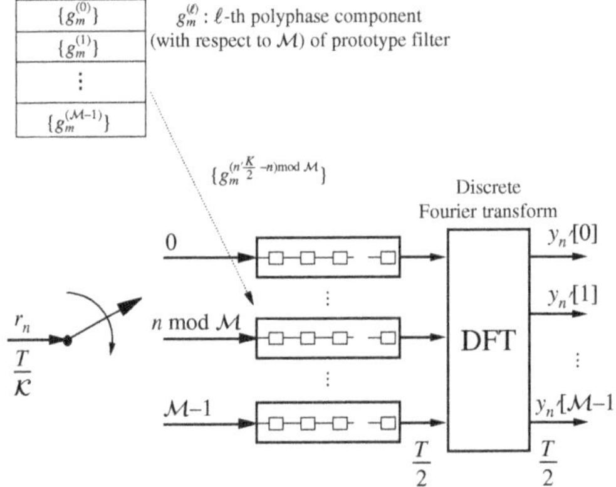

Figure 8.11 Efficient implementation of the receiver of a system employing non-critically sampled filter banks; the filter components are periodically time-varying (see (8.61)).

8.5 Examples of MC systems

We consider two simple examples of critically sampled filter bank modulation systems. For practical applications, equalization techniques and possibly non-critically sampled filter bank realizations are required, as will be discussed in the following sections.

OFDM or DMT

In this text, we refer to orthogonal frequency division multiplexing (OFDM) as a particular technique of MC transmission, which applies to both wired and wireless systems. Other authors use the term discrete

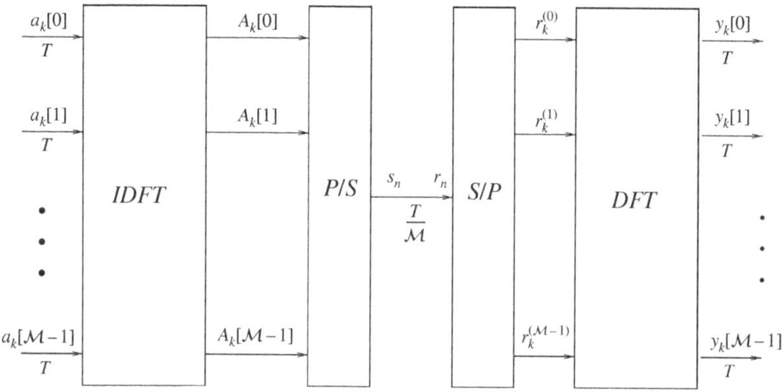

Figure 8.12 Block diagram of an OFDM system with impulse response of the prototype filter given by a rectangular window of length $\mathcal{M}$.

multitone (DMT) modulation for wired transmission systems, whereas they reserve the term OFDM for wireless systems. The transmit and receive filter banks use a prototype filter with impulse response given by [5–8]

$$h_n = \begin{cases} 1 & \text{if } 0 \le n \le \mathcal{M}-1 \\ 0 & \text{otherwise} \end{cases} \tag{8.64}$$

The impulse responses of the polyphase components of the prototype filter are given by

$$\{h_n^{(\ell)}\} = \{\delta_n\}, \qquad \ell = 0, \dots, \mathcal{M}-1 \tag{8.65}$$

and we can easily verify that the orthogonality conditions (8.6) are satisfied.

As shown in Figure 8.12, because the frequency responses of the polyphase components are constant, we obtain directly the transmit signal by applying a P/S conversion at the output of the IDFT. Assuming an ideal channel, at the receiver a S/P converter forms blocks of $\mathcal{M}$ samples, with boundaries between blocks placed so that each block at the output of the IDFT at the transmitter is presented unchanged at the input of the DFT. At the DFT output, the input blocks of $\mathcal{M}$ symbols are reproduced without distortion with a delay equal to T. We note however that the orthogonality conditions are satisfied only if the channel is ideal.

From the frequency response of the prototype filter (see (1.26)):

$$H\left(e^{j2\pi f \frac{T}{\mathcal{M}}}\right) = e^{-j2\pi f \frac{\mathcal{M}-1}{2} \frac{T}{\mathcal{M}}}\, \mathcal{M}\mathrm{sinc}_{\mathcal{M}}(fT) \tag{8.66}$$

using (8.30) the frequency responses of the individual subchannel filters $H_i(z)$ are obtained. Figure 8.13 shows the amplitude of the frequency responses of adjacent subchannel filters, obtained for $f \in (0, 0.06\frac{\mathcal{M}}{T})$ and $\mathcal{M} = 64$. We note that the choice of a rectangular window of length $\mathcal{M}$ as impulse response of the baseband prototype filter leads to a significant overlapping of spectral components of transmitted signals in adjacent subchannels.

Filtered multitone

In filtered multitone (FMT) systems, the transmit and receive filter banks use a prototype filter with frequency response given by [3, 4]

$$H\left(e^{j2\pi f \frac{T}{\mathcal{M}}}\right) = \begin{cases} \left|\frac{1+e^{-j2\pi fT}}{1+\rho e^{-j2\pi fT}}\right| & \text{if } |f| \le \frac{1}{2T} \\ 0 & \text{otherwise} \end{cases} \tag{8.67}$$

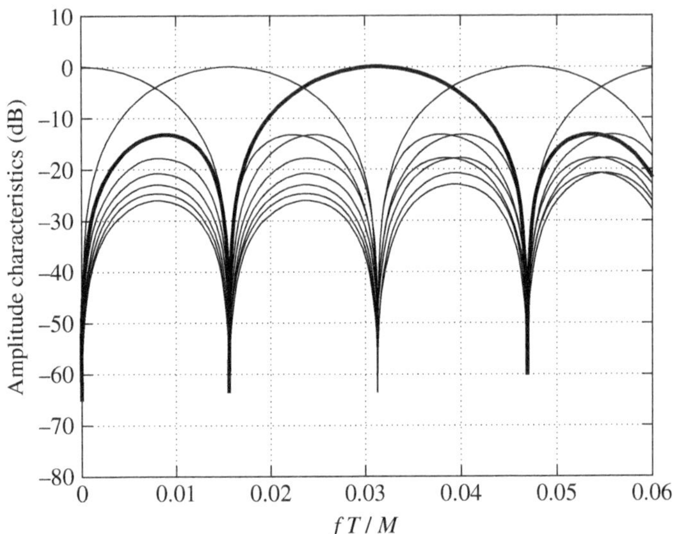

Figure 8.13 Amplitude of the frequency responses of adjacent subchannel filters in OFDM for $f \in (0, 0.06\frac{\mathcal{M}}{T})$ and $\mathcal{M} = 64$. Source: Reproduced with permission from Cherubini et al. [4]. ©2002, IEEE.

where the parameter $0 \leq \rho \leq 1$ controls the spectral roll-off of the filter. The frequency response exhibits spectral nulls at the band edges and, when used as the prototype filter characteristic, leads to transmission free of ICI but with ISI within a subchannel. For $\rho \to 1$, the frequency characteristic of each subchannel exhibits steep roll-off towards the band edge frequencies. On the other hand, for $\rho = 0$ the partial-response class I characteristic is obtained (see (7.634)). In general, it is required that at the output of each subchannel the ICI is negligible with respect to the noise.

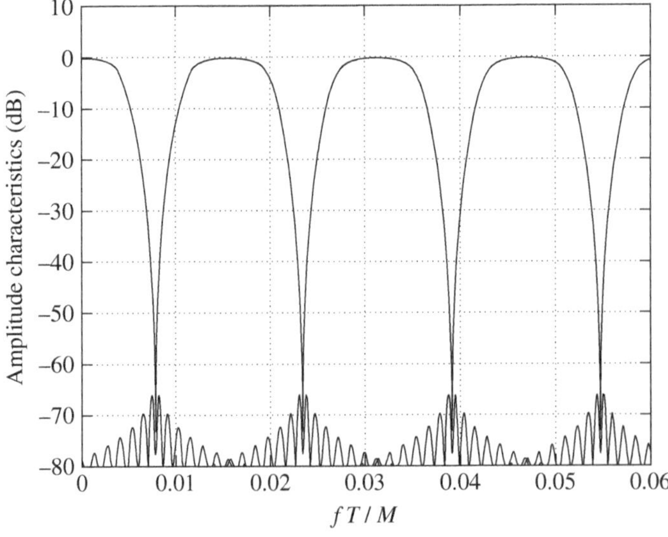

Figure 8.14 Amplitude of the frequency responses of adjacent subchannel filters in an FMT system for $f \in (0, 0.06\frac{\mathcal{M}}{T})$, and design parameters $\mathcal{M} = 64$, $\gamma = 10$, and $\rho = 0.1$. Source: Reproduced with permission from Cherubini et al. [4]. ©2002, IEEE.

The amplitude of the frequency responses of subchannel filters obtained with a minimum-phase prototype FIR filter for $f \in (0, 0.06\frac{\mathcal{M}}{T})$, and design parameters $\mathcal{M} = 64$, $\gamma = 10$, and $\rho = 0.1$ are illustrated in Figure 8.14.

8.6 Analog signal processing requirements in MC systems

In this section, we consider the case of a modulated signal $\{s_n\}$ with a sampling frequency of $\mathcal{M}/T$ that we obtain at the output of the transmit filter bank and must be converted into an analog signal before being sent to a power amplifier and then over the transmission channel with impulse response $g_{Ch}(t)$, $t \in \mathbb{R}$. First, we consider the design criteria of the transmit filters and then we further elaborate on the power amplifier requirements.

8.6.1 Analog filter requirements

Two interpretations of $\{s_n\}$ lead to different filter requirements.

Interpolator filter and virtual subchannels

The first interpretation considers $\{s_n\}$ as a sampled signal.

Assuming $\{s_n\}$ is a real signal, we refer to the scheme of Figure 8.15, where the task of the transmit analog filter g_{Tx}, together with the digital-to-analog-converter (DAC) interpolator filter g_I, is to attenuate the spectral components of $\{s_n\}$ at frequencies higher than $\mathcal{M}/(2T)$. Therefore, aliasing is avoided at the receiver by sampling the signal with rate $\mathcal{M}/T$ at the output of the anti-aliasing filter g_{Rc}. We note that the filter g_{Rc} does not equalize the channel, which would be a difficult task, whenever the channel frequency response presents large attenuations at frequencies within the passband, and is therefore entirely carried out in the digital domain.

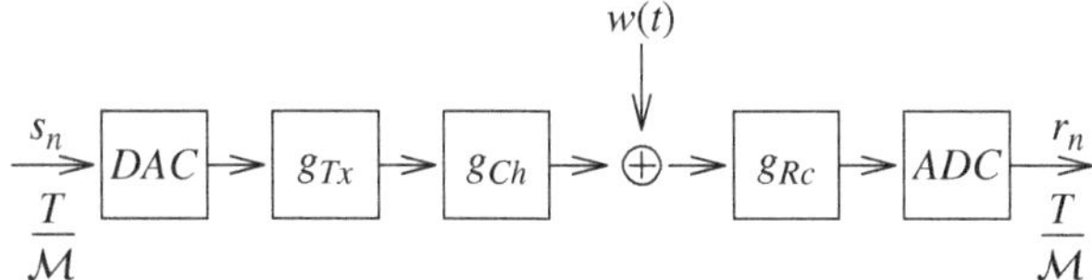

Figure 8.15 Baseband analog transmission of an MC signal.

With reference to Figure 8.15, to simplify the specifications of the transmit analog filter, it is convenient to interpolate the signal $\{s_n\}$ by a digital filter before the DAC.

To further simplify the transmit filters g_I and g_{Tx}, it is desirable to introduce a transition band of $\{s_n\}$ by avoiding transmission on subchannels near the frequency $\mathcal{M}/(2T)$; in other words, we transmit sequences of all zeros, $a_k[i] = 0$, $\forall k$, for $i = (\mathcal{M}/2) - N_{CV}, \dots, (\mathcal{M}/2), \dots, (\mathcal{M}/2) + N_{CV}$. These subchannels are generally called *virtual channels* and their number, $2N_{CV} + 1$, may be a non-negligible percentage of $\mathcal{M}$; this is usually the case in OFDM systems because of the large support of the prototype filter in the frequency domain. In FMT systems, choosing $N_{CV} = 1$ or 2 is sufficient. Typically, a square root raised cosine filter, with Nyquist frequency equal to $\mathcal{M}/(2T)$, is selected to implement the interpolator filter and the anti-aliasing filter.

We recall now the conditions to obtain a real-valued transmitted signal $\{s_n\}$. For OFDM systems with the efficient implementation illustrated in Figure 8.7, it is sufficient that the coefficients of the prototype filter and the samples $A_k[i]$, $\forall k$, $i = 0, \dots, \mathcal{M} - 1$, are real-valued. Observing (8.36), the latter condition implies that the following Hermitian symmetry conditions must be satisfied

$$\begin{aligned} &a_k[0], \quad a_k\left[\frac{\mathcal{M}}{2}\right] \in \mathbb{R} \\ &a_k[i] = a_k^*[\mathcal{M} - i], \qquad i = 1, \dots, \frac{\mathcal{M}}{2} - 1 \end{aligned} \tag{8.68}$$

In this case, the symmetry conditions (8.68) also further reduce the implementation complexity of the IDFT and DFT.

When $\{s_n\}$ is a complex signal, the scheme of Figure 8.16 is adopted, where the filters g_{Tx} and g_{Rc} have the characteristics described above and f_0 is the carrier frequency.

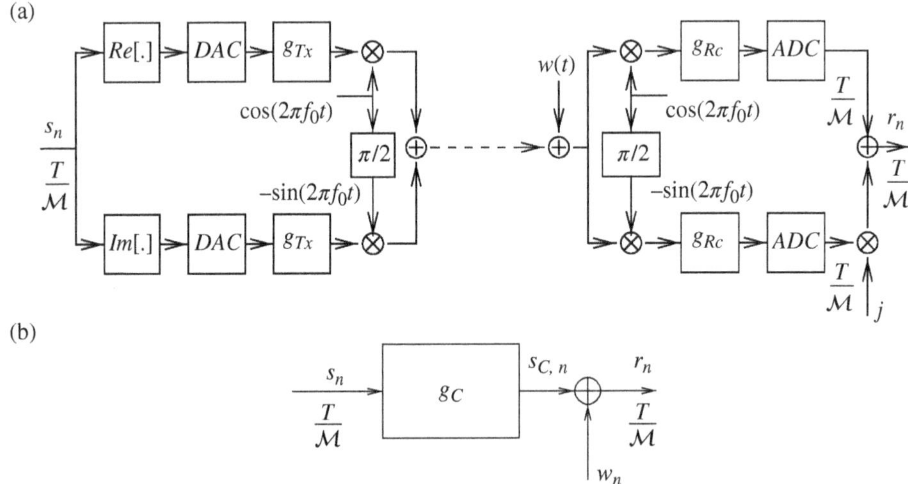

Figure 8.16 (a) Passband analog MC transmission scheme; (b) baseband equivalent scheme.

We note that this scheme is analogous to that of a QAM system, with the difference that the transmit and receive lowpass filters, with impulse responses g_{Tx} and g_{Rc}, respectively, have different requirements. The baseband equivalent scheme shown in Figure 8.16b is obtained from the passband scheme by the method discussed in Chapter 7. Note that g_C is the discrete-time equivalent channel obtained by the cascade of analog transmit filter, channel and receive filter. Since transmit and receive filters are in general designed according to different criteria for SC and MC systems, we use different notations for baseband equivalent channels, namely, g_C for MC systems and h for SC systems. With some approximations, the two pulses may be assumed identical to simplify simulations. As the receive filters approximate an ideal lowpass filter with bandwidth $\mathcal{M}/(2T)$, the signal-to-noise ratio Γ at the channel output is assumed equal to that obtained at the output of the baseband equivalent discrete-time channel, given by

$$\Gamma = \frac{\mathtt{M}_{s_C}}{N_0 \mathcal{M}/T} \tag{8.69}$$

assuming the noise $\{w_n\}$ complex-valued with PSD N_0.

Modulator filter

The second interpretation considers $\{s_n\}$ as a data signal which must be modulated before transmission on the channel. Both schemes of Figures 8.15 and 8.16 hold with different requirements for filters:

g_{Tx} should be a square root Nyquist pulse and typically also g_{Rc}, although this choice is not optimum. On the positive side, no virtual subchannel is needed; however, filters with a higher implementation complexity are required. Moreover, the required channel bandwidth is slightly higher than in the previous interpretation. Overall, this interpretation is only used in discrete-time system simulations but seldom in hardware implementations.

8.6.2 Power amplifier requirements

As seen in Section 4.1.3, the high-power amplifier (HPA) used to amplify the analog signal s before transmission, has a linear *envelope transfer function* for a limited input range. When the envelope A of the input signal is outside that range, the output is not a scaled version with a constant factor of the HPA input, resulting into a signal distortion.

For example, an SC QPSK transmission, and rectangular transmit filter h_{Tx} has at the HPA input the signal

$$s(t) = A_{SC}\cos(2\pi f_0 t + \arg a_k) \tag{8.70}$$

for $t \in [kT, (k+1)T]$, having constant envelope $A(t) = A$ and we can select to work at saturation point with $S = A_{SC}$ and corresponding output back-off (OBO) $OBO = 0$ dB. Incidentally, this choice of h_{Tx} requires a very large channel bandwidth.

For an OFDM signal, instead, the envelope changes not only according to the constellation of the data symbols $\{a_k[i]\}$ but also to the number of subchannel. For example, an OFDM transmission with data symbols $a_k[i]$ still taken from a QPSK constellation, the data block $\boldsymbol{a}_k = [1+j, 1+j, \dots, 1+j]$ has IDFT $\boldsymbol{A}_k = [\mathcal{M}(1+j), 0, \dots, 0]$, and the transmitted signal (still with a rectangular transmit filter) is

$$s(t) = \begin{cases} A_{OFDM}\cos\left(2\pi f_0 t + \frac{\pi}{4}\right) & 0 \le t < \frac{T}{\mathcal{M}} \\ 0 & \frac{T}{\mathcal{M}} \le t \le T \end{cases} \tag{8.71}$$

and the first part of s has envelope $A(t) = A_{OFDM}$, while the remaining *OFDM symbol* has envelope $A(t) = 0$. For the same maximum value of s, the average power of the two configurations yields

$$\mathtt{M}_{Tx,OFDM} = \frac{1}{\mathcal{M}}\mathtt{M}_{Tx,SC} \tag{8.72}$$

Hence, the OBO of OFDM would be much higher than that of SC.

In general, in this set up with rectangular interpolator filters, we may scale $A_M[i]$ to $A'_M[i] = \mathcal{K}A_M[i]$ with $\mathcal{K}$ such that

$$\max|A'_k[i]| \le S \tag{8.73}$$

The corresponding transmitted power would be

$$\mathtt{M}_{Tx,OFDM} = E(|A'_k[i]|^2) \tag{8.74}$$

which, in general, yields a very high OBO. On the other hand, we may wonder what is the probability of that particular data sequence determining the (minimum) value of $\mathcal{K}$ such that (8.73) holds. If this probability is very low, the corresponding high value of OBO may not be justifiable.

In the literature, the *peak to average power ratio* (PAPR), defined as

$$\text{PAPR} = \frac{\max|A_k[i]|^2}{E(|A_k[i]|^2)} \tag{8.75}$$

is used as an estimate of the OBO of the HPA and various methods to limit the PAPR have been proposed in the literature [9]. However, a single parameter is not enough to fully characterize the impact of the

HPA on OFDM performance. For example, more useful performance measures would be to report the probability cumulative distribution of s or s_{Tx} for a given $\mathtt{M}_{Tx}$.

We may realize that most analytical approaches on reducing the OBO are indeed based on reducing the input back-off (IBO) as in (8.75). Evaluation of the OBO requires to introduce the non-linear transformations of the HPA whose output most of the time can be determined only by simulation.

8.7 Equalization

8.7.1 OFDM equalization

We consider the baseband equivalent channel model shown in Figure 8.16.b, where the channel impulse response (CIR) has support $\{0, 1, \dots, N_c - 1\}$, with $N_c > 1$. In this case, the orthogonality conditions for the OFDM system described in Section 8.2 are no longer satisfied; indeed, the transfer matrix $\boldsymbol{S}(z)$, defined by (8.10) and evaluated for a non-ideal channel, has in general elements different from a delay factor along the main diagonal, meaning the presence of ISI for transmission over the individual subchannels, and non-zero elements off the main diagonal, meaning the presence of ICI. A simple method to avoid equalization is based on the concept of circular convolution introduced in Section 1.2, that expresses a convolution in the time domain as product of finite length vectors in the frequency domain (see (1.42)). Using the method indicated as *Relation 2* on page 10, we extend the block of samples $\boldsymbol{A}_k$ by repeating N_{cp} elements with

$$N_{cp} \geq N_c - 1 \tag{8.76}$$

where the repeated part is denoted as *cyclic prefix* (CP) and N_{cp} is called the *CP length*. In this way, we obtain the OFDM system illustrated in Figure 8.17.

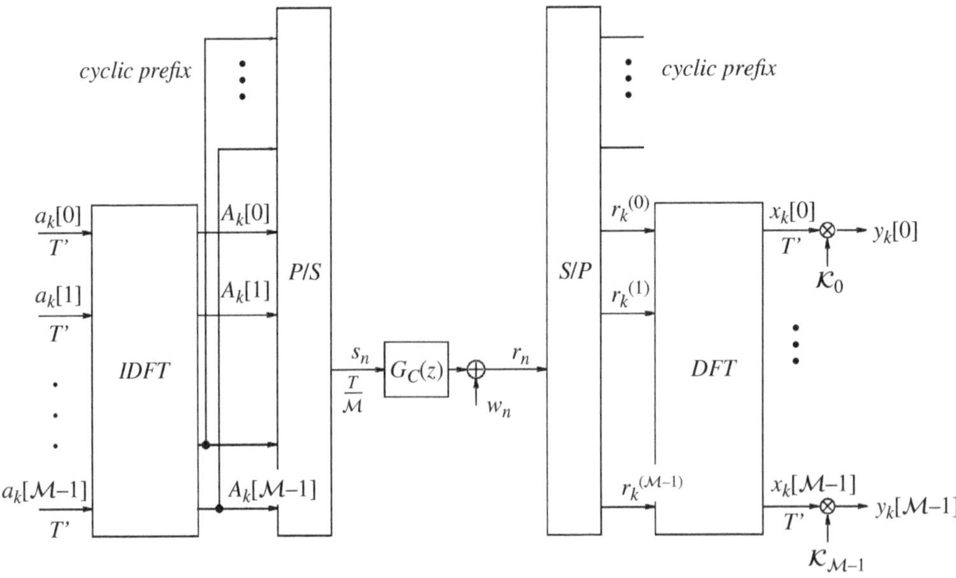

Figure 8.17 Block diagram of an OFDM system with CP and frequency-domain scaling on a per subchannel basis.

For the same channel bandwidth and hence for a given transmission rate $\mathcal{M}/T$, the IDFT (modulation) must be carried out at the rate $\frac{1}{T'} = \frac{\mathcal{M}}{(\mathcal{M}+N_{cp})T} < \frac{1}{T}$. After the modulation, each block of samples is cyclically extended by copying the N_{cp} samples $A_k[\mathcal{M}-N_{cp}],\dots,A_k[\mathcal{M}-1]$ in front of the block, as shown in Figure 8.17. After the P/S conversion, where the N_{cp} samples of the CP are the first to be sent, the $N_{cp}+\mathcal{M}$ samples are transmitted over the channel. At the receiver, blocks of samples of length $N_{cp}+\mathcal{M}$ are taken; the boundaries between blocks are set so that the last $\mathcal{M}$ samples depend only on the elements of only one cyclically extended block of samples. The first N_{cp} samples of a block are discarded.

We now recall the result (1.50). The vector $\boldsymbol{r}_k$ of the last $\mathcal{M}$ samples of the block received at the k-th modulation interval is

$$\boldsymbol{r}_k = \boldsymbol{\Xi}_k\, \boldsymbol{g}_C + \boldsymbol{w}_k \tag{8.77}$$

where $\boldsymbol{g}_C = [g_{C,0},\dots,g_{C,N_c-1},0,\dots,0]^T$ is the $\mathcal{M}$-component vector of the CIR extended with $\mathcal{M}-N_c$ zeros, $\boldsymbol{w}_k=[w_k^{(0)},\dots,w_k^{(\mathcal{M}-1)}]^T$ is a vector of additive white Gaussian noise (AWGN) samples, and $\boldsymbol{\Xi}_k$ is an $\mathcal{M}\times\mathcal{M}$ circulant matrix

$$\boldsymbol{\Xi}_k = \begin{bmatrix} A_k[0] & A_k[\mathcal{M}-1] & \cdots & A_k[1] \\ A_k[1] & A_k[0] & \cdots & A_k[2] \\ \vdots & \vdots & & \vdots \\ A_k[\mathcal{M}-1] & A_k[\mathcal{M}-2] & \cdots & A_k[0] \end{bmatrix} \tag{8.78}$$

Equation (8.77) is obtained by observing that only the elements of the first N_c columns of the matrix $\boldsymbol{\Xi}_k$ contribute to the convolution that determines the vector $\boldsymbol{r}_k$, as the last $\mathcal{M}-N_c$ elements of $\boldsymbol{g}_C$ are equal to zero. The elements of the last $\mathcal{M}-N_c$ columns of the matrix $\boldsymbol{\Xi}_k$ are chosen so that the matrix is circulant, even though they might have been chosen arbitrarily. Moreover, we observe that the matrix $\boldsymbol{\Xi}_k$, being circulant, satisfies the relation

$$\boldsymbol{F}_{\mathcal{M}}\, \boldsymbol{\Xi}_k\, \boldsymbol{F}_{\mathcal{M}}^{-1} = \begin{bmatrix} a_k[0] & 0 & \cdots & 0 \\ 0 & a_k[1] & \cdots & 0 \\ \vdots & \vdots & & \vdots \\ 0 & 0 & \cdots & a_k[\mathcal{M}-1] \end{bmatrix} = \text{diag}\{\boldsymbol{a}_k\} \tag{8.79}$$

Defining the DFT of vector $\boldsymbol{g}_C$ as

$$\boldsymbol{G}_C = [\mathcal{G}_{C,0},\mathcal{G}_{C,1},\dots,\mathcal{G}_{C,\mathcal{M}-1}]^T = \text{DFT}\{\boldsymbol{g}_C\} = \boldsymbol{F}_{\mathcal{M}}\, \boldsymbol{g}_C \tag{8.80}$$

and using (8.79), we find that the demodulator output is given by

$$\boldsymbol{x}_k = \boldsymbol{F}_{\mathcal{M}}\, \boldsymbol{r}_k = \text{diag}\{\boldsymbol{a}_k\}\, \boldsymbol{G}_C + \boldsymbol{W}_k \tag{8.81}$$

where $\boldsymbol{W}_k=[W_k[0],\dots,W_k[\mathcal{M}-1]]^T=\boldsymbol{F}_{\mathcal{M}}\boldsymbol{w}_k$ is given by the DFT of the vector $\boldsymbol{w}_k$. Relation (8.81) is illustrated in Figure 8.18 with

$$x_k[i] = a_k[i]\mathcal{G}_{C,i} + W_k[i], \quad i=0,1,\dots,\mathcal{M}-1 \tag{8.82}$$

Recalling the properties of $\boldsymbol{w}_k$, $\boldsymbol{W}_k$ is a vector of independent Gaussian r.v.s. In particular, if $w_k^{(\ell)}\sim\mathcal{CN}(0,\sigma_w^2)$ with $\sigma_w^2 = N_0\frac{\mathcal{M}}{T}$, then $W_k[i]\sim\mathcal{CN}(0,\sigma_W^2)$ with $\sigma_W^2=\mathcal{M}\sigma_w^2$.

Before detection, the signal $\boldsymbol{x}_k$ (8.81) is multiplied by the diagonal matrix $\boldsymbol{K}$, whose elements on the diagonal are given by[1]

$$\mathcal{K}_i = [\boldsymbol{K}]_{i,i} = \frac{1}{\mathcal{G}_{C,i}}, \qquad i=0,1,\dots,\mathcal{M}-1 \tag{8.83}$$

From (8.81), the input to the data detectors is given by (see Figure 8.19)

$$\boldsymbol{y}_k = \boldsymbol{K}\boldsymbol{x}_k = \boldsymbol{a}_k + \boldsymbol{W}'_k, \quad \boldsymbol{W}'_k = \boldsymbol{K}\boldsymbol{W}_k \tag{8.84}$$

[1] To be precise, the operation indicated by (8.83), rather than equalizing the signal, that is received in the absence of ISI, normalizes the amplitude and adjusts the phase of the desired signal.

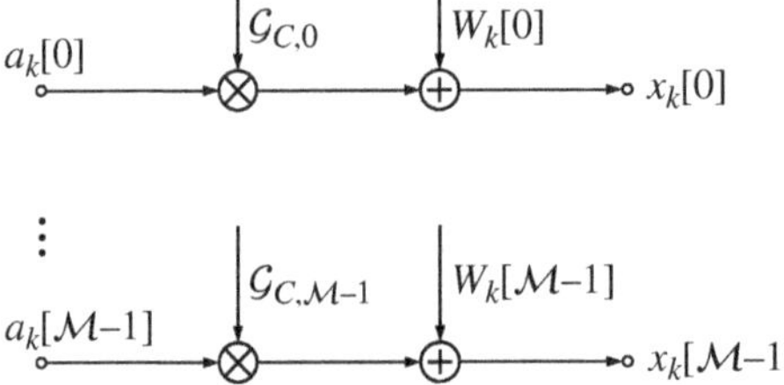

Figure 8.18 OFDM discrete-time model.

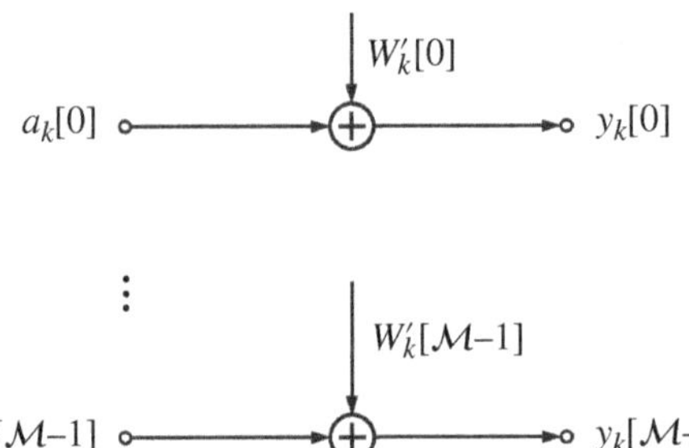

Figure 8.19 OFDM relation between transmitted data and signal at detection point.

i.e.

$$y_k[i] = a_k[i] + W'_k[i] \quad \text{with } W'_k[i] \sim \mathcal{CN}\left(0, \frac{\sigma_W^2}{|\mathcal{G}_{C,i}|^2}\right) \tag{8.85}$$

We assume that the sequence of input symbol vectors $\{\boldsymbol{a}_k\}$ is a sequence of i.i.d. random vectors. Equation (8.84) shows that the sequence $\{\boldsymbol{a}_k\}$ can be detected by assuming transmission over $\mathcal{M}$ independent and orthogonal subchannels in the presence of AWGN.

A drawback of this simple detection scheme is the reduction in the modulation rate by a factor $(\mathcal{M} + N_{cp})/\mathcal{M}$. Therefore, it is essential that the length of the CIR is much smaller than the number of subchannels, so that the reduction of the modulation rate due to the CP can be considered negligible.

To reduce the length of the CIR one approach is to equalize the channel before demodulation [8, 10, 11]. With reference to Figure 8.16b, a linear equalizer with input r_n is used; it is usually chosen as the FF filter of a decision-feedback equalizer (DFE) that is determined by imposing a prefixed length of the feedback filter, smaller than the CP length.

8.7.2 FMT equalization

We analyse three schemes.

Per-subchannel fractionally spaced equalization

We consider an FMT system with non-critically sampled transmit and receive filter banks, so that transmission within individual subchannels with non-zero excess bandwidth is achieved, and subchannel output signals obtained at a sampling rate equal to $2/T$, as discussed in Section 8.4. We recall that the frequency responses of FMT subchannels are characterized by steep roll-off towards the band-edge frequencies, where they exhibit near spectral nulls. This suggests that a per-subchannel DFE is used

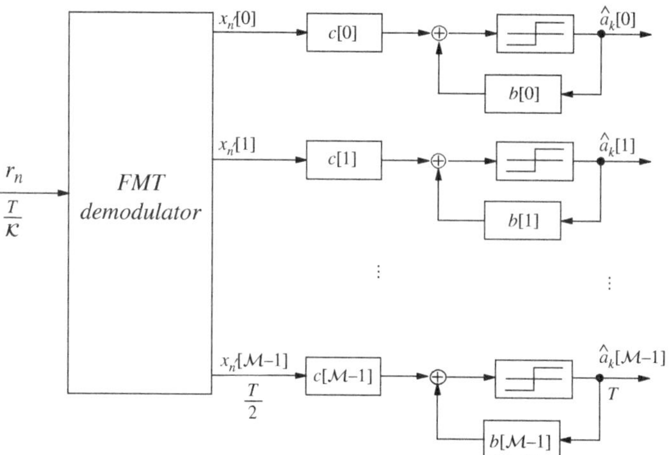

Figure 8.20 Per-subchannel equalization (DFE) for an FMT system with non-critically sampled filter banks.

to remove ISI. The block diagram of an FMT receiver employing per-subchannel fractionally spaced equalization is illustrated in Figure 8.20. Over the i-th subchannel, the DFE is designed for an overall impulse response given by

$$h_{overall,n}[i] = \sum_{n_1=-\infty}^{+\infty} g_{n\mathcal{M}-n_1}[i] \sum_{n_2=0}^{N_c-1} h_{n_1-n_2}[i]\, g_{C,n_2}, \qquad n \in \mathbb{Z} \tag{8.86}$$

In the given scheme, the $\mathcal{M}$ DFEs depend on the transmission channel. If the transmission channel is time variant, each DFE must be able to track the channel variations, or it must be recomputed periodically. Error propagation inherent to decision-feedback equalization can be avoided by resorting to precoding techniques, as discussed in Section 7.7.1.

Per-subchannel T-spaced equalization

We consider now an FMT system with critically sampled filter banks, and subchannel output signals obtained at the sampling rate of $1/T$. The high level of spectral containment of the transmit filters suggests that, if the number of subchannels is sufficiently high, and the group delay in the passband of the transmission channel is approximately constant, the frequency response of every subchannel becomes approximately a constant. In this case, the effect of the transmission channel is that of multiplying every subchannel signal by a complex value. Therefore, as for OFDM systems with CP, equalization of the transmission channel can be performed by choosing a suitable constant for every subchannel. We note however that, whereas for an OFDM system with CP, the model of the transmission channel as a multiplicative constant for each subchannel is exact, if CP is longer than CIR, for an FMT system such model is valid only as an approximation.[2] In FMT, the degree of the approximation depends on the dispersion of the transmission channel and on $\mathcal{M}$, the number of subchannels.

Assuming a constant frequency response for transmission over each subchannel, the equalization scheme is given in Figure 8.21a, where K_i is defined in (8.83), and the DFE is designed to equalize only the cascade of the transmit and receive filters. Using (8.25) and (8.29), we find that the convolution

[2] In fact, while in FMT the channel frequency response is the discrete-time Fourier transform of $g_{C,n}$, $n = 0, 1, \dots, N_c - 1$, say $\mathcal{G}_C(f)$, in OFDM we simply use the DFT of $\{g_{C,n}\}$, say $\mathcal{G}_{C,i}$, $i = 0, 1, \dots, \mathcal{M} - 1$, and from (1.28) $\mathcal{G}_{C,i} = \mathcal{G}_C(f)|_{f=\frac{i}{T}}$.

of transmit and receive subchannel filters is independent of the subchannel index. In fact, we get

$$\sum_{n_1=0}^{\infty} h_{n_1}[i]\, g_{n\mathcal{M}-n_1}[i] = \sum_{n_1=0}^{\infty} h_{n_1}\, g_{n\mathcal{M}-n_1} = h_{eq,n} \tag{8.87}$$

In this case, all DFEs are equal.

Alternative per-subchannel T-spaced equalization

An further simplification is obtained by using the scheme of Figure 8.21b. The idea is that, in the presence of a transmission channel with flat frequency response for each subchannel, a reconstruction of the signal is achieved by designing the ℓ-th polyphase component of the receive prototype filter, $g^{(\ell)}$, to equalize the corresponding polyphase component $h^{(\ell)}$ of the transmit prototype filter. In general, a DFE scheme can be used, where the ℓ-th polyphase components of the receive filters, $g_{FF}^{(\ell)}$ and $g_{FB}^{(\ell)}$, equalize the corresponding ℓ-th polyphase component $h^{(\ell)}$ of the overall subCIR.

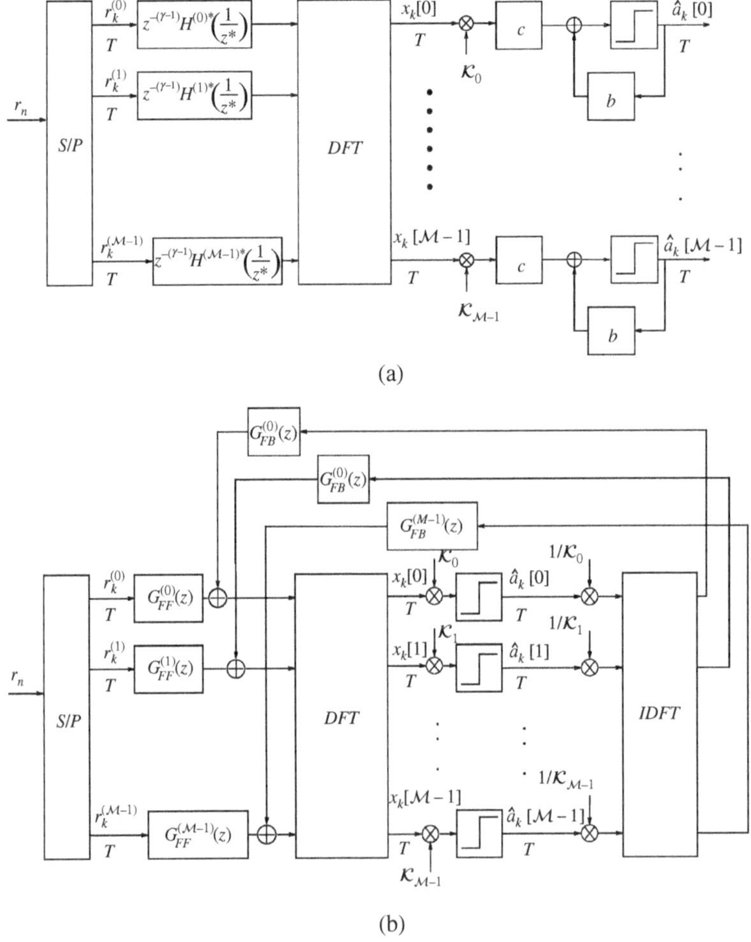

Figure 8.21 (a) Equalization scheme for FMT in the case of approximately constant frequency response for transmission over each subchannel; (b) alternative scheme.

8.8 Orthogonal time frequency space modulation

In OFDM, data symbols are allocated in frequency (subchannel index i), and time (block index k) and *equalization* is performed on a block-by-block basis. To capture the *time diversity* of the channel, the principle of orthogonal time frequency space (OTFS) modulation [12] is to collect the received signal across K blocks, and only then recover data. In fact, it may be simpler to work in the DFT-domain of time (index k), i.e. the Doppler domain with index ℓ, where the channel is less dispersive. Similarly, the channel is less dispersive in the DFT domain of frequency (index i), i.e. the delay domain described by the power delay profile with index m. Here, we should remark that equalization may not be simpler for a more compact channel.

In OTFS, modulation based on OFDM [13] data symbols are spread by a 2D-DFT in time and frequency before being transmitted by OFDM. In particular, let us collect K data blocks of size $\mathcal{M}$ into the $\mathcal{M} \times K$ matrix $\boldsymbol{a}'$. Data are pre-processed (spread) by the symplectic discrete Fourier transform (SDFT) to obtain the signals to be transmitted on frequency i at time k

$$a_k[i] = \sum_{m=0}^{\mathcal{M}-1} \sum_{\ell=0}^{K-1} a'_{m,\ell} e^{-j2\pi\left(\frac{im}{\mathcal{M}} - \frac{k\ell}{K}\right)}, \qquad i = 0, \dots, \mathcal{M}-1, \quad k = 0, \dots, K-1 \tag{8.88}$$

Note that, apart from an irrelevant sign difference in front of $k\ell$, this is simply a 2D-DFT.

OTFS equalization

If $\{s_n\}$ is the OFDM transmitted signal corresponding to data $a_k[i]$ given by (8.88), over a linear *time-variant* dispersive AWGN channel, the received signal is

$$r_n = \sum_{m=0}^{N_c-1} g_C(n,m) s_{n-m} + w_n \tag{8.89}$$

where $g_C(n,m)$, $m = 0, \dots, N_c - 1$, is the CIR at time n.

At the OFDM receiver, after the S/P converter and the DFT, matrix $\boldsymbol{X} = [\boldsymbol{x}_0, \dots, \boldsymbol{x}_{K-1}]$ is formed. Note that in OTFS systems the CP may not be used for a higher spectral efficiency. Next, matrix $\boldsymbol{X}$ goes through the 2D-DFT^{-1} as the inverse of (8.88). Let $\{\mathcal{H}_{\ell,m}\}$ be the equivalent CIR in the Doppler-delay domain, as the DFT of $\{g_C(n,m)\}$ with respect to time index n, after a suitable windowing. The recovered signal can be written as [12]

$$y_{m,\ell} = \sum_{m'=0}^{\mathcal{M}-1} \sum_{\ell'=0}^{K-1} \mathcal{H}_{\ell',m'} a'_{m-m',\ell-\ell'} + \text{noise} \tag{8.90}$$

Although the channel matrix $\mathcal{H}$ is banded, detection of data symbols from (8.90) is a formidable task, as each data symbol is affected by interference from neighbouring (both along Doppler and delay axes) symbols. Iterative cancellation techniques may be used to recover data.

The advantage of OTFS is that the receiver, besides capturing the *time diversity*, shows a reduced variability of the channel across data symbols, i.e. in the Doppler-delay domain. Hence, in OTFS coding is less important.

8.9 Channel estimation in OFDM

From the system model of Figure 8.18, given the vector $\boldsymbol{x}_k$ at the output of the DFT, in order to scale each component of $\boldsymbol{x}_k$ by the DFT of the CIR, we need an estimate of $\boldsymbol{G}_C$. Various methods to estimate $\boldsymbol{G}_C$ are now given. At first, we assume that the first OFDM symbol for $k = 0$ is a known vector of $\mathcal{M}$ symbols, $a_0[i] = \alpha[i]$, $i = 0, 1, \dots, \mathcal{M} - 1$, or $\boldsymbol{a}_0 = \boldsymbol{\alpha}$. In the OFDM literature, a known subchannel symbol is also denoted as *pilot symbol*.

For a simplified notation, we introduce the $\mathcal{M} \times \mathcal{M}$ input data matrix (see also (2.93)):

$$\mathcal{I} = \text{diag}[\alpha[0], \dots, \alpha[\mathcal{M} - 1]] \tag{8.91}$$

In correspondence to the transmission of $\boldsymbol{\alpha}$, the received block $\boldsymbol{x}_0$ is denoted as

$$\boldsymbol{o} = [o[0], \dots, o[\mathcal{M} - 1]]^T \tag{8.92}$$

If $\hat{\boldsymbol{G}}_C$ is an estimate of $\boldsymbol{G}_C$, the channel estimation error vector is defined as

$$\Delta\boldsymbol{G}_C = \hat{\boldsymbol{G}}_C - \boldsymbol{G}_C \tag{8.93}$$

From the decomposition (8.81), we have

$$\begin{aligned} \boldsymbol{o} = \boldsymbol{x}_0 &= \mathcal{I}\boldsymbol{G}_C + \boldsymbol{W}_0 \\ &= \mathcal{I}\hat{\boldsymbol{G}}_C - \mathcal{I}\Delta\boldsymbol{G}_C + \boldsymbol{W}_0 \end{aligned} \tag{8.94}$$

and defined

$$\boldsymbol{R}_{\Delta G_C} = E\,[\Delta\boldsymbol{G}_C^* \Delta\boldsymbol{G}_C^T] \tag{8.95}$$

we measure the estimate quality by the signal-to-noise-ratio (SNR) (see also (7.548))

$$\Lambda_n = \frac{E\,[||\boldsymbol{W}_0||^2]}{\sigma_a^2 E\,[||\Delta\boldsymbol{G}_C||^2]} = \frac{\mathcal{M}\sigma_W^2}{\sigma_a^2 \,\text{tr}\, \boldsymbol{R}_{\Delta G_C}} \tag{8.96}$$

assuming components of $\boldsymbol{W}_0$ are i.i.d., each with variance σ_W^2.

It will be useful to introduce a parameter related to the data constellation

$$\beta = \frac{\dfrac{1}{\mathcal{M}} \displaystyle\sum_{i=0}^{\mathcal{M}-1} \frac{1}{|\alpha[i]|^2}}{\dfrac{1}{\sigma_a^2}} = \frac{E\left[\frac{1}{|a_0[i]|^2}\right]}{\frac{1}{\sigma_a^2}} = \sigma_a^2 E\left[\frac{1}{|a_0[i]|^2}\right] \tag{8.97}$$

Note that for a constant envelope constellation (PSK) $\beta = 1$, while for a 16-QAM constellation $\beta = 17/9$.

We will also make use of the ratio

$$\eta = \frac{\sigma_a^2}{\sigma_W^2} \tag{8.98}$$

Instantaneous estimate or LS method

From (8.94), an estimate of $\boldsymbol{G}_C$ is simply given by

$$\hat{\mathcal{G}}_{C,LS,i} = \frac{o[i]}{\alpha[i]}, \qquad i = 0, 1, \dots, \mathcal{M} - 1 \tag{8.99}$$

or

$$\hat{\boldsymbol{G}}_{C,LS} = \mathcal{I}^{-1}\boldsymbol{o} \tag{8.100}$$

This estimate is also denoted as least squares (LS) because of the following correspondence. Introduced the functional

$$\mathcal{E} = ||\boldsymbol{o} - \hat{\boldsymbol{o}}||^2 = \sum_{i=0}^{\mathcal{M}-1} |o[i] - \hat{o}[i]|^2 \tag{8.101}$$

with

$$\hat{\boldsymbol{o}} = \mathcal{I}\hat{\boldsymbol{G}}_C \tag{8.102}$$

and defined matrix $\mathbf{\Phi}$ and vector $\boldsymbol{\vartheta}$ as

$$\mathbf{\Phi} = \mathcal{I}^H \mathcal{I} = [|\alpha[0]|^2, \dots, |\alpha[\mathcal{M}-1]|^2] \tag{8.103}$$

$$\boldsymbol{\vartheta} = \mathcal{I}^H \boldsymbol{o} = [\alpha^*[0]o[0], \dots, \alpha^*[\mathcal{M}-1]o[\mathcal{M}-1]]^T \tag{8.104}$$

the solution which minimizes (8.101), given by $\mathbf{\Phi}^{-1}\boldsymbol{\vartheta}$, coincides with (8.100).

The statistical parameters associated to the estimate $\hat{G}_{C,LS}$ are easily derived. From (8.104) and (8.94) we have that

$$\boldsymbol{\vartheta} = \mathbf{\Phi} G_C + \mathcal{I}^H W_0 \tag{8.105}$$

By left multiplying by $\mathbf{\Phi}^{-1}$ yields

$$\hat{\boldsymbol{G}}_{C,LS} = \boldsymbol{G}_C + \mathbf{\Phi}^{-1}\mathcal{I}^H \boldsymbol{W}_0 = \boldsymbol{G}_C + \mathcal{I}^{-1}\boldsymbol{W}_0 \tag{8.106}$$

and

$$\Delta \boldsymbol{G}_C = \mathcal{I}^{-1}\boldsymbol{W}_0 \tag{8.107}$$

hence

$$\boldsymbol{R}_{\Delta G_C} = \sigma_W^2 \mathbf{\Phi}^{-1} \tag{8.108}$$

As

$$\text{tr}\,[\mathbf{\Phi}^{-1}] = \sum_{i=0}^{\mathcal{M}-1} \frac{1}{|\alpha[i]|^2} \tag{8.109}$$

it is

$$\Lambda_n = \frac{1}{\beta} \tag{8.110}$$

Note that from (8.108)

$$E\,[|\Delta \mathcal{G}_{C,LS,i}|^2] = \frac{\sigma_W^2}{|\alpha[i]|^2} \tag{8.111}$$

In other words, the channel estimate is affected by an error equal to the channel noise scaled by the symbol value. Note that for a constant envelope constellation $\Lambda_n = 1$.

Two variants of the LS methods are as follows:

1. Transmit pilots at a power level higher than the information symbols. However, this would increase the transmit signal power spectral density which, on the other hand, must be confined within certain bounds;
2. Use more pilots in time, say $\boldsymbol{\alpha}_k$, $k = 0, 1, \dots, N_p - 1$, and

$$\hat{\mathcal{G}}_{C,LS,i} = \frac{1}{N_p} \sum_{k=0}^{N_p-1} \frac{o_k[i]}{\alpha_k[i]}, \qquad i = 0, 1, \dots, \mathcal{M}-1 \tag{8.112}$$

It is seen that the variance of estimate is reduced by a factor N_p and

$$\Lambda_n = \frac{N_p}{\beta} \tag{8.113}$$

However, this solution is seldom used due to its large redundancy.

For an estimate of the channel noise variance, from (8.101) it is

$$\hat{\sigma}_W^2 = \frac{\mathcal{E}}{\mathcal{M}} \tag{8.114}$$

Indeed, this estimate is good only if $\hat{\boldsymbol{G}}_{C,LS} \simeq \boldsymbol{G}_C$.

LMMSE

This solution makes use of the channel correlation matrix R_{G_C}. From (7.584)

$$\hat{G}_{C,LMMSE} = \left(R_o^{-1}R_{oG_C}\right)^T o \tag{8.115}$$

which, from (7.591) becomes

$$\hat{G}_{C,LMMSE} = \left[\sigma_W^2\left(R_{G_C}^*\right)^{-1} + \mathcal{I}^H\mathcal{I}\right]^{-1}\mathcal{I}^H o \tag{8.116}$$

Also, from (8.103) and (8.104)

$$\begin{aligned}\hat{G}_{C,LMMSE} &= \left[\sigma_W^2\left(R_{G_C}^*\right)^{-1} + \Phi\right]^{-1}\vartheta \\ &= R_{G_C}^*[\sigma_W^2\Phi^{-1} + R_{G_C}^*]^{-1}\hat{G}_{C,LS}\end{aligned} \tag{8.117}$$

Note that the linear minimum-mean-square-error (LMMSE) solution introduces a weighting of the LS components $\hat{G}_{C,LS}$ which depends on the ratio between σ_W^2 and the correlation matrix R_{G_C}. In particular if $\sigma_W^2\ (R_{G_C}^*)^{-1}$ is negligible w.r.t. Φ, it turns out that $\hat{G}_{C,LMMSE} = \hat{G}_{C,LS}$. Moreover, from (7.593)

$$R_{\Delta G} = \sigma_W^2\left\{\left[\sigma_W^2\left(R_{G_C}^*\right)^{-1} + \Phi\right]^*\right\}^{-1} \tag{8.118}$$

and

$$\Lambda_n = \frac{1}{\sigma_a^2\frac{1}{\mathcal{M}}\mathrm{tr}\left\{\left[\sigma_W^2\left(R_{G_C}^*\right)^{-1} + \Phi\right]^*\right\}^{-1}} \tag{8.119}$$

An approximated version of (8.117) replaces $\frac{1}{|\alpha[i]|^2}$ in Φ^{-1} with its expectation $E\left[\frac{1}{|\alpha[i]|^2}\right]$, then it is

$$\hat{G}_{C,LMMSE} = R_{G_C}^*\left[\frac{\beta}{\eta}I + R_{G_C}\right]^{-1}\hat{G}_{C,LS} \tag{8.120}$$

where the $\mathcal{M}\times\mathcal{M}$ matrix left multiplying $\hat{G}_{C,LS}$ is supposed to be known. To simplify the algorithm, R_{G_C} is sometimes evaluated for a uniform power-delay profile for a duration equal to the CP length $(N_{cp}+1)$ and a high value of η (e.g. 20 dB). In fact, σ_W^2 is a parameter to be estimated once the channel has been estimated.

Overall, we can say that the LMMSE method exploits the correlation in the frequency domain.

The LS estimate with truncated impulse response

In many simulations, it is seen that what makes the LMMSE method significantly outperforms the LS method is the use of the power delay profile length. So the idea is to take the IDFT of the estimate $\hat{G}_{C,LS}$, as given by (8.99), to obtain:

$$\hat{g}_{C,LS} = \mathrm{IDFT}\ \{\hat{G}_{C,LS}\} \tag{8.121}$$

If $\hat{g}_{C,LS} = [\hat{g}_{C,LS,0},\dots,\hat{g}_{C,LS,\mathcal{M}-1}]^T$, the next step is to keep just the first $N_{cp}+1$ samples to yield an enhanced channel estimate:

$$\hat{g}_{C,LS\text{-}enh,n} = \begin{cases}\hat{g}_{C,LS,n} & n = 0,1,\dots,N_{cp}\\ 0 & n = N_{cp}+1,\dots,\mathcal{M}-1\end{cases} \tag{8.122}$$

with DFT

$$\hat{G}_{C,LS\text{-}enh} = \mathrm{DFT}[g_{C,LS\text{-}enh}] \tag{8.123}$$

Moreover, as $g_{C,LS,n}$, $n = N_{cp}+1,\dots,\mathcal{M}-1$, is just noise, we can estimate the noise variance in time as

$$\hat{\sigma}_w^2 = \frac{1}{\mathcal{M}-N_{cp}-1}\sum_{n=N_{cp}+1}^{\mathcal{M}-1}|\hat{g}_{C,LS,n}|^2 \tag{8.124}$$

and

$$\hat{\sigma_W^2} = \mathcal{M}\hat{\sigma}_w^2 \tag{8.125}$$

It is seen that the noise in estimate $\hat{\boldsymbol{G}}_{C,LS\text{-}enh}$ is reduced by a fraction $\frac{N_{cp}+1}{\mathcal{M}}$ w.r.t. the LS method (8.99) and (see (8.110))

$$\Lambda_n = \frac{\mathcal{M}}{N_{cp}+1}\frac{1}{\beta} \tag{8.126}$$

with (see (8.100))

$$E[|\Delta\mathcal{G}_{C,LS\text{-}enh,i}|^2] = \frac{N_{cp}+1}{\mathcal{M}}\frac{\sigma_W^2}{|\alpha[i]|^2} \tag{8.127}$$

For many channel models, it is seen that the enhanced version of the LS method performs as good as the LMMSE method with a much reduced complexity.

8.9.1 Channel estimate and pilot symbols

We recall that pilots are transmitted symbols which are known at the receiver. They are used for channel estimation and synchronization (tracking). As they introduce overhead, their number should be kept to a minimum value. They are typically arranged in fixed *patterns* over the time-frequency plane.

Pilot patterns depend on the coherence bandwidth and coherence time of the channel. As a practical rule, the pilot frequency spacing should be smaller than the coherence bandwidth, while the pilot time spacing should be smaller than the coherence time.

Some pilots arrangements are shown in Figure 8.22. In the block-type (left of Figure 8.22), an entire OFDM symbol contains pilots. This configuration is repeated after many OFDM symbol periods. This arrangement is particularly recommended for slow-fading channels.

In the comb-type (right of Figure 8.22) for each OFDM symbol, some channels are dedicated to pilots. This arrangement is recommended for fast-fading channels. Other arrangements are given in Figure 8.23.

Once the pilot arrangement is chosen, (polynomial) interpolation, usually first in time next in frequency, is carried out to acquire the channel for every subchannel. Two-dimensional (time and frequency) interpolation techniques may also be used.

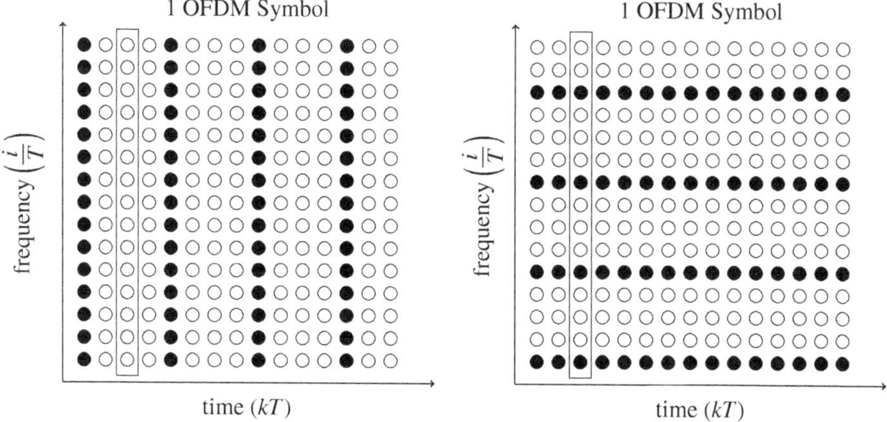

Figure 8.22 Pilot symbols arrangements in the time-frequency plane: the black circles are the pilot symbols, while the white ones are the information.

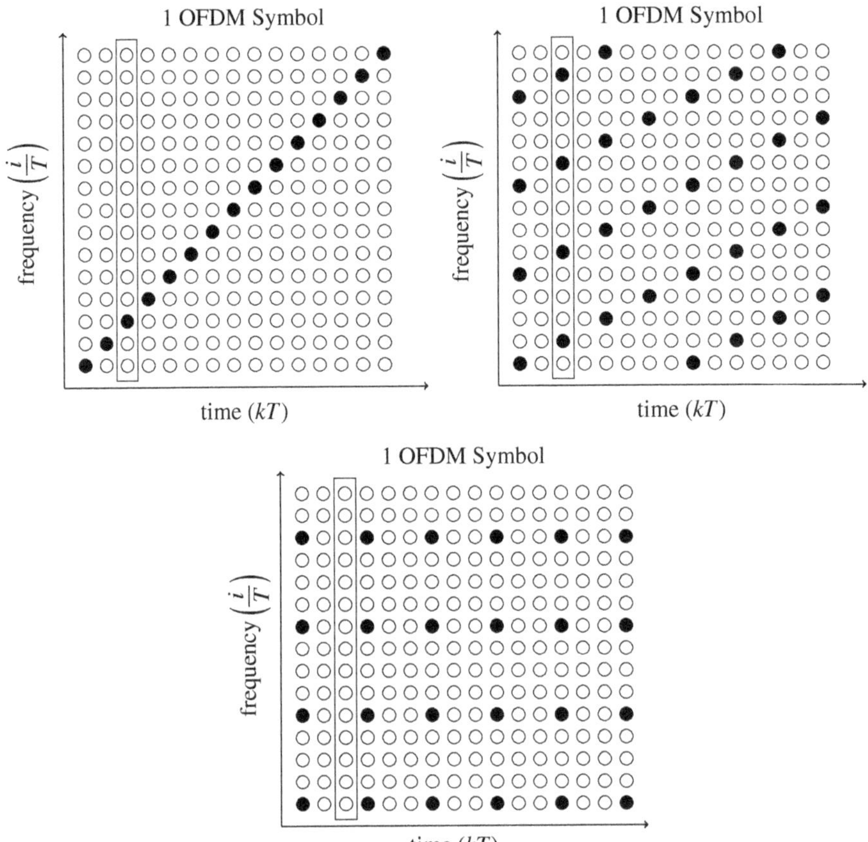

Figure 8.23 Other pilot symbols arrangements in the time-frequency plane.

8.10 Multiuser access schemes

In this section, we introduce two frequency division multiple access (FDMA) schemes that are closely related to OFDM. One solution directly uses OFDM by allocating subchannels to different users, while the other scheme, denoted SC-FDMA (or DFT-spread OFDM), merges advantages of SC and MC modulations.

8.10.1 OFDMA

The orthogonal frequency division multiple access (OFDMA) scheme is a generalization of OFDM to a multiuser context. With OFDMA, each user is assigned a subset of all subchannels. For example, for a transmission by U users, let

$$\mathcal{S}_u \subset \{0, 1, \dots, \mathcal{M} - 1\}, \quad u = 1, 2, \dots, U \tag{8.128}$$

be the subset of subchannels assigned to user u. User u will transmit data symbol on subchannels $\mathcal{S}_u$, while not transmitting anything on other subchannels, i.e. indicating with $a_{u,k}[i]$ the data symbol of user u at OFDM symbol k on subchannel i, we have $a_{u,k}[i] = 0$ for $i \notin \mathcal{S}_u$.

Typically, we have $\mathcal{S}_{u_1} \cap \mathcal{S}_{u_2} = \emptyset$, for $u_1 \neq u_2$, i.e. users transmit on different subchannels in order to avoid interference. Note that the destination of each user transmission may be a single device, such as the base station in the *uplink* of a cellular system, or different devices, including other (typically non-transmitting) users. A receiver interested in data coming from user u will implement a regular OFDM receiver, and extract only the signals of subchannels $\mathcal{S}_u$. Also, a single device (such as a base station in a *downlink* transmission to mobile terminals) can transmit to multiple users by OFDMA.

Note that the subchannel allocation can be flexibly modified, even for each OFDM block, thus providing a very easy resource allocation among the users. Indeed, OFDMA is for example adopted in the downlink of the fourth and fifth generations of the 3GPP cellular standard.

8.10.2 SC-FDMA or DFT-spread OFDM

The power inefficiency of HPAs due to the large dynamic of $A_k[i]$, as discussed in Section 8.6.2 has lead to a number of solutions. Among those, we focus here on the SC-FDMA, which aims at reducing the PAPR, while at the same time reaping the benefits of MC modulation, in particular the flexible resource allocation of OFDMA.

The scheme of SC-FDMA for a single user is shown in Figure 8.24. As for OFDMA, the $\mathcal{M}$ subchannels are assigned to U users according to (8.128). Let $\kappa_u = |\mathcal{S}_u|$, each block of data for user u, $a_{u,k}[l]$, $l = 0, \dots, \kappa_u - 1$, are first pre-processed with a DFT of size κ_u, to obtain

$$a'_{u,k}[i] = \sum_{\ell=0}^{\kappa_u - 1} a_{u,k}[\ell] e^{-j2\pi \frac{\ell i}{\kappa_u}}, \quad i = 0, \dots, \kappa_u - 1 \tag{8.129}$$

Then, $\{a'_{u,k}[i]\}$ are transmitted by an OFDMA scheme using subchannels $\mathcal{S}_u$ out of $\mathcal{M}$.

At the receiver, we apply equalization according to (8.84) to obtain

$$x'_{u,k}[i] = \frac{y_k[i]}{\mathcal{G}_{C,i}} = a'_{u,k}[i] + W'_k[i], \quad i = 0, \dots, \kappa_u - 1 \tag{8.130}$$

where $W'_k[i]$ is defined in (8.85).

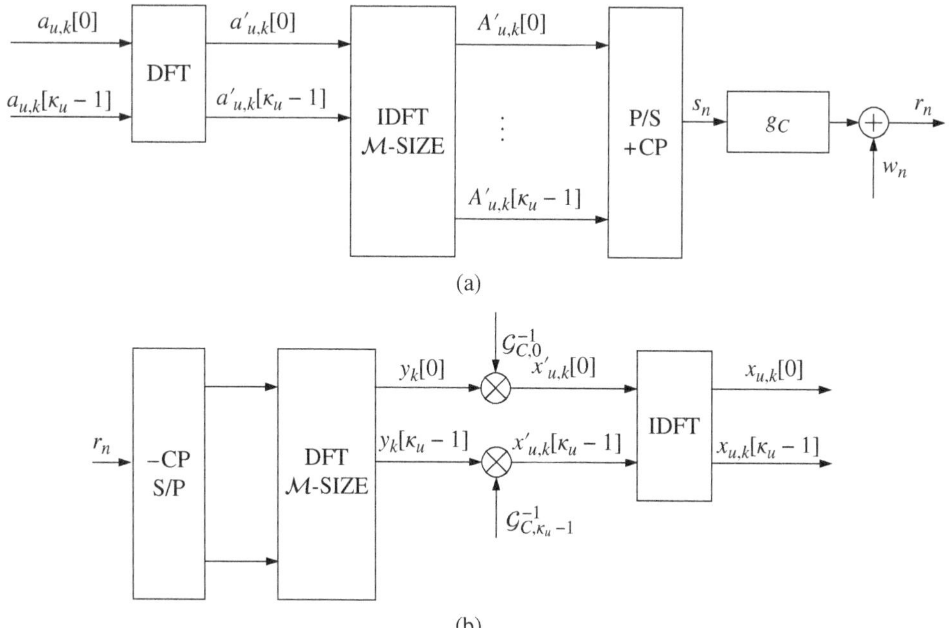

Figure 8.24 SC-FDMA scheme, for user 1: (a) transmitter, (b) receiver.

To simplify notation in Figure 8.24 and (8.130), we assume $\mathcal{S}_u$ given by the first κ_u subchannels of OFDM. Finally, we select the subchannels $\mathcal{S}_u$ on which an IDFT of size κ_u is performed to obtain

$$x_{u,k}[\ell] = \sum_{i=0}^{\kappa_u-1} x'_{u,k}[i]e^{j2\pi\frac{\ell i}{\kappa_u}} = \kappa_u a_{u,k}[\ell] + \tilde{w}_k[\ell], \quad \ell = 0,\dots,\kappa_u - 1 \tag{8.131}$$

Detection of $a_{u,k}[\ell]$ is based on $x_{u,k}[\ell]$ after a suitable scaling. Note that while $W'_k[i]$ is statistically independent across subchannels, even if with a different variance $\sigma_W^2/|\mathcal{G}_{C,i}|^2$, $\tilde{w}_k[\ell]$ turns out to be correlated across index ℓ. Its correlation matrix $\boldsymbol{R}_{\tilde{w}}$ has entry (ℓ_1, ℓ_2) given by

$$[\boldsymbol{R}_{\tilde{w}}]_{\ell_1,\ell_2} = E[\tilde{w}_k[\ell_1]\tilde{w}_k^*[\ell_2]] = \sum_{i=0}^{\kappa_u-1} \frac{\sigma_W^2}{|\mathcal{G}_{C,i}|^2} e^{j2\pi\frac{(\ell_1-\ell_2)i}{\kappa_u}} \tag{8.132}$$

As $\boldsymbol{R}_{\tilde{w}}$ is not diagonal, the above receiver is not optimum. In any case, to summarize

$$\tilde{w}_k[\ell] \sim \mathcal{CN}(0, \sigma_{\tilde{w}_k}^2) \tag{8.133}$$

with

$$\sigma_{\tilde{w}_k}^2 = \sigma_W^2 \sum_{i=0}^{\kappa_u-1} \frac{1}{|\mathcal{G}_{C,i}|^2} \tag{8.134}$$

Note that for a single user using all the subchannels, i.e. $\mathcal{S}_1 = \{0,\dots,\mathcal{M}-1\}$, the DFT precoding cancels out with the transmit IDFT, and we have an SC transmission with the CP, while at the receiver we obtain a ZF linear equalizer (see Observation 7.10) implemented in the frequency domain, exploiting the CP. Hence, performance of this scheme is typically worse than that of OFDM for a frequency selective channel. With multiple users, we can still interpret DFT precoding as the SC transmission of each user on part of the whole spectrum. On one hand, SC-FDMA reduces the PAPR, shifting from an MC to a quasi-SC transmission. On the other hand, clearly SC-FDMA retains the easy allocation of resources of OFDMA, as assigning subchannels to users simply corresponds to selecting the connection between the precoding DFT output and the transmit IDFT (and similarly at the receiver), i.e. choosing sets $\mathcal{S}_u$. SC-FDMA is a particularly useful solution when power efficiency is a concern, and when cheap HPAs are used, in order to reduce the transmitter cost. This is the case for example of smartphones in cellular systems, and indeed both the 4th and the 5th generations of the 3GPP cellular standard adopt SC-FDMA for the uplink.

8.11 Comparison between MC and SC systems

It can be shown that MC and SC systems achieve the same theoretical performance for transmission over ideal AWGN channels [14]. In practice, however, MC systems offer some considerable advantages with respect to SC systems.

- OFDM systems achieve higher spectral efficiency if the channel frequency response exhibits large attenuations at frequencies within the passband. In fact, the band used for transmission can be varied by increments equal to the modulation rate $1/T$, and optimized for each subchannel. Moreover, if the noise exhibits strong components in certain regions of the spectrum, the total band can be subdivided in two or more subbands.

- OFDM systems guarantee a higher robustness with respect to impulse noise. If the average arrival rate of the pulses is lower than the modulation rate, the margin against the impulse noise is of the order of $10 \log_{10} \mathcal{M}$ dB.
- For typical values of $\mathcal{M}$, OFDM systems achieve the same performance as SC systems with a complexity that can be lower.
- In multiple-access systems, the finer granularity of OFDM systems allows a greater flexibility in the spectrum allocation. For example, it is possible to allocate subsets of subchannels to different users, and update the allocation at each OFDM block. The same operations in SC systems would require, for each user spectrum allocation, different transmit filter, whose design and switch (in correspondence of the time fractions of blocks in OFDM) is problematic.

On the other hand, OFDM systems present also a few drawbacks with respect to SC systems.

- In OFDM systems, the transmitted signals exhibit a higher PAPR, that contributes to increase the susceptibility of these systems to non-linear distortion.
- A coding scheme is needed to overcome channels with nulls in the frequency response.
- High sensitivity to carrier frequency offset.
- Because of the block processing of samples, a higher latency is introduced by OFDM systems in the transmission of information.

However, FMT is significantly more robust than OFDM against frequency errors, because of the small spectral overlap between adjacent subchannels. Variants of the basic OFDM scheme are numerous [15], but rarely accepted in recent standards. As an exception, FMT has been adopted by the *European Terrestrial Trunked Radio* (TETRA), and is being considered for the *VHF Data Exchange System* (VDES), a basic method of data exchange in e-Navigation, an application of Internet of things (IoT) in the maritime environment [16].

With significant limitations of the HPA in consumer products, SC-FDMA represents an attractive alternative to OFDM, as it limits PAPR, while providing a simple resource allocation, although with suboptimal equalization.

8.12 Other MC waveforms

We conclude this section by mentioning some MC waveforms that have attracted attention recently. They can see as variants of the OFDM system, with some filtering operation operated either in the frequency or time domain. These solutions have been recently considered for the fifth generation (5G) of mobile communication systems, but up to release 15 only OFDM and SC-FDMA have been included in the 3GPP standard.

The filterbank multicarrier (FBMC) modulation [17] is obtained as the combination of a filter bank with offset QAM (OQAM), resulting into $2\mathcal{M}$ sub-carriers, each carrying a real symbol, having interference limited to two adjacent subchannels.

The generalized frequency division multiplexing (GFDM) [15] can be seen as a FMT system wherein each subchannel signal is filtered by a circular filter, and transmission includes a CP. It can be also seen as a linear-precoded OFDM system, where precoding is performed by a Toeplitz matrix.

The universal filtered MC (UFMC) scheme [18] generalizes the OFDM system by replacing the CP with the ramp-up and ramp-down of suitably designed filters. Therefore, UFMC can be obtained by cascading an OFDM system with a transmit filter operating in time and on each OFDM block.

Bibliography

[1] Vaidyanathan, P.P. (1993). *Multirate Systems and Filter Banks*. Englewood Cliffs, NJ: Prentice-Hall.

[2] Bellanger, M.G., Bonnerot, G., and Coudreuse, M. (1976). Digital filtering by polyphase network: application to sample-rate alteration and filter banks. *IEEE Transactions on Acoustics, Speech, and Signal Processing* ASSP-24: 109–114.

[3] Cherubini, G., Eleftheriou, E., Ölçer, S., and Cioffi, J.M. (2000). Filter bank modulation techniques for very high-speed digital subscriber lines. *IEEE Communications Magazine* 38: 98–104.

[4] Cherubini, G., Eleftheriou, E., and Ölcer, S. (2002). Filtered multitone modulation for very-high-speed digital subscriber lines. *IEEE Journal on Selected Areas in Communications*. 20: (5) 1016–1028.

[5] Weinstein, S.B. and Ebert, P.M. (1971). Data transmission by frequency-division multiplexing using the discrete Fourier transform. *IEEE Transactions on Communications* 19: 628–634.

[6] Bingham, J.A.C. (1990). Multicarrier modulation for data transmission: an idea whose time has come. *IEEE Communications Magazine* 28: 5–14.

[7] Sari, H., Karam, G., and Jeanclaude, I. (1995). Transmission techniques for digital terrestrial TV broadcasting. *IEEE Communications Magazine* 33: 100–109.

[8] Chow, J.S., Tu, J.C., and Cioffi, J.M. (1991). A discrete multitone transceiver system for HDSL applications. *IEEE Journal on Selected Areas in Communications* 9: 895–908.

[9] Rahmatallah, Y. and Mohan, S. (2013). Peak-to-average power ratio reduction in OFDM systems: a survey and taxonomy. *IEEE Communications Surveys Tutorials* 15: 1567–1592.

[10] Melsa, P.J.W., Younce, R.C., and Rohrs, C.E. (1996). Impulse response shortening for discrete multitone transceivers. *IEEE Transactions on Communications* 44: 1662–1672.

[11] Baldemair, R. and Frenger, P. (2001). A time-domain equalizer minimizing intersymbol and intercarrier interference in DMT systems. *Proceedings of GLOBECOM '01*, San Antonio, TX, USA.

[12] Hadani, R., Rakib, S., Tsatsanis, M. et al. (2017). Orthogonal time frequency space modulation. *Proceedings IEEE Wireless Communications and Networking Conference (WCNC)*, pp. 16.

[13] Farhang, A., RezazadehReyhani, A., Doyle, L.E., and Farhang-Boroujeny, B. (2018). Low complexity modem structure for OFDM-based orthogonal time frequency space modulation. *IEEE Wireless Communications Letters* 7: 344–347.

[14] Cioffi, J.M., Dudevoir, G.P., Eyuboglu, M.V., and Forney, G.D. Jr. (1995). MMSE decision-feedback equalizers and coding. Part I and Part II. *IEEE Transactions on Communications* 43: 2582–2604.

[15] Michailow, N., Matthé, M., Gaspar, I.S. et al. (2014). Generalized frequency division multiplexing for 5th generation cellular networks. *IEEE Transactions on Communications* 62: 3045–3061.

[16] Hu, Q., Jing, X., Xu, L. et al. (2018). Performance specifications of filtered multitone modulation for the roll-off factor and filter order in the maritime VHF data exchange system. *2018 IEEE International Conference on Smart Internet of Things (SmartIoT)*, pp. 87–92.

[17] Foch, B.L., Alard, M., and Berrou, C. (1995). Coded orthogonal frequency division multiplex. *Proceedings of the IEEE* 83: 982–996.

[18] Vakilian, V., Wild, T., Schaich, F. et al. (2013). Universal-filtered multi-carrier technique for wireless systems beyond LTE. *2013 IEEE Globecom Workshops (GC Wkshps)*, pp. 223–228.

Chapter 9

Transmission over multiple input multiple output channels

In this chapter, we discuss several transmit/receive techniques over multiple input multiple output (MIMO) channels. The chapter will focus on single carrier (SC) systems over narrowband (NB) MIMO channels. More interesting, the same models hold for a subchannel of a multicarrier (MC) system over a dispersive MIMO channel. As a reference scenario, we will consider wireless links (for which the MIMO channel model was also introduced in Section 4.1.11), where inputs are given by transmitting antennas and outputs are given by receiving antennas. Still, MIMO techniques find applications also in other communication contexts, for example optical fibres and acoustic communications, where antennas are replaced by optical and acoustic devices, respectively. Moreover, a MIMO channel can also model a (wireless) MC system affected by inter-carrier interference, wherein inputs and outputs are subchannels, rather than physical devices.

In the first part of the chapter, we will consider a single transmitter communicating with a single receiver, where each device is equipped with multiple antennas. Therefore, transmitted (received) signals can be jointly processed at the transmitter (receiver). In the second part of the chapter, we will consider instead multiple transmitters (each possibly equipped with multiple antennas) transmitting independent data signals to multiple receivers (again, each possibly equipped with multiple antennas), wherein transmitted (received) signals can be processed only locally, i.e. at a single transmitter (receiver).

9.1 The MIMO NB channel

For a SC transmit signal, we start by developing a *discrete-time* model for the MIMO NB channel. We recall the baseband equivalent model of an SC single input single output (SISO) system given in Figure 7.10, where we assume that h_{Tx} is a square-root Nyquist pulse and g_{Rc} is matched to h_{Tx}. Moreover, the channel is non-dispersive, i.e. flat fading, with channel impulse response

$$g_C(\tau) = \mathcal{G}_C^{(1,1)} \delta(\tau) \tag{9.1}$$

where $\mathcal{G}_C^{(1,1)}$ is the channel gain.

If we denote by $s_k^{(1)}$ the *transmitted signal*, i.e. the digital signal at rate $1/T$ modulated by h_{Tx}, and by $x_k^{(1)}$ the signal at the receiver sampler, their relation is illustrated in Figure 9.1.

Let $\{a_k\}$ be the *generated data* at the output of the bit mapper, in general a quadrature amplitude modulation (QAM) data stream. While in Chapter 7 it was $s_k^{(1)} = a_k$, here the relation between transmitted

Algorithms for Communications Systems and their Applications, Second Edition.
Nevio Benvenuto, Giovanni Cherubini, and Stefano Tomasin.

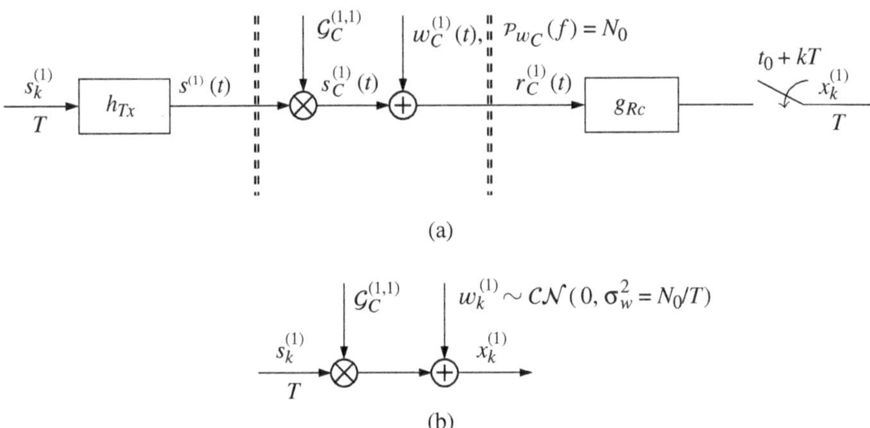

Figure 9.1 (a) Relation between the transmitted digital signal and the discrete-time received signal for a SISO system and flat fading channel, (b) equivalent discrete-time model of a SISO flat fading channel.

samples and generated data may be more complex, as it will be seen. For a SISO system, we define the channel signal-to-noise-ratio (SNR) as (see (6.69))

$$\Gamma = \frac{E\,[|s_C^{(1)}(t)|^2]}{N_0 \frac{1}{T}} = \frac{E\,[|s^{(1)}(t)|^2]}{N_0 \frac{1}{T}} E\,[|\mathcal{G}_C^{(1,1)}|^2] \tag{9.2}$$

where $s^{(1)}$ is the signal at the channel input, and $s_C^{(1)}(t) = \mathcal{G}_C^{(1,1)} s^{(1)}(t)$. As the convolution between h_{Tx} and g_{Rc} yields a Nyquist pulse, for a suitable timing phase t_0 and assuming filters h_{Tx} and g_{Rc} with energy T and $1/T$, respectively (see also (7.106)), it is (see Figure 9.1b)

$$x_k^{(1)} = \mathcal{G}_C^{(1,1)} s_k^{(1)} + w_k^{(1)} \tag{9.3}$$

with

$$w_k^{(1)} \sim \mathcal{CN}\left(0, \sigma_w^2 = \frac{N_0}{T}\right) \tag{9.4}$$

We recall from (1.324) the relation

$$\mathtt{M}_{Tx} = E\,[|s^{(1)}(t)|^2] = E\,[|s_k^{(1)}|^2] \frac{E_{h_{Tx}}}{T} \tag{9.5}$$

Hence, for $E_{h_{Tx}} = T$, the transmit statistical power is equal to the statistical power of $s_k^{(1)}$,

$$\mathtt{M}_{Tx} = E\,[|s_k^{(1)}|^2] \tag{9.6}$$

and

$$\Gamma = \frac{\mathtt{M}_{Tx}}{\sigma_w^2} E\,[|\mathcal{G}_C^{(1,1)}|^2] \tag{9.7}$$

The ratio between $\mathtt{M}_{Tx}$ and σ_w^2

$$\rho = \frac{\mathtt{M}_{Tx}}{\sigma_w^2} \quad \text{with} \quad \Gamma = \rho E\,[|\mathcal{G}_C^{(1,1)}|^2] \tag{9.8}$$

will be useful. Often it is assumed

$$E\,[|\mathcal{G}_C^{(1,1)}|^2] = 1 \tag{9.9}$$

i.e. $\mathcal{G}_C^{(1,1)}$ models the channel fast fading component, while path loss due to distance and shadowing are included in ρ. Moreover, for a Rayleigh fading channel

$$\mathcal{G}_C^{(1,1)} \sim \mathcal{CN}(0,1) \tag{9.10}$$

The extension of Figure 9.1 to a MIMO channel with N_{Tx} transmit antennas and N_{Rc} receive antennas is straightforward, as illustrated in Figure 9.2. Here the channel impulse response between the transmit antenna t and the receive antenna r is given by $\mathcal{G}_C^{(r,t)}\delta(\tau)$. Assuming h_{Tx} and g_{Rc} as in Figure 9.1, the discrete-time model of a MIMO system is reported in Figure 9.2b, where

$$w_k^{(r)} \sim \mathcal{CN}(0,\sigma_w^2), \quad \sigma_w^2 = N_0 \frac{1}{T} \tag{9.11}$$

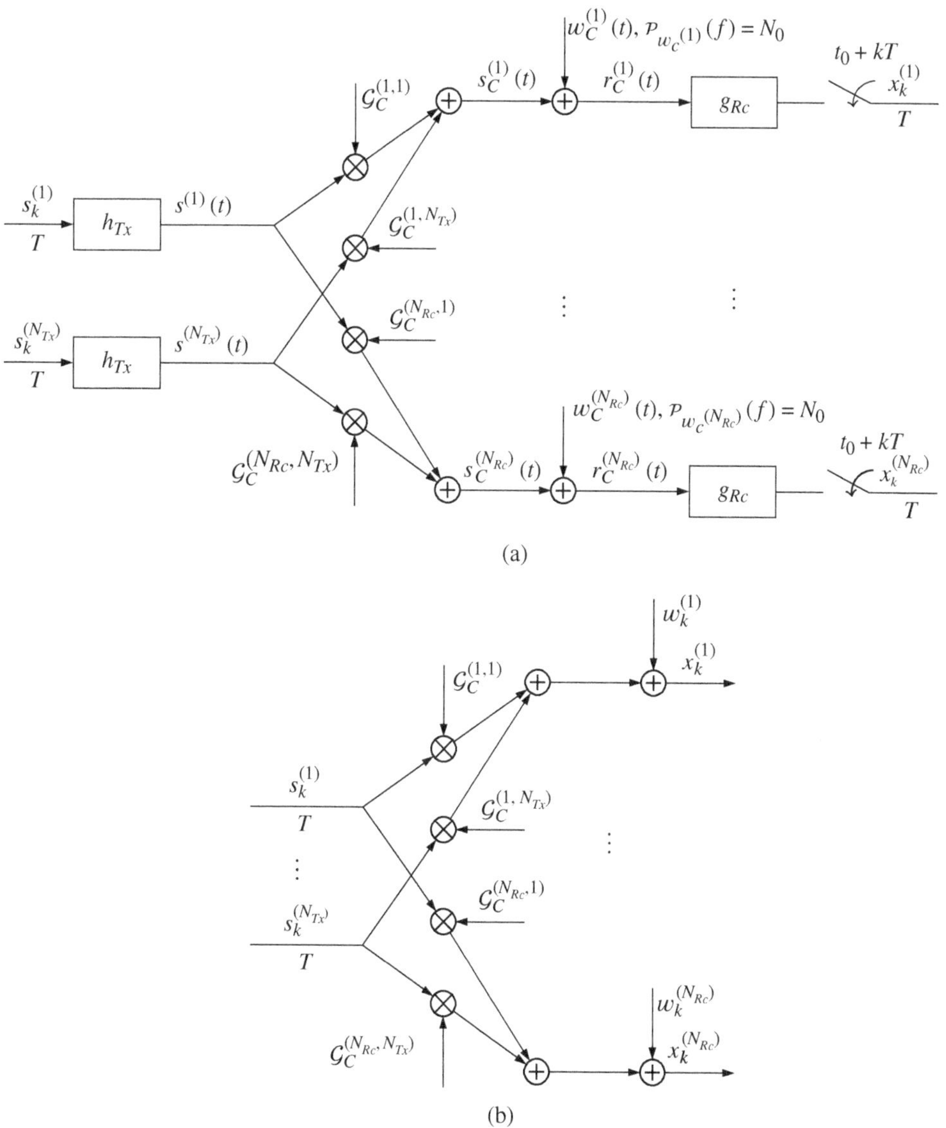

Figure 9.2 (a) A MIMO system and flat fading channel, (b) equivalent discrete-time model.

are uncorrelated across receive antennas, i.e.

$$E[w_k^{(r_1)}(w_k^{(r_2)})^*] = 0 \quad \text{if } r_2 \neq r_1 \tag{9.12}$$

Hence, the correlation matrix of

$$\boldsymbol{w}_k = [w_k^{(1)}, \dots, w_k^{(N_{Rc})}]^T \tag{9.13}$$

is

$$\boldsymbol{R}_w = E[\boldsymbol{w}_k \boldsymbol{w}_k^H] = \sigma_w^2 \boldsymbol{I} \tag{9.14}$$

Let

$$\boldsymbol{s}_k = [s_k^{(1)}, \dots, s_k^{(N_{Tx})}]^T, \quad N_{Tx} \times 1 \tag{9.15}$$

$$\mathcal{G}_C = \begin{bmatrix} \mathcal{G}_C^{(1,1)} & \mathcal{G}_C^{(1,2)} & \dots & \mathcal{G}_C^{(1,N_{Tx})} \\ \mathcal{G}_C^{(2,1)} & \mathcal{G}_C^{(2,2)} & \dots & \mathcal{G}_C^{(2,N_{Tx})} \\ \vdots & \vdots & \ddots & \vdots \\ \mathcal{G}_C^{(N_{Rc},1)} & \mathcal{G}_C^{(N_{Rc},2)} & \dots & \mathcal{G}_C^{(N_{Rc},N_{Tx})} \end{bmatrix}, \quad N_{Rc} \times N_{Tx} \tag{9.16}$$

$$\boldsymbol{x}_k = [x_k^{(1)}, \dots, x_k^{(N_{Rc})}]^T, \quad N_{Rc} \times 1 \tag{9.17}$$

it is

$$\boldsymbol{x}_k = \mathcal{G}_C \boldsymbol{s}_k + \boldsymbol{w}_k \tag{9.18}$$

The transmitted signals across antennas are assumed zero mean and with correlation matrix $\boldsymbol{R}_s = E[\boldsymbol{s}_k \boldsymbol{s}_k^H]$. The total transmit power is given by

$$\mathtt{M}_{Tx} = \text{tr}[\boldsymbol{R}_s] \tag{9.19}$$

Definition of ρ is sill given by (9.8).

Concerning the statistical model of channel matrix $\mathcal{G}_C$, see Chapter 4.

For a MC transmit signal, namely orthogonal frequency division multiplexing (OFDM), and a SISO configuration, we recall the discrete-time model of Figure 8.17. Its extension to a MIMO configuration needs some notation update. Here i is the subchannel index, and, for example, $s_k^{(t)}[i]$ is the transmitted digital signal on antenna t, subchannel i. Similarly to (9.15), (9.16), (9.13), and (9.17), we define $\boldsymbol{s}_k[i]$, $\mathcal{G}_C[i]$, $\boldsymbol{W}_k[i]$, and $\boldsymbol{x}_k[i]$, respectively. It holds (see (8.80))

$$\boldsymbol{x}_k[i] = \mathcal{G}_C[i]\boldsymbol{s}_k[i] + \boldsymbol{W}_k[i], \quad i = 0, \dots, \mathcal{M} - 1 \tag{9.20}$$

where $\mathcal{G}_C[i]$ is the $N_{Rc} \times N_{Tx}$ MIMO channel matrix for subchannel i. In particular, if $g_{C,n}^{(r,t)}$, $n = 0, 1, \dots, N_c^{(r,t)} - 1$, is the impulse response between the transmit antenna t and the receive antenna r, it is

$$\{\mathcal{G}_C^{(r,t)}[i], i = 0, \dots, \mathcal{M} - 1\} = \text{DFT}\{g_{C,n}^{(r,t)}\} \tag{9.21}$$

One word of caution is due concerning the noise

$$\boldsymbol{W}_k[i] = [W_k^{(1)}[i], \dots, W_k^{(N_{Rc})}[i]]^T \tag{9.22}$$

with

$$W_k^{(r)}[i] \sim \mathcal{CN}(0, \sigma_W^2), \qquad \sigma_W^2 = \mathcal{M}\sigma_w^2 \tag{9.23}$$

where σ_w^2 is the channel noise variance in the time domain as in (9.3).

Indicating the statistical power across all transmit antennas, subchannel i, as

$$\mathtt{M}_{Tx}[i] = \sum_{t=1}^{N_{Tx}} \sigma_{s^{(t)}[i]}^2, \qquad i = 0, 1 \dots, \mathcal{M} - 1 \tag{9.24}$$

the total transmit power is

$$\mathtt{M}_{Tx} = \sum_{i=0}^{\mathcal{M}-1} \mathtt{M}_{Tx}[i] \tag{9.25}$$

Depending upon the application, we may use either $\mathtt{M}_{Tx}$ or $\mathtt{M}_{Tx}[i]$, hence (9.8) becomes

$$\rho[i] = \frac{\mathtt{M}_{Tx}[i]}{\sigma_W^2} \quad \text{or} \quad \rho = \frac{\mathtt{M}_{Tx}}{\sigma_W^2} \tag{9.26}$$

To simplify the notation, often we drop both the subchannel index i and the time index k, hence (9.18) and (9.20) merge into the general equation

$$\boldsymbol{x} = \mathcal{G}_C \boldsymbol{s} + \boldsymbol{w} \tag{9.27}$$

We stress the fact that in OFDM the noise in (9.27) is in the frequency domain. For the rest, all considerations about (9.18) hold also about (9.20).

Figure 9.3 shows a block diagram of (9.27), where the double lines describe vector signals. In this figure, we also indicate the size of vector signals, a notation that is drop in the following.

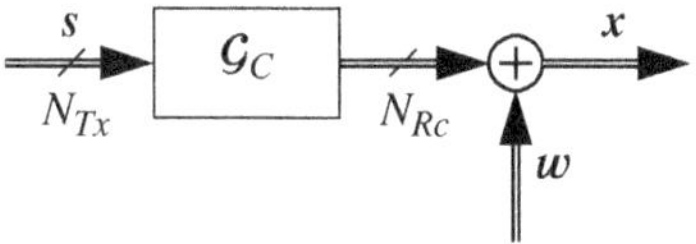

Figure 9.3 NB MIMO channel model.

Spatial multiplexing and spatial diversity

The transmission of $\boldsymbol{s}_k$ is typically associated to N_S data streams, $a_{1,k}, \ldots, a_{N_S,k}$, and this feature goes under the name of *spatial multiplexing*. The generic symbol $a_{n,k}$ takes values in a QAM constellation with minimum distance 2. Hence, the power of $a_{n,k}$, σ_a^2, depends on the constellation size. Moreover, it is $N_S \leq N_{S,max}$, with

$$N_{S,max} = \min\{N_{Tx}, N_{Rc}\} \tag{9.28}$$

This fact will be further analysed in Section 9.2.2. Note that the N_S streams can also be received by multiple users, that turn out to be separated in space (rather than in time or frequency), providing a space-division multiple access (SDMA) scheme. This use of MIMO will be further discussed starting from Section 9.6.

We also need to set a bound on the transmit power $\mathtt{M}_{Tx}$. Here we choose

$$\mathtt{M}_{Tx} \leq N_S \sigma_a^2 \tag{9.29}$$

However, other bounds can be used. This basically corresponds to assigning power σ_a^2 to each stream or equivalently

$$\mathrm{tr}[\boldsymbol{R}_s] \leq N_S \sigma_a^2 \tag{9.30}$$

In a transmission with a single transmit antenna and multiple receive antennas, the signal travels over N_{Rc} channels, and if the distance between receive antennas is large enough and scatterers are numerous, channels are significantly different, if not statistically independent (see Section 4.1.11). Therefore, by using more receive antennas, there is a higher probability of having channels with high gains, thus potentially providing a high SNR (9.8) for some of the receive antennas. Proper operations at the receiver (such as antenna selection, combining of signals from multiple antennas) yield the benefits of multiple channels by providing a high SNR at the detection point. This feature of the MIMO channel goes under the name of *spatial diversity*. In particular, with one transmit antenna and multiple receive antennas (single input multiple output, SIMO), we have *receive diversity*. Similar observations hold when the same data symbol is transmitted by multiple antennas, while being received with a single antenna (multiple input single output, MISO) and we have *transmit diversity*.

The concepts of multiplexing and diversity find also further explanation in Section 6.5.4, where the capacity of a MIMO system is discussed.

Interference in MIMO channels

Transmission over the MIMO channel in general requires suitable signal processing techniques at either or both the transmitter and the receiver. A linear combination of the transmitted samples with complex coefficients given by matrix $\mathcal{G}_C$ is received by each receive antenna. Therefore, if we simply transmit a different data symbol over each transmit antenna, the N_{Rc} received samples will exhibit severe interference. The operations required to avoid interference are tailored to $\mathcal{G}_C$; therefore, it is essential to know the channel state information (CSI) at either the transmitter or the receiver, or at both communication ends. In the following, we consider various MIMO communication techniques, classified according to the device where CSI is available.

9.2 CSI only at the receiver

Various approaches can be adopted when CSI is available at the receiver only. Linear operations, which will be written as multiplications of $\boldsymbol{x}$ by suitably designed matrices, are denoted as *combining* techniques. More in general, non-linear solutions will be denoted as *non-linear detection and decoding*. Note that in the literature combining is also denoted as *receive beamforming* or simply *beamforming*, which however may also refer to transmitter processing (more appropriately denoted as *transmit beamforming*). In order to avoid ambiguities, we will only use *combining* referring to linear operations performed at the receiver. Operations at the transmitter are instead denoted as *precoding*.

9.2.1 SIMO combiner

The simplest MIMO scheme using combining is SIMO, with only one transmit antenna ($N_{Tx} = 1$) and multiple receive antennas ($N_{Rc} > 1$). In this case, we will transmit a single data symbol at each time, thus $s = a$. Moreover, $\mathcal{G}_C$ becomes an N_{Rc}-size column vector and (9.27) is

$$\boldsymbol{x} = \mathcal{G}_C a + \boldsymbol{w} \tag{9.31}$$

Receive antenna r ($r = 1, \dots, N_{Rc}$), sees a scaled and phase-shifted version (by $\mathcal{G}_C^{(r,1)}$) of s.

Applying a combiner means multiplying $\boldsymbol{x}$ by vector $\boldsymbol{C}^T = [C^{(1,1)}, \dots, C^{(1,N_{Rc})}]$, obtaining

$$y = \boldsymbol{C}^T \boldsymbol{x} = \boldsymbol{C}^T \mathcal{G}_C a + \boldsymbol{C}^T \boldsymbol{w} \tag{9.32}$$

Based on y, detection of a follows. Since the transmit signal is in this case a scalar, the combiner output is again a scalar signal. From (9.32), note that the combiner vector $\boldsymbol{C}$ has effects also on the noise. The resulting scheme is shown in Figure 9.4.

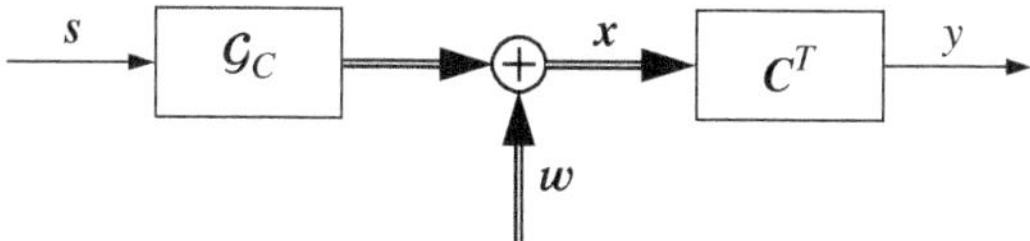

Figure 9.4 SIMO combiner.

From (6.75), the SNR at detection point (which we recall yields the probability of symbol error) can be immediately written as[1]

$$\gamma = \frac{|C^T \mathcal{G}_C|^2}{\frac{1}{2}||C||^2 \sigma_w^2} \tag{9.35}$$

Let

$$\gamma_{MF} = \frac{2}{\sigma_w^2} \tag{9.36}$$

be the SNR at the detection point for an ideal additive white Gaussian noise (AWGN) channel. It is

$$\gamma = \gamma_{MF} \frac{|C^T \mathcal{G}_C|^2}{||C||^2} \tag{9.37}$$

As expected, multiplying C by any non-zero constant does not change γ. On the choice of C various methods are now listed.

1. Selective combining: Only one of the received signals is selected. Let r_{SC} be the antenna corresponding to the received signal with highest power,

$$r_{SC} = \arg \max_{r \in \{1,\dots,N_{Rc}\}} \mathtt{M}_{x^{(r)}} \tag{9.38}$$

where the different powers may be estimated using a training sequence. Based on the decision (9.38), the receiver selects the antenna r_{SC} and consequently extracts the signal $x^{(r_{SC})}$ aligned in phase.

With reference to (9.32), this method is equivalent to setting

$$C^{(1,r)} = \begin{cases} \mathcal{G}_C^{(r,1)*} & r = r_{SC} \\ 0 & r \neq r_{SC} \end{cases} \tag{9.39}$$

and from (9.32)

$$y = |\mathcal{G}_C^{(r_{SC},1)}|^2 \; a + \mathcal{G}_C^{(r_{SC},1)*} \; w^{(r_{SC})} \tag{9.40}$$

At the detection point, we have

$$\gamma_{SC} = \gamma_{MF} |\mathcal{G}_C^{(r_{SC},1)}|^2 \tag{9.41}$$

A variant of this technique provides to select a subset of the signals $\{x^{(r)}\}$; next, their combination takes place.

2. Switched combining: Another antenna is selected only when the statistical power of the received signal drops below a given threshold. Once a new antenna is selected, the signal is processed as in the previous case.

---

[1] If A is an $N \times M$ matrix, $||A||$ denotes its Frobenius norm,

$$||A|| = \left[\sum_{n=1}^{N} \sum_{m=1}^{M} |[A]_{m,n}|^2 \right]^{1/2} \tag{9.33}$$

which extends the classical Euclidean norm of a vector. Also

$$||A||^2 = \text{tr}[AA^H] = \text{tr}[A^* A^T] \tag{9.34}$$

3. Equal gain combining (EGC): In this case, the signals are only aligned in phase; therefore, we have

$$C^{(1,r)} = e^{-j\arg(\mathcal{G}_C^{(r,1)})}, \qquad r = 1, \dots, N_{Rc} \tag{9.42}$$

It results

$$y = a\left(\sum_{r=1}^{N_{Rc}} |\mathcal{G}_C^{(r,1)}|\right) + \sum_{r=1}^{N_{Rc}} e^{-j\arg(\mathcal{G}_C^{(r,1)})}\, w^{(r)} \tag{9.43}$$

which yields

$$\gamma_{EGC} = \gamma_{MF} \frac{\left(\sum_{r=1}^{N_{Rc}} |\mathcal{G}_C^{(r,1)}|\right)^2}{N_{Rc}} \tag{9.44}$$

This technique is often used in receivers of differential phase-shift-keying (DPSK) signals (see Section 16.1.1); in this case, the combining is obtained by summing the differentially demodulated signals on the various receive antennas.

4. Maximal ratio combining (MRC): The maximal ratio combining (MRC) criterion consists in maximizing the SNR at the detection point. From (9.35),

$$\gamma = \gamma_{MF} \frac{\left|\sum_{r=1}^{N_{Rc}} C^{(1,r)}\, \mathcal{G}_C^{(r,1)}\right|^2}{\sum_{r=1}^{N_{Rc}} |C^{(1,r)}|^2} \tag{9.45}$$

Using the Schwarz inequality (see (1.353)), (9.45) is maximized for

$$C^{(1,r)} = \mathcal{K}\, \mathcal{G}_C^{(r,1)*}, \qquad r = 1, \dots, N_{Rc} \tag{9.46}$$

where $\mathcal{K}$ is a constant and

$$\gamma_{MRC} = \gamma_{MF} \sum_{r=1}^{N_{Rc}} |\mathcal{G}_C^{(r,1)}|^2 \tag{9.47}$$

Introducing the SNR of the r-th branch

$$\gamma^{(r)} = \gamma_{MF} |\mathcal{G}_C^{(r,1)}|^2 \tag{9.48}$$

(9.47) can be written as

$$\gamma_{MRC} = \sum_{r=1}^{N_{Rc}} \gamma^{(r)} \tag{9.49}$$

that is, the total SNR is the sum of the SNR's of the individual branches.

5. MMSE optimum combining (OC): The system model (9.31) can be extended with $\boldsymbol{w}$ modelling besides white noise also interference due to other users. If a statistical model of interference is known, the coefficients $\{C^{(1,r)}\}$, $r = 1, \dots, N_{Rc}$, can be determined by the Wiener formulation, which yields the minimum mean square error (MMSE) between y, the sample at the decision point, and a, the desired symbol. Recalling the expression of the autocorrelation matrix

$$\boldsymbol{R} = E[\boldsymbol{x}^* \boldsymbol{x}^T] \tag{9.50}$$

and the crosscorrelation vector

$$\boldsymbol{p} = E[a\, \boldsymbol{x}^*] \tag{9.51}$$

the optimum solution is given by

$$\boldsymbol{C}_{opt}^{T} = \boldsymbol{R}^{-1}\,\boldsymbol{p} \tag{9.52}$$

In the literature, the OC criterion is known as maximizing the ratio between the power σ_a^2 of the desired signal and the power of the residual interference plus noise (signal to interference plus noise ratio, SINR) at detection point. In fact SINR $\propto 1/\text{MMSE}$.

As an alternative, or next to a first solution through the direct method (9.52), the least-mean-square (LMS) or recursive-least-square (RLS) iterative methods may be used (see Chapter 3), that present the advantage of tracking the system variations.

Equalization and diversity

Techniques of the previous section can be extended to a transmission over multipath channels using, instead of a single coefficient $C^{(1,r)}$ per antenna, a filter with N coefficients. Introducing the vector whose elements are the coefficients of the N_{Rc} filters, that is

$$\boldsymbol{C}^{T} = [C_0^{(1,1)}, \dots, C_{N-1}^{(1,1)}, C_0^{(1,2)}, \dots, C_{N-1}^{(1,2)}, \dots, C_0^{(1,N_{Rc})}, \dots, C_{N-1}^{(1,N_{Rc})}] \tag{9.53}$$

and the input vector

$$\boldsymbol{x}_k = [x_k^{(1)}, \dots, x_{k-N+1}^{(1)}, x_k^{(2)}, \dots, x_{k-N+1}^{(2)}, \dots, x_k^{(N_{Rc})}, \dots, x_{k-N+1}^{(N_{Rc})}]^T \tag{9.54}$$

the solution is of the type (9.52).

Also in this case, a decision feedback equalizer (DFE) structure, or better a maximum-likelihood sequence detection (MLSD) receiver, yields improved performance at the cost of a greater complexity. For an in-depth study, we refer to the bibliography [1–18].

We point out that the optimum structure of a *combiner/linear equalizer* is given by a bank of filters, each matched to the overall receive pulse at the antenna output, whose outputs are then summed; the signal thus obtained is equalized by a linear equalizer (LE) or a DFE [19].

9.2.2 MIMO combiner

In the presence of multiple antennas at both the transmitter and the receiver ($N_{Tx} > 1$ and $N_{Rc} > 1$), we obtain a fully MIMO system. We here assume $N_{Rc} \geq N_{Tx}$ and $N_S = N_{Tx}$, with $s^{(t)} = a_t$, $t = 1, \dots, N_{Tx}$, and

$$\boldsymbol{s} = \boldsymbol{a} = [a_1, \dots, a_{N_S}]^T \tag{9.55}$$

For a MIMO system, (9.27) cannot be simplified and a receiver combiner must both collect as much power as possible for each stream and combat the interference due to the multiple streams transmitted over the channel. The vector signal obtained after combining is now

$$\boldsymbol{y} = [y_1, \dots, y_{N_S}]^T = \boldsymbol{C}^T\boldsymbol{x} = \boldsymbol{C}^T\mathcal{G}_C\boldsymbol{a} + \boldsymbol{C}^T\boldsymbol{w} \tag{9.56}$$

where $\boldsymbol{C}^T$ is an $N_S \times N_{Rc}$ matrix and each entry of $\boldsymbol{y}$ now corresponds to a different data stream.

We observe that $\boldsymbol{y}$ is affected by noise $\boldsymbol{C}^T\boldsymbol{w}$, which in general is (i) correlated over the various streams and (ii) with different power on each stream. Moreover, even after combining, some residual interference among streams may be present, unless $\boldsymbol{C}^T\mathcal{G}_C$ is diagonal. However, to simplify implementation, typically the correlation among samples in $\boldsymbol{y}$ is neglected and a separate detector is applied on each element of the combiner output y_r, $r = 1, \dots, N_S$. In other words, the resulting system is a set of N_S parallel channels. The block diagram of the MIMO combiner is shown in Figure 9.5. Note that hard detectors can be replaced by soft detectors, followed by decoders.

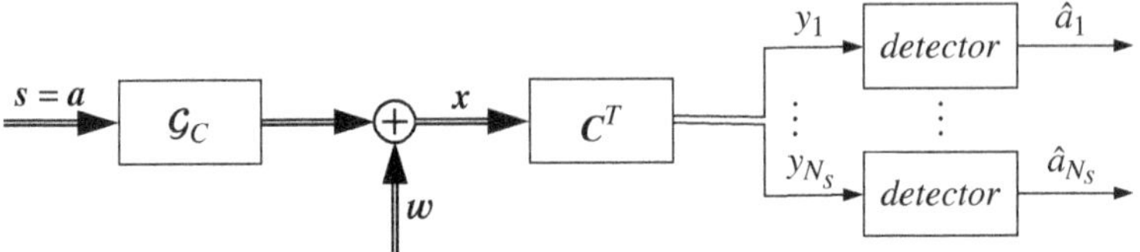

Figure 9.5 MIMO combiner.

Observation 9.1

Since the cascade of the MIMO channel and combiner can be modelled as a set of parallel channels, we can allocate the power to each stream in an optimum way, e.g. in order to maximize the achievable rate of the system. Once we set

$$s = \text{diag}\,([\mathcal{K}_1, \dots, \mathcal{K}_{N_S}]^T)\boldsymbol{a} \tag{9.57}$$

values of $\mathcal{K}_n$, $n = 1, \dots, N_S$, under the constraint (9.29) are obtained by a applying the waterfilling algorithm (see Section 13.1.2). Note however that this operation requires CSI at the transmitter.

Zero-forcing

A first solution for the choice of combiner $\boldsymbol{C}^T$ is denoted *zero-forcing* (ZF) combiner and aims at nulling the interference among streams, i.e.

$$\boldsymbol{C}_{ZF}^T \mathcal{G}_C = \boldsymbol{I} \tag{9.58}$$

where $\boldsymbol{I}$ is the $N_S \times N_S$-size identity matrix. From (9.56), we obtain in this case $\boldsymbol{y} = \boldsymbol{a} + \boldsymbol{C}_{ZF}^T \boldsymbol{w}$.

First note that $\boldsymbol{C}_{ZF}^T$ is readily obtained as the pseudo-inverse of $\mathcal{G}_C$ (see (2.149)), i.e.

$$\boldsymbol{C}_{ZF}^T = [\mathcal{G}_C^H \mathcal{G}_C]^{-1} \mathcal{G}_C^H \tag{9.59}$$

assuming here that the inverse exists, i.e. $\mathcal{G}_C$ is full rank. When the inverse of $\mathcal{G}_C^H \mathcal{G}_C$ does not exist, we cannot decode $N_S = N_{Tx}$ streams.

Second, the resulting SNR at the detection point for stream $n = 1, \dots, N_S$, is

$$\gamma_{ZF,n} = \gamma_{MF} \frac{1}{||[\boldsymbol{C}_{ZF}^T]_{n,\cdot}||^2} \tag{9.60}$$

where $[\boldsymbol{C}_{ZF}^T]_{n,\cdot}$ denotes row n of $\boldsymbol{C}_{ZF}^T$.

MMSE

The ZF combiner may enhance the noise, by only imposing the strict absence of interference among streams, with a resulting potentially high SNR at the detection point. A better trade-off between noise and interference is instead obtained with the *MMSE* combiner aiming at minimizing the average mean-square error (MSE) over the streams, i.e. (see (2.124))

$$J = E[||\boldsymbol{y} - \boldsymbol{a}||^2] \tag{9.61}$$

Assuming independent data streams and noise entries, from (9.27) we obtain the following correlation matrices (see (2.184) and (2.186))

$$\boldsymbol{R}_{\boldsymbol{x}} = E[\boldsymbol{x}^* \boldsymbol{x}^T] = \sigma_a^2 \mathcal{G}_C^* \mathcal{G}_C^T + \sigma_w^2 \boldsymbol{I} \tag{9.62}$$

$$\boldsymbol{R}_{\boldsymbol{x}\boldsymbol{a}} = E[\boldsymbol{x}^* \boldsymbol{a}^T] = \sigma_a^2 \mathcal{G}_C^* \tag{9.63}$$

The MMSE combiner is (see (2.198))

$$C_{MMSE}^T = (R_x^{-1} R_{xa})^T = (\sigma_a^2 \mathcal{G}_C^H \mathcal{G}_C + \sigma_w^2 I)^{-1} \mathcal{G}_C^H \sigma_a^2 \tag{9.64}$$

where we note that $(\sigma_a^2 \mathcal{G}_C^H \mathcal{G}_C + \sigma_w^2 I)$ is always invertible when $\sigma_w^2 \neq 0$. The resulting SNR at detection point for stream $n = 1, \dots, N_S$, is

$$\gamma_{MMSE,n} = \gamma_{MF} \frac{\left| [C_{MMSE}^T \mathcal{G}_C]_{n,n} \right|^2}{\left\| [C_{MMSE}^T]_{n,\cdot} \right\|^2} \tag{9.65}$$

where $[X]_{m,n}$ is the entry (m, n) of matrix X. Note also that as $\sigma_w^2 \to 0$, the MMSE combiner converges to the ZF combiner, i.e. $C_{MMSE} \to C_{ZF}$.

9.2.3 MIMO non-linear detection and decoding

Non-linear processing can be applied on x before stream detection. In general, these schemes are more complex than the combiner, while having the potential to provide better performance, i.e. higher data rates or lower bit error rate (BER). Here we focus on (i) the *vertical Bell labs layered space-time* (V-BLAST) solution [20] and (ii) the spatial modulation.

V-BLAST system

With V-BLAST, we still transmit an independent data symbol on each transmit antenna, according to (9.55), with $N_{Rc} \geq N_{Tx}$. The basic idea of V-BLAST is to combine the received signals, detect a single data stream, remove its contribution from x, and then repeat combining and detection for the next data stream. Combining, detection, and cancellation are iterated N_S times, until all data streams are detected.

Let $x(n)$, $n = 1, \dots, N_S$, be the receive vector of size $N_S = N_{Tx}$ at iteration n, where for the first iteration we have $x(1) = x$. A combiner of size $1 \times N_{Rc}$ is applied on $x(n)$ to obtain stream n, for example designed according to the MMSE criterion (9.64), where $\mathcal{G}_C$ is replaced by $\mathcal{G}_C(n)$, the $N_{Rc} \times (N_S - n + 1)$ sub-matrix of $\mathcal{G}_C$ comprising columns from n to N_{Tx}, i.e.

$$\mathcal{G}_C^{(r,t)}(n) = \mathcal{G}_C^{(r,n+t-1)}, \qquad t = 1, \dots, N_{Tx} - n + 1, \quad r = 1, \dots, N_{Rc} \tag{9.66}$$

and C^T is replaced by the $(N_S - n + 1) \times N_{Rc}$ matrix $C^T(n)$. At iteration n, the combiner output is

$$y_n = [C^T(n)]_{1,\cdot} x(n) \tag{9.67}$$

and detection follows, providing $\hat{a}_n$. Then, the contribution of this and all previously detected data symbols are removed from x, to obtain

$$x(n+1) = x(n) - \mathcal{G}_C^{(\cdot,1)}(n) \hat{a}_n \tag{9.68}$$

with

$$\mathcal{G}_C^{(\cdot,1)}(n) = [\mathcal{G}_C^{(1,n)}, \dots, \mathcal{G}_C^{(N_{Rc},n)}]^T \tag{9.69}$$

We observe that if $\hat{a}_\ell = a_\ell$, $\ell = 1, \dots, n$, we obtain

$$x(n+1) = \mathcal{G}_C(n+1) \begin{bmatrix} a_{n+1} \\ \vdots \\ a_{N_S} \end{bmatrix} + w \tag{9.70}$$

Note that at each iteration the combiner is applied on a channel with smaller dimension and fewer streams, as previously detected streams are removed. The resulting scheme is shown in Figure 9.6.

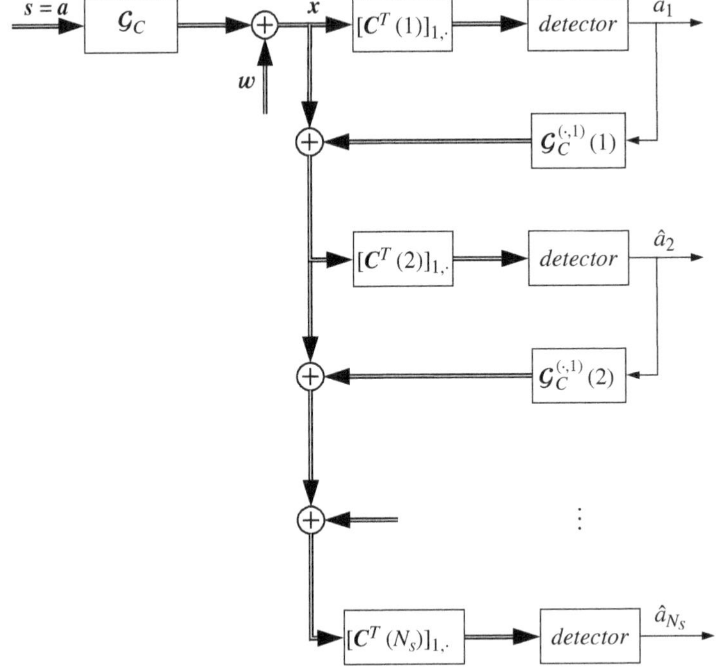

Figure 9.6 V-BLAST system.

The order of detection can be modified, as the best performance is obtained by detecting (and removing) streams with higher SINR first, taking into account the interference of unremoved streams. Note that the detection of data streams is prone to errors that are propagated in forthcoming iterations by the cancellation process. In order to mitigate this problem, each data stream, before the bit mapper, can be encoded with forward-error-correcion (FEC) at the transmitter. FEC decoding is performed on each detected stream. Next, bit mapping follows, then cancellation. This solution however comes at the cost of both increased complexity and higher latency, as a block of data must be decoded before the next stream is detected.

Spatial modulation

Another solution for using the spatial dimension made available by multiple transmit (and receive) antennas is the *spatial modulation* [21]. In this approach, a data symbol is *transmitted with a single antenna* out of the N_{Tx}, thus $N_S = 1$. Besides this information carried by the symbol, also the index of the selected antenna carries an information of $\log_2 N_{Tx}$ [bit]. As $N_S = 1$ in any case, we can also consider the case $N_{Tx} \geq N_{Rc}$.

For example, assume $N_{Tx} = 2, N_{Rc} = 1$, and quadrature phase-shift-keying (QPSK) symbols are transmitted. Given 3 data bits b_1, b_2, and b_3, to be transmitted: b_1 is used to select the antenna used for transmitting the QPSK symbol a, obtained by mapping b_2 and b_3. The received signal will be

$$x = \begin{cases} \mathcal{G}_C^{(1,1)} a + w & b_1 = 1 \\ \mathcal{G}_C^{(1,2)} a + w & b_1 = 0 \end{cases} \tag{9.71}$$

The receiver needs to know the channel with respect to the various antennas (which must be different, i.e. $\mathcal{G}_C^{(1,1)} \neq \mathcal{G}_C^{(1,2)}$). The receiver can apply two minimum-distance decisions on x, one assuming channel

$\mathcal{G}_C^{(1,1)}$, the other assuming channel $\mathcal{G}_C^{(1,2)}$. Let $\hat{a}|_{b_1=1}$ and $\hat{a}|_{b_1=0}$ be the (hard) detected symbols for the two channel cases. Then, the hard decision on bit b_1 is obtained as

$$\hat{b}_1 = \begin{cases} 1 & |x - \mathcal{G}_C^{(1,1)}\hat{a}|_{b_1=1}| < |x - \mathcal{G}_C^{(1,2)}\hat{a}|_{b_1=0}| \\ 0 & |x - \mathcal{G}_C^{(1,1)}\hat{a}|_{b_1=1}| \geq |x - \mathcal{G}_C^{(1,2)}\hat{a}|_{b_1=0}| \end{cases} \tag{9.72}$$

Moreover, if $\hat{b}_1 = 1$ the QPSK detected symbol is $\hat{a} = \hat{a}|_{b_1=1}$, otherwise $\hat{a} = \hat{a}|_{b_1=0}$.

Spatial modulation has the advantage of avoiding spatial interference, since only one transmit antenna is used at any given time; this also reduces the complexity of the transmitter, where only one radio frequency chain is needed. Moreover, the transmission of each QPSK symbol is characterized by a different SNR, as channel changes according to the selected transmit antenna; on the other hand, a higher difference between the gain of the two channels $|\mathcal{G}_C^{(1,1)}|$ and $|\mathcal{G}_C^{(1,2)}|$ reduces the detection errors on b_1.

A further extreme version of spatial modulation is *space shift keying*, where information bits are carried only by the index of the selected transmit antenna. In this case, with reference to (9.71), a is a constant and the receiver only needs to detect from which antenna the transmission occurred. While further simplifying the receiver, this solution further reduces the transmit bit rate.

9.2.4 Space-time coding

The solutions presented until now operate on a single symbol period. We now consider a space modulation scheme operating over multiple symbol periods. Hence, this approach is denoted as *space-time coding*, and it still requires CSI at the receiver only. In particular, we focus here on a MISO transmission, with $N_{Tx} = 2$ and $N_{Rc} = 1$.

First, we extend the channel model (9.27) to include the time index k. If $\boldsymbol{s}_k = [s_k^{(1)}, s_k^{(2)}]^T$ is the transmitted signal at time k,

$$x_k = \mathcal{G}_C \boldsymbol{s}_k + w_k = \mathcal{G}_C^{(1,1)} s_k^{(1)} + \mathcal{G}_C^{(1,2)} s_k^{(2)} + w_k \tag{9.73}$$

at the receiver. Note that we assume that channel vector $\mathcal{G}_C$ is time-invariant.

The Alamouti code

We consider the so-called *Alamouti* code [22], by the name of its inventor, where

$$s_{2k}^{(1)} = \mathcal{K}a_{2k}, \quad s_{2k}^{(2)} = \mathcal{K}a_{2k+1}, \quad s_{2k+1}^{(1)} = -\mathcal{K}a_{2k+1}^*, \quad s_{2k+1}^{(2)} = \mathcal{K}a_{2k}^* \tag{9.74}$$

with $\mathcal{K}$ a suitable scaling factor such that the power constraint (9.29) is satisfied as we transmit a single data stream a_k with two antennas (note that as $N_{Rc} = 1$ it must be $N_S \leq 1$). In particular, as $\boldsymbol{s}_k$ carries two independent symbols in two time periods, or, equivalently, $N_S = 1$ symbol per time period, on average the power constraint (9.29) yields $\mathcal{K} = \frac{1}{\sqrt{2}}$ and

$$\mathtt{M}_{s^{(1)}} = \mathtt{M}_{s^{(2)}} = \frac{\mathtt{M}_{Tx}}{2} \tag{9.75}$$

with $\mathtt{M}_{Tx} = \sigma_a^2$.

Note that symbol a_{2k+1} must be already available for transmission at time k. Note also that the transmitted signal $\boldsymbol{s}_k$ is a non-linear transformation of the data signal a_k due to the transmission of its complex conjugate at time $2k + 1$. From (9.73) and (9.74), at the receiver we obtain

$$x_{2k} = \mathcal{G}_C^{(1,1)} s_{2k}^{(1)} + \mathcal{G}_C^{(1,2)} s_{2k}^{(2)} + w_{2k} = \mathcal{K}\mathcal{G}_C^{(1,1)} a_{2k} + \mathcal{K}\mathcal{G}_C^{(1,2)} a_{2k+1} + w_{2k} \tag{9.76a}$$

$$x_{2k+1} = \mathcal{G}_C^{(1,1)} s_{2k+1}^{(1)} + \mathcal{G}_C^{(1,2)} s_{2k+1}^{(2)} + w_{2k+1} = -\mathcal{K}\mathcal{G}_C^{(1,1)} a_{2k+1}^* + \mathcal{K}\mathcal{G}_C^{(1,2)} a_{2k}^* + w_{2k+1} \tag{9.76b}$$

Now, by performing the following operations on the two received samples

$$\begin{aligned} y_{2k} &= \mathcal{G}_C^{(1,1)*} x_{2k} + \mathcal{G}_C^{(1,2)} x_{2k+1}^* \\ &= \mathcal{K}(|\mathcal{G}_C^{(1,1)}|^2 + |\mathcal{G}_C^{(1,2)}|^2) a_{2k} + \mathcal{G}_C^{(1,1)*} w_{2k} + \mathcal{G}_C^{(1,2)} w_{2k+1}^* \end{aligned} \tag{9.77a}$$

$$\begin{aligned} y_{2k+1} &= \mathcal{G}_C^{(1,2)*} x_{2k} - \mathcal{G}_C^{(1,1)} x_{2k+1}^* \\ &= \mathcal{K}(|\mathcal{G}_C^{(1,1)}|^2 + |\mathcal{G}_C^{(1,2)}|^2) a_{2k+1} + \mathcal{G}_C^{(1,2)*} w_{2k} - \mathcal{G}_C^{(1,1)} w_{2k+1}^* \end{aligned} \tag{9.77b}$$

we obtain equivalent AWGN channels for both symbols a_{2k} and a_{2k+1} with (a) the same channel gain $\mathcal{K}(|\mathcal{G}_C^{(1,1)}|^2 + |\mathcal{G}_C^{(1,2)}|^2)$ and the same noise power $(|\mathcal{G}_C^{(1,1)}|^2 + |\mathcal{G}_C^{(1,2)}|^2)\sigma_w^2$. Moreover, the noise components of the two samples in (9.77) are uncorrelated, hence statistically independent, being Gaussian.

As the two samples carry independent data symbols, and noise samples are independent, two *minimum distance* detectors can be applied separately on y_{2k} and y_{2k+1}. The resulting SNR at the detector input is (from (9.75))

$$\gamma_{AL} = \gamma_{MF} \frac{1}{2} (|\mathcal{G}_C^{(1,1)}|^2 + |\mathcal{G}_C^{(1,2)}|^2) \tag{9.78}$$

Note the loss of 3 dB in γ with respect to the MISO linear precoder (9.109) to be seen in Section 9.3.1, due to the fact that the channel is known only at the receiver. The Alamouti scheme works only if the channel is time-invariant for the duration of two consecutive symbols.

Observation 9.2

The time axis can be replaced by the frequency axis of a MC systems; in other words, a_{2k} is transmitted on a subchannel, and a_{2k+1} on an adjacent subchannel. In order to obtain expressions similar to those derived here, the two adjacent subchannels should have the same gain and phase, a condition typically verified with a small subchannel spacing with respect to the channel coherence bandwidth. The Alamouti code is adopted in various standards, including video data for the European Telecommunications Standards Institute (ETSI) digital video broadcasting standard for terrestrial applications (DVB-T2) [23], and for the master information block (i.e. the broadcast transmission on cell information used for initial connection to the network) of the long-term evolution (LTE) of the 3GPP cellular communication standard [24]. In both cases, the frequency axis is used instead of the time axis.

Observation 9.3

The Alamouti scheme is implementing the maximum-likelihood (ML) detection only for phase-shift-keying (PSK) constellations, where symbol values have the same amplitude [22].

Observation 9.4

The Alamouti scheme can be easily extended to the MIMO case with multiple receive antennas ($N_{Rc} > 1$), and still $N_{Tx} = 2$. Indeed, the MIMO channel can be seen as N_{Rc} parallel MISO (2 input, 1 output) channels. By applying the Alamouti scheme and decoding each of these channels, we obtain N_{Rc} replicas of the transmitted data without interference. In particular, focusing on a_{2k} and receive antenna $r = 1, \dots, N_{Rc}$, (9.77a) can be rewritten as

$$y_{2k}^{(r)} = \mathcal{K}(|\mathcal{G}_C^{(r,1)}|^2 + |\mathcal{G}_C^{(r,2)}|^2) a_{2k} + w_{2k}^{'(r)} \tag{9.79}$$

where $w_{2k}^{'(r)} \sim \mathcal{CN}(0, (|\mathcal{G}_C^{(r,1)}|^2 + |\mathcal{G}_C^{(r,2)}|^2)\sigma_w^2)$. By normalizing the noise power in $y_{2k}^{(r)}$ to σ_w^2, we have

$$\overline{y}_{2k}^{(r)} = \frac{y_{2k}^{(r)}}{\sqrt{|\mathcal{G}_C^{(r,1)}|^2 + |\mathcal{G}_C^{(r,2)}|^2}}, \qquad r = 1, \dots, N_{Rc} \tag{9.80}$$

We thus obtain an equivalent SIMO channel with N_{Rc} outputs $\{\overline{y}_{2k}^{(r)}\}$, each with noise power σ_w^2, and channel gain

$$\mathcal{K}\sqrt{|\mathcal{G}_C^{(r,1)}|^2 + |\mathcal{G}_C^{(r,2)}|^2}, \qquad r = 1, \dots, N_{Rc} \tag{9.81}$$

We then apply combiner (9.46) to merge the output of these channels, here a simple sum as the channel gains (9.81) are real non-negative, providing the following SNR at the detector input

$$\gamma_{AL} = \gamma_{MF}\frac{1}{2}\left[\sum_{t=1}^{2}\sum_{r=1}^{N_{Rc}}|\mathcal{G}_C^{(r,t)}|^2\right] \tag{9.82}$$

The Golden code

The Golden code map two data streams ($N_S = 2$) $a_{1,k}$ and $a_{2,k}$ to the transmitted samples

$$\begin{aligned} s_{2k}^{(1)} &= \frac{\alpha}{\sqrt{5}}(a_{1,2k} + \theta a_{2,2k}), & s_{2k}^{(2)} &= \frac{\alpha}{\sqrt{5}}(a_{1,2k+1} + \theta a_{2,2k+1}) \\ s_{2k+1}^{(1)} &= \frac{j\overline{\alpha}}{\sqrt{5}}(a_{1,2k+1} + \overline{\theta} a_{2,2k+1}), & s_{2k+1}^{(2)} &= \frac{\overline{\alpha}}{\sqrt{5}}(a_{1,2k} + \overline{\theta} a_{2,2k}) \end{aligned} \tag{9.83}$$

where $\theta = \frac{1+\sqrt{5}}{2}$ is the Golden ratio, $\overline{\theta} = 1 - \theta$, $\alpha = 1 + j(1 - \theta)$, and $\overline{\alpha} = 1 + j(1 - \overline{\theta})$. With this code four data symbols $a_{1,2k}$, $a_{2,2k}$, $a_{1,2k+1}$, and $a_{2,2k+1}$, are transmitted over two symbol periods, using two antennas. Therefore the number of symbols per transmit time interval T achieved by this code is twice that of the Alamouti code. The detection of this code requires a more complex ML approach, with respect to the simple operations of the Alamouti receiver (9.77). The Golden code also provides a smaller error probability, in particular for symbol constellations with many points.

We conclude this section mentioning diversity techniques with more than one transmit and receive antenna, called *space-time coding* techniques, whereby data symbols are coded in both space and time [25–27].

9.2.5 MIMO channel estimation

Channel estimation for MIMO refers to the problem of estimating $\mathcal{G}_C$ at the receiver, by exploiting the transmission of symbols already known at the receiver, denoted *training* or *pilot* symbols. Since the channel to be estimated is a $N_{Rc} \times N_{Tx}$ matrix, and at each symbol time we only receive N_{Rc} samples, many symbols must be transmitted in order to estimate the full matrix. Moreover, we will assume that the channel does not change for the duration of the channel estimation.

Let $\boldsymbol{s}_k = [s_k^{(1)}, \dots, s_k^{(N_{Tx})}]^T$ be vector of transmitted symbols at time k, and

$$\boldsymbol{x}_k = [x_k^{(1)}, \dots, x_k^{(N_{Rc})}]^T = \mathcal{G}_C \boldsymbol{s}_k + \boldsymbol{w}_k \tag{9.84}$$

be the vector of received samples at time k, with $\boldsymbol{w}_k = [w_k^{(1)}, \dots, w_k^{(N_{Rc})}]^T$ the AWGN vector. Let us also collect L training symbol vectors into matrix

$$\boldsymbol{S} = [\boldsymbol{s}_0, \dots, \boldsymbol{s}_{L-1}], \qquad N_{Tx} \times L \tag{9.85}$$

Collecting the corresponding L received samples and noise vectors into matrices

$$\boldsymbol{X} = [\boldsymbol{x}_0, \dots, \boldsymbol{x}_{L-1}], \qquad \boldsymbol{W} = [\boldsymbol{w}_0, \dots, \boldsymbol{w}_{L-1}] \tag{9.86}$$

the input–output relation (9.84) can be rewritten as

$$\boldsymbol{X} = \mathcal{G}_C \boldsymbol{S} + \boldsymbol{W}, \qquad N_{Rc} \times L \tag{9.87}$$

The least squares method

We denote by $\hat{\mathcal{G}}_C$ the channel matrix estimate with corresponding output

$$\hat{X} = \hat{\mathcal{G}}_C S \tag{9.88}$$

Defined the error matrix

$$e = X - \hat{X} \tag{9.89}$$

the least squares (LS) estimator aims at minimizing the sum of square output errors

$$\mathcal{E} = ||e||^2 = \sum_{r=1}^{N_{Rc}} \sum_{k=0}^{L-1} |x_k^{(r)} - \hat{x}_k^{(r)}|^2 \tag{9.90}$$

with respect to $\hat{\mathcal{G}}_C$. As $\{x_k^{(r)}\}$ ($\{\hat{x}_k^{(r)}\}$), $k = 0, \dots, L-1$, depends only on row r of $\mathcal{G}_C$ ($\hat{\mathcal{G}}_C$), the above problem can be split into N_{Rc} problems, where the generic function to be minimized with respect to $[\hat{\mathcal{G}}_C]_{r,\cdot}$ is

$$\mathcal{E}^{(r)} = \sum_{k=0}^{L-1} |x_k^{(r)} - \hat{x}_k^{(r)}|^2 \tag{9.91}$$

From (9.88), it is

$$[\hat{X}]_{r,\cdot}^T = S^T [\hat{\mathcal{G}}_C]_{r,\cdot}^T \tag{9.92}$$

hence, using the results of Section 7.8.2, the solution of the problem is (7.545)

$$[\hat{\mathcal{G}}_{C,LS}]_{r,\cdot}^T = (S^* S^T)^{-1} S^* [X]_{r,\cdot}^T \tag{9.93}$$

or

$$[\hat{\mathcal{G}}_{C,LS}]_{r,\cdot} = [X]_{r,\cdot} S^H (S S^H)^{-1} \tag{9.94}$$

Hence, the whole channel matrix estimate is

$$\hat{\mathcal{G}}_{C,LS} = X S^H (S S^H)^{-1} \tag{9.95}$$

assuming $L \geq N_{Tx}$ and S' full rank so that the above inverse exists.

By inserting (9.87) into (9.95), it is

$$\hat{\mathcal{G}}_{C,LS} = \mathcal{G}_C + W S^H (S S^H)^{-1} \tag{9.96}$$

Hence,

$$\Delta \mathcal{G}_C = \hat{\mathcal{G}}_{C,LS} - \mathcal{G}_{C,LS} = W S^H (S S^H)^{-1} \tag{9.97}$$

with corresponding normalized SNR (see (7.548))

$$\Lambda_n = \frac{\sigma_w^2}{\mathtt{M}_{s^{(1)}} \, E\,[||\Delta \mathcal{G}_C||^2]} \tag{9.98}$$

For S orthonormal, $S S^H = I$ and white noise across antennas and time, it turns out

$$\Lambda_n = \frac{L}{N_{Tx} N_{Rc}} \tag{9.99}$$

The LMMSE method

By recalling the notation vec(·), (see (4.148)) as

$$g_C = \text{vec}(\mathcal{G}_C) = [\mathcal{G}_C^{(1,1)}, \ldots, \mathcal{G}_C^{(1,N_{Tx})}, \mathcal{G}_C^{(2,1)}, \ldots, \mathcal{G}_C^{(N_{Rc},N_{Tx})}]^T \tag{9.100}$$

and similarly $\boldsymbol{x} = \text{vec}(\boldsymbol{X})$, $\boldsymbol{w} = \text{vec}(\boldsymbol{W})$, and the $(LN_{Rc}) \times (N_{Tx}N_{Rc})$ matrix

$$\overline{\boldsymbol{S}} = \boldsymbol{S}^T \otimes \boldsymbol{I} \tag{9.101}$$

with $\boldsymbol{I}$ the $N_{Rc} \times N_{Rc}$ identity matrix, we have that (9.84) can be written as

$$\boldsymbol{x} = \overline{\boldsymbol{S}}\boldsymbol{g}_C + \boldsymbol{w} \tag{9.102}$$

The channel estimation problem can be seen as the estimation of channel vector $\boldsymbol{g}_C$, a random vector with correlation matrix $\boldsymbol{R}_{\boldsymbol{g}_C}$, starting from an observation vector $\boldsymbol{x}$, a problem discussed in detail in Section 7.8.2. In particular, assuming $\boldsymbol{w} \sim \mathcal{CN}(\boldsymbol{0}, \sigma_w^2 \boldsymbol{I})$, the linear minimum-mean-square-error (LMMSE) estimator is given by (7.591) as

$$\begin{aligned} \hat{\boldsymbol{g}}_{C,LMMSE} &= (\boldsymbol{R}_{\boldsymbol{x}}^{-1} \boldsymbol{R}_{\boldsymbol{x}\boldsymbol{g}_C})^T \boldsymbol{x} \\ & [\sigma_w^2 (\boldsymbol{R}_{\boldsymbol{g}_C}^*)^{-1} + \overline{\boldsymbol{S}}^H \overline{\boldsymbol{S}}]^{-1} \overline{\boldsymbol{S}}^H \boldsymbol{x} \end{aligned} \tag{9.103}$$

Example 9.2.1

For a MIMO channel with (Gaussian) zero-mean uncorrelated entries $\boldsymbol{R}_{\boldsymbol{g}_C} = \boldsymbol{I}$, identity matrix of size $(N_{Tx}N_{Rc}) \times (N_{Tx}N_{Rc})$, it is

$$\hat{\boldsymbol{g}}_{C,LMMSE} = [\sigma_w^2 \boldsymbol{I} + \overline{\boldsymbol{S}}^H \overline{\boldsymbol{S}}]^{-1} \overline{\boldsymbol{S}}^H \boldsymbol{x} \tag{9.104}$$

Moreover, if $\overline{\boldsymbol{S}}$ has orthonormal columns, we have

$$\overline{\boldsymbol{S}}^H \overline{\boldsymbol{S}} = \boldsymbol{I} \tag{9.105}$$

and

$$\hat{\boldsymbol{g}}_{C,LMMSE} = \frac{1}{1+\sigma_w^2} \overline{\boldsymbol{S}}^H \boldsymbol{x} \tag{9.106}$$

Example 9.2.2

Another example of channel correlation matrix is given by (4.152).

9.3 CSI only at the transmitter

Until now we have considered CSI available only at the receiver, and all the burden of handling the interference introduced by the MIMO channel has been upon the receiver. In this section, we consider instead that CSI is available at the transmitter only, thus the transmitter must pre-elaborate the data symbols before transmission over the MIMO channel. Also in this case, we will start with linear approaches and then consider non-linear solutions. In this section, we will assume $N_{Tx} \geq N_{Rc}$ and $N_S = N_{S,max} = N_{Rc}$.

9.3.1 MISO linear precoding

In a MISO system, multiple transmit antennas ($N_{Tx} > 1$) are employed, while the receiver is equipped with a single antenna ($N_{Rc} = 1$), and therefore, (9.27) becomes $x = \mathcal{G}_C \boldsymbol{s} + w$. At the receiver, a

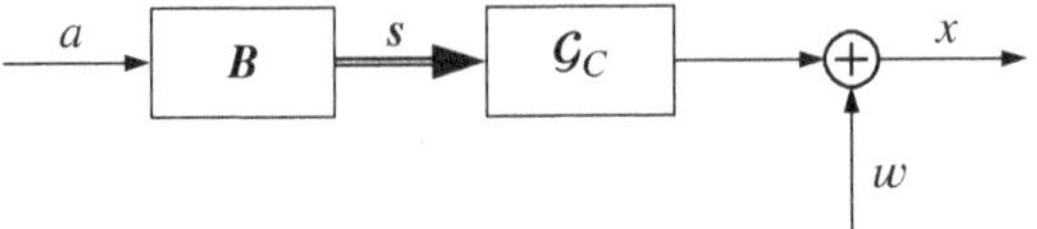

Figure 9.7 MISO linear precoder.

conventional symbol-by-symbol detector is employed. Now, since a single sample x is received for each transmit vector $\boldsymbol{s}$, we can easily conclude that even with $N_{Tx} > 1$ antennas we can transmit a single stream at any given time, $N_S = N_{Rc} = 1$, and the transmit vector can be written as

$$\boldsymbol{s} = \boldsymbol{B}a \tag{9.107}$$

where a is the data symbol of the single stream, while $\boldsymbol{B}$ is an N_{Tx} vector of *linear precoding*, with entries $B^{(t,1)}$, $t = 1, \dots, N_{Tx}$. The resulting scheme is shown in Figure 9.7.

As $\boldsymbol{R}_s = \sigma_a^2 \boldsymbol{B}\boldsymbol{B}^H$ and $N_S = 1$, from the power constraint (9.30), we obtain a constraint on the precoder, namely

$$||\boldsymbol{B}||^2 = \sum_{t=1}^{N_{Tx}} |B^{(t,1)}|^2 = 1 \tag{9.108}$$

The SNR at detection point turns out to be

$$\gamma = \gamma_{MF} |\mathcal{G}_C \boldsymbol{B}|^2 \tag{9.109}$$

The precoder maximizing γ under the constraint (9.108) can be obtained by the Lagrange multipliers method, i.e. finding the maximum of

$$J = ||\mathcal{G}_C \boldsymbol{B}||^2 \sigma_a^2 + \lambda(||\boldsymbol{B}||^2 - 1) \tag{9.110}$$

where λ is the Lagrange multiplier to be optimized to satisfy (9.108). By taking the gradient of J with respect to $\boldsymbol{B}$ and λ, and enforcing (9.108), we obtain[2]

$$\boldsymbol{B} = \frac{\mathcal{G}_C^H}{||\mathcal{G}_C||} \tag{9.111}$$

Note that, similar to MRC (9.46), the MISO linear precoder aligns the signals at the receiver by compensating the phase shift introduced by the channel, and transmits more power on channels having a higher gain $|\mathcal{G}_C^{(1,t)}|$. Correspondingly, we have

$$\gamma_{LP} = \gamma_{MF} \left(\sum_{t=1}^{N_{Tx}} |\mathcal{G}_C^{(1,t)}|^2 \right) \tag{9.112}$$

MISO antenna selection

As illustrated in Figure 9.8, a simple configuration for MISO system is *antenna selection*, wherein one antenna is selected for transmission among the N_{Tx} available, and all others are switched off. As selection criterion, we can choose the antenna having the highest channel gain to the receive antenna, i.e.

$$t_{AS} = \arg\max_t |\mathcal{G}_C^{(1,t)}|^2 \tag{9.113}$$

At detection point, the SNR is

$$\gamma_{AS} = \gamma_{MF} |\mathcal{G}_C^{(1,t_{AS})}|^2 \tag{9.114}$$

[2] Similarly to (9.46), the solution (9.111) can be derived from (9.109) by the Schwarz inequality up to a constant which in turn is determined by the constraint (9.108).

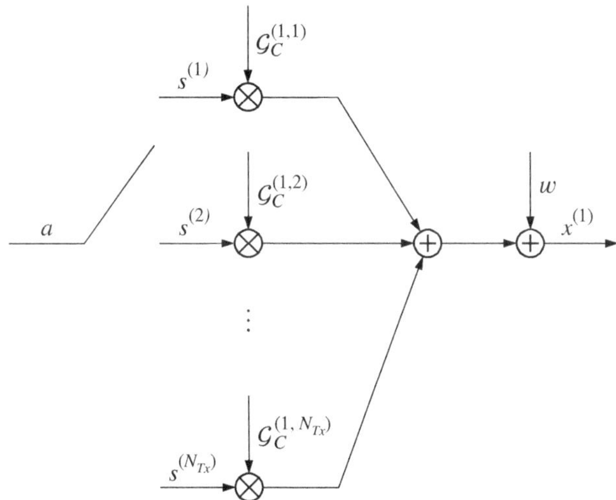

Figure 9.8 MISO antenna selection or switched transmit diversity.

9.3.2 MIMO linear precoding

For a MIMO system ($N_{Tx} > 1$ and $N_{Rc} > 1$), multiple streams can potentially be transmitted, which however interfere with each other at the receiver. Similarly to the combining solutions, two approaches are typically employed in the choice of the linear precoder matrix $\boldsymbol{B}$, namely ZF and MMSE precoding. As we have already seen, a MIMO system can transmit at most N_S independent data streams (see (9.28)). With linear precoding, $\boldsymbol{a} = [a_1, \dots, a_{N_S}]^T$ multiplies the precoder $N_{Tx} \times N_S$ matrix $\boldsymbol{B}$, with entries $B^{(t,n)}$, $t = 1, \dots, N_{Tx}$, $n = 1, \dots, N_S$, to yield the transmitted vector

$$\boldsymbol{s} = \boldsymbol{B}\boldsymbol{a} \tag{9.115}$$

The resulting input–output relation is

$$\boldsymbol{x} = \mathcal{G}_C \boldsymbol{B}\boldsymbol{a} + \boldsymbol{w} \tag{9.116}$$

As the channel is not known at the receiver, detection of the N_S symbols is based independently on each component of $\boldsymbol{x}$. Therefore, also in this case suitable power allocation techniques (e.g. using to the waterfilling algorithm) can be applied across the components of $\boldsymbol{s}$ and under the constraint (see (9.30))

$$\text{tr}[\boldsymbol{B}\boldsymbol{B}^H] = \sum_{n=1}^{N_S} \sum_{t=1}^{N_{Tx}} |B^{(t,n)}|^2 = N_S \tag{9.117}$$

Note that with precoding, the channel matrix $\mathcal{G}_C$ must be already available at the transmitter for the design of $\boldsymbol{B}$. The resulting scheme is similar to Figure 9.7, where now $\boldsymbol{a}$, $\boldsymbol{w}$, and $\boldsymbol{x}$ are vectors.

ZF precoding

Similarly to ZF combining, the *ZF precoder* aims at removing the interference introduced by the channel, operating now only at the transmitter. Let us define the pseudo-inverse matrix

$$\tilde{\boldsymbol{B}} = \mathcal{G}_C^H (\mathcal{G}_C \mathcal{G}_C^H)^{-1} \tag{9.118}$$

assuming that matrix $\mathcal{G}_C \mathcal{G}_C^H$ is invertible. Note that we have $\mathcal{G}_C \tilde{\boldsymbol{B}} = \boldsymbol{I}$ thus $\tilde{\boldsymbol{B}}$ could be the ZF precoder. However, with this choice constraint, (9.117) may not be satisfied.

We now investigate a proper power splitting among components of $\boldsymbol{s}$. To this end, we normalize $\tilde{\boldsymbol{B}}$ such that (9.117) holds, and also $\mathtt{M}_{Tx}$ is equally split among the components of $\boldsymbol{s}$. We leave the power allocation among the components of $\boldsymbol{s}$ in the presence of the channel $\mathcal{G}_C$ as a separate problem to be solved next. From (9.115), the transmit power gain of a_n, due to the precoder $\tilde{\boldsymbol{B}}$, is

$$\tilde{\beta}_n = ||\tilde{\boldsymbol{B}}_{\cdot,n}||^2 \tag{9.119}$$

Therefore, the normalized ZF beamformer

$$\boldsymbol{B}_{ZF} = \tilde{\boldsymbol{B}}\ \mathrm{diag}\left(\left[\sqrt{\frac{N_S}{\tilde{\beta}_1}}, \ldots, \sqrt{\frac{N_S}{\tilde{\beta}_{N_S}}}\right]^T\right) \tag{9.120}$$

satisfies (9.117) and at the same time forces the same transmit power for all data symbols. This precoder, also denoted as *diagonal precoder*, provides at detection point the set of parallel channels

$$x_n = \sqrt{\frac{N_S}{\tilde{\beta}_n}} a_n + w_n, \qquad n = 1, \ldots, N_S \tag{9.121}$$

Once we replace a_n with $\mathcal{K}_n a_n$, with $\mathcal{K}_n$, $n = 1, \ldots, N_S$, suitable scaling factors such that

$$\sum_{n=1}^{N_S} \mathcal{K}_n^2 = 1 \tag{9.122}$$

waterfilling algorithm can be applied (see Section 13.1.2).

A simpler ZF solution is obtained by enforcing the same channel gain for all streams at the receiver. In this case, we have the *regularized zero-forcing* (RZF) precoder

$$\boldsymbol{B}_{RZF} = \kappa \tilde{\boldsymbol{B}} \quad \text{with } \kappa = \sqrt{\frac{N_S}{\mathrm{tr}(\tilde{\boldsymbol{B}}^H \tilde{\boldsymbol{B}})}} \tag{9.123}$$

providing $x_n = \kappa a_n + w_n$. In this case, however, we use a different transmit power per component of $\boldsymbol{s}$, which however has not been optimized, e.g. with waterfilling. The constraint (9.117) is still satisfied.

9.3.3 MIMO non-linear precoding

We may wonder if non-linear precoding is more effective than linear precoding for MIMO systems. First, consider the LQ factorization of channel matrix $\mathcal{G}_C$, a variant of the LU factorization [28], wherein

$$\mathcal{G}_C = \boldsymbol{L}\boldsymbol{Q} \tag{9.124}$$

where the $N_{Tx} \times N_{Tx}$ matrix $\boldsymbol{Q}$ is unitary (i.e. $\boldsymbol{Q}^H\boldsymbol{Q} = \boldsymbol{I}$), and $\boldsymbol{L}$ is a $N_{Rc} \times N_{Tx}$ lower triangular matrix, i.e. having its upper triangular part with only zeros. Note that the diagonal of $\boldsymbol{L}$ has N_S non-zero entries. Then, by transmitting $\boldsymbol{s} = \boldsymbol{Q}^H \tilde{\boldsymbol{s}}$ (where $\tilde{\boldsymbol{s}}$ will be detailed next), we obtain

$$\boldsymbol{x} = \boldsymbol{L}\tilde{\boldsymbol{s}} + \boldsymbol{w} \tag{9.125}$$

We now provide details on how to design $\tilde{\boldsymbol{s}} = [\tilde{s}^{(1)}, \ldots, \tilde{s}^{(N_{Tx})}]^T$. From (9.125), we observe that at the first receive antenna we have $x^{(1)} = L_{1,1}\tilde{s}^{(1)} + w^{(1)}$, without interference from other entries of $\tilde{\boldsymbol{s}}$, and we can set

$$\tilde{s}^{(1)} = a_1 \tag{9.126}$$

The second entry of $\boldsymbol{x}$ instead is affected by interference from $u^{(2)} = L_{2,1}a_1$, as from (9.125)

$$x^{(2)} = L_{2,2}\tilde{s}^{(2)} + u^{(2)} + w^{(2)} = L_{2,2}\tilde{s}^{(2)} + L_{2,1}a_1 + w^{(2)} \tag{9.127}$$

Therefore, by setting

$$\tilde{s}^{(2)} = a_2 - \frac{L_{2,1}}{L_{2,2}} a_1 \tag{9.128}$$

we obtain $x^{(2)} = L_{2,2}a_2 + w^{(2)}$, which again is without interference. This approach can be iterated for the other entries of $\tilde{s}$. Two observations are in place here. First, by a closer look at (9.128), we note that we are transmitting a linear combination of the data symbols a_1 and a_2, thus proceeding with this approach we would re-discover the linear precoding approaches of the previous section. Second, the transmit strategy (9.128) does not satisfy the power constraint (9.117).

We now consider instead non-linear approaches where $\tilde{s}^{(2)}$ is a non-linear function of both a_1 and a_2. Two solutions are investigated, namely *Tomlinson Harashima* (TH) precoding, that operates on a symbol level, which is an extension of the approach encountered for transmission over inter-symbol-interference (ISI) channels (see Section 7.7.1), and dirty paper precoding, which provides the maximum data rate.

We would like first to underline a key difference between MIMO narrowband channels and SISO dispersive channels; in MIMO, the interference occurs among different streams, that are transmitted at the same time, while in dispersive channels interference occurs among symbols transmitted at different times. Therefore, in MIMO, we can use space-time coding techniques that operate on blocks of symbols within each stream in order to avoid interference: per-stream precoding will be smartly exploited in the *dirty paper coding* (DPC) approach.

Dirty paper coding

DPC has been introduced by Costa [29], who focused on a transmission channel affected by interference, which is known at the transmitter but not at the receiver. The *dirt* is present on the paper when an author is writing, as the interference in the considered channel model.

The theoretical paper by Costa only established that DPC is a technique to achieve the MIMO channel capacity, without providing its implementation details. Over the years, it turned out that the implementation of DPC is computationally demanding and the performance promised by theory is reachable at the cost of encoding long data streams, similarly to what happens for a capacity-achieving transmission on the SISO AWGN channel.

An implementation of DPC has been proposed in [30] using linear block codes. In particular, consider the LQ precoded channel model (9.125), with $N_{Tx} = N_{Rc} = N_S = 2$. Let $\{b_{1,\ell}\}$ and $\{b_{2,\ell}\}$ be the two bit streams. We now provide details on how to design $\tilde{\boldsymbol{s}}_k$.

The first bit stream $b_{1,\ell}$ is encoded with any error correcting code with codeword length n_0, and then bit mapped into K symbols $\boldsymbol{a}_1 = [a_{1,1}, \dots, a_{1,K}]^T$. Then, we set

$$\tilde{s}_k^{(1)} = \mathcal{K}_1 a_{1,k} \tag{9.129}$$

where $\mathcal{K}_1$ is a suitable power scaling constant. From (9.127), the resulting interference on $\tilde{\boldsymbol{s}}^{(2)} = [\tilde{s}_1^{(2)}, \dots, \tilde{s}_K^{(2)}]^T$ is

$$\boldsymbol{u}^{(2)} = [L_{2,1}\mathcal{K}_1 a_{1,1}, \dots, L_{2,1}\mathcal{K}_1 a_{1,K}]^T \tag{9.130}$$

Consider now a (k_0, n_0) block code (see Chapter 11), and for a generic bit stream z of length n_0, let $\varsigma(z)$ be its $(n_0 - k_0)$-size *syndrome* (see Section 11.2.1). Data bits of the second stream are not associated to a codeword, but rather to a syndrome; therefore, we transmit one of the $2^{n_0-k_0}$ possible messages, each associated to a different syndrome. Remember that bit sequences of length n_0 having the same syndrome $\varsigma(z)$ belong to the same *coset* (see Section 11.2.1) $C(\varsigma(z))$.

The $(n_0 - k_0)$ bits of the second bit stream $\boldsymbol{b}_2 = [b_{2,1}, \dots, b_{2,n_0-k_0}]^T$ are considered a syndrome, and from its coset $C(\boldsymbol{b}_2)$, we extract the n_0 bit sequence $\boldsymbol{c}_2$ to be mapped into $\boldsymbol{a}_2$ and sent as $\tilde{\boldsymbol{s}}^{(2)}$. In particular, the selected sequence $\boldsymbol{c}_2$ is such that its bit-mapped and scaled symbol sequence $\boldsymbol{a}_2(\boldsymbol{c}_2)$ is closest (in MSE sense) to the interference $\boldsymbol{u}^{(2)}$, i.e.

$$\boldsymbol{c}_2 = \arg\min_{\boldsymbol{z}\in C(\boldsymbol{b}_2)} ||\mathcal{K}_2\boldsymbol{a}_2(\boldsymbol{z}) - \boldsymbol{u}^{(2)}||^2 \tag{9.131}$$

where $\mathcal{K}_2$ is a suitable power scaling constant. In particular, $\boldsymbol{c}_2$ is obtained by applying a decoder of the error correcting code on $\boldsymbol{u}^{(2)}$, constrained to select the decoded codewords from the *reduced set* $C(\boldsymbol{b}_2)$, thus denoted *reduced set code* (RSC) decoder. Lastly,

$$\tilde{s}_k^{(2)} = \frac{\mathcal{K}_2 a_{2,k}(\boldsymbol{c}_2) - u_k^{(2)}}{L_{2,2}}, \qquad k = 1, \ldots, K \tag{9.132}$$

The block diagram of the generation of $\tilde{s}_k^{(2)}$ is shown in Figure 9.9.

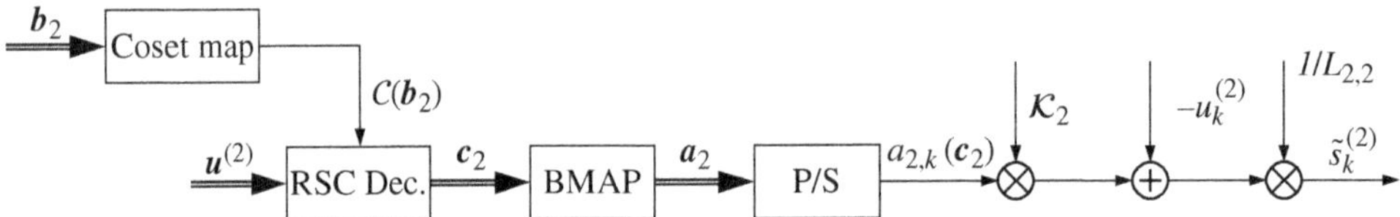

Figure 9.9 Example of generation of the second stream for dirty paper coding.

At the receiver, we have

$$x_k^{(2)} = L_{2,2}\tilde{s}_k^{(2)} + u_k^{(2)} + w_k^{(2)} = \mathcal{K}_2 a_{2,k} + w_k^{(2)} \tag{9.133}$$

Next, detection and inverse bit mapping follow. Last, we compute the syndrome of the detected bit sequence, that in absence of noise is $\boldsymbol{b}_2$.

Note that with this choice, we do not explicitly enforce the power constraint (9.117), which however can be enforced on average by properly choosing $\mathcal{K}_1$ and $\mathcal{K}_2$.

The coding-based solution for DPC can be applied also to other codes than block codes, such as convolutional, turbo, and low-density-parity-check (LDPC) codes [30].

TH precoding

TH precoding removes interference among streams. This approach is in general suboptimal with respect to DPC, as it splits the encoding process into two parts: the error correcting coding and the TH precoding. Still, the TH precoding yields a simple implementation and is effective in reducing interference.

With reference to the channel model (9.125) obtained with LQ precoding, the TH precoder operates as follows. The first stream is transmitted directly, i.e. $\tilde{s}^{(1)} = a_1$. For the second stream, we can apply the TH precoder of Section 7.7.1. Recall that, with reference to Figure 7.58, the constellation alphabet $\mathcal{A}$ is extended to cover the entire plane, and Λ_T denotes the sublattice of Λ_0. Therefore, in the MIMO context, a_2 is transmitted as

$$\tilde{s}^{(2)} = a_2 - \frac{L_{2,1}}{L_{2,2}} a_1 + c \tag{9.134}$$

where c is a point of the sublattice Λ_T chosen so that $\tilde{s}^{(2)}$ belongs to the region $R(\mathcal{A})$ of Figure 7.58.

When comparing with dirty paper, both techniques exploit knowledge of the interference at the transmitter in order to *strengthen* the message rather than being seen as an obstacle. However, notice that in TH precoding, the interference is generated by the transmitter itself, while in dirty paper coding, the known interference is related to a separate stream.

9.3.4 Channel estimation for CSIT

CSI at the transmitter (CSIT) can be obtained in various ways. First, if the two communicating devices are operating in half-duplex alternating transmission and reception in *time division duplexing* (TDD), the channel reciprocity can be exploited. In other words, if $\mathcal{G}_C$ is the channel from the transmitter to the receiver, the channel from the receiver (operating as a transmitter) and the transmitter (operating as a receiver) is $\mathcal{G}_C^T$. Therefore, we first estimate the channel from the receiver to the transmitter, $\hat{\mathcal{G}}_C^T$ and then the transmitter simply takes its transpose to estimate $\hat{\mathcal{G}}_C$.

In *frequency division duplexing* (FDD) systems, where the transmissions in the two directions occur in two different bands, channel reciprocity cannot be exploited to obtain CSIT. In Section 5.6, a solution is described, where the channel matrix $\mathcal{G}_C$ is first estimated at the receiver, it is quantized by vector quantization, and the quantization index is fed back to the transmitter by a feedback signalling channel.

9.4 CSI at both the transmitter and the receiver

We now consider specific techniques when CSI is available at both the transmitter and the receiver.

In this case, we can apply a linear processing at both the transmitter and the receiver. In particular, let

$$N_{S,max} = \text{rank } \mathcal{G}_C \tag{9.135}$$

hence $N_{S,max} \leq \min(N_{Rc}, N_{Tx})$, the singular value decomposition (SVD) of the channel matrix yields (6.114), with $\boldsymbol{\Sigma} = \text{diag}([\xi_1, \dots, \xi_{N_{S,max}}]^T)$ the diagonal matrix of non-zero singular values. We introduce the extended generated data vector

$$\boldsymbol{a}_{ext} = [\boldsymbol{a}^T, \boldsymbol{0}]^T \tag{9.136}$$

of size $N_{Tx} \times 1$. We apply the precoder

$$\boldsymbol{s} = \boldsymbol{B}\boldsymbol{a}_{ext} \tag{9.137}$$

and the combiner $\boldsymbol{C}^T$ to yield

$$\boldsymbol{y}_{ext} = \boldsymbol{C}^T[\mathcal{G}_C\boldsymbol{s} + \boldsymbol{w}] = \boldsymbol{C}^T\mathcal{G}_C\boldsymbol{s} + \tilde{\boldsymbol{w}}_{ext} \tag{9.138}$$

In particular, by choosing as combiner and precoder

$$\boldsymbol{C}^T = \boldsymbol{U}^H, \quad \boldsymbol{B} = \boldsymbol{V} \tag{9.139}$$

we note that we satisfy the transmit power constraint (9.117), as $\boldsymbol{V}\boldsymbol{V}^H = \boldsymbol{I}$. This choice clearly requires CSI at both the transmitter and the receiver. If $\boldsymbol{w}$ is an AWGN vector with i.i.d. entries, $\tilde{\boldsymbol{w}}_{ext}$ has the same statistics of $\boldsymbol{w}$. Denoting by $\boldsymbol{y}$ ($\tilde{\boldsymbol{w}}$) the first $N_{S,max}$ entries of $\boldsymbol{y}_{ext}$ ($\boldsymbol{w}_{ext}$) and removing $N_{Rc} - N_{S,max}$ entries, thanks to the unitary properties of matrices $\boldsymbol{U}$ and $\boldsymbol{V}$, at the detection point we have

$$\boldsymbol{y} = \boldsymbol{\Sigma}\boldsymbol{a} + \tilde{\boldsymbol{w}} \tag{9.140}$$

Therefore, with these precoder and combiner, we have converted the MIMO channel into a set of parallel channels (with independent noise terms)

$$y_n = \xi_n a_n + \tilde{w}_n \tag{9.141}$$

$\tilde{w}_n \sim \mathcal{CN}(0, \sigma_w^2)$, $n = 1, \dots, N_{S,max}$. The waterfilling algorithm can be applied to allocate power to each data stream, considering the power budget.

By comparing this choice of the precoder and combiner with the results obtained on the capacity of a MIMO system (see Section 6.5.4), we will conclude that this solution is capacity-achieving, as long as waterfilling is applied and ideal constellation and coding schemes are used (even separately) for each stream. Therefore, non-linear solutions wherein $\mathcal{G}_C$ is known at both the transmitter and receiver cannot outperform this linear approach.

9.5 Hybrid beamforming

Even simple linear precoding and combining solutions require matrix operations, whose complexity grows with the square of the number of antennas. In particular, for mmWave transmissions (see Section 4.1.11), the number of antennas quickly becomes large, in order to overcome huge attenuations. In this context, a number of solutions have been proposed to reduce the implementation complexity. Two directions have been mostly considered: the use of precoding/combining solutions partially implemented with analog circuits (the so-called *hybrid beamforming*), and the use of analog-to-digital-converters/digital-to-analog-converters (ADCs/DACs) with a small number of bits per sample.

The reduction of the number of bits used in ADC/DAC increases the quantization error at each antenna. Indeed, using a large number of antennas we can average out the noise, so we can use more antennas and ADCs, each with a smaller number of quantization levels, and still achieve good performance.

Let us now focus on the second approach, i.e. the hybrid beamforming. It generically refers to linear operations performed at either the transmitter or the receiver, where some operations are implemented with analog circuits. With reference to a hybrid beamformer used at the receiver, thus operating as a combiner, we obtain the scheme of Figure 9.10. The received vector is first processed by linear operations to provide the vector

$$z = [z_1, \dots, z_{N_{ADC}}]^T = \boldsymbol{C}_A^T \boldsymbol{x} \tag{9.142}$$

where N_{ADC} is the number of ADCs (typically $N_{ADC} \leq N_{Rc}$), and combiner $\boldsymbol{C}_A$ is implemented by an analog circuit. After digital conversion, yielding vector $\tilde{z} = [\tilde{z}_1, \dots, \tilde{z}_{N_{ADC}}]^T$, a further $N_S \times N_{ADC}$ digital combiner $\boldsymbol{C}_D^T$ follows, to obtain

$$\boldsymbol{y} = \boldsymbol{C}_D^T \boldsymbol{z} \tag{9.143}$$

Therefore, ignoring the imperfections of ADC conversion, this scheme implements the overall combiner $\boldsymbol{C}^T = \boldsymbol{C}_D^T \boldsymbol{C}_A^T$.

First note that having fewer ADCs than antennas already limits the resulting combiner $\boldsymbol{C}$, which in general will not be equivalent to one of the solutions studied earlier in this chapter, which require a fully digital implementation.

Then, a second constraint is introduced by the nature of $\boldsymbol{C}_A^T$. In principle, the analog combiner should be designed according to the specific MIMO channel, thus the analog circuit must be reconfigurable. In

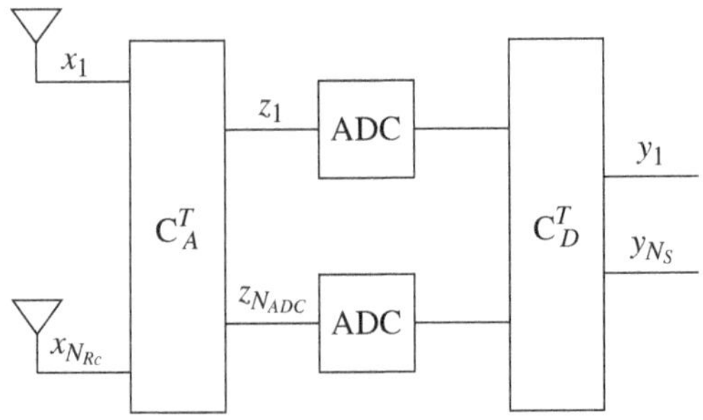

Figure 9.10 Hybrid beamformer operating as a combiner.

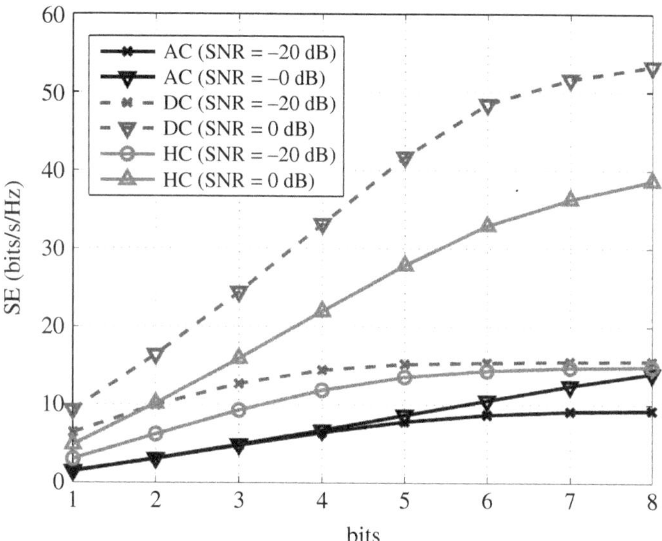

Figure 9.11 Spectral efficiency for analog, digital, and hybrid beamforming structures. Source: Reproduced with permission from Abbas et al. [31]. ©2017, IEEE.

practice, the reconfigurability typically consists in either *switches* connecting ADCs to different circuits or *reconfigurable phase shifters* with a finite set of shifts, as entries of matrix $\boldsymbol{C}_A$. In other words, entries of $\boldsymbol{C}_A$ are constrained to be either zero or $e^{j\phi}$, with $\phi \in \{\phi_0, \phi_1, \dots, \phi_{Q-1}\}$. The choice of these values is typically further constrained by the fact that each ADC may be connected (even through switches) only to a sub-set of antennas, in order to reduce the complexity of the analog circuit.

These solutions provide a trade-off between the complexity of implementing a parametric analog circuits and the flexibility requested by adapting the combiner to the channel. Both simulations and experiments show that even with a small number of ADCs and switches or phase shifts, the performance of hybrid solutions can be very close to the fully digital solution, especially when the number of receive antennas is much larger than the number transmitted data streams (or transmit antennas).

An example of spectral efficiency (SE) (as given by (6.118)) obtained with hybrid beamforming and a fix number of quantization bits per ADC is given in Figure 9.11. AC refers to a purely analog combiner, DC refers to a purely digital combiner, and HC refers to a hybrid combiner. In all cases, ADCs have a finite number of bits per sample. The x-axis indicates the number of bits used for each sample conversion in the ADC. Here $N_{Tx} = 64$, $N_{Rc} = 16$, and $N_{RF} = 4$. The hybrid beamforming architecture comprises analog phase shifters with infinite precision, thus $\boldsymbol{C}_A^T$ is full, with entries $[\boldsymbol{C}_A^T]_{m,p} = e^{j\phi_{m,p}}$, $\phi_{m,p} \in [0, 2\pi)$. According to the small-number-of-paths model (see Section 4.1.11), the mmWave channel is described by a number of clusters from which paths are generated, and the number of clusters is a Poisson random variable with parameter 1.8, while the number of paths per cluster[3] is 20. The two values of channel SNR Γ in the figure refers to non-line-of-sight (LOS) ($\Gamma = -20$ dB) and LOS ($\Gamma = 0$ dB) scenarios. We note that hybrid beamforming structures provide an interesting trade-off between complexity and performance, with respect to the fully digital solution, while significantly outperforming the fully analog solution with phase shifters.

[3] Note that even if the number of paths is large, it turns out that only a few of them are actually relevant (see Appendix 4.A).

Hybrid beamforming and angular domain representation

A special solution is obtained by applying the angular domain representation of mmWave channels (see Appendix 4.A) with hybrid beamforming at both the transmitter and the receiver. Consider that the *virtual channel*, as the 2D discrete Fourier transform (DFT) (4.162) of $\mathcal{G}_C$, can be written as

$$\boldsymbol{\Theta} = \boldsymbol{F}_{M_1} \mathcal{G}_C \boldsymbol{F}_{M_2} \tag{9.144}$$

with $\boldsymbol{F}_{M_i}$ the DFT matrix of size $M_i \times M_i$ (see (1.30)). As $\boldsymbol{\Theta}$ in mmWave channels is sparse (see Appendix 4.A), we can use the following precoder and combiner:

- $\boldsymbol{B}_A = \boldsymbol{F}_{M_2}$ with all columns set to zero, except those corresponding to the non-zero columns of $\boldsymbol{\Theta}$
- $\boldsymbol{C}_A = \boldsymbol{F}_{M_1}$ with all rows set to zero, except those corresponding to the non-zero rows of $\boldsymbol{\Theta}$.

Both matrices can be implemented in analog by phase shifters and switches to active selective elements.

9.6 Multiuser MIMO: broadcast channel

We have already mentioned in Section 9.1 that spatial multiplexing may enable simultaneous transmissions (by SDMA) among multiple devices or *users*, in what is called *multiuser MIMO* (MU-MIMO), in contrast to single-user MIMO (SU-MIMO) systems considered until now in this chapter. In particular, we focus on two relevant MU-MIMO scenarios: the *broadcast* or *downlink* channel and the *multiuser access* or *uplink* channel. The former scenario (treated in this section) occurs for example in a cellular system when a base station transmits to multiple users at the same time on the same frequency, and either or both the transmitter and receivers are equipped with multiple antennas. The latter scenario (treated in Section 9.7) still occurs in a cellular system when multiple users transmit simultaneously on the same frequency to the same base station. Another example of MU-MIMO is the communication between stations and the access point (AP) of a wireless local area network (WLAN).

With reference to the broadcast channel, let the transmitter be equipped with N_{Tx} antennas, and the U users be equipped with N_{Rc} antennas each. [4] Let $\{\mathcal{G}_{C,u}\}$ be the $N_{Rc} \times N_{Tx}$ channel matrices for users $u = 1, 2, \ldots, U$, and let

$$\mathcal{G}_C = [\mathcal{G}_{C,1}^T, \ldots, \mathcal{G}_{C,U}^T]^T \tag{9.145}$$

be the $UN_{Rc} \times N_{Tx}$ overall channel matrix. The input–output relation for user u can be written as

$$\boldsymbol{x}_u = \mathcal{G}_{C,u} \boldsymbol{s} + \boldsymbol{w}_u \tag{9.146}$$

where $\boldsymbol{w}_u$ is the noise vector, and $\boldsymbol{s}$ is the transmitted signal containing data for all users. In general, the transmission of data symbols for different users will generate *multiuser interference* at the receiver, i.e. the reception of a user signal will be interfered by the reception of other users' signals. This can be seen as a generalization of the MIMO interference of SU-MIMO. However, in the broadcast MU-MIMO channel, signals received by different users can only be processed locally, rather than jointly among all receiving antennas as occurs in SU-MIMO systems.

---

[4] For the sake of a simpler presentation, we focus on users with the same number of antennas. Its extension to more general cases is straightforward.

CSI only at the receivers

Consider now a broadcast MU-MIMO channel where CSI is available only at the receivers, i.e. the transmitter does not know the channel. We can apply at each user the MIMO combiners of Section 9.2.2 or the non-linear MIMO detection and decoding schemes of Section 9.2.3. Each user receives (and potentially decodes, e.g. in V-BLAST) data streams of all users and is able to decode its own signal only if $N_{Rc} \geq N_{Tx}$. Note that this condition rarely occurs in wireless systems, such as cellular systems or WLANs, where users have been equipped with a single antenna for a long time, while base stations or access points are equipped with multiple antennas. Only recently the use of multiple antennas at end-user mobile power-supplied devices is spreading.

CSI only at the transmitter

When CSI is available only at the transmitter of a broadcast MU-MIMO channel, all techniques for SU-MIMO systems developed under the hypothesis of having CSI only at the transmitter (see Section 9.3) are immediately applicable, as we still have a single transmitter. As a result of precoding, each user will see an equivalent SISO channel on each receive antenna.

9.6.1 CSI at both the transmitter and the receivers

When CSI is available at both the transmitter and the receivers, we can adopt specific solutions for a broadcast MU-MIMO channel. We assume here that $N_{Tx} \geq UN_{Rc}$, so that we can transmit up to $N_{S,u} = N_{Rc}$ streams per user u, for a total number of $N_S = UN_{Rc}$ streams.

We will consider two linear precoding and combining techniques suited for the broadcast MU-MIMO channel. In particular, we generate the transmitted vector as

$$s = \sum_{u=1}^{U} \boldsymbol{B}_u \boldsymbol{a}_u \tag{9.147}$$

where $\{\boldsymbol{B}_u\}$ are $N_{Tx} \times N_{S,u}$ precoding matrices, one for each user, and $\boldsymbol{a}_u = [a_{u,1}, \dots, a_{u,N_{S,u}}]^T$ is the data vector for user u. We also define the overall precoding matrix

$$\boldsymbol{B} = [\boldsymbol{B}_1, \dots, \boldsymbol{B}_U] \tag{9.148}$$

Block diagonalization

Block diagonalization is a linear precoding technique such that

$$\mathcal{G}_{C,u}\boldsymbol{B}_p = \begin{cases} \boldsymbol{0} & p \neq u \\ \mathcal{G}_{eq,u} & p = u \end{cases} \tag{9.149}$$

Hence, the cascade of precoding and MU-MIMO channel results in a block-diagonal matrix, i.e.

$$\mathcal{G}_C\boldsymbol{B} = \begin{bmatrix} \mathcal{G}_{eq,1} & \boldsymbol{0} & \cdots & \boldsymbol{0} \\ \boldsymbol{0} & \ddots & \ddots & \vdots \\ \vdots & \ddots & \ddots & \boldsymbol{0} \\ \boldsymbol{0} & \cdots & \boldsymbol{0} & \mathcal{G}_{eq,U} \end{bmatrix} \tag{9.150}$$

and $\mathcal{G}_{eq,u}$, $u = 1, \dots, U$, is the $N_{Rc} \times N_{S,u}$ resulting channel matrix for user u. From (9.146) and (9.147), the vector signal received at user u is

$$\boldsymbol{x}_u = \mathcal{G}_{C,u}\left(\sum_{p=1}^{U} \boldsymbol{B}_p \boldsymbol{a}_p\right) + \boldsymbol{w}_u = \mathcal{G}_{eq,u}\boldsymbol{a}_u + \boldsymbol{w}_u \tag{9.151}$$

i.e. we have an equivalent SU-MIMO channel, with input $\boldsymbol{a}_u$ and output $\boldsymbol{x}_u$. For this channel, we apply precoding the techniques (possibly using also further combining) described in the previous sections of this chapter.

In order to obtain (9.149), $\boldsymbol{B}_p$ must be in the right null space of $\mathcal{G}_{C,p}$ for $p \neq u$, or still equivalently, defining the $N_{Rc}(U-1) \times N_{Tx}$ matrix

$$\mathcal{H}_u = [\mathcal{G}_{C,1}^T, \dots, \mathcal{G}_{C,u-1}^T, \mathcal{G}_{C,u+1}^T, \dots, \mathcal{G}_{C,U}^T]^T \tag{9.152}$$

$\boldsymbol{B}_u$ must be in the right null space of $\mathcal{H}_u$, i.e.

$$\mathcal{H}_u \boldsymbol{B}_u = \boldsymbol{0} \tag{9.153}$$

By taking the SVD of $\mathcal{H}_u$, we obtain $N_{nz} = \text{rank}(\mathcal{H}_u)$ non-zero singular values and

$$\mathcal{H}_u = \boldsymbol{U}_{H_u} \text{ diag}([\nu_{H_u,1}, \dots, \nu_{H_u,N_{nz}}, 0, \dots, 0]) \; [\boldsymbol{V}_{H_u,nz} \boldsymbol{V}_{H_u,z}]^H \tag{9.154}$$

where $\boldsymbol{V}_{H_u,z}$ is the $N_{Rc} \times N_{Tx}$ matrix containing the right singular vectors corresponding to the right null space of $\mathcal{H}_u$. Therefore, by setting

$$\boldsymbol{B}_u = \boldsymbol{V}_{H_u,z} \tag{9.155}$$

we satisfy (9.149), and

$$\mathcal{G}_{eq,u} = \mathcal{G}_{C,u} \boldsymbol{V}_{H_u,z} \tag{9.156}$$

Note that $\boldsymbol{V}_{H_u,z}$ has unit norm, thus the power constraint (9.117) is satisfied. Indeed, block diagonalization can be seen as a generalization of ZF, where instead of imposing a diagonal equivalent MIMO channel we impose a block-diagonal equivalent multiuser MIMO channel.

For example, if we use the *capacity-achieving* linear precoder and combiner of Section 9.4 for each user, and denote with $\boldsymbol{B}_{eq,u}$ and $\boldsymbol{C}_{eq,u}$ the precoder and combiner matrices obtained from the SVD of $\mathcal{G}_{eq,u}$, the overall precoder for user u (cascade of block diagonalization and user precoding) is $\boldsymbol{B}_{BC,u} = \boldsymbol{B}_u \boldsymbol{B}_{eq,u}$. By collecting all precoders into matrix $\boldsymbol{B}_{BC} = [\boldsymbol{B}_{BC,1}, \dots, \boldsymbol{B}_{BC,U}]$ and all combiners into matrix $\boldsymbol{C}_{BC}$, block-diagonal with blocks $\boldsymbol{C}_{eq,u}$, $u = 1, \dots, U$, we obtain the overall equivalent channel

$$\boldsymbol{C}_{BC}^T \mathcal{G}_C \boldsymbol{B}_{BC} = \boldsymbol{\Sigma}_{BC} \tag{9.157}$$

where $\boldsymbol{\Sigma}_{BC}$ is a diagonal matrix containing on the diagonal the singular values of matrices $\mathcal{G}_{eq,u}$, $u = 1, \dots, U$.

User selection

Multiuser interference can be avoided by transmitting only to a subset of users that do not interfere with each other. In particular, out of the U users, we select $U' \leq U$ users with indices $\mathcal{U}' = \{u_1', \dots, u_{U'}'\}$, such that their channel matrices are *semi-orthogonal*, i.e.

$$\mathcal{G}_{C,p} \mathcal{G}_{C,q}^H \simeq \boldsymbol{0} \qquad \text{for } p,\, q \in \mathcal{U}' \text{ and } p \neq q \tag{9.158}$$

Let $\mathcal{G}_C' = [\mathcal{G}_{C,u_1}^T, \dots, \mathcal{G}_{C,u_{U'}}^T]^T$ the new overall channel. By choosing the user precoder

$$\boldsymbol{B}_{u_j'} = \frac{\mathcal{G}_{C,u_j'}^H}{||\mathcal{G}_{C,u_j'}||} \tag{9.159}$$

for $j = 1, \dots, U'$, and setting $\boldsymbol{B}' = [\boldsymbol{B}_{u_1'}, \dots, \boldsymbol{B}_{u_{U'}'}]$, from (9.158) we have that $\mathcal{G}_C' \boldsymbol{B}'$ is block-diagonal with blocks of size $N_{Rc} \times N_{S,u_i'}$, thus avoiding multiuser interference. The resulting equivalent SU-MIMO channel for user u is

$$\mathcal{G}_{eq,u} = \frac{\mathcal{G}_{C,u} \mathcal{G}_{C,u}^H}{||\mathcal{G}_{C,u}||} \tag{9.160}$$

for which we can apply precoding/combining SU-MIMO techniques. Note that for single-antenna users ($N_{Rc} = 1$), the precoder is simply the MRC MISO linear precoder (9.111).

Various approaches have been proposed in the literature to select users such that condition (9.158) holds (see [32] for a survey). For example, in [33] a greedy iterative *semi-orthogonal user selection* algorithm has been proposed, where set $\mathcal{U}'$ is built by adding one user at each iteration. In particular, for single-antenna users ($N_{Rc} = 1$), we first select the user having the largest channel gain (set $\mathcal{O}_1 = \{1, \dots, U\}$), with $\mathcal{G}_{C,u}$ $1 \times N_{Tx}$ row vector,

$$u'_1 = \arg\max_{u \in \mathcal{O}_1} ||\mathcal{G}_{C,u}||^2 \tag{9.161}$$

Then, we select the subset of users whose channels are almost orthogonal to $\mathcal{G}_{C,u'_1}$, according to the correlation coefficient

$$\mathtt{r}_{p,u'_1} = \frac{|\mathcal{G}_{C,p}\mathcal{G}^H_{C,u'_1}|}{||\mathcal{G}_{C,p}||\ ||\mathcal{G}_{C,u'_1}||} \tag{9.162}$$

In particular, let $\mathcal{O}_2 = \{p \in \mathcal{O}_1 : \mathtt{r}_{p,u'_1} < \epsilon\}$, with ϵ a small-enough positive number, be the set of users whose channels have a small correlation with that of user u'_1. Among these users, we select the user with the largest channel gain, i.e.

$$u'_2 = \arg\max_{u \in \mathcal{O}_2} ||\mathcal{G}_{C,u}||^2 \tag{9.163}$$

The process is iterated, where at iteration $i > 1$, we extract the user subset

$$\mathcal{O}_{i+1} = \{p \in \mathcal{O}_i : \mathtt{r}_{p,u'_i} < \epsilon\} \tag{9.164}$$

and within $\mathcal{O}_{i+1}$, we select the user u'_{i+1} with the largest channel gain.

Clearly, user selection can also be applied in conjunction with block diagonalization: first, we select almost-orthogonal users and then we apply block diagonalization on the set of selected users to completely remove interference. With this choice, we avoid having to remove interference among users with highly correlated channels, thus saving in transmit power.

Joint spatial division and multiplexing

Both block diagonalization and user selection require the knowledge of CSI $\mathcal{G}_C$ at the transmitter. As discussed in Section 9.3.4, using CSI feedback techniques, the number of feedback bits is proportional to the number of users and increases for time-varying channels, as CSI must be refreshed whenever the channel changes.

The joint spatial division and multiplexing (JSDM) technique [34] provides an alternative solution, reducing the need to have updated CSI from all users at the transmitter, by exploiting the fact that users' channels are typically spatially correlated and their autocorrelation matrices change much more slowly than the instantaneous channel realizations. Also in this case, for the sake of a simpler explanation, let us consider single-antenna users, i.e. $N_{Rc} = 1$. We consider the SVD of the channel autocorrelation matrix of user u (see also the transmit correlation matrix in (4.147))

$$\boldsymbol{R}_u = E[\mathcal{G}_{C,u}\mathcal{G}^H_{C,u}] = \boldsymbol{U}_u\boldsymbol{\Lambda}_u\boldsymbol{U}^H_u \tag{9.165}$$

where $\boldsymbol{\Lambda}_u$ is diagonal with real non-negative entries. For zero-mean Gaussian channel matrices, we can decompose the channel vector $\mathcal{G}_{C,u}$ as (see (4.147) for MIMO channel matrix with correlated Gaussian entries)

$$\mathcal{G}_{C,u} = \boldsymbol{g}^T_u\boldsymbol{\Lambda}^{1/2}_u\boldsymbol{U}^H_u \tag{9.166}$$

with $\boldsymbol{g}_u$ a column vector with $\mathcal{CN}(0, 1)$ independent entries.

Let us now group users, where users in the same group have the same autocorrelation matrix. Let G be the number of groups, and $\boldsymbol{R}_i$, $i = 1, \dots, G$, the corresponding autocorrelation matrices, and $\mathcal{U}_i$ is the set of user indices in group i. We can select a subset of groups $\{i_1, \dots, i_{G'}\}$ such that they have orthogonal crosscorrelation matrices, i.e. it is

$$\boldsymbol{U}_i^H \boldsymbol{U}_j = \boldsymbol{0}, \qquad i \neq j, \quad i,j \in \{i_1, \dots, i_{G'}\} \tag{9.167}$$

Clearly, from (2.142), we also have $\boldsymbol{U}_i^H \boldsymbol{U}_i = \boldsymbol{I}$ for $i \in \{i_1, \dots, i_{G'}\}$. Now, the precoding matrix for user $u \in \mathcal{U}_i$ is the cascade of a suitable *per-user precoding* matrix $\boldsymbol{B}'_u$, to be suitably designed, and the *pre-beamforming* matrix $\boldsymbol{U}_i$, i.e.

$$\boldsymbol{B}_u = \boldsymbol{U}_i \boldsymbol{B}'_u \tag{9.168}$$

With this precoding, the received vector signal can be written as

$$\boldsymbol{x}_u = \mathcal{G}_C \sum_{p=1}^{U} \boldsymbol{B}_p \boldsymbol{s} + \boldsymbol{w}_u = \sum_{p \in \mathcal{U}_i} \boldsymbol{g}_p^T \boldsymbol{\Lambda}_p^{1/2} \boldsymbol{B}'_p \boldsymbol{a}_p + \boldsymbol{w}_u \tag{9.169}$$

Note that the interference of users belonging to other groups has been eliminated by $\boldsymbol{U}_i$, and only the interference of users of the same group remains. For the design of $\boldsymbol{B}'_u$, we can resort to block diagonalization for users in group i. Note that pre-beamforming simplifies the design of the per-user precoder, as it significantly reduces the dimensionality of the precoder from U to $|\mathcal{U}_i|$. Furthermore, the pre-beamformer is designed on slowly changing channel characteristics, thus can be kept constant over multiple transmissions and for all users of the group. Moreover, pre-beamforming is amenable for an analog implementation in a hybrid beamforming scheme (see Section 9.5).

9.6.2 Broadcast channel estimation

The channel estimation at the receivers of a MIMO broadcast channel does not poses major challenges. Indeed, the transmitter can transmit a single sequence of symbols (known to all receivers), from which each receiver estimates its own channel.

When CSIT is needed, we can either estimate the channel from the receiver to the transmitter considering the resulting multiple-access channel (see Section 9.7.2), or (if possible) we can exploit the channel reciprocity, estimating the downlink channel at the receiver and then feeding back a quantized version to the transmitter.

9.7 Multiuser MIMO: multiple-access channel

With reference to the multiple-access MU-MIMO channel, U users are equipped with N_{Tx} antennas each, and the receiver has N_{Rc} antennas.[5] Let $\mathcal{G}_{C,u}$ be the $N_{Rc} \times N_{Tx}$ channel matrix for user $u = 1, 2, \dots, U$, and let

$$\mathcal{G}_C = [\mathcal{G}_{C,1}, \dots, \mathcal{G}_{C,U}] \tag{9.170}$$

be the $N_{Rc} \times UN_{Tx}$ overall channel matrix. The signal vector at the receiver can be written as

$$\boldsymbol{x} = \sum_{u=1}^{U} \mathcal{G}_{C,u} \boldsymbol{s}_u + \boldsymbol{w} \tag{9.171}$$

where $\boldsymbol{s}_u$ is the transmitted vector by user i and $\boldsymbol{w}$ is the noise vector. Clearly, also in this case we have multiuser interference. Let also $\boldsymbol{a}_u = [a_{u,1}, \dots, a_{u,N_{S,u}}]^T$ be the data vector to be transmitted to user u, and $\boldsymbol{a} = [\boldsymbol{a}_1^T, \dots, \boldsymbol{a}_U^T]^T$.

[5] Also in this case, more general solutions with a different number of transmit antennas per user can be easily obtained.

CSI only at the transmitters

We consider here the case wherein CSI is known only at the U transmitters. We also assume that $N_{Rc} = U$, and each user transmits a single data stream to the receiver.

We may linearly precoding the data signal at each user in order to completely eliminate the multiuser interference at the receiver. This can be achieved by ensuring that data signal of user u is received only by antenna u. [6] To this end, let us define the $N_{Rc} \times (N_{Rc} - 1)$ matrix

$$\overline{\mathcal{G}}_{C,u} = [\mathcal{G}_{C,u}^{(1,\cdot)T}, \dots, \mathcal{G}_{C,u}^{(u-1,\cdot)T}, \mathcal{G}_{C,u}^{(u+1,\cdot)T}, \dots, \mathcal{G}_{C,u}^{(N_{Rc},\cdot)T}]^T \tag{9.172}$$

i.e. the user channel matrix $\mathcal{G}_{C,u}$ without the row corresponding to antenna u. We can then apply a procedure similar to block diagonalization, where user u utilizes as precoder $\boldsymbol{B}_u$ the vector of the right null space of $\overline{\mathcal{G}}_{C,u}$. Then, from (9.171), we have

$$x^{(r)} = \sum_{u=1}^{U} \mathcal{G}_{C,u}^{(r,\cdot)} \boldsymbol{B}_u a_u + \boldsymbol{w} = \mathcal{G}_{eq}^{(r)} a_r + w^{(r)} \tag{9.173}$$

where

$$\mathcal{G}_{eq}^{(u)} = \mathcal{G}_{C,u}^{(u,\cdot)} \boldsymbol{B}_u \tag{9.174}$$

Note from (9.173) that we have eliminated the multiuser interference.

CSI only at the receiver

Consider now a multiple-access MU-MIMO channel, where CSI is available only at the receiver. We can apply at the receiver the MIMO combiner of Section 9.2.2 or the non-linear MIMO detection and decoding of Section 9.2.3, seeing antennas of different users as antennas of a single user, wherein each antenna transmits an independent data signal. In this case, it must be $N_{Rc} \geq U N_{Tx}$ for proper detection.

9.7.1 CSI at both the transmitters and the receiver

We now consider the knowledge of CSI at both the transmitters and the receiver for the multiuser access MU-MIMO channel. First note that if the users are equipped with a single antenna each ($N_{Tx} = 1$), CSI at the transmitter is of no use and we can only resort to solutions using CSI at the receiver. In this latter case, we can also apply user selection techniques, as those described for the broadcast channel, in order to select users with semi-orthogonal channels and ease the processing at the receiver. An example is given by cellular systems where the mobile user's terminal has been traditionally equipped with a single antenna.

We now consider the case wherein each user is equipped with multiple antennas, i.e. $N_{Rc} > 1$.

Block diagonalization

For the multiuser access channel with $N_{S,u}$ streams for user u, we aim at designing a block-diagonal precoding matrix $\boldsymbol{B}_{MA}$ (with blocks of size $N_{Tx} \times N_{S,u}$) and a full combining matrix $\boldsymbol{C}_{MA}$ such that

$$\boldsymbol{C}_{MA}^T \mathcal{G}_C \boldsymbol{B}_{MA} = \boldsymbol{\Sigma} \tag{9.175}$$

[6] Note that the association between the users and the receive antennas can be optimized according to instantaneous channel conditions, although it then requires some coordination among users. Here, we consider the simple association of user u with receive antenna u.

with $\boldsymbol{\Sigma}$ a block-diagonal matrix. This is similar to (9.157), obtained by block diagonalization of the broadcast MU-MIMO channel, where however there the combining matrix was block-diagonal while the precoding matrix was full. Now, let us define the equivalent broadcast channel matrix as

$$\mathcal{G}_{C,BC} = \mathcal{G}_C^T \tag{9.176}$$

Note that for uplink and downlink transmissions occurring on the same channel, (9.176) is given by the channel reciprocity. In any case, we will take (9.176) as the definition of matrix $\mathcal{G}_{C,BC}$ without further physical meanings. Then, we can design matrices $\boldsymbol{C}_{BC}$ and $\boldsymbol{B}_{BC}$ according to (9.157), where we assume a broadcast channel for a transmitter with N_{Rc} antennas and receivers with N_{Tx} antennas each. Lastly, if we set

$$\boldsymbol{B}_{MA} = \boldsymbol{C}_{BC}, \qquad \boldsymbol{C}_{MA} = \boldsymbol{B}_{BC} \tag{9.177}$$

we note that we have a block-diagonal precoder, suited for uplink transmission, and a full combining matrix. Moreover, observing that

$$\boldsymbol{\Sigma} = \boldsymbol{\Sigma}^T = \boldsymbol{B}_{BC}^T \mathcal{G}_{C,BC}^T \boldsymbol{C}_{BC} = \boldsymbol{C}_{MA}^T \mathcal{G}_C \boldsymbol{B}_{MA} \tag{9.178}$$

we conclude that with definitions (9.177), condition (9.175) holds. Thus, we have obtained an interference-free MIMO channel with linear precoding and combining, where the pre-coding is performed on a per-user basis. Clearly, this approach requires CSI at both the transmitter and the receiver.

9.7.2 Multiple-access channel estimation

Channel estimation at the receiver in multiple-access channel requires the estimation of the $N_{Rc} \times UN_{Tx}$ channel matrix (9.170) between a transmitter with UN_{Tx} antennas and N_{Rc} antennas. If each user is transmitting different pilot sequences, the receiver jointly estimates all the users' channels. Note that this solution requires at least the transmission of UN_{Tx} symbol vectors by each user, with a sequence length proportional to the number of users. Moreover, we need orthogonal training blocks (see (9.105)).

For CSIT, if we can exploit channel reciprocity, the transmitter can estimate the receiver-transmitters channel at the transmitters by exploiting the broadcast channel estimation techniques of Section 9.6.2. Otherwise, we can estimate the channel at the receiver and then feed back (quantized) estimated channels to the transmitters.

9.8 Massive MIMO

The term *massive MIMO* refers to a MIMO system wherein the number of either or both the transmit and the receive antennas grows to infinity, i.e.

$$N_{Tx} \to \infty \quad \text{and/or} \quad N_{Rc} \to \infty \tag{9.179}$$

Massive MIMO systems were first studied in [35].

Operating under such asymptotic conditions has a number of consequences. Here we indicate two main benefits of massive MIMO systems, namely the channel hardening and the favourable propagation.

9.8.1 Channel hardening

Typically, massive MIMO is referred to channels with uncorrelated Gaussian entries (see (4.154))

$$\boldsymbol{R}_{vec(\mathcal{G}_C)} = \boldsymbol{I} \tag{9.180}$$

This scenario occurs when a large number of scatterers is available. Note that this may not always be the case, such as in the relevant case of mmWave channels, wherein the number of scatterers is low, although a very large number of antennas may be employed.

As a consequence of conditions (9.179) and (9.180), we obtain the *channel hardening phenomenon*. Consider a MISO system, then from (9.180) and the law of large numbers we have

$$\lim_{N_{Tx}\to\infty} \frac{||\mathcal{G}_C||^2}{N_{Tx}} = 1 \tag{9.181}$$

Therefore, the channel gain of any link tends to a constant. For example, using the MISO linear precoding from (9.112), we obtain

$$\gamma_{LP} \to \gamma_{MF} N_{Tx} \tag{9.182}$$

As a consequence of channel hardening, the effects of fading disappears and any receiver experiences always the same SNR at the decoder input. Similar results hold for a SIMO system using a linear combiner matched to the channel, and also for a general massive MIMO system.

Avoiding fading phenomenon significantly simplifies also the resource allocation (power allocation but also subcarrier allocation to multiple users in OFDM systems), since channels conditions are time-invariant and the same for all receivers.

9.8.2 Multiuser channel orthogonality

Consider now a broadcast massive MISO channel, where the transmitter has N_{Tx} antennas and each of the U user has a single ($N_{Rc} = 1$) antenna. Assume that users' channel $\mathcal{G}_{C,u}$, $u = 1, \dots, U$, is zero-mean statistically independent, then we have

$$\lim_{N_{Tx}\to\infty} \mathcal{G}_{C,u_1} \mathcal{G}_{C,u_2}^H = 0, \qquad u_1 \neq u_2, \quad u_1, u_2 = 1, \dots, U \tag{9.183}$$

In this case, users have orthogonal channels and condition (9.158) holds with equality. This condition is also denoted as *favourable propagation*. Therefore, a simple linear precoder matched to each user separately

$$\boldsymbol{B}_u = \mathcal{G}_{C,u}^H \tag{9.184}$$

maximizes the SNR at the detector input, and avoids any multiuser interference.

Bibliography

[1] Jakes, W.C. (1993). *Microwave Mobile Communications*. New York, NY: IEEE Press.

[2] Glance, B. and Greenstein, L.J. (1983). Frequency-selective fading in digital mobile radio with diversity combining. *IEEE Transactions on Communications* 31: 1085–1094.

[3] Kennedy, W.K., Greenstein, L.J., and Clark, M.V. (1994). Optimum linear diversity receivers for mobile communications. *IEEE Transactions on Vehicular Technology* 43: 47–56.

[4] Clark, M.V., Greenstein, L.J., Kennedy, W.K., and Shafi, M. (1992). MMSE diversity combining for wide-band digital cellular radio. *IEEE Transactions on Communications* 40: 1128–1135.

[5] Compton, R.T. (1988). *Adaptive Antennas: Concepts and Performance*. Englewood Cliffs, NJ: Prentice-Hall.

[6] Falconer, D.D., Abdulrahman, M., Lo, N.W.K. et al. (1993). Advances in equalization and diversity for portable wireless systems. *Digital Signal Processing* 3: 148–162.

[7] Foschini, G.J. (1996). Layered space-time architecture for wireless communication in a fading environment when using multi-element antennas. *Bell System Technical Journal* 1: 41–59.

[8] Godara, L.C. (1997). Application of antenna arrays to mobile communications - Part I: Performance improvement, feasibility and system considerations. *IEEE Proceedings* 85: 1031–1060.

[9] Godara, L.C. (1997). Application of antenna arrays to mobile communications - Part II: Beam-forming and direction-of-arrival considerations. *IEEE Proceedings* 85: 1195–1245.

[10] Kirsteins, I.P. and Tufts, D.W. (1994). Adaptive detection using low rank approximation to a data matrix. *IEEE Transactions on Aerospace and Electronic Systems* 30: 55–67.

[11] Kohno, R., Imai, H., Hatori, M., and Pasupathy, S. (1990). Combination of an adaptive array antenna and a canceller of interference for direct-sequence spread-spectrum multiple-access system. *IEEE Journal on Selected Areas in Communications* 8: 675–682.

[12] Lee, W.C.Y. (1973). Effects on correlation between two mobile radio base-station antennas. *IEEE Transactions on Communications* 21: 1214–1224.

[13] Monzingo, R.A. and Miller, T.W. (1980). *Introduction to Adaptive Arrays*. New York, NY: Wiley.

[14] Paulraj, A.J. and Papadias, C.B. (1997). Space-time processing for wireless communications. *IEEE Signal Processing Magazine* 14: 49–83.

[15] Salz, J. and Winters, J.H. (1994). Effect of fading correlation on adaptive arrays in digital wireless communications. *IEEE Transactions on Vehicular Technology* 43: 1049–1057.

[16] Tsoulos, G., Beach, M., and McGeehan, J. (1997). Wireless personal communications for the 21st century: European technological advances in adaptive antennas. *IEEE Communications Magazine* 35: 102–109.

[17] Widrow, B., Mantey, P.E., Griffiths, L.J., and Goode, B.B. (1967). Adaptive antenna systems. *IEEE Proceedings* 55: 2143–2159.

[18] Winters, J.H. (1998). Smart antennas for wireless systems. *IEEE Personal Communications Magazine* 5: 23–27.

[19] Balaban, P. and Salz, J. (1991). Dual diversity combining and equalization in digital cellular mobile radio. *IEEE Transactions on Vehicular Technology* 40: 342–354.

[20] Foschini, G.J. (1977). A reduced-state variant of maximum-likelihood sequence detection attaining optimum performance for high signal-to-noise ratio performance. *IEEE Transactions on Information Theory* 24: 505–509.

[21] Chau, Y.A. and Yu, S.-H. (2001). Space modulation on wireless fading channels. *IEEE 54th Vehicular Technology Conference. VTC Fall 2001. Proceedings (Cat. No.01CH37211)*, Volume 3, pp. 1668–1671.

[22] Alamouti, S.A. (1998). A simple transmit diversity technique for wireless communications. *IEEE Journal on Selected Areas in Communications* 16: 1451–1458.

[23] DVB (2009). Frame structure channel coding and modulation for a second generation digital terrestrial television broadcasting system (DVB-T2), A122.

[24] 3GPP (2008). Evolved universal terrestrial radio access (E-UTRA); physical layer procedures, TS 36.213, v. 8.5.0.

[25] Tarokh, V., Seshadri, N., and Calderbank, A.R. (1998). Space-time codes for high data rate wireless communication: performance criterion and code construction. *IEEE Transactions on Information Theory* 44: 744–765.

[26] Naguib, A., Tarokh, V., Seshadri, N., and Calderbank, A.R. (1998). A space-time coding modem for high data rate wireless communications. *IEEE Journal on Selected Areas in Communications* 16: 1459–1478.

[27] Tarokh, V., Naguib, A., Seshadri, N., and Calderbank, A.R. (1999). Combined array processing and space-time coding. *IEEE Transactions on Information Theory* 45: 1121–1128.

[28] Golub, G.H. and van Loan, C.F. (1989). *Matrix Computations*, 2e. Baltimore, MD and London: The Johns Hopkins University Press.

[29] Costa, M. (1983). Writing on dirty paper (corresp.). *IEEE Transactions on Information Theory* 29: 439–441.

[30] Kim, T., Kwon, K., and Heo, J. (2017). Practical dirty paper coding schemes using one error correction code with syndrome. *IEEE Communications Letters* 21: 1257–1260.

[31] Abbas, W.B., Gomez-Cuba, F., and Zorzi, M. (2017). Millimeter wave receiver efficiency: A comprehensive comparison of beamforming schemes with low resolution ADCs. *IEEE Transactions on Wireless Communications* 16: 8131–8146.

[32] Casta neda, E., Silva, A., Gameiro, A., and Kountouris, M. (2017). An overview on resource allocation techniques for multi-user MIMO systems. *IEEE Communications Surveys Tutorials* 19: 239–284.

[33] Yoo, T. and Goldsmith, A. (2006). On the optimality of multiantenna broadcast scheduling using zero-forcing beamforming. *IEEE Journal on Selected Areas in Communications* 24: 528–541.

[34] Adhikary, A., Nam, J., Ahn, J., and Caire, G. (2013). Joint spatial division and multiplexing–the large-scale array regime. *IEEE Transactions on Information Theory* 59: 6441–6463.

[35] Marzetta, T.L. (2010). Noncooperative cellular wireless with unlimited numbers of base station antennas. *IEEE Transactions on Wireless Communications* 9: 3590–3600.

Chapter 10

Spread-spectrum systems

The term spread-spectrum systems [1–7] was coined to indicate communication systems in which the bandwidth of the signal obtained by a standard modulation method (see Chapter 6) is spread by a certain factor before transmission over the channel, and then despread, by the same factor, at the receiver. The operations of spreading and despreading are the inverse of each other, i.e. for an ideal and noiseless channel the received signal after despreading is equivalent to the transmitted signal before spreading. For transmission over an ideal additive white Gaussian noise (AWGN) channel, these operations do not offer any improvement with respect to a system that does not use spread-spectrum techniques. However, the practical applications of spread-spectrum systems are numerous, for example in multiple-access systems, narrowband interference rejection, and transmission over channels with fading, as discussed in Section 10.2.

10.1 Spread-spectrum techniques

We consider the two most common spread-spectrum techniques: *direct sequence* (DS) and *frequency hopping* (FH).

10.1.1 Direct sequence systems

The baseband equivalent model of a DS system is illustrated in Figure 10.1. We consider the possibility that U users in a multiple-access system simultaneously transmit, using the same frequency band, by *code division multiple access* (CDMA) (see Appendix 18.B).

The sequence of information bits $\{b_\ell^{(u)}\}$ of user u undergoes the following transformations.

Bit-mapper From the sequence of information bits, a sequence of i.i.d. symbols $\{a_k^{(u)}\}$ with statistical power $\mathtt{M}_a$ is generated. The symbols assume values in an M-ary constellation with symbol period T.

Spreading We indicate by the integer N_{SF} the *spreading factor*, and by T_{chip} the *chip period*. These two parameters are related to the symbol period T by the relation

$$T_{chip} = \frac{T}{N_{SF}} \tag{10.1}$$

Algorithms for Communications Systems and their Applications, Second Edition.
Nevio Benvenuto, Giovanni Cherubini, and Stefano Tomasin.

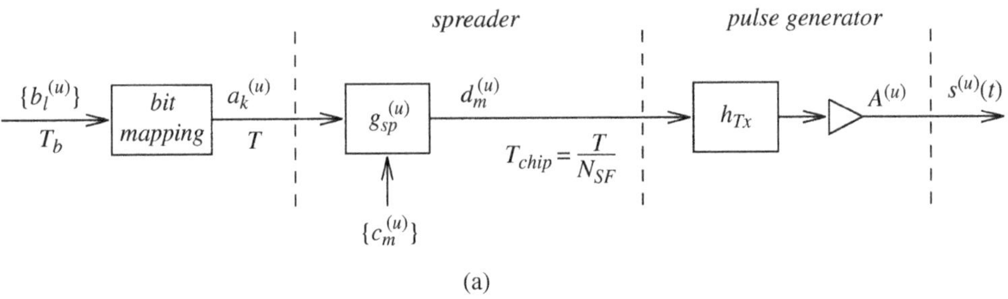

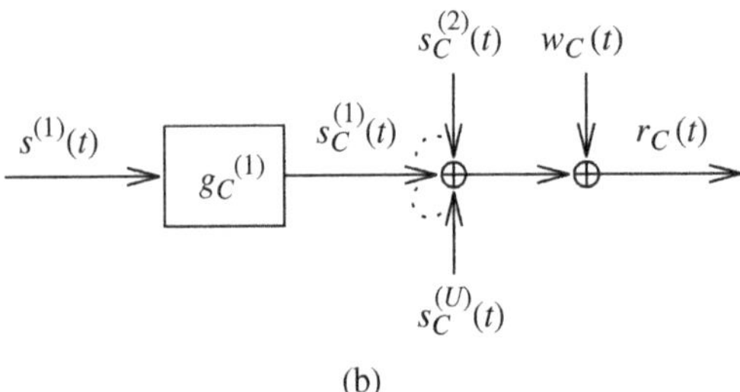

Figure 10.1 Baseband equivalent model of a DS system: (a) transmitter, (b) multiuser channel.

We consider the channelization code $\{c_{Ch,m}^{(u)}\}$, $u \in \{1, \ldots, U\}$, where $c_{Ch,m}^{(u)} \in \{-1, 1\}$ and $|c_{Ch,m}^{(u)}| = 1$. We assume that the sequences of the channelization code are periodic, that is $c_{Ch,m}^{(u)} = c_{Ch,m_{\text{mod } N_{SF}}}^{(u)}$. The channelization codes are orthogonal among users, if

$$\frac{1}{N_{SF}} \sum_{m=0}^{N_{SF}-1} c_{Ch,m}^{(u_1)} \, c_{Ch,m}^{(u_2)*} = \begin{cases} 1 & \text{if } u_1 = u_2 \\ 0 & \text{if } u_1 \neq u_2 \end{cases} \tag{10.2}$$

An example of orthogonal channelization code sequence is given by the Walsh–Hadamard sequences (see Appendix 10.A).

We also introduce the *user code* $\{c_m^{(u)}\}$, also called *signature sequence* or *spreading sequence*, that we initially assume equal to the channelization code, i.e.

$$c_m^{(u)} = c_{Ch,m}^{(u)} \tag{10.3}$$

Consequently, $\{c_m^{(u)}\}$ is also a periodic sequence of period N_{SF}.

The operation of spreading consists of associating to each symbol $a_k^{(u)}$ a sequence of N_{SF} symbols of period T_{chip}, that is obtained as follows. First, each symbol $a_k^{(u)}$ is repeated N_{SF} times with period T_{chip}. As illustrated in Figure 10.2, this operation is equivalent to upsampling $\{a_k^{(u)}\}$, so that $(N_{SF} - 1)$ zeros are inserted between two consecutive symbols, and using a *holder* of N_{SF} values. The obtained sequence is then multiplied by the user code. Formally we have

$$\begin{aligned} \bar{a}_m^{(u)} &= a_k^{(u)}, \qquad m = kN_{SF}, \ldots, kN_{SF} + N_{SF} - 1 \\ d_m^{(u)} &= \bar{a}_m^{(u)} \, c_m^{(u)} \end{aligned} \tag{10.4}$$

If we introduce the filter

$$g_{sp}^{(u)}(i\,T_{chip}) = c_i^{(u)}, \qquad i = 0, \ldots, N_{SF} - 1 \tag{10.5}$$

the correlation of Figure 10.2a can be replaced by the interpolator filter $g_{sp}^{(u)}$, as illustrated in Figure 10.2b.

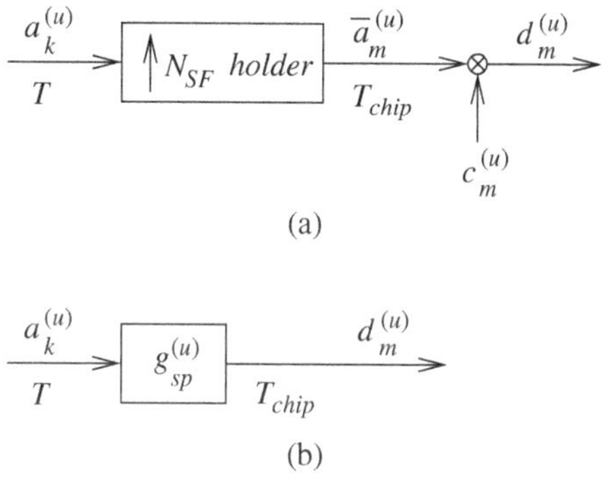

Figure 10.2 Spreading operation: (a) correlator, (b) interpolator filter.

Recalling that $|c_m^{(u)}| = 1$, from (10.4) we get

$$\mathtt{M}_d = \mathtt{M}_a \tag{10.6}$$

Pulse-shaping Let h_{Tx} be the modulation pulse, typically a square root raised cosine function or rectangular window. The baseband equivalent of the transmitted signal of user u is expressed as

$$s^{(u)}(t) = A^{(u)} \sum_{m=-\infty}^{+\infty} d_m^{(u)} \, h_{Tx}(t - m\,T_{chip}) \tag{10.7}$$

where $A^{(u)}$ accounts for the transmit signal power. In fact, if E_h is the energy of h_{Tx} and $\{d_m^{(u)}\}$ is assumed i.i.d., the average statistical power of $s^{(u)}$ is given by (see (1.324))

$$\overline{\mathtt{M}}_{s^{(u)}} = (A^{(u)})^2 \, \mathtt{M}_d \, \frac{E_h}{T_{chip}} \tag{10.8}$$

Using (10.4), an alternative expression for (10.7) is given by

$$s^{(u)}(t) = A^{(u)} \sum_{k=-\infty}^{+\infty} a_k^{(u)} \sum_{\ell=0}^{N_{SF}-1} c_{\ell+kN_{SF}}^{(u)} \, h_{Tx}(t - (\ell + kN_{SF})\,T_{chip}) \tag{10.9}$$

In the scheme of Figure 10.1a we note that, if condition (10.3) holds, then $g_{sp}^{(u)}$ is invariant with respect to the symbol period, and the two filters $g_{sp}^{(u)}$ and h_{Tx} can be combined into one filter (see also (10.9))

$$h_T^{(u)}(t) = \sum_{\ell=0}^{N_{SF}-1} c_\ell^{(u)} \, h_{Tx}(t - \ell\,T_{chip}) \tag{10.10}$$

thus providing

$$s^{(u)}(t) = A^{(u)} \sum_{k=-\infty}^{+\infty} a_k^{(u)} \, h_T^{(u)}(t - kT) \tag{10.11}$$

As shown in Figure 10.3, the cascade of spreader and pulse-shaping filter is still equivalent to the modulator of a quadrature amplitude modulation (QAM). The peculiarity is that the filter $h_T^{(u)}$ has a bandwidth much larger than the Nyquist frequency $1/(2T)$. Therefore, a DS system can be interpreted as a QAM system either with input symbols $\{d_m^{(u)}\}$ and transmit pulse h_{Tx}, or with input symbols $\{a_k^{(u)}\}$ and pulse $h_T^{(u)}$; later both interpretations will be used.

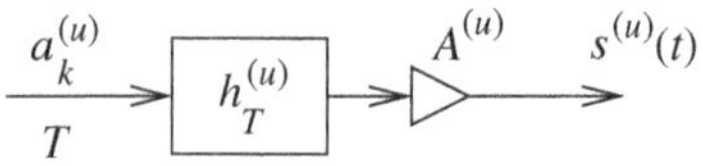

$$a_k^{(u)} \xrightarrow[T]{} \boxed{h_T^{(u)}} \rightarrow A^{(u)} \rightarrow s^{(u)}(t)$$

Figure 10.3 Equivalent scheme of spreader and pulse-shaping filter in a DS system.

Transmission channel Modelling the transmission channel as a filter having impulse response $g_C^{(u)}$, the output signal is given by

$$s_C^{(u)}(t) = (s^{(u)} * g_C^{(u)})(t) \tag{10.12}$$

Multiuser interference At the receiver, say of user 1, we have also the signal of other users (rather than user 1), namely signals $s_C^{(2)}(t), \ldots, s_C^{(U)}(t)$.

The possibility that many users transmit simultaneously over the same frequency band leads to a total signal

$$s_C(t) = \sum_{u=1}^{U} s_C^{(u)}(t) \tag{10.13}$$

We now aim at reconstructing the message $\{a_k^{(1)}\}$ of user $u = 1$, identified as *desired user*. If $\mathtt{M}_{s_C^{(u)}}$ is the statistical power of $s_C^{(u)}$, the following *signal-to-interference ratios* (SIRs) define the relative powers of the user signals:

$$\Gamma_i^{(u)} = \frac{\mathtt{M}_{s_C^{(1)}}}{\mathtt{M}_{s_C^{(u)}}}, \qquad u = 2, 3, \ldots, U \tag{10.14}$$

Noise In Figure 10.1b, the term w_C includes both the noise of the receiver and other interference sources, and it is modelled as white noise with power spectral density (PSD) equal to N_0.

Two signal-to-noise ratios (SNRs) are of interest. To measure the performance of the system in terms of P_{bit}, it is convenient to refer to the SNR defined in Chapter 6 for passband transmission, i.e.

$$\Gamma_s = \frac{\mathtt{M}_{s_C^{(1)}}}{N_0/T} \tag{10.15}$$

We recall that for an uncoded sequence of symbols $\{a_k^{(1)}\}$, the following relation holds:

$$\frac{E_b}{N_0} = \frac{\Gamma_s}{\log_2 M} \tag{10.16}$$

However, there are cases, as for example in the evaluation of the performance of the channel impulse response estimation algorithm, wherein it is useful to measure the power of the noise over the whole transmission bandwidth. Hence, we define

$$\Gamma_c = \frac{\mathtt{M}_{s_C^{(1)}}}{N_0/T_{chip}} = \frac{\Gamma_s}{N_{SF}} \tag{10.17}$$

Receiver The receiver structure varies according to the channel model and number of users. Deferring until Section 10.3, the analysis of more complicated system configurations, here we limit ourselves to an ideal AWGN channel with $g_C^{(u)}(t) = \delta(t)$ and *synchronous users*. The latter assumption implies that the transmitters of the various users are synchronized and transmit at the same instant. For an ideal AWGN channel, this means that at the receiver the optimum timing phase of signals of different users is the same.

With these assumptions, we verify that the optimum receiver is simply given by the matched filter to $h_T^{(1)}$. According to the analog or discrete-time implementation of the matched filters, we get the schemes of Figure 10.4 or Figure 10.5, respectively; note that in Figure 10.5 the receiver front-end comprises an anti-aliasing filter followed by a sampler with sampling period $T_c = T_{chip}/2$. Let t_0 be the optimum timing phase at the matched filter output (see Section 14.7).

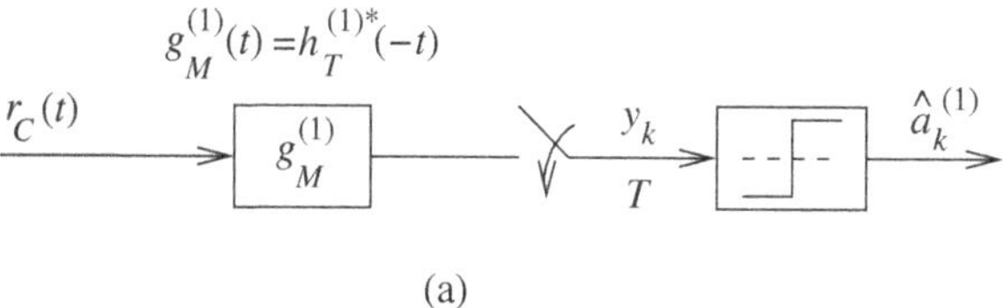

(a)

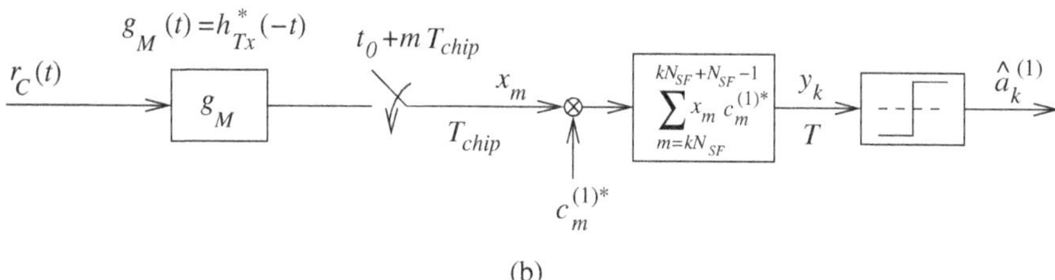

(b)

Figure 10.4 Optimum receiver with analog filters for a DS system with ideal AWGN channel and synchronous users. Two equivalent structures: (a) overall matched filter, (b) matched filter to h_{Tx} and despreading correlator.

For an ideal AWGN channel, it results

$$\begin{aligned} r_C(t) &= \sum_{u=1}^{U} s^{(u)}(t) + w_C(t) \\ &= \sum_{u=1}^{U} A^{(u)} \sum_{i=-\infty}^{+\infty} a_i^{(u)} \sum_{\ell=0}^{N_{SF}-1} c_\ell^{(u)} \, h_{Tx}(t - (\ell + iN_{SF})\,T_{chip}) + w_C(t) \end{aligned} \tag{10.18}$$

In the presence only of the desired user, that is for $U = 1$, it is clear that in the absence of inter-symbol interference (ISI), the structure with the matched filter to $h_T^{(1)}$ is optimum. We verify that the presence of other users is cancelled at the receiver, given that the various user codes are orthogonal.

We assume that the overall analog impulse response of the system is a Nyquist pulse, hence

$$\left.\left(h_{Tx} * g_C^{(u)} * g_{AA} * g_M\right)(t)\right|_{t=t_0+j\,T_{chip}} = (h_{Tx} * g_M)(t)|_{t=t_0+j\,T_{chip}} = E_h\,\delta_j \tag{10.19}$$

We note that, if t_0 is the instant at which the peak of the overall pulse at the output of g_M is observed, then t_0' in Figure 10.5 is given by $t_0' = t_0 - t_{g_M}$, where t_{g_M} is the duration of g_M. Moreover, from (10.19), we get that the noise at the output of g_M, sampled with sampling rate $1/T_{chip}$,

$$\tilde{w}_m = (w_C * g_M)(t)|_{t=t_0+m\,T_{chip}} \tag{10.20}$$

is an i.i.d. sequence with variance $N_0 E_h$.

Hence, from (10.9) and (10.19), the signal at the output of g_M, sampled with sampling rate $1/T_{chip}$, is

$$x_m = E_h \sum_{u=1}^{U} A^{(u)} \sum_{i=-\infty}^{+\infty} a_i^{(u)} \sum_{\ell=0}^{N_{SF}-1} c_{\ell+iN_{SF}}^{(u)} \, \delta_{m-\ell-iN_{SF}} + \tilde{w}_m \tag{10.21}$$

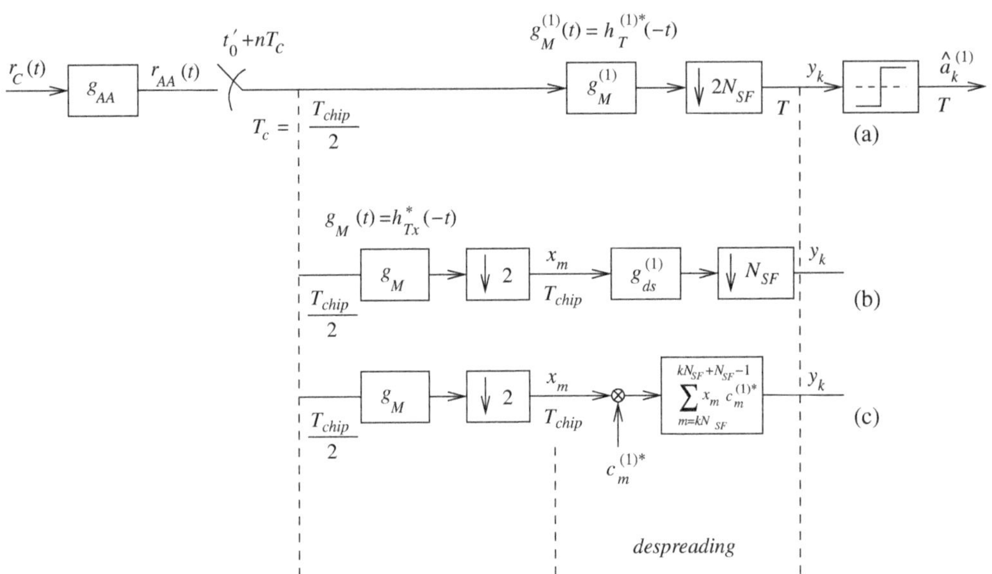

Figure 10.5 Optimum receiver with discrete-time filters for a DS system with ideal AWGN channel and synchronous users. Three equivalent structures: (a) overall matched filter, (b) matched filter to h_{Tx} and despreading filter, and (c) matched filter to h_{Tx} and despreading correlator.

With the change of indices $m = j + kN_{SF}, j = 0, 1, \dots, N_{SF} - 1$, k integer, we get

$$x_{j+kN_{SF}} = E_h \sum_{u=1}^{U} A^{(u)} \, a_k^{(u)} \, c_{j+kN_{SF}}^{(u)} + \tilde{w}_{j+kN_{SF}} \tag{10.22}$$

Despreading We now correlate the sequence of samples $\{x_m\}$, suitably synchronized, with the code sequence of the desired user, and we form the signal

$$y_k = \sum_{j=0}^{N_{SF}-1} x_{j+kN_{SF}} \, c_{j+kN_{SF}}^{(1)*} \tag{10.23}$$

As usual, introducing the filter $g_{ds}^{(u)}$ given by

$$g_{ds}^{(u)}(i\, T_{chip}) = c_{N_{SF}-1-i}^{(1)*}, \qquad i = 0, 1, \dots, N_{SF} - 1 \tag{10.24}$$

the correlation (10.23) is implemented through the filter (10.24), followed by a downsampler, as illustrated in Figure 10.5b.

Replacing (10.22) in (10.23) yields

$$y_k = N_{SF} \, E_h \sum_{u=1}^{U} A^{(u)} \, a_k^{(u)} \, \mathtt{r}_{c^{(u)}\, c^{(1)}}(0) + w_k \tag{10.25}$$

where in general

$$\mathtt{r}_{c^{(u_1)}\, c^{(u_2)}}(n_D) = \frac{1}{N_{SF}-|n_D|} \sum_{j=0}^{N_{SF}-1-|n_D|} c_{j+kN_{SF}+n_D}^{(u_1)} \, c_{j+kN_{SF}}^{(u_2)*}$$
$$n_D = -(N_{SF} - 1), \dots, -1, 0, 1, \dots, N_{SF} - 1 \tag{10.26}$$

is the cross-correlation at lag n_D between the user codes u_1 and u_2.

In the considered case, from (10.3) and (10.2) we get $\mathtt{r}_{c^{(u_1)}\,c^{(u_2)}}(0) = \delta_{u_1-u_2}$. Therefore, (10.25) simply becomes

$$y_k = N_{SF}\,E_h\,A^{(1)}\,a_k^{(1)} + w_k \tag{10.27}$$

where $N_{SF}\,E_h$ is the energy of the pulse associated with $a_k^{(1)}$ (see (10.10)).

In (10.27), the noise is given by

$$w_k = \sum_{j=0}^{N_{SF}-1} \tilde{w}_{j+kN_{SF}}\,c_{j+kN_{SF}}^{(1)*} \tag{10.28}$$

therefore assuming $\{\tilde{w}_m\}$ i.i.d., the variance of w_k is given by

$$\sigma_w^2 = N_{SF}\,\sigma_{\tilde{w}}^2 = N_{SF}\,N_0\,E_h \tag{10.29}$$

Data detector Using a threshold detector, from (10.27) the SNR at the decision point is given by (see (6.75))

$$\gamma = \left(\frac{d_{min}}{2\sigma_I}\right)^2 = \frac{(N_{SF}\,E_h\,A^{(1)})^2}{N_{SF}\,N_0\,E_h/2} = \frac{N_{SF}\,E_h(A^{(1)})^2}{N_0/2} \tag{10.30}$$

On the other hand, from (10.8) and (10.15) we get

$$\Gamma_s = \frac{(A^{(1)})^2\,\mathtt{M}_a\,E_h/T_{chip}}{N_0/(N_{SF}\,T_{chip})} = \frac{N_{SF}\,E_h(A^{(1)})^2\,\mathtt{M}_a}{N_0} = \frac{E_s}{N_0} \tag{10.31}$$

where E_s is the average energy per symbol of the transmitted signal.

In other words, the relation between γ and Γ_s is optimum, as given by (6.80). Therefore, with regard to user $u = 1$, at the decision point the system is equivalent to an M-QAM system. However, as observed before, the transmit pulse $h_T^{(1)}$ has a bandwidth much larger than $1/(2T)$.

Multiuser receiver The derivation of the optimum receiver carried out for user 1 can be repeated for each user. Therefore, we obtain the multiuser receiver of Figure 10.6, composed of a matched filter to the transmit pulse and a despreader bank, where each branch employs a distinct user code.

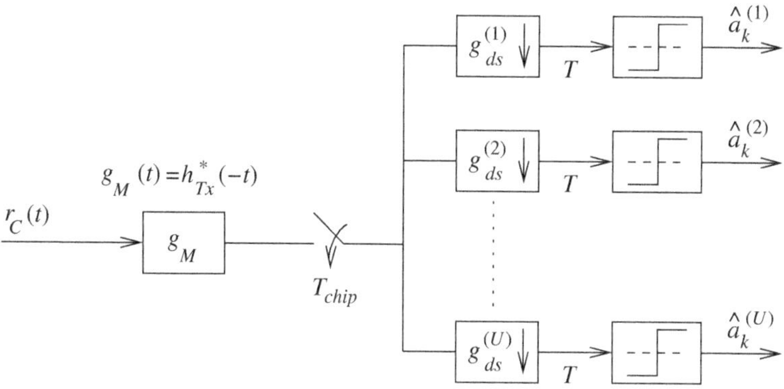

Figure 10.6 Multiuser receiver for a CDMA synchronous system with an ideal AWGN channel.

We observe that for the ideal AWGN channel case, spreading the bandwidth of the transmit signal by a factor U allows the simultaneous transmission of U messages using the same frequency band.

Classification of CDMA systems

Synchronous systems This is the case just examined, in which the user codes are orthogonal and the user signals are time-aligned. In a wireless cellular system, this situation occurs in the forward or downlink transmission from the base station to the mobile stations, e.g. in the 3rd generation of the 3GPP mobile communication standard. From the point of view of each mobile station, all U users share the same channel. Therefore, although the channel impulse response depends on the site of the mobile, we have

$$g_C^{(u)}(t) = g_C(t), \qquad u = 1, \dots, U \tag{10.32}$$

and the residual interference is due to signals originating from adjacent cells in addition to the multipath interference introduced by the channel. In general, interference due to the other users within the same cell is called *multiuser interference* (MUI) or *co-channel interference* (CCI).

Asynchronous systems In this case, the various user signals are not time-aligned. In a wireless cellular system, this situation typically occurs in the reverse or uplink transmission from the mobile stations to the base station.

Because the Walsh–Hadamard codes do not exhibit good cross-correlation properties for lags different from zero, pseudo-noise (PN) scrambling sequences are used (see Appendix 1.C). The user code is then given by

$$c_m^{(u)} = c_{Ch,m}^{(u)} \, c_{scr,m} \tag{10.33}$$

where $\{c_{scr,m}\}$ may be the same for all users in a cell.

It is necessary now to make an important observation. In some systems, the period of $\{c_{scr,m}\}$ is equal to the length of $\{c_{Ch,m}\}$, that is N_{SF}, whereas in other systems, it is much larger than N_{SF}.[1] In the latter case, spreading and despreading operations remain unchanged, even if they are *symbol time varying*, as $\{c_m^{(u)}\}$ changes from symbol to symbol; note that consequently the receiver is also symbol time varying.

Asynchronous systems are characterized by codes with low cross-correlation for non-zero lags; however there is always a residual non-zero correlation among the various user signals. Especially in the presence of multipath channels, the residual correlation is the major cause of interference in the system, which now originates from signals within the cell: for this reason the MUI is usually characterized as *intracell* MUI.

Synchronization

Despreading requires the reproduction at the receiver of a user code sequence synchronous with that used for spreading. Therefore, the receiver must first perform *acquisition*, that is the code sequence $\{c_m^{(u)}\}$ produced by the local generator must be synchronized with the code sequence of the desired user, so that the error in the time alignment between the two sequences is less than one chip interval.

As described in Section 14.7, acquisition of the desired user code sequence is generally obtained by a sequential searching algorithm that, at each step, delays the local code generator by a fraction of a chip, typically half a chip, and determines the correlation between the signals $\{x_m\}$ and $\{c_m^{(u)}\}$; the search terminates when the correlation level exceeds a certain threshold value, indicating that the desired time alignment is attained. Following the acquisition process, a *tracking* algorithm is used to achieve, in the steady state, a time alignment between the signals $\{x_m\}$ and $\{c_m^{(u)}\}$ that has the desired accuracy; the more commonly used tracking algorithms are the *delay-locked loop* and the *tau-dither loop*. The

---

[1] This observation must not be confused with the distinction between the use of short (of period $\simeq 2^{15}$) or long (of period $\simeq 2^{42}$) PN scrambling sequences, which are employed to identify the base stations or the users and to synchronize the system [8].

synchronization method also suggests the use of PN sequences as user code sequences. In practice, the chip frequency is limited to values of the order of hundreds of Mchip/s because of the difficulty in obtaining an accuracy of the order of a fraction of a nanosecond in the synchronization of the code generator. In turn, this determines the limit in the bandwidth of a DS signal.

10.1.2 Frequency hopping systems

The FH spread-spectrum technique is typically used for the spreading of multiple frequency shifting keying (M-FSK) signals. We consider an M-FSK signal [9], with carrier frequency f_0 expressed in complex form, i.e. we consider the analytic signal, as $A\ e^{j2\pi(f_0+\Delta f(t))t}$, where $\Delta f(t) = \sum_{k=-\infty}^{+\infty} a_k\ \mathtt{w}_T(t-kT)$, with $\{a_k\}$ sequence of i.i.d. symbols taken from the alphabet $\mathcal{A} = \{-(M-1), \dots, -1, +1, \dots, M-1\}$, at the symbol rate $1/T$. An FH/M-FSK signal is obtained by multiplying the M-FSK signal by

$$c_{FH}(t) = \sum_{i=-\infty}^{+\infty} e^{j(2\pi f_{0,i}t+\varphi_{0,i})}\ \mathtt{w}_{T_{hop}}(t - iT_{hop}) \tag{10.34}$$

where $\{f_{0,i}\}$ is a pseudorandom sequence that determines shifts in frequency of the FH/M-FSK signal, $\{\varphi_{0,i}\}$ is a sequence of random phases associated with the sequence of frequency shifts, and $\mathtt{w}_{T_{hop}}$ is a rectangular window of duration equal to a hop interval T_{hop}. In an FH/M-FSK system, the transmitted signal is then given by

$$s(t) = Re\ [c_{FH}(t)\ e^{j2\pi(f_0+\Delta f(t))t}] \tag{10.35}$$

In practice, the signal c_{FH} is not generated at the transmitter; the transmitted signal s is obtained by applying the sequence of pseudorandom frequency shifts $\{f_{0,i}\}$ directly to the frequency synthesizer that generates the carrier at frequency f_0. With reference to the implementation illustrated in Figure 10.7, segments of L consecutive chips from a PN sequence, not necessarily disjoint, are applied to a frequency synthesizer that makes the carrier frequency hop over a set of 2^L frequencies. As the band over which the synthesizer must operate is large, it is difficult to maintain the carrier phase coherent between two consecutive hops [10]; if the synthesizer is not equipped with any device to maintain a coherent phase, it is necessary to include a random phase $\varphi_{0,i}$ as in (10.34). In a time interval that is long with respect to T_{hop}, the bandwidth of the signal s, B_{SS}, can be in practice of the order of several GHz. However, in a short time interval during which no FH occurs, the bandwidth of an FH/M-FSK signal is the same as the bandwidth of the M-FSK signal that carries the information, usually much lower than B_{SS}.

Despreading, in this case also called dehopping, is ideally carried out by multiplying the received signal r by a signal $\hat{c}_{FH}$ equal to that used for spreading, apart from the sequence of random phases associated with the frequency shifts. For non-coherent demodulation, the sequence of random phases can be modelled as a sequence of i.i.d. random variables with uniform probability density in $[0, 2\pi)$. The operation of despreading yields the signal x, given by the sum of the M-FSK signal, the noise and possibly interference. The signal x is then filtered by a lowpass filter and presented to the input of the receive section comprising a non-coherent demodulator for M-FSK signals. As in the case of DS systems, the receiver must perform acquisition and tracking of the FH signal, so that the waveform generated by the synthesizer for dehopping reproduces as accurately as possible the signal c_{FH}.

Classification of FH systems

FH systems are traditionally classified according to the relation between T_{hop} and T. *Fast frequency-hopped* (FFH) systems are characterized by one or more frequency hops per symbol interval, that is $T = NT_{hop}$, N integer, and *slow frequency-hopped* (SFH) systems are characterized by the transmission of several symbols per hop interval, that is $T_{hop} = NT$.

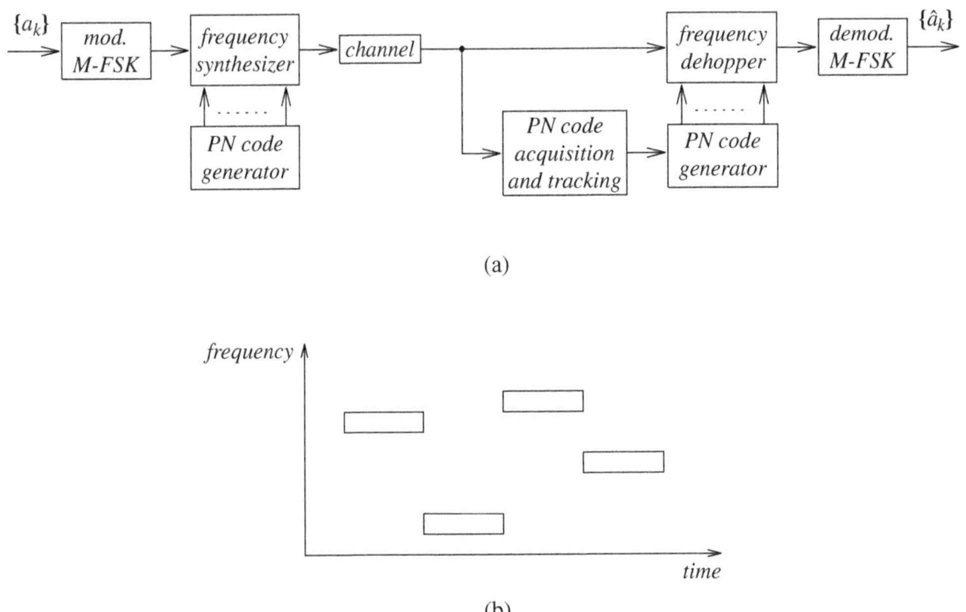

Figure 10.7 (a) Block diagram of an FH/M-FSK system. (b) Time-frequency allocation.

Moreover, a chip frequency F_{chip} is defined also for FH systems, and is given by the largest value among $F_{hop} = 1/T_{hop}$ and $F = 1/T$. Therefore, the chip frequency F_{chip} corresponds to the highest among the clock frequencies used by the system. The frequency spacing between tones of an FH/M-FSK signal is related to the chip frequency and is therefore determined differently for FFH and SFH systems.

SFH systems For SFH systems, $F_{chip} = F$, and the spacing between FH/M-FSK tones is equal to the spacing between the M-FSK tones themselves. In a system that uses a non-coherent receiver for M-FSK signals, orthogonality of tones corresponding to M-FSK symbols is obtained if the frequency spacing is an integer multiple of $1/T$. Assuming the minimum spacing is equal to F, the bandwidth B_{SS} of an FH/M-FSK signal is partitioned into $N_f = B_{SS}/F = B_{SS}/F_{chip}$ subbands with equally spaced centre frequencies; in the most commonly used FH scheme, the N_f tones are grouped into $N_b = N_f/M$ adjacent bands without overlap in frequency, each one having a bandwidth equal to $MF = MF_{chip}$, as illustrated in Figure 10.8. Assuming M-FSK modulation symmetric around the carrier frequency, the centre frequencies of the $N_b = 2^L$ bands represent the set of carrier frequencies generated by the synthesizer, each associated with an L-uple of binary symbols. According to this scheme, each of the N_f tones of the FH/M-FSK signal corresponds to a unique combination of carrier frequency and M-FSK symbol.

In a different scheme, that yields a better protection against an intentional jammer using a sophisticated disturbance strategy, adjacent bands exhibit an overlap in frequency equal to $(M-1)F_{chip}$ as illustrated in Figure 10.9. Assuming that the centre frequency of each band corresponds to a possible carrier frequency, as all N_f tones except $(M-1)$ are available as centre frequencies, the number of carrier frequencies increases from N_f/M to $N_f - (M-1)$, which for $N_f \gg M$ represents an increase by a factor M of the randomness in the choice of the carrier frequency.

FFH systems For FFH systems, where $F_{chip} = F_{hop}$, the spacing between tones of an FH/M-FSK signal is equal to the hop frequency. Therefore, the bandwidth of the spread-spectrum signal is partitioned into a total of $N_f = B_{SS}/F_{hop} = B_{SS}/F_{chip}$ subbands with equally spaced centre frequencies, each

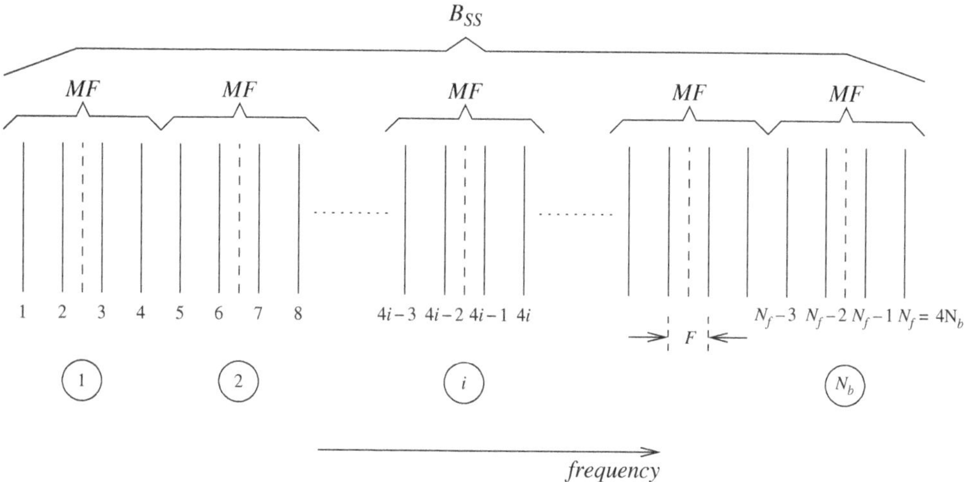

Figure 10.8 Frequency distribution for an FH/4-FSK system with bands non-overlapping in frequency; the dashed lines indicate the carrier frequencies.

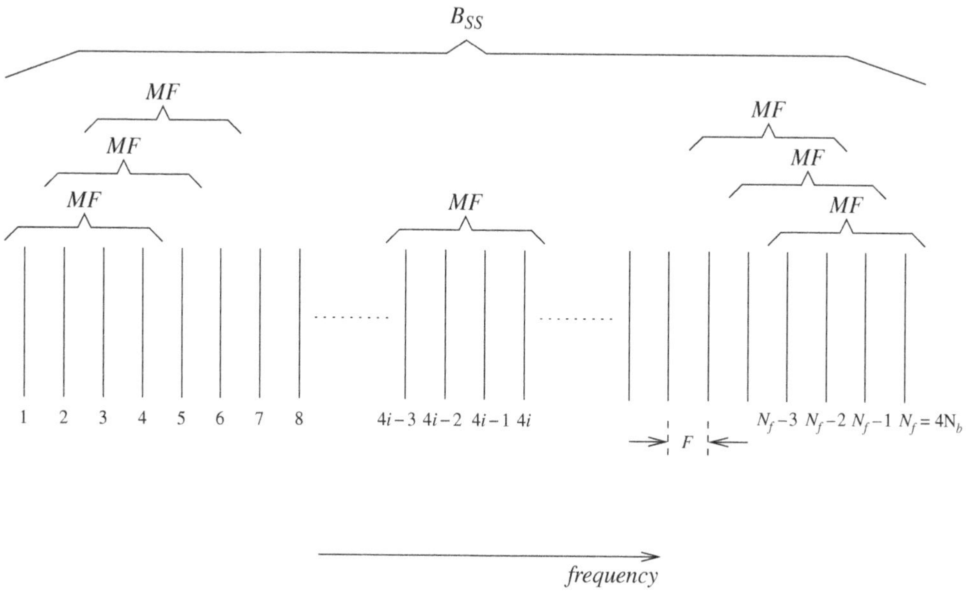

Figure 10.9 Frequency distribution for an FH/4-FSK system with bands overlapping in frequency.

corresponding to a unique L-uple of binary symbols. Because there are F_{hop}/F hops per symbol, the metric used to decide upon the symbol with a non-coherent receiver is suitably obtained by summing F_{hop}/F components of the received signal.

10.2 Applications of spread-spectrum systems

The most common applications of spread-spectrum systems, that will be discussed in the next sections, may be enumerated as follows.

1. *Multiple access.* In alternative to frequency division multiple access (FDMA) and time division multiple access (TDMA) systems, introduced in Appendix 18.B, spread-spectrum systems allow the simultaneous transmission of messages by several users over the channel, as discussed in Section 10.1.1.
2. *Narrowband interference rejection.* We consider the DS case. Because interference is introduced in the channel after signal spreading, at the receiver the despreading operation compresses the bandwidth of the desired signal to the original value, and at the same time it expands by the same factor the bandwidth of the interference, thus reducing the level of the interference PSD. After demodulation, the ratio between the desired signal power and the interference power is therefore larger than that obtained without spreading the signal spectrum.
3. *Robustness against fading.* Widening the signal bandwidth allows the exploitation of multipath diversity of a radio channel affected by fading. Applying a DS spread-spectrum technique, intuitively, has the effect of modifying a channel model that is adequate for transmission of narrowband signals in the presence of flat fading or multipath fading with a few rays, to a channel model with many rays. Using a receiver that combines the desired signal from the different propagation rays, the power of the desired signal at the decision point increases. In an FH system, on the other hand, we obtain diversity in the time domain, as the channel changes from one hop interval to the next. The probability that the signal is affected by strong fading during two consecutive hop intervals is usually low. To recover the transmitted message in a hop interval during which strong fading is experienced, error correction codes with very long interleaver and automatic repeat request (ARQ) schemes are used (see Chapter 11).

10.2.1 Anti-jamming

Narrowband interference We consider the baseband equivalent signals of an M-QAM passband communication system with symbol rate $F = 1/T$, transmitted signal power equal to $\mathtt{M}_s$, and PSD with minimum bandwidth, i.e. $\mathcal{P}_s(f) = E_s \text{rect}(f/F)$, where $E_s F = \mathtt{M}_s$.

We consider now the application of a DS spread-spectrum modulation system. Due to spreading, the bandwidth of the transmitted signal is expanded from F to $B_{SS} = N_{SF}F$. Therefore, for the same transmitted signal power, the PSD of the transmitted signal becomes $\mathcal{P}_{s'}(f) = (E_s/N_{SF})\text{rect}(f/B_{SS})$, where $E_s/N_{SF} = \mathtt{M}_s/B_{SS}$. We note that spreading has decreased the amplitude of the PSD by the factor N_{SF}, as illustrated in Figure 10.10.

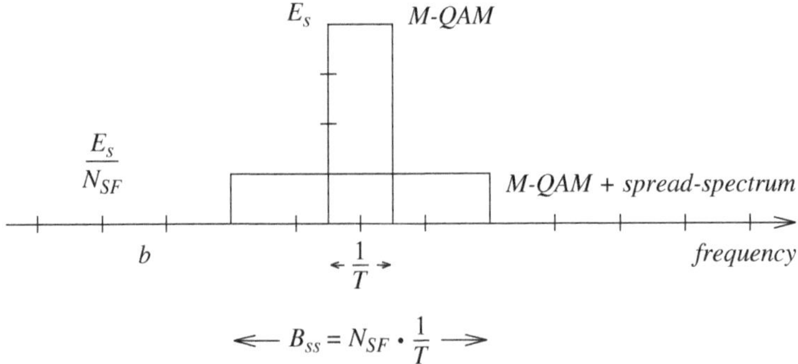

Figure 10.10 Power spectral density of an M-QAM signal with minimum bandwidth and of a spread-spectrum M-QAM signal with spreading factor $N_{SF} = 4$.

In the band of the spread-spectrum signal, in addition to AWGN with PSD N_0, we assume the channel introduces an additive interference signal or jammer with power $\mathtt{M}_{\mathtt{j}}$, uniformly distributed on a bandwidth $B_{\mathtt{j}}$, with $B_{\mathtt{j}} < 1/T$.

With regard to the operation of despreading, we consider the signals after the multiplication by the user code sequence. The interference signal spectrum is expanded and has a PSD equal to $\mathcal{P}_{\mathtt{j}'}(f) = N_{\mathtt{j}}\,\mathrm{rect}(f/B_{SS})$, with $N_{\mathtt{j}} = \mathtt{M}_{\mathtt{j}}/B_{SS}$. The noise, that originally has a uniformly distributed power over all the frequencies, still has PSD equal to N_0, i.e. spreading has not changed the PSD of the noise.

At the output of the despreader, the desired signal exhibits the original PSD equal to $E_s\,\mathrm{rect}(f/F)$. Modelling the despreader filter as an ideal lowpass filter with bandwidth $1/(2T)$, for the SNR γ at the decision point the following relation holds:

$$\left(\frac{1}{2}\,\mathtt{M}_a\right)\gamma = \frac{E_s}{N_0 + N_{\mathtt{j}}} = \frac{\mathtt{M}_s/F}{(N_0 + \mathtt{M}_{\mathtt{j}}/B_{SS})} \tag{10.36}$$

In practice, performance is usually limited by interference and the presence of white noise can be ignored. Therefore, assuming $N_{\mathtt{j}} \gg N_0$, (10.36) becomes

$$\left(\frac{1}{2}\,\mathtt{M}_a\right)\gamma \simeq \frac{E_s}{N_{\mathtt{j}}} = \frac{\mathtt{M}_s/F}{\mathtt{M}_{\mathtt{j}}/B_{SS}} = \frac{\mathtt{M}_s}{\mathtt{M}_{\mathtt{j}}}\,\frac{B_{SS}}{F} \tag{10.37}$$

where $\mathtt{M}_s/\mathtt{M}_{\mathtt{j}}$ is the ratio between the power of the desired signal and the power of the jammer, and B_{SS}/F is the spreading ratio N_{SF} also defined as the *processing gain* of the system.

The above considerations are now defined more precisely in the following case.

Sinusoidal interference We assume that the baseband equivalent received signal is expressed as

$$r_C(t) = s(t) + \mathtt{j}(t) + w_C(t) \tag{10.38}$$

where s is a DS signal given by (10.9) with amplitude $A^{(u)} = 1$, w_C is AWGN with spectral density N_0, and the interferer is given by

$$\mathtt{j}(t) = A_{\mathtt{j}}\,e^{j\varphi} \tag{10.39}$$

In (10.39), $A_{\mathtt{j}} = \sqrt{\mathtt{M}_{\mathtt{j}}}$ is the amplitude of the jammer and φ a random phase with uniform distribution in $[0, 2\pi)$. We also assume a minimum bandwidth transmit pulse, $h_{Tx}(t) = \sqrt{E_h/T_{chip}}\,\mathrm{sinc}(t/T_{chip})$, hence $g_M(t) = h_{Tx}(t)$, and $\mathcal{G}_M(0) = \sqrt{E_h\,T_{chip}}$.

For the coherent receiver of Figure 10.4, at the detection point the sample at instant kT is given by

$$y_k = N_{SF}\,E_h\,a_k + w_k + A_{\mathtt{j}}\,e^{j\varphi}\,\mathcal{G}_M(0)\sum_{j=0}^{N_{SF}-1} c^*_{j+kN_{SF}} \tag{10.40}$$

Modelling the sequence $\{c^*_{kN_{SF}}, c^*_{kN_{SF}+1}, \ldots, c^*_{kN_{SF}+N_{SF}-1}\}$ as a sequence of i.i.d. random variables, the variance of the summation in (10.40) is equal to N_{SF}, and the ratio γ is given by

$$\gamma = \frac{(N_{SF}\,E_h)^2}{(N_{SF}\,N_0\,E_h + \mathtt{M}_{\mathtt{j}}\,E_h\,T_{chip}\,N_{SF})/2} \tag{10.41}$$

Using (10.8) and the relation $E_s = \mathtt{M}_s T$, we get

$$\left(\frac{1}{2}\,\mathtt{M}_a\right)\gamma = \frac{1}{N_0/E_s + \mathtt{M}_{\mathtt{j}}/(N_{SF}\,\mathtt{M}_s)} \tag{10.42}$$

We note that in the denominator of (10.42) the ratio $\mathtt{M}_{\mathtt{j}}/\mathtt{M}_s$ is divided by N_{SF}. Recognizing that $\mathtt{M}_{\mathtt{j}}/\mathtt{M}_s$ is the ratio between the power of the jammer and the power of the desired signal before the despreading operation, and that $\mathtt{M}_{\mathtt{j}}/(N_{SF}\,\mathtt{M}_s)$ is the same ratio after the despreading, we find that, in analogy with the previous case of narrowband interference, also in the case of a sinusoidal jammer, the use of the DS technique reduces the effect of the jammer by a factor equal to the processing gain.

10.2.2 Multiple access

Spread-spectrum multiple-access communication systems represent an alternative to TDMA or FDMA systems and are normally referred to as CDMA systems (see Sections 18.B and 10.1.1). With CDMA, a particular spreading sequence is assigned to each user for accessing the channel; unlike FDMA, where users transmit simultaneously over non-overlapping frequency bands, or TDMA, where users transmit over the same band but in disjoint time intervals, users in a CDMA system transmit simultaneously over the same frequency band.

Because in CDMA systems correlation receivers are usually employed, it is important that the spreading sequences are characterized by low cross-correlation values. We have already observed that CDMA systems may be classified as synchronous or asynchronous. In the first case, the symbol transition instants of all users are aligned; this allows the use of orthogonal sequences as spreading sequences and consequently the elimination of interference caused by one user signal to another; in the second case, the interference caused by multiple access limits the channel capacity, but the system design is simplified.

CDMA has received particular interest for applications in wireless communications systems, for example cellular radio systems, personal communications services (PCS), and wireless local-area networks; this interest is mainly due to performance that spread-spectrum systems achieve over channels characterized by multipath fading.

Other properties make CDMA interesting for application to cellular radio systems, for example the possibility of applying the concept of frequency reuse (see Section 18.1.1). In cellular radio systems based on FDMA or TDMA, to avoid excessive levels of interference from one cell onto neighbouring cells, the frequencies used in one cell are not used in neighbouring cells. In other words, the system is designed so that there is a certain spatial separation between cells that use the same frequencies. For CDMA, this spatial separation is not necessary, making it possible, in principle, to reuse all frequencies.

10.2.3 Interference rejection

Besides the above described properties, that are relative to the application in multiple-access systems, the robustness of spread-spectrum systems in the presence of narrowband interferers is key in other applications, for example in systems where interference is unintentionally generated by other users that transmit over the same channel. We have CCI when a certain number of services are simultaneously offered to users transmitting over the same frequency band. Although in these cases some form of spatial separation among signals interfering with each other is usually provided, for example by using directional antennas, it is often desirable to use spread-spectrum systems for their inherent interference suppression capability. In particular, we consider a scenario in which a frequency band is only partially occupied by a set of narrowband conventional signals; to increase the spectral efficiency of the system, a set of spread-spectrum signals can simultaneously be transmitted over the same band, thus two sets of users can access the transmission channel. Clearly, this scheme can be implemented only if the mutual interference, which a signal set imposes on the other, remains within tolerable limits.

10.3 Chip matched filter and rake receiver

Before introducing a structure that is often employed in receivers for DS spread-spectrum signals, we make the following considerations on the radio channel model introduced in Section 4.1.

Number of resolvable rays in a multipath channel

We want to represent a multipath radio channel with a number of rays having gains modelled as complex valued, Gaussian *uncorrelated* random processes. From (4.64) and (4.65), with some simplification in the notation, a time-invariant channel impulse response is approximated as

$$g_C(\tau) = \sum_{i=0}^{N_{c,\infty}-1} g_i \, \delta(\tau - iT_Q) \tag{10.43}$$

where delays $\tau_i = iT_Q$ are approximated as multiples of a sufficiently small period T_Q.

Hence, from (10.43), the channel output signal s_C is related to the input signal s by

$$s_C(t) = \sum_{i=0}^{N_{c,\infty}-1} g_i \, s(t - iT_Q) \tag{10.44}$$

Now the number of resolvable or uncorrelated rays in (10.44) is generally less than $N_{c,\infty}$ and is related to the bandwidth B of s as the uncorrelated rays are spaced by a delay of the order of $1/B$. Consequently, for a channel with a delay spread τ_{rms} and bandwidth $B \propto 1/T_{chip}$, the number of resolvable rays is

$$N_{c,res} \propto \frac{\tau_{rms}}{T_{chip}} \tag{10.45}$$

Using the notion of channel coherence bandwidth, $B_{coh} \propto 1/\tau_{rms}$, (10.45) may be rewritten as

$$N_{c,res} \propto \frac{B}{B_{coh}} \tag{10.46}$$

We now give an example that illustrates the above considerations. Let $\{g_i\}$ be a realization of the channel impulse response with uncorrelated coefficients having a given power delay profile; the *infinite bandwidth* of the channel will be equal to $B = 1/(2T_Q)$. We now filter $\{g_i\}$ with two filters having, respectively, bandwidth $B = 0.1/(2T_Q)$ and $B = 0.01/(2T_Q)$, and we compare the three pulse shapes given by the input sequence and the two output sequences. We note that the output obtained in correspondence of the filter with the narrower bandwidth has fewer resolvable rays. In fact, in the limit for $B \to 0$ the output is modelled as a single random variable.

Another way to derive (10.45) is to observe that, for t within an interval of duration $1/B$, s does not vary much. Therefore, letting

$$N_{cor} = \frac{N_{c,\infty}}{N_{c,res}} \tag{10.47}$$

(10.44) can be written as

$$s_C(t) = \sum_{j=0}^{N_{c,res}-1} g_{res,j} \, s(t - jN_{cor}T_Q) \tag{10.48}$$

and the gains of the resolvable rays are

$$g_{res,j} \simeq \sum_{i=0}^{N_{cor}-1} g_{i+jN_{cor}} \tag{10.49}$$

In summary, assuming the symbol period T is given and DS spread-spectrum modulation is adopted, the larger N_{SF} the greater the resolution of the radio channel, that is, the channel can be modelled with a larger number of uncorrelated rays, with delays of the order of T_{chip}.

Chip matched filter

We consider the transmission of a *DS signal* (10.9) for $U = 1$ on a *dispersive channel* as described by (10.48). The receiver that maximizes the ratio between the amplitude of the pulse associated with the desired signal sampled with sampling rate $1/T_{chip}$ and the standard deviation of the noise is obtained by the filter matched to the received pulse, named chip matched filter (CMF). We define

$$q_C(t) = (h_{Tx} * g_C * g_{AA})(t) \tag{10.50}$$

and let $g_M(t) = q_C^*(t_0 - t)$ be the corresponding matched filter. In practice, at the output of the filter g_{AA} an estimate of q_C with sampling period $T_c = T_{chip}/2$ is evaluated,[2] which yields the corresponding discrete-time matched filter with sampling period of the input signal equal to T_c and sampling period of the output signal equal to T_{chip} (see Figure 10.11).

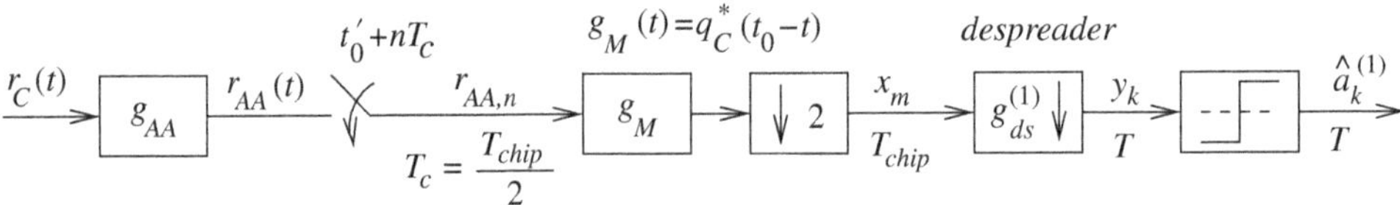

Figure 10.11 Chip matched filter receiver for a dispersive channel.

If q_C is sparse, that is, it has a large support but only a few non-zero coefficients, for the realization of g_M, we retain only the coefficients of q_C with larger amplitude; it is better to set to zero the remaining coefficients because their estimate is usually very noisy (see observation 3 at page 383).

Figure 10.12a illustrates in detail the receiver of Figure 10.11 for a filter g_M with at most N_{MF} coefficients spaced of $T_c = T_{chip}/2$. If we now implement the despreader on every branch of the filter g_M, we obtain the structure of Figure 10.12b. We observe that typically only 3 or 4 branches are active, that is they have a coefficient $g_{M,i}$ different from zero.

Ideally, for an overall channel with N_{res} resolvable paths, we assume

$$q_C(t) = \sum_{i=1}^{N_{res}} q_{C,i}\, \delta(t - \tau_i) \tag{10.51}$$

hence

$$g_M(t) = \sum_{j=1}^{N_{res}} q_{C,j}^*\, \delta(t_0 - t - \tau_j) \tag{10.52}$$

Defining

$$t_{M,j} = t_0 - \tau_j, \qquad j = 1, \dots, N_{res} \tag{10.53}$$

the receiver scheme, analogous to that of Figure 10.12b, is illustrated in Figure 10.13.

To simplify the analysis, we assume that the spreading sequence is a PN sequence with N_{SF} sufficiently large, such that the following approximations hold: (1) the autocorrelation of the spreading sequence is a Kronecker delta and (2) the delays $\{\tau_i\}$ are multiples of T_{chip}.

From (10.51), in the *absence of noise*, the signal r_{AA} is given by

$$r_{AA}(t) = \sum_{n=1}^{N_{res}} q_{C,n} \sum_{i=-\infty}^{+\infty} a_i^{(1)} \sum_{\ell=0}^{N_{SF}-1} c_{\ell+iN_{SF}}^{(1)}\, \delta(t - \tau_n - (\ell + iN_{SF})\, T_{chip}) \tag{10.54}$$

[2] To determine the optimum sampling phase t_0, usually r_{AA} is oversampled with a period T_Q such that $T_c/T_Q = 2$ or 4 for $T_c = T_{chip}/2$; among the 2 or 4 estimates of g_C obtained with sampling period T_c, the one with the largest energy is selected (see Observation 7.6 on page 320).

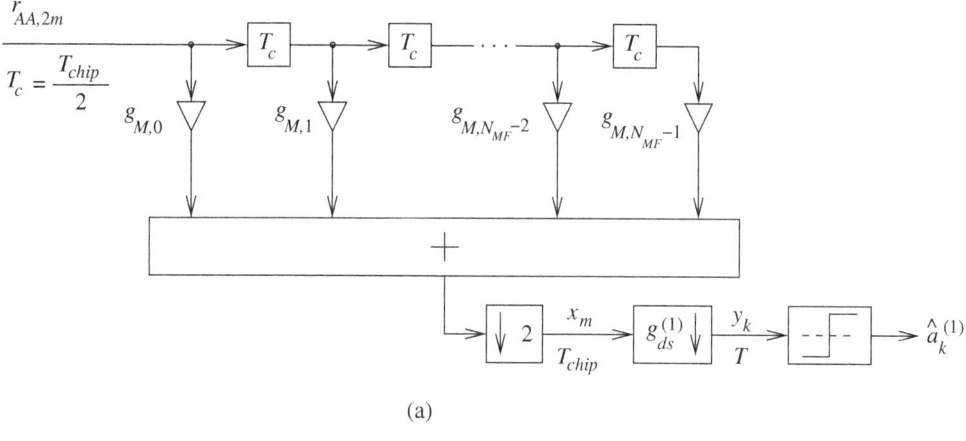

(a)

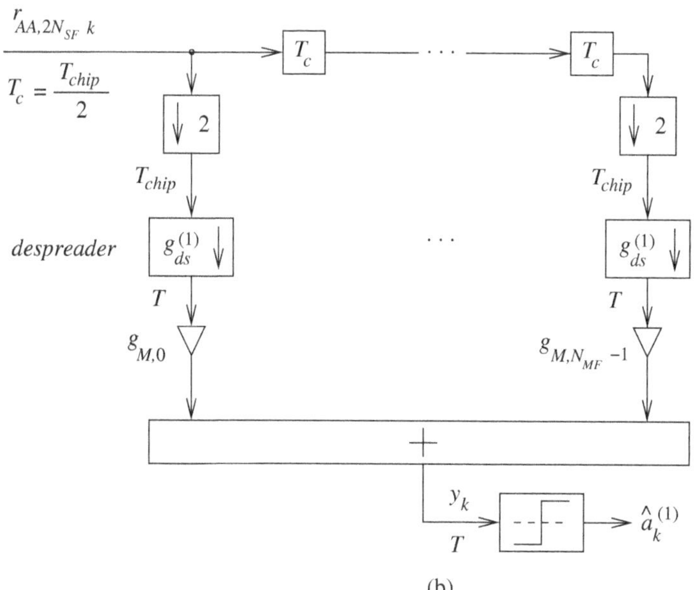

(b)

Figure 10.12 Two receiver structures: (a) chip matched filter with despreader, (b) rake.

and the output of the sampler on branch j is given by[3]

$$x_{j,m} = \sum_{n=1}^{N_{res}} q_{C,n} \sum_{i=-\infty}^{+\infty} a_i^{(1)} \sum_{\ell=0}^{N_{SF}-1} c_{\ell+iN_{SF}}^{(1)} \delta_{m+\frac{\tau_j-\tau_n}{T_{chip}}-(\ell+iN_{SF})} \tag{10.55}$$

Correspondingly the despreader output, assuming $\mathtt{r}_{c^{(1)}}(n_D) = \delta_{n_D}$ and the absence of noise, yields the signal $N_{SF}\, a_k^{(1)}\, q_{C,j}$. The contributions from the various branches are then combined according to the maximum-ratio-combining (MRC) technique (see Section 9.2.1) to yield the sample

$$y_k = \left(N_{SF} \sum_{n=1}^{N_{res}} |q_{C,n}|^2 \right) a_k^{(1)} \tag{10.56}$$

[3] Instead of using the Dirac delta in (10.51), a similar analysis assumes that (1) $g_{AA}(t) = h_{Tx}^*(-t)$ and (2) $\mathtt{r}_{h_{Tx}}$ is a Nyquist pulse. The result is the same as (10.55).

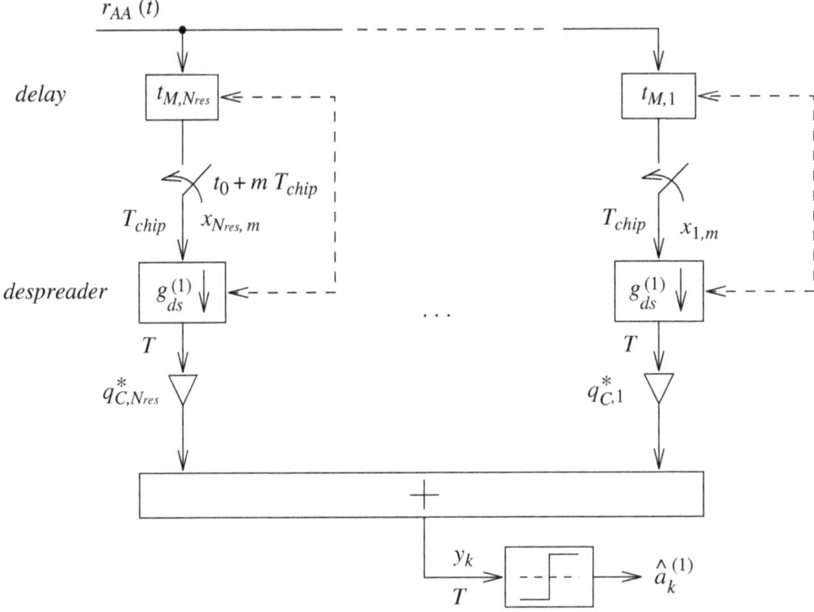

Figure 10.13 Rake receiver for a channel with N_{res} resolvable paths.

where $E_{q_C} = \sum_{n=1}^{N_{res}} |q_{C,n}|^2$ is the energy per chip of the overall channel impulse response.

The name rake originates from the structure of the receiver that is similar to a rake with N_{res} fingers. In practice, near the rake receiver a correlator estimates the delays, with precision $T_{chip}/2$, and the gains of the various channel rays. The rake is initialized with the coefficients of rays with larger gain. The delays and the coefficients are updated whenever a change in the channel impulse response is observed. However, after the initialization has taken place, on each finger of the rake the estimates of the amplitude and of the delay of the corresponding ray may be refined by using the correlator of the despreader, as indicated by the dotted line in Figure 10.13. We note that if the channel is static, the structure of Figure 10.12a with $T_c = T_{chip}/2$ yields a sufficient statistic.

10.4 Interference

For a dispersive channel and in the case of U users, we evaluate the expression of the signal y_k at the decision point using the matched filter receiver of Figure 10.11.

Similarly to (10.50), we define

$$q_C^{(u)}(t) = (h_{Tx} * g_C^{(u)} * g_{AA})(t), \qquad u = 1, \dots, U \tag{10.57}$$

and let

$$g_M^{(v)}(t) = q_C^{(v)*}(t_0 - t), \qquad v = 1, \dots, U \tag{10.58}$$

be the corresponding matched filter. Moreover, we introduce the correlation between $q_C^{(u)}$ and $q_C^{(v)}$, expressed by

$$\mathtt{r}_{q_C^{(u)} q_C^{(v)}}(\tau) = (q_C^{(u)}(t) * q_C^{(v)*}(-t))(\tau) \tag{10.59}$$

Assuming without loss of generality that the desired user signal has the index $u = 1$, we have

$$x_m = \sum_{u=1}^{U} A^{(u)} \sum_{i=-\infty}^{+\infty} a_i^{(u)} \sum_{\ell=0}^{N_{SF}-1} c_{\ell+iN_{SF}}^{(u)} \, \mathtt{r}_{q_C^{(u)} q_C^{(1)}}((m-\ell-iN_{SF})\,T_{chip}) + \tilde{w}_m \tag{10.60}$$

where $\tilde{w}_m$ is given by (10.20).

At the despreader output, we obtain

$$\begin{aligned} y_k &= \sum_{j=0}^{N_{SF}-1} x_{j+kN_{SF}} \, c_{j+kN_{SF}}^{(1)*} + w_k \\ &= \sum_{u=1}^{U} A^{(u)} \sum_{i=-\infty}^{+\infty} a_i^{(u)} \sum_{\ell=0}^{N_{SF}-1} \sum_{j=0}^{N_{SF}-1} c_{\ell+iN_{SF}}^{(u)} \\ &\quad \mathtt{r}_{q_C^{(u)} q_C^{(1)}}((j-\ell+(k-i)\,N_{SF})\,T_{chip}) c_{j+kN_{SF}}^{(1)*} + w_k \end{aligned} \tag{10.61}$$

where w_k is defined in (10.28).

Introducing the change of index $n = \ell - j$ and recalling the definition of cross-correlation between two code sequences (10.26), the double summation in ℓ and j in (10.61) can be written as

$$\begin{aligned} &\sum_{n=-(N_{SF}-1)}^{-1} (N_{SF} - |n|) \mathtt{r}_{q_C^{(u)} q_C^{(1)}}((-n+(k-i)\,N_{SF})T_{chip}) \mathtt{r}^*_{c^{(1)} c^{(u)}}(-n) \\ &+ \sum_{n=0}^{N_{SF}-1} (N_{SF} - |n|) \mathtt{r}_{q_C^{(u)} q_C^{(1)}}((-n+(k-i)\,N_{SF})T_{chip}) \mathtt{r}_{c^{(u)} c^{(1)}}(n) \end{aligned} \tag{10.62}$$

where, to simplify the notation, we have assumed that the user code sequences are periodic of period N_{SF}.

The desired term in (10.61) is obtained for $u = 1$; as $\mathtt{r}^*_{c^{(1)}}(-n) = \mathtt{r}_{c^{(1)}}(n)$, it has the following expression:

$$A^{(1)} \sum_{i=-\infty}^{+\infty} a_i^{(1)} \sum_{n=-(N_{SF}-1)}^{N_{SF}-1} (N_{SF} - |n|) \, \mathtt{r}_{c^{(1)}}(n) \, \mathtt{r}_{q_C^{(1)}}((-n+(k-i)\,N_{SF})\,T_{chip}) \tag{10.63}$$

Consequently, if the code sequences are orthogonal, that is

$$\mathtt{r}_{c^{(1)}}(n) = \delta_n \tag{10.64}$$

and in the absence of ISI, that is

$$\mathtt{r}_{q_C^{(1)}}(iN_{SF}\,T_{chip}) = \delta_i \, E_{q_C^{(1)}} \tag{10.65}$$

where $E_{q_C^{(1)}}$ is the energy per chip of the overall pulse at the output of the filter g_{AA}, then the desired term (10.63) becomes

$$A^{(1)} \, N_{SF} \, E_{q_C} \, a_k^{(1)} \tag{10.66}$$

which coincides with the case of an ideal AWGN channel (see (10.27)). Note that using the same assumptions we find the rake receiver behaves as an MRC (see (10.56)).

If (10.64) is not verified, as it happens in practice, and if

$$\mathtt{r}_{q_C^{(1)}}(nT_{chip}) \neq \delta_n \, E_{q_C} \tag{10.67}$$

the terms for $n \neq 0$ in (10.63) give rise to ISI, in this context also called *inter-path interference* (IPI). Usually, the smaller N_{SF} the larger the IPI. We note however that if the overall pulse at the output of the CMF is a Nyquist pulse, that is

$$\mathtt{r}_{q_C^{(1)}}(nT_{chip}) = \delta_n \, E_{q_C} \tag{10.68}$$

then there is no IPI, even if (10.64) is not verified.

With reference to (10.62) we observe that, in the multiuser case, if $\mathbf{r}_{c^{(u)}\,c^{(1)}}(n) \neq 0$ then y_k is affected by MUI, whose value increases as the cross-correlation between the pulses $q_C^{(u)}$ and $q_C^{(1)}$ increases.

Detection strategies for multiple-access systems

For detection of the user messages in CDMA systems, we make a distinction between two classes of receivers: single-user and multiuser. In the first class, the receivers focus on detecting the data from a single user, and the other user signals are considered as additional noise. In the second class, the receivers seek to simultaneously detect all U messages. The performance of the multiuser receivers is substantially better than that of the single-user receivers, achieved at the expense of a higher computational complexity. Using as front-end a filter bank, where the filters are matched to the channel impulse responses of the U users, the structures of single-user and multiuser receivers are exemplified in Figure 10.14.

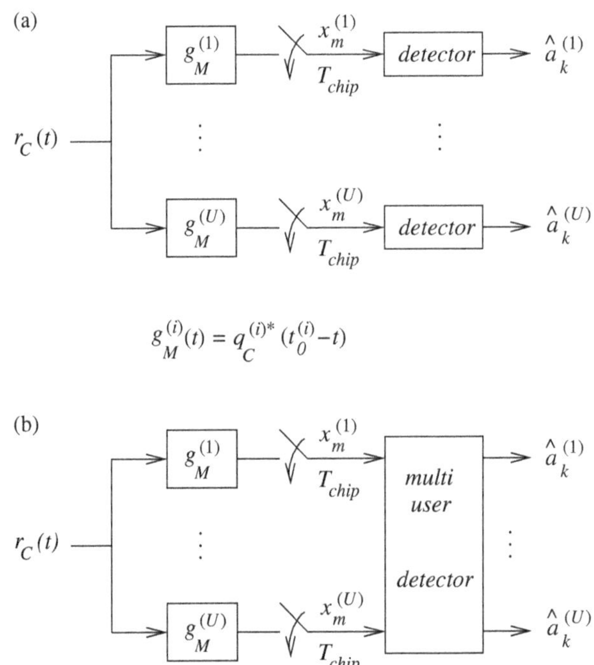

Figure 10.14 (a) Single-user receiver, and (b) multiuser receiver.

10.5 Single-user detection

We consider two equalizers for single-user detection.

Chip equalizer

To mitigate the interference in the signal sampled at the chip rate, after the CMF (see (10.68)) a zero forcing (ZF) or an mean square error (MSE) equalizer (denoted as *chip equalizer*, CE) can be used [11–14]. As illustrated in Figure 10.15, let g_{CE} be the equalizer filter with output $\{\tilde{d}_m\}$. For an MSE criterion, the cost function is given by

$$J = E\,[|\tilde{d}_m - d_m|^2] \tag{10.69}$$

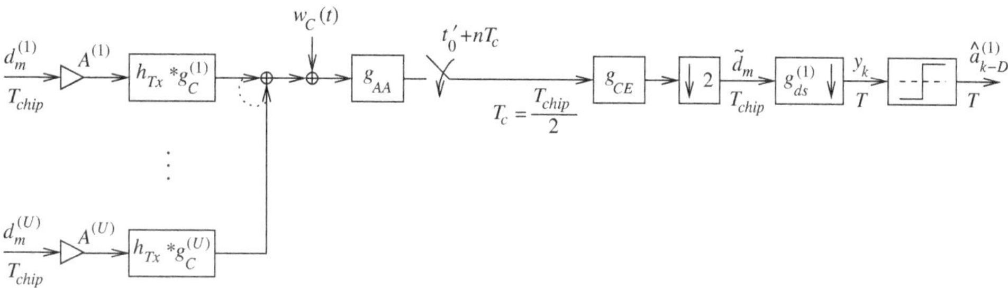

Figure 10.15 Receiver as a fractionally-spaced chip equalizer.

where $\{d_m\}$ is assumed i.i.d.. We distinguish the two following cases.

All code sequences are known This is the case that may occur for downlink transmission in wireless networks. Then, $g_C^{(u)}(t) = g_C(t)$, $u = 1, \ldots, U$, and we assume

$$d_m = \sum_{u=1}^{U} d_m^{(u)} \tag{10.70}$$

that is, for the equalizer design, all user signals are considered as desired signals.

Only the code sequence of the desired user signal is known In this case, we need to assume

$$d_m = d_m^{(1)} \tag{10.71}$$

The other user signals are considered as white noise, with overall PSD N_i, that is added to w_C.

From the knowledge of $q_C^{(1)}$ and the overall noise PSD, the minimum of the cost function defined in (10.69) is obtained by following the same steps developed in Chapter 7. Obviously, if the level of interference is high, the solution corresponding to (10.71) yields a simple CMF, with low performance whenever the residual interference (MUI and IPI) at the decision point is high.

A better structure for single-user detection is obtained by the following approach.

Symbol equalizer

Recalling that we adopt the transmitter model of Figure 10.3, and that we are interested in the message $\{a_k^{(1)}\}$, the optimum receiver with linear filter, symbol equalizer (SE), $g_{SE}^{(1)}$ is illustrated in Figure 10.16.

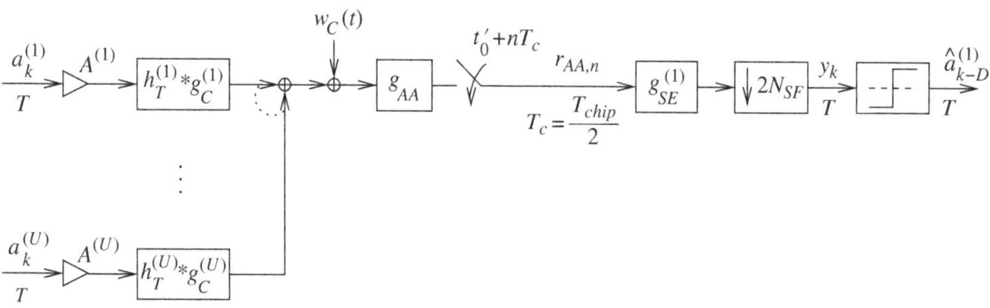

Figure 10.16 Receiver as a fractionally-spaced symbol equalizer.

The cost function is now given by [15–17]

$$J = E[|y_k - a_{k-D}^{(1)}|^2] \tag{10.72}$$

Note that $g_{SE}^{(1)}$, that includes also the function of despreading, depends on the code sequence of the desired user. Therefore, the length of the code sequence is usually not larger than N_{SF}, otherwise we would find a different solution for every symbol period, even if $g_C^{(1)}$ is time invariant. Moreover, in this formulation, the other user signals are seen as interference, and one of the tasks of g_{SE} is to mitigate the MUI.

In an adaptive approach, for example using the least mean square (LMS) algorithm, the solution is simple to determine and does not require any particular a priori knowledge, except the training sequence in $\{a_k^{(1)}\}$ for initial convergence. On the other hand, using a direct approach, we need to identify the autocorrelation of $r_{AA,n}$ and the cross-correlation between $r_{AA,n}$ and $a_{k-D}^{(1)}$. As usual these correlations are estimated directly or, assuming the messages $\{a_k^{(u)}\}$, $u = 1, \dots, U$, are i.i.d. and independent of each other, they can be determined using the knowledge of the various pulses $\{h_T^{(u)}\}$ and $\{g_C^{(u)}\}$, that is the channel impulse responses and code sequences of all users; for the special case of downlink transmission, the knowledge of the code sequences is sufficient, as the channel is common to all user signals.

10.6 Multiuser detection

Multiuser detection techniques are essential for achieving near-optimum performance in communication systems where signals conveying the desired information are received in the presence of ambient noise plus MUI. The leitmotiv of developments in multiuser detection is represented by the reduction in complexity of practical receivers with respect to that of optimal receivers, which is known to increase exponentially with the number of active users and with the delay spread of the channel, while achieving near-optimum performance. A further element that is being recognized as essential to reap the full benefits of interference suppression is the joint application of multiuser detection with other techniques such as spatial-temporal processing and iterative decoding.

10.6.1 Block equalizer

Here we first consider the simplest among multiuser receivers. It comprises a bank of U filters $g_T^{(u)}$, $u = 1, \dots, U$, matched to the impulse responses[4]

$$q_T^{(u)}(t) = \sum_{\ell=0}^{N_{SF}-1} c_\ell^{(u)} \, q_C^{(u)}(t - \ell \, T_{chip}), \qquad u = 1 \dots, U \tag{10.73}$$

where the functions $\{q_C^{(u)}(t)\}$ are defined in (10.57). Decisions taken by threshold detectors on the U output signals, sampled at the symbol rate, yield the detected user symbol sequences. It is useful to introduce this receiver, that we denote as MF, as, substituting the threshold detectors with more sophisticated detection devices, it represents the first stage of several multiuser receivers, as illustrated in general in Figure 10.17.

We introduce the following vector notation. The vector of symbols transmitted by U users in a symbol period T is expressed as

$$\boldsymbol{a}_k = [a_k^{(1)}, \dots, a_k^{(U)}]^T \tag{10.74}$$

[4] We assume that the information on the power of the user signals is included in the impulse responses $g_C^{(u)}$, $u = 1, \dots, U$, so that $A^{(u)} = 1$, $u = 1, \dots, U$.

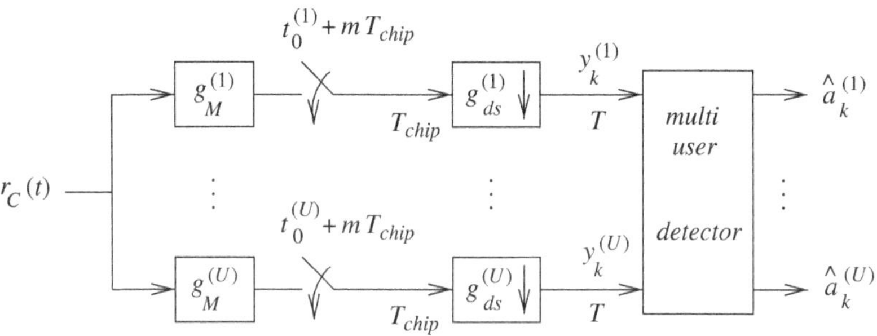

Figure 10.17 Receiver as MF and multiuser detector.

and the vector that carries the information on the codes and the channel impulse responses of the U users is expressed as

$$\boldsymbol{q}_T(t) = [q_T^{(1)}(t), \dots, q_T^{(U)}(t)]^T \tag{10.75}$$

Joint detectors constitute an important class of multiuser receivers. They effectively mitigate both ISI and MUI, exploiting the knowledge of the vector $\boldsymbol{q}_T$. In particular, we consider now *block linear receivers*: as the name suggests, a block linear receiver is a joint detector that recovers the information contained in a window of K symbol periods. Let

$$\begin{aligned}\boldsymbol{a} &= [\boldsymbol{a}_0^T, \dots, \boldsymbol{a}_{K-1}^T]^T \\ &= [a_0^{(1)}, \dots, a_0^{(U)}, \dots, a_{K-1}^{(1)}, \dots, a_{K-1}^{(U)}]^T\end{aligned} \tag{10.76}$$

be the information transmitted by U users and let $\boldsymbol{y}$ be the corresponding vector of KU elements at the MF output. We define the following correlations:

$$\mathtt{r}_{q(u,v)}(k) = \left(q_T^{(u)}(t) * q_T^{(v)*}(-t)\right)(\tau)\Big|_{\tau=kT} \tag{10.77}$$

Assuming

$$\mathtt{r}_{q(u,v)}(k) = 0 \qquad \text{for } |k| > \nu \tag{10.78}$$

with $\nu < K$, and following the approach in [18–21], we introduce the $KU \times KU$ block banded matrix

$$\boldsymbol{T} = \begin{bmatrix} \boldsymbol{R}(0) & \dots & \boldsymbol{R}(-\nu) & \boldsymbol{0} & \dots & \boldsymbol{0} \\ \vdots & \ddots & & \ddots & \ddots & \vdots \\ \vdots & & & & \ddots & \boldsymbol{0} \\ \boldsymbol{R}(\nu) & & & & & \boldsymbol{R}(-\nu) \\ \boldsymbol{0} & \ddots & & & & \vdots \\ \vdots & \ddots & \ddots & & \ddots & \vdots \\ \boldsymbol{0} & \cdots & \boldsymbol{0} & \boldsymbol{R}(\nu) & \cdots & \boldsymbol{R}(0) \end{bmatrix} \tag{10.79}$$

with

$$\boldsymbol{R}(n) = \begin{bmatrix} \mathtt{r}_{q(1,1)}(n) & \dots & \mathtt{r}_{q(1,U)}(n) \\ \vdots & \ddots & \vdots \\ \mathtt{r}_{q(U,1)}(n) & \dots & \mathtt{r}_{q(U,U)}(n) \end{bmatrix} \tag{10.80}$$

Let $\boldsymbol{w}$ be the vector of noise samples at the MF output. It is immediate to verify that its covariance matrix is $N_0\boldsymbol{T}$. Then the matrix $\boldsymbol{T}$ is Hermitian and, assuming that it is definite positive, the Cholesky decomposition (2.138) can be applied

$$\boldsymbol{T} = \boldsymbol{L}^H \boldsymbol{L} \tag{10.81}$$

where $\boldsymbol{L}^H$ is a lower triangular matrix with positive real elements on the main diagonal.

Using (10.76) and (10.79), we find that the vector $\boldsymbol{y}$ satisfies the linear relation

$$\boldsymbol{y} = \boldsymbol{T}\,\boldsymbol{a} + \boldsymbol{w} \tag{10.82}$$

Once the expression (10.82) is obtained, the vector $\boldsymbol{a}$ can be detected by well-known techniques [21].

Applying the zero-forcing criterion, at the decision point we get the vector

$$\begin{aligned} \boldsymbol{z} &= \boldsymbol{T}^{-1}\,\boldsymbol{y} \\ &= \boldsymbol{a} + \boldsymbol{T}^{-1}\,\boldsymbol{w} \end{aligned} \tag{10.83}$$

Equation (10.83) shows that the zero-forcing criterion completely eliminates both ISI and MUI, but it may enhance the noise.

Applying instead the MSE criterion to the signal $\boldsymbol{y}$ leads to the solution (see (2.198))

$$\boldsymbol{z} = (\boldsymbol{T} + N_0\boldsymbol{I})^{-1}\,\boldsymbol{y} \tag{10.84}$$

Both approaches require the inversion of a $KU \times KU$ Hermitian matrix and therefore a large computational complexity. A scheme that is computationally efficient while maintaining comparable performance is described in [22]. A minimum mean square error (MMSE) method with further reduced complexity operates on single output samples, that is for $K = 1$. However, the performance is lower because it does not exploit the correlation among the different observations.

10.6.2 Interference cancellation detector

The block equalizer operates linearly on the received signal. Non-linear techniques, based on detection and interference cancellation can actually provide better results, at the cost of a higher complexity. For the case $K = 1$, an example of non-linear detection is represented by a decision feedback equalizer (DFE) structure (see Section 7.4.4) that yields performance near the optimum maximum-likelihood (ML) receiver. However, it comes at a very higher complexity of both design and implementation. We now introduce two simpler non-linear techniques based on the principle of interference cancellation.

Successive interference cancellation

The first approach, denoted as *successive interference cancellation* (SIC) works at follows. Users' signals are detected in U steps (according to a pre-determined order): suppose here for simplicity that the sequence of user detection is simply $1, \dots, U$, i.e. they are detected according to their index. At the first step, user $u = 1$ is detected, using any single-user detection strategy. Then, its contribution to the received signal is removed from the output of the despreader of user 2, i.e. we compute from (10.61)

$$\begin{aligned} y_k^{(2)} &= y_k - A^{(1)} \sum_{i=-\infty}^{+\infty} \hat{a}_i^{(1)} \sum_{\ell=0}^{N_{SF}-1} \sum_{j=0}^{N_{SF}-1} c^{(1)}_{\ell+iN_{SF}} \\ &\quad \mathtt{r}_{q_C^{(1)}\,q_C^{(2)}}((j-\ell+(k-i)\,N_{SF})\,T_{chip}) c^{(1)*}_{j+kN_{SF}} \\ &= \sum_{u=2}^{U} A^{(u)} \sum_{i=-\infty}^{+\infty} a_i^{(u)} \sum_{\ell=0}^{N_{SF}-1} \sum_{j=0}^{N_{SF}-1} c^{(u)}_{\ell+iN_{SF}} \\ &\quad \mathtt{r}_{q_C^{(u)}\,q_C^{(2)}}((j-\ell+(k-i)\,N_{SF})\,T_{chip}) c^{(2)*}_{j+kN_{SF}} + w_k^{(2)} \end{aligned} \tag{10.85}$$

where the second equation is obtained if detection of the signal of the first user was without errors, thus $\hat{a}_k^{(1)} = a_k^{(1)}$. Moreover, we also have

$$w_k^{(p)} = \sum_{j=0}^{N_{SF}-1} \tilde{w}_{j+kN_{SF}}\; c^{(p)*}_{j+kN_{SF}} \tag{10.86}$$

Then, single user detection is applied to detect the signal of the second user. At the generic step $p = 2, \dots, U$ of SIC, we remove from the output of despreader of user p the contribution of signals of all previously detected users, i.e.

$$\begin{aligned} y_k^{(p)} &= y_k^{(p-1)} - A^{(p-1)} \sum_{i=-\infty}^{+\infty} \hat{a}_i^{(p-1)} \sum_{\ell=0}^{N_{SF}-1} \sum_{j=0}^{N_{SF}-1} c_{\ell+iN_{SF}}^{(p-1)} \\ &\quad \mathtt{r}_{q_C^{(p-1)}\, q_C^{(p)}}((j-\ell+(k-i)\,N_{SF})\,T_{chip}) c_{j+kN_{SF}}^{(p-1)*} \\ &= \sum_{u=p}^{U} A^{(u)} \sum_{i=-\infty}^{+\infty} a_i^{(u)} \sum_{\ell=0}^{N_{SF}-1} \sum_{j=0}^{N_{SF}-1} c_{\ell+iN_{SF}}^{(u)} \\ &\quad \mathtt{r}_{q_C^{(u)}\, q_C^{(p)}}((j-\ell+(k-i)\,N_{SF})\,T_{chip}) c_{j+kN_{SF}}^{(p)*} + w_k^{(p)} \end{aligned} \tag{10.87}$$

and single-user detection is applied.

Note that the order of detection of the users can be optimized, typically starting with the detection users with the highest signal-to-interference-plus-noise ratio (SINR) at the detector input.

The SIC receiver can also be applied on a single symbol. In this case, each single-user detector operates on a single symbol and before detection of symbol k of user u, the interference generated by symbols at time $k' < k$ of all users and of symbols at time k for users $\ell < u$ is removed. With this option we reduce latency, while being in general less effective in cancelling the interference, as in this case

$$\begin{aligned} y_k^{(p)} &= \sum_{u=p}^{U} A^{(u)} \sum_{i=-\infty}^{k} a_i^{(u)} \sum_{\ell=0}^{N_{SF}-1} \sum_{j=0}^{N_{SF}-1} c_{\ell+iN_{SF}}^{(u)} \\ &\quad \mathtt{r}_{q_C^{(u)}\, q_C^{(p)}}((j-\ell+(k-i)\,N_{SF})\,T_{chip}) c_{j+kN_{SF}}^{(p)*} + \\ &\quad + \sum_{u=1}^{U} A^{(u)} \sum_{i=k}^{+\infty} a_i^{(u)} \sum_{\ell=0}^{N_{SF}-1} \sum_{j=0}^{N_{SF}-1} c_{\ell+iN_{SF}}^{(u)} \\ &\quad \mathtt{r}_{q_C^{(u)}\, q_C^{(p)}}((j-\ell+(k-i)\,N_{SF})\,T_{chip}) c_{j+kN_{SF}}^{(p)*} + w_k^{(p)} \end{aligned} \tag{10.88}$$

Note that SIC follows a similar principle to that of the V-BLAST receiver for multiple-input-multiple-output (MIMO) systems (see Section 9.2.3).

Parallel interference cancellation

The second approach, denoted as *parallel interference cancellation (PIC)* works as follows.

PIC operates iteratively on a single symbol for all users. In the first iteration, a single-user detector is applied for all users, providing $\hat{a}_k^{(1,u)}$, $u = 1, \dots, U$. At iteration $i > 1$, MUI is partially removed exploiting detected symbols at iteration $i-1$, to obtain

$$\begin{aligned} y_k^{(i,u)} &= y_k^{(u)'} - \sum_{p=1, p\neq k}^{U} A^{(p)}\, \hat{a}_k^{(i-1,p)} \sum_{\ell=0}^{N_{SF}-1} \sum_{j=0}^{N_{SF}-1} c_{\ell+kN_{SF}}^{(p)} \\ &\quad \mathtt{r}_{q_C^{(p)}\, q_C^{(u)}}((j-\ell)\,T_{chip}) c_{j+kN_{SF}}^{(u)*} \end{aligned} \tag{10.89}$$

where $y_k^{(u)'}$ is the signal at the output of the despreader of user k, where we have removed the interference of symbols $a_\ell^{(p)}$, with $p = 1, \dots, U$, and $\ell < k$. Then, single-user detection is applied for each user, to obtain $\hat{a}_k^{(i,u)}$, $u = 1, \dots, U$. Note that errors in the detection process may lead to imperfect interference cancellation at first iterations. However, in a well-designed system, typically the process converges within few iterations to yield good detection performance.

10.6.3 ML multiuser detector

We now present two implementations of the ML multiuser detector.

Correlation matrix

Using the notation introduced in the previous section, the multiuser signal is

$$r_C(t) = \sum_{u=1}^{U} s_C^{(u)}(t) + w_C(t) \tag{10.90}$$

$$= \sum_{i=0}^{K-1} \boldsymbol{a}_i^T \, \boldsymbol{q}_T(t - iT) + w_C(t) \tag{10.91}$$

The log-likelihood associated with (10.90) is [23]

$$\ell_C = -\int \left| r_C(t) - \sum_{u=1}^{U} s_C^{(u)}(t) \right|^2 dt \tag{10.92}$$

Defining the matrices

$$\boldsymbol{Q}_{k_1-k_2} = \int \boldsymbol{q}_T^*(t - k_1 T) \, \boldsymbol{q}_T^T(t - k_2 T) \, dt \tag{10.93}$$

after several steps, (10.92) can be written as

$$\ell_C = \sum_{k=0}^{K-1} Re \left\{ \boldsymbol{a}_k^H \left[\boldsymbol{Q}_0 \, \boldsymbol{a}_k + \sum_{m=1}^{\nu} 2\boldsymbol{Q}_m \, \boldsymbol{a}_{k-m} - 2\boldsymbol{y}_k \right] \right\} \tag{10.94}$$

where the generic term is the branch metric, having assumed that

$$\boldsymbol{Q}_m = \boldsymbol{0}, \qquad |m| > \nu \tag{10.95}$$

We note that the first two terms within the brackets in (10.94) can be computed off-line.

The sequence $\{\hat{\boldsymbol{a}}_k\}$ that maximizes (10.94) can be obtained using the Viterbi algorithm (see Section 7.5.1); the complexity of this scheme is however exceedingly large, because it requires $O(4^{2U\nu})$ branch metric computations per detected symbol, assuming quadrature phase shift keying (QPSK) modulation.

Whitening filter

We now derive an alternative formulation of the ML multiuser detector; for this reason, it is convenient to express the MF output using the D transform, defined as the z-transform where $D = z^{-1}$ (see Section 1.2) [23]. Defining

$$\boldsymbol{Q}(D) = \sum_{k=-\nu}^{\nu} \boldsymbol{Q}_k \, D^k \tag{10.96}$$

the MF output can be written as

$$\boldsymbol{y}(D) = \boldsymbol{Q}(D) \, \boldsymbol{a}(D) + \boldsymbol{w}(D) \tag{10.97}$$

where $\boldsymbol{w}(D)$ is the noisy term with matrix spectral density $N_0 \boldsymbol{Q}(D)$. Assuming that it does not have poles on the unit circle, $\boldsymbol{Q}(D)$ can be factorized as

$$\boldsymbol{Q}(D) = \boldsymbol{F}^H(D^{-1}) \, \boldsymbol{F}(D) \tag{10.98}$$

where $\boldsymbol{F}(D)$ is minimum phase; in particular $\boldsymbol{F}(D)$ has the form

$$\boldsymbol{F}(D) = \sum_{k=0}^{\nu} \boldsymbol{F}_k \, D^k \tag{10.99}$$

where $\boldsymbol{F}_0$ is a lower triangular matrix. Now let $\boldsymbol{\Gamma}(D) = [\boldsymbol{F}^H(D^{-1})]^{-1}$, an anti-causal filter by construction. Applying $\boldsymbol{\Gamma}(D)$ to $\boldsymbol{y}(D)$ in (10.97), we get

$$\begin{aligned} \boldsymbol{z}(D) &= \boldsymbol{\Gamma}(D)\,\boldsymbol{y}(D) \\ &= \boldsymbol{F}(D)\,\boldsymbol{a}(D) + \boldsymbol{w}'(D) \end{aligned} \tag{10.100}$$

where the noisy term $\boldsymbol{w}'(D)$ is a white Gaussian process. Consequently, in the time domain (10.100) becomes

$$\boldsymbol{z}_k = \sum_{m=0}^{\nu} \boldsymbol{F}_m \, \boldsymbol{a}_{k-n} + \boldsymbol{w}'_k \tag{10.101}$$

With reference to [24], the expression (10.101) is an extension to the multidimensional case of Forney's maximum-likelihood sequence detector (MLSD) approach. In fact, the log-likelihood can be expressed as the sum of branch metrics defined as

$$\begin{aligned} &\left\| \boldsymbol{z}_k - \sum_{m=0}^{\nu} \boldsymbol{F}_m \, \boldsymbol{a}_{k-m} \right\|^2 \\ &= \sum_{u=1}^{U} \left| z_k^{(u)} - \sum_{i=1}^{U} \left(F_0^{(u,i)} \, a_k^{(i)} + \cdots + F_\nu^{(u,i)} \, a_{k-\nu}^{(i)} \right) \right|^2 \end{aligned} \tag{10.102}$$

We note that, as $\boldsymbol{F}_0$ is a lower triangular matrix, the metric has a causal dependence also with regard to the ordering of the users.

For further study on multiuser detection techniques, we refer the reader to [25–27].

10.7 Multicarrier CDMA systems

CDMA can be combined with orthogonal frequency division multiplexing (OFDM) systems in various fashions.

For a downlink transmission where a single base station transmits simultaneously to many users, data symbols are spread by CDMA with different spreading sequence per user, and then chips are allocated over the subchannels. At each user receiver, one-tap OFDM equalization (see (8.83) in Section 8.7.1) is performed, and then despreading follows, which completely removes MUI. While in orthogonal frequency division multiple access (OFDMA) schemes (see Section 8.10.1), each user is allocated a subset of subchannels, in OFDM-CDMA each user transmits over all subchannels, thus increasing the frequency diversity. However, for a channel that is highly selective in frequency and may exhibit a very low gain for some subchannels, the one-tap per subchannel zero-forcing technique considered above may substantially increase the noise level at the detection point, thus deteriorating the system performance. Alternatives, which are obtained for an increased receiver complexity, have been proposed in the literature and include (1) a single-tap MSE per subchannel equalizer, (2) an MSE per user, or even (3) an ML approach in the version MRC [28]. Note that both approaches (1) and (2) introduce MUI.

Using a CDMA-OFDM approach in uplink yields that the spread signal has undergone different channels for the various users generating MUI at the despreader output. Suitable multiuser detection techniques are required [29].

A recent approach [28, 30] denoted index-modulated OFDM-CDMA allocates to each user multiple spreading codes. Data bits are still QAM-mapped, spread, and mapped to OFDM subcarriers. However, the code use for spreading and subcarrier selection is determined again by the data bits. At the receiver, a detector is first used to identify the selected spreading code and subcarriers, before CDMA despreading follows. A particular solution provides that some data bits are used *only* for code or subcarrier selection, thus their detection at the receiver is only associated with the identification of used codes and subcarriers (and not to despreading). Lastly, observe that this approach is similar to *spatial modulation* (see Section 9.2.3), where information is carried also by the selected antenna by a multi-antenna transmitter.

Bibliography

[1] Simon, M.K., Omura, J.K., Scholtz, R.A., and Levitt, B.K. (1994). *Spread Spectrum Communications Handbook*. New York, NY: McGraw-Hill.

[2] Dixon, R.C. (1994). *Spread Spectrum Systems*, 3e. New York, NY: Wiley.

[3] Milstein, L.B. and Simon, M.K. (1996). Spread spectrum communications. In: *The Mobile Communications Handbook*, Chapter 11 (ed. J.D. Gibson), 152–165. New York, NY: CRC/IEEE Press.

[4] Proakis, J.G. (1995). *Digital Communications*, 3e. New York, NY: McGraw-Hill.

[5] Price, R. and Green, P.E. (1958). A communication technique for multipath channels. *IRE Proceedings* 46: 555–570.

[6] Viterbi, A.J. (1995). *CDMA: Principles of Spread-Spectrum Communication*. Reading, MA: Addison-Wesley.

[7] Peterson, R.L., Ziemer, R.E., and Borth, D.E. (1995). *Introduction to Spread Spectrum Communications*. Englewood Cliffs, NJ: Prentice-Hall.

[8] Holma, H. and Toskala, A. (2010). *WCDMA for UMTS: HSPA Evolution and LTE*. Chichester, UK: Wiley.

[9] Benvenuto, N. and Zorzi, M. (2011). *Principles of Communications Networks and Systems*. Wiley.

[10] Cherubini, G. and Milstein, L.B. (1989). Performance analysis of both hybrid and frequency–hopped phase–coherent spread–spectrum system. Part I and Part II. *IEEE Transactions on Communications* 37: 600–622.

[11] Klein, A. (1997). Data detection algorithms specially designed for the downlink of CDMA mobile radio systems. *Proceedings of 1997 IEEE Vehicular Technology Conference*, Phoenix, USA (4–7 May 1997), pp. 203–207.

[12] Li, K. and Liu, H. (1999). A new blind receiver for downlink DS-CDMA communications. *IEEE Communications Letters* 3: 193–195.

[13] Werner, S. and Lilleberg, J. (1999). Downlink channel decorrelation in CDMA systems with long codes. *Proceedings of 1999 IEEE Vehicular Technology Conference*, Houston, USA (16–20 May 1999), pp. 1614–1617.

[14] Hooli, K., Latva-aho, M., and Juntti, M. (1999). Multiple access interference suppression with linear chip equalizers in WCDMA downlink receivers. *Proceedings of 1999 IEEE Global Telecommunications Conference*, Rio de Janeiro, Brazil (5–9 December 1999), pp. 467–471.

[15] Madhow, U. and Honig, M.L. (1994). MMSE interference suppression for direct-sequence spread-spectrum CDMA. *IEEE Transactions on Communications* 42: 3178–3188.

[16] Miller, S.L. (1995). An adaptive direct-sequence code-division multiple-access receiver for multiuser interference rejection. *IEEE Transactions on Communications* 43: 1746–1755.

[17] Rapajic, P.B. and Vucetic, B.S. (1994). Adaptive receiver structures for asynchronous CDMA systems. *IEEE Journal on Selected Areas in Communications* 12: 685–697.

[18] Klein, A. and Baier, P.W. (1993). Linear unbiased data estimation in mobile radio systems applying CDMA. *IEEE Journal on Selected Areas in Communications* 11: 1058–1066.

[19] Blanz, J., Klein, A., Naıhan, M., and Steil, A. (1994). Performance of a cellular hybrid C/TDMA mobile radio system applying joint detection and coherent receiver antenna diversity. *IEEE Journal on Selected Areas in Communications* 12: 568–579.

[20] Kaleh, G.K. (1995). Channel equalization for block transmission systems. *IEEE Journal on Selected Areas in Communications* 13: 110–120.

[21] Klein, A., Kaleh, G.K., and Baier, P.W. (1996). Zero forcing and minimum mean-square-error equalization for multiuser detection in code-division multiple-access channels. *IEEE Transactions on Vehicular Technology* 45: 276–287.

[22] Benvenuto, N. and Sostrato, G. (2001). Joint detection with low computational complexity for hybrid TD-CDMA systems. *IEEE Journal on Selected Areas in Communications* 19: 245–253.

[23] Bottomley, G.E. and Chennakeshu, S. (1998). Unification of MLSE receivers and extension to time-varying channels. *IEEE Transactions on Communications* 46: 464–472.

[24] Duel-Hallen, A. (1995). A family of multiuser decision feedback detectors for asynchronous code-division multiple access channels. *IEEE Transactions on Communications* 43: 421–434.

[25] Verdù, S. (1998). *Multiuser Detection*. Cambridge: Cambridge University Press.

[26] (2001). Multiuser detection techniques with application to wired and wireless communications systems I. *IEEE Journal on Selected Areas in Communications* 19.

[27] (2002). Multiuser detection techniques with application to wired and wireless communications systems II. *IEEE Journal on Selected Areas in Communications* 20.

[28] Li, Q., Wen, M., Basar, E., and Chen, F. (2018). Index modulated OFDM spread spectrum. *IEEE Transactions on Wireless Communications* 17: 2360–2374.

[29] Fazel, K. and Kaiser, S. (2008). *Multi-carrier and spread spectrum systems: from OFDM and MC-CDMA to LTE and WiMAX*. Chichester: Wiley.

[30] Basar, E., Aygölü, U., Panayirci, E., and Poor, H.V. (2013). Orthogonal frequency division multiplexing with index modulation. *IEEE Transactions on Signal Processing* 61: 5536–5549.

Appendix 10.A Walsh Codes

We illustrate a procedure to obtain *orthogonal* binary sequences, with values $\{-1, 1\}$, of length 2^m.

We consider $2^m \times 2^m$ Hadamard matrices $\boldsymbol{A}_m$, with binary elements from the set $\{0, 1\}$.

For the first orders, we have

$$\boldsymbol{A}_0 = [0] \tag{10.103}$$

$$\boldsymbol{A}_1 = \begin{bmatrix} 0 & 0 \\ 0 & 1 \end{bmatrix} \tag{10.104}$$

$$\boldsymbol{A}_2 = \begin{bmatrix} 0 & 0 & 0 & 0 \\ 0 & 1 & 0 & 1 \\ 0 & 0 & 1 & 1 \\ 0 & 1 & 1 & 0 \end{bmatrix} \tag{10.105}$$

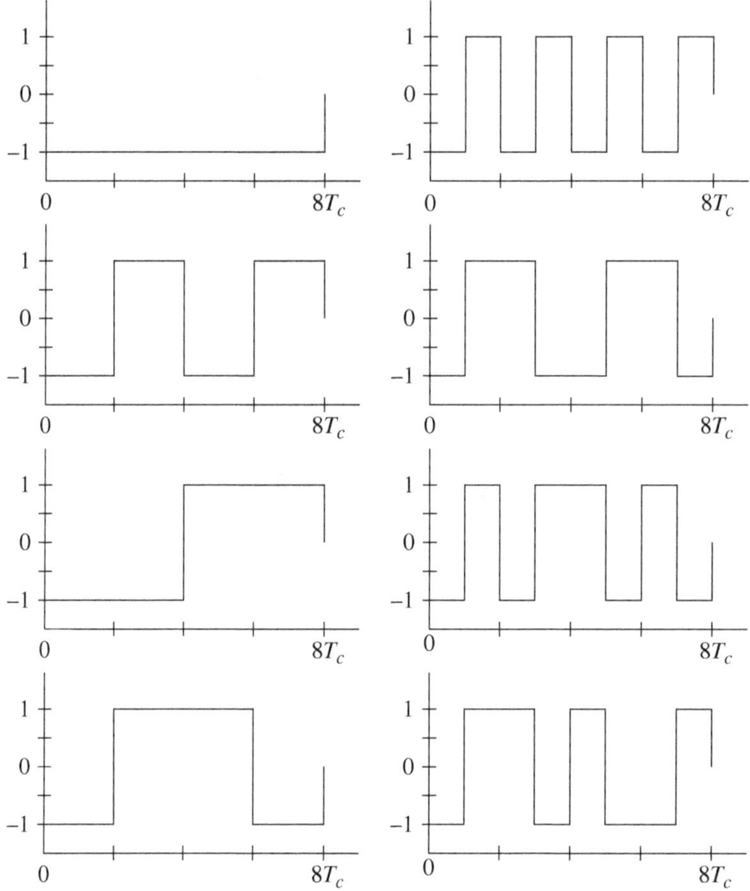

Figure 10.18 Eight orthogonal signals obtained from the Walsh code of length 8.

$$\boldsymbol{A}_3 = \begin{bmatrix} 0 & 0 & 0 & 0 & 0 & 0 & 0 & 0 \\ 0 & 1 & 0 & 1 & 0 & 1 & 0 & 1 \\ 0 & 0 & 1 & 1 & 0 & 0 & 1 & 1 \\ 0 & 1 & 1 & 0 & 0 & 1 & 1 & 0 \\ 0 & 0 & 0 & 0 & 1 & 1 & 1 & 1 \\ 0 & 1 & 0 & 1 & 1 & 0 & 1 & 0 \\ 0 & 0 & 1 & 1 & 1 & 1 & 0 & 0 \\ 0 & 1 & 1 & 0 & 1 & 0 & 0 & 1 \end{bmatrix} \tag{10.106}$$

In general, the construction is recursive

$$\boldsymbol{A}_{m+1} = \begin{bmatrix} \boldsymbol{A}_m & \boldsymbol{A}_m \\ \boldsymbol{A}_m & \overline{\boldsymbol{A}}_m \end{bmatrix} \tag{10.107}$$

where $\overline{\boldsymbol{A}}_m$ denotes the matrix that is obtained by taking the 1's complement of the elements of $\boldsymbol{A}_m$.

A Walsh code of length 2^m is obtained by taking the rows (or columns) of the Hadamard matrix $\boldsymbol{A}_m$ and by mapping 0 into -1. From the construction of Hadamard matrices, it is easily seen that two words of a Walsh code are orthogonal.

Figure 10.18 shows the eight signals obtained with the Walsh code of length 8: the signals are obtained by interpolating the Walsh code sequences by a filter having impulse response

$$\mathtt{w}_{T_c}(t) = \text{rect}\frac{t - T_c/2}{T_c} \tag{10.108}$$

Chapter 11

Channel codes

Forward error correction (FEC) is a widely used technique to achieve reliable data transmission. The redundancy introduced by an encoder for the transmission of data in coded form allows the decoder at the receiver to detect and partially correct errors. An alternative transmission technique, known as *automatic repeat query* or *request* (ARQ), consists in detecting the errors (usually by a *check-sum* transmitted with the data, see page 556) and requesting the retransmission of a data packet whenever it is received with errors.

The FEC technique presents two advantages with respect to the ARQ technique.

1. In systems that make use of the ARQ technique, the data packets do not necessarily have to be retransmitted until they are received without errors; however, for large values of the error probability, the aggregate traffic of the link is higher.
2. In systems that make use of the FEC technique, the receiver does not have to request the retransmission of data packets, thus making possible the use of a simplex link; this feature represents a strong point in many applications like time division multiple access (TDMA) and video satellite links, where a central transmitter broadcasts to receive-only terminals, which are unable to make a possible retransmission request. The FEC technique is also particularly useful in various satellite communication applications, in which the long round-trip delay of the link would cause serious traffic problems whenever the ARQ technique would be used.

We distinguish two broad classes of FEC techniques, each with numerous subclasses, employing *block codes* or *convolutional codes*.

All error correction techniques add redundancy, in the form of additional bits, to the information bits that must be transmitted. Redundancy makes the correction of errors possible and for the classes of codes considered in this chapter represents the coding overhead. The effectiveness of a coding technique is expressed in terms of the *coding gain*, G_{code}, given by the difference between the signal-to-noise ratios (SNRs), in dB, that are required to achieve a certain bit error probability for transmission without and with coding (see Definition 6.2 on page 274). The overhead is expressed in terms of the *code rate*, R_c, given by the ratio between the number of information bits and the number of code bits that are transmitted. The transmission bit rate is inversely proportional to R_c, and is larger than that necessary for uncoded data. If one of the modulation techniques of Chapter 6 is employed, the modulation rate is also larger. In Chapter 12, methods to transmit coded sequences of symbols without an increase in the modulation rate will be discussed.

For further study on the topic of error correcting codes we refer to [1–3].

Algorithms for Communications Systems and their Applications, Second Edition.
Nevio Benvenuto, Giovanni Cherubini, and Stefano Tomasin.

11.1 System model

With reference to the model of a transmission system with coding, illustrated in Figure 6.1, we introduce some fundamental parameters.

A block code is composed of a set of vectors of given length called *code words*; the length of a code word is defined as the number of vector elements, indicated by n_0. The elements of a code word are chosen from an alphabet of q elements: if the alphabet consists of two elements, for example 0 and 1, the code is a *binary code*, and we refer to the elements of each code word as *bits*; if, on the other hand, the elements of a code word are chosen from an alphabet having q elements ($q > 2$), the code is non-binary. It is interesting to note that if q is a power of two, that is $q = 2^b$, where b is a positive integer, each q-ary element has an equivalent binary representation of b bits and therefore a non-binary code word of length N can be mapped to a binary code word of length $n_0 = bN$.

There are 2^{n_0} possible code words in a binary code of length n_0. From these 2^{n_0} possible code words, we choose 2^{k_0} words ($k_0 < n_0$) to form a code. Thus, a block of k_0 information bits is mapped to a code word of length n_0 chosen from a set of 2^{k_0} code words; the resulting block code is indicated as (n_0, k_0) code, and the ratio $R_c = k_0/n_0$ is the code rate. [1]

Observation 11.1

The code rate R_c is related to the *encoder-modulator rate* R_I (6.57) by the following relation:

$$R_I = \frac{k_0}{n_0}\frac{\log_2 M}{I} = R_c \frac{\log_2 M}{I} \tag{11.1}$$

where M is the number of symbols of the I-dimensional constellation adopted by the bit-mapper.

Because the number of bits per unit of time produced by the encoder is larger than that produced by the source, two transmission strategies are possible.

Transmission for a given bit rate of the information message. With reference to Figure 6.1, from the relation

$$k_0\, T_b = n_0\, T_{cod} \tag{11.2}$$

we get

$$\frac{1}{T_{cod}} = \frac{1}{R_c}\frac{1}{T_b} \tag{11.3}$$

note that the bit rate at the modulator input is increased in the presence of the encoder. For a given modulator with M symbols, that is using the same bit mapper, this implies an increase of the modulation rate given by

$$\frac{1}{T'} = \frac{1}{T_{cod}\log_2 M} = \frac{1}{T}\frac{1}{R_c} \tag{11.4}$$

and therefore an increase of the bandwidth of the transmission channel by a factor $1/R_c$. Moreover, for the same transmitted power, from (6.97) in the presence of the encoder the SNR becomes

$$\Gamma' = \Gamma\, R_c \tag{11.5}$$

i.e. it decreases by a factor R_c with respect to the case of transmission of an uncoded message. Therefore, for a given information message bit rate $1/T_b$, the system operates with a lower Γ': consequently, the receiver is prone to introduce more bit errors at the decoder input. In spite of this, for a suitable choice

[1] In this chapter, a block code will be sometimes indicated also with the notation (n, k).

of the code, in many cases, the decoder produces a detected message $\{\hat{b}_\ell\}$ affected by fewer errors with respect to the case of transmission of an uncoded message. We note that the energy per information bit of the encoded message $\{c_m\}$ is equal to that of $\{b_\ell\}$. In fact, for a given bit mapper,

$$L'_b = \frac{k_0}{n_0}\log_2 M = R_c\, L_b \tag{11.6}$$

Assuming the same transmitted power, from (6.61) and (11.4), we get

$$E_{s'_{Ch}} = R_c\, E_{s_{Ch}} \tag{11.7}$$

Therefore, (6.63) yields for the encoded message $\{c_m\}$ an energy per bit of information equal to

$$\frac{E_{s'_{Ch}}}{L'_b} = \frac{E_{s_{Ch}}}{L_b} = E_b \tag{11.8}$$

Since $\Gamma' \neq \Gamma$, a comparison between the performance of the two systems, with and without coding, is made for the same E_b/N_0. In this case the coding gain, in dB, is given by

$$10\,(\log_{10}\Gamma - \log_{10}\Gamma' + \log_{10} R_c) \tag{11.9}$$

Transmission for a given modulation rate. For given transmitted power and given transmission channel bandwidth, Γ remains unchanged in the presence of the encoder. Therefore, there are three possibilities.

1. The bit rate of the information message decreases by a factor R_c and becomes
$$\frac{1}{T'_b} = \frac{1}{T_b}\,R_c \tag{11.10}$$
2. The source emits information bits in packets, and each packet is followed by additional bits generated by the encoder, forming a code word; the resulting bits are transmitted at the rate
$$\frac{1}{T_{cod}} = \frac{1}{T_b} \tag{11.11}$$
3. A block of m information bits is mapped to a transmitted symbol using a constellation with cardinality $M > 2^m$. In this case, transmission occurs without decreasing the bit rate of the information message.

In the first two cases, for the same number of bits of the information message, we have an increase in the duration of the transmission by a factor $1/R_c$.

For a given bit error probability in the sequence $\{\hat{b}_\ell\}$, we expect that in the presence of coding, a smaller Γ is required to achieve a certain error probability as compared to the case of transmission of an uncoded message; this reduction corresponds to the coding gain.

11.2 Block codes

We give the following general definition.[2]

Definition 11.1
The *Hamming distance* between two vectors $\boldsymbol{v}_1$ and $\boldsymbol{v}_2$, $d^H(\boldsymbol{v}_1, \boldsymbol{v}_2)$, is given by the number of elements in which the two vectors differ. □

[2] The material presented in Sections 11.2 and 11.3 is largely based on lectures given at the University of California, San Diego, by Prof. Jack K. Wolf [4], whom the authors gratefully acknowledge.

11.2.1 Theory of binary codes with group structure

Properties

A binary block code of length n is a subset containing M_c of the 2^n possible binary sequences of length n, also called code words. The only requirement on the code words is that they are all of the same length.

Definition 11.2
The *minimum Hamming distance* of a block code, to which we will refer in this chapter simply as the *minimum distance*, is denoted by d^H_{min} and coincides with the smallest number of positions in which any two code words differ. □

An example of a block code with $n = 4, M_c = 4$, and $d^H_{min} = 2$ is given by (11.22).

For the binary symmetric channel model (6.72), assuming that the binary code word $\boldsymbol{c}$ of length n is transmitted, we observe at the receiver[3]

$$z = \boldsymbol{c} \oplus \boldsymbol{e} \tag{11.12}$$

where $\oplus$ denotes the modulo-2 sum of respective vector components; for example $(0111) \oplus (0010) = (0101)$. In (11.12), $\boldsymbol{e}$ is the binary error vector whose generic component is equal to 1 if the channel has introduced an error in the corresponding bit of $\boldsymbol{c}$, and 0 otherwise. We note that z can assume all the 2^n possible combinations of n bits.

With reference to Figure 6.1, the function of the decoder consists in associating with each possible value z of a code word. A commonly adopted criterion is to associate z with the code word $\hat{\boldsymbol{c}}$ that is closest according to the Hamming distance. From this code word, the k_0 information bits, which form the sequence $\{\hat{b}_l\}$, are recovered by inverse mapping.

Interpreting the code words as points in an n-dimensional space, where the distance between points is given by the Hamming distance, we obtain the following properties:

1. A binary block code with minimum distance d^H_{min} can *correct* all patterns of
$$t = \left\lfloor \frac{d^H_{min} - 1}{2} \right\rfloor \tag{11.13}$$
or fewer errors, where $\lfloor x \rfloor$ denotes the integer value of x.
2. A binary block code with minimum distance d^H_{min} can *detect* all patterns of $(d^H_{min} - 1)$ or fewer errors.
3. In a *binary erasure channel*, the transmitted binary symbols are detected using a ternary alphabet $\{0, 1, \textit{erasure}\}$; a symbol is detected as *erasure* if the reliability of a binary decision is low. In the absence of errors, a binary block code with minimum distance d^H_{min} can fill in $(d^H_{min} - 1)$ erasures.
4. Seeking a relation among n, M_c, and d^H_{min}, we find that, for fixed n and odd d^H_{min}, M_c is upper bounded by[4]
$$M_c \leq M_{UB} = \left\lfloor 2^n \Big/ \left\{ 1 + \binom{n}{1} + \binom{n}{2} + \cdots + \binom{n}{\left\lfloor \frac{d^H_{min}-1}{2} \right\rfloor} \right\} \right\rfloor \tag{11.15}$$

[3] In Figure 6.1, z is indicated as $\tilde{\boldsymbol{c}}$.
[4] We recall that the number of binary sequences of length n with m 'ones' is equal to
$$\binom{n}{m} = \frac{n!}{m!(n-m)!} \tag{11.14}$$
where $n! = n(n-1) \cdots 1$.

5. For fixed n and d^H_{min}, it is always possible to find a code with M_c^* words, where

$$M_c^* = \left\lceil 2^n \Big/ \left\{ 1 + \binom{n}{1} + \binom{n}{2} + \cdots + \binom{n}{d^H_{min} - 1} \right\} \right\rceil \tag{11.16}$$

where $\lceil x \rceil$ denotes the smallest integer greater than or equal to x. We will now consider a procedure for finding such a code.

Step 1: choose any code word of length n and exclude from future choices that word and all words that differ from it in $(d^H_{min} - 1)$ or fewer positions. The total number of words excluded from future choices is

$$N_c(n, d^H_{min} - 1) = 1 + \binom{n}{1} + \binom{n}{2} + \cdots + \binom{n}{d^H_{min} - 1} \tag{11.17}$$

Step i: choose a word not previously excluded and exclude from future choices all words previously excluded plus the chosen word and those that differ from it in $(d^H_{min} - 1)$ or fewer positions. Continue this procedure until there are no more words available to choose from. At each step, if still not excluded, at most $N_c(n, d^H_{min} - 1)$ additional words are excluded; therefore, after step i, when i code words have been chosen, at most $i\,N_c(n, d^H_{min} - 1)$ words have been excluded. Then, if $2^n / N_c(n, d^H_{min} - 1)$ is an integer, we can choose at least that number of code words; if it is not an integer, we can choose at least a number of code words equal to the next largest integer.

Definition 11.3

A *binary code with group structure* is a binary block code for which the following conditions are verified:

1. the all zero word is a code word (zero code word);
2. the modulo-2 sum of any two code words is also a code word.

□

Definition 11.4

The *weight* of any binary vector $\boldsymbol{x}$, denoted as $w(\boldsymbol{x})$, is the number of ones in the vector. □

Property 1 of a group code. The minimum distance of the code d^H_{min} is given by

$$d^H_{min} = \min w(\boldsymbol{c}) \tag{11.18}$$

where $\boldsymbol{c}$ can be any non-zero code word.

Proof. The sum of any two distinct code words is a non-zero word. The weight of the resulting word is equal to the number of positions in which the two original words differ. Because two words at the minimum distance differ in d^H_{min} positions, there is a word of weight d^H_{min}. If there were a non-zero word of weight less than d^H_{min}, it would be different from the zero word in less than d^H_{min} positions. □

Property 2 of a group code. If all code words in a group code are written as rows of an $M_c \times n$ matrix, then every column is either zero or consists of half zeros and half ones.

Proof. An all zero column is possible if all code words have a zero in that column. Suppose in column i, there are m 1s and $(M_c - m)$ 0s. Choose one of the words with a 1 in that column and add it to all words that have a 1 in that column, including the word itself: this operation produces m words with a 0 in that column, hence, $(M_c - m) \geq m$. Now, we add that word to each word that has a 0 in that column: this produces $(M_c - m)$ words with a 1 in that column, hence, $(M_c - m) \leq m$. Therefore, $M_c - m = m$ or $m = M_c/2$. □

Corollary 11.1
From Property 2, it turns out that the number of code words M_c must be even for a binary group code. □

Corollary 11.2
Excluding codes of no interest from the transmission point of view, for which all code words have a 0 in a given position, from Property 2 the *average* weight of a code word is equal to $n/2$. □

Parity check matrix

Let $\boldsymbol{H}$ be a binary $r \times n$ matrix, which is called parity check matrix, of the form

$$\boldsymbol{H} = [\boldsymbol{A}\ \boldsymbol{B}] \tag{11.19}$$

where $\boldsymbol{B}$ is an $r \times r$ matrix with $\det[\boldsymbol{B}] \neq 0$, i.e. the columns of $\boldsymbol{B}$ are linearly independent.

A *binary parity check code* is a code consisting of all binary vectors $\boldsymbol{c}$ that are solutions of the equation

$$\boldsymbol{H}\boldsymbol{c} = \boldsymbol{0} \tag{11.20}$$

The matrix product in (11.20) is computed using the modulo-2 arithmetic.

Example 11.2.1
Let the matrix $\boldsymbol{H}$ be given by

$$\boldsymbol{H} = \begin{bmatrix} 1 & 0 & 1 & 1 \\ 0 & 1 & 0 & 1 \end{bmatrix} \tag{11.21}$$

There are four code words in the binary parity check code corresponding to the matrix $\boldsymbol{H}$; they are

$$\boldsymbol{c}_0 = \begin{bmatrix} 0 \\ 0 \\ 0 \\ 0 \end{bmatrix} \quad \boldsymbol{c}_1 = \begin{bmatrix} 1 \\ 0 \\ 1 \\ 0 \end{bmatrix} \quad \boldsymbol{c}_2 = \begin{bmatrix} 0 \\ 1 \\ 1 \\ 1 \end{bmatrix} \quad \boldsymbol{c}_3 = \begin{bmatrix} 1 \\ 1 \\ 0 \\ 1 \end{bmatrix} \tag{11.22}$$

Property 1 of a parity check code. A parity check code is a group code.

Proof. The all zero word is always a code word, as

$$\boldsymbol{H}\boldsymbol{0} = \boldsymbol{0} \tag{11.23}$$

Suppose that $\boldsymbol{c}_1$ and $\boldsymbol{c}_2$ are code words; then $\boldsymbol{H}\boldsymbol{c}_1 = \boldsymbol{0}$ and $\boldsymbol{H}\boldsymbol{c}_2 = \boldsymbol{0}$. It follows that

$$\boldsymbol{H}(\boldsymbol{c}_1 \oplus \boldsymbol{c}_2) = \boldsymbol{H}\boldsymbol{c}_1 \oplus \boldsymbol{H}\boldsymbol{c}_2 = \boldsymbol{0} \oplus \boldsymbol{0} = \boldsymbol{0} \tag{11.24}$$

Therefore, $\boldsymbol{c}_1 \oplus \boldsymbol{c}_2$ is also a code word. □

Property 2 of a parity check code. The code words corresponding to the parity check matrix $\boldsymbol{H} = [\boldsymbol{A}\ \boldsymbol{B}]$ are identical to the code words corresponding to the parity check matrix $\tilde{\boldsymbol{H}} = [\boldsymbol{B}^{-1}\boldsymbol{A}, \boldsymbol{I}] = [\tilde{\boldsymbol{A}}\ \boldsymbol{I}]$, where $\boldsymbol{I}$ is the $r \times r$ identity matrix.

Proof. Let $\boldsymbol{c} = \begin{bmatrix} \boldsymbol{c}_1^{n-r} \\ \boldsymbol{c}_{n-r+1}^{n} \end{bmatrix}$ be a code word corresponding to the matrix $\boldsymbol{H} = [\boldsymbol{A}\ \boldsymbol{B}]$, where $\boldsymbol{c}_1^{n-r}$ are the first $(n-r)$ components of the vector and $\boldsymbol{c}_{n-r+1}^{n}$ are the last r components of the vector. Then

$$\boldsymbol{H}\boldsymbol{c} = \boldsymbol{A}\boldsymbol{c}_1^{n-r} \oplus \boldsymbol{B}\boldsymbol{c}_{n-r+1}^{n} = \boldsymbol{0} \tag{11.25}$$

Multiplying by $\boldsymbol{B}^{-1}$, we get

$$\boldsymbol{B}^{-1}\boldsymbol{A}\boldsymbol{c}_1^{n-r} \oplus \boldsymbol{I}\boldsymbol{c}_{n-r+1}^{n} = \boldsymbol{0} \tag{11.26}$$

or $\tilde{\boldsymbol{H}}\boldsymbol{c} = \boldsymbol{0}$. □

From Property 2, we see that parity check matrices of the form $\tilde{\boldsymbol{H}} = [\tilde{\boldsymbol{A}}\ \boldsymbol{I}]$ are not less general than parity check matrices of the form $\boldsymbol{H} = [\boldsymbol{A}\ \boldsymbol{B}]$, where $\det[\boldsymbol{B}] \neq 0$. In general, we can consider any $r \times n$ matrix as a parity check matrix, provided that some set of r columns has a non-zero determinant. If we are not concerned with the order by which the elements of a code word are transmitted, then such a code would be equivalent to a code formed by a parity check matrix of the form

$$\boldsymbol{H} = [\boldsymbol{A}\ \boldsymbol{I}] \tag{11.27}$$

The form of the matrix (11.27) is called *canonical or systematic form*. We assume that the last r columns of $\boldsymbol{H}$ have a non-zero determinant and Therefore, that the parity check matrix can be expressed in canonical form.

Property 3 of a parity check code. There are exactly $2^{n-r} = 2^k$ code words in a parity check code.

Proof. Referring to the proof of Property 2, we find that

$$\boldsymbol{c}_{n-r+1}^{n} = \boldsymbol{A}\boldsymbol{c}_1^{n-r} \tag{11.28}$$

For each of the $2^{n-r} = 2^k$ possible binary vectors $\boldsymbol{c}_1^{n-r}$, it is possible to compute the corresponding vector $\boldsymbol{c}_{n-r+1}^{n}$. Each of these code words is unique as all of them differ in the first $(n-r) = k$ positions. Assume that there are more than 2^k code words; then at least two will agree in the first $(n-r) = k$ positions. But from (11.28), we find that these two code words also agree in the last r positions and therefore, they are identical. □

The code words have the following structure:

$$\boldsymbol{c} = [m_0 \dots m_{k-1}, p_0 \dots p_{r-1}]^T \tag{11.29}$$

where the first $k = (n-r)$ bits are called *information bits* and the last r bits are called *parity check bits*. As mentioned in Section 11.1, a parity check code that has code words of length n that are obtained by encoding k information bits is an (n, k) code.

Property 4 of a parity check code. A code word of weight w exists if and only if the modulo-2 sum of w columns of $\boldsymbol{H}$ equals $\boldsymbol{0}$.

Proof. $\boldsymbol{c}$ is a code word if and only if $\boldsymbol{H}\boldsymbol{c} = \boldsymbol{0}$. Let $\boldsymbol{h}_i$ be the ith column of $\boldsymbol{H}$ and let c_j be the jth component of $\boldsymbol{c}$. Therefore, if $\boldsymbol{c}$ is a code word, then

$$\sum_{j=1}^{n} \boldsymbol{h}_j c_j = \boldsymbol{0} \tag{11.30}$$

If $\boldsymbol{c}$ is a code word of weight w, then there are exactly w non-zero components of $\boldsymbol{c}$, for example $c_{j_1}, c_{j_2}, \dots, c_{j_w}$. Consequently, $\boldsymbol{h}_{j_1} \oplus \boldsymbol{h}_{j_2} \oplus \dots \oplus \boldsymbol{h}_{j_w} = \boldsymbol{0}$, thus a code word of weight w implies that the sum of w columns of $\boldsymbol{H}$ equals $\boldsymbol{0}$. Conversely, if $\boldsymbol{h}_{j_1} \oplus \boldsymbol{h}_{j_2} \oplus \dots \oplus \boldsymbol{h}_{j_w} = \boldsymbol{0}$, then $\boldsymbol{H}\boldsymbol{c} = \boldsymbol{0}$, where $\boldsymbol{c}$ is a binary vector with elements equal to 1 in positions $j_1, j_2, \dots, j_w$. □

Property 5 of a parity check code. A parity check code has minimum distance d_{min}^H if some modulo-2 sum of d_{min}^H columns of $\boldsymbol{H}$ is equal to $\boldsymbol{0}$, but no modulo-2 sum of fewer than d_{min}^H columns of $\boldsymbol{H}$ is equal to $\boldsymbol{0}$.

Proof. Property 5 follows from Property 1 of a group code and also from Properties 1 and 4 of a parity check code. □

Property 5 may be considered as the fundamental property of parity check codes, as it forms the basis for the design of almost all such codes. An important exception is constituted by low-density parity check codes, which will be discussed in Section 11.8. A bound on the number of parity check bits required for a given block length n and given d_{min}^H derives directly from this property.

Property 6 of a parity check code. A binary parity check code exists of block length n and minimum distance d_{min}^H, having no more than r^* parity check bits, where

$$r^* = \left\lfloor \log_2 \left(\sum_{i=0}^{d_{min}^H - 2} \binom{n-1}{i} \right) \right\rfloor + 1 \tag{11.31}$$

Proof. The proof derives from the following exhaustive construction procedure of the parity check matrix of the code.

Step 1: choose as the first column of $\boldsymbol{H}$ any non-zero vector with r^* components.

Step 2: choose as the second column of $\boldsymbol{H}$ any non-zero vector different from the first.

Step 3: choose as the ith column of $\boldsymbol{H}$ any vector distinct from all vectors obtained by modulo-2 sum of $(d_{min}^H - 2)$ or fewer previously chosen columns.

Clearly, such a procedure will result in a matrix $\boldsymbol{H}$, where no set of $(d_{min}^H - 1)$ or fewer columns of $\boldsymbol{H}$ sum to $\boldsymbol{0}$. However, we must show that we can indeed continue this process for n columns. After applying this procedure for $(n-1)$ columns, there will be at most

$$N_c(n-1, d_{min}^H - 2) = 1 + \binom{n-1}{1} + \binom{n-1}{2} + \cdots + \binom{n-1}{d_{min}^H - 2} \tag{11.32}$$

distinct vectors that are forbidden for the choice of the last column, but there are 2^{r^*} vectors to choose from; observing (11.31) and (11.32) we get $2^{r^*} > N_c(n-1, d_{min}^H - 2)$. Thus, n columns can always be chosen where no set of $(d_{min}^H - 1)$ or fewer columns sums to zero. From Property 5, the code therefore has minimum distance at least d_{min}^H. □

Code generator matrix

Using (11.28), we can write

$$\boldsymbol{c} = \begin{bmatrix} \boldsymbol{c}_1^{n-r} \\ \boldsymbol{c}_{n-r+1}^n \end{bmatrix} = \begin{bmatrix} \boldsymbol{I} \\ \boldsymbol{A} \end{bmatrix} \boldsymbol{c}_1^{n-r} = \boldsymbol{G}^T \boldsymbol{c}_1^{n-r} \tag{11.33}$$

where $\boldsymbol{G} = [\boldsymbol{I}\ \boldsymbol{A}^T]$ is a $k \times n$ binary matrix, and $\boldsymbol{I}$ is the $k \times k$ identity matrix. Taking the transpose of (11.33), we obtain

$$\boldsymbol{c}^T = (\boldsymbol{c}_1^{n-r})^T\, \boldsymbol{G} \tag{11.34}$$

thus the code words, considered now as row vectors, are given as all linear combinations of the rows of $\boldsymbol{G}$, which is called the *generator matrix* of the code. A parity check code can be specified by giving its parity check matrix $\boldsymbol{H}$ or its generator matrix $\boldsymbol{G}$.

Example 11.2.2
Consider the parity check code (7,4) with the parity check matrix

$$H = \begin{bmatrix} 1 & 1 & 0 & 1 & 1 & 0 & 0 \\ 1 & 1 & 1 & 0 & 0 & 1 & 0 \\ 1 & 0 & 1 & 1 & 0 & 0 & 1 \end{bmatrix} = [A \; I] \tag{11.35}$$

Expressing a general code word according to (11.29), to every 4 information bits 3 parity check bits are added, related to the information bits by the equations (see (11.28))

$$\begin{aligned} p_0 &= m_0 \oplus m_1 \oplus m_3 \\ p_1 &= m_0 \oplus m_1 \oplus m_2 \\ p_2 &= m_0 \oplus m_2 \oplus m_3 \end{aligned} \tag{11.36}$$

The generator matrix of this code is given by

$$G = \begin{bmatrix} 1 & 0 & 0 & 0 & 1 & 1 & 1 \\ 0 & 1 & 0 & 0 & 1 & 1 & 0 \\ 0 & 0 & 1 & 0 & 0 & 1 & 1 \\ 0 & 0 & 0 & 1 & 1 & 0 & 1 \end{bmatrix} = [I \; A^T] \tag{11.37}$$

There are 16 code words consisting of all linear combinations of the rows of $\boldsymbol{G}$. By inspection, we find that the minimum weight of a non-zero code word is 3; hence, from (11.18) the code has $d^H_{min} = 3$ and therefore is a single error correcting code.

Decoding of binary parity check codes

Conceptually, the simplest method for decoding a block code is to compare the received block of n bits with each code word and choose that code word that differs from the received word in the minimum number of positions; in case several code words satisfy this condition, choose amongst them at random. Although the simplest conceptually, the described method is out of the question practically because we usually employ codes with very many code words. It is however instructive to consider the application of this method, suitably modified, to decode group codes.

Cosets

The 2^n possible binary sequences of length n are partitioned into 2^r sets, called *cosets*, by a group code with $2^k = 2^{n-r}$ code words; this partitioning is done as follows:

Step 1: choose the first set as the set of code words $\boldsymbol{c}_1, \boldsymbol{c}_2, \dots, \boldsymbol{c}_{2^k}$.

Step 2: choose any vector, say $\boldsymbol{\eta}_2$, that is not a code word; then choose the second set as $\boldsymbol{c}_1 \oplus \boldsymbol{\eta}_2, \boldsymbol{c}_2 \oplus \boldsymbol{\eta}_2, \dots, \boldsymbol{c}_{2^k} \oplus \boldsymbol{\eta}_2$.

Step i: choose any vector, say $\boldsymbol{\eta}_i$, not included in any previous set; choose the ith set, i.e. coset, as $\boldsymbol{c}_1 \oplus \boldsymbol{\eta}_i, \boldsymbol{c}_2 \oplus \boldsymbol{\eta}_i, \dots, \boldsymbol{c}_{2^k} \oplus \boldsymbol{\eta}_i$. The partitioning continues until all 2^n vectors are used.

Note that each coset contains 2^k vectors; if we show that no vector can appear in more than one coset, we will have demonstrated that there are $2^r = 2^{n-k}$ cosets.

Property 1 of cosets. Every binary vector of length n appears in one and only one coset.

Proof. Every vector appears in at least one coset as the partitioning stops only when all vectors are used. Suppose that a vector appeared twice in one coset; then for some value of the index i, we have $\boldsymbol{c}_{j_1} \oplus \boldsymbol{\eta}_i =$

$\boldsymbol{c}_{j_2} \oplus \boldsymbol{\eta}_i$, or $\boldsymbol{c}_{j_1} = \boldsymbol{c}_{j_2}$, that is a contradiction as all code words are unique. Suppose that a vector appears in two cosets; then $\boldsymbol{c}_{j_1} \oplus \boldsymbol{\eta}_{i_1} = \boldsymbol{c}_{j_2} \oplus \boldsymbol{\eta}_{i_2}$, where we assume $i_2 > i_1$. Then $\boldsymbol{\eta}_{i_2} = \boldsymbol{c}_{j_1} \oplus \boldsymbol{c}_{j_2} \oplus \boldsymbol{\eta}_{i_1} = \boldsymbol{c}_{j_3} \oplus \boldsymbol{\eta}_{i_1}$, that is a contradiction as $\boldsymbol{\eta}_{i_2}$ would have appeared in a previous coset, against the hypothesis. □

Example 11.2.3

Consider partitioning the 2^4 binary vectors of length 4 into cosets using the group code with code words 0000, 0011, 1100, 1111, as follows:

$$\begin{array}{rcccc} & 0\,0\,0\,0 & 0\,0\,1\,1 & 1\,1\,0\,0 & 1\,1\,1\,1 \\ \eta_2 = & 0\,0\,0\,1 & 0\,0\,1\,0 & 1\,1\,0\,1 & 1\,1\,1\,0 \\ \eta_3 = & 0\,1\,1\,1 & 0\,1\,0\,0 & 1\,0\,1\,1 & 1\,0\,0\,0 \\ \eta_4 = & 1\,0\,1\,0 & 1\,0\,0\,1 & 0\,1\,1\,0 & 0\,1\,0\,1 \end{array} \tag{11.38}$$

The vectors $\boldsymbol{\eta}_1 = \boldsymbol{0}, \boldsymbol{\eta}_2, \boldsymbol{\eta}_3, \dots, \boldsymbol{\eta}_{2^r}$, are called *coset leaders*; the partitioning (11.38) is called *coset table* or *decoding table*.

Property 2 of cosets. Suppose that instead of choosing $\boldsymbol{\eta}_i$ as the coset leader of the ith coset, we choose another element of that coset; the new coset formed by using this new coset leader contains exactly the same vectors as the old coset.

Proof. Assume that the new coset leader is $\boldsymbol{\eta}_i \oplus \boldsymbol{c}_{j_1}$, and that z is an element of the new coset; then $z = \boldsymbol{\eta}_i \oplus \boldsymbol{c}_{j_1} \oplus \boldsymbol{c}_{j_2} = \boldsymbol{\eta}_i \oplus \boldsymbol{c}_{j_3}$, so z is an element of the old coset. As the new and the old cosets both contain 2^k vectors, and all vectors in a coset are unique, every element of the new coset belongs to the old coset, and vice versa. □

Example 11.2.4

Suppose that in the previous example, we had chosen the third coset leader as 0100; then the table (11.38) would be

$$\begin{array}{rcccc} & 0\,0\,0\,0 & 0\,0\,1\,1 & 1\,1\,0\,0 & 1\,1\,1\,1 \\ \eta_2 = & 0\,0\,0\,1 & 0\,0\,1\,0 & 1\,1\,0\,1 & 1\,1\,1\,0 \\ \eta_3 = & 0\,1\,0\,0 & 0\,1\,1\,1 & 1\,0\,0\,0 & 1\,0\,1\,1 \\ \eta_4 = & 1\,0\,1\,0 & 1\,0\,0\,1 & 0\,1\,1\,0 & 0\,1\,0\,1 \end{array} \tag{11.39}$$

Two conceptually simple decoding methods

Assume that each *coset leader* is chosen as the *minimum weight vector in its coset*; in case several vectors in a coset have the same minimum weight, choose any one of them as the coset leader.

Then a second method of decoding, using the decoding table, is as follows:

Step 1: locate the received vector in the coset table.

Step 2: choose the code word that appears as the first vector in the column containing the received vector.

Proposition 11.1

Decoding using the decoding table decodes to the closest code word to the received word; in case several code words are at the same smallest distance from the received word, it decodes to one of these closest words. □

Proof. Assume that the received word is the *j*th vector in the *i*th coset. The received word, given by $z = \boldsymbol{c}_j \oplus \boldsymbol{\eta}_i$, is corrected to the code word $\boldsymbol{c}_j$ and the distance between the received word and the *j*th code word is $w(\boldsymbol{\eta}_i)$. Suppose that another code word, say $\boldsymbol{c}_k$, is closer to the received vector: then

$$w(\boldsymbol{c}_k \oplus \boldsymbol{c}_j \oplus \boldsymbol{\eta}_i) < w(\boldsymbol{\eta}_i) \tag{11.40}$$

or

$$w(\boldsymbol{c}_\ell \oplus \boldsymbol{\eta}_i) < w(\boldsymbol{\eta}_i) \tag{11.41}$$

but this cannot be as $w(\boldsymbol{\eta}_i)$ is assumed to be the minimum weight vector in its coset and $\boldsymbol{c}_\ell \oplus \boldsymbol{\eta}_i$ is in that coset. □

We note that the coset leaders determine the only error patterns that can be corrected by the code. Coset leaders, moreover, have many other interesting properties: for example if a code has minimum distance d^H_{min}, all binary *n*-tuple of weight less than or equal to $\left\lfloor \frac{d^H_{min}-1}{2} \right\rfloor$ are coset leaders.

Definition 11.5
A code for which coset leaders are all vectors of weight t or less, and no others, is called a *perfect t*-error correcting code. A code for which coset leaders are all vectors of weight t or less, and some vectors of weight $t+1$ but not all, and no others, is called *quasi-perfect t*-error correcting code. □

The *perfect binary codes* are:

1. codes given by the repetition of n bits, with n odd: these codes contain only two code words, 000…0 (all zeros) and 111…1 (all ones), and correct $t = (n-1)/2$ errors ($d^H_{min} = n$);
2. Hamming codes: these codes correct $t = 1$ errors ($d^H_{min} = 3$) and have $n = 2^r - 1$, $k = n - r$, $r > 1$; the columns of the matrix $\boldsymbol{H}$ are given by all non-zero vectors of length r;
3. Golay code: $t = 3$ ($d^H_{min} = 7$), $n = 23$, $k = 12$, $r = 11$.

The following modification of the decoding method dealt with in this section will be useful later on:

Step 1′: locate the received vector in the coset table and identify the coset leader of the coset containing that vector.

Step 2′: add the coset leader to the received vector to find the decoded code word.

Syndrome decoding

A third method of decoding is based on the concept of syndrome. Among the methods described in this section, syndrome decoding is the only method of practical value for a code with a large number of code words.

Definition 11.6
For any parity check matrix $\boldsymbol{H}$, we define the *syndrome* $\boldsymbol{s}(z)$ of a binary vector z of length n as

$$\boldsymbol{s}(z) = \boldsymbol{H}z \tag{11.42}$$

□

We note that the syndrome is a vector of length r, whereas z is a vector of length n. Therefore, many vectors will have the same syndrome. All code words have an all zero Syndrome, and these are the only vectors with this property. This property of the code words is a special case of the following:

Property 3 of cosets. All vectors in the same coset have the same syndrome; vectors in different cosets have distinct syndromes.

Proof. Assume that z_1 and z_2 are in the same coset, say the ith: then $z_1 = \boldsymbol{\eta}_i \oplus \boldsymbol{c}_{j_1}$ and $z_2 = \boldsymbol{\eta}_i \oplus \boldsymbol{c}_{j_2}$. Moreover, $s(z_1) = \boldsymbol{H}z_1 = \boldsymbol{H}(\boldsymbol{\eta}_i \oplus \boldsymbol{c}_{j_1}) = \boldsymbol{H}\boldsymbol{\eta}_i \oplus \boldsymbol{H}\boldsymbol{c}_{j_1} = \boldsymbol{H}\boldsymbol{\eta}_i \oplus \boldsymbol{0} = s(\boldsymbol{\eta}_i)$. Similarly $s(z_2) = s(\boldsymbol{\eta}_i)$, so $s(z_1) = s(z_2) = s(\boldsymbol{\eta}_i)$: this proves the first part of the property.

Now, assume that z_1 and z_2 are in different cosets, say the i_1th and i_2th: then $z_1 = \boldsymbol{\eta}_{i_1} \oplus \boldsymbol{c}_{j_1}$ and $z_2 = \boldsymbol{\eta}_{i_2} \oplus \boldsymbol{c}_{j_2}$, so $s(z_1) = s(\boldsymbol{\eta}_{i_1})$ and $s(z_2) = s(\boldsymbol{\eta}_{i_2})$. If $s(z_1) = s(z_2)$ then $s(\boldsymbol{\eta}_{i_1}) = s(\boldsymbol{\eta}_{i_2})$, which implies $\boldsymbol{H}\boldsymbol{\eta}_{i_1} = \boldsymbol{H}\boldsymbol{\eta}_{i_2}$. Consequently, $\boldsymbol{H}(\boldsymbol{\eta}_{i_1} \oplus \boldsymbol{\eta}_{i_2}) = \boldsymbol{0}$, or $\boldsymbol{\eta}_{i_1} \oplus \boldsymbol{\eta}_{i_2}$ is a code word, say $\boldsymbol{c}_{j_3}$. Then $\boldsymbol{\eta}_{i_2} = \boldsymbol{\eta}_{i_1} \oplus \boldsymbol{c}_{j_3}$, which implies that $\boldsymbol{\eta}_{i_1}$ and $\boldsymbol{\eta}_{i_2}$ are in the same coset, that is a contradiction. Thus, the assumption that $s(z_1) = s(z_2)$ is incorrect. □

From Property 3, we see that there is a one-to-one relation between cosets and syndromes; this leads to the *third method of decoding*, which proceeds as follows:

Step 1″: compute the syndrome of the received vector; this syndrome identifies the coset in which the received vector is. Identify then the leader of that coset.

Step 2″: add the coset leader to the received vector to find the decoded code word.

Example 11.2.5

Consider the parity check matrix

$$\boldsymbol{H} = \begin{bmatrix} 1 & 1 & 0 & 1 & 0 & 0 \\ 1 & 0 & 1 & 0 & 1 & 0 \\ 1 & 1 & 1 & 0 & 0 & 1 \end{bmatrix} \tag{11.43}$$

The coset leaders and their respective syndromes obtained using (11.42) are reported in Table 11.1.

Suppose that the vector $z = 000111$ is received. To decode, we first compute the syndrome $\boldsymbol{H}z = \begin{bmatrix} 1 \\ 1 \\ 1 \end{bmatrix}$, then by Table 11.1 we identify the coset leader as $\begin{bmatrix} 1 \\ 0 \\ 0 \\ 0 \\ 0 \\ 0 \end{bmatrix}$, and obtain the decoded code word (1 0 0 1 1 1).

The advantage of syndrome decoding over the other decoding methods previously described is that there is no need to memorize the entire decoding table at the receiver. The first part of Step 1″, namely computing the syndrome, is trivial. The second part of Step 1″, namely identifying the coset leader corresponding to that syndrome, is the difficult part of the procedure; in general, it requires a RAM with 2^r memory locations, addressed by the syndrome of r bits and containing the coset leaders of n bits. Overall, the memory bits are $n2^r$.

There is also an algebraic method to identify the coset leader. In fact, this problem is equivalent to finding the minimum set of columns of the parity check matrix which sum to the syndrome. In other words, we must find the vector z of minimum weight such that $\boldsymbol{H}z = s$.

For a single error correcting Hamming code, all coset leaders are of weight 1 or 0, so a non-zero syndrome corresponds to a single column of $\boldsymbol{H}$ and the correspondence between syndrome and coset

Table 11.1: Coset leaders and respective syndromes for Example 11.2.5.

Coset leader	Syndrome
000000	000
000001	001
000010	010
000100	100
001000	011
010000	101
100000	111
100001	110

leader is simple. For a code with coset leaders of weight 0, 1, or 2, the syndrome is either 0, a single column of $\boldsymbol{H}$, or the sum of two columns, etc.

For a particular class of codes that will be considered later, the structure of the construction of $\boldsymbol{H}$ identifies the coset leader starting from the syndrome by using algebraic procedures. In general, each class of codes leads to a different technique to perform this task.

Property 7 of parity check codes. There are exactly 2^r correctable error vectors for a parity check code with r parity check bits.

Proof. Correctable error vectors are given by the coset leaders and there are $2^{n-k} = 2^r$ of them, all of which are distinct. On the other hand, there are 2^r distinct syndromes and each corresponds to a correctable error vector. □

For a binary symmetric channel (see Definition 6.1) we should correct all error vectors of weight i, $i = 0, 1, 2, \ldots$, until we exhaust the capability of the code. Specifically, we should try to use a *perfect code* or a *quasi-perfect code*. For a quasi-perfect t-error correcting code, the coset leaders consist of all error vectors of weight $i = 0, 1, 2, \ldots, t$, and some vectors of weight $t + 1$.

Non-binary parity check codes are discussed in Appendix 11.A.

11.2.2 Fundamentals of algebra

The calculation of parity check bits from information bits involves solving linear equations. This procedure is particularly easy for binary codes since we use modulo-2 arithmetic. An obvious question is whether or not the concepts of the previous section generalize to codes with symbols taken from alphabets with a larger cardinality, say alphabets with q symbols. We will see that the answer can be yes or no according to the value of q; furthermore, even if the answer is yes, we might not be able to use modulo-q arithmetic.

Consider the equation for the unknown x

$$ax = b \tag{11.44}$$

Table 11.2: Multiplication table for 3 elements (modulo-3 arithmetic).

·	0	1	2
0	0	0	0
1	0	1	2
2	0	2	1

Table 11.3: Multiplication table for an alphabet with four elements (modulo-4 arithmetic).

·	0	1	2	3
0	0	0	0	0
1	0	1	2	3
2	0	2	0	2
3	0	3	2	1

Table 11.4: Multiplication table for an alphabet with four elements.

·	0	1	2	3
0	0	0	0	0
1	0	1	2	3
2	0	2	3	1
3	0	3	1	2

where a and b are known coefficients, and all values are from the finite alphabet $\{0, 1, 2, \ldots, q-1\}$. First, we need to introduce the concept of multiplication, which is normally given in the form of a *multiplication table*, as the one given in Table 11.2 for the three elements $\{0, 1, 2\}$.

Table 11.2 allows us to solve (11.44) for any values of a and b, except $a = 0$. For example, the solution to equation $2x = 1$ is $x = 2$, as from the multiplication table, we find $2 \cdot 2 = 1$.

Let us now consider the case of an alphabet with four elements. A multiplication table for the four elements $\{0, 1, 2, 3\}$, resulting from the modulo-4 arithmetic, is given in Table 11.3.

Note that the equation $2x = 2$ has two solutions, $x = 1$ and $x = 3$, and equation $2x = 1$ has no solution. It is possible to construct a multiplication table that allows the equation (11.44) to be solved uniquely for x, provided that $a \neq 0$, as shown in Table 11.4.

Note that Table 11.4 is not obtained using modulo 4 arithmetic. For example $2x = 3$ has the solution $x = 2$, and $2x = 1$ has the solution $x = 3$.

modulo-q arithmetic

Consider the elements $\{0, 1, 2, \ldots, q-1\}$, where q is a positive integer larger than or equal to 2. We define two operations for combining pairs of elements from this set. The first, denoted by $\oplus$, is called *modulo-q addition* and is defined as

$$c = a \oplus b = \left\{ \begin{array}{ll} a+b & \text{if } 0 \leq a+b < q \\ a+b-q & \text{if } a+b \geq q \end{array} \right\}. \tag{11.45}$$

Here $a + b$ is the ordinary addition operation for integers that may produce an integer not in the set. In this case, q is subtracted from $a + b$, and $a + b - q$ is always an element in the set $\{0, 1, 2, \dots, q - 1\}$. The second operation, denoted by $\otimes$, is called *modulo-q multiplication* and is defined as

$$d = a \otimes b = \begin{cases} ab & \text{if } 0 \leq ab < q \\ ab - \left\lfloor \frac{ab}{q} \right\rfloor q & \text{if } ab \geq q \end{cases} \tag{11.46}$$

Note that $ab - \left\lfloor \frac{ab}{q} \right\rfloor q$, is the *remainder* or *residue* of the division of ab by q, and is always an integer in the set $\{0, 1, 2, \dots, q - 1\}$. Often, we will omit the notation $\otimes$ and write $a \otimes b$ simply as ab.

We recall that special names are given to sets which possess certain properties with respect to operations. Consider the general set G that contains the elements $\{\alpha, \beta, \gamma, \delta, \dots\}$, and two operations for combining elements from the set. We denote the first operation $\triangle$ (addition), and the second operation $\Diamond$ (multiplication). Often, we will omit the notation $\Diamond$ and write $a \Diamond b$ simply as ab. The properties we are interested in are the followig:

1. *Existence of additive identity*: For every $\alpha \in G$, there exists an element $\emptyset \in G$, called *additive identity*, such that $\alpha \triangle \emptyset = \emptyset \triangle \alpha = \alpha$.
2. *Existence of additive inverse*: For every $\alpha \in G$, there exists an element $\beta \in G$, called *additive inverse of* α, and indicated with $-\alpha$, such that $\alpha \triangle \beta = \beta \triangle \alpha = \emptyset$.
3. *Additive closure*: For every $\alpha, \beta \in G$, not necessarily distinct, $\alpha \triangle \beta \in G$.
4. *Additive associative law*: For every $\alpha, \beta, \gamma \in G$, $\alpha \triangle (\beta \triangle \gamma) = (\alpha \triangle \beta) \triangle \gamma$.
5. *Additive commutative law*: For every $\alpha, \beta \in G$, $\alpha \triangle \beta = \beta \triangle \alpha$.
6. *Multiplicative closure*: For every $\alpha, \beta \in G$, not necessarily distinct, $\alpha \Diamond \beta \in G$.
7. *Multiplicative associative law*: For every $\alpha, \beta, \gamma \in G$, $\alpha \Diamond (\beta \Diamond \gamma) = (\alpha \Diamond \beta) \Diamond \gamma$.
8. *Distributive law*: For every $\alpha, \beta, \gamma \in G$, $\alpha \Diamond (\beta \triangle \gamma) = (\alpha \Diamond \beta) \triangle (\alpha \Diamond \gamma)$ and $(\alpha \triangle \beta) \Diamond \gamma = (\alpha \Diamond \gamma) \triangle (\beta \Diamond \gamma)$.
9. *Multiplicative commutative law*: For every $\alpha, \beta \in G$, $\alpha \Diamond \beta = \beta \Diamond \alpha$.
10. *Existence of multiplicative identity*: For every $\alpha \in G$, there exists an element $I \in G$, called *multiplicative identity*, such that $\alpha \Diamond I = I \Diamond \alpha = \alpha$.
11. *Existence of multiplicative inverse*: For every $\alpha \in G$, except the element $\emptyset$, there exists an element $\delta \in G$, called *multiplicative inverse of* α, and indicated with α^{-1}, such that $\alpha \Diamond \delta = \delta \Diamond \alpha = I$.

Any set G for which Properties 1–4 hold is called a *group with respect to* $\triangle$. If G has a finite number of elements, then G is called *finite group*, and the number of elements of G is called the *order of* G.

Any set G for which Properties 1–5 hold is called an *Abelian group with respect to* $\triangle$.

Any set G for which Properties 1–8 hold is called a *ring with respect to the operations* $\triangle$ *and* $\Diamond$.

Any set G for which Properties 1–9 hold is called a *commutative ring with respect to the operations* $\triangle$ *and* $\Diamond$.

Any set G for which Properties 1–10 hold is called a *commutative ring with identity*.

Any set G for which Properties 1–11 hold is called a *field*.

It can be seen that the set $\{0, 1, 2, \dots, q - 1\}$ is a commutative ring with identity with respect to the operations of addition $\oplus$ defined in (11.45) and multiplication $\otimes$ defined in (11.46). We will show by the next three properties that this set satisfies also Property 11 if and only if q is a prime: in other words, we will show that the set $\{0, 1, 2, \dots, q - 1\}$ is a *field* with respect to the modulo-q addition and modulo-q multiplication if and only if q is a *prime*.

Finite fields are called *Galois fields*; a field of q elements is usually denoted as GF(q).

Property 11a of modulo-q arithmetic. If q is not a prime, each factor of q (less than q and greater than 1) does not have a multiplicative inverse.

Proof. Let $q = ab$, where $1 < a, b < q$; then, observing (11.46), $a \otimes b = 0$. Assume that a has a multiplicative inverse a^{-1}; then $a^{-1} \otimes (a \otimes b) = a^{-1} \otimes 0 = 0$. Now, from $a^{-1} \otimes (a \otimes b) = 0$ it is $1 \otimes b = 0$; this implies $b = 0$, which is a contradiction as $b > 1$. Similarly, we show that b does not have a multiplicative inverse. □

Property 11b of modulo-q arithmetic. If q is a prime and $a \otimes b = 0$, then $a = 0$, or $b = 0$, or $a = b = 0$.

Proof. Assume $a \otimes b = 0$ and $a, b > 0$; then $ab = Kq$, where $K < \min(a, b)$. If $1 < a \leq q - 1$ and a has no factors in common with q, then it must divide K; but this is impossible as $K < \min(a, b)$. The only other possibility is that $a = 1$, but then $a \otimes b \neq 0$ as $ab < q$. □

Property 11c of modulo-q arithmetic. If q is a prime, all non-zero elements of the set $\{0, 1, 2, \dots, q - 1\}$ have multiplicative inverse.

Proof. Assume the converse, that is the element j, with $1 \leq j \leq q - 1$, does not have a multiplicative inverse; then there must be two distinct elements $a, b \in \{0, 1, 2, \dots, q - 1\}$ such that $a \otimes j = b \otimes j$. This is a consequence of the fact that the product $i \otimes j$ can only assume values in the set $\{0, 2, 3, \dots, q - 1\}$, as by assumption $i \otimes j \neq 1$; then

$$(a \otimes j) \oplus (q - (b \otimes j)) = 0 \tag{11.47}$$

On the other hand, $q - (b \otimes j) = (q - b) \otimes j$, and

$$(a \oplus (q - b)) \otimes j = 0 \tag{11.48}$$

But $j \neq 0$ and consequently, by Property 11b, we have $a \oplus (q - b) = 0$. This implies $a = b$, which is a contradiction. □

Definition 11.7
An *ideal* I is a subset of elements of a ring R such that:

1. I is a subgroup of the additive group R, that is the elements of I form a group with respect to the addition defined in R;
2. for any element a of I and any element r of R, ar and ra are in I. □

Polynomials with coefficients from a field

We consider the set of polynomials in one variable; as this set is interesting in two distinct applications, to avoid confusion, we will use a different notation for the two cases.

The first application permits to extend our knowledge of finite fields. We have seen in Section 11.2.2 how to construct a field with a prime number of elements. Polynomials allow us to construct fields in which the number of elements is given by a power of a prime; for this purpose, we will use polynomials in the variable y.

The second application introduces an alternative method to describe code words. We will consider *cyclic codes*, a subclass of parity check codes, and in this context will use polynomials in the single variable x.

Consider any two polynomials with coefficients from the set $\{0, 1, 2, , p-1\}$, where p is a prime:

$$\begin{aligned} g(y) &= g_0 + g_1 y + g_2 y^2 + \cdots + g_m y^m \\ f(y) &= f_0 + f_1 y + f_2 y^2 + \cdots + f_n y^n \end{aligned} \tag{11.49}$$

We assume that $g_m \neq 0$ and $f_n \neq 0$. We define m as *degree of the polynomial* $g(y)$, and we write $m = \deg(g(y))$; in particular, if $g(y) = a$, $a \in \{0, 1, 2, \dots, p-1\}$, we say that $\deg(g(y)) = 0$. Similarly, it is $n = \deg(f(y))$. If $g_m = 1$, we say that $g(y)$ is a *monic polynomial.*

Assume $m \leq n$: then the *addition among polynomials* is defined as

$$f(y) + g(y) = (f_0 \oplus g_0) + (f_1 \oplus g_1)\, y + (f_2 \oplus g_2)\, y^2 + \cdots + (f_m \oplus g_m)\, y^m + \cdots + f_n y^n \tag{11.50}$$

Example 11.2.6
Let $p = 5$, $f(y) = 1 + 3y + 2y^4$, and $g(y) = 4 + 3y + 3y^2$; then $f(y) + g(y) = y + 3y^2 + 2y^4$.

Note that

$$\deg\,(f(y) + g(y)) \leq \max\,(\deg\,(f(y)), \deg(g(y)))$$

Multiplication among polynomials is defined as usual

$$f(y)\, g(y) = d_0 + d_1 y + \cdots + d_{m+n}\, y^{m+n} \tag{11.51}$$

where the arithmetic to perform operations with the various coefficients is modulo p,

$$d_i = (f_0 \otimes g_i) \oplus (f_1 \otimes g_{i-1}) \oplus \ldots \oplus (f_{i-1} \otimes g_1) \oplus (f_i \otimes g_0) \tag{11.52}$$

Example 11.2.7
Let $p = 2$, $f(y) = 1 + y + y^3$, and $g(y) = 1 + y^2 + y^3$; then $f(y)\, g(y) = 1 + y + y^2 + y^3 + y^4 + y^5 + y^6$.

Note that

$$\deg\,(f(y)\, g(y)) = \deg\,(f(y)) + \deg\,(g(y))$$

Definition 11.8
If $f(y)\, g(y) = d(y)$, we say that $f(y)$ divides $d(y)$, and $g(y)$ divides $d(y)$. We say that $p(y)$ is an *irreducible polynomial* if and only if, assuming another polynomial $a(y)$ divides $p(y)$, then $a(y) = a \in \{0, 1, \dots, p-1\}$ or $a(y) = k\, p(y)$, with $k \in \{0, 1, \dots, p-1\}$. □

The concept of an irreducible polynomial plays the same role in the theory of polynomials as does the concept of a prime number in the number theory.

Modular arithmetic for polynomials

We define a modulo arithmetic for polynomials, analogously to the modulo q arithmetic for integers. We choose a polynomial $q(y) = q_0 + q_1 y + \ldots + q_m y^m$ with coefficients that are elements of the field $\{0, 1, 2, \dots, p-1\}$. We consider the set $\mathcal{P}$ of all polynomials of degree less than m with coefficients from the field $\{0, 1, 2, \dots, p-1\}$; this set consists of p^m polynomials.

Example 11.2.8
Let $p = 2$ and $q(y) = 1 + y + y^3$; then the set $\mathcal{P}$ consists of 2^3 polynomials, $\{0, 1, y, y+1, y^2, y^2+1, y^2+y, y^2+y+1\}$.

Example 11.2.9
Let $p = 3$ and $q(y) = 2y^2$; then the set $\mathcal{P}$ consists of 3^2 polynomials, $\{0, 1, 2, y, y+1, y+2, 2y, 2y+1, 2y+2\}$.

We now define two operations among polynomials of the set $\mathcal{P}$, namely modulo $q(y)$ addition, denoted by $\triangle$, and modulo $q(y)$ multiplication, denoted by $\diamondsuit$. Modulo $q(y)$ addition is defined for every pair of polynomials $a(y)$ and $b(y)$ from the set $\mathcal{P}$ as

$$a(y) \triangle b(y) = a(y) + b(y) \tag{11.53}$$

where $a(y) + b(y)$ is defined in (11.50).

The definition of modulo $q(y)$ multiplication requires the knowledge of the *Euclidean division algorithm.*

Euclidean division algorithm. For every pair of polynomials $\alpha(y)$ and $\beta(y)$ with coefficients from some field, and $\deg(\beta(y)) \geq \deg(\alpha(y)) > 0$, there exists a unique pair of polynomials $q(y)$ and $r(y)$ such that

$$\beta(y) = q(y)\ \alpha(y) + r(y) \tag{11.54}$$

where $0 \leq \deg(r(y)) < \deg(\alpha(y))$; polynomials $q(y)$ and $r(y)$ are called, respectively, *quotient polynomial* and *remainder or residue polynomial*. In a notation analogous to that used for integers, we can write

$$q(y) = \left\lfloor \frac{\beta(y)}{\alpha(y)} \right\rfloor \tag{11.55}$$

and

$$r(y) = \beta(y) - \left\lfloor \frac{\beta(y)}{\alpha(y)} \right\rfloor \alpha(y) \tag{11.56}$$

Example 11.2.10
Let $p = 2$, $\beta(y) = y^4 + 1$, and $\alpha(y) = y^3 + y + 1$; then $y^4 + 1 = y(y^3 + y + 1) + y^2 + y + 1$, so $q(y) = y$ and $r(y) = y^2 + y + 1$.

We define modulo $q(y)$ multiplication, denoted by $\diamondsuit$, for polynomials $a(y)$ and $b(y)$ in the set $\mathcal{P}$ as

$$a(y)\diamondsuit b(y) = \begin{cases} a(y)\ b(y) & \text{if } \deg(a(y)\ b(y)) < \deg(q(y)) \\ a(y)\ b(y) - \left\lfloor \dfrac{a(y)\ b(y)}{q(y)} \right\rfloor q(y) & \text{otherwise} \end{cases} \tag{11.57}$$

It is easier to think of (11.57) as a typical multiplication operation for polynomials whose coefficients are given according to (11.52). If in this multiplication, there are terms of degree greater than or equal to $\deg(q(y))$, then we use the relation $q(y) = 0$ to lower the degree.

Example 11.2.11
Let $p = 2$ and $q(y) = 1 + y + y^3$; then $(y^2 + 1)\diamondsuit(y + 1) = y^3 + y^2 + y + 1 = (-1 - y) + y^2 + y + 1 = (1 + y) + y^2 + y + 1 = y^2$.

It can be shown that the set of polynomials with coefficients from some field and degree less than $\deg(q(y))$ is a commutative ring with identity with respect to the operations modulo $q(y)$ addition and modulo $q(y)$ multiplication. We now find under what conditions this set of polynomials and operations forms a field.

Property 11a of modular polynomial arithmetic. If $q(y)$ is not irreducible, then the factors of $q(y)$, of degree greater than zero and less than $\deg(q(y))$, do not have multiplicative inverses.

Proof. Let $q(y) = a(y)\, b(y)$, where $0 < \deg(a(y)), \deg(b(y)) < \deg(q(y))$; then $a(y) \Diamond b(y) = 0$. Assume $a(y)$ has a multiplicative inverse, $a^{-1}(y)$; then, from $a^{-1}(y) \Diamond (a(y) \Diamond b(y)) = a^{-1}(y) \Diamond 0 = 0$ it is $(a^{-1}(y) \Diamond a(y)) \Diamond b(y) = 0$, then $1 \Diamond b(y) = 0$, or $b(y) = 0$. The last equation is a contradiction as by assumption $\deg(b(y)) > 0$. Similarly, we show that $b(y)$ does not have a multiplicative inverse. □

We give without proof the following properties:

Property 11b of modular polynomial arithmetic. If $q(y)$ is irreducible and $a(y) \Diamond b(y) = 0$, then $a(y) = 0$, or $b(y) = 0$, or $a(y) = b(y) = 0$.

Property 11c of modular polynomial arithmetic. If $q(y)$ is irreducible, all non-zero elements of the set of polynomials $\mathcal{P}$ of degree less than $\deg(q(y))$ have multiplicative inverses.

We now have that the set of polynomials with coefficients from some field and degree less than $\deg(q(y))$ forms a field, with respect to the operations of modulo $q(y)$ addition and modulo $q(y)$ multiplication, if and only if $q(y)$ is irreducible.

Furthermore, it can be shown that there exists at least one irreducible polynomial of degree m, for every $m \geq 1$, with coefficients from a generic field $\{0, 1, 2, \ldots, p-1\}$. We now have a method of generating a field with p^m elements.

Example 11.2.12
Let $p = 2$ and $q(y) = y^2 + y + 1$; we have that $q(y)$ is irreducible. Consider the set $\mathcal{P}$ with elements $\{0, 1, y, y+1\}$. The addition and multiplication tables for these elements modulo $y^2 + y + 1$ are given in Table 11.5 and Table 11.6, respectively.

Table 11.5: Modulo $y^2 + y + 1$ addition table for $p = 2$.

$\triangle$	0	1	y	$y+1$
0	0	1	y	$y+1$
1	1	0	$y+1$	y
y	y	$y+1$	0	1
$y+1$	$y+1$	y	1	0

Table 11.6: Modulo $y^2 + y + 1$ multiplication table for $p = 2$.

$\Diamond$	0	1	y	$y+1$
0	0	0	0	0
1	0	1	y	$y+1$
y	0	y	$y+1$	1
$y+1$	0	$y+1$	1	y

Devices to sum and multiply elements in a finite field

For the GF(p^m) obtained by an irreducible polynomial of degree m,

$$q(y) = \sum_{i=0}^{m} q_i\, y^i, \qquad q_i \in \text{GF}(p) \tag{11.58}$$

let $a(y)$ and $b(y)$ be two elements of $\mathcal{P}$:

$$a(y) = \sum_{i=0}^{m-1} a_i\, y^i, \qquad a_i \in \text{GF}(p) \tag{11.59}$$

and

$$b(y) = \sum_{i=0}^{m-1} b_i\, y^i, \qquad b_i \in \text{GF}(p) \tag{11.60}$$

The device to perform the addition(11.53),

$$s(y) = \sum_{i=0}^{m-1} s_i\, y^i = (a(y) + b(y)) \bmod q(y) \tag{11.61}$$

is illustrated in Figure 11.1. The implementation of a device to perform the multiplication is slightly more complicated, as illustrated in Figure 11.2, where T_c is the period of the clock applied to the shift-register (SR) with m elements, and all operations are modulo p.

Let us define

$$d(y) = \sum_{i=0}^{m-1} d_i\, y^i = (a(y)\, b(y)) \bmod q(y) \tag{11.62}$$

The device is based on the following decomposition:

$$\begin{aligned} a(y)\, b(y) \bmod q(y) &= \sum_{i=0}^{m-1} a_i\, y^i\, b(y) \bmod q(y) \\ &= a_0\, b(y) \\ &\quad + a_1 (y\, b(y)) \bmod q(y) \\ &\quad \vdots \\ &\quad + a_{m-1} (y^{m-1}\, b(y)) \bmod q(y) \end{aligned} \tag{11.63}$$

where additions and multiplications are modulo p. Now, using the identity $\sum_{i=0}^{m} q_i\, y^i = 0 \bmod q(y)$, note that the following relation holds:

$$\begin{aligned} y\, b(y) &= b_0 y + b_1 y^2 + \cdots + b_{m-2} y^{m-1} + b_{m-1} y^m \\ &= (-b_{m-1} q_m^{-1} q_0) + (b_0 - b_{m-1} q_m^{-1} q_1)\, y + \cdots + (b_{m-2} - b_{m-1} q_m^{-1} q_{m-1})\, y^{m-1} \end{aligned} \tag{11.64}$$

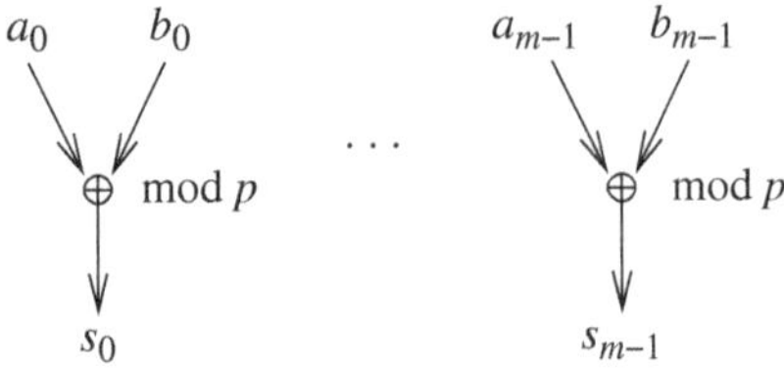

Figure 11.1 Device for the sum of two elements $(a_0, \ldots, a_{m-1})$ and $(b_0, \ldots, b_{m-1})$ of GF(p^m).

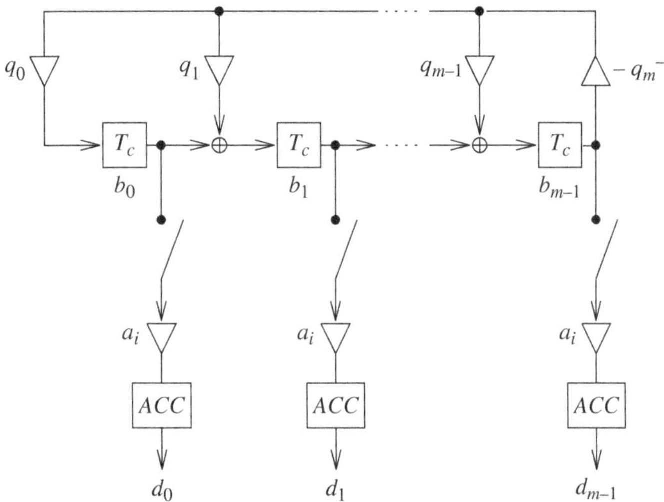

Figure 11.2 Device for the multiplication of two elements $(a_0, \dots, a_{m-1})$ and $(b_0, \dots, b_{m-1})$ of GF(p^m). T_c is the clock period, and ACC denotes an accumulator. All additions and multiplications are modulo p.

The term $(y^i\, b(y)) \bmod q(y)$ is thus obtained by initializing the SR of Figure 11.2 to the sequence $(b_0, \dots, b_{m-1})$, and by applying i clock pulses; the desired result is then contained in the shift register.

Observing (11.63), we find that it is necessary to multiply each element of the SR by a_i and accumulate the result; after multiplications by all coefficients $\{a_i\}$ have been performed, the final result is given by the content of the accumulators. Note that in the binary case, for $p = 2$, the operations of addition and multiplication are carried out by XOR and AND functions, respectively.

Remarks on finite fields

1. We have seen how to obtain finite fields with p (p a prime) elements, given by $\{0, 1, \dots, p-1\}$, or p^m elements, using the Property 11c. These fields are also known as Galois fields and are usually denoted by GF(p) or GF(p^m). It can be shown that there are no other fields with a finite number of elements. Moreover, all fields with the same number of elements are identical, that is all finite fields are generated by the procedures discussed in the previous sections.
2. The field, from which the coefficients of the irreducible polynomial are chosen, is called *ground field*; the field generated using the arithmetic of polynomials is called *extension field.*
3. Every row of the addition table contains each field element once and only once; the same is true for the columns.
4. Every row of the multiplication table, except the row corresponding to the element 0, contains each field element once and only once; the same is true for the columns.
5. If we multiply any non-zero element by itself, we get a non-zero element of the field (perhaps itself). As there are only $(q-1)$ non-zero elements, we must eventually find a situation for

$j > i$ such that an element α multiplied by itself j times will equal α multiplied by itself i times, that is

$$\overbrace{\alpha \otimes \alpha \otimes \alpha \otimes \cdots \otimes \alpha}^{j \text{ times}} = \overbrace{\alpha \otimes \alpha \otimes \cdots \otimes \alpha}^{i \text{ times}} = \beta \tag{11.65}$$

We observe that

$$\overbrace{\alpha \otimes \alpha \otimes \cdots \otimes \alpha}^{j-i \text{ times}} \otimes \overbrace{\alpha \otimes \alpha \otimes \cdots \otimes \alpha}^{i \text{ times}} = \overbrace{\alpha \otimes \alpha \otimes \cdots \otimes \alpha}^{j \text{ times}} \tag{11.66}$$

Substituting (11.65) in (11.66), and observing that β has a multiplicative inverse, we can multiply from the right by this inverse to obtain

$$\alpha^{j-i} = \overbrace{\alpha \otimes \alpha \otimes \cdots \otimes \alpha}^{j-i \text{ times}} = 1 \tag{11.67}$$

Definition 11.9
For every non-zero field element, α, the *order of* α is the smallest integer ℓ such that $\alpha^{\ell} = 1$. □

Example 11.2.13
Consider the field with elements $\{0, 1, 2, 3, 4\}$ and modulo 5 arithmetic. Then

element	order
1	1
2	4
3	4
4	2

(11.68)

Example 11.2.14
Consider the field GF(2^2) with four elements, $\{0, 1, y, y + 1\}$, and addition and multiplication modulo $y^2 + y + 1$. Then

element	order
1	1
y	3
$y+1$	3

(11.69)

6. An element from the field GF(q) is said to be *primitive* if it has order $q - 1$. For fields generated by arithmetic modulo a polynomial $q(y)$, if the field element y is primitive, we say that $q(y)$ is a *primitive irreducible polynomial.*

A property of finite fields that we give without proof is that every finite field has at least one primitive element; we note that once a primitive element has been identified, every other non-zero field element can be obtained by multiplying the primitive element by itself an appropriate number of times. A list of primitive polynomials for the ground field GF(2) is given in Table 11.7.

Example 11.2.15
For the field GF(4) generated by the polynomial arithmetic modulo $q(y) = y^2 + y + 1$, for the ground field GF(2), y is a primitive element (see (11.69)); thus $y^2 + y + 1$ is a primitive polynomial.

Table 11.7: List of primitive polynomials $q(y)$ of degree m for the ground field GF(2).

m		m	
2	$1 + y + y^2$	14	$1 + y + y^6 + y^{10} + y^{14}$
3	$1 + y + y^3$	15	$1 + y + y^{15}$
4	$1 + y + y^4$	16	$1 + y + y^3 + y^{12} + y^{16}$
5	$1 + y^2 + y^5$	17	$1 + y^3 + y^{17}$
6	$1 + y + y^6$	18	$1 + y^7 + y^{18}$
7	$1 + y^3 + y^7$	19	$1 + y + y^2 + y^5 + y^{19}$
8	$1 + y^2 + y^3 + y^4 + y^8$	20	$1 + y^3 + y^{20}$
9	$1 + y^4 + y^9$	21	$1 + y^2 + y^{21}$
10	$1 + y^3 + y^{10}$	22	$1 + y + y^{22}$
11	$1 + y^2 + y^{11}$	23	$1 + y^5 + y^{23}$
12	$1 + y + y^4 + y^6 + y^{12}$	24	$1 + y + y^2 + y^7 + y^{24}$
13	$1 + y + y^3 + y^4 + y^{13}$		

7. The order of every non-zero element of GF(q) must divide $(q - 1)$.

Proof. Every non-zero element β can be written as the power of a primitive element α_p; this implies that there is some $i < (q - 1)$ such that

$$\beta = \overbrace{\alpha_p \otimes \alpha_p \otimes \cdots \otimes \alpha_p}^{i \text{ times}} = \alpha_p^i \tag{11.70}$$

Note that from the definition of a primitive element, we get $\alpha_p^{q-1} = 1$, but $\alpha_p^j \neq 1$ for $j < (q - 1)$; furthermore, there exists an integer ℓ such that $\beta^\ell = \alpha_p^{i\ell} = 1$. Consequently, $(\ell)(i)$ is a multiple of $(q - 1)$, and it is exactly the smallest multiple of i that is a multiple of $(q - 1)$, thus $(i)(\ell) = \text{l.c.m.}(i, q - 1)$, i.e. the least common multiple of i and $(q - 1)$. We recall that

$$\text{l.c.m.}(a, b) = \frac{ab}{\text{g.c.d.}(a, b)} \tag{11.71}$$

where g.c.d.(a, b) is the greatest common divisor of a and b. Thus,

$$(i)(\ell) = \frac{(i)(q - 1)}{\text{g.c.d.}(i, q - 1)} \tag{11.72}$$

and

$$\frac{q - 1}{\text{g.c.d.}(i, q - 1)} = \ell \tag{11.73}$$

□

Example 11.2.16
Let α_p be a primitive element of GF(16); from (11.73), the orders of the non-zero field elements are

field element $\beta = \alpha^i$	g.c.d.$(i, q-1)$	order of field element $\frac{q-1}{\text{g.c.d.}(i, q-1)}$
α_p	1	15
α_p^2	1	15
α_p^3	3	5
α_p^4	1	15
α_p^5	5	3
α_p^6	3	5
α_p^7	1	15
α_p^8	1	15
α_p^9	3	5
α_p^{10}	5	3
α_p^{11}	1	15
α_p^{12}	3	5
α_p^{13}	1	15
α_p^{14}	1	15
α_p^{15}	15	1

(11.74)

8. A ground field can itself be generated as an extension field. For example GF(16) can be generated by taking an irreducible polynomial of degree 4 with coefficients from GF(2), which we would call GF(2^4), or by taking an irreducible polynomial of degree 2 with coefficients from GF(4), which we would call GF(4^2). In either case, we would have the same field, except for the names of the elements.

Example 11.2.17
Consider the field GF(2^3) generated by the primitive polynomial $q(y) = 1 + y + y^3$, with ground field GF(2). As $q(y)$ is a primitive polynomial, each element of GF(2^3), except the zero element, can be expressed as a power of y. Recalling the polynomial representation $\mathcal{P}$, we may associate to each polynomial a vector representation, with m components on GF(p) given by the coefficients of the powers of the variable y. The three representations are reported in Table 11.8.

Roots of a polynomial

Consider a polynomial of degree m with coefficients that are elements of some field. We will use the variable x, as the polynomials are now considered for a purpose that is not that of generating a finite field. In fact, the field of the coefficients may itself have a polynomial representation.

Consider, for example a polynomial in x with coefficients from GF(4). We immediately see that it is not worth using the notation $\{0, 1, y, y+1\}$ to identify the four elements of GF(4), as the notation $\{0, 1, \alpha, \beta\}$ would be much simpler. For example a polynomial of degree three with coefficients from GF(4) is given by $f(x) = \alpha x^3 + \beta x^2 + 1$.

Table 11.8: Three equivalent representations of the elements of GF(2^3).

exponential	polynomial	binary ($y^0y^1y^2$)
0	0	0 0 0
1	1	1 0 0
y	y	0 1 0
y^2	y^2	0 0 1
y^3	$1+y$	1 1 0
y^4	$y+y^2$	0 1 1
y^5	$1+y+y^2$	1 1 1
y^6	$1+y^2$	1 0 1

Given any polynomial $f(x)$, we say that γ is a *root of the equation* $f(x) = 0$ or, more simply, that it is a *root of* $f(x)$, if and only if $f(\gamma) = 0$. The definition is more complicated than it appears, as we must know the meaning of the two members of the equation $f(\gamma) = 0$. For example, we recall that the fundamental theorem of algebra states that every polynomial of degree m has exactly m roots, not necessarily distinct. If we take the polynomial $f(x) = x^2 + x + 1$ with coefficients from $\{0, 1\}$, what are its roots? As $f(0) = f(1) = 1$, we have that neither 0 nor 1 are roots.

Before proceeding, we recall a similar situation that we encounter in ordinary algebra. The polynomial $x^2 + 3$, with coefficients in the field of real numbers, has two roots in the field of complex numbers; however, no roots exist in the field of real numbers; therefore, the polynomial does not have factors whose coefficients are real numbers. Thus, we would say that the polynomial is irreducible, yet even the irreducible polynomial has complex-valued roots and can be factorized.

This situation is due to the fact that, if we have a polynomial $f(x)$ with coefficients from some field, the roots of the polynomial are either from that field or from an extension field of that field. For example, take the polynomial $f(x) = x^2 + x + 1$ with coefficients from GF(2), and consider the extension field GF(4) with elements $\{0, 1, \alpha, \beta\}$ that obey the addition and the multiplication rules given in Tables 11.9 and 11.10, respectively. Then $f(\alpha) = f(x)|_{x=\alpha} = \alpha^2 \bigtriangleup \alpha \bigtriangleup 1 = (\beta \bigtriangleup \alpha) \bigtriangleup 1 = 1 \bigtriangleup 1 = 0$, thus α is a root.

Table 11.9: Addition table for the elements of GF(4).

$\bigtriangleup$	0	1	α	β
0	0	1	α	β
1	1	0	β	α
α	α	β	0	1
β	β	α	1	0

Table 11.10: Multiplication table for the elements of GF(4).

$\Diamond$	0	1	α	β
0	0	0	0	0
1	0	1	α	β
α	0	α	β	1
β	0	β	1	α

Similarly, we find $f(\beta) = f(x)|_{x=\beta} = \beta^2 \triangle \beta \triangle 1 = (\alpha \triangle \beta) \triangle 1 = 1 \triangle 1 = 0$, thus, the two roots of $f(x)$ are α and β. We can factor $f(x)$ into two factors, each of which is a polynomial in x with coefficients from GF(4). For this purpose, we consider $(x \triangle -\alpha)\Diamond(x \triangle -\beta) = (x \triangle \alpha)\Diamond(x \triangle \beta)$; leaving out the notations $\triangle$ and $\Diamond$ for $+$ and $\times$, we get

$$(x \triangle \alpha)\Diamond(x \triangle \beta) = x^2 + (\alpha + \beta)\, x + \alpha\beta = x^2 + x + 1 \tag{11.75}$$

Thus, if we use the operations defined in GF(4), $(x + \alpha)$ and $(x + \beta)$ are factors of $x^2 + x + 1$; it remains that $x^2 + x + 1$ is irreducible as it has no factors with coefficients from GF(2).

Property 1 of the roots of a polynomial. If γ is a root of $f(x) = 0$, then $(x - \gamma)$, that is $(x + (-\gamma))$, is a factor of $f(x)$.

Proof. Using the Euclidean division algorithm, we divide $f(x)$ by $(x - \gamma)$ to get

$$f(x) = Q(x)\,(x - \gamma) + r(x) \tag{11.76}$$

where $\deg(r(x)) < \deg(x - \gamma) = 1$. Therefore,

$$f(x) = Q(x)\,(x - \gamma) + r_0 \tag{11.77}$$

But $f(\gamma) = 0$, so

$$f(\gamma) = 0 = Q(\gamma)\,(\gamma - \gamma) + r_0 = r_0 \tag{11.78}$$

therefore,

$$f(x) = Q(x)\,(x - \gamma) \tag{11.79}$$

□

Property 2 of the roots of a polynomial. If $f(x)$ is an arbitrary polynomial with coefficients from GF(p), p a prime, and β is a root of $f(x)$, then β^p is also a root of $f(x)$.

Proof. We consider the polynomial $f(x) = f_0 + f_1 x + f_2 x^2 + \cdots + f_m x^m$, where $f_i \in$ GF(p), and form the power $(f(x))^p$. It results

$$(f(x))^p = (f_0 + f_1 x + f_2 x^2 + \cdots + f_m x^m)^p = f_0^p + f_1^p x^p + \cdots + f_m^p x^{mp} \tag{11.80}$$

as the cross terms contain a factor p, which is the same as 0 in GF(p). On the other hand, for $f_i \neq 0$, $f_i^p = f_i$, as from Property 7 on page 537 the order of any non-zero element divides $p - 1$; the equation is true also if f_i is the zero element. Therefore,

$$(f(x))^p = f(x^p) \tag{11.81}$$

If β is a root of $f(x) = 0$, then $f(\beta) = 0$, and $f^p(\beta) = 0$. But $f^p(\beta) = f(\beta^p)$, so that $f(\beta^p) = 0$; therefore, β^p is also a root of $f(x)$. □

A more general form of the property just introduced, that we will give without proof, is expressed by the following property.

Property 2a of the roots of a polynomial. If $f(x)$ is an arbitrary polynomial having coefficients from GF(q), with q a prime or a power of a prime, and β is a root of $f(x) = 0$, then β^q is also a root of $f(x) = 0$,

Example 11.2.18
Consider the polynomial $x^2 + x + 1$ with coefficients from GF(2). We already have seen that α, element of GF(4), is a root of $x^2 + x + 1 = 0$. Therefore, α^2 is also a root, but $\alpha^2 = \beta$, so β is a second root. The polynomial has degree 2, thus it has 2 roots, and they are α and β, as previously seen. Note also that β^2 is also a root, but $\beta^2 = \alpha$.

Minimum function

Definition 11.10

Let β be an element of an extension field of GF(q); the *minimum function of* β, $m_\beta(x)$, is the monic polynomial of least degree with coefficients from GF(q) such that $m_\beta(x)|_{x=\beta} = 0$. □

We now list some properties of the minimum function.

1. The minimum function is unique.

Proof. Assume there were two minimum functions, of the same degree and monic, $m_\beta(x)$ and $m'_\beta(x)$. Form the new polynomial $(m_\beta(x) - m'_\beta(x))$ whose degree is less than the degree of $m_\beta(x)$ and $m'_\beta(x)$; but $(m_\beta(x) - m'_\beta(x))|_{x=\beta} = 0$, so we have a new polynomial, whose degree is less than that of the minimum function, that admits β as root. Multiplying by a constant, we can thus find a monic polynomial with this property, but this cannot be since the minimum function is the monic polynomial of least degree for which β is a root. □

2. The minimum function is irreducible.

Proof. Assume the converse were true, that is $m_\beta(x) = a(x)\, b(x)$; then $m_\beta(x)|_{x=\beta} = a(\beta)\, b(\beta) = 0$. Then either $a(\beta) = 0$ or $b(\beta) = 0$, so that β is a root of a polynomial of degree less than the degree of $m_\beta(x)$. By making this polynomial monic, we arrive at a contradiction. □

3. Let $f(x)$ be any polynomial with coefficients from GF(q), and let $f(x)|_{x=\beta} = 0$; then $f(x)$ is divisible by $m_\beta(x)$.

Proof. Use the Euclidean division algorithm to yield

$$f(x) = Q(x)\, m_\beta(x) + r(x) \tag{11.82}$$

where $\deg(r(x)) < \deg(m_\beta(x))$. Then we have that

$$f(\beta) = Q(\beta)\, m_\beta(\beta) + r(\beta) \tag{11.83}$$

but as $f(\beta) = 0$ and $m_\beta(\beta) = 0$, then $r(\beta) = 0$. As $\deg(r(x)) < \deg(m_\beta(x))$, the only possibility is $r(x) = 0$; thus $f(x) = Q(x)\, m_\beta(x)$. □

4. Let $f(x)$ be any irreducible monic polynomial with coefficients from GF(q) for which $f(\beta) = 0$, where β is an element of some extension field of GF(q); then $f(x) = m_\beta(x)$.

Proof. From Property 3 $f(x)$ must be divisible by $m_\beta(x)$, but $f(x)$ is irreducible, so it is only trivially divisible by $m_\beta(x)$, that is $f(x) = K\, m_\beta(x)$: but $f(x)$ and $m_\beta(x)$ are both monic polynomials, therefore, $K = 1$. □

We now introduce some interesting propositions.

1. Let β be an element of GF(q^m), with q prime; then the polynomial $F(x)$, defined as

$$F(x) = \prod_{i=0}^{m-1} (x - \beta^{q^i}) = (x - \beta)\,(x - \beta^q)\,(x - \beta^{q^2}) \cdots (x - \beta^{q^{m-1}}) \tag{11.84}$$

has all its coefficients from GF(q).

Proof. Observing Property 7 on page 537, we have that the order of β divides $q^m - 1$, therefore, $\beta^{q^m} = \beta$. Thus, we can express $F(x)$ as

$$F(x) = \prod_{i=1}^{m} (x - \beta^{q^i}) \tag{11.85}$$

Therefore,

$$F(x^q) = \prod_{i=1}^{m} (x^q - \beta^{q^i}) = \prod_{i=1}^{m} (x - \beta^{q^{i-1}})^q = \prod_{j=0}^{m-1} (x - \beta^{q^j})^q = (F(x))^q \tag{11.86}$$

Consider now the expression $F(x) = \sum_{i=0}^{m} f_i\, x^i$; then

$$F(x^q) = \sum_{i=0}^{m} f_i\, x^{iq} \tag{11.87}$$

and

$$(F(x))^q = \left(\sum_{i=0}^{m} f_i\, x^i \right)^q = \sum_{i=0}^{m} f_i^q\, x^{iq} \tag{11.88}$$

Equating like coefficients in (11.87) and (11.88), we get $f_i^q = f_i$; hence, f_i is a root of the equation $x^q - x = 0$. But on the basis of Property 7 on page 537, the q elements from GF(q) all satisfy the equation $x^q - x = 0$, and this equation only has q roots; therefore, the coefficients f_i are elements from GF(q). □

2. If $g(x)$ is an irreducible polynomial of degree m with coefficients from GF(q), and $g(\beta) = 0$, where β is an element of some extension field of GF(q), then $\beta, \beta^q, \beta^{q^2}, \dots, \beta^{q^{m-1}}$ are all the roots of $g(x)$.

Proof. At least one root of $g(x)$ is in GF(q^m); this follows by observing that, if we form GF(q^m) using the arithmetic modulo $g(y)$, then y will be a root of $g(x) = 0$. From Proposition 11.1, if β is an element from GF(q^m), then $F(x) = \prod_{i=0}^{m-1} (x - \beta^{q^i})$ has all coefficients from GF(q); thus, $F(x)$ has degree m, and $F(\beta) = 0$. As $g(x)$ is irreducible, we know that $g(x) = K\, m_\beta(x)$, but as $F(\beta) = 0$, and $F(x)$ and $g(x)$ have the same degree, then $F(x) = K_1 m_\beta(x)$, and therefore, $g(x) = K_2 F(x)$. As $\beta, \beta^q, \beta^{q^2}, \dots, \beta^{q^{m-1}}$, are all roots of $F(x)$, then they must also be all the roots of $g(x)$. □

3. Let $g(x)$ be a polynomial with coefficients from GF(q) which is also irreducible in this field. Moreover, let $g(\beta) = 0$, where β is an element of some extension field of GF(q); then the degree of $g(x)$ equals the smallest integer k such that

$$\beta^{q^k} = \beta \tag{11.89}$$

Proof. We have that $\deg(g(x)) \geq k$ as $\beta, \beta^q, \beta^{q^2}, \dots, \beta^{q^{k-1}}$, are all roots of $g(x)$ and by assumption are distinct. Assume that $\deg(g(x)) > k$; from Proposition 11.2, we know that β must be at least a double root of $g(x) = 0$, and therefore, $g'(x) = (d/dx)g(x) = 0$ must also have β as a root. As $g(x)$ is irreducible, we have that $g(x) = K\, m_\beta(x)$, but $m_\beta(x)$ must divide $g'(x)$; we get a contradiction because $\deg(g'(x)) < \deg(g(x))$. □

Methods to determine the minimum function

1. Direct calculation.

Example 11.2.19

Consider the field GF(2^3) obtained by taking the polynomial arithmetic modulo the irreducible polynomial $y^3 + y + 1$ with coefficients from GF(2); the field elements are $\{0, 1, y, y + 1, y^2, y^2 + 1, y^2 +$

$y, y^2+y+1\}$. Assume we want to find the minimum function of $\beta = (y+1)$. If $(y+1)$ is a root, also $(y+1)^2 = y^2+1$ and $(y+1)^4 = y^2+y+1$ are roots. Note that $(y+1)^8 = (y+1) = \beta$, thus the minimum function is

$$\begin{aligned} m_{y+1}(x) &= (x-\beta)\,(x-\beta^2)\,(x-\beta^4) \\ &= (x+(y+1))(x+(y^2+1))(x+(y^2+y+1)) \\ &= x^3+x^2+1 \end{aligned} \tag{11.90}$$

2. Solution of the system of the coefficient equations.

Example 11.2.20
Consider the field GF(2^3) of the previous example; as $(y+1)$, $(y+1)^2 = y^2+1$, $(y+1)^4 = y^2+y+1$, $(y+1)^8 = y+1$, the minimum function has degree three; as the minimum function is monic and irreducible, we have

$$m_{y+1}(x) = m_3x^3+m_2x^2+m_1x+m_0 = x^3+m_2x^2+m_1x+1 \tag{11.91}$$

As $m_{y+1}(y+1) = 0$, then

$$(y+1)^3+m_2(y+1)^2+m_1(y+1)+1 = 0 \tag{11.92}$$

that can be written as

$$y^2(1+m_2)+ym_1+(m_2+m_1+1) = 0 \tag{11.93}$$

As all coefficients of the powers of y must be zero, we get a system of equations in the unknown m_1 and m_2, whose solution is given by $m_1 = 0$ and $m_2 = 1$. Substitution of this solution in (11.91) yields

$$m_{y+1}(x) = x^3+x^2+1 \tag{11.94}$$

3. Using the minimum function of the multiplicative inverse.

Definition 11.11
The *reciprocal polynomial* of any polynomial $m_\alpha(x) = m_0+m_1x+m_2x^2+\cdots+m_Kx^K$ is defined by $m^\alpha(x) = m_0x^K+m_1x^{K-1}+\cdots+m_{K-1}x+m_K$. □

We use the following proposition that we give without proof.

The minimum function of the multiplicative inverse of a given element is equal to the reciprocal of the minimum function of the given element. In formulas: let $\alpha\beta = 1$, then $m_\beta(x) = m^\alpha(x)$.

Example 11.2.21
Consider the field GF(2^6) obtained by taking the polynomial arithmetic modulo the irreducible polynomial y^6+y+1 with coefficients from GF(2); the polynomial y^6+y+1 is primitive, thus from Property 7 on page 537 any non-zero field element can be written as a power of the primitive element y. From Proposition 11.2, we have that the minimum function of y is also the minimum function of $y^2, y^4, y^8, y^{16}, y^{32}$, the minimum function of y^3 is also the minimum function of $y^6, y^{12}, y^{24}, y^{48}, y^{33}$, and so forth. We list in Table 11.11 the powers of y that have the same minimum function.

Given the minimum function of y^{11}, $m_{y^{11}} = x^6+x^5+x^3+x^2+1$, we want to find the minimum function of y^{13}. From Table 11.11, we note that y^{13} has the same minimum function as y^{52}; furthermore, we note that y^{52} is the multiplicative inverse of y^{11}, as $(y^{11})(y^{52}) = y^{63} = 1$. Therefore, the minimum function of y^{13} is the reciprocal polynomial of $m_{y^{11}}$, given by $m_{y^{13}} = x^6+x^4+x^3+x+1$.

Table 11.11: Powers of a primitive element in GF(2^6) with the same minimum function.

1	2	4	8	16	32
3	6	12	24	48	33
5	10	20	40	17	34
7	14	28	56	49	35
9	18	36			
11	22	44	25	50	37
13	26	52	41	19	38
15	30	60	57	51	39
21	42				
23	46	29	58	53	43
27	54	45			
31	62	61	59	55	47

Properties of the minimum function

1. Let β be an element of order n in an extension field of GF(q), and let $m_\beta(x)$ be the minimum function of β with coefficients from GF(q); then $x^n - 1 = m_\beta(x)\, b(x)$, but $x^i - 1 \neq m_\beta(x)\, b(x)$ for $i < n$.

Proof. We show that β is a root of $x^n - 1$, as $\beta^n - 1 = 0$, but from Property 3 of the minimum function (see page 541) we know that $m_\beta(x)$ divides any polynomial $f(x)$ such that $f(\beta) = 0$; this proves the first part.

Assume that $x^i - 1 = m_\beta(x)\, b(x)$ for some $i < n$: then

$$x^i - 1|_{x=\beta} = m_\beta(x)\, b(x)|_{x=\beta} = 0 \tag{11.95}$$

so $\beta^i - 1 = 0$ for $i < n$. But from Definition 11.2.9 of the order of β (see page 536), n is the smallest integer such that $\beta^n = 1$, hence, we get a contradiction. □

2. Let $\beta_1, \beta_2, \dots, \beta_L$ be elements of some extension field of GF(q), and let $\ell_1, \ell_2, \dots, \ell_L$ be the orders of these elements, respectively. Moreover, let $m_{\beta_1}(x), m_{\beta_2}(x), \dots, m_{\beta_L}(x)$ be the minimum functions of these elements with coefficients from GF(q), and let $g(x)$ be the smallest monic polynomial with coefficients from GF(q) that has $\beta_1, \beta_2, \dots, \beta_L$ as roots: then

(a) $g(x) = \text{l.c.m.}(m_{\beta_1}(x), m_{\beta_2}(x), \dots, m_{\beta_L}(x))$;
(b) if the minimum functions are all distinct, that is they do not have factor polynomials in common, then $g(x) = m_{\beta_1}(x)\, m_{\beta_2}(x) \cdots m_{\beta_L}(x)$;
(c) if $n = \text{l.c.m.}(\ell_1, \ell_2, \dots, \ell_L)$, then $x^n - 1 = h(x)\, g(x)$, and $x^i - 1 \neq h(x)\, g(x)$ for $i < n$.

Proof.
(a) Noting that $g(x)$ must be divisible by each of the minimum functions, it must be the smallest degree of monic polynomial divisible by $m_{\beta_1}(x), m_{\beta_2}(x), \dots, m_{\beta_L}(x)$, but this is just the definition of the least common multiple.
(b) If all the minimum functions are distinct, as each is irreducible, the least common multiple is given by the product of the polynomials.

(c) As n is a multiple of the order of each element, $\beta_j^n - 1 = 0$, for $j = 1, 2, \dots, L$; then $x^n - 1$ must be divisible by $m_{\beta_j}(x)$, for $j = 1, 2, \dots, L$, and therefore, it must be divisible by the least common multiple of these polynomials. Assume now that $g(x)$ divides $x^i - 1$ for $i < n$; then $\beta_j^i - 1 = 0$ for each $j = 1, 2, \dots, L$, and thus i is a multiple of $\ell_1, \ell_2, \dots, \ell_L$. But n is the smallest integer multiple of $\ell_1, \ell_2, \dots, \ell_L$, hence, we get a contradiction. □

We note that if the extension field is GF(q^k) and $L = q^k - 1 = n$, then $g(x) = x^n - 1$ and $h(x) = 1$.

11.2.3 Cyclic codes

In Section 11.2.1, we dealt with the theory of binary group codes. We now discuss a special class of linear codes. These codes, called *cyclic codes*, are based upon polynomial algebra and lead to particularly efficient implementations for encoding and decoding.

The algebra of cyclic codes

We consider polynomials with coefficients from some field GF(q); in Particular, we consider the polynomial $x^n - 1$, and assume it can be factorized as

$$x^n - 1 = g(x)\, h(x) \tag{11.96}$$

Many such factorizations are possible for a given polynomial $x^n - 1$; we will consider any one of them. We denote the degrees of $g(x)$ and $h(x)$ as r and k, respectively; thus, $n = k + r$. The choice of the symbols n, k and r is intentional, as they assume the same meaning as in the previous sections.

The polynomial arithmetic modulo $q(x) = x^n - 1$ is particularly important in the discussion of cyclic codes.

Proposition 11.2
Consider the set of all polynomials of the form $c(x) = a(x)\, g(x)$ modulo $q(x)$, as $a(x)$ ranges over all polynomials of all degrees with coefficients from GF(q). This set must be finite as there are at most q^n remainder polynomials that can be obtained by dividing a polynomial by $x^n - 1$. Now, we show that there are exactly q^k distinct polynomials. □

Proof. There are at least q^k distinct polynomials $a(x)$ of degree less than or equal to $k - 1$, and each such polynomial leads to a distinct polynomial $a(x)\, g(x)$. In fact, as the degree of $a(x)\, g(x)$ is less than $r + k = n$, no reduction modulo $x^n - 1$ is necessary for these polynomials.

Now, let $a(x)$ be a polynomial of degree greater than or equal to k. To reduce the polynomial $a(x)\, g(x)$ modulo $x^n - 1$, we divide by $x^n - 1$ and keep the remainder; thus

$$a(x)\, g(x) = Q(x)\, (x^n - 1) + r(x) \tag{11.97}$$

where $0 \le \deg(r(x)) < n$. By using (11.96), we can express $r(x)$ as

$$r(x) = (a(x) - h(x)\, Q(x))g(x) = a'(x)\, g(x) \tag{11.98}$$

As $r(x)$ is of degree less than n, $a'(x)$ is of degree less than k, but we have already considered all polynomials of this form; therefore, $r(x)$ is one of the q^k polynomials determined in the first part of the proof. □

Example 11.2.22
Let $g(x) = x + 1$, GF(q) = GF(2), and $n = 4$; then all polynomials $a(x)\ g(x)$ modulo $x^4 - 1 = x^4 + 1$ are given by

$a(x)$	$a(x)\ g(x) \bmod (x^4 - 1)$	code word
0	0	0000
1	$x + 1$	1100
x	$x^2 + x$	0110
$x + 1$	$x^2 + 1$	1010
x^2	$x^3 + x^2$	0011
$x^2 + 1$	$x^3 + x^2 + x + 1$	1111
$x^2 + x$	$x^3 + x$	0101
$x^2 + x + 1$	$x^3 + 1$	1001

(11.99)

We associate with any polynomial of degree less than n and coefficients from GF(q) a vector of length n with components equal to the coefficients of the polynomial, that is

$$f(x) = f_0 + f_1x + f_2x^2 + \cdots + f_{n-1}x^{n-1} \quad \leftrightarrow \quad f = (f_0, f_1, f_2, \ldots, f_{n-1}) \tag{11.100}$$

Note that in the definition, f_{n-1} does not need to be non-zero.

We can now define cyclic codes. The code words will be the vectors associated with a set of polynomials; alternatively, we speak of the polynomials themselves as being code words or code polynomials (see (11.99)).

Definition 11.12
Choose a field GF(q), a positive integer n, and a polynomial $g(x)$ with coefficients from GF(q) such that $x^n - 1 = g(x)\ h(x)$; furthermore, let $\deg(g(x)) = r = n - k$. Words of a cyclic code are the vectors of length n that are associated with all multiples of $g(x)$ reduced modulo $x^n - 1$. In formulas: $c(x) = a(x)\ g(x) \bmod (x^n - 1)$, for $a(x)$ polynomial with coefficients from GF(q).

The polynomial $g(x)$ is called *generator polynomial.* □

Properties of cyclic codes

1. In a cyclic code, there are q^k code words, as shown in the previous section.
2. A cyclic code is a linear code.

Proof. The all zero word is a code word as $0\ g(x) = 0$; any multiple of a code word is a code word, as if $a_1(x)\ g(x)$ is a code word so is $\alpha a_1(x)\ g(x)$. Let $a_1(x)\ g(x)$ and $a_2(x)\ g(x)$ be two code words; then

$$\alpha_1\ a_1(x)\ g(x) + \alpha_2\ a_2(x)\ g(x) = (\alpha_1\ a_1(x) + \alpha_2\ a_2(x))g(x) = a_3(x)\ g(x) \tag{11.101}$$

is a code word. □

3. Every cyclic permutation of a code word is a code word.

Proof. It is enough to show that if $c(x) = c_0 + c_1x + \cdots + c_{n-2}x^{n-2} + c_{n-1}x^{n-1}$ corresponds to a code word, then also $c_{n-1} + c_0x + \cdots + c_{n-3}x^{n-2} + c_{n-2}x^{n-1}$ corresponds to a code word. But if $c(x) = a(x)\ g(x) = c_0 + c_1x + \cdots + c_{n-2}x^{n-2} + c_{n-1}x^{n-1} \bmod(x^n - 1)$, then $xc(x) = xa(x)\ g(x) = c_{n-1} + c_0x + \cdots + c_{n-3}x^{n-2} + c_{n-2}x^{n-1} \bmod(x^n - 1)$. □

Example 11.2.23

Let GF(q) = GF(2), $g(x) = x + 1$, and $n = 4$. From the previous example, we obtain the code words, which can be grouped by the number of cyclic shifts.

$$\begin{array}{ccc} \text{code polynomials} & \text{code words} & \text{cyclic shifts} \\ 0 & 0000 & \} \quad 1 \\ 1+x & 1100 & \\ x+x^2 & 0110 & \\ x^2+x^3 & 0011 & 4 \\ 1+x^3 & 1001 & \\ 1+x+x^2+x^3 & 1111 & \} \quad 1 \\ 1+x^2 & 1010 & \\ x+x^3 & 0101 & 2 \end{array} \tag{11.102}$$

4. $c(x)$ is a code polynomial if and only if $c(x)\,h(x) = 0 \bmod(x^n - 1)$.

Proof. If $c(x)$ is a code polynomial, then $c(x) = a(x)g\,(x) \bmod(x^n - 1)$, but $h(x)\,c(x) = h(x)\,a(x)\,g(x) = a(x)\,(g(x)h(x)) = a(x)(x^n - 1) = 0 \bmod(x^n - 1)$.

Assume now $h(x)\,c(x) = 0 \bmod(x^n - 1)$; then $h(x)\,c(x) = Q(x)(x^n - 1) = Q(x)\,h(x)\,g(x)$, or $c(x) = Q(x)\,g(x)$; therefore, $c(x)$ is a code polynomial. □

5. Let $x^n - 1 = g(x)\,h(x)$, where $g(x) = g_0 + g_1x + \cdots + g_rx^r$ and $h(x) = h_0 + h_1x + \cdots + h_kx^k$; then the code corresponding to all multiples of $g(x)$ modulo $x^n - 1$ has the generator matrix

$$\boldsymbol{G} = \begin{bmatrix} g_0 & g_1 & g_2 & \dots & g_r & 0 & 0 & \dots & 0 \\ 0 & g_0 & g_1 & \dots & g_{r-1} & g_r & 0 & \dots & 0 \\ & & & & & & & & \\ 0 & 0 & 0 & \dots & & & & \dots & g_r \end{bmatrix} \tag{11.103}$$

and parity check matrix

$$\boldsymbol{H} = \begin{bmatrix} 0 & 0 & \dots & 0 & h_k & h_{k-1} & \dots & h_1 & h_0 \\ 0 & 0 & \dots & h_k & h_{k-1} & h_{k-2} & \dots & h_0 & 0 \\ & & & & & & & & \\ h_k & h_{k-1} & \dots\dots & & & & \dots & 0 & 0 \end{bmatrix} \tag{11.104}$$

Proof. We show that $\boldsymbol{G}$ is the generator matrix. The first row of $\boldsymbol{G}$ corresponds to the polynomial $g(x)$, the second to $xg(x)$ and the last row to $x^{k-1}\,g(x)$, but the code words are all words of the form

$$(a_0 + a_1x + \cdots + a_{k-1}x^{k-1})g(x) = a_0\,g(x) + a_1(xg(x)) + \cdots + a_{k-1}(x^{k-1}g(x)) \tag{11.105}$$

But (11.105) expresses all code words as linear combinations of the rows of $\boldsymbol{G}$; therefore, $\boldsymbol{G}$ is the generator matrix of the code.

To show that $\boldsymbol{H}$ is the parity check matrix, we consider the product $c(x)\,h(x)$. If we write

$$c(x) = c_0 + c_1x + \cdots + c_{n-1}x^{n-1} \tag{11.106}$$

and

$$h(x) = h_0 + h_1x + \cdots + h_{k-1}x^{k-1} + h_kx^k + \cdots + h_{n-1}x^{n-1} \tag{11.107}$$

where $h_{k+1} = h_{k+2} = \cdots = h_{n-1} = 0$, we get

$$d(x) = c(x)\,h(x) = d_0 + d_1x + \cdots + d_{2n-2}x^{2n-2} \tag{11.108}$$

where

$$d_i = \left\{ \begin{array}{ll} \sum_{j=0}^{i} c_j \, h_{i-j} & \text{if } 0 \le i \le n-1 \\ \sum_{j=i-(n-1)}^{n-1} c_j \, h_{i-j} & \text{if } n \le i \le 2n-2 \end{array} \right\}. \tag{11.109}$$

We consider reducing $d(x)$ modulo $x^n - 1$, and denote the result as $\hat{d}(x) = \hat{d}_0 + \hat{d}_1 x + \cdots + \hat{d}_{n-1} x^{n-1}$; then $\hat{d}_i = d_i + d_{n+i}$, $i = 0, 1, 2, \ldots, n-1$. If $c(x)\,h(x) = 0 \bmod (x^n - 1)$, then $\hat{d}_i = 0$, $i = 0, 1, 2, \ldots, n-1$, therefore, we get

$$\sum_{j=0}^{i} c_j \, h_{i-j} + \sum_{j=i+1}^{n-1} c_j \, h_{n+i-j} = 0, \qquad i = 0, 1, 2, \ldots, n-1 \tag{11.110}$$

For $i = n-1$, (11.110) becomes

$$\sum_{j=0}^{n-1} c_j \, h_{n-1-j} = 0 \tag{11.111}$$

or $[h_{n-1} \; h_{n-2} \; \ldots \; h_1 \; h_0] \; [c_0 \; c_1 \; \ldots \; c_{n-1}]^T = 0$.

For $i = n-2$, (11.110) becomes

$$\sum_{j=0}^{n-2} c_j \, h_{n-2-j} + c_{n-1} \, h_{n-1} = 0 \tag{11.112}$$

or $[h_{n-2} \; h_{n-3} \; \ldots \; h_0 \; h_{n-1}] \; [c_0 \; c_1 \; \ldots \; c_{n-1}]^T = 0$.

After r steps, for $i = n - r$, (11.110) becomes

$$\sum_{j=0}^{n-r} c_j \, h_{n-r-j} + \sum_{j=n-r+1}^{n-1} c_j \, h_{2n-r-j} = 0 \tag{11.113}$$

or $[h_{n-r} \; h_{n-r-1} \; \cdots \; h_{n-r+2} \; h_{n-r+1}] \; [c_0 \; c_1 \; \cdots \; c_{n-1}]^T = 0$. The r equations can be written in matrix form as

$$\begin{bmatrix} h_{n-1} & h_{n-2} & \ldots & h_1 & h_0 \\ h_{n-2} & h_{n-3} & \ldots & h_0 & h_{n-1} \\ \vdots & & & & \vdots \\ h_{n-r} & h_{n-r-1} & \ldots & h_{n-r+2} & h_{n-r+1} \end{bmatrix} \begin{bmatrix} c_0 \\ c_1 \\ \vdots \\ c_{n-1} \end{bmatrix} = \begin{bmatrix} 0 \\ 0 \\ \vdots \\ 0 \end{bmatrix} \tag{11.114}$$

Therefore, all code words are solutions of the equation $\boldsymbol{Hc} = \boldsymbol{0}$, where $\boldsymbol{H}$ is given by (11.104). It still remains to be shown that all solutions of the equation $\boldsymbol{Hc} = \boldsymbol{0}$ are code words. As $h_{n-1} = h_{n-2} = \cdots = h_{n-r+1} = 0$, and $h_0 \neq 0$, from (11.104) $\boldsymbol{H}$ has rank r, and can be written as $\boldsymbol{H} = [\boldsymbol{A} \; \boldsymbol{B}]$, where $\boldsymbol{B}$ is an $r \times r$ matrix with non-zero determinant; therefore,

$$\begin{bmatrix} c_k \\ c_{k+1} \\ \vdots \\ c_{n-1} \end{bmatrix} = -\boldsymbol{B}^{-1} \boldsymbol{A} \begin{bmatrix} c_0 \\ c_1 \\ \vdots \\ c_{k-1} \end{bmatrix} \tag{11.115}$$

so there are $q^k = q^{n-r}$ solutions of the equation $\boldsymbol{Hc} = \boldsymbol{0}$. As there are q^k code words, all solutions of the equation $\boldsymbol{Hc} = \boldsymbol{0}$ are the code words in the cyclic code. □

Example 11.2.24

Let $q = 2$ and $n = 7$. As $x^7 - 1 = x^7 + 1 = (x^3 + x + 1)(x^3 + x^2 + 1)(x + 1)$, we can choose $g(x) = x^3 + x + 1$ and $h(x) = (x^3 + x^2 + 1)(x + 1) = x^4 + x^2 + x + 1$; thus, the matrices $\boldsymbol{G}$ and $\boldsymbol{H}$ of this code are given by

$$\boldsymbol{G} = \begin{bmatrix} 1\ 1\ 0\ 1\ 0\ 0\ 0 \\ 0\ 1\ 1\ 0\ 1\ 0\ 0 \\ 0\ 0\ 1\ 1\ 0\ 1\ 0 \\ 0\ 0\ 0\ 1\ 1\ 0\ 1 \end{bmatrix} \tag{11.116}$$

$$\boldsymbol{H} = \begin{bmatrix} 0\ 0\ 1\ 0\ 1\ 1\ 1 \\ 0\ 1\ 0\ 1\ 1\ 1\ 0 \\ 1\ 0\ 1\ 1\ 1\ 0\ 0 \end{bmatrix} \tag{11.117}$$

Note that the columns of $\boldsymbol{H}$ are all possible non-zero vectors of length 3, so the code is a Hamming single error correcting (7,4) code.

6. In a code word, any string of r consecutive symbols, even taken cyclically, can identify the check positions.

Proof. From (11.115), it follows that the last r positions can be check positions. Now, if we cyclically permute every code word of m positions, the resultant words are themselves code words; thus the r check positions can be cyclically permuted anywhere in the code words. □

7. As the r check positions can be the first r positions, a simple *encoding method* in canonical form is given by the following steps:

Step 1: represents the k information bits by the coefficients of the polynomial $m(x) = m_0 + m_1x + \cdots + m_{k-1}x^{k-1}$.

Step 2: multiplies $m(x)$ by x^r to obtain $x^r\ m(x)$.

Step 3: divides $x^r\ m(x)$ by $g(x)$ to obtain the remainder $r(x) = r_0 + r_1x + \ldots + r_{r-1}\ x^{r-1}$.

Step 4: forms the code word $c(x) = (x^r\ m(x) - r(x))$; note that the coefficients of $(-r(x))$ are the parity check bits.

Proof. To show that $(x^r\ m(x) - r(x))$ is a code word, we must prove that it is a multiple of $g(x)$: from step 3, we obtain

$$x^r\ m(x) = Q(x)\ g(x) + r(x) \tag{11.118}$$

so that

$$(x^r\ m(x) - r(x)) = Q(x)\ g(x) \tag{11.119}$$

□

Example 11.2.25

Let $g(x) = 1 + x + x^3$, for $q = 2$ and $n = 7$. We report in Table 11.12 the message words $(m_0, \ldots, m_3)$ and the corresponding code words $(c_0, \ldots, c_6)$ obtained by the generator polynomial according to Definition 11.12 on page 546 for $a(x) = m(x)$; the same code in canonical form, obtained by (11.119), is reported in Table 11.13.

Table 11.12: (7,4) binary cyclic code, generated by $g(x) = 1 + x + x^3$.

message $(m_0 m_1 m_2 m_3)$	*code polynomial* $c(x) = m(x)\, g(x) \bmod x^7 - 1$	*code* $(c_0 c_1 c_2 c_3 c_4 c_5 c_6)$
0000	$0g(x) = 0$	0000000
1000	$1g(x) = 1 + x + x^3$	1101000
0100	$xg(x) = x + x^2 + x^4$	0110100
1100	$(1 + x)g(x) = 1 + x^2 + x^3 + x^4$	1011100
0010	$x^2 g(x) = x^2 + x^3 + x^5$	0011010
1010	$(1 + x^2)g(x) = 1 + x + x^2 + x^5$	1110010
0110	$(x + x^2)g(x) = x + x^3 + x^4 + x^5$	0101110
1110	$(1 + x + x^2)g(x) = 1 + x^4 + x^5$	1000110
0001	$x^3 g(x) = x^3 + x^4 + x^6$	0001101
1001	$(1 + x^3)g(x) = 1 + x + x^4 + x^6$	1100101
0101	$(x + x^3)g(x) = x + x^2 + x^3 + x^6$	0111001
1101	$(1 + x + x^3)g(x) = 1 + x^2 + x^6$	1010001
0011	$(x^2 + x^3)g(x) = x^2 + x^4 + x^5 + x^6$	0010111
1011	$(1 + x^2 + x^3)g(x) = 1 + x + x^2 + x^3 + x^4 + x^5 + x^6$	1111111
0111	$(x + x^2 + x^3)g(x) = x + x^5 + x^6$	0100011
1111	$(1 + x + x^2 + x^3)g(x) = 1 + x^3 + x^5 + x^6$	1001011

Table 11.13: (7,4) binary cyclic code in canonical form, generated by $g(x) = 1 + x + x^3$.

message $(m_0 m_1 m_2 m_3)$	*code polynomial* $r(x) = x^r\, m(x) \bmod g(x)$ $c(x) = x^r\, m(x) - r(x)$	*code* $(c_0 c_1 c_2 c_3 c_4 c_5 c_6)$
0000	0	0000000
1000	$1 + x + x^3$	1101000
0100	$x + x^2 + x^4$	0110100
1100	$1 + x^2 + x^3 + x^4$	1011100
0010	$1 + x + x^2 + x^5$	1110010
1010	$x^2 + x^3 + x^5$	0011010
0110	$1 + x^4 + x^5$	1000110
1110	$x + x^3 + x^4 + x^5$	0101110
0001	$1 + x^2 + x^6$	1010001
1001	$x + x^2 + x^3 + x^6$	0111001
0101	$1 + x + x^4 + x^6$	1100101
1101	$x^3 + x^4 + x^6$	0001101
0011	$x + x^5 + x^6$	0100011
1011	$1 + x^3 + x^5 + x^6$	1001011
0111	$x^2 + x^4 + x^5 + x^6$	0010111
1111	$1 + x + x^2 + x^3 + x^4 + x^5 + x^6$	1111111

Encoding by a shift register of length r

We show that the steps of the encoding procedure can be accomplished by a *linear shift register with r stages*. We begin by showing how to divide m_0x^r by the generator polynomial $g(x)$ and obtain the remainder. As

$$g(x) = g_r x^r + g_{r-1}x^{r-1} + \cdots + g_1 x + g_0 \tag{11.120}$$

then

$$x^r = -g_r^{-1}(g_{r-1}x^{r-1} + g_{r-2}x^{r-2} + \cdots + g_1 x + g_0) \bmod g(x) \tag{11.121}$$

and

$$m_0 x^r = -m_0 g_r^{-1}(g_{r-1}x^{r-1} + g_{r-2}x^{r-2} + \cdots + g_1 x + g_0) \bmod g(x) \tag{11.122}$$

is the remainder after dividing m_0x^r by $g(x)$.

We now consider the scheme illustrated in Figure 11.3, where multiplications and additions are in GF(q), and T_c denotes the clock period with which the message symbols $\{m_i\}$, $i = k-1, \ldots, 1, 0$, are input to the shift register. In the *binary case*, the storage elements are flip flops, the addition is the modulo 2 addition, and multiplication by g_i is performed by a switch that is open or closed depending upon whether $g_i = 0$ or 1, respectively. Note that if m_0 is input, the storage elements of the shift register will contain the coefficients of the remainder upon dividing m_0x^r by $g(x)$.

Let us suppose we want to compute the remainder upon dividing $m_1\, x^{r+1}$ by $g(x)$. We could first compute the remainder of the division of $m_1\, x^r$ by $g(x)$, by presenting m_1 at the input, then multiplying the remainder by x, and again reduce the result modulo $g(x)$. But once the remainder of the first division is stored in the shift register, multiplication by x and division by $g(x)$ are obtained simply by clocking the register once with no input. In fact, if the shift register contains the polynomial

$$b(x) = b_0 + b_1 x + \cdots + b_{r-1}x^{r-1} \tag{11.123}$$

and we multiply by x and divide by $g(x)$, we obtain

$$\begin{aligned} x\, b(x) &= b_0 x + b_1 x^2 + \cdots + b_{r-1}x^r \\ &= b_0 x + b_1 x^2 + \cdots + b_{r-2}x^{r-1} \\ &\quad + b_{r-1}(-g_r^{-1}(g_{r-1}x^{r-1} + \cdots + g_1 x + g_0)) \bmod g(x) \\ &= -\, b_{r-1}g_r^{-1}g_0 + (b_0 - b_{r-1}g_r^{-1}g_1)\, x + \cdots \\ &\quad + (b_{r-2} - b_{r-1}g_r^{-1}g_{r-1})\, x^{r-1} \bmod g(x) \end{aligned} \tag{11.124}$$

that is just the result obtained by clocking the register once.

Finally, we note that superposition holds in computing remainders; in other words, if $m_0\, x^r = r_1(x) \bmod g(x)$ and $m_1\, x^{r+1} = r_2(x) \bmod g(x)$, then $m_0\, x^r + m_1\, x^{r+1} = r_1(x) + r_2(x) \bmod g(x)$. Therefore, to compute the remainder upon dividing $m_0\, x^r + m_1\, x^{r+1}$ by $g(x)$ using the scheme of Figure 11.3, we would first input m_1 and then next input m_0 to the shift register.

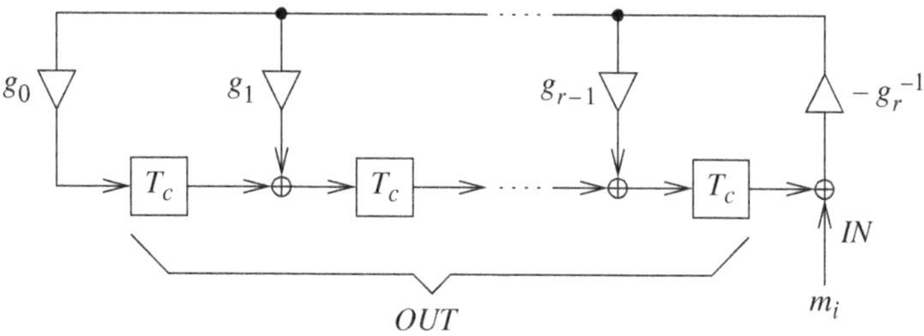

Figure 11.3 Scheme of an encoder for cyclic codes using a shift register with r elements.

Hence, to compute the remainder upon dividing $x^r m(x) = m_0\, x^r + m_1\, x^{r+1} + \cdots + m_{k-1}\, x^{n-1}$ by $g(x)$ we input the symbols $m_{k-1}, m_{k-2}, \ldots, m_1, m_0$ to the device of Figure 11.3; after the last symbol, m_0, enters, the coefficients of the desired remainder will be contained in the storage elements. From (11.119), we note that the parity check bits are the inverse elements (with respect to addition) of the values contained in the register.

In general for an input $z(x)$, polynomial with n coefficients, after n clock pulses the device of Figure 11.3 yields $x^r z(x) \bmod g(x)$.

Encoding by a shift register of length k

It is also possible to accomplish the encoding procedure for cyclic codes by using a shift register with k stages. Again, we consider the first r positions of the code word as the parity check bits, $p_0, p_1, \ldots, p_{r-1}$; utilizing the first row of the parity check matrix, we obtain

$$h_k\, p_{r-1} + h_{k-1}\, m_0 + \cdots + h_1\, m_{k-2} + h_0\, m_{k-1} = 0 \tag{11.125}$$

or

$$p_{r-1} = -h_k^{-1}(h_{k-1}\, m_0 + h_{k-2}\, m_1 + \cdots + h_1\, m_{k-2} + h_0\, m_{k-1}) \tag{11.126}$$

Similarly, using the second row, we obtain

$$p_{r-2} = -h_k^{-1}\ (h_{k-1}\, p_{r-1} + h_{k-2}\, m_0 + \cdots + h_1\, m_{k-3} + h_0\, m_{k-2}) \tag{11.127}$$

and so forth.

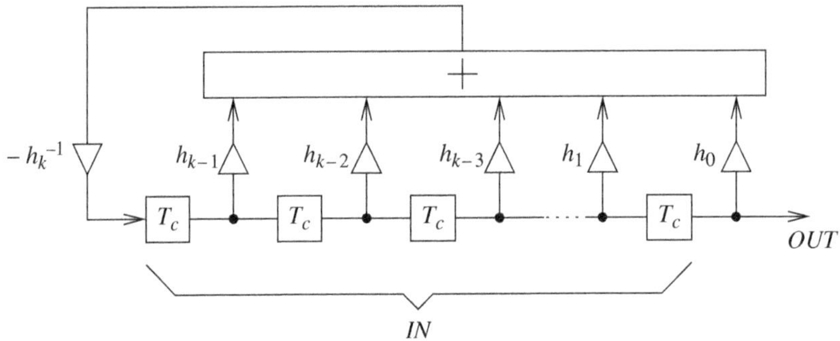

Figure 11.4 Scheme of an encoder for cyclic codes using a shift register with k elements.

Let us consider the scheme of Figure 11.4 and assume that the register initially contains the symbols $m_0, m_1, \ldots, m_{k-1}$. After one clock pulse m_{k-1} will appear at the output, all information symbols will have moved by one place to the right and the parity check symbol p_{r-1} will appear in the first left-most storage element; after the second clock pulse, m_{k-2} will appear at the output, all symbols contained in the storage elements will move one place to the right and the parity check symbol p_{r-2} will appear in the left-most storage element. It is easy to verify that, if we apply n clock pulses to the device, the output will be given by the k message symbols followed by the r parity check bits.

Hard decoding of cyclic codes

We recall (see page 526) that all vectors in the same coset of the decoding table have the same syndrome and that vectors in different cosets have different syndromes.

Proposition 11.3

All polynomials corresponding to vectors in the same coset have the same remainder if they are divided by $g(x)$; polynomials corresponding to vectors in different cosets have different remainders if they are divided by $g(x)$. □

Proof. Let $a_j(x)\ g(x), j = 0, 1, 2, \dots, q^k - 1$, be the code words, and $\eta_i(x), i = 0, 1, 2, \dots, q^r - 1$, be the coset leaders. Assume $z_1(x)$ and $z_2(x)$ are two arbitrary polynomials of degree $n - 1$: if they are in the same coset, say the ith, then

$$z_1(x) = \eta_i(x) + a_{j_1}(x)\ g(x) \tag{11.128}$$

and

$$z_2(x) = \eta_i(x) + a_{j_2}(x)\ g(x) \tag{11.129}$$

As upon dividing $a_{j_1}\ g(x)$ and $a_{j_2}\ g(x)$ by $g(x)$, we get 0 as a remainder, the division of $z_1(x)$ and $z_2(x)$ by $g(x)$ gives the same remainder, namely the polynomial $r_i(x)$, where

$$\eta_i(x) = Q(x)\ g(x) + r_i(x), \qquad \deg(r_i(x)) < \deg(g(x)) = r \tag{11.130}$$

Now, assume $z_1(x)$ and $z_2(x)$ are in different cosets, say the i_1th and i_2th cosets, but have the same remainder, say $r_0(x)$, if they are divided by $g(x)$; then the coset leaders $\eta_{i_1}(x)$ and $\eta_{i_2}(x)$ of these cosets must give the same remainder $r_0(x)$ if they are divided by $g(x)$, i.e.

$$\eta_{i_1}(x) = Q_1(x)\ g(x) + r_0(x) \tag{11.131}$$

and

$$\eta_{i_2}(x) = Q_2(x)\ g(x) + r_0(x) \tag{11.132}$$

Therefore, we get

$$\eta_{i_2}(x) = \eta_{i_1}(x) + (Q_2(x) - Q_1(x))g(x) = \eta_{i_1}(x) + Q_3(x)\ g(x) \tag{11.133}$$

This implies that $\eta_{i_1}(x)$ and $\eta_{i_2}(x)$ are in the same coset, which is a contradiction. □

This result leads to the following decoding method for cyclic codes.

Step 1: compute the remainder upon dividing the received polynomial $z(x)$ of degree $n - 1$ by $g(x)$, for example by the device of Figure 11.5 (see (11.124)), by presenting at the input the sequence of received symbols, and applying n clock pulses. The remainder identifies the coset leader of the coset where the received polynomial is located.

Step 2: subtract the coset leader from the received polynomial to obtain the decoded code word.

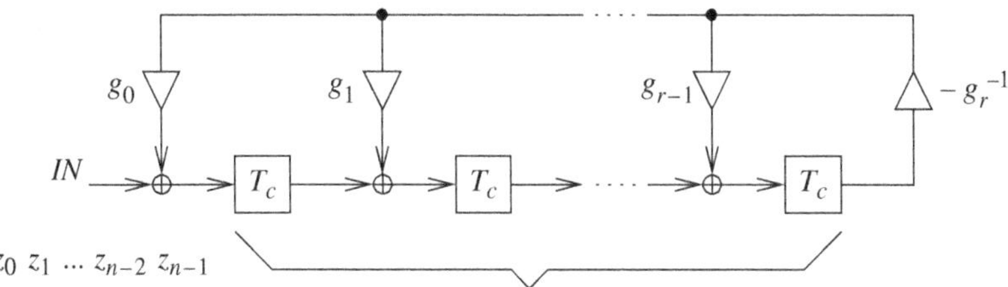

Figure 11.5 Device to compute the division of the polynomial $z(x) = z_0 + z_1x + \dots + z_{n-1}x^{n-1}$ by $g(x)$. After n clock pulses the r storage elements contain the remainder $r_0, r_1, \dots, r_{r-1}$.

Hamming codes

Hamming codes are binary cyclic single error correcting codes.

We consider cyclic codes over GF(2), where $g(x)$ is an irreducible polynomial of degree r such that $g(x)$ divides $x^{2^r-1} - 1$, but not $x^{\ell} - 1$ for $\ell < 2^r - 1$.

To show that $g(x)$ is a primitive irreducible polynomial, we choose $n = 2^r - 1$, thus $x^n - 1 = g(x)\ h(x)$ and the corresponding cyclic code has parameters $n = 2^r - 1$, r, and $k = 2^r - 1 - r$.

Proposition 11.4

This code has minimum distance $d^H_{min} = 3$ and therefore, is a single error correcting code. □

Proof. We first prove that $d^H_{min} \geq 3$ by showing that all single error polynomials have distinct, non-zero remainders if they are divided by $g(x)$.

Assume that $x^i = 0 \bmod g(x)$, for some $0 \leq i \leq n-1$; then $x^i = Q(x)\ g(x)$, which is impossible since $g(x)$ is not divisible by x.

Now, assume that x^i and x^j give the same remainder upon division by $g(x)$ and that $0 \leq i < j \leq n-1$; then

$$x^j - x^i = x^i(x^{j-i} - 1) = Q(x)\ g(x) \tag{11.134}$$

but $g(x)$ does not divide x^i, so it must divide $(x^{j-i} - 1)$. But $0 < j - i \leq n - 1$ and by assumption $g(x)$ does not divide this polynomial. Hence, $d^H_{min} \geq 3$.

By the bound (11.15), we know that for a code with fixed n and k the following inequality holds:

$$2^k \left[1 + \binom{n}{1} + \binom{n}{2} + \cdots + \binom{n}{\left\lfloor \frac{d^H_{min}-1}{2} \right\rfloor} \right] \leq 2^n \tag{11.135}$$

As $n = 2^r - 1$ and $k = n - r$, we have

$$\left[1 + \binom{2^r - 1}{1} + \binom{2^r - 1}{2} + \cdots + \binom{2^r - 1}{\left\lfloor \frac{d^H_{min}-1}{2} \right\rfloor} \right] \leq 2^r \tag{11.136}$$

but

$$1 + \binom{2^r - 1}{1} = 2^r \tag{11.137}$$

and therefore, $d^H_{min} \leq 3$. □

We have seen in the previous section how to implement an encoder for a cyclic code. We consider now the decoder device of Figure 11.6, whose operations are described as follows:

1. Initially, all storage elements of the register contain zeros and the switch SW is in position 0. The received n-bit word $z = (z_0, \ldots, z_{n-1})$ is sequentially clocked into the lower register, with n storage elements, and into the feedback register, with r storage elements, whose content is denoted by $r_0, r_1, \ldots, r_{r-1}$.
2. After n clock pulses, the behaviour of the decoder depends on the value of v: if $v = 0$, the switch SW remains in the position 0 and both registers are clocked once. This procedure is repeated until $v = 1$, which occurs for $r_0 = r_1 = \cdots = r_{r-2} = 0$; then SW moves to position 1 and the content of the last stage of the feedback shift register is added modulo 2 to the content of the last stage of the lower register; both registers are then clocked until the n bits of the entire word are obtained at the output of the decoder. Overall, $2n$ clock pulses are needed.

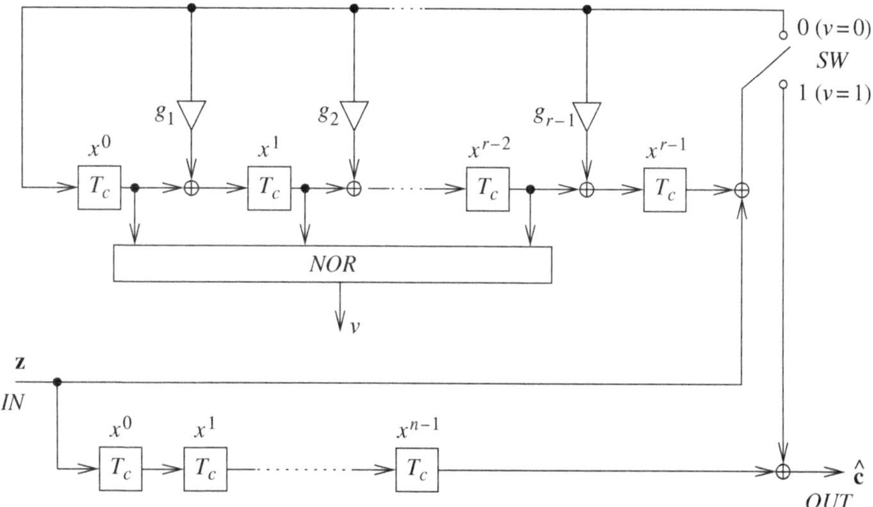

Figure 11.6 Scheme of a decoder for binary cyclic single error correcting codes (Hamming codes). All operations are in GF(2).

We now illustrate the procedure of the scheme of Figure 11.6. First of all we note that for the first n clocks the device coincides with that of Figure 11.3, hence, the content of the shift register is given by

$$r(x) = x^r \, z(x) \bmod g(x) \tag{11.138}$$

We consider two cases.

1. The received word is correct, $z(x) = c(x)$. After the first n clock pulses, from (11.138), we have

$$r(x) = x^r \, c(x) = x^r \, a(x) \, g(x) = 0 \bmod g(x) \tag{11.139}$$

and thus

$$v = 1 \quad \text{and} \quad r_{r-1} = 0 \tag{11.140}$$

In the successive n clock pulses, we have

$$\hat{c}_i = z_i + 0, \qquad i = 0, \dots, n-1 \tag{11.141}$$

therefore, $\hat{\boldsymbol{c}} = \boldsymbol{c}$.

2. The received word is affected by one error, $z(x) = c(x) + x^i$. In other words, we assume that there is a single error in the ith bit, $0 \le i \le n-1$.
After the first n clock pulses, it is

$$r(x) = x^r \, x^i \bmod g(x) \tag{11.142}$$

If $i = n-1$, we have

$$\begin{aligned} r(x) &= x^r \, x^{n-1} \\ &= x^n \, x^{r-1} \\ &= (x^n - 1) \, x^{r-1} + x^{r-1} \\ &= h(x) \, g(x) \, x^{r-1} + x^{r-1} \bmod g(x) \\ &= x^{r-1} \end{aligned} \tag{11.143}$$

and consequently,

$$r_{r-1} = 1 \quad \text{and} \quad r_{r-2} = \cdots = r_0 = 0 \quad (v = 1) \tag{11.144}$$

This leads to switching SW, therefore, during the last n clock pulses, we have

$$\hat{c}_{n-1} = z_{n-1} + 1 \qquad \hat{c}_i = z_i + 0, \qquad i = n-2, \dots, 0 \tag{11.145}$$

Therefore, the bit in the last stage of the buffer is corrected.

If $i = n - j$, then we have

$$r(x) = x^{r-j} \tag{11.146}$$

thus only at the $(n + j - 1)$th clock pulse the condition (11.144) that forces to switch SW from 0 to 1 occurs; therefore, at the next clock pulse, the received bit in error will be corrected.

Burst error detection

We assume that a burst error occurs in the received word and that this burst affects $\ell \leq n - k$ consecutive bits, that is the error pattern is

$$\boldsymbol{e} = (0, 0, 0, \dots, 0, \overset{\text{bit } j}{1}, \dots, \dots, \overset{\text{bit } (j+\ell-1)}{1}, 0, \dots, 0) \tag{11.147}$$

where within the two '1's the values can be either '0' or '1'.

Then we can write the vector $\boldsymbol{e}$ in polynomial form,

$$e(x) = x^j \, B(x) \tag{11.148}$$

where $B(x)$ is a polynomial of degree $\ell - 1 \leq n - k - 1$. Thus, $e(x)$ is divisible by the generator polynomial $g(x)$ if $B(x)$ is divisible by $g(x)$, as x is not a factor of $g(x)$, but $B(x)$ has a degree at most equal to $(n - k - 1)$, lower than the degree of $g(x)$, equal to $n - k$; therefore, $e(x)$ cannot be a code word. We have then that all burst errors of length ℓ less than or equal to $r = n - k$ are detectable by (n, k) cyclic codes. This result leads to the introduction of the *cyclic redundancy check* (CRC) codes.

11.2.4 Simplex cyclic codes

We consider a class of cyclic codes over GF(q) such that the Hamming distance between every pair of distinct code words is a constant; this is equivalent to stating that the weight of all non-zero code words is equal to the same constant. We show that in the binary case, for these codes the non-zero code words are related to the PN sequences of Appendix 1.C.

Let $n = q^k - 1$, and $x^n - 1 = g(x)\, h(x)$, where we choose $h(x)$ as a *primitive* polynomial of degree k; then the resultant code has minimum distance

$$d^H_{min} = (q-1)\, q^{k-1} \tag{11.149}$$

The parameters of some binary codes in this class are listed in Table 11.14.

To show that these codes have minimum distance given by (11.149), first we prove the following:

Table 11.14: Parameters of some simplex binary codes.

n	k	r	d_{min}
7	3	4	4
15	4	11	8
31	5	26	16
63	6	57	32
127	7	120	64

Property

All non-zero code words have the same weight.

Proof. We begin by showing that

$$x^i\, g(x) \neq x^j\, g(x), \qquad \bmod(x^n - 1), \qquad 0 \leq i < j \leq n-1 \tag{11.150}$$

Assume the converse is true, that is $x^i\, g(x) = x^j\, g(x) \bmod(x^n - 1)$; then

$$x^i(x^{j-i} - 1)\, g(x) = Q(x)\, g(x)\, h(x) \tag{11.151}$$

or

$$x^i(x^{j-i} - 1) = Q(x)\, h(x) \tag{11.152}$$

But this is impossible since $h(x)$ is a primitive polynomial of degree k and cannot divide $(x^{j-i} - 1)$, as $(j - i) < n = (q^k - 1)$.

Relation (11.150) implies that all cyclic shifts of the code polynomial $g(x)$ are unique, but there are $n = (q^k - 1)$ cyclic shifts. Furthermore, we know that there are only q^k code words and one is the all-zero word; therefore, all cyclic shifts of $g(x)$ are all the non-zero code words, and they all have the same weight. □

Recall Property 2 of a group code (see page 519), that is if all code words of a linear code are written as rows of a matrix, every column is either formed by all zeros, or it consists of each field element repeated an equal number of times. If we apply this result to a simplex code, we find that no column can be all zero as the code is cyclic, so the sum of the weights of all code words is given by

$$\text{sum of weights} = n(q-1)\,\frac{q^k}{q} = (q^k - 1)\,(q-1)\,q^{k-1} \tag{11.153}$$

But there are $(q^k - 1)$ non-zero code words, all of the same weight; the weight of each word is then given by

$$\text{weight of non-zero code words} = (q-1)\,q^{k-1} \tag{11.154}$$

Therefore, the minimum weight of the non-zero code words is given by

$$d^H_{min} = (q-1)\,q^{k-1} \tag{11.155}$$

Example 11.2.26

Let $q = 2$, $n = 15$, and $k = 4$; hence, $r = 11$, and $d^H_{min} = 8$. Choose $h(x)$ as a primitive irreducible polynomial of degree 4 over GF(2), $h(x) = x^4 + x + 1$.

The generator polynomial $g(x)$ is obtained by dividing $x^{15} - 1$ by $h(x) = x^4 + x + 1$ in GF(2), obtaining

$$g(x) = x^{11} + x^8 + x^7 + x^5 + x^3 + x^2 + x + 1 \tag{11.156}$$

Given an extension field GF(2^k) and $n = 2^k - 1$, from Property 2 on page 544, $x^n - 1$ is given by the l.c.m. of the minimum functions of the elements of the extension field. As $h(x)$ is a primitive polynomial, $g(x)$ is, therefore, given by the l.c.m. of the minimum functions of the elements $1, \alpha^3, \alpha^5, \alpha^7$, from GF($2^4$). By a table similar to Table 11.11, obtained for GF(2^6), and using one of the three methods to determine the minimum function (see page 542), it turns out that the generator polynomial for this code is given by

$$g(x) = (x+1)(x^4 + x^3 + x^2 + x + 1)(x^2 + x + 1)(x^4 + x^3 + 1) \tag{11.157}$$

Relation to PN sequences

We consider a periodic binary sequence of period L, given by $\dots, p(-1), p(0), p(1), \dots$, with $p(\ell) \in \{0,1\}$. We define the normalized autocorrelation function of this sequence as

$$\mathtt{r}_p(m) = \frac{1}{L}\left[L - 2\sum_{\ell=0}^{L-1}(p(\ell) \oplus p(\ell - m))\right] \tag{11.158}$$

Note that with respect to (1.588), now $p(\ell) \in \{0,1\}$ rather than $p(\ell) \in \{-1,1\}$.

Theorem 11.1
If the periodic binary sequence $\{p(\ell)\}$ is formed by repeating any non-zero code word of a simplex binary code of length $L = n = 2^k - 1$, then

$$\mathtt{r}_p(m) = \begin{cases} 1 & m = 0, \pm L, \pm 2L \dots \\ -\dfrac{1}{L} & \text{otherwise} \end{cases} \tag{11.159}$$

□

Proof. We recall that for a simplex binary code all non-zero code words

(a) have weight 2^{k-1},
(b) are cyclic permutations of the same code word.

As the code is linear, the Hamming distance between any code word and a cyclic permutation of this word is 2^{k-1}; this means that for the periodic sequence formed by repeating any non-zero code word, we get

$$\sum_{\ell=0}^{L-1}(p(\ell) \oplus p(\ell - m)) = \begin{cases} 0 & m = 0, \pm L, \pm 2L, \dots \\ 2^{k-1} & \text{otherwise} \end{cases} \tag{11.160}$$

Substitution of (11.160) in (11.158) yields

$$\mathtt{r}_p(m) = \begin{cases} 1 & m = 0, \pm L, \pm 2L, \dots \\ \dfrac{2^k - 1 - 2^k}{2^k - 1} = -\dfrac{1}{2^k - 1} & \text{otherwise} \end{cases} \tag{11.161}$$

□

If we recall the implementation of Figure 11.4, we find that the generation of such sequences is easy. We just need to determine the shift register associated with $h(x)$, load it with anything except all zeros, and let it run. For example choosing $h(x) = x^4 + x + 1$, we get the PN sequence of Figure 1.72, as illustrated in Figure 11.7, where $L = n = 2^4 - 1 = 15$.

11.2.5 BCH codes

An alternative method to specify the code polynomials

Definition 11.13
Suppose we arbitrarily choose L *elements from* GF(q^m) that we denote as $\alpha_1, \alpha_2, \dots, \alpha_L$ (we will discuss later how to select these elements), and we consider polynomials, of degree $n-1$ or less, with *coefficients from* GF(q). A polynomial is a code polynomial if each of the elements $\alpha_1, \alpha_2, \dots, \alpha_L$ is a root of the polynomial. The code then consists of the set of all the code polynomials. □

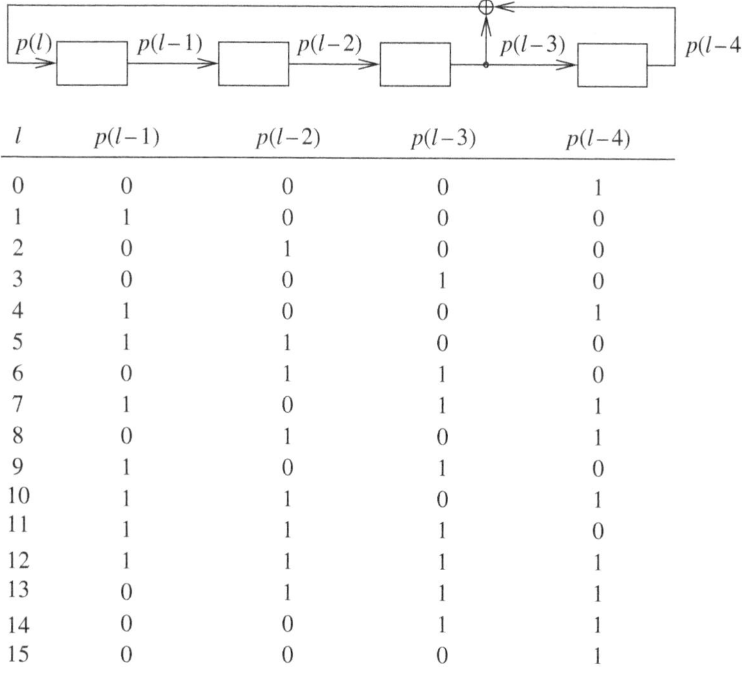

l	$p(l-1)$	$p(l-2)$	$p(l-3)$	$p(l-4)$
0	0	0	0	1
1	1	0	0	0
2	0	1	0	0
3	0	0	1	0
4	1	0	0	1
5	1	1	0	0
6	0	1	1	0
7	1	0	1	1
8	0	1	0	1
9	1	0	1	0
10	1	1	0	1
11	1	1	1	0
12	1	1	1	1
13	0	1	1	1
14	0	0	1	1
15	0	0	0	1

Figure 11.7 Generation of a PN sequence as a repetition of a code word of a simplex code with $L = n = 15$.

Using this method we see that $c(x) = c_0 + c_x + \cdots + c_{n-1}x^{n-1}$ is a code polynomial if and only if $c(\alpha_1) = c(\alpha_2) = \cdots = c(\alpha_L) = 0$; thus

$$\begin{bmatrix} (\alpha_1)^0 & (\alpha_1)^1 & (\alpha_1)^2 & \dots & (\alpha_1)^{n-1} \\ (\alpha_2)^0 & (\alpha_2)^1 & (\alpha_2)^2 & \dots & (\alpha_2)^{n-1} \\ \vdots & & & & \vdots \\ (\alpha_L)^0 & (\alpha_L)^1 & (\alpha_L)^2 & \dots & (\alpha_L)^{n-1} \end{bmatrix} \begin{bmatrix} c_0 \\ c_1 \\ c_2 \\ \vdots \\ c_{n-1} \end{bmatrix} = \begin{bmatrix} 0 \\ 0 \\ \vdots \\ 0 \end{bmatrix} \tag{11.162}$$

All vectors $\boldsymbol{c} = [c_0, c_1, \dots, c_{n-1}]^T$ with elements from GF(q) that are solutions of this set of equations, where operations are performed according to the rules of GF(q^m), are code words. The form of (11.162) resembles equation (11.20), where $\boldsymbol{H}$ is the generalized parity check matrix. One obvious difference is that in (11.20) $\boldsymbol{H}$ and $\boldsymbol{c}$ have elements from the same field, whereas this does not occur for the vector equation (11.162). However, this difference is not crucial as each element from GF(q^m) can be written as a vector of length m with elements from GF(q). Thus, each element $(\alpha_i)^j$ in the matrix is replaced by a column vector with m components. The resultant matrix, with Lm rows and n columns, consists of elements from GF(q) and is, therefore, just a generalized parity check matrix for the considered code.

From the above discussion it appears that, if L roots are specified, the resultant linear code has $r = Lm$ parity check symbols, as the parity check matrix has $r = Lm$ rows. However, not all rows of the matrix are necessarily independent; therefore, the actual number of parity check symbols may be less than Lm.

We now show that if n is properly chosen, the resultant codes are cyclic codes. Let $m_j(x)$ be the minimum function of α_j, $j = 1, 2, \dots, L$, where $\alpha_j \in$ GF(q^m) and $m_j(x)$ has coefficients from GF(q). For Property 3 on page 541, every code polynomial $c(x)$ must be divisible by $m_1(x), m_2(x), \dots, m_L(x)$, and is thus divisible by the least common multiple of such minimum functions, l.c.m.$(m_1(x), m_2(x), \dots, m_L(x))$.

If we define

$$g(x) = \text{l.c.m.}(m_1(x), m_2(x), \dots, m_L(x)) \tag{11.163}$$

then all multiples of $g(x)$ are code words. In particular from Definition 11.2.12, the code is cyclic if

$$x^n - 1 = g(x)\, h(x) \tag{11.164}$$

Let ℓ_i be the order of α_i, $i = 1, 2, \dots, L$, and furthermore, let

$$n = \text{l.c.m.}(\ell_1, \ell_2, \dots, \ell_L) \tag{11.165}$$

From the properties of the minimum function (see page 544, Property 2), we know that $g(x)$ divides $x^n - 1$; thus, the code is cyclic if n is chosen as indicated by (11.165). We note that

$$r = \deg(g(x)) \le mL \tag{11.166}$$

as $\deg(m_i(x)) \le m$. We see that r is equal to mL if all minimum functions are distinct and are of degree m; conversely, $r < mL$ if any minimum function has degree less than m or if two or more minimum functions are identical.

Example 11.2.27

Choose $q = 2$ and let α be a primitive element of GF(2^4); furthermore, let the code polynomials have as roots the elements α, α^2, α^3, α^4. To derive the minimum functions of the chosen elements, we look up for example the Appendix C of [3], where such functions are listed. Minimum functions and orders of elements chosen for this example are given in Table 11.15.

Then

$$\begin{aligned} g(x) &= (x^4 + x + 1)\,(x^4 + x^3 + x^2 + x + 1) \\ n &= \text{l.c.m.}(15, 15, 5, 15) = 15 \end{aligned} \tag{11.167}$$

The resultant code is, therefore, a (15,7) code; later, we will show that $d^H_{min} = 5$.

Table 11.15: Minimum functions and orders of elements α, α^2, α^3, α^4, in GF(2^4).

roots	minimum function	order
α	$x^4 + x + 1$	15
α^2	$x^4 + x + 1$	15
α^3	$x^4 + x^3 + x^2 + x + 1$	5
α^4	$x^4 + x + 1$	15

Bose-Chaudhuri–Hocquenhem codes

The Bose–Chaudhuri–Hocquenhem (BCH) codes are error correcting codes with symbols from GF(q) and roots of code polynomials from GF(q^m).

The basic mathematical fact required to prove the error correcting capability of BCH codes is that if $\alpha_1, \alpha_2, \dots, \alpha_r$ are elements from any field, the determinant of the Vandermonde matrix, given by

$$\det \begin{vmatrix} 1 & 1 & \dots & 1 \\ \alpha_1 & \alpha_2 & \dots & \alpha_r \\ \alpha_1^2 & \alpha_2^2 & \dots & \alpha_r^2 \\ \vdots & & & \vdots \\ \alpha_1^{r-1} & \alpha_2^{r-1} & \dots & \alpha_r^{r-1} \end{vmatrix} \tag{11.168}$$

is non-zero if and only if $\alpha_i \neq \alpha_j$, for all indices $i \neq j$. In particular, we prove the following result:

Lemma. The determinant (11.168) is given by

$$D = \det \begin{vmatrix} 1 & 1 & \dots & 1 \\ \alpha_1 & \alpha_2 & \dots & \alpha_r \\ \alpha_1^2 & \alpha_2^2 & \dots & \alpha_r^2 \\ \vdots & & & \vdots \\ \alpha_1^{r-1} & \alpha_2^{r-1} & \dots & \alpha_r^{r-1} \end{vmatrix} = (-1)^{\frac{r(r+1)}{2}} \prod_{\substack{i,j=1 \\ i<j}}^{r} (\alpha_i - \alpha_j) \tag{11.169}$$

Proof. Consider the polynomial $P(x)$ defined as

$$P(x) = \det \begin{vmatrix} 1 & 1 & \dots & 1 \\ x & \alpha_2 & \dots & \alpha_r \\ x^2 & \alpha_2^2 & \dots & \alpha_r^2 \\ \vdots & & & \vdots \\ x^{r-1} & \alpha_2^{r-1} & \dots & \alpha_r^{r-1} \end{vmatrix} \tag{11.170}$$

so that $D = P(\alpha_1)$. Now, $P(x)$ is a polynomial of degree at most $r-1$ whose zeros are $x = \alpha_2, x = \alpha_3, \dots, x = \alpha_r$, because if $x = \alpha_i$, $i = 2, 3, \dots, r$, the determinant D is equal to zero as two columns of the matrix are identical. Thus,

$$P(x) = k_1(x - \alpha_2)(x - \alpha_3) \cdots (x - \alpha_r) \tag{11.171}$$

and

$$D = P(\alpha_1) = k_1(\alpha_1 - \alpha_2)(\alpha_1 - \alpha_3) \cdots (\alpha_1 - \alpha_r) \tag{11.172}$$

It remains to calculate k_1. The constant k_1 is the coefficient of x^{r-1}; therefore, from (11.170) we get

$$(-1)^r k_1 = \det \begin{vmatrix} 1 & 1 & \dots & 1 \\ \alpha_2 & \alpha_3 & \dots & \alpha_r \\ \vdots & & & \vdots \\ \alpha_2^{r-2} & \alpha_3^{r-2} & \dots & \alpha_r^{r-1} \end{vmatrix} = k_2(\alpha_2 - \alpha_3)(\alpha_2 - \alpha_4) \dots (\alpha_2 - \alpha_r) \tag{11.173}$$

using a result similar to (11.172).

Proceeding we find

$$\begin{aligned} (-1)^{r-1} k_2 &= k_3(\alpha_3 - \alpha_4)(\alpha_3 - \alpha_5) \cdots (\alpha_3 - \alpha_r) \\ (-1)^{r-2} k_3 &= k_4(\alpha_4 - \alpha_5)(\alpha_4 - \alpha_6) \cdots (\alpha_4 - \alpha_r) \\ &\vdots \\ (-1)^2 k_{r-1} &= (-1)(\alpha_{r-1} - \alpha_r) \end{aligned} \tag{11.174}$$

and therefore,

$$D = (-1)^{r+(r-1)+\dots+2+1} \prod_{\substack{i,j=1 \\ i<j}}^{r} (\alpha_i - \alpha_j) = (-1)^{\frac{r(r+1)}{2}} \prod_{\substack{i,j=1 \\ i<j}}^{r} (\alpha_i - \alpha_j) \tag{11.175}$$

□

We now prove the important Bose–Chaudhuri–Hocquenhem theorem.

Theorem 11.2

Consider a code with symbols from GF(q), whose code polynomials have as zeros the elements $\alpha^{m_0}, \alpha^{m_0+1}, \dots, \alpha^{m_0+d-2}$, where α is any element from GF(q^m) and m_0 is any integer. Then the resultant (n, k) cyclic code has the following properties:

a) it has minimum distance $d_{min}^H \geq d$ if the elements $\alpha^{m_0}, \alpha^{m_0+1}, \dots, \alpha^{m_0+d-2}$, are distinct;

b) $n - k \le (d-1)m$; if $q = 2$ and $m_0 = 1$, then $n - k \le \left\lceil \frac{d-1}{2} \right\rceil m$;
c) n is equal to the order of α, unless $d = 2$, in which case n is equal to the order of α^{m_0};
d) $g(x)$ is equal to the least common multiple of the minimum functions of $\alpha^{m_0}, \alpha^{m_0+1}, \ldots, \alpha^{m_0+d-2}$. □

Proof. The proof of part (d) has already been given (see (11.163)); the proof of part (b) then follows by noting that each minimum function is at most of degree m, and there are at most $(d-1)$ distinct minimum functions. If $q = 2$ and $m_0 = 1$, the minimum function of α raised to an even power, for example α^{2i}, is the same as the minimum function of α^i (see Property 2 on page 542), therefore, there are at most $\left(\left\lceil \frac{d-1}{2} \right\rceil m\right)$ distinct minimum functions.

To prove part (c) note that, if $d = 2$, we have only the root α^{m_0}, so that n is equal to the order of α^{m_0}. If there are more than one root, then n must be the least common multiple of the order of the roots. If α^{m_0} and α^{m_0+1} are both roots, then $(\alpha^{m_0})^n = 1$ and $(\alpha^{m_0+1})^n = 1$, so that $\alpha^n = 1$; thus n is a multiple of the order of α. On the other hand, if ℓ is the order of α, $(\alpha^{m_0+i})^{\ell} = (\alpha^{\ell})^{m_0+i} = 1^{m_0+i} = 1$; therefore, ℓ is a multiple of the order of every root. Then n is the least common multiple of numbers all of which divide ℓ, and therefore, $n \le \ell$; thus $n = \ell$.

Finally, we prove part (a). We note that the code words must satisfy the condition

$$\begin{bmatrix} 1 & \alpha^{m_0} & (\alpha^{m_0})^2 & \ldots & (\alpha^{m_0})^{n-1} \\ 1 & \alpha^{m_0+1} & (\alpha^{m_0+1})^2 & \ldots & (\alpha^{m_0+1})^{n-1} \\ \vdots & & & & \vdots \\ 1 & \alpha^{m_0+d-2} & (\alpha^{m_0+d-2})^2 & \ldots & (\alpha^{m_0+d-2})^{n-1} \end{bmatrix} \begin{bmatrix} c_0 \\ c_1 \\ c_2 \\ \vdots \\ c_{n-1} \end{bmatrix} = \begin{bmatrix} 0 \\ 0 \\ \vdots \\ 0 \end{bmatrix} \tag{11.176}$$

We now show that no linear combination of $(d-1)$ or fewer columns is equal to 0. We do this by showing that the determinant of any set of $(d-1)$ columns is non-zero. Choose columns $j_1, j_2, \ldots, j_{d-1}$; then

$$\det \begin{vmatrix} (\alpha^{m_0})^{j_1} & (\alpha^{m_0})^{j_2} & \ldots & (\alpha^{m_0})^{j_{d-1}} \\ (\alpha^{m_0+1})^{j_1} & (\alpha^{m_0+1})^{j_2} & \ldots & (\alpha^{m_0+1})^{j_{d-1}} \\ \vdots & & & \vdots \\ (\alpha^{m_0+d-2})^{j_1} & (\alpha^{m_0+d-2})^{j_2} & \ldots & (\alpha^{m_0+d-2})^{j_{d-1}} \end{vmatrix} \tag{11.177}$$

$$= \alpha^{m_0(j_1+j_2+\cdots+j_{d-1})} \det \begin{vmatrix} 1 & 1 & \ldots & 1 \\ \alpha^{j_1} & \alpha^{j_2} & \ldots & \alpha^{j_{d-1}} \\ \vdots & & & \vdots \\ (\alpha^{j_1})^{d-2} & (\alpha^{j_2})^{d-2} & \ldots & (\alpha^{j_{d-1}})^{d-2} \end{vmatrix} \tag{11.178}$$

$$= \alpha^{m_0(j_1+j_2+\cdots+j_{d-1})} (-1)^{\frac{(d-1)d}{2}} \prod_{\substack{i,k=1 \\ i<k}}^{d-1} (\alpha^{j_i} - \alpha^{j_k}) \neq 0 \tag{11.179}$$

Note that we have proven that $(d-1)$ columns of $\boldsymbol{H}$ are linearly independent even if they are multiplied by elements from GF(q^m). All that would have been required was to show linear independence if the multipliers are from GF(q). □

Binary BCH codes

In this section, we consider binary BCH codes. Choose $m_0 = 1$; then from Property (c) of Theorem 11.2, we get

$$n = \begin{cases} 2^m - 1 & \text{if } \alpha \text{ is a primitive element of GF}(2^m) \\ \dfrac{2^m - 1}{c} & \text{if } \alpha = \beta^c, \text{ where } \beta \text{ is a primitive element of GF}(2^m) \end{cases} \tag{11.180}$$

and $r = n - k$ satisfies the relation (see Property (b))

$$r \le m \left\lceil \frac{d-1}{2} \right\rceil \tag{11.181}$$

Moreover for Property (d)

$$g(x) = \text{l.c.m.}(\text{minimum functions of } \alpha, \alpha^3, \alpha^5, \ldots, \alpha^{d-2}) \text{ with } d \text{ odd number} . \tag{11.182}$$

Example 11.2.28

Consider binary BCH codes of length 63, that is $q = 2$ and $m = 6$. To get a code with design distance d, we choose as roots $\alpha, \alpha^2, \alpha^3, \ldots, \alpha^{d-1}$, where α is a primitive element from GF(2^6). Using Table 11.11 on page 544, we get the minimum functions of the elements from GF(2^6) given in Table 11.16.

Then the roots and generator polynomials for different values of d are given in Table 11.17; the parameters of the relative codes are given in Table 11.18.

Table 11.16: Minimum functions of the elements of GF(2^6).

roots	*minimum function*
$\alpha^1\ \alpha^2\ \alpha^4\ \alpha^8\ \alpha^{16}\ \alpha^{32}$	$x^6 + x + 1$
$\alpha^3\ \alpha^6\ \alpha^{12}\ \alpha^{24}\ \alpha^{48}\ \alpha^{33}$	$x^6 + x^4 + x^2 + x + 1$
$\alpha^5\ \alpha^{10}\ \alpha^{20}\ \alpha^{40}\ \alpha^{17}\ \alpha^{34}$	$x^6 + x^5 + x^2 + x + 1$
$\alpha^7\ \alpha^{14}\ \alpha^{28}\ \alpha^{56}\ \alpha^{49}\ \alpha^{35}$	$x^6 + x^3 + 1$
$\alpha^9\ \alpha^{18}\ \alpha^{36}$	$x^3 + x^2 + 1$
$\alpha^{11}\ \alpha^{22}\ \alpha^{44}\ \alpha^{25}\ \alpha^{50}\ \alpha^{37}$	$x^6 + x^5 + x^3 + x^2 + 1$
$\alpha^{13}\ \alpha^{26}\ \alpha^{52}\ \alpha^{41}\ \alpha^{19}\ \alpha^{38}$	$x^6 + x^4 + x^3 + x + 1$
$\alpha^{15}\ \alpha^{30}\ \alpha^{60}\ \alpha^{57}\ \alpha^{51}\ \alpha^{39}$	$x^6 + x^5 + x^4 + x^2 + 1$
$\alpha^{21}\ \alpha^{42}$	$x^2 + x + 1$
$\alpha^{23}\ \alpha^{46}\ \alpha^{29}\ \alpha^{58}\ \alpha^{53}\ \alpha^{43}$	$x^6 + x^5 + x^4 + x + 1$
$\alpha^{27}\ \alpha^{54}\ \alpha^{45}$	$x^3 + x + 1$
$\alpha^{31}\ \alpha^{62}\ \alpha^{61}\ \alpha^{59}\ \alpha^{55}\ \alpha^{47}$	$x^6 + x^5 + 1$

Example 11.2.29

Let $q = 2$, $m = 6$, and choose as roots α, α^2, α^3, α^4, with $\alpha = \beta^3$, where β is a primitive element of GF(2^6); then $n = \frac{2^m-1}{c} = \frac{63}{3} = 21$, $d^H_{min} \ge d = 5$, and

$$\begin{aligned} g(x) &= \text{l.c.m}(m_{\beta^3}(x), m_{\beta^6}(x), m_{\beta^9}(x), m_{\beta^{12}}(x)) \\ &= m_{\beta^3}(x) m_{\beta^9}(x) \\ &= (x^6 + x^4 + x^2 + x + 1)(x^3 + x^2 + 1) \end{aligned} \tag{11.183}$$

As $r = \deg(g(x)) = 9$, then $k = n - r = 12$; thus, we obtain a (21,12) code.

Example 11.2.30

Let $q = 2$, $m = 4$, and choose as roots α, α^2, α^3, α^4, with α primitive element of GF(2^4); then a (15,7) code is obtained having $d^H_{min} \ge 5$, and $g(x) = (x^4 + x + 1)(x^4 + x^3 + x^2 + x + 1)$.

Example 11.2.31

Let $q = 2$, $m = 4$, and choose as roots α, α^2, α^3, α^4, α^5, α^6, with α primitive element of GF(2^4); then a (15,5) code is obtained having $d^H_{min} \ge 7$, and $g(x) = (x^4 + x + 1)(x^4 + x^3 + x^2 + x + 1)(x^2 + x + 1)$.

Table 11.17: Roots and generator polynomials of BCH codes of length $n = 63 = 2^6 - 1$ for different values of d. α is a primitive element of GF(2^6) (see (11.180)).

d	roots	generator polynomial
3	$\alpha\ \alpha^2$	$(x^6 + x + 1) = g_3(x)$
5	$\alpha\ \alpha^2\ \alpha^3\ \alpha^4$	$(x^6 + x + 1)(x^6 + x^4 + x^2 + x + 1) = g_5(x)$
7	$\alpha\ \alpha^2\ \dots\ \alpha^6$	$(x^6 + x^5 + x^2 + x + 1)\ g_5(x) = g_7(x)$
9	$\alpha\ \alpha^2\ \dots\ \alpha^8$	$(x^6 + x^3 + 1)\ g_7(x) = g_9(x)$
11	$\alpha\ \alpha^2\ \dots\ \alpha^{10}$	$(x^3 + x^2 + 1)\ g_9(x) = g_{11}(x)$
13	$\alpha\ \alpha^2\ \dots\ \alpha^{12}$	$(x^6 + x^5 + x^3 + x^2 + 1)\ g_{11}(x) = g_{13}(x)$
15	$\alpha\ \alpha^2\ \dots\ \alpha^{14}$	$(x^6 + x^4 + x^3 + x + 1)\ g_{13}(x) = g_{15}(x)$
21	$\alpha\ \alpha^2\ \dots\ \alpha^{20}$	$(x^6 + x^5 + x^4 + x^2 + 1)\ g_{15}(x) = g_{21}(x)$
23	$\alpha\ \alpha^2\ \dots\ \alpha^{22}$	$(x^2 + x + 1)\ g_{21}(x) = g_{23}(x)$
27	$\alpha\ \alpha^2\ \dots\ \alpha^{26}$	$(x^6 + x^5 + x^4 + x + 1)\ g_{23}(x) = g_{27}(x)$
31	$\alpha\ \alpha^2\ \dots\ \alpha^{30}$	$(x^3 + x + 1)\ g_{27}(x) = g_{31}(x)$

Table 11.18: Parameters of BCH codes of length $n = 63$.

k	57	51	45	39	36	30	24	18	16	10	7
d	3	5	7	9	11	13	15	21	23	27	31
t	1	2	3	4	5	6	7	10	11	13	15

Reed–Solomon codes

Reed–Solomon codes represent a particular case of BCH codes obtained by choosing $m = 1$; in other words, the field GF(q) and the extension field GF(q^m) coincide. Choosing α as a primitive element of (11.180) we get

$$n = q^m - 1 = q - 1 \tag{11.184}$$

Note that the minimum function with coefficients in GF(q) of an element α^i from GF(q) is

$$m_{\alpha^i}(x) = (x - \alpha^i) \tag{11.185}$$

For $m_0 = 1$, if we choose the roots $\alpha, \alpha^2, \alpha^3, \dots, \alpha^{d-1}$, then

$$g(x) = (x - \alpha)(x - \alpha^2)\dots(x - \alpha^{d-1}) \tag{11.186}$$

so that $r = (d - 1)$; the block length n is given by the order of α. In this case, we show that $d^H_{min} = d$; in fact, for any code, we have $d^H_{min} \leq r + 1$, as we can always choose a code word with every message symbol but one equal to zero and therefore, its weight is at most equal to $r + 1$; from this it follows that $d^H_{min} \leq d$, but from the BCH theorem, we know that $d^H_{min} \geq d$.

Example 11.2.32
Choose α as a primitive element of GF(2^5), and choose the roots α, α^2, α^3, α^4, α^5, and α^6; then the resultant (31,25) code has $d^H_{min} = 7$, $g(x) = (x - \alpha)(x - \alpha^2)(x - \alpha^3)(x - \alpha^4)(x - \alpha^5)(x - \alpha^6)$, and the symbols of the code words are from GF(2^5).

Table 11.19: Addition table for the elements of GF(2^4).

$+$	0	α^{0}	α^{1}	α^{2}	α^{3}	α^{4}	α^{5}	α^{6}	α^{7}	α^{8}	α^{9}	α^{10}	α^{11}	α^{12}	α^{13}	α^{14}
0	0	α^{0}	α^{1}	α^{2}	α^{3}	α^{4}	α^{5}	α^{6}	α^{7}	α^{8}	α^{9}	α^{10}	α^{11}	α^{12}	α^{13}	α^{14}
α^{0}	α^{0}	0	α^{4}	α^{8}	α^{14}	α^{1}	α^{10}	α^{13}	α^{9}	α^{2}	α^{7}	α^{5}	α^{12}	α^{11}	α^{6}	α^{3}
α^{1}	α^{1}	α^{4}	0	α^{5}	α^{9}	α^{0}	α^{2}	α^{11}	α^{14}	α^{10}	α^{3}	α^{8}	α^{6}	α^{13}	α^{12}	α^{7}
α^{2}	α^{2}	α^{8}	α^{5}	0	α^{6}	α^{10}	α^{1}	α^{3}	α^{12}	α^{0}	α^{11}	α^{4}	α^{9}	α^{7}	α^{14}	α^{13}
α^{3}	α^{3}	α^{14}	α^{9}	α^{6}	0	α^{7}	α^{11}	α^{2}	α^{4}	α^{13}	α^{1}	α^{12}	α^{5}	α^{10}	α^{8}	α^{0}
α^{4}	α^{4}	α^{1}	α^{0}	α^{10}	α^{7}	0	α^{8}	α^{12}	α^{3}	α^{5}	α^{14}	α^{2}	α^{13}	α^{6}	α^{11}	α^{9}
α^{5}	α^{5}	α^{10}	α^{2}	α^{1}	α^{11}	α^{8}	0	α^{9}	α^{13}	α^{4}	α^{6}	α^{0}	α^{3}	α^{14}	α^{7}	α^{12}
α^{6}	α^{6}	α^{13}	α^{11}	α^{3}	α^{2}	α^{12}	α^{9}	0	α^{10}	α^{14}	α^{5}	α^{7}	α^{1}	α^{4}	α^{0}	α^{8}
α^{7}	α^{7}	α^{9}	α^{14}	α^{12}	α^{4}	α^{3}	α^{13}	α^{10}	0	α^{11}	α^{0}	α^{6}	α^{8}	α^{2}	α^{5}	α^{1}
α^{8}	α^{8}	α^{2}	α^{10}	α^{0}	α^{13}	α^{5}	α^{4}	α^{14}	α^{11}	0	α^{12}	α^{1}	α^{7}	α^{9}	α^{3}	α^{6}
α^{9}	α^{9}	α^{7}	α^{3}	α^{11}	α^{1}	α^{14}	α^{6}	α^{5}	α^{0}	α^{12}	0	α^{13}	α^{2}	α^{8}	α^{10}	α^{4}
α^{10}	α^{10}	α^{5}	α^{8}	α^{4}	α^{12}	α^{2}	α^{0}	α^{7}	α^{6}	α^{1}	α^{13}	0	α^{14}	α^{3}	α^{9}	α^{11}
α^{11}	α^{11}	α^{12}	α^{6}	α^{9}	α^{5}	α^{13}	α^{3}	α^{1}	α^{8}	α^{7}	α^{2}	α^{14}	0	α^{0}	α^{4}	α^{10}
α^{12}	α^{12}	α^{11}	α^{13}	α^{7}	α^{10}	α^{6}	α^{14}	α^{4}	α^{2}	α^{9}	α^{8}	α^{3}	α^{0}	0	α^{1}	α^{5}
α^{13}	α^{13}	α^{6}	α^{12}	α^{14}	α^{8}	α^{11}	α^{7}	α^{0}	α^{5}	α^{3}	α^{10}	α^{9}	α^{4}	α^{1}	0	α^{2}
α^{14}	α^{14}	α^{3}	α^{7}	α^{13}	α^{0}	α^{9}	α^{12}	α^{8}	α^{1}	α^{6}	α^{4}	α^{11}	α^{10}	α^{5}	α^{2}	0

Observation 11.2

The encoding of Reed–Solomon codes can be done by the devices of Figure 11.3 or Figure 11.4, where the operations are in GF(q). In Tables 11.19 and 11.20 we give, respectively, the tables of additions and multiplications between elements of GF(q) for $q = 2^4$; the conversion of the symbol representation from binary to exponential is implemented using Table 11.21. We note that the encoding operations can be

Table 11.20: Multiplication table for the elements of GF(2^4).

$\cdot$	0	α^{0}	α^{1}	α^{2}	α^{3}	α^{4}	α^{5}	α^{6}	α^{7}	α^{8}	α^{9}	α^{10}	α^{11}	α^{12}	α^{13}	α^{14}
0	0	0	0	0	0	0	0	0	0	0	0	0	0	0	0	0
α^{0}	0	α^{0}	α^{1}	α^{2}	α^{3}	α^{4}	α^{5}	α^{6}	α^{7}	α^{8}	α^{9}	α^{10}	α^{11}	α^{12}	α^{13}	α^{14}
α^{1}	0	α^{1}	α^{2}	α^{3}	α^{4}	α^{5}	α^{6}	α^{7}	α^{8}	α^{9}	α^{10}	α^{11}	α^{12}	α^{13}	α^{14}	α^{0}
α^{2}	0	α^{2}	α^{3}	α^{4}	α^{5}	α^{6}	α^{7}	α^{8}	α^{9}	α^{10}	α^{11}	α^{12}	α^{13}	α^{14}	α^{0}	α^{1}
α^{3}	0	α^{3}	α^{4}	α^{5}	α^{6}	α^{7}	α^{8}	α^{9}	α^{10}	α^{11}	α^{12}	α^{13}	α^{14}	α^{0}	α^{1}	α^{2}
α^{4}	0	α^{4}	α^{5}	α^{6}	α^{7}	α^{8}	α^{9}	α^{10}	α^{11}	α^{12}	α^{13}	α^{14}	α^{0}	α^{1}	α^{2}	α^{3}
α^{5}	0	α^{5}	α^{6}	α^{7}	α^{8}	α^{9}	α^{10}	α^{11}	α^{12}	α^{13}	α^{14}	α^{0}	α^{1}	α^{2}	α^{3}	α^{4}
α^{6}	0	α^{6}	α^{7}	α^{8}	α^{9}	α^{10}	α^{11}	α^{12}	α^{13}	α^{14}	α^{0}	α^{1}	α^{2}	α^{3}	α^{4}	α^{5}
α^{7}	0	α^{7}	α^{8}	α^{9}	α^{10}	α^{11}	α^{12}	α^{13}	α^{14}	α^{0}	α^{1}	α^{2}	α^{3}	α^{4}	α^{5}	α^{6}
α^{8}	0	α^{8}	α^{9}	α^{10}	α^{11}	α^{12}	α^{13}	α^{14}	α^{0}	α^{1}	α^{2}	α^{3}	α^{4}	α^{5}	α^{6}	α^{7}
α^{9}	0	α^{9}	α^{10}	α^{11}	α^{12}	α^{13}	α^{14}	α^{0}	α^{1}	α^{2}	α^{3}	α^{4}	α^{5}	α^{6}	α^{7}	α^{8}
α^{10}	0	α^{10}	α^{11}	α^{12}	α^{13}	α^{14}	α^{0}	α^{1}	α^{2}	α^{3}	α^{4}	α^{5}	α^{6}	α^{7}	α^{8}	α^{9}
α^{11}	0	α^{11}	α^{12}	α^{13}	α^{14}	α^{0}	α^{1}	α^{2}	α^{3}	α^{4}	α^{5}	α^{6}	α^{7}	α^{8}	α^{9}	α^{10}
α^{12}	0	α^{12}	α^{13}	α^{14}	α^{0}	α^{1}	α^{2}	α^{3}	α^{4}	α^{5}	α^{6}	α^{7}	α^{8}	α^{9}	α^{10}	α^{11}
α^{13}	0	α^{13}	α^{14}	α^{0}	α^{1}	α^{2}	α^{3}	α^{4}	α^{5}	α^{6}	α^{7}	α^{8}	α^{9}	α^{10}	α^{11}	α^{12}
α^{14}	0	α^{14}	α^{0}	α^{1}	α^{2}	α^{3}	α^{4}	α^{5}	α^{6}	α^{7}	α^{8}	α^{9}	α^{10}	α^{11}	α^{12}	α^{13}

Table 11.21: Three equivalent representations of the elements of GF(2^4), obtained applying the polynomial arithmetic modulo $x^4 + x + 1$.

exponential	polynomial	binary $(x^0 x^1 x^2 x^3)$
0	0	0 0 0 0
α^0	1	1 0 0 0
α^1	x	0 1 0 0
α^2	x^2	0 0 1 0
α^3	x^3	0 0 0 1
α^4	$1+x$	1 1 0 0
α^5	$x+x^2$	0 1 1 0
α^6	x^2+x^3	0 0 1 1
α^7	$1+x+x^3$	1 1 0 1
α^8	$1+x^2$	1 0 1 0
α^9	$x+x^3$	0 1 0 1
α^{10}	$1+x+x^2$	1 1 1 0
α^{11}	$x+x^2+x^3$	0 1 1 1
α^{12}	$1+x+x^2+x^3$	1 1 1 1
α^{13}	$1+x^2+x^3$	1 0 1 1
α^{14}	$1+x^3$	1 0 0 1

performed by interpreting the field elements as polynomials with coefficients from GF(2), and applying the polynomial arithmetic mod $x^4 + x + 1$ (see Figures 11.1 and 11.2).

Decoding of BCH codes

Suppose we transmit the code polynomial $c_0 + c_1 x + \cdots + c_{n-1} x^{n-1}$, and we receive $z_0 + z_1 x + \cdots + z_{n-1} x^{n-1}$. We define the polynomial $e(x) = (z_0 - c_0) + (z_1 - c_1)x + \ldots + (z_{n-1} - c_{n-1})x^{n-1} = e_0 + e_1 x + \cdots + e_{n-1} x^{n-1}$, where e_i are elements from GF(q^m), the field in which parity control is performed. If we express polynomials as vectors, we obtain $\boldsymbol{z} = \boldsymbol{c} + \boldsymbol{e}$; furthermore, we recall that, defining the matrix

$$\boldsymbol{H} = \begin{bmatrix} 1 & \alpha^{m_0} & (\alpha^{m_0})^2 & \ldots & (\alpha^{m_0})^{n-1} \\ 1 & \alpha^{m_0+1} & (\alpha^{m_0+1})^2 & \ldots & (\alpha^{m_0+1})^{n-1} \\ \vdots & & & & \vdots \\ 1 & \alpha^{m_0+d-2} & (\alpha^{m_0+d-2})^2 & \ldots & (\alpha^{m_0+d-2})^{n-1} \end{bmatrix} \tag{11.187}$$

we obtain

$$\boldsymbol{Hz} = \boldsymbol{He} = \boldsymbol{s} = \begin{bmatrix} s_{m_0} \\ s_{m_0+1} \\ s_{m_0+2} \\ \vdots \\ s_{m_0+d-2} \end{bmatrix} \tag{11.188}$$

where

$$s_j = \sum_{\ell=0}^{n-1} e_\ell (\alpha^j)^\ell, \qquad j = m_0, m_0+1, \ldots, m_0+d-2 \tag{11.189}$$

Assuming there are ν errors, that is $w(\boldsymbol{e}) = \nu$, we can write

$$s_j = \sum_{i=1}^{\nu} \varepsilon_i (\xi_i)^j, \qquad j = m_0, m_0 + 1, \ldots, m_0 + d - 2 \tag{11.190}$$

where the coefficients ε_i are elements from GF(q^m) that represent the values of the errors, and the coefficients ξ_i are elements from GF(q^m) that give the positions of the errors. In other words, if $\xi_i = \alpha^{\ell}$, then an error has occurred in position ℓ, where $\ell \in \{0, 1, 2, \ldots, n-1\}$. The idea of a decoding algorithm is to solve the set of non-linear equations for the unknowns ε_i and ξ_i; then there are 2ν unknowns and $d-1$ equations.

We show that it is possible to solve this set of equations if $2\nu \leq d-1$, assuming $m_0 = 1$; in the case $m_0 \neq 1$, the decoding procedure does not change.

Consider the polynomial in x, also called *error indicator polynomial*,

$$\lambda(x) = \lambda_\nu x^\nu + \lambda_{\nu-1} x^{\nu-1} + \cdots + \lambda_1 x + 1 \tag{11.191}$$

defined as the polynomial that has as zeros the inverse of the elements that locate the positions of the errors, that is ξ_i^{-1}, $i = 1, \ldots, \nu$. Then

$$\lambda(x) = (1 - x\xi_1)(1 - x\xi_2)\cdots(1 - x\xi_\nu) \tag{11.192}$$

If the coefficients of $\lambda(x)$ are known, it is possible to find the zeros of $\lambda(x)$, and thus determine the positions of the errors.

The first step of the decoding procedure consists in evaluating the coefficients $\lambda_1, \ldots, \lambda_\nu$ using the syndromes (11.190). We multiply both sides of (11.191) by $\varepsilon_i \xi_i^{j+\nu}$ and evaluate the expression found for $x = \xi_i^{-1}$, obtaining

$$0 = \varepsilon_i \xi_i^{j+\nu} (1 + \lambda_1 \xi_i^{-1} + \lambda_2 \xi_i^{-2} + \cdots + \lambda_\nu \xi_i^{-\nu}) \tag{11.193}$$

which can be written as

$$\varepsilon_i \, (\xi_i^{j+\nu} + \lambda_1 \xi_i^{j+\nu-1} + \lambda_2 \xi_i^{j+\nu-2} + \cdots + \lambda_\nu \xi_i^{j}) = 0 \tag{11.194}$$

(11.194) holds for $i = 1, \ldots, \nu$, and for every value of j. Adding these equations for $i = 1, \ldots, \nu$, we get

$$\sum_{i=1}^{\nu} \varepsilon_i \, (\xi_i^{j+\nu} + \lambda_1 \xi_i^{j+\nu-1} + \lambda_2 \xi_i^{j+\nu-2} + \ldots + \lambda_\nu \xi_i^{j}) = 0 \qquad \text{for every } j \tag{11.195}$$

or equivalently

$$\sum_{i=1}^{\nu} \varepsilon_i \xi_i^{j+\nu} + \lambda_1 \sum_{i=1}^{\nu} \varepsilon_i \xi_i^{j+\nu-1} + \lambda_2 \sum_{i=1}^{\nu} \varepsilon_i \xi_i^{j+\nu-2} + \cdots + \lambda_\nu \sum_{i=1}^{\nu} \varepsilon_i \xi_i^{j} = 0 \quad \text{for every } j \tag{11.196}$$

As $\nu \leq (d-1)/2$, if $1 \leq j \leq \nu$ the summations in (11.196) are equal to the syndromes (11.190); therefore, we obtain

$$\lambda_1 s_{j+\nu-1} + \lambda_2 s_{j+\nu-2} + \cdots + \lambda_\nu s_j = -s_{j+\nu}, \qquad j = 1, \ldots, \nu \tag{11.197}$$

(11.197) is a system of linear equations that can be written in the form

$$\begin{bmatrix} s_1 & s_2 & s_3 & \cdots & s_{\nu-1} & s_\nu \\ s_2 & s_3 & s_4 & \cdots & s_\nu & s_{\nu+1} \\ \vdots & & & & \vdots & \vdots \\ s_\nu & s_{\nu+1} & s_{\nu+2} & \cdots & s_{2\nu-2} & s_{2\nu-1} \end{bmatrix} \begin{bmatrix} \lambda_\nu \\ \lambda_{\nu-1} \\ \vdots \\ \lambda_1 \end{bmatrix} = \begin{bmatrix} -s_{\nu+1} \\ -s_{\nu+2} \\ \vdots \\ -s_{2\nu} \end{bmatrix} \tag{11.198}$$

Thus, we have obtained ν equations in the unknowns $\lambda_1, \lambda_2, \dots, \lambda_\nu$. To show that these equations are linearly independent we see that, from (11.190), the matrix of the coefficients can be factorized as

$$\begin{bmatrix} s_1 & s_2 & s_3 & \dots & s_{\nu-1} & s_\nu \\ s_2 & s_3 & s_4 & \dots & s_\nu & s_{\nu+1} \\ \vdots & & & & \vdots & \vdots \\ s_\nu & s_{\nu+1} & s_{\nu+2} & \dots & s_{2\nu-2} & s_{2\nu-1} \end{bmatrix} = \begin{bmatrix} 1 & 1 & 1 & \dots & 1 \\ \xi_1 & \xi_2 & \xi_3 & \dots & \xi_\nu \\ \vdots & & & & \vdots \\ \xi_1^{\nu-1} & \xi_2^{\nu-1} & \xi_3^{\nu-1} & \dots & \xi_\nu^{\nu-1} \end{bmatrix} \begin{bmatrix} \varepsilon_1\xi_1 & & & \\ & \varepsilon_2\xi_2 & & 0 \\ & & \ddots & \\ & 0 & & \varepsilon_\nu\xi_\nu \end{bmatrix} \begin{bmatrix} 1 & \xi_1 & \dots & \xi_1^{\nu-1} \\ 1 & \xi_2 & \dots & \xi_2^{\nu-1} \\ \vdots & & & \vdots \\ 1 & \xi_\nu & \dots & \xi_\nu^{\nu-1} \end{bmatrix} \tag{11.199}$$

The matrix of the coefficients has a non-zero determinant if each of the matrices on the right-hand side of (11.199) has a non-zero determinant. We note that the first and third matrix are Vandermonde matrices, hence, they are non-singular if $\xi_i \neq \xi_m$, $i \neq m$; the second matrix is also non-singular, as it is a diagonal matrix with non-zero terms on the diagonal. This assumes we have at most ν errors, with $\nu \leq (d-1)/2$. As ν is arbitrary, we initially choose $\nu = (d-1)/2$ and compute the determinant of the matrix (11.199). If this is non-zero, then we have the correct value of ν; otherwise, if the determinant is zero, we reduce ν by one and repeat the computation. We proceed until we obtain a non-zero determinant, thus finding the number of errors that occurred.

After finding the solution for $\lambda_1, \dots, \lambda_\nu$ (see (11.198)), we obtain the positions of the errors by finding the zeros of the polynomial $\lambda(x)$ (see (11.192)). Note that an exhaustive method to search for the ν zeros of the polynomial $\lambda(x)$ requires that at most n possible roots of the type $\alpha^\ell, \ell \in \{0, 1, \dots, n-1\}$, are taken into consideration. We now compute the value of the non-zero elements ε_i, $i = 1, \dots, \nu$, of the vector $\boldsymbol{e}$. If the code is binary, the components of the vector $\boldsymbol{e}$ are immediately known, otherwise, we solve the system of linear equations (11.190) for the ν unknowns $\varepsilon_1, \varepsilon_2, \dots, \varepsilon_\nu$. The determinant of the matrix of the system of linear equations is given by

$$\det \begin{bmatrix} \xi_1 & \xi_2 & \dots & \xi_\nu \\ \xi_1^2 & \xi_2^2 & \dots & \xi_\nu^2 \\ \vdots & \vdots & & \vdots \\ \xi_1^\nu & \xi_2^\nu & \dots & \xi_\nu^\nu \end{bmatrix} = \xi_1\xi_2 \dots \xi_\nu \det \begin{bmatrix} 1 & 1 & \dots & 1 \\ \xi_1 & \xi_2 & \dots & \xi_\nu \\ \vdots & & & \vdots \\ \xi_1^{\nu-1} & \xi_2^{\nu-1} & \dots & \xi_\nu^{\nu-1} \end{bmatrix} \tag{11.200}$$

The determinant of the Vandermonde matrix in (11.200) is non-zero if ν errors occurred, as the elements $\xi_1, \dots, \xi_\nu$ are non-zero and distinct.

In summary, the original system of non-linear equations (11.190) is solved in the following three steps:

Step 1: find the coefficients $\lambda_1, \lambda_2, \dots, \lambda_\nu$ of the error indicator polynomial by solving a system of linear equations.

Step 2: find the ν roots $\xi_1, \xi_2, \dots, \xi_\nu$ of a polynomial of degree ν.

Step 3: find the values of the errors $\varepsilon_1, \varepsilon_2, \dots, \varepsilon_\nu$ by solving a system of linear equations.

For binary codes, the last step is omitted.

Efficient decoding of BCH codes

The computational complexity for the decoding of BCH codes illustrated in the previous Section, lies mainly in the solution of the systems of linear equations (11.198) and (11.190). For small values of ν, the direct solution of these systems of equations by inverting matrices does not require a high computational complexity; we recall that the number of operations necessary to invert a $\nu \times \nu$ matrix is of the order of

ν^3. However, in many applications, it is necessary to resort to codes that are capable of correcting several errors, and it is thus desirable to find more efficient methods for the solution. The method developed by Berlekamp is based on the observation that the matrix of the coefficients and the known data vector in (11.198) have a particular structure. Assuming that the vector $\boldsymbol{\lambda} = [\lambda_\nu, \lambda_{\nu-1}, \dots, \lambda_1]^T$ is known, then from (11.197) for the sequence of syndromes $s_1, s_2, \dots, s_{2\nu}$, the recursive relation holds

$$s_j = -\sum_{i=1}^{\nu} \lambda_i \, s_{j-i}, \qquad j = \nu+1, \dots, 2\nu \tag{11.201}$$

For a given $\boldsymbol{\lambda}$, (11.201) is the equation of a recursive filter, which can be implemented by a *shift register with feedback*, whose coefficients are given by $\boldsymbol{\lambda}$, as illustrated in Figure 11.8. The solution of (11.198) is thus equivalent to the problem of finding the shift register with feedback of minimum length that, if suitably initialized, yields the sequence of syndromes. This will identify the polynomial $\lambda(x)$ of minimum degree ν, that we recall exists and is unique, as the $\nu \times \nu$ matrix of the original problem admits the inverse.

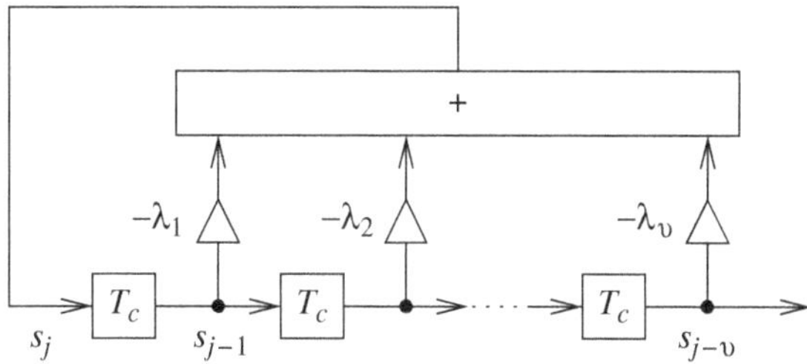

Figure 11.8 Recursive filter to compute syndromes (see (11.201)).

The *Berlekamp–Massey algorithm* to find the recursive filter can be applied in any field and does not make use of the particular properties of the sequence of syndromes $s_1, s_2, \dots, s_{d-1}$. To determine the recursive filter we must find two quantities, that we denote as $(L, \lambda(x))$, where L is the length of the shift register and $\lambda(x)$ is the polynomial whose degree ν must satisfy the condition $\nu \le L$. The algorithm is inductive, that is for each r, starting from $r = 1$, we determine a shift register that generates the first r syndromes. The shift register identified by $(L_r, \lambda^{(r)}(x))$ will then be a shift register of minimum length that generates the sequence $s_1, \dots, s_r$.

Berlekamp–Massey algorithm. Let $s_1, \dots, s_{d-1}$ be a sequence of elements from any field. Assuming the initial conditions $\lambda^{(0)}(x) = 1$, $\beta^{(0)}(x) = 1$, and $L_0 = 0$, we use the following set of recursive equations to determine $\lambda^{(d-1)}(x)$:
for $r = 1, \dots, d-1$,

$$\Delta_r = \sum_{j=0}^{n-1} \lambda_j^{(r-1)} \, s_{r-j} \tag{11.202}$$

$$\delta_r = \begin{cases} 1 & \text{if } \Delta_r \neq 0 \text{ and } 2L_{r-1} \le r-1 \\ 0 & \text{otherwise ,} \end{cases} \tag{11.203}$$

$$L_r = \delta_r (r - L_{r-1}) + (1 - \delta_r)\, L_{r-1} \tag{11.204}$$

$$\begin{bmatrix} \lambda^{(r)}(x) \\ \beta^{(r)}(x) \end{bmatrix} = \begin{bmatrix} 1 & -\Delta_r x \\ \Delta_r^{-1} \delta_r & (1-\delta_r)\, x \end{bmatrix} \begin{bmatrix} \lambda^{(r-1)}(x) \\ \beta^{(r-1)}(x) \end{bmatrix} \tag{11.205}$$

Then $\lambda^{(d-1)}(x)$ is the polynomial of minimum degree such that

$$\lambda_0^{(d-1)} = 1 \tag{11.206}$$

and

$$s_r = -\sum_{j=1}^{n-1} \lambda_j^{(d-1)} \, s_{r-j}, \qquad r = L_{d-1}+1, \dots, d-1 \tag{11.207}$$

Note that Δ_r can be zero only if $\delta_r = 0$; in this case, we assign to $\Delta_r^{-1} \cdot \delta_r$ the value zero. Moreover, we see that the algorithm requires a complexity of the order of d^2 operations, against a complexity of the order of d^3 operations needed by the matrix inversion in (11.198). To prove that the polynomial $\lambda^{(d-1)}(x)$ given by the algorithm is indeed the polynomial of minimum degree with $\lambda_0^{(d-1)} = 1$ that satisfies (11.207), we use the following two lemmas [2.

In Lemma 11.1, we find the relation between the lengths of the shift registers of minimum length obtained in two consecutive iterations, L_r and L_{r-1}. In Lemma 11.2, we use the algorithm to construct a shift register that generates $s_1, \dots, s_r$ starting from the shift register of minimum length that generates $s_1, \dots, s_{r-1}$. We will conclude that the construction yields the shift register of minimum length since it satisfies Lemma 11.1.

Lemma 11.1
We assume that $(L_{r-1}, \lambda^{(r-1)}(x))$ is the shift register of minimum length that generates $s_1, \dots, s_{r-1}$, while $(L_r, \lambda^{(r)}(x))$ is the shift register of minimum length that generates $s_1, \dots, s_{r-1}, s_r$, and $\lambda^{(r)}(x) \neq \lambda^{(r-1)}(x)$; then

$$L_r \geq \max(L_{r-1}, r - L_{r-1}) \tag{11.208}$$

Proof. The inequality (11.208) is the combination of the two inequalities $L_r \geq L_{r-1}$ and $L_r \geq r - L_{r-1}$. The first inequality is obvious, because if a shift register generates a certain sequence it must also generate any initial part of this sequence; the second inequality is obvious if $L_{r-1} \geq r$ because L_r is a non-negative quantity. Thus, we assume $L_{r-1} < r$, and suppose that the second inequality is not satisfied; then $L_r \leq r - 1 - L_{r-1}$, or $r \geq L_{r-1} + L_r + 1$. By assumption we have

$$\begin{cases} s_j = -\sum_{i=1}^{L_{r-1}} \lambda_i^{(r-1)} \, s_{j-i} & j = L_{r-1}+1, \dots, r-1 \\ s_r \neq -\sum_{i=1}^{L_{r-1}} \lambda_i^{(r-1)} \, s_{r-i} \end{cases} \tag{11.209}$$

and

$$s_j = -\sum_{k=1}^{L_r} \lambda_k^{(r)} \, s_{j-k}, \qquad j = L_r + 1, \dots, r \tag{11.210}$$

We observe that

$$s_r = -\sum_{k=1}^{L_r} \lambda_k^{(r)} \, s_{r-k} = \sum_{k=1}^{L_r} \lambda_k^{(r)} \sum_{i=1}^{L_{r-1}} \lambda_i^{(r-1)} \, s_{r-k-i} \tag{11.211}$$

where the expression of s_{r-k} is valid, as $(r-k)$ goes from $r-1$ to $r - L_r$. Hence, it belongs to the set $L_{r-1}+1, \dots, r-1$, as it is assumed $r \geq L_{r-1} + L_r + 1$.

Furthermore,

$$s_r \neq -\sum_{i=1}^{L_{r-1}} \lambda_i^{(r-1)} \, s_{r-i} = \sum_{i=1}^{L_{r-1}} \lambda_i^{(r-1)} \sum_{k=1}^{L_r} \lambda_k^{(r)} \, s_{r-i-k} \tag{11.212}$$

where the expression of s_{r-i} is valid as $(r-i)$ goes from $r-1$ to $r - L_{r-1}$. Hence, it belongs to the set $L_r + 1, \dots, r-1$, as it is assumed $r \geq L_{r-1} + L_r + 1$. The summations on the right-hand side of (11.212) can be exchanged, thus obtaining the right-hand side of (11.211). But this yields s_r of (11.211) different from s_r of (11.212), thus we get a contradiction. □

Lemma 11.2
We assume that $(L_i, \lambda^{(i)}(x))$, $i = 1, \dots, r$, identifies a sequence of shift registers of minimum length such that $\lambda^{(i)}(x)$ generates $s_1, \dots, s_i$. If $\lambda^{(r)}(x) \neq \lambda^{(r-1)}(x)$, then

$$L_r = \max(L_{r-1}, r - L_{r-1}) \tag{11.213}$$

and every shift register that generates $s_1, \dots, s_r$, and has a length that satisfies (11.213), is a shift register of minimum length. The Berlekamp–Massey algorithm yields this shift register.

Proof. From Lemma 11.1, L_r cannot be smaller than the right-hand side of (11.213); thus, if we construct a shift register that yields the given sequence and whose length satisfies (11.213), then it must be a shift register of minimum length. The proof is obtained by induction.

We construct a shift register that satisfies the Lemma at the rth iteration, assuming that shift registers were iteratively constructed for each value of the index k, with $k \leq r - 1$. For each k, $k = 1, \dots, r - 1$, let $(L_k, \lambda^{(k)}(x))$ be the shift register of minimum length that generates $s_1, \dots, s_k$. We assume that

$$L_k = \max(L_{k-1}, k - L_{k-1}), \qquad k = 1, \dots, r - 1 \tag{11.214}$$

Equation (11.214) is verified for $k = 0$, as $L_0 = 0$ and $L_1 = 1$. Let m be the index k at the most recent iteration that required a variation in the length of the shift register. In other words, at the end of the $(r - 1)$th iteration, m is the integer such that

$$L_{r-1} = L_m > L_{m-1} \tag{11.215}$$

From (11.209) and (11.202), we have that

$$s_j + \sum_{i=1}^{L_{r-1}} \lambda_i^{(r-1)} \, s_{j-i} = \sum_{i=0}^{L_{r-1}} \lambda_i^{(r-1)} \, s_{j-i} = \begin{cases} 0 & j = L_{r-1}, \dots, r-1 \\ \Delta_r & j = r \end{cases} \tag{11.216}$$

If $\Delta_r = 0$, then the shift register $(L_{r-1}, \lambda^{(r-1)}(x))$ also generates the first r symbols of the sequence, hence,

$$\begin{cases} L_r = L_{r-1} \\ \lambda^{(r)}(x) = \lambda^{(r-1)}(x) \end{cases} \tag{11.217}$$

If $\Delta_r \neq 0$, then it is necessary to find a new shift register. Recall from (11.215) that there was a variation in the length of the shift register for $k = m$; therefore,

$$s_j + \sum_{i=1}^{L_{m-1}} \lambda_i^{(m-1)} \, s_{j-i} = \begin{cases} 0 & j = L_{m-1}, \dots, m-1 \\ \Delta_m \neq 0 & j = m \end{cases} \tag{11.218}$$

and by induction,

$$L_{r-1} = L_m = \max(L_{m-1}, m - L_{m-1}) = m - L_{m-1} \tag{11.219}$$

as $L_m > L_{m-1}$. We choose now the new polynomial

$$\lambda^{(r)}(x) = \lambda^{(r-1)}(x) - \Delta_r \, \Delta_m^{-1} \, x^{r-m} \, \lambda^{(m-1)}(x) \tag{11.220}$$

and let $L_r = \deg(\lambda^{(r)}(x))$. Then, as $\deg(\lambda^{(r-1)}(x)) \leq L_{r-1}$, and $\deg[x^{r-m}\lambda^{(m-1)}(x)] \leq r - m + L_{m-1}$, we get

$$L_r \leq \max(L_{r-1}, r - m + L_{m-1}) \leq \max(L_{r-1}, r - L_{r-1}) \tag{11.221}$$

Thus, recalling Lemma 11.1, if $\lambda^{(r)}(x)$ generates $s_1, \dots, s_r$, then

$$L_r = \max(L_{r-1}, r - L_{r-1}) \tag{11.222}$$

It remains to prove that the shift register $(L_r, \lambda^{(r)}(x))$ generates the given sequence. By direct computation, we obtain

$$\begin{aligned} s_j &- \left(-\sum_{i=1}^{L_r} \lambda_i^{(r)} \, s_{j-i} \right) \\ &= s_j + \sum_{i=1}^{L_{r-1}} \lambda_i^{(r-1)} \, s_{j-i} - \Delta_r \Delta_m^{-1} \left[s_{j-r+m} + \sum_{i=1}^{L_{m-1}} \lambda_i^{(L_{m-1})} \, s_{j-r+m-i} \right] \\ &= \begin{cases} 0 & j = L_r, L_r+1, \ldots, r-1 \\ \Delta_r - \Delta_r \Delta_m^{-1} \Delta_m = 0 & j = r \end{cases} \end{aligned} \tag{11.223}$$

Therefore, the shift register $(L_r, \lambda^{(r)}(x))$ generates $s_1, \ldots, s_r$. In particular, $(L_{d-1}, \lambda^{(d-1)}(x))$ generates $s_1, \ldots, s_{d-1}$. This completes the proof of Lemma 2. □

We have seen that the computational complexity for the solution of the system of equations (11.198), that yields the error indicator polynomial, can be reduced by the Berlekamp–Massey algorithm. We now consider the system of equations (11.190) that yields the values of the errors. The computation of the inverse matrix can be avoided in the solution of (11.190) by applying the Forney algorithm [2]. We recall the expression (11.191) of the error indicator polynomial $\lambda(x)$, that has zeros for $x = \xi_i^{-1}$, $i = 1, \ldots, \nu$, given by

$$\lambda(x) = \prod_{\ell=1}^{\nu} (1 - x\xi_\ell) \tag{11.224}$$

Define the syndrome polynomial as

$$s(x) = \sum_{j=1}^{d-1} s_j \, x^j = \sum_{j=1}^{d-1} \sum_{j=1}^{\nu} \varepsilon_i \, \xi_i^j \, x^j \tag{11.225}$$

and furthermore, define the *error evaluator polynomial* $\omega(x)$ as

$$\omega(x) = s(x) \, \lambda(x), \qquad \text{mod } x^{d-1} \tag{11.226}$$

Proposition 11.5
The error evaluator polynomial can be expressed as

$$\omega(x) = x \sum_{i=1}^{\nu} \varepsilon_i \, \xi_i \prod_{\substack{j=1 \\ j \neq i}}^{\nu} (1 - \xi_j x) \tag{11.227}$$

□

Proof. From the definition (11.226) of $\omega(x)$, we obtain

$$\begin{aligned} \omega(x) &= \left[\sum_{j=1}^{d-1} \sum_{i=1}^{\nu} \varepsilon_i \, \xi_i^j \, x^j \right] \left[\prod_{\ell=1}^{\nu} (1 - \xi_\ell x) \right], \qquad \text{mod } x^{d-1} \\ &= \sum_{i=1}^{\nu} \varepsilon_i \, \xi_i \, x \left[(1 - \xi_i x) \sum_{j=1}^{d-1} (\xi_i x)^{j-1} \right] \prod_{\substack{\ell=1 \\ \ell \neq i}}^{\nu} (1 - \xi_\ell x), \qquad \text{mod } x^{d-1} \end{aligned} \tag{11.228}$$

By inspection, we see that the term within brackets is equal to $(1 - \xi_i^{d-1} x^{d-1})$; thus,

$$\omega(x) = \sum_{i=1}^{\nu} \varepsilon_i \, \xi_i \, x (1 - \xi_i^{d-1} x^{d-1}) \prod_{\substack{\ell=1 \\ \ell \neq i}}^{\nu} (1 - \xi_\ell x), \qquad \text{mod } x^{d-1} \tag{11.229}$$

We now observe that (11.229), evaluated modulo x^{d-1}, is identical to (11.227). □

Forney algorithm. We introduce the derivative of $\lambda(x)$ given by

$$\lambda'(x) = -\sum_{i=1}^{\nu} \xi_i \prod_{\substack{j=1 \\ j \neq i}}^{\nu} (1 - x\xi_j) \tag{11.230}$$

The values of the errors are given by

$$\varepsilon_\ell = \frac{\omega(\xi_\ell^{-1})}{\prod_{\substack{j=1 \\ j \neq \ell}}^{\nu} (1 - \xi_j \xi_\ell^{-1})} = -\frac{\omega(\xi_\ell^{-1})}{\xi_\ell^{-1} \lambda'(\xi_\ell^{-1})} \tag{11.231}$$

Proof. We evaluate (11.227) for $x = \xi_\ell^{-1}$, obtaining

$$\omega(\xi_\ell^{-1}) = \varepsilon_\ell \prod_{\substack{j=1 \\ j \neq \ell}}^{\nu} (1 - \xi_j \xi_\ell^{-1}) \tag{11.232}$$

which proves the first part of equality (11.231). Moreover, from (11.230), we have

$$\lambda'(\xi_\ell^{-1}) = -\xi_\ell \prod_{\substack{j=1 \\ j \neq \ell}}^{\nu} (1 - \xi_j \xi_\ell^{-1}) \tag{11.233}$$

which proves the second part of equality (11.231). □

Example 11.2.33 (Reed–Solomon (15,9) code with $d = 7$ ($t = 3$), and elements from $GF(2^4)$)
From (11.186), using Tables 11.19 and 11.20, the generator polynomial is given by

$$\begin{aligned} g(x) &= (x - \alpha)(x - \alpha^2)(x - \alpha^3)(x - \alpha^4)(x - \alpha^5)(x - \alpha^6) \\ &= x^6 + \alpha^{10}x^5 + \alpha^{14}x^4 + \alpha^4x^3 + \alpha^6x^2 + \alpha^9x + \alpha^6 \end{aligned} \tag{11.234}$$

Suppose that the code polynomial $c(x) = 0$ is transmitted and that the received polynomial is

$$z(x) = \alpha x^7 + \alpha^5 x^5 + \alpha^{11} x^2 \tag{11.235}$$

In this case $e(x) = z(x)$. From (11.189), using Tables 11.19 and 11.20, the syndromes are

$$\begin{aligned} s_1 &= \alpha\alpha^7 + \alpha^5\alpha^5 + \alpha^{11}\alpha^2 = \alpha^{12} \\ s_2 &= \alpha\alpha^{14} + \alpha^5\alpha^{10} + \alpha^{11}\alpha^4 = 1 \\ s_3 &= \alpha\alpha^{21} + \alpha^5\alpha^{15} + \alpha^{11}\alpha^6 = \alpha^{14} \\ s_4 &= \alpha\alpha^{28} + \alpha^5\alpha^{20} + \alpha^{11}\alpha^8 = \alpha^{13} \\ s_5 &= \alpha\alpha^{35} + \alpha^5\alpha^{25} + \alpha^{11}\alpha^{10} = 1 \\ s_6 &= \alpha\alpha^{42} + \alpha^5\alpha^{30} + \alpha^{11}\alpha^{12} = \alpha^{11} \end{aligned} \tag{11.236}$$

From (11.202) to (11.205), the algorithm develops according to the following $d - 1 = 6$ steps, starting from the initial conditions

$$\begin{aligned} \lambda^{(0)}(x) &= 1 \\ \beta^{(0)}(x) &= 1 \\ L_0 &= 0 \end{aligned} \tag{11.237}$$

Step 1 (r = 1): $\Delta_1 = \alpha^{12}$, and $2L_0 = 0 = r-1$, $\delta_1 = 1$, $L_1 = 1$.

$$\begin{bmatrix} \lambda^{(1)}(x) \\ \beta^{(1)}(x) \end{bmatrix} = \begin{bmatrix} 1 & -\alpha^{12}x \\ \alpha^3 & 0 \end{bmatrix} \begin{bmatrix} 1 \\ 1 \end{bmatrix} = \begin{matrix} 1+\alpha^{12}x \\ \alpha^3 \end{matrix}$$

Step 2 (r = 2): $\Delta_2 = 1+\alpha^9 = \alpha^7$, and $2L_1 = 2 > r-1$, $\delta_2 = 0$, $L_2 = L_1 = 1$.

$$\begin{aligned} \begin{bmatrix} \lambda^{(2)}(x) \\ \beta^{(2)}(x) \end{bmatrix} &= \begin{bmatrix} 1 & -\alpha^7 x \\ 0 & x \end{bmatrix} \begin{bmatrix} 1+\alpha^{12}x \\ \alpha^3 \end{bmatrix} \\ &= \begin{bmatrix} 1+\alpha^{12}x+\alpha^{10}x \\ \alpha^3 x \end{bmatrix} = \begin{bmatrix} 1+\alpha^3 x \\ \alpha^3 x \end{bmatrix} \end{aligned}$$

Step 3 (r = 3): $\Delta_3 = \alpha^{14}+\alpha^3 = 1$, and $2L_2 = 2 = r-1$, $\delta_3 = 1$, $L_3 = 3-1 = 2$.

$$\begin{bmatrix} \lambda^{(3)}(x) \\ \beta^{(3)}(x) \end{bmatrix} = \begin{bmatrix} 1 & -x \\ 1 & 0 \end{bmatrix} \begin{bmatrix} 1+\alpha^3 x \\ \alpha^3 x \end{bmatrix} = \begin{bmatrix} 1+\alpha^3 x+\alpha^3 x^2 \\ 1+\alpha^3 x \end{bmatrix}$$

Step 4 (r = 4): $\Delta_4 = \alpha^{13}+\alpha^3\alpha^{14}+\alpha^3 = 1$, and $2L_3 = 4 > r-1$, $\delta_4 = 0$, $L_4 = L_3 = 2$.

$$\begin{aligned} \begin{bmatrix} \lambda^{(4)}(x) \\ \beta^{(4)}(x) \end{bmatrix} &= \begin{bmatrix} 1 & -x \\ 0 & x \end{bmatrix} \begin{bmatrix} 1+\alpha^3 x+\alpha^3 x^2 \\ 1+\alpha^3 x \end{bmatrix} \\ &= \begin{bmatrix} 1+\alpha^3 x+\alpha^3 x^2+x+\alpha^3 x^2 \\ x+\alpha^3 x^2 \end{bmatrix} = \begin{bmatrix} 1+\alpha^{14}x \\ x+\alpha^3 x^2 \end{bmatrix} \end{aligned}$$

Step 5 (r = 5): $\Delta_5 = 1+\alpha^{14}\alpha^{13} = \alpha^{11}$, and $2L_4 = 4 = r-1$, $\delta_5 = 1$, $L_5 = 5-2 = 3$.

$$\begin{bmatrix} \lambda^{(5)}(x) \\ \beta^{(5)}(x) \end{bmatrix} = \begin{bmatrix} 1 & -\alpha^{11}x \\ \alpha^4 & 0 \end{bmatrix} \begin{bmatrix} 1+\alpha^{14}x \\ x+\alpha^3 x^2 \end{bmatrix} = \begin{bmatrix} 1+\alpha^{14}x+\alpha^{11}x^2+\alpha^{14}x^3 \\ \alpha^4+\alpha^3 x \end{bmatrix}$$

Step 6 (r = 6): $\Delta_6 = \alpha^{11}+\alpha^{14}+\alpha^{11}\alpha^{13}+\alpha^{14}\alpha^{14} = 0$, $\delta_6 = 0$, $L_6 = L_5 = 3$.

$$\begin{aligned} \begin{bmatrix} \lambda^{(6)}(x) \\ \beta^{(6)}(x) \end{bmatrix} &= \begin{bmatrix} 1 & 0 \\ 0 & x \end{bmatrix} \begin{bmatrix} 1+\alpha^{14}x+\alpha^{11}x^2+\alpha^{14}x^3 \\ \alpha^4+\alpha^3 x \end{bmatrix} \\ &= \begin{bmatrix} 1+\alpha^{14}x+\alpha^{11}x^2+\alpha^{14}x^3 \\ \alpha^4 x+\alpha^3 x^2 \end{bmatrix} \end{aligned}$$

The error indicator polynomial is $\lambda(x) = \lambda^{(6)}(x)$. By using the exhaustive method to find the three roots, we obtain

$$\lambda(x) = 1+\alpha^{14}x+\alpha^{11}x^2+\alpha^{14}x^3 = (1-\alpha^7 x)(1-\alpha^5 x)(1-\alpha^2 x) \tag{11.238}$$

Consequently, the three errors are at positions $\ell = 2, 5$, and 7.

To determine the values of the errors, we use the Forney algorithm. The derivative of $\lambda(x)$ is given by

$$\lambda'(x) = \alpha^{14}+\alpha^{14}x^2 \tag{11.239}$$

The error evaluator polynomial is given by

$$\begin{aligned} \omega(x) &= (\alpha^{12}x+x^2+\alpha^{14}x^3+\alpha^{13}x^4+x^5+\alpha^{11}x^6) \\ &\quad (1+\alpha^{14}x+\alpha^{11}x^2+\alpha^{14}x^3) \bmod x^6 \\ &= \alpha^{12}x+\alpha^{12}x^2+\alpha^8 x^3 \end{aligned} \tag{11.240}$$

Thus, the values of the errors are

$$\begin{aligned}\varepsilon_2 &= -\frac{\omega(\alpha^{-2})}{\alpha^{-2}\lambda'(\alpha^{-2})} = \alpha^{11}\\ \varepsilon_5 &= -\frac{\omega(\alpha^{-5})}{\alpha^{-5}\lambda'(\alpha^{-5})} = \alpha^{5}\\ \varepsilon_7 &= -\frac{\omega(\alpha^{-7})}{\alpha^{-7}\lambda'(\alpha^{-7})} = \alpha\end{aligned} \tag{11.241}$$

An alternative approach for the encoding and decoding of Reed–Solomon codes utilizes the concept of Fourier transform on a Galois field [2, 5]. Let α be a primitive element of the field GF(q). The Fourier transform on the field GF(q) (GFFT) of a vector $\boldsymbol{c} = (c_0, c_1, \dots, c_{n-1})$ of n bits is defined as $(C_0, C_1, \dots, C_{n-1})$, where

$$C_j = \sum_{i=0}^{n-1} c_i\, \alpha^{ij}, \qquad j = 0, \dots, n-1 \tag{11.242}$$

Let us consider a code word $\boldsymbol{c}$ of n bits in the *time domain* from a Reed–Solomon cyclic code that corrects up to t errors; then $\boldsymbol{c}$ corresponds to a code polynomial that has as roots $2t = d - 1$ consecutive powers of α. If we take the GFFT of this word, we find that in the 'frequency domain', the transform has $2t$ consecutive components equal to zero. Indeed, from (11.176), specialized to Reed–Solomon codes, and from (11.242), we can show that the two conditions are equivalent, that is a polynomial has $2t$ consecutive powers of α as roots if and only if the transform has $2t$ consecutive components equal to zero. The approach that resorts to the GFFT is, therefore, dual of the approach that uses the generator polynomial. This observation leads to the development of efficient methods for encoding and decoding.

Reed–Solomon codes have found many applications since their original design. Even nowadays, they are used in many fields, including magnetic storage [6–8].

11.2.6 Performance of block codes

In this section, we consider the probability of error in the decoding of block codes, in the case of decoding with *hard* or *soft* input (see Section 6.6 on page 270). For an in-depth study of the subject, we refer the reader, for example, to [9].

With reference to Figure 6.1, let P_{bit} be the bit error probability for the detection of the bits of the binary sequence $\{\tilde{c}_m\}$, or bit error probability of the channel, P_w the error probability for a code word, and $P_{bit}^{(dec)}$ the error probability for a bit of the binary sequence $\{\hat{b}_l\}$ obtained after decoding. For a (n, k) block code with $t = (d_{min}^H - 1)/2$ and hard input decoding, the following inequality holds:

$$P_w \leq \sum_{i=t+1}^{n} \binom{n}{i} P_{bit}^i (1 - P_{bit})^{n-i} \tag{11.243}$$

which, under the condition $nP_{bit} \ll 1$, can be approximated as

$$P_w \simeq \binom{n}{t+1} P_{bit}^{t+1} (1 - P_{bit})^{n-t-1} \tag{11.244}$$

The inequality (11.243) follows from the channel model (11.12) assuming errors that are i.i.d., and from the consideration that the code may not be perfect (see page 525), and therefore, it could correct also some received words with more than t errors.

If a word error occurs, the most probable event is that the decoder decides for a code word with distance $d^H_{min} = 2t + 1$ from the transmitted code word, thus making d^H_{min} bit errors in the sequence $\{\hat{c}_m\}$. As $\boldsymbol{c}$ is formed of n bits, we have that at the decoder output, the bit error probability is

$$P^{(dec)}_{bit} \simeq \frac{2t+1}{n} P_w \tag{11.245}$$

Example 11.2.34
For a (5,1) repetition code with $d^H_{min} = 5$ (see page 525), decoding with hard input yields

$$P_w = \binom{5}{3} P^3_{bit}(1 - P_{bit})^2 + \binom{5}{4} P^4_{bit}(1 - P_{bit}) + P^5_{bit} \tag{11.246}$$

Example 11.2.35
For an (n, k) Hamming code with $d^H_{min} = 3$ (see page 525), (11.243) yields

$$\begin{aligned} P_w &\leq \sum_{i=2}^{n} \binom{n}{i} P^i_{bit}(1 - P_{bit})^{n-i} \\ &= 1 - [(1 - P_{bit})^n + nP_{bit}(1 - P_{bit})^{n-1}] \end{aligned} \tag{11.247}$$

For example, for a (15,11) code, if $P_{bit} = 10^{-3}$, then $P_w \simeq 10^{-4}$, and from (11.245), we get $P^{(dec)}_{bit} \simeq 2\ 10^{-5}$.

The decoders that have been considered so far are classified as hard input decoders, as the demodulator output is quantized to the values of the coded symbols before decoding. In general, other decoding algorithms with soft input may be considered, that directly process the demodulated signal, and consequently, the decoder input is real valued (see Section 11.3.2).

In the case of antipodal binary signals and soft input decoding, we get

$$P_w \simeq (2^k - 1)\, Q\left(\sqrt{\frac{R_c\, d^H_{min}\, 2E_b}{N_0}}\right) \tag{11.248}$$

11.3 Convolutional codes

Convolutional codes are a subclass of the class of *tree codes*, so named because their code words are conveniently represented as sequences of nodes in a tree. Tree codes are of great interest because decoding algorithms have been found that are easy to implement and can be applied to the entire class of tree codes, in contrast to decoding algorithms for block codes, each designed for a specific class of codes, as for example BCH codes.

Several approaches have been used in the literature for describing convolutional codes; here we will illustrate these approaches by first considering a specific example.

Example 11.3.1
Consider a rate 1/2 binary convolutional code, obtained by the encoder illustrated in Figure 11.9a. For each bit b_k that enters the encoder, two output bits, $c^{(1)}_k$ and $c^{(2)}_k$, are transmitted. The first output $c^{(1)}_k$ is obtained if the switch at the output is in the upper position, and the second output $c^{(2)}_k$ is obtained if the switch is in the lower position; the two previous input bits, b_{k-1} and b_{k-2}, are stored in the memory of the encoder. As the information bit is not presented directly to one of the outputs, we say that the code is non-systematic. The two coded bits are generated as linear combinations of the

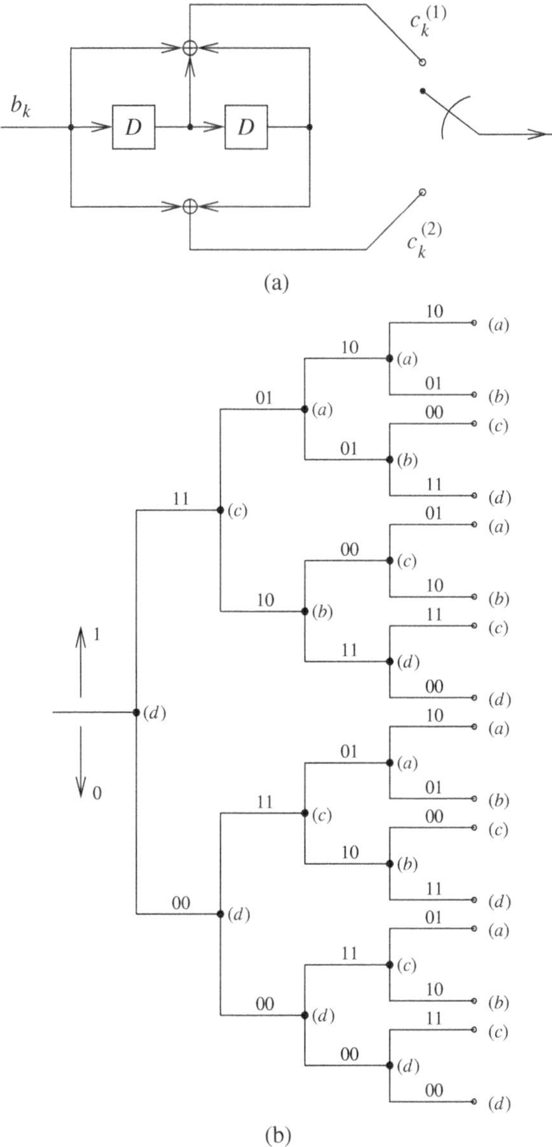

Figure 11.9 (a) Encoder and (b) tree diagram for the convolutional code of Example 11.3.1.

bits of the message; denoting the input sequence as $\dots, b_0, b_1, b_2, b_3, \dots$, and the output sequence as $\dots, c_0^{(1)}, c_0^{(2)}, c_1^{(1)}, c_1^{(2)}, c_2^{(1)}, c_2^{(2)}, c_3^{(1)}, c_3^{(2)}, \dots$, then the following relations hold:

$$
\begin{aligned}
c_k^{(1)} &= b_k \oplus b_{k-1} \oplus b_{k-2} \\
c_k^{(2)} &= b_k \oplus b_{k-2}
\end{aligned}
\tag{11.249}
$$

A convolutional code may be described in terms of a tree, trellis, or state diagram; for the code defined by (11.249), these descriptions are illustrated in Figures 11.9b and 11.10a,b, respectively.

With reference to the tree diagram of Figure 11.9b, we begin at the left (root) node and proceed to the right by choosing an upper path if the input bit is equal to 1 and a lower path if the input bit is 0. We

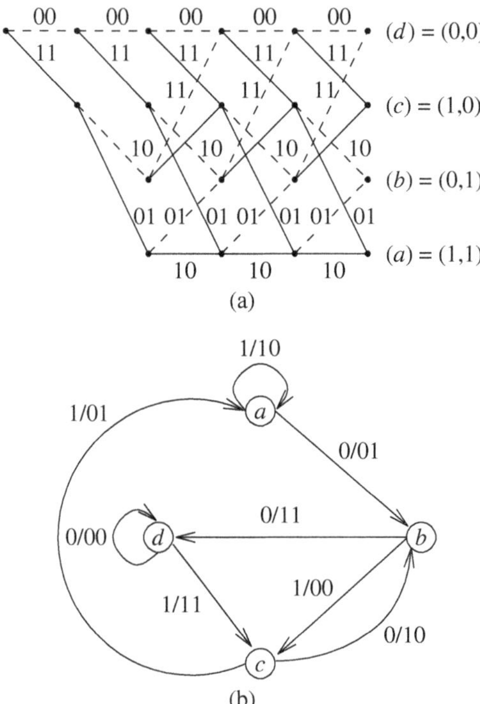

Figure 11.10 (a) Trellis diagram and (b) state diagram for the convolutional code of Example 11.3.1.

output the two bits represented by the label on the branch that takes us to the next node and then repeat this process at the next node. The nodes or states of the encoder are labeled with the letters *a*, *b*, *c*, and *d*, which indicate the relation with the four possible values assumed by the two bits stored in the encoder, according to the table:

$$\begin{array}{ccc} b_{k-1} & b_{k-2} & \text{label} \\ 0 & 0 & d \\ 1 & 0 & c \\ 0 & 1 & b \\ 1 & 1 & a \end{array} \tag{11.250}$$

If for example we input the sequence $b_0, b_1, b_2, b_3, \ldots = 1\ 1\ 0\ 1\ \ldots$, we would, then output the sequence $c_0^{(1)}, c_0^{(2)}, c_1^{(1)}, c_1^{(2)}, c_2^{(1)}, c_2^{(2)}, c_3^{(1)}, c_3^{(2)}, \ldots = 1\ 1\ 0\ 1\ 0\ 1\ 0\ 0\ \ldots$. As, at any depth in the tree, nodes with the same label will have the same tree growing from them, we can superimpose these nodes on a single node. This results in the trellis diagram represented in Figure 11.10a, where solid and dashed lines correspond to transitions determined by input bits equal to 1 and 0, respectively.

The state diagram for the encoder is illustrated in Figure 11.10b. The four states (a, b, c, d) correspond to the four possible combinations of bits stored in the encoder. If the encoder is in a certain state, a transition to one of two possible states occurs, depending on the value of the input bit. Possible transitions between states are represented as arcs, on which an arrow indicates the direction of the transition; with each arc is associated a label that indicates the value assumed by the input bit and also the value of the resulting output bits. The description of the encoder by the state diagram is convenient for analysing the properties of the code, as we will see later.

It is also convenient to represent code sequences in terms of the D transform, as

$$\begin{aligned} b(D) &= b_0 + b_1 D + b_2\, D^2 + b_3\, D^3 + \cdots \\ c^{(1)}(D) &= c_0^{(1)} + c_1^{(1)} D + c_2^{(1)}\, D^2 + c_3^{(1)}\, D^3 + \cdots = g^{(1,1)}(D)\, b(D) \\ c^{(2)}(D) &= c_0^{(2)} + c_1^{(2)} D + c_2^{(2)}\, D^2 + c_3^{(2)}\, D^3 + \cdots = g^{(2,1)}(D)\, b(D) \end{aligned} \tag{11.251}$$

where $g^{(1,1)}(D) = 1 + D + D^2$, and $g^{(2,1)}(D) = 1 + D^2$.

11.3.1 General description of convolutional codes

In general, we consider convolutional codes with symbols from GF(q); assuming the encoder produces n_0 output code symbols for every k_0 input message symbols, the code rate is equal to k_0/n_0. It is convenient to think of the message sequence as being the interlaced version of k_0 different message sequences, and to think of the code sequence as the interlaced version of n_0 different code sequences. In other words, given the information sequence $\{b_\ell\}$, we form the k_0 subsequences

$$b_k^{(i)} = b_{kk_0+i-1}, \qquad i = 1, \ldots, k_0 \tag{11.252}$$

that have D transform defined as

$$\begin{aligned} b_0^{(1)} b_1^{(1)} b_2^{(1)} \cdots &\Longleftrightarrow b^{(1)}(D) = b_0^{(1)} + b_1^{(1)} D + b_2^{(1)} D^2 + \cdots \\ b_0^{(2)} b_1^{(2)} b_2^{(2)} \cdots &\Longleftrightarrow b^{(2)}(D) = b_0^{(2)} + b_1^{(2)} D + b_2^{(2)} D^2 + \cdots \\ \vdots \qquad & \qquad\qquad \vdots \\ b_0^{(k_0)} b_1^{(k_0)} b_2^{(k_0)} \cdots &\Longleftrightarrow b^{(k_0)}(D) = b_0^{(k_0)} + b_1^{(k_0)} D + b_2^{(k_0)} D^2 + \cdots \end{aligned} \tag{11.253}$$

Let $c^{(1)}(D), c^{(2)}(D), \ldots, c^{(n_0)}(D)$ be the D transforms of the n_0 output sequences; then

$$c^{(j)}(D) = \sum_{i=1}^{k_0} g^{(j,i)}(D)\, b^{(i)}(D), \qquad j = 1, 2, \ldots, n_0 \tag{11.254}$$

An (n_0, k_0) convolutional code is then specified by giving the coefficients of all the polynomials $g^{(j,i)}(D)$, $i = 1, 2, \ldots, k_0$, $j = 1, 2, \ldots, n_0$.

If for all $j = 1, 2, \ldots, k_0$, we have

$$g^{(j,i)}(D) = \begin{cases} 1 & j = i \\ 0 & j \neq i \end{cases} \tag{11.255}$$

then the code is systematic and k_0 of the n_0 output sequences are just the message sequences.

An encoder for a convolutional code needs storage elements. Let ν be the *constraint length* of the code,[5]

$$\nu = \max_{j,i}(\deg g^{(j,i)}(D)) \tag{11.256}$$

Therefore, the encoder of a convolutional code must store ν previous blocks of k_0 message symbols to form a block of n_0 output symbols.

The general structure of an encoder for a code with $k_0 = 1$ and $n_0 = 2$ is illustrated in Figure 11.11; for such an encoder, νk_0 storage elements are necessary. If the code is systematic, then the encoder can be implemented with $\nu(n_0 - k_0)$ storage elements, as illustrated in Figure 11.12 for $k_0 = 2$ and $n_0 = 3$.

If we interpret the sequence $\{c_k\}$ as the output of a sequential finite-state machine (see Appendix 7.B), at instant k the trellis of a non-systematic code is defined by the three signals:

1. Input

$$[b_k^{(1)}, b_k^{(2)}, \ldots, b_k^{(k_0)}] \tag{11.257}$$

[5] Many authors define the constraint length as $\nu + 1$, where ν if given by (11.256).

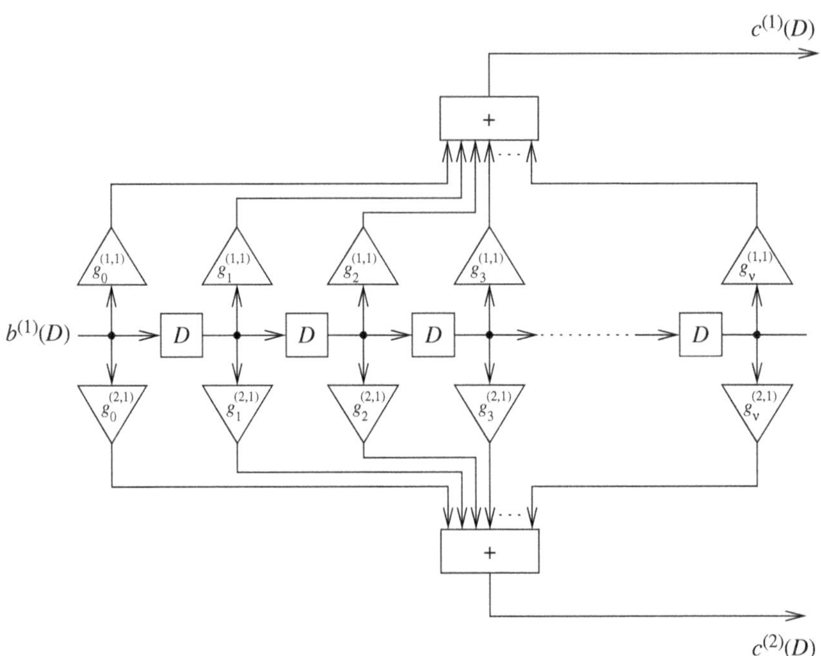

Figure 11.11 Block diagram of an encoder for a convolutional code with $k_0 = 1$, $n_0 = 2$, and constraint length ν.

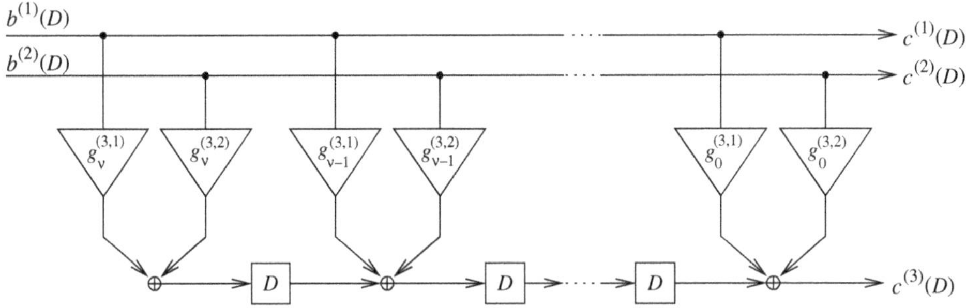

Figure 11.12 Block diagram of an encoder for a systematic convolutional code with $k_0 = 2$, $n_0 = 3$, and constraint length ν.

2. State

$$[b_k^{(1)}, \dots, b_k^{(k_0)}, \dots, b_{k-(\nu-1)}^{(1)}, \dots, b_{k-(\nu-1)}^{(k_0)}] \tag{11.258}$$

3. Output

$$[c_k^{(1)}, c_k^{(2)}, \dots, c_k^{(n_0)}] \tag{11.259}$$

where $c_k^{(j)}, j = 1, \dots, n_0$, is given by (11.254). Then there are $q^{k_0\nu}$ states in the trellis. There are q^{k_0} branches departing from each state and q^{k_0} branches merging into a state. The output vector consists of n_0 q-ary symbols.

Parity check matrix

A semi-infinite parity check matrix can be defined in general for convolutional codes; however, we note that it is only in the case of systematic codes that we can easily express the elements of this matrix in terms of the coefficients of the generator polynomials $g^{(j,i)}(D)$. We write the coefficients of the generator polynomials in the form

$$\begin{array}{cccc} g_0^{(1,1)} g_0^{(2,1)} \cdots g_0^{(n_0,1)} & g_1^{(1,1)} \cdots g_1^{(n_0,1)} & \cdots & g_\nu^{(1,1)} g_\nu^{(2,1)} \cdots g_\nu^{(n_0,1)} \\ g_0^{(1,2)} g_0^{(2,2)} \cdots g_0^{(n_0,2)} & g_1^{(1,2)} \cdots g_1^{(n_0,2)} & \cdots & g_\nu^{(1,2)} g_\nu^{(2,2)} \cdots g_\nu^{(n_0,2)} \\ \vdots & & & \vdots \\ g_0^{(1,k_0)} g_0^{(2,k_0)} \cdots g_0^{(n_0,k_0)} & g_1^{(1,k_0)} \cdots g_1^{(n_0,k_0)} & \cdots & g_\nu^{(1,k_0)} g_\nu^{(2,k_0)} \cdots g_\nu^{(n_0,k_0)} \end{array} \tag{11.260}$$

If the code is systematic, the parity matrix of the generator polynomials can be written as

$$\boldsymbol{I}\,\boldsymbol{P}_0\;\boldsymbol{0}\,\boldsymbol{P}_1\;\cdots\;\boldsymbol{0}\,\boldsymbol{P}_\nu \tag{11.261}$$

where $\boldsymbol{I}$ and $\boldsymbol{0}$ are $k_0 \times k_0$ matrices and $\boldsymbol{P}_i$, $i = 0, \ldots, \nu$, are $k_0 \times (n_0 - k_0)$ matrices.

The semi-infinite parity check matrix is then

$$\boldsymbol{H}_\infty = \begin{bmatrix} -\boldsymbol{P}_0^T & \boldsymbol{I} & \boldsymbol{0} & \boldsymbol{0} & \boldsymbol{0} & \boldsymbol{0} \ldots \\ -\boldsymbol{P}_1^T & \boldsymbol{0} & -\boldsymbol{P}_0^T & \boldsymbol{I} & \boldsymbol{0} & \boldsymbol{0} \ldots \\ \ldots & & \ldots & & \ldots & \ldots \\ -\boldsymbol{P}_\nu^T & \boldsymbol{0} & -\boldsymbol{P}_{\nu-1}^T & \boldsymbol{0} & -\boldsymbol{P}_{\nu-2}^T & \boldsymbol{0} \ldots \\ \boldsymbol{0} & \boldsymbol{0} & -\boldsymbol{P}_\nu^T & \boldsymbol{0} & -\boldsymbol{P}_{\nu-1}^T & \boldsymbol{0} \ldots \\ \ldots & & \ldots & & \ldots & \ldots \end{bmatrix} \tag{11.262}$$

Thus, for any code word $\boldsymbol{c}$ of infinite length, $\boldsymbol{H}_\infty \boldsymbol{c} = \boldsymbol{0}$. Often, rather than considering the semi-infinite matrix $\boldsymbol{H}_\infty$, we consider the finite matrix $\boldsymbol{H}$ defined as

$$\boldsymbol{H} = \begin{bmatrix} -\boldsymbol{P}_0^T & \boldsymbol{I} & \boldsymbol{0} & \boldsymbol{0} \ldots & \boldsymbol{0} & \boldsymbol{0} \\ -\boldsymbol{P}_1^T & \boldsymbol{0} & -\boldsymbol{P}_0^T & \boldsymbol{I} \ldots & \boldsymbol{0} & \boldsymbol{0} \\ \ldots & & \ldots & \ldots & \ldots & \\ -\boldsymbol{P}_\nu^T & \boldsymbol{0} & -\boldsymbol{P}_{\nu-1}^T & \boldsymbol{0} \ldots & -\boldsymbol{P}_0^T & \boldsymbol{I} \end{bmatrix} \tag{11.263}$$

The bottom row of matrices of the matrix $\boldsymbol{H}$ is called the *basic parity check matrix*. From it, we can see that the parity symbols in a block are given by the linear combination of information bits in that block, corresponding to non-zero terms in $-\boldsymbol{P}_0^T$, in the immediately preceding block, corresponding to non-zero terms in $-\boldsymbol{P}_1^T$, and so on until the νth preceding block, corresponding to non-zero terms in $-\boldsymbol{P}_\nu^T$.

Generator matrix

From (11.260), we introduce the matrices

$$\boldsymbol{g}_i = \begin{bmatrix} g_i^{(1,1)} & g_i^{(2,1)} & \ldots & g_i^{(n_0,1)} \\ \vdots & & & \vdots \\ g_i^{(1,k_0)} & g_i^{(2,k_0)} & \ldots & g_i^{(n_0,k_0)} \end{bmatrix}, \qquad i = 0, \ldots, \nu \tag{11.264}$$

Hence, the generator matrix is of the form

$$\boldsymbol{G}_\infty = \begin{bmatrix} \boldsymbol{g}_0 & \boldsymbol{g}_1 & \ldots & \boldsymbol{g}_\nu & \boldsymbol{0} & \ldots \\ \boldsymbol{0} & \boldsymbol{g}_0 & \ldots & \boldsymbol{g}_{\nu-1} & \boldsymbol{g}_\nu & \ldots \\ \ldots & \ldots & \ldots & \ldots & \ldots & \ldots \end{bmatrix} \tag{11.265}$$

Some examples of convolutional codes with the corresponding encoders and generator matrices are illustrated in Figure 11.13.

Transfer function

An important parameter of a convolutional code is d^H_{free}, that determines the performance of the code (see Section 11.3.3)

Definition 11.14

Let $\boldsymbol{e}(D) = [e^{(n_0)}(D), \dots, e^{(1)}(D)]$ be any error sequence between two code words $\boldsymbol{c}_1(D) = [c_1^{(n_0)}(D), \dots, c_1^{(1)}(D)]$ and $\boldsymbol{c}_2(D) = [c_2^{(n_0)}(D), \dots, c_2^{(1)}(D)]$, that is $\boldsymbol{c}_1(D) = \boldsymbol{c}_2(D) + \boldsymbol{e}(D)$, and $\boldsymbol{e}_k = [e_k^{(n_0)}, \dots, e_k^{(1)}]$ denotes the kth element of the sequence. We define the free Hamming distance of the code as

$$d^H_{free} = \min_{\boldsymbol{e}(D)} \sum_{k=0}^{\infty} w(\boldsymbol{e}_k) \tag{11.266}$$

where w is introduced in Definition 11.4 on page 519. As the code is linear, d^H_{free} corresponds to the minimum number of symbols different from zero in a non-zero code word. □

Next, we consider a method to compute the weights of all code words in a convolutional code; to illustrate the method, we examine the simple binary encoder of Figure 11.9a.

We begin by reproducing the trellis diagram of the code in Figure 11.14, where each path is now labeled with the weight of the output bits corresponding to that path. We consider all paths that diverge from state (d) and return to state (d) for the first time after a number of steps j. By inspection, we find one such path of weight 5 returns to state (d) after 3 steps; moreover, we find two distinct paths of weight 6, one that returns to state (d) after 4 steps and another after 5 steps. Hence, we find that this code has $d^H_{free} = 5$.

We now look for a method that enables us to find the weights of all code words as well as the lengths of the paths that give origin to the code words with these weights. Consider the state diagram for this code, redrawn in Figure 11.15 with branches labeled as D^2, D, or $D^0 = 1$, where the exponent corresponds to the weight of the output bits corresponding to that branch. Next, we split node (0,0) to obtain the state diagram of Figure 11.16, and we compute a *generating function* for the weights. The generating function is the transfer function of a signal flow graph with unit input. From Figure 11.16, we obtain this transfer function by solving the system of equations

$$\begin{aligned} \beta &= D^2\alpha + 1\gamma \\ \gamma &= D\beta + D\delta \\ \delta &= D\beta + D\delta \\ \eta &= D^2\gamma \end{aligned} \tag{11.267}$$

Then, we get

$$t(D) = \frac{\eta}{\alpha} = \frac{D^5}{1-2D} = D^5 + 2D^6 + 4D^7 + \cdots + 2^i D^{i+5} + \dots \tag{11.268}$$

From inspection of $t(D)$, we find there is one code word of weight 5, two of weight 6, four of weight 7, ... Equation (11.268) holds for code words of infinite length.

If we want to find code words that return to state (d) after j steps, we refer to the state diagram of Figure 11.17. The term L introduced in the label on each branch keeps track of the length of the sequence, as the exponent of L is augmented by 1 every time a transition occurs. Furthermore, we introduce the term I in the label on a branch if the corresponding transition is due to an information bit equal to 1; for each path on the trellis diagram, we can compute the corresponding number of information bits equal

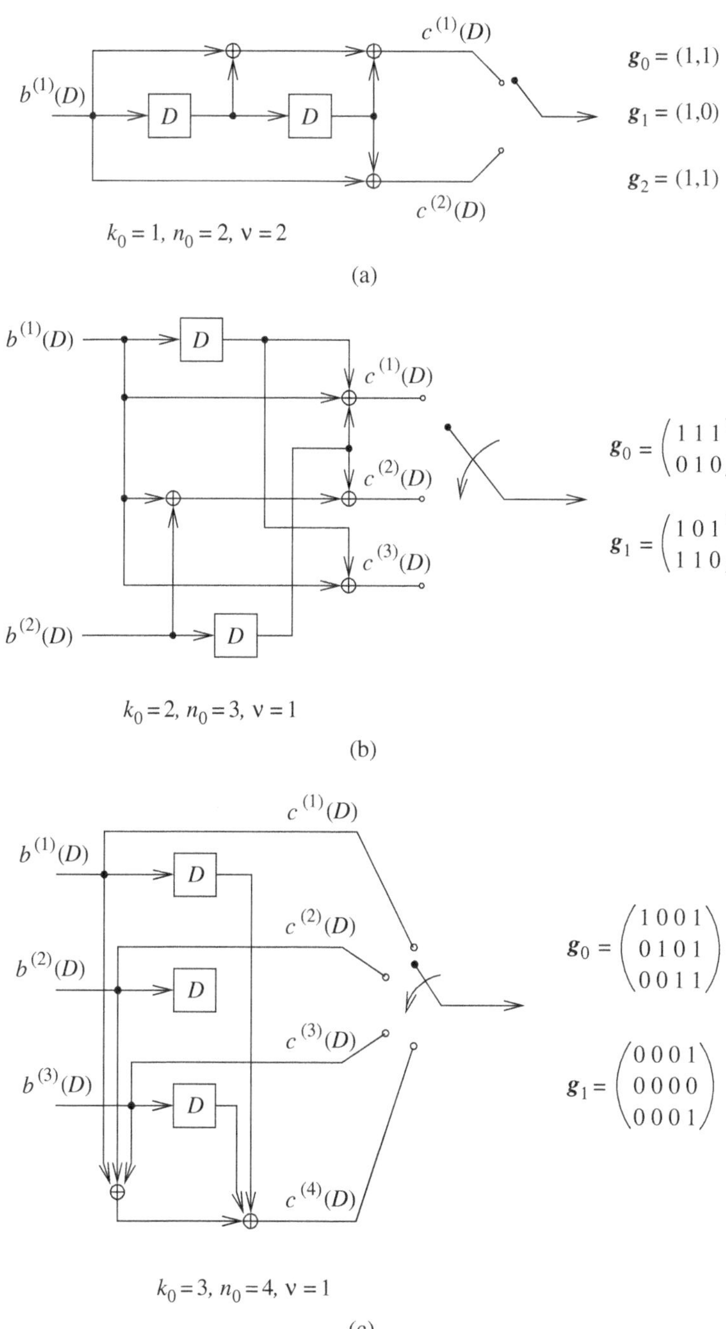

Figure 11.13 Examples of encoders for three convolutional codes.

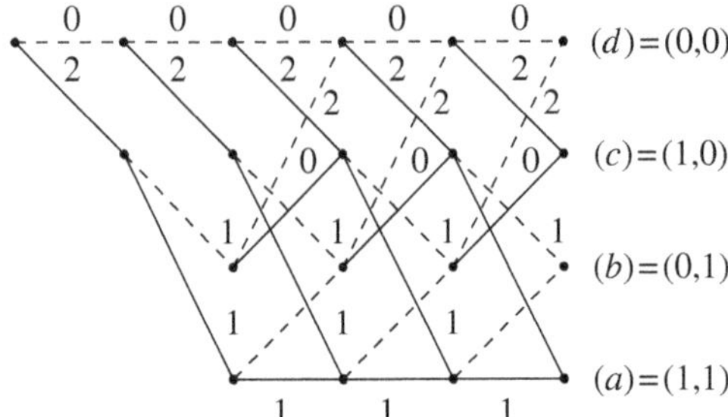

Figure 11.14 Trellis diagram of the code of Example 11.3.1; the labels represent the Hamming weight of the output bits.

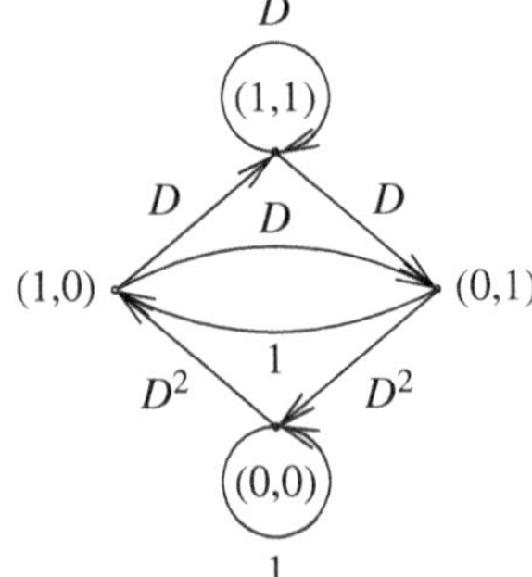

Figure 11.15 State diagram of the code of Example 11.3.1; the labels represent the Hamming weight of the generated bits.

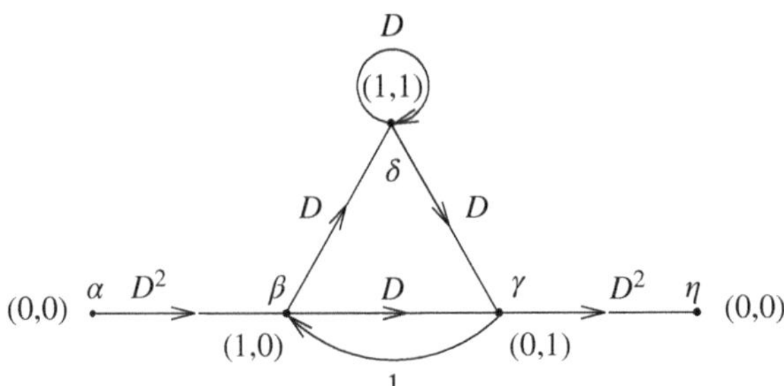

Figure 11.16 State diagram of the code of Example 11.3.1; node (0,0) is split to compute the transfer function of the code.

to 1. The augmented transfer function is given by

$$\begin{aligned} t(D,L,I) &= \frac{D^5L^3I}{1-DL(1+L)I} \\ &= D^5L^3I + D^6L^4(1+L)I^2 + D^7L^5(1+L)^2I^3 + \cdots \\ &\quad + D^{5+i}L^{3+i}(1+L)^iI^{1+i} + \cdots \end{aligned} \tag{11.269}$$

Thus, we see that the code word of weight 5 is of length 3 and is originated by a sequence of information bits that contains one bit equal to 1, there are two code words of weight 6, one of length 4 and the

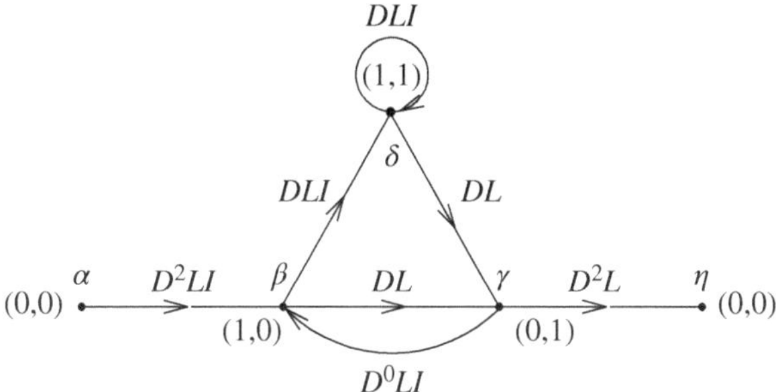

Figure 11.17 State diagram of the code of Example 36; node (0,0) is split to compute the augmented transfer function.

other of length 5, both of which are originated by a sequence of information bits that contain two bits equal to 1, ...

Catastrophic error propagation

For certain codes, a finite number of channel errors may lead to an infinite number of errors in the sequence of decoded bits. For example, consider the code with encoder and state diagram illustrated in Figure 11.18a,b, respectively. Note that in the state diagram, the self loop at state (1,1) does not increment the weight of the code word so that a code word corresponding to a path passing through the states (0,0),(1,0),(1,1),(1,1), ...,(1,1),(0,1),(0,0) is of weight 6, independently of the number of times it passes through the self-loop at state (1,1). In other words, long sequences of coded bits equal to zero may be obtained by remaining in the state $(0, 0)$ with a sequence of information bits equal to zero, or by

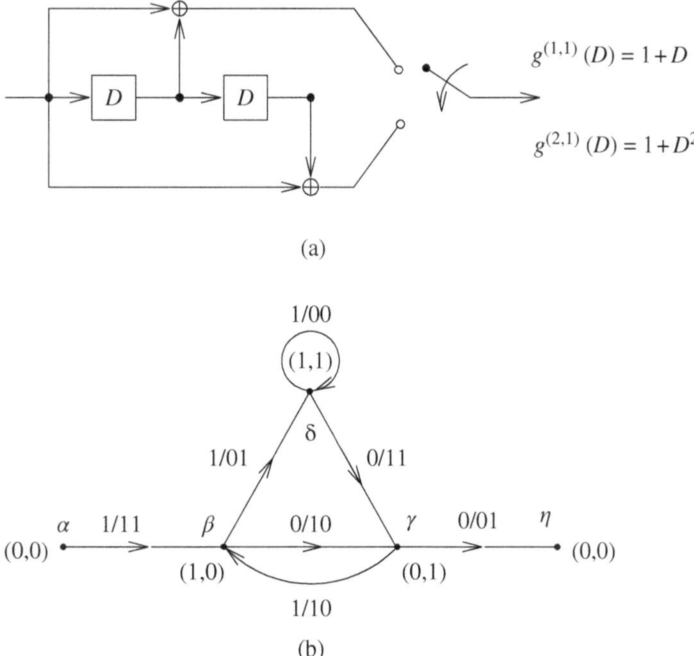

Figure 11.18 (a) Encoder and (b) state diagram for a catastrophic convolutional code.

remaining in the state (1, 1) with a sequence of information bits equal to one. Therefore, a limited number of channel errors, in this case 6, can cause a large number of errors in the sequence of decoded bits.

Definition 11.15
A convolutional code is *catastrophic* if there exists a closed loop in the state diagram that has all branches with zero weight. □

For codes with rate $1/n_0$, it has been shown that a code is catastrophic if and only if all generator polynomials have a common polynomial factor. In the above example, the common factor is $1 + D$. This can be proven using the following argument: suppose that $g^{(1,1)}(D), g^{(2,1)}(D), \dots, g^{(k_0,1)}(D)$ all have the common factor $g_c(D)$, so that

$$g^{(i,1)}(D) = g_c(D)\ \tilde{g}^{(i,1)}(D) \tag{11.270}$$

Suppose the all zero sequence is sent, $b^{(1)}(D) = 0$ and that the finite error sequence $\tilde{g}^{(i,1)}(D)$, equal to that defined in (11.270), occurs in the ith subsequence output, for $i = 1, 2, \dots, n_0$, as illustrated in Figure 11.19a. The same output sequence is obtained if the sequence of information bits with infinite length $b^{(1)}(D) = 1/g_c(D)$ is sent, and no channel errors occur, as illustrated in Figure 11.19b. Thus, a finite number of errors yields a decoded sequence of information bits that differ from the transmitted sequence in an infinite number of positions.

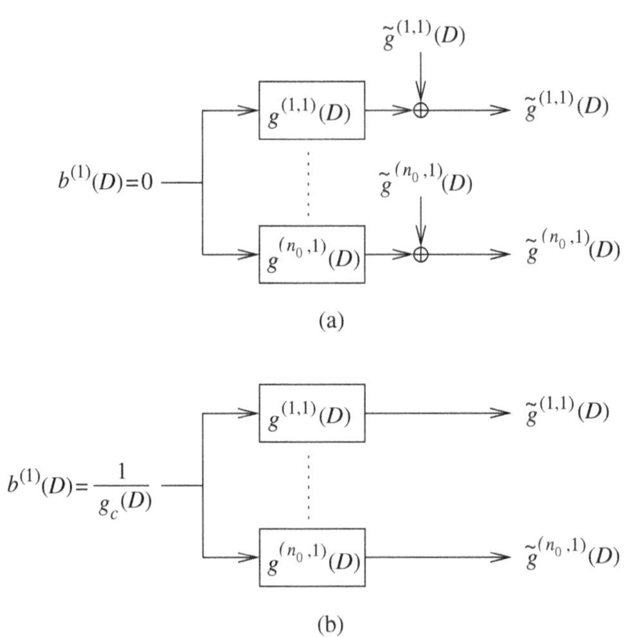

Figure 11.19 Two distinct infinite sequences of information bits that produce the same output sequence with a finite number of errors.

11.3.2 Decoding of convolutional codes

Various algorithms have been developed for the decoding of convolutional codes. One of the first decoding methods was algebraic decoding, which is similar to the methods developed for the decoding of block codes. However, this method has the disadvantages that it is applicable only to a limited number of codes having particular characteristics and exhibits performance that is lower as compared to decoding methods based on the observation of the whole received sequence. The latter methods, also called *probabilistic decoding methods*, include the Viterbi algorithm (VA), the sequential decoding algorithm by Fano [9], and the forward–backward algorithm (FBA) by Bahl–Cocke–Jelinek–Raviv (BCJR).

Before illustrating the various decoding methods, we consider an important function.

Interleaving

The majority of block codes as well as convolutional codes is designed by assuming that the errors introduced by the noisy channel are statistically independent. This assumption is not always true in practice. To make the channel errors, at least approximately, statistically independent it is customary to resort to an interleaver, as described in Section 6.1.

Two decoding models

We consider a binary convolutional code with $k_0 = 1$, $n_0 = 2$, and constraint length ν. In general, from (11.254), we write the code sequence as a function of the message sequence as

$$\begin{aligned} c_k^{(1)} &= g^{(1,1)}(b_k, \dots, b_{k-\nu}) \\ c_k^{(2)} &= g^{(2,1)}(b_k, \dots, b_{k-\nu}) \end{aligned} \tag{11.271}$$

At the receiver, two models may be adopted.

Model with hard input. With reference to the transmission system of Figure 6.1, we consider the sequence at the output of the binary channel. In this case, the demodulator has already detected the transmitted symbols, for example by a threshold detector, and the inverse bit mapper provides the binary sequence $\{z_m = \tilde{c}_m\}$ to the decoder, from which we obtain the interlaced binary sequences

$$\begin{aligned} z_k^{(1)} &= z_{2k} = c_k^{(1)} \oplus e_k^{(1)} \\ z_k^{(2)} &= z_{2k+1} = c_k^{(2)} \oplus e_k^{(2)} \end{aligned} \tag{11.272}$$

where the errors $e_k^{(i)} \in \{0, 1\}$, for a memoryless binary symmetric channel, are i.i.d. (see (6.73)).

From the description on page 579, introducing the state of the encoder at instant k as the vector with ν elements

$$\boldsymbol{s}_k = [b_k, \dots, b_{k-(\nu-1)}] \tag{11.273}$$

the desired sequence in (11.272), that coincides with the encoder output, can be written as (see (11.271) and (11.273))

$$\begin{aligned} c_k^{(1)} &= f^{(1)}(\boldsymbol{s}_k, \boldsymbol{s}_{k-1}) \\ c_k^{(2)} &= f^{(2)}(\boldsymbol{s}_k, \boldsymbol{s}_{k-1}) \end{aligned} \tag{11.274}$$

Model with soft input. Again with reference to Figure 6.1, at the decision point of the receiver, the signal can be written as (see (7.334))

$$z_k = u_k + w_k \tag{11.275}$$

where we assume w_k is white Gaussian noise with variance $\sigma_w^2 = 2\sigma_I^2$, and u_k is given by (7.335)

$$u_k = \sum_{n=-L_1}^{L_2} \eta_n a_{k-n} \tag{11.276}$$

where $\{a_k\}$ is the sequence of symbols at the output of the bit mapper. Note that in (11.276), the symbols $\{a_k\}$ are in general not independent, as the input of the bit mapper is a code sequence according to the law (11.274).

The relation between u_k and the bits $\{b_\ell\}$ depends on the intersymbol interference in (11.276), the type of bit mapper and the encoder (11.271). We consider the case of absence of inter-symbol interference (ISI), that is (see also (6.13))

$$u_k = a_k \tag{11.277}$$

and a 16-pulse-amplitude-modulation (PAM) system where, without interleaving, four consecutive code bits $c_{2k}^{(1)}, c_{2k}^{(2)}, c_{2k-1}^{(1)}, c_{2k-1}^{(2)}$ are mapped into a symbol of the constellation. For an encoder with constraint length v, we have

$$\begin{aligned} u_k &= \tilde{f}(a_k) = \tilde{f}[\text{BMAP}\,\{[c_{2k}^{(1)}, c_{2k}^{(2)}, c_{2k-1}^{(1)}, c_{2k-1}^{(2)}]\}] \\ &= \tilde{f}[\text{BMAP}\,\{g^{(1,1)}\,(b_{2k}, \dots, b_{2k-v}), g^{(2,1)}\,(b_{2k}, \dots, b_{2k-v}), \\ &\quad g^{(1,1)}\,(b_{2k-1}, \dots, b_{2k-1-v}), g^{(2,1)}(b_{2k-1}, \dots, b_{2k-1-v})\}] \end{aligned} \tag{11.278}$$

In other words, let

$$\boldsymbol{s}_k = [b_{2k}, \dots, b_{2k-v+1}] \tag{11.279}$$

we can write

$$u_k = f(\boldsymbol{s}_k, \boldsymbol{s}_{k-1}) \tag{11.280}$$

We observe that in this example each state of the trellis admits four possible transitions. As we will see in Chapter 12, better performance is obtained by jointly optimizing the encoder and the bit mapper.

Decoding by the Viterbi algorithm

The VA, described in Section 7.5.1, is a probabilistic decoding method that implements the maximum-likelihood (ML) criterion, which minimizes the probability of detecting a sequence that is different from the transmitted sequence.

VA with hard input. The trellis diagram is obtained by using the definition (11.273), and the branch metric is the Hamming distance between $\boldsymbol{z}_k = [z_k^{(1)}, z_k^{(2)}]^T$ and $\boldsymbol{c}_k = [c_k^{(1)}, c_k^{(2)}]^T$, (see Definition 11.1),

$$d^H(\boldsymbol{z}_k, \boldsymbol{c}_k) = \text{number of positions where } \boldsymbol{z}_k \text{ differs from } \boldsymbol{c}_k, \tag{11.281}$$

where $\boldsymbol{c}_k$ is generated according to the rule (11.274).

VA with soft input. The trellis diagram is now obtained by using the definition (11.279), and the branch metric is the Euclidean distance between z_k and u_k,

$$|z_k - u_k|^2 \tag{11.282}$$

where u_k, in the case of the previous example of absence of ISI and 16-PAM transmission, is given by (11.280).

As an alternative to the VA, we can use the forward–backward algorithm (FBA) of Section 7.5.2.

Decoding by the forward-backward algorithm

The previous approach, which considers joint detection in the presence of ISI and convolutional decoding, requires a computational complexity that in many applications may turn out to be exceedingly large. In fact the state (11.279), that takes into account both encoding and the presence of ISI, usually is difficult to define and is composed of several bits of the sequence $\{b_\ell\}$. An approximate solution is obtained by considering the detection and the decoding problems separately, however, assuming that the detector passes the soft information on the detected bits to the decoder. This is also the principle of bit-interleaved coded modulation [10].

Soft output detection by FBA. For a channel model at detection point with ISI, as by (11.276), extraction of soft information (through log-likelihood ratios, LLRs) associated to code bits follows the derivation of (7.436) where we use the FBA of page 348 to extract the state metrics $V_k(i)$, $i = 1, \ldots, N_s$. In the particular case of no ISI, extraction of LLRs associated to code bits is simplified, as outlined in Section 6.2.2.

An alternative to the FBA is obtained by modifying the VA to yield a soft output Viterbi algorithm (SOVA), as discussed in the next section.

Convolutional decoding with soft input The decoder for the convolutional code typically uses the VA with branch metric (associated with a cost function to be minimized) given by

$$\| \boldsymbol{\ell}_k^{(in)} - \boldsymbol{c}_k \|^2 \tag{11.283}$$

where $\boldsymbol{c}_k$ is given by (11.274) for a code with $n_0 = 2$, and $\boldsymbol{\ell}_k^{(in)} = [\ell_k^{(in,1)}, \ell_k^{(in,2)}]$ are the LLRs associated, respectively, with $c_k^{(1)}$ and $c_k^{(2)}$. As $|c_k^{(j)}|^2 = 1$, (11.283) can be rewritten as

$$\begin{aligned}(\ell_k^{(in,1)} - c_k^{(1)})^2 &+ (\ell_k^{(in,2)} - c_k^{(2)})^2 \\ &= (\ell_k^{(in,1)})^2 + (\ell_k^{(in,2)})^2 + 2 - 2c_k^{(1)}\ell_k^{(in,1)} - 2c_k^{(2)}\ell_k^{(in,2)}\end{aligned} \tag{11.284}$$

Leaving out the terms that do not depend on $\boldsymbol{c}_k$ and extending the formulation to a convolutional code with rate k_0/n_0, the branch metric (associated with a cost function to be maximized) is expressed as (see also Ref. [11])

$$2 \sum_{j=1}^{n_0} c_k^{(j)} \ell_k^{(in,j)} \tag{11.285}$$

where the factor 2 can be omitted.

Observation 11.3

As we have previously stated, best system performance is obtained by jointly designing the encoder and the bit mapper. However, in some systems, typically radio, an interleaver is used between the encoder and the bit mapper. In this case, joint detection and decoding are impossible to implement in practice. Detection with soft output followed by decoding with soft input remains a valid approach, obviously after re-ordering the LLR as determined by the deinterleaver.

In applications that require a soft output (see Section 11.7), the decoder, that is called in this case *soft-input soft-output*, can use one of the versions of the FBA or the SOVA. [6]

[6] An extension of soft-input soft-output decoders for the decoding of block codes is found in [12, 13].

Sequential decoding

Sequential decoding of convolutional codes represented the first practical algorithm for ML decoding. It has been employed, for example, for the decoding of signals transmitted by deep-space probes, such as the Pioneer, 1968 [14]. There exist several variants of sequential decoding algorithms that are characterized by the search of the optimum path in a tree diagram (see Figure 11.9b), instead of along a trellis diagram, as considered, e.g. by the VA.

Sequential decoding is an attractive technique for the decoding of convolutional codes and trellis codes if the number of states of the encoder is large [15]. In fact, as the implementation complexity of ML decoders such as the Viterbi decoder grows exponentially with the constraint length of the code, ν, the complexity of sequential decoding algorithms is essentially independent of ν. On the other hand, sequential decoding presents the drawback that the number of computations N_{op} required for the decoding process to advance by one branch in the decoder tree is a random variable with a Pareto distribution, i.e.

$$P[N_{op} > N] = AN^{-\rho} \tag{11.286}$$

where A and ρ are constants that depend on the channel characteristics and on the specific code and the specific version of sequential decoding used. Real-time applications of sequential decoders require buffering of the received samples. As practical sequential decoders can perform only a finite number of operations in a given time interval, re-synchronization of the decoder must take place if the maximum number of operations that is allowed for decoding without incurring buffer saturation is exceeded.

Sequential decoders exhibit very good performance, with a reduced complexity as compared to the VA, if the constraint length of the code and the SNR are sufficiently large. Otherwise, the average number of operations required to produce one symbol at the decoder output is very large.

The Fano Algorithm In this section, we consider sequential decoding of trellis codes, a class of codes that will be studied in detail in Chapter 12. However, the algorithm can be readily extended to convolutional codes. At instant k, the k_0 information bits $\boldsymbol{b}_k = [b_k^{(1)}, \dots, b_k^{(k_0)}]$ are input to a rate $k_0/(k_0+1)$ convolutional encoder with constraint length ν that outputs the coded bits $\boldsymbol{c}_k = [c_k^{(0)}, \dots, c_k^{(k_0)}]$. The k_0+1 coded bits select from a constellation with $M = 2^{k_0+1}$ elements a symbol a_k, which is transmitted over an additive white Gaussian noise (AWGN) channel. Note that the encoder tree diagram has 2^{k_0} branches that correspond to the values of $\boldsymbol{b}_k$ stemming from each node. Each branch is labelled by the symbol a_k selected by the vector $\boldsymbol{c}_k$. The received signal is given by (see (11.275))

$$z_k = a_k + w_k \tag{11.287}$$

The received signal sequence is input to a sequential decoder. Using the notation of Section 7.5.1, in the absence of ISI and assuming $\boldsymbol{a} = \boldsymbol{\alpha}$, the ML metric to be maximized can be written as [9]

$$\Gamma(\boldsymbol{\alpha}) = \sum_{k=0}^{K-1} \left[\log_2 \frac{P_{z_k|a_k}(\rho_k|\alpha_k)}{\sum\limits_{\alpha_i \in \mathcal{A}} P_{z_k|a_k}(\rho_k|\alpha_i) P_{a_k}(\alpha_i)} - B \right] \tag{11.288}$$

where B is a suitable constant that determines a trade-off between computational complexity and performance and is related to the denominator in (11.288). Choosing $B = k_0$ and $P_{a_k}(\alpha_i) = \frac{1}{M} = 2^{-(k_0+1)}$,

$\alpha_i \in \mathcal{A}$, we obtain

$$\Gamma(\boldsymbol{\alpha}) = \sum_{k=0}^{K-1} \left[\log_2 \frac{e^{-\frac{|\rho_k - \alpha_k|^2}{2\sigma_I^2}}}{\sum_{\alpha_i \in \mathcal{A}} e^{-\frac{|\rho_k - \alpha_i|^2}{2\sigma_I^2}}} + 1 \right] \tag{11.289}$$

Various algorithms have been proposed for sequential decoding [16–18]. We will restrict our attention here to the Fano algorithm [9, 15].

The Fano algorithm examines only one path of the decoder tree at any time using the metric in (11.288). The considered path extends to a certain node in the tree and corresponds to a segment of the entire code sequence $\boldsymbol{\alpha}$, up to symbol α_k.

Three types of moves between consecutive nodes on the decoder tree are allowed: forward, lateral, and backward. On a *forward move*, the decoder goes one branch to the right in the decoder tree from the previously hypothesized node. This corresponds to the insertion of a new symbol α_{k+1} in (11.288). On a *lateral move*, the decoder goes from a path on the tree to another path differing only in the last branch. This corresponds to the selection of a different symbol α_k in (11.288). The ordering among the nodes is arbitrary, and a lateral move takes place to the next node in order after the current one. A *backward move* is a move one branch to the left on the tree. This corresponds to the removal of the symbol α_k from (11.288).

To determine which move needs to be made after reaching a certain node, it is necessary to compute the metric Γ_k of the current node being hypothesized and consider the value of the metric Γ_{k-1} of the node one branch to the left of the current node, as well as the current value of a threshold Th, which can assume values that are multiple of a given constant Δ. The transition diagram describing the Fano algorithm is illustrated in Figure 11.20. Typically, Δ assumes values that are of the order of the minimum distance between symbols.

As already mentioned, real-time applications of sequential decoding require buffering of the input samples with a buffer of size S. Furthermore, the depth of backward search is also finite and is usually chosen at least five times the constraint length of the code. To avoid erasures of output symbols in case of buffer saturation, in [19] a *buffer looking algorithm* (BLA) is proposed. The buffer is divided into L sections, each with size S_j, $j = 1, \dots, L$. A conventional sequential decoder (primary decoder) and $L - 1$ secondary decoders are used. The secondary decoders employ fast algorithms, such as the M-algorithm [20], or variations of the Fano algorithm that are obtained by changing the value of the bias B in the metric (11.288).

Example 11.3.2 (Sequential decoding of a 512-state 16-PAM trellis code)

We illustrate sequential decoding with reference to a 512-state 16-PAM trellis code specified for *single-line high-bit-rate digital subscriber line* (SHDSL) transmission (see Section 18.6). The encoder for this trellis code comprises a rate $1/2$ non-systematic non-recursive convolutional encoder with constraint length $\nu = 9$ and a bit mapper as specified in Figure 11.21. The symbol error probabilities versus SNR Γ obtained by sequential decoding with infinite buffer size and depth of backward search of 64 and 128 symbols, and by a 512-state VA decoding with length of the path memory of 64 and 128 symbols are shown in Figure 11.22. Also shown for comparison are the error probabilities obtained for uncoded 8-PAM and 16-PAM transmission.

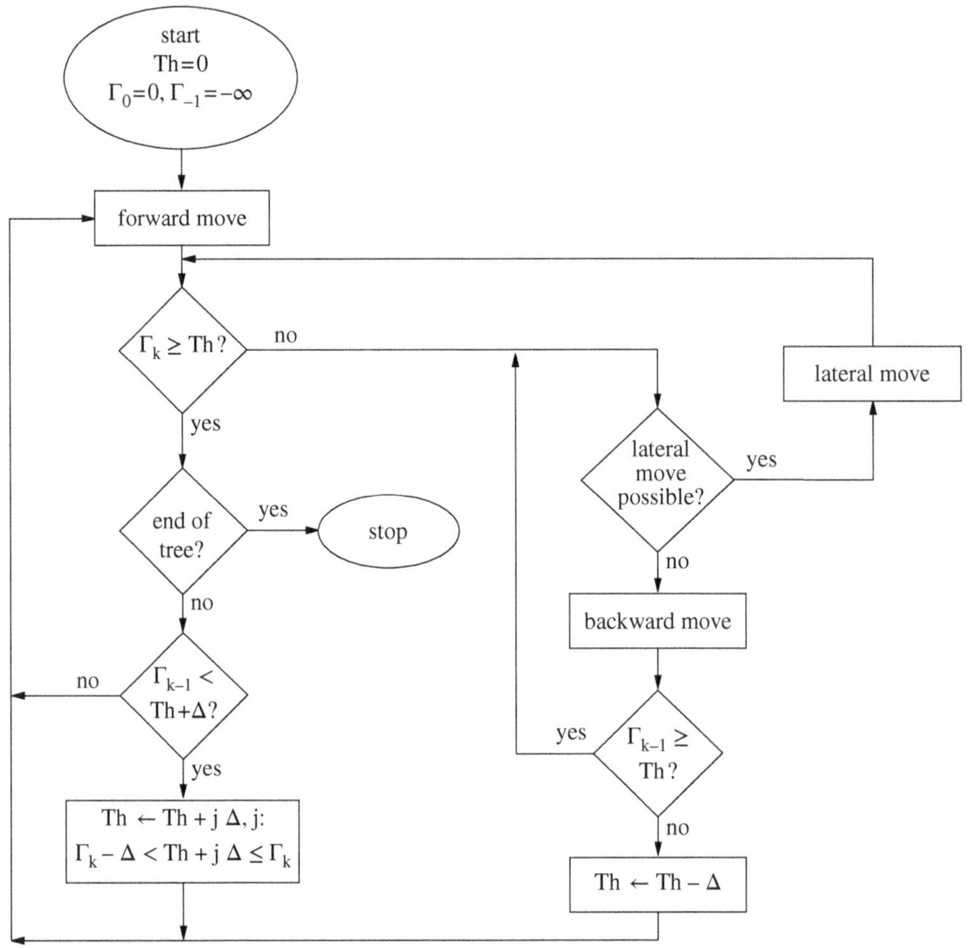

Figure 11.20 Transition diagram of the Fano algorithm.

11.3.3 Performance of convolutional codes

For binary convolutional codes with free distance d^H_{free} and bit error probability of the channel equal to P_{bit}, decoding with *hard input* yields

$$P^{(dec)}_{bit} \simeq A\, 2^{-d^H_{free}}\, P_{bit} \tag{11.290}$$

and decoding with *soft input*, for a system with antipodal signals, yields

$$P^{(dec)}_{bit} \simeq A\, Q\left(\sqrt{\frac{R_c\, d^H_{free}\, 2E_b}{N_0}}\right) \tag{11.291}$$

where A is a constant [21].

In particular, we consider BPSK transmission over an ideal AWGN channel. Assuming an encoder with rate $R_c = 1/2$ and constraint length $\nu = 6$, the coding gain for a soft Viterbi decoder is about 3.5 dB for $P_{bit} = 10^{-3}$; it becomes about 4.6 dB for $P_{bit} = 10^{-5}$. Note that a soft decoder provides a gain of about 2.4 dB with respect to a hard decoder, for $P_{bit} < 10^{-3}$.

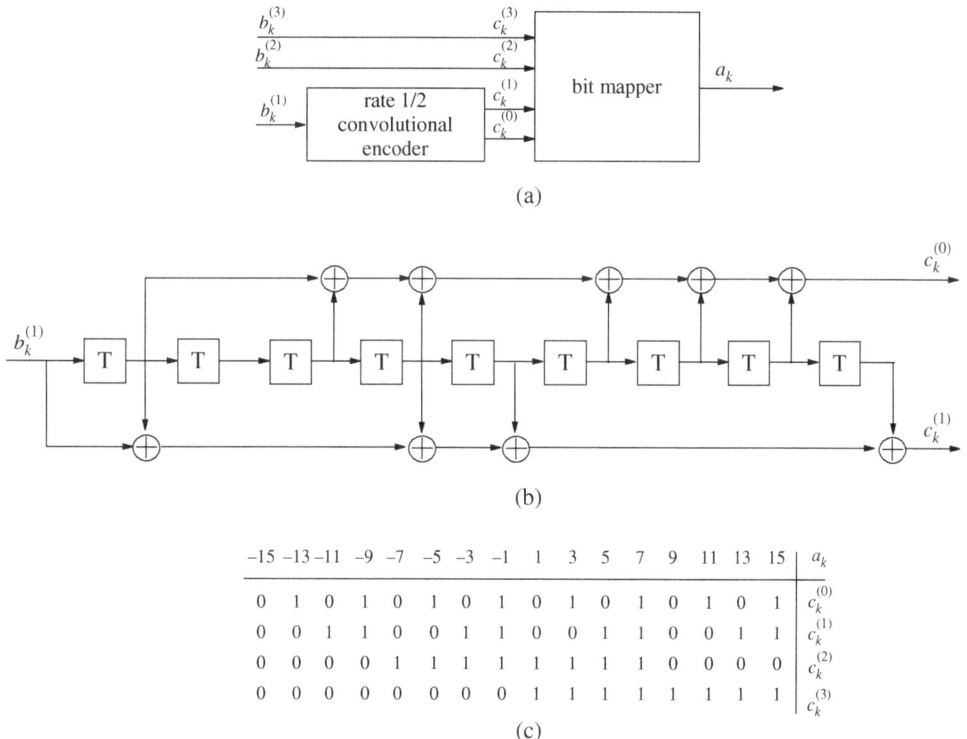

−15	−13	−11	−9	−7	−5	−3	−1	1	3	5	7	9	11	13	15	a_k
0	1	0	1	0	1	0	1	0	1	0	1	0	1	0	1	$c_k^{(0)}$
0	0	1	1	0	0	1	1	0	0	1	1	0	0	1	1	$c_k^{(1)}$
0	0	0	0	1	1	1	1	1	1	1	1	0	0	0	0	$c_k^{(2)}$
0	0	0	0	0	0	0	0	1	1	1	1	1	1	1	1	$c_k^{(3)}$

Figure 11.21 (a) Block diagram of the encoder and bit mapper for a trellis code for 16-PAM transmission, (b) structure of the rate-1/2 convolutional encoder, and (c) bit mapping for the 16-PAM format.

11.4 Puncturing

This operation takes place after the encoder, before the interleaver. By puncturing we omit to transmit some code bits, generated by (11.254). Obviously, these positions are known at the receiver, which inserts a LLR equal to zero before decoding. Hence, puncturing provides a higher code rate without modifying substantially the encoder and the decoder.

11.5 Concatenated codes

Concatenated coding is usually introduced to achieve an improved error correction capability [22]. Interleaving is also commonly used in concatenated coding schemes, as illustrated in Figure 11.23 so that the decoding processes of the two codes (inner and outer) can be considered approximately independent. The first decoding stage is generally utilized to produce soft decisions on the information bits that are passed to the second decoding stage. While the FBA directly provides a soft output, the VA must be slightly modified.

The soft-output Viterbi algorithm

We have seen that the FBA in the original maximum-a-posteriori (MAP) version directly yields a soft output, at the expense of a large computational complexity. The Max-Log-MAP criterion has a reduced

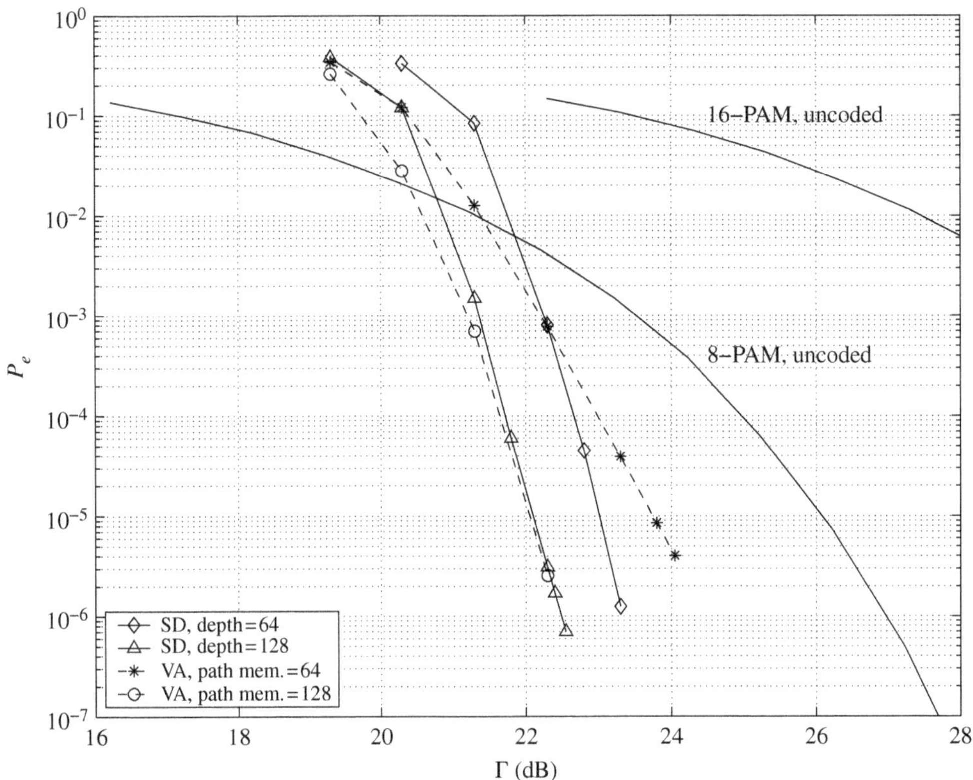

Figure 11.22 Symbol error probabilities for the 512-state 16-PAM trellis code with sequential decoding (depth of search limited to 64 or 128) and 512-state Viterbi decoding (length of path memory limited to 64 or 128). Symbol error probabilities for uncoded 8-PAM and 16-PAM transmission are also shown.

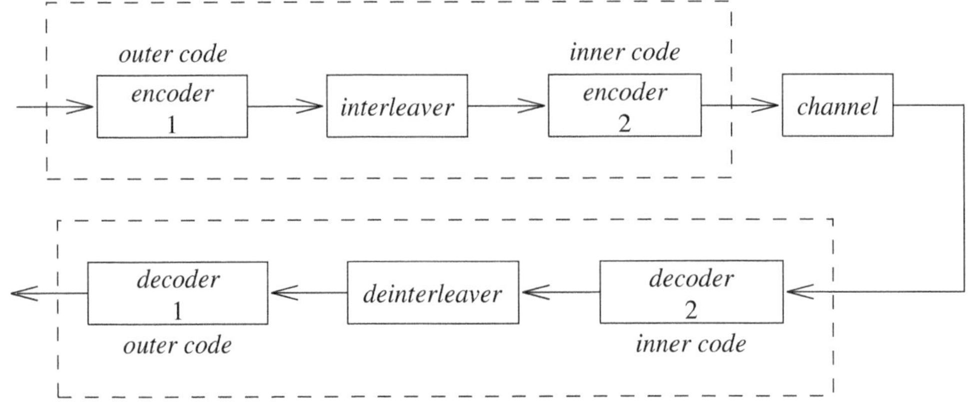

Figure 11.23 Transmission scheme with concatenated codes and interleaver.

complexity, but there remains the problem of performing the two procedures, forward and backward. We now illustrate how to modify the VA to obtain a soft output equivalent to that produced by the Max-Log-MAP.

In this section, we consider the different methods to generate the soft output.

The difference metric algorithm (DMA). Figure 11.24 shows a section of a trellis diagram with four states, where we assume $\boldsymbol{s}_k = (b_k, b_{k-1})$. We consider two states at instant $k-1$ that differ in the least significant bit b_{k-2} of the binary representation, that is $\boldsymbol{s}_{k-1}^{(0)} = (00)$ and $\boldsymbol{s}_{k-1}^{(1)} = (01)$. A transition from each of these two states to state $\boldsymbol{s}_k = (00)$ at instant k is allowed. According to the VA, we choose the survivor sequence minimizing the metric

$$\{\Gamma_{k-1}(\boldsymbol{s}_{k-1}^{(i)}) + \gamma_k^{\boldsymbol{s}_{k-1}^{(i)} \to \boldsymbol{s}_k}\} \tag{11.292}$$

where $\Gamma_{k-1}(\boldsymbol{s}_{k-1}^{(i)})$ is the path metric associated with the survivor sequence up to state $\boldsymbol{s}_{k-1}^{(i)}$ at instant $k-1$, and $\gamma_k^{\boldsymbol{s}_{k-1}^{(i)} \to \boldsymbol{s}_k}$ denotes the branch metric associated with the transition from state $\boldsymbol{s}_{k-1}^{(i)}$ to state $\boldsymbol{s}_k$. Let $A_k = \Gamma_{k-1}(00) + \gamma_k^{00\to 00}$ and $B_k = \Gamma_{k-1}(01) + \gamma_k^{01\to 00}$, then we choose the upper or the lower transition according to whether $\Delta_k = A_k - B_k$ is smaller or larger than zero, respectively. Note that $|\Delta_k|$ is a reliability measure of the selection of a certain sequence as survivor sequence.

In other words, if $|\Delta_k|$ is small, there is a non-negligible probability that the bit b'_{k-2} associated with the transition from state $\boldsymbol{s}_{k-1}^{(i)}$ to $\boldsymbol{s}_k$ on the survivor sequence is in error. The difference $|\Delta_k| = \lambda_k$ yields the value of the soft decision for b_{k-2}, in case the final sequence chosen by the VA includes the state $\boldsymbol{s}_k$; conversely, this information is disregarded. Thus, the difference metric algorithm (DMA) can be formulated as follows:

For each state $\boldsymbol{s}_k$ of the trellis diagram at instant k, the metric $\Gamma_k(\boldsymbol{s}_k)$ and the most recent $(K_d + 1)$ bits of the survivor sequence $\boldsymbol{b}'(\boldsymbol{s}_k) = \{b'_k, \dots, b'_{k-K_d}\}$ are stored, where K_d denotes the path memory depth of the VA. Furthermore, the reliability measures $\lambda(\boldsymbol{s}_k) = \{\lambda_k, \dots, \lambda_{k-K_d}\}$ associated with the bits $\boldsymbol{b}'(\boldsymbol{s}_k)$ are also stored. Interpreting b_k and $\hat{b}_k$ as binary symbols in the alphabet $\{-1, 1\}$, the soft output associated with b_k is given by

$$\tilde{\ell}_k = \hat{b}_k \ \lambda_{k+2} \tag{11.293}$$

where $\{\hat{b}_k\}$ is the sequence of information bits associated with the ML sequence.

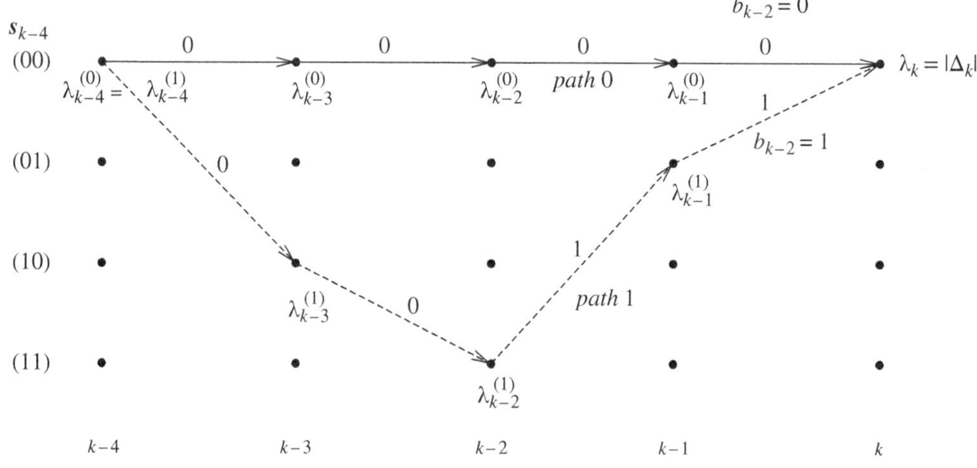

Figure 11.24 Soft-output VA.

The soft-output VA (SOVA). As for the DMA, the SOVA determines the difference between the metrics of the survivor sequences on the paths that converge to each state of the trellis and updates at every instant k the reliability information $\lambda(\boldsymbol{s}_k)$ for each state of the trellis. To perform this update, the sequences on the paths that converge to a certain state are compared to identify the positions at which the information bits of the two sequences differ. With reference to Figure 11.24, we denote the two paths that converge to the state (00) at instant k as *path* 0 and *path* 1. Without loss of generality, we assume that the sequence associated with path 0 is the survivor sequence, and thus the sequence with the smaller cost; furthermore, we define $\lambda(\boldsymbol{s}_k^{(0)}) = \{\lambda_k^{(0)}, \dots, \lambda_{k-K_d}^{(0)}\}$ and $\lambda(\boldsymbol{s}_k^{(1)}) = \{\lambda_k^{(1)}, \dots, \lambda_{k-K_d}^{(1)}\}$ as the two reliability vectors associated with the information bits of two sequences. If one information bit *along path 0* differs from the corresponding information bit *along path 1*, then its reliability is updated according to the rule, for $i = k - K_d, \dots, k-1$,

$$\lambda_i = \min\left(|\Delta_k|, \lambda_i^{(0)}\right) \qquad \text{if } b_{i-2}^{(0)} \neq b_{i-2}^{(1)} \tag{11.294}$$

With reference to Figure 11.24, the two sequences on path 0 and on path 1 diverge from state $\boldsymbol{s}_k = (00)$ at instant $k-4$. The two sequences differ in the associated information bits at the instants k and $k-1$; therefore, only λ_{k-1} will be updated.

Modified SOVA. In the modified version of the SOVA, the reliability of an information bit along the survivor path is also updated if the information bit is the same, according to the rule, for $i = k - K_d, \dots, k-1$

$$\lambda_i = \begin{cases} \min\left(|\Delta_k|, \lambda_i^{(0)}\right) & \text{if } b_{i-2}^{(0)} \neq b_{i-2}^{(1)} \\ \min\left(|\Delta_k| + \lambda_i^{(1)}, \lambda_i^{(0)}\right) & \text{if } b_{i-2}^{(0)} = b_{i-2}^{(1)} \end{cases} \tag{11.295}$$

Note that (11.294) is still used to update the reliability if the information bits differ; this version of the SOVA gives a better estimate of λ_i. As proven in [23], if the VA is used as decoder, the modified SOVA is equivalent to Max–Log–MAP decoding. An example of how the modified SOVA works is illustrated in Figure 11.25.

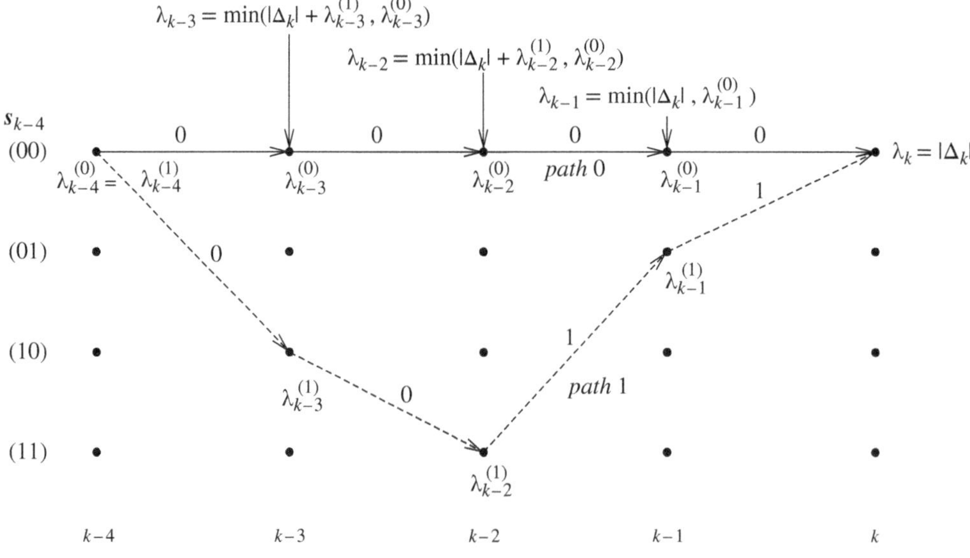

Figure 11.25 Modified soft-output VA.

11.6 Turbo codes

Turbo codes, proposed in 1993 by Berrou et al. [24, 25], are an evolution of concatenated codes, in the form of parallel concatenated convolutional codes (PCCC), and allow reliable transmission of information at rates near the Shannon limit [24–26]. As will be discussed in this section, the term turbo, even though it is used to qualify these codes, is rather tied to the decoder, whose principle is reminiscent of that of turbo engines.

Encoding

For the description of turbo codes, we refer to the first code of this class that appeared in the scientific literature [24, 25]. A sequence of information bits is encoded by a simple *recursive systematic convolutional* (RSC) binary encoder with code rate $1/2$, to produce a first sequence of parity bits, as illustrated in Figure 11.26. The same sequence of information bits is permuted by a long interleaver and then encoded by a second recursive systematic convolutional encoder with code rate $1/2$ to produce a second sequence of parity bits. Then the sequence of information bits and the two sequences of parity bits are transmitted. Note that the resulting code has rate $R_c = 1/3$. Higher code rates R_c are obtained by transmitting only some of the parity bits (puncturing). For example, for the turbo code in [24, 25], a code rate equal to 1/2 is obtained by transmitting only the bits of the parity sequence 1 with odd indices and the bits of the parity sequence 2 with even indices. A specific example of turbo encoder is reported in Figure 11.27.

The exceptional performance of turbo codes is due to one particular characteristic. We can think of a turbo code as being a block code for which an input word has a length equal to the interleaver length, and a code word is generated by initializing to zero the memory elements of the convolutional encoders before the arrival of each input word. This block code has a group structure. As for the usual block codes, the asymptotic performance, for large values of the SNR, is determined by the code words of minimum weight and by their number. For low values of the SNR, also the code words of non-minimum weight and their multiplicity need to be taken into account. Before the introduction of turbo codes, the focus on designing codes was mainly on asymptotic performance, and thus on maximizing the minimum

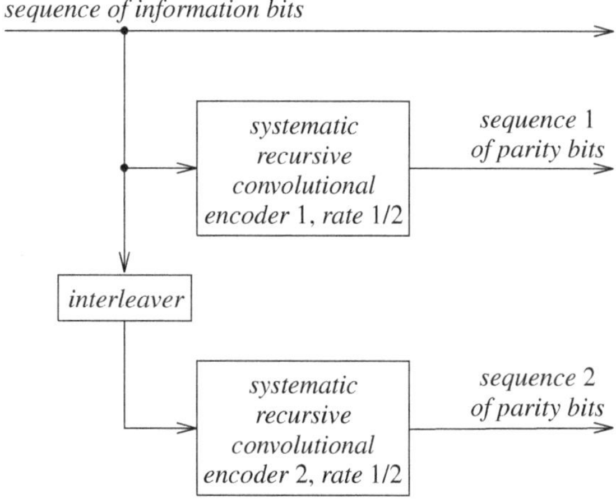

Figure 11.26 Encoder of a turbo code with code rate $R_c = \frac{1}{3}$.

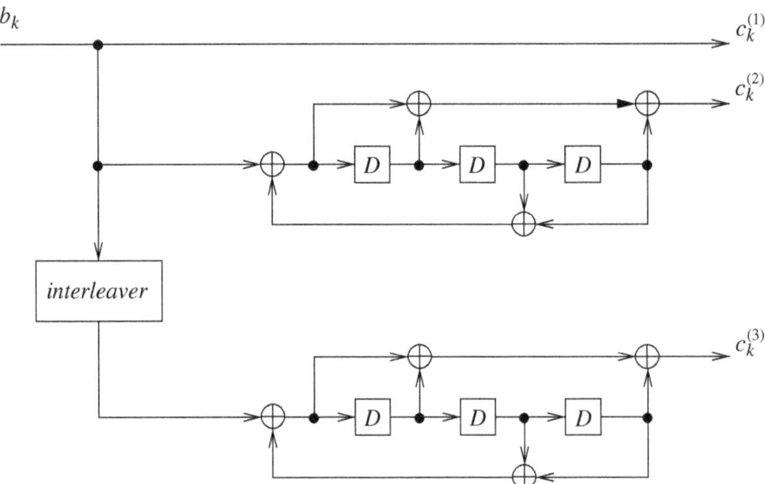

Figure 11.27 Turbo encoder adopted by the UMTS standard.

distance. With turbo codes, the approach is different. Because of the large ensemble of code words, the performance curve, in terms of bit error probability as a function of the SNR, rapidly decreases for low values of the SNR. For P_{bit} lower than 10^{-5}, where performance is determined by the minimum distance between code words, the bit error probability curve usually exhibits a reduction in the value of slope.

The two encoders that compose the scheme of Figure 11.26 are called *component encoders*, and they are usually identical. As mentioned above, Berrou and Glavieux proposed two recursive systematic convolutional encoders as component encoders. Later, it was shown that it is not necessary to use systematic encoders [21, 27].

Recursive convolutional codes are characterized by the property that the code bits at a given instant do not depend only on the information bits at the present instant and the previous ν instants, where ν is the constraint length of the code, but on all the previous information bits, as the encoder exhibits a structure with feedback.

Starting from a non-recursive, non-systematic convolutional encoder for a code with rate $1/n_0$, it is possible to obtain in a very simple way a recursive systematic encoder for a code with the same rate and the same code words, and hence, with the same free distance d^H_{free}. Obviously, for a given input word, the output code words will be different in the two cases. Consider, for example a non-recursive, non-systematic convolutional encoder for a code with code rate $1/2$. The code bits can be expressed in terms of the information bits as (see (11.254))

$$\begin{aligned} c'^{(1)}(D) &= g^{(1,1)}(D)\, b(D) \\ c'^{(2)}(D) &= g^{(2,1)}(D)\, b(D) \end{aligned} \tag{11.296}$$

The corresponding recursive systematic encoder is obtained by dividing the polynomials in (11.296) by $g^{(1,1)}(D)$ and implementing the functions

$$c^{(1)}(D) = b(D) \tag{11.297}$$

$$c^{(2)}(D) = \frac{g^{(2,1)}(D)}{g^{(1,1)}(D)}\, b(D) \tag{11.298}$$

Let us define

$$d(D) = \frac{c^{(2)}(D)}{g^{(2,1)}(D)} = \frac{b(D)}{g^{(1,1)}(D)} \tag{11.299}$$

then the code bits can be expressed as a function of the information bits and the bits of the sequence $\{d_k\}$ as

$$c_k^{(1)} = b_k \tag{11.300}$$

$$c_k^{(2)} = \sum_{i=0}^{\nu} g_i^{(2,1)} d_{k-i} \tag{11.301}$$

where, using the fact that $g_0^{(1,1)}(D) = 1$, from (11.299), we get

$$d_k = b_k + \sum_{i=1}^{\nu} g_i^{(1,1)} d_{k-i} \tag{11.302}$$

We recall that the operations in the above equations are in GF(2). Another recursive systematic encoder that generates a code with the same free distance is obtained by exchanging the role of the polynomials $g^{(1,1)}(D)$ and $g^{(2,1)}(D)$ in the above equations.

One recursive systematic encoder corresponding to the non-recursive, non-systematic encoder of Figure 11.9a is illustrated in Figure 11.28.

The 16-state component encoder for a code with code rate 1/2 used in the turbo code of Berrou and Glavieux [24, 25], is shown in Figure 11.29. The encoder in Figure 11.27, with an 8-state component encoder for a code with code rate 1/2, is adopted in the standard for third-generation *universal mobile telecommunications systems* (UMTS) [28]. Turbo codes are also used in *digital video broadcasting* (DVB) [29] standards and in space telemetry applications as defined by the *Consultative Committee for Space Data Systems* (CCSDS) [30]. In [31] are listed generator polynomials of recursive systematic convolutional encoders for codes with rates 1/2, 1/3, 1/4, 2/3, 3/4, 4/5, and 2/4, that can be used for the construction of turbo codes.

Another fundamental component in the structure of turbo codes is represented by a *non-uniform* interleaver. We recall that a uniform[7] interleaver, as that described in Section 11.3.2, operates by writing input bits in a matrix by rows and reading them by columns. In practice, a non-uniform interleaver determines a permutation of the sequence of input bits so that adjacent bits in the input sequence are separated, after the permutation, by a number of bits that varies with the position of the bits in the input sequence. The

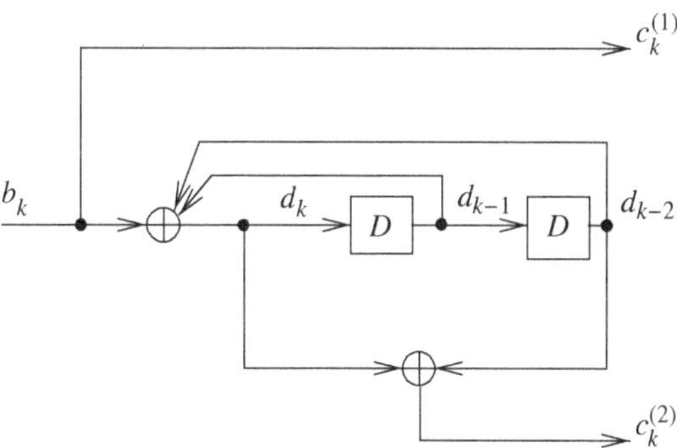

Figure 11.28 Recursive systematic encoder that generates a code with the same free distance as the non-recursive, non-systematic encoder of Figure 11.9a.

[7] The adjective *uniform*, referred to an interleaver, is used with a different meaning in [27].

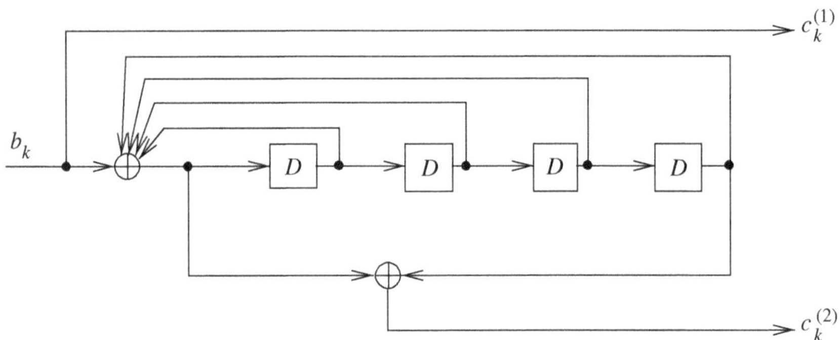

Figure 11.29 A 16-state component encoder for the turbo code of Berrou and Glavieux.

interleaver determines directly the minimum distance of the code and therefore, performance for high values of the SNR. Nevertheless, the choice of the interleaver is not critical for low values of the SNR. Beginning with the interleaver originally proposed in [24, 25], various interleavers have since been proposed (see Ref. [32] and references contained therein).

One of the interleavers that yields better performance is the so-called spread interleaver [33]. Consider a block of M_1 input bits. The integer numbers that indicate the position of these bits after the permutation are randomly generated with the following constraint: each integer randomly generated is compared with the S_1 integers previously generated; if the distance from them is shorter than a prefixed threshold S_2, the generated integer is discarded and another one is generated until the condition is satisfied. The two parameters S_1 and S_2 must be larger than the memory of the two-component encoders. If the two-component encoders are equal, it is convenient to choose $S_1 = S_2$. The computation time needed to generate the interleaver increases with S_1 and S_2, and there is no guarantee that the procedure terminates successfully. Empirically, it has been verified that, choosing both S_1 and S_2 equal to the closest integer to $\sqrt{M_1/2}$, it is possible to generate the interleaver in a finite number of steps.

Many variations of the basic idea of turbo codes have been proposed. For example, codes generated by serial concatenation of two convolutional encoders, connected by means of a non-uniform interleaver [34]. Parallel and serial concatenation schemes were then extended to the case of multi-level constellations to obtain coded modulation schemes with high spectral efficiency (see [35] and references therein).

The basic principle of iterative decoding

The presence of the interleaver in the scheme of Figure 11.26 makes an encoder for a turbo code have a very large memory even if very simple component encoders are used. Therefore, the optimum maximum-likelihood sequence detection (MLSD) decoder would require a Viterbi decoder with an exceedingly large number of states and it would not be realizable in practice. For this reason, we resort to a suboptimum iterative decoding scheme with a much lower complexity, that however, as it was verified empirically, exhibits near optimum performance [27]. The decoder for the turbo encoder of Figure 11.26 is illustrated in Figure 11.30. The received sequences corresponding to the sequence of information bits and the first sequence of parity bits are decoded using a soft input soft output decoder, corresponding to the first convolutional encoder. Thus, this decoder provides a soft decision for each information bit; these soft decisions are then used by a second decoder corresponding to the second convolutional encoder, together with the received sequence corresponding to the second sequence of parity bits. Soft decisions so obtained are taken back to the input of the first decoder for a new iteration, where the additional

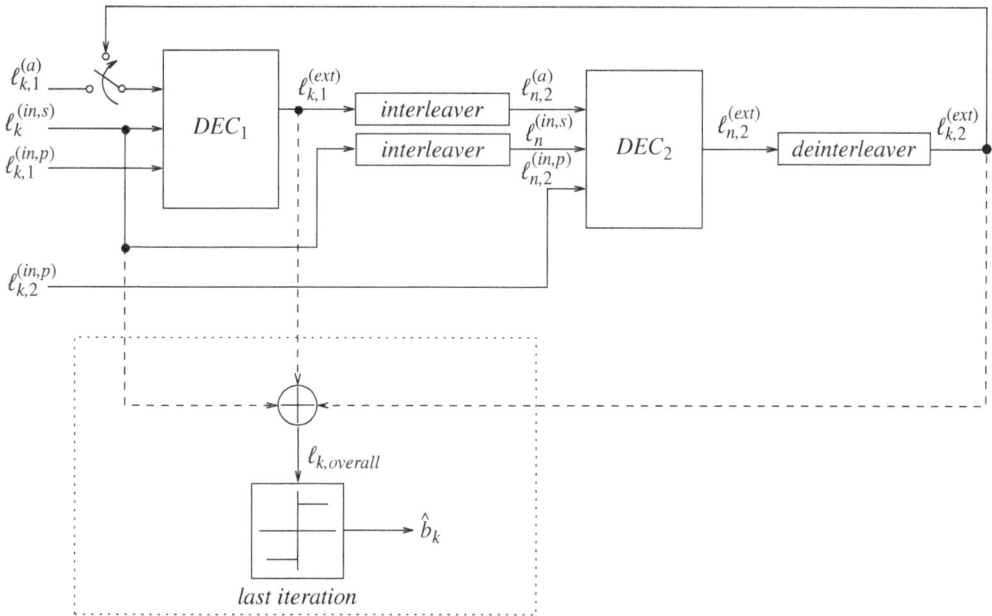

Figure 11.30 Principle of the decoder for a turbo code with rate 1/3.

information obtained by the second decoder is used to produce more reliable soft decisions. The procedure continues iteratively for about 10–20 cycles until final decisions are made on the information bits.

The two component decoders of the scheme in Figure 11.30 are soft input, soft output decoders that produce estimates on the reliability of the decisions. Therefore, they implement the SOVA or the FBA (or one of its simplified realizations). The basic principle of iterative decoding is the following: each component decoder uses the *hints* of the other to produce more reliable decisions. In the next sections, we will see in detail how this is achieved, and in particular how the reliability of the decisions is determined.

The algorithms for iterative decoding introduced with the turbo codes were also immediately applied in wider contexts. In fact, this iterative procedure may be used every time the transmission system includes multiple processing elements with memory interconnected by an interleaver. Iterative decoding procedures may be used, for example for detection in the presence of intersymbol interference, also called turbo equalization or turbo detection [36] (see Section 11.7), for non-coherent decoding [37, 38], and for joint detection and decoding in the case of transmission over channels with fading [39].

Before discussing in detail iterative decoding, it is useful to revisit the FBA.

FBA revisited

The formulation of the FBA presented here is useful for the decoding of recursive systematic convolutional codes [40].

We consider a binary recursive systematic convolutional encoder for a code with rate k_0/n_0 and constraint length ν. Let the encoder input be given by a sequence of K vectors, each composed of k_0 binary components. As described on page 579, each information vector to be encoded is denoted by (see (11.252))

$$\boldsymbol{b}_k = [b_k^{(1)}, b_k^{(2)}, \dots, b_k^{(k_0)}], \quad b_k^{(i)} \in \{-1, 1\}, \quad k = 0, 1, \dots K-1 \tag{11.303}$$

where k_0 can be seen either as the number of encoder inputs or as the length of an information vector. As the convolutional encoder is systematic, at instant k the state of the convolutional encoder is given by the vector (see extension of (11.302))

$$s_k = [d_k^{(k_0+1)}, \dots, d_{k-\nu+1}^{(k_0+1)}, d_k^{(k_0+2)}, \dots, d_{k-\nu+1}^{(k_0+2)}, \dots, d_k^{(n_0)}, \dots, d_{k-\nu+1}^{(n_0)}] \tag{11.304}$$

which has a number of components $N_2 = \nu \cdot (n_0 - k_0)$, equal to the number of the encoder memory elements. The set of states $\mathcal{S}$, that is the possible values assumed by s_k, is given by

$$s_k \in \mathcal{S} = \{\sigma_1, \sigma_2, \dots, \sigma_{N_s}\} \tag{11.305}$$

where $N_s = 2^{N_2}$ is the number of encoder states. It is important to observe that the convolutional encoder can be seen as a sequential finite-state machine with i.i.d. input b_k, and state transition function $s_k = f_s(b_k, s_{k-1})$. Hence, for a given information vector b_k, the transition from state $s_{k-1} = \sigma_i$ to state $s_k = \sigma_j$ is unique, in correspondence of which a code vector is generated, that is expressed as

$$c_k = [c_k^{(1)}, c_k^{(2)}, \dots, c_k^{(k_0)}, c_k^{(k_0+1)}, \dots, c_k^{(n_0)}] = [\, c_k^{(s)}, c_k^{(p)} \,] \tag{11.306}$$

where the superscript$^{(s)}$ denotes systematic bits, and$^{(p)}$ denotes parity check bits. Then $c_k^{(s)} = b_k$, and from (11.301), we can express $c_k^{(p)}$ as a function of s_k and s_{k-1} as

$$c_k^{(p)} = f^{(p)}(s_k, s_{k-1}) \tag{11.307}$$

The values assumed by the code vectors are indicated by

$$\boldsymbol{\beta} = [\beta^{(1)}, \beta^{(2)}, \dots, \beta^{(k_0)}, \beta^{(k_0+1)}, \dots, \beta^{(n_0)}] = [\, \boldsymbol{\beta}^{(s)}, \boldsymbol{\beta}^{(p)} \,] \tag{11.308}$$

The actual input to the decoder is the LLR $\ell_k^{(in)}$ associated with c_k and whose expression is recalled in page 589. Moreover, we *model the decoder input* as if the code binary symbols were transmitted by a binary modulation scheme over an AWGN channel. In this case, at the decoder input, we get the signal (see (11.275)),

$$z_k = c_k + w_k \tag{11.309}$$

where $c_k \in \{-1, 1\}$ denotes a code bit, and $\{w_k\}$ is a sequence of real-valued i.i.d. Gaussian noise samples with (unknown) variance σ_{dec}^2. It is useful to organize the samples $\{z_k\}$ into subsequences that follow the structure of the code vectors (11.306). Then we introduce the vectors

$$z_k = [z_k^{(1)}, z_k^{(2)}, \dots, z_k^{(k_0)}, z_k^{(k_0+1)}, \dots, z_k^{(n_0)}] = [\, z_k^{(s)}, z_k^{(p)} \,] \tag{11.310}$$

As usual, we denote as ρ_k an observation of z_k,

$$\rho_k = [\rho_k^{(1)}, \rho_k^{(2)}, \dots, \rho_k^{(k_0)}, \rho_k^{(k_0+1)}, \dots, \rho_k^{(n_0)}] = [\rho_k^{(s)}, \rho_k^{(p)}] \tag{11.311}$$

We recall from Section 7.5 that the FBA yields the *detection of the single information vector* b_k, $k = 0, 1, \dots, K-1$, expressed as

$$\hat{b}_k = [\hat{b}_k^{(1)}, \hat{b}_k^{(2)}, \dots, \hat{b}_k^{(k_0)}] \tag{11.312}$$

through the computation of the a posteriori probability. We also recall that in general, for a sequence $a = [a_0, \dots, a_k, \dots, a_{K-1}]$, with the notation a_l^m, we indicate the subsequence formed by the components $[a_l, a_{l+1}, \dots, a_m]$.

Defining the likelihood of the generic information vector (see (6.26)),

$$\mathrm{L}_k(\boldsymbol{\beta}^{(s)}) = P\,[b_k = \boldsymbol{\beta}^{(s)} \mid z_0^{K-1} = \rho_0^{K-1}] \tag{11.313}$$

detection by the MAP criterion is expressed as

$$\hat{b}_k = \arg\max_{\boldsymbol{\beta}^{(s)}} \mathrm{L}_k(\boldsymbol{\beta}^{(s)}), \qquad k = 0, 1, \dots, K-1 \tag{11.314}$$

We note that the likelihood associated with the individual bits of the information vector $\boldsymbol{b}_k$ are obtained by suitably adding (11.313), as

$$\begin{aligned} \mathrm{L}_{k,i}(\alpha) &= P\,[b_k^{(i)} = \alpha \mid \boldsymbol{z}_0^{K-1} = \boldsymbol{\rho}_0^{K-1}] \\ &= \sum_{\substack{\boldsymbol{\beta}^{(s)} \in \{-1,1\}^{k_0} \\ \beta^{(i)} = \alpha}} \mathrm{L}_k(\boldsymbol{\beta}^{(s)}), \quad \alpha \in \{-1, 1\} \end{aligned} \tag{11.315}$$

In a manner similar to the analysis of page 347, we introduce the following quantities:

1. The state transition probability $\Pi(j \mid i) = P\,[\boldsymbol{s}_k = \boldsymbol{\sigma}_j \mid \boldsymbol{s}_{k-1} = \boldsymbol{\sigma}_i]$, that assumes non-zero values only if there is a transition from the state $\boldsymbol{s}_{k-1} = \boldsymbol{\sigma}_i$ to the state $\boldsymbol{s}_k = \boldsymbol{\sigma}_j$ for a certain input $\boldsymbol{\beta}^{(s)}$, and we write
$$\Pi(j \mid i) = P\,[\boldsymbol{b}_k = \boldsymbol{\beta}^{(s)}] = \mathrm{L}_k^{(a)}(\boldsymbol{\beta}^{(s)}) \tag{11.316}$$
$\mathrm{L}_k^{(a)}(\boldsymbol{\beta}^{(s)})$ is called the *a priori information* on the information vector $\boldsymbol{b}_k = \boldsymbol{\beta}^{(s)}$, and is one of the soft inputs.
2. For an AWGN channel, the channel transition probability $p_{\boldsymbol{z}_k}(\boldsymbol{\rho}_k \mid j, i)$ can be separated into two contributions, one due to the systematic bits and the other to the parity check bits,
$$\begin{aligned} p_{\boldsymbol{z}_k}(\boldsymbol{\rho}_k \mid j, i) &= P\,[\boldsymbol{z}_k = \boldsymbol{\rho}_k \mid \boldsymbol{s}_k = \boldsymbol{\sigma}_j, \boldsymbol{s}_{k-1} = \boldsymbol{\sigma}_i] \\ &= P\,[\boldsymbol{z}_k^{(s)} = \boldsymbol{\rho}_k^{(s)} \mid \boldsymbol{s}_k = \boldsymbol{\sigma}_j, \boldsymbol{s}_{k-1} = \boldsymbol{\sigma}_i] \\ &\quad P\,[\boldsymbol{z}_k^{(p)} = \boldsymbol{\rho}_k^{(p)} \mid \boldsymbol{s}_k = \boldsymbol{\sigma}_j, \boldsymbol{s}_{k-1} = \boldsymbol{\sigma}_i] \\ &= P\,[\boldsymbol{z}_k^{(s)} = \boldsymbol{\rho}_k^{(s)} \mid \boldsymbol{c}_k^{(s)} = \boldsymbol{\beta}^{(s)}]\, P\,[\boldsymbol{z}_k^{(p)} = \boldsymbol{\rho}_k^{(p)} \mid \boldsymbol{c}_k^{(p)} = \boldsymbol{\beta}^{(p)}] \\ &= \left(\left(\frac{1}{\sqrt{2\pi\sigma_I^2}} \right)^{k_0} e^{-\frac{1}{2\sigma_I^2} \|\boldsymbol{\rho}_k^{(s)} - \boldsymbol{\beta}^{(s)}\|^2} \right) \\ &\quad \left(\left(\frac{1}{\sqrt{2\pi\sigma_I^2}} \right)^{n_0 - k_0} e^{-\frac{1}{2\sigma_I^2} \|\boldsymbol{\rho}_k^{(p)} - \boldsymbol{\beta}^{(p)}\|^2} \right) \end{aligned} \tag{11.317}$$
3. We merge (11.316) and (11.317) into one variable (see(7.390)),
$$\begin{aligned} C_k(j \mid i) &= P\,[\boldsymbol{z}_k = \boldsymbol{\rho}_k, \boldsymbol{s}_k = \boldsymbol{\sigma}_j \mid \boldsymbol{s}_{k-1} = \boldsymbol{\sigma}_i] = p_{\boldsymbol{z}_k}(\boldsymbol{\rho}_k \mid j, i)\, \Pi(j \mid i) \\ &= \left(\frac{1}{\sqrt{2\pi\sigma_I^2}} \right)^{n_0} C_k^{(s)}(j \mid i)\, C_k^{(p)}(j \mid i) \end{aligned} \tag{11.318}$$
where
$$C_k^{(s)}(j \mid i) = e^{-\frac{1}{2\sigma_I^2} \|\boldsymbol{\rho}_k^{(s)} - \boldsymbol{\beta}^{(s)}\|^2}\, \mathrm{L}_k^{(a)}(\boldsymbol{\beta}^{(s)}) \tag{11.319}$$
$$C_k^{(p)}(j \mid i) = e^{-\frac{1}{2\sigma_I^2} \|\boldsymbol{\rho}_k^{(p)} - \boldsymbol{\beta}^{(p)}\|^2} \tag{11.320}$$
The two previous quantities are related, respectively, to the systematic bits and the parity check bits of a code vector. Observe that the exponential term in (11.319) represents the reliability of a certain a priori information $\mathrm{L}_k^{(a)}(\boldsymbol{\beta}^{(s)})$ associated with $\boldsymbol{\beta}^{(s)}$.
4. The computation of the forward and backward metrics is carried out as in the general case.
   - *Forward metric*, for $k = 0, 1, \ldots, K-1$:
$$F_k(j) = \sum_{\ell=1}^{N_s} C_k(j \mid \ell)\, F_{k-1}(\ell), \quad j = 1, \ldots, N_s \tag{11.321}$$

– *Backward metric*, for $k = K-1, K-2, \dots, 0$:

$$B_k(i) = \sum_{m=1}^{N_s} B_{k+1}(m)\; C_{k+1}(m \mid i), \qquad i = 1, \dots, N_s \tag{11.322}$$

Suitable initializations are obtained, respectively, through (7.398) and (7.405).

5. By using the total probability theorem, the likelihood (11.313) can be written as

$$\mathrm{L}_k(\boldsymbol{\beta}^{(s)}) = \mathrm{A} \sum_{\substack{i=1 \\ \sigma_j = f_s(\boldsymbol{\beta}^{(s)}, \sigma_i)}}^{N_s} P\,[\boldsymbol{s}_{k-1} = \boldsymbol{\sigma}_i, \boldsymbol{s}_k = \sigma_j, \boldsymbol{z}_0^{K-1} = \boldsymbol{\rho}_0^{K-1}] \tag{11.323}$$

where f_s is the state transition function, and the multiplicative constant $\mathrm{A} = 1/P[\boldsymbol{z}_0^{K-1} = \boldsymbol{\rho}_0^{K-1}]$ is irrelevant for vector detection as can be seen from (11.314). We note that the summation in (11.323) is over all transitions from the general state $\boldsymbol{s}_{k-1} = \boldsymbol{\sigma}_i$ to the state $\boldsymbol{s}_k = \boldsymbol{\sigma}_j = f_s(\boldsymbol{\beta}^{(s)}, \sigma_i)$ generated by the information vector $\boldsymbol{b}_k = \boldsymbol{\beta}^{(s)}$. On the other hand, the probability in (11.323) can be written as

$$\begin{aligned}
P\,&[\boldsymbol{s}_{k-1} = \sigma_i, \boldsymbol{s}_k = \sigma_j, \boldsymbol{z}_0^{K-1} = \boldsymbol{\rho}_0^{K-1}] \\
&= P\,[\boldsymbol{z}_{k+1}^{K-1} = \boldsymbol{\rho}_{k+1}^{K-1} \mid \boldsymbol{s}_{k-1} = \sigma_i, \boldsymbol{s}_k = \sigma_j, \boldsymbol{z}_0^{k} = \boldsymbol{\rho}_0^{k}] \\
&\quad P\,[\boldsymbol{s}_k = \sigma_j, \boldsymbol{z}_k = \boldsymbol{\rho}_k \mid \boldsymbol{s}_{k-1} = \sigma_i, \boldsymbol{z}_0^{k-1} = \boldsymbol{\rho}_0^{k-1}] \\
&\quad P\,[\boldsymbol{s}_{k-1} = \sigma_i, \boldsymbol{z}_0^{k-1} = \boldsymbol{\rho}_0^{k-1}] \\
&= P\,[\boldsymbol{z}_{k+1}^{K-1} = \boldsymbol{\rho}_{k+1}^{K-1} \mid \boldsymbol{s}_k = \sigma_j] \\
&\quad P\,[\boldsymbol{s}_k = \sigma_j, \boldsymbol{z}_k = \boldsymbol{\rho}_k \mid \boldsymbol{s}_{k-1} = \sigma_i]\; P\,[\boldsymbol{s}_{k-1} = \sigma_i, \boldsymbol{z}_0^{k-1} = \boldsymbol{\rho}_0^{k-1}] \\
&= B_k(j)\; C_k(j|i)\; F_{k-1}(i)
\end{aligned} \tag{11.324}$$

Substituting for $C_k(j|i)$ the expression in (11.318), the likelihood becomes

$$\mathrm{L}_k(\boldsymbol{\beta}^{(s)}) = \mathrm{B}\; \mathrm{L}_k^{(a)}(\boldsymbol{\beta}^{(s)})\; \mathrm{L}_k^{(int)}(\boldsymbol{\beta}^{(s)})\; \mathrm{L}_k^{(ext)}(\boldsymbol{\beta}^{(s)}) \tag{11.325}$$

where $\mathrm{B} = \mathrm{A}/(2\pi\sigma_I^2)$ is an irrelevant constant,

$$\mathrm{L}_k^{(int)}(\boldsymbol{\beta}^{(s)}) = e^{-\frac{1}{2\sigma_I^2}\|\boldsymbol{\rho}_k^{(s)} - \boldsymbol{\beta}^{(s)}\|^2} \tag{11.326}$$

and

$$\mathrm{L}_k^{(ext)}(\boldsymbol{\beta}^{(s)}) = \sum_{\substack{i=1 \\ \sigma_j = f_s(\boldsymbol{\beta}^{(s)}, \sigma_i)}}^{N_s} B_k(j)\; C_k^{(p)}(j|i)\; F_{k-1}(i) \tag{11.327}$$

Observing each term in (11.325), we make the following considerations.

i. $\mathrm{L}_k^{(a)}(\boldsymbol{\beta}^{(s)})$ represents the *a priori information* on the information vector $\boldsymbol{b}_k = \boldsymbol{\beta}^{(s)}$.

ii. $\mathrm{L}_k^{(int)}(\boldsymbol{\beta}^{(s)})$ depends on the received samples associated with the information vector and on the channel characteristics.

iii. $\mathrm{L}_k^{(ext)}(\boldsymbol{\beta}^{(s)})$ represents the *extrinsic information* extracted from the received samples associated with the parity check bits. This is the incremental information on the information vector obtained by the decoding process.

6. Typically, it is easier to work with the logarithm of the various likelihoods.
We associate with each bit of the code vector $\boldsymbol{c}_k$ a LLR

$$\boldsymbol{\ell}_k^{(in)} = [\ell_k^{(in,1)}, \dots, \ell_k^{(in,n_0)}] = [\boldsymbol{\ell}_k^{(in,s)}, \boldsymbol{\ell}_k^{(in,p)}] \tag{11.328}$$

For binary modulation, from (11.309), ρ_k and $\ell_k^{(in)}$ are related through (6.39)

$$\ell_k^{(in)} = \frac{2}{\sigma_{dec}^2}\rho_k \tag{11.329}$$

where ρ_k is the observation at the instant k. Moreover, we can see that σ_{dec}^2 is not relevant since it is a common scaling factor not relevant for final detection.

We define now two quantities that are related, respectively, to the systematic bits and the parity check bits of the code vector, as

$$\ell_k^{(s)}(\boldsymbol{\beta}^{(s)}) = \frac{1}{2}\sum_{m=1}^{k_0} \ell_k^{(in,m)} \beta^{(m)} \tag{11.330}$$

and

$$\ell_k^{(p)}(j, i) = \frac{1}{2}\sum_{m=k_0+1}^{n_0} \ell_k^{(in,m)} \beta^{(m)} \tag{11.331}$$

where by (11.307) and (11.308), we have

$$\boldsymbol{\beta}^{(p)} = [\beta^{(k_0+1)}, \dots, \beta^{(n_0)}] = f^{(p)}(\boldsymbol{\sigma}_j, \boldsymbol{\sigma}_i) \tag{11.332}$$

Expressing (11.319) and (11.320) as a function of the likelihoods (11.330) and (11.331), apart from factors that do not depend on $\{\beta^{(m)}\}, m = 1, \dots, n_0$, we get

$$C_k^{'(s)}(j \mid i) = e^{\ell_k^{(s)}(\boldsymbol{\beta}^{(s)})} e^{\ell_k^{(a)}(\boldsymbol{\beta}^{(s)})} \tag{11.333}$$

and

$$C_k^{'(p)}(j \mid i) = e^{\ell_k^{(p)}(j,i)} \tag{11.334}$$

To compute the forward and backward metrics, we use, respectively, (11.321) and (11.322), where the variable $C_k(j|i)$ is replaced by $C_k'(j \mid i) = C_k^{'(s)}(j \mid i)\, C_k^{'(p)}(j \mid i)$. Similarly in (11.327) $C_k^{(p)}(j \mid i)$ is replaced by $C_k^{'(p)}(j \mid i)$. Taking the logarithm of (11.327), we obtain the extrinsic component $\ell_k^{(ext)}(\boldsymbol{\beta}^{(s)})$.

Finally, from (11.325), by neglecting non-essential terms, the log-likelihood associated with the information vector $\boldsymbol{b}_k = \boldsymbol{\beta}^{(s)}$ is given by

$$\ell_k(\boldsymbol{\beta}^{(s)}) = \ell_k^{(a)}(\boldsymbol{\beta}^{(s)}) + \ell_k^{(int)}(\boldsymbol{\beta}^{(s)}) + \ell_k^{(ext)}(\boldsymbol{\beta}^{(s)}) \tag{11.335}$$

where $\ell_k^{(int)}(\boldsymbol{\beta}^{(s)}) = \ell_k^{(s)}(\boldsymbol{\beta}^{(s)})$ is usually called the intrinsic component.

Expression (11.335) suggests an alternative method to (11.327) to obtain $\ell_k^{(ext)}(\boldsymbol{\beta}^{(s)})$, which uses the direct computation of $\ell_k(\boldsymbol{\beta}^{(s)})$ by (11.323) and (11.324), where $C_k(j \mid i)$ is replaced by $C_k'(j \mid i)$, whose factors are given in (11.333) and (11.334). From the known a priori information $\ell_k^{(a)}(\boldsymbol{\beta}^{(s)})$ and the intrinsic information (11.330), from (11.335), we get

$$\ell_k^{(ext)}(\boldsymbol{\beta}^{(s)}) = \ell_k(\boldsymbol{\beta}^{(s)}) - \ell_k^{(a)}(\boldsymbol{\beta}^{(s)}) - \ell_k^{(int)}(\boldsymbol{\beta}^{(s)}) \tag{11.336}$$

Going back to the expression (11.335), detection of the vector $\boldsymbol{b}_k$ is performed according to the rule

$$\hat{\boldsymbol{b}}_k = \arg\max_{\boldsymbol{\beta}^{(s)}} \ell_k(\boldsymbol{\beta}^{(s)}) \tag{11.337}$$

Note that to compute $\ell_k^{(ext)}(\boldsymbol{\beta}^{(s)})$ ($\ell_k(\boldsymbol{\beta}^{(s)})$) by the logarithm of (11.327) (or (11.323) and (11.324)), we can use the Max-Log-MAP method introduced in (6.43) and page 352.

Example 11.6.1 (Systematic convolutional code with rate 1/2)
For a convolutional code with rate $R_c = 1/2$, the information vector $\boldsymbol{b}_k = [b_k]$ is composed of only one bit ($k_0 = 1$), like the systematic part and the parity check part of $\boldsymbol{c}_k = [c_k^{(s)}, c_k^{(p)}]$.
In this case, it is sufficient to determine the log-likelihoods $\ell_k(-1)$ and $\ell_k(1)$, or better the LLR

$$\ell_k = \ln \frac{\mathrm{L}_k(1)}{\mathrm{L}_k(-1)} = \ell_k(1) - \ell_k(-1) \tag{11.338}$$

Detection of the information bit is performed according to the rule

$$\hat{b}_k = \mathrm{sgn}(\ell_k) \tag{11.339}$$

The a priori information at the decoder input is given by

$$\ell_k^{(a)} = \ln \frac{P[b_k = 1]}{P[b_k = -1]} \tag{11.340}$$

from which we derive the a priori probabilities

$$P[b_k = \beta^{(s)}] = e^{\ell_k^{(a)}(\beta^{(s)})} = \frac{e^{\beta^{(s)}\ell_k^{(a)}}}{1 + e^{\beta^{(s)}\ell_k^{(a)}}} = \frac{e^{-\frac{1}{2}\ell_k^{(a)}}}{1 + e^{-\ell_k^{(a)}}}\, e^{\frac{1}{2}\beta^{(s)}\ell_k^{(a)}}, \quad \beta^{(s)} \in \{-1, 1\} \tag{11.341}$$

By using LLRs, (11.330) yields

$$\ell_k^{(int)} = \ell_k^{(int)}(1) - \ell_k^{(int)}(-1) = \ell_k^{(in,1)} = \ell_k^{(in,s)} \tag{11.342}$$

In turn (11.333) and (11.334) for $k_0 = 1$ and $n_0 = 2$ simplify into

$$C_k^{\prime(s)}(j \mid i) = e^{\frac{1}{2}\,\beta^{(s)}\,(\ell_k^{(in,s)} + \ell_k^{(a)})} \tag{11.343}$$

$$C_k^{\prime(p)}(j \mid i) = e^{\frac{1}{2}\,\beta^{(p)}\,\ell_k^{(in,p)}} \tag{11.344}$$

The extrinsic component is obtained starting from (11.327) and using the above variables

$$\ell_k^{(ext)} = \ln \frac{\mathrm{L}_k^{(ext)}(1)}{\mathrm{L}_k^{(ext)}(-1)} = \ell_k^{(ext)}(1) - \ell_k^{(ext)}(-1) \tag{11.345}$$

From (11.335), apart from irrelevant terms, the LLR associated with the information bit b_k can be written as

$$\ell_k = \ell_k^{(a)} + \ell_k^{(int)} + \ell_k^{(ext)} \tag{11.346}$$

where the meaning of each of the three contributions is as follows:

- *A priori information* $\ell_k^{(a)}$: It is an a priori reliability measure on the bit b_k. This value can be extracted either from the known statistic of the information sequence or, in the case of iterative decoding of turbo codes, from the previous analysis.
- *Channel information* $\ell_k^{(int)} = \ell_k^{(in,s)}$: As it is evident from the case of binary modulation, where $\ell_k^{(in,s)} = \frac{2}{\sigma_{dec}^2}\rho_k^{(s)}$, if the noise variance is low, the contribute of $\ell_k^{(int)}$ usually dominates with respect to the other two terms; in this case bit detection is simply obtained by the sign of $\rho_k^{(s)}$.
- *Extrinsic information* $\ell_k^{(ext)}$: It is a reliability measure that is determined by the redundancy in the transmitted sequence. This contribute improves the reliability of transmission over a noisy channel using the parity check bits.

The decomposition (11.346) forms the basis for the iterative decoding of turbo codes.

Example 11.6.2 (Non-systematic code and LLR associated with the code bits)
Consider the case of a non-systematic code. If the code is also non- recursive, we need to use in place of (11.304) the state definition (11.273).

Now, all bits are parity check bits and (11.306) and (11.310) become, respectively,

$$\boldsymbol{c}_k = \boldsymbol{c}_k^{(p)} = [c_k^{(1)}, \ldots, c_k^{(n_0)}] \tag{11.347}$$

$$\boldsymbol{z}_k = \boldsymbol{z}_k^{(p)} = [z_k^{(1)}, \ldots, z_k^{(n_0)}] \tag{11.348}$$

However, the information vector is still given by $\boldsymbol{b}_k = [b_k^{(1)}, \ldots, b_k^{(k_0)}]$ with values $\boldsymbol{\alpha} = [\alpha^{(1)}, \ldots, \alpha^{(k_0)}]$, $\alpha^{(i)} \in \{-1, 1\}$. The likelihood (11.313) is given by

$$\mathrm{L}_k(\boldsymbol{\alpha}) = P\,[\boldsymbol{b}_k = \boldsymbol{\alpha} \mid \boldsymbol{z}_0^{K-1} = \boldsymbol{\rho}_0^{K-1}] \tag{11.349}$$

The various terms with superscript $^{(s)}$ of the previous analysis vanish by setting $k_0 = 0$. Therefore, (11.330) and (11.331) become

$$\ell_k^{(int)}(\boldsymbol{\beta}^{(s)}) = \ell_k^{(s)}(\boldsymbol{\beta}^{(s)}) = 0 \tag{11.350}$$

and

$$\ell_k^{(p)}(j, i) = \frac{1}{2} \sum_{m=1}^{n_0} \ell_k^{(in,m)} \beta^{(m)} \tag{11.351}$$

where $\boldsymbol{\beta} = \boldsymbol{\beta}^{(p)} = [\beta^{(1)}, \ldots, \beta^{(n_0)}] = f(\boldsymbol{\sigma}_j, \boldsymbol{\sigma}_i)$ is the code vector associated with the transition from state $\boldsymbol{\sigma}_i$ to state $\boldsymbol{\sigma}_j$. Note that, apart from irrelevant factors, (11.351) coincides with (11.285).

For $k_0 = 1$, it is convenient to use LLRs; in particular, (11.346) yields a LLR associated with the information bit b_k that is given by

$$\ell_k = \ell_k^{(a)} + \ell_k^{(ext)} \tag{11.352}$$

where $\ell_k^{(ext)}$ can be obtained directly using (11.345), (11.334), and (11.327).

In some applications, it is useful to associate a LLR with the encoded bit $c_k^{(q)}, q = 1, \ldots, n_0$, rather than to the information bit b_k. We define

$$\overline{\ell}_{k,q} = \ln \frac{P\,[c_k^{(q)} = 1 \mid \boldsymbol{z}_0^{K-1} = \boldsymbol{\rho}_0^{K-1}]}{P\,[c_k^{(q)} = -1 \mid \boldsymbol{z}_0^{K-1} = \boldsymbol{\rho}_0^{K-1}]} \tag{11.353}$$

Let $\overline{\ell}_{k,q}^{(a)}$ be the a priori information on the code bits. The analysis is similar to the previous case, but now, with respect to the encoder output, $c_k^{(q)}$ is regarded as an information bit, while the remaining bits $c_k^{(m)}, m = 1, \ldots, n_0,\ m \neq q$, are regarded as parity check bits. Equations (11.330), (11.331), (11.343), and (11.344), are modified as follows:

$$\overline{\ell}_{k,q}^{(s)}(\beta^{(q)}) = \frac{1}{2} \ell_k^{(in,q)} \beta^{(q)} \tag{11.354}$$

$$\overline{\ell}_{k,q}^{(p)}(j, i) = \frac{1}{2} \sum_{\substack{m=1 \\ m \neq q}}^{n_0} \ell_k^{(in,m)} \beta^{(m)} \tag{11.355}$$

and

$$C_{k,q}^{\prime(s)}(j \mid i) = e^{\frac{1}{2}\,\beta^{(q)}\,(\ell_k^{(in,q)} + \overline{\ell}_{k,q}^{(a)})} \tag{11.356}$$

$$C_{k,q}^{\prime(p)}(j \mid i) = e^{\overline{\ell}_{k,q}^{(p)}(j,i)} \tag{11.357}$$

Associated with (11.357), we obtain $\overline{\ell}^{(ext)}_{k,q}$ by using (11.345) and (11.327). The overall result is given by

$$\overline{\ell}_{k,q} = \overline{\ell}^{(a)}_{k,q} + \ell^{(in,q)}_{k} + \overline{\ell}^{(ext)}_{k,q}, \quad q = 1, \ldots, n_0 \tag{11.358}$$

Example 11.6.3 (Systematic code and LLR associated with the code bits)
With reference to the previous example, if the code is systematic, whereas (11.346) holds for the systematic bit $c^{(1)}_k$, for the parity check bits $c^{(q)}_k$, the following relations hold [41]. For $k_0 = 1$, let α be the value of the information bit b_k, $b_k = \alpha$, with $\alpha \in \{-1, 1\}$, associated with the code vector

$$\boldsymbol{c}_k = \boldsymbol{\beta} = [\alpha, \beta^{(2)}, \ldots, \beta^{(n_0)}] \tag{11.359}$$

where we assume $\beta^{(1)} = \alpha$. For $q = 2, \ldots, n_0$, we get

$$\overline{\ell}^{(s)}_{k,q}(\beta^{(q)}) = \frac{1}{2}\ell^{(in,q)}_{k}\beta^{(q)} \tag{11.360}$$

$$\overline{\ell}^{(p)}_{k,q}(j,i) = \frac{1}{2}\sum_{\substack{m=1 \\ m \neq q}}^{n_0} \ell^{(in,m)}_{k}\beta^{(m)} + \frac{1}{2}\ell^{(a)}_{k}\alpha \tag{11.361}$$

where $\ell^{(a)}_k$ is the a priori information of b_k. Furthermore,

$$C'^{(s)}_{k,q}(j \mid i) = e^{\frac{1}{2}\,\beta^{(q)}\,\ell^{(in,q)}_{k}} \tag{11.362}$$

$$C'^{(p)}_{k,q}(j \mid i) = e^{\overline{\ell}^{(p)}_{k,q}(j,i)} \tag{11.363}$$

From (11.363), we get $\overline{\ell}^{(ext)}_{k,q}$ using (11.345) and (11.327). The overall result is given by

$$\overline{\ell}_{k,q} = \ell^{(in,q)}_{k} + \overline{\ell}^{(ext)}_{k,q} \tag{11.364}$$

Iterative decoding

In this section, we consider the iterative decoding of turbo codes with $k_0 = 1$. In this case, as seen in Example 11.6.1, using the LLRs simplifies the procedure. In general, for $k_0 > 1$ we should refer to the formulation (11.335).

We now give a step-by-step description of the decoding procedure of a turbo code with rate 1/3, of the type shown in Figure 11.27, where each of the two component decoders DEC_1 and DEC_2 implements the FBA for recursive systematic convolutional codes with rate 1/2. The decoder scheme is shown in Figure 11.30, where the subscript in LLR corresponds to the component decoder. In correspondence to the information bit b_k, the turbo code generates the vector

$$\boldsymbol{c}_k = [c^{(1)}_k, c^{(2)}_k, c^{(3)}_k] \tag{11.365}$$

where $c^{(1)}_k = b_k$. We now introduce the following notation for the observation vector $\boldsymbol{\ell}^{(in)}_k$ that relates to the considered decoder:

$$\boldsymbol{\ell}^{(in)}_k = [\ell^{(in,s)}_{k}, \ell^{(in,p)}_{k,1}, \ell^{(in,p)}_{k,2}] \tag{11.366}$$

where $\ell^{(in,s)}_{k}$ corresponds to the systematic part, and $\ell^{(in,p)}_{k,1}$ and $\ell^{(in,p)}_{k,2}$ correspond to the parity check parts generated by the first and second convolutional encoder, respectively.

If some parity check bits are punctured to increase the code rate, at the receiver, the corresponding LLRs $\ell^{(in,m)}_{k}$ are set to zero.

1. *First iteration*:
   1.1 *Decoder DEC_1*: If the statistic of the information bits is unknown, then the bits of the information sequence are considered i.i.d. and the a priori information is zero,
$$\ell_{k,1}^{(a)} = \ln \frac{P[b_k = 1]}{P[b_k = -1]} = 0 \tag{11.367}$$
For $k = 0, 1, 2, \dots, K - 1$, observed $\ell_k^{(in,s)}$ and $\ell_{k,1}^{(in,p)}$, we compute according to (11.343) and (11.344) the variables $C_k^{\prime(s)}$ and $C_k^{\prime(p)}$, and from these the corresponding forward metric $F_k(j)$ (11.321). After the entire sequence has been received, we compute the backward metric $B_k(i)$ (11.322) and, using (11.327), we find $\mathrm{L}_{k,1}^{(ext)}(1)$ and $\mathrm{L}_{k,1}^{(ext)}(-1)$. The decoder soft output is the extrinsic information obtained by the LLR
$$\ell_{k,1}^{(ext)} = \ln \frac{\mathrm{L}_{k,1}^{(ext)}(1)}{\mathrm{L}_{k,1}^{(ext)}(-1)} \tag{11.368}$$
   1.2 *Interleaver*: Because of the presence of the interleaver, the parity check bit $c_n^{(3)}$ is obtained in correspondence to a transition of the convolutional encoder state determined by the information bit b_n, where n depends on the interleaver pattern. In decoding, the extrinsic information $\ell_{k,1}^{(ext)}$, extracted from DEC_1, and the systematic observation $\ell_k^{(in,s)}$ are scrambled by the turbo code interleaver and associated with the corresponding observation $\ell_{n,2}^{(in,p)}$ to form the input of the second component decoder.
   1.3 *Decoder DEC_2*: The extrinsic information generated by DEC_1 is set as the a priori information $\ell_{n,2}^{(a)}$ to the component decoder DEC_2,
$$\ell_{n,2}^{(a)} = \ln \frac{P[b_n = 1]}{P[b_n = -1]} = \ell_{n,1}^{(ext)} \tag{11.369}$$
The basic idea consists in supplying DEC_2 only with the extrinsic part $\ell_{n,1}^{(ext)}$ of $\ell_{n,1}$, in order to minimize the correlation between the a priori information and the observations used by DEC_2. Ideally, the a priori information should be an independent estimate.
As done for DEC_1, we extract the extrinsic information $\ell_{n,2}^{(ext)}$.
   1.4 *Deinterleaver*: The deinterleaver realizes the inverse function of the interleaver, so that the extrinsic information extracted from DEC_2, $\ell_{n,2}^{(ext)}$, is synchronized with the systematic part $\ell_k^{(in,s)}$ and the parity check part $\ell_{k,1}^{(in,p)}$ of the observation of DEC_1. By a feedback loop, the a posteriori information $\ell_{k,2}^{(ext)}$ is placed at the input of DEC_1 as a priori information $\ell_{k,1}^{(a)}$, giving origin to an iterative structure.
2. *Successive iterations*: Starting from the second iteration, each component decoder has at its input an a priori information. The information on the bits become more reliable as the a priori information stabilizes in sign and increases in amplitude.
3. *Last iteration*: When the decoder achieves convergence, the iterative process can stop and form the overall LLR (11.346),
$$\ell_{k,overall} = \ell_k^{(in,s)} + \ell_{k,1}^{(ext)} + \ell_{k,2}^{(ext)}, \quad k = 0, 1, \dots, K - 1 \tag{11.370}$$
and detection of the information bits b_k is obtained by
$$\hat{b}_k = \mathrm{sgn}(\ell_{k,overall}) \tag{11.371}$$

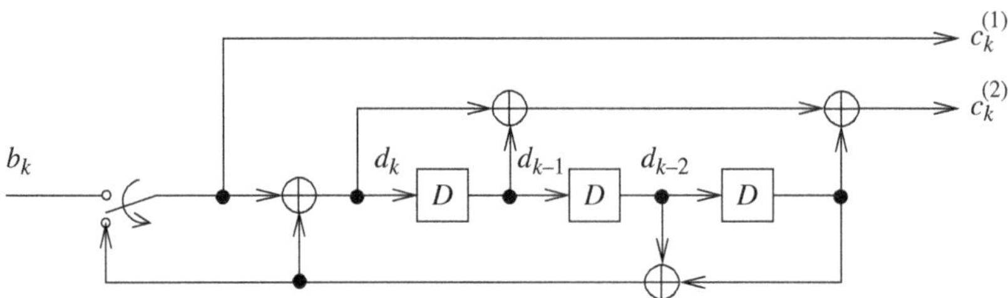

Figure 11.31 Termination of trellis.

To make decoding more reliable, the final state of each component decoder is set to zero, thus enabling an initialization of the backward metric as in (7.405). As illustrated in Figure 11.31, at the instant following the input of the last information bit, that is for $k = K$, the commutator is switched to the lower position, and therefore, we have $d_k = 0$; after ν clock intervals the zero state is reached. The bits $c_k^{(1)}$ and $c_k^{(2)}$, for $k = K, K+1, \dots, K+\nu-1$, are appended at the end of the code sequence to be transmitted.

Performance evaluation

Performance of the turbo code with the encoder of Figure 11.27 is evaluated in terms of error probability and convergence of the iterative decoder implemented by the FBA. For the memoryless AWGN channel, error probability curves versus E_b/N_0 are plotted in Figure 11.32 for a sequence of information bits of length $K = 640$, and various numbers of iterations of the iterative decoding process. Note that performance improves as the number of iterations increases; however, the gain between consecutive iterations becomes smaller as the number of iterations increases.

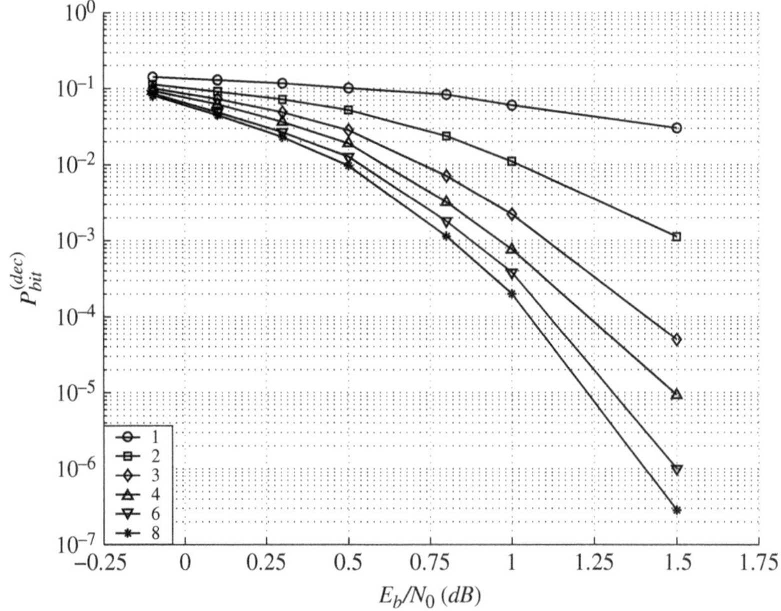

Figure 11.32 Performance of the turbo code defined by the UMTS standard, with length of the information sequence K=640, and various numbers of iterations of the iterative decoding process.

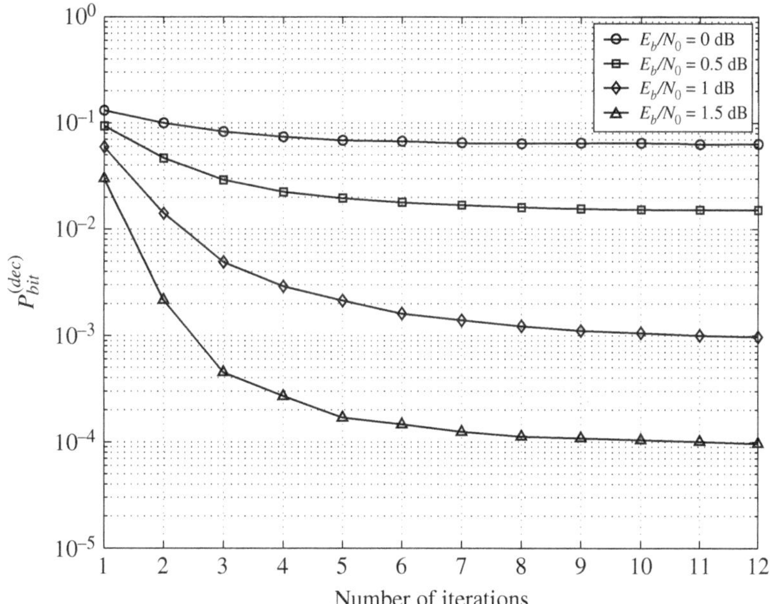

Figure 11.33 Curves of convergence of the decoder for the turbo code defined by the UMTS standard, for $K = 320$ and various values of E_b/N_0.

In Figure 11.33, the error probability $P_{bit}^{(dec)}$ is given as a function of the number of iterations, for fixed values of E_b/N_0, and $K = 320$. From the behaviour of the error probability, we deduce possible criteria for stopping the iterative decoding process at convergence [40]. A timely stop of the iterative decoding process leads to a reduction of the decoding delay and of the overall computational complexity of the system. Note, however, that convergence is not always guaranteed.

The performance of the code depends on the length K of the information sequence. Figure 11.34 illustrates how the bit error probability decreases by increasing K, for a constant E_b/N_0. A higher value of K corresponds to an interleaver on longer sequences and thus the assumption of independence among the inputs of each component decoder is better satisfied. Moreover, the burst errors introduced by the channel are distributed over all the original sequence, increasing the correction capability of the decoder. As the length of the interleaver grows also the latency of the system increases.

11.7 Iterative detection and decoding

We consider the transmitter of Figure 11.35 composed of a convolutional encoder, interleaver, bit mapper, and modulator for 16-PAM. Interpreting the channel as a finite-state machine, the overall structure may be interpreted as a *serial concatenated convolutional code* (SCCC). The procedure of soft-input, soft-output detection and soft input (SI) decoding of page 589 can be made iterative by applying the principles of the previous section, by including a soft-input, soft-output decoding stage. With reference to Figure 11.36, a step-by-step description follows:

0. *Initialization*: Suppose we have no information on the a priori probability of the code bits, therefore, we associate with c_n a zero LLR,

$$\ell_{n,det}^{(a)} = 0 \tag{11.372}$$

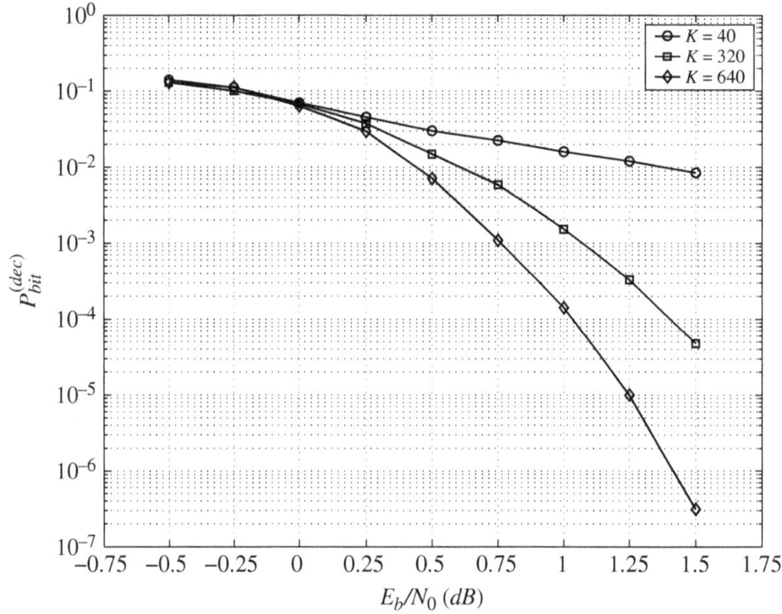

Figure 11.34 Performance of the turbo code defined by the UMTS standard achieved after 12 iterations, for K=40, 320, and 640.

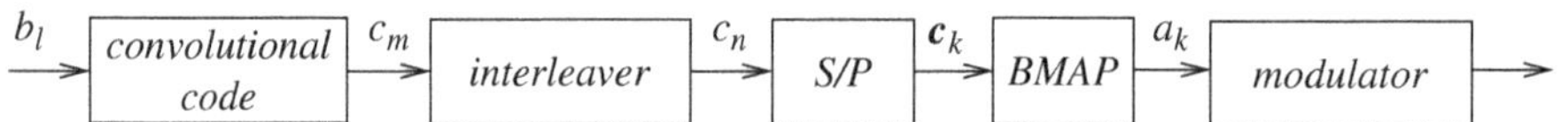

Figure 11.35 Encoder structure, bit mapper, and modulator; for 16-PAM: $\boldsymbol{c}_k = [c_{4k}, c_{4k-1}, c_{4k-2}, c_{4k-3}]$.

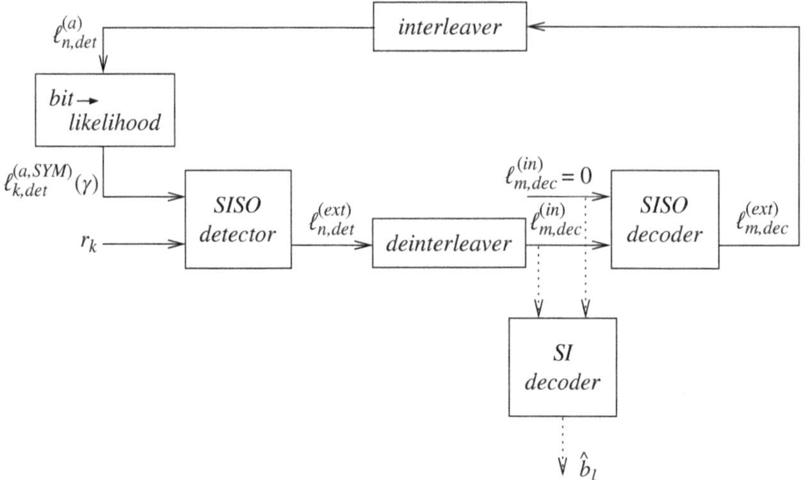

Figure 11.36 Iterative detection and decoding.

1. *Detector*: First, we associate a log-likelihood with the two possible values of $c_n = \eta$, $\eta \in \{-1, 1\}$, according to the rule

$$\ell^{(a)}_{n,det}(1) = \ell^{(a)}_{n,det}, \qquad \ell^{(a)}_{n,det}(-1) = 0 \tag{11.373}$$

Then we express the symbol a_k as a function of the bits $\{c_n\}$ according to the bit mapper, for example for 16-PAM,

$$a_k = \text{BMAP}\{\boldsymbol{c}_k = [c_{4k}, c_{4k-1}, c_{4k-2}, c_{4k-3}]\} \tag{11.374}$$

Assuming the sequence $\{c_n\}$ is a sequence of i.i.d. binary symbols, we associate with each value of the symbol a_k the a priori information expressed by the log-likelihood

$$\ell^{(a,SYM)}_{k,det}(\gamma) = \sum_{t=0}^{3} \ell^{(a)}_{4k-t,det}(\eta_t), \quad \gamma \in \mathcal{A} \tag{11.375}$$

where $\gamma = \text{BMAP}\{[\eta_0, \dots, \eta_3]\}$, $\eta_t \in \{-1, 1\}$.
For multilevel transmission over a channel with ISI, the FBA of Section 7.5.2 provides a log-likelihood for each value of a_k. The new feature is that now in (7.385), we take into account the a priori information on the various values of a_{k+L_1}; then (7.390) becomes

$$C_k(j \mid i) = \frac{1}{\sqrt{2\pi\sigma_I^2}} e^{-\frac{1}{2\sigma_I^2}|\rho_k - u_k|^2} e^{\ell^{(a,SYM)}_{k+L_1,det}(\gamma)} \tag{11.376}$$

where $\gamma = f(\sigma_j, \sigma_i) \in \mathcal{A}$ is the symbol associated with the transition from the state σ_i to the state σ_j on the trellis determined by the ISI.
If $V_k(i)$, $i = 1, \dots, N_s$, denotes the metric corresponding to the various states of the trellis, we associate with each value of the code bits $\{c_n\}$ the following likelihood:

$$\text{L}_{4(k+L_1)-t,det}(\alpha) = \sum_{\substack{i=1 \\ \sigma_i \text{ such that} \\ c_{4(k+L_1)-t} = \alpha}}^{N_s} V_k(i), \quad \alpha \in \{-1, 1\}, \quad t = 0, \dots, 3 \tag{11.377}$$

Taking the logarithm of (11.377), we obtain the LLR

$$\ell_{n,det} = \ell_{n,det}(1) - \ell_{n,det}(-1) \tag{11.378}$$

To determine the extrinsic information associated with $\{c_n\}$, we subtract the a priori information from (11.378),

$$\ell^{(ext)}_{n,det} = \ell_{n,det} - \ell^{(a)}_{n,det} \tag{11.379}$$

Note that in this application, the detector considers the bits $\{c_n\}$ as information bits and the log-likelihood associated with c_n at the detector output is due to the channel information[8] in addition to the a priori information.
In [42], the quantity $\ell^{(a)}_{n,det}$ in (11.379) is weighted by a coefficient, which is initially chosen small, when the a priori information is not reliable, and is increased after each iteration.
2. *Deinterleaver*: The metrics $\ell^{(ext)}_{n,det}$ are re-ordered according to the deinterleaver to provide the sequence $\ell^{(ext)}_{m,det}$.

[8] For the iterative decoding of turbo codes, this information is defined as intrinsic.

3. *Soft-input, soft-output decoder*: As input LLR, we use

$$\ell^{(in)}_{m,dec} = \ell^{(ext)}_{m,det} \tag{11.380}$$

and we set

$$\ell^{(a)}_{m,dec} = 0 \tag{11.381}$$

in the lack of an a priori information on the code bits $\{c_m\}$. Indeed, we note that in the various formulas, the roles of $\ell^{(a)}_{m,dec}$ and $\ell^{(in)}_{m,dec}$ can be interchanged.
Depending on whether the code is systematic or not, we use the soft-input, soft-output decoding procedure reported in Examples 11.6.1 and 11.6.2, respectively. In both cases, we associate with the encoded bits c_m the quantity

$$\ell^{(ext)}_{m,dec} = \ell_{m,dec} - \ell^{(in)}_{m,dec} \tag{11.382}$$

that is passed to the soft-input soft-output detector as a priori information, after reordering by the interleaver.

4. *Last iteration*: After a suitable number of iterations, the various metrics stabilize, and from the LLRs $\{\ell^{(in)}_{m,dec}\}$ associated with $\{c_m\}$, the SI decoding of bits $\{b_l\}$ is performed, using the procedure of Example 11.6.1.

11.8 Low-density parity check codes

Low-density parity check (LDPC) codes were introduced by Gallager [9] as a family of linear block codes with parity check matrices containing mostly zeros and only a small number of ones. The *sparsity* of the parity check matrices defining LDPC codes is the key for the efficient decoding of these codes by a message-passing procedure also known as the *sum-product algorithm*. LDPC codes and their efficient decoding were *reinvented* by MacKay and Neal [43, 44] in the mid 1990s, shortly after Berrou and Glavieux introduced the turbo-codes. Subsequently, LDPC codes have generated interest from a theoretical as well as from a practical viewpoint and many new developments have taken place.

It is today well acknowledged that LDPC codes are as good as turbo codes, or even better. Also the decoding techniques used for both methods can be viewed as different realizations of the same fundamental decoding process. However, the soft input soft output FBA, or suboptimal versions of it, used for turbo decoding is rather complex, whereas the sum-product algorithm used for LDPC decoding lends itself to parallel implementation and is computationally simpler. LDPC codes, on the other hand, may lead to more stringent requirements in terms of storage.

11.8.1 Representation of LDPC codes

Matrix representation

From Section 11.2.1, we recall that a linear (n_0, k_0) block code, where n_0 and k_0 denote the transmitted block length and the source block length, respectively, can be described in terms of a parity check matrix $\boldsymbol{H}$, such that the equation

$$\boldsymbol{Hc} = \boldsymbol{0} \tag{11.383}$$

is satisfied for all code words $\boldsymbol{c}$. Each row of the $r_0 \times n_0$ parity check matrix, where $r_0 = n_0 - k_0$ is the number of parity check bits, defines a parity check equation that is satisfied by each code word $\boldsymbol{c}$. For example, the (7,4) Hamming code is defined by the following parity check equations:

$$\begin{bmatrix} 1 & 1 & 1 & 0 & 1 & 0 & 0 \\ 1 & 1 & 0 & 1 & 0 & 1 & 0 \\ 1 & 0 & 1 & 1 & 0 & 0 & 1 \end{bmatrix} \begin{bmatrix} c_1 \\ c_2 \\ c_3 \\ c_4 \\ c_5 \\ c_6 \\ c_7 \end{bmatrix} = \mathbf{0} \leftrightarrow \begin{array}{ll} c_5 = c_1 \oplus c_2 \oplus c_3 & \text{(check 1)} \\ c_6 = c_1 \oplus c_2 \oplus c_4 & \text{(check 2)} \\ c_7 = c_1 \oplus c_3 \oplus c_4 & \text{(check 3)} \end{array} \tag{11.384}$$

LDPC codes differ in major ways with respect to the above simple example; they usually have long block lengths n_0 in order to achieve near Shannon-limit performance, their parity check matrices are defined in non-systematic form and exhibit a number of ones that is much less than $r_0 \cdot n_0$. Traditionally, the parity-check matrices of LDPC codes have been defined pseudo-randomly subject to the requirement that $\boldsymbol{H}$ be sparse, and code construction of binary LDPC codes involves randomly assigning a small number of the values in all-zero matrix to be 1. So the lack of any obvious algebraic structure in randomly constructed LDPC codes is another reason that sets them apart from traditionally parity-check codes. Moreover, they have a decoding algorithm whose complexity is linear in the block length of the code, which allows the decoding of large codes.

A parity check matrix for a (J, K)*-regular LDPC code* has exactly J ones in each of its columns and K ones in each of its rows. If $\boldsymbol{H}$ is low density, but the number of 1s in each column or row is not constant, then the code is an *irregular LDPC code*.

Importantly, an error correction code can be described by more than one parity check matrix. It is not required that two parity-check matrices, for the same code, have the same number of rows, but they need the rank over GF(2) of both be the same, since the number of information bits, k_0, in a binary code it is given by

$$k_0 = n_0 - \text{rank}_2(\boldsymbol{H}) \tag{11.385}$$

where $rank_2(\boldsymbol{H})$ is the number of rows $\boldsymbol{H}$ which are linearly independent over GF(2). The pair (J, K) together with the code length n_0 specifies an *ensemble* of codes, rather than any particular code.

Gallager showed that the minimum distance of a *typical* regular LDPC code increases linearly with n_0, provided $J \geq 3$. Therefore, regular LDPC codes are constructed with J on the order of 3 or 4, subject to the above constraint. For large block lengths, the random placement of 1s in $\boldsymbol{H}$ such that each row has exactly K 1s and each column has exactly J 1s requires some effort and systematic methods for doing this have been developed.

We emphasize that the performance of an LDPC code depends on the random realization of the parity check matrix $\boldsymbol{H}$. Hence, these codes form a constrained random code ensemble.

Graphical representation

A parity check matrix can generally be represented by a bipartite graph, also called Tanner graph, with two types of nodes: the *bit nodes* and the *parity check nodes* (or check nodes) [45]. A bit node n, representing the code bit c_n, is connected by an *edge* to the check node m only if the element (m, n) of the parity check matrix is equal to 1. No bit (check) node is connected to a bit (check) node. In a graph, the number of edges incident upon a node is called the *degree* of the node. For example, the (7,4) Hamming code can be represented by the graph shown in Figure 11.37.

We note in this specific case that because the parity check matrix is given in systematic form, bit nodes c_5, c_6, and c_7 in the associated graph are connected to single distinct check nodes. The parity check matrix

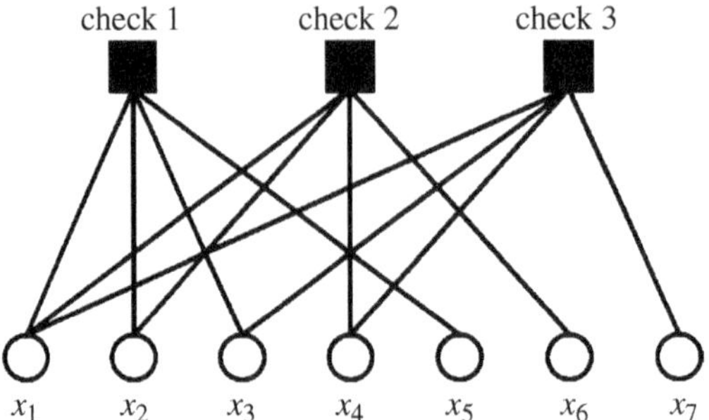

Figure 11.37 Tanner graph corresponding to the parity check matrix of the (7,4) Hamming code.

of a (J, K)-regular LDPC code leads to a graph where every bit node is connected to precisely J check nodes, and every check node is connected to precisely K bit nodes.

A *cycle* in a Tanner graph is a sequence of connected vertices which starts and ends at the same vertex in the graph, and which contains other vertices no more than once. The length of a cycle is the number of edges it contains, and the *girth* of a graph is the size of its smallest cycle. The shortest possible cycle in a bipartite graph is clearly a length-4 cycle, and such cycles manifest themselves in the $\boldsymbol{H}$ matrix as four 1s that lie on the corners of a submatrix of $\boldsymbol{H}$. We are interested in cycles, particularly short cycles, because they degrade the performance of the iterative decoding algorithm used for LDPC codes.

Graphical representations of LDPC codes are useful for deriving and implementing the iterative decoding procedure introduced in [9]. Gallager decoder is a message-passing decoder, in a sense to be made clear below, based on the so-called sum-product algorithm, which is a general decoding algorithm for codes defined on graphs.[9]

11.8.2 Encoding

Encoding procedure

From (11.34), encoding is performed by multiplying the vector of k_0 information bits $\boldsymbol{b}$ by the generator matrix $\boldsymbol{G}$ of the LDPC code:

$$\boldsymbol{c}^T = \boldsymbol{b}^T \boldsymbol{G} \tag{11.386}$$

where the operations are in GF(2). Hence, the generator and parity check matrices satisfy the relation

$$\boldsymbol{H}\boldsymbol{G}^T = \boldsymbol{0} \tag{11.387}$$

The parity check matrix in systematic form is $\tilde{\boldsymbol{H}} = [\tilde{\boldsymbol{A}}, \boldsymbol{I}]$, where $\boldsymbol{I}$ is the $r_0 \times r_0$ identity matrix and $\tilde{\boldsymbol{A}}$ is a binary matrix. Recall also that any other $r_0 \times n_0$ matrix $\boldsymbol{H}$ whose rows span the same space as $\tilde{\boldsymbol{H}}$ is a valid parity check matrix.

Given the block length n_0 of the transmitted sequence and the block length k_0 of the information sequence, we select a column weight J, greater than or equal to 3. To define the code, we generate a rectangular $r_0 \times n_0$ matrix $\boldsymbol{H} = [\boldsymbol{A}\ \boldsymbol{B}]$ at random with exactly J ones per column and, assuming a proper choice of n_0 and k_0, exactly K ones per row. The $r_0 \times k_0$ matrix $\boldsymbol{A}$ and the square $r_0 \times r_0$ matrix $\boldsymbol{B}$ are very sparse. If the rows of $\boldsymbol{H}$ are independent, which is usually true with high probability if J is odd [44],

[9] A wide variety of other algorithms (e.g. the VA, the FBA, the iterative turbo decoding algorithm, the fast Fourier transform,...) can also be derived as specific instances of the sum-product algorithm [46].

by Gaussian elimination and reordering of columns we determine an equivalent parity check matrix $\tilde{\boldsymbol{H}}$ in systematic form. We obtain the generator matrix in systematic form as

$$\boldsymbol{G}^T = \begin{bmatrix} \boldsymbol{I} \\ \tilde{\boldsymbol{A}} \end{bmatrix} = \begin{bmatrix} \boldsymbol{I} \\ \boldsymbol{B}^{-1}\boldsymbol{A} \end{bmatrix} \tag{11.388}$$

where I is the $k_0 \times k_0$ identity matrix.

One of the most relevant problem of this encoding procedure is its quadratic complexity in the block length. In order to reduce this complexity, either linear or quadratic, Richardson and Urbanke (RU) took advantage of the sparsity of the $\boldsymbol{H}$ matrix. The RU algorithm consists of two steps: a preprocessing and the actual encoding [47].

11.8.3 Decoding

The key to extracting maximal benefit from LDPC codes is *soft decision* decoding. In general, they are iterative decoding algorithms which are based on the code's Thanner graph, where *message passing* is an essential part of them. In particular, we imagine that bit nodes are a sort of processors of one type, check nodes processors of another type, and the edges represent message paths.

Hard decision decoder

To illustrate the process of iterative decoding, we first show an intuitive algorithm which uses an hard decision (0 or 1) assessment of each received bit. For this algorithm, the messages are simple: each bit node n sends a message to each of the check nodes to which it is connected, declaring if it is 1 or 0, and each check node i sends a message to each of the bit nodes to which it is connected declaring what value the bit node n should have in order to satisfy the check equation i, given the value of the bit nodes (other than n) involved in it. This algorithm operates as follows:

Initialization: Each bit node is assigned the bit value received, by hard decision, from the channel and send messages to check nodes to which it is connected indicating its value.

First step: Using the messages from the bit nodes, each check node determines which value should have each bit node, in order to satisfy the parity check equation. If the sum (modulo 2) of the bit nodes (other than the message receiver) is 1, the left bit node will receive the message 1, otherwise, if the sum is 0, the message sent will be 0.

Second step: If the majority of the messages received by each bit node is different from the value with it associated, the bit node flips its current value, in order to satisfies the greater part of check equations, otherwise, the value is retained.

Third step: If $\boldsymbol{H}\hat{\boldsymbol{c}} = \boldsymbol{0}$, where $\hat{\boldsymbol{c}}$ is the updated received code word, decoding is stopped; if the maximum number of allowed iterations is reached, the algorithm terminates and a failure to converge is reported; otherwise, each bit node sends new messages to the check nodes to which it is connected, indicating its value, and the algorithm returns to the *First step.*

Example 11.8.1

Consider the following (6,2) code with parity check matrix

$$\boldsymbol{H} = \begin{bmatrix} 1 & 1 & 0 & 1 & 0 & 0 \\ 0 & 1 & 1 & 0 & 1 & 0 \\ 1 & 0 & 0 & 0 & 1 & 1 \\ 0 & 0 & 1 & 1 & 0 & 1 \end{bmatrix} \tag{11.389}$$

with $J = 2$ and $K = 3$. The corresponding Thanner graph is shown in Figure 11.38:

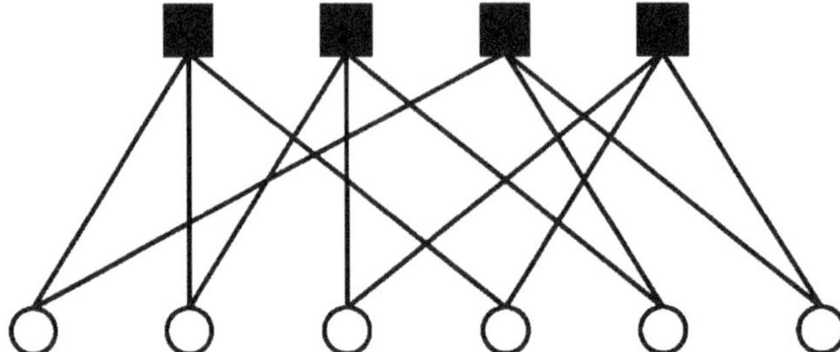

Figure 11.38 Tanner graph corresponding to the parity check matrix of code (11.389).

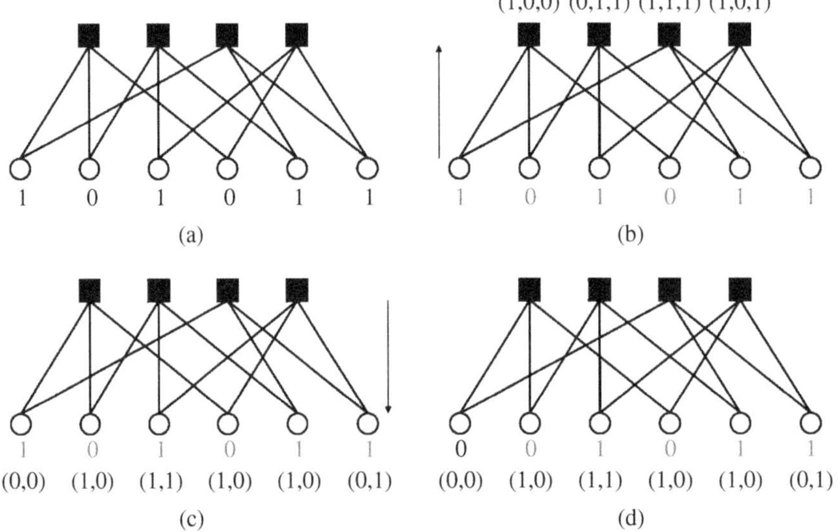

Figure 11.39 Illustrative example of iterative decoding: each sub-figure show the messages flow of the algorithm at each step. The arrow indicates the messages direction.

Assume that the code word sent is $\boldsymbol{c} = [0\ 0\ 1\ 0\ 1\ 1]$ and that the word received by hard decision from the channel is $\tilde{\boldsymbol{c}} = [1\ 0\ 1\ 0\ 1\ 1]$. The steps required to decode this received word are shown in Figure 11.39.

First of all the bit values are initialized to be 1, 0, 1, 0, 1, and 1, respectively (Figure 11.39a), and messages are sent to the check nodes indicating these values (Figure 11.39b). Then each check node sends the desired value, in order to satisfy the related equation, to the bit nodes connected with it (Figure 11.39c). Finally, we can see that the first bit has the majority of its messages indicating a value different from the initial one and so it flips from 1 to 0. Now, the word received is a code word, so the algorithm halts and returns $\tilde{\boldsymbol{c}} = [0\ 0\ 1\ 0\ 1\ 1]$.

The existence of cycles in the Thanner graph of a code reduces the effectiveness of the iterative decoding process. To illustrate the effect of a 4-cycle, we can take into consideration the code which leads to the representations on Figure 11.40, where the sent code word is $\boldsymbol{c} = [0\ 0\ 1\ 0\ 0\ 1]$, while the word received from the channel is $\tilde{\boldsymbol{c}} = [1\ 0\ 1\ 0\ 0\ 1]$.

As a result of the 4-cycle each of the first two code word bits are involved in the same two parity-check equations, and so when neither of the parity check equations is satisfied, it is not possible to determine which bit is causing the error.

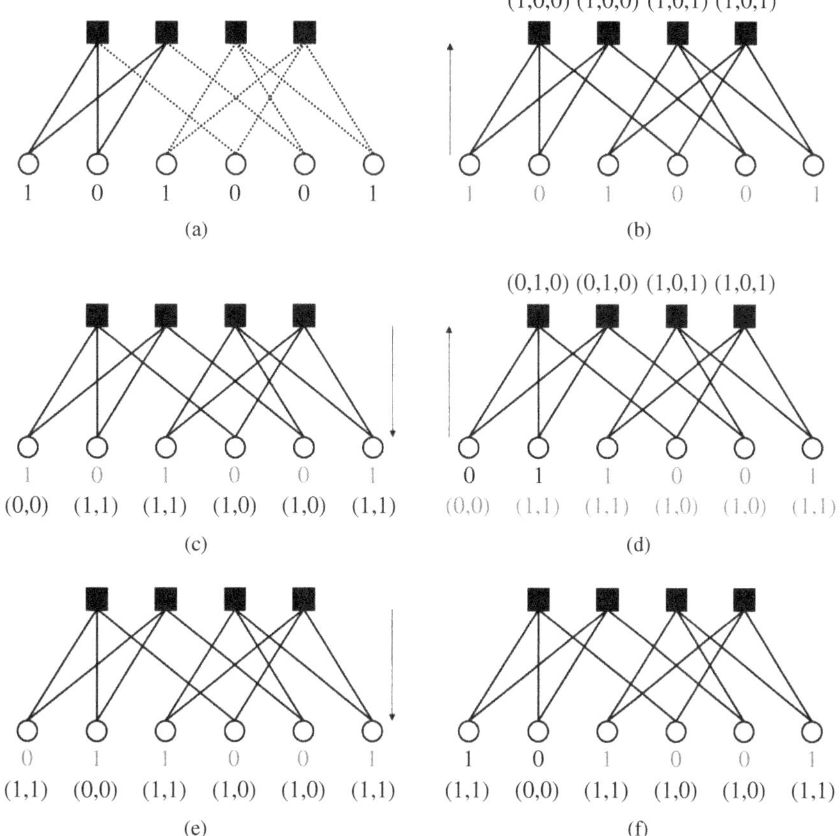

Figure 11.40 Illustrative example of a 4-cycle Thanner graph. The cycle is shown in solid lines in the first subfigure.

The sum-product algorithm decoder

The above description of hard input decoding is mainly to get an idea about the decoder procedure. Now, we introduce the *sum-product algorithm* (SPA), also denoted as *belief propagation algorithm* (BPA) or *message passing algorithm*: the underlying idea is exactly the same as in hard input decoding, but with messages representing now probabilities.

Given the vector of input LLRs $\boldsymbol{\ell}^{(in)} = [\ell_1^{(in)}, \dots, \ell_{n_0}^{(in)}]^T$ associated to code words $\boldsymbol{c} = [c_1, \dots c_{n_0}]^T$, the decoder problem is to determine the most likely binary vector $\boldsymbol{x}$ such that (see (11.383))

$$\boldsymbol{s} = \boldsymbol{H}\boldsymbol{x} = \boldsymbol{0} \tag{11.390}$$

A word on notation: when we use parity check matrices or equations, we assume the $\{0, 1\}$ alphabet, while we use the $\{-1, 1\}$ alphabet on detected bits of decoder output. First, we give some general definitions:

- *checks* s_i, $i = 1, \dots, r_0$: elements of the vector $\boldsymbol{s}$. They are represented by the check nodes in the corresponding Tanner graph.
- $L(i) = \{n : H_{i,n} = 1\}$, $i = 1, \dots, r_0$: set of the bit nodes that participate in the check with index i.
- $L(i)\setminus\tilde{n}$: set $L(i)$ from which the element with index $\tilde{n}$ has been removed.

- $M(n) = \{i : H_{i,n} = 1\}$, $n = 1, \dots, n_0$: set of the check nodes in which the bit with index n participates.
- $M(n)\setminus\tilde{i}$: set $M(n)$ from which the element with index $\tilde{i}$ has been removed.

Then the aim is to compute the *a posteriori probability* (APP) that a given bit in code word $\boldsymbol{c}$ equals β, $\beta \in \{0, 1\}$, given input vector $\boldsymbol{\ell}^{(in)}$. Without loss of generality, let us focus on the decoding of bit x_n, so that we are interested in computing the APP

$$q_n^\beta = P[x_n = \beta \mid \boldsymbol{s} = \boldsymbol{0}] \tag{11.391}$$

For the sake of a simpler notation in the above and following expressions, we drop the condition on the input LLRs $\boldsymbol{\ell}^{(in)}$.

The algorithm consists of two alternating steps, illustrated in Figure 11.41, in which two quantities $q_{i,n}^\beta$ and $r_{i,n}^\beta$, associated with each non-zero element of the matrix $\boldsymbol{H}$, are iteratively updated. The quantity $q_{i,n}^\beta$ denotes the probability that $x_n = \beta$, $\beta \in \{0, 1\}$, given the check equations (other than check i) involving x_n are satisfied:

$$q_{i,n}^\beta = P[x_n = \beta \mid \{s_{i'} = 0, i' \in M(n)\setminus i\}] \tag{11.392}$$

On the other hand, the quantity $r_{i,n}^\beta$ denotes the probability of check i being satisfied and the bits (other than x_n) having a known distribution (given by the probabilities $\{q_{i,n'}^\beta : n' \in L(i)\setminus n,\ \beta \in \{0, 1\}\}$), given $x_n = \beta$, $\beta \in \{0, 1\}$:

$$r_{i,n}^\beta = P[s_i = 0, \{x_{n'}, n' \in L(i)\setminus n\} \mid x_n = \beta] \tag{11.393}$$

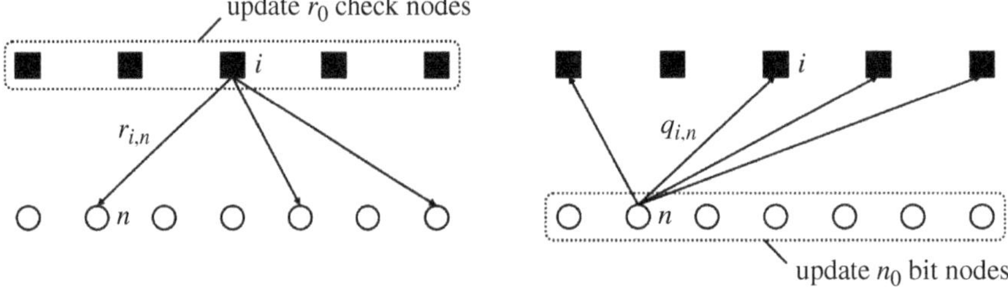

Figure 11.41 Message-passing decoding.

In the first step, the quantities $r_{i,n}^\beta$ associated with check node i are updated and passed as messages to the bit nodes checked by check node i. This operation is performed for all check nodes. In the second step, quantities $q_{i,n}^\beta$ associated with bit node n are updated and passed as messages to the check nodes that involve bit node n. This operation is performed for all bit nodes.

From (11.398) and (11.401), we have that

$$\begin{aligned} r_{i,n}^0 &= 1 - P[s_i = 1, \{x_{n'}, n' \in L(i)\setminus n\} \mid x_n = 0] \\ &= 1 - P[s_i = 0, \{x_{n'}, n' \in L(i)\setminus n\} \mid x_n = 1] \\ &= 1 - r_{i,n}^1 \end{aligned} \tag{11.394}$$

The algorithm is described as follows:

Initialization: Let $p_n^0 = P[x_n = 0]$ denote the probability that $x_n = 0$ given the observation $\boldsymbol{\ell}^{(in)}$, and $p_n^1 = P[x_n = 1] = 1 - p_n^0$. From the definition (6.37) of LLR $\ell_n^{(in)}$

$$p_n^0 = \frac{1}{1 + e^{\ell_n^{(in)}}}, \qquad p_n^1 = \frac{1}{1 + e^{-\ell_n^{(in)}}} \tag{11.395}$$

Let $q_{i,n}^\beta = p_n^\beta$, $n \in L(i)$, $i = 1, \dots, r_0$, $\beta \in \{0, 1\}$. Then all bit nodes sends their $q_{i,n}^\beta$ messages to the connected check nodes.

First step: We run through the checks, and for the ith check we compute for each $n \in L(i)$ the probability $r^0_{i,n}$ that, given $x_n = 0$, $s_i = 0$ and the other bits $\{x_{n'} : n' \neq n\}$ have a distribution $\{q^0_{i,n'}, q^1_{i,n'}\}$.
In order to evaluate $r^0_{i,n}$, we introduce the important result (due to Gallager, [48]): consider the sequence of K independent bits $\{d_{n'}\}$, $n' = 1, \dots, K$, for which $P[d_{n'} = 1] = p_n$. Then the probability that the sequence $\{d_{n'}\}$, $n' = 1, \dots, K$, contains an even number of 1s is

$$\frac{1}{2} + \frac{1}{2}\prod_{n'=1}^{K}(1 - 2p_{n'}) \tag{11.396}$$

The proof is by induction on K. In view of this result, from (11.394), we obtain

$$r^0_{i,n} = \frac{1}{2} + \frac{1}{2}\prod_{n' \in L(i)\setminus n}(1 - 2q^1_{i,n'}) \tag{11.397}$$

since, given $x_n = 0$, check equation i must contain an even number of 1s in order to be satisfied. Moreover, from (11.402), $r^1_{i,n} = 1 - r^0_{i,n}$.

Second step: After computing $r^0_{i,n}$ and $r^1_{i,n}$, we update the values of the probabilities $q^0_{i,n}$ and $q^1_{i,n}$. From (11.400) we obtain

$$q^\beta_{i,n} = \frac{P[x_n = \beta, \{s_{i'} = 0, i' \in M(n)\setminus i\}]}{P[\{s_{i'} = 0, i' \in M(n)\setminus i\}]} \tag{11.398}$$

Lumping in $\alpha_{i,n}$, the contribution of the terms that do not depend on β, we get

$$\begin{aligned} q^\beta_{i,n} &= \alpha_{i,n}\, P[s_{i'} = 0, i' \in M(n)\setminus i \mid x_n = \beta]\, p^\beta_n \\ &= \alpha_{i,n}\, p^\beta_n \prod_{i' \in M(n)\setminus i} r^\beta_{i',n} \end{aligned} \tag{11.399}$$

where we used Bayes rule twice to obtain the first line and the independence assumption on code bits to obtain the second line. The constant $\alpha_{i,n}$ is chosen such that $q^0_{i,n} + q^1_{i,n} = 1$. Taking into account the information from all check nodes, from (11.399), we can also compute the *pseudo a posteriori probabilities* q^0_n and q^1_n at this iteration,

$$q^0_n = \alpha_n\, p^0_n \prod_{i \in M(n)} r^0_{i,n} \tag{11.400}$$

$$q^1_n = \alpha_n\, p^1_n \prod_{i \in M(n)} r^1_{i,n} \tag{11.401}$$

where α_n is chosen such that $q^0_n + q^1_n = 1$.
At this point, it is possible to detect a code word $\hat{\boldsymbol{c}}$ by the log-MAP criterion; for $n = 1, \dots, n_0$

$$\hat{c}_n = \mathrm{sgn}\left(\ln \frac{q^1_n}{q^0_n}\right) \tag{11.402}$$

Third step: If $\boldsymbol{H}\hat{\boldsymbol{c}} = \boldsymbol{0}$, where $\hat{\boldsymbol{c}}$ is the updated received code word, decoding is stopped; if the maximum number of allowed iterations is reached, the algorithm terminates and a failure to converge is reported; otherwise, each bit node sends their new $q^\beta_{i,n}$ messages to the check nodes to which it is connected, and the algorithm returns to the *First step*.

The algorithm would produce the exact a posteriori probabilities of all the bits if the bipartite graph defined by $\boldsymbol{H}$ contained no cycles [49].

We remark that the sum-product algorithm for the decoding of LDPC codes has been derived under the assumption that the check nodes s_i, $i = 1, \dots, r_0$, are statistically independent given the bit nodes x_n, $n = 1, \dots, n_0$, and vice versa, i.e. the variables of the vectors $\boldsymbol{s}$ and $\boldsymbol{x}$ form a Markov field [46]. Although this assumption is not strictly true, it turns out that the algorithm yields very good performance with low computational complexity. However, 4-cycle Tanner graphs would introduce non-negligible statistical dependence between nodes. A general method for constructing Tanner graphs with large girth is described in [50].

The LR-SPA decoder

Actually, the SPA requires many multiplications which are more costly to implement than additions; beside multiplications of probabilities could become numerically unstable. The result will come very close to zero for large block lengths. To prevent this problem, likelihood ratios (LRs) or LLRs can be used. First, we introduce the SPA using LRs

$$\begin{aligned} \mathrm{L}_n^{(a)} &= \frac{p_n^1}{p_n^0} \\ \mathrm{L}_{i,n}^{(r)} &= \frac{r_{i,n}^1}{r_{i,n}^0} = \frac{1-r_{i,n}^0}{r_{i,n}^0} \\ \mathrm{L}_{i,n}^{(q)} &= \frac{q_{i,n}^1}{q_{i,n}^0} \\ \mathrm{L}_n &= \frac{q_n^1}{q_n^0} \end{aligned} \tag{11.403}$$

Using these ratios, the SPA can be rewritten as follows:

Initialization: Given the a priori LR $L_n^{(a)} = e^{\ell_n^{(in)}}$, we assign $\mathrm{L}_{i,n}^{(q)} = \mathrm{L}_n^{(a)}$, $n \in L(i)$, $i = 1, \dots, r_0$. Next, all bit nodes send their $\mathrm{L}_{i,n}^{(q)}$ messages to the check nodes to which they are connected.

First step: Using the information from their bit nodes, each check node computes

$$\begin{aligned} \mathrm{L}_{i,n}^{(r)} &= \frac{1 - \frac{1}{2} - \frac{1}{2} \prod_{n' \in L(i) \setminus n} (1 - 2q_{i,n'}^1)}{\frac{1}{2} + \frac{1}{2} \prod_{n' \in L(i) \setminus n} (1 - 2q_{i,n'}^1)} \\ &= \frac{1 - \prod_{n' \in L(i) \setminus n} (1 - 2q_{i,n'}^1)}{1 + \prod_{n' \in L(i) \setminus n} (1 - 2q_{i,n'}^1)} \end{aligned} \tag{11.404}$$

Now, from definition of $\mathrm{L}_{i,n'}^{(q)}$ it is

$$1 - 2q_{i,n'}^1 = \frac{1 - \mathrm{L}_{i,n'}^{(q)}}{1 + \mathrm{L}_{i,n'}^{(q)}} \tag{11.405}$$

and we can write

$$\mathrm{L}_{i,n}^{(r)} = \frac{1 - \prod_{n' \in L(i) \setminus n} \left(\frac{1 - \mathrm{L}_{i,n'}^{(q)}}{1 + \mathrm{L}_{i,n'}^{(q)}} \right)}{1 + \prod_{n' \in L(i) \setminus n} \left(\frac{1 - \mathrm{L}_{i,n'}^{(q)}}{1 + \mathrm{L}_{i,n'}^{(q)}} \right)} \tag{11.406}$$

Second step: Using (11.407), now each bit node updates $\mathrm{L}_{i,n}^{(q)}$ by simply dividing $q_{i,n}^1$ by $q_{i,n}^0$, obtaining

$$\begin{aligned} \mathrm{L}_{i,n}^{(q)} &= \frac{\alpha_{i,n}\, p_n^1 \prod_{i' \in M(n) \setminus i} r_{i',n}^1}{\alpha_{i,n}\, p_n^0 \prod_{i' \in M(n) \setminus i} r_{i',n}^0} \\ &= \mathrm{L}_n^{(a)} \prod_{i' \in M(n) \setminus i} \mathrm{L}_{i',n}^{(r)} \end{aligned} \tag{11.407}$$

while from (11.408) and (11.409)

$$\mathrm{L}_n = \mathrm{L}_n^{(a)} \prod_{i \in M(n)} \mathrm{L}_{i,n}^{(r)} \tag{11.408}$$

and detection becomes, for $n = 1, \dots, n_0$

$$\hat{c}_n = \mathrm{sgn}(\ln(\mathrm{L}_n)) \tag{11.409}$$

Third step: If $\boldsymbol{H}\hat{\boldsymbol{c}} = \boldsymbol{0}$, where $\hat{\boldsymbol{c}}$ is the updated received code word, decoding is stopped; if the maximum number of allowed iterations is reached, the algorithm terminates and a failure to converge is reported; otherwise, each bit node sends their new $\mathrm{L}_{i,n}^{(q)}$ messages to the check nodes to which it is connected, and the algorithm returns to the *First step*.

The LLR-SPA or log-domain SPA decoder

The log-domain SPA is now presented. We first define

$$\begin{aligned} \ell_n^{(a)} &= \ln \frac{p_n^1}{p_n^0} = \ell_n^{(in)} \\ \ell_{i,n}^{(r)} &= \ln \frac{r_{i,n}^1}{r_{i,n}^0} \\ \ell_{i,n}^{(q)} &= \ln \frac{q_{i,n}^1}{q_{i,n}^0} \\ \ell_n &= \ln \frac{q_n^1}{q_n^0} \end{aligned} \tag{11.410}$$

Initialization: Assign $\ell_{i,n}^{(q)} = \ell_n^{(a)}$, $n \in L(i)$, $i = 1, \dots, r_0$. Now, all bit nodes send their $\ell_{i,n}^{(q)}$ message to the connected check nodes.

First step: We replace $r_{i,n}^0$ with $1 - r_{i,n}^1$ and from (11.405) we obtain

$$1 - 2r_{i,n}^1 = \prod_{n' \in L(i) \setminus n} (1 - 2q_{i,n'}^1) \tag{11.411}$$

Using the fact that if v_1 and v_0 are probabilities with $v_0 + v_1 = 1$, $\tanh\left[\frac{1}{2}\ln(v_1/v_0)\right] = v_1 - v_0 = 1 - 2v_0$, we can rewrite (11.419) as

$$\tanh\left(\frac{1}{2}\ell_{i,n}^{(r)}\right) = -\prod_{n' \in L(i) \setminus n} \tanh\left[\frac{1}{2}(-\ell_{i,n'}^{(q)})\right] \tag{11.412}$$

However, we are still left with products and the tanh function. We can remedy as follows: First, we factorize $-\ell_{i,n'}^{(q)}$ into its sign and amplitude. Let

$$s_{i,n'} = \mathrm{sgn}\,(-\ell_{i,n'}^{(q)}), \quad M_{i,n'} = |\ell_{i,n'}^{(q)}| \tag{11.413}$$

it is

$$-\ell_{i,n'}^{(q)} = s_{i,n'}\, M_{i,n'} \tag{11.414}$$

Hence, (11.420) can be rewritten as

$$\tanh\left(\frac{1}{2}\ell_{i,n}^{(r)}\right) = \left(-\prod_{n' \in L(i) \setminus n} s_{i,n'}\right) \prod_{n' \in L(i) \setminus n} \tanh\left(\frac{1}{2}M_{i,n'}\right) \tag{11.415}$$

from which we derive

$$\begin{aligned}
\ell_{i,n}^{(r)} &= \left(- \prod_{n' \in L(i) \setminus n} s_{i,n'} \right) 2 \tanh^{-1} \left(\prod_{n' \in L(i) \setminus n} \tanh \left(\frac{1}{2} M_{i,n'} \right) \right) \\
&= \left(- \prod_{n' \in L(i) \setminus n} s_{i,n'} \right) 2 \tanh^{-1} \ln^{-1} \ln \left(\prod_{n' \in L(i) \setminus n} \tanh \left(\frac{1}{2} M_{i,n'} \right) \right) \\
&= \left(- \prod_{n' \in L(i) \setminus n} s_{i,n'} \right) 2 \tanh^{-1} \ln^{-1} \left(\sum_{n' \in L(i) \setminus n} \ln \left(\tanh \left(\frac{1}{2} M_{i,n'} \right) \right) \right)
\end{aligned} \tag{11.416}$$

We now define the function

$$\phi(x) = -\ln \left[\tanh \left(\frac{x}{2} \right) \right] = \ln \left(\frac{e^x + 1}{e^x - 1} \right) \tag{11.417}$$

As shown in Figure 11.42, the function is fairly well behaved and so it may be implemented by a look-up table. Using the fact that $\phi^{-1}(x) = \phi(x)$ for $x > 0$, (11.416) becomes

$$\ell_{i,n}^{(r)} = \left(- \prod_{n' \in L(i) \setminus n} s_{i,n'} \right) \phi \left(\sum_{n' \in L(i) \setminus n} \phi(M_{i,n'}) \right) \tag{11.418}$$

Second step: From (11.407) it is

$$\ell_{i,n}^{(q)} = \ell_n^{(a)} + \sum_{i' \in M(n) \setminus i} \ell_{i',n}^{(r)} \tag{11.419}$$

while (11.408) yields

$$\ell_n = \ell_n^{(a)} + \sum_{i \in M(n)} \ell_{i,n}^{(r)} \tag{11.420}$$

Detection is performed as, for $n = 1, \dots, n_0$,

$$\hat{c}_n = \text{sgn}\,(\ell_n) \tag{11.421}$$

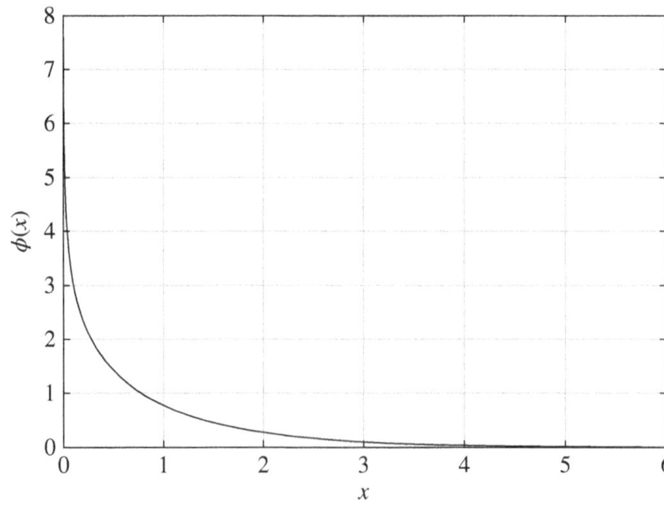

Figure 11.42 $\phi(x)$ function.

Third step: If $\boldsymbol{H}\hat{\boldsymbol{c}} = \boldsymbol{0}$, where $\hat{\boldsymbol{c}}$ is the updated received code word, decoding is stopped; if the maximum number of allowed iterations is reached the algorithm terminates and a failure to converge is reported; otherwise, each bit node sends their new $\ell_{i,n}^{(q)}$ messages to the check nodes to which it is connected, and the algorithm returns to the *First step*, or, better, equation (11.414).

The min-sum decoder

Decoders of even lower complexity than LLR-SPA can be devised at the cost of a small performance degradation, typically on the order of 0.5 dB which, however, is a function of the code and the channel.

Considering the shape of the $\phi(x)$ function, it issues that the term corresponding to the smallest $M_{i,n}$ in the summation of (11.418) dominates, so that we approximate

$$\phi\left(\sum_{n'\in L(i)\setminus n} \phi\,(M_{i,n'})\right) \simeq \phi(\phi\,(\min_{n'\in L(i)\setminus n} M_{i,n'})) \\ = \min_{n'\in L(i)\setminus n} M_{i,n'} \tag{11.422}$$

Thus, the min-sum algorithm is simply the LLR-SPA where the *First step* is replaced by

$$\ell_{i,n}^{(r)} = \left(-\prod_{n'\in L(i)\setminus n} s_{i,n'}\right)\min_{n'\in L(i)\setminus n} M_{i,n'} \tag{11.423}$$

Other decoding algorithms

Despite of being the simplest method to avoid calculating the hyper tangent function, the min-sum algorithm cannot approach the performance of the LLR-SPA decoder.

In [51], two simplified versions of the SPA, devised to achieve a faster iterative decoding, are proposed. Both algorithms are implemented with real additions only, and this simplifies the decoding complexity.

To reduce decoding complexity, also the concept of *forced convergence* has been introduced. Restricting the message passing to the nodes that still significantly contribute to the decoding result, this approach substantially reduces the complexity with negligible performance loss. If the channel is good (high SNR), some $\ell_{i,n}^{(q)}$ tend to have high magnitude, hence, their corresponding ϕ values is close to zero. In this case, their magnitude contribution in (11.418) can be neglected. However, their sign must be considered. The min-sum algorithm is the extreme case where the bit node with minimum magnitude among the connected bit nodes is used.

Other simplified approaches are given in [52–54], while for a flexible encoder hardware implementation we refer to [55].

11.8.4 Example of application

We study in this section the application of LDPC codes to two-dimensional QAM transmission over an AWGN channel [56].

Performance and coding gain

Recall from (6.82), the expression of the error probability for uncoded M-QAM transmission,

$$P_e \simeq 4Q\left(\sqrt{\frac{3}{M-1}\Gamma}\right) \tag{11.424}$$

where Γ is the SNR given by (6.97). In general, the relation between M and the rate of the encoder-modulator from (11.1) is given by

$$R_I = \frac{k_0}{n_0} \frac{\log_2 M}{2} \tag{11.425}$$

Recall also, from (6.71), that the SNR per dimension is given by

$$\Gamma_I = \Gamma = 2R_I \frac{E_b}{N_0} \tag{11.426}$$

Using (6.129), we introduce the normalized SNR

$$\overline{\Gamma}_I = \frac{\Gamma_I}{2^{2R_I} - 1} = \frac{2R_I}{2^{2R_I} - 1} \frac{E_b}{N_0} \tag{11.427}$$

Then for an uncoded M-QAM system we express (11.424) as

$$P_e \simeq 4Q\,(\sqrt{3\overline{\Gamma}_I}) \tag{11.428}$$

As illustrated in Figure 6.13, the curve of P_e versus $\overline{\Gamma}_I$ indicates that the *gap to capacity* for uncoded QAM with $M \gg 1$ is equal to $\overline{\Gamma}_{gap,dB} \simeq 9.8$ dB at a symbol error probability of 10^{-7}. We, therefore, determine the value of the normalized SNR $\overline{\Gamma}_I^c$ needed for the coded system to achieve a symbol error probability of 10^{-7}, and compute the coding gain at that symbol error probability as

$$G_{code,dB} = 9.8 - 10 \log_{10} \overline{\Gamma}_I^c \tag{11.429}$$

From Figure 6.13, as for large SNRs the Shannon limit cannot be approached to within less than 1.53 dB without shaping, we note that an upper limit to the coding gain measured in this manner is about 8.27 dB. Simulation results for three high-rate $(n_0,\ k_0)$ binary LDPC codes are specified in Table 11.22 in terms of the coding gains obtained at a symbol error probability of 10^{-7} for transmission over an AWGN channel for 16, 64, and 4096-QAM modulation formats. Transmitted QAM symbols are obtained from coded bits via Gray labelling. To measure error probabilities, one code word is decoded using the message-passing (sum-product) algorithm for given maximum number of iterations. Figure 11.43 shows the effect on performance of the maximum number of iterations allowed in the decoding process for code 2 specified in Table 11.22 and 16-QAM transmission.

Table 11.22: LDPC codes considered for the simulation and coding gains achieved at a symbol error probability of 10^{-7} for different QAM constellations. The spectral efficiencies ν are also indicated.

	k_0	n_0	code ratek_0/n_0	16-QAM	64-QAM	4096-QAM
Code 1	433	495	0.8747	4.9 dB	4.6 dB	3.5 dB
				(3.49 bit/s/Hz)	(5.24 bit/s/Hz)	(10.46 bit/s/Hz)
Code 2	1777	1998	0.8894	6.1 dB	5.9 dB	4.8 dB
				(3.55 bit/s/Hz)	(5.33 bit/s/Hz)	(10.62 bit/s/Hz)
Code 3	4095	4376	0.9358	6.2 dB	6.1 dB	5.6 dB
				(3.74 bit/s/Hz)	(5.61 bit/s/Hz)	(11.22 bit/s/Hz)

The codes given in Table 11.22 are due to MacKay and have been obtained by a random construction method. The results of Table 11.22 indicate that LDPC codes offer net coding gains that are similar to those that have been reported for turbo codes. Furthermore, LDPC codes achieve asymptotically an excellent performance without exhibiting *error floors* and admit a wide range of trade-offs between performance and decoding complexity.

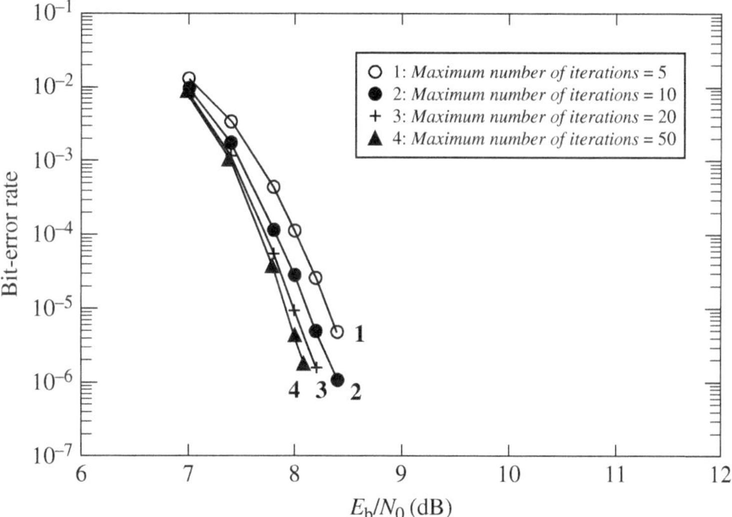

Figure 11.43 Performance of LDPC decoding with Code 2 and 16-QAM for various values of the maximum number of iterations.

11.8.5 Comparison with turbo codes

LDPC codes are capable of near capacity performance while admitting a simple decoder. The advantages that the LDPC codes have over turbo codes are the following: (i) they allow parallelizable decoder; (ii) they are more amenable to high code rates; (iii) they generally posses a lower error-rate floor (for the same length and rate); (iv) they possess superior performance in the presence of burst errors (due to interference, fading, and so on); (v) they require no interleavers in the encoder and decoder; and (vi) a single LDPC code can be universally good aver a collection of channels. Among their disadvantages are the following (i) most LDPC codes have somewhat complex encoders; (ii) the connectivity among the decoder component processors can be large and unwieldy; and (iii) turbo codes can often perform better when the code length is short.

In Figure 11.44 a comparison between (i) regular LDPC, (ii) irregular LDPC, (iii) turbo codes, and (iv) Shannon bound, is represented. The LDPC curves have been obtained by letting the decoder run for enough iterations to get the best possible performance. We should note the great performance of irregular LDPC codes, which are within a fraction of dB, in terms of E_b/N_0, from the Shannon bound, and outperform turbo codes of the same length.

11.9 Polar codes

Polar codes are a new class of channel codes, introduced in 2008 by Arıkan in his seminal work [58].[10] By specific encoding and decoding, the unreliable communication channel is turned into virtual bit-channels of which one portion is highly reliable and the other is highly unreliable. When the code length tends to infinity, the highly reliable channels become error free, and the unreliable channels become fully random – this phenomenon is referred to as *channel polarization*. Furthermore, the proportion of the

[10] The material presented in this section was written by Dr Ingmar Land, whom the authors gratefully acknowledge.

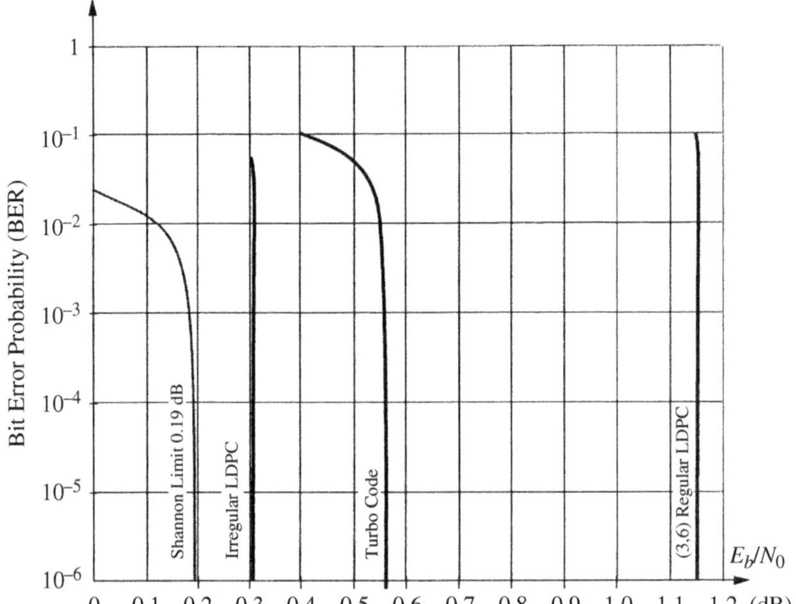

Figure 11.44 Simulation results for a regular (3,6) LDPC, an optimized irregular LDPC and a turbo code. All three codes are of rate $R_c = 1/2$ and have block length 10^6. Source: Reproduced with permission from Richardson et al. [57]. ©2001, IEEE.

reliable channels tends to the capacity of the communication channel, and thus polar coding achieves capacity. Remarkably, the encoding and decoding complexity of this capacity-achieving coding scheme has only order $O(n_0 \ln n_0)$ in the code length n_0.

As opposed to other modern codes like turbo codes or LDPC codes, which are iteratively decoded, polar coding is based on successive cancellation (SC) decoding. While in theory, plain SC decoding is sufficient to achieve capacity, successive cancellation list (SCL) decoding is applied in the short-length to medium-length regime [59–61]. The performance of polar codes is comparable to that of turbo codes or LDPC codes; depending on the application, code length and code rate, and implementation constraints, polar codes may be superior [60, 62]. Polar codes (under SCL decoding) have been adopted for 5G in 2016, namely as channel coding for uplink and downlink control information for the enhanced mobile broadband (eMBB) communication service [63, 64].

11.9.1 Encoding

The structure of a polar code encoder is shown in Figure 11.45. In the following, we will go through the blocks and their functionalities.

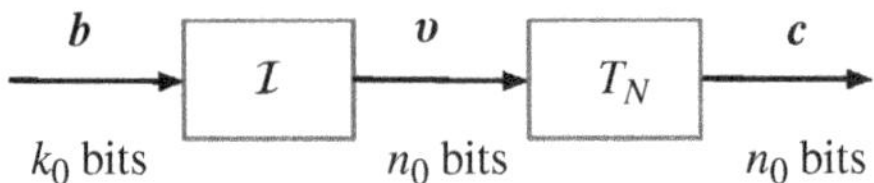

Figure 11.45 Polar code encoding.

A polar code $P(n_0, k_0)$ of information length k_0, code length $n_0 = 2^{\epsilon_0}$, $\epsilon_0 \in \mathbb{N}$, and code rate $R_c = k_0/n_0$ is defined by the *transformation matrix*

$$\boldsymbol{T}_{n_0} = \boldsymbol{T}_2^{\otimes \epsilon_0} \tag{11.430}$$

given by the ϵ_0-fold Kronecker product (see (4.152)) of the *kernel* matrix

$$\boldsymbol{T}_2 = \begin{bmatrix} 1 & 0 \\ 1 & 1 \end{bmatrix} \tag{11.431}$$

and by an *information set* $\mathcal{I} \subset \mathcal{S}$ of size k_0, where $\mathcal{S} = \{0, 1, \dots, n_0 - 1\}$. The set $\mathcal{F} = \mathcal{S} \setminus \mathcal{I}$ of size $(n_0 - k_0)$ is the complementary set of $\mathcal{I}$ and is called the *frozen set*. Note that the information set and the frozen set are index sets.

Consider the encoding of an *information word* $\boldsymbol{b}$ of length k_0 bits into a codeword $\boldsymbol{c}$ of length n_0 bits. $\boldsymbol{b}$ is first mapped to *input vector* $\boldsymbol{v}$ of length n_0, as

$$\begin{aligned} \boldsymbol{v}_{\mathcal{I}} &= \boldsymbol{b} \\ \boldsymbol{v}_{\mathcal{F}} &= \boldsymbol{0} \end{aligned} \tag{11.432}$$

i.e. the elements of $\boldsymbol{v}$ indexed by the information set are set to the values of the information bits, and the remaining elements, indexed by the frozen set, are set (*frozen*) to zero.[11] This input vector is then multiplied by the transformation matrix to obtain the *codeword*, (see also (11.34))

$$\boldsymbol{c}^T = \boldsymbol{v}^T \boldsymbol{T}_{n_0} \tag{11.433}$$

This construction[12] yields polar codes of lengths n_0 that are powers of 2. Polar codes of other lengths may be obtained by puncturing or shortening, see Section 11.9.5. The choice of the indices in the information and frozen sets is a part of the polar code design, see Section 11.9.4.

Example 11.9.1

Consider a polar code $P(8, 5)$ of length $n_0 = 8$ and information length $k_0 = 5$. With $\epsilon_0 = \log_2 n_0 = 3$, the transformation matrix is given by

$$\begin{aligned} \boldsymbol{T}_8 &= \boldsymbol{T}_2^{\otimes 3} = \boldsymbol{T}_2 \otimes \boldsymbol{T}_2 \otimes \boldsymbol{T}_2 \\ &= \begin{bmatrix} 1 & 0 \\ 1 & 1 \end{bmatrix} \otimes \begin{bmatrix} 1 & 0 \\ 1 & 1 \end{bmatrix} \otimes \begin{bmatrix} 1 & 0 \\ 1 & 1 \end{bmatrix} = \begin{bmatrix} 1&0&0&0&0&0&0&0 \\ 1&1&0&0&0&0&0&0 \\ 1&0&1&0&0&0&0&0 \\ 1&1&1&1&0&0&0&0 \\ 1&0&0&0&1&0&0&0 \\ 1&1&0&0&1&1&0&0 \\ 1&0&1&0&1&0&1&0 \\ 1&1&1&1&1&1&1&1 \end{bmatrix} \end{aligned} \tag{11.434}$$

The information set $\mathcal{I}$ has size $k_0 = 5$ and the frozen set $\mathcal{F}$ has size $(n_0 - k_0) = 3$; we assume the sets

$$\begin{aligned} \mathcal{I} &= \{2, 3, 5, 6, 7\} \\ \mathcal{F} &= \{0, 1, 4\} \end{aligned}$$

---

[11] Typically, the frozen bits are chosen to be zero and then the polar code is linear; in general, however, the frozen bits may be set to arbitrary known values without affecting the performance.

[12] Note that Arıkan used a bit-reversal permutation in the encoding of the polar code of his original work. Doing so leads to an equivalent code and has no further effect on code properties, decoding algorithms, or performance.

For encoding the information word $\boldsymbol{b}^T = [b_0 \ \dots \ b_4]$ of length $k_0 = 5$, its elements are first mapped into the input vector,

$$\boldsymbol{v}^T = [0\ 0\ b_0\ b_1\ 0\ b_2\ b_3\ b_4] \tag{11.435}$$

i.e. the information bits $\{b_n\}$ are placed on the positions of $\boldsymbol{v}$ indexed by the information set, and the remaining positions of $\boldsymbol{v}$ are set (*frozen*) to the known values zero. The codeword $\boldsymbol{c}^T = [c_0 \ \dots \ c_7]$ is obtained by multiplying $\boldsymbol{v}$ by the transformation matrix,

$$[c_0 \ \dots \ c_7]^T = [0\ 0\ b_0\ b_1\ 0\ b_2\ b_3\ b_4]^T \begin{bmatrix} 1&0&0&0&0&0&0&0 \\ 1&1&0&0&0&0&0&0 \\ 1&0&1&0&0&0&0&0 \\ 1&1&1&1&0&0&0&0 \\ 1&0&0&0&1&0&0&0 \\ 1&1&0&0&1&1&0&0 \\ 1&0&1&0&1&0&1&0 \\ 1&1&1&1&1&1&1&1 \end{bmatrix} \tag{11.436}$$

Note that the frozen positions effectively eliminate the rows of $\boldsymbol{T}_8$ indexed by the frozen set $\mathcal{F}$. Thus, the remaining rows represent the generator matrix of the polar code.

Internal CRC

A CRC is typically used to detect decoding failures. In polar coding, an *internal CRC* may be used for different purpose, namely to significantly increase the error-rate performance. The encoder structure is shown in Figure 11.46.

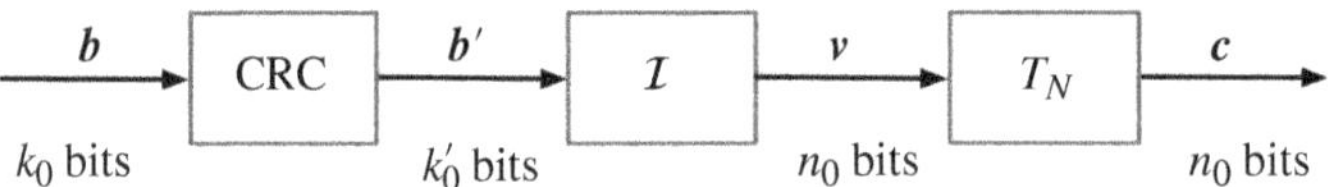

Figure 11.46 Polar code encoding with CRC.

The information word $\boldsymbol{b}$ of length k_0 is first extended by a CRC $\boldsymbol{b}_{CRC}$ of length $k_{0,CRC}$ to the vector

$$\boldsymbol{b}' = \begin{bmatrix} \boldsymbol{b} \\ \boldsymbol{b}_{CRC} \end{bmatrix} \tag{11.437}$$

of length $k'_0 = k_0 + k_{0,CRC}$. The information set $\mathcal{I}$ of size k'_0 and the frozen set $\mathcal{F}$ of size $(n_0 - k'_0)$ are then used to map the elements of $\boldsymbol{b}'$ to the input vector $\boldsymbol{v}$ of length n_0,

$$\begin{aligned} \boldsymbol{v}_{\mathcal{I}} &= \boldsymbol{b}' \\ \boldsymbol{v}_{\mathcal{F}} &= \boldsymbol{0} \end{aligned} \tag{11.438}$$

The input vector $\boldsymbol{v}$ is finally encoded into the codeword $\boldsymbol{c}$ as before, according to (11.441). Note that the rate[13] of the polar code is still $R_c = k_0/n_0$, but there are fewer frozen bits.

The internal CRC typically improves the distance spectrum of the polar code. Starting from the code $P(n_0, k'_0)$ without CRC, the CRC constraints remove some of the codewords, and thus in general also reduce the number of minimum distance codewords. If the CRC is long enough (and well chosen), it

[13] Some authors refer to k_0/n_0 as the overall rate (CRC and polar code), and to k'_0/n_0 as the rate of the polar code (excluding the CRC).

may even remove all minimum-distance codewords, such that the code $P(n_0, k_0)$ obtains an *increased minimum distance*. Under SC decoding, these effects cannot be exploited; under SCL decoding, however, this internal CRC leads to a significant performance improvement and is of paramount importance in the short-length and medium-length regime [59, 60].

LLRs associated to code bits

The receiver, by soft detection methods (see page 589), provides to the decoder the LLR ℓ_n in correspondence of each code bit c_n. Moreover, as in previous codes, for analysis purposes, we *model the decoder input* as (see (11.309))

$$z_n = a_n + w_n \tag{11.439}$$

where codebit c_n is mapped into the binary symbol a_n: $0 \to +1$, $1 \to -1$. Note that this mapping is different from that used in LDPC codes. In (11.439), w_n is AWGN with variance σ^2_{dec}. Hence, set $z = [z_0, \dots, z_{n_0-1}]$, if the observation is $z = \rho$, it is

$$\ell_n = \ln \frac{P[c_n = 0|z = \rho]}{P[c_n = 1|z = \rho]} \tag{11.440}$$

For the above *model* (11.439), from (6.39), it is

$$\ell_n = \frac{2}{\sigma^2_{dec}} \rho_n \tag{11.441}$$

11.9.2 Tanner graph

The relationship between the input bits $\{v_n\}$ and the output bits (code bits) $\{c_n\}$ is defined by the transformation matrix $\boldsymbol{T}_{n_0}$, and it may be represented by a *Tanner graph*, similar to the graph of LDPC codes. The decoding algorithms for polar codes, as discussed in Section 11.9.3, operate on this graph by passing messages along the edges, however, in a way different from LDPC message passing algorithms. This section describes the construction of the Tanner graph. We start with the example of $\boldsymbol{T}_8$ and then generalize the procedure.

Example 11.9.2

Consider $\boldsymbol{T}_8$ with $\epsilon_0 = \log_2 8 = 3$, as depicted in Figure 11.47. The graph consists of $n_0 = 8$ variable nodes at the left, corresponding to the 8 input bits $\{v_n\}$, and $n_0 = 8$ variable nodes at the right, corresponding to the 8 code bits $\{c_n\}$, $n = 0, 1, \dots, 7$. In between there are $\epsilon_0 = 3$ stages, connected by intermediate variable nodes $\{v_n^{(s)}\}$, $s = 1, 2$. For convenience, we define $\{v_n^{(0)} = v_n\}$ and $\{v_n^{(3)} = c_n\}$; this way for stage s, the input bits are $\{v_n^{(s-1)}\}$ and the output bits are $v_n^{(s)}$, see also Figure 11.48. This gives an overall of $n_0(\epsilon_0 + 1) = 8 \cdot 4$ variables nodes in the graph. We now have a closer look at the relations defined by each stage.

In stage 1, pairs of input bits are processed as

$$[v_{2i}^{(0)} \; v_{2i+1}^{(0)}] \cdot \begin{bmatrix} 1 & 0 \\ 1 & 1 \end{bmatrix} = [v_{2i}^{(1)} \; v_{2i+1}^{(1)}]$$

for $i = 0, 1, 2$, and 3: the first output bit is equal to the sum of the two input bits, and the second output bit is equal to the second input bit.

In stage 2, pairs of blocks of two input bits are processed. Consider first the two input-blocks $[v_0^{(1)} \; v_1^{(1)}]$ and $[v_2^{(1)} \; v_3^{(1)}]$, and the two output-blocks $[v_0^{(2)} \; v_1^{(2)}]$ and $[v_2^{(2)} \; v_3^{(2)}]$. Their relation is given by

$$[v_0^{(2)} \; v_1^{(2)}] = [v_0^{(1)} \; v_1^{(1)}] \oplus [v_2^{(1)} \; v_3^{(1)}]$$
$$[v_2^{(2)} \; v_3^{(2)}] = [v_2^{(1)} \; v_3^{(1)}]$$

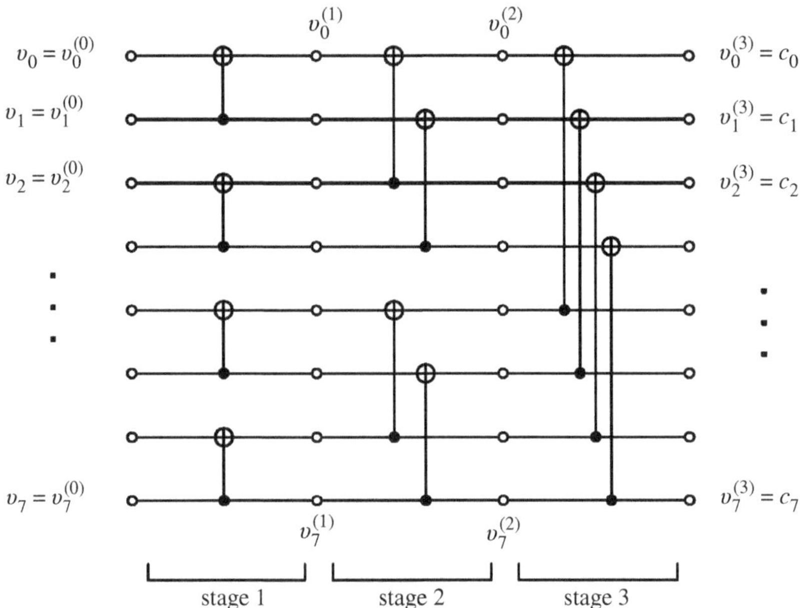

Figure 11.47 Tanner graph for transformation matrix $\boldsymbol{T}_8$.

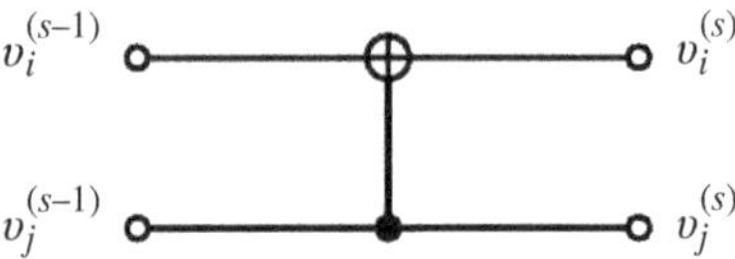

Figure 11.48 One $\boldsymbol{T}_2$-block at stage s of the Tanner graph.

where the addition is element-wise. This can equivalently be formulated as the application of $\boldsymbol{T}_2$ to the first bits of the two input blocks,

$$[v_0^{(1)} \; v_2^{(1)}] \cdot \begin{bmatrix} 1 & 0 \\ 1 & 1 \end{bmatrix} = [v_0^{(2)} \; v_2^{(2)}]$$

and its application to the second bits of the two input blocks,

$$[v_1^{(1)} \; v_3^{(1)}] \cdot \begin{bmatrix} 1 & 0 \\ 1 & 1 \end{bmatrix} = [v_1^{(2)} \; v_3^{(2)}]$$

In the same way, the four input bits $v_4^{(1)}$, $v_5^{(1)}$, $v_6^{(1)}$, and $v_7^{(1)}$, are transformed to obtain the four output bits $v_4^{(2)}$, $v_5^{(2)}$, $v_6^{(2)}$, and $v_7^{(2)}$.

In stage 3, two blocks of four input bits each is processed. Similarly to stage 2, the operation may be written using vectors,

$$\begin{aligned}
&[v_0^{(3)} \; v_1^{(3)} \; v_2^{(3)} \; v_3^{(3)}] = [v_0^{(2)} \; v_1^{(2)} \; v_2^{(2)} \; v_3^{(2)}] \oplus [v_4^{(2)} \; v_5^{(2)} \; v_6^{(2)} \; v_7^{(2)}] \\
&[v_4^{(3)} \; v_5^{(3)} \; v_6^{(3)} \; v_7^{(3)}] = [v_4^{(2)} \; v_5^{(2)} \; v_6^{(2)} \; v_7^{(2)}]
\end{aligned}$$

or equivalently using $\boldsymbol{T}_2$-blocks,

$$[v_i^{(2)} \; v_{i+4}^{(2)}] \cdot \begin{bmatrix} 1 & 0 \\ 1 & 1 \end{bmatrix} = [v_i^{(3)} \; v_{i+4}^{(3)}] \tag{11.442}$$

for $i = 0, 1, 2$, and 3: the first bits of the two input blocks are combined, the second bits of the two input blocks are combined, etc.

Going through the graph from right to left reveals the recursive nature of the overall transform. At stage 3, the output vector of length 8 is obtained by combining two input vectors of length 4. At stage 2, each output vector of length 4 is obtained by combining two input vectors of length 2. And at stage 1, each output vector of length 2 is obtained by combining single input bits (vectors of length 1). This recursive structure of the Tanner graph reflects the Kronecker product in the transformation matrix $\boldsymbol{T}_8$, as given in (11.442), each stage of the graph corresponding to one of the kernels $\boldsymbol{T}_2$.

The example motivates the general construction.[14] Assume that we want to construct the graph of $\boldsymbol{T}_{n_0}$: this graph has $\epsilon_0 = \log_2 n_0$ stages and an overall of $n_0(\epsilon_0 + 1)$ variable nodes, denoted by $\{v_n^{(s)}\}$, $n = 0, 1, \dots, n_0 - 1$, $s = 0, 1, \dots, \epsilon_0$.

At stage s, there are $2^{\epsilon_0 - s}$ output blocks of length 2^s bits $\{v_n^{(s)}\}$, and each output block is formed by combining two input blocks of length 2^{s-1} bits $\{v_n^{(s-1)}\}$. For each output block, the two respective input blocks are combined as given in the example, namely the first bits of the two input blocks are combined according to $\boldsymbol{T}_2$, the second bits of the two input blocks are combined according to $\boldsymbol{T}_2$, etc.

This property is used to formulate the algorithm to build the Tanner graph. The relation between input bits and output bits is given by

$$[v_i^{(s-1)} \; v_j^{(s-1)}] \cdot \begin{bmatrix} 1 & 0 \\ 1 & 1 \end{bmatrix} = [v_i^{(s)} \; v_j^{(s)}] \tag{11.443}$$

for

$$\begin{aligned} i &= l + t \cdot 2^s \\ j &= l + t \cdot 2^s + 2^{s-1} \end{aligned} \tag{11.444}$$

where $l = 0, 1, \dots, 2^{s-1} - 1$ denotes the index within the input block, and $t = 0, 1, \dots, 2^{\epsilon_0 - s} - 1$ denotes the index of the output block. This $\boldsymbol{T}_2$-block of the Tanner graph is shown in Figure 11.48.

The Tanner graph, according to (11.443), may be used to efficiently compute the transformation $\boldsymbol{T}_{n_0}$, for a given input vector, as the computations can be highly parallelized. Further, the Tanner graph provides the basis for the decoding algorithms of polar codes.

11.9.3 Decoding algorithms

The following sections explain the principle and algorithm of SC decoding and how to perform the more powerful SCL decoding.

Consider a polar code $P(n_0, k_0)$ with information set $\mathcal{I}$ and frozen set $\mathcal{F}$. The decoding algorithms aim at maximum-likelihood (ML) decoding, i.e. finding among all codewords in the polar code $P(n_0, k_0)$ the one with the largest likelihood:

$$\hat{\boldsymbol{c}}_{\text{ML}} = \underset{\boldsymbol{C} \in P}{\text{argmax}} \; P[\boldsymbol{z} = \boldsymbol{\rho} | \boldsymbol{c} = \boldsymbol{C}] \tag{11.445}$$

Equivalently, we may search over all valid input sequences $\boldsymbol{v}$, i.e. the input vectors with the frozen bits being zero, and minimize the negative log-likelihood:

$$\hat{\boldsymbol{v}}_{\text{ML}} = \underset{\boldsymbol{V}:\boldsymbol{V}_{\mathcal{F}} = \boldsymbol{0}}{\text{argmin}} - \ln P[\boldsymbol{z} = \boldsymbol{\rho} | \boldsymbol{v} = \boldsymbol{V}] \tag{11.446}$$

To simplify the following derivations, we assume that all input vectors $\boldsymbol{v}$ are equiprobable (including the ones with frozen bits not being zero), and thus that all input bits $\{v_n\}$ are equiprobable (even the

[14] The proof is direct from the recursive structure and the details are omitted.

frozen bits). As the decoder searches only over the valid input vectors, i.e. $\boldsymbol{v}$ with $\boldsymbol{v}_{\mathcal{F}} = 0$, this assumption does not change the decoding outcome. Using this assumption, we can write the decoding rule above as

$$\begin{aligned} \hat{\boldsymbol{v}}_{\text{ML}} &= \underset{V:V_{\mathcal{F}}=\mathbf{0}}{\text{argmin}} - \ln P[\boldsymbol{v} = V | z = \boldsymbol{\rho}] \\ &= \underset{V:V_{\mathcal{F}}=\mathbf{0}}{\text{argmin}} \sum_{n=0}^{n_0-1} - \ln P[v_n = V_n | z = \boldsymbol{\rho}; v_0 = V_0, v_1 = V_1, \dots, v_{n-1} = V_{n-1}] \end{aligned} \tag{11.447}$$

Bayes' rule was applied in the last step to expand the original expression into a chain. Note that (11.447) coincides with the MAP criterion.

Practical decoding algorithms approximate this optimal decoding rule in order to reduce the decoding complexity. In SC decoding, the individual terms of the sum are minimized successively; in SC list decoding, the full expression is used, but the search space is reduced.

Successive cancellation decoding – the principle

The schematic of SC decoding [58] is depicted in Figure 11.49. The principle is as follows: To detect the first bit, v_0, given the channel LLRs (the LLRs on the code bits), $\boldsymbol{\ell} = [\ell_0, \ell_1, \dots, \ell_{n_0-1}]$, we compute the a posteriori LLR associated to v_0,

$$\lambda_0 = \ell_{v_0}(\boldsymbol{\ell}) \tag{11.448}$$

and a hard-decision is made to obtain the detected bit $\hat{v}_0$. If the index 0 is in the frozen set, i.e. if $0 \in \mathcal{F}$, $\hat{v}_0$ is set to the known value zero. Then $\hat{v}_0$ is fed back into the decoding process. To detect the second bit, v_1, again, all channel LLRs and the decoded bit $\hat{v}_0$ are used to compute the a posteriori LLR

$$\lambda_1 = \ell_{v_1}(\boldsymbol{\ell}; \hat{v}_0) \tag{11.449}$$

and a hard-decision is made to obtain the detected bit $\hat{v}_1$. Again, if $1 \in \mathcal{F}$, $\hat{v}_1$ is set to zero; then $\hat{v}_1$ is fed back into the decoding process. Similarly, decoding continues for all other input bits.

In general, successively the a posteriori LLRs are computed for all input bits $\{v_n\}$, $n = 0, 1, \dots, n_0 - 1$,

$$\lambda_n = \ell_{v_n}(\boldsymbol{\ell}; \hat{v}_0, \dots, \hat{v}_{n-1}) \tag{11.450}$$

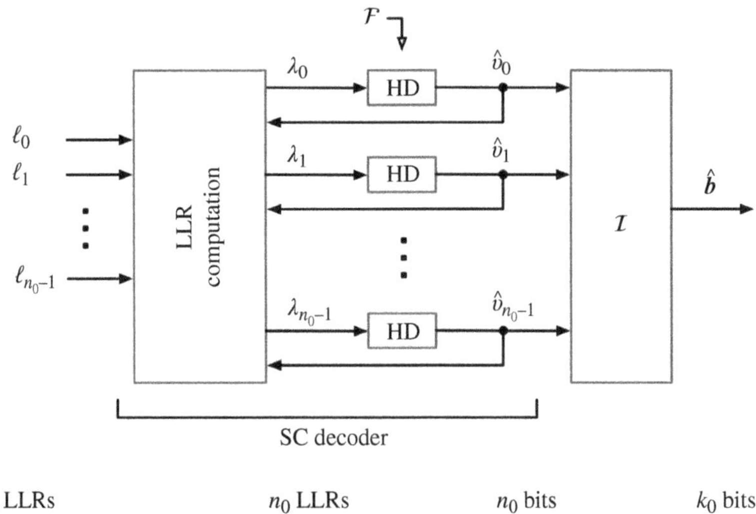

Figure 11.49 Structure of the polar code SC decoder.

and the detected bits are determined by a hard decision, subject to the frozen set, as

$$\hat{v}_n = \begin{cases} 0 & \text{if } \lambda_n \geq 0 \text{ and } n \in \mathcal{I} \\ 1 & \text{if } \lambda_n < 0 \text{ and } n \in \mathcal{I} \\ 0 & \text{if } n \in \mathcal{F}; \end{cases} \tag{11.451}$$

and then the detected bits are fed back into the decoding process.

From the detected input bits, the detections of the information bits $\{\hat{b}_n\}$ are extracted via the information set $\mathcal{I}$,

$$\hat{\boldsymbol{b}} = \hat{\boldsymbol{v}}_{\mathcal{I}} \tag{11.452}$$

This operation corresponds to (11.440) on the encoder side. If an internal CRC is present, then in addition, the $k_{0,CRC}$ CRC bits have to be removed, corresponding to (11.447) and (11.446) on the encoder side. The specific use of the CRC bits in SCL decoding will be discussed later on.

During the internal computations of the a posteriori LLRs, the feedback of detected input bits is equivalent to cancelling the contribution of this input bit to the channel LLRs. Therefore, this decoding principle has been termed *successive cancellation decoding*.

SC decoding is prone to error propagation, as with a single wrongly detected bit $\hat{v}_n$ fed back into the decoding process, the whole codeword will be in error, and thus decoding cannot recover from any wrong decision on input bits. There are two ways to deal with this problem: (i) Positions that are likely to be in error should be in the frozen set such that their values are known by the decoder; this idea is considered in polar code design, see Section 11.9.4. (ii) Instead of feeding back the detected bit into the decoding process, both hypotheses may be used for a while and the actual decision may be delayed, which allows for recovery from some wrong decisions; this idea is employed in SC list decoding, as addressed later on.

Successive cancellation decoding – the algorithm

SC decoding requires the efficient computation of the a posteriori LLRs given detection of the previous input bits. This can be performed on the Tanner graph of the polar code. We will use the Tanner graph as determined in Section 11.9.2, with input bits at the left and code bits at the right. Thus, the channel LLRs $\{\ell_n\}$, corresponding to the code bits $\{c_n\}$, go into the graph at the right, and the a posteriori LLRs $\{\lambda_n\}$, corresponding to the input bits $\{v_n\}$ leave the graph at the left. We will first describe the general algorithm and then specify it to the example of $\boldsymbol{T}_8$.

The algorithm on the graph is based on message passing: LLRs travel through the graph from right to left; detected bit and their sums, termed partial sums, travel through the graph from left to right. To each intermediate bit value $\{v_n^{(s)}\}$ in the graph, we associate its detected value $\{\hat{v}_n^{(s)}\}$ and its LLR $\{\ell_n^{(s)}\}$, see Figure 11.50. The LLRs for the last stage ϵ_0 are initialized as $\ell_n^{(\epsilon_0)} = \ell_n$, $n = 0, 1, \dots, n_0 - 1$. The a posteriori LLRs are given by the LLRs of the first stage, $\lambda_n = \ell_n^{(0)}$, $n = 0, 1, \dots, n_0 - 1$. The core of the algorithm is the operation on a single $\boldsymbol{T}_2$-block at stage s, $s = 0, 1, \dots, \epsilon_0$, as shown in Figure 11.48, with indices i and j as specified in (11.458).

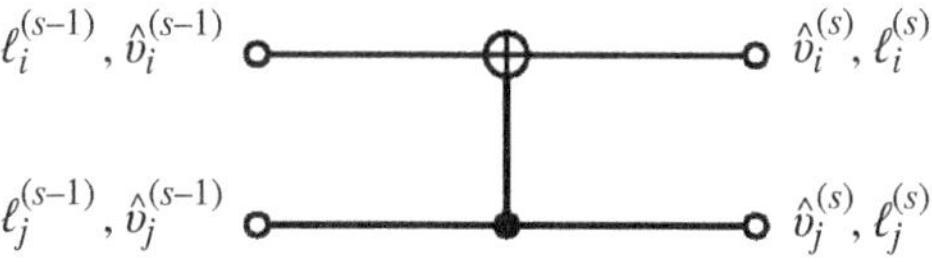

Figure 11.50 $\boldsymbol{T}_2$-block at stage s with LLRs and detected bits.

For application to decoding, we rewrite the input–output constraint (11.457) as

$$v_i^{(s-1)} = v_i^{(s)} \oplus v_j^{(s)} \tag{11.453}$$

$$v_j^{(s-1)} = v_i^{(s-1)} \oplus v_i^{(s)} = v_j^{(s)} \tag{11.454}$$

There are three kinds of operations on the values of this $\boldsymbol{T}_2$-block [58, 61]:

Computation of first LLR/f-function (right-to-left): Assume that LLRs $\ell_i^{(s)}$ and $\ell_j^{(s)}$, corresponding to the *output bits* $v_i^{(s)}$ and $v_j^{(s)}$ are available. Given the check constraint (11.467), the LLR $\ell_i^{(s-1)}$, corresponding to the *input bit* $v_i^{(s-1)}$, can be computed as

$$\begin{aligned}\ell_i^{(s-1)} &= 2\tanh^{-1}\left(\tanh\frac{\ell_i^{(s)}}{2}\cdot\tanh\frac{\ell_j^{(s)}}{2}\right)\\ &\simeq \operatorname{sgn}\ell_i^{(s)}\cdot\operatorname{sgn}\ell_j^{(s)}\cdot\min\{|\ell_i^{(s)}|,|\ell_j^{(s)}|\} &(11.455)\\ &= f(\ell_i^{(s)},\ell_j^{(s)}) &(11.456)\end{aligned}$$

Replacing the exact operation by the f-function has no significant impact on performance but heavily reduces the complexity. This operation on LLRs is the same as decoding a degree-3 check node of LDPC codes.

Computation of second LLR/g-function (right-to-left): Assume that the LLRs $\ell_i^{(s)}$ and $\ell_j^{(s)}$ are available, as well as the detected input bit $\hat{v}_i^{(s-1)}$. Given the equality constraint (11.468), the LLR $\ell_j^{(s-1)}$, corresponding to the input bit $v_j^{(s-1)}$ can be computed as

$$\ell_j^{(s-1)} = (-1)^{\hat{v}_i^{(s-1)}}\cdot\ell_i^{(s)} + \ell_j^{(s)} = g(\ell_i^{(s)},\ell_j^{(s)};\hat{v}_i^{(s-1)}) \tag{11.457}$$

This operation corresponds to decoding a degree-3 variable node of LDPC codes, with the difference that the sign of $\ell_i^{(s)}$ needs to be corrected according to the value of the bit $\hat{v}_i^{(s-1)}$.

Computation of partial sums (left-to-right): Assume that the detected input bits $\hat{v}_i^{(s-1)}$ and $\hat{v}_j^{(s-1)}$ are available. Then the two detected output bits can be computed as

$$[\hat{v}_i^{(s-1)}\ \hat{v}_j^{(s-1)}]\cdot\begin{bmatrix}1 & 0\\ 1 & 1\end{bmatrix} = [\hat{v}_i^{(s)}\ \hat{v}_j^{(s)}] \tag{11.458}$$

corresponding to (11.457). These values are called *partial sums*.

Given these three kinds of operations for all $(n_0\epsilon_0/2)$ $\boldsymbol{T}_2$-blocks in the Tanner graph, there is only one possible scheduling. We will explain this schedule based on an example for $\boldsymbol{T}_8$ and generalize it afterwards.

Example 11.9.3

Continuing the previous example with $\boldsymbol{T}_8$, we label the Tanner graph from Figure 11.47 with the intermediate LLRs, as shown in Figure 11.51. Note that we keep information set $\mathcal{I} = \{2,3,5,6,7\}$ and frozen set $\mathcal{F} = \{0,1,4\}$ as before.

The LLRs of the code bits at the right are initialized as $\{\ell_n^{(3)} = \ell_n\}$, $n = 0, 1, \ldots, 7$. Then the f-functions and g-functions are applied with the schedule as indicated by the numbers with the hashes; the computations of the partial sums are not indicated in the graph.

Input bit 0: First, the four f-functions of stage 3 (#1, #2, #3, #4) produce the LLRs $\ell_0^{(2)}$, $\ell_1^{(2)}$, $\ell_2^{(2)}$, $\ell_3^{(2)}$. Then the two upper f-functions of stage 2 (#5, #6) are called to obtain $\ell_0^{(1)}$, $\ell_1^{(1)}$. And, last, the uppermost f-function of stage 1 (#7) produces the a posteriori LLR $\lambda_0 = \ell_0^{(0)}$. Position 0 is a frozen position, therefore, we set $\hat{v}_0 = 0$, independently of λ_0.

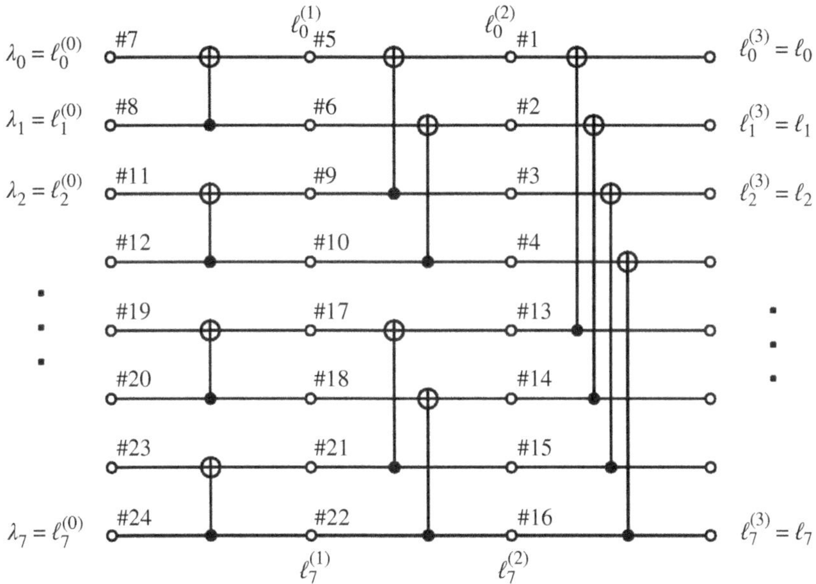

Figure 11.51 Tanner graph for $\boldsymbol{T}_8$, as used for decoding.

Input bit 1: The uppermost g-function (#8) uses $\hat{v}_0 = 0$ to compute λ_1 and $\hat{v}_1$. Position 1 is also a frozen position, therefore, we set $\hat{v}_1 = 0$. From $\hat{v}_0^{(0)}$ and $\hat{v}_1^{(0)}$, the partial sums $\hat{v}_0^{(1)}$ and $\hat{v}_1^{(1)}$ are computed.

Input bit 2: The two upper g-functions of stage 2 (#9, #10) take $\hat{v}_0^{(1)}$ and $\hat{v}_1^{(1)}$ and produce the LLRs $\ell_2^{(1)}$, $\ell_3^{(1)}$. The second f-function of stage 1 (#11) computes then λ_2. As this is an information position, the detected bit $\hat{v}_2$ is set to the hard decision of λ_2.

Input bit 3: The second g-function of stage 2 (#12) uses $\hat{v}_2$ to compute λ_3. This is again an information position, so $\hat{v}_3$ is set to the hard decision of λ_3. Now the partial sums $\hat{v}_2^{(1)}$ and $\hat{v}_3^{(1)}$ can be computed, and further $\hat{v}_0^{(2)}$, $\hat{v}_1^{(2)}$, $\hat{v}_2^{(2)}$, and $\hat{v}_3^{(2)}$.

Input bits 4–7: The four g-functions of stage 3 (#13, #14, #15, #16) produce the LLRs $\ell_4^{(2)}$, $\ell_5^{(2)}$, $\ell_6^{(2)}$, and $\ell_7^{(2)}$. From here on, the procedure is the same as for the upper half of the graph in stages 2 and 1.

Note that all three operations, i.e. f-function, g-function, and partial-sum computation, need to wait until all required input values are available and then each of them is activated exactly once. This defines a unique scheduling.

The example shows in detail the scheduling for $\boldsymbol{T}_8$. For the general case, the schedule for the LLR computation (and computation of partial sums) can be determined in a similar way. For fast decoding, the schedule may be pre-computed offline and stored, rather than determined during decoding.

The algorithm for SC decoding may be improved in terms of complexity or speed. Here are a few examples: (i) Instead of starting with propagating LLRs from right to left, the algorithm may start with propagating partial sums due to frozen bits from left to right; this way, a posteriori LLRs of frozen bits are not computed. (ii) Instead of making decisions on input bits individually, decisions may be made on blocks of input bits where all bits are frozen or all bits are not frozen [65]. (iii) Extending this idea, small codes may be identified that can efficiently be decoded in a single step, e.g. [66]. Further enhancements can be found in literature.

Successive cancellation list decoding

The SC decoder cannot recover from a wrong decision on an input bit. To overcome this deficiency and improve performance, SC list decoding may be used [59–61].

The principle is as follows: Whenever a detected input-bit is to be fed back into the decoding process, two instances of the decoder continue from this point onwards, each using one of the bit values. Further a path metric is associated with each of these paths. The number of paths grows exponentially in the information positions, and thus needs to be limited to a certain list length L. When the number of paths in the list exceeds the list length, only the L paths with the best metric are kept and the others are discarded. Once decoding has reached the end and the paths have length n_0 bits, the path with the best metric is selected. If a CRC is present, we first discard all paths that do not fulfill the CRC; then among the remaining paths we select the one with the best metric.

We look now into the path metric in detail. For convenient notation, we denote a partial path as

$$\boldsymbol{v}_{[n]} = [v_0, v_1, \dots, v_n] \tag{11.459}$$

with value $\boldsymbol{V}_{[n]} = [V_0, V_1, \dots, V_n]$. The metric to be used for path selection is as given in the decoding rule (11.461). For a complete path $\boldsymbol{v}_{[n_0-1]} = \boldsymbol{V}_{[n_0-1]}$, the *path metric* is given by

$$\begin{aligned} \Gamma(\boldsymbol{V}_{[n_0-1]}) &= \sum_{n=0}^{n_0-1} -\ln P[v_n = V_n | z = \rho; \boldsymbol{v}_{[n-1]} = \boldsymbol{V}_{[n-1]}] \\ &= \sum_{n=0}^{n_0-1} \ln(1 + e^{-\lambda_n (-1)^{V_n}}) \end{aligned} \tag{11.460}$$

The first line is taken from (11.461). The second line uses the definition of λ_n in (11.464) and the way to express a probability by the corresponding LLR, as given by (11.395).

To simplify notation, the value of v_n, V_n, will still be denoted as v_n.

The individual terms of the sum in (11.474) are called the *path metric increments*. For those, we can distinguish the case where the sign of the LLR corresponds to the bit value, i.e. $\text{sgn}\, \lambda_n = (-1)^{v_n}$, and the case where it does not, i.e. $\text{sgn}\, \lambda_n \neq (-1)^{v_n}$. For each case, we then use only the absolute value of the LLR to determine the value of the path metric increment,

$$\ln(1 + e^{-\lambda_n (-1)^{v_n}}) = \begin{cases} \ln(1 + e^{|\lambda_n|}) = |\lambda_n| + \ln(1 + e^{-|\lambda_n|}) & \text{for sgn}\, \lambda_n \neq (-1)^{v_n} \\ \ln(1 + e^{-|\lambda_n|}) & \text{for sgn}\, \lambda_n = (-1)^{v_n} \end{cases} \tag{11.461}$$

or with some approximations

$$\ln(1 + e^{-\lambda_n (-1)^{v_n}}) \simeq \begin{cases} |\lambda_n| & \text{for sgn}\, \lambda_n \neq (-1)^{v_n} \\ 0 & \text{for sgn}\, \lambda_n = (-1)^{v_n} \end{cases} \tag{11.462}$$

which provides for a simple low-complexity method to compute the path metric.

Assume that the partial path $\boldsymbol{v}_{[n-1]}$ has path metric $\Gamma(\boldsymbol{v}_{[n-1]})$, the LLR of the current bit position is λ_n. Then the metric of the extended path $\boldsymbol{v}_{[i]} = [\boldsymbol{v}_{[n-1]}, v_n]$ can be written as

$$\Gamma(\boldsymbol{v}_{[n]}) = \Gamma(\boldsymbol{v}_{[n-1]}) + \begin{cases} |\lambda_n| & \text{for sgn}\, \lambda_n \neq (-1)^{v_n} \\ 0 & \text{for sgn}\, \lambda_n = (-1)^{v_n} \end{cases} \tag{11.463}$$

This means that not following the path as suggested by the sign of the LLR leads to a penalization, and the penalty corresponds to the absolute value of the LLR; both make sense intuitively.

Following the principle of SC list decoding as outlined above and using the path metric and its computation, we can formally describe the SCL decoding algorithm. Assume a polar code $P(n_0, k_0)$. Assume further an SCL decoder with list length L; each list entry consists of a partial path $\boldsymbol{v}_{[n]}^{(l)}$ and its metric $\Gamma(\boldsymbol{v}_{[n]}^{(l)})$, $l = 0, 1, \ldots, L-1$. The SCL decoder operates as follows:

Initialization: Take as input the LLRs $\{\ell_n\}$ from the channel, $n = 0, 1, \ldots, n_0 - 1$. Initialize the path list $\mathcal{L}$ of size L to the empty set, $\mathcal{L} = \emptyset$.

Iteration: For input bit position $n = 0, 1, \ldots, n_0 - 1$, perform the following:

LLR calculation and list expansion: If the path list is empty (if and only if $n = 0$), run the SC decoder to compute λ_0. Otherwise, for each path $\boldsymbol{v}_{[n-1]}^{(l)}$ in the path list, $l = 0, 1, \ldots, |\mathcal{L}|$, take the path out of the list, run the SC decoder, and compute $\lambda_n(\boldsymbol{v}_{[n-1]}^{(l)})$. If position n is a frozen bit, i.e. if $n \in \mathcal{F}$, extend the path to $[\boldsymbol{v}_{[n-1]}^{(l)}, 0]$, compute the path metric $\Gamma\,([\boldsymbol{v}_{[n-1]}^{(l)}, 0])$, and put the new path including its metric in the list $\mathcal{L}$. If position n is an information bit, $n \in \mathcal{I}$, extend the path to $[\boldsymbol{v}_{[n-1]}^{(l)}, 0]$ and $[\boldsymbol{v}_{[n-1]}^{(l)}, 1]$, compute the path metrics $\Gamma([\boldsymbol{v}_{[n-1]}^{(l)}, 0])$ and $\Gamma([\boldsymbol{v}_{[n-1]}^{(l)}, 1])$, and put the two new paths including their metrics in the list $\mathcal{L}$. The new path metrics are computed according to (11.477).

List pruning: If the list contains more than L paths, keep the L paths with the smallest path metrics and discard the others.

Decision: If a CRC is present, discard all paths in $\mathcal{L}$ that do not fulfill the CRC. Then select the path with the smallest path metric. Finally, extract the information bits from the selected input vector and remove the CRC bits if a CRC is present.

A few comments on this decoding algorithm are appropriate:

1. At the iteration for input–bit position n, the list size remains constant if $n \in \mathcal{F}$, and it is doubled if $n \in \mathcal{I}$. The doubling happens as each existing path l is extended by both possible hypotheses $v_n^{(l)} = 0$ and $v_n^{(l)} = 1$. The step with the list pruning keeps the list at maximum size L.
2. Every path in the list is associated with a decoder memory storing all intermediate LLRs and all partial sums. After the list-pruning step, it may be that both extensions of a previous path are kept; in this case, the memory needs to be duplicated. More efficient memory managements and implementations have been addressed in literature.
3. In the decision step, the CRC acts as a kind of genie, i.e. the CRC is likely to pick the transmitted codeword from the list, if the transmitted codeword is in the list and has not been discarded before.
4. If a CRC is present and none of the paths in the list fulfill the CRC at the end, there are two options: (i) the decoder may output the information bits corresponding to the best path, accepting decoding errors; (ii) the decoder may declare a decoding failure. The choice depends on the application.

The complexity of SCL decoding is about L times larger than that of SC decoding, i.e. of order $O(L\, n_0 \ln n_0)$, as an SC decoder has to be run for each of the L paths in the list. The error-rate performance, however, is largely improved.

Other decoding algorithms

Various other decoding algorithms have been proposed for polar codes, most based on SC decoding, among which are the following three: *SC stack decoding* is similar to SC list decoding, however, not all paths in the list are extended in every step, and thus the list in general contains paths of different lengths [67]. In *SC flip decoding*, the CRC is used to check after SC decoding if decoding was successful; if not, SC decoding is run again, however, with a different decision on the least reliable position, thus leading to a new decoding outcome [68]. *Belief-propagation decoding*, as for LDPC codes, may be directly applied to the Tanner graph of the polar code, allowing for a high degree of parallelism [69].

11.9.4 Frozen set design

A polar code of length $n_0 = 2^{\epsilon_0}$ and information length k_0 is defined by its information set $\mathcal{I}$ of size k_0 and its frozen set $\mathcal{F}$ of size $(n_0 - k_0)$. Polar code design is thus concerned with the choice of which $n_0 - k_0$ of the n_0 input-bit positions to put into the frozen set and which k_0 positions to put into the information set. The SC decoder cannot make wrong decisions on frozen bits, as they are known to be zero, and we want that wrong decisions on information bits are unlikely; therefore, unreliable positions should be frozen and reliable positions should be used to transmit information. This section deals with an algorithm to determine the reliability of the input-bit positions. A received signal *model* as in (11.439) is assumed.

Genie-aided SC decoding

The frozen-set design is based on *genie-aided SC decoding* [58], as follows: we denote the transmitted input vector by $\check{\boldsymbol{v}}$ and the detected value (after decoding) by $\hat{\boldsymbol{v}}$, as before. Arbitrary input vectors $\check{v}$, without any frozen bits, are assumed to be transmitted over an AWGN channel with a chosen *design SNR*. For each input-bit position n, $n = 0, 1, \dots, n_0 - 1$, decoding is performed and its reliability is determined:

1. The SC decoder runs as shown in Figure 11.49, and the a posteriori LLRs $\{\lambda_n\}$ are computed, as explained above.
2. The hard-decisions $\{\hat{v}_n\}$ of $\{\lambda_n\}$ are used to count errors and determine the bit-error rate p_n for position n. Note that no bits are frozen.
3. The correct bit values $\check{v}_n$, i.e. the actually transmitted ones, are fed back into the decoding process. This is the genie-aided part.

According to this decoding model, the a posteriori LLRs are now

$$\lambda_n = \ell_{v_n}(\boldsymbol{\ell}; \check{v}_0, \dots, \check{v}_{n-1}) \tag{11.464}$$

with *perfect feedback* (genie-aided and thus error-free). The hard-decisions are

$$\hat{v}_i = \begin{cases} 0 & \text{if } \lambda_n > 0 \\ 1 & \text{if } \lambda_n < 0 \\ \text{randomly chosen} & \text{if } \lambda_n = 0 \end{cases} \tag{11.465}$$

and bit errors are counted to determine the *position-wise bit-error rates* (BERs),

$$p_n = P[\hat{v}_n \neq \check{v}_n] \tag{11.466}$$

$n = 0, 1, \dots, n_0 - 1$.

We aim at minimizing the *union bound for the word error rate under SC decoding*,

$$P_e = P[\hat{\boldsymbol{v}} \neq \check{\boldsymbol{v}}] \leq \sum_{n \in \mathcal{I}} p_n \tag{11.467}$$

(The inequality is due to possibly correlated error events.) To do so, we collect the $(n_0 - k_0)$ positions with the largest BER p_n into the frozen set and the k_0 positions with the smallest BER p_n into the information set:

$$\mathcal{F} = \{n \in \{0, 1, \dots, n_0 - 1\} \ : \ \{p_n\} \text{ are the } (n_0 - k_0) \text{ largest values}\} \tag{11.468}$$

$$\mathcal{I} = \{n \in \{0, 1, \dots, n_0 - 1\} \ : \ \{p_n\} \text{ are the } k_0 \text{ smallest values}\} \tag{11.469}$$

The required position-wise BERs $\{p_n\}$ may be determined by Monte Carlo simulation. Note that due to binary modulation model and the symmetry of the AWGN channel, it is sufficient to transmit the all-zero input codevector. Though precise, this is method is time-consuming, as the values $\{p_n\}$ need to be about as accurate as the overall target BER of the decoder.

Design based on density evolution

A faster method with low complexity is based on *density evolution using a Gaussian approximation* (DE/GA). Density evolution and its variant with the Gaussian approximation were originally suggested for LDPC code design [70] and later proposed to polar code design [71]; further refinements for polar codes have been proposed in literature.

The concept of DE/GA is as follows. We assume that the all-zero input vector is transmitted, without loss of generality as explained above; thus all partial sums and all code bits are zero as well. Then on the decoder side, the LLRs $\{\ell_n^{(s)}\}$ are considered as random variables with a conditional probability density given that the corresponding bit value is zero. The goal is to start with the Gaussian pdf of $\{\ell_n\}$ to compute the intermediate densities of $\{\ell_n^{(s)}\}$, and finally the densities of the a posteriori LLRs $\{\lambda_n\}$; this process is referred to as *density evolution*. To simplify the computations, we approximate the LLRs as Gaussian random variables. As Gaussian LLRs have the property that their *variance is twice their mean* value [70], we can fully characterize each distribution by its mean value. Thus, for DE/GA, it is sufficient to track the mean values of the LLRs.

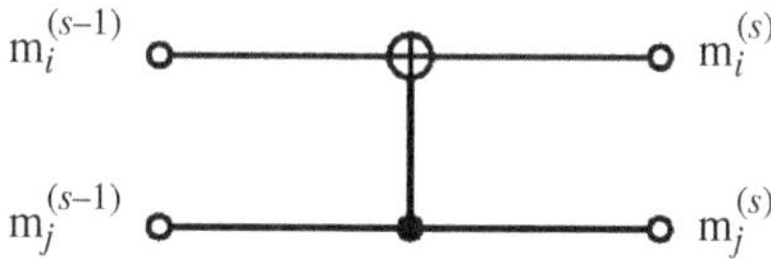

Figure 11.52 T_2-block at stage s with mean values of LLRs.

For the initialization, we need the mean values of the channel LLRs. Based on the channel model (11.439), the channel LLRs are computed according to (11.450). Under the assumption of the all-zero-codeword, from the bit-mapping (11.439), the channel output z_n has mean $+1$, and thus the channel LLR ℓ_n has mean $\mathtt{m}_n = 2/\sigma_{dec}^2$ for all positions n. Thus, we initialize the DE/GA with

$$\mathtt{m}_n^{(\epsilon_0)} = \mathtt{m}_n = \frac{2}{\sigma_{dec}^2} \tag{11.470}$$

for $n = 0, 1, \dots, n_0 - 1$.

The computation of the LLRs in the Tanner graph is based on T_2-blocks, and so is the density evolution. With reference to Figure 11.50, we have the corresponding mean values of the LLRs as shown in Figure 11.52. Assume a T_2-block at stage s with the mean values $\mathtt{m}_i^{(s)}$ and $\mathtt{m}_j^{(s)}$ (at the right) are available. Our goal is to compute the mean values $\mathtt{m}_i^{(s-1)}$ and $\mathtt{m}_j^{(s-1)}$ (at the left).

The LLR $\ell_n^{(s-1)}$ is computed by the f-function, corresponding to a check-node in LDPC coding, see (11.470). Correspondingly, the DE/GA for a check node may be applied,

$$\mathtt{m}_i^{(s-1)} = \varphi^{-1}(1 - (1 - \varphi(\mathtt{m}_i^{(s)})) \cdot (1 - \varphi(\mathtt{m}_j^{(s)}))) \tag{11.471}$$

The function φ relates the mean value $\mathtt{m} = E[\ell]$ of a Gaussian LLR to the expectation $E[\tanh(\ell/2)]$, $\varphi : E[\ell] \mapsto E[\tanh(\ell/2)]$, and it is formally defined as

$$\varphi(\mathtt{m}) = \begin{cases} 1 - \frac{1}{\sqrt{4\pi\mathtt{m}}} \int_{-\infty}^{\infty} \tanh(\frac{a}{2})\, e^{-\frac{(a-\mathtt{m})^2}{2\mathtt{m}}} da & \text{for } \mathtt{m} > 0 \\ 1 & \text{for } \mathtt{m} = 0 \end{cases}$$

the function $\varphi^{-1}(\mathtt{m})$ is its inverse.

To simplify the computations, the function φ may be approximated by

$$\varphi(\mathtt{m}) \simeq \begin{cases} e^{a\mathtt{m}^2+b\mathtt{m}} & \text{for } 0 \le \mathtt{m} < \mathtt{m}_0 \\ e^{\alpha\mathtt{m}^\gamma+\beta} & \text{for } \mathtt{m}_0 \le \mathtt{m} \end{cases} \tag{11.472}$$

with $\alpha = -0.4527$, $\beta = 0.0218$, $\gamma = 0.86$, $a = 0.0564$, $b = -0.48560$, $\mathtt{m}_0 = 0.867861$ [70, 72], where the parameters have been determined by curve fitting; note that both approximations can be analytically inverted. Refined approximations of $\varphi(\mathtt{m})$ have been proposed in literature, particularly for improved performance of long codes under SC decoding, which is very sensitive to the choice of the frozen positions. SCL decoding is typically more robust to the frozen set design.

The LLR $\ell_j^{(s-1)}$ is computed by the g-function, corresponding to a variable-node in LDPC coding, see (11.471). As all feedback bits are zero, all partial sums are zero, and therefore the detected bit is $\hat{v}_i^{(s-1)} = 0$. As for decoding, we then simply add two Gaussian LLRs, the mean of $\ell_j^{(s-1)}$ is given by the sum of the two mean values on the right,

$$\mathtt{m}_j^{(s-1)} = \mathtt{m}_i^{(s)} + \mathtt{m}_j^{(s)} \tag{11.473}$$

Equations (11.485) and (11.488) are referred to as the *density-evolution equations*. Applying them to the Tanner graph from right to left, we can perform the full DE/GA.

When this DE is finished, we set the mean values μ_n of the a posteriori LLRs λ_n to $\mu_n = \mathtt{m}_n^{(0)}$. From the means of the a posteriori LLRs and the Gaussian approximation, we can determine the position-wise BERs as

$$p_n \simeq \mathrm{Q}(\sqrt{\mu_n/2}) \tag{11.474}$$

As the BER is continuous in the mean, we may directly use the mean values to determine the frozen set and the information set:

$$\mathcal{F} = \{n \in \{0, 1, \dots, n_0 - 1\} : \{\mu_n\} \text{ are the } (n_0 - k_0) \text{ smallest values}\} \tag{11.475}$$

$$\mathcal{I} = \{n \in \{0, 1, \dots, n_0 - 1\} : \{\mu_n\} \text{ are the } k_0 \text{ largest values}\} \tag{11.476}$$

This provides a fast and convenient method for the frozen set design.

We will illustrate the approach for the example of the transformation $\boldsymbol{T}_8$.

Example 11.9.4

The LLRs involved in the DE are shown in Figure 11.53. Assume the initial channel LLRs are $\mathtt{m}_n = 2.1$. Then we set $\mathtt{m}_n^{(3)} = \mathtt{m}_n = 2.1$, $n = 0, 1, \dots, 7$.

Consider stage 3 and first the $\boldsymbol{T}_2$-block with the index pair $(i,j) = (0, 4)$. We have $\mathtt{m}_0^{(3)} = \mathtt{m}_4^{(3)} = 2.1$, and we compute the DE as

$$\begin{aligned} \mathtt{m}_0^{(2)} &= \varphi^{-1}(1 - (1 - \varphi(\mathtt{m}_0^{(3)})) \cdot (1 - \varphi(\mathtt{m}_4^{(3)}))) \\ &= \varphi^{-1}(1 - (1 - \varphi(2.1)) \cdot (1 - \varphi(2.1))) = 0.89 \\ \mathtt{m}_4^{(2)} &= \mathtt{m}_0^{(3)} + \mathtt{m}_4^{(3)} = 2.1 + 2.1 = 4.2 \end{aligned} \tag{11.477}$$

The calculations for the other three $\boldsymbol{T}_2$-blocks with index pairs (i,j) being $(1, 5)$, $(2, 6)$, and $(3, 7)$ are exactly the same, as the input values are the same.

At stage 2 for the $\boldsymbol{T}_2$-block with indices $(i,j) = (0, 2)$, we have $\mathtt{m}_0^{(2)} = \mathtt{m}_2^{(2)} = 0.89$. The DE is computed as

$$\begin{aligned} \mathtt{m}_0^{(1)} &= \varphi^{-1}(1 - (1 - \varphi(\mathtt{m}_0^{(2)})) \cdot (1 - \varphi(\mathtt{m}_2^{(2)}))) \\ &= \varphi^{-1}(1 - (1 - \varphi(0.89)) \cdot (1 - \varphi(0.89))) = 0.23 \\ \mathtt{m}_2^{(1)} &= \mathtt{m}_0^{(2)} + \mathtt{m}_2^{(2)} = 0.89 + 0.89 = 1.77 \end{aligned} \tag{11.478}$$

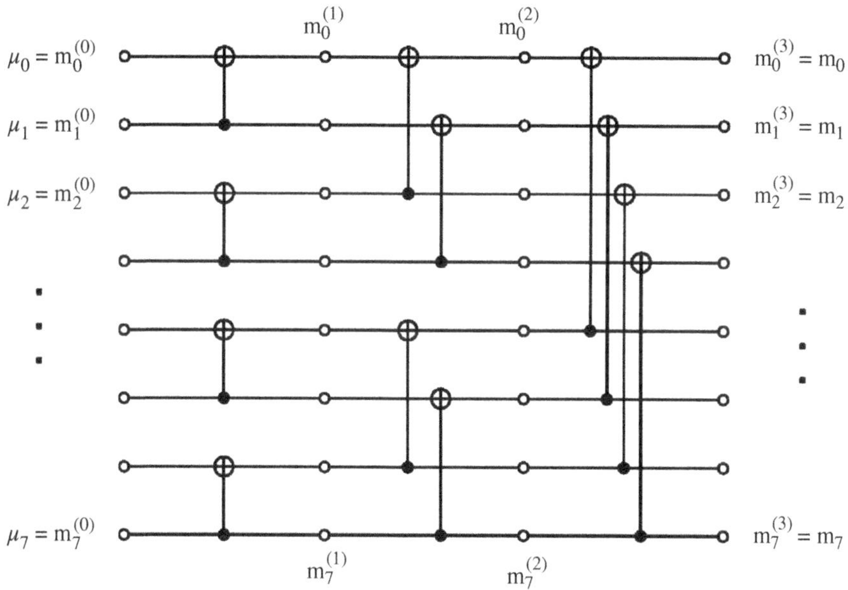

Figure 11.53 Tanner graph for $\boldsymbol{T}_8$, as used for DE/GA.

For the $\boldsymbol{T}_2$-block with indices $(i,j) = (1, 3)$, we obtain the same result as for the one with $(i,j) = (0, 2)$, as the input values are the same. Similarly, we continue for the remaining two $\boldsymbol{T}_2$-blocks at stage 2.

For stage 1, the DE has to be computed for all four $\boldsymbol{T}_2$-blocks individually, as their input values are different; for each $\boldsymbol{T}_2$-block, however, the two input values are identical. Note that input bit v_7 has the highest reliability, with $\mathtt{m}_0^{(7)} = 8 \cdot 2.1 = 16.8$.

Using $\mu_n = \mathtt{m}_n^{(1)}$, we obtain the mean values of the a-posteriori LLRs as

$$[\mu_0, \dots, \mu_7] = [0.04, 0.47, 0.69, 3.55, 1.11, 4.89, 6.15, 16.8] \tag{11.479}$$

Ranking the indices in increasing order of the mean values, and thus in increasing reliability, we get the ordered index sequence

$$[0, 1, 2, 4, 3, 5, 6, 7] \tag{11.480}$$

If we want a polar code with information length $k_0 = 4$, we collect the $n_0 - k_0 = 8 - 4 = 4$ first indices of this sequence into the frozen set and the last $k_0 = 4$ indices into the information set:

$$\mathcal{F} = \{0, 1, 2, 4\}$$
$$\mathcal{I} = \{3, 5, 6, 7\}$$

This completes the frozen and information set design.

For a general polar code of length $n_0 = 2^{\epsilon_0}$ and information length k_0, the procedure is similar to this example, only comprising more stages.

Channel polarization

The channels from an input bit v_n to its corresponding a posteriori LLR λ_n are termed bit-channels, $n = 0, 1, \dots, n_0$. The method for the frozen set design gives rise to two theoretical questions: When the code length n_0 tends to infinity, (i) how good and how bad are the bit-channels, and (ii) how many of the bit-channels are good? The answer was given by Arıkan in his seminal work [58].

For $n_0 \to \infty$, the bit-channels tend to become fully reliable (BER p_n being zero), or fully unreliable (BER p_n being $1/2$); this phenomenon is referred to as *channel polarization*. Further, denoting the capacity of the communication channel with C, the number of reliable channels tend to Cn_0.

Thus, asymptotically, $k_0 = Cn_0$ positions can be used for the information bits and the probability of making wrong decisions on these positions during SC decoding tends to zero; the other positions are frozen. The resulting code rate tends to $R = k_0/n_0 = Cn_0/n_0 = C$, i.e. to the channel capacity. This gives the fundamental result that polar codes under SC decoding are *capacity-achieving*.

11.9.5 Puncturing and shortening

Puncturing and shortening are methods to change the length and rate of a given code, called the *mother code*, while using the encoder and decoder of the original code. As from Section 11.4, in *puncturing*, certain code bits of the mother code are not transmitted; at the receiver side, the corresponding LLRs are set to zero, as the probabilities for 0 and 1 are both $1/2$. In *shortening*, encoding of the mother code is done such that certain code bits are always zero, and these are not transmitted; at the receiver side, the corresponding LLRs are set to plus infinity, as the bits are known to be zero.

Polar codes based on the kernel $\boldsymbol{T}_2$ can be constructed for any information word length k_0 but only for code lengths $n_0 = 2^{\epsilon_0}$ that are powers of two. Therefore, puncturing and shortening are important methods in polar coding to obtain code lengths different from powers of two. Further, due to the way polar codes are encoded and decoded, the frozen set design requires special attention when puncturing or shortening are applied. This section explains the underlying principles. Further details may be found in [64, 73, 74].

Puncturing

Consider a polar code $P(n_0, k_0)$ of length $n_0 = 2^{\epsilon_0}$, information length k_0, and rate $R_c = k_0/n_0$ to be used as the mother polar code. By puncturing, we reduce the code length to $n'_0 < n_0$ and keep the information length k_0; thus, we obtain a code of rate $R'_c = k_0/n'_0 > R_c$, which is higher than the rate R_c of the mother code.

The encoding is performed for the mother polar code and then we puncture $p_0 = n_0 - n'_0$ code bits, selected by the index set $\mathcal{A}_p \subset \{0, 1, \dots, n_0 - 1\}$ of size p_0, called the *puncturing pattern*; i.e. the code bits, $\{c_n\}, n \in \mathcal{A}_p$, are not transmitted. At the receiver side, the LLRs $\{\ell_n\}$ for the unpunctured positions, i.e. for $n \notin \mathcal{A}_p$, are computed as usual from the observations $\{y_n\}$. For the punctured positions, the LLRs are set to zero, $\{\ell_n = 0\}$ for $n \in \mathcal{A}_p$. SC-decoding or SC-based decoding is then performed as usual for the mother polar code.

SC decoding reacts in a special way to puncturing: every channel LLR with value zero produces one a posteriori LLR with value zero. Instead of a proof, we motivate this by an example. Figures 11.54 and 11.55 show the Tanner graph of the transform $\boldsymbol{T}_4 = \boldsymbol{T}_2 \otimes \boldsymbol{T}_2$, consisting of two stages, the former with the bits for encoding, the latter with the LLRs for decoding. Assume that c_2 was punctured, such that $\ell_2^{(2)} = \ell_2 = 0$. When running the decoder, $\ell_0^{(1)}$ will become zero since $\ell_2^{(2)} = 0$; on the other hand, $\ell_2^{(1)}$ will be non-zero according to $\ell_0^{(2)}$ (unless this LLR is zero as well); see (11.470) and (11.471). For the same reason, we obtain $\ell_0^{(0)} = 0$. Thus, puncturing position 2 leads to the a posteriori LLR $\lambda_0 = 0$ in position 0.

Therefore, the positions with zero a posteriori LLR, induced by the puncturing patterns $\mathcal{A}_p$, need to be in the frozen set $\mathcal{F}$. Two special structures of puncturing patterns have the property that the index

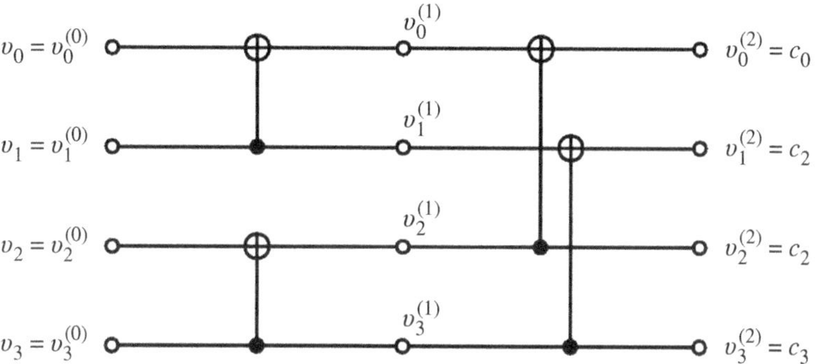

Figure 11.54 Tanner graph of T_4 for encoding.

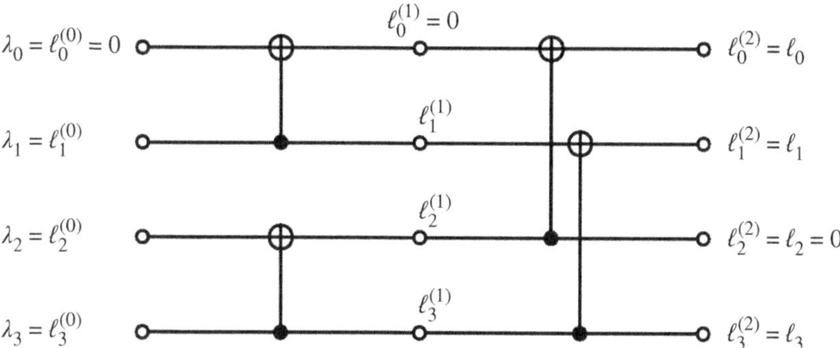

Figure 11.55 Tanner graph of T_4 for decoding with punctured position.

pattern of the zero a posterior LLRs is identical to the puncturing pattern, which is a property convenient for polar code design. The first structure is called *block puncturing*, where the first p_0 positions are punctured,

$$\mathcal{A}_p = \{0, 1, \dots, p_0 - 1\} \tag{11.481}$$

The second structure is called *bit-reversal puncturing*. The bit-reversal permutation is a permutation of length $n_0 = 2^{\epsilon_0}$, where the index n with binary representation $(d_0, d_1, \dots, d_{\epsilon_0 - 1})$ is mapped to index π_n with binary representation $(d_{\epsilon_0-1}, d_{\epsilon_0-2}, \dots, d_0)$ [58]. The bit-reversal puncturing uses the puncturing pattern

$$\mathcal{A}_p = \{\pi_0, \pi_1, \dots, \pi_{p_0-1}\} \tag{11.482}$$

For those two puncturing schemes, we have then the simple requirement that $\mathcal{A}_p \subset \mathcal{F}$.

Shortening

Consider, as before, a polar code $P(n_0, k_0)$ of length $n_0 = 2^{\epsilon_0}$, information length k_0 and rate $R_c = k_0/n_0$, to be used as the mother polar code. By shortening, we reduce the code length to $n_0' = n_0 - n_S$, and we reduce the information length to $k_0' = k_0 - n_S$, shortening by n_S positions; thus, we obtain a code of rate $R_c' = k_0'/n_0' < R_c$, which is lower than the rate R_c of the mother code.

For shortening a systematically encoded code, $k_0 - n_S$ information bits are used to actually transmit message bits, and the other n_S information bits are set to zero; the n_S zero information bits are not transmitted, and only the $k_0 - n_S$ message-carrying information bits and the parity bits are transmitted. For systematically encoded polar codes (not discussed here), this method may directly be applied.

Non-systematically encoded polar codes (as discussed above and often used) require a different approach. The encoding is performed for the mother polar code, however, n_S additional input bits are frozen in a way such that after encoding, n_S code bits are always zero. The index set of these zero code bits, denoted by $\mathcal{A}_s \subset \{0, 1, \dots, n_0 - 1\}$, is called the *shortening pattern* of size $n_S = |\mathcal{A}_s|$. These code bits are not transmitted. At the receiver side, the LLRs ℓ_n for the positions that are not shortened, i.e. for $n \notin \mathcal{A}_s$, are computed as usual. For the shortened positions, the LLRs are set to be plus infinity, $\ell_n = +\infty$ for $n \in \mathcal{A}_s$; the reason is that the values of these bits are known to be zero, thus the probabilities for 0 and 1 are 1 and 0, respectively, and correspondingly the LLR is plus infinity. SC-decoding or SC-based decoding is then performed as usual for the mother polar code.

Similarly to the case of puncturing, shortening has a special effect on SC decoding: every channel LLR with value infinity produces one a posteriori LLR with value infinity. Again, we provide an example instead of a proof, and we use Figures 11.54 and 11.56 for illustration. Assume that c_3 was shortened, such that $\ell_3^{(2)} = \ell_3 = +\infty$. When running the decoder, $\ell_3^{(1)}$ will become infinity due to $\ell_3^{(2)}$, independent of $\ell_1^{(2)}$; on the other hand, $\ell_1^{(1)}$ will be non-zero due to $\ell_1^{(2)}$; see (11.470) and (11.471). For similar reasons, we obtain $\ell_3^{(0)} = +\infty$. Thus, shortening position 3 leads to the a posteriori LLR $\lambda_3 = +\infty$ in position 3. Note that this is the input bit position used to make sure the code bit c_3 is zero, such that it can be shortened.

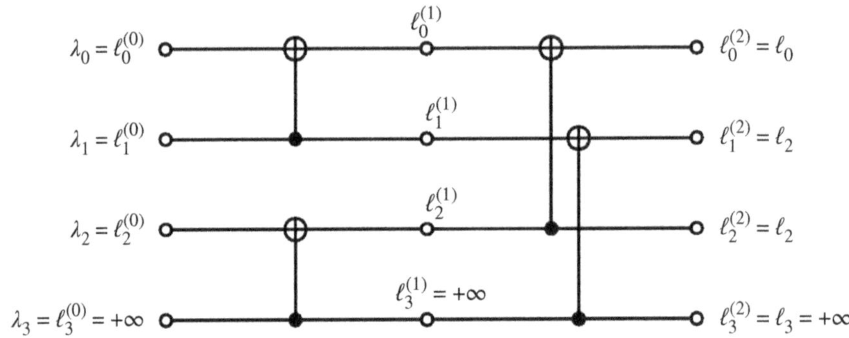

Figure 11.56 Tanner graph of $\boldsymbol{T}_4$ for decoding with shortened position.

Similar to puncturing, there are two special structures of shortening patterns $\mathcal{A}_s$ with the property that the index set of the additionally frozen input bits is identical to the shortening pattern. The structure, called *block shortening*, is obtained by the shortening the last n_S bits, i.e.

$$\mathcal{A}_s = \{n_0 - n_S, n_0 - n_S - 1, \dots, n_0 - 2, n_0 - 1\} \tag{11.483}$$

The second, called *bit-reversal shortening*, is obtained by using the last n_S positions of the bit-reversal permutation, i.e.

$$\mathcal{A}_s = \{\pi_{n_0-n_S}, \pi_{n_0-n_S-1}, \dots, \pi_{n_0-2}, \pi_{n_0-1}\} \tag{11.484}$$

As for puncturing, this gives the simple requirement that $\mathcal{A}_s \subset \mathcal{F}$.

Note that there are other options for puncturing and shortening patterns with the desired property, e.g. patterns composed recursively of blocked and bit-reversal patterns.

Frozen set design

We have seen above that for puncturing and shortening of polar codes, we have the constraints that the puncturing and the shortening pattern for the block and bit-reversal schemes have to be within the frozen sets,

$$\mathcal{A}_p \subset \mathcal{F}, \qquad \mathcal{A}_s \subset \mathcal{F} \tag{11.485}$$

Considering these properties, there are two ways to determine the overall frozen set. First, the puncturing or shortening pattern may be fixed, and then DE/GA may be run, using mean value zero for the punctured positions and mean value infinity for the shortened positions. This leads to the best performance but needs to be redone for every puncturing or shortening pattern. Second, the frozen set $\mathcal{F}$ determined for the mother code may be used, and then the union of this frozen set with the puncturing or shortening pattern be used as the overall frozen set, i.e. $\mathcal{F}_p = \mathcal{F} \cup \mathcal{A}_p$ or $\mathcal{F}_s = \mathcal{F} \cup \mathcal{A}_s$, respectively. This requires DE/GA only to be performed for the mother code but typically leads to a performance loss.

11.9.6 Performance

We consider a polar code $P(1024, 512)$ of length $n_0 = 1024$ and rate $R_c = 512/1024 = 1/2$, first *without inner CRC*. The information set and frozen set $\mathcal{I}$ and $\mathcal{F}$ are designed with DE/GA, as described in Section 11.9.4. The code is transmitted over a 2-PAM $(-1, 1)$ AWGN channel with noise variance σ_{dec}^2 (see (11.439)); the SNR is given in $\frac{E_b}{N_0} = \frac{1}{R_c}\frac{1}{2\sigma_{dec}^2}$. The LLRs of the code bits are computed as given in (11.450). For decoding, we apply plain SC decoding and SCL decoding with list lengths $L = 2, 4$, and 8.

Figure 11.57 shows the BER. Going from SC decoding to SCL decoding with list length $L = 2$ significantly improves the performance. Further increase of the list length lowers the error rates at low SNR. In the high SNR regime, the BER is dominated by the minimum distance of the code, which is not particularly good, and thus increasing the list length is not beneficial.

Next, we consider a polar code *with internal CRC*. Code length and information length are again $n_0 = 1024$ and $k_0 = 512$, the CRC has length $k_{0,CRC} = 11$ bits. As discussed in Section 11.9.1, given $k_0 = 512$ information bits and $k_{0,CRC} = 11$ CRC bits, the information set $\mathcal{I}$ is now of size $k_0' = 512 + 11 = 523$. As

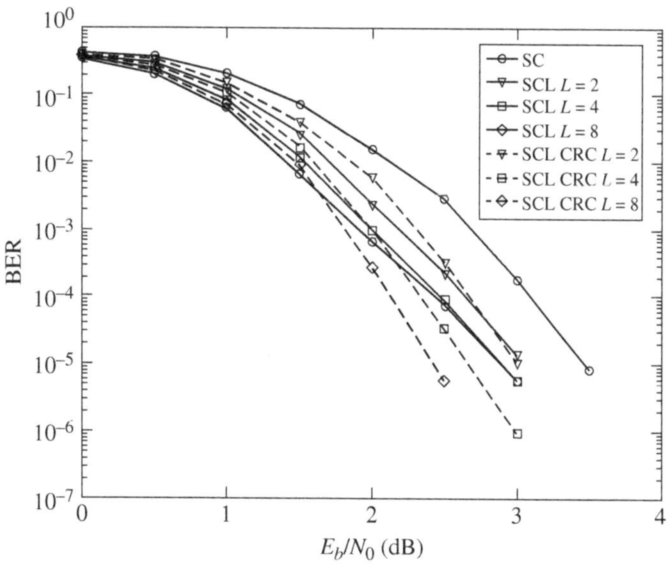

Figure 11.57 Performance of a polar code of length $n_0 = 1024$ and rate $R = 1/2$ under SC and SCL decoding, without inner CRC (solid lines) and with inner CRC (dashed lines).

before, the information set and frozen set $\mathcal{I}$ and $\mathcal{F}$ are designed with DE/GA. We apply SCL decoding with list lengths $L = 2, 4$, and 8, and the CRC is exploited in the decoder to select the detected codeword from the list, as discussed in Section 11.9.3.

Figure 11.57 shows also the BER for this CRC-aided SCL decoding algorithm. For list length $L = 2$, the CRC leads to a worse performance than without CRC. This is due to the fact, that with CRC the number of frozen bits is smaller, and the short list length does not allow to exploit the benefits of the CRC. For larger list lengths, however, the error rates are clearly lower. In particular, the curves with CRC show a steeper slope than the curves without CRC, caused by the increased minimum distance of the overall code, see Section 11.9.1. To take full advantage of this improved distance properties, the list length has to be long enough. The joint selection of CRC length and list length for SCL decoding is an important part of the polar code design.

In general, polar codes under SCL decoding achieve a performance comparable to LDPC codes and turbo codes in the short-length and medium-length regime. Depending on the particular application and constraints on decoding (like complexity, area, power consumption, latency), polar codes may be preferable to other codes [62].

11.10 Milestones in channel coding

Figure 11.58 shows the spectral efficiency versus E_b/N_0 for various codes proposed over time, showing the slow but inexorable shift towards the Shannon bound.

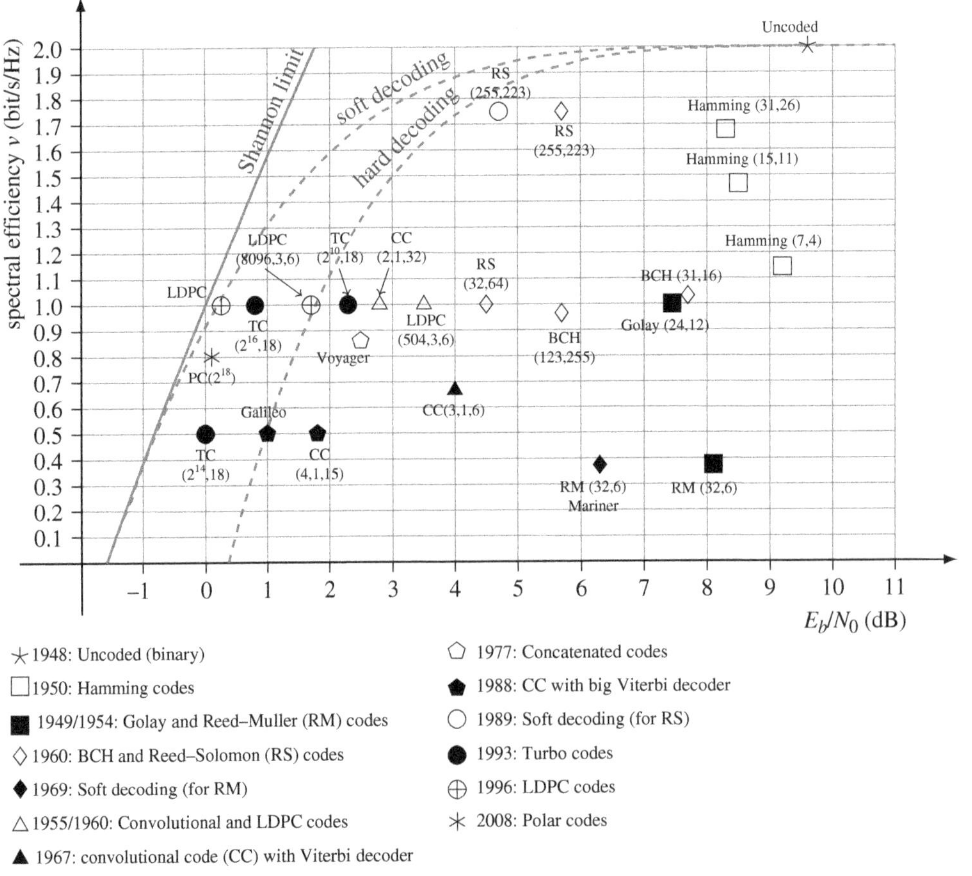

Figure 11.58 Spectral efficiency versus E_b/N_0 for various codes proposed over time. Source: Courtesy of Prof. Matthieu Bloch.

Bibliography

[1] Lin, S. and Costello, D.J. Jr. (1983). *Error Control Coding*. Englewood Cliffs, NJ: Prentice-Hall.

[2] Blahut, R.E. (1983). *Theory and Practice of Error Control Codes*. Reading, MA: Addison-Wesley.

[3] Peterson, W.W. and Weldon, E.J. Jr. (1972). *Error-Correcting Codes*, 2e. Cambridge, MA: MIT Press.

[4] Wolf, J.K. (1980). *Lecture Notes*. University of California: San Diego, CA.

[5] Wicker, S.B. and Bhargava, V.K. (eds.) (1994). *Reed-Solomon Codes and Their Applications*. Piscataway, NJ: IEEE Press.

[6] Oh, J., Ha, J., Park, H., and Moon, J. (2015). RS-LDPC concatenated coding for the modern tape storage channel. *IEEE Transactions on Communications* 64: 59–69.

[7] Cideciyan, R., Furrer, S., and Lantz, M. (2016). Product codes for data storage on magnetic tape. *IEEE Transactions on Magnetics* 53: 1–10.

[8] Furrer, S., Lantz, M., Reininger, P. et al. (2017). 201 Gb/in^2 recording areal density on sputtered magnetic tape. *IEEE Transactions on Magnetics* 54: 1–8.

[9] Gallager, R. (1968). *Information Theory and Reliable Communication*. New York, NY: Wiley.

[10] Caire, G., Taricco, G., and Biglieri, E. (1998). Bit-interleaved coded modulation. *IEEE Transactions on Information Theory* 44: 927–946.

[11] Hagenauer, J. and Hoeher, P. (1989). A Viterbi algorithm with soft-decision output and its applications. *Proceedings of GLOBECOM '89*, Dallas, Texas, USA, pp. 2828–2833.

[12] Fossorier, M.P.C. and Lin, S. (1995). Soft-decision decoding of linear block codes based on ordered statistics. *IEEE Transactions on Information Theory* 41: 1379–1396.

[13] Fossorier, M.P.C. and Lin, S. (1998). Soft-input soft-output decoding of linear block codes based on ordered statistics. *Proceedings of GLOBECOM '98*, Sidney, Australia, pp. 2828–2833.

[14] Costello, D.J. Jr., Hagenauer, J., Imai, H., and Wicker, S.B. (1998). Applications of error-control coding. *IEEE Transactions on Information Theory* 44: 2531–2560.

[15] Fano, R.M. (1963). A heuristic discussion on probabilistic decoding. *IEEE Transactions on Information Theory* 9: 64–74.

[16] Zigangirov, K. (1966). Some sequential decoding procedures. *Problemy Peredachi Informatsii* 2: 13–25.

[17] Jelinek, F. (1969). An upper bound on moments of sequential decoding effort. *IEEE Transactions on Information Theory* 15: 140–149.

[18] Jelinek, F. (1969). Fast sequential decoding algorithm using a stack. *IBM Journal of Research and Development* 13: 675–685.

[19] Wang, F.Q. and Costello, D.J. (1994). Erasure-free sequential decoding of trellis codes. *IEEE Transactions on Information Theory* 40: 1803–1817.

[20] Lin, C.F. and Anderson, J.B. (1986). M-algorithm decoding of channel convolutional codes. *Proc., Princeton Conf. Inform. Sci. Syst.*, Princeton, NJ, USA, pp. 362–365.

[21] Benedetto, S. and Biglieri, E. (1999). *Principles of Digital Transmission with Wireless Applications*. New York, NY: Kluwer Academic Publishers.

[22] Forney, G.D. Jr. (1966). *Concatenated Codes*. Cambridge, MA: MIT Press.

[23] Fossorier, M.P.C., Burkert, F., Lin, S., and Hagenauer, J. (1998). On the equivalence between SOVA and max-log-MAP decodings. *IEEE Communications Letters* 2: 137–139.

[24] Berrou, C., Glavieux, A., and Thitimajshima, P. (1993). Near Shannon limit error-correcting coding and decoding: turbo codes. *Proceedings of IEEE International Conference on Communications*, Geneva, Switzerland (23–26 May 1993), pp. 1064–1070.

[25] Berrou, C. and Glavieux, A. (1996). Near optimum error-correcting coding and decoding: turbo-codes. *IEEE Transactions on Communications* 44: 1261–1271.

[26] Sklar, B. (1997). A primer on turbo code concepts. *IEEE Communications Magazine* 35: 94–101.

[27] Benedetto, S. and Montorsi, G. (1996). Unveiling turbo codes: some results on parallel concatenated coding schemes. *IEEE Transactions on Information Theory* 42: 409–428.

[28] 3-rd Generation Partnership Project (3GPP) (2000). Technical Specification Group (TSG), Radio Access Network (RAN), Working Group 1 (WG1), Multiplexing and channel coding (TDD).

[29] International Telecommunication Union (ITU), Radiocommunication Study Groups (1998). A guide to digital terrestrial television broadcasting in the VHF/UHF bands.

[30] Consultative Committee for Space Data Systems (CCSDS) (1999). Telemetry Systems (Panel 1), Telemetry channel coding.

[31] Benedetto, S., Garello, R., and Montorsi, G. (1998). A search for good convolutional codes to be used in the construction of turbo codes. *IEEE Transactions on Communications* 46: 1101–1105.

[32] Takeshita, O.Y. and Costello, D.J. Jr. (2000). New deterministic interleaver designs for turbo codes. *IEEE Transactions on Information Theory* 46: 1988–2006.

[33] Divsalar, D. and Pollara, F. (1995). Turbo codes for PCS applications. *Proceedings of 1995 IEEE International Conference on Communications*, Seattle, Washington, USA, pp. 54–59.

[34] Benedetto, S., Divsalar, D., Montorsi, G., and Pollara, F. (1998). Serial concatenation of interleaved codes: performance analysis, design, and iterative decoding. *IEEE Transactions on Information Theory* 45: 909–926.

[35] Benedetto, S. and Montorsi, G. (2000). Versatile bandwidth-efficient parallel and serial turbo-trellis-coded modulation. *Proceedings of 2000 International Symposium on Turbo Codes & Related Topics*, Brest, France, pp. 201–208.

[36] Douillard, C., Jezequel, M., Berrou, C. et al. (1995). Iterative correction of intersymbol interference turbo-equalization. *European Transactions on Telecommunications (ETT)* 6: 507–511.

[37] Colavolpe, G., Ferrari, G., and Raheli, R. (2000). Noncoherent iterative (turbo) decoding. *IEEE Transactions on Communications* 48: 1488–1498.

[38] Bauch, G., Khorram, H., and Hagenauer, J. (1997). Iterative equalization and decoding in mobile communications system. *Proceedings of European Personal Mobile Communications Conference*, Bristol, UK, pp. 307–312.

[39] Hoeher, P. and Lodge, J. (1999). ŞTurbo-DPSKŤ: iterative differential PSK demodulation. *IEEE Transactions on Communications* 47: 837–843.

[40] Hagenauer, J., Offer, E., and Papke, L. (1996). Iterative decoding of binary block and convolutional codes. *IEEE Transactions on Information Theory* 42: 429–445.

[41] Bauch, G., Khorram, H., and Hagenauer, J. (1997). Iterative equalization and decoding in mobile communications systems. *Proceedings of EPMCC*, pp. 307–312.

[42] Picart, A., Didier, P., and Glavieux, G. (1997). Turbo-detection: a new approach to combat channel frequency selectivity. *Proceedings of IEEE International Conference on Communications*, pp. 1498–1502.

[43] MacKay, D. and Neal, R. (1996). Near Shannon limit performance of low density parity check codes. *Electronics Letters* 32: 1645–1646.

[44] MacKay, D. (1999). Good error-correcting codes based on very sparse matrices. *IEEE Transactions on Information Theory* 45: 399–431.

[45] Tanner, R.M. (1981). A recursive approach to low complexity codes. *IEEE Transactions on Information Theory* 27: 533–547.

[46] Frey, B.J. (1998). *Graphical Models for Machine Learning and Digital Communications*. Cambridge, MA: MIT Press.

[47] Richardson, T.J. and Urbanke, R.L. (2001). Efficient encoding of low-density parity-check codes. *IEEE Transactions on Information Theory* 47: 638–656.

[48] Gallager, R. (1963). *Low Density Parity Check Codes*. Cambridge, MA: MIT Press.

[49] Pearl, J. (1988). *Probabilistic Reasoning in Intelligent Systems: Networks of Plausible Interference.* Morgan Kaufmann.

[50] Hu, X.-Y., Eleftheriou, E., and Arnold, D.M. (2001). Progressive edge-growth Tanner graphs. *Proceedings of GLOBECOM '01*, San Antonio, Texas, USA.

[51] Fossorier, M.P.C., Mahaljevic, M., and Imai, H. (1999). Reduced complexity iterative decoding of low-density parity check codes based on belief propagation. *IEEE Transactions on Computers* 47: 673–680.

[52] Zimmermann, E., Rave, W., and Fettweis, G. (2005). Forced convergence decoding of LDPC codes - EXIT chart analysis and combination with node complexity reduction techniques. *11th European Wireless Conference (EW'2005)*, Nicosia, Cyprus.

[53] Zimmermann, E., Pattisapu, P., and Fettweis, G. (2005). Bit-flipping post-processing for forced convergence decoding of LDPC codes. *13th European Signal Processing Conference*, Antalya, Turkey.

[54] Shin, K. and Lee, J. (2007). Low complexity LDPC decoding techniques with adaptive selection of edges. *2007 IEEE 65th Vehicular Technology Conference - VTC2007-Spring*, pp. 2205–2209.

[55] Lee, D.-U., Luk, W., Wang, C., and Jones, C. (2004). A flexible hardware encoder for low-density parity-check codes. *12th Annual IEEE Symposium on Field-Programmable Custom Computing Machines*, pp. 101–111.

[56] Cherubini, G., Eleftheriou, E., and Ölcer, S. (2000). On advanced signal processing and coding techniques for digital subscriber lines. Records of the Workshop ŞWhat is *next* in xDSL?Ť, Vienna, Austria.

[57] Richardson, T.J., Shokrollahi, M.A., and Urbanke, R.L. (2001). Design of capacity-approaching irregular low-density parity-check codes. *IEEE Transactions on Information Theory* 47 2: 619–637.

[58] Arıkan, E. (2009). Channel polarization: a method for constructing capacity-achieving codes for symmetric binary-input memoryless channels. *IEEE Transactions on Information Theory* 55: 3051–3073.

[59] Tal, I. and Vardy, A. (2015). List decoding of polar codes. *IEEE Transactions on Information Theory* 61: 2213–2226.

[60] Niu, K., Chen, K., Lin, J., and Zhang, Q.T. (2014). Polar codes: primary concepts and practical decoding algorithms. *IEEE Communications Magazine* 52: 60–69.

[61] Balatsoukas-Stimming, A. and Parizi, M.B. (2015). LLR-based successive cancellation list decoding of polar codes. *IEEE Transactions on Signal Processing* 63: 5165–5179.

[62] Balatsoukas-Stimming, A., Giard, P., and Burg, A. (2017). Comparison of polar decoders with existing low-density parity-check and turbo decoders. *Proceedings of IEEE Wireless Communications and Networking Conference*, San Francisco, CA, USA.

[63] Hui, D., Sandberg, S., Blankenship, Y. et al. (2018). Channel coding in 5G new radio: a tutorial overview and performance comparison with 4G LTE. *IEEE Vehicular Technology Magazine* 13: 60–69.

[64] Bioglio, V., Condo, C., and Land, I. (2019). Design of polar codes in 5G new radio, *arXiv*.

[65] Alamdar-Yazdi, A. and Kschischang, F.R. (2011). A simplified successive- cancellation decoder for polar codes. *IEEE Communications Letters* 15: 1378–1380.

[66] Sarkis, G., Giard, P., Vardy, A. et al. (2014). Fast polar decoders: algorithm and implementation. *IEEE Journal on Selected Areas in Communications* 32: 946–957.

[67] Chen, K., Niu, K., and Lin, J. (2013). Improved successive cancellation decoding of polar codes. *IEEE Transactions on Communications* 61: 3100–3107.

[68] Afisiadis, O., Balatsoukas-Stimming, A., and Burg, A. (2014). A low-complexity improved successive cancellation decoder for polar codes. *Proceedings of Asilomar Conference on Signals, Systems, and Computers*, Pacific Grove, California, USA.

[69] Cammerer, S., Ebada, M., Elkelesh, A., and ten Brink, S. (2018). Sparse graphs for belief propagation decoding of polar codes. *Proceedings of IEEE International Symposium on Information Theory (ISIT)*, Vail, Colorado, USA.

[70] Chung, S.-Y., Richardson, T.J., and Urbanke, R.L. (2001). Analysis of sum-product decoding of low-density parity-check codes using a Gaussian approximation. *IEEE Transactions on Information Theory* 47: 657–670.

[71] Trifonov, P. (2012). Efficient design and decoding of polar codes. *IEEE Transactions on Communications* 60: 3221–3227.

[72] Ha, J., Kim, J., and McLaughlin, S.W. (2004). Rate-compatible puncturing of low-density parity-check codes. *IEEE Transactions on Information Theory* 50: 2824–2836.

[73] Zhang, L., Zhang, Z., Wang, X. et al. (2014). On the puncturing patterns for punctured polar codes. *Proceedings of IEEE International Symposium on Information (ISIT)*, Honolulu, Hawaii, USA.

[74] Miloslavskaya, V. (2015). Shortened polar codes. *IEEE Transactions on Information Theory* 61: 4852–4865.

Appendix 11.A Non-binary parity check codes

Assume that code words are sequences of symbols from the finite field GF(q) (see Section 11.2.2), all of length n. As there are q^n possible sequences, the introduction of redundancy in the transmitted sequences is possible if the number of code words M_c is less than q^n.

We denote by $\boldsymbol{c}$ a transmitted sequence of n symbols taken from GF(q). We also assume that the symbols of the received sequence $\boldsymbol{z}$ are from the same alphabet. We define the *error sequence* $\boldsymbol{e}$ by the equation (see (11.12) for the binary case)

$$\boldsymbol{z} = \boldsymbol{c} + \boldsymbol{e} \tag{11.486}$$

where + denotes a component by component addition of the vectors in accordance with the rules of addition in the field GF(q).

Definition 11.16

The number of non-zero components of a vector $\boldsymbol{x}$ is defined as the weight of the vector, denoted by $w(\boldsymbol{x})$. □

Then $w(\boldsymbol{e})$ is equal to the number of errors occurred in transmitting the code word.

Definition 11.17

The *minimum distance* of a code, denoted d^H_{min}, is equal to the minimum Hamming distance between all pairs of code words; i.e. it is the same as for binary codes. □

We will give without proof the following propositions, similar to those for binary codes on page 518.

1. A non-binary block code with minimum distance d^H_{min} can correct all error sequences of weight $\left\lfloor \frac{d^H_{min}-1}{2} \right\rfloor$ or less.
2. A non-binary block code with minimum distance d^H_{min} can detect all error sequences of weight $(d^H_{min} - 1)$ or less.

As in the binary case, we ask for a relation among the parameters of a code: n, M_c, d^H_{min}, and q. It can be proven that for a block code with length n and minimum distance d^H_{min}, M_c must satisfy the inequality

$$M_c \leq \left\lfloor q^n / \left\{ 1 + \binom{n}{1}(q-1) + \binom{n}{2}(q-1)^2 + \cdots + \binom{n}{\left\lfloor \frac{d^H_{min}-1}{2} \right\rfloor}(q-1)^{\left\lfloor \frac{d^H_{min}-1}{2} \right\rfloor} \right\} \right\rfloor \tag{11.487}$$

Furthermore, for n and d^H_{min} given, it is always possible to find a code with M_c^* words, where

$$M_c^* = \left\lceil q^n / \left\{ 1 + \binom{n}{1}(q-1) + \binom{n}{2}(q-1)^2 + \cdots + \binom{n}{d^H_{min}-1}(q-1)^{d^H_{min}-1} \right\} \right\rceil \tag{11.488}$$

Linear codes

Definition 11.18

A *linear code* is a block code with symbols from GF(q) for which:

(a) the all zero word is a code word;
(b) any multiple of a code word is a code word;
(c) any linear combination of any two code words is a code word.

□

Example 11.A.1

A binary group code is a linear code with symbols from GF(2).

Example 11.A.2

Consider a block code of length 5 having symbols from GF(3) with code words

$$\begin{matrix} 0&0&0&0&0 \\ 1&0&0&2&1 \\ 0&1&1&2&2 \\ 2&0&0&1&2 \\ 1&1&1&1&0 \\ 2&1&1&0&1 \\ 0&2&2&1&1 \\ 1&2&2&0&2 \\ 2&2&2&2&0 \end{matrix} \tag{11.489}$$

It is easily verified that this code is a linear code.

We give the following two properties of a linear code.

1. The minimum distance of the code, d^H_{min}, is given as

$$d^H_{min} = \min w(\bar{c}) \tag{11.490}$$

where $\bar{c}$ can be any non-zero code word.

Proof. By definition of the Hamming distance between two code words, we get

$$d^H(\boldsymbol{c}_1, \boldsymbol{c}_2) = w(\boldsymbol{c}_1 + (-\boldsymbol{c}_2)) \tag{11.491}$$

By Property (b), $(-\boldsymbol{c}_2)$ is a code word if $\boldsymbol{c}_2$ is a code word; by Property (c), $\boldsymbol{c}_1 + (-\boldsymbol{c}_2)$ must also be a code word. As two code words differ in at least d^H_{min} positions, there is a code word of weight d^H_{min}; if there were a code word of weight less than d^H_{min}, this word would be different from the zero word in fewer than d^H_{min} positions. □

2. If all code words in a linear code are written as rows of an $M_c \times n$ matrix, every column is composed of all zeros, or contains all elements of the field, each repeated M_c/q times.

Parity check matrix

Let $\boldsymbol{H}$ be an $r \times n$ matrix with coefficients from GF(q), expressed as

$$\boldsymbol{H} = [\boldsymbol{A}\ \boldsymbol{B}] \tag{11.492}$$

where the $r \times r$ matrix $\boldsymbol{B}$ is such that $\det[B] \neq 0$.

A generalized non-binary parity check code is a code composed of all vectors $\boldsymbol{c}$ of length n, with elements from GF(q), that are the solutions of the equation

$$\boldsymbol{Hc} = \boldsymbol{0} \tag{11.493}$$

The matrix $\boldsymbol{H}$ is called the *generalized parity check matrix.*

Propriety 1 of non-binary generalized parity check codes. A non-binary generalized parity check code is a linear code.

Proof.

(a) The all zero word is a code word, as $\boldsymbol{H0} = \boldsymbol{0}$.

(b) Any multiple of a code word is a code word, because if $\boldsymbol{c}$ is a code word, then $\boldsymbol{Hc} = \boldsymbol{0}$. But $\boldsymbol{H}(\alpha\boldsymbol{c}) = \alpha\boldsymbol{Hc} = \boldsymbol{0}$, and therefore $\alpha\boldsymbol{c}$ is a code word; here α is any element from GF(q).

(c) Any linear combination of any two code words is a code word, because if $\boldsymbol{c}_1$ and $\boldsymbol{c}_2$ are two code words, then $\boldsymbol{H}(\alpha\boldsymbol{c}_1 + \beta\boldsymbol{c}_2) = \alpha\boldsymbol{Hc}_1 + \beta\boldsymbol{Hc}_2 = \alpha\boldsymbol{0} + \beta\boldsymbol{0} = \boldsymbol{0}$, and therefore $\alpha\boldsymbol{c}_1 + \beta\boldsymbol{c}_2$ is a code word. □

Property 2 of non-binary generalized parity check codes. The code words corresponding to the matrix $\boldsymbol{H} = [A\ \boldsymbol{B}]$ are identical to the code words corresponding to the parity check matrix $\tilde{\boldsymbol{H}} = [\boldsymbol{B}^{-1}\boldsymbol{A}, I]$.

Proof. Same as for the binary case. □

The matrices in the form $[\boldsymbol{A}\ \boldsymbol{I}]$ are said to be in *canonical or systematic form.*

Property 3 of non-binary generalized parity check codes. A code consists of exactly $q^{n-r} = q^k$ code words.

Proof. Same as for the binary case (see Property 3 on page 521). □

The first $k = n - r$ symbols are called *information symbols*, and the last r symbols are called *generalized parity check symbols.*

Property 4 of non-binary generalized parity check codes. A code word of weight w exists if and only if some linear combination of w columns of the matrix $\boldsymbol{H}$ is equal to $\boldsymbol{0}$.

Proof. $\boldsymbol{c}$ is a code word if and only if $\boldsymbol{H}\boldsymbol{c} = \boldsymbol{0}$. Let c_j be the jth component of $\boldsymbol{c}$ and let $\boldsymbol{h}_i$ be the ith column of $\boldsymbol{H}$; then if $\boldsymbol{c}$ is a code word we have

$$\sum_{j=1}^{n} \boldsymbol{h}_j\, c_j = \boldsymbol{0} \tag{11.494}$$

If $\boldsymbol{c}$ is a code word of weight w, there are exactly w non-zero components of $\boldsymbol{c}$, say $c_{j_1}, c_{j_2}, \ldots, c_{j_w}$; then

$$c_{j_1}\boldsymbol{h}_{j_1} + c_{j_2}\boldsymbol{h}_{j_2} + \cdots + c_{j_w}\boldsymbol{h}_{j_w} = \boldsymbol{0} \tag{11.495}$$

thus a linear combination of w columns of $\boldsymbol{H}$ is equal to $\boldsymbol{0}$. Conversely, if (11.495) is true, then $\boldsymbol{H}\boldsymbol{c} = \boldsymbol{0}$, where $\boldsymbol{c}$ is a vector of weight w with non-zero components $c_{j_1}, c_{j_2}, \ldots, c_{j_w}$. □

Property 5 of non-binary generalized parity check codes. A code has minimum distance d_{min}^H if some linear combination of d_{min}^H columns of $\boldsymbol{H}$ is equal to $\boldsymbol{0}$, but no linear combination of fewer than d_{min}^H of columns of $\boldsymbol{H}$ is equal to $\boldsymbol{0}$.

Proof. This property is obtained by combining Property 1 of a linear code and Properties 1 and 4 of a nonbinary generalized parity check code. □

Property 5 is fundamental for the design of non-binary codes.

Example 11.A.3
Consider the field GF (4), and let α be a primitive element of this field; Moreover, consider the generalized parity check matrix

$$\boldsymbol{H} = \begin{bmatrix} 1 & 1 & 1 & 1 & 0 \\ 1 & \alpha & \alpha^2 & 0 & 1 \end{bmatrix} \tag{11.496}$$

We find that no linear combination of two columns is equal to $\boldsymbol{0}$. However, there are many linear combinations of three columns that are equal to $\boldsymbol{0}$, for example $\boldsymbol{h}_1 + \boldsymbol{h}_4 + \boldsymbol{h}_5 = \boldsymbol{0}$, $\alpha\boldsymbol{h}_2 + \alpha\boldsymbol{h}_4 + \alpha^2\boldsymbol{h}_5 = \boldsymbol{0}, \ldots$; hence, the minimum distance of this code is $d_{min}^H = 3$.

Code generator matrix

We assume that the parity check matrix is in canonical form; then

$$\boldsymbol{c}_{n-r+1}^{n} = -\boldsymbol{A}\boldsymbol{c}_1^{n-r} \tag{11.497}$$

and

$$\boldsymbol{c} = \begin{bmatrix} \boldsymbol{c}_1^{n-r} \\ \boldsymbol{c}_{n-r+1}^{n} \end{bmatrix} = \begin{bmatrix} \boldsymbol{I} \\ -\boldsymbol{A} \end{bmatrix} \boldsymbol{c}_1^{n-r} = \boldsymbol{G}^T\, \boldsymbol{c}_1^{n-r} \tag{11.498}$$

The matrix $\boldsymbol{G}$ is called the *generator matrix* of the code and is expressed as

$$\boldsymbol{G} = [\boldsymbol{I}, -\boldsymbol{A}^T] \tag{11.499}$$

so that

$$\boldsymbol{c}^T = (\boldsymbol{c}_1^{n-r})^T\, \boldsymbol{G} \tag{11.500}$$

Thus, the code words, considered as row vectors, are given as all linear combinations of the rows of the matrix $\boldsymbol{G}$. A non-binary generalized parity check code can be specified by giving its generalized parity check matrix or its generator matrix.

Example 11.A.4

Consider the field GF (4) and let α be a primitive element of this field; Moreover, consider the generalized parity check matrix (11.496). The generator matrix of this code is given by

$$G = \begin{bmatrix} 1 & 0 & 0 & 1 & 1 \\ 0 & 1 & 0 & 1 & \alpha \\ 0 & 0 & 1 & 1 & \alpha^2 \end{bmatrix} \tag{11.501}$$

There are 64 code words corresponding to all linear combinations of the rows of the matrix $\boldsymbol{G}$.

Decoding of non-binary parity check codes

Methods for the decoding of non-binary generalized parity check codes are similar to those for the binary case. Conceptually, the simplest method consists in comparing the received block of n symbols with each code word and choosing that code word that differs from the received word in the fewest positions. An equivalent method for a linear code consists in partitioning the q^n possible sequences into q^r sets.

The partitioning is done as follows:

Step 1: choose the first set as the set of $q^{n-r} = q^k$ code words, $\boldsymbol{c}_1, \boldsymbol{c}_2, \dots, \boldsymbol{c}_{q^k}$.

Step 2: choose any vector, say $\boldsymbol{\eta}_2$, that is not a code word; then choose the second set as $\boldsymbol{c}_1 + \boldsymbol{\eta}_2, \boldsymbol{c}_2 + \boldsymbol{\eta}_2, \dots, \boldsymbol{c}_{q^k} + \boldsymbol{\eta}_2$.

Step i: choose any vector, say $\boldsymbol{\eta}_i$, not included in any previous set; choose the ith set as $\boldsymbol{c}_1 + \boldsymbol{\eta}_i, \boldsymbol{c}_2 + \boldsymbol{\eta}_i, \dots, \boldsymbol{c}_{q^k} + \boldsymbol{\eta}_i$.

The partitioning continues until all q^n vectors are used; each set is called a *coset*, and the vectors $\boldsymbol{\eta}_i$ are called *coset leaders*. The all zero vector is the *coset leader* for the first set.

Coset

We give the following properties of the cosets omitting the proofs.

1. Every one of the q^n vectors occurs in one and only one coset.
2. Suppose that, instead of choosing $\boldsymbol{\eta}_i$ as coset leader of the ith coset, we choose another element of that coset as the coset leader; then the coset formed by using the new coset leader contains exactly the same vectors as the old coset.
3. There are q^r cosets.

Two conceptually simple decoding methods

We now form a *coset table* by choosing as coset leader for each coset the vector of minimum weight in that coset. The table consists of an array of vectors, with the ith row in the array being the ith coset; the coset leaders make up the first column, and the jth column consists of the vectors $\boldsymbol{c}_j, c_j + \boldsymbol{\eta}_2, \boldsymbol{c}_j + \boldsymbol{\eta}_3, \dots, c_j + \boldsymbol{\eta}_{q^r}$.

A method for decoding consists of the following steps:

Step 1: locate the received vector in the coset table.

Step 2: choose the code word that appears as the first vector in the column containing the received vector.

This decoding method decodes to the closest code word to the received word and the coset leaders are the correctable error patterns.

A modified version of the described decoding method is

Step 1′: locate the received vector in the coset table and then identify the coset leader of the coset containing this vector.

Step 2′: subtract the coset leader from the received vector to find the decoded code word.

Syndrome decoding

Another method of decoding is the syndrome decoding. For any generalized parity, check matrix $\boldsymbol{H}$ and all vectors z of length n, we define the syndrome of z, $\boldsymbol{s}(z)$, as

$$s(z) = \boldsymbol{H}z \tag{11.502}$$

We can show that all vectors in the same coset have the same syndrome and vectors in different cosets have different syndromes. This leads to the following decoding method:

Step 1″: compute the syndrome of the received vector, as this syndrome identifies the coset in which the received vector is in, and so identifies the leader of that coset.

Step 2″: subtract the coset leader from the received vector to find the decoded code word.

The difficulty with this decoding method is in the second part of step 1″, that is identifying the coset leader that corresponds to the computed syndrome; this step is equivalent to finding a linear combination of the columns of $\boldsymbol{H}$ which is equal to that syndrome, using the smallest number of columns. Exploiting the algebraic structure of the generalized parity check matrix for certain classes of codes, the syndrome provides the coset leader by algebraic means.

Chapter 12

Trellis coded modulation

During the 1980s, an evolution in the methods to transmit data over channels with limited bandwidth took place, giving origin to techniques for joint coding and modulation that are generally known with the name of *trellis coded modulation* (TCM). The main characteristic of TCM lies in joint design of modulation mapping and coding. The first article on TCM appeared in 1976 by Ungerboeck; later, a more detailed publication by the same author on the principles of TCM [1] spurred considerable interest in this topic [2–8], leading to a full development of the theory of TCM.

TCM techniques use multilevel modulation with a set of signals from a one, two, or multi-dimensional space. The choice of the signals that generate a code sequence is determined by a finite-state encoder. In TCM, the set of modulation signals is expanded with respect to the set used by an uncoded, i.e. without redundancy, system; in this manner, it is possible to introduce redundancy in the transmitted signal without widening the bandwidth. At the receiver, the signals in the presence of additive noise and channel distortion are decoded by a maximum-likelihood sequence decoder. By simple TCM techniques using a four-state encoder, it is possible to obtain a coding gain of 3 dB with respect to conventional uncoded modulation; with more sophisticated TCM techniques, coding gains of 6 dB or more can be achieved (see Chapter 6).

Errors in the decoding of the received signal sequence are less likely to occur if the waveforms, which represent the code sequences, are easily distinguishable from each other; in mathematical terms, the signal sequences, represented in the Euclidean multidimensional space, need to be separated by large distances. The novelty of TCM is in postulating the expansion of the set of symbols[1] in order to provide the redundancy necessary for the encoding process. The construction of modulation code sequences that are characterized by a *free distance*, i.e. the minimum Euclidean distance between code sequences, that is much larger than the minimum distance between uncoded modulation symbols, with the same information bit rate, and the same bandwidth and power of the modulated signal, is obtained by the joint design of encoder and bit mapper. The term trellis derives from the similarity between state transition diagrams of a TCM encoder and trellis diagrams of binary convolutional codes; the difference lies in the fact that, in TCM schemes, the branches of the trellis are labelled with modulation symbols rather than binary symbols.

[1] In the first part of this chapter, we mainly use the notion of *symbols* of an alphabet with cardinality M, although the analysis could be conducted by referring to vectors in the signal space as *modulation signals*. We will use the term *signals* instead of *symbols* only in the multidimensional case.

Algorithms for Communications Systems and their Applications, Second Edition.
Nevio Benvenuto, Giovanni Cherubini, and Stefano Tomasin.

The IEEE 802.3ab standard for 1 Gbit/s Ethernet [9] adopts four-dimensional (4-D) 5-pulse-amplitude -modulation (PAM) trellis-coded modulation (TCM), which uses signal-constellation expansion in conjunction with set partitioning to perform modulation and coding jointly, thus achieving coding gains for improved system robustness. The emerging IEEE P802.3bs standard for 400 Gbit/s Ethernet over eight lanes enables both 2-PAM and 4-PAM operation, while targeting a data rate of 50 Gbit/s per lane for chip-to-chip and chip-to-module applications. Design techniques for high-speed low-complexity eight-state 4-D 5-PAM TCM decoding, which exhibit the potential for application to future Ethernet standards achieving the required reliability by concatenated Reed Solomon (RS) coding and 4-D 5-PAM TCM, have been presented in [10].

12.1 Linear TCM for one- and two-dimensional signal sets

12.1.1 Fundamental elements

Consider the transmission system illustrated in Figure 6.1, that consists of the modulator, transmission channel, demodulator, and data detector. Errors occasionally occur in the symbol detection, and $\hat{a}_k \neq a_k$. Usually, the simplest data detector is a threshold detector that takes an instantaneous hard decision on the value $\hat{a}_k$ of the transmitted symbol, based on the observation of the sample z_k at the demodulator output. Detection is of the nearest-neighbour type, i.e. the detector decides for the symbol of the constellation that is at the minimum Euclidean distance from the received sample z_k. The objective of traditional channel coding techniques consists in detecting and/or correcting the errors present in the detected sequence of bits $\{\tilde{c}_m\}$.

In the approach followed in Chapter 11, a binary encoder was used to map k_0 information binary symbols $\{b_\ell\}$ in n_0 code binary symbols $\{c_m\}$. As mentioned in Section 11.1, we note that, if we want to maintain the effective rate $1/T_b$ of the information message and at the same time the modulation rate $1/T$ of the system, we need to increase the cardinality of the modulation alphabet. If we do not consider the joint design of encoder and bit mapper, however, a reduction of the bit error probability cannot be efficiently achieved, as we see from the following example.

Example 12.1.1

Consider an uncoded 4-phase-shift-keying (PSK) system and an 8-PSK system that uses a binary error correcting code with rate 2/3; both systems transmit two information bits per modulation interval, which corresponds to a spectral efficiency of 2 bit/s/Hz. If the 4-PSK system works with an error probability of 10^{-5}, for a given signal-to-noise ratio Γ, the 8-PSK system works with an error probability larger than 10^{-2}, due to the smaller Euclidean distance between signals of the 8-PSK system. We must use an error correcting code with minimum Hamming distance $d^H_{min} \geq 7$ to reduce the error probability to the same value of the uncoded 4-PSK system. A binary convolutional code with rate 2/3 and constraint length 6 has the required value of $d^H_{free} = 7$. Decoding requires a decoder with 64 states that implements the Viterbi algorithm (VA). However, even after increasing the complexity of the 8-PSK system, we have obtained an error probability only equal to that of the uncoded 4-PSK system.

Two problems determine the unsatisfactory result obtained with the traditional approach. The first is originated by the use of independent hard decisions taken by the detector before decoding; hard input decoding leads to an irreversible loss of information; the remedy is the use of soft decoding (see page 586), whereby the decoder directly operates on the samples at the demodulator output. The second derives from the independent design of encoder and bit mapper.

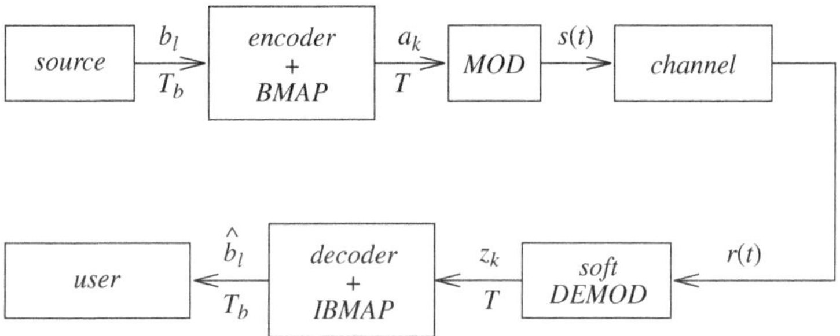

Figure 12.1 Block diagram of a transmission system with trellis coded modulation.

We now consider the transmission system of Figure 12.1, where the transmitted symbol sequence $\{a_k\}$ is produced by a finite-state machine (FSM) having the information bit sequence $\{b_\ell\}$ as input, possibly with a number of information bits per modulation interval larger than one. We denote by $1/T$ the modulation rate and by $\mathcal{A}$ the alphabet of a_k. For an additive white Gaussian noise (AWGN) channel, at the decision point the received samples in the absence of inter-symbol-interference (ISI) are given by (see (7.334))

$$z_k = a_k + w_k \tag{12.1}$$

where $\{w_k\}$ is a sequence of white Gaussian noise samples. Maximum-likelihood sequence detection (MLSD) represents the optimum strategy for decoding a sequence transmitted over a dispersive noisy channel. The decision rule consists in determining the sequence $\{\hat{a}_k\}$ closest to the received sequence in terms of Euclidean distance (see (6.25)) in the set $\mathcal{S}$ of all possible code symbol sequences. MLSD is efficiently implemented by the VA, provided that the generation of the code symbol sequences follows the rules of a FSM.

In relation to (7.355), we define as *free distance*, d_{free}, the minimum Euclidean distance between two code symbol sequences $\{\alpha_k\}$ and $\{\beta_k\}$, that belong to the set $\mathcal{S}$, given by

$$d_{free}^2 = \min_{\{\alpha_k\}\neq\{\beta_k\}} \sum_k |\alpha_k - \beta_k|^2, \qquad \{\alpha_k\}, \{\beta_k\} \in \mathcal{S} \tag{12.2}$$

The most probable error event is determined by two code symbol sequences of the set $\mathcal{S}$ at the minimum distance. The assignment of symbol sequences using a code that is optimized for Hamming distance does not guarantee an acceptable structure in terms of Euclidean distance, as in general the relation between the Hamming distance and the Euclidean distance is not monotonic.

Encoder and modulator must then be jointly designed for the purpose of assigning to symbol sequences waveforms that are separated in the Euclidean signal space by a distance equal to at least d_{free}, where d_{free} is greater than the minimum distance between the symbols of an uncoded system.

At the receiver, the demodulator–decoder does not make errors if the received signal in the Euclidean signal space is at a distance smaller than $d_{free}/2$ from the transmitted sequence.

Basic TCM scheme

The objective of TCM is to obtain an error probability lower than that achievable with uncoded modulation, for the same bit rate of the system, channel bandwidth, transmitted signal power, and noise power spectral density.

The generation of code symbol sequences by a sequential FSM sets some constraints on the symbols of a sequence, thus introducing interdependence among them (see Appendix 7.B). The transmitted symbol at instant kT depends not only on the information bits generated by the source at the same instant, as in the case of memoryless modulation, but also on the previous symbols.

We define as $\boldsymbol{b}_k$ the vector of $\log_2 M$ information bits at instant kT. We recall that for M-ary uncoded transmission there exists a one to one correspondence between $\boldsymbol{b}_k$ and the symbol $a_k \in \mathcal{A}$. We also introduce the state $\boldsymbol{s}_k$ at the instant kT. According to the model of Appendix 7.B, the generation of a sequence of encoded symbols is obtained by the two functions

$$\begin{aligned} a_k &= f(\boldsymbol{b}_k, \boldsymbol{s}_{k-1}) \\ \boldsymbol{s}_k &= g(\boldsymbol{b}_k, \boldsymbol{s}_{k-1}) \end{aligned} \tag{12.3}$$

For an input vector $\boldsymbol{b}_k$ and a state $\boldsymbol{s}_{k-1}$, the first equation describes the choice of the transmitted symbol a_k from a certain constellation, the second the choice of the next state $\boldsymbol{s}_k$.

Interdependence between the symbols $\{a_k\}$ is introduced without a reduction of the bit rate by increasing the cardinality of the alphabet. For example, for a length K of the sequence of input vectors, if we change $\mathcal{A}$ of cardinality M with $\mathcal{A}' \supset \mathcal{A}$ of cardinality $M' > M$, and we select M^K sequences as a subset of $(\mathcal{A}')^K$, a better separation of the code sequences in the Euclidean space may be obtained. Hence, we can obtain a minimum distance d_{free} between any two sequences larger than the minimum distance between signals in $\mathcal{A}^K$.

Note that this operation may cause an increase in the average symbol energy from $E_{s,u}$ for uncoded transmission to $E_{s,c}$ for coded transmission, and hence a loss in efficiency given by $E_{s,c}/E_{s,u}$.

Furthermore, we define as N_{free} the number of sequences that a code sequence has, *on average*, at the distance d_{free} in the Euclidean multidimensional space.

Example

Suppose we want to transmit two bits of information per symbol. Instead of using quadrature phase-shift-keying (QPSK) modulation, we can use the scheme illustrated in Figure 12.2.

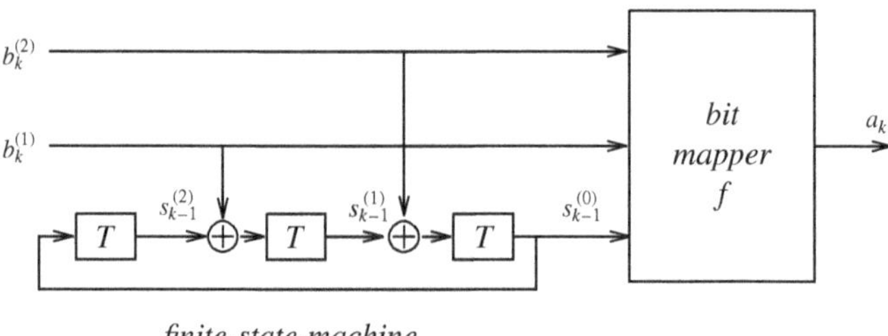

Figure 12.2 Eight-state trellis encoder and bit mapper for the transmission of 2 bits per modulation interval by 8-PSK.

The scheme has two parts. The first is a finite-state sequential machine with 8 states, where the state $\boldsymbol{s}_k$ is defined by the content of the memory cells $\boldsymbol{s}_k = [s_k^{(2)}, s_k^{(1)}, s_k^{(0)}]$. The two bits $\boldsymbol{b}_k = [b_k^{(2)}, b_k^{(1)}]$ are input to the FSM, which undergoes a transition from state $\boldsymbol{s}_{k-1}$ to one of four next possible states, $\boldsymbol{s}_k$, according to the function g. The second part is the bit mapper, which maps the two information bits and one bit that depends on the state, i.e. the three bits $[b_k^{(2)}, b_k^{(1)}, s_{k-1}^{(0)}]$, in one of the symbols of an eight-ary constellation according to the function f, for example an 8-PSK constellation using the map of Figure 12.5. Note that the transmission of two information bits per modulation interval is achieved. Therefore, the constellation of the system is expanded by a factor 2 with respect to uncoded QPSK transmission. Recall from the discussion in Section 6.5 that most of the achievable coding gain for transmission over an ideal AWGN channel of two bits per modulation interval can be obtained by doubling the cardinality of the constellation from four to eight symbols. We will see that TCM using the simple scheme of Figure 12.2 achieves a coding gain of 3.6 dB.

For the graphical representation of the functions f and g, it is convenient to use a trellis diagram; the nodes of the trellis represent the FSM states, and the branches represent the possible transitions between states. For a given state $\boldsymbol{s}_{k-1}$, a branch is associated with each possible vector $\boldsymbol{b}_k$ by the function g, that reaches a next state $\boldsymbol{s}_k$. Each branch is labelled with the corresponding value of the transmitted symbol a_k. For the encoder of Figure 12.2 and the map of Figure 12.5, the corresponding trellis is shown in Figure 12.3, where the trellis is terminated by forcing the state of the FSM to zero at the instant $k = 4$. For a general representation of the trellis, see Figure 12.13.

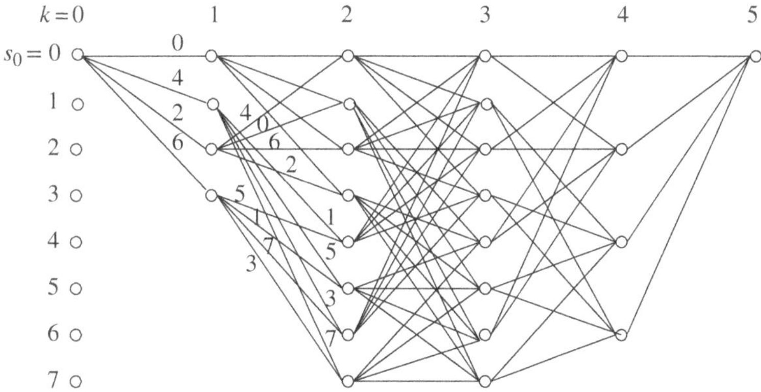

Figure 12.3 Trellis diagram for the encoder of Figure 12.2 and the map of Figure 12.5. Each branch is labelled by the corresponding value of a_k.

Each path of the trellis corresponds to only one message sequence $\{b_\ell\}$ and is associated with only one sequence of code symbols $\{a_k\}$. The optimum decoder searches the trellis for the most probable path, given the received sequence $\{z_k\}$ is observed at the output of the demodulator. This search is usually realized by the VA (see Section 7.5.1). Because of the presence of noise, the chosen path may not coincide with the correct one, but diverge from it at the instant $k = i$ and rejoin it at the instant $k = i + L$; in this case, we say that an *error event of length L* has occurred, as illustrated in the example in Figure 12.4 for an error event of length three (see Definition 7.3 on page 358).

Note that in a trellis diagram more branches may connect the same pair of nodes. In this case we speak of parallel transitions, and by the term free distance of the code we denote the minimum among the distances between symbols on parallel transitions and the distances between code sequences associated with pairs of paths in the trellis that originate from a common node and merge into a common node after L transitions, $L > 1$.

By utilizing the sequence of samples $\{z_k\}$, the decoding of a TCM signal is done in two phases. In the first phase, called *subset decoding*, within each subset of symbols assigned to the parallel transitions in the trellis diagram, the receiver determines the symbol closest to the received sample; these symbols are then memorized together with their square distances from the received sample. In the second phase, we apply the VA to find the code sequence $\{\hat{a}_k\}$ along the trellis such that the sum of the square distances between the code sequence and the sequence $\{z_k\}$ is minimum. Recalling that the signal is obtained at the output of the demodulator in the presence of AWGN with variance σ_I^2 per dimension, the probability of an error event for large values of the signal-to-noise ratio is approximated by (see (7.356))

$$P_e \simeq N_{free} Q\left(\frac{d_{free}}{2\sigma_I}\right) \tag{12.4}$$

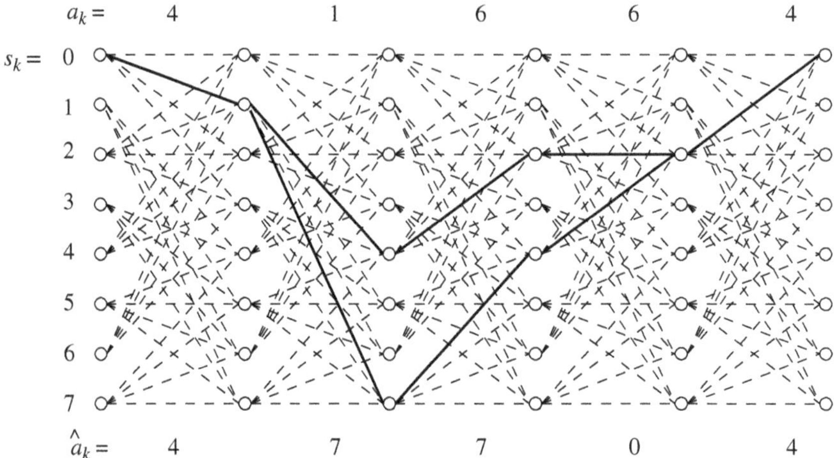

Figure 12.4 Section of the trellis for the decoder of an eight-state trellis code. The two continuous lines indicate two possible paths relative to two 8-PSK signal sequences, $\{a_k\}$ and $\{\hat{a}_k\}$.

where d_{free} is defined in (12.2).

From the Definition 6.2 on page 274 and the relation (12.4) between Euclidean distance and error probability, we give the definition of *asymptotic coding gain*, G_{code},[2] as the ratio between the minimum distance, d_{free}, between code sequences and the minimum Euclidean distance for uncoded sequences, equal to the minimum distance between symbols of the constellation of an uncoded system, $\tilde{\Delta}_0$, normalized by the ratio between the average energy of the coded sequence, $E_{s,c}$, and the average energy of the uncoded sequence, $E_{s,u}$. The coding gain is then expressed in dB as

$$G_{code} = 10 \log_{10} \frac{d_{free}^2 / \tilde{\Delta}_0^2}{E_{s,c}/E_{s,u}} \tag{12.5}$$

12.1.2 Set partitioning

The design of trellis codes is based on a method called mapping by set partitioning. This method requires that the bit mapper assigns symbol values to the input binary vectors so that the minimum Euclidean distance between possible code sequences $\{a_k\}$ is maximum. For a given encoder, the search of the optimum assignment is made by taking into consideration subsets of the symbol set $\mathcal{A}$. These subsets are obtained by successive partitioning of the set $\mathcal{A}$, and are characterized by the property that the minimum Euclidean distance between symbols in a subset corresponding to a certain level of partitioning is larger than or equal to the minimum distance obtained at the previous level.

Consider the symbol alphabet $\mathcal{A} = \mathcal{A}_0$ with 2^n elements, that corresponds to level zero of partitioning. At the first level of partitioning, that is characterized by the index $q = 1$, the set $\mathcal{A}_0$ is subdivided into two disjoint subsets $\mathcal{A}_1(0)$ and $\mathcal{A}_1(1)$ with 2^{n-1} elements each. Let $\Delta_1(0)$ and $\Delta_1(1)$ be the minimum Euclidean distances between elements of the subsets $\mathcal{A}_1(0)$ and $\mathcal{A}_1(1)$, respectively; define Δ_1 as the minimum between the two Euclidean distances $\Delta_1(0)$ and $\Delta_1(1)$; we choose a partition for which Δ_1 is maximum. At the level of partitioning characterized by the index $q > 1$, each of the 2^{q-1} subsets

[2] To emphasize the dependence of the asymptotic coding gain on the choice of the symbol constellations of the coded and uncoded systems, sometimes the information on the considered modulation schemes is included as a subscript in the symbol used to denote the coding gain, e.g. $G_{8PSK/4PSK}$ for the introductory example.

$\mathcal{A}_{q-1}(\ell)$, $\ell = 0, 1, \ldots, 2^{q-1} - 1$, is subdivided into two subsets, thus originating 2^q subsets. During the procedure, it is required that the minimum Euclidean distance at the q-th level of partitioning,

$$\Delta_q = \min_{\ell \in \{0,1,\ldots,2^q-1\}} \Delta_q(\ell) \quad \text{with } \Delta_q(\ell) = \min_{\substack{\alpha_i,\, \alpha_m \in \mathcal{A}_q(\ell) \\ \alpha_i \neq \alpha_m}} |\alpha_i - \alpha_m| \tag{12.6}$$

is maximum. At the n-th level of partitioning, the subsets $\mathcal{A}_n(\ell)$ consist of only one element each; to subsets with only one element we assign the minimum distance $\Delta_n = \infty$; at the end of the procedure, we obtain a tree diagram of binary partitioning for the symbol set. At the q-th level of partitioning, to the two subsets obtained by a subset at the $(q-1)$-th level we assign the binary symbols $y^{(q-1)} = 0$ and $y^{(q-1)} = 1$, respectively; in this manner, an n-tuple of binary symbols $\boldsymbol{y}_i = (y_i^{(n-1)}, \ldots, y_i^{(1)}, y_i^{(0)})$ is associated to each element α_i found at an end node of the tree diagram.[3]

Therefore, the Euclidean distance between two elements of $\mathcal{A}$, α_i and α_m, indicated by the binary vectors $\boldsymbol{y}_i$ and $\boldsymbol{y}_m$ that are equal in the first q components, satisfies the relation

$$|\alpha_i - \alpha_m| \geq \Delta_q \quad \text{for } y_i^{(p)} = y_m^{(p)}, \quad p = 0, \ldots, q-1, \quad i \neq m \tag{12.7}$$

In fact, because of the equality of the components in the positions from (0) up to $(q-1)$, we have that the two elements are in the same subset $\mathcal{A}_q(\ell)$ at the q-th level of partitioning. Therefore, their Euclidean distance is at least equal to Δ_q.

Example 12.1.2

The partitioning of the set $\mathcal{A}_0$ of symbols with statistical power $E[|a_k|^2] = 1$ for an 8-PSK system is illustrated in Figure 12.5. The minimum Euclidean distance between elements of the set $\mathcal{A}_0$ is given by $\Delta_0 = 2\sin(\pi/8) = 0.765$. At the first level of partitioning the two subsets $\mathcal{B}_0 = \{(y^{(2)}, y^{(1)}, 0); y^{(i)} = 0, 1\}$

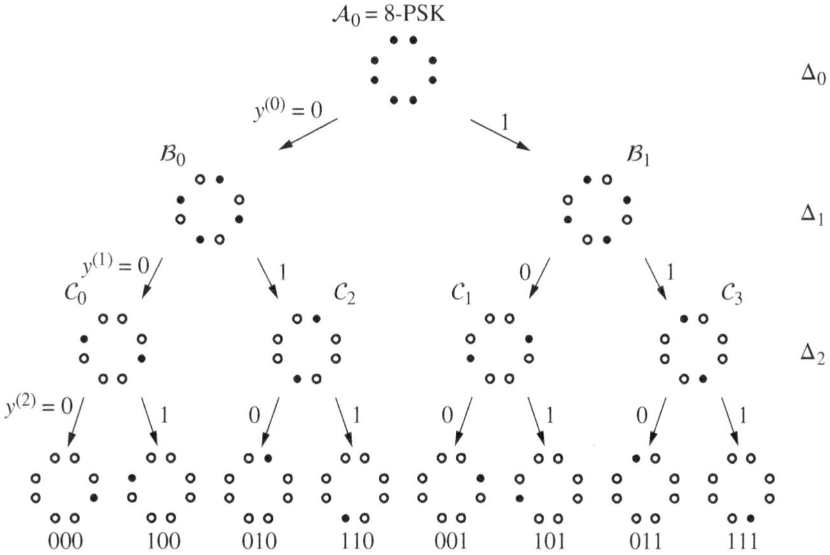

Figure 12.5 Partitioning of the symbol set for an 8-PSK system. Source: Reproduced with permission from Ungerboeck [1]. ©1982, IEEE.

[3] For TCM encoders, the n-tuples of binary code symbols will be indicated by $\boldsymbol{y} = (y^{(n-1)}, \ldots, y^{(0)})$ rather than by the notation $\boldsymbol{c}$ employed in Chapter 11.

and $\mathcal{B}_1 = \{(y^{(2)}, y^{(1)}, 1); y^{(i)} = 0, 1\}$ are found, with four elements each and minimum Euclidean distance $\Delta_1 = \sqrt{2}$. At the second level of partitioning, four subsets $C_0 = \{(y^{(2)}, 0, 0); y^{(2)} = 0, 1\}$, $C_2 = \{(y^{(2)}, 1, 0); y^{(2)} = 0, 1\}$, $C_1 = \{(y^{(2)}, 0, 1); y^{(2)} = 0, 1\}$, and $C_3 = \{(y^{(2)}, 1, 1); y^{(2)} = 0, 1\}$ are found with two elements each and minimum Euclidean distance $\Delta_2 = 2$. Finally, at the last level eight subsets $D_0, \dots, D_7$, are found, with one element each and minimum Euclidean distance $\Delta_3 = \infty$.

Example 12.1.3

The partitioning of the set $\mathcal{A}_0$ of symbols with statistical power $E[|a_k|^2] = 1$ for a 16-quadrature amplitude modulation (QAM) system is illustrated in Figure 12.6. The minimum Euclidean distance between the elements of $\mathcal{A}_0$ is given by $\Delta_0 = 2/\sqrt{10} = 0.632$. Note that at each successive partitioning level, the minimum Euclidean distance among the elements of a subset increases by a factor $\sqrt{2}$. Therefore at the third level of partitioning, the minimum Euclidean distance between the elements of each of the subsets D_i, $i = 0, 1, \dots, 7$, is given by $\Delta_3 = \sqrt{8}\Delta_0$.

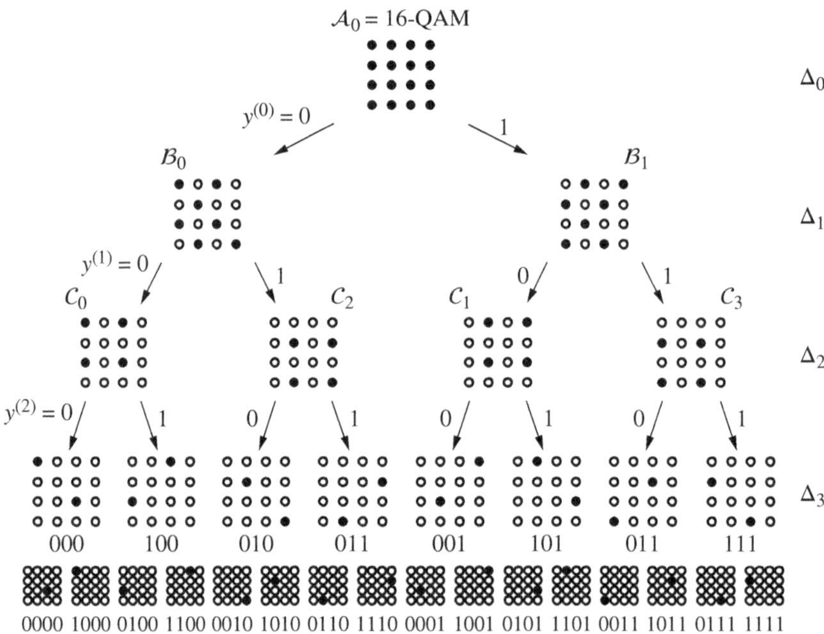

Figure 12.6 Partitioning of the symbol set for a 16-QAM system. Source: Reproduced with permission from Ungerboeck [1]. ©1982, IEEE.

12.1.3 Lattices

Several constellations and the relative partitioning can be effectively described by lattices; furthermore, as we will see in the following sections, the formulation based on lattices is particularly convenient in the discussion on multidimensional trellis codes.

In general, let $\mathbf{Z}_D = \mathbf{Z}^D$, where $\mathbf{Z} = \mathbb{Z}$ denotes the set of integers; a lattice Λ in $\mathbb{R}^D$ is defined by the relation

$$\Lambda = \left\{ (i_1, \dots, i_D)\, \boldsymbol{G} \mid (i_1, \dots, i_D) \in \mathbf{Z}_D \right\} \tag{12.8}$$

where $\boldsymbol{G}$ is a non-singular $D \times D$ matrix, called *lattice generator matrix*, by means of which we obtain a correspondence $\boldsymbol{Z}_D \to \Lambda$. The vectors given by the rows of $\boldsymbol{G}$ form a basis for the lattice Λ; the vectors of the basis define a parallelepiped whose volume $V_0 = |\det(\boldsymbol{G})|$ represents the *characteristic volume* of the lattice. The volume V_0 is equivalent to the volume of a Voronoi cell associated with an element or *point of lattice* Λ and defined as the set of points in $\mathbb{R}^D$ whose distance from a given point of Λ is smaller than the distance from any other point of Λ. The set of Voronoi cells associated with the points of Λ is equivalent to the space $\mathbb{R}^D$. A lattice is characterized by two parameters:

1. d_{min}, defined as the minimum distance between points of the lattice;
2. the *kissing number*, defined as the number of lattice points at minimum distance from a given point.

We obtain a subgroup $\Lambda_q(0)$ if points of the lattice Λ are chosen as basis vectors in a matrix $\boldsymbol{G}_q$, such that they give rise to a characteristic volume $V_q = |\det(\boldsymbol{G}_q)| > V_0$.

Example 12.1.4 ($\boldsymbol{Z}_p$ lattice)
In general, as already mentioned, the notation $\boldsymbol{Z}_p$ is used to define a lattice with an infinite number of points in the p-dimensional Euclidean space with coordinates given by integers. The generator matrix $\boldsymbol{G}$ for the lattice $\boldsymbol{Z}_p$ is the $p \times p$ identity matrix; the minimum distance is $d_{min} = 1$ and the kissing number is equal to $2p$. The $\boldsymbol{Z}_2$ type constellations (see Figure 12.7a) for QAM systems are finite subsets of $\boldsymbol{Z}_2$, with centre at the origin and minimum Euclidean distance equal to Δ_0.

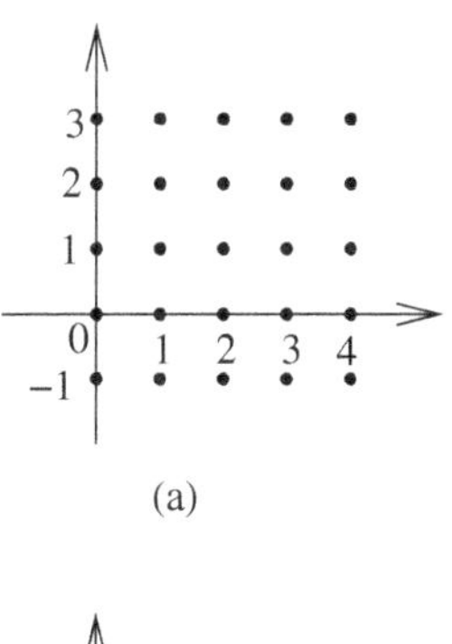

(a)

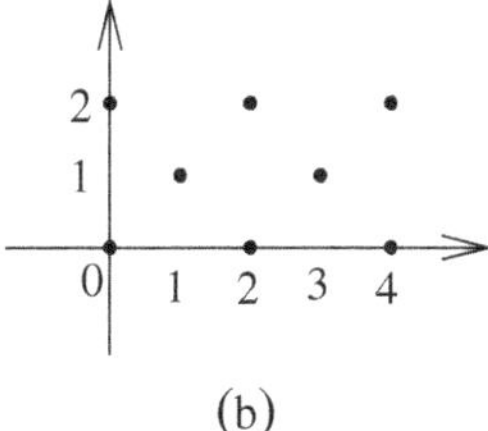

(b)

Figure 12.7 (a) $\boldsymbol{Z}_2$ lattice; (b) $\boldsymbol{D}_2$ lattice.

Example 12.1.5 ($\boldsymbol{D}_n$ lattice)
$\boldsymbol{D}_n$ is the set of all n-dimensional points whose coordinates are integers that sum to an even number; it may be regarded as a version of the $\boldsymbol{Z}_n$ lattice from which the points whose coordinates are integers that sum to an odd number were removed. The minimum distance is $d_{min} = \sqrt{2}$ and the kissing number is $2n(n-1)$. The lattice $\boldsymbol{D}_2$ is represented in Figure 12.7b.

$\boldsymbol{D}_4$, called the *Schläfli lattice*, constitutes the densest lattice in $\mathbb{R}^4$; this means that if four-dimensional spheres with centres given by the points of the lattice are used to fill $\mathbb{R}^4$, then $\boldsymbol{D}_4$ is the lattice having the largest number of spheres per unit of volume.

Example 12.1.6 ($\boldsymbol{E}_8$ lattice)
$\boldsymbol{E}_8$ is given by points

$$\left\{ (x_1, \dots, x_8) \mid \forall i : x_i \in \boldsymbol{Z} \text{ or } \forall i : x_i \in \boldsymbol{Z} + \frac{1}{2} \sum_{i=1}^{8} x_i = 0 \bmod 2 \right\} \tag{12.9}$$

In other words, $\boldsymbol{E}_8$ is the set of eight-dimensional points whose components are all integers, or all halves of odd integers, that sum to an even number. $\boldsymbol{E}_8$ is called the *Gosset lattice*.

We now discuss set partitioning with the aid of lattices. First we recall the properties of subsets obtained by partitioning.

If the set $\mathcal{A}$ has a group structure with respect to a certain operation (see page 529), the partitioning can be done so that the sequence of subsets $\mathcal{A}_0, \mathcal{A}_1(0), \dots, \mathcal{A}_n(0)$, with $\mathcal{A}_q(0) \subset \mathcal{A}_{q-1}(0)$, form a *chain of subgroups* of $\mathcal{A}_0$; in this case, the subsets $\mathcal{A}_q(\ell)$, $\ell \in \{1, \dots, 2^q - 1\}$, are called *cosets* of the subgroup $\mathcal{A}_q(0)$ with respect to $\mathcal{A}_0$ (see page 523), and are obtained from the subgroup $\mathcal{A}_q(0)$ by translations. The distribution of Euclidean distances between elements of a coset $\mathcal{A}_q(\ell)$ is equal to the distribution of the Euclidean distances between elements of the subgroup $\mathcal{A}_q(0)$, as the *difference* between two elements of a coset yields an element of the subgroup; in particular, for the minimum Euclidean distance in subsets at a certain level of partitioning it holds

$$\Delta_q(\ell) = \Delta_q \qquad \forall \ell \in \{0, \dots, 2^q - 1\} \tag{12.10}$$

The lattice Λ in $\mathbb{R}^D$ defined as in (12.8) has a *group structure with respect to the addition.* With a suitable translation and normalization, we obtain that the set $\mathcal{A}_0$ for PAM or QAM is represented by a subset of $\boldsymbol{Z}$ or $\boldsymbol{Z}_2$. To get a QAM constellation from $\boldsymbol{Z}_2$, we define for example the translated and normalized lattice $\boldsymbol{Q} = c(\boldsymbol{Z}_2 + \{1/2, 1/2\})$, where c is an arbitrary scaling factor, generally chosen to normalize the statistical power of the symbols to 1. Figure 12.8 illustrates how QAM constellations are obtained from $\boldsymbol{Z}_2$.

If we apply binary partitioning to the set $\boldsymbol{Z}$ or $\boldsymbol{Z}_2$, we still get infinite lattices in $\mathbb{R}$ or $\mathbb{R}^2$, in which the minimum Euclidean distance increases with respect to the original lattice. Formally, we can assign the binary representations of the tree diagram obtained by partitioning to the lattices; for transmission, a symbol is chosen as representative for each lattice at an end node of the tree diagram.

Definition 12.1
The notation $\boldsymbol{X}/\boldsymbol{X}'$ denotes the set of subsets obtained from the decomposition of the group $\boldsymbol{X}$ in the subgroup $\boldsymbol{X}'$ and its cosets. The set $\boldsymbol{X}/\boldsymbol{X}'$ forms in turn a group, called the *quotient group* of $\boldsymbol{X}$ with respect to $\boldsymbol{X}'$. It is called *binary* if the number of elements is a power of two. □

PAM In this case, the subgroups of the lattice $\boldsymbol{Z}$ are expressed by

$$\mathcal{A}_q(0) = \{2^q i \mid i \in \boldsymbol{Z}\} = 2^q \boldsymbol{Z} \tag{12.11}$$

Let

$$t(\ell) = \ell \in \{0, \dots, 2^q - 1\} \tag{12.12}$$

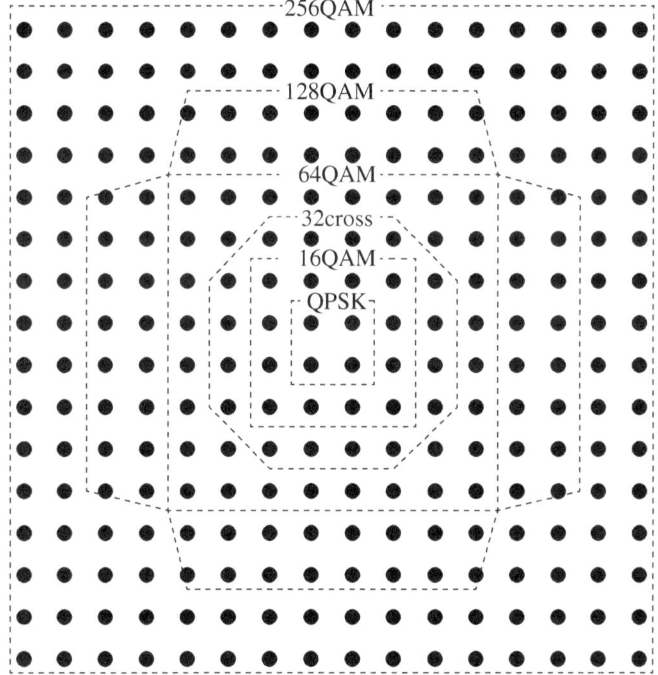

Figure 12.8 The integer lattice Z_2 as template for QAM constellations.

then the cosets of a subgroup, expressed as

$$\mathcal{A}_q(\ell) = \{a + t(\ell) \mid a \in \mathcal{A}_q(0)\} \tag{12.13}$$

are obtained by translations of the subgroup. The sequence

$$\mathbf{Z} \;/\; 2\mathbf{Z} \;/\; 2^2\mathbf{Z} \;/\; \dots \;/\; 2^q\mathbf{Z} \;/\; \dots \tag{12.14}$$

forms a binary *partitioning chain* of the lattice $\mathbf{Z}$, with increasing minimum distances given by

$$\Delta_q = 2^q \tag{12.15}$$

QAM The first subgroup in the binary partitioning chain of the two-dimensional lattice $\mathbf{Z}_2$ is a lattice that is obtained from $\mathbf{Z}_2$ by rotation of $\pi/4$ and multiplication by $\sqrt{2}$. The matrix of this linear transformation is

$$\boldsymbol{R} = \begin{pmatrix} 1 & 1 \\ -1 & 1 \end{pmatrix} \tag{12.16}$$

Successive subgroups in the binary partitioning chain are obtained by repeated application of the linear transformation $\boldsymbol{R}$,

$$\mathcal{A}_q(0) = \mathbf{Z}_2\boldsymbol{R}^q \tag{12.17}$$

Let

$$i\mathbf{Z}_2\boldsymbol{R}^q = \left\{(ik, im)\boldsymbol{R}^q \mid k, m \in \mathbf{Z}\right\}, \qquad i \in \mathbb{N} \tag{12.18}$$

then the sequence

$$\mathbf{Z}_2 \;/\; \mathbf{Z}_2\boldsymbol{R} \;/\; \mathbf{Z}_2\boldsymbol{R}^2 \;/\; \mathbf{Z}_2\boldsymbol{R}^3 \;/\; \dots = \mathbf{Z}_2 \;/\; \mathbf{Z}_2\boldsymbol{R} \;/\; 2\mathbf{Z}_2 \;/\; 2\mathbf{Z}_2\boldsymbol{R} \;/\; \dots \tag{12.19}$$

forms a binary partitioning chain of the lattice $\mathbf{Z}_2$, with increasing minimum distances given by

$$\Delta_q = 2^{q/2} \tag{12.20}$$

This binary partitioning chain is illustrated in Figure 12.9.

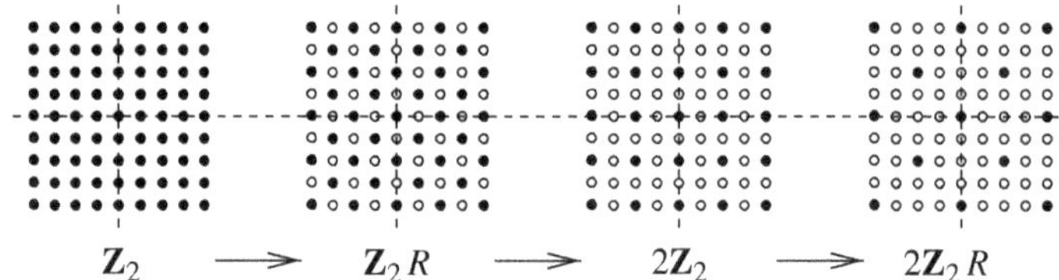

Figure 12.9 Binary partitioning chain of the lattice $\mathbf{Z}_2$.

The cosets $\mathcal{A}_q(\ell)$ are obtained by translations of the subgroup $\mathcal{A}_q(0)$ as in the one-dimensional case (12.13), with

$$\mathbf{t}(\ell) = (i, m) \tag{12.21}$$

where, as it can be observed in Figure 12.10 for the case $q = 2$ with $\mathcal{A}_2(0) = 2\mathbf{Z}_2$,

$$\begin{aligned} &i \in \left\{0, \dots, 2^{\frac{q}{2}} - 1\right\}, \quad m \in \left\{0, \dots, 2^{\frac{q}{2}} - 1\right\}, \quad \ell = 2^{\frac{q}{2}} i + m \quad q \text{ even} \\ &i \in \left\{0, \dots, 2^{\frac{q+1}{2}} - 1\right\}, \ m \in \left\{0, \dots, 2^{\frac{q-1}{2}} - 1\right\}, \ \ell = 2^{\frac{q-1}{2}} i + m \quad q \text{ odd} \end{aligned} \tag{12.22}$$

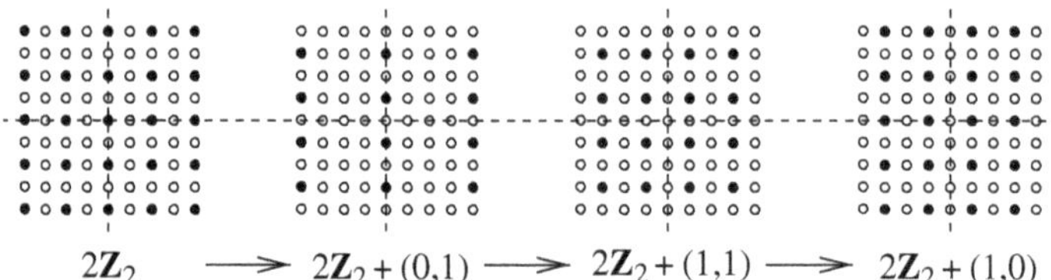

Figure 12.10 The four cosets of $2\mathbf{Z}_2$ in the partition $\mathbf{Z}_2/2\mathbf{Z}_2$.

M-PSK In this case, the set of symbols $\mathcal{A}_0 = \{e^{j2\pi(k/M)} \mid k \in \mathbf{Z}\}$ on the unit circle of the complex plane forms a *group with respect to the multiplication*. If the number of elements M is a power of two, the sequence

$$\mathcal{A}_0 \ / \ \mathcal{A}_1(0) \ / \ \mathcal{A}_2(0) \ / \ \dots \ / \ \mathcal{A}_{\log_2 M}(0) \qquad \text{with } \mathcal{A}_q(0) = \left\{ e^{j2\pi \frac{2^q}{M} k} \mid k \in \mathbf{Z} \right\} \tag{12.23}$$

forms a binary partitioning chain of the set $\mathcal{A}_0$, with increasing minimum distances

$$\Delta_q = \begin{cases} 2 \sin\left(\pi \dfrac{2^q}{M}\right) & \text{for } 0 \le q < \log_2 M \\ \infty & \text{for } q = \log_2 M \end{cases} \tag{12.24}$$

The cosets of the subgroups $\mathcal{A}_q(0)$ are given by

$$\mathcal{A}_q(\ell) = \left\{ a\, e^{j2\pi \frac{\ell}{M}} \mid a \in \mathcal{A}_q(0) \right\}, \qquad \ell \in \{0, \dots, 2^q - 1\} \tag{12.25}$$

12.1.4 Assignment of symbols to the transitions in the trellis

As discussed in the previous sections, an encoder can be modelled as a finite-state machine with a given number of states and well-defined state transitions. If the encoder input consists of m binary symbols per modulation interval,[4] then there are 2^m possible transitions from a state to the next state; there may be parallel transitions between pairs of states; furthermore, for reasons of symmetry, we take into consideration only encoders with a uniform structure. After selecting a diagram having the desired characteristics in terms of state transitions, the design of a code is completed by assigning symbols to the state transitions, such that d_{free} is maximum. Following the indications of information theory (see Section 6.5), the symbols are chosen from a redundant set $\mathcal{A}$ of 2^{m+1} elements.

Example 12.1.7 (Uncoded transmission of 2 bits per modulation interval by 8-PSK)
Consider an uncoded 4-PSK system as reference system. The uncoded transmission of 2 bits per modulation interval by 4-PSK can be viewed as the result of the application of a trivial encoder with only one state, and trellis diagram with four parallel transitions, as illustrated in Figure 12.11. A distinct symbol of the alphabet of a 4-PSK system is assigned to each parallel transition. Note from Figure 12.5 that the alphabet of a 4-PSK system is obtained choosing the subset $\mathcal{B}_0$ (or $\mathcal{B}_1$) obtained by partitioning the alphabet $\mathcal{A}_0$ of an 8-PSK system. Therefore, $d_{min} = \Delta_1 = \sqrt{2}$. In the trellis diagram, any sequence on the trellis represents a possible symbol sequence. The optimum receiver decides for the symbol of the subset $\mathcal{B}_0$ (or $\mathcal{B}_1$) that is found at the minimum distance from the received signal.

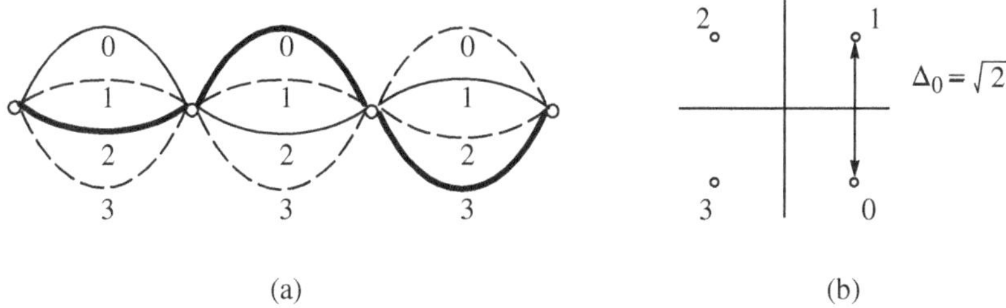

Figure 12.11 Uncoded transmission of 2 bits per modulation interval by 4-PSK. (a) Trellis diagram with a single state and (b) 4-PSK constellation.

Now consider the two-state trellis diagram of Figure 12.12. The symbols of the subsets $\mathcal{B}_0 = \mathcal{C}_0 \cup \mathcal{C}_2$ and $\mathcal{B}_1 = \mathcal{C}_1 \cup \mathcal{C}_3$ are assigned to the transitions that originate from the first and second state, respectively; this guarantees that the minimum d_{free} between the code symbol sequences is at least equal to that obtained for uncoded 4-PSK system. With a trellis diagram with only two states, it is impossible to have only signals of $\mathcal{B}_0$ or $\mathcal{B}_1$ assigned to the transitions that originate from a certain state, and also to all transitions that lead to the same state; therefore, we find that the minimum Euclidean distance between code symbol sequences is in this case equal to $d_{free} = \sqrt{\Delta_1^2 + \Delta_0^2} = 1.608$, greater than that obtained for uncoded 4-PSK transmission. As $E_{s,c} = E_{s,u}$ and $\tilde{\Delta}_0 = \Delta_1$, the coding gain is $G_{8PSK/4PSK} = 20\log_{10}\left(\sqrt{\Delta_1^2 + \Delta_0^2}/\sqrt{\Delta_1^2}\right) = 1.1$ dB. Furthermore, $N_{free} = 2$.

---

[4] In this chapter, encoding and bit mapping are jointly optimized and m represents the number of information bits L_b per modulation interval (see (6.57)). For example, for QAM the rate of the encoder–modulator is $R_I = \frac{m}{2}$.

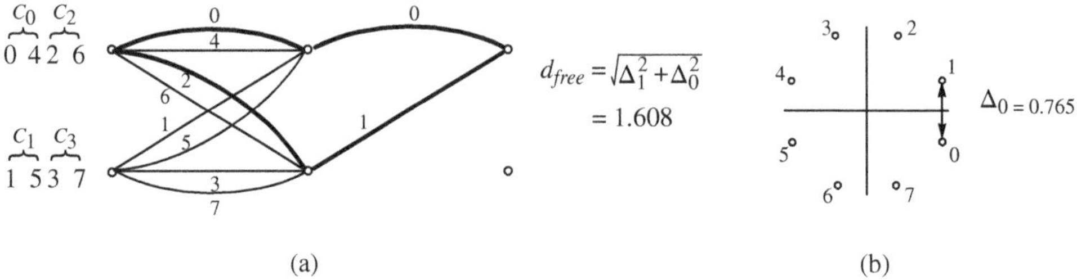

Figure 12.12 Transmission of 2 bits per modulation interval using (a) a two-state trellis code and (b) 8-PSK constellation. For each state, the values of the symbols assigned to the transitions that originate from that state are indicated.

High coding gains for the transmission of 2 bits per modulation interval by 8-PSK are obtained by the codes represented in Figure 12.13, with trellis diagrams having 4, 8, and 16 states. For the heuristic design of these codes with moderate complexity, we resort to the following rules proposed by Ungerboeck:

1. All symbols of the set $\mathcal{A}_0$ must be assigned equally likely to the state transitions in the trellis diagram, using criteria of regularity and symmetry;
2. To transitions that originate from the same state are assigned symbols of the subset $\mathcal{B}_0$, or symbols of the subset $\mathcal{B}_1$;
3. To transitions that merge to the same state are assigned symbols of the subset $\mathcal{B}_0$, or symbols of the subset $\mathcal{B}_1$;
4. To parallel transitions between two states we assign the symbols of one of the subsets C_0, C_1, C_2, or C_3.

Rule 1 intuitively points to the fact that good trellis codes exhibit a regular structure. Rules 2, 3, and 4 guarantee that the minimum Euclidean distance between code symbol sequences that differ in one or more elements is at least twice the minimum Euclidean distance between uncoded 4-PSK symbols, so that the coding gain is greater than or equal to 3 dB, as we will see in the next examples.

Example 12.1.8 (Four-state trellis code for the transmission of 2 bit/s/Hz by 8-PSK)
Consider the code with four states represented in Figure 12.13. Between each pair of code symbol sequences in the trellis diagram that diverge at a certain state and merge after more than one transition, the Euclidean distance is greater than or equal to $\sqrt{\Delta_1^2 + \Delta_0^2 + \Delta_1^2} = 2.141$. For example, this distance exists between sequences in the trellis diagram labelled by the symbols 0–0–0 and 2–1–2; on the other hand, the Euclidean distance between symbols assigned to parallel transitions is equal to $\Delta_2 = 2$. Therefore, the minimum Euclidean distance between code symbol sequences is equal to 2; hence, with a four-state trellis code, we obtain a gain equal to 3 dB over uncoded 4-PSK transmission. Note that, as the minimum distance d_{free} is determined by parallel transitions, the sequence at minimum distance from a transmitted sequence differs only by one element that corresponds to the transmitted symbol rotated by 180°.

A possible implementation of the encoder/bit-mapper for a four-state trellis code is illustrated in Figure 12.14.

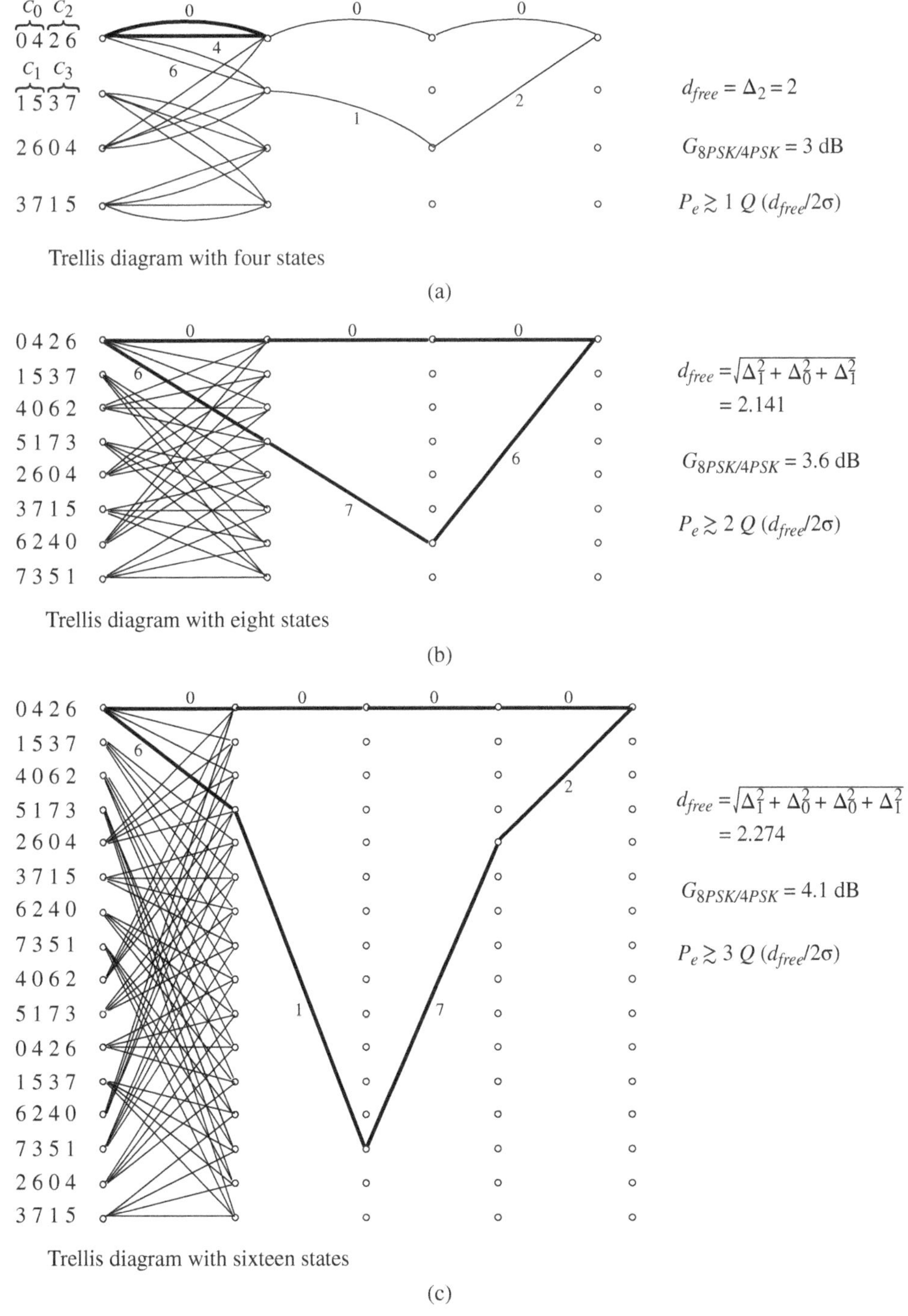

Figure 12.13 Trellis codes with (a) 4, (b) 8, and (c) 16 states for transmission of 2 bits per modulation interval by 8-PSK. Source: Reproduced with permission from Ungerboeck [1]. ©1982, IEEE.

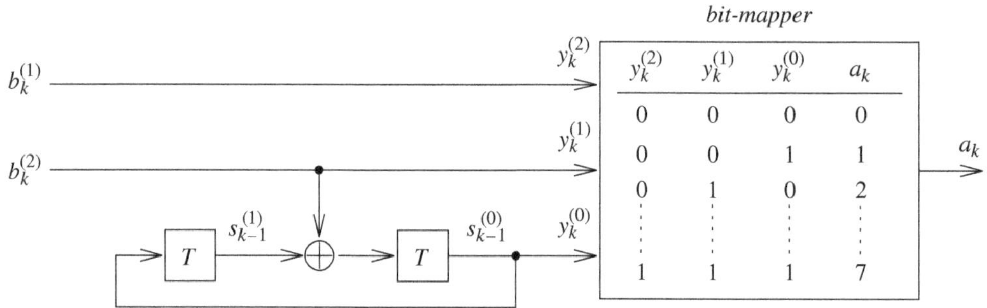

$y_k^{(2)}$	$y_k^{(1)}$	$y_k^{(0)}$	a_k
0	0	0	0
0	0	1	1
0	1	0	2
⋮	⋮	⋮	⋮
1	1	1	7

Figure 12.14 Encoder/bit-mapper for a 4-state trellis code for the transmission of 2 bits per modulation interval by 8-PSK.

The 4-state trellis code for transmission with spectral efficiency of 2 bit/s/Hz by 8-PSK was described in detail as an introductory example. In Figure 12.13, the values of d_{free} and N_{free} for codes with 4, 8, and 16 states are reported.

Consider now a 16-QAM system for the transmission of 3 bits per modulation interval; in this case, the reference system uses uncoded 8-PSK or 8-amplitude modulation (AM)-phase modulation (PM), as illustrated in Figure 12.15.

Example 12.1.9 (Eight-state trellis code for the transmission of 3 bit/s/Hz by 16-QAM)
The partitioning of a symbol set with unit statistical power for a 16-QAM system is shown in Figure 12.6. For the assignment of symbols to the transitions on the trellis consider the subsets of the symbol set $\mathcal{A}_0$ denoted by $D_0, D_1, \dots, D_7$, that contain two elements each. The minimum Euclidean distance between the signals in $\mathcal{A}_0$ is $\Delta_0 = 2/\sqrt{10} = 0.632$; the minimum Euclidean distance between elements of a subset D_i, $i = 0, 1, \dots, 7$, is $\Delta_3 = \sqrt{8}\Delta_0$.

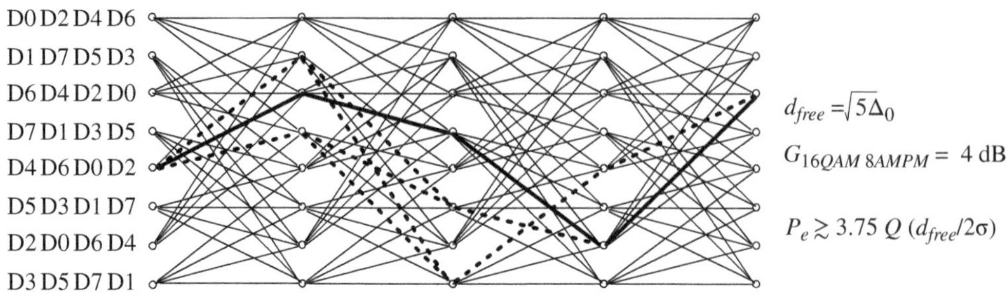

Figure 12.15 Eight-state trellis code for the transmission of 3 bits per modulation interval by 16-QAM.

In the 8-state trellis code illustrated in Figure 12.15, four transitions diverge from each state and four merge to each state. To each transition one of the subsets D_i, $i = 0, 1, \dots, 7$, is assigned; therefore, a transition in the trellis corresponds to a pair of parallel transitions. The assignment of subsets to the transitions satisfies Ungerboeck rules. The subsets D_0, D_4, D_2, D_6, or D_1, D_5, D_3, D_7 are assigned to the four transitions from or to the same state. In evaluating d_{free}, this choice guarantees a square Euclidean distance equal to at least $2\Delta_0^2$ between sequences that diverge from a state and merge after L transitions, $L > 1$. The square distance between sequences that diverge from a state and merge after two transitions is equal to $6\Delta_0^2$. If two sequences diverge and merge again after three or more transitions, at least one intermediate transition contributes to an incremental square Euclidean distance equal to Δ_0^2; thus, the minimum Euclidean distance between code symbol sequences that do not differ only for one symbol is given by $\sqrt{5}\Delta_0$. As the Euclidean distance between symbols assigned to parallel transitions is equal to $\sqrt{8}\Delta_0$, the free distance of the code is $d_{free} = \sqrt{5}\Delta_0$. Because the minimum Euclidean distance for an uncoded 8-AM-PM reference system with the same average symbol energy is $\tilde{\Delta}_0 = \sqrt{2}\Delta_0$, the coding gain is $G_{16QAM/8AM-PM} = 20\log_{10}\{\sqrt{5/2}\} = 4$ dB.

In the trellis diagram of Figure 12.15, four paths are shown that represent error events at minimum distance from the code sequence, having symbols taken from the subsets $D_0 - D_0 - D_3 - D_6$; the sequences in error diverge from the same state and merge after three or four transitions. It can be shown that for each code sequence and for each state, there are two paths leading to error events of length three and two of length four. The number of code sequences at the minimum distance depends on the code sequence being considered, hence N_{free} represents a mean value. The proof is simple in the case of uncoded 16-QAM, where the number of symbols at the minimum distance from a symbol that is found at the centre of the constellation is larger than the number of symbols at the minimum distance from a symbol found at the edge of the constellation; in this case, we obtain $N_{free} = 3$. For the eight-state trellis code of Figure 12.15, we get $N_{free} = 3.75$. For constellations of type $\mathbf{Z}_2$ with a number of signals that tends to infinity, the limit is $N_{free} = 4$ for uncoded modulation, and $N_{free} = 16$ for coded modulation with an eight-state code.

12.1.5 General structure of the encoder/bit-mapper

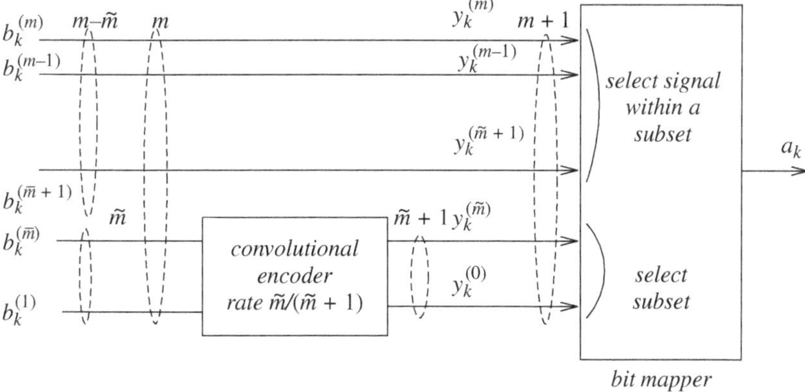

Figure 12.16 General structure of the encoder/bit-mapper for TCM.

The structure of trellis codes for TCM can be described in general by the combination of a convolutional encoder and a special bit mapper, as illustrated in Figure 12.16. A code symbol sequence is generated as follows. Of the m information bits $[b_k^{(m)}, \dots, b_k^{(1)}]$ that must be transmitted during a cycle of the encoder/bit-mapper operations, $\tilde{m} \le m$ are input to a convolutional encoder with rate $R_c = \tilde{m}/(\tilde{m}+1)$; the $\tilde{m}+1$ bits at the encoder output, $\tilde{\mathbf{y}}_k = [y_k^{(\tilde{m})}, \dots, y_k^{(0)}]^T$, are used to *select one of the* $2^{\tilde{m}+1}$ *subsets* of the symbol set $\mathcal{A}$ with 2^{m+1} elements, and to determine the next state of the encoder. The remaining $(m-\tilde{m})$ uncoded bits *determine which of the* $2^{m-\tilde{m}}$ *symbols* of the selected subset must be transmitted. For example in the encoder/bit-mapper for a four-state code illustrated in Figure 12.14, the two bits $y_k^{(0)}$ and $y_k^{(1)}$ select one of the four subsets C_0, C_2, C_1, C_3 of the set $\mathcal{A}_0$ with eight elements. The uncoded bit $y_k^{(2)}$ determines which of the two symbols in the subset C_i is transmitted.

Let $\mathbf{y}_k = [y_k^{(m)}, \dots, y_k^{(1)}, y_k^{(0)}]^T$ be the $(m+1)$-dimensional binary vector at the input of the bit-mapper at the k-th instant, then the selected symbol is expressed as $a_k = a[\mathbf{y}_k]$. Note that in the trellis the symbols of each subset are associated with $2^{m-\tilde{m}}$ parallel transitions.

The free Euclidean distance of a trellis code, given by (12.2), can be expressed as

$$d_{free} = \min\left\{ \Delta_{\tilde{m}+1}, d_{free}(\tilde{m}) \right\} \tag{12.26}$$

where $\Delta_{\tilde{m}+1}$ is the minimum distance between symbols assigned to parallel transitions and $d_{free}(\tilde{m})$ denotes the minimum distance between code sequences that differ in more than one symbol. In the particular case $\tilde{m} = m$, each subset has only one element and therefore there are no parallel transitions; this occurs, for example, for the encoder/ bit-mapper for an 8-state code illustrated in Figure 12.2.

From Figure 12.16, observe that the vector sequence $\{\tilde{\mathbf{y}}_k\}$ is the output sequence of a convolutional encoder. Recalling (11.263), for a convolutional code with rate $\tilde{m}/(\tilde{m}+1)$ and constraint length ν, we have the following constraints on the bits of the sequence $\{\tilde{\mathbf{y}}_k\}$:

$$\sum_{i=0}^{\tilde{m}} h_\nu^{(i)} y_{k-\nu}^{(i)} \otimes h_{\nu-1}^{(i)} y_{k-\nu+1}^{(i)} \otimes \cdots \otimes h_0^{(i)} y_k^{(i)} = 0 \qquad \forall k \tag{12.27}$$

where $\{h_j^{(i)}\}, 0 \le j \le \nu, 0 \le i \le \tilde{m}$, are the parity check binary coefficients of the encoder. For an encoder having ν binary memory cells, a trellis diagram is generated with 2^ν states.

Note that (12.27) defines only the constraints on the code bits, but not the input/output relation of the encoder.

Using polynomial notation, for the binary vector sequence $\mathbf{y}(D)$ (12.27) becomes

$$\left[y^{(m)}(D), \dots, y^{(1)}(D), y^{(0)}(D)\right] \left[h^{(m)}(D), \dots, h^{(1)}(D), h^{(0)}(D)\right]^T = 0 \tag{12.28}$$

where $h^{(i)}(D) = h_\nu^{(i)} D^\nu + h_{\nu-1}^{(i)} D^{\nu-1} + \cdots + h_1^{(i)} D + h_0^{(i)}$, for $i = 0, 1, \dots, \tilde{m}$, and $h^{(i)}(D) = 0$ for $\tilde{m} < i \le m$. From (12.28), we observe that the code sequences $\mathbf{y}(D)$ can be obtained by a systematic encoder with feedback as[5]

$$\begin{bmatrix} y^{(m)}(D) \\ \vdots \\ y^{(1)}(D) \\ y^{(0)}(D) \end{bmatrix} = \begin{bmatrix} & 0 \\ & \vdots \\ & 0 \\ \mathbf{I}_m & h^{(\tilde{m})}(D)/h^{(0)}(D) \\ & \vdots \\ & h^{(1)}(D)/h^{(0)}(D) \end{bmatrix}^T \begin{bmatrix} b^{(m)}(D) \\ \vdots \\ b^{(1)}(D) \end{bmatrix} \tag{12.29}$$

The rational functions $h^{(i)}(D)/h^{(0)}(D)$, $i = 1, \dots, \tilde{m}$, are realizable if the following condition is satisfied [11]:

$$h_0^{(i)} = h_\nu^{(i)} = \left\{ \begin{matrix} 0 & i \ne 0 \\ 1 & i = 0 \end{matrix} \right\}, \qquad \nu \ge 2 \tag{12.30}$$

[5] Note that (12.29) is analogous to (11.298); here the parity check coefficients are used.

The implementation of a systematic encoder with feedback having ν binary memory elements is illustrated in Figure 12.17.

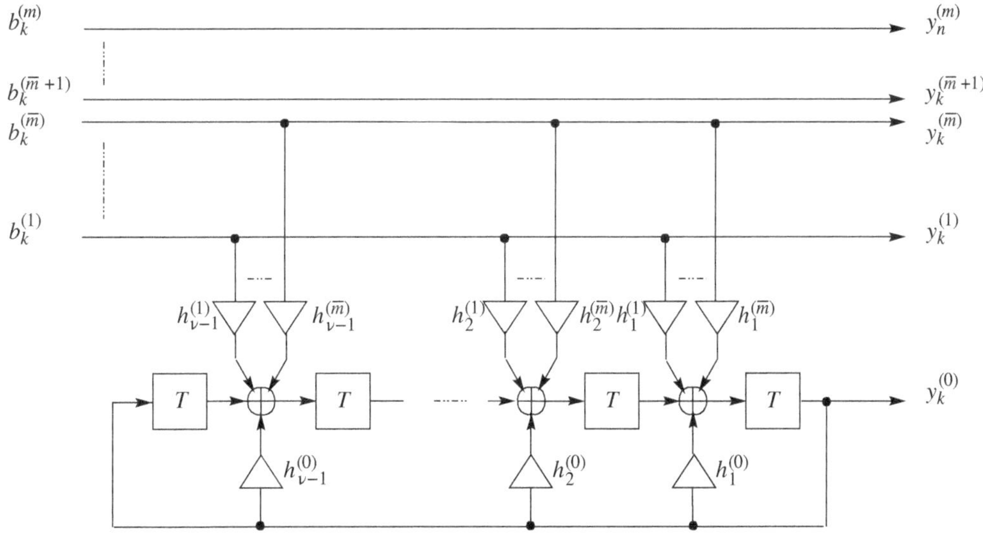

Figure 12.17 Block diagram of a systematic convolutional encoder with feedback. Source: Reproduced with permission from Ungerboeck [1]. ©1982, IEEE.

Computation of d_{free}

Consider the two code sequences

$$\boldsymbol{y}_1(D) = \left[y_1^{(m)}(D), \dots, y_1^{(0)}(D)\right]^T \tag{12.31}$$

$$\boldsymbol{y}_2(D) = \left[y_2^{(m)}(D), \dots, y_2^{(0)}(D)\right]^T \tag{12.32}$$

related by $\boldsymbol{y}_2(D) = \boldsymbol{y}_1(D) \otimes \boldsymbol{e}(D)$, as the code is linear. Let $\boldsymbol{e}_k = [e_k^{(m)}, \dots, e_k^{(0)}]$, then the error sequence $\boldsymbol{e}(D)$ is given by

$$\boldsymbol{e}(D) = \boldsymbol{e}_k D^k + \boldsymbol{e}_{k+1} D^{k+1} + \cdots + \boldsymbol{e}_{k+L} D^{k+L} \tag{12.33}$$

where $\boldsymbol{e}_i, \boldsymbol{e}_{i+L} \neq 0$, and $L > 0$. Note that $\boldsymbol{e}(D)$ is also a valid code sequence as the code is linear. To find a lower bound on the Euclidean distance between the code symbol sequences $a_1(D) = a[\boldsymbol{y}_1(D)]$ and $a_2(D) = a[\boldsymbol{y}_2(D)]$ obtained from $\boldsymbol{y}_1(D)$ and $\boldsymbol{y}_2(D)$, we define the function $d[\boldsymbol{e}_k] = \min_{z_k} d(a[z_k], a[z_k \otimes \boldsymbol{e}_k])$, where minimization takes place in the space of the binary vectors $z_k = [z_k^{(m)}, \dots, z_k^{(1)}, z_k^{(0)}]^T$, and $d(\cdot, \cdot)$ denotes the Euclidean distance between specified symbols. For the square distance between the sequences $a_1(D)$ and $a_2(D)$ then the relation holds

$$\sum_{k=i}^{i+L} d^2\left(a[\boldsymbol{y}_k], a[\boldsymbol{y}_k \otimes \boldsymbol{e}_k]\right) \geq \sum_{k=i}^{i+L} d^2[\boldsymbol{e}_k] = d^2[\boldsymbol{e}(D)] \tag{12.34}$$

We give the following fundamental theorem [1].

Theorem 12.1
For each sequence $\boldsymbol{e}(D)$ there exists a pair of symbol sequences $a_1(D)$ and $a_2(D)$ for which relation (12.34) is satisfied with the equal sign. □

Proof. Due to the symmetry in the subsets of symbols obtained by partitioning, $d[\boldsymbol{e}_k] = \min_{z_k} d(a[z_k], a[z_k \otimes \boldsymbol{e}_k])$ can arbitrarily be obtained by letting the component $z_k^{(0)}$ of the vector $\boldsymbol{z_k}$ equal to 0 or to 1 and performing the minimization only with respect to components $[z_k^{(m)}, \dots, z_k^{(1)}]$. As encoding does not impose any constraint on the component sequence $[y_k^{(m)}, \dots, y_k^{(1)}]$,[6] for each sequence $\boldsymbol{e}(D)$ a code sequence $\boldsymbol{y}(D)$ exists such that the relation (12.34) is satisfied as equality for every value of the index k. □

The free Euclidean distance between code symbol sequences can therefore be determined by a method similar to that used to find the free Hamming distance between binary code sequences $\boldsymbol{y}(D)$ (see (11.266)). We need to find an efficient algorithm to examine all possible error sequences $\boldsymbol{e}(D)$ (12.33) and to substitute the square Euclidean distance $d^2[\boldsymbol{e}_k]$ to the Hamming weight of $\boldsymbol{e}_k$; thus

$$d^2_{free}(\tilde{m}) = \min_{\boldsymbol{e}(D)=\boldsymbol{y}_2(D)-\boldsymbol{y}_1(D)\neq\boldsymbol{0}} \sum_{k=i}^{i+L} d^2[\boldsymbol{e}_k] \tag{12.35}$$

Let $q(\boldsymbol{e}_k)$ be the number of consecutive components equal to zero in the vector $\boldsymbol{e}_k$, starting with component $e_k^{(0)}$. For example, if $\boldsymbol{e}_k = [e_k^{(m)}, \dots, e_k^{(3)}, 1, 0, 0]^T$, then $q(\boldsymbol{e}_k) = 2$. From the definition of the indices assigned to symbols by partitioning, we obtain that $d[\boldsymbol{e}_k] \geq \Delta_{q(\boldsymbol{e}_k)}$; moreover, this relation is satisfied as equality for almost all vectors $\boldsymbol{e}_k$. Note that $d[\boldsymbol{0}] = \Delta_{q(\boldsymbol{0})} = 0$. Therefore,

$$d^2_{free}(\tilde{m}) \geq \min_{\boldsymbol{e}(D)\neq\boldsymbol{0}} \sum_{k=i}^{i+L} \Delta^2_{q(\boldsymbol{e}_k)} = \Delta^2_{free}(\tilde{m}) \tag{12.36}$$

If we assume $d_{free}(\tilde{m}) = \Delta_{free}(\tilde{m})$, the risk of committing an error in evaluating the free distance of the code is low, as the minimum is usually reached by more than one error sequence. By the definition of free distance in terms of the minimum distances between elements of the subsets of the symbol set, the computation of $d_{free}(\tilde{m})$ will be independent of the particular assignment of the symbols to the binary vectors with $(m + 1)$ components, provided that the values of the minimum distances among elements of the subsets are not changed.

At this point, it is possible to identify a further important consequence of the constraint (12.30) on the binary coefficients of the systematic convolutional encoder. We can show that an error sequence $\boldsymbol{e}(D)$ begins with $\boldsymbol{e}_i = (e_i^{(m)}, \dots, e_i^{(1)}, 0)$ and ends with $\boldsymbol{e}_{i+L} = (e_{i+L}^{(m)}, \dots, e_{i+L}^{(1)}, 0)$. It is therefore guaranteed that all transitions that originate from the same state and to the transitions that merge at the same state are assigned signals of the subset $\mathcal{B}_0$ or those of the subset $\mathcal{B}_1$. The square Euclidean distance associated with an error sequence is therefore greater than or equal to $2\Delta_1^2$. The constraint on the parity check coefficients allows however to determine only a lower bound for $d_{free}(\tilde{m})$. For a given sequence $\Delta_0 \leq \Delta_1 \leq \dots \leq \Delta_{\tilde{m}+1}$ of minimum distances between elements of subsets and a code with constraint length ν, a convolutional code that yields the maximum value of $d_{free}(\tilde{m})$ is usually found by a computer program for code search. The search of the $(\nu - 1)(\tilde{m} + 1)$ parity check binary coefficients is performed by means such that the explicit computation of $d_{free}(\tilde{m})$ is often avoided.

Tables 12.1 and 12.2 report the optimum codes for TCM with symbols of the type $\mathbf{Z}_1$ and $\mathbf{Z}_2$, respectively [2]. For 8-PSK, the optimum codes are given in Table 12.3 [2]. Parity check coefficients are specified in octal notation; for example, the binary vector $[h_6^{(0)}, \dots, h_0^{(0)}] = [1, 0, 0, 0, 1, 0, 1]$ is represented by $\boldsymbol{h}^{(0)} = 105_8$. In the tables, an asterisk next to the value d_{free} indicates that the free distance is determined by the parallel transitions, that is $d_{free}(\tilde{m}) > \Delta_{\tilde{m}+1}$.

[6] From the parity equation (12.29), we observe that a code sequence $\{z_k\}$ can have arbitrary values for each m-tuple $[z_k^{(m)}, \dots, z_k^{(1)}]$.

Table 12.1: Codes for one-dimensional modulation.

2^ν	$\tilde{m}$	h^1	h^0	d^2_{free}/Δ^2_0	$G_{4AM/2AM}$ ($m=1$)	$G_{8AM/4AM}$ ($m=2$)	G_{code} ($m \to \infty$)	N_{free} ($m \to \infty$)
4	1	2	5	9.0	2.55	3.31	3.52	4
8	1	04	13	10.0	3.01	3.77	3.97	4
16	1	04	23	11.0	3.42	4.18	4.39	8
32	1	10	45	13.0	4.15	4.91	5.11	12
64	1	024	103	14.0	4.47	5.23	5.44	36
128	1	126	235	16.0	5.05	5.81	6.02	66

Source: Reproduced with permission from Ungerboeck [1]. ©1982, IEEE.

Table 12.2: Codes for two-dimensional modulation.

2^ν	$\tilde{m}$	h^2	h^1	h^0	d^2_{free}/Δ^2_0	$G_{16QAM/8PSK}$ ($m=3$)	$G_{32QAM/16QAM}$ ($m=4$)	$G_{64QAM/32QAM}$ ($m=5$)	G_{code} ($m \to \infty$)	N_{free} ($m \to \infty$)
4	1	—	2	5	4.0*	4.36	3.01	2.80	3.01	4
8	2	04	02	11	5.0	5.33	3.98	3.77	3.98	16
16	2	16	04	23	6.0	6.12	4.77	4.56	4.77	56
32	2	10	06	41	6.0	6.12	4.77	4.56	4.77	16
64	2	064	016	101	7.0	6.79	5.44	5.23	5.44	56
128	2	042	014	203	8.0	7.37	6.02	5.81	6.02	344
256	2	304	056	401	8.0	7.37	6.02	5.81	6.02	44
512	2	0510	0346	1001	8.0*	7.37	6.02	5.81	6.02	4

Source: Reproduced with permission from Ungerboeck [2]. ©1987, IEEE.

Table 12.3: Codes for 8-PSK.

2^ν	$\tilde{m}$	h^2	h^1	h^0	d^2_{free}/Δ^2_0	$G_{8PSK/4PSK}$ ($m=2$)	N_{free}
4	1	—	2	5	4.0*	3.01	1
8	2	04	02	11	4.586	3.60	2
16	2	16	04	23	5.172	4.13	≃2.3
32	2	34	16	45	5.758	4.59	4
64	2	066	030	103	6.343	5.01	≃5.3
128	2	122	054	277	6.586	5.17	≃0.5

Source: Reproduced with permission from Ungerboeck [2]. ©1987, IEEE.

12.2 Multidimensional TCM

So far we have dealt with TCM schemes that use two-dimensional (2D) constellations, that is to send m information bits per modulation interval we employ a 2D constellation of $2^{(m+1)}$ points; the intrinsic cost is represented by doubling the cardinality of the 2D constellation with respect to uncoded schemes, as a bit of redundancy is generated at each modulation interval. Consequently, the minimum distance within points of the constellation is reduced, for the same average power of the transmitted signal; without this cost, the coding gain would be 3 dB higher.

An advantage of using a multidimensional constellation to generate code symbol sequences is that doubling the cardinality of a constellation does not lead to a 3 dB loss; in other words, the signals are spaced by a larger Euclidean distance d_{min} and therefore the margin against noise is increased.

A simple way to generate a multidimensional constellation is obtained by time division. If for example ℓ two-dimensional symbols are transmitted over a time interval of duration T_s, and each of them has a duration T_s/ℓ, we may regard the ℓ 2D symbols as an element of a 2ℓ-dimensional constellation. Therefore in practice, multidimensional signals can be transmitted as sequences of one or two-dimensional symbols.

In this section, we describe the construction of 2ℓ-dimensional TCM schemes for the transmission of m bits per 2D symbol, and thus $m\ell$ bits per 2ℓ-dimensional signal. We maintain the principle of using a redundant signal set, with a number of elements doubled with respect to that used for uncoded modulation; therefore, the 2ℓ-dimensional TCM schemes use sets of $2^{m\ell+1}$ 2ℓ-dimensional signals. With respect to two-dimensional TCM schemes, this implies a lower redundancy in the two-dimensional component sets. For example, in the 4D case doubling the number of elements causes an expansion of the 2D component constellations by a factor $\sqrt{2}$; this corresponds to a half bit of redundancy per 2D component constellation. The cost of the expansion of the signal set is reduced by 1.5 dB in the 4D case and by 0.75 dB in the 8D case. A further advantage of multidimensional constellations is that the design of schemes invariant to phase rotation is simplified (see Section 12.3).

Encoding

Starting with a constellation $\mathcal{A}_0$ of the type $\mathbf{Z}_1$ or $\mathbf{Z}_2$ in the one or two-dimensional signal space, we consider multidimensional trellis codes where the signals to be assigned to the transitions in the trellis diagram come from a constellation $\mathcal{A}_I^0$, $I > 2$, in the multidimensional space $\mathbb{R}^I$. In practice, if $(m+1)$ binary output symbols of the finite state encoder determine the assignment of modulation signals, it is possible to associate with these binary symbols a sequence of ℓ modulation signals, each in the constellation $\mathcal{A}_0$, transmitted during ℓ modulation intervals, each of duration T; this sequence can be considered as an element of the space $\mathbb{R}^I$, as signals transmitted in different modulation intervals are assumed orthogonal. If we have a constellation $\mathcal{A}_0$ with M elements, the relation $M^\ell \geq 2^{m+1}$ holds.

Possible sequences of ℓ symbols of a constellation $\mathcal{A}_0$ give origin to a block code B of length ℓ in the space $\mathbb{R}^I$. Hence, we can consider the multidimensional TCM as an encoding method in which the binary output symbols of the finite state encoder represent the information symbols of the block code. Part of the whole coding gain is obtained by choosing a multidimensional constellation $\mathcal{A}_I^0$ with a large minimum Euclidean distance, that is equivalent to the choice of an adequate block code; therefore, it is necessary to identify block codes in the Euclidean space $\mathbb{R}^I$ that admit a partitioning of code sequences in subsets such that the minimum Euclidean distance among elements of a subset is the largest possible. For a linear block code, it can be shown that this partitioning yields as subsets a subgroup of the code and its cosets. Consider the trellis diagram of a code for the multidimensional TCM where each state has $2^{\tilde{m}+1}$ transitions to adjacent states, and between pairs of adjacent states there exist $2^{m-\tilde{m}}$ parallel transitions. Then, the construction of the trellis code requires that the constellation $\mathcal{A}_I^0$ is partitioned into $2^{\tilde{m}+1}$ block codes $B^j_{\tilde{m}+1}$, $j \in \{0, \dots, 2^{\tilde{m}+1}-1\}$, each of length ℓ and *rate* equal to $(m-\tilde{m})/I$ bits/dimension. In any period equivalent to ℓ modulation intervals, $(\tilde{m}+1)$ binary output symbols of the finite state encoder select one of the code blocks, and the remaining $(m-\tilde{m})$ binary symbols determine the sequence to be transmitted among those belonging to the selected code. From (6.67), for a constellation $\mathcal{A}_0$ of type $\mathbf{Z}_2$, $B_{min}T_s = \ell$, and $M^\ell = 2^{\ell m+1}$, the spectral efficiency of a system with multidimensional TCM is equal to

$$\nu = \frac{m}{\ell} = \frac{\ell \log_2 M - 1}{\ell} = \log_2 M - \frac{1}{\ell} \text{ bit/s/Hz} \tag{12.37}$$

The optimum constellation in the space $\mathbb{R}^I$ is obtained by solving the problem of finding a lattice such that, given the minimum Euclidean distance d_{min} between two points, the number of points per unit of volume is maximum; for $I = 2$, the solution is given by the hexagonal lattice. For a number of constellation points $M \gg 1$, the ratio between the statistical power of signals of a constellation of the type $\mathbf{Z}_2$ and that of signals of a constellation chosen as a subset of the hexagonal lattice is equal to 0.62 dB; hence, a constellation subset of the hexagonal lattice yields a coding gain equal to 0.62 dB with respect to a constellation of the type $\mathbf{Z}_2$.

For $I = 4$ and $I = 8$, the solutions are given by the Schläfli lattice $\boldsymbol{D}_4$ and the Gosset lattice $\boldsymbol{E}_8$, respectively, defined in Examples 12.1.5 and 12.1.6 of page 668. Note that to an increase in the number of dimensions and in the density of the optimum lattice also corresponds an increase of the number of lattice points with minimum distance from a given point.

To design codes with a set of modulation signals whose elements are represented by symbol sequences, we have to address the problem of partitioning a lattice in a multidimensional space.

In the multidimensional TCM, if $\tilde{m} + 1$ binary output symbols of the finite state encoder determine the next state to a given state, the lattice Λ must be partitioned into $2^{\tilde{m}+1}$ subsets $\Lambda_{\tilde{m}+1}(j), j = 0, \ldots, 2^{\tilde{m}+1} - 1$, such that the minimum Euclidean distance $\Delta_{\tilde{m}+1}$ between the elements of each subset is maximum; hence, the problem consists in determining the subsets of Λ so that the density of the points of $\Lambda_{\tilde{m}+1}(j)$, $j = 0, \ldots, 2^{\tilde{m}+1} - 1$, is maximum. In the case of an I-dimensional lattice Λ that can be expressed as the Cartesian product of ℓ terms all equal to a lattice in the space $\mathbb{R}^{I/\ell}$, the partitioning of Λ can be derived from the partitioning of the I/ℓ-dimensional lattice.

Example 12.2.1 (Partitioning of the lattice $\mathbf{Z}_4$)
The lattice $\mathcal{A}_4^0 = \mathbf{Z}_4$ can be expressed in terms of the Cartesian product of the lattice $\mathbf{Z}_2$ with itself, $\mathbf{Z}_4 = \mathbf{Z}_2 \otimes \mathbf{Z}_2$; the subsets of the lattice $\mathcal{A}_4^0$ are therefore characterized by two sets of signals belonging to a $\mathbf{Z}_2$ lattice and by their subsets. The partitioning of a signal set belonging to a $\mathbf{Z}_2$ lattice is represented in Figure 12.6. Thus,

$$\begin{aligned} \mathbf{Z}_4 &= \mathcal{A}_4^0 = \mathcal{A}_0 \otimes \mathcal{A}_0 = (\mathcal{B}_0 \cup \mathcal{B}_1) \otimes (\mathcal{B}_0 \cup \mathcal{B}_1) \\ &= (\mathcal{B}_0 \otimes \mathcal{B}_0) \cup (\mathcal{B}_0 \otimes \mathcal{B}_1) \cup (\mathcal{B}_1 \otimes \mathcal{B}_0) \cup (\mathcal{B}_1 \otimes \mathcal{B}_1) \end{aligned} \tag{12.38}$$

At the first level of partitioning, the two optimum subsets are

$$\mathcal{B}_4^0 = (\mathcal{B}_0 \otimes \mathcal{B}_0) \cup (\mathcal{B}_1 \otimes \mathcal{B}_1) \tag{12.39}$$

$$\mathcal{B}_4^1 = (\mathcal{B}_0 \otimes \mathcal{B}_1) \cup (\mathcal{B}_1 \otimes \mathcal{B}_0) \tag{12.40}$$

In terms of the four-dimensional TCM, the assignment of pairs of two-dimensional modulation signals to the transitions in the trellis diagram is such that in the first modulation interval all the points of the component QAM constellation are admissible. If two pairs assigned to adjacent transitions are such that the signals in the first modulation interval are separated by the minimum Euclidean distance Δ_0, the signals in the second modulation interval will also have Euclidean distance at least equal to Δ_0; in this way the minimum Euclidean distance among points of $\mathcal{B}_4^0$ or $\mathcal{B}_4^1$ is equal to $\Delta_1 = \sqrt{2}\Delta_0$. The subset $\mathcal{B}_4^0$ is the Schläfli lattice $\boldsymbol{D}_4$, that is the densest lattice in the space $\mathbb{R}^4$; the subset $\mathcal{B}_4^1$ is instead the coset of $\boldsymbol{D}_4$ with respect to $\mathbf{Z}_4$. The next partitioning in the four subsets

$$\mathcal{B}_0 \otimes \mathcal{B}_0, \qquad \mathcal{B}_1 \otimes \mathcal{B}_1, \qquad \mathcal{B}_0 \otimes \mathcal{B}_1, \qquad \mathcal{B}_0 \otimes \mathcal{B}_1 \tag{12.41}$$

does not yield any increase in the minimum Euclidean distance. Note that the four subsets at the second level of partitioning differ from the $\mathbf{Z}_4$ lattice only by the position with respect to the origin, direction, and scale; therefore, the subsets at successive levels of partitioning are obtained by iterating the same procedure described for the first two. Thus, we have the following partitioning chain

$$\boldsymbol{Z}_4 \ / \ \boldsymbol{D}_4 \ / \ (\boldsymbol{Z}_2\boldsymbol{R})^2 \ / \ 4\boldsymbol{D}_4 \ / \ \dots \tag{12.42}$$

where $\boldsymbol{R}$ is the 2×2 matrix given by (12.16). The partitioning of the $\boldsymbol{Z}_4$ lattice is illustrated in Figure 12.18 [2].

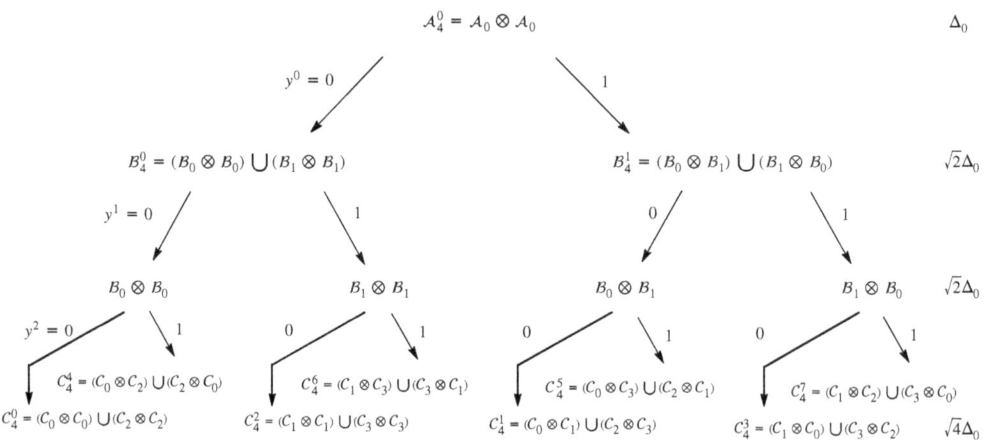

Figure 12.18 Partitioning of the lattice $\mathcal{A}_4^0 = \boldsymbol{Z}_4$. Source: Reproduced with permission from Ungerboeck [2]. ©1987, IEEE.

Optimum codes for the multidimensional TCM are found in a similar manner to that described for the one and two-dimensional TCM codes. Codes and relative asymptotic coding gains for the four-dimensional TCM, with respect to uncoded modulation with signals of the type $\boldsymbol{Z}_2$, are reported in Table 12.4 [2]. These gains are obtained for signal sets with a large number of elements that, in the signal space, take up the same volume of elements of the signal set used for uncoded modulation; then, the comparison is made for the same statistical power and the same peak power of the two-dimensional signals utilized for uncoded modulation.

Decoding

The decoding of signals generated by multidimensional TCM is achieved by a sequence of operations that is the inverse with respect to the encoding procedure described in the previous section. The first stage of decoding consists in determining, for each modulation interval, the Euclidean distance between the received sample and all M signals of the constellation $\mathcal{A}_0$ of symbols and also, within each subset $\mathcal{A}_q(i)$

Table 12.4: Codes for four-dimensional modulation.

2^ν	$\tilde{m}$	$\boldsymbol{h}^4$	$\boldsymbol{h}^3$	$\boldsymbol{h}^2$	$\boldsymbol{h}^1$	$\boldsymbol{h}^0$	d^2_{free}/Δ^2_0	G_{code} $(m \to \infty)$	N_{free} $(m \to \infty)$
8	2	—	—	04	02	11	4.0	4.52	88
16	2	—	—	14	02	21	4.0	4.52	24
32	3	—	30	14	02	41	4.0	4.52	8
64	4	050	030	014	002	101	5.0	5.48	144
128	4	120	050	022	006	203	6.0	6.28	

Source: Reproduced with permission from Ungerboeck [2]. ©1987, IEEE.

of the constellation $\mathcal{A}_0$, the signal $\hat{a}_k(i)$ that has the minimum Euclidean distance from the received sample. The second stage consists in the decoding, by a maximum-likelihood decoder, of $2^{\tilde{m}+1}$ block codes $B_{\tilde{m}+1}(\ell')$. Due to the large number M^{ℓ} of signals in the multidimensional space $\mathbb{R}^I$, in general the block codes have a number of elements such that the complexity of a maximum-likelihood decoder that should compute the metric for each element would result excessive. The task of the decoder can be greatly simplified thanks to the method followed for the construction of block codes $B_{\tilde{m}+1}(\ell')$. Block codes are identified by the subsets of the multidimensional constellation $\mathcal{A}_I^0$, which are expressed in terms of the Cartesian product of subsets $\mathcal{A}_q(i)$ of the one- or two-dimensional constellation $\mathcal{A}_0$. Decoding of the block codes is jointly carried out by a trellis diagram defined on a finite number ℓ of modulation intervals, where Cartesian products of subsets of the constellation $\mathcal{A}_0$ in different modulation intervals are represented as sequences of branches, and the union of subsets as the union of branches. Figure 12.19 illustrates the trellis diagram for the decoding of block codes obtained by the partitioning into two, four, and eight subsets of a constellation $\mathcal{A}_4^0 \subset \mathbf{Z}_4$.

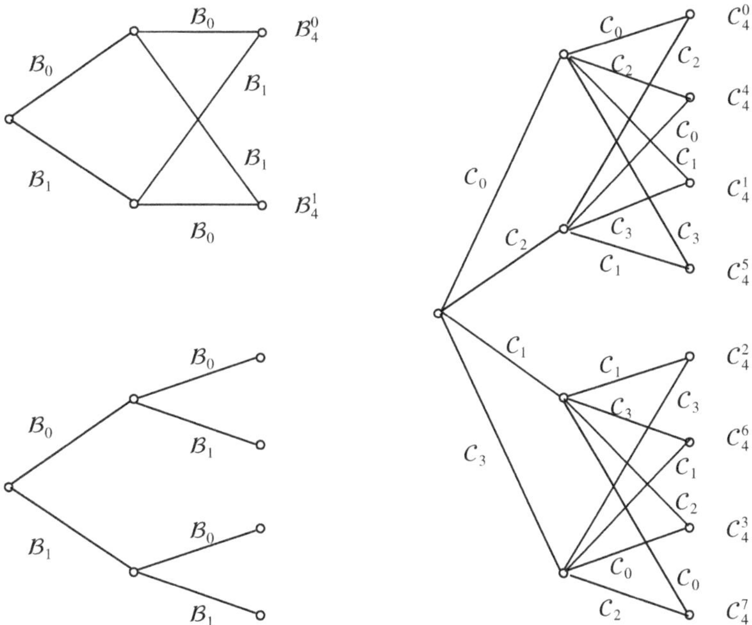

Figure 12.19 Trellis diagram for the decoding of block codes obtained by the partitioning the lattice $\mathcal{A}_4^0 = \mathbf{Z}_4$ into two, four, and eight subsets.

As the decisions taken in different modulation intervals are independent, the branch metrics of the trellis diagram is the square Euclidean distance $d^2(z_k, \hat{a}_k(i))$ between the received sample during the k-th modulation interval and the signal $\hat{a}_k(i)$ of the subset $\mathcal{A}_q(i)$ at the minimum Euclidean distance from z_k. The maximum-likelihood decoder for block codes can then be implemented by the VA, which is applied for ℓ iterations from the initial node of the trellis diagram to the $2^{\tilde{m}+1}$ terminal nodes; the procedure to decode block codes by a trellis diagram is due to Wolf [12]. The third and last decoding stage consists in using the metrics, obtained at the $2^{\tilde{m}+1}$ terminal nodes of the trellis diagram for the decoding of the block codes, in the Viterbi decoder for the trellis code, that yields the sequence of detected binary vectors $(\hat{b}_k^{(m)}, \dots, \hat{b}_k^{(1)})$.

To evaluate the complexity of each iteration in multidimensional TCM decoding, it is necessary to consider the overall number of branches in the trellis diagrams considered at the different decoding stages; this number includes the ℓM branches for decisions on signals within the subsets $\mathcal{A}_q(i)$ in ℓ modulation intervals, the N_R branches in the trellis diagram for the decoding of the block codes, and the $2^{\nu+\tilde{m}}$ branches in the trellis diagram for the decoding of the trellis code. For a code with efficiency equal to $(\log_2 M - 1/\ell)$ bits per modulation interval, the complexity expressed as the number of branches in the trellis diagram per transmitted information bit is thus given by

$$\frac{\ell M + N_R + 2^{\nu+\tilde{m}}}{\ell \log_2 M - 1} \tag{12.43}$$

The number N_R can be computed on the basis of the particular choice of the multidimensional constellation partitioning chain; for example, in the case of partitioning of the lattice $\mathbf{Z}_4$ into four or eight subsets, for the four-dimensional TCM we obtain $N_R = 20$. Whereas in the case of two-dimensional TCM, the decoding complexity essentially lies in the implementation of the VA to decode the trellis code, in the four-dimensional TCM most of the complexity is due to the decoding of the block codes.

In multidimensional TCM, it is common to find codes with a large number N_{free} of error events at the minimum Euclidean distance; this characteristic is due to the fact that in dense multidimensional lattices, the number of points at the minimum Euclidean distance rapidly increases with the increase in the number of dimensions. In this case, the minimum Euclidean distance is not sufficient to completely characterize code performance, as the difference between asymptotic coding gain and effective coding gain, for values of interest of the error probability, cannot be neglected.

12.3 Rotationally invariant TCM schemes

In PSK or QAM transmission schemes with coherent demodulation, an ambiguity occurs at the receiver whenever we ignore the value of the carrier phase used at the transmitter. In binary phase-shift-keying (BPSK), for example, there are two equilibrium values of the carrier phase synchronization system at the receiver, corresponding to a rotation of the phase equal to 0° and 180°, respectively; if the rotation of the phase is equal to 180°, the binary symbols at the demodulator output result inverted. In QAM, instead, because of the symmetry of the $\mathbf{Z}_2$ lattice points, there are four possible equilibrium values. The solution to the problem is represented by the insertion in the transmitted sequence of symbols for the synchronization, or by differential encoding (see Section 16.1.2). We recall that with differential encoding the information symbols are assigned to the phase difference between consecutive elements in the symbol sequence; in this way, the absolute value of phase in the signal space becomes irrelevant to the receiver.

With TCM, we obtain code signal sequences that, in general, do not present symmetries with respect to the phase rotation in the signal space. This means that, for a code symbol sequence, after a phase rotation determined by the phase synchronization system, we may obtain a sequence that does not belong to the code; therefore, trellis coding not invariant to phase rotation can be seen as a method for the construction of signal sequences that allow the recovery of the absolute value of the carrier phase. Then, for the demodulation, we choose the value of the carrier phase corresponding to an equilibrium value of the synchronization system, for which the Euclidean distance between the received sequence and the code sequence is minimum.

For fast carrier phase synchronization, various rotationally invariant trellis codes have been developed; these codes are characterized by the property that code signal sequences continue to belong to the code even after the largest possible number of phase rotations. In this case, with differential encoding of the information symbols, the independence of demodulation and decoding from the carrier phase recovered by the synchronization system is guaranteed. A differential decoder is then applied to the sequence of binary symbols $\{\tilde{b}_\ell\}$ at the output of the decoder for the trellis code to obtain the desired detection of the sequence of information symbols $\{\hat{b}_\ell\}$. Rotationally invariant trellis codes were initially proposed by Wei [13, 14]. In general, the invariance to phase rotation is more easily obtained with multidimensional TCM; in the case of two-dimensional TCM for PSK or QAM systems, it is necessary to use non-linear codes in GF(2).

In the case of TCM for PSK systems, the invariance to phase rotation can be directly obtained using PSK signals with differential encoding. The elements of the symbol sets are not assigned to the binary vectors of the convolutional code but rather to the phase differences relative to the previous symbols.

Bibliography

[1] Ungerboeck, G. (1982). Channel coding with multilevel/phase signals. *IEEE Transactions on Information Theory* 28: 55–67.

[2] Ungerboeck, G. (1987). Trellis coded modulation with redundant signal sets. Part I and Part II. *IEEE Communications Magazine* 25: 6–21.

[3] Pietrobon, S.S., Deng, R.H., Lafanechere, A. et al. (1990). Trellis-coded multidimensional phase modulation. *IEEE Transactions on Information Theory* 36: 63–89.

[4] Biglieri, E., Divsalar, D., McLane, P.J., and Simon, M.K. (1991). *Introduction to Trellis-Coded Modulation with Applications*. New York, NY: Macmillan Publishing Company.

[5] Pietrobon, S.S. and Costello, D.J. Jr. (1993). Trellis coding with multidimensional QAM signal sets. *IEEE Transactions on Information Theory* 39: 325–336.

[6] Huber, J. (1992). *Trelliscodierung*. Heidelberg: Springer-Verlag.

[7] Schlegel, C. (1997). *Trellis Coding*. New York, NY: IEEE Press.

[8] Forney, G.D. Jr. and Ungerboeck, G. (1998). Modulation and coding for linear Gaussian channels. *IEEE Transactions on Information Theory* 44: 2384–2415.

[9] IEEE Standard 802.3ab (2015). *Physical layer parameters and specifications for 1000 Mb/s operation over 4-pair of category 5 balanced copper cabling, type 1000BASE-T*. IEEE.

[10] Yueksel, H., Braendli, M., Burg, A. et al. (2018). Design techniques for high-speed multi-level Viterbi detectors and trellis-coded-modulation decoders. *IEEE Transactions on Circuits and Systems I: Regular Papers* 65: 3529–3542.

[11] Forney, G.D. Jr. (1970). Convolutional codes I: algebraic structure. *IEEE Transactions on Information Theory* IT–16: 720–738.

[12] Wolf, J.K. (1978). Efficient maximum likelihood decoding of linear block codes using a trellis. *IEEE Transactions on Information Theory* IT–24: 76–80.

[13] Wei, L.F. (1987). Trellis-coded modulation with multidimensional constellations. *IEEE Transactions on Information Theory* 33: 483–501.

[14] Wei, L.F. (1989). Rotationally invariant trellis-coded modulations with multidimensional *M*-PSK. *IEEE Journal on Selected Areas in Communications* 7: 1281–1295.

Chapter 13

Techniques to achieve capacity

In this chapter, we introduce practical solutions that provide bit rates close to capacity. First, we consider the *parallel channel*, comprising a set of independent narrowband (NB) additive white Gaussian noise (AWGN) subchannels: such a channel is found when considering multicarrier (MC) single-input-single-output (SISO) transmissions over a dispersive AWGN channel, or when applying a suitable precoder and combiner to a multiple-input-multiple-output (MIMO) NB AWGN channel. Then, we provide solutions tailored to single-carrier SISO transmissions, combined with joint precoding and coding techniques.

13.1 Capacity achieving solutions for multicarrier systems

In the following, we discuss techniques to achieve capacity, with reference to the orthogonal frequency division multiplexing (OFDM) system modelled in (8.83), comprising $\mathcal{M}$ subchannels, with AWGN with power σ_W^2 and gain $\mathcal{G}_{C,i}$ on subchannel i, $i = 0, 1, \dots, \mathcal{M} - 1$. The considered techniques can also be applied in other scenarios, e.g. MIMO transmissions over NB AWGN channels, providing $N_{S,max}$ equivalent subchannels with gains $\{\xi_n\}$, $n = 1, 2, \dots, N_{S,max}$ (see (9.141)).

13.1.1 Achievable bit rate of OFDM

For an OFDM system (see Section 8.7.1), the signal-to-noise-ratio (SNR), normalized to unit transmit power, over a given subchannel i is (see (8.83)))

$$\Theta[i] = \frac{|\mathcal{G}_{C,i}|^2}{\sigma_W^2}, \quad i = 0, 1, \dots, \mathcal{M} - 1 \tag{13.1}$$

while the SNR is ($\mathtt{M}[i] = E[|a_k[i]|^2]$)

$$\Gamma[i] = \mathtt{M}[i]\Theta[i], \quad i = 0, 1, \dots, \mathcal{M} - 1 \tag{13.2}$$

For a quadrature amplitude modulation (QAM) transmission, from (6.90), the capacity of the SISO NB AWGN subchannel i is (see also (6.103))

$$C_{[bit/s]}[i] = \frac{1}{T'}\log_2(1 + \mathtt{M}[i]\Theta[i]) \tag{13.3}$$

Algorithms for Communications Systems and their Applications, Second Edition.
Nevio Benvenuto, Giovanni Cherubini, and Stefano Tomasin.

where powers $\{\mathtt{M}[i]\}$ are chosen in order to *maximize* the overall capacity

$$C_{[bit/s]} = \sum_{i=0}^{\mathcal{M}-1} C_{[bit/s]}[i] \tag{13.4}$$

under the total power constraint

$$\sum_{i=0}^{\mathcal{M}-1} \mathtt{M}[i] \leq \mathtt{M}_{tot} \tag{13.5}$$

By applying the Lagrange multiplier method, the optimal power allocation is given by

$$\mathtt{M}[i] = \left(L - \frac{1}{\Theta[i]}\right)^{+} \tag{13.6}$$

where L is chosen in order to satisfy (13.5) with equality.

13.1.2 Waterfilling solution

From (13.6), the optimum solution is to allocate a higher power to subchannels with a higher gain, where we assume that subchannels have decreasing normalized SNRs, i.e.

$$\Theta[i+1] < \Theta[i], \qquad i = 1, \dots, \mathcal{M}-2 \tag{13.7}$$

With reference to the illustrative example of Figure 13.1, we can see that $\frac{1}{\Theta[\mathcal{M}-1]} > L$ and $\frac{1}{\Theta[\mathcal{M}-2]} > L$. Hence, in this case $\mathtt{M}_{opt}[i] = 0$, for $i = \mathcal{M}-1, \mathcal{M}-2$, i.e. the subchannels $\mathcal{M}-1$ and $\mathcal{M}-2$ are not active for transmission. Instead, for the other subchannels, the gap between the constant L and the factor $\frac{1}{\Theta[i]}$ is the power assigned for the subchannel. This illustrates the concept of water filling, where for a lake with depth $1/\Theta[i]$ at location i, $i = 0, \dots, \mathcal{M}-1$, the water (power) has the same level L in the whole lake.

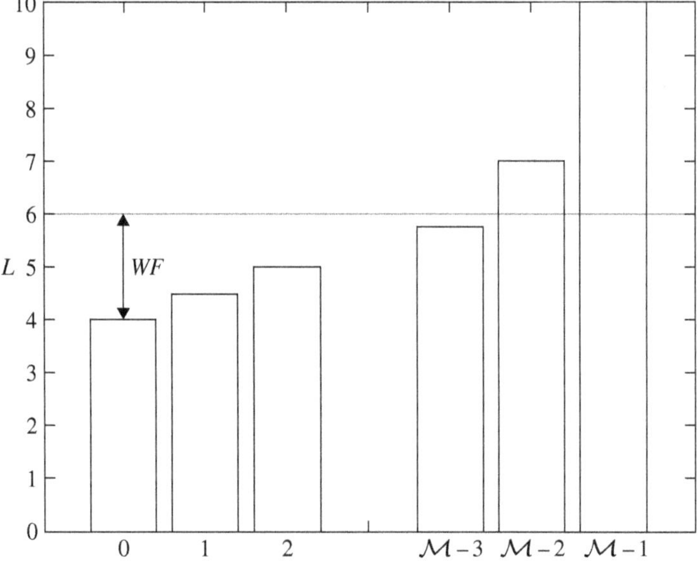

Figure 13.1 Increasing order of factors $\frac{1}{\Theta[i]}$.

Iterative solution

We now show an iterative solution for the computation of the water level L, and consequently the optimal power allocation. First, we assume that (13.7) holds.

Let N denote the number of *active* subchannels, with $\frac{1}{\Theta[i]}$ below level L. In general, for a given N, from

$$\mathtt{M}[i] = L - \frac{1}{\Theta[i]} \qquad i = 0, \dots, N-1 \tag{13.8}$$

and the power constraint

$$\mathtt{M}_{tot} = \sum_{i=0}^{N-1} \mathtt{M}[i] = N\,L - \sum_{i=0}^{N-1} \frac{1}{\Theta[i]} \tag{13.9}$$

we obtain the following iterative algorithm.

0. Initialize N to the number of available subchannels $\mathcal{M}$.
1. From (13.9) calculate the level

$$L = \frac{1}{N}\left[\mathtt{M}_{tot} + \sum_{i=0}^{N-1} \frac{1}{\Theta[i]}\right] \tag{13.10}$$

2. From (13.8) calculate the transmit power on subchannel i

$$\mathtt{M}[i] = L - \frac{1}{\Theta[i]} \qquad i = 0, \dots, N-1 \tag{13.11}$$

4. If $\mathtt{M}[i] > 0$, for $i = 0, \dots, N-1$, (i.e. check if $\mathtt{M}[N-1] > 0$) we are done. Otherwise decrease N by one and go to step 1.

13.1.3 Achievable rate under practical constraints

In practice, due to constraints imposed by both signal constellations and available codes, the achievable bit rate on subchannel i is smaller than (13.3), and is determined starting from the desired bit error probability. As capacity and achievable bit rate are proportional to $1/T'$, we will drop this factor in the following, referring instead to the achievable rate, or *throughput*, in bit/MCsymbol, as given by

$$\mathtt{b} = \sum_{i=0}^{N-1} \mathtt{b}[i] \text{ [bit/MCsymbol]} \tag{13.12}$$

where N is the number of active subchannels, and $\mathtt{b}[i]$ is the achievable rate for subchannel i.

The following derivation takes into account the fact that practical codes do not achieve the capacity. In particular, we refer to M-QAM for subchannel i, and to a bit error rate for which the symbol error probability in the uncoded case is given by (6.82), neglecting the factor term $(1 - 1/M)$ for simplicity,

$$P_e[i] = 4Q\left(\sqrt{\frac{3}{M-1}\Gamma[i]}\right) = 4Q(\sqrt{3\bar{\Gamma}[i]}) \tag{13.13}$$

where (see (6.131))

$$\bar{\Gamma}[i] = \frac{\Gamma[i]}{2^{\mathtt{b}[i]} - 1} \tag{13.14}$$

with $\mathtt{b}[i] \leq \log_2 M$ (equality holds in the uncoded case). For a given $P_e[i]$, we have a corresponding value of

$$\bar{\Gamma}_{gap}[i] = \bar{\Gamma}[i] \tag{13.15}$$

which provides $\mathtt{b}[i] = \mathtt{b}$. Therefore, by inverting (13.14) we have

$$\mathtt{b}[i] = \log_2\left(1 + \frac{\Gamma[i]}{\overline{\Gamma}_{gap}[i]}\right) = \log_2\left(1 + \Gamma[i] \cdot 10^{-\frac{\overline{\Gamma}_{gap,dB}[i]}{10}}\right) \tag{13.16}$$

For example, for $P_e[i] = 2 \times 10^{-7}$, in the uncoded case the SNR gap from capacity is $\overline{\Gamma}_{gap,dB}[i] = 9.8$ dB. Introducing a system margin $\Gamma_{mar,dB}$ for unforeseen channel impairments ($\Gamma_{mar,dB} = 6$ dB for subscriber loop applications) and a code with gain $G_{code,dB}[i]$, we define

$$\Gamma_{gap,dB}[i] = \overline{\Gamma}_{gap,dB}[i] + \Gamma_{mar,dB}[i] - G_{code,dB}[i] \tag{13.17}$$

and $\mathtt{b}[i]$ is computed from (13.16) replacing $\overline{\Gamma}_{gap,dB}[i]$ with $\Gamma_{gap,dB}[i]$. When $\Gamma_{gap} = 1$ (i.e. $\Gamma_{gap,dB} = 0$), (13.16) is equal to the Shannon capacity (13.3) (apart from the scaling factor $1/T'$). Any reliable and implementable system must transmit at a rate below capacity, and hence, Γ_{gap} is a measure of loss with respect to theoretically optimum performance.

Effective SNR and system margin in MC systems

Analogously to (6.111) for a linear dispersive channel, we define an effective SNR for OFDM. Assuming the same $\Gamma_{gap,dB}$ across subchannels, since from (13.12) and (13.16), it is

$$\mathtt{b} = \sum_{i=0}^{N-1} \log_2\left(1 + \frac{\Gamma[i]}{\Gamma_{gap}}\right) = \log_2\left[\prod_{i=0}^{N-1}\left(1 + \frac{\Gamma[i]}{\Gamma_{gap}}\right)\right] \tag{13.18}$$

we define the effective SNR as

$$\frac{\Gamma_{eff}}{\Gamma_{gap}} = \sqrt[N]{\prod_{i=0}^{N-1}\left(1 + \frac{\Gamma[i]}{\Gamma_{gap}}\right)} - 1 \tag{13.19}$$

Then, from (13.18) we have

$$\mathtt{b} = N\log_2\left(1 + \frac{\Gamma_{eff}}{\Gamma_{gap}}\right) \quad \text{[bit/MCsymbol]} \tag{13.20}$$

If +1 and -1 are negligible in (13.19), we can approximate Γ_{eff} as the geometric mean of the subchannel SNRs, i.e.

$$\Gamma_{eff} = \sqrt[N]{\prod_{i=0}^{N-1} \Gamma[i]} \tag{13.21}$$

Moreover, from (13.17) and (13.20), we have the relation as

$$\Gamma_{mar,dB} = 10\log_{10}\frac{\Gamma_{eff}}{2^{\frac{\mathtt{b}}{N}} - 1} + G_{code,dB} - \overline{\Gamma}_{gap,dB} \tag{13.22}$$

Uniform power allocation and minimum rate per subchannel

In order to maximize the achievable rate (13.18), we can apply the waterfilling algorithm using $\Theta[i]/\Gamma_{gap}$ as normalized SNR. Here we consider a simpler solution used in some applications, to simplify computation. We will enforce a uniform power allocation among active subchannels, and we will assume that a subchannel must transmit at least ρ bits of information when active: more on this assumption can be found in Section 13.1.4 on *transmission modes*.

First, once the N active subchannels have been selected, for a uniform power allocation strategy we set

$$\mathtt{M}[i] = \begin{cases} \frac{\mathtt{M}_{tot}}{N} & i = 0, \dots, N-1 \\ 0 & i = N, \dots, \mathcal{M}-1 \end{cases} \tag{13.23}$$

It is seen that the optimum (waterfilling) and uniform power distributions yield a similar achievable rate. In other words, it is more important to select the active subchannels over which transmission takes place rather than allocate optimum powers.

A simple algorithm to determine the active subchannels is the following:

0. Set N to the number of subchannels $\mathcal{M}$.
1. Evaluate the power per subchannel

$$\mathtt{M}[i] = \frac{\mathtt{M}_{tot}}{N}, \qquad i = 0, \dots, N-1 \tag{13.24}$$

   We recall that $\Gamma[i] = \mathtt{M}[i]\Theta[i]$

2. For $i = 0, \dots, N-1$, evaluate

$$\mathtt{b}[i] = \log_2\left(1 + \frac{\Gamma[i]}{\Gamma_{gap}}\right) \tag{13.25}$$

   If $\mathtt{b}[i] < \rho$ drop subchannel i, as it cannot transmit the minimum number of information bits ρ.

3. Evaluate the number of active subchannels N'.
4. If $N = N'$, we are done and N is the number of active subchannels. Otherwise we reduce the number of active subchannels, i.e. set $N = N'$, and repeat from step 1.

13.1.4 The bit and power loading problem revisited

Unfortunately, the above derivations yield $\mathtt{b}[i]$, the number of information bits per subchannel, as a real number, while in practice $\mathtt{b}[i]$ may assume only a finite number of values depending upon the used code and the constellation. Two problems can be approached:

- Maximize the achievable rate with a given total transmit power.
- Minimize the total transmit power for a given achievable rate.

Transmission modes

A *transmission mode* consists of a given QAM constellation and a coding scheme, and in general, only a finite set of K *transmission modes* is available for subchannel i, and we write $\mathtt{m}[i] \in \{1, \dots, K\}$. To our concern, mode k is characterized by the couple

$$(\mathtt{b}_k, \Gamma_k), \qquad k = 1, \dots, K \tag{13.26}$$

where $\mathtt{b}_k$ is number of information bits/QAM-symbol carried by mode k and Γ_k is the minimum SNR Γ required to transmit with mode k while ensuring a given P_{bit}.

Without loss of generality, we assume

$\mathtt{b}_k < \mathtt{b}_{k+1}$ (number of information bits/QAM symbol ordered in increasing order)

$\mathtt{b}_0 = 0$ (subchannel can be turned off)

A transmission mode $\mathtt{m}[i]$ and a transmit power

$$\mathtt{M}[i] = \frac{\Gamma_{\mathtt{m}[i]}}{\Theta[i]} \tag{13.27}$$

have been assigned to each subchannel, $i = 0, \dots, \mathcal{M}-1$.

A possible allocation of modes $\boldsymbol{m} = [\mathtt{m}[0], \dots, \mathtt{m}[\mathcal{M}-1]]$, with corresponding bits $\boldsymbol{b} = [\mathtt{b}_{\mathtt{m}[0]}, \dots, \mathtt{b}_{\mathtt{m}[\mathcal{M}-1]}]$ and powers $\boldsymbol{M} = [\mathtt{M}[0], \dots, \mathtt{M}[\mathcal{M}-1]]$ given by (13.27) is called *bit and power loading*.

Example 13.1.1

Table 13.1 reports an example of transmission modes. SNRs Γ_k, $k = 1, \dots, 7$, correspond to a bit error probability $P_{bit} = 10^{-3}$ and have been obtained through system simulation (soft decoding has been used).

Table 13.1: Example of transmission modes.

	Code bits per QAM symbol	Code rate	Info. bits per QAM symbol	Γ_k (dB)
Mode 0	0	0	0	—
Mode 1	1 (BPSK)	1/2	0.5	−0.39
Mode 2	1 (BPSK)	3/4	0.75	2.55
Mode 3	2 (QPSK)	1/2	1	2.68
Mode 4	2 (QPSK)	3/4	1.5	5.50
Mode 5	4 (16-QAM)	9/16	2.25	8.87
Mode 6	4 (16-QAM)	3/4	3	11.65
Mode 7	6 (64-QAM)	3/4	4.5	17

Problem formulation

- ***Throughput maximization problem (TMP)***

  In this problem, we aim at distributing the total available power among the $\mathcal{M}$ subchannels, such that the achievable rate is maximized. The TMP is stated as follows.

  $$\arg \max_{\mathtt{m}[0],\mathtt{m}[1],\dots,\mathtt{m}[\mathcal{M}-1]} \mathtt{b} = \sum_{i=0}^{\mathcal{M}-1} \mathtt{b}_{\mathtt{m}[i]} \tag{13.28}$$

  subject to (13.5)

  $$\sum_{i=0}^{\mathcal{M}-1} \mathtt{M}[i] \le \mathtt{M}_{tot} \tag{13.29}$$

  Note that we consider again all $\mathcal{M}$ subchannels here, as we aim at determining which will be active (whose total number will be N).

- ***Margin maximization problem (MMP)***

  The goal is to determine the bit allocation that requires the least amount of power. The MMP is stated as follows.

  $$\arg \min_{\mathtt{m}[0],\mathtt{m}[1],\dots,\mathtt{m}[\mathcal{M}-1]} \mathtt{M}_{tot} = \sum_{i=0}^{\mathcal{M}-1} \mathtt{M}[i] \tag{13.30}$$

  subject to

  $$\sum_{i=0}^{\mathcal{M}-1} \mathtt{b}_{\mathtt{m}[i]} = \mathtt{b}_{tot} \tag{13.31}$$

  where $\mathtt{b}_{tot}$ is the target achievable rate.

Some simplifying assumptions

Most algorithms developed to solve above problems work under one or more of the following hypotheses: *Convex hypothesis*

$$\frac{\mathtt{b}_{k+1} - \mathtt{b}_k}{\Gamma_{k+1} - \Gamma_k} < \frac{\mathtt{b}_k - \mathtt{b}_{k-1}}{\Gamma_k - \Gamma_{k-1}} \tag{13.32}$$

i.e. going from $\mathtt{b}_k$ to $\mathtt{b}_{k+1}$ more power is required than going from $\mathtt{b}_{k-1}$ to $\mathtt{b}_k$.

Constant granularity hypothesis

$$\mathtt{b}_k = k\rho \tag{13.33}$$

where ρ is a constant called *granularity*. Usually $\rho = 1$; however, ρ may not be an integer, as some coding schemes can yield a fractional number of information bits/symbol.

The constant gap approximation

The constant gap approximation assumes that $\overline{\Gamma}_{gap}[i]$ of (13.15) is only function of P_e, i.e. it is a constant with respect to $\mathtt{b}[i]$. Indeed many coding schemes verify the gap approximation quite well, but only for very small values of P_e. Thus, under the constant gap approximation, all transmission modes have the same gap $\overline{\Gamma}_{gap}$.

On loading algorithms

We can represent all possible loadings on the $(\mathtt{M}, \mathtt{b})$ plane where $\mathtt{M} = \mathtt{M}_{tot} = \sum_{i=0}^{\mathcal{M}-1} \mathtt{M}[i]$ and $\mathtt{b} = \mathtt{b}_{tot} = \sum_{i=0}^{\mathcal{M}-1} \mathtt{b}[i]$.

From Figure 13.2, we can see that loadings are arranged in rows. For a given row, the same achievable rate can be obtained with many values of the total power. The left-most loadings of each rows are called *efficient loadings*, because they demand the minimum power. In the convex hypothesis, efficient loadings draw a convex curve

The optimum loading is obviously an efficient loading. An efficient loading has the following properties:

1. No other loading exists with the same achievable rate and a lower total power (or a higher achievable rate with the same power).
2. Under convexity and constant granularity hypotheses, if we take ρ bits away from any subchannel and add them to any other subchannel, the new loading (with the same achievable rate) has increased total power.

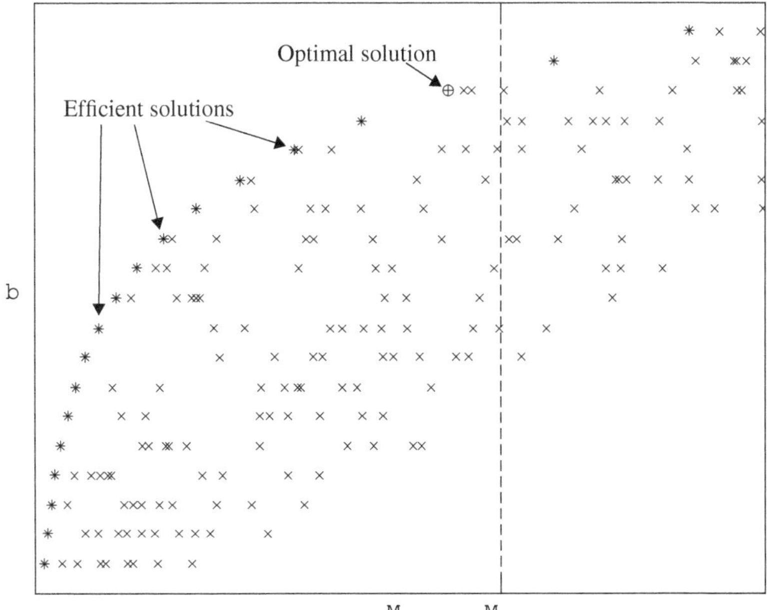

Figure 13.2 Representation of all possible loadings.

Some bit and power loading algorithms are *greedy*, i.e. they allocate one bit at a time to the subchannel that will do the most good for the current partial allocation. In other words, the algorithm maximizes the achievable rate increase at each step, regardless to the global effects of this choice.

The Hughes-Hartogs algorithm

For MMP, the Hughes-Hartogs (HH) [1] algorithm successively assigns the bits to the subchannels until the target achievable rate $\mathtt{b}_{tot}$ is reached. Thereby, at each steep, the selected subchannel is such that the transmission of an additional bit can be done with the smallest additional transmit power. The HH algorithm is *optimal* when the *hypotheses of convexity* and *constant granularity* hold, and it has a complexity $O(\mathcal{M}\mathtt{b}_{tot}/\rho)$. As shown in Figure 13.3, it visits all efficient loadings in increasing order until the optimum one is found.

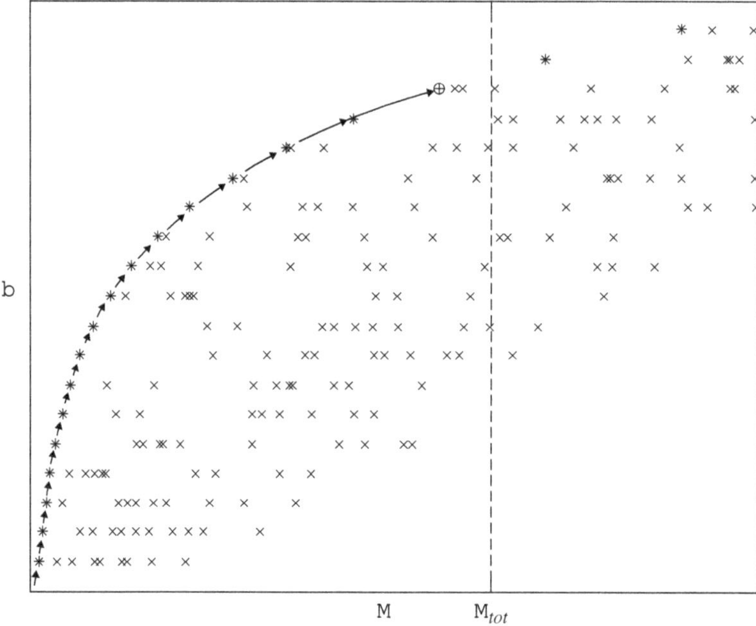

Figure 13.3 The Hughes-Hartogs algorithm.

The Krongold–Ramchandran–Jones algorithm

The Krongold–Ramchandran–Jones (KRJ) algorithm [2] makes use of the convexity hypothesis and exploits a function which, given a parameter $\lambda > 0$, returns an efficient loading $\boldsymbol{L}_\lambda = (\mathtt{M}_\lambda, b_\lambda)$. It has the following property: the total power $\mathtt{M}_\lambda$ and achievable rate $\mathtt{b}_\lambda$ of loading $\boldsymbol{L}_\lambda$ are non-increasing function of λ. Let a plane wave of slope λ propagate down-rightwards on the $(\mathtt{M}, \mathtt{b})$ plane. The first loading point $\boldsymbol{L}_\lambda$ hit by the plane wave is efficient (see Figure 13.4). Moreover, hit points have equal or smaller values of $\mathtt{M}$ and $\mathtt{b}$ for increasing values of λ.

$\boldsymbol{L}_\lambda$ is built by applying the principle of the plane wave to each subchannel. For each subchannel i, let us represent the set of modes on the $(\mathtt{M}[i], \mathtt{b}[i])$ plane; each mode corresponds to the point $(\mathtt{M}[i], \mathtt{b}_{\mathtt{m}[i]})$, with

$$\mathtt{M}[i] = \frac{\Gamma_{\mathtt{m}[i]}}{\Theta[i]} \tag{13.34}$$

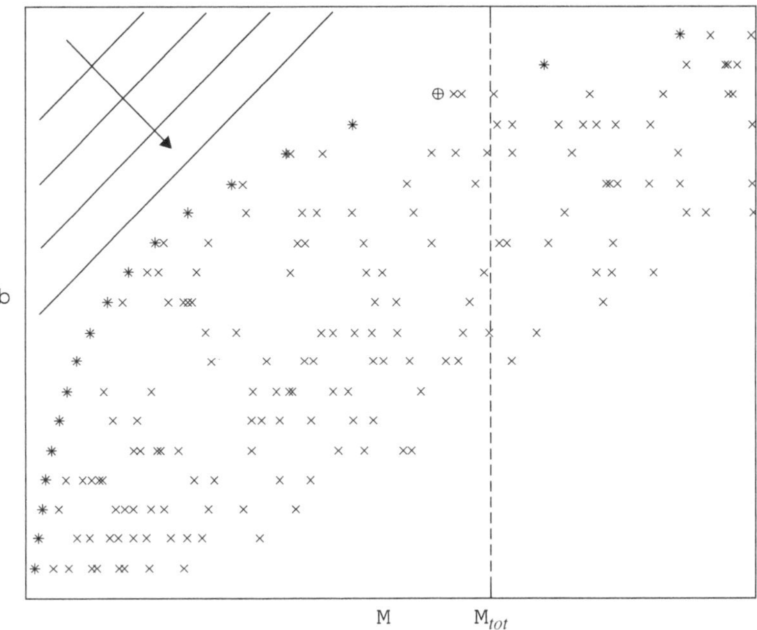

Figure 13.4 Description of the plane wave used by the KRJ algorithm.

Each representation includes K points. Indeed from (13.34) the set of points for subchannel i is obtained by the horizontal scaling of the points $(\mathtt{b}_k, \Gamma_k)$, $k = 1, \dots, K$, by a factor $\Theta[i]$.

From Figure 13.4, if we connect all adjacent efficient modes with segments, we obtain a collection of slopes, where slope λ is associated to the segment edge point with lower power $\mathtt{m}_\lambda[i] = (\mathtt{M}_\lambda[i], \mathtt{b}_\lambda[i])$, for an overall loading $\boldsymbol{L}_\lambda$. Since $\mathtt{b}_\lambda$ and $\mathtt{M}_\lambda$ are non-increasing functions of λ, the algorithm can be summarized as follows.

The TMP and MMP algorithms:

We will give the algorithm for TMP, as shown in Figure 13.5, and, when different, in square brackets the MMP.

1. Start with two slopes $\lambda_{lo} = +\infty$ and $\lambda_{hi} = 0$; build the plane wave solutions $\boldsymbol{L}_{lo} = \boldsymbol{L}_{\lambda_{lo}}$ and $\boldsymbol{L}_{hi} = \boldsymbol{L}_{\lambda_{hi}}$ and compute their corresponding total power and achievable rate $\mathtt{M}_{lo} = \mathtt{M}(\boldsymbol{L}_{lo})$, $\mathtt{M}_{hi} = \mathtt{M}(\boldsymbol{L}_{hi})$, $\mathtt{b}_{lo} = \mathtt{b}(\boldsymbol{L}_{lo})$, and $\mathtt{b}_{hi} = \mathtt{b}(\boldsymbol{L}_{hi})$.
2. Compute the new slope
$$\lambda_{new} = \frac{\mathtt{b}_{hi} - \mathtt{b}_{lo}}{\mathtt{M}_{hi} - \mathtt{M}_{lo}} \tag{13.35}$$
3. Build the plane wave solution $\boldsymbol{L}_{new} = \boldsymbol{L}_{\lambda_{new}}$ and compute the achievable rate $\mathtt{b}_{new}$ and power $\mathtt{M}_{new}$.
4. *If* $\mathtt{M}_{new} = \mathtt{M}_{lo}$, *or* $\mathtt{M}_{new} = \mathtt{M}_{hi}$, *or* $\mathtt{M}_{new} = \mathtt{M}_{tot}$, [*If* $\mathtt{b}_{new} = \mathtt{b}_{tot}$]
   then go to step 5;
   else if $\mathtt{M}_{new} < \mathtt{M}_{tot}$, [*else if* $\mathtt{b}_{new} < \mathtt{b}_{tot}$]
   then $\lambda_{low} \leftarrow \lambda_{new}$, $\mathtt{M}_{low} \leftarrow \mathtt{M}_{new}$, $\mathtt{b}_{low} \leftarrow \mathtt{b}_{new}$; go to step 2.
   else if $\mathtt{M}_{new} > \mathtt{M}_{tot}$, [*else if* $\mathtt{b}_{new} > \mathtt{b}_{tot}$]
   then $\lambda_{hi} \leftarrow \lambda_{new}$, $\mathtt{M}_{hi} \leftarrow \mathtt{M}_{new}$, $\mathtt{b}_{hi} \leftarrow \mathtt{b}_{new}$; go to step 2.
5. The optimal solution has been found. If $\mathtt{M}_{new} = \mathtt{M}_{tot}$ (which is really very unlikely), then the optimal solution is $\boldsymbol{L}_{new}$, otherwise the optimal solution is $\boldsymbol{L}_{lo}$. (The optimal solution is simply $\boldsymbol{L}_{new}$.)

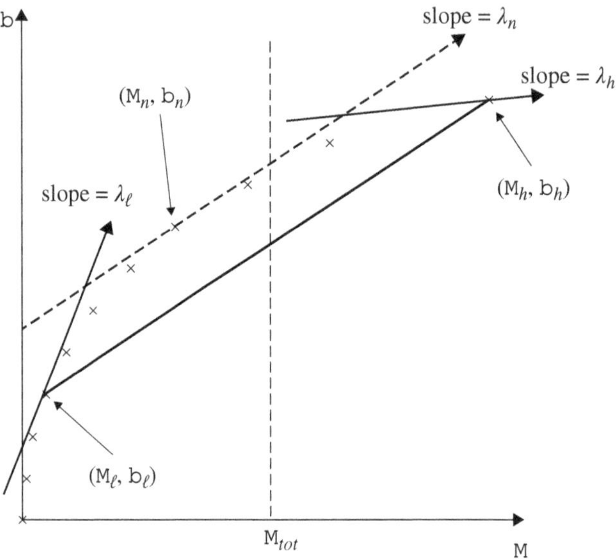

Figure 13.5 Bisection search algorithm.

This algorithm is in practice a bisection search among efficient solutions, which are known to yield a power and number of bits/symbol that are a decreasing function of λ. We recall that, when building a plane wave solution $\boldsymbol{L}_\lambda$, λ can be thought of as the slope of the curve that connects the efficient solutions, at the point corresponding to $\boldsymbol{L}_\lambda$. This is why step 2 yields a solution that is midway between $\boldsymbol{L}_{low}$ and $\boldsymbol{L}_{hi}$.

Because the total number of efficient solutions is $\mathcal{M}K$, the bisection search finds the optimal solution in approximately $O(\log_2(\mathcal{M}K))$ iterations. For each subchannel, $i = 0, \dots, \mathcal{M} - 1$, finding $\mathtt{m}_\lambda[i]$ would require $O(K)$ comparisons with a pre-computed table which stores the map between slopes and efficient modes. The overall complexity is $O(\mathcal{M}K\log_2(\mathcal{M}K))$.

The KRJ algorithm is *optimal*.

The Chow–Cioffi–Bingham algorithm

The Chow–Cioffi–Bingham (CCB) algorithm [3] uses the *hypotheses of convexity*, *constant granularity*, and *constant gap approximation*. It would be suboptimal even if the constant gap hypothesis were satisfied exactly. It estimates the subchannels to be turned off, and distributes power almost equally among remaining subchannels. The key assumption is that performance is mostly determined by the choice of subchannels to be turned off. We provide here its description for the MMP.

1. We start by assigning the same power to all subchannels:

$$\mathtt{M}[i] = \frac{1}{\mathcal{M}}\mathtt{M}_{tot}, \qquad i = 0, \dots, \mathcal{M} - 1 \tag{13.36}$$

   We also calculate $\Gamma[i] = \mathtt{M}[i]\Theta[i]$, $i = 0, \dots, \mathcal{M} - 1$. Let N be the number of active subchannels, initialized to $N = \mathcal{M}$. Then we assume a starting margin $\gamma = 1$.
2. For each subchannel we compute:

$$\mathtt{b}'[i] = \log_2\left(1 + \frac{\Gamma[i]}{\Gamma_{gap}\gamma}\right) \tag{13.37}$$

If we assign $\mathtt{b}'[i]$ bits/symbol to subchannel i, we would actually obtain the same margin on all subchannels, with the total power equal to $\mathtt{M}_{tot}$. However, $\mathtt{b}'[i]/\rho$ in general is not an integer, so we have to choose the nearest available mode. Hence we assign[1]:

$$\mathtt{m}[i] = \max\left[\text{round}\left(\frac{\mathtt{b}'[i]}{\rho}\right), 0\right] \tag{13.38}$$

For future use, we also define

$$d[i] = \mathtt{b}'[i] - \mathtt{b}_{\mathtt{m}[i]} \tag{13.39}$$

We also perform the following operation for $i = 0, \dots, \mathcal{M} - 1$

$$\text{if } \mathtt{m}[i] = 0, \text{then } N \leftarrow N - 1 \tag{13.40}$$

3. We compute the number of bits/symbol obtained with the current loading: $\mathtt{b} = \sum_{i=1}^{N} \mathtt{b}_{\mathtt{m}[i]}$. The obtained solution yields *approximately* the same margin γ for all subchannels, although γ may not be optimal, and $\mathtt{b}$ may not be the maximum achievable rate, since we have arbitrarily assigned uniformly distributed power at the beginning.
4. The new margin γ obtained when lowering the achievable rate by $\mathtt{b} - \mathtt{b}_{tot}$ is estimated. To this end, we consider an equivalent *average* subchannel with achievable rate $\mathtt{b}/N$, margin γ, and unknown SNR Γ. For such an equivalent subchannel, it would be

$$\gamma = \frac{\Gamma \overline{\Gamma}_{gap}}{2^{\mathtt{b}/N}} \tag{13.41}$$

Setting the achievable rate of the equivalent subchannel to $\mathtt{b}_{tot}/N$ yields the new margin

$$\frac{\gamma_{new}}{\gamma} = \frac{2^{\mathtt{b}/N} - 1}{2^{\mathtt{b}_{tot}/N} - 1} \simeq 2^{\frac{\mathtt{b}-\mathtt{b}_{tot}}{N}} \tag{13.42}$$

Then, we update γ as

$$\gamma \leftarrow \gamma \cdot 2^{\frac{\mathtt{b}-\mathtt{b}_{tot}}{N}} \tag{13.43}$$

5. If $\mathtt{b} = \mathtt{b}_{tot}$, we stop iterating and we go to the last step.
6. If this is the first iteration, then we go back to step 2, because we have to recompute the combination of modes according to the new margin. Otherwise, if it is not the first iteration, we check if N has changed since the last iteration. If it has not, it is not useful to go on iterating, because the difference between $\mathtt{b}$ and $\mathtt{b}_{tot}$ is only due to rounding, and not to subchannel exclusions. Thus, if N has changed we go back to step 2, otherwise we go on to the next step.
7. Now we fine-tune the bit distribution to get exactly $\mathtt{b} = \mathtt{b}_{tot}$. We have to add (or subtract) $\mathtt{b} - \mathtt{b}_{tot}$ bits. We add (subtract) ρ bits at a time to the subchannels that have received less (more) bits than they should, due to rounding. Thus we do
   - *While* $\mathtt{b} < \mathtt{b}_{tot}$
     $i^* = \text{argmax}_i d[i]$
     $\mathtt{m}[i^*] \leftarrow \mathtt{m}[i^*] + 1,\ d[i^*] \leftarrow d[i^*] - \rho,\ \mathtt{b} \leftarrow \mathtt{b} + 1$
   - *While* $\mathtt{b} > \mathtt{b}_{tot}$
     $i^* = \text{argmax}_{i:\mathtt{m}[i]>0} d[i]$
     $\mathtt{m}[i^*] \leftarrow \mathtt{m}[i^*] - 1,\ d[i^*] \leftarrow d[i^*] + \rho,\ \mathtt{b} \leftarrow \mathtt{b} - 1$

   Actually, we can do this more efficiently with a complexity proportional to the number of active channels [3].

[1] round(x) denotes the nearest integer to x.

8. Lastly, we compute the correct powers according to the combination of selected modes:

$$\mathtt{M}[i] = (2^{\mathtt{b}_{\mathtt{m}[i]}} - 1)\frac{\overline{\Gamma}_{gap}}{\Theta[i]}, \qquad i = 0, \ldots, N-1 \tag{13.44}$$

and we scale by

$$\mathtt{M}'_{tot} = \sum_{i=0}^{N-1} \mathtt{M}[i] \tag{13.45}$$

so that the total power is $\mathtt{M}_{tot}$ as:

$$\mathtt{M}[i] \leftarrow \mathtt{M}[i]\frac{\mathtt{M}_{tot}}{\mathtt{M}'_{tot}}, \qquad i = 0, \ldots, N-1 \tag{13.46}$$

The solution we have obtained ensures that all subchannels have exactly the same margin, and hence if $\gamma > 1$ it guarantees that the error probability constraint is satisfied; it also uses a total power $\mathtt{M}_{tot}$ and yields an achievable rate $\mathtt{b}_{tot}$. However, it is not guaranteed to be the optimal solution, that is the margin may be slightly lower than it could be.

In general, the CCB algorithm may be slightly suboptimal with respect to the HH algorithm, while having lower complexity $O(\mathcal{M})$.

Comparison

A comparison between the various algorithms examined is reported in Table 13.2. For the CCB algorithm, N_{iter} is the number of iterations. Note that the HH algorithm is very complex in some applications like asymmetric digital subscribers line (ADSL), where a large number of modes are available and $\mathcal{M}$ is large.

Table 13.2: Comparison between four loading algorithms.

Algorithm	optimal	Hypothesis				Complexity
		ρ	convex	Γ_{gap}	Other	
HH	Yes	x	x			$O(\mathcal{M}\mathtt{b}_{tot}/\rho)$
KRJ	Yes		x			$O(\mathcal{M}K\log_2(\mathcal{M}K))$
CCB	No	x	x	x		$O(\mathcal{M}N_{iter})$

13.2 Capacity achieving solutions for single carrier systems

If the passband $\mathcal{B}$ consists of only one frequency interval, as an alternative to OFDM transmission the capacity of a linear dispersive channel can be achieved by single carrier transmission. For a comparison of the characteristics of the two systems, we refer to page 444.

First we examine the equivalence (in terms of capacity) between a continuous-time channel and a discrete-time channel with inter-symbol interference (ISI); this equivalence is obtained by referring to a transmit filter that shapes the spectrum of the transmit signal as indicated by water pouring, and a receiver that implements a matched filter (MF) or a whitened matched filter (WMF), as illustrated in Section 7.4.6 (Figure 7.35). By the WMF, we obtain a canonical form of the discrete-time channel with *trailing ISI*, that is ISI due only to postcursors, and AWGN (see Figure 13.6). As will be shown in this section, the contribution of the ISI of the resulting discrete-time channel to the capacity of the

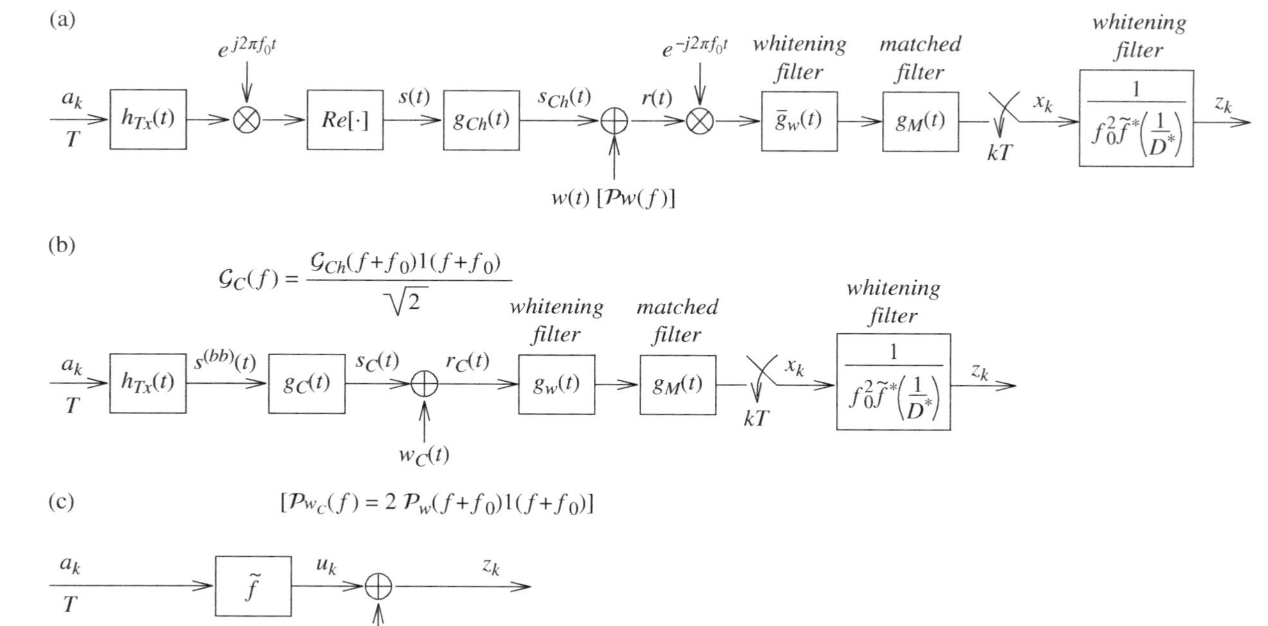

Figure 13.6 Equivalence between a continuous time system, (a) passband model, (b) baseband equivalent model with $g_w(t) = \frac{1}{\sqrt{2}}\bar{g}_w(t)$, and (c) a discrete-time system, for transmission over a linear dispersive channel.

channel becomes negligible for high SNRs. This suggests that capacity can be achieved by combining ISI cancellation techniques in the discrete-time domain with channel coding and shaping.

Assume that the passband $\mathcal{B}$ that allows achieving capacity consists of only one frequency interval $[f_1, f_2]$, with $0 < f_1 < f_2$ and bandwidth $B = f_2 - f_1$. If $\mathcal{B}$ consists of several intervals, then the same procedure can be separately applied to each interval, although in this case MC transmission is preferable.

Consider passband transmission with modulation interval T and with minimum bandwidth of a complex-valued symbol sequence, that is we choose $T = 1/B$.

We recall the signal analysis that led to the scheme of Figure 7.35, extending it to the case of noise $w(t)$ with power spectral density (PSD) $\mathcal{P}_w(f)$ not necessarily constant. With reference to Figure 13.6, from (7.26) the transmitted signal is given by

$$s(t) = Re\left[\sum_k a_k h_{Tx}(t - kT)\, e^{j2\pi f_0 t}\right] \tag{13.47}$$

where $\{a_k\}$ is modelled as a sequence of complex-valued i.i.d. symbols with *Gaussian* distribution and variance σ_a^2, and $h_{Tx}(t)$ denotes the transmit filter impulse response, with Fourier transform $\mathcal{H}_{Tx}(f)$.

The transmit filter is chosen such that the PSD of $s(t)$, given by (7.24) and (7.25), is equal to $\mathcal{P}_{s,opt}(f)$, that is

$$\frac{\sigma_a^2}{4T}\, |\mathcal{H}_{Tx}(f)|^2 = \mathcal{P}_{s,opt}(f + f_0)\; 1(f + f_0) \tag{13.48}$$

Note that this requires the adaptation of the transmit filter to the channel impulse response, a solution rarely adopted in practice, where the transmit filter is typically fixed, and implemented in part by analog circuitry.

At the receiver, the signal $r(t)$ is first demodulated and filtered by a filter with impulse response $\overline{g}_w$ that *suppresses* the signal components around $-2f_0$ and *whitens* the noise. As an alternative we could use a passband whitening filter phase splitter that suppresses the signal components with negative frequency and whitens the noise in the passband, see Section 7.4.9.

Consider the baseband equivalent model of Figure 13.6b, where $\mathcal{G}_C(f) = \frac{1}{\sqrt{2}}\, \mathcal{G}_{Ch}(f + f_0)\; 1(f + f_0)$ and $\mathcal{P}_{w_C}(f) = 2\mathcal{P}_w(f + f_0)\; 1(f + f_0)$; we define

$$\mathcal{B}_0 = [f_1 - f_0, f_2 - f_0] \tag{13.49}$$

as the new passband of the desired signal at the receiver. The whitening filter $g_w(t) = \frac{1}{\sqrt{2}}\overline{g}_w(t)$ has frequency response given by

$$|\mathcal{G}_w(f)|^2 = \begin{cases} \dfrac{1}{\mathcal{P}_{w_C}(f)} & f \in \mathcal{B}_0 \\ 0 & \text{elsewhere} \end{cases} \tag{13.50}$$

From the scheme of Figure 7.35, the whitening filter is then followed by a MF g_M with frequency response

$$\mathcal{G}_M(f) = [\mathcal{H}_{Tx}(f)\; \mathcal{G}_C(f)\; \mathcal{G}_w(f)]^* \tag{13.51}$$

The cascade of whitening filter g_w and MF g_M yields the composite filter $g_{Rc}^{MF}(f)$ with frequency response

$$\begin{aligned} \mathcal{G}_{Rc}^{MF}(f) &= \mathcal{H}_{Tx}^*(f)\; \mathcal{G}_C^*(f)\; |\mathcal{G}_w(f)|^2 \\ &= \frac{\mathcal{H}_{Tx}^*(f)\; \mathcal{G}_C^*(f)}{\mathcal{P}_{w_C}(f)} \end{aligned} \tag{13.52}$$

The overall QAM pulse q at the output of the MF has frequency response given by

$$\mathcal{Q}(f) = \mathcal{H}_{Tx}(f)\; \mathcal{G}_C(f)\; \mathcal{G}_{Rc}^{MF}(f) = \frac{|\mathcal{H}_{Tx}(f)\; \mathcal{G}_C(f)|^2}{\mathcal{P}_{w_C}(f)} \tag{13.53}$$

Note that $\mathcal{Q}(f)$ has the properties of a PSD with passband $\mathcal{B}_0$; in particular, $\mathcal{Q}(f)$ is equal to the noise PSD $\mathcal{P}_{w_R}(f)$ at the MF output, as

$$\mathcal{P}_{w_R}(f) = \mathcal{P}_{w_C}(f)\ |\mathcal{G}_{Rc}^{MF}(f)|^2 = \frac{|\mathcal{H}_{Tx}(f)\ \mathcal{G}_C(f)|^2}{\mathcal{P}_{w_C}(f)} = \mathcal{Q}(f) \tag{13.54}$$

Therefore, the sequence of samples at the MF output can be expressed as

$$x_k = \sum_{i=-\infty}^{+\infty} a_i\ h_{k-i} + \tilde{w}_k \tag{13.55}$$

where the coefficients $h_i = q(iT)$ are given by the samples of the overall impulse response $q(t)$, and

$$\tilde{w}_k = w_R(kT) = [w_C(t') * g_{Rc}^{MF}(t')](t)|_{t=kT} \tag{13.56}$$

In general, the Fourier transform of the discrete-time response $\{h_i\}$ is given by

$$\mathcal{H}^{(dis)}(f) = \frac{1}{T}\sum_{\ell=-\infty}^{+\infty} \mathcal{Q}\left(f - \frac{\ell}{T}\right) \tag{13.57}$$

In this case, because $\mathcal{Q}(f)$ is limited to the passband $\mathcal{B}_0$ with bandwidth $B = 1/T$, there is no aliasing; the function $\mathcal{H}^{(dis)}(f)$, periodic of period $1/T$, is therefore equal to $(1/T)\mathcal{Q}(f)$ in the band $\mathcal{B}_0$. As $\mathcal{P}_{w_R}(f) = \mathcal{Q}(f)$, $\{\tilde{w}_k\}$ is a sequence of Gaussian noise samples with autocorrelation sequence $\{\mathtt{r}_{\tilde{w}_k}(n) = h_n\}$. Note moreover that $\{h_i\}$ satisfies the Hermitian property, as $\mathcal{H}^{(dis)}$ is real valued.

We have thus obtained a discrete-time equivalent channel that can be described using the D transform as

$$x(D) = a(D)h(D) + \tilde{w}(D) \tag{13.58}$$

where $h(D)$ has Hermitian symmetry.

We now proceed to develop an alternative model of the discrete-time equivalent channel with causal, monic, and minimum-phase response $\tilde{f}(D)$, and AWGN $w(D)$. With this regard, we recall the theorem of spectral factorization for discrete time systems (see page 36). If $\mathcal{H}^{(dis)}(f)$ satisfies the Paley–Wiener condition, then the function $h(D)$ can be factorized as follows:

$$h(D) = \tilde{f}^*\left(\frac{1}{D^*}\right) f_0^2 \tilde{f}(D) \tag{13.59}$$

$$\mathcal{H}^{(dis)}(f) = \tilde{\mathcal{F}}^*(f)\, f_0^2\ \tilde{\mathcal{F}}(f) \tag{13.60}$$

where the function $\tilde{f}(D) = 1 + \tilde{f}_1 D + \cdots$ is associated with a causal ($\tilde{f}_i = 0$ for $i < 0$), monic and minimum-phase sequence $\tilde{f}_i$, and $\tilde{\mathcal{F}}(f) = \tilde{f}(e^{-j2\pi fT})$ is the Fourier transform of the sequence $\{\tilde{f}_i\}$. The factor f_0^2 is the geometric mean of $\mathcal{H}^{(dis)}(f)$ over an interval of measure $1/T$, that is

$$\log f_0^2 = T\int_{1/T} \log \mathcal{H}^{(dis)}(f)\, df \tag{13.61}$$

where logarithms may have any common base.

Then, (13.58) can be written as

$$x(D) = a(D)\, f_0^2\, \tilde{f}(D)\, \tilde{f}^*\left(\frac{1}{D^*}\right) + w'(D)\, f_0 \tilde{f}^*\left(\frac{1}{D^*}\right) \tag{13.62}$$

where $w'(D)$ is a sequence of i.i.d. Gaussian noise samples with unit variance. Filtering $x(D)$ by a filter having transfer function $1/[f_0^2\tilde{f}^*(1/D^*)]$, we obtain the discrete-time equivalent canonical model of the dispersive channel

$$z(D) = a(D)\tilde{f}(D) + w(D) \tag{13.63}$$

where $w(D)$ is a sequence of i.i.d. Gaussian noise samples with variance $1/f_0^2$. Equation (13.63) is obtained under the assumption that $\tilde{f}(D)$ has a stable reciprocal function, and hence, $\tilde{f}^*(1/D^*)$ has an anti-causal stable reciprocal function; this condition is verified if $h(D)$ has no spectral zeros.

However, to obtain the reciprocal of $\tilde{f}(D)$ does not represent a problem, as $z(D)$ can be indirectly obtained from the sequence of samples at the output of the WMF. The transfer function of the composite filter that consists of the whitening filter g_w and the WMF has a transfer function given by

$$\mathcal{G}_{Rc}^{WMF}(f) = \frac{\mathcal{G}_{Rc}^{MF}(f)}{f_0^2 \tilde{\mathcal{F}}^*(f)} = \frac{\mathcal{H}_{Tx}^*(f)\, \mathcal{G}_C^*(f)}{\mathcal{P}_{w_C}(f)} \frac{\tilde{\mathcal{F}}(f)}{\mathcal{H}^{(dis)}(f)} \tag{13.64}$$

The only condition for the stability of the filter (13.64) is given by the Paley–Wiener criterion.

Achieving capacity

Note that the model (13.63) expresses the output sequence as the sum of the noiseless sequence $a(D)\,\tilde{f}(D)$ and AWGN $w(D)$.

From (13.63), if $a(D)$ is an uncoded sequence with symbols taken from a finite constellation, and $\tilde{f}(D)$ has finite length, then the received sequence in the absence of noise $a(D)\,\tilde{f}(D)$ can be viewed as the output of a finite state machine, and the sequence $a(D)$ can be optimally detected by the Viterbi algorithm, as discussed in Chapter 7. As an alternative, maximum-likelihood sequence detection (MLSD) can be directly performed by considering the MF output sequence $x(D)$, using a trellis of the same complexity but with a different metric (see Section 7.5.3).

In fact the MF output sequence $x(D)$ can be obtained from the WMF output sequence $z(D)$ by filtering $z(D)$ with a stable filter having transfer function $f_0^2 \tilde{f}^*(1/D^*)$. As $x(D)$ is a sufficient statistic[2] (see Definition 7.2 on page 354) for the detection of $a(D)$, also $z(D)$ is a sufficient statistic; therefore, the capacity of the overall channel including the MF or the WMF is equal to the capacity $C_{[bit/s]}$ given by (6.105). Therefore, capacity can be achieved by coding in combination with the cancellation of ISI.

We now evaluate the capacity. Using (13.53), (13.48), and (6.102) with the definitions of $\mathcal{G}_C$ and $\mathcal{P}_{w_C}$, yields

$$\mathcal{Q}(f) = \frac{|\mathcal{H}_{Tx}(f)\, \mathcal{G}_C(f)|^2}{\mathcal{P}_{w_C}(f)} = \frac{4T}{\sigma_a^2}\, \mathcal{P}_{s,opt}(f+f_0)\, \frac{1}{4}\, \Theta(f+f_0)\, 1(f+f_0) \tag{13.65}$$

and capacity can be expressed as

$$\begin{aligned} C_{[bit/s]} &= \int_{\mathcal{B}} \log_2[1 + \mathcal{P}_{s,opt}(f)\, \Theta(f)]\, df \\ &= \int_{\mathcal{B}_0} \log_2[1 + \mathcal{P}_{s,opt}(f+f_0)\, \Theta(f+f_0)]\, df \\ &= \int_{\mathcal{B}_0} \log_2\left[1 + \frac{\sigma_a^2}{T}\, \mathcal{Q}(f)\right] df \end{aligned} \tag{13.66}$$

Recall from (13.57) that $\mathcal{H}^{(dis)}(f)$, periodic of period $1/T$, is equal to $(1/T)\mathcal{Q}(f)$ in the band $\mathcal{B}_0$; therefore, using (13.61) for $B = 1/T$, the capacity $C_{[bit/s]}$ and its approximation for large values of the SNR Γ can be expressed as

$$C_{[bit/s]} = \int_{1/T} \log_2\left(1 + \sigma_a^2\, \mathcal{H}^{(dis)}(f)\right) df \simeq \int_{1/T} \log_2\left(\sigma_a^2\, \mathcal{H}^{(dis)}(f)\right) df = B \log_2(\sigma_a^2 f_0^2) \tag{13.67}$$

[2] In particular, we have a sufficient statistic when projecting the received signal on a basis of the desired (useful) signals. For example, considering the basis $\{e^{\pm j2\pi ft}, t \in \mathbb{R}, f \in \mathcal{B}\}$ to represent real-valued signals with passband $\mathcal{B}$ in the presence of additive noise, that is the Fourier transform of the noisy signal filtered by an ideal filter with passband $\mathcal{B}$, we are able to reconstruct the noisy signals within the passband of the desired signals; therefore, the noisy signal filtered by a filter with passband $\mathcal{B}$ is a sufficient statistic.

Assume that the tail of the impulse response that causes ISI can be in some way eliminated, so that at the receiver we observe the sequence $a(D) + w(D)$ rather than the sequence (13.63). The SNR of the resultant ideal AWGN channel becomes $\Gamma_{ISI\text{-}free} = \sigma_a^2 f_0^2$; thus, from (6.96) the capacity of the ISI-free channel and its approximation for large values of Γ become

$$C_{ISI\text{-}free\ [bit/s]} = B \log_2(1 + \sigma_a^2 f_0^2) \simeq B \log_2(\sigma_a^2 f_0^2) \tag{13.68}$$

Comparing (13.67) and (13.68), we finally obtain

$$C_{[bit/s]} \simeq C_{ISI\text{-}free\ [bit/s]} \tag{13.69}$$

Price was the first to observe that for large values of Γ we obtain (13.69), that is for high SNRs the capacity $C_{[bit/s]}$ of the linear dispersive channel is approximately equal to the capacity of the ideal ISI-free channel obtained assuming that the residual ISI in the discrete-time channel can be in some way eliminated from the sequence $x(D)$; in other words, the residual ISI of the discrete channel does not significantly contribute to the capacity [4]. Note however again that this results leverages the adaptation of the transmit filter to channel conditions, in order to obtain (13.48), a mostly theoretical setting rarely adopted in practice.

It is interesting to observe that in faster-than-Nyquist systems, where ISI is predominant, the capacity is indeed higher than (13.69) (see references in Section 7.9).

Bibliography

[1] Hughes-Hartogs, D. (1987). Ensemble modem structure for imperfect transmission media. US Patent No. 4, 679, 227, July 1987; 4, 731, 816, March 1988, and 4, 833, 796, May 1989.

[2] Krongold, B.S., Ramchandran, K., and Jones, D.L. (2000). Computationally efficient optimal power allocation algorithms for multicarrier communication systems. *IEEE Transactions on Communications* 48: 23–27.

[3] Chow, J.S., Tu, J.C., and Cioffi, J.M. (1991). A discrete multitone transceiver system for HDSL applications. *IEEE Journal on Selected Areas in Communications* 9: 895–908.

[4] Price, R. (1972). Nonlinearly feedback-equalized PAM versus capacity for noisy filter channels. In: *Proceedings of 1972 International Conference on Communications*, pp. 22.12–22.17.

Chapter 14

Synchronization

In quadrature amplitude modulation (QAM) systems, coherent detection requires the generation of a carrier signal at the frequency of the transmitted modulated signal. Although the nominal frequency of the transmitter and the receiver are the same, imperfections of the components or Doppler shift effects determine an unknown (significant) difference between the receiver oscillator frequency and the carrier frequency of the received signal.

In this chapter, we will discuss methods for carrier phase and frequency recovery, as well as algorithms to estimate the timing phase. To avoid ambiguity, we refer to the latter as timing recovery algorithms, dropping the term phase. These algorithms are developed for application in the pulse amplitude modulation (PAM) and QAM transmission systems of Chapter 7, and the spread spectrum systems of Chapter 10. The problem of carrier frequency synchronization for orthogonal frequency division multiplexing (OFDM) systems will be addressed in Section 14.9.

Carrier recovery must be performed using one of the following two strategies.

1. The first consists in multiplexing, usually in the frequency domain, a special signal, called *pilot signal*, at the transmitter. This allows extracting the carrier at the receiver and therefore synchronizing the receive oscillator in phase and frequency with the transmit oscillator. If the pilot signal consists of a non-modulated carrier, carrier recovery is obtained by the *phase-locked loop* (PLL) described in Section 14.2.
2. The second consists in getting the carrier directly from the modulated signal; this approach presents the advantage that all the transmitted power is allocated for the transmission of the signal carrying the desired information. Some structures that implement this strategy are reported in Section 14.3.

14.1 The problem of synchronization for QAM systems

As a generalization of the receiver block diagram shown in Figure 7.5, the representation of the analog front end for a passband QAM system is illustrated in Figure 14.1. The received signal r is multiplied by a complex-valued carrier generated by a local oscillator, then filtered by an anti-imaging filter, g_{AI}, that extracts the complex representation of r, r_C (see Section 1.4). Often the function of the anti-imaging filter is performed by other filters, however, here g_{AI} is considered only as a model for the analysis and is assumed to be non-distorting.

In Figure 14.1, the reconstructed carrier, in complex form, has the expression

$$\hat{v}_C(t) = \exp\{-j(2\pi f_1 t + \varphi_1)\} \tag{14.1}$$

Algorithms for Communications Systems and their Applications, Second Edition.
Nevio Benvenuto, Giovanni Cherubini, and Stefano Tomasin.

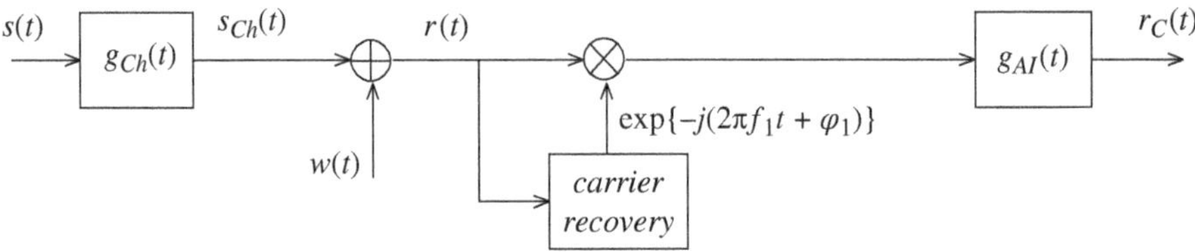

Figure 14.1 Analog front end for passband QAM systems.

Therefore, if the carrier generated at the transmitter, as observed at the receiver (*received carrier*) is given by $\exp\{j(2\pi f_0 t + \varphi_0)\}$,[1] in general we have a reconstruction error of the carrier phase φ_{PA} given by

$$\varphi_{PA}(t) = (2\pi f_1 t + \varphi_1) - (2\pi f_0 t + \varphi_0) \tag{14.2}$$

By defining

$$\begin{aligned} \Omega &= 2\pi(f_0 - f_1) \\ \theta &= (\varphi_0 - \varphi_1) \end{aligned} \tag{14.3}$$

(14.2) can be rewritten as

$$\varphi_{PA}(t) = -\Omega t - \theta \tag{14.4}$$

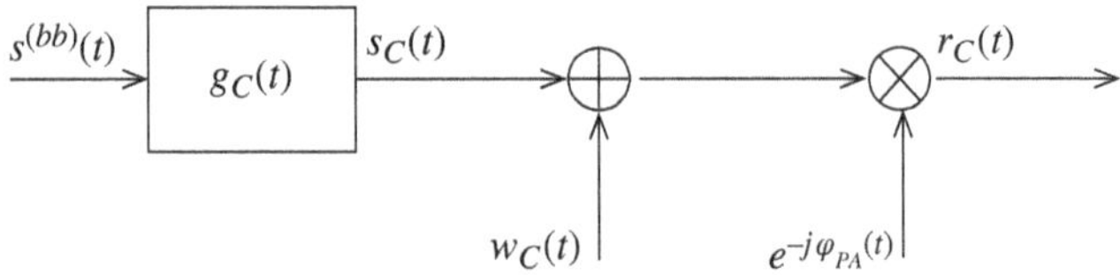

Figure 14.2 Baseband equivalent model of the channel and analog front end for a QAM system.

With reference to the notation of Figure 7.11, observing that now the phase offset is not included in the baseband equivalent channel impulse response (CIR), we have

$$g_C(t) = \frac{1}{2\sqrt{2}}\, g_{Ch}^{(bb)}(t)\, e^{-j \arg \mathcal{G}_{Ch}(f_0)} \tag{14.5}$$

The resulting baseband equivalent scheme of a QAM system is given in Figure 14.2. We assume that the anti-imaging filter frequency response is flat within the frequency interval

$$|f| \leq B + \frac{\Omega_{max}}{2\pi} \tag{14.6}$$

where B is the bandwidth of the signal s_C and Ω_{max} is the maximum value of $|\Omega|$.

The received signal r_C is affected by a frequency offset Ω and a phase offset θ; moreover, both the transmit filter h_{Tx} and the channel filter g_C introduce a transmission delay t_0. To simplify the analysis, this delay is assumed known with an error in the range $(-T/2, T/2)$. This coarse timing estimate can be obtained, for example, by a correlation method with known input (see (7.603)). This corresponds to assuming the overall pulse q_C non-causal, with peak at the origin.

Once set $t_0 = \varepsilon T$, with $|\varepsilon| \leq 1/2$, the signal r_C can be written as

$$r_C(t) = e^{j(\Omega t + \theta)} \sum_{k=-\infty}^{+\infty} a_k\, q_C(t - kT - \varepsilon T) + w_{C_\varphi}(t) \tag{14.7}$$

[1] In this chapter, φ_0 is given by the sum of the phase of the transmitted carrier and the channel phase at $f = f_0$, equal to $\arg \mathcal{G}_{Ch}(f_0)$.

where

$$\begin{aligned} q_C(t) &= (h_{Tx} * g_C)(t + \varepsilon T) \\ w_{C_\varphi}(t) &= w_C(t)\, e^{-j\varphi_{PA}(t)} \end{aligned} \tag{14.8}$$

Furthermore, the receiver clock is independent of the transmitter clock, consequently the receiver clock period, that we denote as T_c, is different from the symbol period T at the transmitter: we assume that the ratio $F_0 = T/T_c$ is not necessary a rational number.

The synchronization process consists in recovering the carrier, in the presence of phase offset θ and frequency offset Ω, and the timing phase or time shift εT.

14.2 The phase-locked loop

We assume now that at the transmitter an unmodulated carrier is superimposed to the data signal, in order to ease synchronization at the receiver.

We also assume that the transmitted sinusoidal component v has been isolated from the signal at the receiver by a suitable narrowband filter. Now the problem consists in generating by a local oscillator, for example using a PLL, a signal v_{VCO} with the same frequency and phase of v, apart from a known offset.

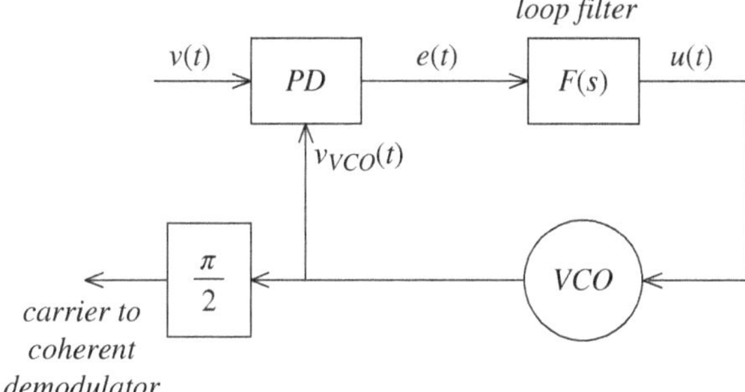

Figure 14.3 Block diagram of a PLL.

A PLL, whose block diagram is shown in Figure 14.3, is a control system used to automatically regulate the phase of a locally generated signal v_{VCO} so that it coincides with that of an input signal v. We assume that the two signals are given by

$$v(t) = A_1 \sin[\omega_0 t + \varphi_0(t)], \qquad v_{VCO}(t) = A_2 \cos[\omega_1 t + \varphi_1(t)] \tag{14.9}$$

where the phase φ_0 is a slowly time varying function with respect to ω_0, or

$$\left|\frac{d\varphi_0(t)}{dt}\right| \ll \omega_0 \tag{14.10}$$

We write the instantaneous phase of v_{VCO} as follows:

$$\omega_1 t + \varphi_1(t) = \omega_0 t + (\omega_1 - \omega_0)\, t + \varphi_1(t) \tag{14.11}$$

and we let

$$\hat{\varphi}_0(t) = (\omega_1 - \omega_0)\, t + \varphi_1(t) \tag{14.12}$$

thus

$$v_{VCO}(t) = A_2 \cos[\omega_0 t + \hat{\varphi}_0(t)] \tag{14.13}$$

and $\hat{\varphi}_0$ then represents the estimate of the phase φ_0 obtained by the PLL.

We define the *phase error* as the difference between the instantaneous phases of the signals v and v_{VCO}, that is

$$\phi(t) = (\omega_0 t + \varphi_0(t)) - (\omega_1 t + \varphi_1(t)) = \varphi_0(t) - \hat{\varphi}_0(t) \tag{14.14}$$

As illustrated in Figure 14.3, a PLL comprises:

- a *phase detector* (PD), that yields an output signal e given by the sine of the difference between the instantaneous phases of the two input signals, that is

$$e(t) = K_D \sin[\phi(t)] \tag{14.15}$$

  where K_D denotes the phase detector gain; we observe that e is an odd function of ϕ; therefore, the PD produces a signal having the same sign as the *phase error*, at least for values of ϕ between $-\pi$ and $+\pi$;
- a lowpass filter $F(s)$, called *loop filter*, whose output u is equal to

$$u(t) = f * e(t) \tag{14.16}$$

- a *voltage controlled oscillator* (VCO), which provides a periodic output signal v_{VCO} whose phase $\hat{\varphi}_0$ satisfies the relation

$$\frac{d\hat{\varphi}_0(t)}{dt} = K_0\, u(t) \tag{14.17}$$

  called VCO *control law*, where K_0 denotes the VCO gain.

In practice, the PD is often implemented by a simple multiplier; then the signal e is proportional to the product of v and v_{VCO}. If K_m denotes the multiplier gain and we define

$$K_D = \frac{1}{2}\, A_1\, A_2\, K_m \tag{14.18}$$

then we obtain

$$\begin{aligned} e(t) &= K_m\, v(t)\, v_{VCO}(t) \\ &= K_m\, A_1 \sin[\omega_0 t + \varphi_0(t)]\, A_2 \cos[\omega_1 t + \varphi_1(t)] \\ &= K_D \sin[\phi(t)] + K_D \sin[2\omega_0 t + \varphi_0(t) + \hat{\varphi}_0(t)] \end{aligned} \tag{14.19}$$

Note that, with respect to the signal e defined in (14.15), there is now an additional term with radian frequency $2\omega_0$. However, as from (14.10), φ_0 is slowly varying in comparison with the term at frequency $2\omega_0$, the high frequency components are eliminated by the lowpass filter $F(s)$ or, in case the lowpass filter is not implemented, by the VCO that has a lowpass frequency response. Therefore, the two schemes, with a PD or with a multiplier, may be viewed as equivalent; because of its simplicity, the latter will be considered in the following analysis.

14.2.1 PLL baseband model

We now derive a baseband equivalent model of the PLL. From (14.16), we have

$$u(t) = \int_0^t f(t-\xi)\, e(\xi)\, d\xi \tag{14.20}$$

Substitution of (14.20) in (14.17) yields

$$\frac{d\hat{\varphi}_0(t)}{dt} = K_0 \int_0^t f(t-\xi)\, e(\xi)\, d\xi \tag{14.21}$$

By this relation, we derive the baseband scheme of Figure 14.4. Subtraction of the phase estimate $\hat{\varphi}_0$ from the phase φ_0 yields the phase error ϕ that, transformed by the non-linear block $K_D \sin(\cdot)$, in turn

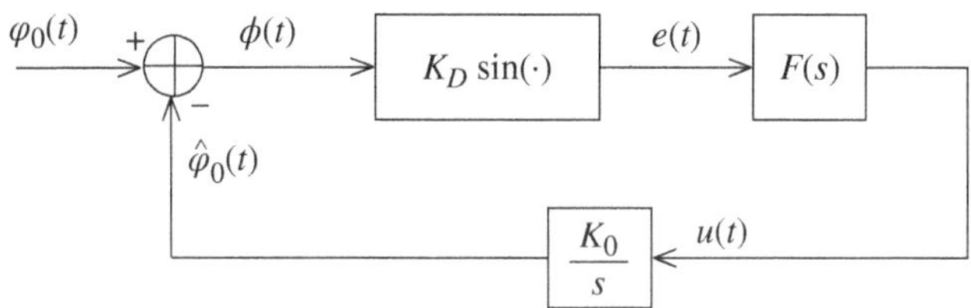

Figure 14.4 Baseband model of a PLL.

gives the signal e. The signal e is input to the loop filter $F(s)$, which outputs the control signal u. The integration block, with gain K_0, integrates the signal u and yields the estimate $\hat{\varphi}_0$, thus closing the loop.

Replacing $\hat{\varphi}_0$ by $\varphi_0(t) - \phi$ in (14.21), and e by (14.15) we have

$$\frac{d\phi(t)}{dt} = \frac{d\varphi_0(t)}{dt} - K_D\, K_0 \int_0^t f(t-\xi) \sin[\phi(\xi)]\, d\xi \tag{14.22}$$

This equation represents the integro-differential equation that governs the dynamics of the PLL. Later, we will study this equation for particular expressions of the phase φ_0, and only for the case $\phi(t) \simeq 0$, i.e. assuming the PLL is in the steady state or in the so-called *lock condition*; the transient behaviour, that is for the case $\phi(t) \neq 0$, is difficult to analyse and we refer to [1] for further study.

Linear approximation

Assume that the phase error ϕ is small, or $\phi(t) \simeq 0$; then, the following approximation holds

$$\sin[\phi(t)] \simeq \phi(t) \tag{14.23}$$

and (14.22) simplifies into

$$\frac{d\phi(t)}{dt} = \frac{d\varphi_0(t)}{dt} - K_D\, K_0 \int_0^t f(t-\xi)\, \phi(\xi)\, d\xi \tag{14.24}$$

In this way, the non-linear block $K_D \sin(\cdot)$ of Figure 14.4 becomes a multiplier by the constant K_D, and the whole structure is linear, as illustrated in the simplified block diagram of Figure 14.5.

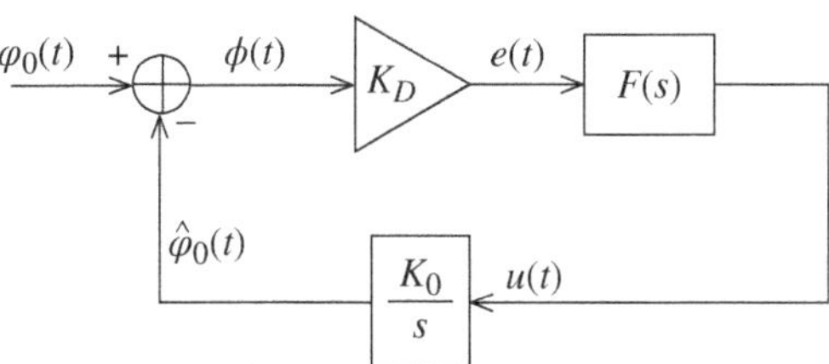

Figure 14.5 Linearized baseband model of the PLL.

We denote by $P_\phi(s)$ the Laplace transform of ϕ; by taking the Laplace transform of (14.24) and assuming $\hat{\varphi}_0(0) = 0$, we obtain

$$sP_\phi(s) = s\Phi_0(s) - K_D\, K_0\, F(s)\, P_\phi(s) \tag{14.25}$$

Substituting $P_\phi(s)$ with $\Phi_0(s) - \hat{\Phi}_0(s)$, we derive the *loop transfer function* as

$$H(s) = \frac{\hat{\Phi}_0(s)}{\Phi_0(s)} = \frac{KF(s)}{s + KF(s)}, \qquad K = K_D\, K_0 \tag{14.26}$$

Table 14.1: Three expressions of φ_0 and corresponding Laplace transforms.

$\varphi_0(t)$	$\Phi_0(s)$
$\varphi_s\ 1(t)$	$\frac{\varphi_s}{s}$
$\omega_s t\ 1(t)$	$\frac{\omega_s}{s^2}$
$\omega_r \frac{t^2}{2}\ 1(t)$	$\frac{\omega_r}{s^3}$

Then from (14.26), we get the following two relations:

$$P_\phi(s) = \Phi_0(s) - \hat{\Phi}_0(s) = [1 - H(s)]\ \Phi_0(s) \tag{14.27}$$

$$\frac{P_\phi(s)}{\Phi_0(s)} = \frac{1}{1 + [KF(s)/s]} \tag{14.28}$$

We define as *steady state error* ϕ_∞ the limit for $t \to \infty$ of ϕ; recalling the final value theorem, and using (14.28), ϕ_∞ can be computed as follows:

$$\phi_\infty = \lim_{t\to\infty} \phi(t) = \lim_{s\to 0} sP_\phi(s) = \lim_{s\to 0} s\Phi_0(s)\ \frac{1}{1 + KF(s)/s} \tag{14.29}$$

We compute now the value of ϕ_∞ for the three expressions of φ_0 given in Table 14.1 along with the corresponding Laplace transforms.

- *Phase step*: $\varphi_0(t) = \varphi_s\ 1(t)$;

$$\begin{aligned} \phi_\infty &= \lim_{s\to 0} \left[s\ \frac{\varphi_s}{s}\ \frac{1}{1 + [KF(s)/s]} \right] \\ &= \lim_{s\to 0} \left[\frac{\varphi_s s}{s + KF(s)} \right] \end{aligned} \tag{14.30}$$

  thus we obtain

$$\phi_\infty = 0 \ \Leftrightarrow\ F(0) \neq 0 \tag{14.31}$$

  Observe that (14.31) holds even if $F(s) = 1$, i.e. in case the loop filter is absent.

- *Frequency step*: $\varphi_0(t) = \omega_s t\ 1(t)$;

$$\phi_\infty = \lim_{s\to 0} \left[s\ \frac{\omega_s}{s^2}\ \frac{1}{1 + [KF(s)/s]} \right] = \lim_{s\to 0} \left[\frac{\omega_s}{s + KF(s)} \right] \tag{14.32}$$

  If we choose

$$F(s) = s^{-k}\ F_1(s), \text{with } k \geq 1 \text{ and } 0 < |F_1(0)| < \infty \tag{14.33}$$

  then $\phi_\infty = 0$.

- *Frequency ramp*: $\varphi_0(t) = (\omega_r t^2/2)\ 1(t)$

$$\phi_\infty = \lim_{s\to 0} \left[s\ \frac{\omega_r}{s^3}\ \frac{1}{1 + [KF(s)/s]} \right] = \lim_{s\to 0} \left[\frac{\omega_r}{s^2 + KF(s)s} \right] \tag{14.34}$$

  If we use a loop filter of the type (14.33) with $k = 1$, i.e. with one pole at the origin, then we obtain a steady state error ϕ_∞ given by

$$\phi_\infty = \frac{\omega_r}{KF_1(0)} \neq 0 \tag{14.35}$$

As a general rule we can state that, in the presence of an input signal having Laplace transform of the type s^{-k} with $k \geq 1$, to get a steady state error $\phi_\infty = 0$, a filter with at least $(k-1)$ poles at the origin is needed.

The choice of the above elementary expressions of the phase φ_0 for the analysis is justified by the fact that an arbitrary phase φ_0 can always be approximated by a Taylor series expansion truncated to the second order, and therefore as a linear combination of the considered functions.

14.2.2 Analysis of the PLL in the presence of additive noise

We now extend the PLL baseband model and relative analysis to the case in which white noise w with spectral density $N_0/2$ is added to the signal v. Introducing the in-phase and quadrature components of w, from (1.96) we get the relation

$$w(t) = w_I(t)\cos(\omega_0 t) - w_Q(t)\sin(\omega_0 t) \tag{14.36}$$

where w_I and w_Q are two uncorrelated random processes having spectral density in the desired signal band given by

$$\mathcal{P}_{w_I}(f) = \mathcal{P}_{w_Q}(f) = N_0 \tag{14.37}$$

Letting

$$K_w = \frac{1}{2} A_2 K_m \tag{14.38}$$

the multiplier output signal e assumes the expression

$$\begin{aligned} e(t) &= K_m[v(t) + w(t)]\, v_{VCO}(t) \\ &= K_D \sin[\phi(t)] + K_w\, w_Q(t)\sin[\hat{\varphi}_0(t)] + K_w\, w_I(t)\cos[\hat{\varphi}_0(t)] \\ &\quad + K_D \sin[2\omega_0 t + \varphi_0(t) + \hat{\varphi}_0(t)] - K_w\, w_Q(t)\sin[2\omega_0 t + \hat{\varphi}_0(t)] \\ &\quad + K_w\, w_I(t)\cos[2\omega_0 t + \hat{\varphi}_0(t)] \end{aligned} \tag{14.39}$$

Neglecting the high-frequency components in (14.39), (14.20) becomes

$$u(t) = \int_0^t f(t-\xi)\{K_D \sin[\phi(\xi)] + K_w\, w_Q(\xi)\sin[\hat{\varphi}_0(\xi)] + K_w\, w_I(\xi)\cos[\hat{\varphi}_0(\xi)]\}\, d\xi \tag{14.40}$$

Defining the noise signal

$$w_e(t) = K_w[w_I(t)\sin\hat{\varphi}_0(t) + w_Q(t)\cos\hat{\varphi}_0(t)] \tag{14.41}$$

from (14.21) we get the integro-differential equation that describes the dynamics of the PLL in the presence of noise, expressed as

$$\frac{d\phi(t)}{dt} = \frac{d\varphi_0(t)}{dt} - K_0 \int_0^t f(t-\xi)\{K_D \sin[\phi(\xi)] + w_e(\xi)\}\, d\xi \tag{14.42}$$

From (14.42), we obtain the PLL baseband model illustrated in Figure 14.6.

Noise analysis using the linearity assumption

In the case $\phi(t) \simeq 0$, we obtain the linearized PLL baseband model shown in Figure 14.7.

We now determine the contribution of the noise w to the phase error ϕ in terms of variance of the phase error, σ_ϕ^2, assuming that the phase of the desired input signal is zero, or $\varphi_0(t) = 0$. From (14.14), we obtain

$$\hat{\varphi}_0(t) = -\phi(t) \tag{14.43}$$

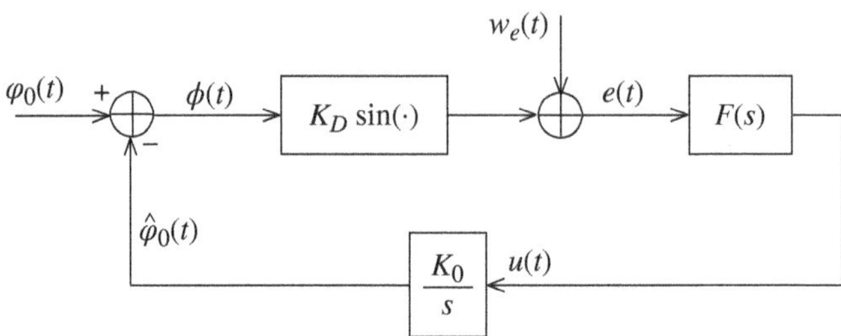

Figure 14.6 PLL baseband model in the presence of noise.

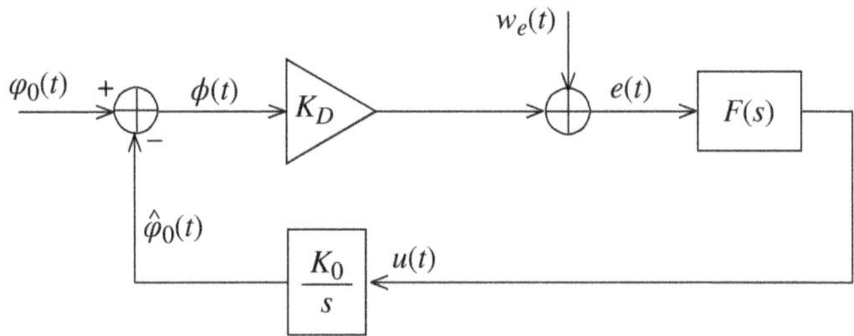

Figure 14.7 Linearized PLL baseband model in the presence of additive noise.

Recalling that the transfer function of a filter that has w_e as input and $\hat{\varphi}_0$ as output is given by $(1/K_D)\, H(s)$ (see (14.26)), the spectral density of ϕ is given by

$$\mathcal{P}_\phi(f) = \mathcal{P}_{\hat{\varphi}_0}(f) = \frac{1}{K_D^2}\, |\mathcal{H}(f)|^2\, \mathcal{P}_{w_e}(f) \tag{14.44}$$

where

$$\mathcal{H}(f) = H(j2\pi f) \tag{14.45}$$

To obtain $\mathcal{P}_{w_e}(f)$, we use (14.41). Assuming w_I and w_Q are uncorrelated white random processes with autocorrelation

$$\mathtt{r}_{w_I}(\tau) = \mathtt{r}_{w_Q}(\tau) = N_0\, \delta(\tau) \tag{14.46}$$

and using the property of the Dirac function $\delta(\tau)f(t+\tau) = \delta(\tau)f(t)$, the autocorrelation of w_e turns out to be

$$\begin{aligned} \mathtt{r}_{w_e}(t, t-\tau) &= K_w^2\, \mathtt{r}_{w_I}(\tau)\{\sin[\hat{\varphi}_0(t)]\sin[\hat{\varphi}_0(t-\tau)] + \cos[\hat{\varphi}_0(t)]\cos[\hat{\varphi}_0(t-\tau)]\} \\ &= K_w^2\, N_0\, \delta(\tau)\{\sin^2[\hat{\varphi}_0(t)] + \cos^2[\hat{\varphi}_0(t)]\} \\ &= K_w^2\, N_0\, \delta(\tau) \end{aligned} \tag{14.47}$$

Taking the Fourier transform of (14.47), we get

$$\mathcal{P}_{w_e}(f) = K_w^2\, N_0 \tag{14.48}$$

Therefore using (14.18) and (14.38), from (14.44) we get the variance of the phase error, given by

$$\sigma_\phi^2 = \int_{-\infty}^{+\infty} \frac{1}{K_D^2}\, |\mathcal{H}(f)|^2\, \mathcal{P}_{w_e}(f)\, df = \frac{N_0}{A_1^2} \int_{-\infty}^{+\infty} |\mathcal{H}(f)|^2\, df \tag{14.49}$$

From (1.74) we now define the *equivalent noise bandwidth of the loop filter* as

$$B_L = \frac{\int_0^{+\infty} |\mathcal{H}(f)|^2\, df}{|\mathcal{H}(0)|^2} \tag{14.50}$$

Then (14.49) can be written as

$$\sigma_{\phi}^2 = \frac{2N_0 B_L}{A_1^2} \tag{14.51}$$

where $A_1^2/2$ is the statistical power of the desired input signal, and $N_0 B_L = (N_0/2)2B_L$ is the input noise power evaluated over a bandwidth B_L.

In Table 14.2, the expressions of B_L for different choices of the loop filter $F(s)$ are given.

Table 14.2: Expressions of B_L for different choices of the loop filter $F(s)$.

Loop order	$F(s)$	$H(s)$	B_L
First	1	$\frac{K}{s+K}$	$\frac{K}{4}$
Second	$\frac{s+a}{s}$	$\frac{K(s+a)}{s^2+Ks+Ka}$	$\frac{K+a}{4}$
Sec.-imperfect	$\frac{s+a}{s+b}$	$\frac{K(s+a)}{s^2+(K+b)s+Ka}$	$\frac{K(K+a)}{4(K+b)}$
Third	$\frac{s^2+as+b}{s^2}$	$\frac{K(s^2+as+b)}{s^3+Ks^2+aKs+bK}$	$\frac{K(aK+a^2-b)}{4(aK-b)}$

14.2.3 Analysis of a second-order PLL

In this section, we analyse the behaviour of a second-order PLL, using the linearity assumption. In particular, we find the expression of the phase error ϕ for the input signals given in Table 14.1, and we evaluate the variance of the phase error σ_{ϕ}^2.

From Table 14.2, the transfer function of a second-order loop is given by

$$H(s) = \frac{K(s+a)}{s^2+Ks+Ka} \tag{14.52}$$

We define the *natural radian frequency*, ω_n, and the *damping factor of the loop*, ζ, as

$$\omega_n = \sqrt{Ka}, \qquad \zeta = \frac{1}{2a}\sqrt{Ka} \tag{14.53}$$

As $K = 2\zeta\omega_n$ and $a = \omega_n/(2\zeta)$, (14.52) can be expressed as

$$H(s) = \frac{2\zeta\,\omega_n s + \omega_n^2}{s^2 + 2\zeta\,\omega_n s + \omega_n^2} \tag{14.54}$$

Once the expression of φ_0 is known, ϕ can be obtained by (14.27) and finding the inverse transform of $P_{\phi}(s)$. The relation is simplified if in place of s we introduce the normalized variable $\tilde{s} = s/\omega_n$; in this case, we obtain

$$P_{\phi}(\tilde{s}) = \frac{\tilde{s}^2}{\tilde{s}^2 + 2\zeta\tilde{s} + 1}\,\Phi_0(\tilde{s}) \tag{14.55}$$

which depends only on the parameter ζ.

In Figures 14.8–14.10, we show the plots of ϕ, with ζ as a parameter, for the three inputs of Table 14.1, respectively. Note that for the first two inputs ϕ converges to zero, while for the third input it converges to a non-zero value (see (14.35)), because $F(s)$ has only one pole at the origin, as can be seen from Table 14.2. We note that if ϕ is a phase step the speed of convergence increases with increasing ζ, whereas if ϕ is a frequency step, the speed of convergence is maximum for $\zeta = 1$.

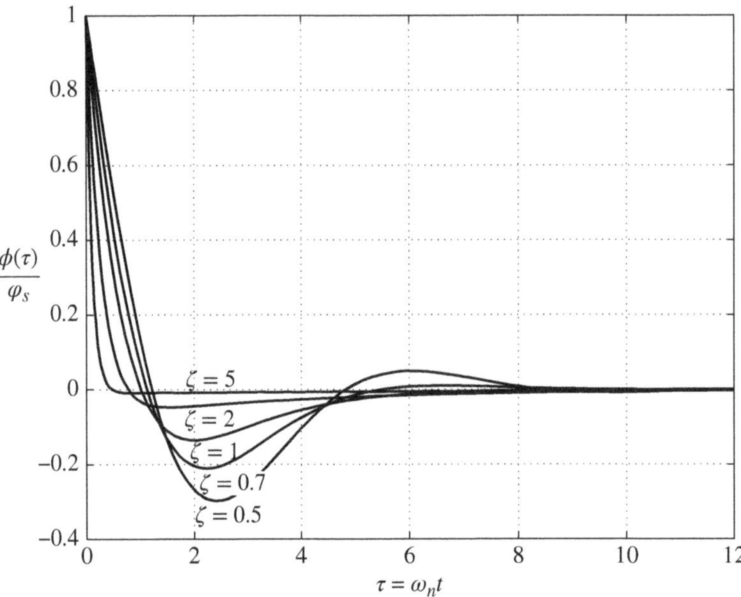

Figure 14.8 Plots of $\phi(\tau)$ as a function of $\tau = \omega_n t$, for a second-order loop filter with a *phase step input signal*: $\varphi_0(t) = \varphi_s 1(t)$.

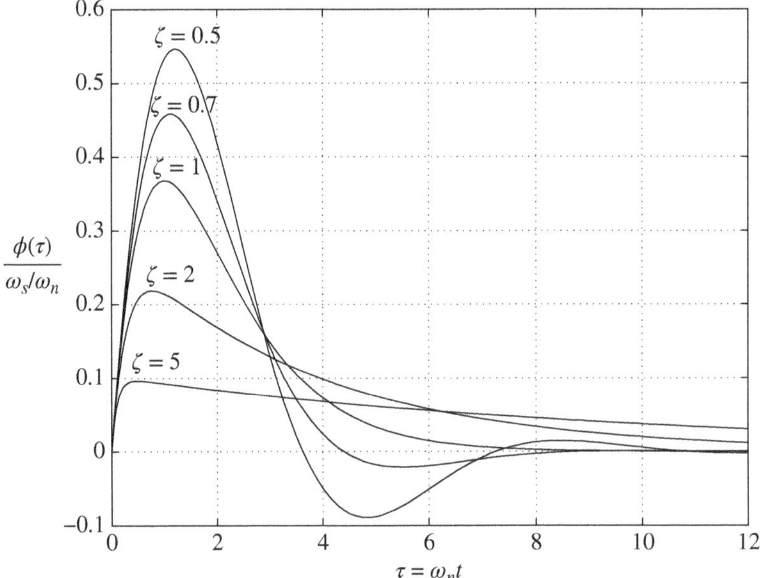

Figure 14.9 Plots of $\phi(\tau)$ as a function of $\tau = \omega_n t$, for a second-order loop filter with a *frequency step input signal*: $\varphi_0(t) = \omega_s t\ 1(t)$.

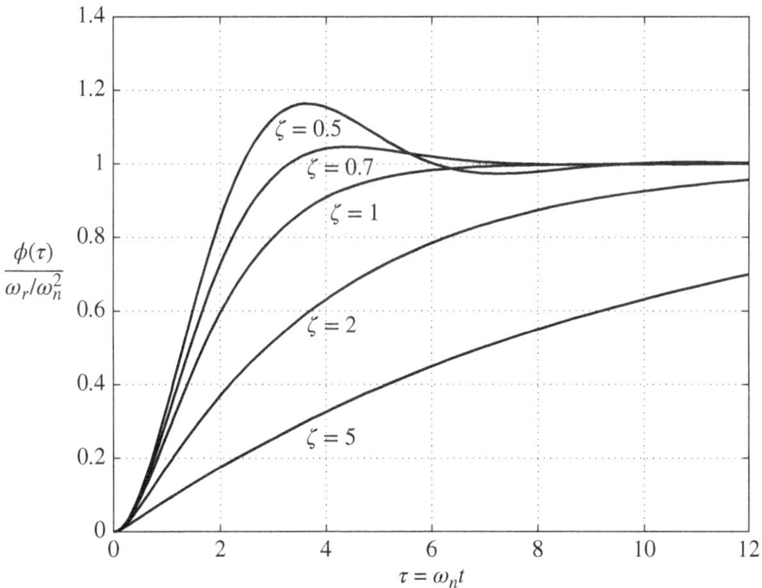

Figure 14.10 Plots of $\phi(\tau)$ as a function of $\tau = \omega_n t$, for a second-order loop filter with a *frequency ramp input signal*: $\varphi_0(t) = \omega_r(t^2/2)\ 1(t)$.

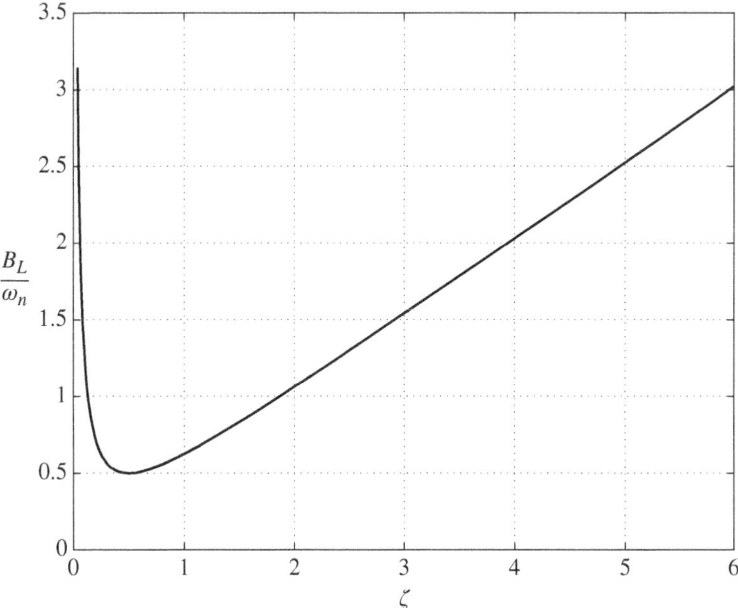

Figure 14.11 Plot of B_L as function of ζ for a second-order loop.

In Figure 14.11, the plot of B_L is shown as a function of ζ. As

$$B_L = \omega_n \left(\frac{\zeta}{2} + \frac{1}{8\zeta}\right) \tag{14.56}$$

we note that B_L has a minimum for $\zeta = 0.5$, and that for $\zeta > 0.5$ it increases as $(1/2)\ \zeta$; the choice of ζ is therefore critical and represents a trade-off between the variance of the phase error and the speed of convergence.

For a detailed analysis of the second- and third-order loops, we refer to [1].

14.3 Costas loop

In the previous section, the PLL was presented as a structure capable of performing carrier recovery for signals of the type

$$s_{Ch}(t) = \sin(2\pi f_0 t + \varphi_0) \tag{14.57}$$

and, in general, for signals that contain periodic components of period n/f_0, with n positive integer.

We now discuss carrier recovery schemes for both PAM-double-side-band (DSB) (see (1.123) where a is a PAM signal as given by (7.2)) and QAM (see Section 7.1.3) signals; these signals do not contain periodic components, but are *cyclostationary* and hence have periodic statistical moments.

We express the generic received signal s_{Ch}, of PAM-DSB or QAM type, in terms of the complex envelope $s_{Ch}^{(bb)}$, that for simplicity we denote by a, as

$$s_{Ch}(t) = Re[a(t)\ e^{j(2\pi f_0 t + \varphi_0)}] \tag{14.58}$$

If in the reference carrier of the complex envelope we include also the phase φ_0, (7.36) becomes equal to (14.5), and

$$s_C(t) = \frac{1}{\sqrt{2}}\ s_{Ch}^{(bb)}(t) = \frac{a(t)}{\sqrt{2}} \tag{14.59}$$

The expression of s_C, apart from the delay εT and the phase offset $e^{-j\arg \mathcal{G}_{Ch}(f_0)}$, is given by (7.39).

The autocorrelation of s_{Ch} is given by (see also (1.228))

$$\begin{aligned} \mathtt{r}_{s_{Ch}}(t, t-\tau) &= \frac{1}{2}\ Re[\mathtt{r}_a(t, t-\tau)\ e^{j2\pi f_0 \tau}] \\ &+ \frac{1}{2}\ Re[\mathtt{r}_{aa^*}(t, t-\tau)\ e^{j[2\pi f_0(2t-\tau)+2\varphi_0]}] \end{aligned} \tag{14.60}$$

from which the statistical power is obtained as

$$\mathtt{M}_{s_{Ch}}(t) = \mathtt{r}_{s_{Ch}}(t,t) = \frac{1}{2}\ E[\ |a(t)|^2] + \frac{1}{2}\ Re\{E[a^2(t)]\ e^{j(4\pi f_0 t + 2\varphi_0)}\} \tag{14.61}$$

14.3.1 PAM signals

We assume that the channel frequency response $\mathcal{G}_C(f)$, obtained from (14.5), is Hermitian[2]; then g_C and a are real valued, hence

$$|a(t)|^2 = [a(t)]^2 \tag{14.62}$$

and (14.61) becomes

$$\mathtt{M}_{s_{Ch}}(t) = \frac{1}{2}\ \mathtt{M}_a(t)[1 + \cos(4\pi f_0 t + 2\varphi_0)] \tag{14.63}$$

[2] In practice, it is sufficient that $\mathcal{G}_C(f)$ is Hermitian in a small interval around $f = 0$ (see page 17).

As a is a cyclostationary random process with period T (see Example 1.7.9 on page 49), $\mathtt{M}_a$ is periodic of period T; therefore, $\mathtt{M}_{s_{Ch}}$ is also periodic and, assuming $1/T \ll f_0$, which is often verified in practice, its period is equal to $1/(2f_0)$.

Suppose the signal s_{Ch} is input to a device (*squarer*) that computes the square of the signal. The output of the squarer has a mean value (deterministic component) equal to the statistical power of s_{Ch}, given by (14.63); if the squarer is cascaded with a narrow passband filter $\mathcal{H}_N(f)$ (see Figure 14.12), with $\mathcal{H}_N(2f_0) = 1$, then the mean value of the filter output is a sinusoidal signal with frequency $2f_0$, phase $2\varphi_0$ (in practice we need to sum also the phase introduced by the filter), and amplitude $(1/2)\ \mathtt{M}_a$.

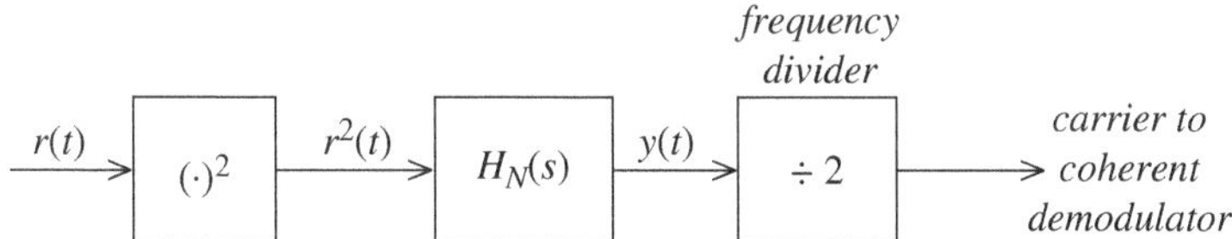

Figure 14.12 Carrier recovery in PAM-DSB systems.

Assuming that $\mathcal{H}_N(f)$ completely suppresses the components at low frequencies, the output filter signal is given by

$$y(t) = \frac{1}{2}\ Re[(h_N^{(bb)} * a^2)(t)\ e^{j[4\pi f_0 t + 2\varphi_0]}] \tag{14.64}$$

This expression is obtained from the scheme of Figure 14.13 in which the product of two generic passband signals, y_1 and y_2, is expressed as a function of their complex envelopes, and decomposed into a baseband component z_1 and a passband component z_2 centred at $\pm 2f_0$.

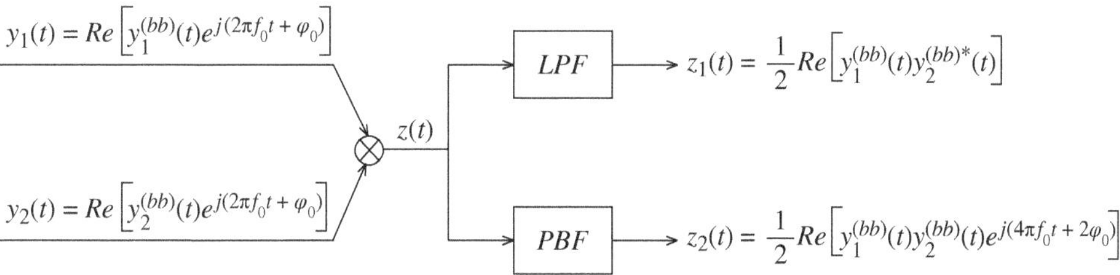

Figure 14.13 Baseband and passband components of the product of two generic passband signals, y_1 and y_2, as a function of their complex envelopes.

If $\mathcal{H}_N(f)$ is Hermitian around the frequency $2f_0$, then $(h_N^{(bb)} * a^2)$ is real valued,[3] and y can be written as

$$y(t) = \frac{(h_N^{(bb)} * a^2)(t)}{2}\ \cos(4\pi f_0 t + 2\varphi_0) \tag{14.65}$$

Thus, we have obtained a sinusoidal signal with frequency $2f_0$, phase $2\varphi_0$, and slowly varying amplitude, function of the bandwidth of $\mathcal{H}_N(f)$.

The carrier can be reconstructed by passing the signal y through a limiter, which eliminates the dependence on the amplitude, and then to a *frequency divider* that returns a sinusoidal signal with frequency and phase equal to half those of the square wave.

[3] In PAM-single-side-band (SSB) and PAM-vestigial-side-band (VSB) systems, $(h_N^{(bb)} * a^2)(t)$ will also contain a quadrature component.

In the case of a time-varying phase φ_0, the signal y can be sent to a PLL with a VCO that operates at frequency $2f_0$, and generates a reference signal equal to

$$v_{VCO}(t) = -A\sin(4\pi f_1 t + 2\varphi_1(t)) = -A\sin(4\pi f_0 t + 2\hat{\varphi}_0(t)) \tag{14.66}$$

The signal v_{VCO} must then be sent to a frequency divider to obtain the desired carrier. The block diagram of this structure is illustrated in Figure 14.14. Observe that the passband filter $\mathcal{H}_N(f)$ is substituted by the lowpass filter $\mathcal{H}_{LPF}(f)$, with $\mathcal{H}_{LPF}(0) = 1$, inserted in the feedback (FB) loop of the PLL; this structure is called the *squarer/PLL*.

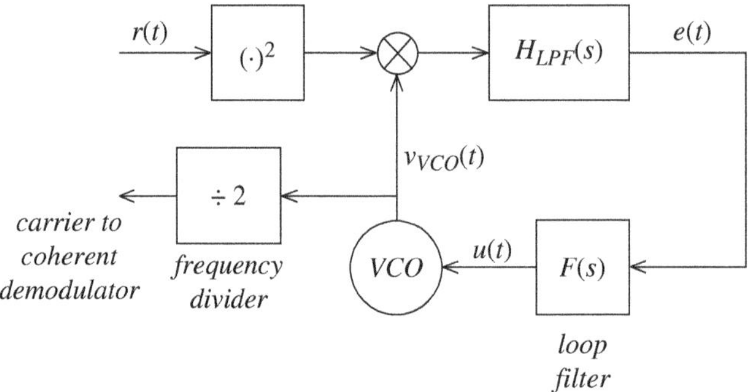

Figure 14.14 Squarer/PLL for carrier recovery in PAM-DSB systems.

An alternative tracking structure to the squarer/PLL, called *Costas loop*, is shown in Figure 14.15. In a Costas loop, the signal e is obtained by multiplying the I and Q components of the signal r; the VCO directly operates at frequency f_0, thus eliminating the frequency divider, and generates the reconstructed carrier

$$v_{VCO}(t) = \sqrt{2A}\cos(2\pi f_0 t + \hat{\varphi}_0(t)) \tag{14.67}$$

By the equivalences of Figure 14.13, we find that the input of the loop filter is identical to that of the squarer/PLL and is given by

$$e(t) = \frac{A}{4}\,a^2(t)\sin[2\phi(t)] \tag{14.68}$$

where ϕ is the phase error (14.14).

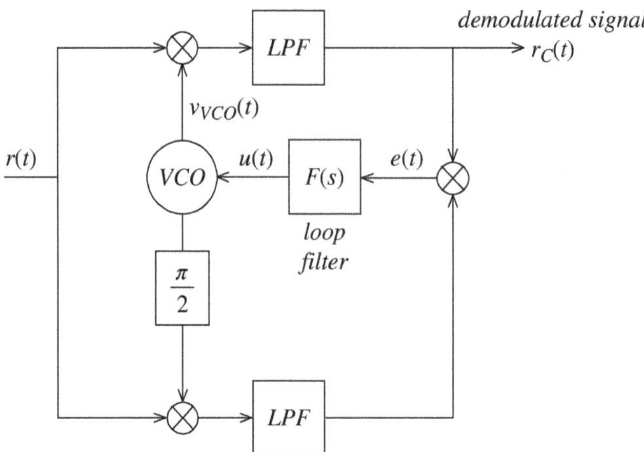

Figure 14.15 Costas loop for PAM-DSB systems.

14.3.2 QAM signals

The schemes of Figures 14.12 and 14.14 cannot be directly applied to QAM systems. Indeed, the symmetry of the constellation of a QAM system usually leads to $E[a^2(t)] = 0$ (see (1.328)); therefore, the periodic component in (14.61) is suppressed.

We compute now the fourth power of s_{Ch}; after a few steps, we obtain

$$\begin{aligned} s^4_{Ch}(t) = &\frac{1}{8}\ Re[a^4(t)\exp(j8\pi f_0 t + j4\varphi_0)] \\ &+ Re[|a(t)|^2\ a^2(t)\exp(j4\pi f_0 t + j2\varphi_0)] \\ &+ \frac{3}{8}\ |a(t)|^4 \end{aligned} \tag{14.69}$$

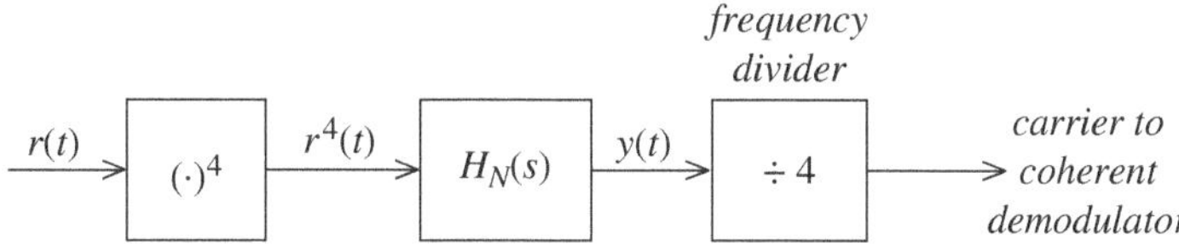

Figure 14.16 Carrier recovery in QAM systems.

Filtering s^4_{Ch} by a passband filter centred at $\pm 4f_0$ (see Figure 14.16), eventually followed by a PLL in the case of a time-varying phase, we obtain a signal y having a mean value given by

$$E[y(t)] = \frac{1}{4}\ \{E[a_I^4(t)] - \mathtt{M}^2_{a_I}\}\cos(8\pi f_0 t + 4\varphi_0), \qquad a_I(t) = Re[a(t)] = s^{(bb)}_{Ch,I} \tag{14.70}$$

which is a periodic signal with period $1/(4f_0)$ and phase $4\varphi_0$.

In the case of M-phase-shift-keying (PSK) signals, there exists a variant of the Costas loop called *extended Costas loop*; the scheme for a quadrature phase-shift-keying (QPSK) system is illustrated in Figure 14.17.

In the presence of additive noise, a passband filter centred at $\pm f_0$ is placed in front of all schemes to limit the noise without distorting the desired signal. For the performance analysis, we refer to [2, 3], in which similar conclusions to those described in Section 14.2 for the PLL are reached.

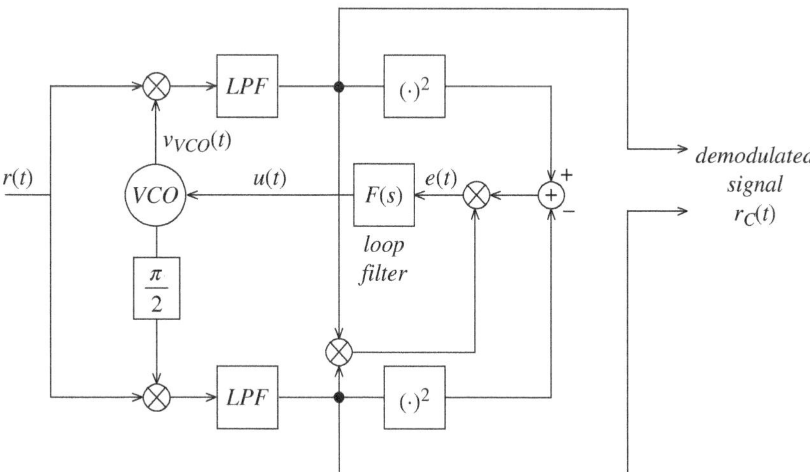

Figure 14.17 Extended Costas loop for QPSK systems.

14.4 The optimum receiver

With reference to the signal model (14.7), once the carrier has been recovered by one of the two methods described in the previous sections, after demodulation, the three parameters $(\theta, \varepsilon, \Omega)$ need to be estimated. In this section, the optimum receiver is obtained using the maximum-likelihood (ML) criterion discussed in Chapter 6 (see also (2.167)). The next two sections synthetically describe various estimation methods given in [4, 5].

From (14.7), the received signal r_C can be expressed as

$$\begin{aligned} r_C(t) &= e^{j(\Omega t+\theta)}\, s_C(t) + w_{C_\varphi}(t) \\ &= s_C(t;\theta,\varepsilon,\Omega) + w_{C_\varphi}(t) \end{aligned} \tag{14.71}$$

where s_C is given by[4]

$$s_C(t) = \sum_{k=-\infty}^{+\infty} a_k\, q_C(t - kT - \varepsilon T) \tag{14.72}$$

We express (14.71) using the vector notation, that is

$$\boldsymbol{r} = \boldsymbol{s} + \boldsymbol{w} \tag{14.73}$$

with the following assumptions:

1. the phase offset θ is equal to z,
2. the time shift εT is equal to eT,
3. the frequency offset Ω is equal to o,
4. the transmitted data sequence $\boldsymbol{a}$ is equal to $\boldsymbol{\alpha}$;

then the probability density function of $\boldsymbol{r}$ is given by (see (1.303))

$$p_{\boldsymbol{r}|\theta,\varepsilon,\Omega,\boldsymbol{a}}(\boldsymbol{\rho} \mid z, e, o, \boldsymbol{\alpha}) = \mathcal{K} \exp\left(-\frac{1}{N_0}\ \| \boldsymbol{\rho} - \boldsymbol{s}\|^2\right) \tag{14.74}$$

The quantities $\|\boldsymbol{\rho}\|^2$ and $\|\boldsymbol{s}\|^2$ are constants[5] [3]; therefore, (14.74) is proportional to the likelihood

$$\mathrm{L}_{\theta,\varepsilon,\Omega,\boldsymbol{a}}(z, e, o, \boldsymbol{\alpha}) = \exp\left\{\frac{2}{N_0}\ Re[\boldsymbol{\rho}^T \boldsymbol{s}^*]\right\} \tag{14.75}$$

Referring to the transmission of K symbols or to a sufficiently large observation interval $T_K = KT$, (14.75) can be written as

$$\mathrm{L}_{\theta,\varepsilon,\Omega,\boldsymbol{a}}(z, e, o, \boldsymbol{\alpha}) = \exp\left\{Re\left[\frac{2}{N_0}\int_{T_K} \rho(t)\ s_C^*(t)\ e^{-j(ot+z)}\ dt\right]\right\} \tag{14.76}$$

Inserting the expression (14.72) of s_C, limited to the transmission of K symbols, in (14.76) and interchanging the operations of summation and integration, we obtain

$$\mathrm{L}_{\theta,\varepsilon,\Omega,\boldsymbol{a}}(z, e, o, \boldsymbol{\alpha}) = \exp\left\{\frac{2}{N_0}\sum_{k=0}^{K-1} Re\left[\alpha_k^*\ e^{-jz}\int_{T_K} \rho(t)\ e^{-jot}\ q_C^*(t - kT - eT)\ dt\right]\right\} \tag{14.77}$$

We introduce the matched filter[6]

$$g_M(t) = q_C^*(-t) \tag{14.78}$$

[4] The phasor $e^{-j\arg \mathcal{G}_{Ch}(f_0)}$ is included in q_C.

[5] Here we are not interested in the detection of $\boldsymbol{a}$, as in the formulation of Section 7.5, but rather in the estimate of the parameters θ, ε, and Ω; in this case, if the observation is sufficiently long, we can assume that $\|s\|^2$ is invariant with respect to the different parameters.

[6] In this formulation, the filter g_M is anticausal; in practice, a delay equal to the duration of q_C must be taken into account.

and assume that the pulse $(q_C * g_M)$ is a Nyquist pulse; therefore, there exists a suitable sampling phase for which inter-symbol-interference (ISI) is avoided.

Last, if we denote the integral in (14.77) by $x(kT + eT, o)$, that is

$$x(kT + eT, o) = (r_C(\tau)\, e^{-jo\tau} * g_M(\tau))(t)|_{t=kT+eT} \tag{14.79}$$

(14.77) becomes

$$\mathrm{L}_{\theta,\varepsilon,\Omega,\boldsymbol{a}}(z, e, o, \boldsymbol{\alpha}) = \exp\left\{\frac{2}{N_0}\sum_{k=0}^{K-1} Re[\alpha_k^*\, x(kT + eT, o)\, e^{-jz}]\right\} \tag{14.80}$$

Let us now suppose that the optimum values of z, e, o that maximize (14.80), i.e. the *estimates* of z, e, o have been determined in some manner.

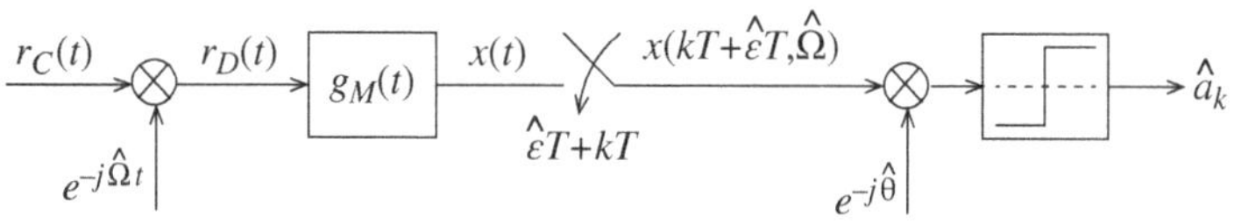

Figure 14.18 Analog receiver for QAM systems.

The structure of the optimum receiver derived from (14.80) is illustrated in Figure 14.18. The signal r_C is multiplied by $\exp\{-j\hat{\Omega}t\}$, where $\hat{\Omega}$ is an estimate of Ω, to remove the frequency offset, then filtered by the matched filter g_M and sampled at the sampling instants $kT + \hat{\varepsilon}T$, where $\hat{\varepsilon}$ is an estimate of ε. The samples $x(kT + \hat{\varepsilon}T, \hat{\Omega})$ are then multiplied by $\exp\{-j\hat{\theta}\}$ to remove the phase offset. Last, the data detector decides on the symbol $\hat{a}_k$ that, in the absence of ISI, maximizes the k-th term of the summation in (14.80) evaluated for $(z, e, o) = (\hat{\theta}, \hat{\varepsilon}, \hat{\Omega})$:

$$\hat{a}_k = \arg\max_{\alpha_k}\ Re[\alpha_k^*\, x(kT + \hat{\varepsilon}T, \hat{\Omega})\, e^{-j\hat{\theta}}] \tag{14.81}$$

The digital version of the scheme of Figure 14.18 is illustrated in Figure 14.19; it uses an anti-aliasing filter and a sampler with period T_c such that (recall the sampling theorem on page 17)

$$\frac{1}{2T_c} \geq B + \frac{\Omega_{max}}{2\pi} \tag{14.82}$$

Observation 14.1

To simplify the implementation of the digital receiver, the ratio $F_0 = T/T_c$ is chosen as an integer; in this case, for $F_0 = 4$ or 8, the interpolator filter may be omitted and the timing after the matched filter has a precision of $T_c = T/F_0$. This approach is usually adopted in radio systems for the transmission of packet data.

To conclude this section, we briefly discuss the algorithms for timing and carrier phase recovery.

Timing recovery

Ideally, at the output of the anti-aliasing filter, the received signal should be sampled at the instants $t = kT + \varepsilon T$; however, there are two problems:

1. the value of ε is not known;
2. the clock at the receiver allows sampling at multiples of T_c, not at multiples of T, and the ratio T/T_c is not necessarily a rational number.

Therefore, time synchronization methods are usually composed of two basic functions.

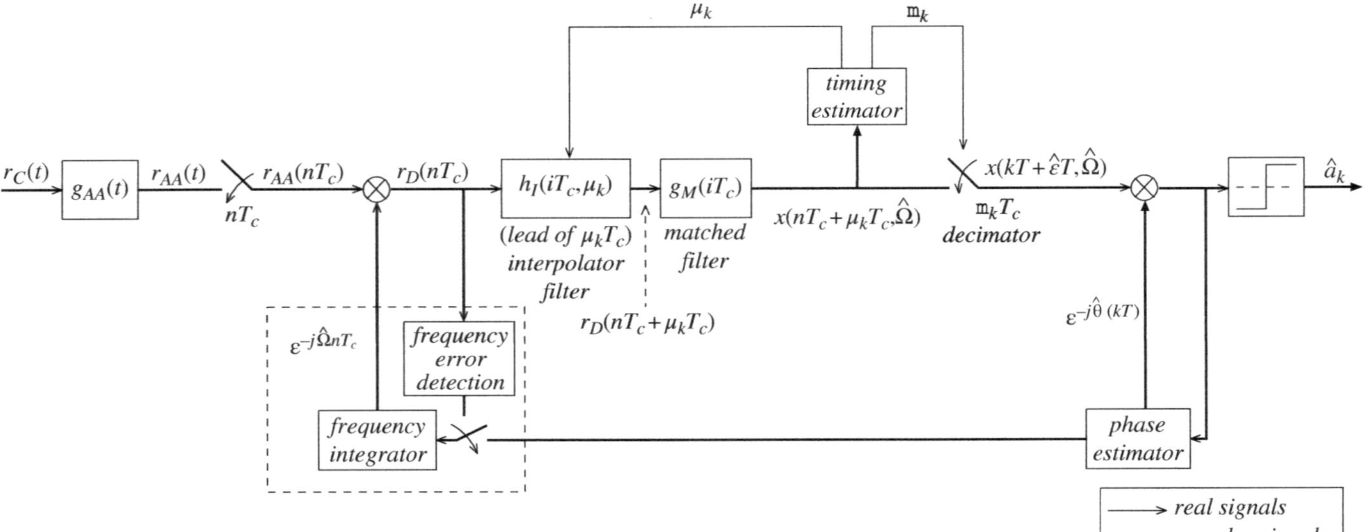

Figure 14.19 Digital receiver for QAM systems. Source: Reproduced with permission from Meyr et al. [4]. ©1998, IEEE.

Timing estimate The first function gives an estimate $\hat{\varepsilon}$ of ε.

Interpolation and decimation The sampling instants $t = kT + \hat{\varepsilon}T$ can be written as

$$kT + \hat{\varepsilon}T = \left[k\,\frac{T}{T_c} + \hat{\varepsilon}\,\frac{T}{T_c}\right] T_c \tag{14.83}$$

The expression within brackets admits the following decomposition:

$$\begin{aligned}\left[k\,\frac{T}{T_c} + \hat{\varepsilon}\,\frac{T}{T_c}\right] &= \left\lfloor\left(k\,\frac{T}{T_c} + \hat{\varepsilon}\,\frac{T}{T_c}\right)\right\rfloor + \mu_k \\ &= \mathtt{m}_k + \mu_k\end{aligned} \tag{14.84}$$

Given a real number a, $\lfloor a \rfloor$ denotes the largest integer smaller than or equal to a (floor), and $\mu = a - \lfloor a \rfloor$ is the fractional part that we denote as $[a]_F$. Suppose now that the estimate of ε is time varying; we denote it by $\hat{\varepsilon}_k$. Consider the $(k+1)$-th sampling instant, expressed as

$$(k+1)\,T + \hat{\varepsilon}_{k+1}T \tag{14.85}$$

By summing and subtracting $\hat{\varepsilon}_k T$, (14.85) can be rewritten as follows:

$$(k+1)\,T + \hat{\varepsilon}_{k+1}T = kT + \hat{\varepsilon}_k T + T + (\hat{\varepsilon}_{k+1} - \hat{\varepsilon}_k)\,T \tag{14.86}$$

Substituting $kT + \hat{\varepsilon}_k T$ with $\mathtt{m}_k T_c + \mu_k T_c$, we obtain

$$\begin{aligned}(k+1)\,T + \hat{\varepsilon}_{k+1}T &= \mathtt{m}_k T_c + \mu_k T_c + T + (\hat{\varepsilon}_{k+1} - \hat{\varepsilon}_k)\,T \\ &= \left[\mathtt{m}_k + \mu_k + \frac{T}{T_c}(1 + (\hat{\varepsilon}_{k+1} - \hat{\varepsilon}_k))\right] T_c\end{aligned} \tag{14.87}$$

Recalling that $\mathtt{m}_k$ is a positive integer, and μ_k is real valued and belongs to the interval [0,1), from (14.87) the following recursive expressions for $\mathtt{m}_k$ and μ_k are obtained:

$$\begin{aligned}\mathtt{m}_{k+1} &= \mathtt{m}_k + \left\lfloor \mu_k + \frac{T}{T_c}(1 + (\hat{\varepsilon}_{k+1} - \hat{\varepsilon}_k)) \right\rfloor \\ \mu_{k+1} &= \left[\mu_k + \frac{T}{T_c}(1 + (\hat{\varepsilon}_{k+1} - \hat{\varepsilon}_k))\right]_F\end{aligned} \tag{14.88}$$

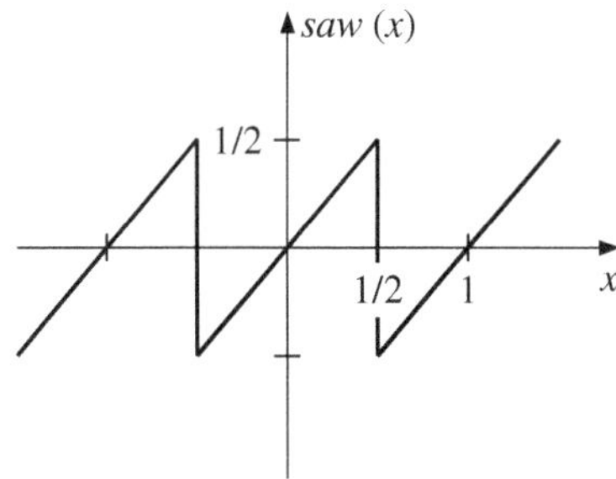

Figure 14.20 Plot of the saw-tooth function saw(x).

The quantity saw$(\hat{\varepsilon}_{k+1} - \hat{\varepsilon}_k)$ is often substituted for $(\hat{\varepsilon}_{k+1} - \hat{\varepsilon}_k)$, where saw($x$) is the saw-tooth function illustrated in Figure 14.20. Thus, the difference between two successive estimates belongs to the interval $[-1/2, 1/2]$; this choice reduces the effects that a wrong estimate of ε would have on the value of the pair $(\mathtt{m}_k, \mu_k)$.

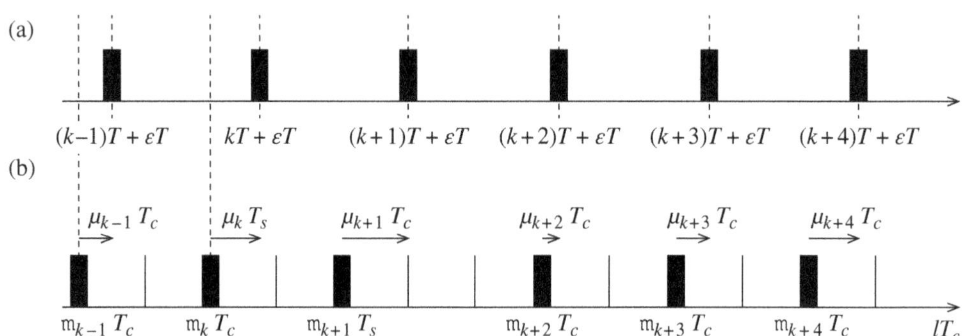

Figure 14.21 (a) Transmitter time scale; (b) receiver time scale with $T_c < T$, for the ideal case $\hat{\varepsilon} = \varepsilon$. Source: Reproduced with permission from Meyr et al. [4]. ©1998, IEEE.

Figure 14.21 illustrates the graphic representation of (14.84) in the ideal case $\hat{\varepsilon} = \varepsilon$. The transmitter time scale, defined by multiples of T, is shifted by a constant quantity equal to εT. The receiver time scale is defined by multiples of T_c. The fact that the ratio T/T_c may be a non-rational number has two consequences: first, the time shift $\mu_k T_c$ is time varying even if εT is a constant; second, the instants $\mathtt{m}_k T_c$ form a non-uniform subset of the receiver time axis, such that on average the considered samples are separated by an interval T.

With reference to Figure 14.19, to obtain the samples of the signal x at the instants $(\mathtt{m}_k + \mu_k)\, T_c$, we can proceed as follows:

(a) implement a digital interpolator filter that provides samples of the received signal at the instants $(n + \mu_k)\, T_c$, starting from samples at the instants nT_c (see Section 1.A.5);
(b) implement a downsampler that yields samples at the instants $(\mathtt{m}_k + \mu_k)\, T_c = kT + \hat{\varepsilon}T$.

With regard to the digital interpolator filter, consider a signal r_D with bandwidth $B_{r_D} \leq 1/(2T_c)$. From the sampling theorem, the signal r_D can be reconstructed from its samples $r_D(iT_c)$ using the relation

$$r_D(t) = \sum_{i=-\infty}^{+\infty} r_D(iT_c)\,\text{sinc}\left(\frac{t - iT_c}{T_c}\right) \tag{14.89}$$

This expression is valid for all t; in particular, it is valid for $t = t_1 + \mu_k T_c$, thus yielding the signal $r_D(t_1 + \mu_k T_c)$. Sampling this signal at $t_1 = nT_c$, we obtain

$$r_D(nT_c + \mu_k T_c) = \sum_{i=-\infty}^{+\infty} r_D(iT_c)\,\text{sinc}\,(n + \mu_k - i) \tag{14.90}$$

Observe that the second member of (14.90) is a discrete-time convolution; in fact, introducing the interpolator filter with impulse response h_I and parameter μ_k,

$$h_I(iT_c; \mu_k) = \text{sinc}(i + \mu_k), \qquad i = -\infty, \dots, +\infty \tag{14.91}$$

(14.90) can be rewritten as

$$r_D(nT_c + \mu_k T_c) = [r_D(iT_c) * h_I(iT_c; \mu_k)]\ (nT_c) \tag{14.92}$$

In other words, to obtain from samples of r_D at instants nT_c the samples at $nT_c + \mu_k T_c$, we can use a filter with impulse response $h_I(iT_c; \mu_k)$.[7]

[7] In practice, the filter impulse response $h_I(\cdot; \mu_k)$ must have a finite number N of coefficients. The choice of N depends on the ratio T/T_c and on the desired precision; for example for $T/T_c = 2$ and a normalized mean-square error (MSE),

With regard to the cascade of the matched filter $g_M(iT_c)$ and the decimator at instants $\mathtt{m}_k T_c$, we point out that a more efficient solution is to implement a filter with input at instants nT_c that generates output samples only at instants $\mathtt{m}_k T_c$.

We conclude this section by recalling that, if after the matched filter g_M, or directly in place of g_M, there is an equalizer filter c with input signal having sampling period equal to T_c, the function of the filter h_I is performed by the filter c itself (see Section 7.4.3).

Carrier phase recovery

An offset of the carrier phase equal to θ has the effect of rotating the complex symbols by $\exp(j\theta)$; this error can be corrected by multiplying the matched filter output by $\exp(-j\hat{\theta})$, where $\hat{\theta}$ is an estimate of θ.

Carrier phase recovery consists of three basic functions:

Phase estimate In the scheme of Figure 14.19, phase estimation is performed after the matched filter, using samples with sampling period equal to the symbol period T. In this scheme, timing recovery is implemented before phase recovery, and must operate in one of the following modes:

(a) with an arbitrary phase offset;
(b) with a phase estimate anticipating the multiplication by $e^{-j\hat{\theta}}$ after the decimator;
(c) jointly recovering phase and timing.

Phase rotation
(a) The samples $x(kT + \hat{\varepsilon}T, \hat{\Omega})$ are multiplied by the complex signal $\exp(-j\hat{\theta}(kT))$ (see Figure 14.19); a possible residual frequency offset $\Delta\Omega$ can be corrected by a time-varying phase given by $\hat{\theta}(kT) = \hat{\theta} + kT\widehat{\Delta\Omega}$.
(b) The samples $x(kT + \hat{\varepsilon}T, \hat{\Omega})\, e^{-j\hat{\theta}}$ are input to the data detector, assuming $(\hat{\theta}, \hat{\varepsilon})$ are the true values of (θ, ε).

Frequency synchronization A first coarse estimate of the frequency offset needs to be performed in the analog domain. In fact, algorithms for timing and phase recovery only work in the presence of a small residual frequency offset. A second block provides a fine estimate of this residual offset that is used for frequency offset compensation.

14.5 Algorithms for timing and carrier phase recovery

In this section, we discuss digital algorithms to estimate the time shift and the carrier phase offset under the assumption of absence of frequency offset, or $\Omega = 0$. Thus, the output samples of the decimator of Figure 14.19 are expressed as $x(kT + \hat{\varepsilon}T, 0)$; they will be simply denoted as $x(kT + \hat{\varepsilon}T)$, or in compact notation as $x_k(\hat{\varepsilon})$.

---

given by

$$J = 2T \int_0^1 \int_{-1/(2T)}^{1/(2T)} |e^{j2\pi f\mu} - \mathcal{H}_I(f;\mu)|^2 \, df \, d\mu \tag{14.93}$$

where

$$\mathcal{H}_I(f;\mu) = \sum_{i=-(N/2)+1}^{N/2} h_I(iT_c;\mu)\, e^{-j2\pi f i T_c} \tag{14.94}$$

equal to −50 dB, it turns out $N \simeq 5$. Of course, more efficient interpolator filters than that defined by (14.91) can be utilized.

14.5.1 ML criterion

The expression of the likelihood is obtained from (14.80) assuming $o = 0$, that is

$$\begin{aligned} \mathrm{L}_{\theta,\varepsilon,\boldsymbol{a}}(z, e, \boldsymbol{\alpha}) &= \exp\left\{\frac{2}{N_0}\sum_{k=0}^{K-1} Re[\alpha_k^* \; x_k(e)\; e^{-jz}]\right\} \\ &= \prod_{k=0}^{K-1} \exp\left\{\frac{2}{N_0} Re[\alpha_k^* \; x_k(e)\; e^{-jz}]\right\} \end{aligned} \tag{14.95}$$

Assumption of slow time varying channel

In general, both the time shift and the phase offset are time varying; from now on, we assume that the rate at which these parameters vary is much lower than the symbol rate $1/T$. Thus, it is useful to consider two time scales: one that refers to the symbol period T, for symbol detection and estimation of synchronization parameters, and the other that refers to a period much larger than T, for the variation of the synchronization parameters.

14.5.2 Taxonomy of algorithms using the ML criterion

Synchronization algorithms are obtained by the ML criterion (see (14.77) or equivalently (14.74)) averaging the probability density function $p_{r|\theta,\varepsilon,\boldsymbol{a}}(\boldsymbol{\rho} \mid z, e, \boldsymbol{\alpha})$ with respect to parameters that do not need to be estimated. Then, we obtain the following likelihood functions:

- for the joint estimate of (θ, ε),

$$p_{r|\theta,\varepsilon}(\boldsymbol{\rho} \mid z, e) = \sum_{\boldsymbol{\alpha}} p_{r|\theta,\varepsilon,\boldsymbol{a}}(\boldsymbol{\rho} \mid z, e, \boldsymbol{\alpha}) \; P[\boldsymbol{a} = \boldsymbol{\alpha}] \tag{14.96}$$

- for the estimate of the phase,

$$p_{r|\theta}(\boldsymbol{\rho} \mid z) = \int \left[\sum_{\boldsymbol{\alpha}} p_{r|\theta,\varepsilon,\boldsymbol{a}}(\boldsymbol{\rho} \mid z, e, \boldsymbol{\alpha}) \; P[\boldsymbol{a} = \boldsymbol{\alpha}] \; p_\varepsilon(e)\right] de \tag{14.97}$$

- for the estimate of timing,

$$p_{r|\varepsilon}(\boldsymbol{\rho} \mid e) = \int \left[\sum_{\boldsymbol{\alpha}} p_{r|\theta,\varepsilon,\boldsymbol{a}}(\boldsymbol{\rho} \mid z, e, \boldsymbol{\alpha}) \; P[\boldsymbol{a} = \boldsymbol{\alpha}] \; p_\theta(z)\right] dz \tag{14.98}$$

With the exception of some special cases, the above functions cannot be computed in close form. Consequently, we need to develop appropriate approximation techniques.

A first classification of synchronization algorithms is based on whether knowledge of the data sequence is available or not; in this case, we distinguish two classes:

1. *decision-directed* (DD) or *data-aided* (DA);
2. *non-data-aided* (NDA).

If the data sequence is known, for example by sending a training sequence $\boldsymbol{a} = \boldsymbol{\alpha}_0$ during the acquisition phase, we speak of *DA* algorithms. As the sequence $\boldsymbol{a}$ is known, in the sum in the expression of the likelihood function, only the term for $\boldsymbol{\alpha} = \boldsymbol{\alpha}_0$ remains. For example, the joint estimate of (θ, ε) reduces to the maximization of the likelihood $p_{r|\theta,\varepsilon,\boldsymbol{a}}(\boldsymbol{\rho} \mid z, e, \boldsymbol{\alpha}_0)$, and we get

$$(\hat{\theta}, \hat{\varepsilon})_{DA} = \arg\max_{z,e} \; p_{r|\theta,\varepsilon,\boldsymbol{a}}(\boldsymbol{\rho} \mid z, e, \boldsymbol{\alpha}_0) \tag{14.99}$$

On the other hand, whenever we use the detected sequence $\hat{\boldsymbol{a}}$ as if it were the true sequence $\boldsymbol{a}$, we speak of *data-directed* algorithms. If there is a high probability that $\hat{\boldsymbol{a}} = \boldsymbol{a}$, again in the sum in the expression of the likelihood function only one term remains. Taking again the joint estimate of (θ, ε) as an example, in (14.96) the sum reduces to

$$\sum_{\boldsymbol{\alpha}} p_{r|\theta,\varepsilon,\boldsymbol{a}}(\rho \mid z, e, \boldsymbol{\alpha})\, P[\boldsymbol{a} = \boldsymbol{\alpha}] \simeq p_{r|\theta,\varepsilon,\boldsymbol{a}}(\rho \mid z, e, \hat{\boldsymbol{a}}) \tag{14.100}$$

as $P[\boldsymbol{a} = \hat{\boldsymbol{a}}] \simeq 1$. The joint estimate (θ, ε) is thus given by

$$(\hat{\theta}, \hat{\varepsilon})_{DD} = \arg\max_{z,e}\; p_{r|\theta,\varepsilon,\boldsymbol{a}}(\rho \mid z, e, \hat{\boldsymbol{a}}) \tag{14.101}$$

NDA algorithms apply instead, in an exact or approximate fashion, the averaging operation with respect to the data sequence.

A second classification of synchronization algorithms is made based on the synchronization parameters that must be eliminated; then, we have four cases

1. DD&Dε: *data and timing directed,*

$$p_{r|\theta}(\rho \mid z) = p_{r|\theta,\varepsilon,\boldsymbol{a}}(\rho \mid z, \hat{\varepsilon}, \hat{\boldsymbol{a}}) \tag{14.102}$$

2. DD, *timing independent,*

$$p_{r|\theta}(\rho \mid z) = \int p_{r|\theta,\varepsilon,\boldsymbol{a}}(\rho \mid z, e, \hat{\boldsymbol{a}})\, p_{\varepsilon}(e)\, de \tag{14.103}$$

3. DD&Dθ: *data and phase directed,*

$$p_{r|\varepsilon}(\rho \mid e) = p_{r|\theta,\varepsilon,\boldsymbol{a}}(\rho \mid \hat{\theta}, e, \hat{\boldsymbol{a}}) \tag{14.104}$$

4. DD, *phase independent or non-coherent* (NC),

$$p_{r|\varepsilon}(\rho \mid e) = \int p_{r|\theta,\varepsilon,\boldsymbol{a}}(\rho \mid z, e, \hat{\boldsymbol{a}})\, p_{\theta}(z)\, dz \tag{14.105}$$

A further classification is based on the method for obtaining the timing and phase estimates:

1. *Feedforward (FF)*: The FF algorithms directly estimate the parameters (θ, ε) without using signals that are modified by the estimates; this implies using signals before the interpolator filter for timing recovery and before the phase rotator (see Figure 14.19) for carrier phase recovery.
2. *Feedback (FB)*: The FB algorithms estimate the parameters (θ, ε) using also signals that are modified by the estimates; in particular, they yield an estimate of the errors $e_\theta = \hat{\theta} - \theta$ and $e_\varepsilon = \hat{\varepsilon} - \varepsilon$, which are then used to control the interpolator filter and the phase rotator, respectively. In general, FB structures are able to track slow changes of parameters.

Next, we give a brief description of FB estimators, with emphasis on the fundamental blocks and on input–output relations.

Feedback estimators

In Figure 14.22, the block diagrams of a FB phase (FBθ) estimator and of a FB timing (FBε) estimator are illustrated. These schemes can be easily extended to the case of a FB frequency offset estimator. The two schemes only differ in the first block. In the case of the FBθ estimator, the first block is a phase rotator that, given the input signal $s(kT)$ ($x(kT + \hat{\varepsilon}T)$ in Figure 14.19), yields $s(kT)\exp(-j\hat{\theta}_k)$, where $\hat{\theta}_k$ is the estimate of θ at instant kT; in the case of the FBε estimator, it is an interpolator filter that, given the input signal $s(kT)$ ($r_D(nT_c)$ in Figure 14.19), returns $s(kT + \hat{\varepsilon}_k T)$ ($r_D(kT + \hat{\varepsilon}T)$ in Figure 14.19), where $\hat{\varepsilon}_k$ is the estimate of ε at instant kT.

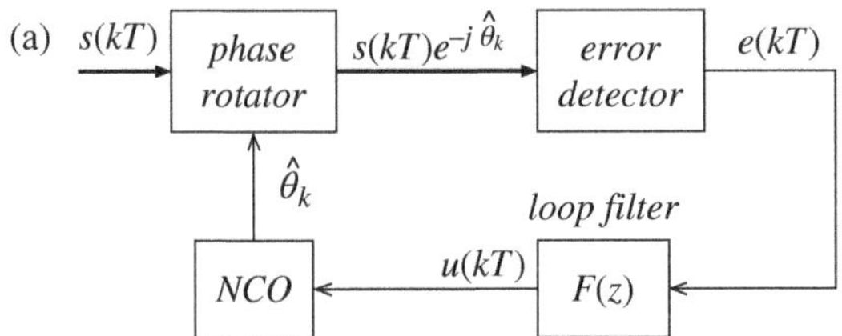

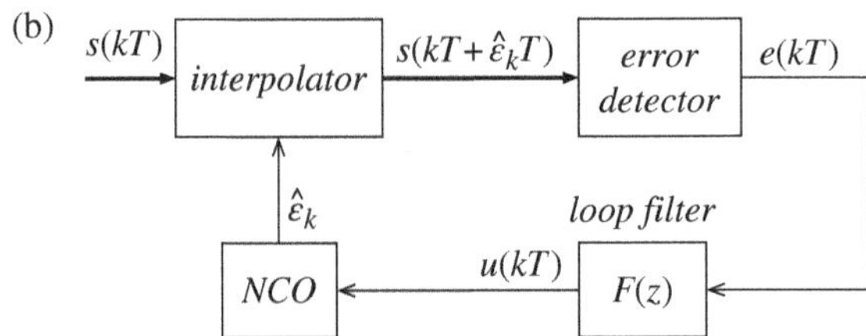

Figure 14.22 (a) Feedback estimator of θ; (b) feedback estimator of ε.

We analyse only the FBθ estimator, as the FBε estimator is similar. The error detector block is the fundamental block and has the function of generating a signal $e(kT)$, called *error signal*, whose *mean value is written as*

$$E[e(kT)] = g_\theta(\theta - \hat{\theta}_k) \tag{14.106}$$

where $g_\theta(\cdot)$ is an odd function. Then, the error signal $e(kT)$ admits the following decomposition

$$e(kT) = g_\theta(\theta - \hat{\theta}_k) + \eta_\theta(kT; \theta; \hat{\theta}_k) \tag{14.107}$$

where $\eta_\theta(\cdot)$ is a disturbance term called *loop noise*; the loop filter $F(z)$ is a lowpass filter that has two tasks: it regulates the speed of convergence of the estimator and mitigates the effects of the loop noise. The loop filter output, $u(kT)$, is input to the *numerically controlled oscillator* (NCO), that updates the phase estimate θ according to the following recursive relation:

$$\hat{\theta}_{k+1} = \hat{\theta}_k + \mu_\theta\, u(kT) \tag{14.108}$$

where μ_θ denotes the NCO gain.

In most FB estimators, the error signal $e(kT)$ is obtained as the derivative of the likelihood, or of its logarithm, with respect to the parameter to be estimated, evaluated using the most recent estimates $\hat{\theta}_k$, $\hat{\varepsilon}_k$. In particular, we have the two cases

$$\begin{aligned} e(kT) &\propto \frac{\partial}{\partial e}\, \mathrm{L}_{\theta,\varepsilon,\boldsymbol{a}}(\hat{\theta}_k, e, \boldsymbol{\alpha})\Big|_{e=\hat{\varepsilon}_k} \\ e(kT) &\propto \frac{\partial}{\partial z}\, \mathrm{L}_{\theta,\varepsilon,\boldsymbol{a}}(z, \hat{\varepsilon}_k, \boldsymbol{\alpha})\Big|_{z=\hat{\theta}_k} \end{aligned} \tag{14.109}$$

We note that in FB estimators, the vector $\boldsymbol{a}$ represents the transmitted symbols from instant 0 up to the instant corresponding to the estimates $\hat{\theta}_k$, $\hat{\varepsilon}_k$.

Early-late estimators

Early-late estimators constitute a subclass of FB estimators, where the error signal is computed according to (14.109), and the derivative operation is approximated by a finite difference [3], i.e. given a signal p, its derivative is computed as follows:

$$\frac{dp(t)}{dt} \simeq \frac{p(t+\delta) - p(t-\delta)}{2\delta} \tag{14.110}$$

where δ is a positive real number.

Consider, for example, a DD&Dθ timing estimator and denote by $G_{\rm L}[\cdot]$ the function that, given the input signal $x_k(\hat{\varepsilon}_k)$, yields the likelihood $\mathrm{L}_\varepsilon(e)$ evaluated at $e = \hat{\varepsilon}_k$. From (14.110), the error signal is given by

$$e(kT) \propto G_{\rm L}[x_k(\hat{\varepsilon}_k + \delta)] - G_{\rm L}[x_k(\hat{\varepsilon}_k - \delta)] \tag{14.111}$$

The block diagram of Figure 14.22b is modified into that of Figure 14.23. Observe that in the upper branch the signal $x_k(\hat{\varepsilon}_k)$ is anticipated of δT, while in the lower branch, it is delayed of $-\delta T$. Hence, the name of *early-late estimator*.

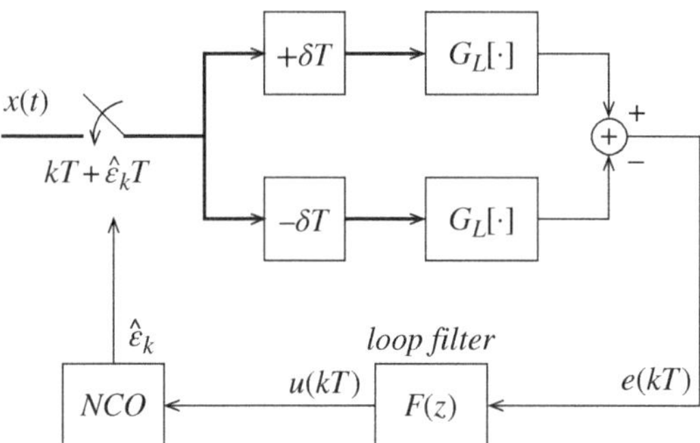

Figure 14.23 FB early-late timing estimator.

14.5.3 Timing estimators

Non-data aided

The estimate of ε can be obtained from (14.95) by eliminating the dependence on the parameters $\boldsymbol{\alpha}$ and z that are not estimated. To remove this dependence, we take the expectation with respect to $\boldsymbol{a}$; assuming i.i.d. symbols, we obtain the likelihood

$$\mathrm{L}_{\theta,\varepsilon}(z, e) = \prod_{k=0}^{K-1} E_{a_k}\left[\exp\left\{\frac{2}{N_0} Re[a_k^* \, x_k(e) \, e^{-jz}]\right\}\right] \tag{14.112}$$

where E_{a_k} denotes the expectation with respect to a_k.

For an M-PSK signal, with $M > 2$, we approximate a_k as $e^{j\varphi_k}$, where φ_k is a uniform random variable (r.v.) on $(-\pi; \pi]$; then, (14.112) becomes

$$\mathrm{L}_{\theta,\varepsilon}(z, e) = \prod_{k=0}^{K-1} \int_{-\pi}^{+\pi} \exp\left\{\frac{2}{N_0} Re[e^{-jv_k} \, x_k(e) \, e^{-jz}] \, \frac{dv_k}{2\pi}\right\} \tag{14.113}$$

If we use the definition of the Bessel function (4.58), (14.113) is independent of the phase θ and we get

$$\mathrm{L}_\varepsilon(e) = \prod_{k=0}^{K-1} I_0\left(\frac{|x_k(e)|}{N_0/2}\right) \tag{14.114}$$

On the other hand, if we take the expectation of (14.95) only with respect to the phase θ, we get

$$\mathrm{L}_{\varepsilon,\boldsymbol{a}}(e,\boldsymbol{\alpha}) = \prod_{k=0}^{K-1} I_0\left(\frac{|x_k(e)\ \alpha_k^*|}{N_0/2}\right) \tag{14.115}$$

We observe that, for M-PSK, $\mathrm{L}_{\varepsilon,\boldsymbol{a}}(e,\boldsymbol{\alpha}) = \mathrm{L}_{\varepsilon}(e)$, as $|\alpha_k|$ is a constant, while this does not occur for M-QAM.

To obtain estimates from the two likelihood functions just obtained, if the signal-to-noise ratio (SNR) Γ is sufficiently high, we utilize the fact that $I_0(\cdot)$ can be approximated as

$$I_0(\zeta) \simeq 1 + \frac{\zeta^2}{2} \qquad \text{for } |\zeta| \ll 1 \tag{14.116}$$

Taking the logarithm of the likelihood and eliminating non-relevant terms, we obtain the following NDA estimator and DA estimator.

$$\text{NDA}: \qquad \hat{\varepsilon} = \arg\max_e \ln\{\mathrm{L}_{\varepsilon}(e)\} \simeq \arg\max_e \sum_{k=0}^{K-1} |x_k(e)|^2 \tag{14.117}$$

$$\text{DA}: \qquad \hat{\varepsilon} = \arg\max_e \ln\{\mathrm{L}_{\varepsilon,\boldsymbol{a}}(e,\boldsymbol{\alpha})\} \simeq \arg\max_e \sum_{k=0}^{K-1} |x_k(e)|^2\ |\alpha_k|^2 \tag{14.118}$$

On the other hand, if the SNR $\Gamma \ll 1$, (14.112) can be approximated using a power series expansion of the exponential function. Taking the logarithm of (14.112), using the hypothesis of i.i.d. symbols, with $E[a_n] = 0$, and eliminating non-relevant terms, we obtain the following log-likelihood:

$$\ell_{\theta,\varepsilon}(z,e) = E[|a_n|^2] \sum_{k=0}^{K-1} |x_k(e)|^2 + Re\left[E[a_n^2] \sum_{k=0}^{K-1} (x_k^*(e))^2\ e^{j2z}\right] \tag{14.119}$$

Averaging with respect to θ, we obtain the following phase independent log-likelihood:

$$\ell_{\varepsilon}(e) = \sum_{k=0}^{K-1} |x_k(e)|^2 \tag{14.120}$$

which yields the same NDA estimator as (14.117).

For a modulation technique characterized by $E[a_n^2] \neq 0$, (14.119) may be used to obtain an NDA joint estimate of phase and timing. In fact for a phase estimate given by

$$\hat{\theta} = -\frac{1}{2}\ \arg\left\{E[a_n^2] \sum_{k=0}^{K-1} (x_k^*(e))^2\right\} \tag{14.121}$$

the second term of (14.119) is maximized. Substitution of (14.121) in (14.119) yields a new estimate $\hat{\varepsilon}$ given by

$$\hat{\varepsilon} = \arg\max_e\ E[|a_n|^2] \sum_{k=0}^{K-1} |x_k(e)|^2 + \left|E[a_n^2] \sum_{k=0}^{K-1} x_k^2(e)\right| \tag{14.122}$$

The block diagram of the joint estimator is shown in Figure 14.24, where P values of the time shift ε, $\varepsilon^{(m)}$, $m = 1, \dots, P$, equally spaced in $[-1/2, 1/2]$ are considered; usually, the resolution obtained with $P = 8$ or 10 is sufficient. For each time shift $\varepsilon^{(m)}$, the log-likelihood (14.122) is computed and the value of $\varepsilon^{(m)}$ associated with the largest value of the log-likelihood is selected as the timing estimate. Furthermore, we observe that in the generic branch m, filtering by the matched filter $g_M(iT_c + \varepsilon^{(m)}T)$

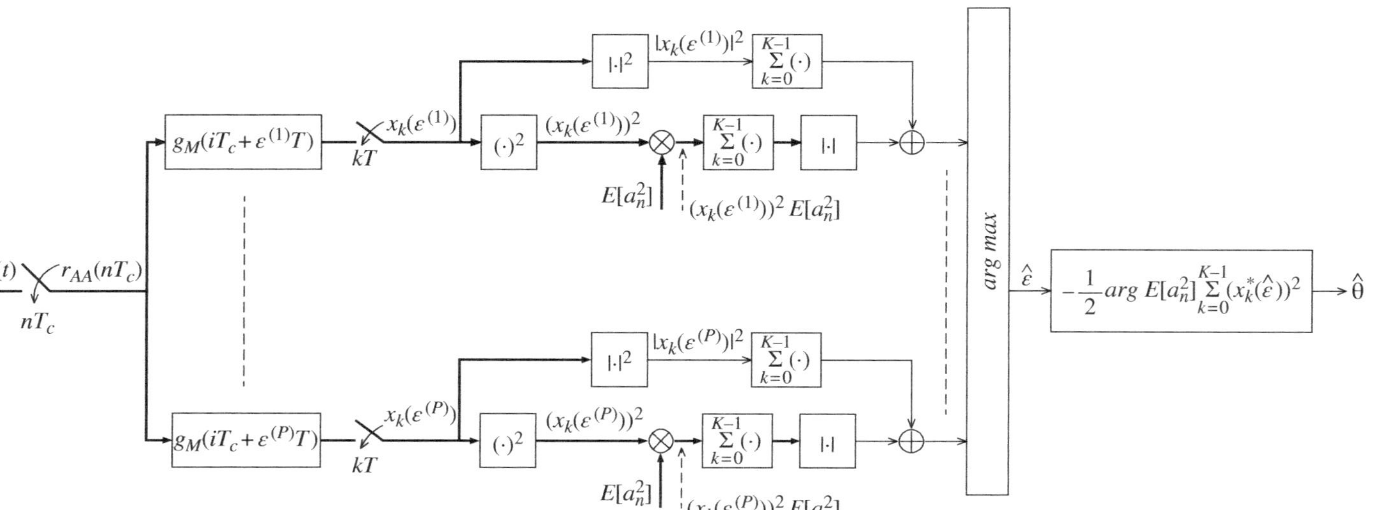

Figure 14.24 NDA joint timing and phase (for $E[a_n^2] \neq 0$) estimator. Source: Reproduced with permission from Meyr et al. [4]. ©1998, IEEE.

and sampling at the instants kT can be implemented by the cascade of an interpolator filter $h_I(iT_c; \mu^{(m)})$ (where $\mu^{(m)}$ depends on $\varepsilon^{(m)}$) and a filter $g_M(iT_c)$, followed by a decimator that provides samples at the instants $\mathtt{m}_k T_c$, as illustrated in Figure 14.19 and described in Section 14.4.

NDA synchronization via spectral estimation

Let us consider the log-likelihood (14.120) limited to a symmetric observation time interval $[-LT, LT]$. Thus, we get

$$\ell_\varepsilon(e) = \sum_{k=-L}^{L} |x(kT + eT)|^2 \tag{14.123}$$

Now, as x is a QAM signal, the process $|x(kT + eT)|^2$ is approximately cyclostationary in e of period 1 (see Section 7.1.2). We introduce the following Fourier series representation

$$|x(kT + eT)|^2 = \sum_{i=-\infty}^{+\infty} c_i^{(k)} \, e^{j2\pi ie} \tag{14.124}$$

where the coefficients $\{c_i^{(k)}\}$ are r.v.s. given by

$$c_i^{(k)} = \int_0^1 |x(kT + eT)|^2 \, e^{-j2\pi ie} \, de \tag{14.125}$$

Now (14.123) is equal to the average of the cyclostationary process $|x(kT + eT)|^2$ in the interval $[-L; L]$. Defining

$$c_i = \sum_{k=-L}^{L} c_i^{(k)} \tag{14.126}$$

it results [4] that only c_0 and c_1 have non-zero mean, and (14.123) can be written as

$$\ell_\varepsilon(e) = c_0 + 2Re[c_1 \, e^{j2\pi e}] + \underbrace{\sum_{|i|\geq 2} 2Re[c_i \, e^{j2\pi ie}]}_{\substack{\text{disturbance with zero mean} \\ \text{for each value of } e}} \tag{14.127}$$

As c_0 and $|c_1|$ are independent of e, the maximum of $\ell_\varepsilon(e)$ yields

$$\hat{\varepsilon} = -\frac{1}{2\pi} \arg c_1 \tag{14.128}$$

However, the coefficient c_1 is obtained by integration, which in general is hard to implement in the digital domain; on the other hand, if the bandwidth of $|x(lT)|^2$ satisfies the relation

$$B_{|x|^2} = \frac{1}{T}(1 + \rho) < \frac{1}{2T_c} \tag{14.129}$$

where ρ is the roll-off factor of the matched filter, then c_1 can be computed by discrete Fourier transform (DFT). Let $F_0 = T/T_c$, then we get

$$c_1 = \sum_{k=-L}^{L} \left[\frac{1}{F_0} \sum_{l=0}^{F_0-1} |x([kF_0 + l] \, T_c)|^2 \, e^{-j(2\pi/F_0)l} \right] \tag{14.130}$$

A simple implementation of the estimator is possible for $F_0 = 4$; in fact, in this case no multiplications are needed. As $e^{-j(2\pi/4)l} = (-j)^l$, (14.130) simplifies into

$$c_1 = \sum_{k=-L}^{L} \left[\frac{1}{4} \sum_{l=0}^{3} |x([4k + l] \, T_c)|^2 \, (-j)^l \right] \tag{14.131}$$

Figure 14.25 illustrates the implementation of the estimator for $F_0 = 4$.

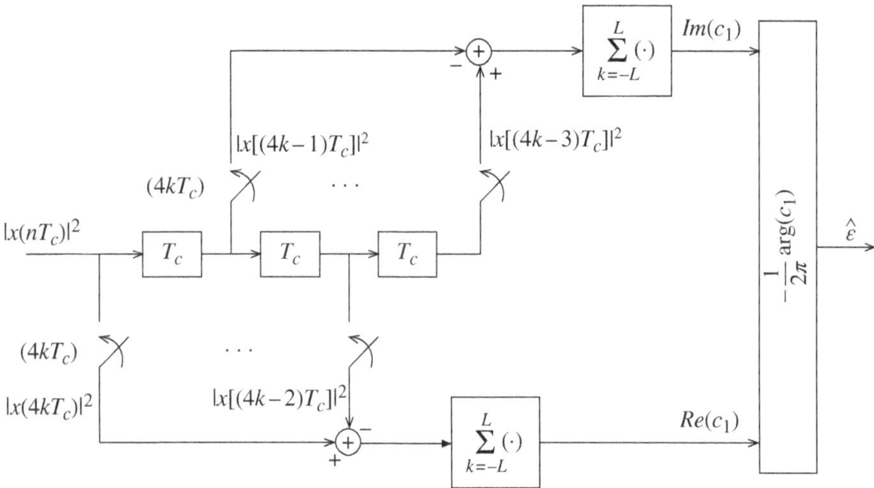

Figure 14.25 NDA timing estimator via spectral estimation for the case $F_0 = 4$. Source: Reproduced with permission from Meyr et al. [4]. ©1998, IEEE.

Data aided and data directed

If in (14.95) we substitute the parameters α_k and z with their estimates, we obtain the phase independent DA (DD) log-likelihood

$$\mathrm{L}_\varepsilon(e) = \exp\left\{\frac{2}{N_0}\,Re\left[\sum_{k=0}^{K-1}\hat{a}_k^*\,x_k(e)\,e^{-j\hat{\theta}}\right]\right\} \tag{14.132}$$

from which we immediately derive the estimate

$$\hat{\varepsilon} = \arg\max_{e}\ \mathrm{L}_\varepsilon(e) \tag{14.133}$$

The block diagram of the estimator is shown in Figure 14.26. Note that this algorithm can only be used in the case phase recovery is carried out before timing recovery.

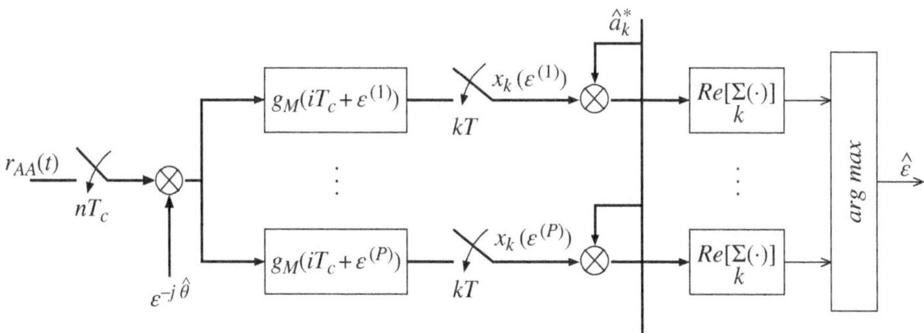

Figure 14.26 Phase independent DA (DD) timing estimator. Source: Reproduced with permission from Meyr et al. [4]. ©1998, IEEE.

For a joint phase and timing estimator, from (14.95) we get

$$\mathrm{L}_{\theta,\varepsilon}(z,e) = \exp\left\{\frac{2}{N_0}\,Re\left[\sum_{k=0}^{K-1}\hat{a}_k^*\,x_k(e)\,e^{-jz}\right]\right\} \tag{14.134}$$

Defining

$$\mathtt{r}(e) = \sum_{k=0}^{K-1} \hat{a}_k^* \, x_k(e) \tag{14.135}$$

the estimation algorithm becomes

$$\begin{aligned}(\hat{\theta}, \hat{\varepsilon}) &= \arg \max_{z,e} \; Re[\mathtt{r}(e)\, e^{-jz}] \\ &= \arg \max_{z,e} \; |\mathtt{r}(e)| \; Re[e^{-j(z-\arg(\mathtt{r}(e)))}]\end{aligned} \tag{14.136}$$

The two-variable search of the maximum reduces to a single-variable search; as a matter of fact, once the value of e that maximizes $|\mathtt{r}(e)|$ is obtained, which is independent of z, the second term

$$Re[e^{-j(z-\arg(\mathtt{r}(e)))}] \tag{14.137}$$

is maximized by $z = \arg(\mathtt{r}(e))$. Therefore, the joint estimation algorithm is given by

$$\begin{aligned}\hat{\varepsilon} &= \arg \max_{e} \; |\mathtt{r}(e)| = \arg \max_{e} \left| \sum_{k=0}^{K-1} \hat{a}_k^* \, x_k(e) \right| \\ \hat{\theta} &= \arg \mathtt{r}(\hat{\varepsilon}) = \arg \sum_{k=0}^{K-1} \hat{a}_k^* \, x_k(\hat{\varepsilon})\end{aligned} \tag{14.138}$$

Figure 14.27 illustrates the implementation of this second estimator. Note that this scheme is a particular case of (7.603).

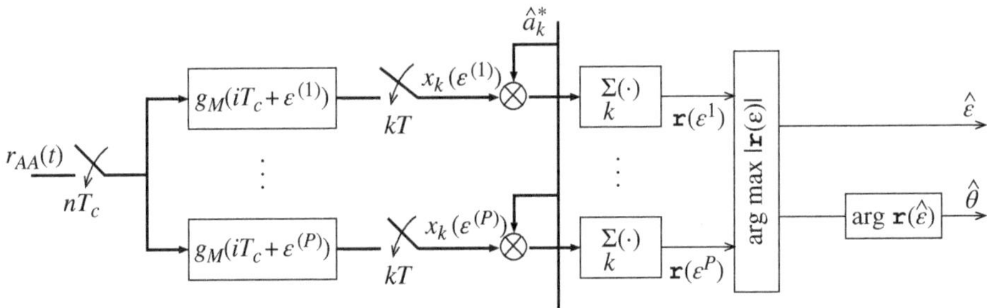

Figure 14.27 DA (DD) joint phase and timing estimator. Source: Reproduced with permission from Meyr et al. [4]. ©1998, IEEE.

For both estimators, estimation of the synchronization parameters is carried out every K samples, according to the assumption of slow parameter variations made at the beginning of the section.

Observation 14.2

When the channel is not known, to implement the matched filter g_M, we need to estimate the overall impulse response q_C. Then, the estimation of q_C, for example by one of the methods presented in Appendix 1.C, and of timing can be performed jointly.

Let $F_0 = T/T_c$ and $Q_0 = T/T_Q$ be integers, with $Q_0 \geq F_0$. From the signal $\{r_{AA}(qT_Q)\}$, obtained by oversampling r_{AA} or by interpolation of $\{r_{AA}(nT_c)\}$, and the knowledge of the training sequence $\{a_k\}$, $k = 0, \ldots, L_{TS} - 1$, we estimate either q_C with sampling period T_Q or equivalently its Q_0/F_0 polyphase components with sampling period T_c (see Observation 7.6). Limiting the estimate to the more significant consecutive samples around the peak, the estimating of the timing phase with precision T_Q coincides with selecting the polyphase component with the largest energy among the Q_0/F_0 polyphase components. This determines the optimum filter g_M with sampling period T_c. Typically, for radio systems $F_0 = 2$, and $Q_0 = 4$ or 8.

Data and phase directed with feedback: differentiator scheme

Differentiating the log-likelihood (14.95) with respect to e, neglecting non-relevant terms, and evaluating the result at $(\hat{\theta}, e, \hat{\boldsymbol{a}})$, we obtain

$$\frac{\partial}{\partial e} \ln\{\mathrm{L}_\varepsilon(e)\} \propto Re\left[\sum_{k=0}^{K-1} \hat{a}_k^* \frac{\partial}{\partial e} x(kT + eT)\, e^{-j\hat{\theta}}\right] \tag{14.139}$$

With reference to the scheme of Figure 14.22, if we suppose that the sum in (14.139) is approximated by the filtering operation by the loop filter $F(z)$, the error signal $e(kT)$ results

$$e(kT) = Re\left[\hat{a}_k^* \frac{\partial}{\partial e} x(kT + eT)|_{e=\hat{\varepsilon}_k}\, e^{-j\hat{\theta}}\right] \tag{14.140}$$

The partial derivative of $x(kT + eT)$ with respect to e can be carried out in the digital domain by a differentiator filter with an ideal frequency response given by

$$\mathcal{H}_d(f) = j2\pi f, \qquad |f| \leq \frac{1}{2T_c} \tag{14.141}$$

In practice, if $T/T_c \geq 2$, it is simpler to implement a differentiator filter by a finite difference filter having a symmetric impulse response given by

$$h_d(iT_c) = \frac{1}{2T_c}\,(\delta_{i+1} - \delta_{i-1}) \tag{14.142}$$

Figure 14.28 illustrates the block diagram of the estimator, where the compact notation $\dot{x}$ is used in place of $(dx(t)/dt)$; moreover, based on the analysis of Section 14.5.2, if $u(kT)$ is the loop filter output, the estimate of ε is given by

$$\hat{\varepsilon}_{k+1} = \hat{\varepsilon}_k + \mu_\varepsilon u(kT) \tag{14.143}$$

where μ_ε is a suitable constant. Applying (14.88) to the value of $\hat{\varepsilon}_{k+1}$, we obtain the values of μ_{k+1} and $\mathtt{m}_{k+1}$.

Data and phase directed with feedback: Mueller and Muller scheme

The present algorithm gets its name from Mueller and Muller, who first proposed it in 1976 [6]. Consider the estimation error $e_\varepsilon = \hat{\varepsilon} - \varepsilon$ and the pulse $q_R(t) = q_C * g_M(t)$. The basic idea consists in generating an *error signal* whose mean value assumes one of the following two expressions:

$$\text{Type A:} \qquad E[e(kT)] = Re\left\{\frac{1}{2}[q_R(e_\varepsilon T + T) - q_R(e_\varepsilon T - T)]\right\} \tag{14.144}$$

$$\text{Type B:} \qquad E[e(kT)] = Re\{q_R(e_\varepsilon T + T)\} \tag{14.145}$$

Observe that, under the assumptions of Section 14.4, q_R is a Nyquist pulse; moreover, we assume that in the absence of channel distortion, q_R is an even function.

Note that the signal (14.144) is an odd function of the estimation error e_ε for $e_\varepsilon \in (-\infty, \infty)$, whereas the signal (14.145) is an odd function of e_ε only around $e_\varepsilon = 0$. Under lock conditions, i.e. for $e_\varepsilon \to 0$, the two versions of the algorithm exhibit a similar behaviour. However, the type A algorithm outperforms the type B algorithm in transient conditions because the mean value of the error signal for the type B algorithm is not symmetric. Moreover, the type A algorithm turns out to be effective also in the presence of signal distortion.

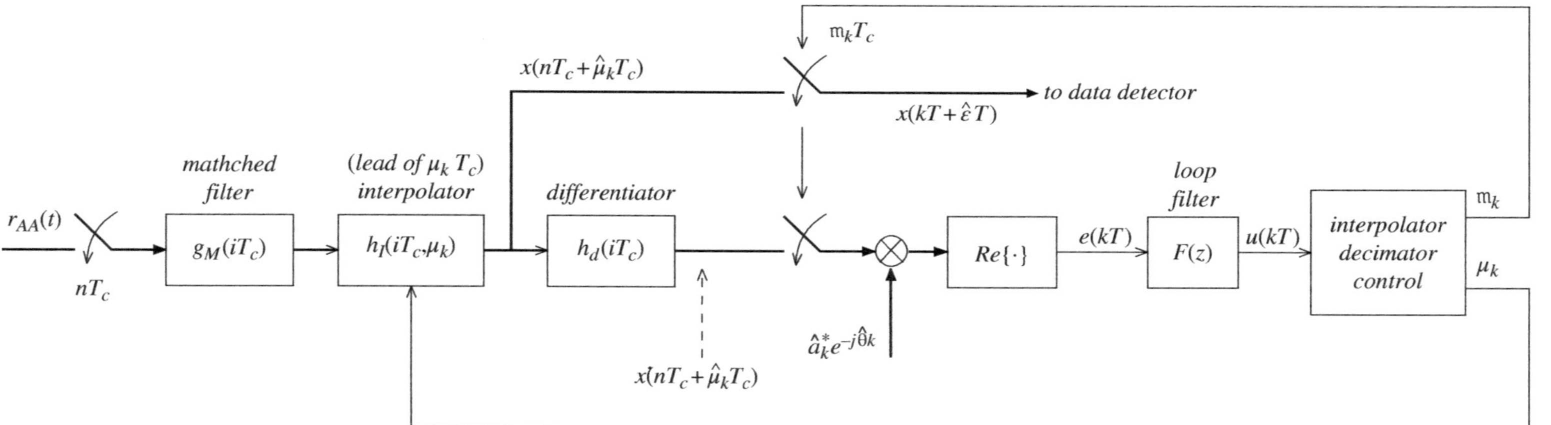

Figure 14.28 DD&Dθ-FB timing estimator. Source: Reproduced with permission from Meyr et al. [4]. ©1998, IEEE.

The error signal for the type A algorithm is chosen equal to

$$e(kT) = \kappa \; Re[[\hat{a}_{k-1}^* \; x_k(\hat{\varepsilon}) - \hat{a}_k^* \; x_{k-1}(\hat{\varepsilon})] \; e^{-j\hat{\theta}}] \tag{14.146}$$

where κ is a suitable constant whose value is discussed below. Assuming that $\hat{a}_{k-1} = a_{k-1}$, $\hat{a}_k = a_k$, and $\hat{\theta} = \theta$, from (14.71) and (14.79) for $o = \Omega = 0$ (14.146) can be written as

$$\begin{aligned} e(kT) = \kappa \; Re \Bigg\{ \Bigg[& a_{k-1}^* \sum_{i=-\infty}^{+\infty} a_i \; q_R(kT + e_\varepsilon T - iT) + \\ & - a_k^* \sum_{i=-\infty}^{+\infty} a_i \; q_R((k-1)T + e_\varepsilon T - iT) \Bigg] \\ & + a_{k-1}^* \; \tilde{w}_k - a_k^* \; \tilde{w}_{k-1} \end{aligned} \tag{14.147}$$

where $\tilde{w}_k$ is the decimated noise signal at the matched filter output. We define

$$q_m(e_\varepsilon) = q_R(mT + e_\varepsilon T) \tag{14.148}$$

then with a suitable change of variables, (14.147) becomes

$$e(kT) = \kappa Re \left\{ \left[a_{k-1}^* \sum_{m=-\infty}^{+\infty} a_{k-m} q_m(e_\varepsilon) - a_k^* \sum_{m=-\infty}^{+\infty} a_{k-1-m} q_m(e_\varepsilon) \right] + a_k^* \tilde{w}_k - a_k^* \tilde{w}_{k-1} \right\} \tag{14.149}$$

Taking the mean value of $e(kT)$, we obtain

$$\begin{aligned} E[e(kT)] &= \kappa \; Re\{(E[|a_k|^2] - |\mathtt{m}_a|^2)[q_1(e_\varepsilon) - q_{-1}(e_\varepsilon)]\} \\ &= \kappa \; Re\{(E[|a_k|^2] - |\mathtt{m}_a|^2)[q_R(e_\varepsilon T + T) - q_R(e_\varepsilon T - T)]\} \end{aligned} \tag{14.150}$$

For

$$\kappa = \frac{1}{2(E[|a_k|^2] - |\mathtt{m}_a|^2)} \tag{14.151}$$

we obtain (14.144). Similarly, in the case of the type B algorithm, the error signal assumes the expression

$$e(kT) = \frac{1}{(E[|a_k|^2] - |\mathtt{m}_a|^2)} \; [\hat{a}_{k-1} - \mathtt{m}_a]^* \; x_k(\hat{\varepsilon}) \; e^{-j\hat{\theta}} \tag{14.152}$$

Figure 14.29 illustrates the block diagram of the direct section of the type A estimator. The constant κ is included in the loop filter and is not explicitly shown.

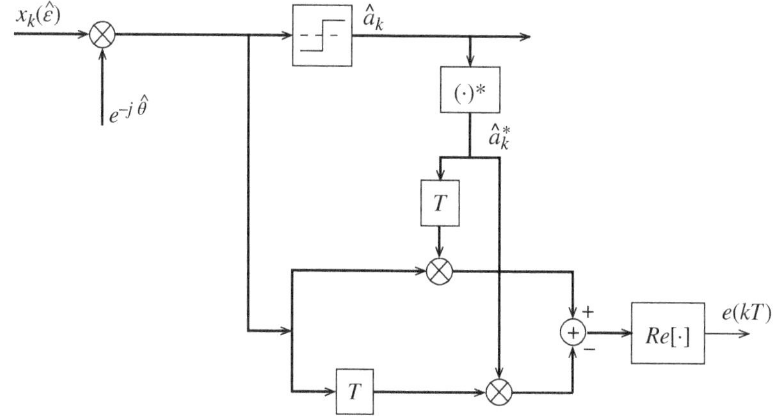

Figure 14.29 Mueller and Muller type A timing estimator.

Non-data aided with feedback

We consider the log-likelihood (14.120), obtained for a NDA estimator, and differentiate it with respect to e to get

$$\begin{aligned}\frac{\partial \ell_{\varepsilon}(e)}{\partial e} &= \frac{\partial}{\partial e}\sum_{k=0}^{K-1}|x(kT+eT)|^2 \\ &= \sum_{k=0}^{K-1} 2Re[x(kT+eT)\,\dot{x}^*(kT+eT)]\end{aligned} \tag{14.153}$$

If we assume that the sum is carried out by the loop filter, the error signal is given by

$$e(kT) = Re[x(kT+\hat{\varepsilon}_k T)\,\dot{x}^*(kT+\hat{\varepsilon}_k T)] \tag{14.154}$$

14.5.4 Phasor estimators

Data and timing directed

We discuss an algorithm that directly yields the phasor $\exp(j\hat{\theta})$ in place of the phase $\hat{\theta}$. Assuming that an estimate of $\hat{\boldsymbol{a}}$ and $\hat{\varepsilon}$ is available, the likelihood (14.95) becomes

$$\mathrm{L}_{\theta}(z) = \exp\left\{\frac{2}{N_0}\,Re\left[\sum_{k=0}^{K-1}\hat{a}_k^*\,x_k(\hat{\varepsilon})\,e^{-jz}\right]\right\} \tag{14.155}$$

and is maximized by

$$e^{j\hat{\theta}} = e^{j\arg\sum_{k=0}^{K-1}\hat{a}_k^*\,x_k(\hat{\varepsilon})} \tag{14.156}$$

Figure 14.30 shows the implementation of the estimator (14.156).

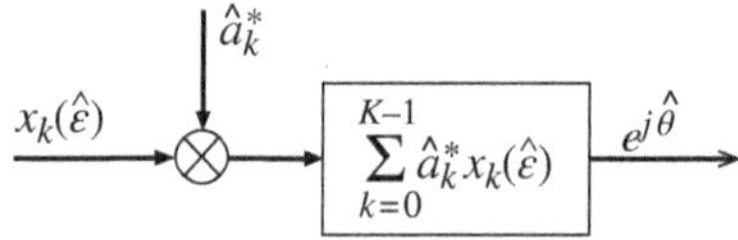

Figure 14.30 DD&Dε estimator of the phasor $e^{j\theta}$.

Non-data aided for M-PSK signals

In an M-PSK system, to remove the data dependence from the decimator output signal in the scheme of Figure 14.19, we raise the samples $x_k(\hat{\varepsilon})$ to the M-th power. Assuming absence of ISI, we get

$$x_k^M(\hat{\varepsilon}) = [a_k e^{j\theta} + \tilde{w}_k]^M = a_k^M e^{jM\theta} + w_{M,k} \tag{14.157}$$

where $\tilde{w}_k$ represents the decimator output noise, and $w_{M,k}$ denotes the overall disturbance.

As $a_k^M = (e^{j2\pi l/M})^M = 1$, (14.157) becomes

$$x_k^M(\hat{\varepsilon}) = e^{jM\theta} + w_{M,k} \tag{14.158}$$

From (14.95), we substitute $(x_k(\hat{\varepsilon}))^M$ for $\alpha_k^* x_k(\hat{\varepsilon})$ obtaining the likelihood

$$\mathrm{L}_{\theta}(z) = \exp\left\{\frac{2}{N_0}\,Re\left[\sum_{k=0}^{K-1}(x_k(\hat{\varepsilon}))^M e^{-jzM}\right]\right\} \tag{14.159}$$

which is maximized by the phasor

$$\exp(j\hat{\theta}M) = \exp\left[j \arg \sum_{k=0}^{K-1} (x_k(\hat{\varepsilon}))^M\right] \tag{14.160}$$

We note that raising $x_k(\hat{\varepsilon})$ to the M-th power causes a phase ambiguity equal to a multiple of $(2\pi)/M$; in fact, if $\hat{\theta}$ is a solution to (14.160), also $(\hat{\theta} + 2\pi l/M)$ for $l = 0, \ldots, M-1$, are solutions. This ambiguity can be removed, for example, by differential encoding (see Section 16.1.2). The estimator block diagram is illustrated in Figure 14.31.

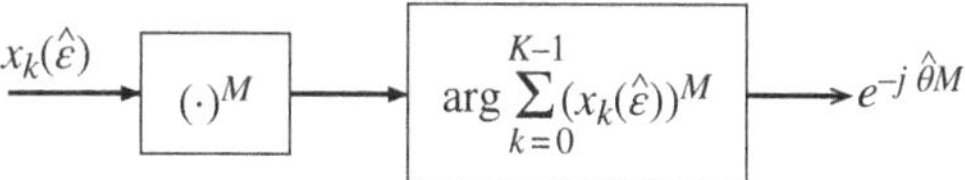

Figure 14.31 NDA estimator of the phasor $e^{j\theta}$ for M-PSK.

Data and timing directed with feedback

Consider the likelihood (14.155) obtained for the DD&Dε estimator of the phasor $e^{j\theta}$. Taking the logarithm, differentiating it with respect to z, and neglecting non-relevant terms, we obtain the error signal

$$e(kT) = Im[\hat{a}_k^* \, x_k(\hat{\varepsilon}) \, e^{-j\hat{\theta}_k}] \tag{14.161}$$

Observe that, in the absence of noise, $x_k(\hat{\varepsilon}) = a_k \, e^{j\theta}$, and (14.161) becomes

$$e(kT) = |\hat{a}_k|^2 \sin(\theta - \hat{\theta}_k) \tag{14.162}$$

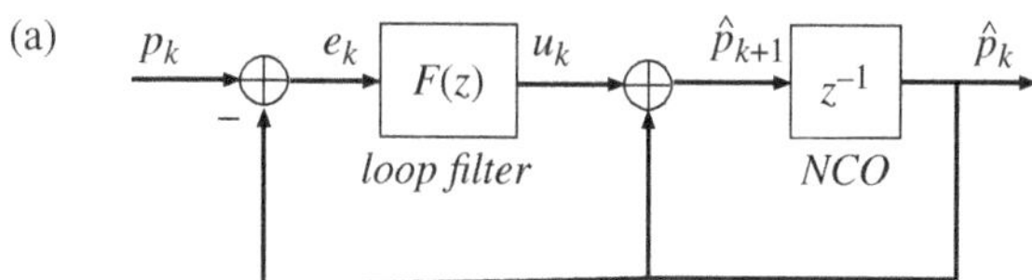

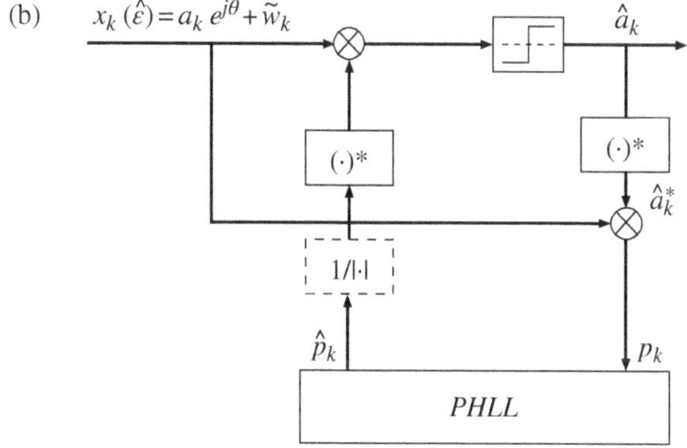

Figure 14.32 (a) PHLL; (b) DD&Dε-FB phasor estimator. Source: Reproduced with permission from Meyr et al. [4]. ©1998, IEEE.

Hence, we can use a digital version of the PLL to implement the estimator. However, the error signal (14.161) introduces a phase ambiguity; in fact it assumes the same value if we substitute $(\hat{\theta}_k - \pi)$ for $\hat{\theta}_k$. An alternative to the digital PLL is given by the *phasor-locked loop* (PHLL), that provides an estimate of the phasor $e^{j\theta}$, rather than the estimate of θ, thus eliminating the ambiguity.

The block diagram of the PHLL is illustrated in Figure 14.32a; it is a FB structure with the phasor $p_k = e^{j\theta_k}$ as input and the estimate $\hat{p}_k = e^{j\hat{\theta}_k}$ as output. The error signal e_k is obtained by subtracting the estimate $\hat{p}_k$ from p_k; then e_k is input to the loop filter $F(z)$ that yields the signal u_k, which is used to update the phasor estimate according to the recursive relation

$$\hat{p}_{k+1} = \hat{p}_k + (f * u)(k) \tag{14.163}$$

Figure 14.32b illustrates the block diagram of a DD&Dε phasor estimator that implements the PHLL. Observe that the input phasor p_k is obtained by multiplying $x_k(\hat{\varepsilon})$ by $\hat{a}_k^*$ to remove the dependence on the data. The dashed block normalizes the estimate $\hat{p}_k$ in the QAM case.

14.6 Algorithms for carrier frequency recovery

As mentioned in Section 14.4, phase and timing estimation algorithms work correctly only if the frequency offset is small. Therefore, the frequency offset must be compensated before the estimate of the other two synchronization parameters takes place. Hence, the algorithms that we will present are mainly NDA and *non-clock aided* (NCA); timing-directed algorithms are possible only in the case the frequency offset has a magnitude much smaller than $1/T$.

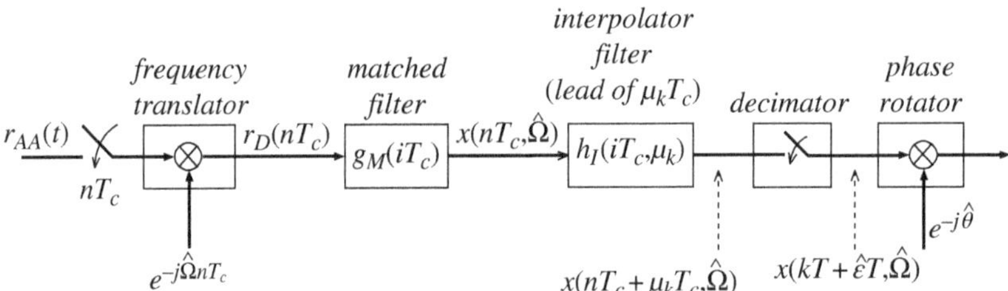

Figure 14.33 Receiver of Figure 14.19 with interpolator and matched filter interchanged. Source: Reproduced with permission from Meyr et al. [4]. ©1998, IEEE.

In Figure 14.33, we redraw part of the digital receiver scheme of Figure 14.19. Observe that the position of the matched filter is interchanged with that of the interpolator filter. In this scheme, the samples $r_{AA}(nT_c)$ are multiplied by $\exp(-j\hat{\Omega}nT_c)$ to remove the frequency offset.

In [4] it is shown that, whenever Ω satisfies the condition

$$\left|\frac{\Omega T}{2\pi}\right| \leq 0.15 \tag{14.164}$$

the following approximation holds:

$$x(kT + eT, o) \simeq e^{-jokT}\, x_k(e) \tag{14.165}$$

Then, the likelihood (14.80) can be written as

$$\mathrm{L}_{\theta,\varepsilon,\Omega,a}(z, e, o, \boldsymbol{\alpha}) = \exp\left\{ \frac{2}{N_0} \sum_{k=0}^{K-1} Re[\alpha_k^*\, x_k(e)\, e^{-jokT}\, e^{-jz}] \right\} \tag{14.166}$$

Therefore, in the schemes of Figures 14.19 and 14.33, the frequency translator may be moved after the decimator, together with the phase rotator.

14.6.1 Frequency offset estimators

Non-data aided

Suppose the receiver operates with a low SNR Γ. Similarly to (14.123) the log-likelihood for the joint estimate of (ε, Ω) in the observation interval $[-LT, LT]$ is given by

$$\ell_{\varepsilon,\Omega}(e, o) = \sum_{k=-L}^{L} |x(kT + eT, o)|^2 \tag{14.167}$$

By expanding $\ell_{\varepsilon,\Omega}(e, o)$ in Fourier series and using the notation introduced in the previous section, we obtain

$$\ell_{\varepsilon,\Omega}(e, o) = c_0 + 2Re[c_1\, e^{j2\pi e}] + \underbrace{\sum_{|i|\geq 2} 2Re[c_i\, e^{j2\pi ie}]}_{\text{disturbance}} \tag{14.168}$$

Now the mean value of c_0, $E[c_0]$, depends on o, but is independent of e and furthermore is maximized for $o = \hat{\Omega}$. Hence

$$\hat{\Omega} = \arg\max_{o} c_0 \tag{14.169}$$

As we did for the derivation of (14.131), starting with (14.169) and assuming the ratio $F_0 = T/T_c$ is an integer, we obtain the following joint estimate of (Ω, ε) [4]:

$$\begin{aligned} \hat{\Omega} &= \arg\max_{o} \sum_{n=-LF_0}^{LF_0-1} |x(nT_c, o)|^2 \\ \hat{\varepsilon} &= \arg \sum_{n=-LF_0}^{LF_0-1} |x(nT_c, o)|^2\, e^{-j2\pi n/F_0} \end{aligned} \tag{14.170}$$

The implementation of the estimator is illustrated in Figure 14.34. Observe that the signal $x(nT_c, o)$ can be rewritten as

$$\begin{aligned} x(nT_c, o) &= \sum_i r_{AA}(iT_c)\, e^{-joiT_c}\, g_M(nT_c - iT_c) \\ &= e^{-jonT_c} \sum_i r_{AA}(iT_c)\, e^{-jo(i-n)T_c}\, g_M(nT_c - iT_c) \\ &= e^{-jonT_c} \sum_i r_{AA}(iT_c)\, g_M^{(pb)}(nT_c - iT_c; o) \end{aligned} \tag{14.171}$$

where the expression of the filter

$$g_M^{(pb)}(iT_c; o) = g_M(iT_c)\, e^{joiT_c} \tag{14.172}$$

depends on the offset o. Defining

$$x_o(nT_c) = \sum_i r_{AA}(iT_c)\, g_M^{(pb)}(nT_c - iT_c; o) \tag{14.173}$$

we note that $|x(nT_c, o)| = |x_o(nT_c)|$, and hence in the m-th branch of Figure 14.34, the cascade of the frequency translator and the filter can be substituted with a simple filter with impulse response $g_M^{(pb)}(iT_c; \Omega^{(m)})$.

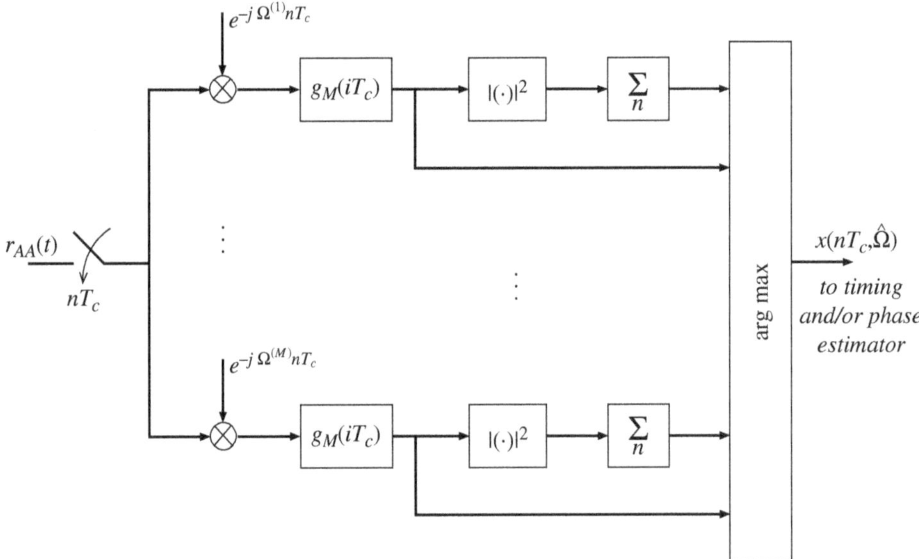

Figure 14.34 NDA frequency offset estimator. Source: Reproduced with permission from Meyr et al. [4]. ©1998, IEEE.

Non-data aided and timing independent with feedback

Differentiating the log-likelihood defined by (14.170), equal to $c_0(\Omega)$, with respect to Ω, we obtain the error signal

$$e(nT_c) = 2Re\left[x(nT_c, o)\,\frac{\partial}{\partial o}\,x^*(nT_c, o)\right]\Bigg|_{o=\hat{\Omega}_n} \tag{14.174}$$

Observe that, as $c_0(\Omega)$ is independent of e, then also $e(nT_c)$ is independent of e. From the first of (14.171), the partial derivative of $x(nT_c, o)$ with respect to o is given by

$$\frac{\partial}{\partial o}\,x(nT_c, o) = \sum_{i=-\infty}^{+\infty} (-jiT_c)\; r_{AA}(iT_c)\; e^{-joiT_c}\; g_M(nT_c - iT_c) \tag{14.175}$$

We define the *frequency matched filter* as

$$g_{FM}(iT_c) = (jiT_c)\cdot g_M(iT_c) \tag{14.176}$$

Observe now that, if the signal $r_D(nT_c) = r_{AA}(nT_c)\,e^{-jonT_c}$ is input to the filter $g_{FM}(iT_c)$, then from (14.175), the output is given by

$$x_{FM}(nT_c) = g_{FM} * r_D(nT_c) = \frac{\partial}{\partial o}\,x(nT_c, o) + jnT_c\,x(nT_c, o) \tag{14.177}$$

from which we obtain

$$\frac{\partial}{\partial o}\,x(nT_c, o) = x_{FM}(nT_c) - jnT_c\,x(nT_c, o) \tag{14.178}$$

Therefore, the expression of the error signal (14.174) becomes

$$e(nT_c) = 2Re[x(nT_c, \hat{\Omega}_n)\,x^*_{FM}(nT_c, \hat{\Omega}_n)] \tag{14.179}$$

The block diagram of the resultant estimator is shown in Figure 14.35. The loop filter output $u(kT_c)$ is sent to the NCO that yields the frequency offset estimate according to the recursive equation

$$\hat{\Omega}_{n+1}\,T_c = \hat{\Omega}_n\,T_c + \mu_\Omega\,u(nT_c) \tag{14.180}$$

where μ_Ω is the NCO gain.

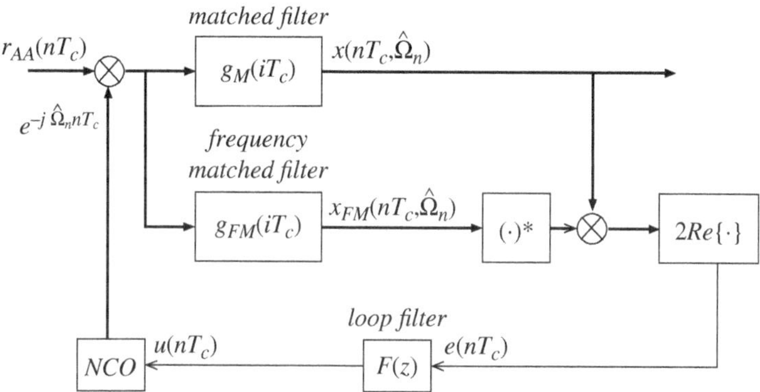

Figure 14.35 NDA-NDε-FB frequency offset estimator. Source: Reproduced with permission from Meyr et al. [4]. ©1998, IEEE.

Non-data aided and timing directed with feedback

Consider the log-likelihood (14.167). To get the Dε estimator, we substitute e with the estimate $\hat{\varepsilon}$ obtaining the log-likelihood

$$\ell_\Omega(o) = \sum_{k=0}^{K-1} |x(kT + \hat{\varepsilon}T, o)|^2 \tag{14.181}$$

Proceeding as in the previous section, we get the block diagram illustrated in Figure 14.36.

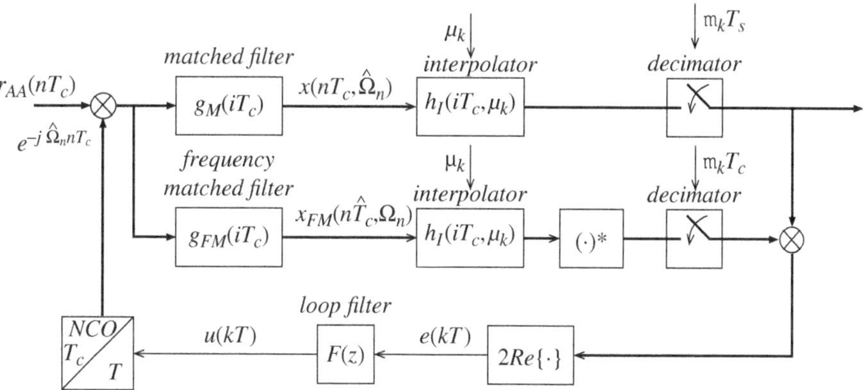

Figure 14.36 NDA-Dε-FB frequency offset estimator.

14.6.2 Estimators operating at the modulation rate

As mentioned at the beginning of Section 14.6, whenever the condition (14.164) is satisfied, the frequency translator can be moved after the decimator and consequently frequency offset estimation may take place after timing estimation, thus obtaining Dε algorithms. The likelihood (14.166) for an observation interval $[-(K-1)/2, (K-1)/2]$, with K odd, evaluated at $e = \hat{\varepsilon}$ becomes

$$\mathrm{L}_{\theta,\Omega,\boldsymbol{a}}(z, o, \boldsymbol{\alpha}) = \exp\left\{\frac{2}{N_0}\sum_{k=-(K-1)/2}^{(K-1)/2} Re[\alpha_k^*\, x_k(\hat{\varepsilon})\, e^{-jokT}\, e^{-jz}]\right\} \tag{14.182}$$

Data aided and data directed

We assume a known training sequence $\boldsymbol{a} = \boldsymbol{\alpha}_0$ is transmitted during the acquisition phase, and denote as $\alpha_{0,k}$ the k-th symbol of the sequence. The log-likelihood yields the joint estimate of (Ω, θ) as

$$(\hat{\Omega}, \hat{\theta}) = \arg\max_{z,o} \sum_{k=-(K-1)/2}^{(K-1)/2} Re[\alpha_{0,k}^* \, x_k(\hat{\varepsilon}) \, e^{-jokT} \, e^{-jz}] \tag{14.183}$$

Note that the joint estimate is computed by finding the maximum of a function of one variable; in fact, defining

$$\mathtt{r}(o) = \sum_{k=-(K-1)/2}^{(K-1)/2} \alpha_{0,k}^* \, x_k(\hat{\varepsilon}) \, e^{-jokT} \tag{14.184}$$

(14.183) can be rewritten as

$$(\hat{\Omega}, \hat{\theta}) = \arg\max_{z,o} \; |\mathtt{r}(o)| \; Re[e^{-j(z-\arg\{\mathtt{r}(o)\})}] \tag{14.185}$$

The maximum (14.183) is obtained by first finding the value of o that maximizes $|\mathtt{r}(o)|$,

$$\hat{\Omega} = \arg\max_{o} \; |\mathtt{r}(o)| \tag{14.186}$$

and then finding the value of z for which the term within brackets in (14.185) becomes real valued,

$$\hat{\theta} = \arg\{\mathtt{r}(\hat{\Omega})\} \tag{14.187}$$

We now want to solve (14.186) in close form; a necessary condition to get a maximum is that the derivative of $|\mathtt{r}(o)|^2 = \mathtt{r}(o) \, \mathtt{r}^*(o)$ with respect to o is equal to zero for $o = \hat{\Omega}$. Defining

$$b_k = \left[\frac{K^2 - 1}{4} - k(k+1)\right] \tag{14.188}$$

we get

$$\hat{\Omega}T = \arg\left\{ \sum_{k=-(K-1)/2}^{(K-1)/2} b_k \, \frac{\alpha_{0,k+1}^*}{\alpha_{0,k}^*} \, [x_{k+1}(\hat{\varepsilon}) \, x_k^*(\hat{\varepsilon})] \right\} \tag{14.189}$$

The DD estimator is obtained by substituting $\boldsymbol{\alpha}_0$ with the estimate $\hat{\boldsymbol{a}}$.

Non-data aided for M-PSK

By raising $(\alpha_{0,k+1}^* / \alpha_{0,k}^*)(x_{k+1}(\hat{\varepsilon}) \, x_k^*(\hat{\varepsilon}))$ to the M-th power, we obtain the NDA version of the DA (DD) algorithm for M-PSK signals as

$$\hat{\Omega}T = \arg\left\{ \sum_{k=-(K-1)/2}^{(K-1)/2} b_k \, [x_{k+1}(\hat{\varepsilon}) \, x_k^*(\hat{\varepsilon})]^M \right\} \tag{14.190}$$

14.7 Second-order digital PLL

In FB systems, the recovery of the phase θ and the frequency Ω can be jointly performed by a second-order *digital phase-locked loop* (DPLL), given by

$$\hat{\theta}_{k+1} = \hat{\theta}_k + \mu_\theta \, e_\theta(kT) + \hat{\Omega}_k \tag{14.191}$$

$$\hat{\Omega}_{k+1} = \hat{\Omega}_k + \mu_{\Omega,1} \, e_\theta(kT) + \mu_{\Omega,2} \, e_\Omega(kT) \tag{14.192}$$

where e_θ and e_Ω are estimates of the phase error and of the frequency error, respectively, and μ_θ, $\mu_{\Omega,1}$, and $\mu_{\Omega,2}$ are suitable constants. Typically, (see Example 15.6.4 on page 783) $\mu_{\Omega,1} \simeq \mu_{\Omega,2} \simeq \sqrt{\mu_\theta}$.

Observe that (14.191) and (14.192) form a digital version of the second-order analog PLL illustrated in Figure 14.7.

14.8 Synchronization in spread-spectrum systems

In this section, we discuss early-late FB schemes, named *delay-locked loops* (DLLs) [7], for the timing recovery in spread-spectrum systems (see Chapter 10).

14.8.1 The transmission system

Transmitter

In Figure 14.37 (see Figures 10.2 and 10.3), the (baseband equivalent) scheme of a spread-spectrum transmitter is illustrated. The symbols $\{a_k\}$, generated at the modulation rate $1/T$, are input to a holder, which outputs the symbols $\bar{a}_m$ at the *chip* rate $1/T_{chip}$. The *chip* period T_{chip} is given by

$$T_{chip} = \frac{T}{N_{SF}} \tag{14.193}$$

where N_{SF} denotes the spreading factor. Then, the symbols $\{a_k\}$ are multiplied by the spreading code $\{c_m\}$ and input to a filter with impulse response h_{Tx} that includes the digital-to-analog-converter (DAC). The (baseband equivalent) transmitted signal s is expressed as

$$s(t) = \sum_{m=-\infty}^{+\infty} \bar{a}_m \, c_m \, h_{Tx}(t - mT_{chip}) \tag{14.194}$$

In terms of the symbols $\{a_k\}$, we obtain the following alternative representation that will be used next:

$$s(t) = \sum_{k=-\infty}^{+\infty} a_k \sum_{m=kN_{SF}}^{kN_{SF}+N_{SF}-1} c_m \, h_{Tx}(t - mT_{chip}) \tag{14.195}$$

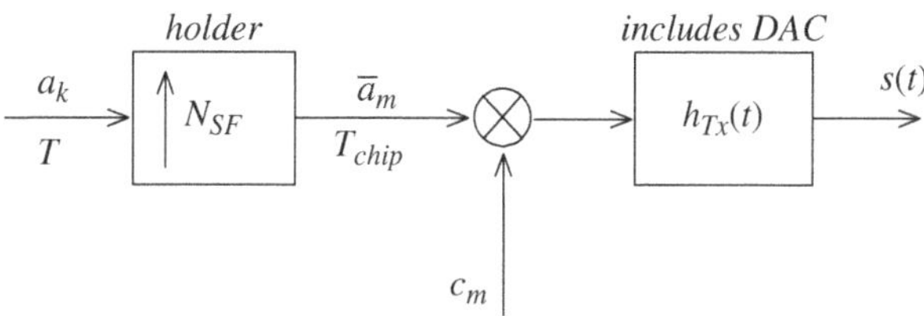

Figure 14.37 Baseband transmitter for spread-spectrum systems.

Optimum receiver

Under the same assumptions of Section 14.1 and for the transmission of K symbols, the received signal r_C is expressed as

$$r_C(t) = \sum_{k=0}^{K-1} a_k \sum_{m=kN_{SF}}^{kN_{SF}+N_{SF}-1} c_m \, q_C(t - mT_{chip} - \varepsilon T_{chip}) \, e^{j(\Omega t+\theta)} + w_{C_\varphi}(t) \tag{14.196}$$

The likelihood $\mathtt{L}_{ss} = \mathtt{L}_{\theta,\varepsilon,\Omega,\boldsymbol{a}}(z, e, o, \boldsymbol{\alpha})$ can be computed as in (14.77); after a few steps, we obtain

$$\mathtt{L}_{ss} = \exp\left\{\frac{2}{N_0}\sum_{k=0}^{K-1} Re\left[\alpha_k^*\, e^{-jz} \sum_{m=kN_{SF}}^{kN_{SF}+N_{SF}-1} c_m^* \int_{T_K} \rho(t)\, e^{-jot}\, g_M(mT_{chip} + eT_{chip} - t)\, dt\right]\right\} \tag{14.197}$$

Defining the two signals

$$\begin{aligned} x(t, o) &= \int_{T_K} \rho(\tau)\, e^{-jo\tau}\, g_M(t - \tau)\, d\tau \\ y(kT, e, o) &= \sum_{m=kN_{SF}}^{kN_{SF}+N_{SF}-1} c_m^*\, x(mT_{chip} + eT_{chip}, o) \end{aligned} \tag{14.198}$$

the likelihood becomes

$$\mathtt{L}_{\theta,\varepsilon,\Omega,\boldsymbol{a}}(z, e, o, \boldsymbol{\alpha}) = \exp\left\{\frac{2}{N_0}\sum_{k=0}^{K-1} Re[\alpha_k^*\, y(kT, e, o)\, e^{-jz}]\right\} \tag{14.199}$$

To obtain the samples $y(kT, e, o)$, we can proceed as follows:

1. obtain the samples

$$y(lT_{chip}, e, o) = \sum_{m=l-N_{SF}+1}^{l} c_m^*\, x(mT_{chip} + eT_{chip}, o) \tag{14.200}$$

2. decimate $y(lT_{chip}, e, o)$ at $l = (k+1)N_{SF} - 1$, i.e. evaluate $y(lT_{chip}, e, o)$ for $l = N_{SF} - 1, 2N_{SF} - 1, \dots, K \cdot N_{SF} - 1$.

By (14.199), it is possible to derive the optimum digital receiver. In particular, up to the decimator that outputs samples at instants $\{\mathtt{m}_m T_c\}$, the receiver is identical to that of Figure 14.19[8]; then in Figure 14.38, only part of the receiver is shown.

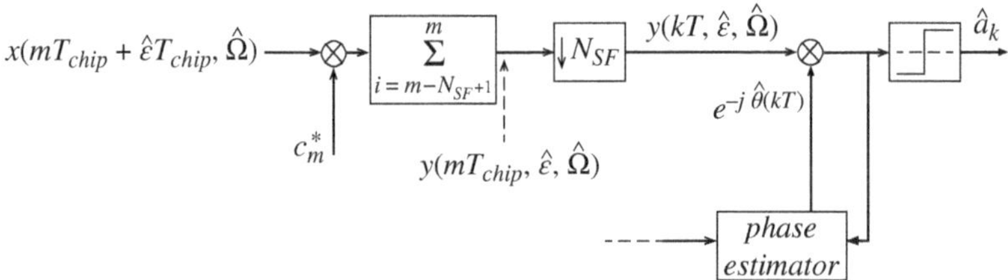

Figure 14.38 Digital receiver for spread-spectrum systems.

14.8.2 Timing estimators with feedback

Assume there is no frequency offset, i.e. $\Omega = 0$. The likelihood is obtained by letting $o = 0$ in (14.199) to yield

$$\mathtt{L}_{\theta,\varepsilon,\boldsymbol{a}}(z, e, \boldsymbol{\alpha}) = \exp\left\{\frac{2}{N_0}\sum_{k=0}^{K-1} Re[\alpha_k^*\, y_k(e)\, e^{-jz}]\right\} \tag{14.201}$$

[8] Note that now $T_c \simeq T_{chip}/2$, and the sampling instants at the decimator are such that $mT_{chip} + \hat{\varepsilon}\, T_{chip} = \mathtt{m}_m T_c + \mu_m T_c$. The estimate $\hat{\varepsilon}$ is updated at every symbol period $T = T_{chip} \cdot N_{SF}$.

where we use the compact notation

$$y_k(e) = y(kT, e, 0) \tag{14.202}$$

The early-late FB estimators are obtained by approximating the derivative of (14.201) with respect to e with a finite difference, and evaluating it for $e = \hat{\varepsilon}_k$.

Non-data aided: non-coherent DLL

In the NDA case, the log-likelihood is obtained from (14.120). Using $y_k(e)$ instead of $x_k(e)$ yields

$$\ell_\varepsilon(e) = \sum_{k=0}^{K-1} |y_k(e)|^2 \tag{14.203}$$

From (14.110), the derivative of $\ell_\varepsilon(e)$ is approximated as

$$\frac{\partial \ell_\varepsilon(e)}{\partial e} \simeq \frac{1}{2\delta} \sum_{k=0}^{K-1} [|y_k(e+\delta)|^2 - |y_k(e-\delta)|^2] \tag{14.204}$$

By including the constant $1/(2\delta)$ in the loop filter and also assuming that the sum is performed by the loop filter, we obtain the error signal

$$e(kT) = |y_k(\hat{\varepsilon}_k + \delta)|^2 - |y_k(\hat{\varepsilon}_k - \delta)|^2 \tag{14.205}$$

The block diagram of the estimator is shown in Figure 14.39. Note that the lag and the lead equal to δT_{chip} are implemented by interpolator filters operating at the sampling period T_c (see (14.91)) with parameter μ equal, respectively, to $-\tilde{\delta}$ and $+\tilde{\delta}$, where $\tilde{\delta} = \delta(T_{chip}/T_c)$. The estimator is called non-coherent digital DLL [8, 9], as the dependence of the error signal on the pair $(\theta, \boldsymbol{a})$ is eliminated without computing the estimates.

Non-data aided modified code tracking loop

We compute now the derivative of (14.203) in exact form,

$$\frac{\partial \ell_\varepsilon(e)}{\partial e} = 2Re\left[\sum_{k=0}^{K-1} y_k(e)\, \frac{d}{de}\, y_k^*(e)\right] \tag{14.206}$$

Approximating the derivative of $y_k^*(e)$ as in (14.110), we obtain

$$\frac{\partial \ell_\varepsilon(e)}{\partial e} \simeq \frac{1}{\delta} \sum_{k=0}^{K-1} Re[y_k(e)(y_k(e+\delta) - y_k(e-\delta))^*] \tag{14.207}$$

Assuming the loop filter performs the multiplication by $1/\delta$ and the sum, the error signal is given by

$$e(kT) = Re[y_k(\hat{\varepsilon}_k)(y_k(\hat{\varepsilon}_k + \delta) - y(\hat{\varepsilon}_k - \delta))^*] \tag{14.208}$$

The block diagram of the estimator is shown in Figure 14.40 and is called *modified code tracking loop* (MCTL) [10]; also in this case, the estimator is non-coherent.

Data and phase directed: coherent DLL

In case the estimates $\hat{\theta}$ and $\hat{\boldsymbol{a}}$ are given, the log-likelihood is expressed as

$$\ell_\varepsilon(e) = \sum_{k=0}^{K-1} Re[\hat{a}_k^*\, y_k(e)\, e^{-j\hat{\theta}}] \tag{14.209}$$

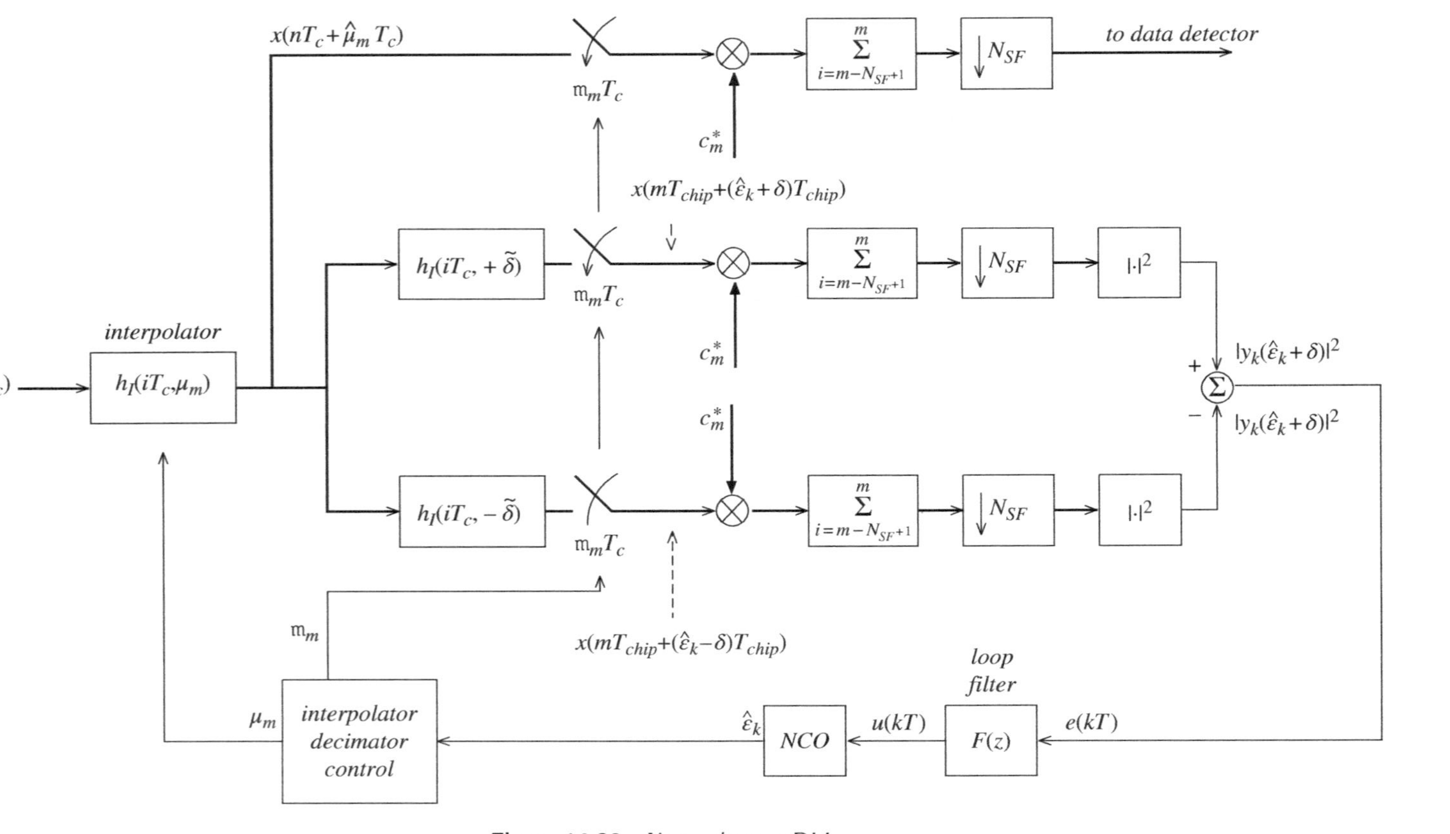

Figure 14.39 Non-coherent DLL.

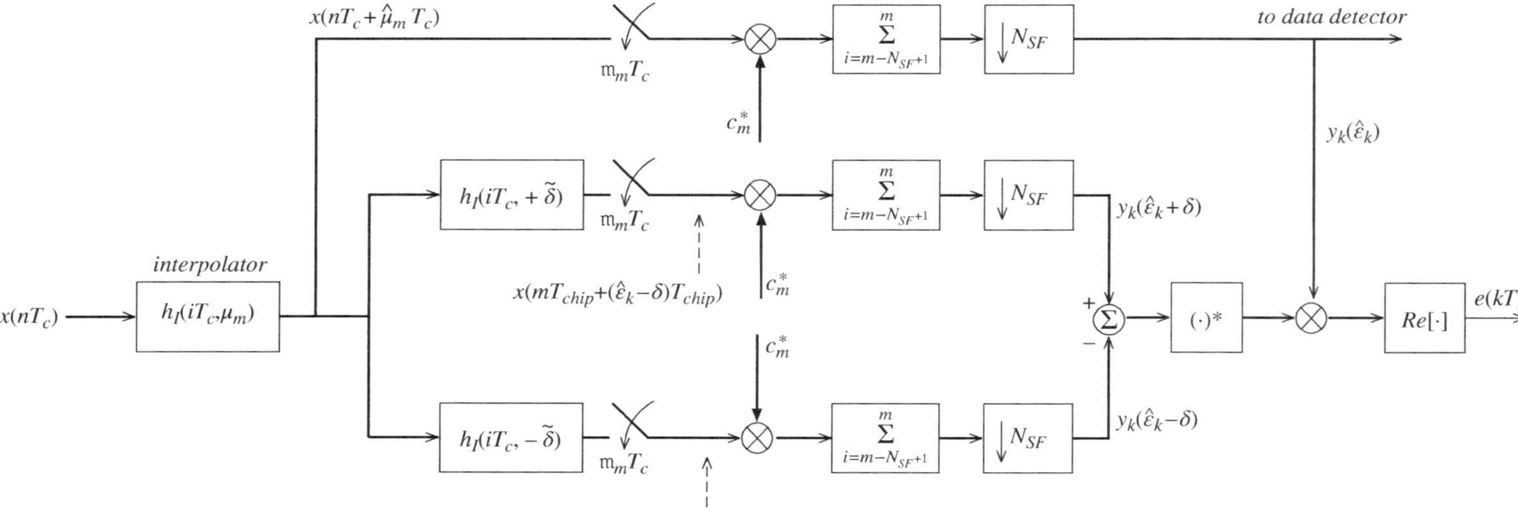

Figure 14.40 Direct section of the non-coherent MCTL.

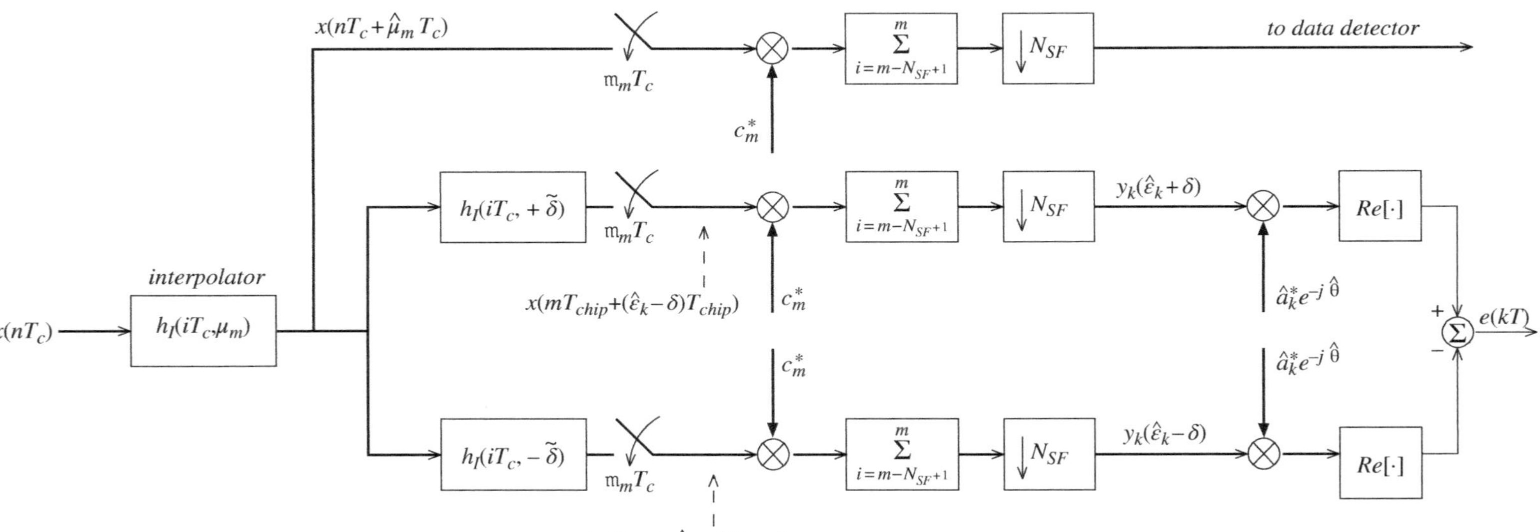

Figure 14.41 Direct section of the coherent DLL.

Approximating the derivative as in (14.110) and including both the multiplicative constant and the summation in the loop filter, the error signal is given by

$$e(kT) = Re[\hat{a}_k^* \, e^{-j\hat{\theta}}[y_k(\hat{\varepsilon}_k + \delta) - y_k(\hat{\varepsilon}_k - \delta)]] \tag{14.210}$$

Figure 14.41 illustrates the block diagram of the estimator, which is called coherent DLL [9–11], as the error signal is obtained by the estimates $\hat{\theta}$ and $\hat{\boldsymbol{a}}$.

In the three schemes of Figures 14.39–14.41, the direct section of the DLL gives estimates of $\mathtt{m}_m$ and μ_m at every symbol period T, whereas the FB loop may operate at the chip period T_{chip}. Observe that by removing the decimation blocks the DLL is able to provide timing estimates at every chip period.

14.9 Synchronization in OFDM

Various algorithms may be applied to achieve synchronization of multicarrier systems for transmission over dispersive channels, depending on the system and on the type of equalization adopted.

For multicarrier systems, two synchronization processes are identified: synchronization of the clock of the analog-to-digital converter (ADC) at the receiver front-end, or *clock synchronization*, and synchronization of the vector $\boldsymbol{r}_k$ of (8.77) at the output of the serial-to-parallel (S/P) element, or *frame synchronization*. Clock synchronization guarantees alignment of the timing phase at the receiver with that at the transmitter; frame synchronization, on the other hand, extracts from the sequence of received samples blocks of $\mathcal{M}$ samples that form the sequence of vectors $\boldsymbol{r}_k$ that are presented at the input of the DFT.

For filtered multitone (FMT) systems with non-critically sampled filter banks and fractionally-spaced equalization, the synchronization is limited to clock recovery (see Section 7.4.3). For OFDM systems, we provide now a more in depth description of a relevant synchronization method.

14.9.1 Frame synchronization

With regards to frame synchronization, in this first section we will discuss the effects of OFDM *symbol time offset* (STO) and how it may be resolved or attenuated, in particular through the *Schmidl and Cox* method [12]. Later, we will focus on the effects of carrier frequency offset (CFO) and its estimate.

Effects of STO

The effect of STO depends on the starting position of the block over which vector $\boldsymbol{r}_k$ is defined and DFT is taken. Four different cases of timing offset (TO) must be considered, as shown in Figure 14.42, where each case is identified by a delay δ in the start of $\boldsymbol{r}_k$, with respect to the end of the cyclic prefix (CP). For notation simplicity, we ignore the noise. Note that $\delta = 0$ corresponds to (8.77).

- *Case I*: The begin of $\boldsymbol{r}_k$ is before the end of the CIR, i.e. $-N_{cp} \le \delta \le -N_{cp} + N_c - 1$. It is

$$r_k^{(\ell)} = [\boldsymbol{r}_k]_\ell = \begin{cases} \sum_{m=N_c-(N_c-N_{cp}-1-\delta)+\ell}^{N_c-1} A_{k-1}[\mathcal{M} - 1 - (m- \\ \quad (N_c - (N_c - N_{cp} - 1 - \delta) + \ell))] g_{C,m} + \\ \quad + \sum_{m=0}^{N_c-1-(N_c-N_{cp}-1-\delta)+\ell} A_k[(\ell + \delta - m) \bmod \mathcal{M}] g_{C,m} \\ \quad \ell = 0, \dots, (N_c - N_{cp} - 1 - \delta) - 1 \\ \sum_{m=0}^{N_c-1} A_k[(\ell + \delta - m) \bmod \mathcal{M}] g_{C,m} \\ \quad \ell = (N_c - N_{cp} - 1 - \delta), \dots, \mathcal{M} - 1 \end{cases} \tag{14.211}$$

hence we have interference from the previous block $k - 1$.

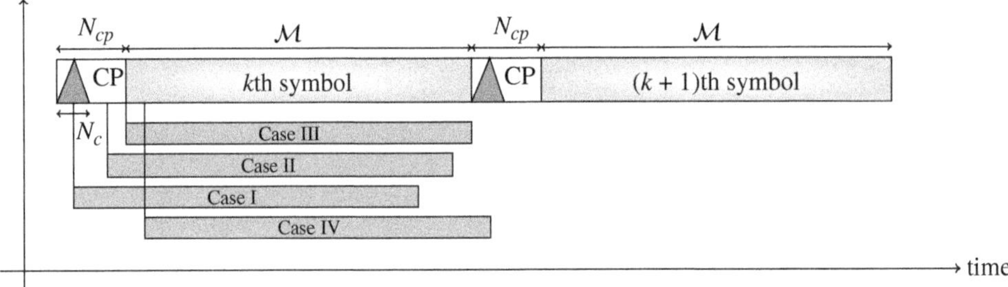

Figure 14.42 The four different cases on the location of the timing offset.

- *Case II*: The begin of $\boldsymbol{r}_k$ is after the end of the CIR and before the end of the CP, $-N_{cp} + N_c - 1 \leq \delta \leq -1$. It is

$$r_k^{(\ell)} = \sum_{m=0}^{N_c-1} A_k[(\ell + \delta - m) \bmod \mathcal{M}] g_{C,m}, \qquad \ell = 0, 1, \dots, \mathcal{M} - 1 \tag{14.212}$$

  thus we do not have interference from other blocks, and the delay will become an additional phase shift of $i\delta T$ on the channel frequency response, i.e. In particular, we have

$$x_k[i] = e^{j2\pi i\delta/\mathcal{M}} \mathcal{G}_C[i] a_k[i], \qquad i = 0, 1, \dots \mathcal{M} - 1 \tag{14.213}$$

  Note that for $\delta = -\lceil (N_{cp} - N_c)/2 \rceil$, the block $\boldsymbol{r}_k$ begins in the middle point between the end of the CIR and the end of the CP. In this case, possible uncontrolled variations in synchronization still keep the system in Case II, which we have already seen that does not introduce inter-block interference. Therefore, this particular choice of δ is particularly robust.
- *Case III*: The begin of $\boldsymbol{r}_k$ is at the end of the CIR, i.e. $\delta = 0$. This has been the synchronization case of (8.77). This case is critical, since a small synchronization error lead to Case IV, which we will see to be affected by interference.
- *Case IV*: The begin of $\boldsymbol{r}_k$ is after the end of the CP, i.e. $0 < \delta < N_c$. It is

$$r_k^{(\ell)} = \begin{cases} \sum_{m=0}^{N_c-1} A_k[\ell + \delta - m \bmod \mathcal{M}] g_{C,m} & \ell = 0, \dots, \mathcal{M} - 1 - \delta \\ \sum_{m=\ell-(\mathcal{M}-\delta)+1}^{N_c-1} A_k[(\ell + \delta - m) \bmod \mathcal{M}] g_{C,m} & \\ \quad + \sum_{m=0}^{\ell-(\mathcal{M}-\delta)} A_{k+1}[\ell + \delta - N_{cp} - m] g_{C,m} & \ell = \mathcal{M} - \delta, \dots, \mathcal{M} - 1 \end{cases} \tag{14.214}$$

  Therefore, $\boldsymbol{r}_k$ is affected by interference from the next block $k + 1$.

Schmidl and Cox algorithm

We consider now the Schmidl and Cox algorithm for the estimation of δ, i.e. for OFDM symbol synchronization. It is based on the transmission of an OFDM symbol, whose time domain block $\boldsymbol{A}_k$ (see Section 8.7.1) has the second half identical to the first half (see Figure 14.43). This can be obtained by transmitting a pseudonoise (PN) sequence on the even subchannels and zeros on the odd subchannels of OFDM symbol $k = 0$ (see $\{a_0[i]\}$ in Table 14.3).

Therefore, in the absence of noise and for an ideal channel, we have from Figure 8.17

$$r_n = r_{n+\mathcal{M}/2}, \qquad n = N_{cp}, \dots, N_{cp} + \mathcal{M}/2 - 1 \tag{14.215}$$

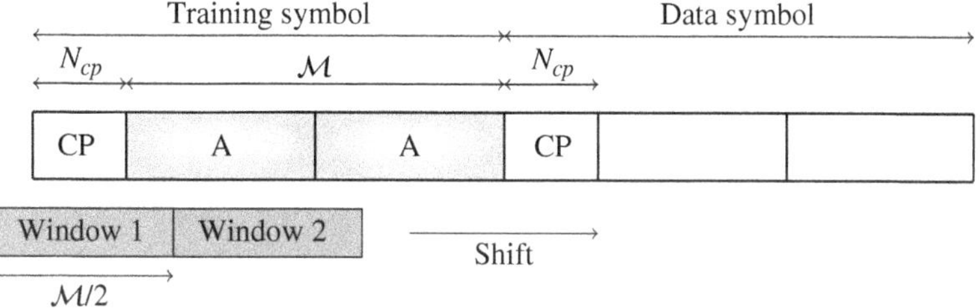

Figure 14.43 Example of payload using repetitive training symbol for STO estimation.

Table 14.3: Illustration of use of PN sequences for training [72].

Subchannel i	$a_0[i]$	$a_1[i]$	$v[i]$
0	$7+7j$	$5-5j$	$-j$
1	0	$-5-5j$	
2	$-7+7j$	$-5-5j$	j
3	0	$-5+5j$	
4	$7+7j$	$-5-5j$	-1
5	0	$5+5j$	
6	$7-7j$	$-5+5j$	-1
7	0	$5-5j$	
8	$7+7j$	$5+5j$	1

With this in mind, two metrics can be defined:

- The correlation of r at lead n, evaluated on $\mathcal{M}/2$ samples, i.e.

$$P(n) = \sum_{i=0}^{\mathcal{M}/2-1} r_{n+i+N_{cp}} r^*_{n+i+N_{cp}+\mathcal{M}/2} \tag{14.216}$$

- The energy of r at lead n, evaluated on $\mathcal{M}/2$ samples (see Figure 14.43, window 2)

$$E(n) = \sum_{i=0}^{\mathcal{M}/2-1} |r_{n+i+N_{cp}+\mathcal{M}/2}|^2 \tag{14.217}$$

With reference to a channel introducing only a delay D_0 (see Figure 8.17), we observe that $P(n)$ reaches the maximum value for $n = D_0$, since the two windows will contain the same values hence the correlation is maximized. These two quantities can then be used to define the timing metric

$$c(n) = \frac{|P(n)|}{E(n)} \tag{14.218}$$

whose maximization yields the estimate of the STO [12] as

$$\hat{\delta} = \arg\max_{\delta} c(\delta) \tag{14.219}$$

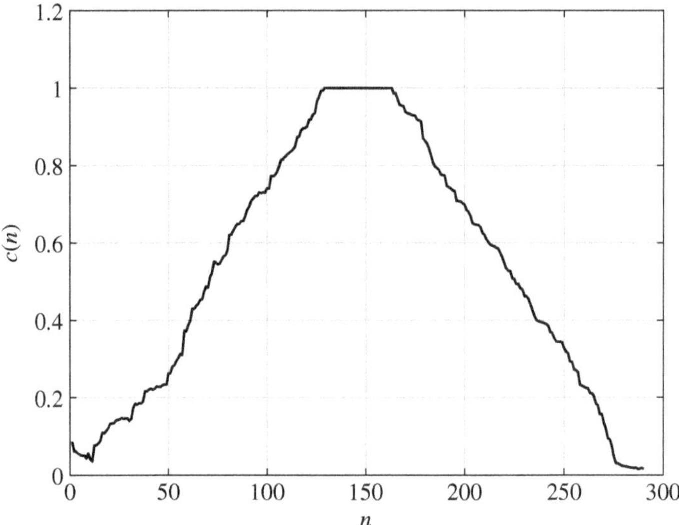

Figure 14.44 Example of timing metric for an AWGN channel.

When studying $c(n)$ as a function of n, (see Figure 14.44), we observe a plateau, or flat interval, of extension given by the difference between the guard interval duration and the CIR length, i.e. $N_{cp} + 1 - N_c$. Precisely, the nature of the channel determines the length of this plateau:

- for an ideal additive white Gaussian noise (AWGN) channel with $N_c = 1$, the plateau extension is $N_{cp} + 1$, since for $c(n)$ is maximum for $n = -N_{cp}, \dots, 0$; indeed, in (14.216) we find the same values of the CP in the first half of the OFDM symbol.
- for a frequency selective channel, instead, the plateau has duration $N_{cp} + 2 - N_c$, since, in this case, only after the end of the CIR we find the same values of the CP in the first half of the OFDM symbol.

In order to achieve the synchronization of Case II in Figure 14.42, we can first find the value of the plateau as

$$c_{max} = \max_n c(n) \tag{14.220}$$

then the edges of the plateau can be found by setting a level slightly below the maximum (e.g. $\eta = 0.9$, i.e. 90% of the maximum) and finding δ_1 and δ_2 such that

$$\begin{aligned} c(\delta_1) &= \eta\, c_{max} \\ &= c(\delta_2) \end{aligned} \tag{14.221}$$

and $c(n) < c(\delta_1)$ for $n < \delta_1$, $c(n) < c(\delta_2)$ for $n > \delta_2$ Then, synchronization according to Case II is obtained by choosing the middle point of the plateau support

$$\hat{\delta} = \frac{\delta_1 + \delta_2}{2} \tag{14.222}$$

14.9.2 Carrier frequency synchronization

The CFO $\Omega/(2\pi) = f_0 - f_1$ is originated either by a difference between the transmit (f_0) and receive (f_1) oscillator frequencies or by a Doppler shift (see Section 4.1.6). In a way symmetric to STO, the CFO yields a phase offset of the time-domain signal.

In ideal channels with $g_{C,0} \neq 0$, and $g_{C,n} = 0$, for $n \neq 0$, it results

$$r_n = g_{C,0} s_n e^{j2\pi(f_0 - f_1)nT/\mathcal{M}} + w_n \tag{14.223}$$

where w_n is AWGN with power σ_w^2.

Introducing the normalized CFO to twice the inter-carrier spacing

$$\Delta = \frac{f_0 - f_1}{2/T} = (f_0 - f_1)\frac{T}{2} \tag{14.224}$$

from (14.216) and (14.223), we have

$$P(\hat{\delta}) = e^{-j2\pi\Delta} \sum_{i=0}^{\mathcal{M}/2-1} |r_{\hat{\delta}+i+N_{cp}}|^2 = e^{-j2\pi(\Delta_I + \Delta_F)} \sum_{i=0}^{\mathcal{M}/2-1} |r_{\hat{\delta}+i+N_{cp}}|^2 \tag{14.225}$$

where $\Delta_I = \lfloor \Delta \rfloor$ and $\Delta_F = \Delta - \Delta_F$. An estimate of the fractional part of Δ, Δ_F, can be obtained as

$$\hat{\Delta}_F = -\arg \frac{\hat{P}(\hat{\delta})}{2\pi} \tag{14.226}$$

On the other hand, from (14.223) the integer part of Δ, Δ_I, yields a shift of data on subchannels,[9] i.e.

$$x_k[i] = a_k[i - 2\Delta_I] \tag{14.227}$$

The Schmidl and Cox algorithm, after sending the first OFDM symbol ($k = 0$) for frame synchronization, sends a second OFDM block ($k = 1$) that contains a different PN sequence (with respect to that of the first symbol) on the odd frequencies, see for example Table 14.3. Let us define

$$v[2i] = \sqrt{2}\frac{a_1[2i]}{a_0[2i]} \tag{14.228}$$

for $i = 0, \ldots, \mathcal{M}/2 - 1$. Then, we estimate the number of even shifted positions by exploiting the fact that the phase shift is the same for each pair of frequencies in both OFDM symbols, [12], thus

$$\hat{\Delta}_I = \arg\max_{\xi} \frac{\left|\sum_{i=0}^{\mathcal{M}/2-1} x_0^*[2i + 2\xi] v^*[2i] x_1[2i + 2\xi]\right|^2}{\left(\sum_{i=0}^{\mathcal{M}/2-1} |x_1[2i]|^2\right)^2} \tag{14.229}$$

Estimator performance

We now assess the performance of the Δ_F estimator (14.226).

Following the derivations of [12], we can show that the estimator is unbiased and, assuming that $|s_n| = 1$, the variance of the estimate is

$$\text{var}(\hat{\Delta}) = \frac{2\sigma_w^2}{(2\pi)^2 \mathcal{M} |g_{C,0}|^2} \tag{14.230}$$

When we compare this variance with the minimum variance of an unbiased estimator provided by the Cramér-Rao lower bound (see Example 2.4.6), we conclude that the estimate (14.226) achieves the minimum error variance among all possible estimators.

Other synchronization solutions

Various frequency synchronization methods have been proposed for OFDM systems that include both DA or training-assisted schemes [13–16], and NDA or blind schemes [17–20]. NDA methods use the inherent structure of the OFDM block for synchronization. However, these algorithms usually require

[9] It should be kept in mind that the cyclic shifting of the subcarrier frequencies actually holds only in the discrete-time case, where the DFT of a finite-time signal provides a periodic signal. This does not hold in the continuous-time case, where the subcarrier signals at margins is lost.

a large number of OFDM blocks to achieve satisfactory performance. DA synchronization schemes employ either the auto-correlation of received training blocks or its cross-correlation with a local copy at the receiver. Methods resorting to the auto-correlation for synchronization use training symbols with repetitions that offer robustness to large CFO and multipath channel. To facilitate synchronization and cell search procedures in OFDM systems specified by the long-term evolution (LTE) standard, constant-amplitude zero auto-correlation (CAZAC) sequences, for which circularly shifted copies of a sequence are uncorrelated, have been proposed as training symbols. CAZAC sequences, also known as *Zadoff–Chu* sequences, have been introduced in Appendix 1.C and are ideal candidates for uplink synchronization because different users can transmit CAZAC sequences with different shifts, and timing offset (TO) estimation can be performed efficiently using cross-correlation with the primary CAZAC sequence. Joint timing and frequency synchronization using CAZAC sequences have been proposed for downlink transmission as well [21].

14.10 Synchronization in SC-FDMA

A word of caution is needed on synchronization of single-carrier frequency division multiple access (SC-FDMA), see Section 8.10.2: in order to avoid interblock and intersymbol interference at the receiver, signals coming from all users must satisfy the orthogonality principle, which for SC-FDMA (as for OFDM) corresponds to the fact that the users' channels fall within the CP.

Consider first a single-user transmission over two CIRs, g_C, and g'_C that coincide apart from a delay $D_0 > 0$, i.e.

$$g'_{C,n} = \begin{cases} g_{C,n-D_0} & n = D_0, \dots, D_0 + N_c \\ 0 & \text{otherwise} \end{cases} \tag{14.231}$$

Clearly, a single-user SC-FDMA receiver can use in both cases a CP of length $N_{cp} = N_c - 1$, where for channel g'_C the begin of block $\boldsymbol{r}_k$ will be delayed by D_0 with respect to the signal going through channel g_C. In a multiuser system instead, where a user is transmitting over g_C and another over g'_C, we must use a CP of length $N_c - 1 + D_0$ in order to avoid interference.

When signals from many users are collected at a single receiver (such as in the uplink of a cellular system), an alternative solution provides that transmission by different users is suitably delayed in order to have the impulse responses of the users to be overlapping at the receiver. In our example, we delay by D_0 the transmission of user going through channel g_C, so that the correct block synchronization for each signal individually is the same, and then a CP of size $N_{cp} = N_c - 1$ ensures orthogonality for both signals. This solution goes under the name of *time advance* in 3GPP cellular systems.

With multiple receiving devices in different positions, thus experiencing different delays, avoiding the use of longer CPs may be problematic.

Bibliography

[1] Meyr, H. and Ascheid, G. (1990). *Synchronization in Digital Communications*, vol. 1. New York, NY: Wiley.

[2] Franks, L.E. (1980). Carrier and bit synchronization in data communication - a tutorial review. *IEEE Transactions on Communications* 28: 1107–1120.

[3] Proakis, J.G. (1995). *Digital Communications*, 3e. New York, NY: McGraw-Hill.

[4] Meyr, H., Moeneclaey, M., and Fechtel, S.A. (1998). *Digital Communication Receivers*. New York, NY: Wiley.

[5] Mengali, U. and D'Andrea, A.N. (1997). *Synchronization Techniques for Digital Receivers*. New York, NY: Plenum Press.

[6] Mueller, K.H. and Muller, M.S. (1976). Timing recovery in digital synchronous data receivers. *IEEE Transactions on Communications* 24: 516–531.

[7] Simon, M.K., Omura, J.K., Scholtz, R.A., and Levitt, B.K. (1994). *Spread Spectrum Communications Handbook*. New York, NY: McGraw-Hill.

[8] De Gaudenzi, R., Luise, M., and Viola, R. (1993). A digital chip timing recovery loop for band-limited direct-sequence spread-spectrum signals. *IEEE Transactions on Communications* 41: 1760–1769.

[9] De Gaudenzi, R. (1999). Direct-sequence spread-spectrum chip tracking loop in the presence of unresolvable multipath components. *IEEE Transactions on Vehicular Technology* 48: 1573–1583.

[10] Yost, R.A. and Boyd, R.W. (1982). A modified PN code tracking loop: its performance analysis and comparative evaluation. *IEEE Transactions on Communications* 30: 1027–1036.

[11] De Gaudenzi, R. and Luise, M. (1991). Decision-directed coherent delay-lock tracking loop for DS-spread spectrum signals. *IEEE Transactions on Communications* 39: 758–765.

[12] Schmidl, T.M. and Cox, D.C. (1997). Robust frequency and timing synchronization for OFDM. *IEEE Transactions on Communications* 45: 1613–1621.

[13] Speth, M., Fechtel, S.A., Fock, G., and Meyr, H. (1999). Optimum receiver design for wireless broad-band systems using OFDM. I. *IEEE Transactions on Communications* 47: 1668–1677.

[14] Minn, H., Bhargava, V.K., and Letaief, K.B. (2003). A robust timing and frequency synchronization for OFDM systems. *IEEE Transactions on Wireless Communications* 2: 822–839.

[15] Ren, G., Chang, Y., Zhang, H., and Zhang, H. (2005). Synchronization method based on a new constant envelop preamble for OFDM systems. *IEEE Transactions on Broadcasting* 51: 139–143.

[16] Wang, C. and Wang, H. (2009). On joint fine time adjustment and channel estimation for OFDM systems. *IEEE Transactions on Wireless Communications* 8: 4940–4944.

[17] Bolcskei, H. (2001). Blind estimation of symbol timing and carrier frequency offset in wireless OFDM systems. *IEEE Transactions on Communications* 49: 988–999.

[18] Luise, M., Marselli, M., and Reggiannini, R. (2002). Low-complexity blind carrier frequency recovery for OFDM signals over frequency-selective radio channels. *IEEE Transactions on Communications* 50: 1182–1188.

[19] Park, B., Cheon, H., Ko, E. et al. (2004). A blind OFDM synchronization algorithm based on cyclic correlation. *IEEE Signal Processing Letters* 11: 83–85.

[20] Yao, Y. and Giannakis, G.B. (2005). Blind carrier frequency offset estimation in SISO, MIMO, and multiuser OFDM systems. *IEEE Transactions on Communications* 53: 173–183.

[21] Gul, M.M.U., Ma, X., and Lee, S. (2015). Timing and frequency synchronization for OFDM downlink transmissions using Zadoff–Chu sequences. *IEEE Transactions on Wireless Communications* 14: 1716–1729.

Chapter 15

Self-training equalization

By the expression *self-training equalization* we refer to channel equalization techniques by which we obtain the initial convergence of the parameters of an adaptive equalizer without resorting to the transmission of a training sequence. Although these techniques generally achieve suboptimum performance at convergence, they are applied in many cases of practical interest, where the transmission of known training sequences is not viable, as for example in broadcast systems, or may result in an undesirable increase of system complexity [1–6]. In these cases, it is necessary to consider self-training of the adaptive equalizer, where the terms for adjusting the coefficient parameters are obtained by processing the received signal. Usually, the receiver performs self-training of the equalizer by referring to the knowledge of the input signal characteristics, for example the probability distribution of the input symbols.

The subject of self-training equalization in communications systems has received considerable attention since the publication of the article by Sato [7] in 1975; the proposed algorithms have been alternatively called *self-recovering*, *self-adaptive*, or *blind*.

15.1 Problem definition and fundamentals

With reference to the discrete-time equivalent scheme of Figure 7.15 shown in Figure 15.1, we consider a real-valued discrete-time system given by the cascade of a linear channel $\{h_i\}$, in general with non-minimum phase transfer function[1] H, and an equalizer $C \leftrightarrow \{c_i\}$; in general, the impulse responses $\{h_i\}$ and $\{c_i\}$ are assumed with unlimited duration. The overall system is given by $\Psi = H \cdot C \leftrightarrow \{\psi_i\}$, where $\psi_i = \sum_{\ell=-\infty}^{\infty} c_\ell h_{i-\ell}$. The channel input symbol sequence $\{a_k\}$ is modelled as a sequence of i.i.d. random variables with symmetric probability density function $p_a(\alpha)$. The channel output sequence is $\{x_k\}$. In this section, additive noise introduced by the channel is ignored.

We note that a non-minimum phase rational transfer function (see Definition 1.5 on page 12) can be expressed as

$$H(z) = H_0 \, \frac{P_1(z)\, P_2(z)}{P_3(z)} \tag{15.1}$$

where H_0 denotes the gain, $P_3(z)$ and $P_1(z)$ are monic polynomials with zeros inside the unit circle, and $P_2(z)$ is a monic polynomial with zeros outside the unit circle. We introduce the inverse functions with respect to the polynomials $P_1(z)$ and $P_2(z)$ given by $P_1^{-1}(z) = 1/P_1(z)$ and $P_2^{-1}(z) = 1/P_2(z)$,

[1] In this chapter, to simplify notation, often the argument of the z-transform is not explicitly indicated.

Algorithms for Communications Systems and their Applications, Second Edition.
Nevio Benvenuto, Giovanni Cherubini, and Stefano Tomasin.

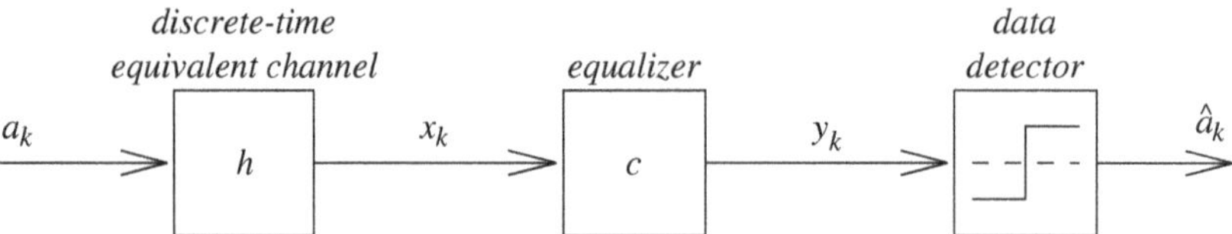

Figure 15.1 Discrete-time equivalent channel and equalizer filter.

respectively. By expanding $P_1^{-1}(z)$ and $P_2^{-1}(z)$ in Laurent series we obtain, apart from lag factors,

$$\begin{aligned} P_1^{-1}(z) &= \sum_{n=0}^{+\infty} c_{1,n}\, z^{-n} \\ P_2^{-1}(z) &= \sum_{n=-\infty}^{0} c_{2,n}\, z^{-n} \end{aligned} \tag{15.2}$$

both converging in a ring that includes the unit circle. Therefore, we have

$$H^{-1}(z) = \frac{1}{H_0}\, P_3(z) \left(\sum_{n=0}^{+\infty} c_{1,n}\, z^{-n} \right) \left(\sum_{n=-\infty}^{0} c_{2,n}\, z^{-n} \right) \tag{15.3}$$

In general, from (15.3) we note that, as the system is non-minimum phase, the inverse system cannot be described by a finite number of parameters; in other words, the reconstruction of a transmitted symbol at the equalizer output at a certain instant requires the knowledge of the entire received sequence.

In practice, to obtain an implementable system, the series in (15.3) are truncated to N terms, and an approximation of H^{-1} with a lag equal to $(N_1 - 1)$ modulation intervals is given by

$$H^{-1}(z) \simeq \frac{1}{H_0}\, z^{-(N_1-1)}\, P_3(z) \left(\sum_{n=0}^{N-1} c_{1,n}\, z^{-n} \right) \left(\sum_{n=-(N-1)}^{0} c_{2,n}\, z^{-n} \right) \tag{15.4}$$

Therefore, the inverse system can only be defined apart from a lag factor.

The problem of self-training equalization is formulated as follows: from the knowledge of the probability distribution of the channel input symbols $\{a_k\}$ and from the observation of the channel output sequence $\{x_k\}$, we want to find an equalizer C such that the overall system impulse response is inter-symbol interference (ISI) free.

Observe that, if the channel H is minimum phase,[2] both H and H^{-1} are causal and stable; in this case, the problem of channel identification, and hence the determination of the sequence $\{a_k\}$, can be solved by whitening (see page 117) the observation $\{x_k\}$ using known procedures that are based on the second order statistical description of signals. If, as it happens in the general case, the channel H is non-minimum phase, by the second order statistical description (1.187), it is possible to identify only the amplitude characteristic of the channel transfer function, but not its phase characteristic.

In particular, if the probability density function of the input symbols is Gaussian, the output $\{x_k\}$ is also Gaussian and the process is completely described by a second order analysis; therefore, the above observations are valid and in general the problem of self-training equalization cannot be solved for Gaussian processes. Note that we have referred to a system model in which the sampling frequency is equal to the symbol rate; a solution to the problem using a second order description with reference to an oversampled model is obtained in [8].

Furthermore we observe that, as the probability density function of the input symbols is symmetric, the sequence $\{-a_k\}$ has the same statistical description as the sequence $\{a_k\}$; consequently, it is not

[2] We recall that all auto-regressive (AR) models (1.445) are minimum phase.

possible to distinguish the desired equalizer H^{-1} from the equalizer $-H^{-1}$. Hence, the inverse system can only be determined apart from the sign and a lag factor. Therefore, the solution to the problem of self-training equalization is given by $C = \pm H^{-1}$, which yields the overall system $\Psi = \pm I$, where I denotes the identity, with the exception of a possible lag.

In this chapter, we will refer to the following theorem, demonstrated in [9].

Theorem 15.1

Assuming that the probability density function of the input symbols is non-Gaussian, then $\Psi = \pm I$ if the output sample

$$y_k = \sum_{n=-\infty}^{+\infty} c_n \, x_{k-n} \tag{15.5}$$

has a probability density function $p_{y_k}(b)$ equal to the probability density function of the symbols $\{a_k\}$. □

Therefore, using this theorem to obtain the solution, it is necessary to determine an algorithm for the adaptation of the coefficients of the equalizer C such that the probability distribution of y_k converges to the distribution of a_k.

We introduce the cost function

$$J = E[\Phi(y_k)] \tag{15.6}$$

where y_k is given by (15.5) and Φ is an even, real-valued function that must be chosen so that the optimum solution, determined by

$$C_{opt}(z) = \arg\min_{C(z)} J \tag{15.7}$$

is found at the points $\pm H^{-1}$, apart from a lag factor.

In most applications, minimization is done by equalizers of finite length N having coefficients at instant k given by

$$\boldsymbol{c}_k = [c_{0,k}, c_{1,k}, \dots, c_{N-1,k}]^T \tag{15.8}$$

Let

$$\boldsymbol{x}_k = [x_k, x_{k-1}, \dots, x_{k-(N-1)}]^T \tag{15.9}$$

then (15.5) becomes

$$y_k = \sum_{n=0}^{N-1} c_{n,k} \, x_{k-n} = \boldsymbol{c}_k^T \, \boldsymbol{x}_k \tag{15.10}$$

Then also (15.7) simplifies into

$$\boldsymbol{c}_{opt} = \arg\min_{\boldsymbol{c}} J \tag{15.11}$$

If Θ is the derivative of Φ, the gradient of J with respect to $\boldsymbol{c}$ is given by (see (2.18))

$$\nabla_{\boldsymbol{c}} J = E[\boldsymbol{x}_k \, \Theta(y_k)] \tag{15.12}$$

For the minimization of J, we use a stochastic gradient algorithm for which (see (3.40))

$$\boldsymbol{c}_{k+1} = \boldsymbol{c}_k - \mu \, \Theta(y_k) \, \boldsymbol{x}_k \tag{15.13}$$

where μ is the adaptation gain. Note that the convergence of C to H^{-1} or to $-H^{-1}$ depends on the initial choice of the coefficients.

Before tackling the problem of choosing the function Φ, we consider the following problem.

Minimization of a special function

We consider an overall system ψ with impulse response having unlimited duration. We want to determine the values of the sequence $\{\psi_i\}$ that minimize the following cost function:

$$\mathcal{V} = \sum_{i=-\infty}^{+\infty} |\psi_i| \tag{15.14}$$

subject to the constraint

$$\sum_{i=-\infty}^{+\infty} \psi_i^2 = 1 \tag{15.15}$$

The function $\mathcal{V}$ characterizes the *peak* amplitude of the system *output signal* $\{y_k\}$ and the constraint can be interpreted as a requirement that the solution $\{\psi_i\}$ belong to the sphere with centre the origin and radius $r = 1$ in the parameter space $\{\psi_i\}$.

Letting $\boldsymbol{\psi} = [\dots, \psi_0, \psi_1, \dots]^T$, it results $\nabla_{\psi} \mathcal{V} = [\dots, \text{sgn}(\psi_0), \text{sgn}(\psi_1), \dots]^T$; then, if ψ_{max} is the maximum value assumed by ψ_i, the cost function (15.14) presents stationary points for $\psi_i = \pm\psi_{max}$. Now, taking into account the constraint (15.15), it is easily seen that the minimum of (15.14) is reached only when one element of the sequence $\{\psi_i\}$ is different from zero (with a value equal to ψ_{max}) and the others are all zeros. In other words, the only points of minimum are given by $\Psi = \pm I$, and the other stationary points correspond to saddle points. Figure 15.2 illustrates the cost function $\mathcal{V}$ along the unit circle for a system with two parameters.

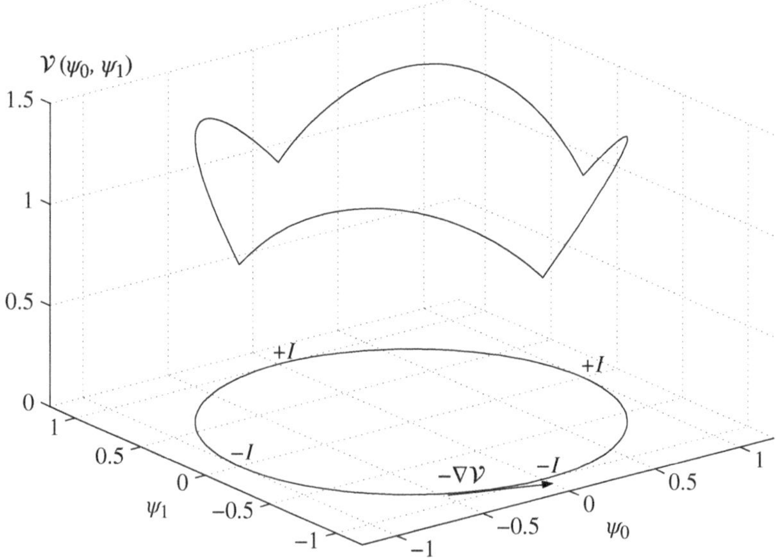

Figure 15.2 Illustration of the cost function $\mathcal{V}$ for the system $\Psi \leftrightarrow \{\psi_0, \psi_1\}$ and of the gradient $-\nabla\mathcal{V}$ projected onto the straight line tangent to the unit circle.

The minimization can also be obtained recursively, updating the parameters $\{\psi_i\}$ as indicated by the direction of steepest descent[3] of the cost function $\mathcal{V}$.

[3] The direction is defined by the gradient vector.

The projection of the gradient $\mathcal{V}$ onto the plane tangent to the unit sphere at the point $\boldsymbol{\psi}$ is given by

$$\nabla_{\psi}\mathcal{V} - [(\nabla_{\psi}\mathcal{V})^T\boldsymbol{\psi}]\boldsymbol{\psi} \tag{15.16}$$

then the recursive equation yields

$$\psi'_{i,k+1} = \psi_{i,k} - \mu\left[\text{sgn}(\psi_{i,k}) - \psi_{i,k}\sum_{\ell=-\infty}^{+\infty}|\psi_{\ell,k}|\right] \tag{15.17}$$

$$\psi_{i,k+1} = \frac{\psi'_{i,k+1}}{\sqrt{\sum_{\ell=-\infty}^{+\infty}(\psi'_{\ell,k+1})^2}} \tag{15.18}$$

Note that, if the term $\psi_{i,k}\sum_{\ell}|\psi_{\ell,k}|$ is omitted in (15.17), with good approximation the direction of the steepest descent is still followed, provided that the adaptation gain μ is sufficiently small.

From (15.17), for each parameter ψ_i the updating consists of a correction towards zero of a fixed value and a correction in the opposite direction of a value proportional to the parameter amplitude. Assuming that the initial point is not a saddle point, by repeated iterations of the algorithm, one of the parameters ψ_i approaches the value one, while all others converge to zero, as shown in Figure 15.3.

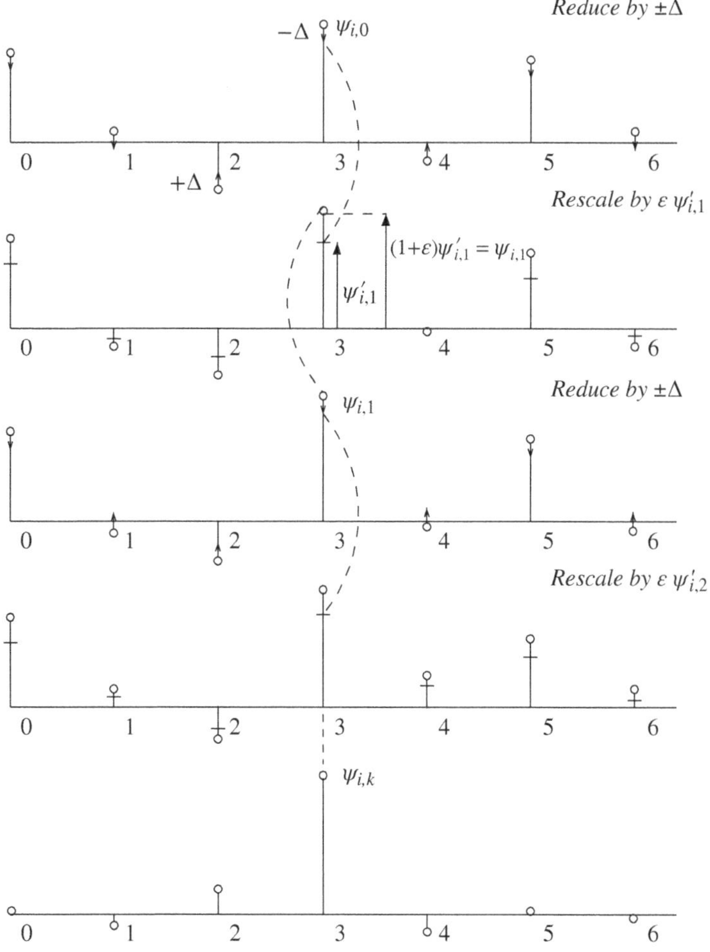

Figure 15.3 Update of the parameters $\{\psi_i\}$, $i = 0, \dots, 6$.

We now want to obtain the same adaptation rule for the parameters $\{\psi_i\}$ using the output signal y_k. We define $\boldsymbol{\psi}_k$ as the vector of the parameters of $\boldsymbol{\psi}$ at instant k,

$$\boldsymbol{\psi}_k = [\ldots, \psi_{0,k}, \psi_{1,k}, \ldots]^T \tag{15.19}$$

and

$$\boldsymbol{a}_k = [\ldots, a_k, a_{k-1}, \ldots]^T \tag{15.20}$$

Therefore,

$$y_k = \boldsymbol{\psi}_k^T \, \boldsymbol{a}_k \tag{15.21}$$

Assume that at the beginning of the adaptation process the overall system Ψ satisfies the condition $\| \boldsymbol{\psi} \|^2 = 1$, but it deviates significantly from the system identity; then the equalizer output signal y_k will occasionally assume positive or negative values which are much larger in magnitude than $\alpha_{max} = \max a_k$. The peak value of y_k is given by $\pm\alpha_{max} \sum_i |\psi_{i,k}|$, obtained with symbols $\{a_{k-i}\}$ equal to $\pm\alpha_{max}$, and indicates that the distortion is too large and must be reduced. In this case, a correction of a fixed value towards zero is obtained using the error signal

$$e_k = y_k - \alpha_{max} \, \text{sgn}(y_k) \tag{15.22}$$

and applying the stochastic gradient algorithm

$$\boldsymbol{\psi}_{k+1} = \boldsymbol{\psi}_k - \mu \, e_k \, \boldsymbol{a}_k \tag{15.23}$$

If the coefficients are scaled so that the condition $\| \boldsymbol{\psi}_k \|^2 = 1$ is satisfied at every k, we obtain a coefficient updating algorithm that approximates algorithm (15.17)–(15.18).

Obviously, algorithm (15.23) cannot be directly applied, as the parameters of the overall system Ψ are not available. However, observe that if the linear transformation H is non-singular, then formally $C = H^{-1}\Psi$. Therefore, the overall minima of $\mathcal{V}$ at the points $\Psi = \pm I$ are mapped into overall minima at the points $C = \pm H^{-1}$ of a cost function J that is the image of $\mathcal{V}$ under the transformation given by H^{-1}, as illustrated in Figure 15.4. Furthermore, it is seen that the direction of steepest descent of $\mathcal{V}$ has not been modified by this transformation. Thus, the updating terms for the equalizer coefficients are still given by (15.22)–(15.23), if symbols a_k are replaced by the channel output samples x_k.

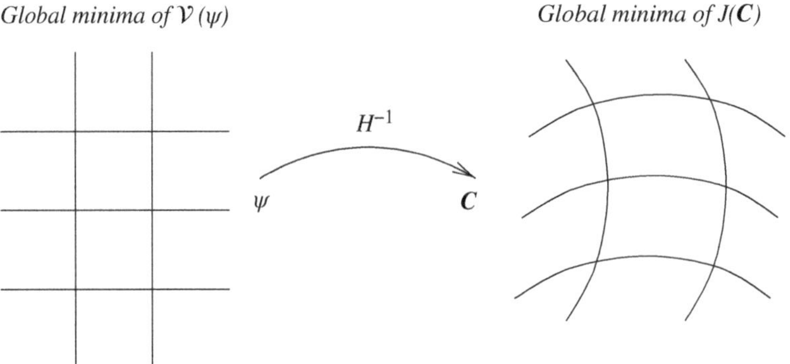

Figure 15.4 Illustration of the transformation $C = H^{-1}\Psi$.

Then, a general algorithm that converges to the desired solution $C = H^{-1}$ can be formulated as follows:

- observe the equalizer output signal $\{y_k\}$ and determine its peak value;

- whenever a peak value occurs, update the coefficients according to the algorithm

$$\begin{aligned} e_k &= y_k - \alpha_{max}\,\mathrm{sgn}(y_k) \\ \boldsymbol{c}_{k+1} &= \boldsymbol{c}_k - \mu\, e_k\, \boldsymbol{x}_k \end{aligned} \tag{15.24}$$

- scale the coefficients so that the statistical power of the equalizer output samples is equal to the statistical power of the channel input symbols.

We observe that it is not practical to implement an algorithm that requires computing the peak value of the equalizer output signal and updating the coefficients only when a peak value is observed. In the next sections, we describe algorithms that allow the updating of the coefficients at every modulation interval, thus avoiding the need of computing the peak value of the signal at the equalizer output and of scaling the coefficients.

15.2 Three algorithms for PAM systems

The Sato algorithm

The Sato cost function is defined as

$$J = E\left[\frac{1}{2}\, y_k^2 - \gamma_S\, |y_k|\right] \tag{15.25}$$

where

$$\gamma_S = \frac{E[a_k^2]}{E[|a_k|]} \tag{15.26}$$

The gradient of J is given by

$$\nabla_c J = E[\boldsymbol{x}_k (y_k - \gamma_S\,\mathrm{sgn}(y_k))] \tag{15.27}$$

We introduce the signal

$$\epsilon_{S,k} = y_k - \gamma_S\,\mathrm{sgn}(y_k) \tag{15.28}$$

which assumes the meaning of *pseudo error* that during self-training replaces the error used in the least-mean-square (LMS) decision-directed algorithm. We recall that for the LMS algorithm the error signal is[4]

$$e_k = y_k - \hat{a}_k \tag{15.29}$$

where $\hat{a}_k$ is the detection of the symbol a_k, obtained by a threshold detector from the sample y_k. Figure 15.5 shows the pseudo error $\epsilon_{S,k}$ as a function of the value of the equalizer output sample y_k.

Therefore, the Sato algorithm for the coefficient updating of an adaptive equalizer assumes the expression

$$\boldsymbol{c}_{k+1} = \boldsymbol{c}_k - \mu\, \epsilon_{S,k}\, \boldsymbol{x}_k \tag{15.30}$$

It was proven [9] that, if the probability density function of the output symbols $\{a_k\}$ is sub-Gaussian,[5] then the Sato cost function (15.25) admits as unique points of minimum the systems $C = \pm H^{-1}$, apart from a possible lag; however, note that the uniqueness of the points of minimum of the Sato cost function is obtained by assuming a continuous probability distribution of input symbols. In the case of a discrete

[4] Note that in this chapter, the error signal is defined with opposite sign with respect to the previous chapters.

[5] A probability density function $p_{a_k}(\alpha)$ is sub-Gaussian if it is uniform or if $p_{a_k}(\alpha) = K\exp\{-g(\alpha)\}$, where $g(\alpha)$ is an even function such that both $g(\alpha)$ and $\frac{1}{\alpha}\frac{dg}{d\alpha}$ are strictly increasing in the domain $[0, +\infty)$.

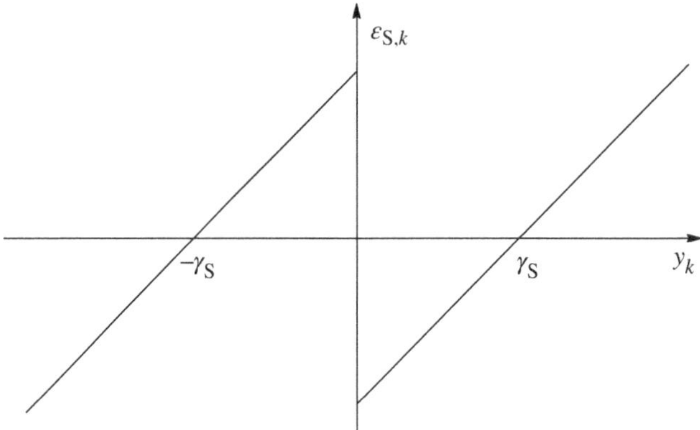

Figure 15.5 Characteristic of the pseudo error $\epsilon_{S,k}$ as a function of the equalizer output.

probability distribution with an alphabet $\mathcal{A} = \{\pm 1, \pm 3, \dots, \pm(M-1)\}$, the convergence properties of the Sato algorithm are not always satisfactory.

Another undesirable characteristic of the Sato algorithm is that the pseudo error $\epsilon_{S,k}$ is not equal to zero for $C = \pm H^{-1}$, unless we consider binary transmission. In fact only the gradient of the cost function given by (15.27) is equal to zero for $C = \pm H^{-1}$; moreover, we find that the variance of the pseudo error may assume non-negligible values in the neighbourhood of the desired solution.

Benveniste–Goursat algorithm

To mitigate the above mentioned inconvenience, we observe that the error e_k used in the LMS algorithm in the absence of noise becomes zero for $C = \pm H^{-1}$. It is possible to combine the two error signals obtaining the pseudo error proposed by Benveniste and Goursat [10], given by

$$\epsilon_{G,k} = \kappa_1 \, e_k + \kappa_2 \, |e_k| \, \epsilon_{S,k} \tag{15.31}$$

where κ_1 and κ_2 are constants. If the distortion level is high, $|e_k|$ assumes high values and the second term of (15.31) allows convergence of the algorithm during the self-training. Near convergence, for $C \simeq \pm H^{-1}$, the second term has the same order of magnitude as the first and the pseudo error assumes small values in the neighbourhood of $C = \pm H^{-1}$.

Note that an algorithm that uses the pseudo error (15.31) allows a smooth transition of the equalizer from the self-training mode to the decision-directed mode. In the case of a sudden change in channel characteristics, the equalizer is found working again in self-training mode. Thus, the transitions between the two modes occur without control on the level of distortion of the signal $\{y_k\}$ at the equalizer output.

Stop-and-go algorithm

The stop-and-go algorithm [11] can be seen as a variant of the Sato algorithm that achieves the same objectives of the Benveniste–Goursat algorithm with better convergence properties. The pseudo error for the stop-and-go algorithm is formulated as

$$\epsilon_{P,k} = \begin{cases} e_k & \text{if } \operatorname{sgn}(e_k) = \operatorname{sgn}(\epsilon_{S,k}) \\ 0 & \text{otherwise} \end{cases} \tag{15.32}$$

where $\epsilon_{S,k}$ is the Sato pseudo error given by (15.28), and e_k is the error used in the decision-directed algorithm given by (15.29). The basic idea is that the algorithm converges if updating of the equalizer coefficients is turned off with sufficiently high probability every time the sign of error (15.29) differs from the sign of error $e_{id,k} = y_k - a_k$, that is $\text{sgn}(e_k) \neq \text{sgn}(e_{id,k})$. As $e_{id,k}$ is not available in a self-training equalizer, with the stop-and-go algorithm coefficient updating is turned off whenever the sign of error e_k is different from the sign of Sato error $\epsilon_{S,k}$. Obviously, in this way we also get a non-zero probability that coefficient updating is inactive when the condition $\text{sgn}(e_k) = \text{sgn}(e_{id,k})$ occurs, but this does not usually bias the convergence of the algorithm.

Remarks

At this point, we can make the following observations.

- Self-training algorithms based on the minimization of a cost function that includes the term $E[|y_k|^p]$, $p \geq 2$, can be explained referring to the algorithm (15.24), because the effect of raising to the p-th power the amplitude of the equalizer output sample is that of emphasizing the contribution of samples with large amplitude.
- Extension of the Sato cost function (15.25) to quadrature amplitude modulation (QAM) systems, which we discuss in Section 15.5, is given by

$$J = E\left[\frac{1}{2}\, |y_k|^2 - \gamma_S\, |y_k|\right] \tag{15.33}$$

  where $\gamma_S = E[|a_k|^2]/E[|a_k|]$. In general this term guarantees that, at convergence, the statistical power of the equalizer output samples is equal to the statistical power of the input symbols.
- In the algorithm (15.24), the equalizer coefficients are updated only when we observe a peak value of the equalizer output signal. As the peak value decreases with the progress of the equalization process, updating of the equalizer coefficients ideally depends on a threshold that varies depending on the level of distortion in the overall system impulse response.

15.3 The contour algorithm for PAM systems

The algorithm (15.24) suggests that the equalizer coefficients are updated when the equalizer output sample reaches a threshold value, which in turn depends on the level of distortion in the overall system impulse response. In practice, we define a threshold at instant k as $T_k = \alpha_{max} + \gamma_{CA,k}$, where the term $\gamma_{CA,k} \geq 0$ represents a suitable measure of distortion. Each time the absolute value of the equalizer output sample reaches or exceeds the threshold T_k, the coefficients are updated so that the peak value of the equalizer output is *driven* towards the constellation boundary $\pm\alpha_{max}$; at convergence of the equalizer coefficients, $\gamma_{CA,k}$ vanishes. Figure 15.6 illustrates the evolution of the *contour* T_k for a two dimensional constellation.

The updating of coefficients described above can be obtained by a stochastic gradient algorithm that is based on a cost function $E[\Phi(y_k)]$ defined on the parameter space $\{\psi_i\}$. We assume that the overall system Ψ initially corresponds to a point on a sphere of arbitrary radius r. With updating terms that on the average exhibit the same sign as the terms found in the general algorithm (15.24), the point on the sphere of radius r moves in such a way to reduce distortion. Moreover, if the radial component of the gradient, i.e. the component that is orthogonal to the surface of the sphere, is positive for $r > 1$, negative for $r < 1$ and vanishes on the sphere of radius $r = 1$, it is not necessary to scale the coefficients and the convergence will take place to the point of global minimum $\Psi = \pm I$. Clearly, the derivative function of $\Phi(y_k)$ with respect to y_k that defines the pseudo error can be determined in various ways. A suitable

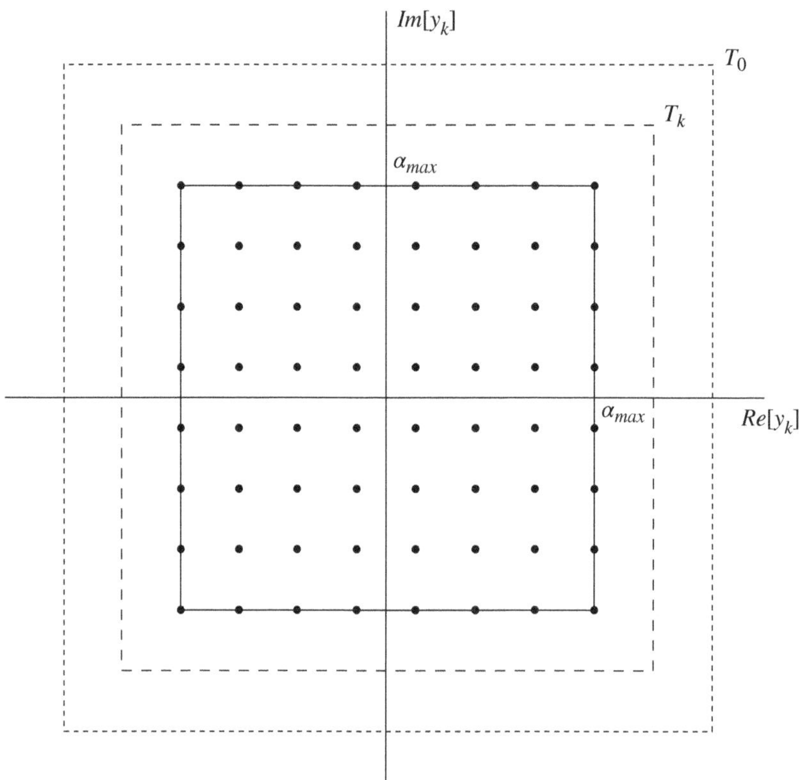

Figure 15.6 Evolution in time of the *contour* T_k for a two-dimensional constellation.

definition is given by

$$\Theta(y_k) = \begin{cases} y_k - [\alpha_{max} + \gamma_{CA,k}]\,\text{sgn}(y_k) & |y_k| \geq \alpha_{max} \\ -\gamma_{CA,k}\,\text{sgn}(y_k) & \text{otherwise} \end{cases} \tag{15.34}$$

For a self-training equalizer, the updating of coefficients as indicated by (15.8) is obtained by

$$\boldsymbol{c}_{k+1} = \boldsymbol{c}_k - \mu\, \Theta(y_k)\, \boldsymbol{x}_k \tag{15.35}$$

In (15.35), to avoid the computation of the threshold T_k at each iteration, the amplitude of the equalizer output signal is compared with α_{max} rather than with T_k; note that the computation of $\Theta(y_k)$ depends on the event that y_k falls inside or outside the constellation boundary. In the next section, for a two dimensional constellation, we define in general the *constellation boundary* as the contour line that connects the outer points of the constellation; for this reason, we refer to this algorithm as the *contour algorithm* [12].

To derive the algorithm (15.34)–(15.35) from the algorithm (15.24), several approximations are introduced; consequently, the convergence properties cannot be directly derived from those of the algorithm (15.24). In the Appendix 15.A, we show how γ_{CA} should be defined to obtain the desired behaviour of the algorithm (15.34)–(15.35) in the case of systems with input symbols having a uniform continuous distribution.

An advantage of the functional introduced in this section with respect to the Sato cost function is that the variance of the pseudo error vanishes at the points of minimum $\Psi = \pm I$; this means that it is possible to obtain the convergence of the mean square error (MSE) to a steady state value that is close to the achievable minimum value. Furthermore, the radial component of the gradient of $E[\Phi(y_k)]$ vanishes at every point on the unit sphere, whereas the radial component of the gradient in the Sato cost function vanishes on the unit sphere only at the points $\Psi = \pm I$. As the direction of steepest descent does not intersect the unit sphere, the contour algorithm avoids overshooting of the convergence trajectories observed using the Sato algorithm; in other words, the stochastic gradient yields a coefficient updating that is made in the correct direction more often than in the case of the Sato algorithm. Therefore, substantially better convergence properties are expected for the contour algorithm even in systems with a discrete probability distribution of input symbols.

The complexity of the algorithm (15.34)–(15.35) can be deemed prohibitive for practical implementations, especially for self-training equalization in high-speed communication systems, as the parameter $\gamma_{CA,k}$ must be estimated at each iteration. In the next section, we discuss a simplified algorithm that allows implementation with low complexity; we will see later how the simplified formulation of the contour algorithm can be extended to self-training equalization of partial response (PR) and QAM systems.

Simplified realization of the contour algorithm

We assume that the input symbols $\{a_k\}$ form a sequence of i.i.d. random variables with a uniform discrete probability density function. From the scheme of Figure 15.1, in the presence of noise, the channel output signal is given by

$$x_k = \sum_{i=-\infty}^{\infty} h_i \, a_{k-i} + w_k \tag{15.36}$$

where $\{w_k\}$ denotes additive white Gaussian noise. The equalizer output is given by $y_k = \boldsymbol{c}_k^T \boldsymbol{x}_k$. To obtain an algorithm that does not require the knowledge of the parameter γ_{CA}, the definition (15.34) suggests introducing the pseudo error

$$\epsilon_{CA,k} = \begin{cases} y_k - \alpha_{max} \operatorname{sgn}(y_k) & \text{if } |y_k| \geq \alpha_{max} \\ -\delta_k \operatorname{sgn}(y_k) & \text{otherwise} \end{cases} \tag{15.37}$$

where δ_k is a non-negative parameter that is updated at every iteration as follows:

$$\delta_{k+1} = \begin{cases} \delta_k - \dfrac{M-1}{M} \Delta & \text{if } |y_k| \geq \alpha_{max} \\ \delta_k + \dfrac{1}{M} \Delta & \text{otherwise} \end{cases} \tag{15.38}$$

and Δ is a positive constant. The initial value δ_0 is not a critical system parameter and can be, for example, chosen equal to zero; the coefficient updating algorithm is thus given by

$$\boldsymbol{c}_{k+1} = \boldsymbol{c}_k - \mu \, \epsilon_{CA,k} \, \boldsymbol{x}_k \tag{15.39}$$

In comparison to (15.34), now δ_k does not provide a measure of distortion as γ_{CA}. The definition (15.37) is justified by the fact that the term $y_k - [\alpha_{max} + \gamma_{CA}]\operatorname{sgn}(y_k)$ in (15.34) can be approximated as $y_k - \alpha_{max}\operatorname{sgn}(y_k)$, because if the event $|y_k| \geq \alpha_{max}$ occurs the pseudo error $y_k - \alpha_{max}\operatorname{sgn}(y_k)$ can be used for coefficient updating. Therefore, δ_k should increase in the presence of distortion only in the case the event $|y_k| < \alpha_{max}$ occurs more frequently than expected. This behaviour of the parameter δ_k is obtained by applying (15.38). Moreover, (15.38) guarantees that δ_k assumes values that approach zero at the convergence of the equalization process; in fact, in this case the probabilities of the events $\{|y_k| < \alpha_{max}\}$ and $\{|y_k| \geq \alpha_{max}\}$ assume approximately the values $(M-1)/M$ and $1/M$, respectively, that correspond to the probabilities of such events for a noisy PAM signal correctly equalized. Figure 15.7 shows the pseudo error $\epsilon_{CA,k}$ as a function of the value of the equalizer output sample y_k.

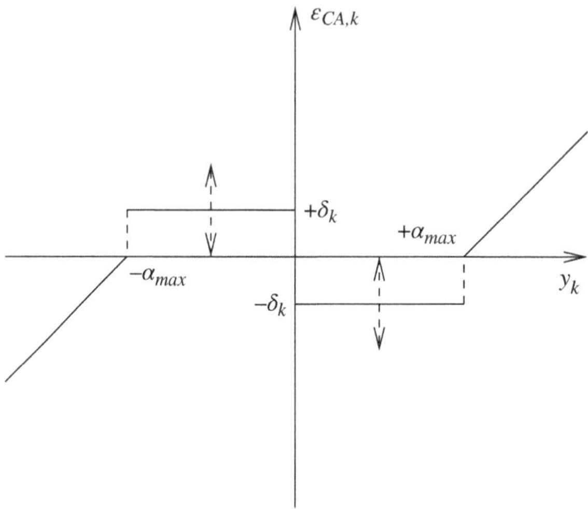

Figure 15.7 Characteristic of the pseudo error $\epsilon_{CA,k}$ as a function of the equalizer output.

The contour algorithm has been described for the case of uniformly distributed input symbols; however, this assumption is not necessary. In general, if $\{y_k\}$ represents an equalized signal, the terms $(M-1)/M$ and $1/M$ in (15.38) are, respectively, substituted with

$$p_0 = P[|a_k| < \alpha_{max}] + \frac{1}{2}\,P[|a_k| = \alpha_{max}] \simeq P[|y_k| < \alpha_{max}] \tag{15.40}$$

and

$$p_1 = \frac{1}{2}\,P[|a_k| = \alpha_{max}] \simeq P[|y_k| \geq \alpha_{max}] \tag{15.41}$$

We note that the considered receiver makes use of signal samples at the symbol rate; the algorithm can also be applied for the initial convergence of a fractionally spaced equalizer (see Section 7.4.3) in case a sampling rate higher than the symbol rate is adopted.

15.4 Self-training equalization for partial response systems

Self-training adaptive equalization has been mainly studied for full response systems; however, self-training equalization methods for partial response systems have been proposed for linear and non-linear equalizers [7, 13, 14]. In general, a self-training equalizer is more difficult to implement for partial response systems (see Appendix 7.C), especially if the input symbol alphabet has more than two elements. Moreover, as self-training is a slow process, the accuracy achieved in the recovery of the timing of the received signal before equalization plays an important role. In fact, if timing recovery is not accomplished, because of the difference between the sampling rate and the modulation rate we have that the sampling phase varies with respect to the timing phase of the remote transmitter clock; in this case, we speak of *phase drift* of the received signal; self-training algorithms fail if the phase drift of the received signal is not sufficiently small. In this section, we discuss the extension to partial response systems of the algorithms for PAM systems presented in the previous section.

The Sato algorithm

We consider a multilevel partial response class IV (PR-IV) system, also called modified duobinary system. Using the D transform, the desired transfer function of the overall system is given by

$\psi(D) = (1 - D^2)$. The objective of an adaptive equalizer for a PR-IV system consists in obtaining an equalized signal of the form

$$y_k = (a_k - a_{k-2}) + w_{y,k} = u_k + w_{y,k} \tag{15.42}$$

where $w_{y,k}$ is a disturbance due to noise and residual distortion. We consider the case of quaternary modulation. Then, the input symbols a_k are from the set $\{-3, -1, +1, +3\}$, and the output signal $u_k = a_k - a_{k-1}$, for an ideal channel in the absence of noise, can assume one of the seven values $\{-6, -4, -2, 0, +2, +4, +6\}$.[6]

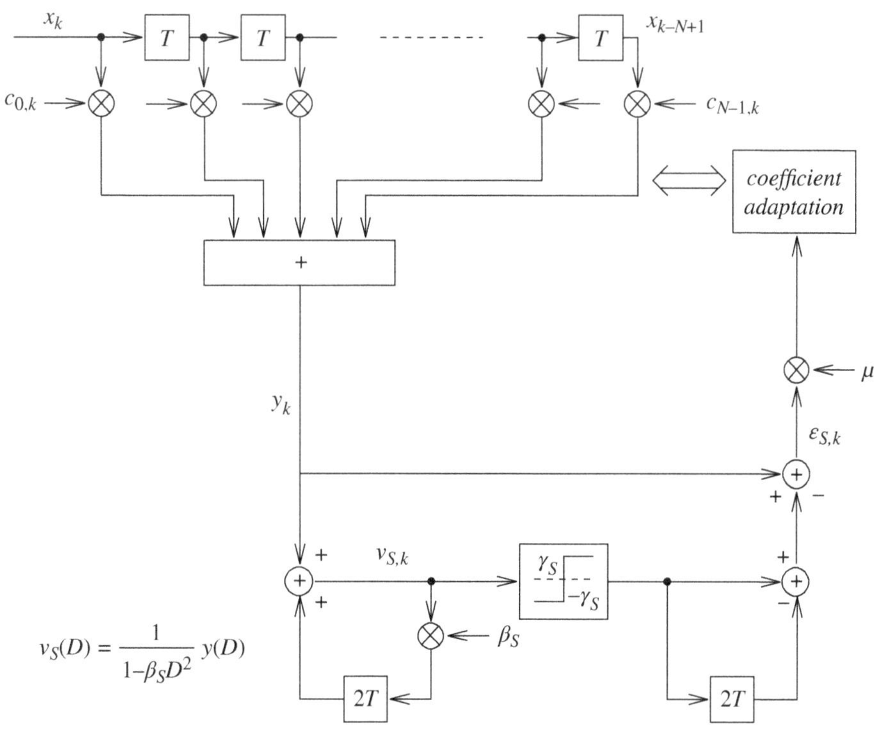

Figure 15.8 Block diagram of a self-training equalizer for a PR-IV system using the Sato algorithm.

As illustrated in Figure 15.8, to obtain a pseudo error to be employed in the equalizer coefficient updating algorithm, the equalizer output signal $\{y_k\}$ is transformed into a full-response signal $v_{S,k}$ by the linear transformation

$$v_{S,k} = y_k + \beta_S \, v_{S,k-2} \tag{15.43}$$

where β_S is a constant that satisfies the condition $0 < \beta_S < 1$. Then, the signal $v_{S,k}$ is quantized by a quantizer with two levels corresponding to $\pm\gamma_S$, where γ_S is given by (15.26). The obtained signal is again transformed into a partial response signal that is subtracted from the equalizer output to generate the pseudo error

$$\epsilon_{S,k} = y_k - \gamma_S[\text{sgn}(v_{S,k}) - \text{sgn}(v_{S,k-2})] \tag{15.44}$$

Then, the Sato algorithm for partial response systems is expressed as

$$\boldsymbol{c}_{k+1} = \boldsymbol{c}_k - \mu \, \epsilon_{S,k} \, \boldsymbol{x}_k \tag{15.45}$$

[6] In general, for an ideal PR-IV channel in the absence of noise, if the alphabet of the input symbols is $\mathcal{A} = \{\pm 1, \pm 3, \dots, \pm(M-1)\}$, the output symbols assume one of the $(2M-1)$ values $\{0, \pm 2, \dots, \pm 2(M-1)\}$.

The contour algorithm

In this case, first the equalizer output is transformed into a full response signal and a pseudo error is computed. Then, the error to compute the terms for coefficient updating is formed.

The method differs in two ways from the Sato algorithm described above. First, the channel equalization, carried out to obtain the full response signals, is obtained by combining linear and non-linear feedback, whereas in the case of the Sato algorithm, it is carried out by a linear filter. Second, the knowledge of the statistical properties of the input symbols is used to determine the pseudo error, as suggested by the contour algorithm.

As mentioned above, to apply the contour algorithm the equalizer output signal is transformed into a full response signal using a combination of linear feedback and decision feedback, that is we form the signal

$$v_k = y_k + \beta_{CA}\, v_{k-2} + (1 - \beta_{CA})\, \hat{a}_{k-2} \tag{15.46}$$

Equation (15.46) and the choice of the parameter β_{CA} are justified in the following way. If $\beta_{CA} = 0$ is selected, we obtain an equalization system with decision feedback, that presents the possibility of significant error propagation. The effect of the choice $\beta_{CA} = 1$ is easily seen using the D transform. From (15.42) and assuming correct decisions, (15.46) can be expressed as

$$v(D) = a(D) + \frac{w_y(D)}{1 - \beta_{CA}\, D^2} \tag{15.47}$$

Therefore with $\beta_{CA} = 1$, we get the linear inversion of the PR-IV channel with infinite noise enhancement at frequencies $f = 0$ and $\pm 1/(2T)$ Hz. The value of β_{CA} is chosen in the interval $0 < \beta_{CA} < 1$, to obtain the best trade-off between linear feedback and decision feedback.

We now apply the contour algorithm (15.37)–(15.39) using the signal v_k rather than the equalizer output signal y_k. The pseudo error is defined as

$$\epsilon^v_{CA,k} = \begin{cases} v_k - \alpha_{max}\,\text{sgn}(v_k) & \text{if } |v_k| \geq \alpha_{max} \\ -\delta^v_k\,\text{sgn}(v_k) & \text{otherwise} \end{cases} \tag{15.48}$$

where δ^v_k is a non-negative parameter that is updated at each iteration as

$$\delta^v_{k+1} = \begin{cases} \delta^v_k - \dfrac{M-1}{M}\,\Delta & \text{if } |v_k| \geq \alpha_{max} \\ \delta^v_k + \dfrac{1}{M}\,\Delta & \text{otherwise} \end{cases} \tag{15.49}$$

The stochastic gradient must be derived taking into consideration the channel equalization performed with linear feedback and decision feedback. We define the error on the M-ary symbol a_k after channel equalization as

$$e^v_k = v_k - a_k \tag{15.50}$$

Assuming correct decisions, by (15.46) it is possible to express the equalizer output as

$$y_k = (a_k - a_{k-2}) + e^v_k - \beta_{CA}\, e^v_{k-2} \tag{15.51}$$

Equation (15.51) shows that an estimate of the term $e^v_k - \beta_{CA}\, e^v_{k-2}$ must be included as error signal in the expression of the stochastic gradient. After initial convergence of the equalizer coefficients, the estimate $\hat{e}^v_k = v_k - \hat{a}_k$ is reliable. Therefore, decision directed coefficient updating can be performed according to the algorithm

$$\boldsymbol{c}_{k+1} = \boldsymbol{c}_k - \mu_{dd}(\hat{e}^v_k - \beta_{CA}\, \hat{e}^v_{k-2})\, \boldsymbol{x}_k \tag{15.52}$$

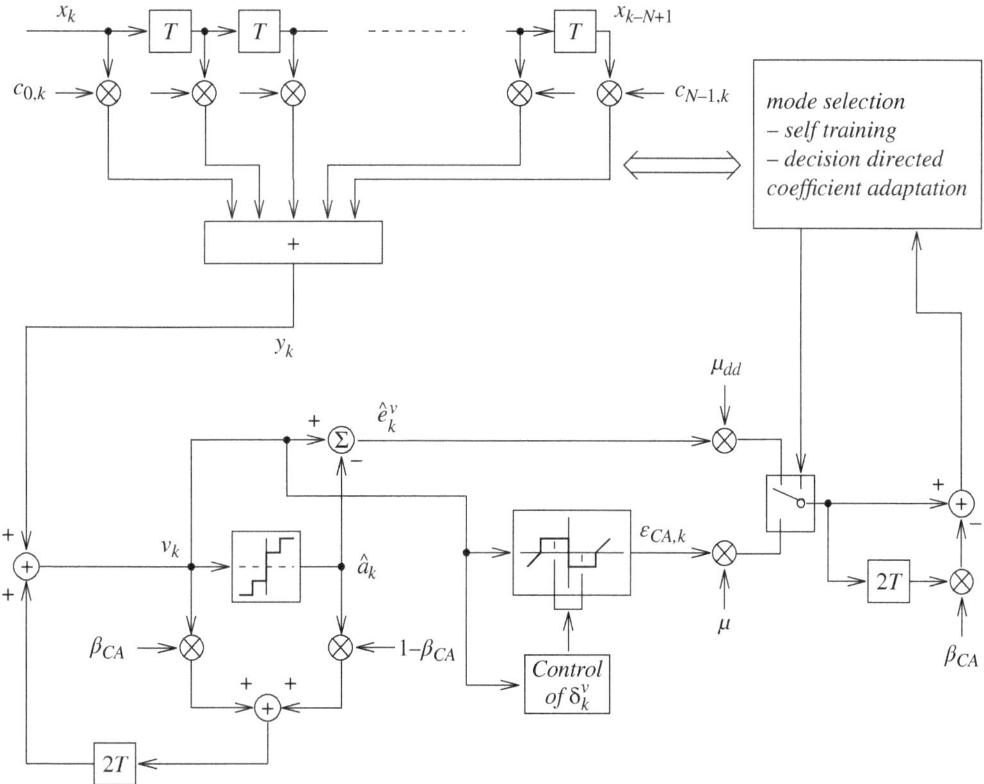

Figure 15.9 Block diagram of a self-training equalizer with the contour algorithm for a QPR-IV system.

The contour algorithm for coefficient updating during self-training is obtained by substituting the decision directed error signal e_k^v with the pseudo error $\epsilon_{CA,k}^v$ (15.48),

$$\boldsymbol{c}_{k+1} = \boldsymbol{c}_k - \mu(\epsilon_{CA,k}^v - \beta_{CA}\, \epsilon_{k-2}^v)\, \boldsymbol{x}_k \tag{15.53}$$

During self-training, satisfactory convergence behaviour is usually obtained for $\beta_{CA} \simeq 1/2$. Figure 15.9 shows the block diagram of an equalizer for a PR-IV system with a quaternary alphabet (QPR-IV), with the generation of the error signals to be used in decision directed and self-training mode.

In the described scheme, the samples of the received signal can be initially filtered by a filter with transfer function $1/(1 - aD^2)$, $0 < a < 1$, to reduce the correlation among samples. The obtained signal is then input to the equalizer delay line.

15.5 Self-training equalization for QAM systems

We now describe various self-training algorithms for passband transmission systems that employ a two-dimensional constellation.

The Sato algorithm

Consider a QAM transmission system with constellation $\mathcal{A}$ and a sequence $\{a_k\}$ of i.i.d. symbols from $\mathcal{A}$ such that $a_{k,I} = Re[a_k]$ and $a_{k,Q} = Im[a_k]$ are independent and have the same probability distribution. We

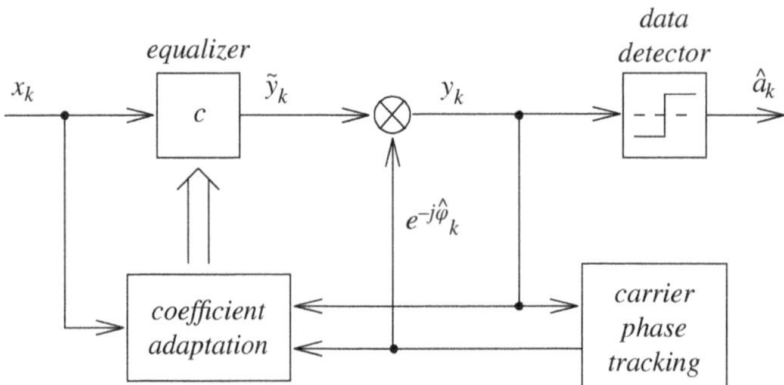

Figure 15.10 Block diagram of a self-training equalizer with the Sato algorithm for a QAM system.

assume a receiver in which sampling of the received signal occurs at the symbol rate and tracking of the carrier phase is carried out at the equalizer output, as shown in Figure 15.10 (see scheme of Figure 7.41 with a baseband equalizer filter where $\{x_k\}$ is already demodulated and $\theta_k = \hat{\varphi}_k$). If $\tilde{y}_k = \boldsymbol{c}^T \boldsymbol{x}_k$ is the equalizer filter output, the sample at the decision point is then given by

$$y_k = \tilde{y}_k \, e^{-j\hat{\varphi}_k} \tag{15.54}$$

We let

$$y_{k,I} = Re[y_k] \qquad \text{and} \qquad y_{k,Q} = Im[y_k] \tag{15.55}$$

and introduce the Sato cost function for QAM systems,

$$J = E[\Phi(y_{k,I}) + \Phi(y_{k,Q})] \tag{15.56}$$

where

$$\Phi(v) = \frac{1}{2} v^2 - \gamma_S \, |v| \tag{15.57}$$

and

$$\gamma_S = \frac{E[a_{k,I}^2]}{E[|a_{k,I}|]} = \frac{E[a_{k,Q}^2]}{E[|a_{k,Q}|]} \tag{15.58}$$

The gradient of (15.56) with respect to $\boldsymbol{c}$ yields (see also (7.321))

$$\nabla_{\boldsymbol{c}} J = \nabla_{Re[\boldsymbol{c}]} J + j \, \nabla_{Im[\boldsymbol{c}]} J = E[e^{j\hat{\varphi}_k} \, \boldsymbol{x}_k^* (\Theta(y_{k,I}) + j \, \Theta(y_{k,Q}))] \tag{15.59}$$

where

$$\Theta(v) = \frac{d}{dv} \, \Phi(v) = v - \gamma_S \, \mathrm{sgn}(v) \tag{15.60}$$

The partial derivative with respect to the carrier phase estimate is given by (see also (7.326))

$$\frac{\partial}{\partial \hat{\varphi}} \, J = E[Im(y_k (\Theta(y_{k,I}) + j \, \Theta(y_{k,Q}))^*)] \tag{15.61}$$

Defining the Sato pseudo error for QAM systems as

$$\epsilon_{S,k} = y_k - \gamma_S \, \mathrm{sgn}(y_k) \tag{15.62}$$

and observing that

$$Im(y_k \, \epsilon_{S,k}^*) = -Im(\gamma_S \, y_k \, \mathrm{sgn}(y_k^*)) \tag{15.63}$$

the equalizer coefficient updating and carrier phase estimate are given by

$$\boldsymbol{c}_{k+1} = \boldsymbol{c}_k - \mu \, \epsilon_{S,k} \, e^{j\hat{\varphi}_k} \, \boldsymbol{x}_k^* \tag{15.64}$$

$$\hat{\varphi}_{k+1} = \hat{\varphi}_k + \mu_\varphi \, Im[\gamma_S \, y_k \, \text{sgn}(y_k^*)] \tag{15.65}$$

where μ_φ is the adaptation gain of the carrier phase tracking loop. Equations (15.64) and (15.65) are analogous to (7.323) and (7.328) for the decision directed case.

The same observations made for self-training equalization of PAM systems using the Sato algorithm hold for QAM systems. In particular, assuming that the algorithm converges to a point of global minimum of the cost function, we recall that the variance of the pseudo error assumes high values in the neighbourhood of the point of convergence. Therefore, in the steady state it is necessary to adopt a decision directed algorithm. To obtain smooth transitions between the self-training mode and the decision directed mode without the need of a further control of the distortion level, in QAM systems we can use extensions of the Benveniste–Goursat and stop-and-go algorithms considered for self-training of PAM systems.

15.5.1 Constant-modulus algorithm

In the constant-modulus algorithm (CMA) for QAM systems proposed by Godard [15], self-training equalization is based on the cost function

$$J = E[(|\tilde{y}_k|^p - R_p)^2] = E[(|y_k|^p - R_p)^2] \tag{15.66}$$

where p is a parameter that usually assumes the value $p = 1$ or $p = 2$. We note that, as J depends on the absolute value of the equalizer output raised to the p-th power, the CMA does not require the knowledge of the carrier phase estimate. Figure 15.11 shows the block diagram of a receiver using the CMA.

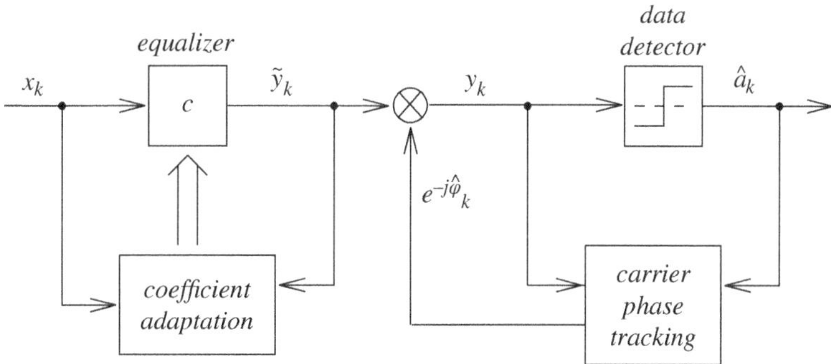

Figure 15.11 Block diagram of a self-training equalizer using the CMA.

The gradient of (15.66) is given by

$$\nabla_c J = 2p \; E[(|\tilde{y}_k|^p - R_p)|\tilde{y}_k|^{p-2} \; \tilde{y}_k \; \boldsymbol{x}_k^*] \tag{15.67}$$

The constant R_p is chosen so that the gradient is equal to zero for a perfectly equalized system; therefore, we have

$$R_p = \frac{E[|a_k|^{2p}]}{E[|a_k|^p]} \tag{15.68}$$

For example, for a 64-QAM constellation, we obtain $R_1 = 6.9$ and $R_2 = 58$. The 64-QAM constellation and the circle of radius $R_1 = 6.9$ are illustrated in Figure 15.12.

By using (15.67), we obtain the equalizer coefficient updating law

$$\boldsymbol{c}_{k+1} = \boldsymbol{c}_k - \mu(|\tilde{y}_k|^p - R_p)|\tilde{y}_k|^{p-2} \; \tilde{y}_k \; \boldsymbol{x}_k^* \tag{15.69}$$

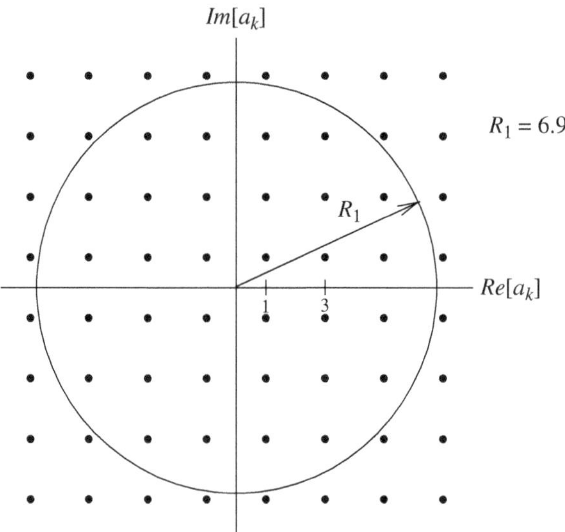

Figure 15.12 The 64-QAM constellation and the circle of radius $R_1 = 6.9$.

For $p = 1$, (15.69) becomes

$$\boldsymbol{c}_{k+1} = \boldsymbol{c}_k - \mu(|\tilde{y}_k| - R_1)\frac{\tilde{y}_k}{|\tilde{y}_k|}\,\boldsymbol{x}_k^* \tag{15.70}$$

We note that the Sato algorithm, introduced in Section 15.2, can then be viewed as a particular case of the CMA.

The contour algorithm

Let us consider a receiver in which the received signal is sampled at the symbol rate and the carrier phase recovery is ideally carried out before the equalizer. The scheme of Figure 15.1 is still valid, and the complex-valued baseband equivalent channel output is given by

$$x_k = \sum_{i=-\infty}^{+\infty} h_i\, a_{k-i} + w_k \tag{15.71}$$

The equalizer output is expressed as $y_k = \boldsymbol{c}^T \boldsymbol{x}_k$. To generalize the notion of pseudo error of the contour algorithm introduced in Section 15.3 for PAM systems, we define a contour line C that connects the outer points of the constellation. For simplicity, we assume a square constellation with $L \times L$ points, as illustrated in Figure 15.13 for the case $L = 8$.

Let S be the region of the complex plane enclosed by the contour line C and let $C \notin S$ by definition. We denote by y_k^C the closest point to y_k on C every time that $y_k \notin S$, that is every time the point y_k is found outside the region enclosed by C. The pseudo error (15.37) is now extended as follows:

$$\epsilon_{CA,k} = \begin{cases} y_k - y_k^C & \text{if } y_k \notin S \\ \left.\begin{array}{ll} -\delta_k \operatorname{sgn}(y_{k,I}) & \text{if } |y_{k,I}| \geq |y_{k,Q}| \\ -j\,\delta_k \operatorname{sgn}(y_{k,Q}) & \text{if } |y_{k,I}| < |y_{k,Q}| \end{array}\right\} & \text{if } y_k \in S \end{cases} \tag{15.72}$$

Also in this case, δ_k is a non-negative parameter, updated at each iteration as

$$\delta_{k+1} = \begin{cases} \delta_k - p_S \Delta & \text{if } y_k \notin S \\ \delta_k + (1 - p_S)\,\Delta & \text{if } y_k \in S \end{cases} \tag{15.73}$$

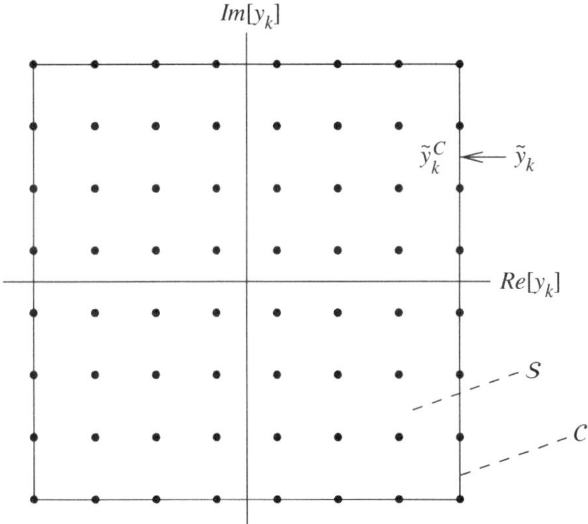

Figure 15.13 Illustration of the contour line and surface S for a 64-QAM constellation.

where, in analogy with (15.40), the probability $p_S \simeq P[y_k \in S]$ is computed assuming that y_k is an equalized signal in the presence of additive noise.

Let α_{max} be the maximum absolute value of the real and imaginary parts of the square $L \times L$ symbol constellation. If $|y_{k,I}| \geq \alpha_{max}$ or $|y_{k,Q}| \geq \alpha_{max}$, but not both, the projection of the sample y_k on the contour line C yields a non-zero pseudo error along one dimension and a zero error in the other dimension. If both $|y_{k,I}|$ and $|y_{k,Q}|$ are larger than α_{max}, y_k^C is chosen as the corner point of the constellation closest to y_k; in this case, we obtain a non-zero pseudo error in both dimensions.

Thus, the equalizer coefficients are updated according to the algorithm

$$\boldsymbol{c}_{k+1} = \boldsymbol{c}_k - \mu \, \epsilon_{CA,k} \, \boldsymbol{x}_k^* \tag{15.74}$$

Clearly, the contour algorithm can also be applied to systems that use non-square constellations. In any case, the robust algorithm for carrier phase tracking that is described in the next section requires that the shape of the constellation is non-circular.

Joint contour algorithm and carrier phase tracking

We now apply the idea of generating an error signal with respect to the contour line of a constellation to the problem of carrier phase recovery and frequency offset compensation [12]. With reference to the scheme of Figure 15.10, we denote as $\hat{\varphi}_k$ the carrier phase estimate used for the received signal demodulation at instant k. If carrier recovery follows equalization, the complex equalizer output signal is given by

$$y_k = \tilde{y}_k \, e^{-j\hat{\varphi}_k} \tag{15.75}$$

As for equalizer coefficient updating, reliable information for updating the carrier phase estimate $\hat{\varphi}_k$ is only available if y_k falls outside of the region S. As illustrated in Figure 15.14, the phase estimation error can then be computed as (see also (15.65))

$$\Delta\varphi_k \simeq Im(y_k \, y_k^{C*}) = -Im[y_k(y_k^* - y_k^{C*})] \tag{15.76}$$

If y_k falls within S, the phase error is set to zero, that is $\Delta\varphi_k = 0$.

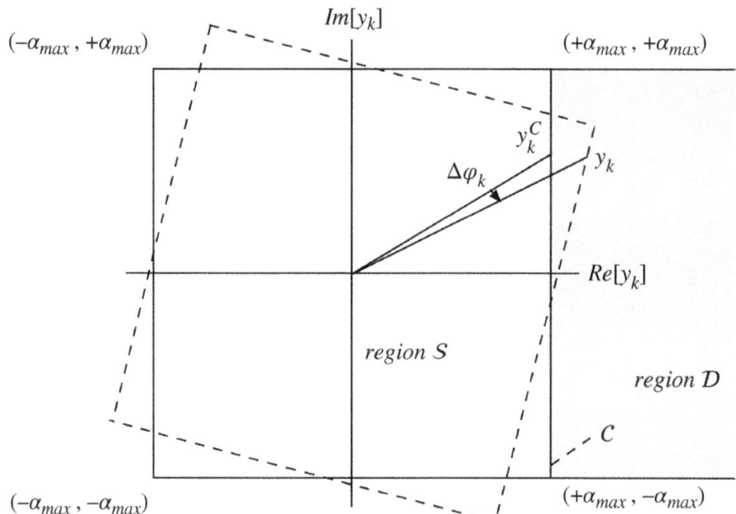

Figure 15.14 Illustration of the rotation of the symbol constellation in the presence of a phase error, and definition of $\Delta\varphi_k$.

From Figure 15.14, we note that the phase error $\Delta\varphi_k$ is invariant with respect to a rotation of y_k equal to an integer multiple of $\pi/2$. Then to determine $\Delta\varphi_k$, we can first rotate y_k as

$$y'_k = y_k\, e^{j\ell\pi/2} \tag{15.77}$$

where ℓ is chosen such that $Re[y'_n] > |Im[y'_n]|$ (shaded region in Figure 15.14). Furthermore, we observe that the information on the phase error obtained by samples of the sequence $\{y_k\}$ that fall in the corner regions, where $|y_{k,I}| > \alpha_{max}$ and $|y_{k,Q}| > \alpha_{max}$, is not important. Thus we calculate a phase error only if y_k is outside of S, but not in the corner regions, that is if $y'_k \in D$, with $D = \{y'_k \,:\, Re[y'_k] > \alpha_{max}, |Im[y'_k]| < \alpha_{max}\}$. Then, (15.76) becomes

$$\Delta\varphi_k = \begin{cases} -Im[y'_k][Re[y'_k] - \alpha_{max}] & \text{if } y'_k \in D \\ 0 & \text{otherwise} \end{cases} \tag{15.78}$$

In the presence of a frequency offset equal to $\Omega/(2\pi)$, the probability distribution of the samples $\{y_k\}$ rotates at a rate of $\Omega/(2\pi)$ revolutions per second. For large values of $\Omega/(2\pi)$, the phase error $\Delta\varphi_k$ does not provide sufficient information for the carrier phase tracking system to achieve a lock condition; therefore, the update of $\hat{\varphi}_k$ must be made by a second-order phase-locked loop (PLL), where in the update of the second-order term a factor that is related to the value of the frequency offset must be included (see Section 14.7).

The needed information is obtained observing the statistical behaviour of the term $Im[y'_k]$, conditioned by the event $y'_k \in D$. At instants in which the sampling distribution of y_k is aligned with S, the distribution of $Im[y'_k]$ is uniform in the range $[-\alpha_{max}, \alpha_{max}]$. Between these instants, the distribution of $Im[y'_k]$ exhibits a time varying behaviour with a downward or upward trend depending on the sign of the frequency offset, with a minimum variance when the corners of the rotating probability distribution of y_k, which we recall rotates at a rate of $\Omega/(2\pi)$ revolutions per second, cross the coordinate axes. Defining

$$Q(v) = \begin{cases} v & \text{if } |v| < \alpha_{max} \\ 0 & \text{otherwise} \end{cases} \tag{15.79}$$

from the observation of Figure 15.14, a simple method to extract information on $\Omega/(2\pi)$ consists in evaluating

$$\Delta\ Im[y'_k] = Q\{Im[y'_k] - Im[y'_m]\}, \qquad y'_k \in \mathcal{D} \tag{15.80}$$

where $m < k$ denotes the last time index for which $y'_k \in \mathcal{D}$. In the mean, $\Delta\ Im[y'_k]$ exhibits the sign of the frequency offset.

The equations for the updating of the parameters of a second-order PLL for the carrier phase recovery then become

$$\begin{cases} \hat{\varphi}_{k+1} = \hat{\varphi}_k + \mu_\varphi\ \Delta\varphi_k + \Delta\hat{\varphi}_{c,k} \\ \Delta\hat{\varphi}_{c,k+1} = \Delta\hat{\varphi}_{c,k} + \mu_{f_1}\ \Delta\varphi_k + \mu_{f_2}\ \Delta\ Im[y'_k] \end{cases} \quad \text{if } y'_k \in \mathcal{D}$$
$$\begin{cases} \hat{\varphi}_{k+1} = \hat{\varphi}_k \\ \Delta\hat{\varphi}_{c,k+1} = \Delta\hat{\varphi}_{c,k} \end{cases} \quad \text{otherwise} \tag{15.81}$$

where μ_φ, μ_{f_1}, and μ_{f_2} are suitable adaptation gains; typically, μ_φ is in the range 10^{-4} to 10^{-3}, $\mu_{f_1} = (1/4)\mu_\varphi^2$, and $\mu_{f_2} \simeq \mu_{f_1}$.

The rotation of y_k given by (15.77) to obtain y'_k also has the advantage of simplifying the error computation for self-training equalizer coefficient adaptation with the contour algorithm.

With no significant effect on performance, we can introduce a simplification similar to that adopted to update the carrier phase, and let the pseudo error equal zero if y_k is found in the corner regions, that is $\epsilon_{CA,k} = 0$ if $Im[y'_k] > \alpha_{max}$. By using (15.72) and (15.73) to compute the pseudo error, the coefficient updating equation (15.74) becomes (see (7.323))

$$\boldsymbol{c}_{k+1} = \boldsymbol{c}_k - \mu\ \epsilon_{CA,k}\ e^{j\hat{\varphi}_k}\ \boldsymbol{x}_k^* \tag{15.82}$$

15.6 Examples of applications

In this section, we give examples of applications that illustrate the convergence behaviour and steady state performance of self-training equalizers, with particular regard to the contour algorithm.

We initially consider self-training equalization for PAM transmission systems over unshielded twisted-pair (UTP) cables with frequency response given by

$$\mathcal{G}_{Ch}(f) = 10^{-\frac{1.2}{20}}\, e^{-(0.00385\sqrt{jf}+0.00028f)L} \tag{15.83}$$

where f is expressed in MHz and L in meters.

Example 15.6.1

As a first example, we consider a 16-PAM system ($M = 16$) with a uniform probability distribution of the input symbols and symbol rate equal to 25 MBaud; the transmit and receive filters are designed to yield an overall raised cosine channel characteristic for a cable length of 50 m. In the simulations, the cable length is chosen equal to 100 m, and the received signal is disturbed by additive white Gaussian noise. The signal-to-noise ratio at the receiver input is equal to $\Gamma = 36$ dB. Self-training equalization is achieved by a fractionally spaced equalizer having $N = 32$ coefficients, and input signal sampled with sampling period equal to $T/2$. Figure 15.15 shows the convergence of the contour algorithm (15.37)–(15.38) for $\delta_0 = 0$ and $\boldsymbol{c}_0$ chosen equal to the zero vector. The results are obtained for a cable with attenuation $\alpha(f)|_{f=1} = 3.85 \times 10^{-6}$ [m^{-1} Hz$^{-1/2}$], parameters of the self-training equalizer given by $\mu = 10^{-5}$ and $\Delta = 2.5 \times 10^{-4}$, and ideal timing recovery.

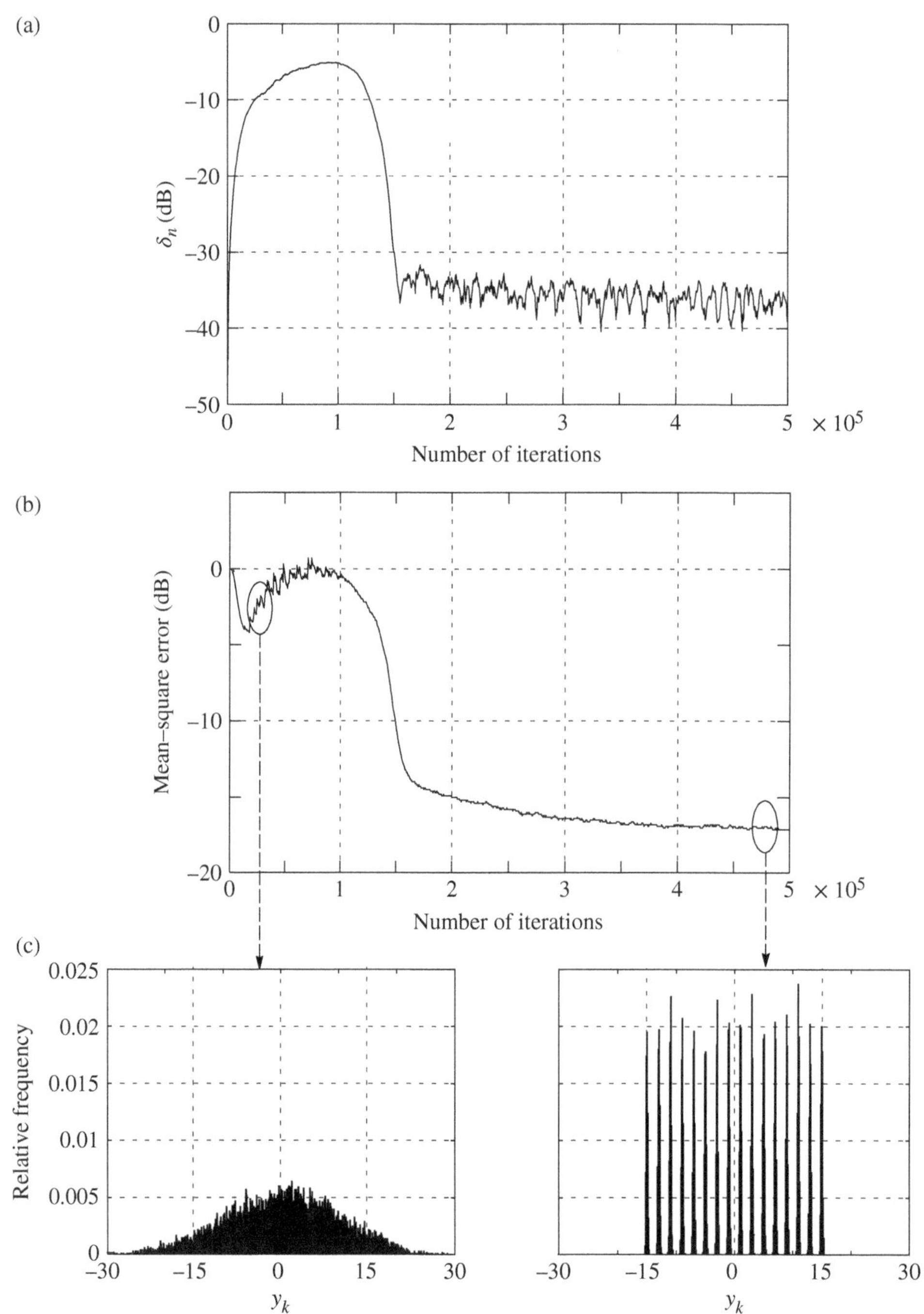

Figure 15.15 Illustration of the convergence of the contour algorithm for a 16-PAM system: (a) behaviour of the parameter δ_n, (b) MSE convergence, and (c) relative frequency of equalizer output samples at the beginning and the end of the convergence process.

Example 15.6.2

We consider self-training equalization for a baseband quaternary partial response class IV system ($M = 4$) for transmission at 125 Mbit/s over UTP cables; a very-large-scale integration (VLSI) transceiver implementation for this system will be described in Chapter 19. We compare the performance of the contour algorithm, described in Section 15.3, with the Sato algorithm for partial response systems. Various realizations of the MSE convergence of a self-training equalizer with $N = 16$ coefficients are shown in Figures 15.16 and 15.17 for the Sato algorithm and the contour algorithm, respectively. The curves are parameterized by $t = \Delta T/T$, where $T = 16$ ns, and ΔT denotes the difference between the sampling phase of the channel output signal and the optimum sampling phase that yields the minimum MSE; we note that the contour algorithm has a faster convergence with respect to the Sato algorithm and yields significantly lower values of MSE in the steady state. The Sato algorithm can be applied only if timing recovery is achieved prior to equalization; note that the convergence characteristics of the contour algorithm makes self-training equalization possible even in the presence of considerable distortion and phase drift of the received signal.

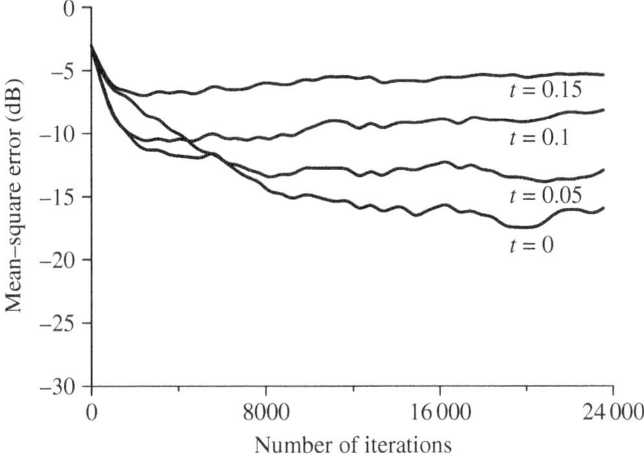

Figure 15.16 MSE convergence with the Sato algorithm for a QPR-IV system.

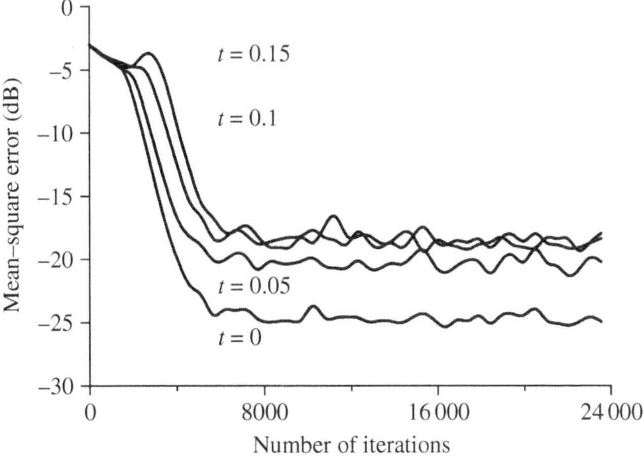

Figure 15.17 MSE convergence with the contour algorithm for a QPR-IV system.

Example 15.6.3

We now examine self-training equalization for a 256-QAM transmission system having a square constellation with $L = 16$ ($M = 256$), and symbol rate equal to 6 MBaud. Along each dimension, symbols ± 3, ± 1 have probability $2/20$, and symbols ± 15, ± 13, ± 11, ± 9, ± 7, and ± 5 have probability $1/20$. The overall baseband equivalent channel impulse response is illustrated in Figure 15.18. The received signal is disturbed by additive white Gaussian noise. The signal-to-noise ratio at the receiver input is equal to $\Gamma = 39$ dB. Signal equalization is obtained by a fractionally spaced equalizer having $N = 32$ coefficients, and input signal sampled with sampling period equal to $T/2$.

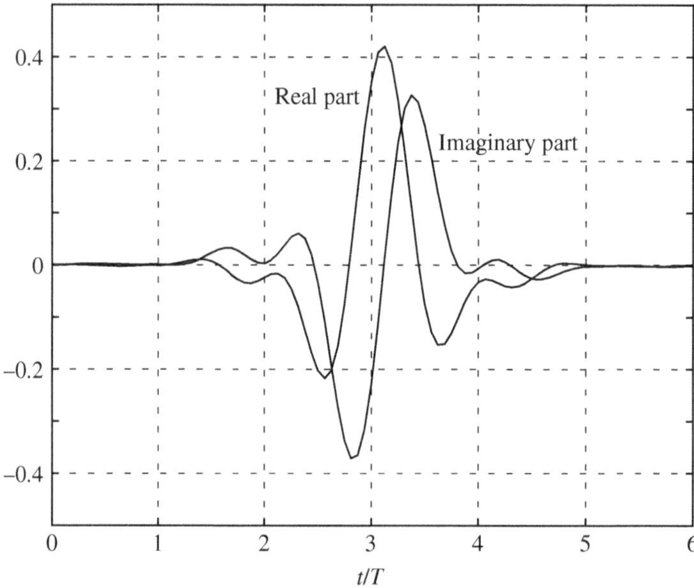

Figure 15.18 Overall baseband equivalent channel impulse response for simulations of a 256-QAM system. Source: Reproduced with permission from Cherubini et al. [12]. ©1998, IEEE.

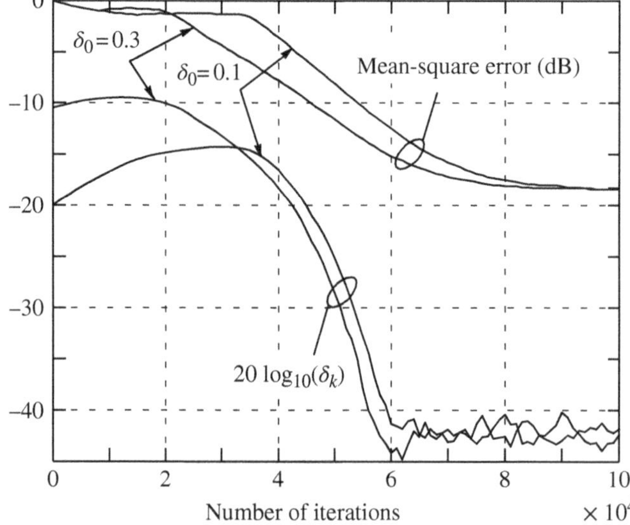

Figure 15.19 Convergence behaviour of MSE and parameter δ_k using the contour algorithm for a 256-QAM system with non-uniform distribution of input symbols. Source: Reproduced with permission from Cherubini et al. [12]. ©1998, IEEE.

Figure 15.19 shows the convergence of the contour algorithm and the behaviour of the parameter δ_k for $p_S = 361/400$, various initial values of δ_0, and $\boldsymbol{c}_0$ given by a vector with all elements equal to zero except for one element. Results are obtained for $\mu = 10^{-4}$, $\Delta = 10^{-4}$, and ideal timing and carrier phase recovery.

Example 15.6.4

With reference to the previous example, we examine the behaviour of the carrier phase recovery algorithm, assuming ideal timing recovery. Figure 15.20 illustrates the behaviour of the MSE and of the second order term $\Delta\hat{\varphi}_{c,k}$ for an initial frequency offset of +2.5 kHz, $\mu_\varphi = 4 \times 10^{-4}$, $\mu_{f_1} = 8 \times 10^{-8}$, and $\mu_{f_2} = 2 \times 10^{-8}$.

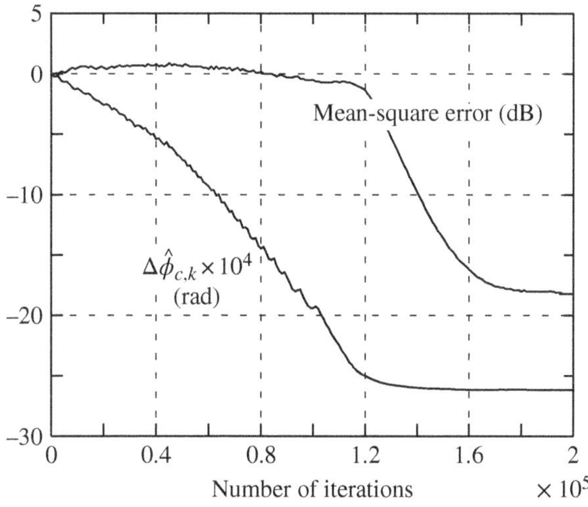

Figure 15.20 Illustration of the convergence behaviour of MSE and second order term $\Delta\hat{\varphi}_{c,k}$ using the contour algorithm in the presence of an initial frequency offset equal to 500 ppm for a 256-QAM system with non-uniform distribution of input symbols. Source: Reproduced with permission from Cherubini et al. [12]. ©1998, IEEE.

Bibliography

[1] Ichikawa, H., Sango, J., and Murase, T. (1987). 256 QAM multicarrier 400 Mb/s microwave radio system field tests. *Proceedings of 1987 IEEE International Conference on Communications*, Philadelphia, PA, pp. 1803–1808.

[2] Ross, F.J. and Taylor, D.P. (1991). An enhancement to blind equalization algorithms. *IEEE Transactions on Communications* 39: 636–639.

[3] Proakis, J.G. and Nikias, C.L. (1991). Blind equalization. *Proceedings of SPIE Adaptive Signal Processing*, San Diego, CA, Volume 1565 (22–24 July 1991), pp. 76–87.

[4] Bellini, S. (1991). Blind equalization and deconvolution. *Proc. SPIE Adaptive Signal Processing*, San Diego, CA, Volume 1565 (22–24 July 1991), pp. 88–101.

[5] Benvenuto, N. and Goeddel, T.W. (1995). Classification of voiceband data signals using the constellation magnitude. *IEEE Transactions on Communications* 43: 2759–2770.

[6] Liu, R. and Tong, L. (eds.) (1998). Special issue on blind systems identification and estimation. *IEEE Proceedings* 86.

[7] Sato, Y. (1975). A method of self-recovering equalization for multilevel amplitude-modulation systems. *IEEE Transactions on Communications* 23: 679–682.

[8] Tong, L., Xu, G., Hassibi, B., and Kailath, T. (1995). Blind channel identification based on second-order statistics: a frequency-domain approach. *IEEE Transactions on Information Theory* 41: 329–334.

[9] Benveniste, A., Goursat, M., and Ruget, G. (1980). Robust identification of a nonminimum phase system: blind adjustment of a linear equalizer in data communications. *IEEE Transactions on Automatic Control* 25: 385–399.

[10] Benveniste, A. and Goursat, M. (1984). Blind equalizers. *IEEE Transactions on Communications* 32: 871–883.

[11] Picchi, G. and Prati, G. (1987). Blind equalization and carrier recovery using a 'Stop-and-Go' decision directed algorithm. *IEEE Transactions on Communications* 35: 877–887.

[12] Cherubini, G., Ölçer, S., and Ungerboeck, G. (1998). The contour algorithm for self-training equalization. In: *Proceedings of Broadband Wireless Communications, 9th Tyrrhenian International Workshop on Digital Communications*, Lerici, Italy (7–10 September 1997) (ed. M. Luise and S. Pupolin), 58–69. Berlin: Springer-Verlag.

[13] Cherubini, G., Ölçer, S., and Ungerboeck, G. (1995). Self-training adaptive equalization for multilevel partial-response transmission systems. *Proceedings of 1995 IEEE International Symposium on Information Theory*, Whistler, Canada (17–22 September 1995), p. 401.

[14] Cherubini, G. (1994). Nonlinear self-training adaptive equalization for partial-response systems. *IEEE Transactions on Communications* 42: 367–376.

[15] Godard, D.N. (1980). Self recovering equalization and carrier tracking in two-dimensional data communication systems. *IEEE Transactions on Communications* 28: 1867–1875.

Appendix 15.A On the convergence of the contour algorithm

Given $y = y_0 = \sum_{i=-\infty}^{+\infty} c_i\, x_{-i}$, we show that the only minima of the cost function $J = E[\Phi(y)]$ correspond to the equalizer settings $C = \pm H^{-1}$, for which we obtain $\Psi = \pm I$, except for a possible delay. If the systems H and C have finite energy, then also Ψ has finite energy and $E[\Phi(y)]$ may be regarded as a functional $J(\Psi) = E[\Phi(y)]$, where y is expressed as $y = \sum_{i=-\infty}^{+\infty} \psi_i\, a_{-i}$; thus, we must prove that the only minima of $\mathcal{V}(\Psi)$ are found at points $\Psi = \pm I$. We consider input symbols with a probability density function $p_{a_k}(\alpha)$ uniform in the interval $[-\alpha_{max}, \alpha_{max}]$.

For the analysis, we express the system Ψ as $\Psi = r\overline{\Psi}$, where $r \geq 0$, and $\overline{\Psi} \leftrightarrow \{\overline{\psi}_i\}$, with $\sum_i \overline{\psi}_i^2 = 1$, denotes the normalized overall system. We consider the cost function J as a functional $\mathcal{V}(\Psi) = E[\Phi(r\overline{y})]$, where $\overline{y} = \overline{y}_0 = \sum_{i=-\infty}^{+\infty} \overline{\psi}_i a_{-i}$ denotes the output of the system $\overline{\Psi}$, and Φ has derivative Θ given by (15.34). Let

$$\tilde{\Theta}(x) = \begin{cases} x - \alpha_{max}\,\mathrm{sgn}(x) & \text{if } |x| \geq \alpha_{max} \\ 0 & \text{otherwise} \end{cases} \tag{15.84}$$

and $p_{\overline{y}_k}(x)$ denote the probability density function of $\overline{y}$.

We express the parameter γ_{CA} as

$$\begin{aligned} \gamma_{CA} &= \gamma_{\overline{\Psi}} + \gamma_r \\ \gamma_{\overline{\Psi}} &= \frac{\int b\, \tilde{\Theta}(b)\, p_{\overline{y}_k}(b)\, db}{\int |b|\, p_{\overline{y}_k}(b)\, db} \\ \gamma_r &= \begin{cases} 1 - r & \text{if } r \leq 1 \\ 0 & \text{otherwise} \end{cases} \end{aligned} \tag{15.85}$$

Examine the function $\mathcal{V}(\overline{\Psi})$ on the unit sphere $\mathcal{S}$. To claim that the only minima of $\mathcal{V}(\overline{\Psi})$ are found at points $\Psi = \pm I$, we apply Theorem 3.5 of [9]. Consider a pair of indices (i,j), $i \neq j$, and a fixed system with coefficients $\{\overline{\psi}_\ell\}_{\ell \neq i,j}$, such that $R^2 = 1 - \sum_{\ell \neq i,j} \overline{\psi}_\ell^2 > 0$. For $\varphi \in [0, 2\pi)$, let $\overline{\Psi}_\varphi \in \mathcal{S}$ be the system with coefficients $\{\overline{\psi}_\ell\}_{\ell \neq i,j}$, $\overline{\psi}_i = R\cos\varphi$, and $\overline{\psi}_j = R\sin\varphi$; moreover let $(\partial/\partial\varphi)\mathcal{V}(\overline{\Psi}_\varphi)$ be the derivative of $\mathcal{V}(\overline{\Psi}_\varphi)$ with respect to φ at point $\overline{\Psi} = \overline{\Psi}_\varphi$. As $p_{a_k}(\alpha)$ is sub-Gaussian, it can be shown that

$$\frac{\partial}{\partial\varphi}\mathcal{V}(\overline{\Psi}_\varphi) = 0 \qquad \text{for } \varphi = k\,\frac{\pi}{4} \qquad k \in \mathbb{Z} \tag{15.86}$$

and

$$\frac{\partial}{\partial\varphi}\mathcal{V}(\overline{\Psi}_\varphi) > 0 \qquad \text{for } 0 < \varphi < \frac{\pi}{4} \tag{15.87}$$

From the above equations we have that the stationary points of $\mathcal{V}(\overline{\Psi}_\varphi)$ correspond to systems characterized by the property that all non-zero coefficients have the same absolute value. Furthermore, using symmetries of the problem, we find the only minima are at $\pm I$, except for a possible delay, and the other stationary points of $\mathcal{V}(\overline{\Psi}_\varphi)$ are saddle points.

The study of the functional $\mathcal{V}$ is then extended to the entire parameter space. As the results obtained for the restriction of $\mathcal{V}$ to $\mathcal{S}$ are also valid on a sphere of arbitrary radius r, we need to study only the radial derivatives of $\mathcal{V}$. For this reason, we consider the function $\tilde{\mathcal{V}}(r) = \mathcal{V}(r\overline{\Psi})$, whose first and second derivatives are

$$\tilde{\mathcal{V}}'(r) = \int b\,\tilde{\Theta}(rb)\,p_{\bar{y}_k}(b)\,db - (\gamma_{\overline{\Psi}} + \gamma_r)\int |b|\,p_{\bar{y}_k}(b)\,db \tag{15.88}$$

and

$$\tilde{\mathcal{V}}''(r) = \int b^2\,\tilde{\Theta}'(rb)\,p_{\bar{y}_k}(b)\,db - \gamma_r'\int |b|\,p_{\bar{y}_k}(b)\,db \tag{15.89}$$

where $\tilde{\Theta}'$ and γ_r' denote derivatives.

Recalling the expressions of $\gamma_{\overline{\Psi}}$ and γ_r given by (15.85), we obtain $\tilde{\mathcal{V}}'(0) < 0$ and $\tilde{\mathcal{V}}''(r) > 0$. Therefore, there exists a radius r_0 such that the radial component of the gradient is negative for $r < r_0$ and positive for $r > r_0$. For a fixed point $\overline{\Psi} \in \mathcal{S}$, r_0 is given by the solution of the equation

$$\int b\,\tilde{\Theta}(rb)\,p_{\bar{y}_k}(b)\,db - (\gamma_{\overline{\Psi}} + \gamma_r)\int |b|\,p_{\bar{y}_k}(b)\,db = 0 \tag{15.90}$$

Substituting the expressions of $\gamma_{\overline{\Psi}}$ and γ_r in (15.90), we obtain $r_0 = 1$, $\forall\overline{\Psi} \in \mathcal{S}$. Therefore the only minima of $\mathcal{V}$ are at $\pm I$. Furthermore, as the radial component of the gradient vanishes on $\mathcal{S}$, the steepest descent lines of $\mathcal{V}$ do not cross the unit sphere. Using the same argument given in [9], we conclude that the points $\pm I$ are the only *stable attractors* of the steepest descent lines of the function $\mathcal{V}$, and that the unique stable attractors of the steepest descent lines of J are $\pm H^{-1}$.

Note that the parameter $\gamma_{\overline{\Psi}}$ is related to the distortion of the distribution of the input sequence filtered by a normalized system; this parameter varies along the trajectories of the stochastic gradient algorithm and vanishes as a point of minimum is reached. Moreover, note that the parameter γ_r indicates the deviation of the overall system gain from the desired unit value; if the gain is too small, the gradient is augmented with an additional driving term.

Chapter 16

Low-complexity demodulators

In this chapter, we discuss modulation and demodulation schemes that are well suited for applications to mobile radio systems because of their simplicity, low-power consumption, and robustness against disturbances introduced by the transmission channel.

16.1 Phase-shift keying

In phase-shift keying (PSK), values of a_k are given by (6.8). A PSK transmitter for $M = 8$ is shown in Figure 16.1. The bit mapper (BMAP) maps a sequence of $\log_2 M$ bits to a constellation point represented by a_k. The quadrature components $a_{k,I}$ and $\alpha_{k,Q}$ are input to interpolator filters h_{Tx}. The filter output signals are multiplied by the carrier signal, $\cos(2\pi f_0 t)$, and by the carrier signal phase-shifted by $\pi/2$, for example by a Hilbert filter, $\sin(2\pi f_0 t)$, respectively. The transmitted signal is obtained by adding the two components.

An implementation of the receiver is illustrated in Figure 16.2, which is similar to that of Figure 7.5. We note that the decision regions are angular sectors with phase $2\pi/M$. For $M = 2, 4$, and 8, simple decision rules can be defined. For $M > 8$, detection can be made by observing the phase v_k of the received sample y_k.

Curves of P_{bit} as a function of Γ are shown in Figure 6.10.

For $M = 2$ (binary phase-shift keying, BPSK), we get $\varphi_1 = \varphi_0$ and $\varphi_2 = \pi + \varphi_0$, where φ_0 is an arbitrary phase. P_e is given in (6.84).

The transmitter and the receiver for a BPSK system are shown in Figure 16.3 and have a very simple implementation. The BMAP of the transmitter maps *0* in *−1* and *1* in *+1*. At the receiver, the decision element implements the *sign* function to detect binary data. The inverse bit-mapper (IBMAP) to recover the bits of the information message is straightforward.

16.1.1 Differential PSK

We assume now that the receiver recovers the carrier signal, except for a phase offset φ_a. In particular, with reference to the scheme of Figure 16.2, the reconstructed carrier is $\cos(2\pi f_0 t - \varphi_a)$. Consequently, it is as if the constellation at the receiver were rotated by φ_a. To prevent this problem, there are two strategies. By the *coherent (CO) method*, a receiver estimates φ_a from the received signal, and considers the original constellation for detection, using the received signal multiplied by a phase offset, i.e. $y_k\, e^{-j\hat{\varphi}_a}$,

Algorithms for Communications Systems and their Applications, Second Edition.
Nevio Benvenuto, Giovanni Cherubini, and Stefano Tomasin.

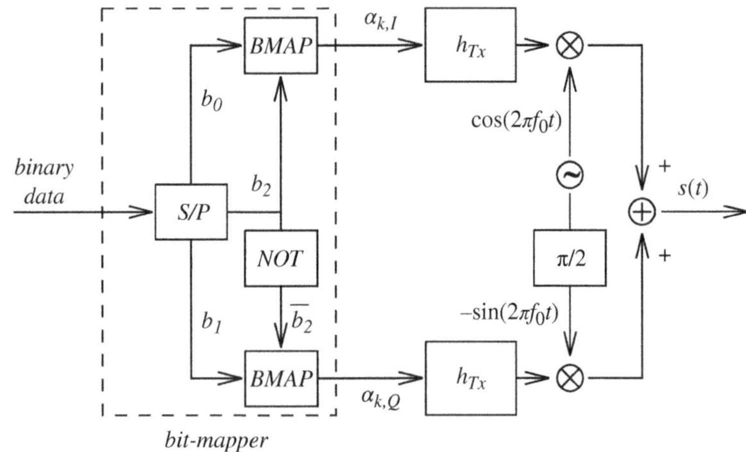

Figure 16.1 Transmitter of an 8-PSK system.

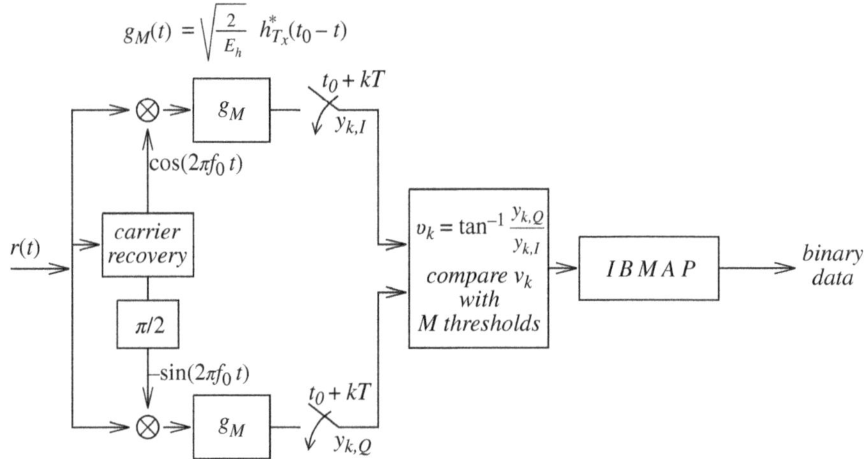

Figure 16.2 Receiver of an M-PSK system. Thresholds are set at $(2\pi/M)n$, $n = 1, \dots, M$.

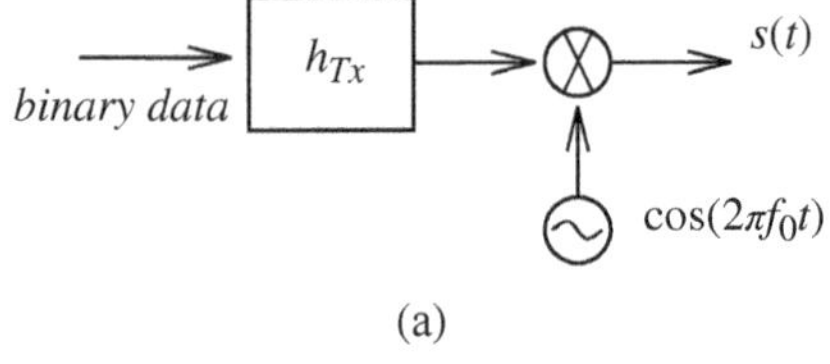

(a)

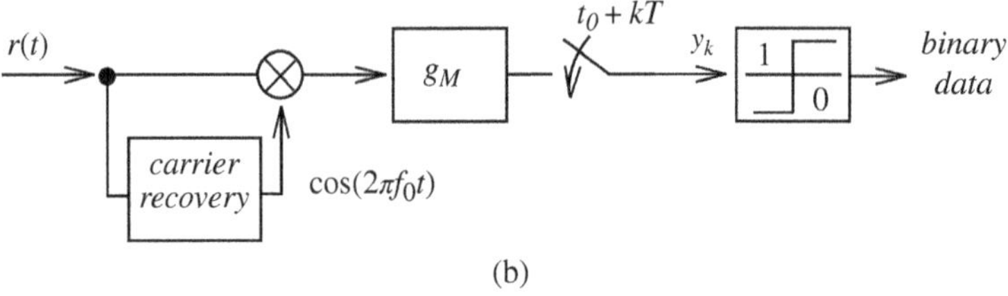

(b)

Figure 16.3 Schemes of (a) transmitter and (b) receiver for a BPSK system with $\varphi_0 = 0$.

where $\hat{\varphi}_a$ is the estimate of φ_a. By the *differential non-coherent (NC) method*, a receiver detects the data using the difference between the phases of signals at successive sampling instants. In other words

- for M-PSK, the phase of the transmitted signal at instant kT is given by (6.10), with
$$\theta_k \in \left\{ \frac{\pi}{M}, \frac{3\pi}{M}, \ldots, \frac{(2M-1)\pi}{M} \right\} \tag{16.1}$$
- for M-differential phase-shift keying (M-DPSK),[1] the transmitted phase as instant kT is given by
$$\psi'_k = \psi'_{k-1} + \theta_k, \qquad \theta_k \in \left\{ 0, \frac{2\pi}{M}, \ldots, \frac{2\pi}{M}(M-1) \right\} \tag{16.2}$$
that is, the phase associated with the transmitted signal at instant kT is equal to that transmitted at the previous instant $(k-1)T$ plus the increment θ_k, which can assume one of M values. We note that the decision thresholds for θ_k are now placed at $(\pi/M)(2n-1)$, $n = 1, \ldots, M$.

For a phase offset equal to φ_a introduced by the channel, the phase of the signal at the detection point becomes
$$\psi_k = \psi'_k + \varphi_a \tag{16.3}$$
In any case,
$$\psi_k - \psi_{k-1} = \theta_k \tag{16.4}$$
and the ambiguity of φ_a is removed. For phase-modulated signals, three differential non-coherent receivers that determine an estimate of (16.4) are discussed in Section 16.2.

Error probability of M-DPSK

For $\Gamma \gg 1$, using the definition of the Marcum function $Q_1(\cdot,\cdot)$ (see Appendix 6.B), it can be shown that the error probability is approximated by the following bound [1, 2]
$$\begin{aligned} P_e \lessapprox 1 &+ Q_1\left(\sqrt{\Gamma\left(1 - \sin\frac{\pi}{M}\right)}, \sqrt{\Gamma\left(1 + \sin\frac{\pi}{M}\right)}\right) \\ &- Q_1\left(\sqrt{\Gamma\left(1 + \sin\frac{\pi}{M}\right)}, \sqrt{\Gamma\left(1 - \sin\frac{\pi}{M}\right)}\right) \end{aligned} \tag{16.5}$$
Moreover, if M is large, the approximation (6.160) can be used and we get
$$\begin{aligned} P_e &\simeq 2Q\left[\sqrt{\Gamma}\left(\sqrt{1 + \sin\frac{\pi}{M}} - \sqrt{1 - \sin\frac{\pi}{M}}\right)\right] \\ &\simeq 2Q\left(\sqrt{\Gamma}\sin\frac{\pi}{M}\right) \end{aligned} \tag{16.6}$$
For Gray labelling of the values of θ_k in (16.2), the bit error probability is given by
$$P_{bit} = \frac{P_e}{\log_2 M} \tag{16.7}$$
For $M = 2$, the exact formula of the error probability is [1, 2]
$$P_{bit} = P_e = \frac{1}{2} e^{-\Gamma} \tag{16.8}$$
For $M = 4$, the exact formula is [1, 2]
$$P_e = 2Q_1(a, b) - I_0(ab)\, e^{-0.5(a^2+b^2)} \tag{16.9}$$

[1] Note that we consider a differential non-coherent receiver to which is associated a differential symbol encoder at the transmitter (see (16.2) or (16.14)). However, as we will see in the next section, a differential encoder and a coherent receiver can be used.

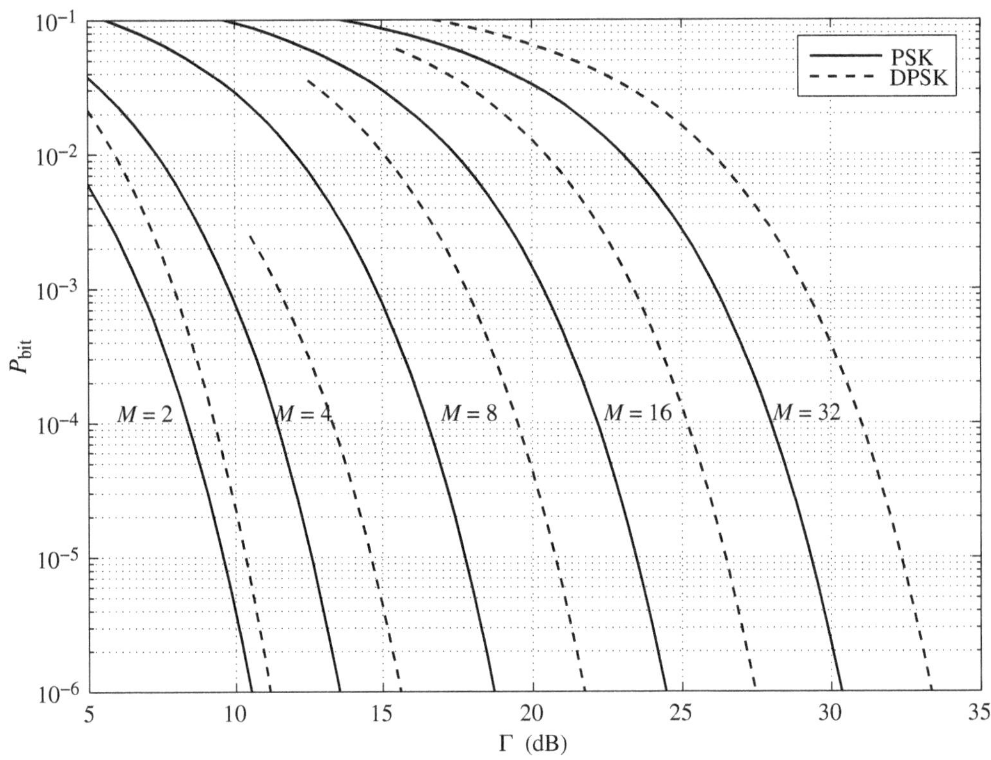

Figure 16.4 Comparison between PSK and DPSK.

where

$$a = \sqrt{\Gamma(1 - \sqrt{1/2})}, \qquad b = \sqrt{\Gamma(1 + \sqrt{1/2})} \tag{16.10}$$

and where the function I_0 is defined in (4.56).

Using the previous results, a comparison in terms of P_{bit} between DPSK (16.6) and PSK (6.83) is given in Figure 16.4. We note that, for $P_{bit} = 10^{-3}$, DPSK presents a loss of only 1.2 dB in Γ for $M = 2$, that increases to 2.3 dB for $M = 4$, and to 3 dB for $M > 4$.

As a DPSK receiver is simpler as compared to a coherent PSK receiver, in that it does not require recovery of the carrier phase, for $M = 2$ DPSK is usually preferred to PSK.

Note that, if the previously received sample is used as a reference, DPSK gives lower performance with respect to PSK, especially for $M \geq 4$, because both the current sample and the reference sample are corrupted by noise. This drawback can be mitigated if the reference sample is constructed by using more than one previously received sample [3]. In this way, we establish a gradual transition between differential phase demodulation and coherent demodulation. In particular, if the reference sample is constructed using the samples received in the two previous modulation intervals, DPSK and PSK yield similar performance [3]. PSK, especially in its differential form, is used in satellite communications. This is due to the fact that satellite transmission undergoes very sudden changes in signal attenuation, which do not affect the performance of a DPSK receiver as would be the case for a quadrature-amplitude-modulation (QAM) receiver. As an application example, we recall that 8-PSK is used in satellite digital video broadcasting (DVB-S) [4].

16.1.2 Differential encoding and coherent demodulation

If φ_a is a multiple of $2\pi/M$, at the receiver the phase difference can be formed between the phases of two consecutive coherently detected symbols, instead of between the phases of two consecutive samples. In this case, symbols are differentially encoded before modulation.

Differentially encoded BPSK

Let b_k be the value of the information bit at instant kT, $b_k \in \{0, 1\}$.

BPSK system without differential encoding The phase $\theta_k \in \{0, \pi\}$ is associated with b_k by the bit map of Table 16.1.

Differential encoder For any $c_{-1} \in \{0, 1\}$, we encode the information bits as

$$c_k = c_{k-1} \oplus b_k, \qquad b_k \in \{0, 1\}, \quad k \geq 0 \tag{16.11}$$

where $\oplus$ denotes the modulo 2 sum; therefore, $c_k = c_{k-1}$ if $b_k = 0$, and[2] $c_k = \bar{c}_{k-1}$ if $b_k = 1$. For the bit map of Table 16.2, we have that $b_k = 1$ causes a phase transition, and $b_k = 0$ causes a phase repetition.

Decoder If $\{\hat{c}_k\}$ are the detected coded bits at the receiver, the information bits are recovered by

$$\hat{b}_k = \hat{c}_k \oplus (-\hat{c}_{k-1}) = \hat{c}_k \oplus \hat{c}_{k-1} \tag{16.12}$$

We note that a phase ambiguity $\varphi_a = \pi$ does not alter the recovered sequence $\{\hat{b}_k\}$: in fact. In this case, $\{\hat{c}_k\}$ becomes $\{\hat{c}'_k = \hat{c}_k \oplus 1\}$ and we have

$$(\hat{c}_k \oplus 1) \oplus (\hat{c}_{k-1} \oplus 1) = \hat{c}_k \oplus \hat{c}_{k-1} = \hat{b}_k \tag{16.13}$$

Multilevel case

Let $\{d_k\}$ be a multilevel information sequence, with $d_k \in \{0, 1, \ldots, M-1\}$. In this case, we have

$$c_k = c_{k-1} \underset{M}{\oplus} d_k \tag{16.14}$$

Table 16.1: Bit map for a BPSK system

b_k	Transmitted phase θ_k (rad)
0	0
1	π

Table 16.2: Bit map for a differentially encoded BPSK system

c_k	Transmitted phase ψ_k (rad)
0	0
1	π

[2] $\bar{c}$ denotes the one's complement of c: $\bar{1} = 0$ and $\bar{0} = 1$.

where $\underset{M}{\oplus}$ denotes the modulo M sum. Because $c_k \in \{0, 1, \dots, M-1\}$, the phase associated with the bit map is $\psi_k \in \{\pi/M, 3\pi/M, \dots, (2M-1)\pi/M\}$. This encoding and bit-mapping scheme are equivalent to (16.2).

At the receiver, the information sequence is recovered by

$$\hat{d}_k = \hat{c}_k \underset{M}{\oplus} (-\hat{c}_{k-1}) \tag{16.15}$$

It is easy to see that an offset equal to $j \in \{0, 1, \dots, (M-1)\}$ in the sequence $\{\hat{c}_k\}$, corresponding to a phase offset equal to $\{0, 2\pi/M, \dots, (M-1)2\pi/M\}$ in $\{\psi_k\}$, does not cause errors in $\{\hat{d}_k\}$. In fact,

$$\left(\hat{c}_k \underset{M}{\oplus} j\right) \underset{M}{\oplus} \left[-\left(\hat{c}_{k-1} \underset{M}{\oplus} j\right)\right] = \hat{c}_k \underset{M}{\oplus} (-\hat{c}_{k-1}) = \hat{d}_k \tag{16.16}$$

Performance of a PSK system with differential encoding and *coherent demodulation* by the scheme of Figure 16.2 is worse as compared to a system with absolute phase encoding. However, for small P_e, up to values of the order of 0.1, we observe that an error in $\{\hat{c}_k\}$ causes two errors in $\{\hat{d}_k\}$. Approximately, P_e increases by a factor 2,[3] which causes a negligible loss in terms of Γ.

To combine Gray labelling of values of c_k with the differential encoding (16.14), a two-step procedure is adopted:

1. represent the values of d_k with a Gray labelling using a combinatorial table, as illustrated for example in Table 16.3 for $M = 8$;
2. determine the differentially encoded symbols according to (16.14).

Example 16.1.1 (Differential encoding 2B1Q)
We consider a differential encoding scheme for a four-level system that makes the reception insensitive to a possible change of sign of the transmitted sequence. For $M = 4$, this implies insensitivity to a phase rotation equal to π in a 4-PSK signal or to a change of sign in a 4-pulse-amplitude-modulation (PAM) signal.

For $M = 4$, we give the law between the binary representation of $d_k = (d_k^{(1)}, d_k^{(0)})$, $d_k^{(i)} \in \{0, 1\}$, and the binary representation of $c_k = (c_k^{(1)}, c_k^{(0)})$, $c_k^{(i)} \in \{0, 1\}$:

$$\begin{aligned} c_k^{(1)} &= d_k^{(1)} \oplus c_{k-1}^{(1)} \\ c_k^{(0)} &= d_k^{(0)} \oplus c_k^{(1)} \end{aligned} \tag{16.19}$$

Table 16.3: Gray coding for $M = 8$

Three information bits			values of d_k
0	0	0	0
0	0	1	1
0	1	1	2
0	1	0	3
1	1	0	4
1	1	1	5
1	0	1	6
1	0	0	7

[3] If we indicate with $P_{e,Ch}$ the channel error probability, then the error probability after decoding is given by [1]

Binary case $$P_{bit} = 2P_{bit,Ch}\,[1 - P_{bit,Ch}] \tag{16.17}$$

Quaternary case $$P_e = 4P_{e,Ch} - 8P_{e,Ch}^2 + 8P_{e,Ch}^3 - 4P_{e,Ch}^4 \tag{16.18}$$

Table 16.4: Bit map for the differential encoder 2B1Q

$c_k^{(1)}$	$c_k^{(0)}$	transmitted symbol a_k
0	0	−3
0	1	−1
1	0	1
1	1	3

The bit map is given in Table 16.4.

The equations of the differential decoder are

$$\begin{aligned} \hat{d}_k^{(1)} &= \hat{c}_k^{(1)} \oplus \hat{c}_{k-1}^{(1)} \\ \hat{d}_k^{(0)} &= \hat{c}_k^{(0)} \oplus \hat{c}_k^{(1)} \end{aligned} \tag{16.20}$$

16.2 (D)PSK non-coherent receivers

We introduce three non-coherent receivers to demodulate phase modulated signals.

16.2.1 Baseband differential detector

For a continuous M-DPSK transmission with symbol period T, the transmitted signal is given by

$$s(t) = \sum_{k=-\infty}^{+\infty} Re\left[e^{j\psi_k} h_{Tx}(t-kT) e^{j2\pi f_0 t}\right] \tag{16.21}$$

where ψ_k is the phase associated with the transmitted symbol at instant kT given by the recursive equation (16.2).

At the receiver, the signal r is a version of s, filtered by the transmission channel and corrupted by additive noise. We denote as g_A the cascade of passband filters used to amplify the desired signal and partially remove noise. As shown in Figure 16.5, let x be the passband received signal, centred around

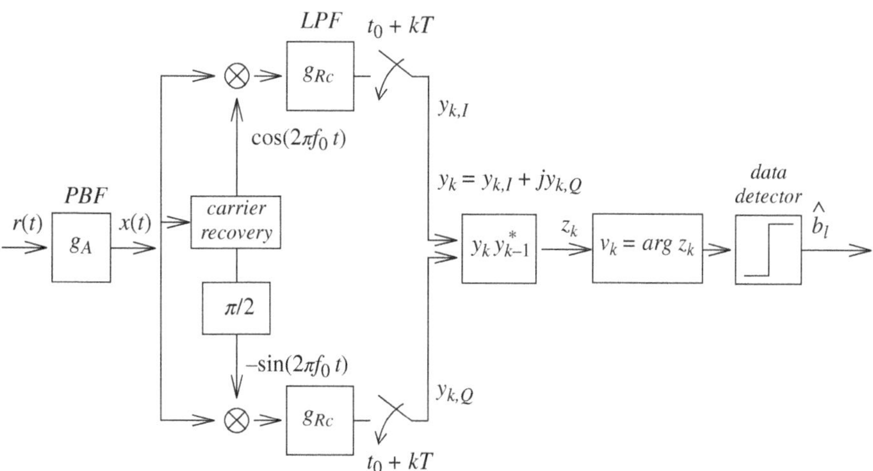

Figure 16.5 Non-coherent baseband differential receiver. Thresholds are set at $(2n-1)\pi/M$, $n = 1, 2, \dots, M$.

the frequency f_0, equal to the carrier of the transmitted signal

$$x(t) = Re\left[x^{(bb)}(t)e^{j2\pi f_0 t}\right] \tag{16.22}$$

where $x^{(bb)}$ is the complex envelope of x with respect to f_0. Using the polar notation $x^{(bb)}(t) = M_x(t)$ $e^{j\Delta\varphi_x(t)}$, (16.22) can be written as

$$x(t) = M_x(t)\cos\left(2\pi f_0 t + \Delta\varphi_x(t)\right) \tag{16.23}$$

where $\Delta\varphi_x$ is the instantaneous *phase deviation* of x with respect to the carrier phase (see (1.141)).

In the ideal case of absence of distortion and noise, sampling at suitable instants yields $\Delta\varphi_x(t_0 + kT) = \psi_k$. Then, for the recovery of the phase θ_k we can use the receiver scheme of Figure 16.5, which, based on signal x, determines the baseband component y as

$$y(t) = y_I(t) + jy_Q(t) = \frac{1}{2}x^{(bb)} * g_{Rc}(t) \tag{16.24}$$

The phase variation of the sampled signal $y_k = y(t_0 + kT)$ between two consecutive symbol instants is obtained by means of the signal

$$z_k = y_k y_{k-1}^* \tag{16.25}$$

Always in the ideal case and assuming that g_{Rc} does not distort the phase of $x^{(bb)}$, z_k turns out to be proportional to $e^{j\theta_k}$. The simplest data detector is the threshold detector (TD) based on the value of

$$v_k = \arg z_k = \Delta\varphi_x(t_0 + kT) - \Delta\varphi_x\left(t_0 + (k-1)T\right) \tag{16.26}$$

Note that a possible phase offset $\Delta\varphi_0$ and a frequency offset Δf_0, introduced by the receive mixer, yields a signal y given by

$$y(t) = \left[\frac{1}{2}x^{(bb)}(t)e^{-j(\Delta\varphi_0 + 2\pi\Delta f_0 t)}\right] * g_{Rc}(t) \tag{16.27}$$

Assuming that g_{Rc} does not distort the phase of $x^{(bb)}$, the signal v_k becomes

$$v_k = \Delta\varphi_x(t_0 + kT) - \Delta\varphi_x(t_0 + (k-1)T) - 2\pi\Delta f_0 T \tag{16.28}$$

which shows that the phase offset $\Delta\varphi_0$ does not influence v_k, while a frequency offset must be compensated by the data detector, summing the constant phase $2\pi\Delta f_0 T$.

The baseband equivalent scheme of the baseband differential receiver is given in Figure 16.6.

The choice of h_{Tx}, $g_A^{(bb)}$, and g_{Rc} is governed by the same considerations as in the case of a QAM system; for an ideal channel, the convolution of these elements must be a Nyquist pulse.

16.2.2 IF-band (1 bit) differential detector

The scheme of Figure 16.5 is introduced only to illustrate the basic principle, as its implementation complexity is similar to that of a coherent scheme. Indeed, this scheme has a reduced complexity because specifications on the carrier recovery can be less stringent; moreover, it does not need phase recovery.

An alternative scheme, denoted as 1-bit differential detector (1BDD), that does not use carrier recovery is illustrated in Figure 16.7. In this case, the signal is first delayed of a symbol period T and then

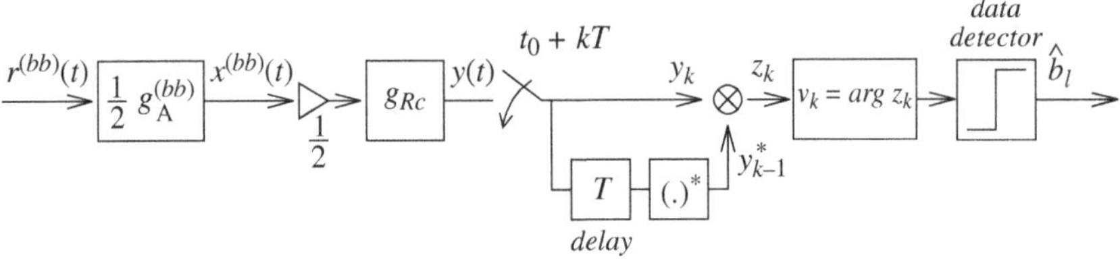

Figure 16.6 Baseband equivalent scheme of Figure 16.5.

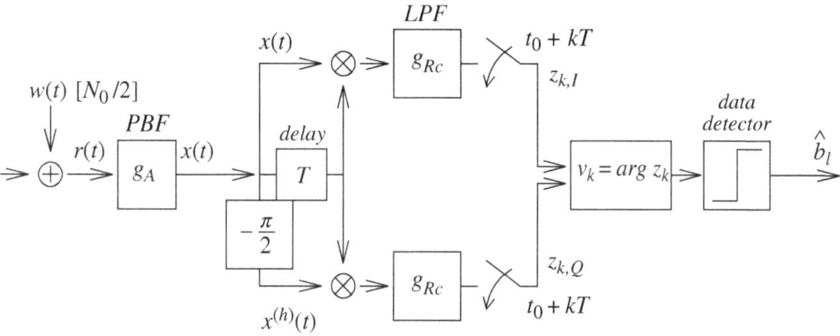

Figure 16.7 Non-coherent 1 bit differential detector.

multiplied with itself (I branch) and with its $-\pi/2$ phase shifted version (Q branch) by a Hilbert filter. On the two branch, the signals are given by

$$\begin{aligned} I &: x(t)x(t-T) \\ &= M_x(t)\cos\left[2\pi f_0 t + \Delta\varphi_x(t)\right] M_x(t-T)\cos\left[2\pi f_0(t-T) + \Delta\varphi_x(t-T)\right] \\ Q &: x^{(h)}(t)x(t-T) \\ &= M_x(t)\sin\left[2\pi f_0 t + \Delta\varphi_x(t)\right] M_x(t-T)\cos\left[2\pi f_0(t-T) + \Delta\varphi_x(t-T)\right] \end{aligned} \tag{16.29}$$

The filter g_{Rc} removes the components around $2f_0$; the sampled filter outputs are then given by

$$\begin{aligned} I &: z_{k,I} = M_x(t_0+kT)M_x\left(t_0+(k-1)T\right) \\ &\qquad \frac{1}{2}\cos\left[2\pi f_0 T + \Delta\varphi_x(t_0+kT) - \Delta\varphi_x(t_0+(k-1)T)\right] \\ Q &: z_{k,Q} = M_x(t_0+kT)M_x\left(t_0+(k-1)T\right) \\ &\qquad \frac{1}{2}\sin\left[2\pi f_0 T + \Delta\varphi_x(t_0+kT) - \Delta\varphi_x(t_0+(k-1)T)\right] \end{aligned} \tag{16.30}$$

If $f_0 T = n$, n an integer, or by removing this phase offset by phase shifting x or z_k, it results

$$v_k = \tan^{-1}\frac{z_{k,Q}}{z_{k,I}} = \Delta\varphi_x(t_0+kT) - \Delta\varphi_x\left(t_0+(k-1)T\right) \tag{16.31}$$

as in (16.26).

The baseband equivalent scheme is shown in Figure 16.8, where, assuming that g_{Rc} does not distort the desired signal,

$$z_k = \frac{1}{2}x^{(bb)}(t_0+kT)x^{(bb)*}\left(t_0+(k-1)T\right) \tag{16.32}$$

Typically, in this modulation system, the transmit filter h_{Tx} is a rectangular pulse or a Nyquist pulse; instead, g_A is a narrow band filter to eliminate out of band noise.

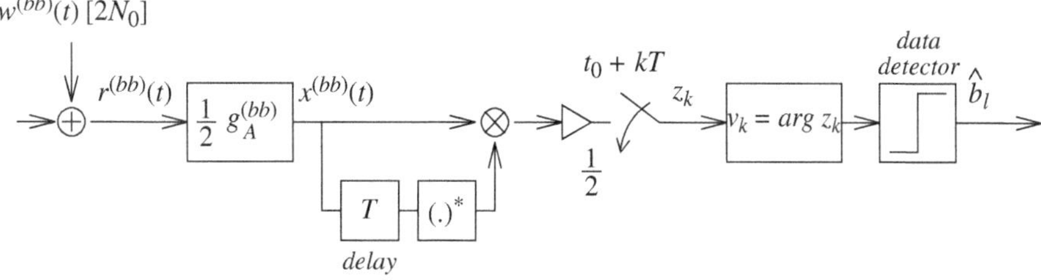

Figure 16.8 Baseband equivalent scheme of Figure 16.7.

For a simple differential binary phase-shift-keying (DBPSK) modulation, with $\theta_k \in \{0, \pi\}$, we only consider the I branch, and $v_k = z_{k,I}$ is compared with a threshold set to 0.

Signal at detection point

With reference to the transmitted signal (16.21), we consider the isolated pulse k

$$s(t) = Re[e^{j\psi_k} h_{Tx}(t - kT) e^{j2\pi f_0 t}] \tag{16.33}$$

where

$$\psi_k = \psi_{k-1} + \theta_k, \qquad \theta_k \in \left\{0, \frac{2\pi}{M}, \ldots, \frac{2\pi(M-1)}{M}\right\} \tag{16.34}$$

In general, referring to the scheme of Figure 16.8, the filtered signal is given by

$$x^{(bb)}(t) = u^{(bb)}(t) + w_R^{(bb)}(t) \tag{16.35}$$

where $u^{(bb)}$ is the desired signal at the demodulator input (isolated pulse k)

$$u^{(bb)}(t) = e^{j\psi_k} h_{Tx} * \frac{1}{2} g_{Ch}^{(bb)} * \frac{1}{2} g_A^{(bb)}(t - kT) \tag{16.36}$$

as $(1/2)g_{Ch}^{(bb)}$ is the complex envelope of the impulse response of the transmission channel, and $w_R^{(bb)}$ is zero mean additive complex Gaussian noise with variance $\sigma^2 = 2N_0 B_{rn}$, where

$$\begin{aligned} B_{rn} &= \int_{-\infty}^{+\infty} |\mathcal{G}_A(f)|^2 \, df = \int_{-\infty}^{+\infty} |g_A(t)|^2 \, dt \\ &= \int_{-\infty}^{+\infty} \frac{1}{4} \left|\mathcal{G}_A^{(bb)}(f)\right|^2 \, df = \int_{-\infty}^{+\infty} \frac{1}{4} \left|g_A^{(bb)}(t)\right|^2 \, dt \end{aligned} \tag{16.37}$$

is the equivalent bilateral noise bandwidth.

We assume the following configuration: the transmit filter impulse response is given by

$$h_{Tx}(t) = \sqrt{\frac{2E_s}{T}} \operatorname{rect} \frac{t - T/2}{T} \tag{16.38}$$

the channel introduces only a phase offset

$$g_{Ch}^{(bb)}(t) = 2e^{j\varphi_a} \delta(t) \tag{16.39}$$

and the receive filter is matched to the transmit filter,

$$g_A^{(bb)}(t) = \sqrt{\frac{2}{T}} \operatorname{rect} \frac{t - T/2}{T} \tag{16.40}$$

At the sampling intervals $t_0 + kT = T + kT$, let

$$w_k = w_R^{(bb)}(t_0 + kT) \tag{16.41}$$

and

$$A = \sqrt{\frac{2E_s}{T}} \frac{1}{2} \sqrt{\frac{2}{T}} T = \sqrt{E_s} \tag{16.42}$$

then

$$x^{(bb)}(t_0 + kT) = A e^{j(\psi_k + \varphi_a)} + w_k \tag{16.43}$$

with

$$E\left[|w_k|^2\right] = \sigma^2 = 2N_0 B_{rn} = N_0 \tag{16.44}$$

since $B_{rn} = 1/2$. Moreover, it results

$$\begin{aligned} z_k &= \frac{1}{2} x^{(bb)}(T + kT) x^{(bb)*}(T + (k-1)T) \\ &= \frac{1}{2} \left[A e^{j(\psi_k + \varphi_a)} + w_k\right] \left[A e^{-j(\psi_{k-1} + \varphi_a)} + w_{k-1}^*\right] \\ &= \frac{1}{2} A \left[\underbrace{A e^{j\theta_k}}_{\text{desired term}} + \underbrace{e^{j(\psi_k + \varphi_a)} w_{k-1}^* + e^{-j(\psi_{k-1} + \varphi_a)} w_k + \frac{w_k w_{k-1}^*}{A}}_{\text{disturbance}} \right] \end{aligned} \tag{16.45}$$

The desired term is similar to that obtained in the coherent case, M-phases on a circle of radius $A = \sqrt{E_s}$. The variance of w_k is equal to N_0, and $\sigma_I^2 = N_0/2$. However, even neglecting the term $w_k w_{k-1}^*/A$, if w_k and w_{k-1} are statistically independent we have

$$E\left[\left| e^{j(\psi_k + \varphi_a)} w_{k-1}^* + e^{-j(\psi_{k-1} + \varphi_a)} w_k \right|^2 \right] = 2\sigma^2 = 2N_0 \tag{16.46}$$

There is an asymptotic penalty, that is for $E_s/N_0 \to \infty$, of 3 dB with respect to the coherent receiver case. Indeed, for a 4-DPSK, a more accurate analysis demonstrates that the penalty is only 2.3 dB for higher values of E_s/N_0.

16.2.3 FM discriminator with integrate and dump filter

The scheme of Figure 16.9 makes use of a limiter discriminator (LD), as for *frequency modulated* (FM) signals, followed by an integrator filter over a symbol period, or *integrate and dump* (I&D) filter. This is limiter discriminator integrator (LDI) scheme. Ideally, the discriminator output provides the instantaneous frequency deviation of x, i.e.

$$\Delta f(t) = \frac{d}{dt} \Delta\varphi_x(t) \tag{16.47}$$

Then, integrating (16.47) over a symbol period, we have

$$\int_{t_0+(k-1)T}^{t_0+kT} \Delta f(\tau)\, d\tau = \Delta\varphi_x(t_0 + kT) - \Delta\varphi_x\left(t_0 + (k-1)T\right) + 2n\pi \tag{16.48}$$

that coincides with (16.31) taking mod 2π. An implementation of the limiter–discriminator is given in Figure 16.10, while the baseband equivalent scheme is given in Figure 16.11, which employs the general relation

$$\Delta f(t) = \frac{Im\left[\dot{x}^{(bb)}(t) x^{(bb)*}(t)\right]}{2\pi |x^{(bb)}(t)|^2} \tag{16.49}$$

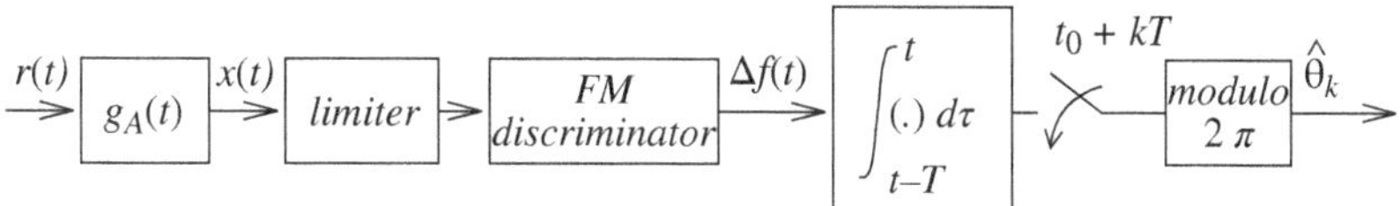

Figure 16.9 FM discriminator and integrate and dump filter.

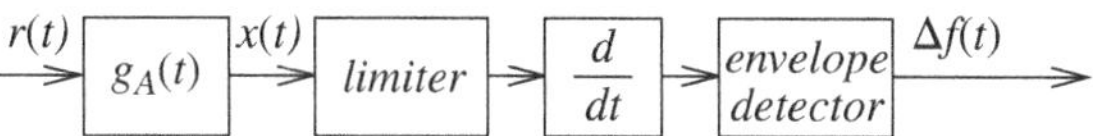

Figure 16.10 Implementation of a limiter–discriminator.

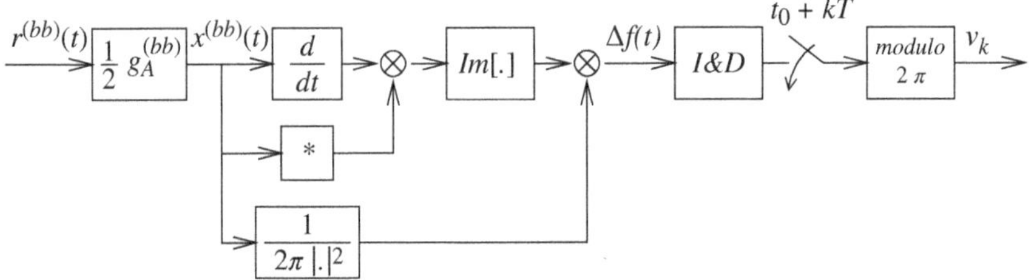

Figure 16.11 Baseband equivalent scheme of a FM discriminator followed by an integrate and dump filter.

In conclusion, all three schemes (baseband differential detector, differential detector, and FM discriminator) yield the same output.

16.3 Optimum receivers for signals with random phase

In this section, we ignore inter-symbol interference (ISI) and consider transmit and receiver structures for some modulation formats.

Let us consider transmission over an ideal additive white Gaussian noise (AWGN) channel of one of the signals

$$s_n(t) = Re\left[s_n^{(bb)}(t)\, e^{j2\pi f_0 t}\right], \quad n = 1,2,\dots,M, \quad 0 < t < T \tag{16.50}$$

where $s_n^{(bb)}$ is the complex envelope of s_n, relative to the carrier frequency f_0, with support $(0, t_0)$, $t_0 \le T$. Hence, the received signal is

$$r(t) = s_n(t) + w(t) \tag{16.51}$$

with w AWGN with power spectral density (PSD) $N_0/2$.

If in (16.50) every signal $s_n^{(bb)}$ has a bandwidth smaller than f_0, then the energy of s_n is given by

$$E_n = \int_0^{t_0} s_n^2(t)\, dt = \int_0^{t_0} \frac{1}{2}\, \left|s_n^{(bb)}(t)\right|^2\, dt \tag{16.52}$$

At the receiver, we observe the signal

$$r(t) = s_n(t;\varphi) + w(t) \tag{16.53}$$

where

$$\begin{aligned} s_n(t;\varphi) &= Re\left[s_n^{(bb)}(t) e^{j\varphi} e^{j2\pi f_0 t}\right] \\ &= Re\left[s_n^{(bb)*}(t) e^{-j\varphi} e^{-j2\pi f_0 t}\right], \quad n = 1,2,\dots,M \end{aligned} \tag{16.54}$$

In other words, at the receiver we assume the carrier is known, except for a phase φ that we assume to be a uniform r.v. in $[-\pi, \pi)$. Receivers, which do not rely on the knowledge of the carrier phase, are called *non-coherent receivers*.

We give three examples of signalling schemes that employ non-coherent receivers.

Example 16.3.1 (Non-coherent binary FSK)

For a non coherent binary frequency shift keying (FSK) transmission, the useful part of the received signals are expressed as (see also Ref. [5]):

$$\begin{aligned} s_1(t;\varphi_1) &= A\cos(2\pi f_1 t + \varphi_1), \quad 0 < t < T \\ s_2(t;\varphi_2) &= A\cos(2\pi f_2 t + \varphi_2), \quad 0 < t < T \end{aligned} \tag{16.55}$$

where, if T is the symbol period and E_s the average signalling energy,

$$A = \sqrt{\frac{2E_s}{T}}, \qquad \varphi_1, \varphi_2 \sim U[-\pi, \pi) \tag{16.56}$$

and

$$f_1 = f_0 - f_d, \qquad f_2 = f_0 + f_d \tag{16.57}$$

where f_d is the frequency deviation with respect to the carrier f_0. We recall that if

$$f_1 + f_2 = k_1 \frac{1}{T} \; (k_1 \text{ integer}) \qquad \text{or else} \qquad f_0 \gg \frac{1}{T} \tag{16.58}$$

and if

$$2 f_d T = k \; (k \text{ integer}) \tag{16.59}$$

then $s_1(t, \varphi_1)$ and $s_2(t, \varphi_2)$ are orthogonal.

The minimum value of f_d is given by

$$(f_d)_{min} = \frac{1}{2T} \tag{16.60}$$

which is twice the value we find for the coherent demodulation case [5].

Example 16.3.2 (On–off keying)
On–off keying (OOK) is a binary modulation scheme where, for example,

$$s_1(t; \varphi) = A \cos(2\pi f_0 t + \varphi), \qquad 0 < t < T \tag{16.61}$$

and

$$s_2(t; \varphi) = 0 \tag{16.62}$$

where $A = \sqrt{4E_s/T}$.

Example 16.3.3 (Double side band (DSB) modulated signalling with random phase)
We consider an M-ary baseband signalling scheme, $\{s_n^{(bb)}(t)\}$, $n = 1, \dots, M$, that is modulated in the passband by the *double sideband* technique (see Example 1.5.3 on page 39).

The useful part of the received signals are expressed as

$$s_n(t; \varphi) = s_n^{(bb)}(t) \cos(2\pi f_0 t + \varphi), \qquad n = 1, \dots, M \tag{16.63}$$

ML criterion

Given $\varphi = p$, that is for known φ, the maximum-likelihood (ML) criterion to detect the transmitted signal starts from the following likelihood function [5]

$$\mathrm{L}_n[p] = \exp\left(\frac{2}{N_0} \int_0^{t_0} r(t)\, s_n(t; p)\, dt - \frac{1}{N_0} \int_0^{t_0} s_n^2(t; p)\, dt \right) \tag{16.64}$$

Using the result

$$\int_0^{t_0} s_n^2(t; \varphi)\, dt = E_n \tag{16.65}$$

we have

$$\mathrm{L}_n[p] = \exp\left(-\frac{E_n}{N_0}\right) \exp\left(\frac{2}{N_0} \int_0^{t_0} r(t)\, s_n(t; p)\, dt \right), \qquad n = 1, \dots, M \tag{16.66}$$

Given $\varphi = p$, the ML criterion yields the decision rule

$$\hat{a}_0 = \arg\max_n \mathrm{L}_n[p] \tag{16.67}$$

The dependency on the r.v. φ is removed by taking the expectation of $\mathrm{L}_n[p]$ with respect to φ[4]:

$$\begin{aligned}\mathrm{L}_n &= \int_{-\pi}^{\pi} \mathrm{L}_n[p]\, p_\varphi(p)\, dp \\ &= e^{-\frac{E_n}{N_0}} \frac{1}{2\pi} \int_{-\pi}^{\pi} \exp\left(\frac{2}{N_0} Re\left[\int_0^{t_0} r(t)\, s_n^{(bb)*}(t)\, e^{-j(p+2\pi f_0 t)}\, dt\right]\right) dp\end{aligned} \tag{16.68}$$

using (16.54). We define

$$L_n = \frac{1}{\sqrt{E_n}} \int_0^{t_0} r(t)[s_n^{(bb)}(t)\, e^{j2\pi f_0 t}]^*\, dt \tag{16.69}$$

Introducing the polar notation $L_n = |L_n| e^{j \arg L_n}$, (16.68) becomes

$$\begin{aligned}\mathrm{L}_n &= e^{-\frac{E_n}{N_0}} \frac{1}{2\pi} \int_{-\pi}^{\pi} e^{\frac{2\sqrt{E_n}}{N_0} Re[L_n e^{-jp}]}\, dp \\ &= e^{-\frac{E_n}{N_0}} \frac{1}{2\pi} \int_{-\pi}^{\pi} e^{\frac{2\sqrt{E_n}}{N_0} |L_n| \cos(p - \arg L_n)}\, dp\end{aligned} \tag{16.70}$$

We recall that the Bessel functions (4.56) $I_0(x)$ is monotonic increasing for $x > 0$. Then (16.70) becomes

$$\mathrm{L}_n = e^{-\frac{E_n}{N_0}} I_0\left(\frac{2\sqrt{E_n}}{N_0} |L_n|\right), \qquad n = 1, \dots, M \tag{16.71}$$

Taking the logarithm, we obtain the *log-likelihood function*

$$\ell_n = \ln I_0\left(\frac{2\sqrt{E_n}}{N_0} |L_n|\right) - \frac{E_n}{N_0} \tag{16.72}$$

If the signals have all the same energy, and considering that both ln and I_0 are monotonic functions, the ML decision criterion can be expressed as

$$\hat{a}_0 = \arg\max_n |L_n| \tag{16.73}$$

Implementation of a non-coherent ML receiver

The scheme that implements the criterion (16.73) is illustrated in Figure 16.12 for the case of all E_n equal. Polar notation is adopted for the complex envelope:

$$s_n^{(bb)}(t) = |s_n^{(bb)}(t)| e^{j\Phi_n(t)} \tag{16.74}$$

From (16.69), the scheme first determines the real and the imaginary parts of L_n starting from $s_n^{(bb)}$, and then determines the square magnitude. Note that the available signal is $s_n^{(bb)}(t)\, e^{j\varphi_0}$, where φ_0 is a constant, rather than $s_n^{(bb)}$. This however does not modify the magnitude of L_n. As shown in Figure 16.13, the generic branch of the scheme in Figure 16.12, composed of the I branch and the Q branch, can be implemented by a complex-valued passband filter (see (16.69)); the bold line denotes a complex-valued signal.

[4] Averaging with respect to the phase φ cannot be considered for PSK and QAM systems, where information is also carried by the phase of the signal.

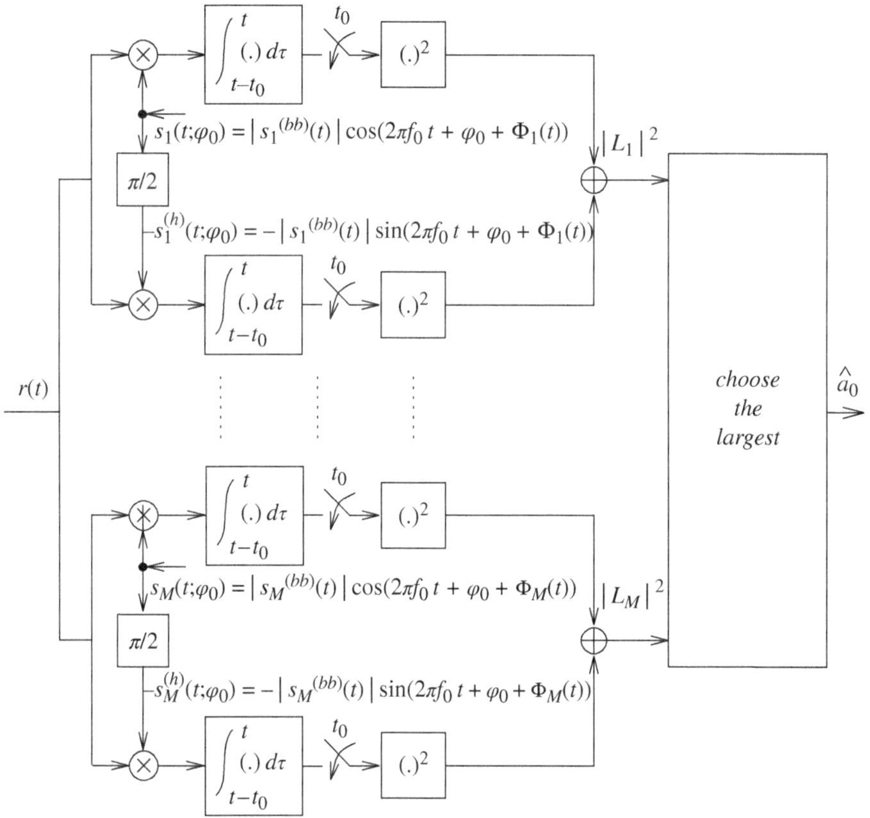

Figure 16.12 Non-coherent ML receiver of the type *square-law detector*.

Figure 16.13 Implementation of a branch of the scheme of Figure 16.12 by a complex-valued passband matched filter.

Alternatively, the matched filter (MF) can be real valued if it is followed by a *phase-splitter*. In this case, the receiver is illustrated in Figure 16.14. For the generic branch, the desired value $|L_n|$ coincides with the absolute value of the output signal of the phase-splitter at instant t_0,

$$|L_n| = \left| y_n^{(a)}(t) \right|_{t=t_0} \tag{16.75}$$

The cascade of the phase-splitter and the *modulo* transformation is called *envelope detector* of the signal y_n (see (1.130) and (1.136)).

A simplification arises if the various signals $s_n^{(bb)}$ have a bandwidth B much lower than f_0. In this case, recalling (1.136), if $y_n^{(bb)}$ is the complex envelope of y_n, at the matched filter output the following relation holds

$$\begin{aligned} y_n(t) &= Re\left[y_n^{(bb)}(t) e^{j2\pi f_0 t} \right] \\ &= \left| y_n^{(bb)}(t) \right| \cos\left(2\pi f_0 t + \arg y_n^{(bb)}(t) \right) \end{aligned} \tag{16.76}$$

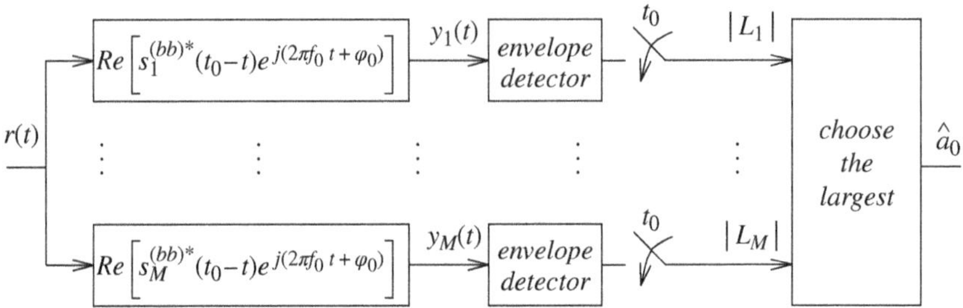

Figure 16.14 Non-coherent ML receiver of the type *envelope detector*, using passband matched filters.

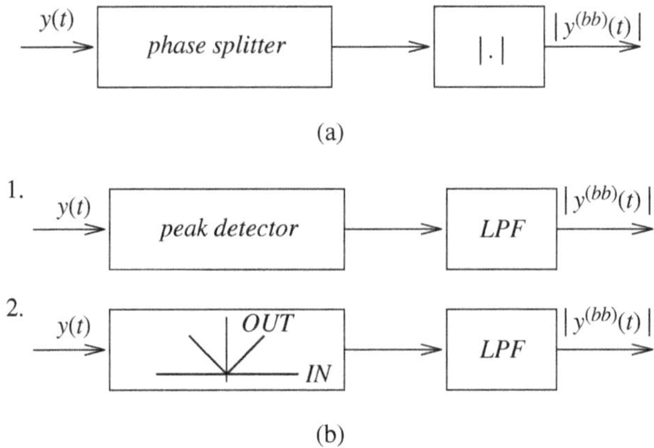

Figure 16.15 (a) Ideal implementation of an envelope detector, and (b) two simpler approximate implementations.

Moreover, from (16.75) and (1.133) we have

$$|L_n| = \left| y_n^{(bb)}(t) \right|_{t=t_0} \tag{16.77}$$

Now, if $f_0 \gg B$, to determine the amplitude $|y_n^{(bb)}(t)|$, we can use one of the schemes of Figure 16.15.

Example 16.3.4 (Non-coherent binary FSK)
We show in Figure 16.16 two alternative schemes of the ML receiver, where $\mathtt{w}_T$ is a rectangular window (see (1.400)) of length T, for the modulation system considered in Example 16.3.1.

Example 16.3.5 (On–off keying)
We illustrate in Figure 16.17 the receiver for the modulation system of Example 16.3.2 where, recalling (16.71), we have

$$U_{Th} = \frac{N_0}{2\sqrt{E_1}} I_0^{-1}(e^{E_1/N_0}) \tag{16.78}$$

where

$$E_1 = \frac{A^2 T}{2} \tag{16.79}$$

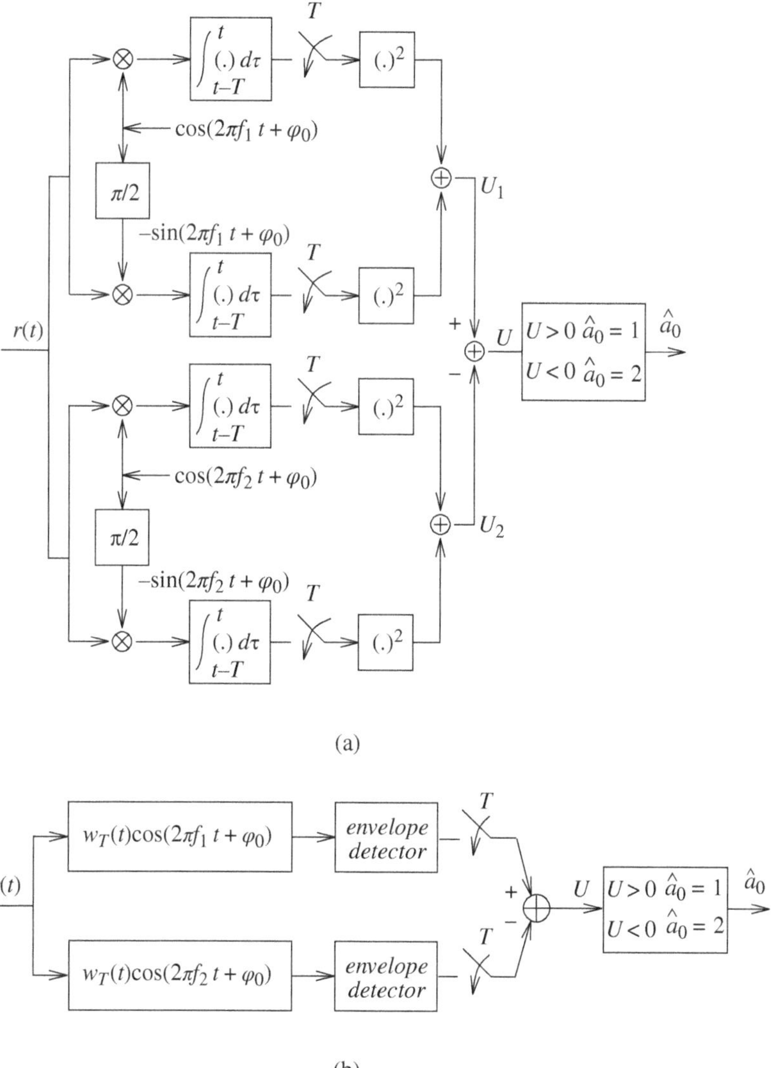

Figure 16.16 Two ML receivers for a non-coherent 2-FSK system. (a) Square-law detector and (b) envelope detector.

Figure 16.17 Envelope detector receiver for an on–off keying system.

Example 16.3.6 (DSB modulated signalling with random phase)
With reference to the Example 16.3.3, we show in Figure 16.18 the receiver for a baseband M-ary signalling scheme that is DSB modulated with random phase. Depending upon the signalling type, further simplifications arise by extracting functions that are common to the different branches.

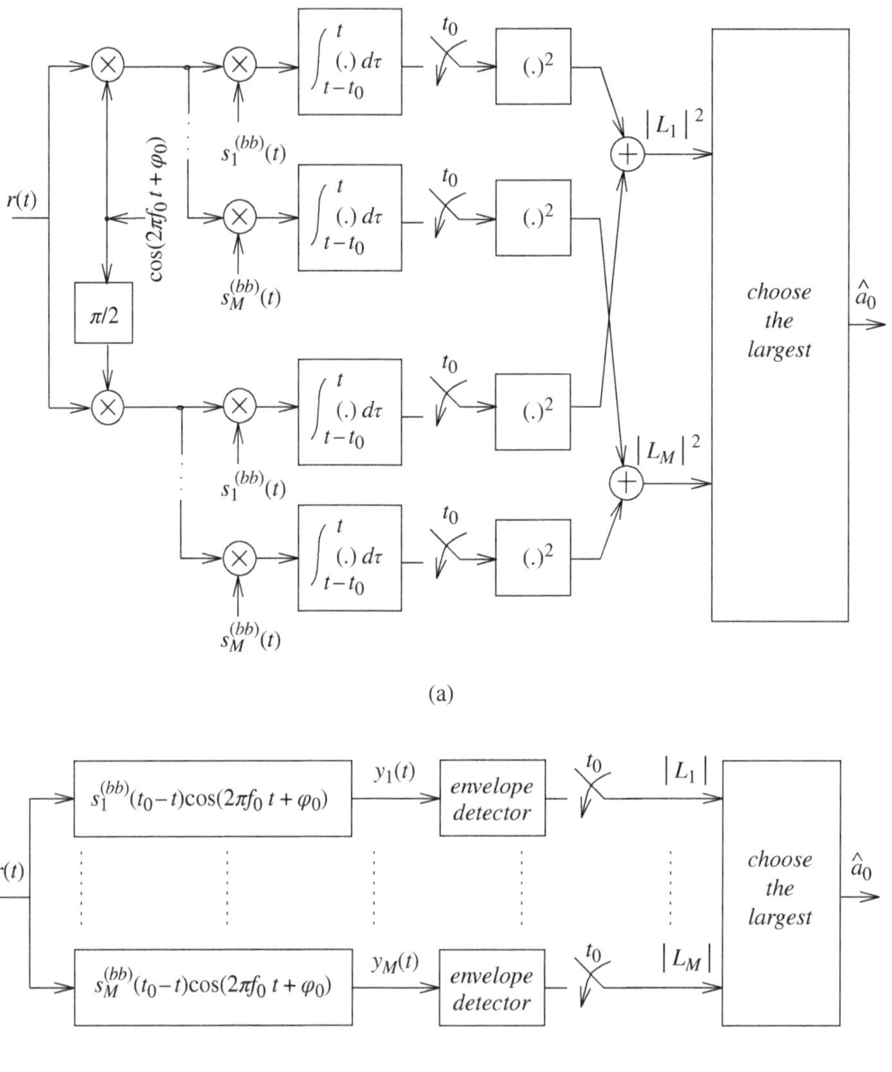

Figure 16.18 Two receivers for a DSB modulation system with M-ary signalling and random phase. (a) Square-law detector and (b) envelope detector.

Error probability for a non-coherent binary FSK system

We now derive the error probability of the system of Example 16.3.4. We assume that s_1 is transmitted. Then,

$$r(t) = s_1(t) + w(t) \tag{16.80}$$

and

$$P_{bit} = P[U_1 < U_2 \mid s_1] \tag{16.81}$$

Equivalently, if we define

$$V_1 = \sqrt{U_1} \quad \text{and} \quad V_2 = \sqrt{U_2} \tag{16.82}$$

we have

$$P_{bit} = P[V_1 < V_2 \mid s_1] \tag{16.83}$$

Now, recalling assumption (16.59), that is $s_1(t, \varphi_1) \perp s_2(t, \varphi_2)$, we have

$$\begin{aligned} U_2 &= \left(\int_0^T r(t)\cos(2\pi f_2 t + \varphi_0)\,dt \right)^2 + \left(\int_0^T r(t)\sin(2\pi f_2 t + \varphi_0)\,dt \right)^2 \\ &= w_{2,c}^2 + w_{2,s}^2 \end{aligned} \tag{16.84}$$

where

$$\begin{aligned} w_{2,c} &= \int_0^T w(t)\cos(2\pi f_2 t + \varphi_0)\,dt \\ w_{2,s} &= \int_0^T w(t)\sin(2\pi f_2 t + \varphi_0)\,dt \end{aligned} \tag{16.85}$$

If we define

$$\begin{aligned} w_{1,c} &= \int_0^T w(t)\cos(2\pi f_1 t + \varphi_0)\,dt \\ w_{1,s} &= \int_0^T w(t)\sin(2\pi f_1 t + \varphi_0)\,dt \end{aligned} \tag{16.86}$$

we have

$$\begin{aligned} U_1 &= \left(\int_0^T r(t)\cos(2\pi f_1 t + \varphi_0)\,dt \right)^2 + \left(\int_0^T r(t)\sin(2\pi f_1 t + \varphi_0)\,dt \right)^2 \\ &= \left(\frac{AT}{2}\cos(\varphi_0 - \varphi_1) + w_{1,c} \right)^2 + \left(\frac{AT}{2}\sin(\varphi_0 - \varphi_1) + w_{1,s} \right)^2 \end{aligned} \tag{16.87}$$

where from (16.56) we also have

$$\frac{AT}{2} = \sqrt{\frac{E_s T}{2}} \tag{16.88}$$

As w is a white Gaussian random process with zero mean, $w_{2,c}$ and $w_{2,s}$ are two jointly Gaussian r.v.s. with

$$\begin{aligned} E[w_{2,c}] &= E[w_{2,s}] = 0 \\ E[w_{2,c}^2] &= E[w_{2,s}^2] = \frac{N_0}{2}\,\frac{T}{2} \\ E[w_{2,c}\,w_{2,s}] &= \int_0^T \int_0^T \frac{N_0}{2}\,\delta(t_1 - t_2)\cos(2\pi f_2 t_1 + \varphi_0)\sin(2\pi f_2 t_2 + \varphi_0)\,dt_1\,dt_2 = 0 \end{aligned} \tag{16.89}$$

Similar considerations hold for $w_{1,c}$ and $w_{1,s}$.

Therefore V_2, with statistical power $2(N_0T/4)$, has a Rayleigh probability density

$$p_{V_2}(v_2) = \frac{v_2}{N_0T/4}\, e^{-\frac{v_2^2}{2(N_0T/4)}}\, 1(v_2) \tag{16.90}$$

whereas V_1 has a Rice probability density function

$$p_{V_1}(v_1) = \frac{v_1}{N_0T/4}\, e^{-\frac{(v_1^2 + (AT/2)^2)}{2(N_0T/4)}}\, I_0\left(\frac{v_1(AT/2)}{N_0T/4} \right) 1(v_1) \tag{16.91}$$

Consequently, (16.83) becomes[5]

$$\begin{aligned} P_{bit} &= \int_0^{+\infty} P[V_1 < v_2 \mid V_2 = v_2]\, p_{V_2}(v_2)\, dv_2 \\ &= \int_0^{+\infty} \left(\int_0^{v_2} p_{V_1}(v_1)\, dv_1 \right) p_{V_2}(v_2)\, dv_2 \\ &= \frac{1}{2}\, e^{-\frac{1}{2}\Gamma} \end{aligned} \tag{16.93}$$

It can be shown that this result is not limited to frequency shift keying (FSK) systems and is valid for any pair of non-coherent orthogonal signals with energy E_s.

Performance comparison of binary systems

The received signals are given by (16.55), where f_d satisfies the constraint (16.59). The two signals are orthogonal ($\rho = 0$); hence from (16.93), we have ($\Gamma = E_s/N_0$)

$$\text{FSK(NC):} \qquad P_{bit} = \frac{1}{2}\, e^{-\frac{1}{2}\Gamma} \tag{16.94}$$

A comparison with a non-coherent binary system with differentially encoded bits, such as DBPSK, is illustrated in Figure 16.19. The differential receiver for DBPSK directly gives the original uncoded bits,

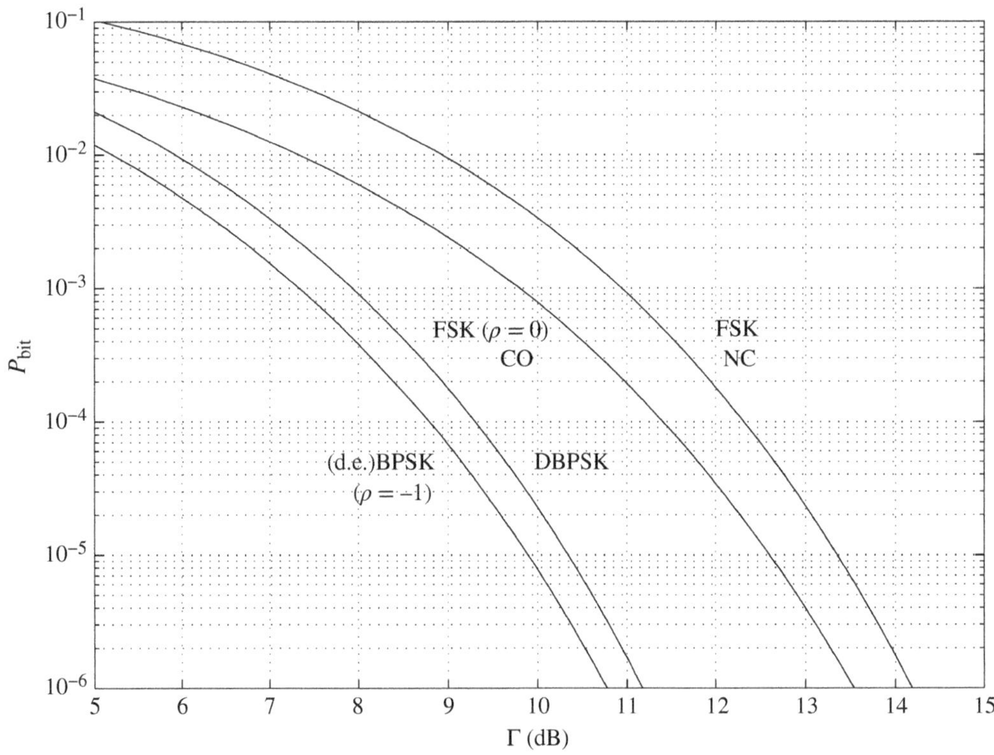

Figure 16.19 Bit error probability as a function of Γ for BPSK and binary FSK systems, with coherent (CO) and non-coherent (NC) detection. Here ρ denotes the correlation coefficient between transmit signals.

---

[5] To compute the following integrals, we recall the Weber–Sonine formula:

$$\int_0^{+\infty} x\, e^{-\alpha(x^2/2)}\, I_0(\beta x)\, dx = \frac{1}{\alpha}\, e^{\beta^2/2\alpha} \tag{16.92}$$

thus from (16.8) we have

$$\text{DBPSK:} \qquad P_{bit} = \frac{1}{2}e^{-\Gamma} \tag{16.95}$$

A comparison between (16.94) and (16.95) indicates that DBPSK is better than FSK by about 3 dB in Γ, for the same P_{bit}.

The performance of a binary frequency shift keying (BFSK) system and that of a differentially encoded BPSK system with coherent detection are compared in Figure 16.19. In particular, from Ref. [5] it follows

$$\text{FSK(CO):} \qquad P_{bit} = Q\left(\sqrt{\Gamma}\right) \tag{16.96}$$

Taking into account differential decoding, from (6.84) we have[6]

$$\text{(d.e.)BPSK:} \qquad P_{bit} \simeq 2Q\left(\sqrt{2\Gamma}\right) \tag{16.97}$$

We observe that the difference between coherent FSK and non-coherent FSK is less than 2 dB for $P_e \leq 10^{-3}$, and becomes less than 1 dB for $P_e \leq 10^{-5}$. We also note that because of the large bandwidth required, FSK systems with $M > 2$ are not widely used.

16.4 Frequency-based modulations

16.4.1 Frequency shift keying

The main advantage of FSK modulators consists in generating a signal having a constant envelope, therefore the distortion introduced by a high-power amplifier (HPA) is usually negligible. However, they present two drawbacks:

- wider spectrum of the modulated signal as compared to amplitude and/or phase modulated systems,
- complexity of the optimum receiver in the presence of non-ideal channels.

As already expressed in (16.55), a BFSK modulator maps the information bits in frequency deviations ($\pm f_d$) around a carrier with frequency f_0; the possible transmitted waveforms are then given by

$$\begin{aligned} s_1(t) &= A\cos\left(2\pi(f_0 - f_d)t + \varphi_1\right) \\ s_2(t) &= A\cos\left(2\pi(f_0 + f_d)t + \varphi_2\right), \qquad kT < t < (k+1)T \end{aligned} \tag{16.98}$$

where, if we denote by E_s the average energy of an isolated pulse, $A = \sqrt{2E_s/T}$.

Figure 16.20 illustrates the generation of the above signals by two oscillators, with frequency $f_1 = f_0 - f_d$ and $f_2 = f_0 + f_d$, selected at instants kT by the variable $a_k \in \{1, 2\}$ related to the information bits. A particular realization of s_1 and s_2 is shown in Figure 16.21a. The resultant signal is given by

$$s(t) = \sum_{k=-\infty}^{+\infty} s_{a_k}(t - kT), \qquad a_k \in \{1, 2\} \tag{16.99}$$

A realization of s is shown in Figure 16.21b.

We consider both coherent and non-coherent receivers.

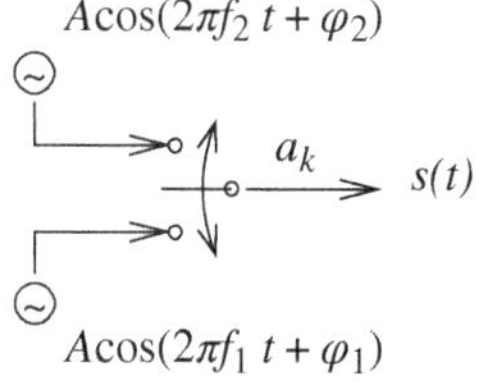

Figure 16.20 Generation of a binary (non-coherent) FSK signal by two oscillators.

[6] For a more accurate evaluation of the probability of error see footnote 3 on page 792.

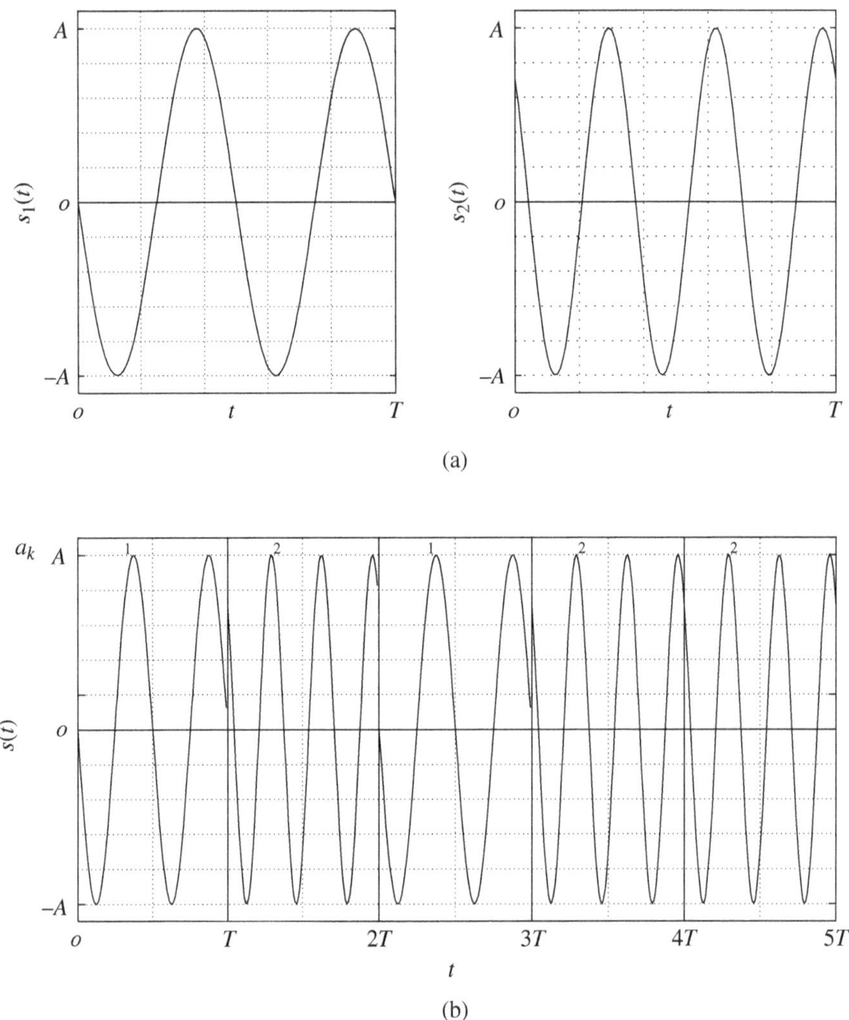

Figure 16.21 (a) Binary FSK waveforms for $f_1 = 2/T$, $f_2 = 3/T$, $\varphi_1 = \pi/2$, and $\varphi_2 = \pi/4$ and (b) transmitted signal for a particular sequence $\{a_k\}$.

Coherent demodulator

A coherent demodulator is used when the transmitter provides all M signals with a well-defined phase; at the receiver, there must be a circuit for the recovery of the phase of the various carriers. In practice, this coherent receiver is rarely used because of its implementation complexity.

In any case, for orthogonal signals we can adopt the scheme of Figure 16.22 for the binary case and the bit error probability is given by (16.96).

We note that the signals are orthogonal if

$$(2f_d)_{min} = \frac{1}{2T} \tag{16.100}$$

Non-coherent demodulator

The transmitted waveforms can now have a different phase, and in any case unknown to the receiver (see Example 16.3.1 on page 799). For orthogonal signals, from the general scheme of Figure 16.16, we

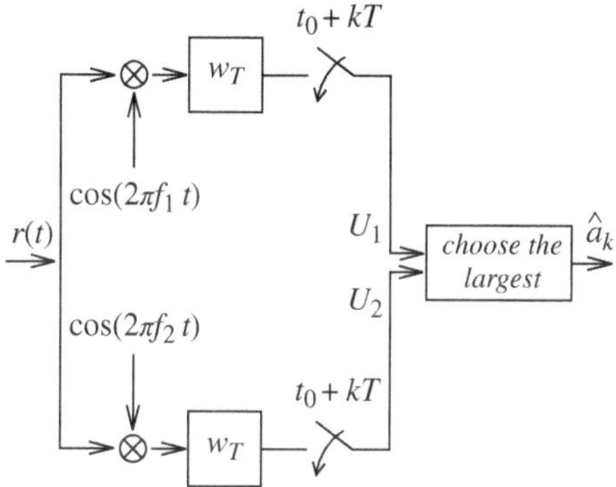

Figure 16.22 Coherent demodulator for orthogonal binary FSK.

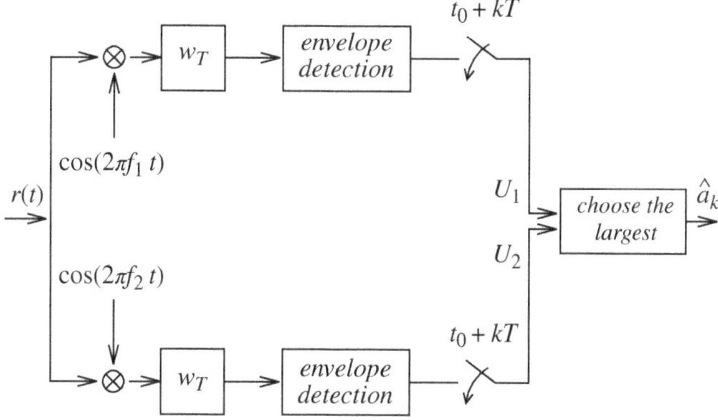

Figure 16.23 Non-coherent demodulator for orthogonal binary FSK.

give in Figure 16.23 a possible non-coherent demodulator. In this case, the bit error probability is given by (16.94),

From (16.60), we note that, in the non-coherent case, signals are orthogonal if

$$(2f_d)_{min} = \frac{1}{T} \tag{16.101}$$

Therefore, a non-coherent FSK system needs a double frequency deviation and has slightly lower performance (see Figure 16.19) as compared to coherent FSK; however, it does not need to acquire the carrier phase.

Limiter–discriminator FM demodulator

A simple non-coherent receiver, with good performance, for $f_1 T \gg 1, f_2 T \gg 1$ and sufficiently high Γ, is depicted in Figure 16.24.

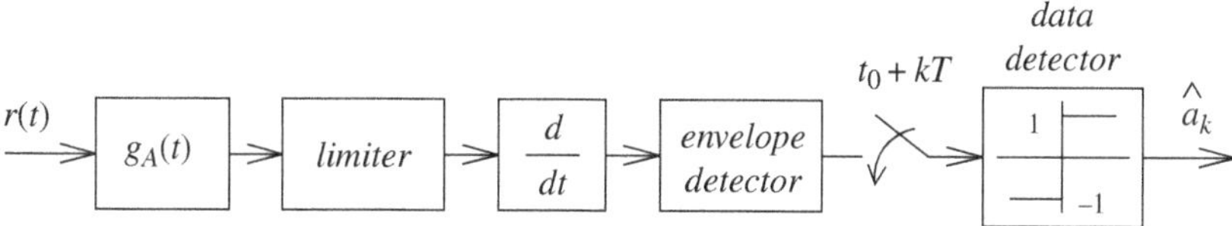

Figure 16.24 Limiter–discriminator FSK demodulator.

After a passband filter to partially eliminate noise, a classical FM demodulator is used to extract the instantaneous frequency deviation of r, equal to $\pm f_d$. Sampling at instants $t_0 + kT$ and using a threshold detector, we get the detected symbol $\hat{a}_k \in \{-1, 1\}$.

In general, for $2f_dT = 1$, the performance of this scheme is very similar to that of a non-coherent orthogonal FSK.

16.4.2 Minimum-shift keying

We recall the following characteristics of an ideal FSK signal.

- In order to avoid high-frequency components in the PSD of s, the phase of an FSK signal should be continuous and not as represented in Figure 16.21. For signals (16.98), with $\varphi_1 = \varphi_2$, this implies that it must be $f_nT = I_n$, $n = 1, 2$, I_n an integer.
- To minimize the error probability, it must be $s_n \perp s_m$, $m \neq n$.
- To minimize the required bandwidth, the separation between the various frequencies must be minimum.

For signals as given by (16.98), choosing $|I_1 - I_2| = 1$ we have $f_2 = f_1 + 1/T$. In this case, s_1 and s_2 are as shown in Figure 16.25.

Moreover from (16.101) it is easy to prove that the signals are orthogonal. The result is that the frequency deviation is such that $2f_d = |f_2 - f_1| = 1/T$, and the carrier frequency is $f_0 = (f_1 + f_2)/2 = f_1 + 1/(2T)$.

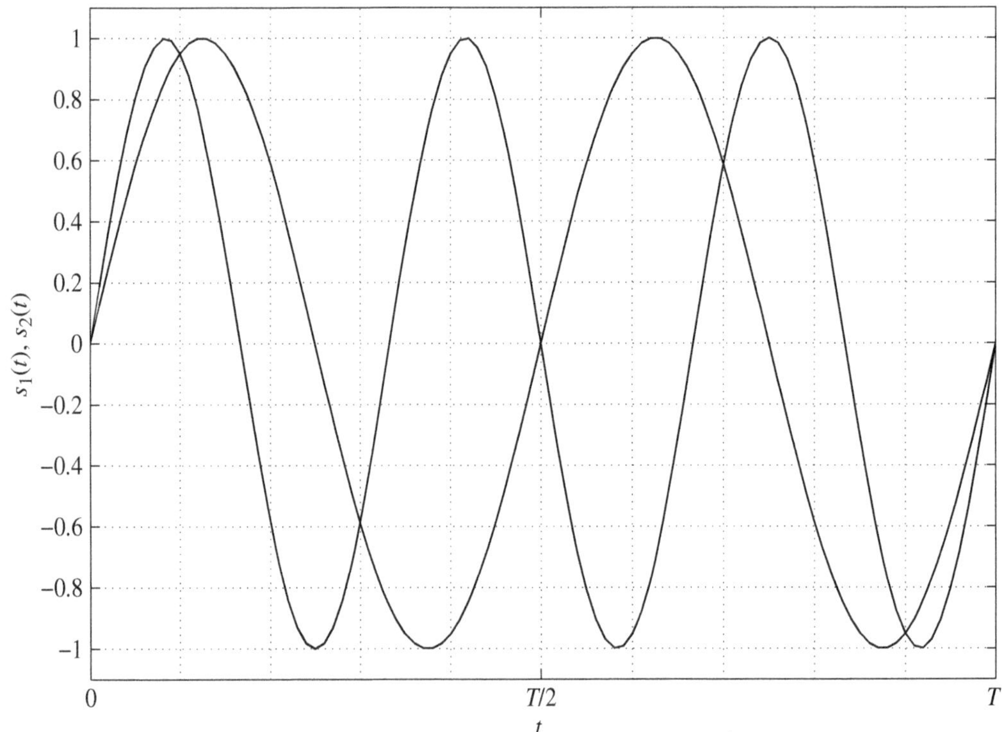

Figure 16.25 Waveforms as given by (16.98) with $A = 1$, $f_1 = 2/T$, $f_2 = 3/T$, and $\varphi_1 = \varphi_2 = 0$.

In a digital implementation of an FSK system, the following observation is very useful. Independent of f_1 and f_2, a method to obtain a phase continuous FSK signal given by

$$s(t) = A \cos \varphi(t) \tag{16.102}$$

where

$$\varphi(t) = 2\pi f_0 t + a_k 2\pi f_d t + \eta_k, \qquad a_k \in \{-1, 1\}\ kT \le t < (k+1)T \tag{16.103}$$

is to employ a single oscillator, whose phase satisfies the constraint $\varphi((k+1)T^{(-)}) = \varphi((k+1)T^{(+)})$; thus, it is sufficient that at the beginning of each symbol interval η_{k+1} is set equal to

$$\eta_{k+1} = ((a_k - a_{k+1})2\pi f_d(k+1)T + \eta_k) \bmod 2\pi \tag{16.104}$$

An alternative method is given by the scheme of Figure 16.26, in which the sequence $\{a_k\}$, with binary elements in $\{-1, 1\}$, is filtered by g to produce a PAM signal

$$x_f(t) = \sum_{k=-\infty}^{+\infty} a_k g(t - kT), \qquad a_k \in \{-1, 1\} \tag{16.105}$$

The signal x_f is input to a voltage controlled oscillator (VCO), whose output is given by

$$s(t) = A \cos \left(2\pi f_0 t + 2\pi h \int_{-\infty}^{t} x_f(\tau)\, d\tau \right) \tag{16.106}$$

In (16.106), f_0 is the carrier, $h = 2f_d T$ is the modulation index and

$$\varphi(t) = 2\pi f_0 t + 2\pi h \int_{-\infty}^{t} x_f(\tau)\, d\tau \tag{16.107}$$

represents the phase of the modulated signal.

Choosing for g a rectangular pulse

$$g(t) = \frac{1}{2T} \mathrm{w}_T(t) \tag{16.108}$$

we obtain a modulation scheme called *continuous phase* frequency shift keying (CPFSK). In turn, CPFSK is a particular case of the *continuous phase modulation* described in Appendix 16.A.

We note that the area of g is equal to 0.5; in this case, the information is in the instantaneous frequency of s, in fact

$$f_s(t) = \frac{1}{2\pi} \frac{d\varphi(t)}{dt} = f_0 + h x_f(t) = f_0 + \frac{h a_k}{2T}, \qquad kT < t < (k+1)T \tag{16.109}$$

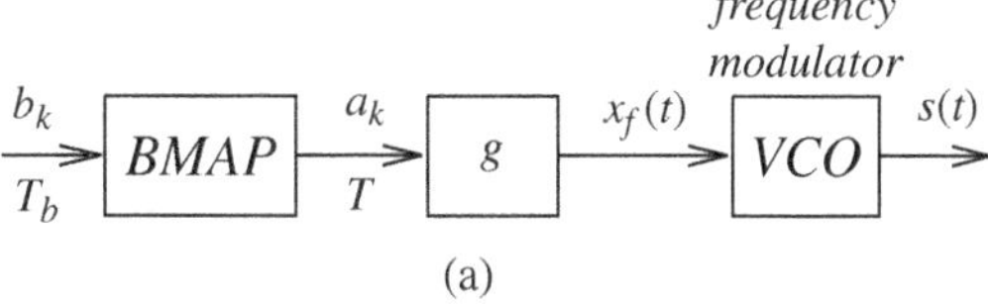

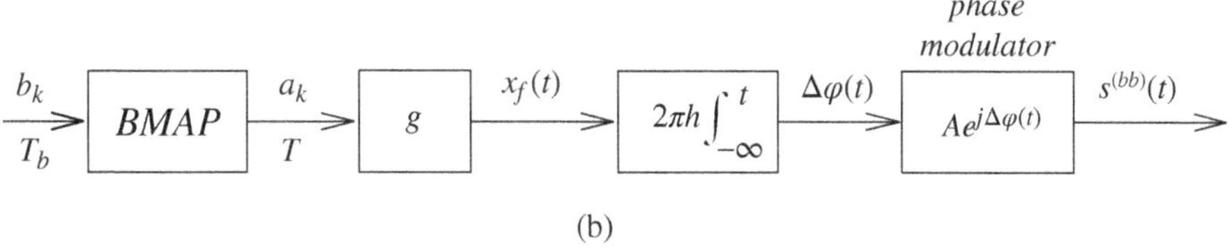

Figure 16.26 CPFSK modulator. (a) Passband model and (b) baseband equivalent model.

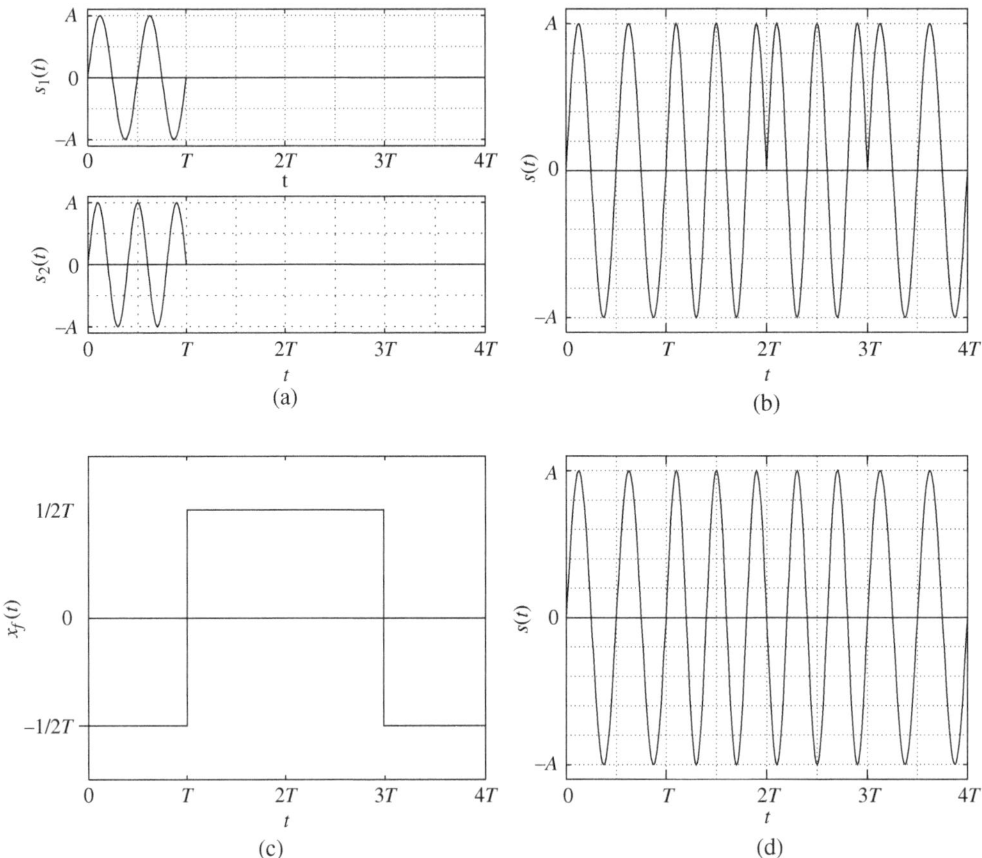

Figure 16.27 Comparison between FSK and MSK signals. (a) FSK waveforms, (b) FSK signal, (c) signal $x_f(t)$, and (d) MSK signal.

The *minimum shift keying* (MSK) modulation is equivalent to CPFSK with modulation index $h = 0.5$, that is for $f_d = 1/(4T)$. Summarizing: an MSK scheme is a BFSK scheme ($T_b = T$) in which, besides having $f_d = 1/(4T)$, the modulated signal has continuous phase.

Figure 16.27 illustrates a comparison between an FSK signal with $f_d = 1/(4T)$ as given by (16.98), and an MSK signal for a binary sequence $\{a_k\}$ equal to $\{-1, 1, 1, -1\}$. Note that in the MSK case, it is like having four waveforms $s_1, -s_1, s_2, -s_2$, each pair related to $a_k = 1$ and $a_k = -1$, respectively. Signal selection within the pair is done in a way to guarantee the phase continuity of s.

Power spectral density of CPFSK

From the expressions given in [1], the behaviour of the continuous part of the PSD is represented in Figure 16.28. Note that for higher values of h, a peak will tend to emerge around $f_d = h/(2T)$, showing the instantaneous frequency of waveforms.

It is seen that phase continuity implies a lower bandwidth, at least if we use the definition of bandwidth based on a given percentage of signal power.

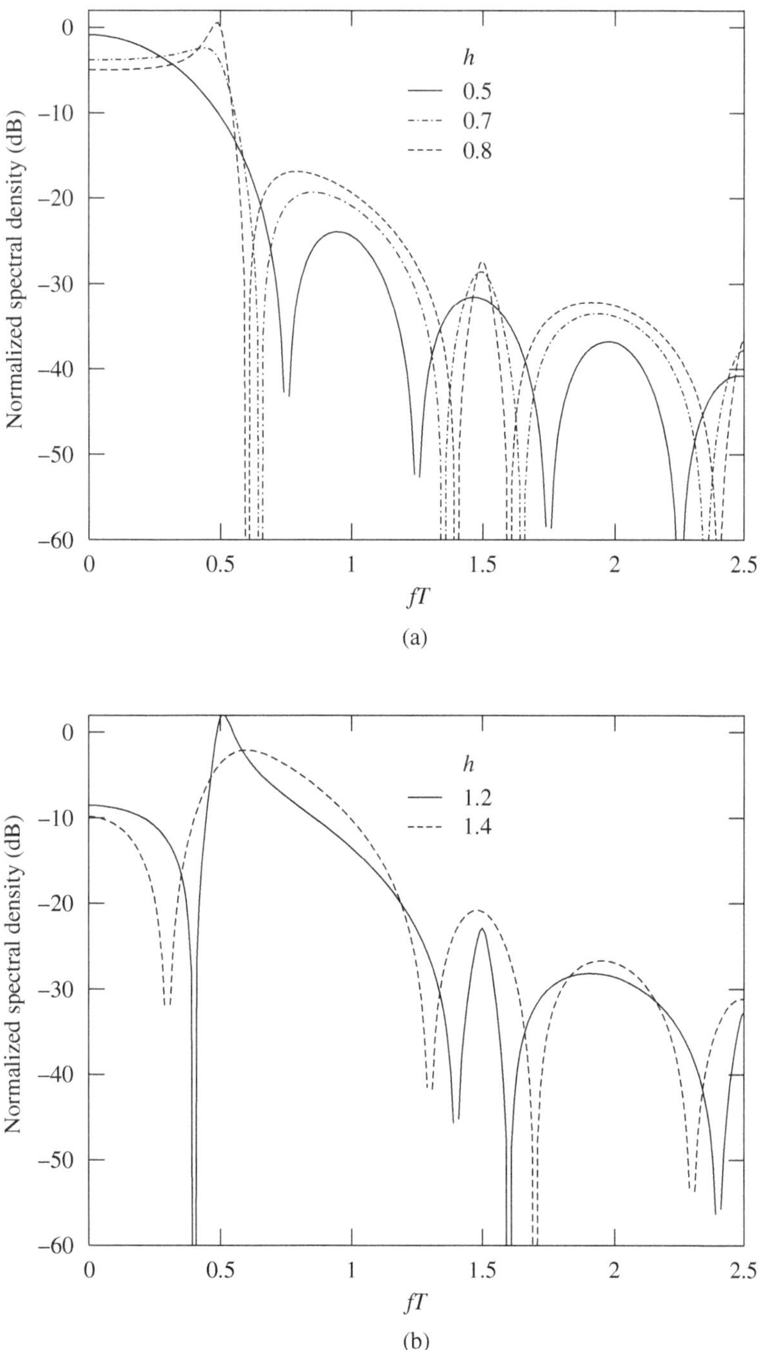

Figure 16.28 Continuous part of the PSD of a CPFSK signal for five values of the modulation index. (a) $h < 1$ and (b) $h > 1$.

Performance

Among the different demodulation schemes listed in Figure 16.29, we show in Figure 16.30 a differential 1BDD non-coherent demodulator. This scheme is based on the fact that the phase deviation of an MSK signal can vary of $\pm\pi/2$ between two suitably instants spaced of T_b. In any case, the performance for an ideal AWGN channel is that of a DBPSK scheme, but with half the phase variation; hence from (16.8), with E_s that becomes $E_s/2$, we get

$$P_{bit} = \frac{1}{2}e^{-\frac{1}{2}\Gamma} \tag{16.110}$$

Note that this is also the performance of a non-coherent orthogonal BFSK scheme (see (16.94)).

In Figure 16.32, we illustrate a coherent demodulator that at alternate instants on the I branch and on the Q branch is of the BPSK type. In this case, from (6.84), the error probability for decisions on the symbols $\{c_k\}$ is given by

$$P_{bit,Ch} = Q(\sqrt{2\Gamma}) \tag{16.111}$$

As it is as if the bits $\{c_k\}$ were differentially encoded, to obtain the bit error probability for decisions on the symbols $\{a_k\}$, we use (16.18):

$$P_{bit} = 4P_{bit,Ch} - 8P^2_{bit,Ch} + 8P^3_{bit,Ch} - 4P^4_{bit,Ch} \tag{16.112}$$

In Figure 16.32, we show error probability curves for various receiver types. From the graph we note that to obtain an error probability of 10^{-3}, going from a coherent system to a non-coherent one, it is necessary to increase the value of Γ as indicated in Table 16.5.

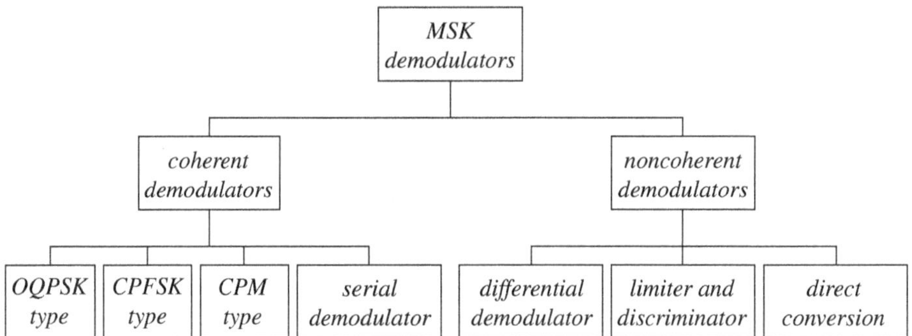

Figure 16.29 MSK demodulator classification.

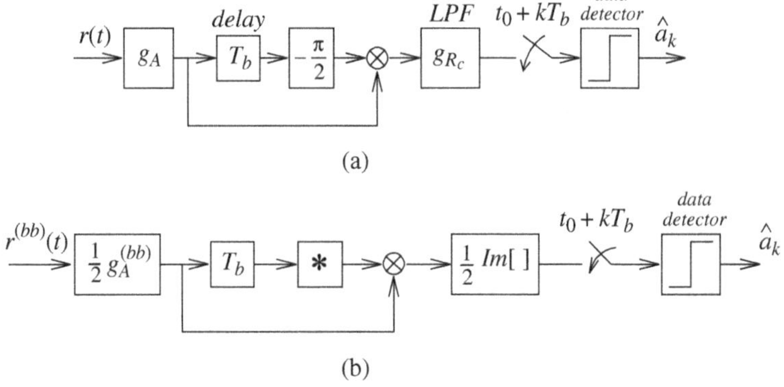

Figure 16.30 Differential (1BDD) non-coherent MSK demodulator. (a) Passband model and (b) baseband equivalent model.

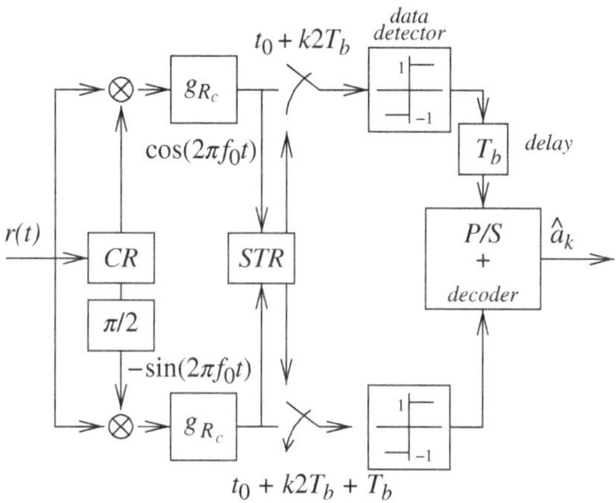

Figure 16.31 Coherent MSK demodulator.

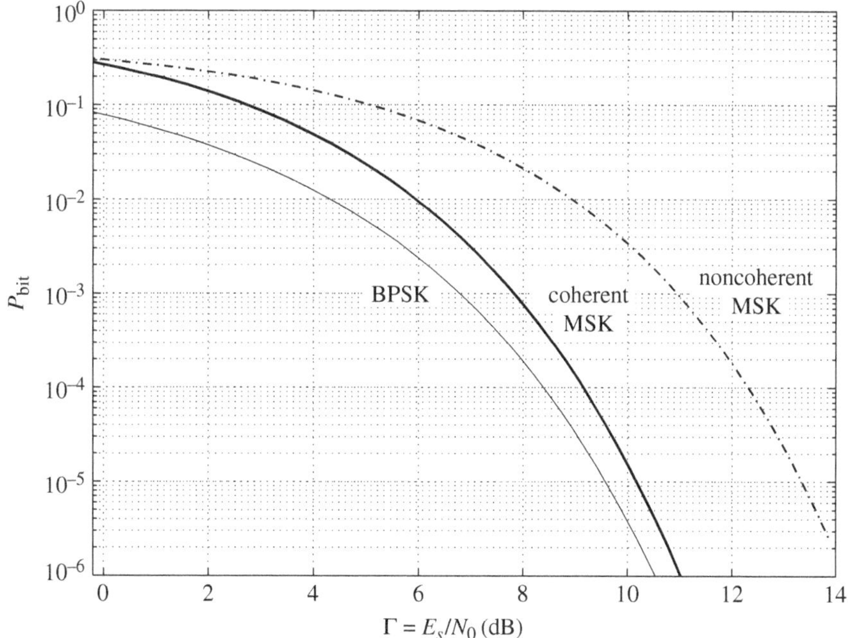

Figure 16.32 Comparison among various error probabilities.

Table 16.5: Increment of Γ (in dB) for an MSK scheme with respect to a coherent BPSK demodulator for $P_{bit} = 10^{-3}$

	coherent BPSK	coherent MSK	non-coherent MSK
Γ	6.79	+1.1	+4.2

MSK with differential precoding

In the previously considered coherent scheme, the transmitted symbols $\{a_k\}$ are obtained from non-run zero (NRZ) mapping of bits $\{b_k\}$. Note that the transformation that maps the bits of $\{a_k\}$ into $\{c_k\}$

corresponds to a *differential encoder* given by[7]

$$c_k = c_{k-1} a_k, \qquad a_k \in \{-1, 1\} \tag{16.113}$$

with $c_k \in \{-1, 1\}$. If we now use a differential (pre)coding, where $\tilde{a}_k \in \{0, 1\}$, $\tilde{a}_k = b_k \oplus b_{k-1}$ and $a_k = 1 - 2\tilde{a}_k = 1 - 2(b_k \oplus b_{k-1})$, then from (16.113) with $\tilde{c}_k \in \{0, 1\}$ and $c_k = 1 - 2\tilde{c}_k$, we get

$$\begin{aligned} \tilde{c}_k &= \tilde{c}_{k-1} \oplus \tilde{a}_k \\ &= \tilde{a}_0 \oplus \tilde{a}_1 \oplus \cdots \oplus \tilde{a}_k \\ &= b_{-1} \oplus b_k \end{aligned} \tag{16.114}$$

In other words, the symbol c_k is directly related to the information bit b_k; the performance loss due to (16.112) is thus avoided and we obtain

$$P_{bit} = P_{bit,Ch} \tag{16.115}$$

We emphasize that this differential (pre)coding scheme should be avoided if differential non-coherent receivers are employed, because one error in $\{\hat{a}_k\}$ generates a long error sequence in $\{\hat{b}_k = \hat{\tilde{a}}_k \oplus \hat{b}_{k-1}\}$.

16.4.3 Remarks on spectral containment

Modulating both BPSK and quadrature phase-shift keying (QPSK) with h_{Tx} given by a rectangular pulse with duration equal to the symbol period, a comparison between the various PSDs is illustrated in Figure 16.33. We note that for limited bandwidth channels, it is convenient to choose h_{Tx} of the raised cosine or square root raised cosine type. However, in some radio applications, the choice of a rectangular pulse may be appropriate, as it generates a signal with lower a peak/average power ratio and therefore more suitable for being amplified with a power amplifier that operates near saturation.

Two observations on Figure 16.33 follow.

- For the same T_b, the main lobe of QPSK extends up to $1/T = 0.5/T_b$, whereas that of MSK extends up to $1/T = 0.75/T_b$; thus, the lobe of MSK is 50% wider than that of QPSK, consequently requiring larger bandwidth.
- At high frequencies, the PSD of MSK decays as $1/f^4$, whereas the PSD of QPSK decays as $1/f^2$.

16.5 Gaussian MSK

The Gaussian minimum shift keying (GMSK) is a variation of MSK in which, to reduce the bandwidth of the modulated signal s, the PAM signal x_f is filtered by a Gaussian filter.

Consider the scheme illustrated in Figure 16.34, in which we have the following filters:

- interpolator filter

$$g_I(t) = \frac{1}{2T} \mathrm{w}_T(t) \tag{16.116}$$

- shaping filter

$$g_G(t) = \frac{K}{\sqrt{2\pi}} e^{-K^2 t^2/2} \qquad \text{with} \qquad K = \frac{2\pi B_t}{\sqrt{\ln(2)}} \qquad (B_t \text{ is the 3 dB bandwidth}) \tag{16.117}$$

- overall filter

$$g(t) = g_I * g_G(t) \tag{16.118}$$

[7] It corresponds to the exclusive OR (16.11) if $a_k \in \{0, 1\}$.

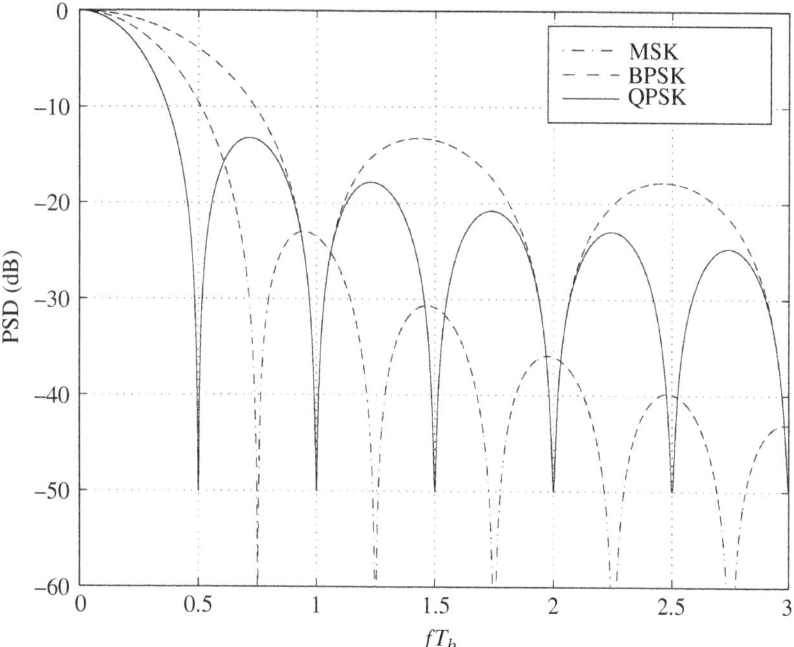

Figure 16.33 Normalized PSD of the complex envelope of signals obtained by three modulation schemes.

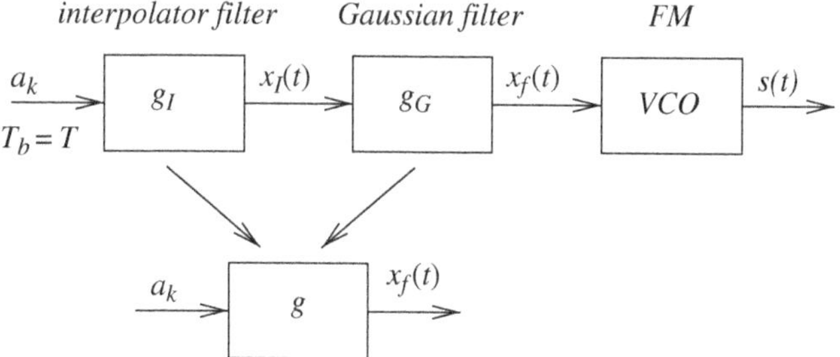

Figure 16.34 GMSK modulator.

Considering the signals we have:

- transmitted binary symbols

$$a_k \in \{-1, 1\} \tag{16.119}$$

- interpolated signal

$$x_I(t) = \sum_{k=-\infty}^{+\infty} a_k g_I(t - kT) = \sum_{k=-\infty}^{+\infty} a_k \frac{1}{2T} \mathtt{w}_T(t - kT) \tag{16.120}$$

- PAM signal

$$x_f(t) = \sum_{k=-\infty}^{+\infty} a_k g(t - kT) \tag{16.121}$$

- modulated signal

$$s(t) = A\cos\left(2\pi f_0 t + 2\pi h \int_{-\infty}^{t} x_f(\tau)\, d\tau\right) = A\cos\left(2\pi f_0 t + \Delta\varphi(t)\right) \tag{16.122}$$

where h is the modulation index, nominally equal to 0.5, and $A = \sqrt{2E_s/T}$.

From the above expressions, it is clear that the GMSK signal is a FM signal with phase deviation $\Delta\varphi(t) = \pi \int_{-\infty}^{t} x_f(\tau)\, d\tau$.

An important parameter is the *3dB bandwidth*, B_t, of the Gaussian filter. However, a reduction in B_t, useful in making prefiltering more selective, corresponds to a broadening of the PAM pulse with a consequent increase in the intersymbol interference, as can be noted in the plots of Figure 16.35. Thus, a trade-off between the two requirements is necessary.

The product B_tT was chosen equal to 0.3 in the Global System for Mobile Communications (GSM) standard, and equal to 0.5 in the Digital Enhanced Cordless Telecommunications (DECT) standard (see Chapter 18). The case $B_tT = \infty$, i.e. without the Gaussian filter, corresponds to MSK.

Analysing g in the frequency domain, we have

$$g(t) = g_I * g_G(t) \quad \stackrel{\mathcal{F}}{\longleftrightarrow} \quad G(f) = G_I(f) \cdot G_G(f) \tag{16.123}$$

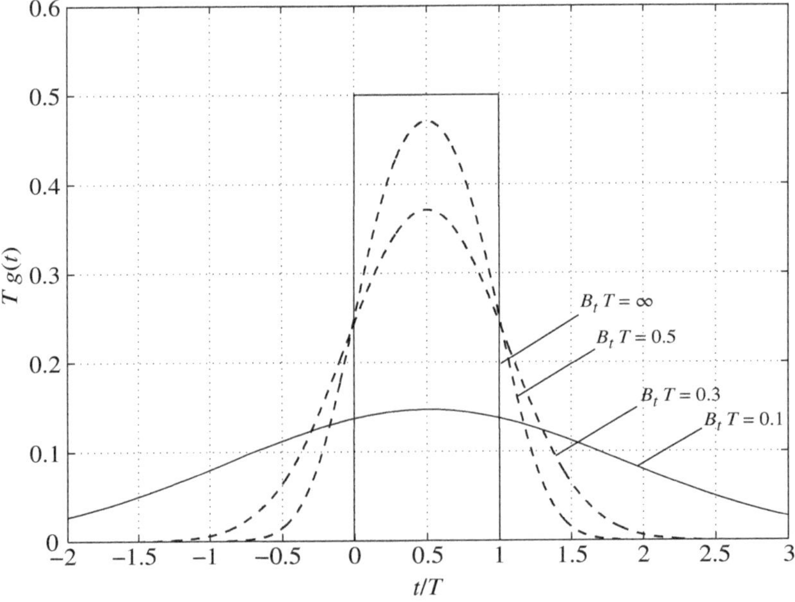

Figure 16.35 Overall pulse $g(t) = g_I * g_G(t)$, with amplitude normalized to $1/T$, for various values of the product B_tT.

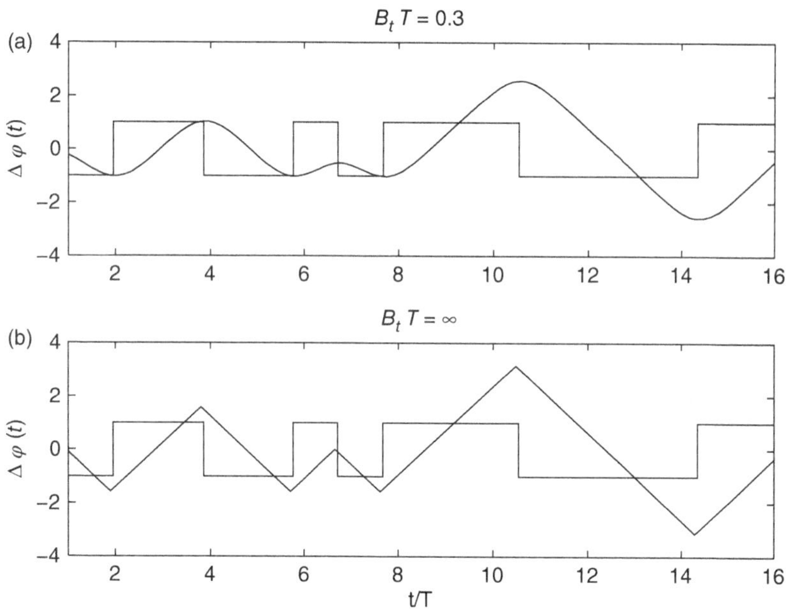

Figure 16.36 Phase deviation $\Delta\varphi$ of (a) GMSK signal for $B_tT = 0.3$, and (b) MSK signal.

As

$$G_I(f) = \frac{1}{2}\text{sinc}(fT)e^{-j\pi fT}$$
$$G_G(f) = e^{-2\pi^2(f/K)^2} \tag{16.124}$$

it follows that

$$G(f) = \frac{1}{2}\text{sinc}(fT)e^{-2\pi^2(f/K)^2}e^{-j\pi fT} \tag{16.125}$$

In Figure 16.36, the behaviour of the phase deviation of a GMSK signal with $B_tT = 0.3$ is compared with the phase deviation of an MSK signal; note that in both cases the phase is continuous, but for GMSK we get a smoother curve, without discontinuities in the slope.

Possible trajectories of the phase deviation for $B_tT = 0.3$ and $B_tT = \infty$ are illustrated in Figure 16.37.

PSD of GMSK

GMSK is a particular case of phase modulation which performs a non-linear transformation on the message, thus, it is very difficult to evaluate analytically the PSD of a GMSK signal. Hence, we resort to a discrete time spectral estimate, for example the Welch method (see Section 1.9). The estimate is made with reference to the baseband equivalent of a GMSK signal, that is using the complex envelope of the modulated signal given by $s^{(bb)}(t) = Ae^{j\Delta\varphi(t)}$.

The result of the spectral estimate is illustrated in Figure 16.38. Note that the central lobe has an extension up to $0.75/T$; therefore, the sampling frequency, F_Q, to simulate the baseband equivalent scheme of Figure 16.34 may be chosen equal to $F_Q = 4/T$ or $F_Q = 8/T$.

16.5.1 Implementation of a GMSK scheme

For the scheme of Figure 16.34, three possible configurations are given, depending on the position of the digital-to-analog converter (DAC).

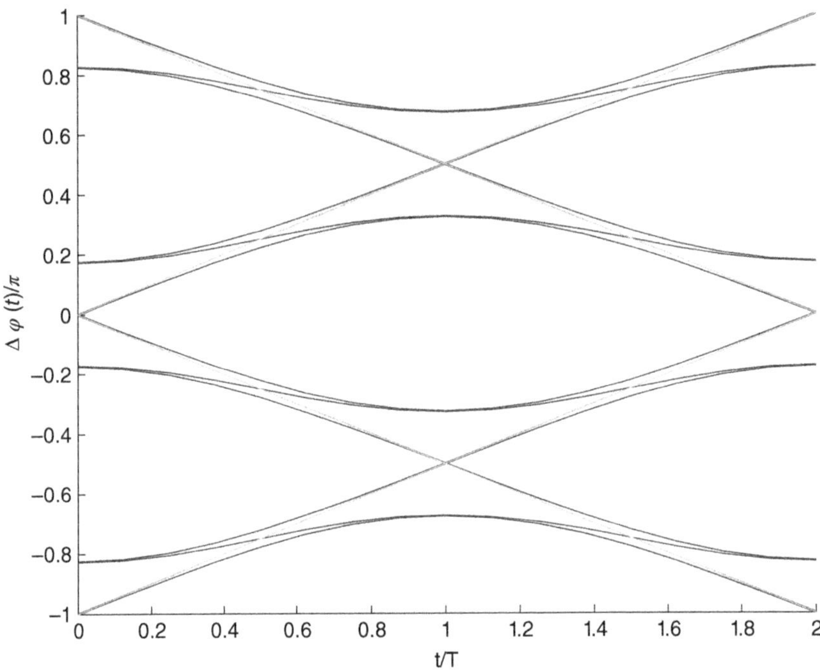

Figure 16.37 Trajectories of the phase deviation of a GMSK signal for $B_tT = 0.3$ (solid line) and $B_tT = \infty$ (dotted line).

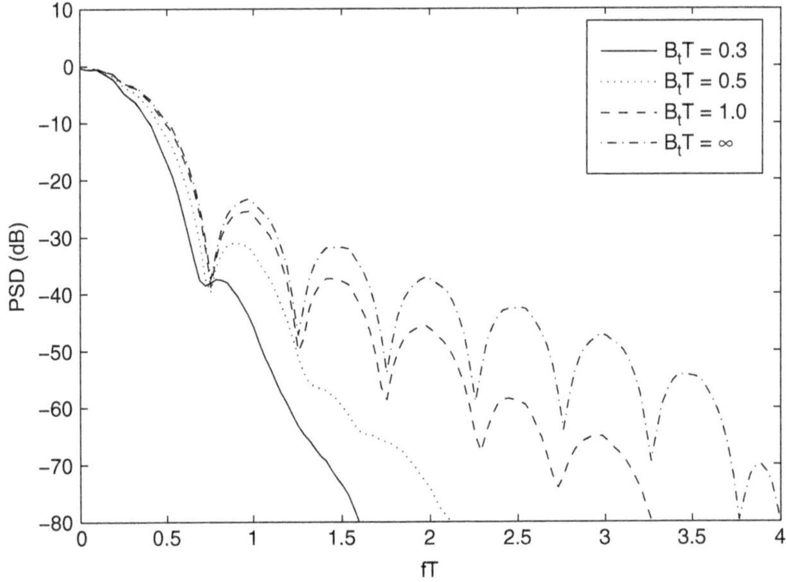

Figure 16.38 Estimate of the PSD of a GMSK signal for various values of B_tT.

Configuration I

The first configuration, illustrated in Figure 16.39, is in the analog domain; in particular, an analog low-pass Gaussian filter is employed. As shown in [6], it is possible to implement a good approximation of the Gaussian filter by resorting to simple devices, such as lattice filters. However, the weak point of this scheme is represented by the VCO, because in an open loop, the voltage/frequency VCO characteristic is non-linear, and as a consequence, the modulation index can vary even of a factor 10 in the considered frequency range.

Configuration II

The second configuration is represented in Figure 16.40. The digital filter $g(nT_Q)$ that approximates the analog filter g is designed by the window method [[7], p. 444]. For an oversampling factor of $Q_0 = 8$, letting $T_Q = T/Q_0$, we consider four filters that are obtained by windowing the pulse g to intervals $(0, T)$, $(-T/2, 3T/2)$, $(-T, 2T)$, and $(-3T/2, 5T/2)$, respectively; the coefficients of the last filter are listed in Table 16.6, using the fact that $g(nT_Q)$ has even symmetry with respect to the peak at $4T_Q = T/2$.

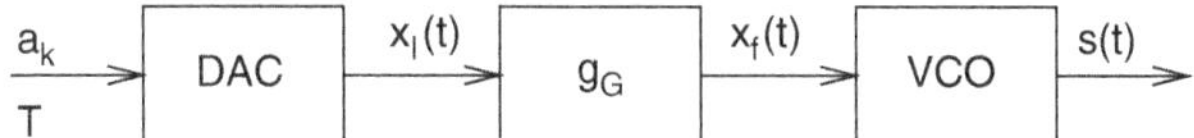

Figure 16.39 GMSK modulator: configuration I.

Figure 16.40 GMSK modulator: configuration II.

Table 16.6: Digital filter coefficients obtained by windowing g; $T_Q = T/8$

$g(nT_Q)$	*value*
$g(4T_Q)$	0.371 19
$g(5T_Q)$	0.361 77
$g(6T_Q)$	0.334 78
$g(7T_Q)$	0.293 81
$g(8T_Q)$	0.244 11
$g(9T_Q)$	0.191 58
$g(10T_Q)$	0.141 68
$g(11T_Q)$	0.098 50
$g(12T_Q)$	0.064 23
$g(13T_Q)$	0.039 21
$g(14T_Q)$	0.022 36
$g(15T_Q)$	0.011 89
$g(16T_Q)$	0.005 89
$g(17T_Q)$	0.002 71
$g(18T_Q)$	0.001 16
$g(19T_Q)$	0.000 46

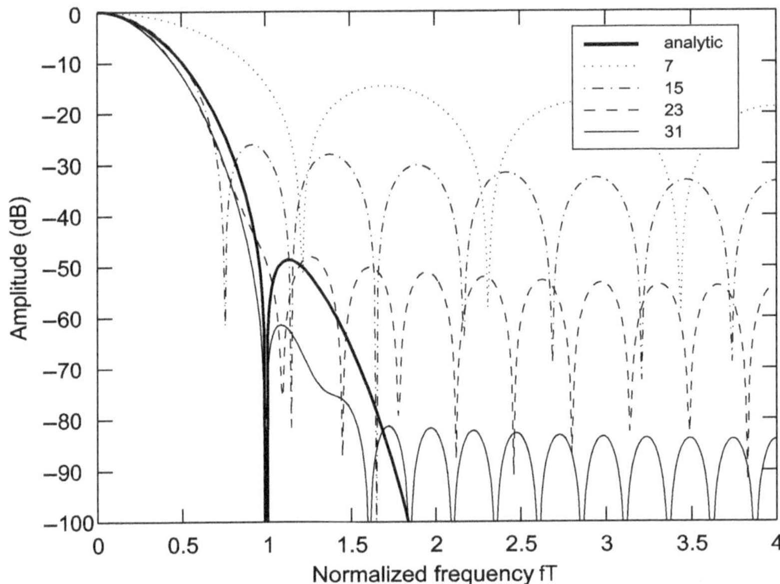

Figure 16.41 Frequency responses of g and $g(nT_Q)$, for $T_Q = T/8$, and various lengths of the FIR filters.

A comparison among the frequency responses of the four discrete-time filters and the continuous-time filter is illustrated in Figure 16.41. For a good approximation to the analog filter, the possible choices are limited to the two finite-impulse-response (FIR) filters with 23 and 31 coefficients, with support $(-T, 2T)$ and $(-3T/2, 5T/2)$, respectively. From now on, we will refer to the filter with 31 coefficients.

We note from Figure 16.35 that, for $B_tT \geq 0.3$, most of the pulse energy is contained within the interval $(-T, 2T)$; therefore, the effect of interference does not extend over more than three symbol periods. With reference to Figure 16.40, the filter g is an interpolator filter from T to T/Q_0, that can be efficiently implemented by using the polyphase representation of the impulse response (see Appendix 1.A). Then, recognizing that $x_{kQ_0+\ell} = f^{(\ell)}(a_{k+2}, a_{k+1}, a_k, a_{k-1})$, the input–output relation can be memorized in a RAM or look up table, and let the input vector $[a_{k+2}, a_{k+1}, a_k, a_{k-1}]$ address one of the possible $2^4 = 16$ output values; this must be repeated for every phase $\ell \in \{0, 1, \dots, Q_0 - 1\}$. Therefore, there are Q_0 RAMs, each with 16 memory locations.

Configuration III

The weak point of the previous scheme is again represented by the analog VCO; thus, it is convenient to partially implement in the digital domain also the frequency modulation stage.

Real-valued scheme The real-valued scheme is illustrated in Figure 16.42. The samples $\{u_n\}$ are given by

$$u_n = u(nT_Q) = A\cos\left(2\pi f_1 nT_Q + \Delta\varphi(nT_Q)\right) \tag{16.126}$$

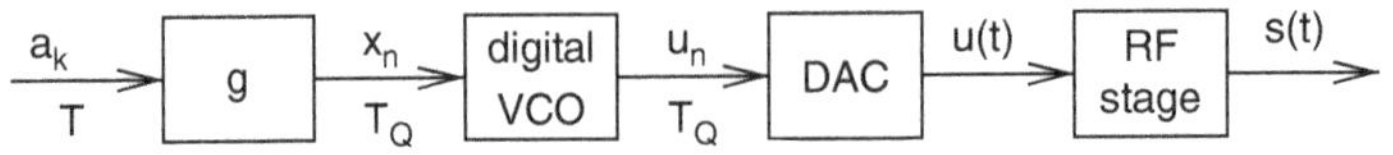

a_k (T) → g → x_n (T_Q) → digital VCO → u_n (T_Q) → DAC → u(t) → RF stage → s(t)

Figure 16.42 GMSK modulator: configuration III.

where f_1 is an intermediate frequency smaller than the sampling rate,

$$f_1 = \frac{N_1}{N_2}\frac{1}{T_Q} \tag{16.127}$$

where N_1 and N_2 are relatively prime numbers, with $N_1 < N_2$. Thus $f_1 T_Q = N_1/N_2$, and

$$u_n = A\cos\left(2\pi\frac{N_1}{N_2}n + \Delta\varphi_n\right) = A\cos\varphi_n \tag{16.128}$$

where

$$\Delta\varphi_n = \Delta\varphi(nT_Q) = \pi\int_{-\infty}^{nT_Q} x(\tau)\,d\tau \simeq \pi\sum_{i=-\infty}^{n} T_Q x(iT_Q) \tag{16.129}$$

More simply, let $x_n = x_f(nT_Q)$ and $X_n = X_{n-1} + x_n$; then it follows that $\Delta\varphi_n = \pi T_Q X_n$. Therefore in (16.128), φ_n, with carrier frequency f_1, becomes

$$\varphi_n^{(f_1)} = 2\pi\frac{N_1}{N_2}n + \pi T_Q X_n = \varphi_{n-1}^{(f_1)} + 2\pi\frac{N_1}{N_2} + \pi T_Q x_n \tag{16.130}$$

that is the value $\varphi_n^{(f_1)}$ is obtained by suitably scaling the accumulated values of x_n. To obtain u_n, we map the value of $\varphi_n^{(f_1)}$ into the memory address of a RAM, which contains values of the cosine function (see Figure 16.43).

Obviously, the size of the RAM depends on the accuracy with which u_n and φ_n are quantized.[8] We note that $u(nT_Q)$ is a real-valued passband signal, with PSD centred around the frequency f_1; the choice of f_1 is constrained by the bandwidth of the signal $u(nT_Q)$, equal to about $1.5/T$, and also by the sampling period, chosen in this example equal to $T/8$; then, it must be

$$-\frac{3}{4}\frac{1}{T} + f_1 > 0, \qquad \frac{3}{4}\frac{1}{T} + f_1 < \frac{4}{T} \tag{16.131}$$

or $3/(4T) < f_1 < 13/(4T)$. A possible choice is $f_1 = 1/(4T_Q)$ assuming $N_1 = 1$ and $N_2 = 4$. With this choice, we have a *image spacing/signal bandwidth* ratio equal to 4/3. Moreover, $\cos\varphi_n = \cos(2(\pi/4)n + \Delta\varphi_n)$ becomes $\cos((\pi/2)n + \Delta\varphi_n)$, which in turn is equal to $\pm\cos(\Delta\varphi_n)$ for n even and $\pm\sin(\Delta\varphi_n)$ for n odd. Therefore, the scheme of Figure 16.42 can be further simplified.

Complex-valued scheme Instead of digitally shifting the signal to an intermediate frequency, it is possible to process it at baseband, thus simplifying the implementation of the DAC equalizer filter.

Consider the scheme of Figure 16.44, where at the output of the exponential block we have the sampled signal $s^{(bb)}(nT_Q) = Ae^{j\Delta\varphi(nT_Q)} = s_I^{(bb)}(nT_Q) + js_Q^{(bb)}(nT_Q)$, where $s_I^{(bb)}$ and $s_Q^{(bb)}$ are the in phase and quadrature components. Then, we get

$$\begin{aligned} s_I^{(bb)}(nT_Q) &= A\cos(\Delta\varphi_n) \\ s_Q^{(bb)}(nT_Q) &= A\sin(\Delta\varphi_n) = A\cos\left(\tfrac{\pi}{2} - \Delta\varphi_n\right) \end{aligned} \tag{16.132}$$

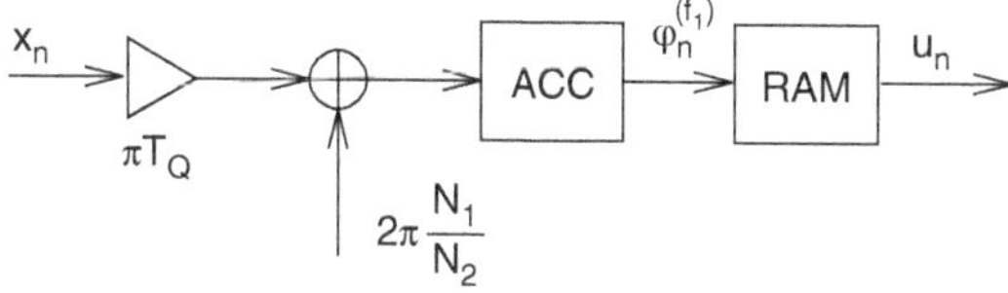

Figure 16.43 Digital implementation of the VCO.

[8] To avoid quantization effects, the number of bits used to represent the accumulated values is usually much larger than the number of bits used to represent φ_n. In practice, φ_n coincides with the most significant bits of the accumulated values.

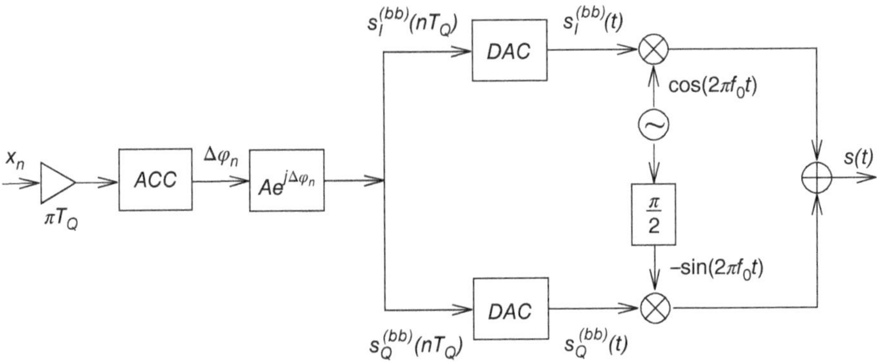

Figure 16.44 GMSK modulator with a complex-valued digital VCO.

Once the two components have been interpolated by the two DACs, the signal s can be reconstructed as

$$s(t) = s_I^{(bb)}(t)\cos(2\pi f_0 t) - s_Q^{(bb)}(t)\sin(2\pi f_0 t) \tag{16.133}$$

With respect to the real-valued scheme, we still have a RAM which stores the values of the cosine function, but two DACs are now required.

16.5.2 Linear approximation of a GMSK signal

According to Laurent [8], if $B_t T \geq 0.3$, a GMSK signal can be approximated by a QAM signal given by

$$\begin{aligned} s^{(bb)}(t) &= \sum_{k=-\infty}^{+\infty} e^{j\frac{\pi}{2}\sum_{i=-\infty}^{k} a_i} h_{Tx}(t - kT_b) \\ &= \sum_{k=-\infty}^{+\infty} j^{k+1}\, c_k\, h_{Tx}(t - kT_b), \qquad c_k = c_{k-1} a_k \end{aligned} \tag{16.134}$$

where h_{Tx} is a suitable real-valued pulse that depends on the parameter $B_t T$ and has a support equal to $(L+1)T_b$, if LT_b is the support of g. For example, we show in Figure 16.45 the plot of h_{Tx} for a GMSK signal with $B_t T = 0.3$.

The linearization of $s^{(bb)}$, that leads to interpreting GMSK as a QAM extension of MSK with a different transmit pulse, is very useful for the design of the optimum receiver, which is the same as for QAM systems. Figure 16.46 illustrates the linear approximation of the GMSK model.

As for MSK, also for GMSK it is useful to differentially (pre)code the data $\{a_k\}$ if a coherent demodulator is employed.

Performance of GMSK

We consider the performance of a GMSK system for an ideal AWGN channel, and compare it with the that of $\pi/4$-DQPSK, where the transmitted symbol is $a_k = e^{j\psi_k}$, with

$$\psi_k = \psi_{k-1} + \theta_k \tag{16.135}$$

where now $\theta_k \in \{\pi/4, 3\pi/4, 5\pi/4, 7\pi/4\}$. Note that for k even, ψ_k assumes values in $\{0, \pi/2, \pi, 3\pi/2\}$, and for k odd $\psi_k \in \{\pi/4, 3\pi/4, 5\pi/4, 7\pi/4\}$. Phase variations between two consecutive instants are $\pm\pi/4$ and $\pm 3\pi/4$, with a good peak/average power ratio of the modulated signal.

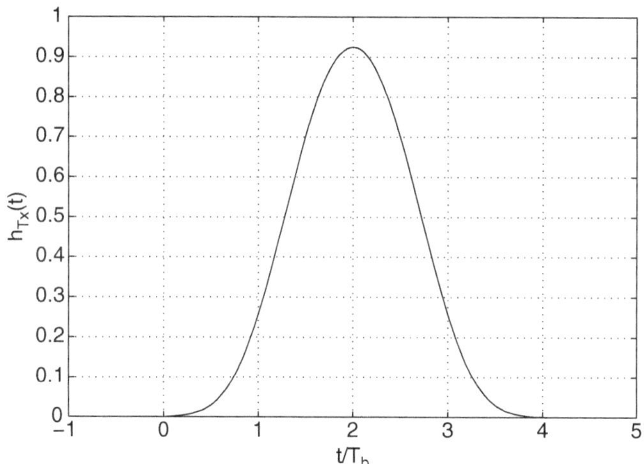

Figure 16.45 Pulse h_{Tx} for a GMSK signal with $B_t T = 0.3$.

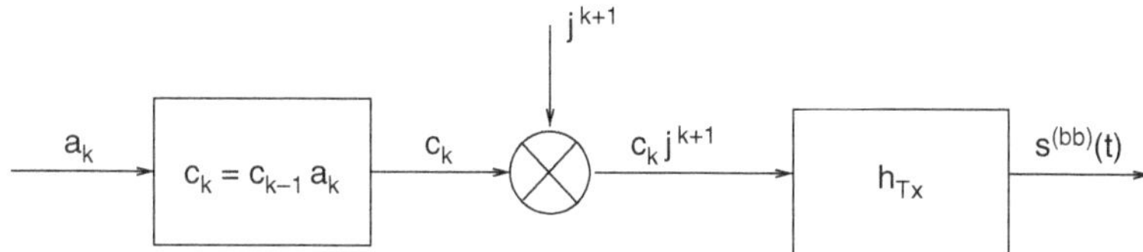

Figure 16.46 Linear approximation of a GMSK signal. h_{Tx} is a suitable pulse which depends on the value of $B_t T$.

Coherent demodulator Assuming a coherent receiver, as illustrated in Figure 16.31, performance of the optimum receiver according to the maximum-*a posteriori* (MAP) criterion, evaluated on the basis of the minimum distance of the received signals, is approximated by the following relation [9]

$$P_{bit,Ch} = Q(\sqrt{c\Gamma}) \tag{16.136}$$

where the coefficient c assumes the values given in Table 16.7 for four modulation systems. The plots of $P_{bit,Ch}$ for the various cases are illustrated in Figure 16.48.

As usual, if the data $\{a_k\}$ are differentially (pre)coded, $P_{bit} = P_{bit,Ch}$ holds, otherwise the relation (16.112) holds. From now on we assume that a differential precoder is employed in the presence of coherent demodulation.

Table 16.7: Values of coefficient c as a function of the modulation system

Modulation system	c
MSK	2.0
GMSK,$B_t T = 0.5$	1.93
GMSK,$B_t T = 0.3$	1.78
$\pi/4$-DQPSK, $\forall\rho$	1.0

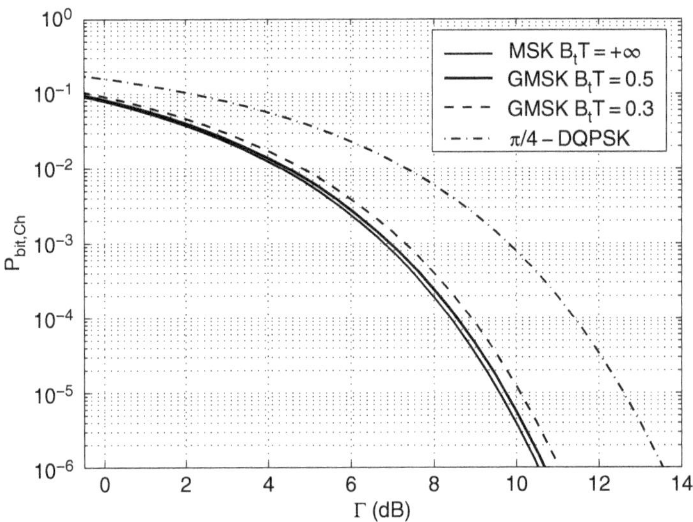

Figure 16.47 $P_{bit,Ch}$ as a function of Γ for the four modulation systems of Table 16.7.

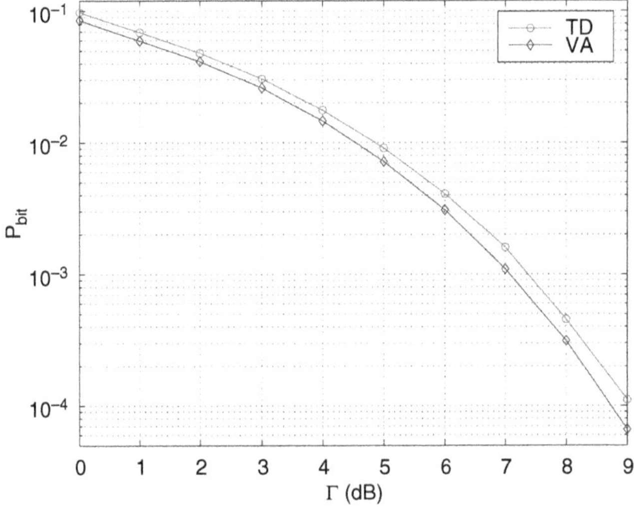

Figure 16.48 P_{bit} as a function of Γ for a coherently demodulated GMSK ($B_tT = 0.3$), for an ideal channel and Gaussian receive filter with $B_rT = 0.3$. Two detectors are compared: (1) four-state VA and (2) threshold detector.

For the case $B_tT = 0.3$, and for an ideal AWGN channel, in Figure 16.48 we also give the performance obtained for a receive filter g_{Rc} of the Gaussian type [10, 11], whose impulse response is given in (16.117), where the 3 dB bandwidth is now denoted by B_r. Clearly, the optimum value of B_rT depends on the modulator type and in particular on B_tT. System performance is evaluated using a 4-state Viterbi algorithm (VA) or a threshold detector (TD). The VA uses an estimated overall system impulse response obtained by the linear approximation of GMSK. The Gaussian receive filter is characterized by $B_rT = 0.3$, chosen for best performance. We observe that the VA gains a fraction of dB as compared to the TD; furthermore, the performance is slightly better than the approximation (16.136).

Non-coherent demodulator (1BDD) For a non-coherent receiver, as illustrated in Figure 16.30, without including the receive filter g_A, we illustrate in Figure 16.49 the eye diagram at the decision point of a GMSK system, for three values of B_tT.

We note that, for decreasing values of B_tT, the system exhibits an increasing level of ISI, and the eye tends to shut; including the receive filter g_A with finite bandwidth, this phenomenon is further emphasized.

Simulation results obtained by considering a receive Gaussian filter g_A, and a Gaussian baseband equivalent system, whose bandwidth is optimized for each different modulation, are shown in Figure 16.50.

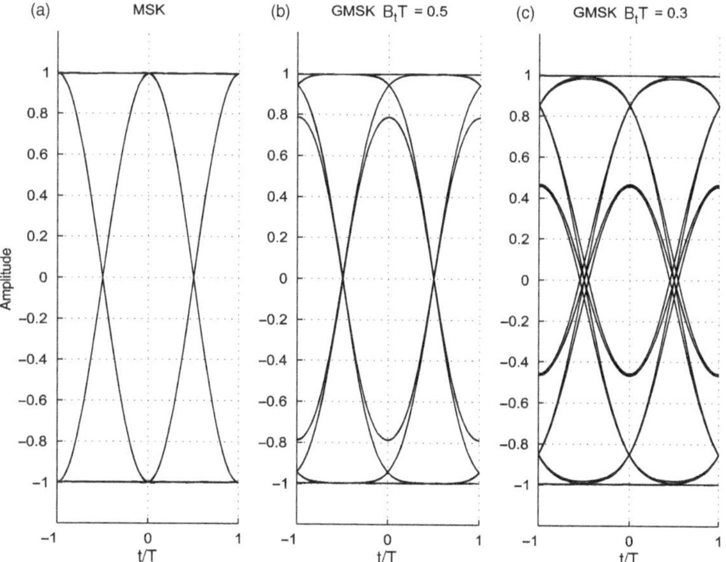

Figure 16.49 Eye diagram at the decision point of the 1BDD for a GMSK system, for an ideal channel and without the filter g_A: (a) $B_tT = +\infty$, (b) $B_tT = 0.5$, (c) $B_tT = 0.3$.

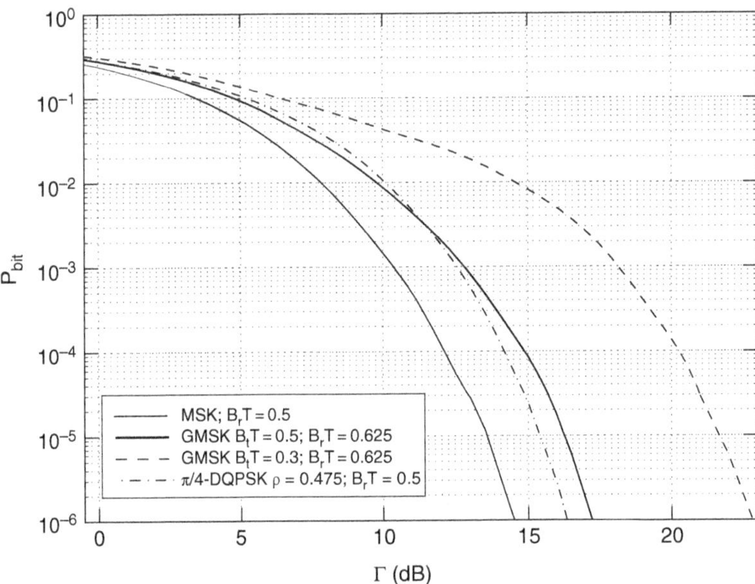

Figure 16.50 P_{bit} as a function of Γ obtained with the 1BDD for GMSK, for an ideal channel and Gaussian receive filter having a normalized bandwidth B_rT.

The sensitivity of the 1BDD to the parameter B_tT of GMSK is higher than that of a coherent demodulator. With decreasing B_tT, we observe a considerable worsening of performance; in fact, to achieve a $P_{bit} = 10^{-3}$ the GMSK with $B_tT = 0.5$ requires a signal-to-noise ratio Γ that is 2.5 dB higher with respect to MSK, whereas GMSK with $B_tT = 0.3$ requires an increment of 7.8 dB. In Figure 16.50, we also show for comparison purposes the performance of $\pi/4$-DQPSK with a receive Gaussian filter.

The performance of another widely used demodulator, LDI (see Section 16.2.3),is quite similar to that of 1BDD, showing a substantial equivalence between the two non-coherent demodulation techniques applied to GMSK and $\pi/4$-DQPSK.

Comparison Always for an ideal AWGN channel, a comparison among the various modulators and demodulators is given in Table 16.8. As a first observation, we note that a coherent receiver with an optimized Gaussian receive filter provides the same performance, evaluated by the approximate relation (16.136), of the MAP criterion. Furthermore, we note that for GMSK with $B_tT \leq 0.5$, because of strong ISI, a differential receiver undergoes a substantial penalty in terms of Γ to achieve a given P_{bit} with respect to a coherent receiver; this effect is mitigated by cancelling ISI by suitable equalizers [12].

Another method is to detect the signal in the presence of ISI, in part due to the channel and in part to the differential receiver, by the VA. Substantial improvements with respect to the simple threshold detector are obtained, as shown in [13].

In the previous comparison between amplitude modulation ($\pi/4$-DQPSK) and phase modulation (GMSK) schemes, we did not take into account the non-linearity introduced by the power amplifier (see Chapter 4), which leads to (1) signal distortion and (2) spectral spreading that creates interference in adjacent channels. Usually, the latter effect is dominant and is controlled using a HPA with a back-off that can be even of several dB. In some cases, signal predistortion before the HPA allows a decrease of the output back-off (OBO).

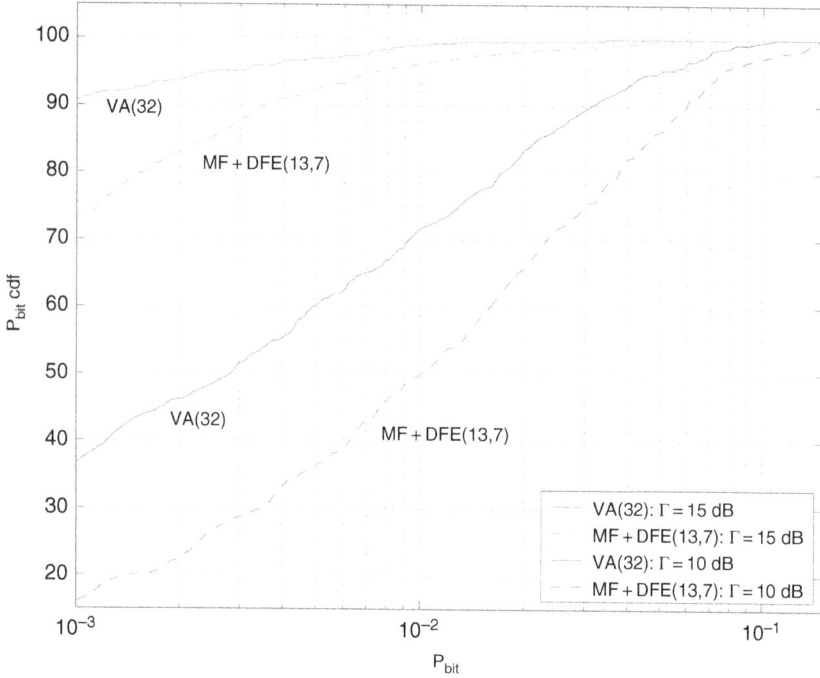

Figure 16.51 Comparison between Viterbi algorithm and DFE preceded by a MF for the multipath channel EQ6, in terms of bit error rate (BER) cdf.

Table 16.8: Required values of Γ, in dB, to achieve a $P_{bit} = 10^{-3}$ for various modulation and demodulation schemes

Modulation	*Demodulation*		
	coherent (MAP)	coherent (g_A Gauss. + TD)	differential (g_A Gauss. + 1BDD)
$\pi/4$-DQPSK or QPSK ($\rho = 0.3$)	9.8	9.8 ($\rho = 0.3$)	12.5 ($B_rT = 0.5$)
MSK	6.8	6.8 ($B_rT = 0.25$)	10.3 ($B_rT = 0.5$)
GMSK ($B_tT = 0.5$)	6.9	6.9 ($B_rT = 0.3$)	12.8 ($B_rT = 0.625$)
GMSK ($B_tT = 0.3$)	7.3	7.3 ($B_rT = 0.3$)	18.1 ($B_rT = 0.625$)

Overall, the best system is the one that achieves, for the same P_{bit}, the smaller value of

$$(\Gamma)_{dB} + (\text{OBO})_{dB} \tag{16.137}$$

In other words (16.137), for the same P_{bit}, additive channel noise and transmit HPA, selects the system for which the transmitted signal has the lowest power. Obviously in (16.137), Γ depends on the OBO; at high frequencies, where the HPA usually introduces large levels of distortion, the OBO for a linear modulation scheme may be so large that a phase modulation scheme may be the best solution in terms of (16.137).

Performance in the presence of multipath

We conclude this section giving the performance in terms of P_{bit} cumulative distribution function (cdf) (see Appendix 7.A) of a GMSK scheme with $B_tT = 0.3$ for transmission over channels in the

Table 16.9: Power delay profiles for the analysed channels

Coefficient	*Relative delay*	*Power delay profile*			
		EQ	HT	UA	RA
0	0	1/6	0.02	0.005 66	0.000 86
1	T	1/6	0.75	0.017 25	0.997 62
2	$2T$	1/6	0.08	0.802 56	0.001 09
3	$3T$	1/6	0.02	0.152 64	0.000 26
4	$4T$	1/6	0.01	0.016 68	0.000 11
5	$5T$	1/6	0.12	0.005 21	0.000 06

presence of frequency selective Rayleigh fading (see Chapter 4). The receive filter is Gaussian with $B_rT = 0.3$, implemented as a FIR filter with 24 $T/8$-spaced coefficients; the detector is a 32-state VA or a decision-feedback equalizer (DFE) with 13 coefficients of the T-spaced feedforward (FF) filter and seven coefficients of the feedback (FB) filter. For the DFE, a substantial performance improvement is observed by placing before the FF filter a matched filter (MF) that, to save computational complexity, operates at T; the improvement is mainly due to a better acquisition of the timing phase.

The channel model is also obtained by considering a T-spaced impulse response; in particular, the performance is evaluated for the four models given in Table 16.9: *equal gain* (EQ), *hilly terrain* (HT), *urban area* (UA), and *rural area* (RA). The difference in performance between the two receivers is higher for the EQ channel; this is the case shown in Figure 16.51 for two values of the signal-to-noise ratio Γ at the receiver.

Bibliography

[1] Benedetto, S. and Biglieri, E. (1999). *Principles of Digital Transmission with Wireless Applications.* New York, NY: Kluwer Academic Publishers.

[2] Proakis, J.G. (1995). *Digital Communications*, 3e. New York, NY: McGraw-Hill.

[3] Divsalar, D., Simon, M.K., and Shahshahani, M. (1990). The performance of trellis-coded MDPSK with multiple symbol detection. *IEEE Transactions on Communications* 38: 1391–1403.

[4] European DVB Project (1993). Digital broadcasting system for television, sound and data services; framing structure, channel coding and modulation for 11/12 GHz satellite services.

[5] Benvenuto, N. and Zorzi, M. (2011). *Principles of Communications Networks and Systems.* Wiley.

[6] Razavi, B. (1997). *RF Microelectronics.* Englewood Cliffs, NJ: Prentice-Hall.

[7] Oppenheim, A.V. and Schafer, R.W. (1989). *Discrete-Time Signal Processing.* Englewood Cliffs, NJ: Prentice-Hall.

[8] Laurent, P. (1986). Exact and approximate construction of digital phase modulations by superposition of amplitude modulated pulses (AMP). *IEEE Transactions on Communications* 34: 150–160.

[9] Ohmori, S., Wakana, H., and Kawase, S. (1998). *Digital Communications Technologies.* Boston, MA: Artech House.

[10] Murota, K. and Hirade, K.H. (1991). GMSK modulation for digital mobile radio telephony. *IEEE Transactions on Communications* 29: 1045.

[11] Simon, M.K. and Wang, C.C. (1984). Differential detection of Gaussian MSK in a mobile radio environment. *IEEE Transactions on Vehicular Technology* 33: 311–312.

[12] Benvenuto, N., Bisaglia, P., Salloum, A., and Tomba, L. (2000). Worst case equalizer for non-coherent HIPERLAN receivers. *IEEE Transactions on Communications* 48: 28–36.

[13] Benvenuto, N., Bisaglia, P., and Jones, A.E. (1999). Complex noncoherent receivers for GMSK signals. *IEEE Journal on Selected Areas in Communications* 17: 1876–1885.

Appendix 16.A Continuous phase modulation

A signal with constant envelope can be defined by its passband version as

$$s(t) = \sqrt{\frac{2E_s}{T}} \cos\left(2\pi f_0 t + \Delta\varphi(t, \boldsymbol{a})\right) \tag{16.138}$$

where E_s is the energy per symbol, T the symbol period, f_0 the carrier frequency, and $\boldsymbol{a}$ denotes the symbol message $\{a_k\}$ at the modulator input. For a continuous phase modulation (CPM) scheme, the phase deviation $\Delta\varphi(t, \boldsymbol{a})$ can be expressed as

$$\Delta\varphi(t, \boldsymbol{a}) = 2\pi h \int_{-\infty}^{t} x_f(\tau)\, d\tau \tag{16.139}$$

with

$$x_f(t) = \sum_{k=-\infty}^{+\infty} a_k g(t - kT) \tag{16.140}$$

where g is called *instantaneous frequency pulse*. In general, the pulse g satisfies the following properties:

$$\text{limited duration} \quad g(t) = 0 \qquad \text{for } t < 0 \text{ and } t > LT \tag{16.141}$$

$$\text{symmetry} \quad g(t) = g(LT - t) \tag{16.142}$$

$$\text{normalization} \quad \int_{0}^{LT} g(\tau)\, d\tau = \frac{1}{2} \tag{16.143}$$

Alternative definition of CPM

From (16.140) and (16.139), we can redefine the phase deviation as follows

$$\begin{aligned} \Delta\varphi(t, \boldsymbol{a}) &= 2\pi h \int_{-\infty}^{t} \sum_{i=-\infty}^{k} a_i g(\tau - iT)\, d\tau = 2\pi h \sum_{i=-\infty}^{k} a_i \int_{-\infty}^{t-iT} g(\tau)\, d\tau, \\ &\qquad kT \le t < (k+1)T \end{aligned} \tag{16.144}$$

or

$$\Delta\varphi(t, \boldsymbol{a}) = 2\pi h \sum_{i=-\infty}^{k} q(t - iT), \qquad kT \le t < (k+1)T \tag{16.145}$$

with

$$q(t) = \int_{-\infty}^{t} g(\tau)\, d\tau \tag{16.146}$$

The pulse q is called *phase response pulse* and represents the most important part of the CPM signal because it indicates to what extent each information symbol contributes to the overall phase deviation. In Figure 16.52, the phase response pulse q is plotted for PSK, CPFSK, and BFSK. In general, the maximum value of the slope of the pulse q is related to the width of the main lobe of the PSD of the modulated signal s, and the number of continuous derivatives of q influences the shape of the secondary lobes.

In general, information symbols $\{a_k\}$ belong to an M-ary alphabet $\mathcal{A}$, that for M even is given by $\{\pm 1, \pm 3, \dots, \pm(M-1)\}$. The constant h is called modulation index and determines, together with the dimension of the alphabet, the maximum phase variation in a symbol period, equal to $(M-1)h\pi$. By changing q (or g), h, and M, we can generate several continuous phase modulation schemes. The modulation index h is always given by the ratio of two integers, $h = \ell/p$, because this implies that the phase deviation, evaluated modulo 2π, assumes values in a finite alphabet. In fact, we can write

$$\Delta\varphi(t, \boldsymbol{a}) = 2\pi h \sum_{i=k-L+1}^{k} a_i q(t - iT) + \psi_{k-L}, \qquad kT \le t < (k+1)T \tag{16.147}$$

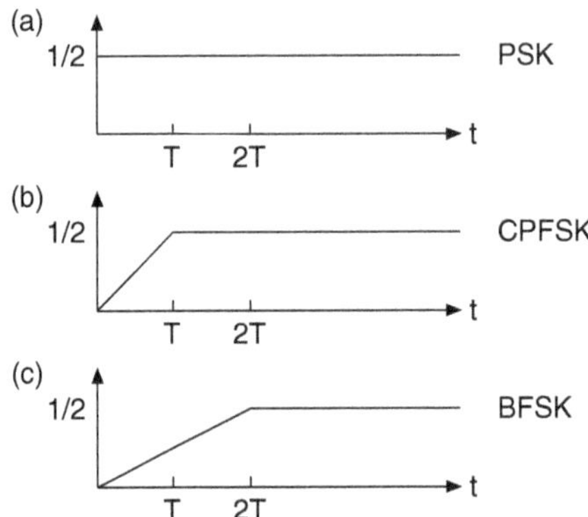

Figure 16.52 Three examples of phase response pulses for CPM: (a) PSK, (b) CPFSK, (c) BFSK.

with *phase state*

$$\psi_{k-L} = \left[\pi \frac{\ell}{p} \sum_{i=-\infty}^{k-L} a_i\right]_{mod\ 2\pi} \tag{16.148}$$

The phase state represents the overall contribution given by the symbols $\ldots, a_{-1}, a_0, a_1, \ldots, a_{k-L}$ to the phase duration in the interval $[kT, (k+1)T)$, and can only assume $2p$ distinct values. The first term in (16.145) is called *corrective state* and, because it depends on L symbols $a_{k-L+1}, \ldots, a_k$, at a certain instant t it can only assume M^{L-1} distinct values. The phase deviation is therefore characterized by a total number of values equal to $2pM^{L-1}$.

CPM schemes with $L = 1$, i.e. CPFSK, are called *full response schemes* and have a reduced complexity with a number of states equal to $2p$; schemes with $L \neq 1$ are instead called *partial response schemes*. Because of the memory in the modulation process a partial response scheme allows a trade-off between the error probability at the receiver and the shaping of the modulated-signal PSD. However, this advantage is obtained at the expense of a greater complexity of the receiver. For this reason, the modulation index is usually a simple rational number as $1, 1/2, 1/4$, and $1/8$.

Advantages of CPM

The advantage of the continuous phase modulation technique derives first of all from the constant envelope property of the CPM signal; in fact a signal with a constant envelope allows very efficient power amplifiers. In the case of linear modulation techniques, as QAM or orthogonal frequency division multiplexing (OFDM), it is necessary to compensate for the non-linearity of the amplifier by predistortion or to decrease the average power in order to work in linear conditions.

Before the introduction of trellis code modulation (TCM), it was believed that source and/or channel coding would improve performance only at the expense of a loss in transmission efficiency, and hence would require a larger bandwidth; CPM provides both good performance and highly efficient transmission. However, one of the drawbacks of the CPM is the implementation complexity of the optimum receiver, especially in the presence of dispersive channels.

Chapter 17

Applications of interference cancellation

The algorithms and structures discussed in this chapter can be applied to both wired and wireless systems, even though transmission systems over twisted-pair cables will be considered to describe examples of applications. Full-duplex data transmission over a single twisted-pair cable permits the simultaneous flow of information in two directions using the same frequency band. Examples of applications of this technique are found in digital communications systems that operate over the telephone network. In a digital subscriber loop, at each end of the full-duplex link, a circuit called hybrid separates the two directions of transmission. To avoid signal reflections at the near and far-end hybrid, a precise knowledge of the line impedance would be required. As the line impedance depends on line parameters that, in general, are not exactly known, an attenuated and distorted replica of the transmit signal leaks to the receiver input as an echo signal. Data-driven adaptive echo cancellation mitigates the effects of impedance mismatch.

A similar problem is caused by crosstalk in transmission systems over voice-grade unshielded twisted-pair (UTP) cables for local-area network applications, where multipair cables are used to physically separate the two directions of transmission. Crosstalk is a statistical phenomenon due to randomly varying differential capacitive and inductive coupling between adjacent two-wire transmission lines (see Ref. [1]). At the rates of several megabit-per-second that are usually considered for local-area network applications, *near-end crosstalk* (NEXT) represents the dominant disturbance; hence adaptive NEXT cancellation must be performed to ensure reliable communications.

A different problem shows up in frequency-division duplexing (FDD) transmission, where different frequency bands are used in the two directions of transmission. In such a case, *far-end crosstalk* (FEXT) interference is dominant. This situation occurs, for instance, in *very high speed digital subscriber line* (VDSL) systems, where multiple users are connected to a central station via UTPs located in the same cable binder.

In voiceband data modems, the model for the echo channel is considerably different from the echo channel model adopted for baseband transmission. The transmitted signal is a passband quadrature amplitude modulation (QAM) signal, and the far-end echo may exhibit significant carrier-phase jitter and carrier-frequency shift, which are caused by signal processing at intermediate points in the telephone network. Therefore, a digital adaptive echo canceller for voiceband modems needs to embody algorithms that account for the presence of such additional impairments.

In the first three sections of this chapter,[1] we describe the echo channel models and adaptive echo canceller structures for various digital communications systems, which are classified according to

[1] The material presented in Sections 17.1–17.3 is reproduced with permission from Cherubini [2].

Algorithms for Communications Systems and their Applications, Second Edition.
Nevio Benvenuto, Giovanni Cherubini, and Stefano Tomasin.

the employed modulation techniques. We also address the trade-offs between complexity, speed of adaptation, and accuracy of cancellation in adaptive echo cancellers. In Section 17.4, we address the problem of FEXT interference cancellation for upstream transmission in a VDSL system. The system is modelled as a *multi-input multi-output* (MIMO) system, to which detection techniques of Section 9.2.3 can be applied.

17.1 Echo and near-end crosstalk cancellation for PAM systems

The model of a full-duplex baseband pulse amplitude modulation (PAM) data transmission system employing adaptive echo cancellation is shown in Figure 17.1. To describe system operations, we consider one end of the full-duplex link. The configuration of an echo canceller for a PAM transmission system (see Section 2.5.5) is shown in Figure 17.2. The transmitted data consist of a sequence $\{a_k\}$ of i.i.d. real-valued symbols from the M-ary alphabet $\mathcal{A} = \{\pm 1, \pm 3, \dots, \pm(M-1)\}$. The sequence $\{a_k\}$ is filtered by a digital transmit filter, whose output is converted into an analog signal by a digital-to-analog converter (DAC) with frequency response $H_{D/A}(f)$, and T denotes the modulation interval. The DAC output is filtered by the analog transmit filter and is input to the channel through the hybrid (see also Appendix 7.D).

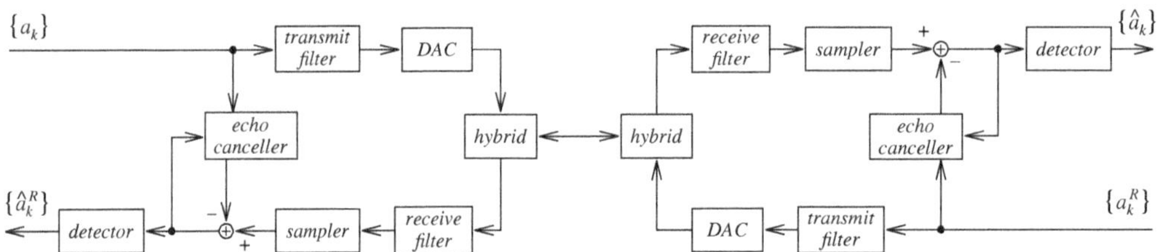

Figure 17.1 Model of a full-duplex PAM transmission system.

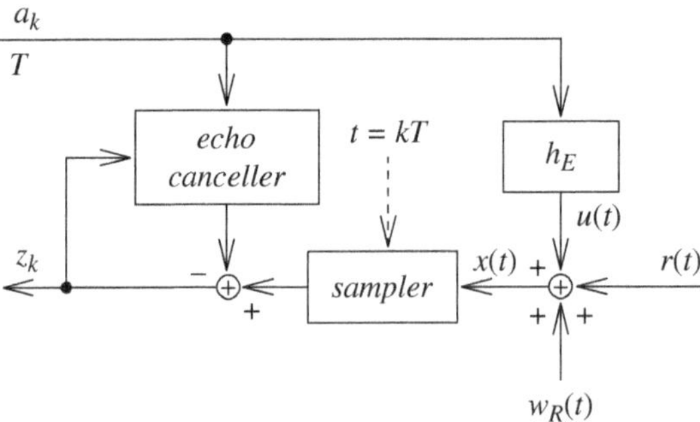

Figure 17.2 Configuration of an echo canceller for a PAM transmission system.

The signal x at the output of the low-pass analog receive filter has three components, namely, the signal from the far-end transmitter r, the echo u, and additive Gaussian noise w_R. The signal x is given by

$$x(t) = r(t) + u(t) + w_R(t) = \sum_{k=-\infty}^{+\infty} a_k^R \, h(t-kT) + \sum_{k=-\infty}^{+\infty} a_k \, h_E(t-kT) + w_R(t) \tag{17.1}$$

where $\{a_k^R\}$ is the sequence of symbols from the remote transmitter, and h and $h_E(t) = \{h_{D/A} * g_E\}(t)$ are the impulse responses of the overall channel and of the echo channel, respectively. In the expression of $h_E(t)$, the function $h_{D/A}$ is the inverse Fourier transform of $H_{D/A}(f)$. The signal obtained after echo cancellation is processed by a detector that outputs the sequence of detected symbols $\{\hat{a}_k^R\}$.

Crosstalk cancellation and full-duplex transmission

In the case of full-duplex PAM data transmission over multi-pair cables for local-area network applications, where NEXT represents the main disturbance, the configuration of a digital NEXT canceller is also obtained as shown in Figure 17.2, with the echo channel replaced by the crosstalk channel. For these applications, however, instead of *mono-duplex* transmission, where one pair is used to transmit only in one direction and the other pair to transmit only in the reverse direction, *dual-duplex* transmission may be adopted. Bi-directional transmission at rate R_b over two pairs is then accomplished by full-duplex transmission of data streams at $R_b/2$ over each of the two pairs. The lower modulation rate and/or spectral efficiency required per pair for achieving an aggregate rate equal to R_b represents an advantage of dual-duplex over mono-duplex transmission. Dual-duplex transmission requires two transmitters and two receivers at each end of a link, as well as separation of the simultaneously transmitted and received signals on each pair, as illustrated in Figure 17.3. In dual-duplex transceivers, it is therefore necessary to suppress echoes originated by reflections at the hybrids and at impedance discontinuities in the cable, as well as self NEXT, by adaptive digital echo and NEXT cancellation. Although a dual-duplex scheme might appear to require higher implementation complexity than a mono-duplex scheme, it turns out that the two schemes are equivalent in terms of the number of multiply-and-add operations per second that are needed to perform the various filtering operations.

One of the transceivers in a full-duplex link will usually employ an externally provided reference clock for its transmit and receive operations. The other transceiver will extract timing from the receive signal, and use this timing for its transmitter operations. This is known as *loop timing*, also illustrated in Figure 17.3. If signals were transmitted in opposite directions with independent clocks, signals received from the remote transmitter would generally shift in phase relative to the also received echo signals. To cope with this effect, some form of interpolation (see Chapter 14) would be required.

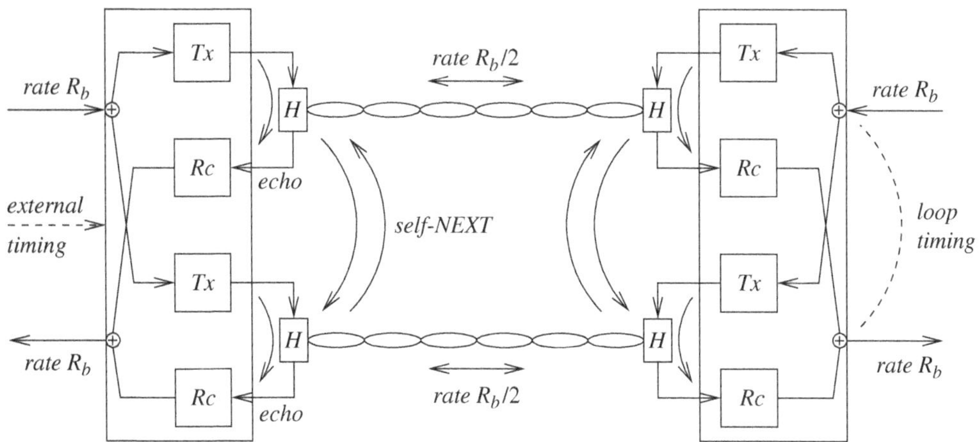

Figure 17.3 Model of a dual-duplex transmission system.

Polyphase structure of the canceller

In general, we consider baseband signalling techniques such that the signal at the output of the overall channel has non-negligible excess bandwidth, i.e. non-negligible spectral components at frequencies larger than half of the modulation rate, $|f| \geq 1/(2T)$. Therefore, to avoid aliasing, the signal x is sampled at twice the modulation rate or at a higher sampling rate. Assuming a sampling rate equal to F_0/T, $F_0 > 1$, the i-th sample during the k-th modulation interval is given by

$$\begin{aligned} x\left((kF_0+i)\frac{T}{F_0}\right) &= x_{kF_0+i} = r_{kF_0+i} + u_{kF_0+i} + w_{kF_0+i}, \qquad i = 0, \ldots, F_0-1 \\ &= \sum_{n=-\infty}^{+\infty} h_{nF_0+i}\, a^R_{k-n} + \sum_{n=-\infty}^{+\infty} h_{E,nF_0+i}\, a_{k-n} + w_{kF_0+i} \end{aligned} \tag{17.2}$$

where $\{h_{kF_0+i}, i = 0, \ldots, F_0-1\}$ and $\{h_{E,kF_0+i}, i = 0, \ldots, F_0-1\}$ are the discrete-time impulse responses of the overall channel and the echo channel, respectively, and $\{w_{kF_0+i}, i = 0, \ldots, F_0-1\}$ is a sequence of Gaussian noise samples with zero mean and variance σ_w^2. Equation (17.2) suggests that the sequence of samples $\{x_{kF_0+i},\ i = 0, \ldots, F_0-1\}$ be regarded as a set of F_0 interlaced sequences, each with a sampling rate equal to the modulation rate. Similarly, the sequence of echo samples $\{u_{kF_0+i}, i = 0, \ldots, F_0-1\}$ can be regarded as a set of F_0 interlaced sequences that are output by F_0 independent echo channels with discrete-time impulse responses $\{h_{E,kF_0+i}\}, i = 0, \ldots, F_0-1$, and an identical sequence $\{a_k\}$ of input symbols (see also Figures 7.83 and 7.63). Hence, echo cancellation can be performed by F_0 interlaced echo cancellers, as shown in Figure 17.4. As the performance of each canceller is independent of the other F_0-1 units, in the remaining part of this section we will consider the operations of a single echo canceller.

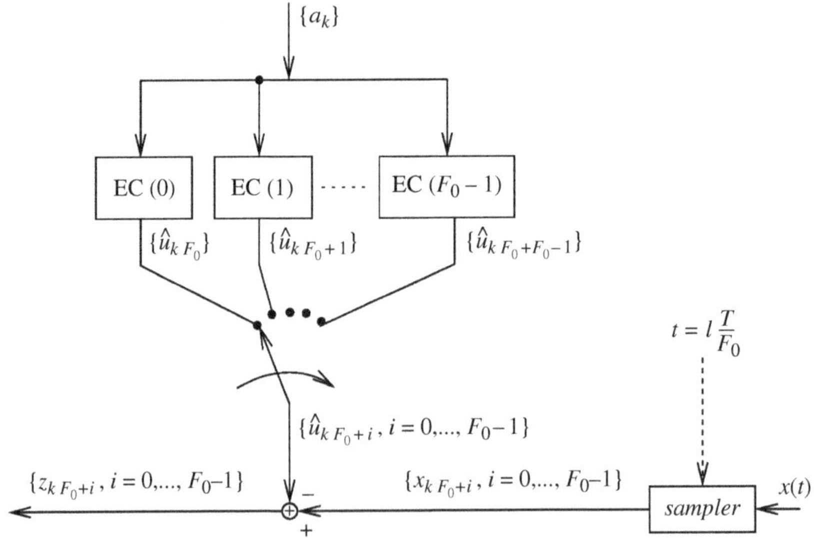

Figure 17.4 A set of F_0 interlaced echo cancellers (EC).

Canceller at symbol rate

The echo canceller generates an estimate $\hat{u}_k$ of the echo signal. If we consider a transversal filter implementation, $\hat{u}_k$ is obtained as the vector product of the vector of filter coefficients at time $t = kT$,

$\boldsymbol{c}_k = [c_{0,k}, \dots, c_{N-1,k}]^T$, and the vector of signals stored in the echo canceller delay line at the same instant, $\boldsymbol{a}_k = [a_k, \dots, a_{k-N+1}]^T$, expressed by

$$\hat{u}_k = \boldsymbol{c}_k^T \, \boldsymbol{a}_k = \sum_{n=0}^{N-1} c_{n,k} \, a_{k-n} \tag{17.3}$$

The estimate of the echo is subtracted from the received signal. The result is defined as the cancellation error signal

$$z_k = x_k - \hat{u}_k = x_k - \boldsymbol{c}_k^T \, \boldsymbol{a}_k \tag{17.4}$$

The echo attenuation that must be provided by the echo canceller to achieve proper system operation depends on the application. For example, for the integrated services digital network (ISDN) U-Interface transceiver, the echo attenuation must be larger than 55 dB [3]. It is then required that the echo signals outside of the time span of the echo canceller delay line be negligible, i.e. $h_{E,n} \approx 0$ for $n < 0$ and $n > N - 1$. As a measure of system performance, we consider the mean-square error J_k at the output of the echo canceller at time $t = kT$, defined by

$$J_k = E[z_k^2] \tag{17.5}$$

For a particular coefficient vector $\boldsymbol{c}_k$, substitution of (17.4) into (17.5) yields (see (2.17))

$$J_k = E[x_k^2] - 2\boldsymbol{c}_k^T \, \boldsymbol{p} + \boldsymbol{c}_k^T \, \boldsymbol{R}\boldsymbol{c}_k \tag{17.6}$$

where $\boldsymbol{p} = E[x_k \, \boldsymbol{a}_k]$ and $\boldsymbol{R} = E[\boldsymbol{a}_k \, \boldsymbol{a}_k^T]$. With the assumption of i.i.d. transmitted symbols, the correlation matrix $\boldsymbol{R}$ is diagonal. The elements on the diagonal are equal to the variance of the transmitted symbols, $\sigma_a^2 = (M^2 - 1)/3$. From (2.35), the minimum mean-square error is given by

$$J_{min} = E[x_k^2] - \boldsymbol{c}_{opt}^T \, \boldsymbol{R}\boldsymbol{c}_{opt} \tag{17.7}$$

where the optimum coefficient vector is $\boldsymbol{c}_{opt} = \boldsymbol{R}^{-1}\boldsymbol{p}$.

We note that proper system operation is achieved only if the transmitted symbols are uncorrelated with the symbols from the remote transmitter. If this condition is satisfied, the optimum filter coefficients are given by the values of the discrete-time echo channel impulse response, i.e. $c_{opt,n} = h_{E,n}, n = 0, \dots, N - 1$ (see Section 2.5.1).

Adaptive canceller

By the least mean square (LMS) algorithm, the coefficients of the echo canceller converge in the mean to $\boldsymbol{c}_{opt}$. The LMS algorithm (see Section 3.1.2) for an N-tap adaptive linear transversal filter is formulated as follows:

$$\boldsymbol{c}_{k+1} = \boldsymbol{c}_k + \mu \, z_k \, \boldsymbol{a}_k \tag{17.8}$$

where μ is the adaptation gain.

The block diagram of an adaptive transversal filter echo canceller is shown in Figure 17.5.

If we define the vector $\Delta\boldsymbol{c}_k = \boldsymbol{c}_k - \boldsymbol{c}_{opt}$, the mean-square error can be expressed by (2.35)

$$J_k = J_{min} + \Delta\boldsymbol{c}_k^T \, \boldsymbol{R} \, \Delta\boldsymbol{c}_k \tag{17.9}$$

where the term $\Delta\boldsymbol{c}_k^T \, \boldsymbol{R} \, \Delta \, \boldsymbol{c}_k$ represents an excess mean-square error due to the misadjustment of the filter settings. Under the assumption that the vectors $\Delta\boldsymbol{c}_k$ and $\boldsymbol{a}_k$ are statistically independent, the dynamics of the mean-square error are given by (see (3.196))

$$J_k = \sigma_0^2 \left[1 - \mu\sigma_a^2(2 - \mu N\sigma_a^2)\right]^k + \frac{2J_{min}}{2 - \mu N\sigma_a^2} \tag{17.10}$$

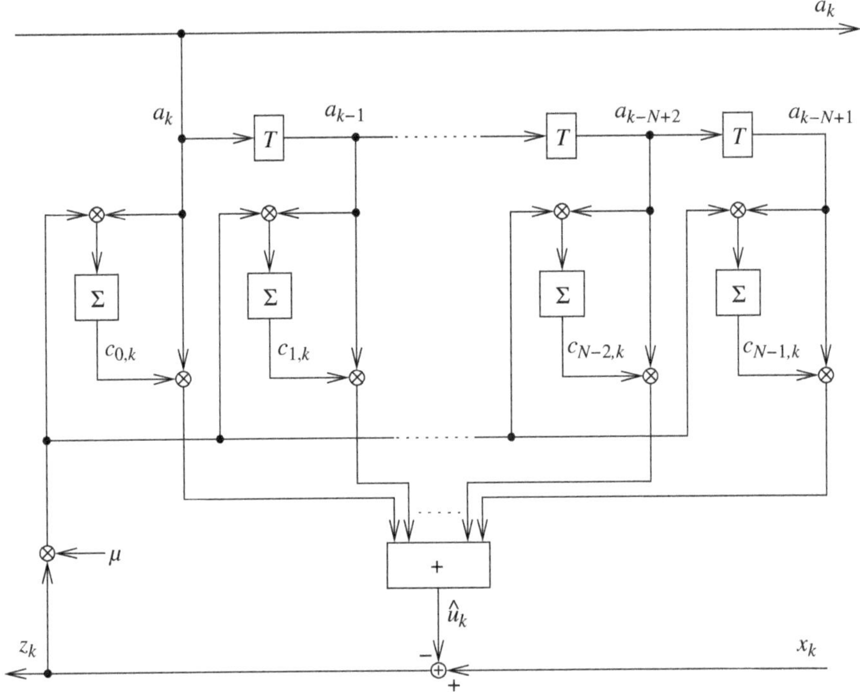

Figure 17.5 Block diagram of an adaptive transversal filter echo canceller.

where σ_0^2 is determined by the initial conditions. The mean-square error converges to a finite steady-state value J_∞ if the stability condition $0 < \mu < 2/(N\sigma_a^2)$ is satisfied. The optimum adaptation gain that yields fastest convergence at the beginning of the adaptation process is $\mu_{opt} = 1/(N\sigma_a^2)$. The corresponding time constant and asymptotic mean-square error are $\tau_{opt} = N$ and $J_\infty = 2J_{min}$, respectively.

We note that a fixed adaptation gain equal to μ_{opt} could not be adopted in practice, as after echo cancellation the signal from the remote transmitter would be embedded in a residual echo having approximately the same power. If the time constant of the convergence mode is not a critical system parameter, an adaptation gain smaller than μ_{opt} will be adopted to achieve an asymptotic mean-square error close to J_{min}. On the other hand, if fast convergence is required, a variable adaptation gain will be chosen.

Several techniques have been proposed to increase the speed of convergence of the LMS algorithm. In particular, for echo cancellation in data transmission, the speed of adaptation is reduced by the presence of the signal from the remote transmitter in the cancellation error. To mitigate this problem, the data signal can be adaptively removed from the cancellation error by a decision-directed algorithm [4].

Modified versions of the LMS algorithm have been also proposed to reduce system complexity. For example, the sign algorithm suggests that only the sign of the error signal be used to compute an approximation of the gradient [5] (see Section 3.1.4). An alternative method to reduce the implementation complexity of an adaptive echo canceller consists in the choice of a filter structure with a lower computational complexity than the transversal filter.

Canceller structure with distributed arithmetic

At high rates, very large scale integration (VLSI) technology is needed for the implementation of transceivers for full-duplex data transmission. High-speed echo cancellers and NEXT cancellers that do not require multiplications represent an attractive solution because of their low complexity. As an example of an architecture suitable for VLSI implementation, we consider echo cancellation by a distributed-arithmetic filter, where multiplications are replaced by table look-up and shift-and-add operations [6]. By segmenting the echo canceller into filter sections of shorter lengths, various trade-off concerning the number of operations per modulation interval and the number of memory locations needed to store the look-up tables are possible. Adaptivity is achieved by updating the values stored in the look-up tables by the LMS algorithm. To describe the principles of operations of a distributed-arithmetic echo canceller, we assume that the number of elements in the alphabet of the input symbols is a power of two, $M = 2^W$. Therefore, each symbol is represented by the vector $[a_k^{(0)}, \dots, a_k^{(W-1)}]$, where $a_k^{(i)} \in \{0, 1\}$, $i = 0, \dots, W-1$, are independent binary random variables, i.e.

$$a_k = \sum_{w=0}^{W-1} (2a_k^{(w)} - 1)\, 2^w = \sum_{w=0}^{W-1} b_k^{(w)}\, 2^w \tag{17.11}$$

where $b_k^{(w)} = (2a_k^{(w)} - 1) \in \{-1, +1\}$.

By substituting (17.11) into (17.3) and segmenting the delay line of the echo canceller into L sections with $K = N/L$ delay elements each, we obtain

$$\hat{u}_k = \sum_{\ell=0}^{L-1} \sum_{w=0}^{W-1} 2^w \left[\sum_{m=0}^{K-1} b_{k-\ell K-m}^{(w)}\, c_{\ell K+m,k} \right] \tag{17.12}$$

Note that the summation within parenthesis in (17.12) may assume at most 2^K distinct real values, one for each binary sequence $\{a_{k-\ell K-m}^{(w)}\}$, $m = 0, \dots, K-1$. If we pre-compute these 2^K values and store them in a look-up table addressed by the binary sequence, we can substitute the real time summation by a simple reading from the table.

Equation (17.12) suggests that the filter output can be computed using a set of $L2^K$ values that are stored in L tables with 2^K memory locations each. The binary vectors $\boldsymbol{a}_{k,\ell}^{(w)} = [a_{k-\ell K}^{(w)}, \dots, a_{k-\ell K-K+1}^{(w)}]$, $w = 0, \dots, W-1$, $\ell = 0, \dots, L-1$, determine the addresses of the memory locations where the values that are needed to compute the filter output are stored. The filter output is obtained by WL table look-up and shift-and-add operations.

We observe that $\boldsymbol{a}_{k,\ell}^{(w)}$ and its binary complement $\overline{\boldsymbol{a}}_{k,\ell}^{(w)}$ select two values that differ only in their sign. This symmetry is exploited to halve the number of values to be stored. To determine the output of a distributed-arithmetic filter with reduced memory size, we reformulate (17.12) as

$$\hat{u}_k = \sum_{\ell=0}^{L-1} \sum_{w=0}^{W-1} 2^w\, b_{k-\ell K}^{(w)} \left[c_{\ell K,k} + b_{k-\ell K}^{(w)} \sum_{m=1}^{K-1} b_{k-\ell K-m}^{(w)}\, c_{\ell K+m,k} \right] \tag{17.13}$$

Then the binary symbol $b_{k-\ell K}^{(w)}$ determines whether a selected value is to be added or subtracted. Each table has now 2^{K-1} memory locations, and the filter output is given by

$$\hat{u}_k = \sum_{\ell=0}^{L-1} \sum_{w=0}^{W-1} 2^w\, b_{k-\ell K}^{(w)}\, d_k(i_{k,\ell}^{(w)}, \ell) \tag{17.14}$$

where

$$d_k(n, \ell) = c_{\ell K,k} + b_{k-\ell K}^{(w)} \sum_{m=1}^{K-1} b_{k-\ell K-m}^{(w)}\, c_{\ell K+m,k}, \qquad n = i_{k,\ell}^{(w)}$$

and

$$i_{k,\ell}^{(w)} = \begin{cases} \sum_{m=1}^{K-1} a_{k-\ell K-m}^{(w)} \; 2^{m-1} & a_{k-\ell K}^{(w)} = 1 \\ \sum_{m=1}^{K-1} \overline{a}_{k-\ell K-m}^{(w)} \; 2^{m-1} & a_{k-\ell K}^{(w)} = 0 \end{cases} \tag{17.15}$$

Both $d_k(n, \ell)$ and $i_{k,\ell}^{(w)}$ values can be stored in look-up tables.

We note that, as long as (17.12) and (17.13) hold for some coefficient vector $[c_{0,k}, \ldots, c_{N-1,k}]$, a distributed-arithmetic filter emulates the operation of a linear transversal filter. For arbitrary values $d_k(n, \ell)$, however, a non-linear filtering operation results.

The expression of the LMS algorithm to update the values of a distributed-arithmetic echo canceller is derived as in (3.204). To simplify the notation, we set

$$\hat{u}_k(\ell) = \sum_{w=0}^{W-1} 2^w \; b_{k-\ell K}^{(w)} \; d_k(i_{k,\ell}^{(w)}, \ell) \tag{17.16}$$

hence (17.14) can be written as

$$\hat{u}_k = \sum_{\ell=0}^{L-1} \hat{u}_k(\ell) \tag{17.17}$$

We also define the vector of the values in the ℓ-th look-up table as

$$\boldsymbol{d}_k(\ell) = [d_k(0, \ell), \ldots, d_k(2^{K-1} - 1, \ell)]^T \tag{17.18}$$

indexed by the variable (17.15).

The values $\boldsymbol{d}_k(\ell)$ are updated according to the LMS criterion, i.e.

$$\boldsymbol{d}_{k+1}(\ell) = \boldsymbol{d}_k(\ell) - \frac{1}{2} \mu \, \nabla_{\boldsymbol{d}_k(\ell)} \, z_k^2 \tag{17.19}$$

where

$$\nabla_{\boldsymbol{d}_k(\ell)} \, z_k^2 = 2 z_k \, \nabla_{\boldsymbol{d}_k(\ell)} \, z_k = -2 z_k \, \nabla_{\boldsymbol{d}_k(\ell)} \, \hat{u}_k = -2 z_k \, \nabla_{\boldsymbol{d}_k(\ell)} \, \hat{u}_k(\ell) \tag{17.20}$$

The last expression has been obtained using (17.17) and the fact that only $\hat{u}_k(\ell)$ depends on $\boldsymbol{d}_k(\ell)$. Defining

$$\boldsymbol{y}_k(\ell) = [y_k(0, \ell), \ldots, y_k(2^{K-1} - 1, \ell)]^T = \nabla_{\boldsymbol{d}_k(\ell)} \, \hat{u}_k(\ell)$$

(17.19) becomes

$$\boldsymbol{d}_{k+1}(\ell) = \boldsymbol{d}_k(\ell) + \mu \, z_k \, \boldsymbol{y}_k(\ell) \tag{17.21}$$

For a given value of k and ℓ, we assign the following values to the W addresses (17.15):

$$I^{(w)} = i_{k,\ell}^{(w)}, \qquad w = 0, 1, \ldots, W-1$$

From (17.16) we get

$$y_k(n, \ell) = \sum_{w=0}^{W-1} 2^w \; b_{k-\ell K}^{(w)} \; \delta_{n-I^{(w)}} \tag{17.22}$$

In conclusion, in (17.21) for every instant k and for each value of the index $w = 0, 1, \ldots, W-1$, the product $2^w \; b_{k-\ell K}^{(w)} \; \mu z_k$ is added to the memory location indexed by $I^{(w)}$. The complexity of the implementation can be reduced by updating, at every iteration k, only the values corresponding to the addresses given by the most significant bits of the symbols in the filter delay line. In this case, (17.22) simplifies into

$$y_k(n, \ell) = \begin{cases} 2^{W-1} \; b_{k-\ell K}^{(W-1)} & n = I^{(W-1)} \\ 0 & n \neq I^{(W-1)} \end{cases} \tag{17.23}$$

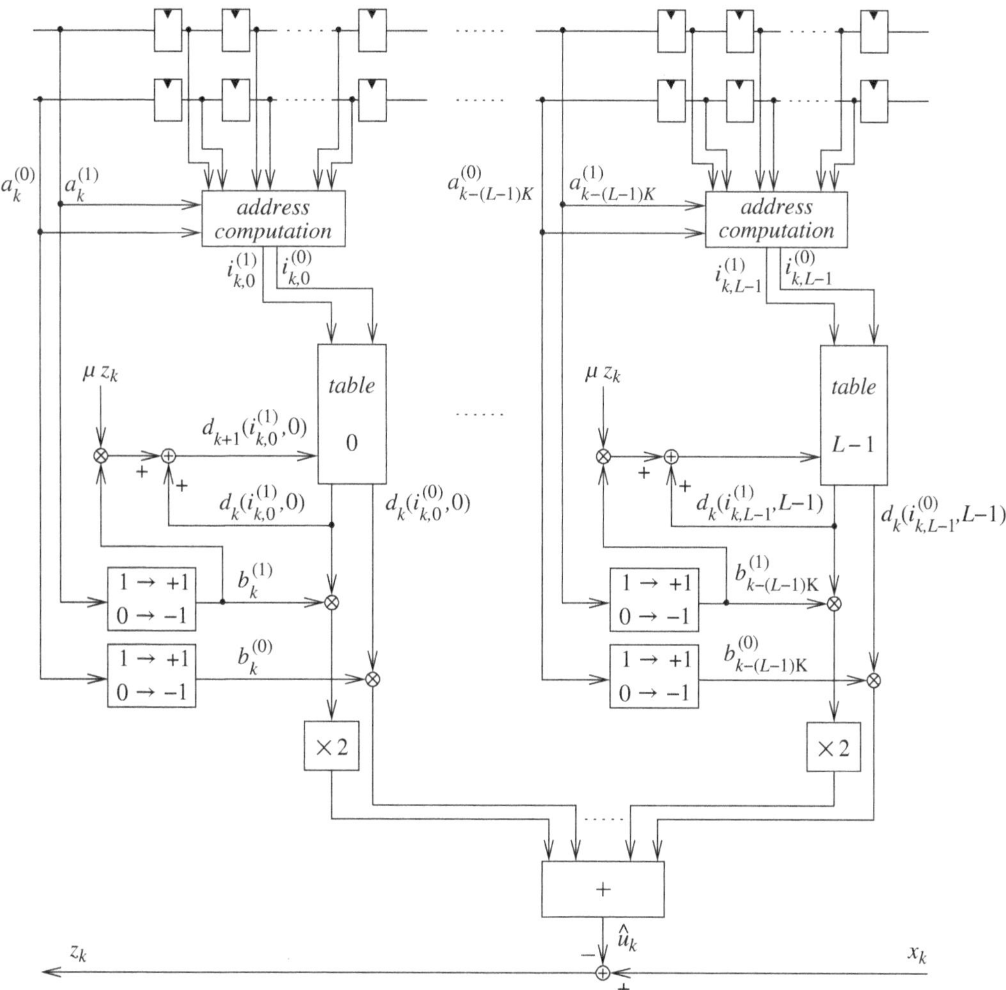

Figure 17.6 Block diagram of an adaptive distributed-arithmetic echo canceller.

The block diagram of an adaptive distributed-arithmetic echo canceller with input symbols from a quaternary alphabet is shown in Figure 17.6.

The analysis of the mean-square error convergence behaviour and steady-state performance can be extended to adaptive distributed-arithmetic echo cancellers [7]. The dynamics of the mean-square error are in this case given by

$$J_k = \sigma_0^2\left[1 - \frac{\mu\sigma_a^2}{2^{K-1}}\,(2 - \mu L\sigma_a^2)\right]^k + \frac{2J_{min}}{2 - \mu L\sigma_a^2} \tag{17.24}$$

The stability condition for the echo canceller is $0 < \mu < 2/(L\sigma_a^2)$. For a given adaptation gain, echo canceller stability depends on the number of tables and on the variance of the transmitted symbols. Therefore, the time span of the echo canceller can be increased without affecting system stability, provided that the number L of tables is kept constant. In that case, however, mean-square error convergence will be slower. From (17.24), we find that the optimum adaptation gain that provides the fastest mean-square error convergence at the beginning of the adaptation process is $\mu_{opt} = 1/(L\sigma_a^2)$. The time constant of the convergence mode is $\tau_{opt} = L2^{K-1}$. The smallest achievable time constant is therefore proportional to the total number of values. As mentioned above, the implementation of a distributed-arithmetic echo canceller can be simplified by updating at each iteration only the values that are addressed by the most significant bits of the symbols stored in the delay line. The complexity required for adaptation can thus be reduced at the price of a slower rate of convergence.

17.2 Echo cancellation for QAM systems

Although most of the concepts presented in the preceding section can be readily extended to echo cancellation for communications systems employing QAM, the case of full-duplex transmission over a voice-band data channel requires a specific discussion. We consider the system model shown in Figure 17.7. The transmitter generates a sequence $\{a_k\}$ of i.i.d. complex-valued symbols from a two-dimensional constellation $\mathcal{A}$, that are modulated by the carrier $e^{j2\pi f_0 kT}$, where T and f_0 denote the modulation interval and the carrier frequency, respectively. The discrete-time signal at the output of the transmit passband phase splitter filter may be regarded as an analytic signal, which is generated at the rate of Q_0/T samples/s, $Q_0 > 1$. The real part of the analytic signal is converted into an analog signal by a DAC and input to the channel. We note that by transmitting the real part of a complex-valued signal, positive and negative-frequency components become folded. The attenuation in the image band of the transmit filter thus determines the achievable echo suppression. In fact, the receiver cannot extract aliasing image-band components from desired passband frequency components, and the echo canceller is able to suppress only echo arising from transmitted passband components.

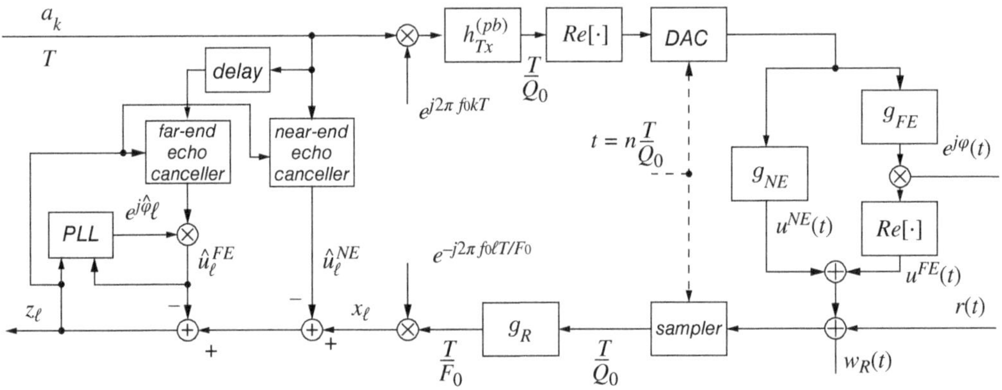

Figure 17.7 Configuration of an echo canceller for a QAM transmission system.

The output of the echo channel is represented as the sum of two contributions. The near-end echo u^{NE} arises from the impedance mismatch between the hybrid and the transmission line, as in the case of baseband transmission. The far-end echo u^{FE} represents the contribution due to echoes that are generated at intermediate points in the telephone network. These echoes are characterized by additional impairments, such as jitter and frequency shift, which are accounted for by introducing a carrier-phase rotation equal to φ in the model of the far-end echo.

At the receiver, samples of the signal at the channel output are obtained synchronously with the transmitter timing, at the sampling rate of Q_0/T samples/s. The discrete-time received signal is converted to a complex-valued baseband signal $\{x_{kF_0+i}\}$, $i = 0, \dots, F_0 - 1$, at the rate of F_0/T samples/s,

$1 < F_0 < Q_0$, through filtering by the receive phase splitter filter, decimation, and demodulation. From delayed transmit symbols, estimates of the near and far-end echo signals after demodulation, $\{\hat{u}^{NE}_{kF_0+i}\}$, $i = 0, \dots, F_0 - 1$, and $\{\hat{u}^{FE}_{kF_0+i}\}$, $i = 0, \dots, F_0 - 1$, respectively, are generated using F_0 interlaced near and far-end echo cancellers. The cancellation error is given by

$$z_\ell = x_\ell - (\hat{u}^{NE}_\ell + \hat{u}^{FE}_\ell) \tag{17.25}$$

A different model is obtained if echo cancellation is accomplished before demodulation. In this case, two equivalent configurations for the echo canceller may be considered. In one configuration, the modulated symbols are input to the transversal filter, which approximates the passband echo response. Alternatively, the modulator can be placed after the transversal filter, which is then called a baseband transversal filter [8].

In the considered implementation, the estimates of the echo signals after demodulation are given by

$$\hat{u}^{NE}_{kF_0+i} = \sum_{n=0}^{N_{NE}-1} c^{NE}_{nF_0+i,k}\, a_{k-n}, \qquad i = 0, \dots, F_0 - 1 \tag{17.26}$$

$$\hat{u}^{FE}_{kF_0+i} = \left(\sum_{n=0}^{N_{FE}-1} c^{FE}_{nF_0+i,k}\, a_{k-n-D_E} \right) e^{j\hat{\varphi}_{kF_0+1}}, \qquad i = 0, \dots, F_0 - 1 \tag{17.27}$$

where $[c^{NE}_{0,k}, \dots, c^{NE}_{F_0 N_{NE}-1,k}]$ and $[c^{FE}_{0,k}, \dots, c^{FE}_{F_0 N_{FE}-1,k}]$ are the coefficients of the F_0 interlaced near and far-end echo cancellers, respectively, $\{\hat{\varphi}_{kF_0+i}\}$, $i = 0, \dots, F_0 - 1$, is the sequence of far-end echo phase estimates, and D_{FE} denotes the bulk delay accounting for the round-trip delay from the transmitter to the point of echo generation. To prevent overlap of the time span of the near-end echo canceller with the time span of the far-end echo canceller, the condition $D_{FE} > N_{NE}$ must be satisfied. We also note that, because of the different nature of near and far-end echo generation, the time span of the far-end echo canceller needs to be larger than the time span of the near-end canceller, i.e. $N_{FE} > N_{NE}$.

Adaptation of the filter coefficients in the near and far-end echo cancellers by the LMS algorithm leads to

$$\begin{aligned} c^{NE}_{nF_0+i,k+1} &= c^{NE}_{nF_0+i,k} + \mu\, z_{kF_0+i} (a_{k-n})^*, \\ & \qquad n = 0, \dots, N_{NE} - 1, \quad i = 0, \dots, F_0 - 1 \end{aligned} \tag{17.28}$$

and

$$\begin{aligned} c^{FE}_{nF_0+i,k+1} &= c^{FE}_{nF_0+i,k} + \mu\, z_{kF_0+i} (a_{k-n-D_{FE}})^*\, e^{-j\hat{\varphi}_{kF_0+i}}, \\ & \qquad n = 0, \dots, N_{FE} - 1, \quad i = 0, \dots, F_0 - 1 \end{aligned} \tag{17.29}$$

respectively.

The far-end echo phase estimate is computed by a second-order phase-lock loop algorithm (see Section 14.7), where the following gradient approach is adopted:

$$\begin{cases} \hat{\varphi}_{\ell+1} = \hat{\varphi}_\ell - \dfrac{1}{2}\, \mu_\varphi\, \nabla_{\hat{\varphi}}\, |z_\ell|^2 + \Delta\varphi_\ell \qquad (\text{mod } 2\pi) \\ \Delta\varphi_{\ell+1} = \Delta\varphi_\ell - \dfrac{1}{2}\, \mu_f\, \nabla_{\hat{\varphi}}\, |z_\ell|^2 \end{cases} \tag{17.30}$$

where $\ell = kF_0 + i$, $i = 0, \dots, F_0 - 1$, μ_φ and μ_f are parameters of the loop, and

$$\nabla_{\hat{\varphi}}\, |z_\ell|^2 = \frac{\partial |z_\ell|^2}{\partial \hat{\varphi}_\ell} = -2\, Im\{z_\ell (\hat{u}^{FE}_\ell)^*\} \tag{17.31}$$

We note that the algorithm (17.30) requires F_0 iterations per modulation interval, i.e. we cannot resort to interlacing to reduce the complexity of the computation of the far-end echo phase estimate.

17.3 Echo cancellation for OFDM systems

We discuss echo cancellation for orthogonal frequency division multiplexing (OFDM) with reference to a discrete-multitone (DMT) system (see Chapter 8), as shown in Figure 17.8. Let $\{h_i\}$, $i = 0, \dots, N_c - 1$, be the channel impulse response, sampled with period $T/\mathcal{M}$, with length $N_c \ll \mathcal{M}$ and $\{h_{E,i}\}$, $i = 0, \dots, N-1$, be the discrete-time echo impulse response whose length is $N < \mathcal{M}$, where $\mathcal{M}$ denotes the number of subchannels of the DMT system. To simplify the notation, the length of the cyclic prefix will be set to $L = N_c - 1$. Recall that in a DMT transmitter the block of $\mathcal{M}$ samples at the inverse discrete Fourier transform (IDFT) output in the k-th modulation interval, $[A_k[0], \dots, A_k[\mathcal{M}-1]]$, is cyclically extended by copying the last L samples at the beginning of the block. After a P/S conversion, wherein the L samples of the cyclic extension are the first to be output, the $L + \mathcal{M}$ samples of the block are transmitted into the channel. At the receiver, the sequence is split into blocks of length $L + \mathcal{M}$, $[x_{k(\mathcal{M}+L)}, \dots, x_{(k+1)(\mathcal{M}+L)-1}]$. These blocks are separated in such a way that the last $\mathcal{M}$ samples depend only on a single cyclically extended block, then the first L samples are discarded.

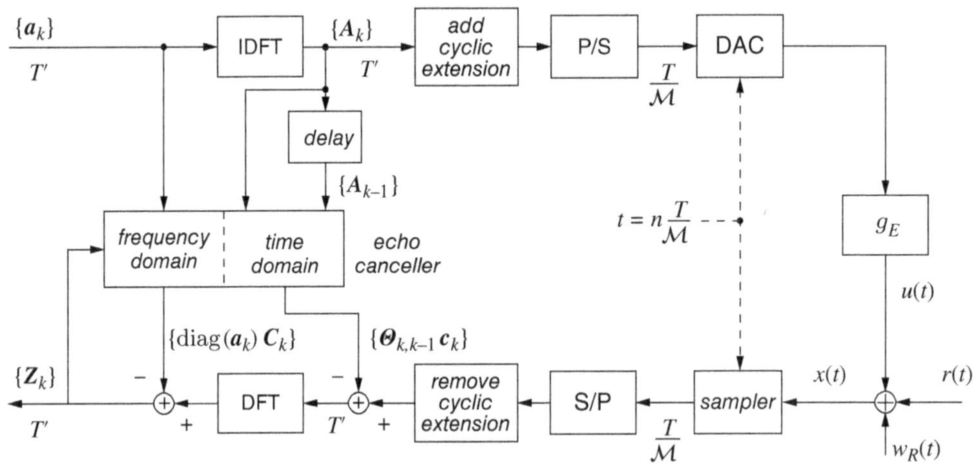

Figure 17.8 Configuration of an echo canceller for a DMT transmission system.

Case $N \leq L + 1$ We initially assume that the length of the echo channel impulse response is $N \leq L + 1$. Furthermore, we assume that the boundaries of the received blocks are placed such that the last $\mathcal{M}$ samples of the k-th received block are expressed by the vector (see (8.77))

$$\boldsymbol{x}_k = \boldsymbol{\Xi}_k^R \, \boldsymbol{h} + \boldsymbol{\Xi}_k \, \boldsymbol{h}_E + \boldsymbol{w}_k \tag{17.32}$$

where $\boldsymbol{h} = [h_0, \dots, h_L, 0, \dots, 0]^T$ is the vector of the overall channel impulse response extended with $\mathcal{M} - L - 1$ zeros, $\boldsymbol{h}_E = [h_{E,0}, \dots, h_{E,N-1}, 0, \dots, 0]^T$ is the vector of the overall echo impulse response extended with $\mathcal{M} - N$ zeros, $\boldsymbol{w}_k$ is a vector of additive white Gaussian noise (AWGN) samples, $\boldsymbol{\Xi}_k^R$ is the circulant matrix with elements given by the signals from the remote transmitter

$$\boldsymbol{\Xi}_k^R = \begin{bmatrix} A_k^R[0] & A_k^R[\mathcal{M}-1] & \dots & A_k^R[1] \\ A_k^R[1] & A_k^R[0] & \dots & A_k^R[2] \\ \vdots & \vdots & & \vdots \\ A_k^R[\mathcal{M}-1] & A_k^R[\mathcal{M}-2] & \dots & A_k^R[0] \end{bmatrix} \tag{17.33}$$

and Ξ_k is the circulant matrix with elements generated by the local transmitter

$$\Xi_k = \begin{bmatrix} A_k[0] & A_k[\mathcal{M}-1] & \dots & A_k[1] \\ A_k[1] & A_k[0] & \dots & A_k[2] \\ \vdots & \vdots & & \vdots \\ A_k[\mathcal{M}-1] & A_k[\mathcal{M}-2] & \dots & A_k[0] \end{bmatrix} \tag{17.34}$$

In the frequency domain, the echo is expressed as

$$U_k = \text{diag}(a_k)\, H_E \tag{17.35}$$

where H_E denotes the discrete Fourier transform (DFT) of the vector h_E. In this case, the echo canceller provides an echo estimate that is given by

$$\hat{U}_k = \text{diag}(a_k)\, C_k \tag{17.36}$$

where C_k denotes the DFT of the vector c_k of the N coefficients of the echo canceller filter extended with $\mathcal{M} - N$ zeros.

In the time domain, (17.36) corresponds to the estimate

$$\hat{u}_k = \Xi_k\, c_k \tag{17.37}$$

Case $N > L + 1$ In practice, however, we need to consider the case $N > L + 1$. The expression of the cancellation error is then given by

$$x_k = \Xi_k^R\, h + \Psi_{k,k-1}\, h_E + w_k \tag{17.38}$$

where $\Psi_{k,k-1}$ is a circulant matrix given by

$$\Psi_{k,k-1} = \begin{bmatrix} A_k[0] & A_k[\mathcal{M}-1] & \dots & A_k[\mathcal{M}-L] & A_{k-1}[\mathcal{M}-1] & \dots & A_{k-1}[L+1] \\ A_k[1] & A_k[0] & \dots & A_k[\mathcal{M}-L+1] & A_k[\mathcal{M}-L] & \dots & A_{k-1}[L+2] \\ \vdots & & & & & & \\ A_k[\mathcal{M}-1] & A_k[\mathcal{M}-2] & \dots & A_k[\mathcal{M}-L-1] & A_k[\mathcal{M}-L-2] & \dots & A_k[0] \end{bmatrix}. \tag{17.39}$$

From (17.38), the expression of the cancellation error in the time domain is then given by

$$z_k = x_k - \Psi_{k,k-1}\, c_k \tag{17.40}$$

We now introduce the Toeplitz triangular matrix

$$\Theta_{k,k-1} = \Psi_{k,k-1} - \Xi_k \tag{17.41}$$

Substitution of (17.41) into (17.40) yields

$$z_k = x_k - \Theta_{k,k-1}\, c_k - \Xi_k\, c_k \tag{17.42}$$

In the frequency domain, (17.42) can be expressed as

$$Z_k = F_{\mathcal{M}}(x_k - \Theta_{k,k-1}\, c_k) - \text{diag}(a_k)\, C_k \tag{17.43}$$

Equation (17.43) suggests a computationally efficient, two-part echo cancellation technique. First, in the time domain, a short convolution is performed and the result subtracted from the received signals to compensate for the insufficient length of the cyclic extension. Second, in the frequency domain, cancellation of the residual echo is performed over a set of $\mathcal{M}$ independent echo subchannels. Observing that (17.43) is equivalent to

$$Z_k = X_k - \tilde{\Psi}_{k,k-1}\, C_k \tag{17.44}$$

where $\tilde{\boldsymbol{\Psi}}_{k,k-1} = \boldsymbol{F}_{\mathcal{M}} \boldsymbol{\Psi}_{k,k-1} \boldsymbol{F}_{\mathcal{M}}^{-1}$, the echo canceller adaptation by the LMS algorithm in the frequency domain takes the form

$$\boldsymbol{C}_{k+1} = \boldsymbol{C}_k + \mu \, \tilde{\boldsymbol{\Psi}}_{k,k-1}^H \, \boldsymbol{Z}_k \tag{17.45}$$

where μ is the adaptation gain.

We note that, alternatively, echo canceller adaptation may also be performed by the simplified algorithm [9]

$$\boldsymbol{C}_{k+1} = \boldsymbol{C}_k + \mu \, \text{diag}(\boldsymbol{a}_k^*) \, \boldsymbol{Z}_k \tag{17.46}$$

which entails a substantially lower computational complexity than the LMS algorithm, at the price of a slower rate of convergence.

In DMT systems, it is essential that the length of the channel impulse response be much smaller than the number of subchannels, so that the reduction in rate due to the cyclic extension may be considered negligible. Therefore, time-domain equalization is adopted in practice to shorten the length of the channel impulse response. From (17.43), however, we observe that transceiver complexity depends on the relative lengths of the echo and of the channel impulse responses. To reduce the length of the cyclic extension as well as the computational complexity of the echo canceller, various methods have been proposed to shorten both the channel and the echo impulse responses jointly [10].

17.4 Multiuser detection for VDSL

In this section, we address the problem of multiuser detection for upstream VDSL transmission (see Chapter 18), where FEXT signals at the input of a VDSL receiver are viewed as interferers that share the same channel as the remote user signal [11].

We assume knowledge of the FEXT responses at the central office and consider a decision-feedback equalizer (DFE) structure with cross-coupled linear feedforward (FF) equalizers and feedback (FB) filters for crosstalk suppression. DFE structures with cross-coupled filters have also been considered for interference suppression in wireless code-division multiple-access (CDMA) communications [12], see also Appendix 19.A. Here we determine the optimum DFE coefficients in a minimum mean-square error (MMSE) sense assuming that each user adopts OFDM modulation for upstream transmission. A system with reduced complexity may be considered for practical applications, in which for each user and each subchannel only the most significant interferers are suppressed.

To obtain a receiver structure for multiuser detection that exhibits moderate complexity, we assume that each user adopts filtered multitone (FMT) modulation with $\mathcal{M}$ subchannels for upstream transmission (see Chapter 8). Hence, the subchannel signals exhibit non-zero excess bandwidth as well as negligible spectral overlap. Assuming upstream transmission by U users, the system illustrated in Figure 17.9 is considered. In general, the sequences of subchannel signal samples at the multicarrier demodulator output are obtained at a sampling rate equal to a rational multiple F_0 of the modulation rate $1/T$. To simplify the analysis, here an integer $F_0 \geq 2$ is assumed.

We introduce the following definitions:

1. $\{a_k^{(u)}[i]\}$, sequence of i.i.d. complex-valued symbols from a QAM constellation $\mathcal{A}^{(u)}[i]$ transmitted by user u over subchannel i, with variance $\sigma^2_{a^{(u)}[i]}$;
2. $\{\hat{a}_k^{(u)}[i]\}$, sequence of detected symbols of user u at the output of the decision element of subchannel i;
3. $h_n^{(u)}[i]$, overall impulse response of subchannel i of user u;
4. $h_{\text{FEXT},n}^{(u,v)}[i]$, overall FEXT response of subchannel i, from user v to user u;

5. $G^{(u)}[i]$, gain that determines the power of the signal of user u transmitted over subchannel i;
6. $\{\tilde{w}_n^{(u)}[i]\}$, sequence of additive Gaussian noise samples with correlation function $\mathtt{r}_{w^{(u)}[i]}(m)$.

At the output of subchannel i of the user-u demodulator, the complex baseband signal is given by

$$x_n^{(u)}[i] = G^{(u)}[i] \sum_{k=-\infty}^{\infty} h_{n-kF_0}^{(u)}[i]\, a_k^{(u)}[i] + \sum_{\substack{v=1 \\ v \neq u}}^{U} G^{(v)}[i] \sum_{k=-\infty}^{\infty} h_{\mathrm{FEXT},n-kF_0}^{(u,v)}[i] a_k^{(v)}[i] + \tilde{w}_n^{(u)}[i] \tag{17.47}$$

For user u, symbol detection at the output of subchannel i is achieved by a DFE structure such that the input to the decision element is obtained by combining the output signals of U linear filters and U FB filters from all users, as illustrated in Figure 17.9.

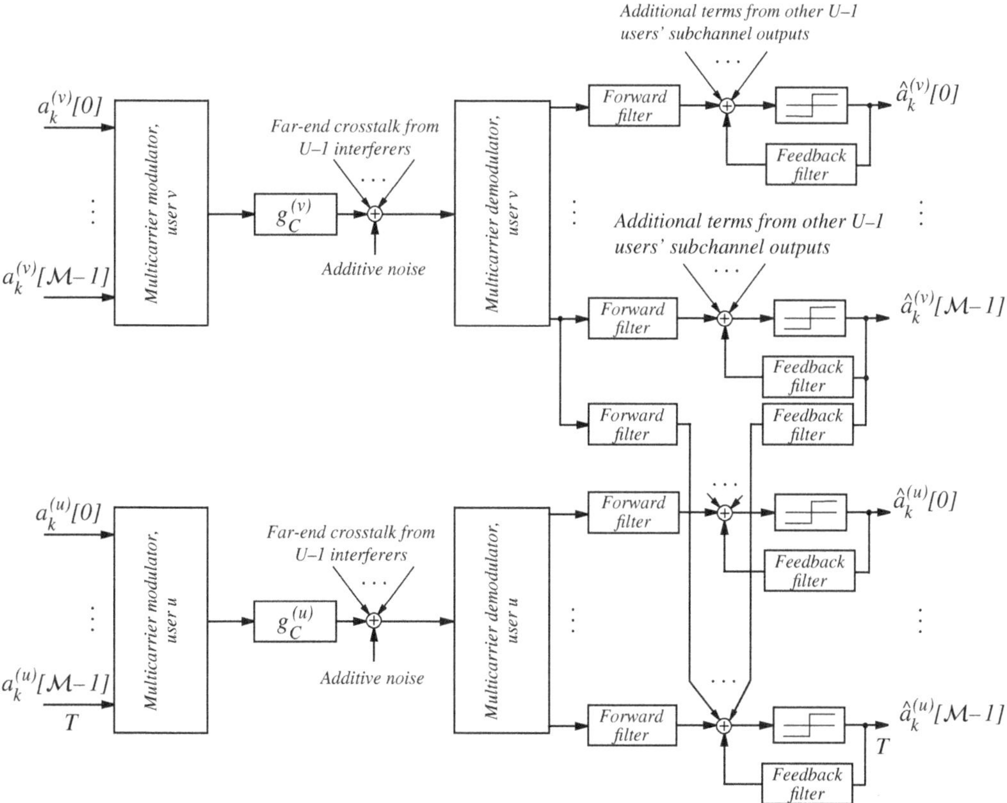

Figure 17.9 Block diagram of transmission channel and DFE structure. Source: Reproduced with permission from Cherubini [11]. ©2001, IEEE.

In practice, to reduce system complexity, for each user only a subset of all other user signals (interferers) is considered as an input to the DFE structure [11]. The selection of the subset signals is based on the power and the number of the interferers. This strategy, however, results in a loss of performance, as some strong interferers may not be considered. This effect is similar to the near–far problem in CDMA systems. To alleviate this problem, it is necessary to introduce power control of the transmitted signals; a suitable method will be described in the next section.

To determine the DFE filter coefficients, we assume M_1 and M_2 coefficients for each FF and FB filter, respectively. We define the following vectors:

1. $\boldsymbol{x}^{(u)}_{kF_0}[i] = [x^{(u)}_{kF_0}[i], \dots, x^{(u)}_{kF_0-M_1+1}[i]]^T$, signal samples stored at instant kF_0 in the delay lines of the FF filters with input given by the demodulator output of user u at subchannel i;
2. $\boldsymbol{c}^{(u,v)}[i] = [c^{(u,v)}_0[i], \dots, c^{(u,v)}_{M_1-1}[i]]^T$, coefficients of the FF filter from the demodulator output of user v to the decision element input of user u at subchannel i;
3. $\boldsymbol{b}^{(u,v)}[i] = [b^{(u,v)}_1[i], \dots, b^{(u,v)}_{M_2}[i]]^T$, coefficients of the FB filter from the decision element output of user v to the decision element input of user u at subchannel i;
4. $\hat{\boldsymbol{a}}^{(u)}_k[i] = \left[\hat{a}^{(u)}_{k-1-\left\lceil \frac{D^{(u)}[i]}{F_0}\right\rceil}[i], \dots, \hat{a}^{(u)}_{k-M_2-\left\lceil \frac{D^{(u)}[i]}{F_0}\right\rceil}[i]\right]^T$, symbol decisions stored at instant k in the delay lines of the FB filters with input given by the decision element output of user u at subchannel i, where $D^{(u)}[i]$ is a suitable integer delay related to the DFE; we assume no decision errors, that is $\hat{a}^{(u)}_k[i] = a^{(u)}_k[i]$.

The input to the decision element of user u at subchannel i at instant k is then expressed by (see Section 7.4.4)

$$
\begin{aligned}
y^{(u)}_k[i] &= \boldsymbol{c}^{(u,u)^T}[i]\,\boldsymbol{x}^{(u)}_{kF_0}[i] + \boldsymbol{b}^{(u,u)^T}[i]\,\boldsymbol{a}^{(u)}_k[i] \\
&\quad + \sum_{\substack{v=1 \\ v\neq u}}^{U} \{\boldsymbol{c}^{(u,v)^T}[i]\,\boldsymbol{x}^{(v)}_{kF_0}[i] + \boldsymbol{b}^{(u,v)^T}[i]\,\boldsymbol{a}^{(v)}_k[i]\}
\end{aligned} \tag{17.48}
$$

and the error signal is given by

$$
e^{(u)}_k[i] = y^{(u)}_k[i] - a^{(u)}_{k-\left\lceil \frac{D^{(u)}[i]}{F_0}\right\rceil} \tag{17.49}
$$

Without loss of generality, we extend the technique developed in Section 7.4.4 for the single-user fractionally spaced DFE to determine the optimum coefficients of the DFE structure for user $u = 1$. We introduce the following vectors and matrices:

1. $\boldsymbol{h}^{(u)}_m[i] = G^{(u)}[i]\left[h^{(u)}_{mF_0+M_1-1+D^{(u)}[i]}[i], \dots, h^{(u)}_{mF_0+D^{(u)}[i]}[i]\right]^T$, vector of M_1 samples of the impulse response of subchannel i of user u;
2. $\boldsymbol{h}^{(u,v)}_{\text{FEXT},m}[i] = G^{(v)}[i]\left[h^{(u,v)}_{\text{FEXT},mF_0+M_1-1+D^{(u,v)}[i]}[i], \dots, h^{(u,v)}_{\text{FEXT},mF_0+D^{(u,v)}[i]}[i]\right]^T$, vector of M_1 samples of the FEXT impulse response of subchannel i from user v to user u; assuming the differences between the propagation delays of the signal of user u and of the crosstalk signals originated by the other users are negligible, we have $D^{(u,v)}[i] = D^{(u)}[i]$;
3.

$$
\begin{aligned}
\boldsymbol{R}^{(1,1)}[i] = E[\boldsymbol{x}^{(1)^*}_k[i]\,\boldsymbol{x}^{(1)^T}_k[i]] - \sum_{m=1}^{M_2}\Bigg(&\sigma^2_{a^{(1)}[i]}\boldsymbol{h}^{(1)*}_m[i]\,\boldsymbol{h}^{(1)^T}_m[i] \\
&+ \sum_{v=2}^{V}\sigma^2_{a^{(v)}[i]}\boldsymbol{h}^{(1,v)^*}_{\text{FEXT},m}[i]\,\boldsymbol{h}^{(1,v)^T}_{\text{FEXT},m}[i]\Bigg)
\end{aligned} \tag{17.50}
$$

$$
\begin{aligned}
\boldsymbol{R}^{(l,l)}[i] &= E[\boldsymbol{x}^{(l)^*}_k[i]\,\boldsymbol{x}^{(l)^T}_k[i]] \\
&\quad - \sum_{m=1}^{M_2}(\sigma^2_{a^{(l)}[i]}\boldsymbol{h}^{(l)^*}_m[i]\,\boldsymbol{h}^{(l)^T}_m[i] + \sigma^2_{a^{(1)}[i]}\boldsymbol{h}^{(l,1)^*}_{\text{FEXT},m}[i]\,\boldsymbol{h}^{(l,1)^T}_{\text{FEXT},m}[i]), \\
&\quad l = 2, \dots, U
\end{aligned} \tag{17.51}
$$

$$\begin{aligned}\boldsymbol{R}^{(l,1)}[i] &= E[\boldsymbol{x}_k^{(l)^*}[i]\,\boldsymbol{x}_k^{(1)^T}[i]]\\ &-\sum_{m=1}^{M_2}\left(\sigma^2_{a^{(1)}[i]}\boldsymbol{h}^{(l,1)^*}_{\text{FEXT},m}[i]\,\boldsymbol{h}_m^{(1)^T}[i]+\sigma^2_{a^{(l)}[i]}\boldsymbol{h}^{(l)^*}_{\text{FEXT},m}[i]\,\boldsymbol{h}^{(l,1)^T}_{\text{FEXT},m}[i]\right.\\ &\left.+\sum_{\substack{p=2\\p\neq l}}^{U}\sigma^2_{a^{(p)}[i]}\boldsymbol{h}^{(l,p)^*}_{\text{FEXT},m}[i]\,\boldsymbol{h}^{(1,p)^T}_{\text{FEXT},m}[i]\right),\\ &l=2,\ldots,U\end{aligned} \tag{17.52}$$

$$\begin{aligned}\boldsymbol{R}^{(l,j)}[i] &= E[\boldsymbol{x}_k^{(l)^*}[i]\,\boldsymbol{x}_k^{(j)^T}[i]]\\ &-\sum_{m=1}^{M_2}(\sigma^2_{a^{(1)}[i]}\boldsymbol{h}^{(l,1)^*}_{\text{FEXT},m}[i]\,\boldsymbol{h}^{(j,1)^T}_{\text{FEXT},m}[i]+\sigma^2_{a^{(j)}[i]}\boldsymbol{h}^{(l,j)^*}_{\text{FEXT},m}[i]\,\boldsymbol{h}_m^{(j)^T}[i]),\\ &1<j<l\le U\end{aligned} \tag{17.53}$$

where all above matrices are $M_1 \times M_1$ square matrices;

4.

$$\boldsymbol{R}^{(1)}[i]=\begin{bmatrix}\boldsymbol{R}^{(1,1)}[i] & \boldsymbol{R}^{(1,2)}[i] & \ldots & \boldsymbol{R}^{(1,U)}[i]\\ \boldsymbol{R}^{(2,1)}[i] & \boldsymbol{R}^{(2,2)}[i] & \ldots & \boldsymbol{R}^{(2,U)}[i]\\ \vdots & \vdots & \ddots & \vdots\\ \boldsymbol{R}^{(U,1)}[i] & \boldsymbol{R}^{(U,2)}[i] & \ldots & \boldsymbol{R}^{(U,U)}[i]\end{bmatrix} \tag{17.54}$$

where $\boldsymbol{R}^{(1)}[i]$ in general is a positive semi-definite Hermitian matrix, for which we assume here the inverse exists;

5.

$$\boldsymbol{p}^{(1)}[i]=\sigma^2_{a^{(1)}[i]}[\boldsymbol{h}_0^{(1)}[i],\boldsymbol{h}^{(2,1)}_{\text{FEXT},0}[i],\ldots,\boldsymbol{h}^{(U,1)}_{\text{FEXT},0}[i]]^T \tag{17.55}$$

Defining the vectors $\boldsymbol{c}^{(1)}[i]=[\boldsymbol{c}^{(1,1)^T}[i],\boldsymbol{c}^{(1,2)^T}[i],\ldots,\boldsymbol{c}^{(1,U)^T}[i]]^T$, and $\boldsymbol{b}^{(1)}[i]=[\boldsymbol{b}^{(1,1)^T}[i],\ldots,\boldsymbol{b}^{(1,U)^T}[i]]^T$, the optimum coefficients are given by

$$\boldsymbol{c}^{(1)}_{opt}[i]=[\boldsymbol{R}^{(1)}[i]]^{-1}\boldsymbol{p}^{(1)}[i] \tag{17.56}$$

and

$$\boldsymbol{b}^{(1)}_{opt}[i]=\begin{bmatrix}\boldsymbol{h}_1^{(1)^T}[i],\boldsymbol{h}^{(2,1)^T}_{\text{FEXT},1}[i],\ldots,\boldsymbol{h}^{(U,1)^T}_{\text{FEXT},1}[i]\ldots\\ \boldsymbol{h}_{M_1}^{(1)^T}[i],\boldsymbol{h}^{(2,1)^T}_{\text{FEXT},M_1}[i],\ldots,\boldsymbol{h}^{(U,1)^T}_{\text{FEXT},M_1}[i]\\ \boldsymbol{h}^{(1,2)^T}_{\text{FEXT},1}[i],\boldsymbol{h}_1^{(2)^T}[i],\ldots,\boldsymbol{h}_1^{(U,2)^T}[i]\ldots\\ \boldsymbol{h}^{(1,2)^T}_{\text{FEXT},M_1}[i],\boldsymbol{h}^{(2)^T}_{\text{FEXT},M_1}[i],\ldots,\boldsymbol{h}^{(U,2)^T}_{\text{FEXT},M_1}[i]\ldots\\ \boldsymbol{h}^{(1,U)^T}_{\text{FEXT},1}[i],\boldsymbol{h}^{(2,U)^T}_{\text{FEXT},1}[i],\ldots,\boldsymbol{h}_1^{(U)^T}[i]\ldots\\ \boldsymbol{h}^{(1,U)^T}_{\text{FEXT},M_1}[i],\boldsymbol{h}^{(2,U)^T}_{\text{FEXT},M_1}[i],\ldots,\boldsymbol{h}^{(U)^T}_{\text{FEXT},M_1}[i]\end{bmatrix}\boldsymbol{c}^{(1)}_{opt}[i] \tag{17.57}$$

The MMSE value at the decision point of user 1 on subchannel i is thus given by

$$J^{(1)}_{min}[i]=\sigma^2_{a^{(1)}[i]}-\boldsymbol{p}^{(1)^H}[i]\boldsymbol{c}^{(1)}_{opt}[i] \tag{17.58}$$

The performance of an OFDM system is usually measured in terms of *achievable bit rate* for given channel and crosstalk characteristics (see Chapter 13). The number of bits per modulation interval that can be loaded with a bit-error probability of 10^{-7} on subchannel i is given by (see (13.16))

$$\mathsf{b}^{(1)}[i]=\log_2\left(1+\frac{\sigma^2_{a^{(1)}[i]}}{J^{(1)}_{min}[i]}\,10^{(G_{code,dB}-\overline{\Gamma}_{gap,dB})/10}\right) \tag{17.59}$$

where $G_{code,dB}$ is the coding gain assumed to be the same for all users and all subchannels. The achievable bit rate for user 1 is therefore given by

$$R_b^{(1)} = \frac{1}{T} \sum_{i=0}^{\mathcal{M}-1} \mathrm{b}^{(1)}[i] \text{ bit/s} \tag{17.60}$$

17.4.1 Upstream power back-off

Upstream *power back-off* (PBO) methods are devised to allow remote users in a VDSL system to achieve a fair distribution of the available capacity in the presence of FEXT [13]. The upstream VDSL transmission rates, which are achievable with PBO methods, usually depend on parameters, for example a reference length or the integral of the algorithm of the received signal power spectral density, that are obtained as the result of various trade-offs between services to be offered and allowed maximum line length. However, the application of such PBO methods results in a suboptimum allocation of the signal power for upstream transmission. It is desirable to devise a PBO algorithm with the following characteristics:

- for each individual user, the transmit signal power spectral density (PSD) is determined by taking into account the distribution of known target rates and estimated line lengths of users in the network, and
- the total power for upstream signals is kept to a minimum to reduce interference with other services in the same cable binder.

In the preceding section, we found the expression (17.60) of the achievable upstream rate for a user in a VDSL system with U users, assuming perfect knowledge of FEXT impulse responses and multiuser detection. However, PBO may be applied by assuming only the knowledge of the statistical behaviour of FEXT coupling functions with no attempt to cancel interference. In this case, the achievable bit rate of user u is given by (see also Section 6.105)

$$R_b^{(u)} = \int_B \log_2 \left[1 + \frac{\mathcal{P}^{(u)}(f)|\mathcal{H}^{(u)}(f)|^2}{\sum_{\substack{v=1 \\ v\neq u}}^{U} \mathcal{P}^{(v)}(f)|\mathcal{H}_{\text{FEXT}}^{(u,v)}(f)|^2 + N_0} 10^{(G_{code,dB}-\Gamma_{gap,dB})/10} \right] df \tag{17.61}$$

where $\mathcal{P}^{(u)}(f)$ denote the PSD of the signal transmitted by user u, $\mathcal{H}^{(u)}(f)$ is the frequency response of the channel for user u, $\mathcal{H}_{\text{FEXT}}^{(u,v)}(f)$ is the FEXT frequency response from user v to user u, and N_0 is the PSD of AWGN. The expression of the average FEXT power coupling function is given by

$$|\mathcal{H}_{\text{FEXT}}^{(u,v)}(f)|^2 = k_t f^2 \min(L_u, L_v)|\mathcal{H}^{(v)}(f)|^2 \tag{17.62}$$

where L_u and L_v denote the lengths of the lines of user u and v, respectively, and k_t is a constant.

Assuming that the various functions are constant within each subchannel band for OFDM modulation, we approximate (17.61) as

$$R_b^{(u)} = \sum_{i=0}^{\mathcal{M}-1} \frac{1}{T} \log_2 \left[1 + \frac{\mathcal{P}^{(u)}(f_i)|\mathcal{H}^{(u)}(f_i)|^2}{\sum_{\substack{v=1 \\ v\neq u}}^{U} \mathcal{P}^{(v)}(f_i)|\mathcal{H}_{\text{FEXT}}^{(u,v)}(f_i)|^2 + N_0} 10^{(G_{code,dB}-\Gamma_{gap,dB})/10} \right] \tag{17.63}$$

where f_i denotes the centre frequency of subchannel i.

Let $\overline{\mathcal{P}}^{(u)}(f_i)$ denote the PSD of the signal transmitted by user u on subchannel i with gain $G^{(u)}[i] = 1$. Then, the PBO problem can be formulated as follows: find the minimum of the function

$$\sum_{u=1}^{U} \int_B \mathcal{P}^{(u)}(f)\, df \simeq \sum_{u=1}^{U} \sum_{i=0}^{\mathcal{M}-1} \frac{1}{T} (G^{(u)}[i])^2 \overline{\mathcal{P}}^{(u)}(f_i) \tag{17.64}$$

subject to the constraints:

1.
$$0 \leq (G^{(u)}[i])^2 \overline{\mathcal{P}}^{(u)}(f_i) \leq \mathcal{P}_{max}, \qquad u = 1, \dots, U, \qquad i = 0, \dots, \mathcal{M} - 1 \tag{17.65}$$

and

2.
$$R_b^{(u)} \geq R_{b,\text{target}}^{(u)}, \qquad u = 1, \dots, U \tag{17.66}$$

where $\mathcal{P}_{max}$ is a constant maximum PSD value and $R_{b,\text{target}}^{(u)}$ is the target rate for user u.

In (17.66), $R_b^{(u)}$ is given by (17.60) or (17.63), depending on the receiver implementation.

Finding the optimum upstream transmit power distribution for each user is therefore equivalent to solving a non-linear programming problem in the $U\mathcal{M}$ parameters $G^{(u)}[i]$, $u = 1, \dots, U$, $i = 0, \dots, \mathcal{M} - 1$. The optimum values of these parameters that minimize (17.64) can be found by simulated annealing [14, 15].

17.4.2 Comparison of PBO methods

The European Telecommunication Standard Institute (ETSI) has defined two modes of operation, named A and B, for PBO [13]. For a scenario using upstream VDSL transmission of two adjacent links with unequal lengths, mode A states that the signal-to-noise ratio degradation to either link shall not exceed 3 dB relative to the equal-length FEXT case. Mode B requires that the signal-to-noise ratio on the longer line shall not be degraded relative to the equal-length FEXT case; furthermore, degradation to the signal-to-noise ratio on the shorter line shall be bounded such that the shorter line can support at least the upstream rate supported on the longer line. Several methods compliant with either mode A or B have been proposed. PBO methods are also classified into methods that allow shaping of the PSD of the transmitted upstream VDSL signal, e.g. the *equalized FEXT* method, and methods that lead to an essentially flat PSD of the transmitted signal over each individual upstream band, e.g. the *average log* method. Both the equalized FEXT and the average log method, which are described below, comply with mode B.

The equalized FEXT method requires that the PSD of user u be computed as [13]

$$\mathcal{P}^{(u)}(f) = \min \left[\frac{L_{ref} |\mathcal{H}_{ref}(f)|^2}{L_v |\mathcal{H}^{(u)}(f)|^2} \mathcal{P}_{max}, \mathcal{P}_{max} \right] \tag{17.67}$$

where L_{ref} and $\mathcal{H}_{ref}$ denote a reference length and a reference channel frequency response, respectively.

The average log method requires that, for an upstream channel in the frequency band (f_1, f_2), user u adopt a constant PSD given by [16]

$$\mathcal{P}^{(u)}(f) = \mathcal{P}_{(f_1, f_2)}^{(u)}, \qquad f \in (f_1, f_2) \tag{17.68}$$

where $\mathcal{P}_{(f_1, f_2)}^{(u)}$ is a constant PSD level chosen such that it satisfies the condition

$$\int_{f_1}^{f_2} \log_2 [\mathcal{P}_{(f_1, f_2)}^{(u)} |\mathcal{H}^{(u)}(f)^2|] \, df = K_{(f_1, f_2)} \tag{17.69}$$

where $K_{(f_1, f_2)}$ is a constant.

In this section, the achievable rates of VDSL upstream transmission using the optimum algorithm (17.64) and the average log method are compared for various distances and services. The numerical results presented in this section are derived assuming a 26-gauge telephone twisted-pair cable. The noise models for the alien-crosstalk disturbers at the line termination and at the network termination are taken as specified in [17] for the fibre-to-the-exchange case. AWGN with a PSD of −140 dBm/Hz is assumed. We consider upstream VDSL transmission of $U = 40$ users over the frequency band given by the union of $\mathcal{B}_1 = (2.9, 5.1\,\text{MHz})$ and $\mathcal{B}_2 = (7.05, 12.0\,\text{MHz})$, similar to those specified in [13]. The maximum PSD value is $\mathcal{P}_{max} = -60$ dBm/Hz. FEXT power coupling functions are determined according to (17.62), where $k_t = 6.65 \times 10^{-21}$. Upstream transmission is assumed to be based on FMT modulation with bandwidth of the individual subchannels equal to 276 kHz and excess bandwidth of 12.5%; for an efficient implementation, a frequency band of (0, 17.664 MHz) is assumed, with $\mathcal{M} = 64$ subchannels, of which only 26 are used. For the computation of the achievable rates, for an error probability of 10^{-7} a signal-to-noise ratio gap to capacity equal to $\overline{\Gamma}'_{gap,dB} = \overline{\Gamma}_{gap,dB} + 6 = 15.8$ dB, which includes a 6 dB margin against additional noise sources that may be found in the digital-subscriber-line (DSL) environment [18], and $G_{code} = 5.5$ dB are assumed.

For each of the methods and for given target rates, we consider two scenarios: the users are (i) all the same distance L from the central office and (ii) uniformly distributed at 10 different nodes, having distances $jL_{max}/10, j = 1, \dots, 10$, from the central office. To assess the performance of each method, the maximum line length L_{max} is found, such that all users can reliably achieve a given target rate $R_{b,\text{target}} =$ 13 MBit/s. The achievable rates are also computed for the case that all users are at the same distance from the central office and no PBO is applied.

For the optimum algorithm, the achievable rates are computed using (17.63). Furthermore, different subchannel gains may be chosen for the two bands, but transmission gains within each band are equal. Figure 17.10 shows the achievable rates for each group of four users with the optimum algorithm for the given target rate. The maximum line length L_{max} for scenario (ii) turns out to be 950 m. For application to scenario (ii), the optimum algorithm requires the computation of 20 parameters. Note that for all users at the same distance from the central office, i.e. scenario (i), the optimum algorithm requires the computation of two gains equal for all users. For scenario (i), the achievable rate is equal to the target rate up to a certain characteristic length L_{max}, which corresponds to the length for which the target rate is achieved without applying any PBO. Also note that L_{max} for scenario (ii) is larger than the characteristics length found for scenario (i).

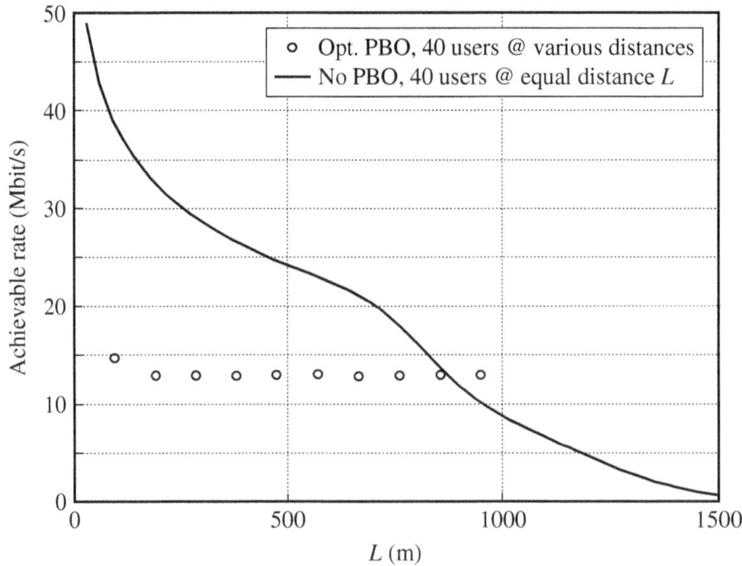

Figure 17.10 Achievable rates of individual users versus cable length using the optimum upstream PBO algorithm for a target rate of 13 Mbit/s. Source: Reproduced with permission from Cherubini [11]. ©2001, IEEE.

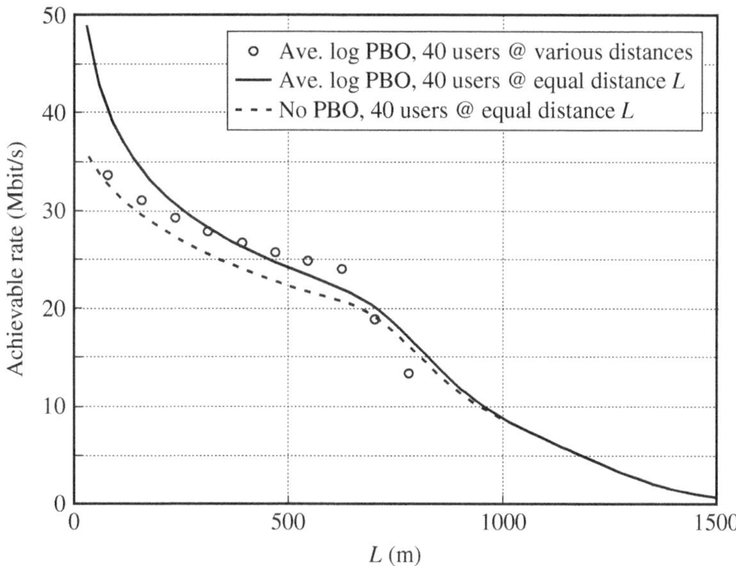

Figure 17.11 Achievable rates of individual users versus cable length using the average log upstream PBO method for a target rate of 13 Mbit/s. Source: Reproduced with permission from Cherubini [11]. ©2001, IEEE.

Figure 17.11 illustrates the achievable rates with the average log algorithm (17.69). Joint optimization of the two parameters K_{B_1} and K_{B_2} for maximum reach under scenario (ii) yields $K_{B_1} = 0.02$ mW, $K_{B_2} = 0.05$ mW, and $L_{max} = 780$ m. By comparison with Figure 17.10, we note that for the VDSL transmission spectrum plan considered, optimum upstream PBO leads to an increase in the maximum reach of up to 20%. This increase depends on the distribution of target rates and line lengths of the users in the network.

At this point, further observations can be made on the application of PBO.

- Equal upstream services have been assumed for all users. The optimum algorithm described is even better suited for mixed-service scenarios.
- The application of PBO requires the transmit PSDs of the individual user signals to be recomputed at the central office whenever one or more users join the network or drop from the network.

To illustrate system performance achievable with multiuser detection, we consider $U = 20$ users, uniformly distributed at 10 different nodes having distances $jL_{max}/10$, $j = 1, \dots, 10$, from the central office, where identification of FEXT impulse responses is performed. For the computation of the achievable rates of individual users, the FEXT impulse responses are generated by a statistical model, and $L_{max} = 500$ m is assumed. Furthermore, to assess the relative merits of multiuser detection and coding, the achievable rates are computed for the two cases of uncoded and coded transmission. For coded transmission, a powerful coding technique yielding 8.5 dB coding gain for an error probability of 10^{-7} is assumed. Figure 17.12 illustrates the achievable rates for perfect suppression of all interferers, which corresponds to the single-user bound. For comparison, the achievable rates for the case that all users are the same distance away from the central office and neither multiuser detection nor coding are applied are also given. Figure 17.13 illustrates the achievable rates for perfect suppression of the 10 worst interferers and no application of PBO. We observe that, without PBO, the partial application of multiuser detection does not lead to a significant increase of achievable rates for all users, even assuming large coding gains. Finally, Figure 17.14 depicts the achievable rates obtained for perfect suppression of the 10 worst interferers and application of the optimum PBO algorithm with target rate of $R_{b,target} = 75$ Mbit/s for all

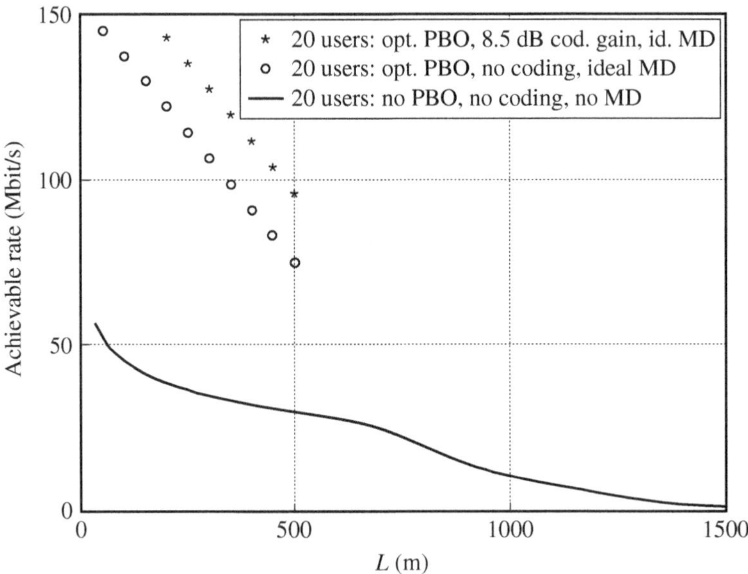

Figure 17.12 Achievable rates of individual users versus cable length with all interferers suppressed. Source: Reproduced with permission from Cherubini [11]. ©2001, IEEE.

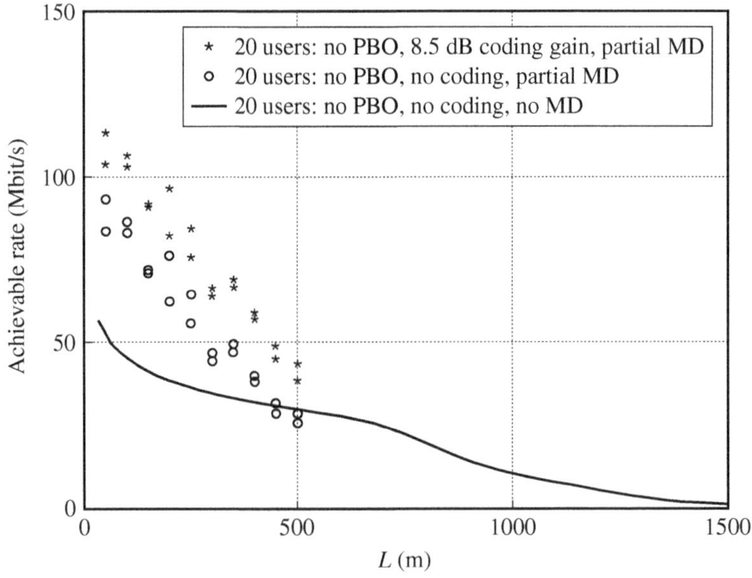

Figure 17.13 Achievable rates of individual users versus cable length with 10 interferers suppressed and no PBO applied. Source: Reproduced with permission from Cherubini 11. ©2001, IEEE.

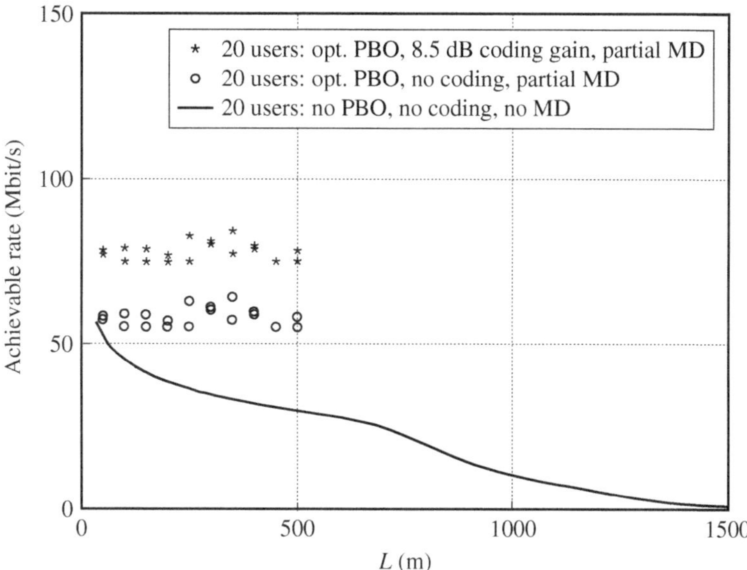

Figure 17.14 Achievable rates of individual users versus cable length with 10 interferers suppressed and optimum PBO applied for a target rate of 75 Mbit/s. Source: Reproduced with permission from Cherubini [11]. ©2001, IEEE.

users. The target rate is achieved by all users with the joint application of multiuser detection, coding, and power back-off.

To summarize the results of this section, a substantial increase in performance with respect to methods that do not require the identification of FEXT responses is achieved by resorting to reduced-complexity multiuser detection (MD) in conjunction with power back-off. This approach yields a performance close to the single-user bound shown in Figure 17.12.

Bibliography

[1] Benvenuto, N. and Zorzi, M. (2011). *Principles of Communications Networks and Systems*. Wiley.

[2] Cherubini, G. (1999). Echo cancellation. In: *The Mobile Communications Handbook*, Chapter 7, 2e (ed. J.D. Gibson), 7.1–7.15. Boca Raton, FL: CRC Press.

[3] Messerschmitt, D.G. (1986). Design issues for the ISDN U-Interface transceiver. *IEEE Journal on Selected Areas in Communications* 4: 1281–1293.

[4] Falconer, D.D. (1982). Adaptive reference echo-cancellation. *IEEE Transactions on Communications* 30: 2083–2094.

[5] Duttweiler, D.L. (1982). Adaptive filter performance with nonlinearities in the correlation multiplier. *IEEE Transactions on Acoustics, Speech, and Signal Processing* 30: 578–586.

[6] Smith, M.J., Cowan, C.F.N., and Adams, P.F. (1988). Nonlinear echo cancelers based on transposed distributed arithmetic. *IEEE Transactions on Circuits and Systems* 35: 6–18.

[7] Cherubini, G. (1993). Analysis of the convergence behavior of adaptive distributed-arithmetic echo cancelers. *IEEE Transactions on Communications* 41: 1703–1714.

[8] Weinstein, S.B. (1977). A passband data-driven echo-canceler for full-duplex transmission on two-wire circuits. *IEEE Transactions on Communications* 25: 654–666.

[9] Ho, M., Cioffi, J.M., and Bingham, J.A.C. (1996). Discrete multitone echo cancellation. *IEEE Transactions on Communications* 44: 817–825.

[10] Melsa, P.J.W., Younce, R.C., and Rohrs, C.E. (1996). Impulse response shortening for discrete multitone transceivers. *IEEE Transactions on Communications* 44: 1662–1672.

[11] Cherubini, G. (2001). Optimum upstream power back-off and multiuser detection for VDSL. *Proceedings of GLOBECOM '01*, San Antonio, TX, USA.

[12] Duel-Hallen, A. (1993). Decorrelating decision-feedback multiuser detector for synchronous code-division multiple-access channel. *IEEE Transactions on Communications* 41: 285–290.

[13] (2000). Access transmission systems on metallic access cables; Very high speed Digital Subscriber Line (VDSL); Part 2: Transceiver specification. *ETSI Technical Specification 101 270-2 V1.1.1*, May 2000.

[14] Kirkpatrik, S., Gelatt, C.D. Jr., and Vecchi, M.P. (1983). Optimization by simulated annealing approach. *Science* 220: 671–680.

[15] Vanderbilt, D. and Louie, S. (1984). A Monte Carlo simulated annealing approach to optimization over continuous variables. *Journal of Computational Physics* 56: 259–271.

[16] (1999). Constant average log: robust new power back-off method. *Contribution D.815 (WP1/15), ITU-T SG 15, Question 4/15*, April 1999.

[17] ETSI VDSL specifications (Part 1) functional requirements. *Contribution D.535 (WP1/15), ITU-T SG 15, Question 4/15*. June 21 V- July 2, 1999.

[18] Starr, T., Cioffi, J.M., and Silverman, P.J. (1999). *Digital Subscriber Line Technology*. Upper Saddle River, NJ: Prentice-Hall.

Chapter 18

Examples of communication systems

In this chapter, we provide relevant examples of communication systems, where many of the considered techniques and algorithms presented in previous chapters are used. First, the fifth generation of cellular wireless systems, as defined by *release 15* of the 3rd Generation Partnership Project (3GPP) is introduced. Then, an overview of the GSM standard is provided. Although this belongs to the second generation of cellular standards, it is still used for its ability to efficiently cover large area and provide basic data services suitable for the Internet of things (IoT). We conclude the overview of wireless systems with the wireless local area network (WLAN) and the digital enhanced cordless telecommunications (or digital European cordless telecommunications, DECT) standards. As example of transmission over unshielded twisted pairs, we provide an overview of digital subscriber lines (DSL) and Ethernet technologies. Lastly, an example of hybrid fibre and coax communication system is discussed.

18.1 The 5G cellular system

We now provide an overview of the physical layer of the cellular communication standard defined by 3GPP. In particular, we refer to the *release 15*, which aims at defining the 5th generation (5G) of cellular systems.

18.1.1 Cells in a wireless system

A fundamental concept when we speak of wireless systems is that of a *cell*. The coverage area of a certain system is not *served* by a single transmitter but by numerous transmitters, called *base stations* (BSs) or *base station transceivers* (BTSs): each one of them can *cover* only a small part of this area, called cell. To each cell, a set of carrier frequencies completely different from those of the neighbouring cells is assigned, so that *co-channel interference* is as low as possible.

The traditional method of frequency reuse, illustrated in Figure 18.1, is related to the original concept of cell. To two cells separated by an adequate distance, the same set of carrier frequencies can be assigned with minimum co-channel interference; then, there is a periodic repetition in assigning the frequencies, as seen in Figure 18.1. *Cluster* is a set of cells that subdivides the available bandwidth of the system. Obviously, each BS will interfere with the other BSs, especially with those in the same cell and in adjacent cells, as there is always a certain percentage of irradiated power outside of the nominal bandwidth of the assigned channel; hence, the strategy in the choice of carrier frequencies assigned to the various cells is of fundamental importance.

Algorithms for Communications Systems and their Applications, Second Edition.
Nevio Benvenuto, Giovanni Cherubini, and Stefano Tomasin.

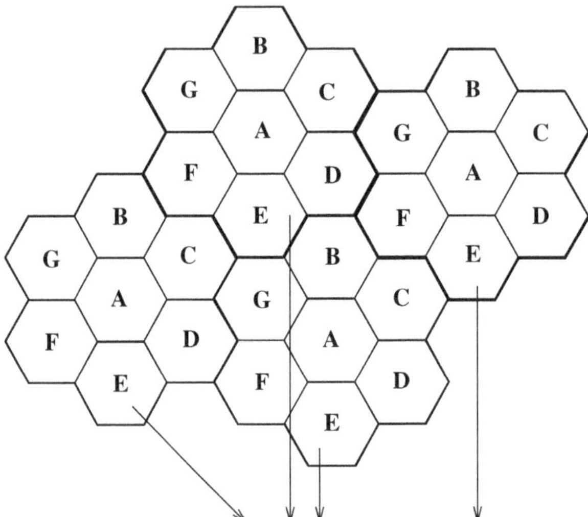

Figure 18.1 Illustration of cell and frequency reuse concepts; cells with the same letter are assigned the same carrier frequencies.

Related to the concept of cell is also the concept of *hand-off* or *hand-over*. When a user moves from one cell to another, the communication cannot be interrupted; hence, a mechanism must be present that, at every instant, keeps track of which BTS sends the *strongest* signal and possibly is capable of changing the BTS to which the mobile terminal is *linked*.

Most recent cellular systems overcome the concept of frequency reuse, for a more flexible cell concept (also exploiting spatial diversity), assigning to each cell the whole spectrum.

18.1.2 The release 15 of the 3GPP standard

The 3GPP activity of 5G started in 2015, with the definition of 74 use cases that would characterize the new standard. With respect to the previous generation, the envisioned 5G scenario included a significantly higher data rate (in accordance with a rapidly growing use of mobile data over the past years), but also a much larger number of connected devices, including those of the IoT.

Among the applications, virtual reality, autonomous driving, heath solutions, emergency communications, and tactile Internet were the most promising and still challenging. For some applications such as virtual reality and tactile Internet, a key factor is a low latency, i.e. a short time for a message to go from end-to-end at the application level. A typical example is provided by a person interacting with a moving object (e.g. playing with a ball by remote through a robot). If the visual feedback on the effects of the actions is not fast enough, the actions on the ball will not be effective, e.g. the player will believe to hit the ball, but the ball will be in another position in the meantime. Another important factor is reliability, in particular for critical applications where connection interruptions may significantly disrupt the application. Therefore, low end-to-end latency and high reliability were also included in the requirements.

The 74 initial cases have been later summarized into four categories:

- *Enhanced mobile broadband*: This category targets the spectral efficiency enhancement, by providing higher data rates (up to Gbit/s) to the end users. It follows the evolution over the past years of mobile service usage.

- *Critical communications*: This category addresses new applications where timely and reliable communications are important. It includes self-driving cars, mission critical broadband applications, augmented reality, industrial and vehicular automation.
- *Massive machine-type communications*: This category collects a new group of envisioned applications, related to the IoT in various realms.
- *Network operations*: This category does not focus on performance requirements of specific services, but more on operational requirements that 5G networks have to fulfil.

The first three categories are somehow related to similar categories defined by the international telecommunication union (ITU) that set the rules for the 5G of cellular systems.

The 3GPP upgrades the cellular standard by *releases* that are provided every 6–12 months. The standardization of 5G has started with *release 15*. The release defines protocols and interfaces at the various communication layers, and here we focus on the physical layer (layer 1), as described by the technical specification (TS) documents of the series 38.200 (see Ref. [1] for an overview). Not all the requirements and features of 5G are implemented in release 15, as some have been left for next releases.[1] The evolution of the standard, with access to all technical specifications, can be followed on the 3GPP web site https://www.3gpp.org.

18.1.3 Radio access network

The radio access network (RAN) comprises multiple cells, where each cell is a region in the space where mobile users are served by a single fixed base station. According to the 5G 3GPP nomenclature, the base station is denoted as new generation node base (gNB), while the mobile user is denoted as user equipment (UE). Uplink transmissions refer to communications from the UEs to the gNB, while downlink transmissions refer to the communications from the gNB to the UE. gNB is typically equipped with multiple antennas, while UE devices typically have a single antenna, although the standard provides the possibility of using more antennas. Therefore, the resulting network is in general a multiuser multiple-input-multiple-output (MIMO) system.

The new radio (NR) is the air interface developed by 3GPP for the physical and medium access control (MAC) layers of the communications between gNB and UE. This will be the focus of the rest of this section.

Time-frequency plan

NR has been designed to work with two frequency bands: the *sub-6* GHz, from 450 MHz to 6 GHz, and *millimetre wave* (mmWave), from 24.25 GHz to 52.6 GHz.

Each cell is assigned a specific band, that is used by the cell continuously over time. In release 15 of the 3GPP standard, time is divided in *radio frames* of the duration of 10 ms, and within each radio frame, both uplink and downlink transmissions occur. Each radio frame is split into 10 *subframes* of the same duration, each in turn split into *slots*. The adopted modulation format is orthogonal frequency division multiplexing (OFDM), possibly with discrete Fourier transform (DFT) spreading (see Section 8.10.2) in uplink, denoted as *transform precoding* in the 3GPP standard. Therefore, each slot is finally split into OFDM *symbols*. Therefore, we have the multiple-access schemes orthogonal frequency division multiple access (OFDMA) and single-carrier frequency division multiple access (SC-FDMA).

Five configurations, denoted with index $\mu = 0, \dots, 4$, are provided by 3GPP [2], with subframes of $N_{slot}^{subframe,\mu}$ slots each, for a total of $N_{slot}^{frame,\mu}$ slots per frame. In any case, we have $N_{symb}^{slot} = 14$ symbols per

[1] Note that in the bibliography of this chapter, we provide references to release documents as of April, 2019.

slot. Moreover, it is possible to transmit data blocks smaller than a slot, to facilitate low-rate low-latency communications. Table 18.1 shows the parameters for the five configurations, while Figure 18.2 shows an example of radio frame configuration for $\mu = 1$.

Table 18.1: Radio frame configurations.

μ	$N_{slot}^{frame,\mu}$	$N_{slot}^{subframe,\mu}$	N_{symb}^{slot}
0	10	1	14
1	20	2	14
2	40	4	14
3	80	8	14
4	160	16	14

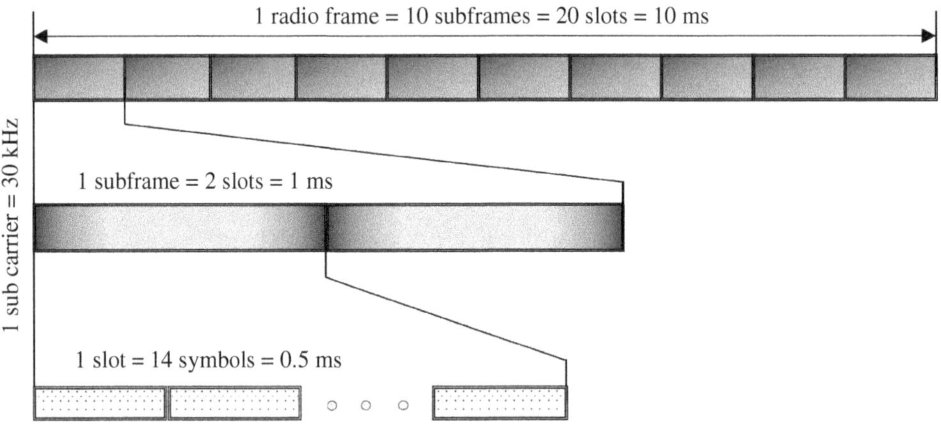

Figure 18.2 Example of radio frame configuration with index $\mu = 1$.

In time-division duplexing (TDD) mode, within each sub-frame, some slots are used for uplink and others for downlink, according to predetermined patterns. In frequency-division duplexing (FDD) mode, the cell band is split into two sub-bands, each exclusively and continuously used for either uplink or downlink.

Users are served over different tiles of the time-frequency plan defined by OFDMA (SC-FDMA). The basic tile assigned to a user is a *resource block* (RB) spanning 12 subcarriers and a duration of 1 ms (sub-frame duration), as shown in Figure 18.3. The mapping to RB is done first in frequency then in time to enable early decoding at the receiver, thus reducing latency.

The NR includes also the possibility of performing *carrier aggregation*, i.e. assign different portions of the spectrum to the same user. For example, it is possible to assign to a user subcarriers in both the sub-6 GHz and the mmWave bands. Indeed, working at mmWave with a huge bandwidth is well suited for small cells, and having also a communication link at lower frequencies is of great help to assist handover (i.e. the mobility of an UE) from a small cell to another.

The synchronization of transmission is established by the gNB, which sets the time-frequency boundaries of the radio frame and all its parts. UEs are required to align their transmission to this general timing. In the following, we will also describe the techniques used to achieve alignment.

The channel coding includes both low-density parity check (LDPC) codes (for the data channel) and polar codes (for the control channel). Moreover, dedicated coding solutions are defined for data blocks of size between 1 and 11 bits. Further details on coding in NR standard are provided in [3].

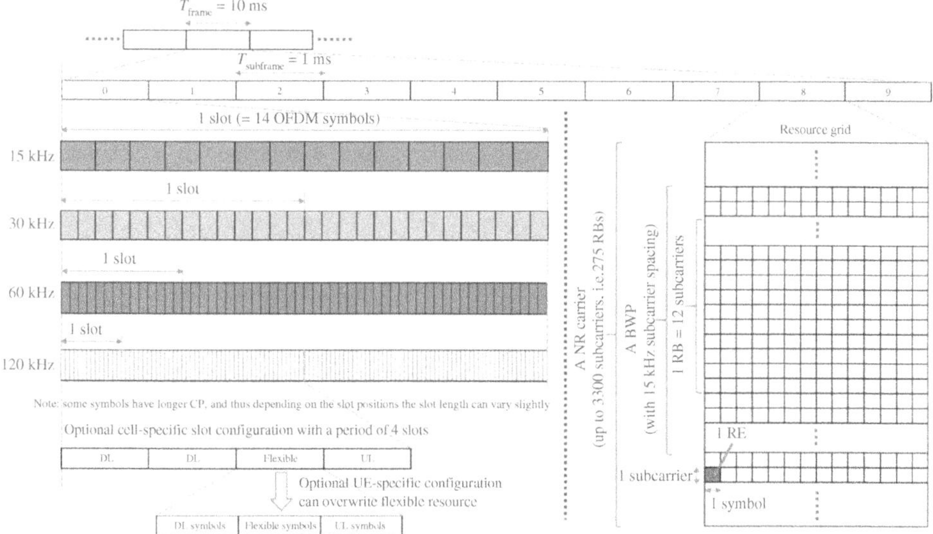

Figure 18.3 5G NR frame structure and basic terminologies. Source: Lin et al. 2019 [4]. Reproduced with permission from IEEE.

NR data transmission chain

Data bits are coded and scrambled (see Chapter 6) before modulation. In downlink, the complex symbols are assigned (by the layer map block) to streams (denoted *layers* by 3GPP), where for details on mapping the reader is referred to [2, Table 7.3.1.3-1].

In order to exploit possibly available 1, 2, or 4 antennas, the streams are precoded in space (see Section 9.3.2) before OFDM modulation.

In uplink, it is possible to use spreading by DFT and mapping its output to a subset of the OFDM subcarriers (see Section 18.1.5).

OFDM numerology

The parameters of the OFDM transmission are common for both uplink and downlink of release 15 [2]. The sample period is $T_c = 1/(\Delta f_{\max} N_f)$, where $\Delta f_{\max} = 480 \cdot 10^3$ Hz and $N_f = 4096$ resulting in $T_c = 0.5$ ns.

Various configurations of the number of subcarriers and length of the cyclic prefix (CP) are available, in what is denoted as the modulation *numerology*. A key parameter is the *subcarrier spacing configuration* (depending on the configuration index μ), that defines the subcarrier spacing as $\Delta f = 2^{\mu} \cdot 15$ kHz. The number of subcarriers is then defined as $\mathcal{M} = 2048\ \kappa\ 2^{-\mu}$, where $\kappa = 64$, thus providing a number of subcarriers between $2^{13} = 8192$ and $2^{17} = 131072$. Note that multiplying the number of subcarriers by their spacing provides the same bandwidth of about 2 GHz for all modes. Moreover, a smaller subcarrier spacing is more suitable for lower frequencies (sub-6 GHz band), where larger cells and longer delay spreads are expected. On the other hand, a larger subcarrier spacing is adequate for higher frequencies (as in mmWave).

The CP length is defined by the configuration index μ and by the CP mode, that can be normal or extended. For the normal mode, the CP length (in samples) is $144\ \kappa\ \cdot 2^{-\mu} + 16\ \kappa$ for OFDM symbols 0 and $7 \cdot 2^{\mu}$, within each sub-frame, and $144\ \kappa \cdot 2^{-\mu}$ for other OFDM symbols. For the extended mode instead the CP length is $512\ \kappa \cdot 2^{-\mu}$ for all OFDM symbols. Also in this case, the dependency of the CP length by the configuration index μ (thus the subcarrier spacing) ensures a fixed minimum duration (in time) of the CP: for example, the normal CP has a minimum duration of $(144 \cdot 64/16)T_c = 0.29$ μs.

Channel estimation

In NR, the training signals used for channel estimation are denoted as *reference signals*. Some reference signals are present both in uplink and downlink transmissions. In particular, these are as follows:

- Demodulation reference signal (DMRS),
- Phase tracking reference signal (PTRS),

and they will be described in more details in the following.

In most cases, the reference signals in NR are transmitted on demand, rather than on a regular basis, following a lean design principle. Their position in the time-frequency plan is extremely flexible, providing an energy reduction with respect to previous generations, wherein reference signals were transmitted also when not needed.

18.1.4 Downlink

The downlink signal comprises different logical signals, which can either carry data and control bits coming from higher communication layers, and are denoted as *physical channels*, or carry signalling information generated and used only at the physical layer, and are denoted as *physical signals*.

The *physical channels* defined in the downlink are as follows:

- Physical downlink shared channel (PDSCH),
- Physical downlink control channel (PDCCH), and
- Physical broadcast channel (PBCH).

The PDSCH channel carries the data, and it is encoded by LDPC codes, scrambled and modulated as described in Section 18.1.3. It can be transmitted on up to 4 MIMO streams (layers for 3GPP nomenclature), where the precoder is left to the operator and is not specified by the standard.

The PDCCH contains control information such as downlink and uplink scheduling assignments. PDCCH includes the DMRS for UE-specific precoding of the control channel. PDCCH is transmitted in specific configurable control resource set (CORESET), spanning from 1 to 3 consecutive OFDM symbols, and 12 resource elements (REs) of one OFDM symbol in one RB are organized into a RE group (REG). For the structure of PDCCH, see Figure 18.4a. Codewords are bit mapped into complex symbols, as described in Section 18.1.3, with quadrature phase shift keying (QPSK), 16 quadrature amplitude modulation (QAM), 64 QAM, and 256 QAM.

The PBCH carries the synchronization signal.

The *physical signals* defined in the downlink are as follows:

- Demodulation reference signals (DMRSs),
- Phase tracking reference signals (PTRSs),
- Channel-state information reference signal (CSIRS),
- Primary synchronization signal (PSS),
- Secondary synchronization signal (SSS).

The primary and secondary synchronization signals are named together synchronization signals (SS) and are used for synchronization purposes. CSIRS, DMRS, and PTRS are used for channel estimation and tracking purposes.

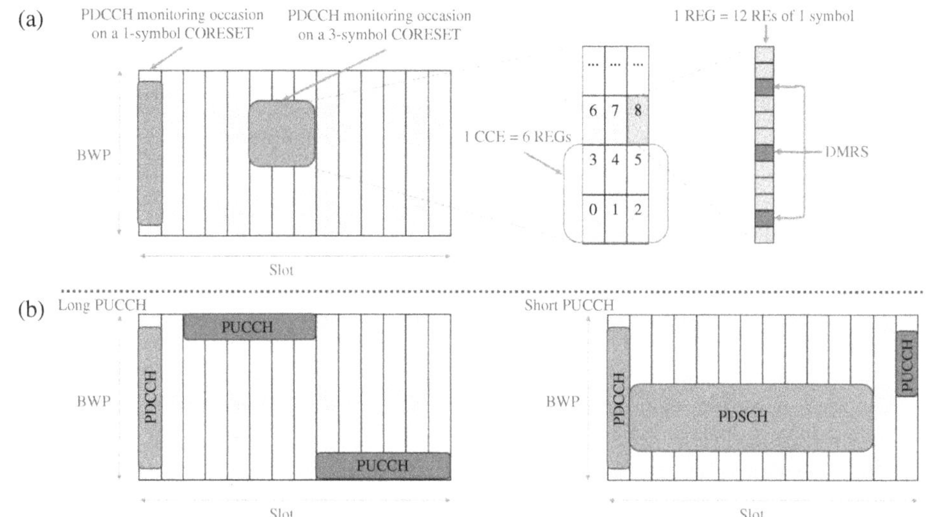

Figure 18.4 5g NR (a) PDCCH and (b) PUCCH. Source: Lin et al. 2019 [4]. Reproduced with permission from IEEE.

Synchronization

The synchronization of the UEs to the gNB adjusts the reference time at the UE, by exploiting the synchronization signals SSS and PSS transmitted in downlink by gNB. In order to access the network, a UE first detects the synchronization signal block (SSB), which comprises 1 PSS, 1 SSS, and 2 PBCH symbols.

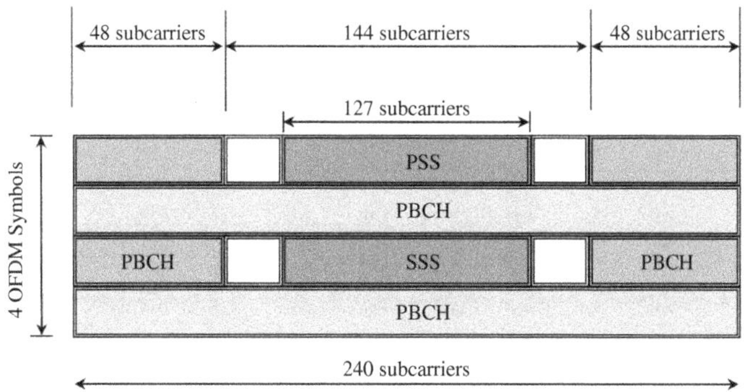

Figure 18.5 The NR SSB.

The SSB is transmitted over 4 OFDM symbols and 240 contiguous subcarriers (20 RBs), as shown in Figure 18.5. Each SSB comprises a primary and a secondary synchronization signal, and by receiving this signal, the UE obtains the cell identity, gets synchronous with downlink transmissions in both time and frequency domains, and acquires the frame timing of the PBCH. The PSS uses binary phase shift keying (BPSK)-modulated m-sequence of length 127, while SSS is obtained as a BPSK-modulated Gold sequence of length 127.

Initial access or beam sweeping

In mmWaves, an adequate signal-to-noise ratio (SNR) at the receiver can be achieved only by precoding and multiple transmit antennas, in order to overcome the strong attenuation incurred at those frequencies (see also Sections 4.1.11 and 9.5). A special issue occurs when a UE enters a cell. In this case, the gNB is not aware of the UE, therefore cannot precode any control signal to it (to start data transmission), and similarly, the UE does not know the spatial direction of the MIMO channel to the gNB and may even have problems in receiving control signals from the gNB. To this end, an *initial access* or *beam sweeping* procedure is provided.

In particular, the SSB can be precoded to specific spatial directions, and multiple SSBs can be transmitted consequently in a *SS burst set* with a maximum duration of 5 ms. The burst is used to transmit SSBs in multiple directions, as illustrated in Figure 18.6. Since each SSB is identified by a unique index, the UE infers the suitable direction for good reception, and then uses it for uplink transmission to the gNB or even feedback the index to gNB, where it is going to be used for downlink precoding. In particular, the feedback is obtained by selecting specific resources of physical random access channel (PRACH) for random access message transmission (see Section 18.1.5), corresponding to the selected SSB index. Release 15 defines various numbers of SSBs per burst, depending on the operating frequency: up to 4 when operating below 3 GHz, 8 when operating between 3 GHz and 6 GHz, and 64 when operating between 6 GHz and 52.6 GHz.

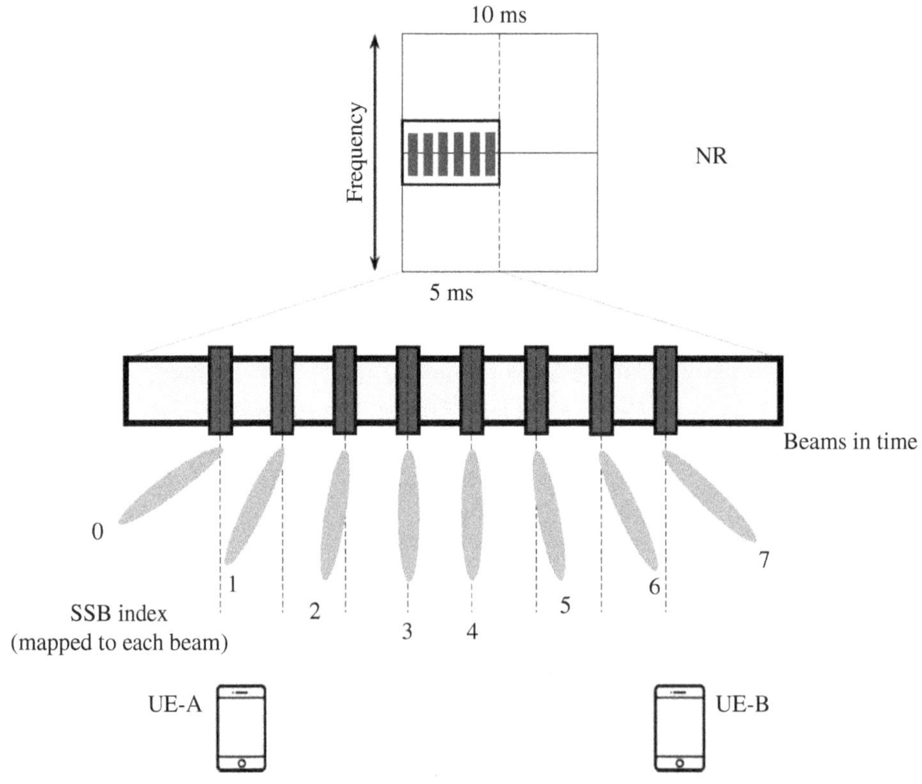

Figure 18.6 The beam sweeping procedure in NR.

Channel estimation

Channel estimation is performed at the UE by exploiting specific training signals transmitted in the downlink by the gNB, and known at the UE, namely the DMRS and the CSIRS.

DMRS is specific for each physical channel, and for each UE. It is transmitted on demand. The density of DMRS in the time-frequency plane is flexible, and is organized in a comb-type structure (see Section 8.9.1). It can be integrated with MIMO supporting 12 different precoding and is QPSK-modulated and based on Gold sequences. DMRS is used in idle mode, i.e. when the UE is not connected.

CSIRS is transmitted by the gNB so that the UE can estimate the downlink channel. Moreover, it also provides ways to measure the received signal power, to track precoding and time-frequency offsets. CSIRS is specific for each UE, but multiple users can share the same resource. The allocation of CSIRS in the time-frequency plan is very flexible. When used for time frequency tracking, CSIRS can be periodic or aperiodic. CSIRS is used when the UE is connected.

The PTRS is used to track the phase of the local oscillators (see phase noise in Section 4.1.3). It has a low density in the frequency domain (as the phase noise is white), but high density in the time domain for proper tracking. PTRS is always transmitted together with DMRS.

Channel state information reporting

The channel estimate obtained by CSIRS can be reported back to gNB. The feedback includes

1. the index of a quantized version of the precoding matrix (in the transmit precoding matrix indicator, TPMI) (see Section 5.6),
2. the number of suggested streams, denoted as rank indicator (RI) of the channel matrix (see (9.135)), and
3. the SNR (channel quality indicator, CQI).

Various fixed TPMI codebooks are defined. For example, for a single-layer transmission using two antenna ports[2] the precoding codebook is

$$\left\{ \frac{1}{\sqrt{2}}\begin{bmatrix}1\\0\end{bmatrix}, \frac{1}{\sqrt{2}}\begin{bmatrix}0\\1\end{bmatrix}, \frac{1}{\sqrt{2}}\begin{bmatrix}1\\1\end{bmatrix}, \begin{bmatrix}1\\-1\end{bmatrix}, \frac{1}{\sqrt{2}}\begin{bmatrix}1\\j\end{bmatrix}, \frac{1}{\sqrt{2}}\begin{bmatrix}1\\-j\end{bmatrix} \right\} \tag{18.1}$$

while for two two-layers over two antennas is

$$\left\{ \frac{1}{\sqrt{2}}\begin{bmatrix}1 & 0\\0 & 1\end{bmatrix}, \frac{1}{\sqrt{2}}\begin{bmatrix}1 & 1\\1 & -1\end{bmatrix}, \frac{1}{\sqrt{2}}\begin{bmatrix}1 & 1\\j & -j\end{bmatrix} \right\} \tag{18.2}$$

The complete set of codebooks is provided in [2].

18.1.5 Uplink

The *physical channels* defined in the uplink are as follows:

- Physical random access channel (PRACH),
- Physical uplink shared channel (PUSCH), and
- Physical uplink control channel (PUCCH).

[2] NR standard refers to *antenna ports* rather than to antennas.

The PRACH is used to let a UE initiate a communication with the gNB, and carries control signals. The PUSCH carries the UE data, and its transmit chain is described in Section 18.1.3, where transform precoding can be optionally used. The PUCCH carries control information such as the feedback hybrid automatic repeat request (HARQ), channel state information, and scheduling requests. PUCCH includes both control information and DMRS. For a structure of the PUCCH, see Figure 18.4b.

The *physical signals* defined in the uplink are as follows:

- DMRSs,
- PTRSs, and
- Sounding reference signal (SRS).

DMRS and PTRS are similar to those available in downlink, and they provide information only for the RBs used by the UE for data transmission. The SRS provides information on the whole bandwidth and is used by the gNB to schedule uplink (and downlink when TDD is used) transmissions in the forthcoming frames.

Transform precoding numerology

OFDM with DFT spreading (see Section 8.10.2) is used on QPSK-modulated symbols in the uplink PUSCH. Spreading (*transform precoding*) is obtained by a DFT of size $N = 12 \cdot \overline{N}$, with $\overline{N} = 2^{\alpha_2} \cdot 3^{\alpha_3} \cdot 5^{\alpha_5}$, where α_2, α_3, and α_5 are non-negative integers. The resulting possible values of the DFT size are reported in Table 18.2. Since DFT sizes are product of powers of 2, 3, and 5, an efficient fast Fourier transform (FFT) implementation is available (see Section 1.2).

Table 18.2: DFT size for transform-precoding.

$\overline{N}$	1	2	3	4	5	6	8	9	10	12	15	16
N	12	24	36	48	60	72	96	108	120	144	180	192
$\overline{N}$	18	20	24	25	27	30	32	36	40	45	48	50
N	216	240	288	300	324	360	384	432	480	540	576	600
$\overline{N}$	54	60	64	72	75	80	81	90	96	100		
N	648	720	768	864	900	960	972	1080	1152	1200		

Channel estimation

The DMRS in uplink is similar to that of downlink. However, in this case, it can be generated either starting from a Gold sequence or from a constant-amplitude zero autocorrelation (CAZAC) (also known as Zadoff-Chu) sequence.

SRS provides uplink channel estimation at the gNB, when in connected mode. The estimated channel is used for equalization in uplink but also, when channel reciprocity can be exploited, to design downlink precoding. The SRS is specific for each UE, with a variety of sequence design. Moreover, since it must cover the whole communication band (not only the RB used for data transmission), frequency hopping mechanisms are defined to obtain a channel estimate on different frequencies over various slots.

Synchronization

The PRACH is used to access the network by the UE. In particular, to start the transmission the UE first sends, within the PRACH, a random-access preamble containing CAZAC sequences, of either 139 or 839 bits, for large and small cells, respectively. For the small-cell configuration, it is also possible

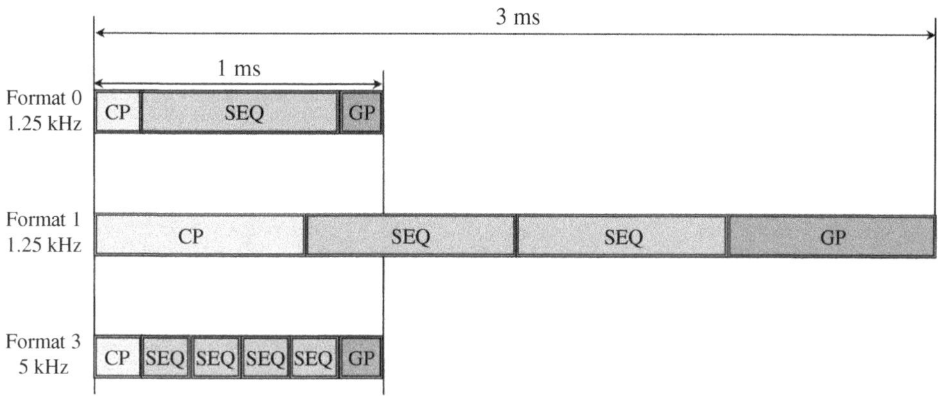

Figure 18.7 PRACH formats (long sequence).

to transmit special OFDM symbols whose last part operates as CP of the next OFDM symbol, and the length of the OFDM preamble equals the length of OFDM data. With this arrangement, the gNB uses the same DFT for data and random-access preamble detection. Moreover, by adding the various replicas of the preamble, synchronization is more robust to time varying channels and frequency errors. An example of PRACH format is shown in Figure 18.7. Note that the structure is similar to that defined for the Schmidl and Cox algorithm (see Section 14.9.1) for time synchronization.

Upon transmission of the preamble, the UE can apply different antenna precoders, thus the same preamble can be received from different angular direction (*beams*) by the gNB. For mmWave with beam sweeping, within PRACH the preamble is an indication of the precoding index to be used by the gNB in downlink.

Timing advance

Once downlink synchronization is achieved, the UE knows the timing of each slot. However, a message transmitted at time t by the UE will reach gNB at time $t + t_0$, with t_0 being the propagation time from the UE to gNB. Note that each UE will have its own propagation time, therefore the scheduled time for uplink transmission will be completely modified.

In order to solve this issue, the *timing advance* procedure is used, by which the gNB measures the effective time $t + t_0$ of reception of the message, and compares it with the expected time t. Then, it sends a signalling message to the UE with a request to anticipate transmission by t_0 in the future. For next uplink transmissions scheduled at time t', the UE will start transmitting at time $t' - t_0$ so that it is received by gNB at the correct time t'.

The timing advance value is measured upon reception of messages on the PRACH channel, wherein the UE indicates the transmission time.

18.1.6 Network slicing

Network slicing is a key concept of 3GPP 5G system. It refers to the capability of the cellular network (including both the RAN and the core network) to support simultaneous services with different qualities. Each service is associated to a *slice*, and the 3GPP currently defines three *slice types*: enhanced mobile broadband (eMBB), massive machine type communications (mMTC), and ultra reliable low latency communications (URLLC). They differ in terms of data rate, number of users per unit area that can be served, latency, and other metrics. A slice describes whatever is needed in terms of network resources and configuration to provide the requested performance metrics.

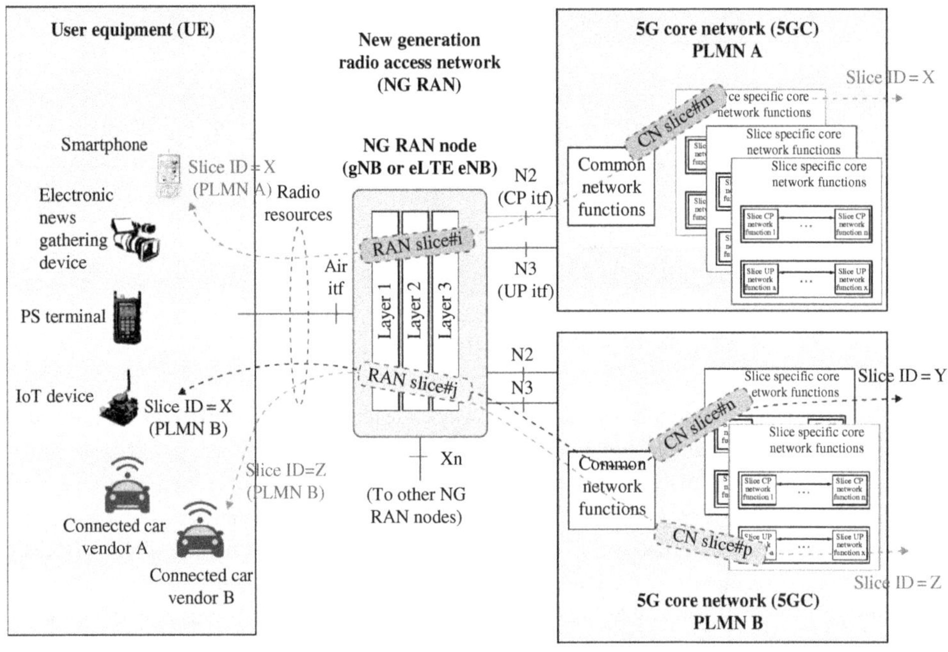

Figure 18.8 Illustration of the network slicing architecture. Source: Ferrus et al. 2018 [5]. Reproduced with permission from IEEE.

A single end-user device may run multiple services, thus being associated to different slices. For example, a smartphone may be equipped with sensors that are performing machine-to-machine communication, thus being connected by an mMTC slice to the cellular network, while at the same time, a video streaming on the same smartphone will be connected by an eMBB slice.

Figure 18.8 shows an example of network slicing architecture. We note that a slice affects both the RAN and the core network, as interventions at all layers and end-to-end are needed in general to ensure the requested quality of service. The figure shows a NR RAN node serving two public land mobile networks (PLMNs), thus two (virtual) operators. Network slicing is enabled by an evolution toward software control of the network (software defined networking) and by the possibility to make available specific network functions to the operator, regardless of the specific hardware/software implementation, in what is called network function virtualization (NFV).

At the time of writing, 3GPP has not yet defined the network slicing functionalities for the RAN. They will include a list of capabilities of gNBs, and lists of offered services, together with entities that manage the activation of slices at gNBs and their mobility functions.

18.2 GSM

The global system for mobile communications (GSM) was started in the early 1990s with the objective of providing one standard within Europe, and belongs to the so-called second generation of cellular standards. Although various generations followed, GSM has been widely in use for its ability to efficiently cover large area and provide basic data services suitable for the IoT.

The services that the GSM system offers are as follows [6]:

- *Telephony service*: Digital telephone service with guaranty of service to users that move at a speed of up to 250 km/h;
- *Data service*: Can realize the transfer of data packets with bit rates in the range from 300 to 9600 bit/s.
- *ISDN service*: Some services, as the identification of a user that sends a call and the possibility of sending *short messages* (SMSs), are realized by taking advantage of the *integrated services digital network* (ISDN), whose description can be found in [7].

A characteristic of the GSM system is the use of the *subscriber identity module* (SIM) card together with a four-digit *number* (ID); inserting the card in any mobile terminal, it identifies the subscriber who wants to use the service. Important is also the protection of privacy offered to the subscribers of the system.

Figure 18.9 represents the structure of a GSM system, that can be subdivided into three subsystems.

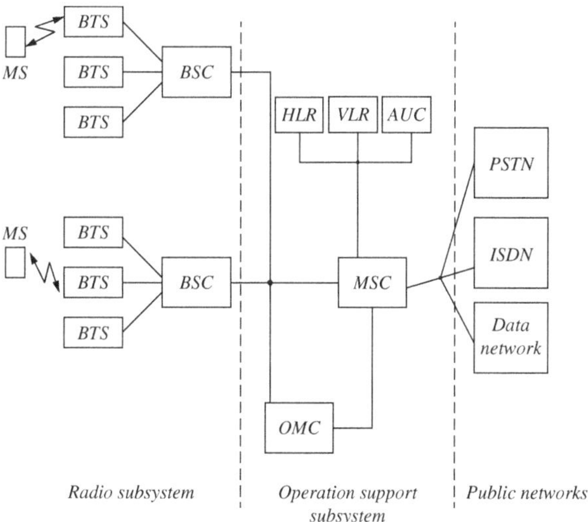

Figure 18.9 GSM system structure.

The first subsystem, composed of the set of BTSs and mobile terminals or *mobile stations* (MSs), is called *radio subsystem*; it allows the communication between the MSs and the *mobile switching centre* (MSC), that coordinates the calls and also other system control operations. To an MSC are linked many *base station controllers* (BSCs); each BSC is linked up to several hundreds BTSs, each of which identifies a cell and directly realizes the link with the mobile terminal. The hand-off procedure between two BTSs is assisted by the mobile terminal in the sense that it is a task of the MS to establish at any instant which BTS is sending the *strongest* signal. In the case of the hand-over between two BTS linked to the same BSC, the entire procedure is handled by the BSC itself and not by the MSC; in this way, the MSC can save many operations.

The second subsystem is the *network switching subsystem* (NSS), that in addition to the MSC includes:

- *Home location register* (HLR): It is a database that contains information regarding subscribers that reside in the same geographical area as the MSC;
- *Visitor location register* (VLR): It is a database that contains information regarding subscribers that are temporarily under the control of the MSC, but do not reside in the same geographical area;

- *Authentication centre* (AUC): It controls codes and other information for correct communications management;
- *Operation maintenance centres* (OMCs): They take care of the proper functioning of the various blocks of the structure.

Lastly, the MSC is directly linked to the public networks: public switched telephone network (PSTN) for telephone services, ISDN for particular services as SMS, and data network for the transmission of data packets. Note that GSM offers a rigid network structure, while 5G system has converted all its subsystems into flexible software network functions to be used dynamically.

Radio subsystem

We now give some details with regard to the radio subsystem. The total bandwidth allocated for the system is 50 MHz; frequencies that go from 890 to 915 MHz are reserved for MS-BTS communications, whereas the bandwidth 935–960 MHz is for communications in the opposite direction.[3] In this way, a full-duplex communication by *FDD* is realized. Within the total bandwidth, there are 248 carriers allocated, that identify as many frequency channels called absolute radio frequency channel number (ARFCN); of these, 124 are for uplink communications and 124 for downlink communications. The separation between two adjacent carriers is 200 kHz; the bandwidth subdivision is illustrated in Figure 18.10. Full-duplex communication is achieved by assigning two carriers to the user, one for transmission and one for reception, such that they are about 45 MHz apart.

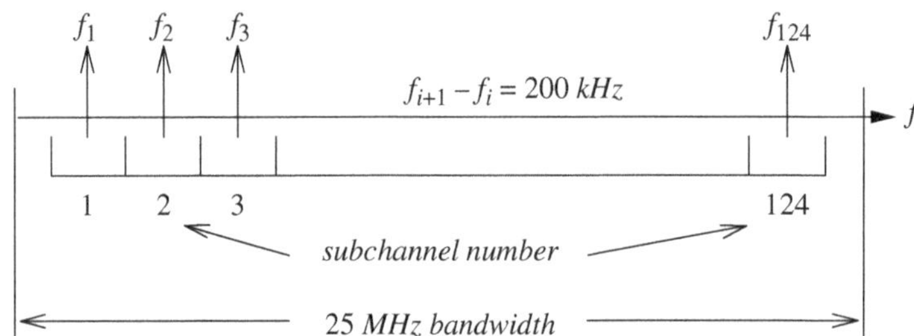

Figure 18.10 Bandwidth allocation of the GSM system.

Each carrier is used for the transmission of an overall bit rate R_b of 270.833 kbit/s, corresponding to a bit period $T_b = 3.692$ μs. The system employs Gaussian minimum shift-keying (GMSK) with parameter $B_t T$ equal to 0.3; the aim is to have a power efficient system. However, the spectral efficiency is not very high; in fact we have

$$\nu = \frac{270.833}{200} \simeq 1.354 \quad \text{bit/s/Hz} \tag{18.3}$$

which is smaller than that of other systems.

Besides this frequency division multiplexing (FDM) structure, there is also a time division multiple access (TDMA) structure; each transmission is divided into eight time intervals, or *time slots*, that identify the time division multiplexing (TDM) frame. Figure 18.11 shows the structure of a frame as well as that of a single time slot.

As a slot is composed of 156.25 bits (not an integer number because of the guard time equal to 8.25 bits), its duration is about 576.92 μs; therefore, the frame duration is about 4.615 ms. In the figure, it is

[3] There also exists a version of the same system that operates at around the frequency of 1.8 GHz (in the United States of America the frequency is 1.9 GHz).

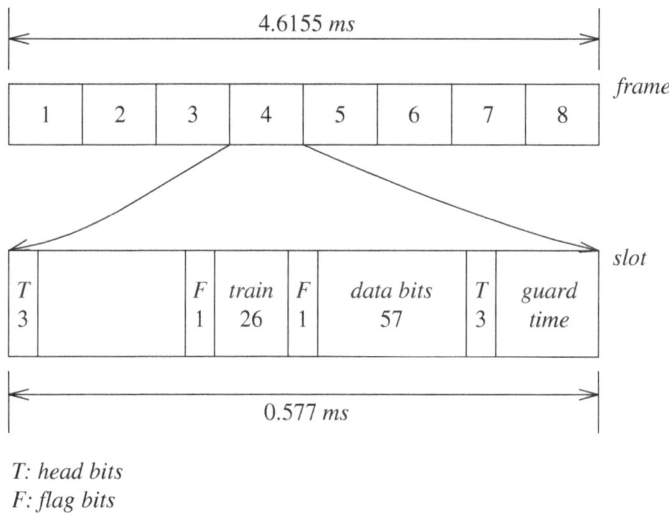

Figure 18.11 TDM frame structure and slot structure of the GSM system.

important to note the training sequence of 26 bits, used to analyse the channel by the MS or BS. The *flag bits* signal if the 114 information bits are for voice transmission or for control of the system. Lastly, the *tail bits* indicate the beginning and the end of the frame bits.

Although transmissions in the two directions occur over different carriers, to each communication is dedicated a pair of time slots spaced 4 slots apart (one for the transmit station and one for the receive station); for example the first and fifth or the second and sixth etc. Considering the sequence of 26 consecutive frames, that have a duration of about 120 ms, the 13th and 26th frames are used for control; then in 120 ms, a subscriber can transmit (or receive) $114 \cdot 24 = 2736$ bits, which corresponds to a bit rate of 22.8 kbit/s. Indeed the net bit rate of the message can be 2.4, 4.8, 9.6, or 13 kbit/s. Redundancy bits are introduced by the channel encoder for protection against errors, so that we get a bit rate of 22.8 kbit/s in any case.

The original speech encoder chosen for the system was a residual-excited predictor, improved by a *long-term predictor*, with a rate of 13 kbit/s. The use of a *voice activity detector* reduces the rate to a minimum value during silence intervals. For channel coding, a convolutional encoder with code rate 1/2 is used. In [7], the most widely used speech and channel encoders are described, together with the data interleavers.

In Figure 18.12, a scheme is given that summarizes the protection mechanism against errors used by the GSM system. The speech encoder generates, in 20 ms, 260 bits; as the bit rate per subscriber is 22.8 kbit/s, in 20 ms 456 bits must then be generated, introducing a few redundancy bits.[4]

To achieve reliable communications in the presence of multipath channels with delay spread up to 16 μs, at the receiver equalization by a decision feedback equalizer (DFE) and/or detection by the Viterbi algorithm are implemented.

In [7, 8], the calling procedure followed by the GSM system and other specific details are fully described.

---

[4] The described slot structure and coding scheme refer to the transmission of user information, namely speech. Other types of communications, as for the control and management of the system, use different coding schemes and different time slot structures (see [8]).

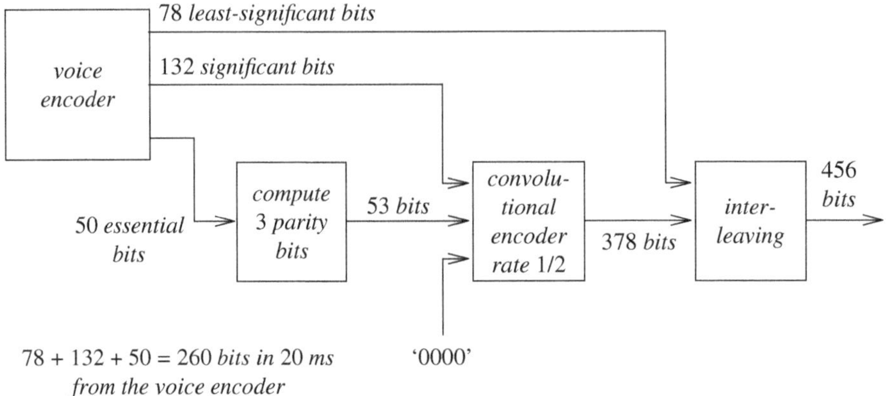

Figure 18.12 Channel coding for the GSM system.

18.3 Wireless local area networks

For the transmission at bit rates of the order of 1–10 Mbit/s, WLANs normally use the *industrial, scientific, and medical* (ISM) frequency bands defined by the United States *Federal Communications Commission* (FCC), that is 902–928 MHz, 2400–2483.5 MHz, and 5725–5850 MHz. Specifications for physical and MAC layers of WLANs are developed by various standardization organizations, among which we cite the IEEE 802.11 Working Group and the *European telecommunications standard institute* (ETSI).

In the United States, WLANs are allowed to operate in the industrial, scientific, and medical (IMS) frequency bands without the need of a license from the authorities, which, however, sets restrictions on the power of the radio signal that must be less than 1 W and specifies that spread-spectrum technology (see Chapter 10) must be used whenever the signal power is larger than 50 mW. Most WLANs employ direct sequence or frequency hopping spread-spectrum systems; WLANs that use narrowband modulation systems usually operate in the band around 5.8 GHz, with a transmitted signal power lower than 50 mW, in compliance with FCC regulations. The coverage radius of WLANs is typically of the order of a few hundred metres.

Medium access control protocols

Unlike cabled LANs, WLANs operate over channels with multipath fading, and channel characteristics typically vary over short distances. Channel monitoring to determine whether other stations are transmitting requires a larger time interval than that required by a similar operation in cabled LANs; this translates into an efficiency loss of the carrier sense multiple access (CSMA) protocols, whenever they are used without modifications. The MAC layer specified by the IEEE 802.11 Working Group is based on the CSMA protocol with *collision avoidance* (CSMA/CA), in which four successive stages are foreseen for the transmission of a data packet, as illustrated in Figure 18.13 [9].

The CSMA/CA principle is simple. All mobile stations that have packets to transmit compete for channel access by sending *request to send* (RTS) messages by a CSMA protocol. If the base stations recognize a RTS message sent by a mobile station, it sends a *clear to send* (CTS) message to the same mobile station and this one transmits the packet; if reception of this packet occurs correctly, then the base station sends an *acknowledgement* (ACK) message to the mobile station. With CSMA/CA, the only possibility of collision will occur during the RTS stage of the protocol; however, we note that also

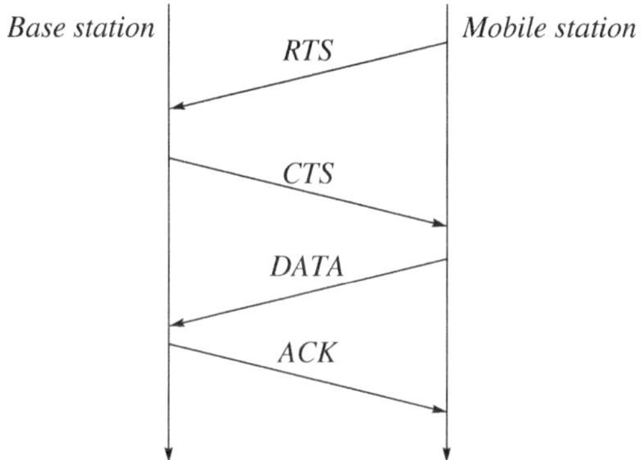

Figure 18.13 Illustration of the CSMA/CA protocol.

the efficiency of this protocol is reduced with respect to that of the simple CSMA/CD because of the presence of the RTS and CTS stages.

As mobility is allowed, the number of mobile stations that are found inside a cell can change at any instant. Therefore, it is necessary that each station informs the others of its presence as it moves around. A protocol used to solve this problem is the so-called hand-off, or hand-over, protocol, which can be described as follows:

- a switching station, or all base stations with a coordinated operation, registers the information relative to the signal levels of all mobile stations inside each cell;
- if a mobile station M is serviced by base station B1, but the signal level of station M becomes larger if received by another base station B2, the switching station proceeds to a *hard hand-off* operation whose final result is that mobile station M is considered part of the cell covered by base station B2.

18.4 DECT

The *digital European cordless telecommunications* (DECT) system is a cordless digital system used in Europe for speech transmission. Unlike cellular systems that use cells with radius of the order of a few kilometres, the DECT system is mainly employed indoor and the cell radius is at most of a few tens of meters (typically 100 m). The main characteristics of DECT are summarized in Table 18.3.

Table 18.3: Table summarizing the main characteristics of DECT.

Frequency range	1880–1900 MHz
RF channel spacing	1728 kHz
Modulation	GMSK
Transmission bit rate	1152 kbit/s
Voice encoding method	32 kbit/s ADPCM
Access method	FDMA/TDMA/TDD
Frame duration	10 ms (24 time slots)
Subscriber TX peak power	250 mW
Radius of service	100–150 m
Frequency planning	Dynamic channel allocation

While originally the hand-off problem was not considered, as each MS corresponded to one BS only, now even for DECT we speak of hand-off assisted by the mobile terminal, such that the system configuration is similar to that of a cellular system (see Figure 18.9). An interesting characteristic is the use of the *dynamic channel selection* (DCS) algorithm, that allows the portable to know at every moment which channel (frequency) is the best (with the lowest level of interference) for communication and select it. We briefly illustrate the calling procedure followed by the system:

1. when a mobile terminal wishes to make a call, it first measures the received signals from the various BS[5] and selects the one which yields the best signal level;
2. by the DCS algorithm, the mobile terminal selects the best free channels of the selected BS;
3. MS sends a message, called *access request*, over the least interfered channel;
4. BS sends (or not) an answer: *access granted*;
5. if the MS receives this message, in turn it transmits the *access confirm* message and the communication starts;
6. if the MS does not receive the *access granted* signal on the selected channel, it abandons this channel and selects the second least interfered channel, repeating the procedure; after failing on five channels, the MS selects another BS and repeats all operations.

The total band allocated to the system goes from 1880 to 1900 MHz and is subdivided into ten *subbands*, each with a width of 1.728 MHz; this is the frequency division multiple access (FDMA) structure of DECT represented in Figure 18.14. Each channel has an overall bit rate of 1.152 Mbit/s, that corresponds to a bit period of about 868 ns.

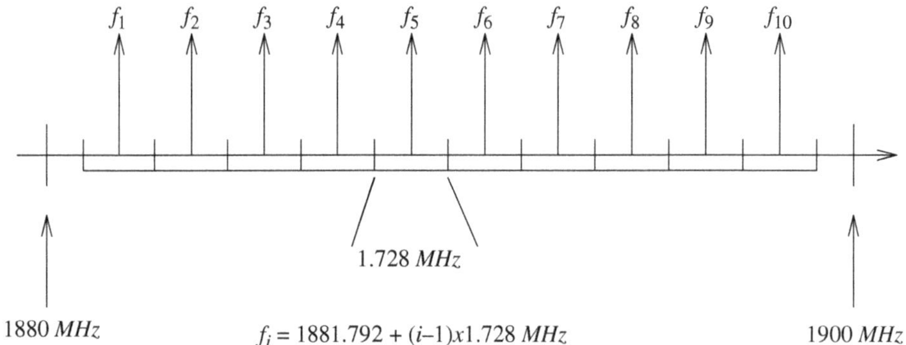

Figure 18.14 FDMA structure of the DECT system.

Similarly to the other systems, TDMA is used. The TDM frame, given in Figure 18.15, is composed of 24 slots. The first 12 are used for the communication BS-MS, the other 12 for the reverse communication; thus, we realize a full-duplex communication by TDD. In this DECT differs from all above considered wireless systems which use FDD; DECT allocates half of the frame for transmission and the other half for reception.

In Figure 18.15, the slot structure is also shown; it is composed of 480 bits of which the first 32 are fixed and correspond to a synchronization word, the successive 388 are information bits, and the remaining 60 constitute the guard time. The frame has a duration of $480 \cdot 24/1152 = 10$ ms. The field of 388 bits reserved for information bits is subdivided into two subfields A and B. The first (64 bits)

[5] For DECT, the acronyms *portable handset* (PH) in place of MS and *radio fixed part* (RFP) in place of BS, are often used.

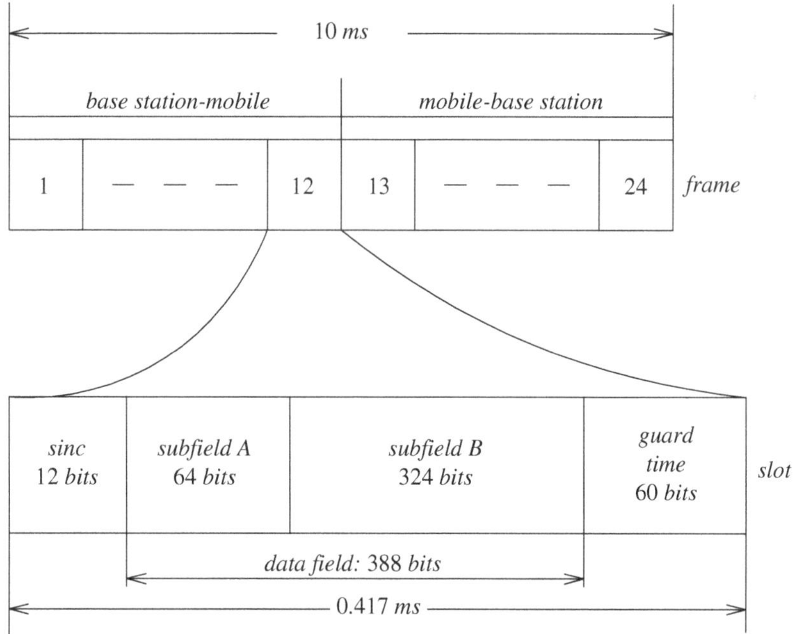

Figure 18.15 TDM frame structure and slot structure for the DECT system.

contains information for signalling and control of the system, the second (324 bits) contains user data. If the signal is speech, 4 of these 324 bits are parity bits, which translates into a net bit rate of 320 bits in a frame interval of 10 ms, and therefore in a bit rate of 32 kbit/s; an adaptive differential pulse code modulation voice encoder at 32 kbit/s is used and no channel coding is provided.

For transmission, GMSK with parameter $B_t T$ of 0.5 is adopted. At the MS receiver, a pair of antennas are often used to realize switched antenna diversity. With this mechanism, fading and interference are mitigated. We summarize in Table 18.4 the characteristics of the various standards so far discussed.

18.5 Bluetooth

The Bluetooth standard [10, 11] is designed for data transmission. It operates around the frequency of 2.45 GHz, in the unlicensed and open ISM band. Bluetooth uses a frequency hopping (FH) /TDD scheme with Gaussian shaped frequency-shift-keying (FSK) ($h \leq 0.25$). The modulation rate is 1 MBaud with a slot of 625 μs; a different hop channel is used for each slot. This gives a nominal hop rate of 1600 hops/s. There are 79 hop carriers spaced by 1 MHz. The maximum bit rate of Bluetooth 1.0 is of 723.2 kbit/s in one direction and 57.6 kbit/s in the reverse direction. Evolved versions of Bluetooth achieve data rates up to 25 Mbit/s.

18.6 Transmission over unshielded twisted pairs

We now provide two examples of transmission systems over unshielded twisted pairs (UTPs), namely the DSL, for communication in the customer service area, and the Ethernet, for communications in local area networks, technologies.

Table 18.4: Summary of characteristics of the standards GSM and DECT.

System	GSM	DECT
Multiple access	TDMA/FDMA	TDMA/FDMA
Band (MHz)		
Downlink	935–960 1805–1880	1880–1900
Uplink	890–915 1710–1785	1880–1900
Duplexing	FDD	TDD
Spacing between carriers	200 kHz	1728 kHz
Modulation	GMSK 0.3	GMSK 0.5
Bit rate per carrier (kbit/s)	270.833	1152
Speech encoder	RPE-LTP 1 3	ADPCM bit rate (kbit/s) 32
Convolutional encoder[a)] code rate	1/2 1/3 (R)	none
Frame duration (ms)	4.615	10

a) All these standards use a code redundancy check (CRC) code, possibly together with a convolutional code.

18.6.1 Transmission over UTP in the customer service area

In customer service areas, UTP cables also represent a low-cost alternative to optical fibres for links that allow data transmission at a considerably higher bit rates than those achievable by modems for transmission over the telephone channel, over distances that can reach 6 km; in fact, although optical fibres have substantially better transmission characteristics (see Chapter 4), a reliable link over a local loop is preferable in many cases given the large number of already installed cables [12–14].

The various *DSL* technologies (in short xDSL) allow full-duplex transmission between user and central office at bit rates that may be different in the two directions [15, 16]. For example, the *high bit rate digital subscriber line* (HDSL) offers a solution for full-duplex transmission at a bit rate of 1.544 Mbit/s, also called T1 rate (see Section 18.8), over two twisted pairs and up to distances of 4500 m; the *single-line high-speed DSL* (SHDSL) provides full duplex transmission at rates up to 2.32 Mbit/s over a single twisted pair, up to distances of 2000 m.

A third example is given by the *asymmetric digital subscriber line* (ADSL) technology (see Figure 18.16); originally, ADSL was proposed for the transmission of video-on-demand signals; later it emerged as a technology capable of providing a large number of services. For example, ADSL-3 is designed for downstream transmission of four compressed video signals, each having a bit rate of 1.5 Mbit/s, in addition to the full-duplex transmission of a signal with a bit rate of 384 kbit/s, a control signal with a bit rate of 16 kbit/s, and an analog telephone signal, up to distances of 3600 m.

A further example is given by the *very high-speed DSL* (VDSL) technology, mainly designed for the *fibre-to-the-curb* (FTTC) architecture. The considered data rates are up to 26 Mbit/s downstream i.e. from the central office or optical network unit to the remote terminal, and 4.8 Mbit/s upstream for asymmetric transmission, and up to 14 Mbit/s for symmetric transmission, up to distances not exceeding a few hundred meters. Figure 18.17 illustrates the FTTC architecture, where the link between the user

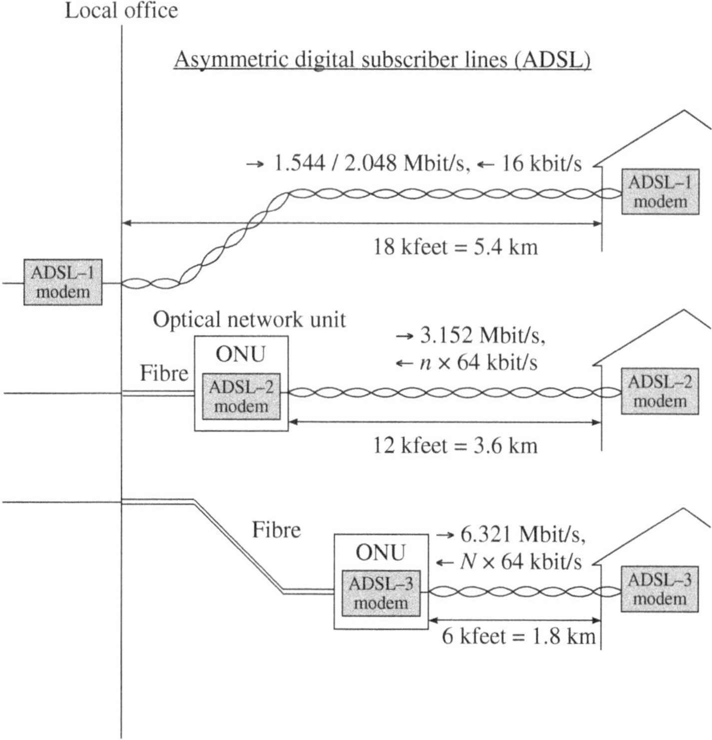

Figure 18.16 Illustration of links between ADSL modems over the subscriber line.

and an *optical network unit* (ONU) is obtained by a UTP cable with maximum length of 300 m, and the link between the ONU and a local central office is realized by optical fibre; in the figure, links between the user and the ONU that are realized by coaxial cable or optical fibre are also indicated, as well as the direct link between the user and the local central office using optical fibre with a *fibre-to-the-home* (FTTH) architecture.

Different baseband and passband modulation techniques are considered for high-speed transmission over UTP cables in the customer service area. For example, the Study Group T1E1.4 of *Committee T1* chose 2B1Q quaternary pulse amplitude modulation (PAM), see Example 16.1.1 on page 793, for HDSL, and DMT modulation (see Chapter 8) for ADSL. Among the organizations that deal with the standardization of DSL technologies we also mention the Study Group TM6 of ETSI and the Study Group 15 of the ITU-T. A table summarizing the characteristics of DSL technologies is given in Table 18.5; spectral allocations of the various signals are illustrated in Figure 18.18.

We now discuss more in detail the VDSL technology. Reliable and cost effective VDSL transmission at a few tens of Megabit per second is made possible by the use of FDD (see Appendix 18.A), which avoids signal disturbance by near-end crosstalk (NEXT), a particularly harmful form of interference at VDSL transmission frequencies. Ideally, using FDD, transmissions on neighbouring pairs within a cable binder couple only through far-end crosstalk (FEXT) (see also Section 17.4), the level of which is significantly below that of NEXT. In practice, however, other forms of signal coupling come into play because upstream and downstream transmissions are placed spectrally as close as possible to each other in order to avoid wasting useful spectrum. Closely packed transmission bands exacerbate interband interference by echo and NEXT from similar systems (self-NEXT), possibly leading to severe performance degradation. Fortunately, it is possible to design modulation schemes that make efficient use of the available

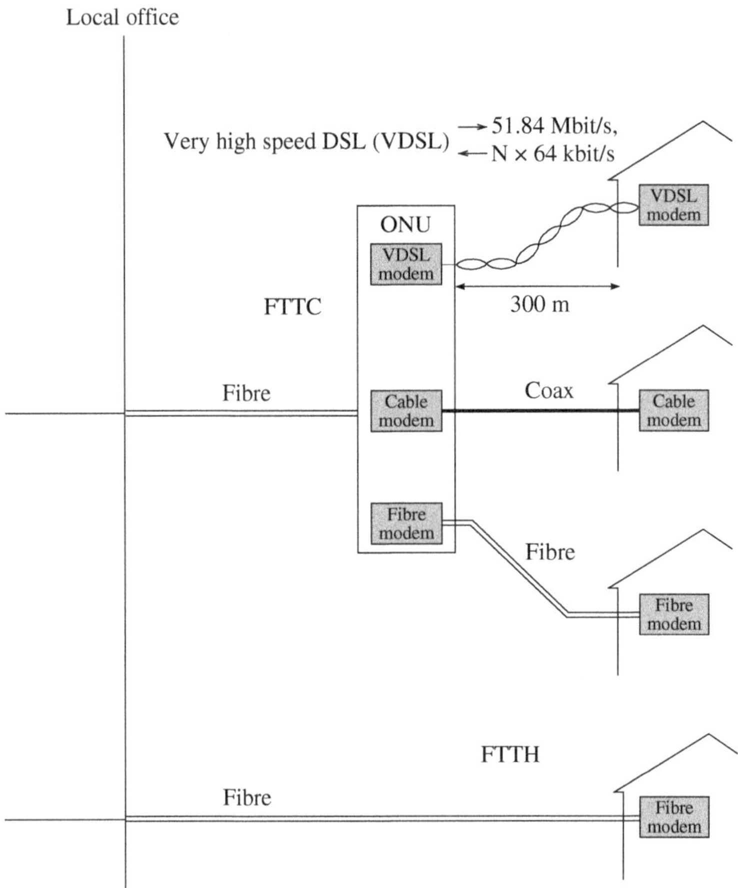

Figure 18.17 Illustration of FTTC and FTTH architectures.

Table 18.5: Characteristics of DSL technologies.

acronym	standard	modulation	bit rate (Mbit/s)	distance (m)
basic rate ISDN		2B1Q	0.144	≤ 6000
HDSL	G.991.1	2B1Q	1.544, 2.048	≤ 4000
SHDSL	G.shdsl	TC-PAM	0.192–2.32	≤ 2000
ADSL	G.992.1	DMT	*downstream* 6.144 *upstream* 0.640	≤ 3600
ADSL lite	G.992.1	DMT	*downstream* 1.5 *upstream* 0.512	*best effort service*
VDSL			*downstream* ≤ 26 *upstream* ≤ 14	≤ 1500

spectrum and simultaneously achieve a sufficient degree of separation between transmissions in opposite directions by relying solely on digital signal processing techniques. This form of FDD is sometimes referred to as digital duplexing.

The concept of *divide and conquer* has been used many times to facilitate the solution of very complex problems; therefore, it appears unavoidable that digital duplexing for VDSL will be realized by the

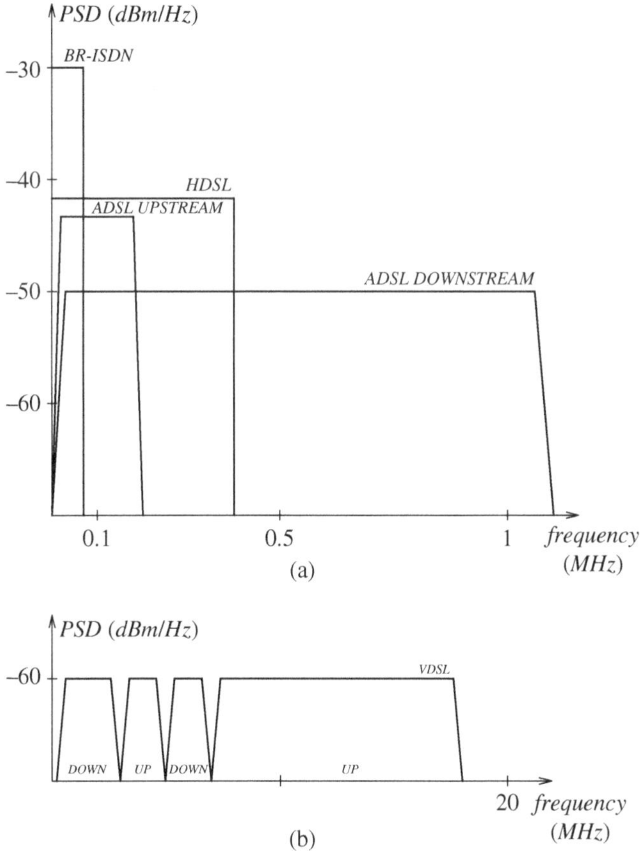

Figure 18.18 Spectral allocation of signals for (a) HDSL and ADSL, and (b) VDSL.

sophisticated version of this concept represented by multicarrier transmission, although single carrier methods have been also proposed. As discussed in Section 8.5, various variants of multicarrier transmission exist. The digital duplexing method for VDSL known as Zipper [17] is based on discrete multitone (DMT) modulation; here, we consider filtered multitone (FMT) modulation (see Section 8.2), which involves a different set of trade-offs for achieving digital duplexing in VDSL and offers system as well as performance advantages over DMT [18, 19].

The key advantages of FMT modulation for VDSL can be summarized as follows. First, there is flexibility to adapt to a variety of spectrum plans for allocating bandwidth for upstream and downstream transmission by proper assignment of the subchannels. This feature is also provided by DMT modulation, but not as easily by single-carrier modulation systems. Second, FMT modulation allows a high-level of subchannel spectral containment and thereby avoids disturbance by echo and self-NEXT. Furthermore, disturbance by a narrowband interferer, e.g. from amplitude modulation (AM) or amateur radio sources, does not affect neighbouring subchannels as the side lobe filter characteristics are significantly attenuated. Third, FMT modulation does not require synchronization of the transmissions at both ends of a link or at the binder level, as is sometimes needed for DMT modulation.

As an example of system performance, we consider the bit rate achievable for different values of the length of a twisted pair, assuming symmetrical transmission at 22.08 MBaud and full-duplex transmission by FDD based on FMT modulation. The channel model is obtained by considering a line with attenuation equal to 11.1 dB/100 m at 11.04 MHz, with 49 near-end crosstalk interference signals, 49 far-end crosstalk interference signals, and additive white Gaussian noise with a power spectral density (PSD) of −140 dBm/Hz. The transmitted signal power is assumed equal to 10 dBm. The FMT system considered here employs the same linear-phase prototype filter for the realization of transmit and receive polyphase filter banks, designed for $\mathcal{M} = 256$, $K = 288$, and polyphase filter length $\gamma = 10$; we recall that with these values of $\mathcal{M}$ and K the excess bandwidth within each subchannel is equal to 12.5%. Per-subchannel equalization is obtained by a Tomlinson–Harashima precoder (THP) (see Section 7.7.1) with 9 coefficients at the transmitter and a fractionally spaced linear equalizer with 26 $T/2$ spaced coefficients at the receiver. With these parameter values, and using the bit loading technique of Chapter 13, the system achieves bit rates of 24.9, 10.3, and 6.5 Mbit/s for the three lengths of 300, 1000, and 1400 m, respectively [20]. We refer to Section 17.4 for a description of the general case where users are connected at different distances from the central office and power control is applied.

18.6.2 High-speed transmission over UTP in local area networks

High-speed transmission over UTP cables installed in buildings is studied by different standardization organizations. For example, the asynchronous transfer mode (ATM) Forum and the Study Group 13 of the ITU-T consider transmission at bit rates of 155.52 Mbit/s and above for the definition of the *ATM* interface between the user and the network. The IEEE 802.3 Working Group investigates the transmission at 1 Gbit/s over four twisted pairs, type UTP-5, for Ethernet (1000BASE-T) networks.

UTP cables were classified by the electronic industries association and telecommunications industries association (EIA/TIA) according to the transmission characteristics (see Chapter 4). We recall that UTP-3, or voice-grade, cables exhibit a signal attenuation and a crosstalk coupling much greater than that of UTP-5, or data-grade, cables. For local area network (LAN) applications, the maximum cable length for a link between stations is 100 m. Existing cabling systems use bundles of twisted pairs, usually 4 or 25 pairs, and signals may cross line discontinuities, represented by connectors. For transmission over UTP-3 cables, in order to meet limits on emitted radiation, the signal band must be confined to frequencies below 30 MHz and sophisticated signal processing techniques are required to obtain reliable transmission.

Standards for Ethernet networks that use the *carrier sense multiple access with collision detection* (CSMA/CD) protocol are specified by the IEEE 802.3 Working Group for different transmission media and bit rates. With the CSMA/CD protocol, a station can transmit a data packet only if no signal from other stations is being transmitted on the transmission medium. As the probability of collision between messages cannot be equal to zero because of the signal propagation delay, a transmitting station must continuously monitor the channel; in the case of a collision, it transmits a special signal called jam signal to inform the other stations of the event, and then stops transmission. Retransmission takes place after a random delay. The 10BASE-T standard for operations at 10 Mbit/s over two unshielded twisted pairs of category 3 or higher defines one of the most widely used implementations of Ethernet networks; this standard considers conventional *mono duplex* (see Section 17.1) transmission, where each pair is utilized to transmit only in one direction using simple Manchester line coding (see Appendix 7.C) to transmit data packets, as shown in Figure 18.19. Transmitters are not active outside of the packets transmission intervals, except for transmission of a signal called *link beat* that is occasionally sent to assure the link connection.

The request for transmission speeds higher than 10 Mbit/s motivated the IEEE 802.3 Working Group to define standards for fast Ethernet that maintain the CSMA/CD protocol and allow transmission at 100 Mbit/s and above. For example, the 100BASE-FX standard defines a physical layer (PHY) for Ethernet networks over optical fibres. The 100BASE-TX standard instead considers conventional mono

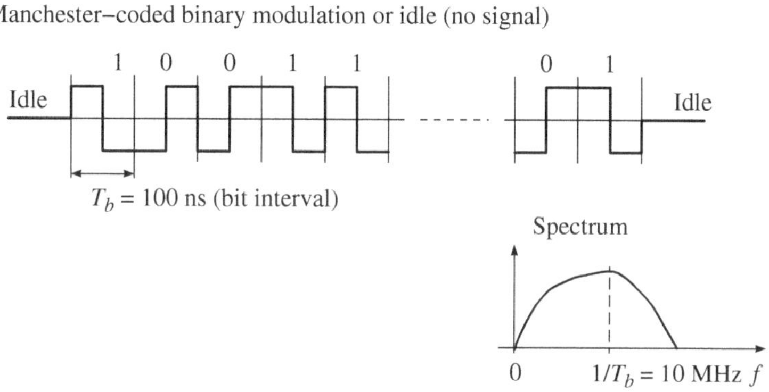

Figure 18.19 Illustration of 10BASE-T signal characteristics.

duplex transmission over two twisted pairs of category 5; the bit rate of 100 Mbit/s is obtained by transmission with a modulation rate of 125 MBaud and multilevel transmission (MLT-3) line coding combined with a channel code with rate 4/5 and scrambling, as illustrated in Figure 18.20. We also mention the 100BASE-T4 standard, which considers transmission over four twisted pairs of category 3; the bit rate of 100 Mbit/s is obtained in the following way by using an 8B6T code with a modulation rate equal to 25 MBaud. On the first two pairs, data are transmitted at a bit rate of 33.3 Mbit/s in half duplex fashion, while on the two remaining pairs, data are transmitted at 33.3 Mbit/s in mono duplex fashion.

A further version of fast Ethernet is represented by the 100BASE-T2 standard, which allows users of the 10BASE-T technology to increase the bit rate from 10 to 100 Mbit/s without modifying the cabling from category 3 to 5, or using four pairs for a link over UTP-3 cables. The bit rate of 100 Mbit/s is achieved with *dual duplex* transmission over two twisted pairs of category 3, where each pair is used to transmit in the two directions (see Figure 18.21). The 100BASE-T2 standard represents the most advanced technology for high-speed transmission over UTP-3 cables in LANs; the transceiver design for 100BASE-T2 will be illustrated in Chapter 19.

Other examples of LANs are the *token ring* and the *fibre distributed data interface* (FDDI) networks; standards for token ring networks were specified by the IEEE 802.5 Working Group, and standards for FDDI networks were specified by the American National Standards Institute (ANSI). In these networks, the access protocol is based on the circulation of a token. A station is allowed to transmit a data packet only after having received the token; once the transmission has been completed, the token is passed to the next station. The IEEE 802.5 standard specifies operations at 16 Mbit/s over two UTPs of category 3 or higher, with mono duplex transmission. For the transmission of data packets, Manchester line coding is adopted; the token is indicated by coding violations, as illustrated in Figure 18.22. The ANSI standard for FDDI networks specifies a physical layer for mono duplex transmission at 100 Mbit/s over two UTPs of category 5 identical to that adopted for the Ethernet 100BASE-TX standard.

Table 18.6 summarizes the characteristics of existing standards for high-speed transmission over UTP cables.

18.7 Hybrid fibre/coaxial cable networks

A *hybrid fibre/coax* (HFC) network is a multiple-access network, in which a *head-end controller* (HC) broadcasts data and information for *MAC* over a set of channels in the downstream direction to a certain number of stations, and these stations send information to the HC over a set of shared channels

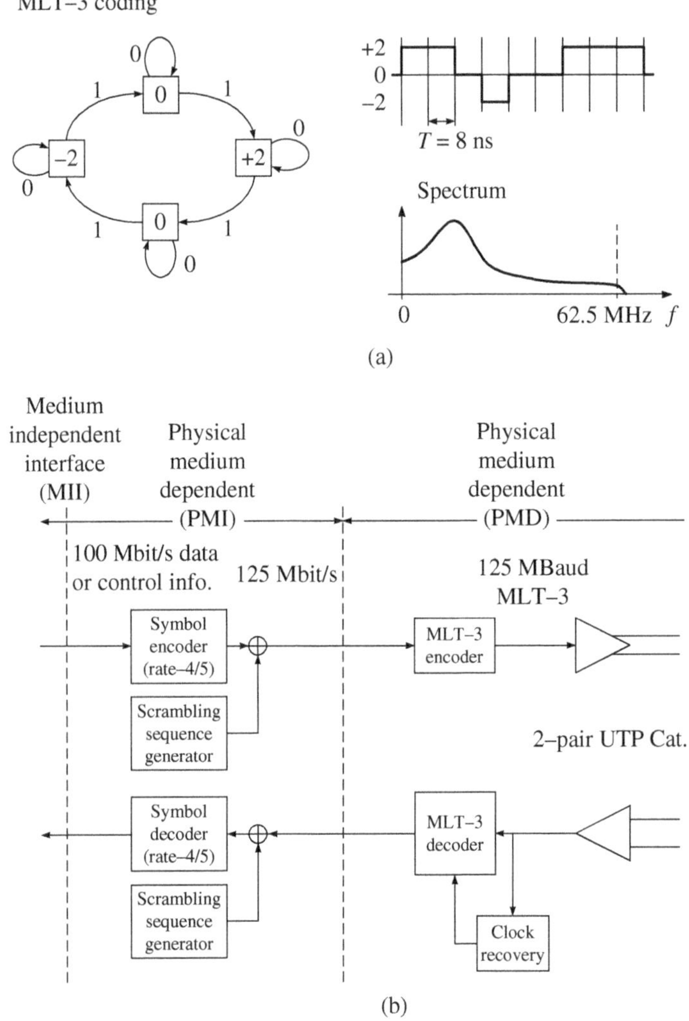

Figure 18.20 Illustration of (a) 100BASE-TX signal characteristics and (b) 100BASE-TX transceiver block diagram.

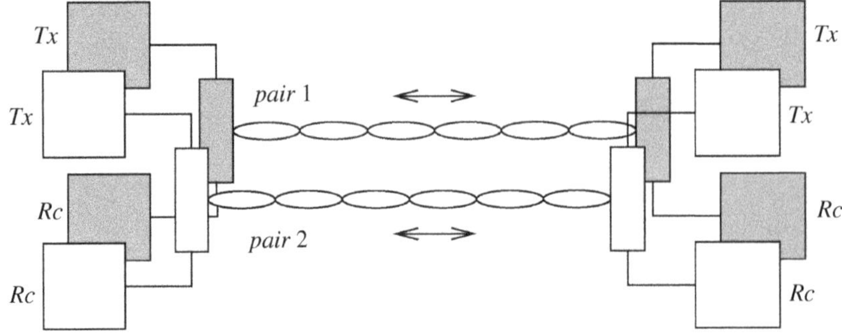

Figure 18.21 Illustration of a dual duplex transmission system.

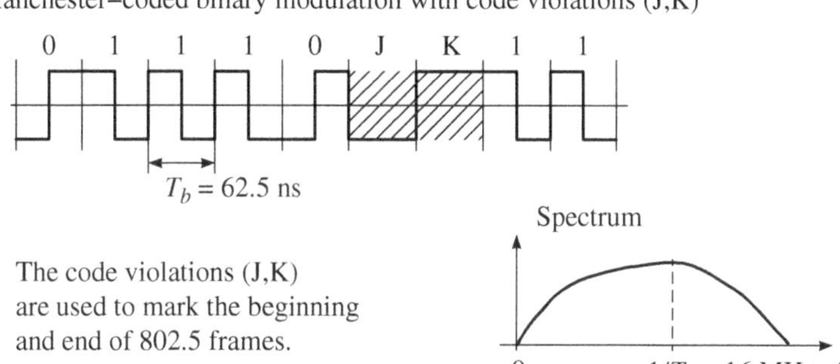

Figure 18.22 Illustration of signal characteristics for transmission over token ring networks.

Table 18.6: Scheme summarizing characteristics of standards for high-speed transmission over UTP cables.

acronym	standard	bit rate (Mbit/s)	cable type
Legacy LANs			
10BASE-T Ethernet	IEEE 802.3	10	2-pair UTP-3
16TR Token Ring	IEEE 802.5	16	2-pair UTP-3
Fibre Distributed Data Interface			
FDDI	ANSI X3T9.5	100	2-pair UTP-5
Fast Ethernet			
100BASE-TX	IEEE 802.3	100	2-pair UTP-5
100BASE-T4	IEEE 802.3	100	4-pair UTP-3
100BASE-T2	IEEE 802.3	100	2-pair UTP-3
1000BASE-T4	IEEE 802.3	1000	4-pair UTP-5
AnyLAN			
100 VG	IEEE 802.12	100	4-pair UTP-3
ATM User-Network Interfaces			
ATM-25	ATM Forum	25.6	2-pair UTP-3
ATM-51	ATM Forum	51.84	2-pair UTP-3
ATM-100	ATM Forum	100	2-pair UTP-5
ATM-155[a]	ATM Forum	155.52	2-pair UTP-3
[a] Cannot operate over UTP-3 in the presence of *alien-NEXT* interference			

in the upstream direction [21, 22]. The topology of an HFC network is illustrated in Figure 18.23: an HFC network is a point-to-multipoint, tree and branch access network in the downlink, with downstream frequencies in the 50–860 MHz band, and a multipoint-to-point, bus access network in the uplink, with upstream frequencies in the 5–42 MHz band. Examples of frequency allocations are shown in Figure 18.24. The maximum round-trip delay between the HC and a station is of the order of 1 ms. The IEEE 802.14 Working Group is one of the standardization bodies working on the specifications for the PHY and MAC layers of HFC networks. A set of PHY and MAC layer specifications adopted by North American cable operators is described in [23].

In the downstream direction, transmission takes place in broadcast made over channels with bandwidth equal to 6 or 8 MHz, characterized by low distortion and high SNR, typically $\geq$ 42 dB. The J.83

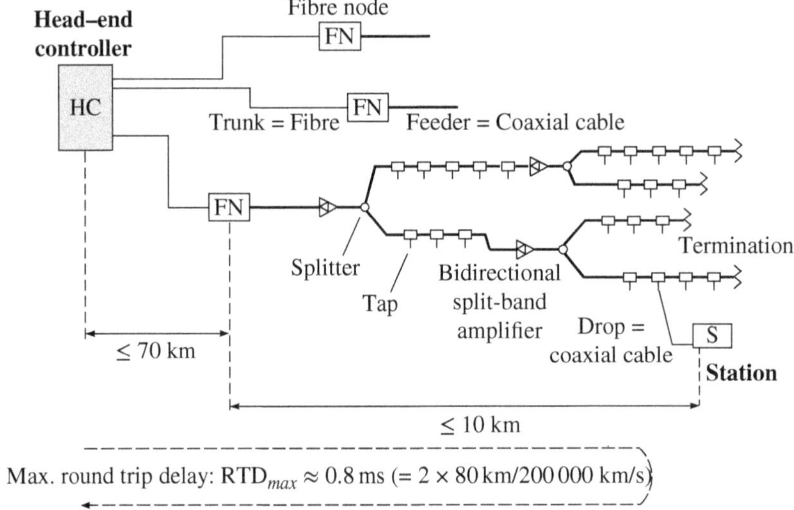

Figure 18.23 Illustration of the HFC network topology.

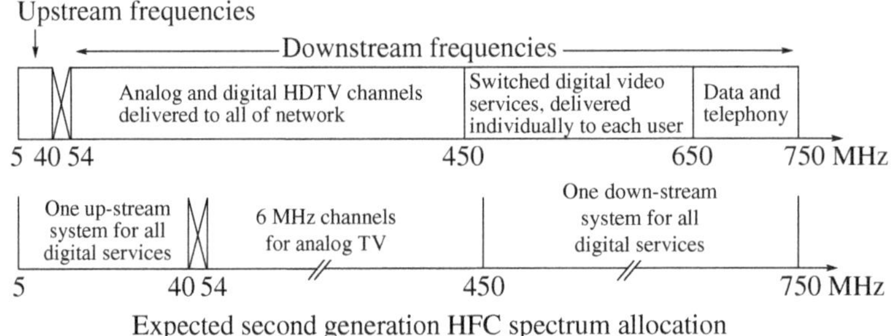

Figure 18.24 Examples of frequency allocations in HFC networks.

document of the ITU-T defines two QAM transmission schemes for transmission in the downstream direction, with a bit rate in the range from 30 to 45 Mbit/s [24]; by these schemes, spectral efficiencies of 5–8 bit/s/Hz are therefore obtained.

In the upstream direction, the implementation of PHY and MAC layers is considerably more difficult than in the downstream. In fact, we can make the following observations:

- signals are transmitted in bursts from the stations to the HC; therefore, it is necessary for the HC receiver to implement fast synchronization algorithms;
- signals from individual stations must be received by the HC at well-defined instants of arrival and power levels; therefore, procedures are required for the determination of the round-trip delay between the HC and each station, as well as for the control of the power of the signal transmitted by each station, as channel attenuation in the upstream direction may present considerable variations, of the order of 60 dB;
- the upstream channel is usually disturbed by impulse noise and narrowband interference signals; moreover, the distortion level is much higher than in the downstream channel.

Interference signals in the upstream channel are caused by domestic appliances and high frequency (HF) radio stations; these signals accumulate along the paths from the stations to the HC and exhibit time varying characteristics; they are usually called *ingress noise*. Because of the high level of disturbance signals, the spectral efficiency of the upstream transmission is limited to 2–4 bit/s/Hz.

The noise PSD suggests that the upstream transmission is characterized by the possibility of changing the frequency band of the transmitted signal (*frequency agility*), and of selecting different modulation rates and spectral efficiencies. In [23], a QAM scheme for upstream transmission that uses a 4 or 16 point constellation and a maximum modulation rate of 2.560 MBaud is defined; the carrier frequency, modulation rate, and spectral efficiency are selected by the HC and transmitted to the stations as MAC information.

Ranging and power adjustment in OFDMA systems

In some multiple-access systems, a HC broadcasts data and MAC information over a set of downlink channels to several stations. These stations send information to the HC over a set of shared uplink channels. Examples of systems exhibiting these characteristics are the two-way HFC systems. This technology has been adopted the data over cable service interface specification (DOCSIS), originally developed for cable modems. DOCSIS most recent version specifies OFDM downstream transmission and OFDMA upstream transmission to achieve efficient use of the spectrum in both uplink and downlink. In the uplink, implementation of physical (PHY) layer transmission and MAC layer functions pose considerable technical challenges. In fact, individual station signals must be received at the HC at defined arrival times and power levels. Therefore, it is important to determine the round-trip delay between the HC and each individual station (ranging) as well as the individual transmit power control for each station to compensate for widely varying attenuations in the uplink direction. In the following, we illustrate an algorithm for initial ranging and power adjustment in OFDMA systems [25], which however may be adapted also for SC systems.

At the beginning of operations, for the realization of the ranging and power adjustment process, a station first tunes its receiver to a downlink channel that conveys MAC information and acquires the global timing reference provided by the HC. Thereafter, the station sends a ranging-request message in the specified subchannel, which results in negligible interference in adjacent subchannels. If a ranging-response message is not received, subsequent ranging-request messages are sent with increasing transmit power, e.g. incremented by 1 dB steps. For ranging-request messages, one may consider a transmission format with a preamble containing a CAZAC sequence [26] of length L, e.g. $L = 16$, which is repeated R times, e.g. $R = 8$, followed by a special start-of-message (SOM) sequence.

Upon detection of the CAZAC sequence, the HC performs subchannel identification using, for example, a least-squares algorithm, see Section 7.8. From the amplitude and phase characteristics of the identified subchannel response, a transmit power level adjustment and a timing phase in $[0, T]$ are derived. The timing phase is needed to compute the round-trip delay compensation so that the MC signal is received in proper synchronism. The detection of the signature sequence provides the HC with further timing information that is used to determine the total round-trip delay from the head-end node to the station, and hence the round-trip delay compensation.

After determining the transmit power level adjustment and the round-trip delay compensation, the HC sends this information to the station as part of a ranging-response message. The station then waits for a message from the HC with an individual station maintenance information and sends at the specified time a ranging-request message using the power level and timing corrections. The HC receives the ranging-request message in proper synchronism at the filter bank output. The HC returns another ranging-response message to the station with information about any additional fine-tuning required. The ranging request/response steps are repeated until the response contains a *Ranging Successful* notification.

Ranging and power adjustment for uplink transmission

We describe now the registration procedure of a cable modem at the HC. The time base relative to the upstream transmission is divided into intervals by the mechanism adopted by the HC for the allocation of resources to the stations. Each interval is constituted by an integer number of subintervals usually called *mini-slots*; a mini-slot then represents the smallest time interval for the definition of an opportunity for upstream transmission.

In [23], a TDMA scheme is considered where uplink transmission is divided into a stream of mini-slots. Each mini-slot is numbered relative to a master reference clock maintained by the HC. The HC distributes timing information to the cable modems by means of time synchronization messages, which include time stamps. From these time stamps, the stations establish a local time base locked to the time base of the HC. For uplink transmission, access to the mini-slots is controlled by *allocation map* (MAP) messages, which describe transmission opportunities on available uplink channels. A MAP message includes a variable number of *information elements* (IE), each of which defines the modality of access to a range of mini-slots in an uplink channel, as illustrated in Figure 18.25. Each station has a unique address of 48 bits; with each active station is also associated, for service request, a 14-bit *service identifier* (SID).

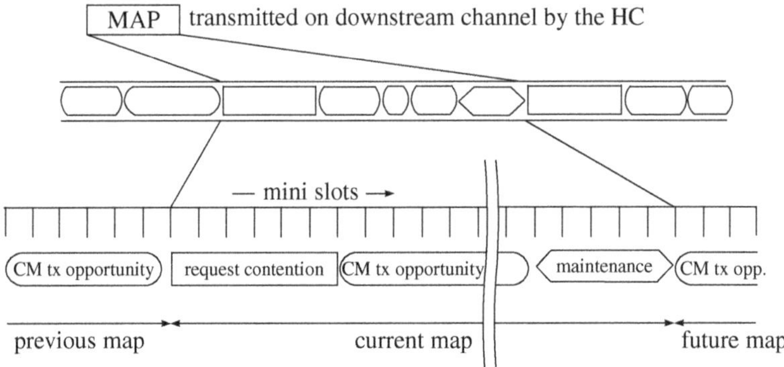

Figure 18.25 Example of a MAP message. Source: Reproduced with permission of Cable Television Laboratories, Inc.

At the beginning of the registration procedures, a station tunes its receiver to the downstream channel on which it receives SYNC messages from the HC; the acquired local timing is delayed with respect to the HC timing due to the signal propagation delay. The station monitors the downstream channel until it receives a MAP message with an IE of initial maintenance, which specifies the time interval during which new stations may send a *ranging request* (RNG-REQ) message to join the network. The duration of the initial maintenance interval is equivalent to the maximum round-trip delay plus the transmission time of a RNG-REQ message. At the instant specified in the MAP message, the station sends a first RNG-REQ message using the lowest power level of the transmitted signal, and is identified by a SID equal to zero as a station that requests to join the network.

If the station does not receive a response within a pre-established time, it means that a collision occurred between RNG-REQ messages sent by more than one station, or that the power level of the transmitted signal was too low; to reduce the probability of repeated collisions, a collision resolution protocol is used with random back-off. After the back-off time interval, of random duration, is elapsed, the station waits for a new MAP message containing an IE of initial maintenance and at the specified instant re-transmits a RNG-REQ message with a higher power level of the transmitted signal. These steps are repeated until the HC detects a RNG-REQ message, from which it can determine the round-trip delay and the correction of the power level that the station must apply for future transmissions. In particular, the compensation for the round-trip delay is computed so that, once applied, the transmitted signals from the station arrive at the HC at well-defined time instants. Then, the HC sends to the station, in a *ranging response* (RNG-RSP) message, the information on round-trip delay compensation and power level correction to be used for future transmissions; this message also includes a temporary SID.

The station awaits for a MAP message containing an IE of station maintenance, individually addressed to it by its temporary SID, and in turn responds through a RNG-REQ message, signing it with its temporary SID and using the specified round-trip delay compensation and power level correction; next the HC

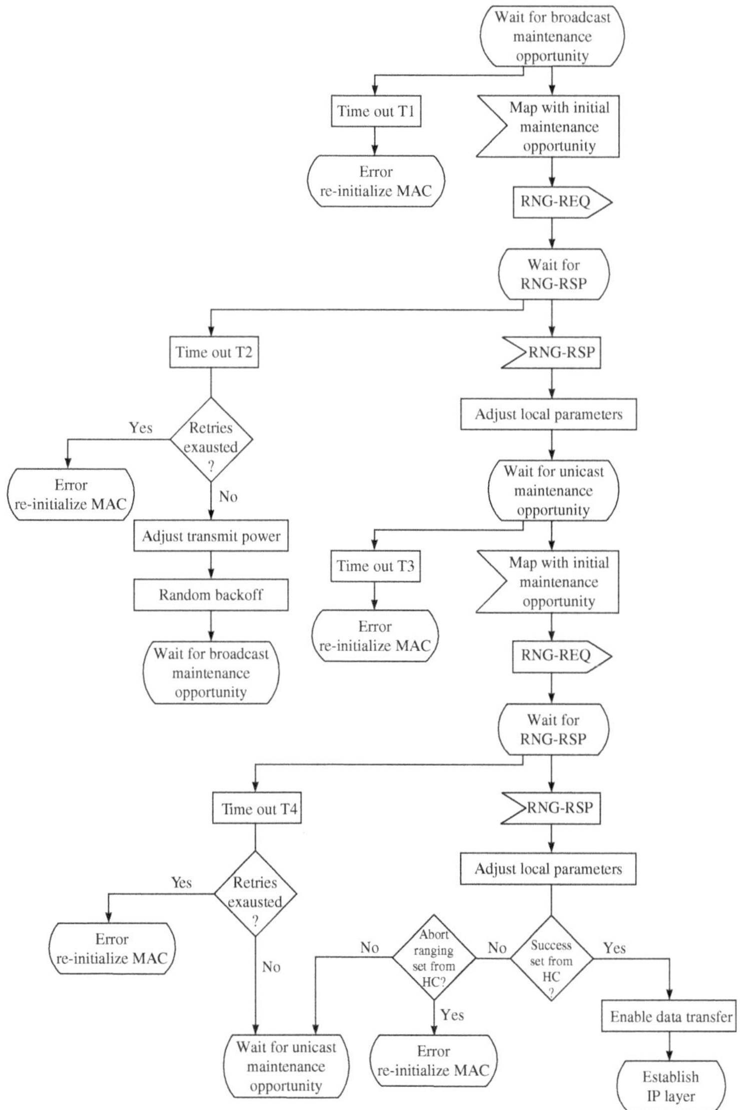

Figure 18.26 Finite state machine used by the cable modem for the registration procedure. Source: Reproduced with permission of Cable Television Laboratories, Inc.

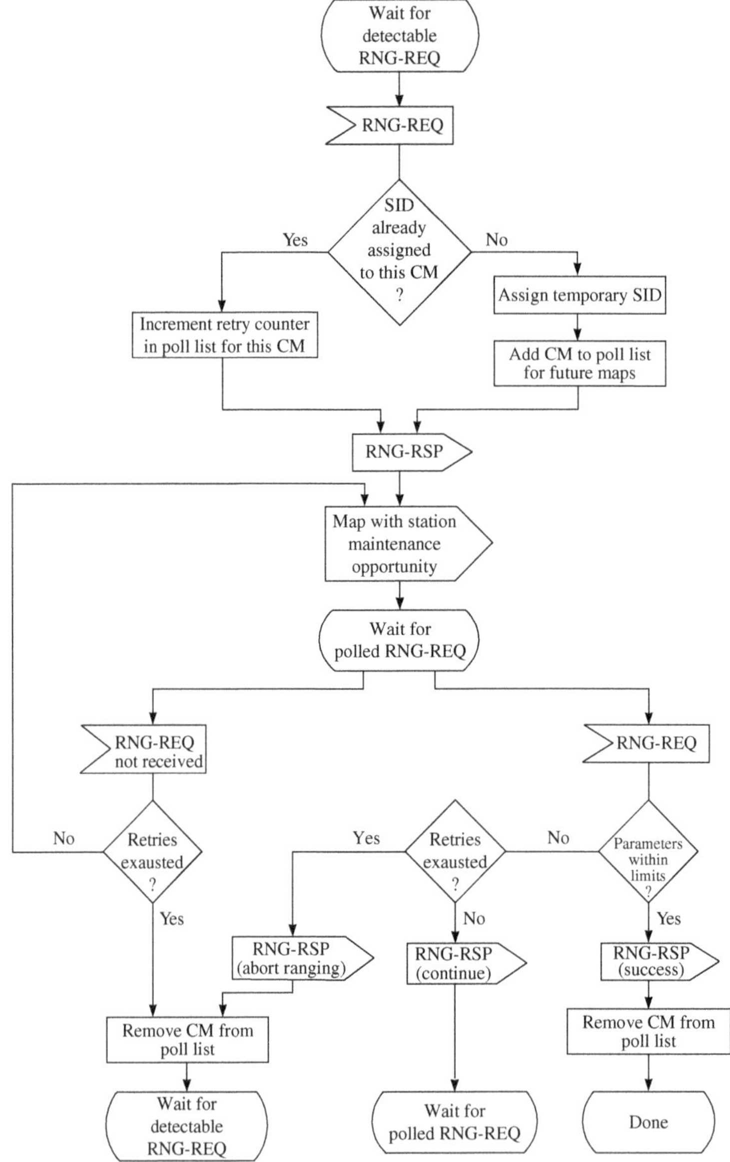

Figure 18.27 Finite state machine used by the HC for the registration procedure. Source: Reproduced with permission of Cable Television Laboratories, Inc.

sends another RNG-REQ message to the station with information for a further refinement of round-trip delay compensation and power level correction. The steps of ranging request/response are repeated until the HC sends a ranging successful message; at this point, the station can send a registration request (REG-REQ) message, to which the HC responds with a registration response (REG-RSP) message confirming the registration and specifying one or more SID that the cable modem must use during the following transmissions. Finite state machines for the ranging procedure and the regulation of the power level for the cable modem and for the HC are shown in Figures 18.26 and 18.27, respectively.

Bibliography

[1] 3GPP Technical Specification Group Radio Access Network (2017). NR; physical layer; general description 12 2017. TS 38.201 V15.0.0.

[2] 3GPP Technical Specification Group Radio Access Network (2018). NR; physical channels and modulation 6 2018. TS 38.211 V15.2.0.

[3] 3GPP Technical Specification Group Radio Access Network (2018). NR; multiplexing and channel coding 6 2018. TS 38.212 V15.2.0.

[4] Lin, X., Li, J., Baldemair, R. et al. (2019). 5G new radio: unveiling the essentials of the next generation wireless access technology. *IEEE Communications Standards Magazine* 3(3): 30–37.

[5] Ferrus, R., Sallent, O., Perez-Romero, J., and Agusti, R. (2018). On 5G radio access network slicing: radio interface protocol features and configuration. *IEEE Communications Magazine* 56: 184–192.

[6] Rahnema, M. (1993). Overview of the GSM system and protocol architecture. *IEEE Communications Magazine* 31: 92–100.

[7] Rappaport, T.S. (1996). *Wireless Communications: Principles and Practice*. Englewood Cliffs, NJ: Prentice-Hall.

[8] Goodman, D.J. (1997). *Wireless Personal Communication Systems*. Reading, MA: Addison-Wesley.

[9] Singh, S. (1996). Wireless LANs. In: *The Mobile Communications Handbook*, Chapter 4 (ed. J.D. Gibson), 540–552. New York, NY: CRC/IEEE Press.

[10] Miller, B.A. and Bisdikian, C. (2001). *Bluetooth Revealed*. Upper Saddle River, NJ: Prentice-Hall.

[11] Haartsen, J.C. and Mattison, S. (2000). Bluetooth — a new low–power radio interface providing short–range connectivity. *Proceedings of the IEEE* 88: 1651–1661.

[12] Ahamed, S.V., Gruber, P.L., and Werner, J.-J. (1995). Digital subscriber line (HDSL and ADSL) capacity of the outside loop plant. *IEEE Journal on Selected Areas in Communications* 13: 1540–1549.

[13] Chen, W.Y. and Waring, D.L. (1994). Applicability of ADSL to support video dial tone in the copper loop. *IEEE Communications Magazine* 32: 102–109.

[14] Hawley, G.T. (1997). Systems considerations for the use of xDSL technology for data access. *IEEE Communications Magazine* 35: 56–60.

[15] Chen, W.Y. and Waring, D.L. (1994). Applicability of ADSL to support video dial tone in the copper loop. *IEEE Communications Magazine* 32: 102–109.

[16] Starr, T., Cioffi, J.M., and Silverman, P.J. (2000). *Understanding Digital Subscriber Line Technology*. Englewood Cliffs, NJ: Prentice-Hall.

[17] Sjoberg, F., Isaksson, M., Nilsson, R. et al. (1999). Zipper: a duplex method for WDSL based on DMT. *IEEE Transactions on Communications* 47: 1245–1252.

[18] Cherubini, G., Eleftheriou, E., Ölçer, S., and Cioffi, J.M. (2000). Filter bank modulation techniques for very high-speed digital subscriber lines. *IEEE Communications Magazine* 38: 98–104.

[19] Cherubini, G., Eleftheriou, E., and Ölcer, S. (2002). Filtered multitone modulation for very-high-speed digital subscriber lines. *IEEE Journal on Selected Areas in Communications* 20 5: 1016–1028.

[20] Cherubini, G., Eleftheriou, E., and Ölcer, S. (1999). Filtered mutitone modulation for VDSL. *Proceedings of IEEE GLOBECOM '99*, Rio de Janeiro, Brazil, pp. 1139–1144.

[21] Eldering, C.A., Himayat, N., and Gardner, F.M. (1995). CATV return path characterization for reliable communications. *IEEE Communications Magazine* 33: 62–69.

[22] Bisdikian, C., Maruyama, K., Seidman, D.I., and Serpanos, D.N. (1996). Cable access beyond the hype: on residential broadband data services over HFC networks. *IEEE Communications Magazine* 34: 128–135.

[23] Fellows, D. and Jones, D. (2001). DOCSIS$^{(\mathrm{tm})}$. *IEEE Communications Magazine* 39: 202–209.

[24] ITU-T Study Group 9 (1995). Digital multi-programme systems for television sound and data services for cable distribution. ITU-T Recommendation J.83, 24 October 1995.

[25] Cherubini, G. (2003). Hybrid TDMA/CDMA based on filtered multitone modulation for uplink transmission in HFC networks. *IEEE Communications Magazine* 41: 108–115.

[26] Chu, D.C. (1972). Polyphase codes with good periodic correlation properties. *IEEE Transactions on Information Theory* 18: 531–532.

[27] Abramson, N. (ed.) (1993). *Multiple Access Communications: Foundations for Emerging Technologies*. Piscataway, NJ: IEEE Press.

Appendix 18.A Duplexing

A transmission link between two users of a communication network may be classified as

(a) *Full duplex*, when two users A and B can send information to each other simultaneously, not necessarily by using the same transmission channels in the two directions.
(b) *Half duplex*, when two users A and B can send information in only one direction at a time, from A to B or from B to A, alternatively.
(c) *Simplex*, when only A can send information to B, that is the link is unidirectional.

Three methods

In the following, we give three examples of transmission methods which are used in practice.

(a) *Frequency-division duplexing* (FDD): In this case, the two users are assigned different transmission bands using the same transmission medium, thus allowing full-duplex transmission. Examples of FDD systems are the GSM, which uses a radio channel (see Section 18.2), and the VDSL, which uses a twisted pair cable (see Section 18.6.1).
(b) *Time-division duplexing* (TDD): In this case, the two users are assigned different slots in a time frame (see Appendix 18.B). If the duration of one slot is small with respect to that of the message, we speak of *full-duplex* TDD systems. Examples of TDD systems are the DECT, which uses a radio channel (see Section 18.4), and the *ping-pong* BR-ISDN, which uses a twisted pair cable.
(c) *Full-duplex systems over a single band*: In this case, the two users transmit simultaneously in two directions using the same transmission band; examples are the HDSL (see Section 18.6.1), and in general high-speed transmission systems over twisted-pair cables for LAN applications (see Section 18.6.2). The two directions of transmission are separated by a hybrid; the receiver eliminates echo signals by echo cancellation techniques. We note that *full-duplex* transmission over a single band is possible also over radio channels, but in practice, alternative methods are still preferred because of the complexity required by echo cancellation.

Appendix 18.B Deterministic access methods

We distinguish three cases for channel access by N users:

1. Subdivision of the channel passband into N_B separate subbands that may be used for transmission (see Figure 18.28a).

2. Subdivision of a sequence of modulation intervals into adjacent subsets called *frames*, each in turn subdivided into N_S adjacent subsets called *slots*. Within a *frame*, each *slot* is identified by an index i, $i = 0, \dots, N_S - 1$ (see Figure 18.28b).
3. Signalling by N_0 orthogonal signals (see for example Figure 10.18).

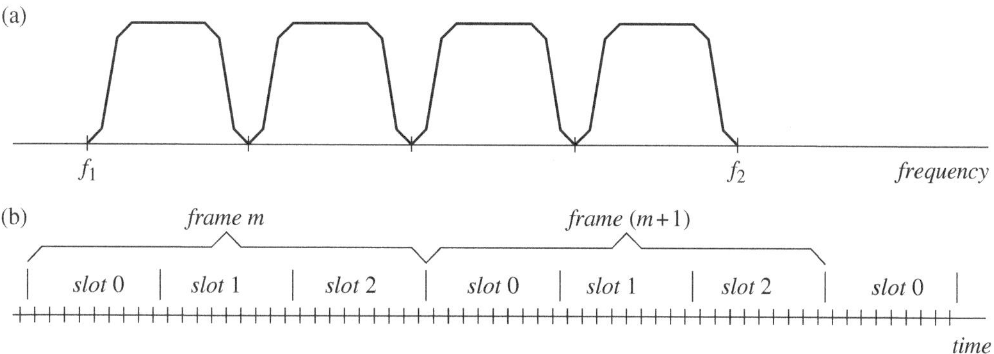

Figure 18.28 Illustration of (a) FDMA, and (b) TDMA.

N users may share the channel using one of the following methods.[6]

1. *Frequency division multiple access* (FDMA): To each user is assigned one of the N_B subbands.
2. *Time division multiple access* (TDMA): To each user is assigned one of the N_S time sequences (slots), whose elements identify the modulation intervals.
3. *Code division multiple access* (CDMA): To each user is assigned a modulation scheme that employs one of the N_0 orthogonal signals, preserving the orthogonality between modulated signals of the various users. For example, for the case $N_0 = 8$, to each user may be assigned one orthogonal signal of those given in Figure 10.18; for binary modulation, within a modulation interval each user then transmits the assigned orthogonal signal or its antipodal version.

We give an example of implementation of the TDMA principle.

Example 18.B.1 (Time-division multiplexing)
Time-division multiplexing (TDM) is the interlacing of several digital messages into one digital message with a higher bit rate; as an example we illustrate the generation of the European base group, called E1, at 2.048Mbit/s, that is obtained by multiplexing 30 pulse-code-modulation (PCM) coded speech signals (or channels) at 64kbit/s. As shown in Figure 18.29, each 8-bit sample of each channel is inserted into a pre-assigned *slot* of a frame composed of 32 slots, equivalent to $32 \cdot 8 = 256$ bits. The frame structure must contain information bits to identify the beginning of a frame (channel ch0) by 8 known framing bits; 8bits are employed for signalling between central offices (channel ch16). The remaining 30 channels are for the transmission of signals. As the duration of a frame is of 125 μs, equal to the interval between two PCM samples of the same channel, the overall digital message has a bit rate of 256 bit/125 μs = 2.048 Mbit/s; we note however that of the 256 bits of the frame only $30 \cdot 8 = 240$ bits carry information related to signals.

[6] The access methods discussed in this section are deterministic, as each user knows exactly at which point in time the channel resources are reserved for transmission; an alternative approach is represented by random access techniques, e.g. ALOHA, CSMA/CD, collision resolution protocols [27].

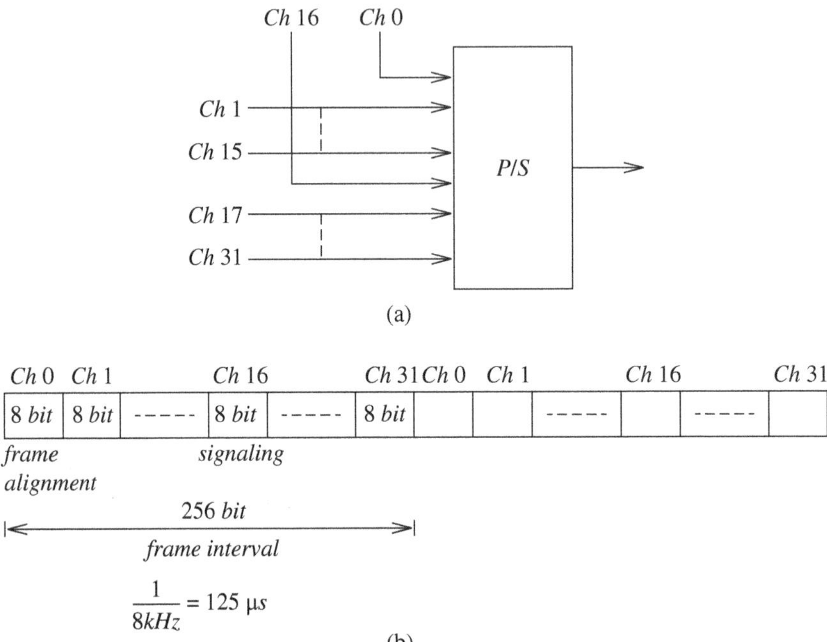

Figure 18.29 TDM in the European base group at 2.048 Mbit/s. (a) Basic scheme. (b) Frame structure.

In the United States, Canada, and Japan, the base group, analog to E1, is called *T1 carrier system* and has a bit rate of 1.544 Mbit/s, obtained by multiplexing 24 PCM speech coded signals at 64 kbit/s. In this case, the frame is such that 1bit per channel is employed for signalling. This bit is *robbed* from the least important bit of the 8-bit PCM sample, thus making it a 7-bit code word per sample; there is then only 1bit for the synchronization of the whole frame. The entire frame is formed of $24 \cdot 8 + 1 = 193$ bits.

Chapter 19

High-speed communications over twisted-pair cables

In this chapter, we describe the design of two high-speed data transmission systems over unshielded twisted-pair (UTP) cables [1, 2].

19.1 Quaternary partial response class-IV system

Figure 19.1 shows the block diagram of a transceiver for a *quaternary partial response class-IV* (QPR-IV), or quaternary modified duobinary (see Appendix 7.C), system for data transmission at 125 Mbit/s over UTP cables [1]. In the transmitter, information bits are first scrambled and then input to a 2B1Q differential encoder that yields output symbols belonging to the quaternary alphabet $\mathcal{A} = \{-3, -1, +1, +3\}$ (see Example 16.1.1 on page 793); differential encoding makes the transmission of information insensitive to the polarity of the received signals. Signal shaping into partial response class-IV (PR-IV) form is accomplished by the cascade of the following elements: the digital-to-analog converter (DAC), the analog transmit filter (ATF), the cable, the analog receive filter (ARF) with automatic gain control (AGC), the analog-to-digital converter (ADC), and, in the digital domain, the fixed decorrelation filter (DCF) and the adaptive equalizer. After equalization, the sequence of transmitted quaternary symbols is detected by a Viterbi algorithm. Since the Viterbi algorithm introduces a delay in the detection of the transmitted sequence, as an alternative it is possible to use a threshold detector that individually detects transmitted symbols with a negligible delay. If the mean-square error (MSE) is below a fixed value, threshold detection, or symbol-by-symbol detection (SSD), is selected as the corresponding symbol decisions are then sufficiently reliable. If, instead, the MSE exceeds the fixed value, the Viterbi algorithm is employed. Finally, differential 1Q2B decoding and descrambling are performed.

Analog filter design

Figure 19.2 depicts the overall analog channel considered for the design of ATF and ARF. The impulse response is denoted by $h(t, L, u_c)$, where L is the cable length and u_c is the control signal used for AGC. To determine ATF and ARF with low implementation complexity, the transfer functions of these filters are first expressed in terms of poles and zeros; the pole-zero configurations are then jointly optimized by simulated annealing as described in [3]. The cost function that is used for the optimization reflects two criteria: (i) the MSE between the impulse response $h(t, L, u_c)$ for a cable length equal to 50 m, and the ideal PR-IV response must be below a certain value and (ii) spectral components of the transmitted

Algorithms for Communications Systems and their Applications, Second Edition.
Nevio Benvenuto, Giovanni Cherubini, and Stefano Tomasin.

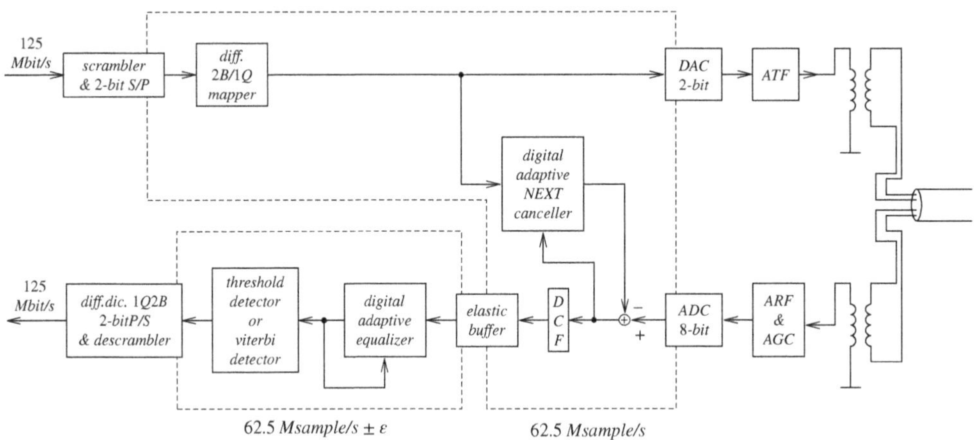

Figure 19.1 Block diagram of a QPR-IV transceiver. Source: Reproduced with permission from Cherubini et al. [1]. ©1995, IEEE.

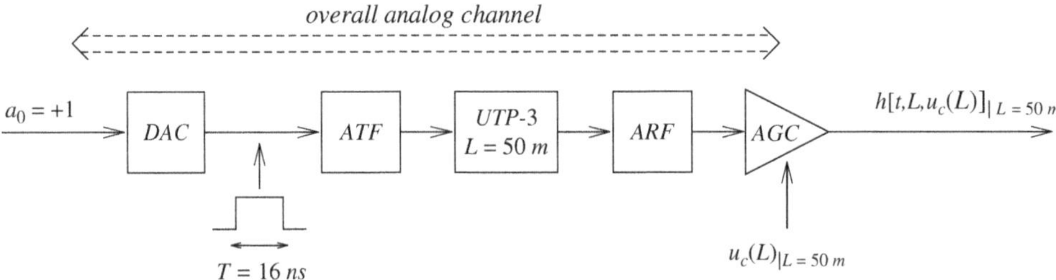

Figure 19.2 Overall analog channel considered for the joint optimization of the analog transmit (ATF) and receive (ARF) filters. Source: Reproduced with permission from Cherubini et al. [1]. ©1995, IEEE.

signal above 30 MHz should be well suppressed for achieving compliance with regulations on radiation limits. Investigations with various orders of the respective transfer functions have shown that a good approximation of an ideal PR-IV response is obtained with 5 poles and 3 zeros for the ATF, and 3 poles for the ARF.

Received signal and adaptive gain control

The received signal at the input of the ADC is expressed as

$$x(t) = \sum_{k=-\infty}^{+\infty} a_k \, h[t - kT, L, u_c(L)] + \sum_{k=-\infty}^{+\infty} a_k^N \, h_N[t - kT, L, u_c(L)] + w_R(t) \tag{19.1}$$

where $\{a_k\}$ and $\{a_k^N\}$ are the sequences of quaternary symbols generated by the local transmitter and remote transmitter, respectively, $h_N[t, L, u_c(L)]$ is the near-end crosstalk (NEXT) channel response and w_R is additive Gaussian noise. The signal x is sampled by the ADC that operates synchronously with the DAC at the modulation rate of $1/T = 62.5$ MBaud. The adjustment of the AGC circuit is computed digitally, using the sampled signal $x_k = x(kT)$, such that the ADC output signal achieves a constant average statistical power $\mathtt{M}_R$, i.e.

$$E[x_k^2]|_{u_c=u_c(L)} = \frac{1}{T}\int_0^T E[x^2(t)]\,dt\Bigg|_{u_c=u_c(L)} = \mathtt{M}_R \tag{19.2}$$

This ensures that the received signal is converted with optimal precision independently of the cable length; moreover, a controlled level of the signal at the adaptive digital equalizer input is required for achieving optimal convergence properties.

Near-end crosstalk cancellation

As illustrated in Chapter 17, NEXT cancellation is achieved by storing the transmit symbols a_k^N in a delay line and computing an estimated NEXT signal $\hat{u}_k^N$, which is then subtracted from the received signal x_k, that is

$$\tilde{x}_k = x_k - \hat{u}_k^N = x_k - \sum_{i=0}^{N_N-1} c_{i,k}^N\, a_{k-i}^N \tag{19.3}$$

where $\{c_{i,k}^N\}$, $i = 0, \dots, N_N - 1$, are the coefficients of the adaptive NEXT canceller. Using the minimization of $E[\tilde{x}_k^2]$ as a criterion for updating the NEXT canceller coefficients (see Section 17.1), leads to the least mean square (LMS) algorithm

$$c_{i,k+1}^N = c_{i,k}^N + \mu_N\, \tilde{x}_k\, a_{k-i}^N, \qquad 0 \le i \le N_N - 1 \tag{19.4}$$

where μ_N is the adaptation gain.

As discussed in Chapter 17, high-speed full-duplex transmission over two separate wire pairs with NEXT cancellation and full-duplex transmission over a single pair with echo cancellation pose similar challenges. In the latter case, a hybrid is included to separate the two directions of transmission; a QPR-IV transceiver with NEXT cancellation can then be used also for full-duplex transmission over a single pair, as in this case the NEXT canceller acts as an echo canceller.

Decorrelation filter

After NEXT cancellation, the signal is filtered by a DCF, which is used to improve the convergence properties of the adaptive digital equalizer by reducing the correlation between the samples of the sequence $\{\tilde{x}_k\}$. The filtering operation performed by the DCF represents an approximate inversion of the PR-IV frequency response. The DCF has frequency response $1/(1 - \beta\, e^{-j4\pi fT})$, with $0 < \beta < 1$, and provides at its output the signal

$$z_k = \tilde{x}_k + \beta z_{k-2} \tag{19.5}$$

Adaptive equalizer

The samples $\{z_k\}$ are stored in an elastic buffer, from which they are transferred into the equalizer delay line. Before describing this operation in more detail, we make some observations about the adaptive equalizer. As mentioned above, the received signal x is sampled in synchronism with the timing of the local transmitter. Due to the frequency offset between local and remote transmitter clocks, the phase of the remote transmitter clock will drift in time relative to the sampling phase. As the received signal is bandlimited to one half of the modulation rate, signal samples taken at the symbol rate are not affected by aliasing; hence, a fractionally spaced equalizer is not required for a QPR-IV system. Furthermore, as the signal value x can be reconstructed from the T-spaced samples $\{z_k\}$, an equalizer of sufficient length acts also as an interpolator. The adaptive equalizer output signal is given by

$$y_k = \sum_{i=0}^{N_E-1} c_{i,k}^E \, z_{k-i} \tag{19.6}$$

where $\{c_{i,k}^E\}$, $i = 0, \ldots, N_E - 1$, denote the filter coefficients.

Compensation of the timing phase drift

The effect of timing phase drift can be compensated by continuously adjusting the equalizer coefficients. As a result of these adjustments, for a positive frequency offset of $+\delta/T$ Hz, i.e. for a frequency of the local transmitter clock larger than the frequency of the remote transmitter clock, the value of the i-th coefficient at time k is approximately equal to the value assumed by the $(i+1)$-th coefficient $1/\delta$ modulation intervals earlier. In other words, the coefficients move to the left by one position relative to the equalizer delay line after $1/\delta$ modulation intervals; conversely, for a negative frequency offset of $-\delta/T$ Hz, the coefficients move to the right by one position relative to the equalizer delay line after $1/\delta$ modulation intervals. Hence, the centre of gravity of the filter coefficients drifts. Proper operation of a finite-length equalizer requires that the coefficients be re-centred; this is accomplished as follows:

- Normally, for each equalizer output y_k, one new signal z_k is transferred from the buffer into the equalizer delay line;
- Periodically, after a given number of modulation intervals, the sums of the magnitudes of N' first and last coefficients are compared; if the first coefficients are too large, the coefficients are shifted by one position towards the end of the delay line (right shift) and two new signals z_k and z_{k+1} are transferred from the buffer into the delay line; if the last coefficients are too large, the coefficients are shifted by one position towards the beginning of the delay line (left shift) and no new signal is retrieved from the buffer.

To prevent buffer overflow or underflow, the rate of the remaining receiver operations is controlled so that the elastic buffer is kept half full on average, thus performing indirect timing recovery. The control algorithm used to adjust the voltage-controlled oscillator (VCO) providing the timing signal for the remaining receiver operations is described in Chapter 14.

Adaptive equalizer coefficient adaptation

Overall system complexity is reduced if self-training adaptive equalization is employed, as in this case a start-up procedure with the transmission of a known training sequence is not needed. We describe here the operation of the adaptive equalizer and omit the presentation of the self-training algorithm, discussed in Section 15.4.

The MSE at the equalizer output is continuously monitored using seven-level tentative decisions. If the MSE is too large, self-training adaptive equalization is performed during a fixed time period T_{ST}. At the end of the self-training period, if the MSE is sufficiently small, equalizer operation is continued with the decision directed LMS algorithm

$$c_{i,k+1}^E = c_{i,k}^E - \mu_E \, z_{k-i} \, \hat{e}_k, \qquad 0 \le i \le N_E - 1 \tag{19.7}$$

where $\hat{e}_k = y_k - (\hat{a}_k - \hat{a}_{k-2})$ is the error obtained using tentative decisions $\hat{a}_k$ on the transmitted quaternary symbols, and μ_E is the adaptation gain.

Convergence behaviour of the various algorithms

We resort to computer simulations to study the convergence behaviour of the adaptive digital NEXT canceller and of the adaptive equalizer. The length of the NEXT canceller is chosen equal to $N_N = 48$ to ensure that, in the worst case signal attenuation, the power of the residual NEXT is with high probability more than 30 dB below the power of the signal from the remote transmitter. In Figure 19.3, the residual NEXT statistical power at the canceller output is plotted versus the number of iterations for a worst-case cable length of $L = 100$ m and an adaptation gain $\mu_N = 2^{-18}$. In the same figure, the statistical power in dB of the portion of the NEXT signal that cannot be cancelled due to finite NEXT-canceller length is also indicated. For the simulations, the NEXT canceller was assumed to be realized in the distributed-arithmetic form (see Section 17.1).

The convergence of the MSE at the output of the equalizer in the absence of timing phase drift is shown in Figure 19.4a,b for best and worst-case sampling phase, respectively, and a value $\Gamma_{Tx} = 2E_{Tx}/N_0$ of 43 dB, where E_{Tx} is the average energy per modulation interval of the transmitted signal. The NEXT canceller is assumed to have converged to the optimum setting. An equalizer length of $N_E = 24$ is chosen, which guarantees that the mean-square interpolation error with respect to an ideal QPR-IV signal is less than −25 dB for the worst-case sampling phase. The self-training period is $T_{ST} \approx 400$ μs, corresponding to approximately 25000T. The adaptation gains for self-training and decision directed adjustment have the same value $\mu_E = 2^{-9}$, that is chosen for best performance in the presence of a worst-case timing phase drift $\delta = 10^{-4}$. Figure 19.5 shows the MSE convergence curves obtained for $\delta = 10^{-4}$.

19.1.1 VLSI implementation

Adaptive digital NEXT canceller

As shown in Chapter 17, for the very large-scale integration (VLSI) implementation of the adaptive digital NEXT canceller, a distributed-arithmetic filter presents significant advantages over a transversal

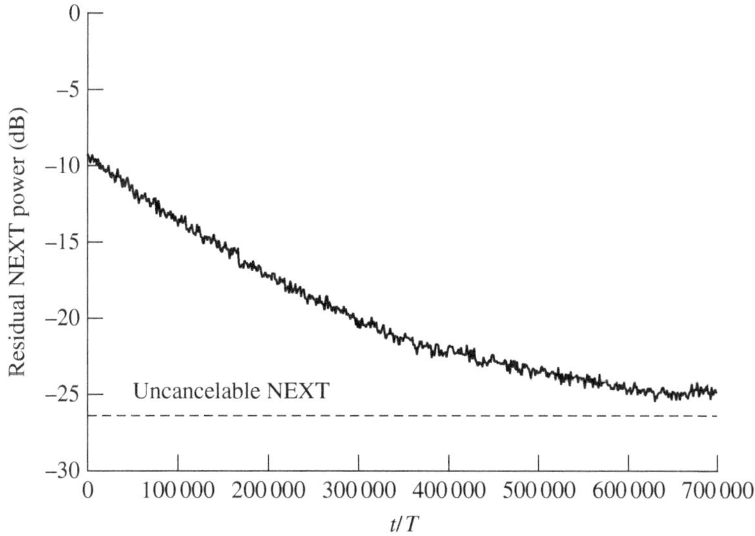

Figure 19.3 Convergence of the adaptive NEXT canceller. Source: Reproduced with permission from Cherubini et al. [1]. ©1995, IEEE.

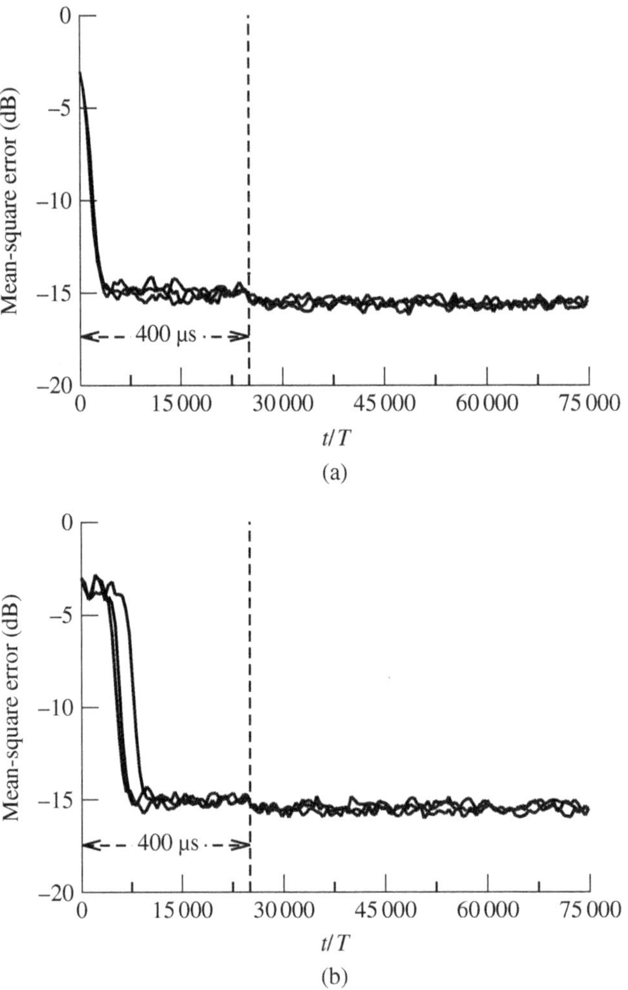

Figure 19.4 Convergence of the adaptive equalizer for (a) best-case sampling phase and (b) worst-case sampling phase. Source: Reproduced with permission from Cherubini et al. [1]. ©1995, IEEE.

filter in terms of implementation complexity. In a NEXT canceller distributed-arithmetic filter, the partial products that appear in the expression of a transversal filter output are not individually computed; evaluation of partial products is replaced by table look-up and shift-and-add operations of binary words. To compute the estimate of the NEXT signal to be subtracted from the received signal, look-up values are selected by the bits in the NEXT canceller delay line and added by a carry-save adder. By segmenting the delay line of the NEXT canceller into sections of shorter lengths, a trade-off concerning the number of operations per modulation interval and the number of memory locations that are needed to store the look-up values is possible. The convergence of the look-up values to the optimum setting is achieved by an LMS algorithm.

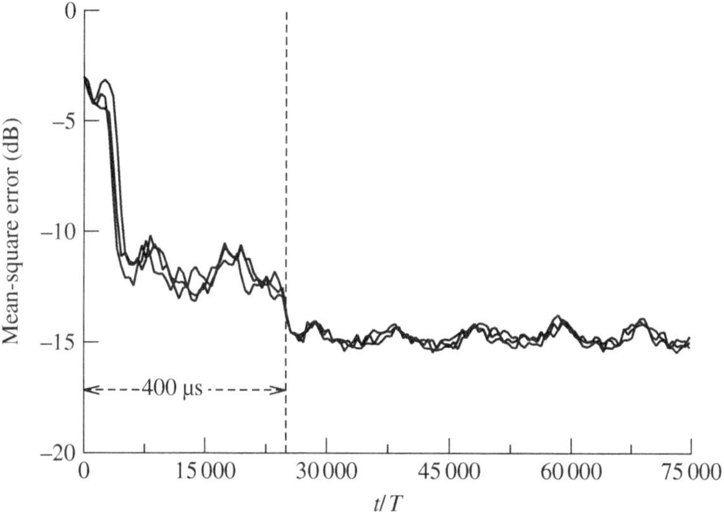

Figure 19.5 Convergence of the adaptive equalizer for worst-case timing phase drift. Source: Reproduced with permission from Cherubini et al. [1]. ©1995, IEEE.

If the delay line of the NEXT canceller is segmented into L sections with $K = N_N/L$ delay elements each, the NEXT canceller output signal is given by

$$\hat{u}_k^N = \sum_{i=0}^{N_N-1} a_{k-i}^N \, c_{i,k}^N = \sum_{\ell=0}^{L-1} \sum_{m=0}^{K-1} a_{k-\ell K-m}^N \, c_{\ell K+m,k}^N \tag{19.8}$$

In a distributed-arithmetic implementation, the quaternary symbol a_k^N is represented by the binary vector $[a_k^{(0)}, a_k^{(1)}]$, that is

$$a_k^N = \sum_{w=0}^{1} (2a_k^{(w)} - 1)2^w = \sum_{w=0}^{1} b_k^{(w)} 2^w \tag{19.9}$$

where $b_k^{(w)} = (2a_k^{(w)} - 1) \in \{-1, +1\}$. Introducing (19.9) into (19.8), we obtain (see (17.12))

$$\hat{u}_k^N = \sum_{\ell=0}^{L-1} \sum_{w=0}^{1} 2^w \left[\sum_{m=0}^{K-1} b_{k-\ell K-m}^{(w)} \, c_{\ell K+m,k}^N \right] \tag{19.10}$$

Equation (19.10) suggests that the filter output can be computed using a set of $L2^K$ look-up values that are stored in L look-up tables with 2^K memory locations each. Extracting the term $b_{k-\ell K}^{(w)}$ out of the square bracket in (19.10), to determine the output of a distributed-arithmetic filter with reduced memory size $L2^{K-1}$, we rewrite (19.10) as (see (17.13))

$$\hat{u}_k^N = \sum_{\ell=0}^{L-1} \sum_{w=0}^{1} 2^w \, b_{k-\ell K}^{(w)} \, d_k^N(i_{k,\ell}^{(w)}, \ell) \tag{19.11}$$

where $\{d_k^N(n, \ell)\}$, $n = 0, \ldots, 2^{K-1} - 1$, $\ell = 0, \ldots, L-1$, are the look-up values, and $i_{k,\ell}^{(w)}$ denotes the selected look-up address that is computed as follows:

$$i_{k,\ell}^{(w)} = \begin{cases} \sum_{m=1}^{K-1} a_{k-\ell K-m}^{(w)} \ 2^{m-1} & \text{if } a_{k-\ell K}^{(w)} = 1 \\ \sum_{m=1}^{K-1} \overline{a}_{k-\ell K-m}^{(w)} \ 2^{m-1} & \text{if } a_{k-\ell K}^{(w)} = 0 \end{cases} \tag{19.12}$$

where $\overline{a}_n$ is the one's complement of a_n. The expression of the LMS algorithm to update the look-up values of a distributed-arithmetic NEXT canceller takes the form

$$\begin{gathered} d_{k+1}^N(n,\ell) = d_k^N(n,\ell) + \mu_N \ \tilde{x}_k \ \sum_{w=0}^{1} 2^w \ b_{k-\ell K}^{(w)} \ \delta_{n-i_{k,\ell}^{(w)}}, \\ n = 0, \ldots, 2^{K-1} - 1, \quad \ell = 0, \ldots, L-1 \end{gathered} \tag{19.13}$$

where δ_n is the Kronecker delta. We note that at each iteration only those look-up values that are selected to generate the filter output are updated. The implementation of the NEXT canceller is further simplified by updating at each iteration only the look-up values that are addressed by the most significant bits of the symbols, i.e. those with index $w = 1$, stored in the delay line (see (17.23)). The block diagram of an adaptive distributed-arithmetic NEXT canceller is shown in Figure 19.6. In the QPR-IV transceiver, for the implementation of a NEXT canceller with a time span of $48T$, $L = 16$ segments with $K = 3$ delay elements each are employed. The look-up values are stored in 16 tables with four 16-bit registers each.

Adaptive digital equalizer

As discussed A in the previous section, the effect of the drift in time of the phase of the remote transmitter clock relative to the sampling phase is compensated by continuously updating and occasionally re-centring the coefficients of the digital equalizer. The need for fast equalizer adaptation excludes a distributed-arithmetic approach for the implementation of the digital equalizer. An efficient solution for occasional re-centring of the N_E equalizer coefficients is obtained by the following structure, which is based on an approach where N_E *multiply-accumulate* (MAC) units are employed.

According to (19.6), assuming that the coefficients are not time varying, at a given time instant k, each MAC unit is presented with a different tap coefficient and carries out the multiplication of this tap coefficient with the signal sample z_k. The result is a partial product that is added to a value stored in the MAC unit, which represents the sum of partial products up until time instant k. The MAC unit that has accumulated N_E partial products provides the equalizer output signal at time instant k, $y_k = \sum_{i=0}^{N_E-1} c_i^E \ z_{k-i}$, and its memory is cleared to allow for the accumulation of the next N_E partial products. At this time instant, the MAC unit that has accumulated the result of $(N_E - 1)$ partial products has stored the term $\sum_{i=1}^{N_E-1} c_i^E \ z_{k-(i-1)}$. At time instant $(k + 1)$, this unit computes the term $c_0^E \ z_{k+1}$ and provides the next equalizer output signal

$$y_{k+1} = c_0^E \ z_{k+1} + \sum_{i=1}^{N_E-1} c_i^E \ z_{k-(i-1)} \tag{19.14}$$

This MAC unit is then reset and its output will be considered again N_E time instants later.

Figure 19.7 depicts the implementation of the digital equalizer. The N_E coefficients $\{c_i^E\}$, $i = 0, \ldots, N_E - 1$, normally circulate in the delay line shown at the top of the figure. Except when re-centring of the equalizer coefficients is needed, N_E coefficients in the delay line are presented each to a different MAC unit, and the signal sample z_k is input to all MAC units. At the next time instant, the coefficients are cyclically shifted by one position, and the new signal sample z_{k+1} is input to all the units. The multiplexer shown at the bottom of the figure selects in turn the MAC unit that provides the equalizer output signal.

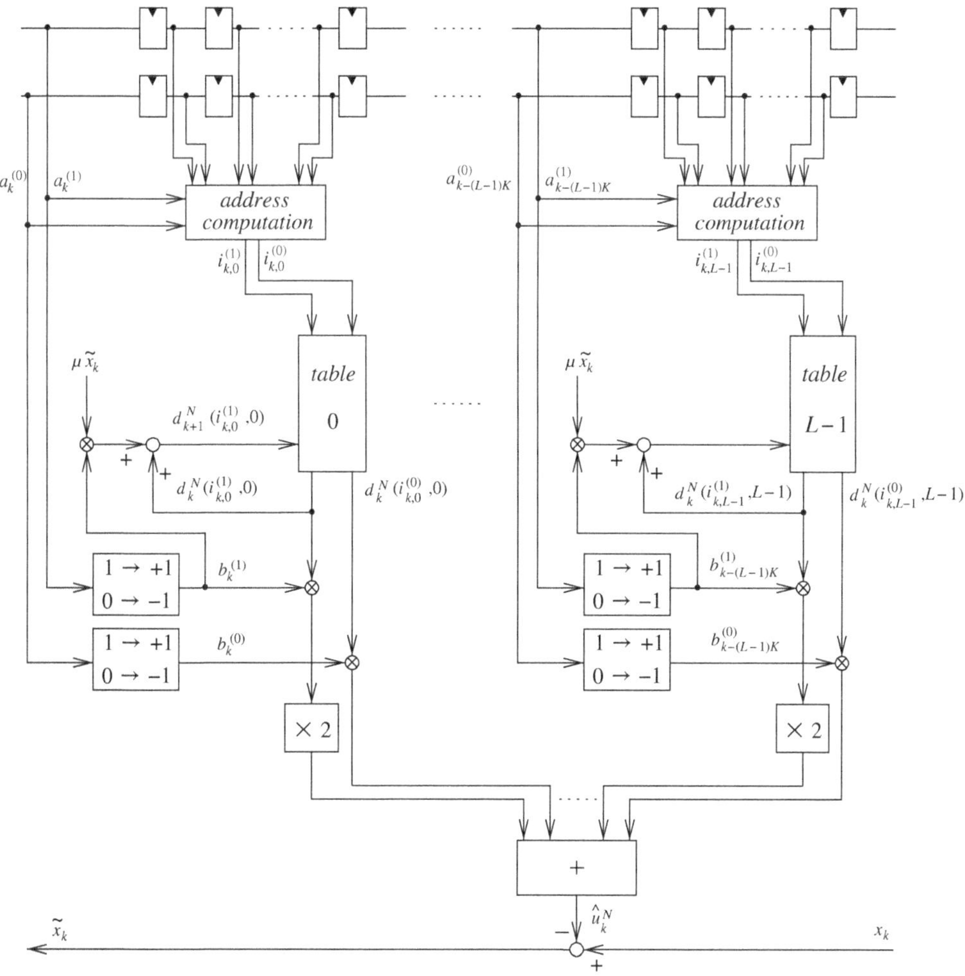

Figure 19.6 Adaptive distributed-arithmetic NEXT canceller. Source: Reproduced with permission from Cherubini et al. [1]. ©1995, IEEE.

To explain the operations for re-centring of the equalizer coefficients, we consider as an example a simple equalizer with $N_E = 4$ coefficients; in Figure 19.8, where for simplicity the coefficients are denoted by $\{c_0, c_1, c_2, c_3\}$, the coefficients and signal samples at the input of the 4 MAC units are given as a function of the time instant. At time instant k, the output of the MAC unit 0 is selected, at time instant $k + 1$, the output of the MAC unit 1 is selected, and so on. For a negative frequency offset between the local and remote transmitter clocks, a re-centring operation corresponding to a left shift of the equalizer coefficients occasionally occurs as illustrated in the upper part of Figure 19.8. We note that as a result of this operation, a new coefficient c_4, initially set equal to zero, is introduced. We also note that signal samples with proper delay need to be input to the MAC units. A similar operation occurs for a right shift of the equalizer coefficients, as illustrated in the lower part of the figure; in this case, a new coefficient c_{-1} , initially set equal to zero, is introduced. In the equalizer implementation shown in Figure 19.7, the control operations to select the filter coefficients and the signal samples are implemented by the multiplexer MUXC at the input of the delay line and by the multiplexers MUXS(0), …, MUXS(N_E − 1) at the input of the MAC units, respectively. A left or right shift of the equalizer coefficients is completed

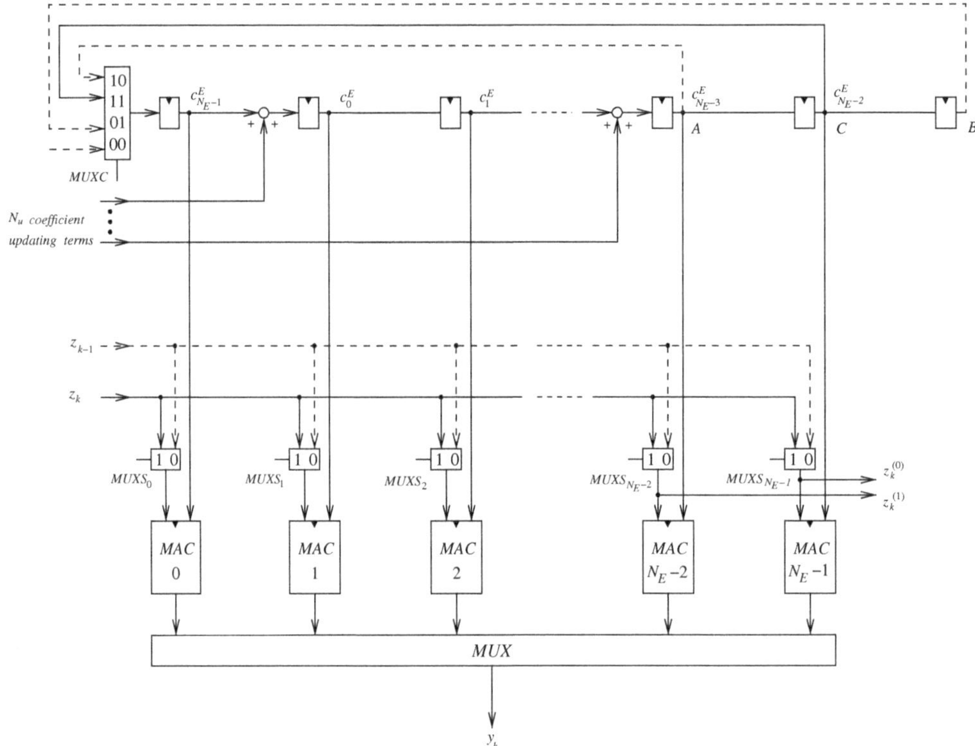

Figure 19.7 Digital adaptive equalizer: coefficient circulation and updating, and computation of output signal. Source: Reproduced with permission from Cherubini et al. [1]. ©1995, IEEE.

in N_E cycles. To perform a left shift, in the first cycle the multiplexer MUXC is controlled so that a new coefficient $c^E_{N_E} = 0$ is inserted into the delay line. During the following $(N_E - 1)$ cycles, the input of the delay line is connected to point B. After inserting the coefficient c^E_1 at the N_E-th cycle, the input of the delay line is connected to point C and normal equalizer operations are restored. For a right shift, the multiplexer MUXC is controlled so that during the first $N_E - 1$ cycles the input of the delay line is connected to point A. A new coefficient $c^E_{-1} = 0$ is inserted into the delay line at the N_E-th cycle and normal equalizer operations are thereafter restored. At the beginning of the equalizer operations, the equalizer coefficients are initialized by inserting the sequence $\{0, \dots, 0, +1, 0, -1, 0, \dots, 0\}$ into the delay line.

The adaptation of the equalizer coefficients in decision-directed mode is performed according to the LMS algorithm (19.7). However, to reduce implementation complexity, equalizer coefficients are not updated at every cycle; during normal equalizer operations, each coefficient is updated every N_E/N_U cycles by adding correction terms at N_U equally spaced fixed positions in the delay line, as shown in Figure 19.9. The architecture adopted for the computation of the correction terms is similar to the architecture for the computation of the equalizer output signal. The gradient components $-\mu_E\, z_{k-i}\, \hat{e}_k$, $i = 0, \dots, N_E - 1$, are accumulated in N_E MAC units, as illustrated in the figure. The delay line stores the signal samples $z_k^{(0)}$ and $z_k^{(1)}$. The inputs to each MAC unit are given by the error signal and a signal from the delay line. The multiplexers at the input of each register in the delay line allow selecting the appropriate inputs to the MAC units in connection with the re-centring of the equalizer coefficients. At

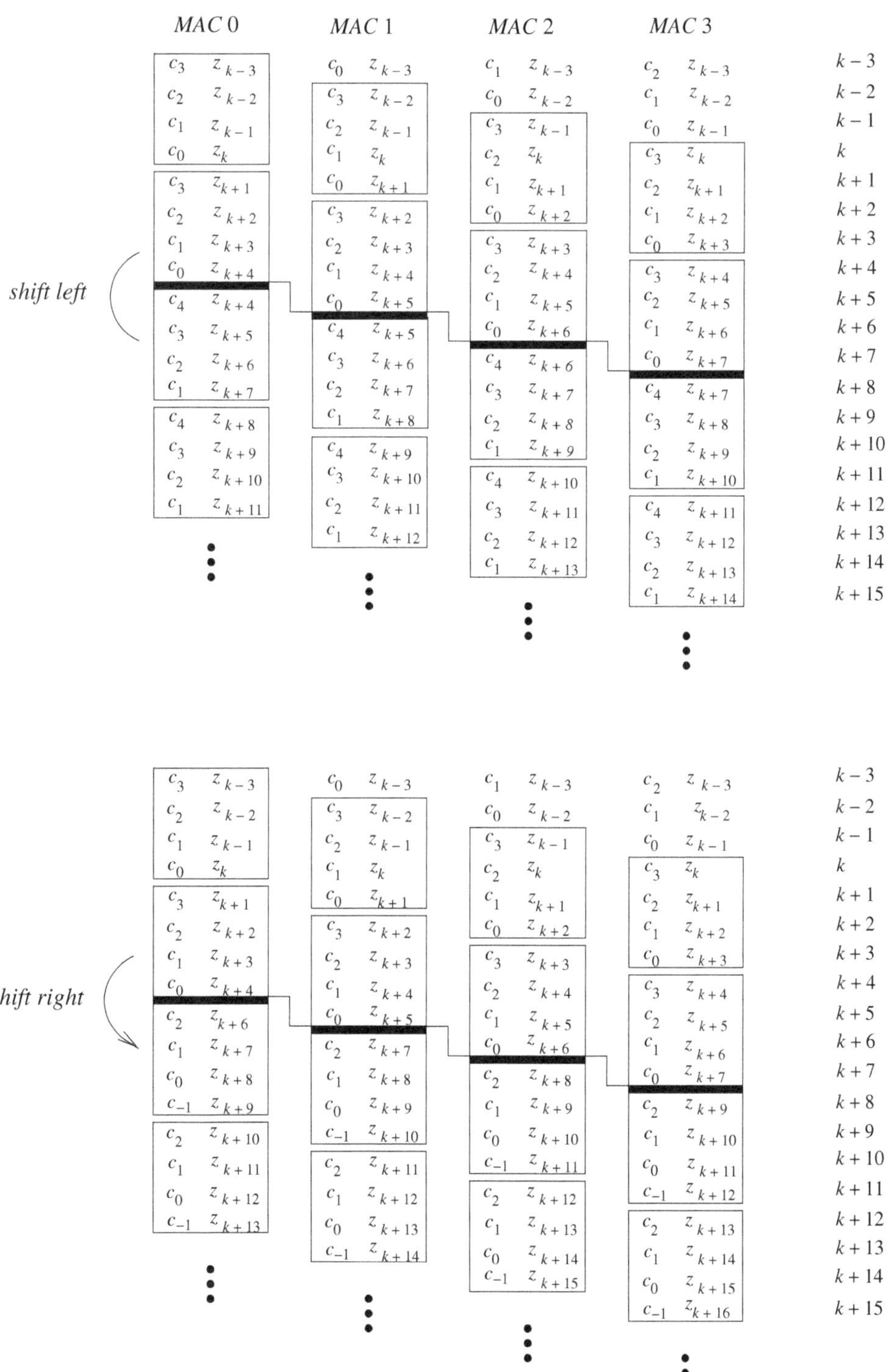

Figure 19.8 Coefficients and signals at the input of the multiply-accumulate (MAC) units during coefficient shifting. Source: Reproduced with permission from Cherubini et al. [1]. ©1995, IEEE.

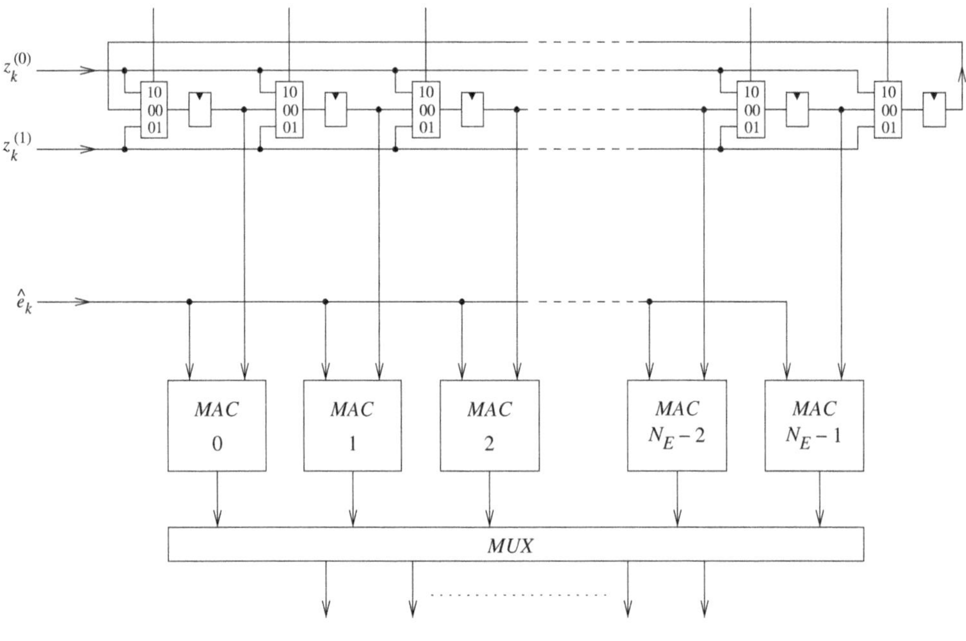

Figure 19.9 Adaptive digital equalizer: computation of coefficient adaptation. Source: Reproduced with permission from Cherubini et al. [1]. ©1995, IEEE.

each cycle, the output of N_U MAC units are selected and input to the N_U adders in the delay line where the equalizer coefficients are circulating, as illustrated in Figure 19.7.

Timing control

An elastic buffer is provided at the boundary between the transceiver sections that operate at the transmit and receive timings. The signal samples at the output of the DCF are obtained at a rate that is given by the transmit timing and stored at the same rate into the elastic buffer. Signal samples from the elastic buffer are read at the same rate that is given by the receive timing. The VCO that generates the receive timing signal is controlled in order to prevent buffer underflow or overflow.

Let WP_k and RP_k denote the values of the two pointers that specify the write and read addresses, respectively, for the elastic buffer at the k-th cycle of the receiver clock. We consider a buffer with eight memory locations, so that $\text{WP}_k, \text{RP}_k \in \{0, 1, 2, \dots, 7\}$. The write pointer is incremented by one unit at every cycle of the transmitter clock, while the read pointer is also incremented by one unit at every cycle of the receiver clock. The difference pointer,

$$\text{DP}_k = \text{WP}_k - \text{RP}_k \quad (\text{mod } 8) \tag{19.15}$$

is used to generate a binary control signal $\Delta_k \in \{\pm 1\}$ that indicates whether the frequency of the VCO must be increased or decreased:

$$\Delta_k = \begin{cases} +1 & \text{if } DP_k = 4, 5 \\ -1 & \text{if } DP_k = 2, 3 \\ \Delta_{k-1} & \text{otherwise} \end{cases} \tag{19.16}$$

The signal Δ_k is input to a digital loop filter which provides the control signal to adjust the VCO. If the loop filter comprises both a proportional and an integral term, with corresponding gains of μ_τ and $\mu_{\Delta\tau}$, respectively, the resulting second order phase-locked loop is described by (see Section 14.7)

$$\begin{aligned} \tau_{k+1} &= \tau_k + \mu_\tau \Delta_k + \Delta\tau_k \\ \Delta\tau_{k+1} &= \Delta\tau_k + \mu_{\Delta\tau} \Delta_k \end{aligned} \tag{19.17}$$

where τ_k denotes the difference between the phases of the transmit and receive timing signals. With a proper setting of the gains μ_τ and $\mu_{\Delta\tau}$, the algorithm (19.17) allows for correct initial frequency acquisition of the VCO and guarantees that the write and read pointers do not overrun each other during steady-state operations.

For every time instant k, the two consecutive signal samples stored in the memory locations with the address values RP_k and $(\mathrm{RP}_k - 1)$ are read and transferred to the equalizer. These signal samples are denoted by z_k and z_{k-1} in Figure 19.10. When a re-centring of the equalizer coefficients has to take place,

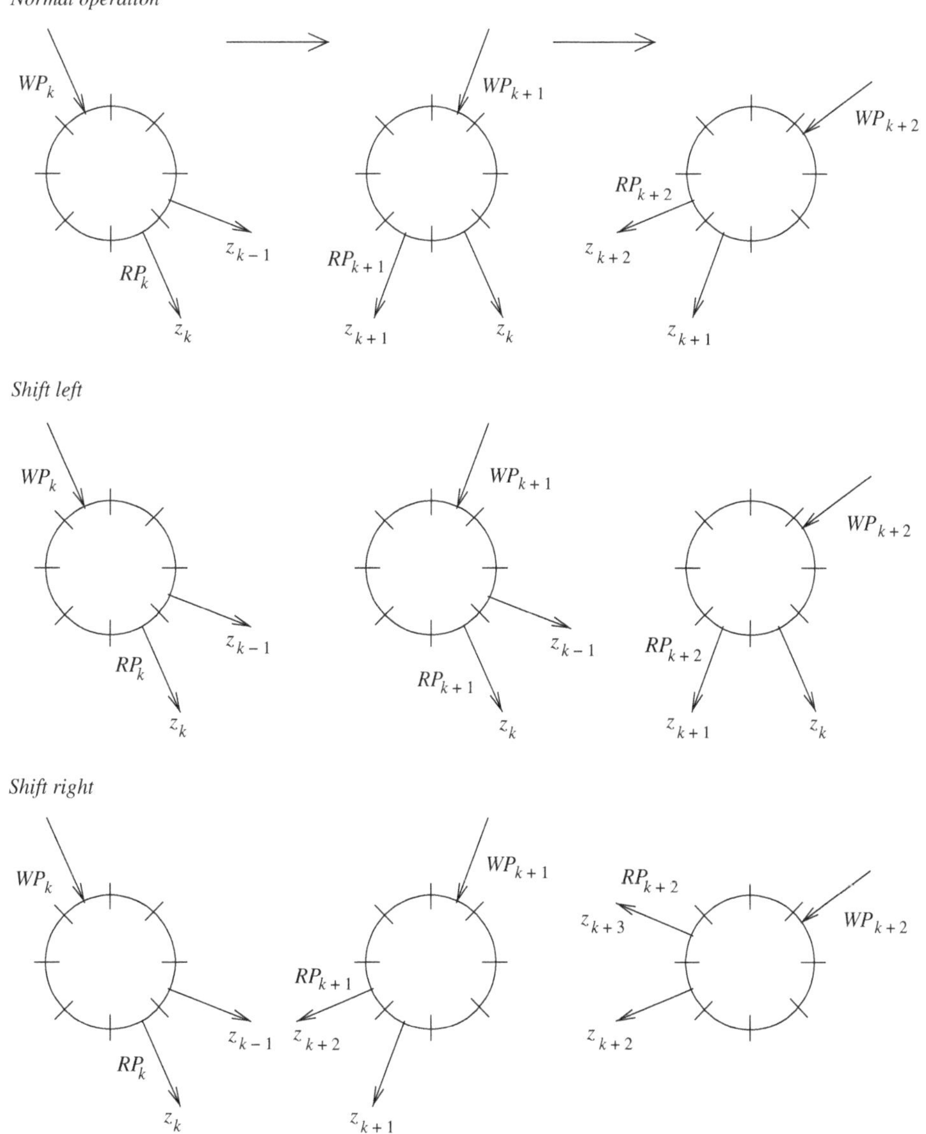

Figure 19.10 Elastic buffer: control of the read pointer.

for one cycle of the receiver clock the read pointer is either not incremented (left shift), or incremented by two units (right shift). These operations are illustrated in the figure, where the elastic buffer is represented as a circular memory. We note that by the combined effect of the timing control scheme and the re-centring of the adaptive equalizer coefficients the frequency of the receive timing signal equals on average the modulation rate at the remote transceiver.

Viterbi detector

For the efficient implementation of near maximum likelihood sequence detection (MLSD) of QPR-IV signals, we consider the reduced-state Viterbi detector of Example 7.5.3 on page 365. In other words, the signal samples at the output the of $(1 - D^2)$ partial response channel are viewed as being generated by two interlaced $(1 - D')$ dicode channels, where $D' = D^2$ corresponds to a delay of two modulation intervals, $2T$. The received signal samples are hence deinterlaced into even and odd time-indexed sequences. The Viterbi algorithm using a 2-state trellis is performed independently for each sequence. This reduced-state Viterbi algorithm retains at any time instant k only the two states with the smallest and second smallest metrics and their survivor sequences, and propagates the difference between these metrics instead of two metrics. Because the minimum distance error events in the partial-response trellis lead to quasi-catastrophic error propagation, a sufficiently long path memory depth is needed. A path memory depth of $64T$ has been found to be appropriate for this application.

19.2 Dual-duplex system

We now describe the 100BASE-T2 system for fast Ethernet mentioned in Section 18.6.2 [2], [4].

Dual-duplex transmission

Figure 19.11 indicates that near-end crosstalk (NEXT) represents the main disturbance for transmission at high data rates over UTP-3 cables. As illustrated in Figure 19.12, showing the principle of dual-duplex transmission (see also Section 17.1), *self NEXT* is defined as NEXT from the transmitter output to the receiver input of the same transceiver, and can be cancelled by adaptive filters as discussed in Chapter 17. *Alien NEXT* is defined instead as NEXT from the transmitter output to the receiver input of another transceiver; this is generated in the case of simultaneous transmission over multiple links within one multi-pair cable, typically with 4 or 25 pairs. Suppression of alien NEXT from other transmissions in multi-pair cables and far-end crosstalk (FEXT), although normally not very significant, requires specific structures (see for example Section 17.4).

To achieve best performance for data transmission over UTP-3 cables, signal bandwidth must be confined to frequencies not exceeding 30 MHz. As shown in Chapter 18, this restriction is further mandated by the requirement to meet Federal Communications Commission (FCC) and Comité européen de normalisation en électronique et en électrotechnique (CENELEC) class B limits on emitted radiation from communication systems. These limits are defined for frequencies above 30 MHz. Twisted pairs used in UTP-3 cables have fewer twists per unit of length and generally exhibit a lower degree of homogeneity than pairs in UTP-5 cables; therefore, transmission over UTP-3 cables produces a higher level of radiation than over UTP-5 cables. Thus, it is very difficult to comply with the class B limits if signals containing spectral components above 30 MHz are transmitted over UTP-3 cables.

As illustrated in Figure 19.12, for 100BASE-T2 a *dual-duplex* baseband transmission concept was adopted. Bidirectional 100 Mbit/s transmission over two pairs is accomplished by full-duplex transmission of 50 Mbit/s streams over each of two wire pairs. The lower modulation rate and/or spectral

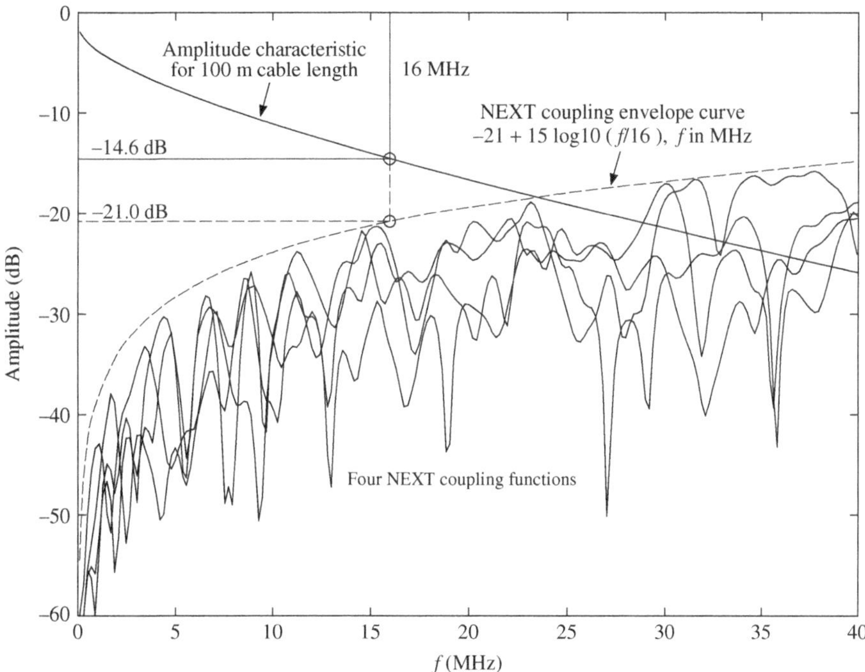

Figure 19.11 Amplitude of the frequency response for a voice-grade twisted-pair cable with length equal to 100 m, and four realizations of NEXT coupling function. Source: Reproduced with permission from Cherubini et al. [2]. ©1997, IEEE.

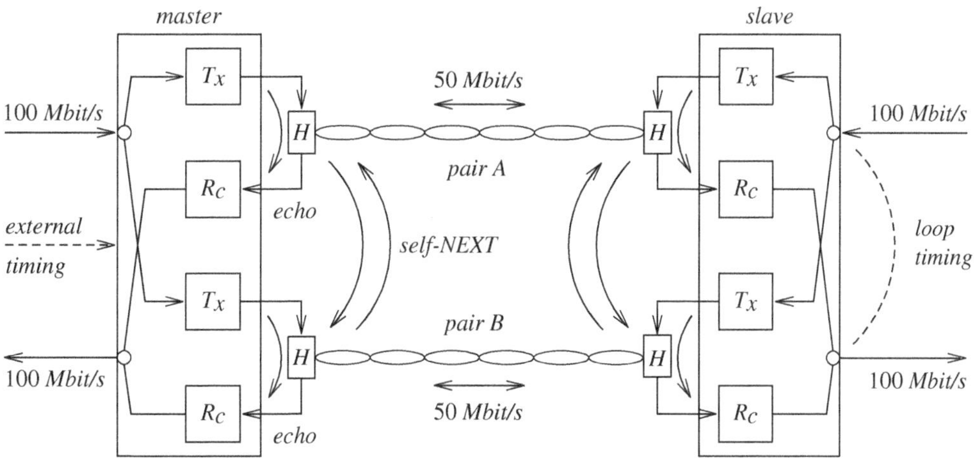

Figure 19.12 Dual-duplex transmission over two wire pairs. Source: Reproduced with permission from Cherubini et al. [2]. ©1997, IEEE.

modulation efficiency required per pair for achieving the 100 Mbit/s aggregate rate represents an obvious advantage over *mono-duplex* transmission, where one pair would be used to transmit only in one direction and the other to transmit only in the reverse direction. Dual-duplex transmission requires two transmitters and two receivers at each end of a link, as well as separation of the simultaneously transmitted and received signals on each wire pair. Sufficient separation cannot be accomplished by analog hybrid circuits only. In 100BASE-T2 transceivers, it is necessary to suppress residual echoes returning from the hybrids and impedance discontinuities in the cable as well as self NEXT by adaptive digital echo and NEXT cancellation. Furthermore, by sending transmit signals with nearly 100% excess bandwidth, received 100BASE-T2 signals exhibit spectral redundancy that can be exploited to mitigate the effect of alien NEXT by adaptive digital equalization. It will be shown later in this chapter that, for digital NEXT cancellation and equalization as well as echo cancellation in the case of dual-duplex transmission, dual-duplex and mono-duplex schemes require a comparable number of multiply-add operations per second.

The dual transmitters and receivers of a 100BASE-T2 transceiver will henceforth be referred to simply as transmitter and receiver. Signal transmission in 100BASE-T2 systems takes place in an uninterrupted fashion over both wire pairs in order to maintain timing synchronization and the settings of adaptive filters at all times. Quinary pulse-amplitude baseband modulation at the rate of 25 MBaud is employed for transmission over each wire pair. The transmitted quinary symbols are randomized by *side-stream scrambling*. The redundancy of the quinary symbol sets is needed to encode 4-bit data *nibbles*, to send between data packets an idle sequence that also conveys information about the status of the local receiver, and to insert special delimiters marking the beginning and end of data packets.

Physical layer control

The diagram in Figure 19.13 shows in a simplified form the operational states defined for the 100BASE-T2 physical layer. Upon power-up or following a request to re-establish a link, an auto-negotiation process is executed during which two stations connected to a link segment advertise their transmission capabilities by a simple pulse-transmission technique. While auto-negotiation is in

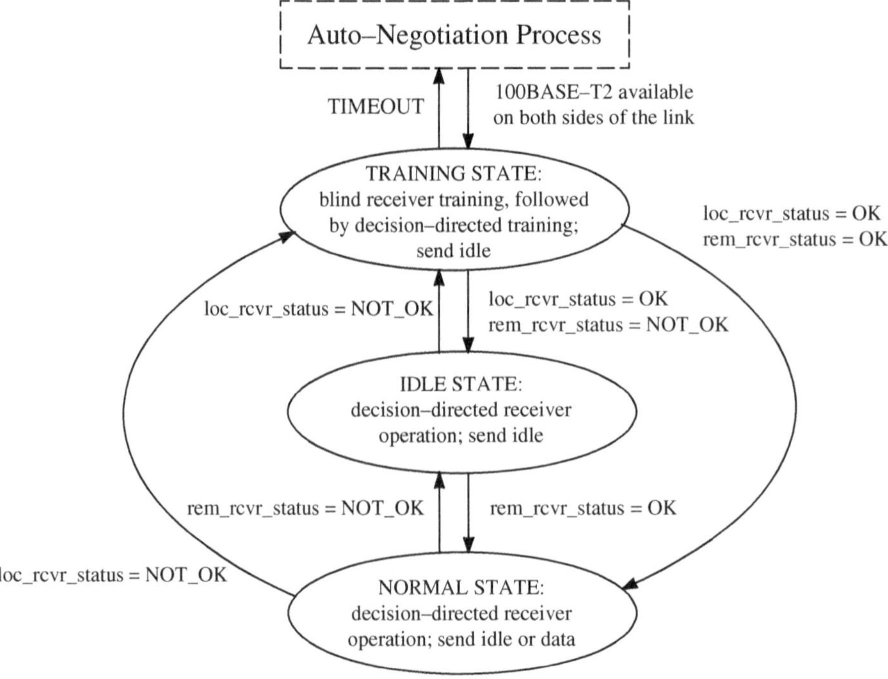

Figure 19.13 State diagram of 100BASE-T2 physical layer control. Source: Reproduced with permission from Cherubini et al. [2]. ©1997, IEEE.

progress, the 100BASE-T2 transmitters remain silent. If the physical layers of both stations are capable of 100BASE-T2 operation, the auto-negotiation process further determines a master/slave relation between the two 100BASE-T2 transceivers: the master transceiver will employ an externally provided reference clock for its transmit and receive operations. The slave transceiver will extract timing from the received signal and use this timing for its transmitter operations. This operation is usually referred to as *loop timing*. If signals were transmitted in opposite directions with independent clocks, signals received from the remote transmitter would generally shift in phase relative to the also-received echo and self-NEXT signals, as discussed in the previous section. To cope with this effect, some form of interpolation would be required, which can significantly increase the transceiver complexity.

After auto-negotiation is completed, both 100BASE-T2 transceivers enter the TRAINING state. In this state, a transceiver expects to receive an idle sequence and also sends an idle sequence, which indicates that its local receiver is not yet trained (`loc_rcvr_status = NOT_OK`). When proper local receiver operation has been achieved by blind training and then by further decision-directed training, a transition to the IDLE state occurs. In the IDLE state, a transceiver sends an idle sequence expressing normal operation at its receiver (`loc_rcvr_status = OK`) and waits until the received idle sequence indicates correct operation of the remote receiver (`rem_rcvr_status = OK`). At this time, a transceiver enters the NORMAL state, during which data nibbles or idle sequences are sent and received as demanded by the higher protocol layers. The remaining transitions shown in the state diagram of Figure 19.13 mainly define recovery functions.

The *medium independent interface* (MII) between the 100BASE-T2 physical layer and higher protocol layers is the same as for the other 10/100 Mbit/s IEEE 802.3 physical layers. If the control line `TX_EN` is inactive, the transceiver sends an idle sequence. If `TX_EN` is asserted, 4-bit data nibbles TXD(3:0) are transferred from the MII to the transmitter at the transmit clock rate of 25 MHz. Similarly, reception of data results in transferring 4-bit data nibbles RXD(3:0) from the receiver to the MII at the receive clock of 25 MHz. Control line `RX_DV` is asserted to indicate valid data reception. Other control lines, such as CRS (carrier sense) and COL (collision), are required for carrier-sense-multiple-access/collision-avoidance (CSMA/CD) specific functions.

Coding and decoding

The encoding and decoding rules for 100BASE-T2 are now described. During the k-th modulation interval, symbols a_k^A and a_k^B from the quinary set $\{-2, -1, 0, +1, +2\}$ are sent over pair A and pair B, respectively. The encoding functions are designed to meet the following objectives:

- the symbols $-2, -1, 0, +1, +2$ occur with probabilities 1/8, 1/4, 1/4, 1/4, 1/8, respectively;
- idle sequences and data sequences exhibit identical power spectral densities;
- reception of an idle sequence can rapidly be distinguished from reception of data;
- scrambler state, pair A and pair B assignment, and temporal alignment and polarities of signals received on these pairs can easily be recovered from a received idle sequence.

At the core of idle sequence generation and side-stream scrambling is a binary *maximum-length shift-register* (MLSR) sequence $\{p_k\}$ (see Appendix 1.C) of period $2^{33} - 1$. One new bit of this sequence is produced at every modulation interval. The transmitters in the master and slave transceivers generate the sequence $\{p_k\}$ using feedback polynomials $g_M(x) = 1 + x^{13} + x^{33}$ and $g_S(x) = 1 + x^{20} + x^{33}$, respectively. The encoding operations are otherwise identical for the master and slave transceivers. From delayed elements $\{p_k\}$, four *derived bits* are obtained at each modulation interval as follows:

$$\begin{aligned} x_k &= p_{k-3} \oplus p_{k-8} \\ y_k &= p_{k-4} \oplus p_{k-6} \\ a_k &= p_{k-1} \oplus p_{k-5} \\ b_k &= p_{k-2} \oplus p_{k-12} \end{aligned} \tag{19.18}$$

where $\oplus$ denotes modulo 2 addition. The sequences $\{x_k\}$, $\{y_k\}$, $\{a_k\}$, and $\{b_k\}$ represent shifted versions of $\{p_k\}$, that differ from $\{p_k\}$ and from each other only by large delays. When observed in a constrained time window, the five sequences appear as mutually uncorrelated sequences. Figures 19.14 and 19.15 illustrate the encoding process for the idle mode and data mode, respectively. Encoding is based in both cases on the generation of pairs of two-bit vectors $(\boldsymbol{X}_k, \boldsymbol{Y}_k)$, $(\boldsymbol{S}_k^a, \boldsymbol{S}_k^b)$, and $(\boldsymbol{T}_k^a, \boldsymbol{T}_k^b)$, and Gray-code labelling of $(\boldsymbol{T}_k^a, \boldsymbol{T}_k^b)$ into symbol pairs (D_k^a, D_k^b), where $D_k^\gamma \in \{-2, -1, 0, +1\}$, $\gamma = a, b$. The generation of these quantities is determined by the sequences $\{x_k\}$, $\{y_k\}$ and $\{p_k\}$, the even/odd state of the time index k (equal to $2n$ or $2n+1$), and the local receiver status. Finally, pairs of transmit symbols (a_k^A, a_k^B) are obtained by scrambling the signs of (D_k^a, D_k^b) with the sequences $\{a_k\}$ and $\{b_k\}$. In Figure 19.16, the symbol pairs transmitted in the idle and data modes are depicted as two-dimensional signal points.

We note that, in idle mode, if $p_k = 1$ then symbols $a_k^A \in \mathcal{A}_x = \{-1, +1\}$ and $a_k^B \in \mathcal{A}_y = \{-2, 0, +2\}$ are transmitted; if $p_k = 0$ then $a_k^A \in \mathcal{A}_y$ and $a_k^B \in \mathcal{A}_x$ are transmitted. This property enables a receiver to recover a local replica of $\{p_k\}$ from the two received quinary symbol sequences.

The associations of the two sequences with pair A and pair B can be checked, and a possible temporal shift between these sequences can be corrected. Idle sequences have the further property that in every two-symbol interval with even and odd time indices k, two symbols ± 1, one symbol 0, and one symbol ± 2 occur. The signs depend on the receiver status of the transmitting transceiver and on the elements

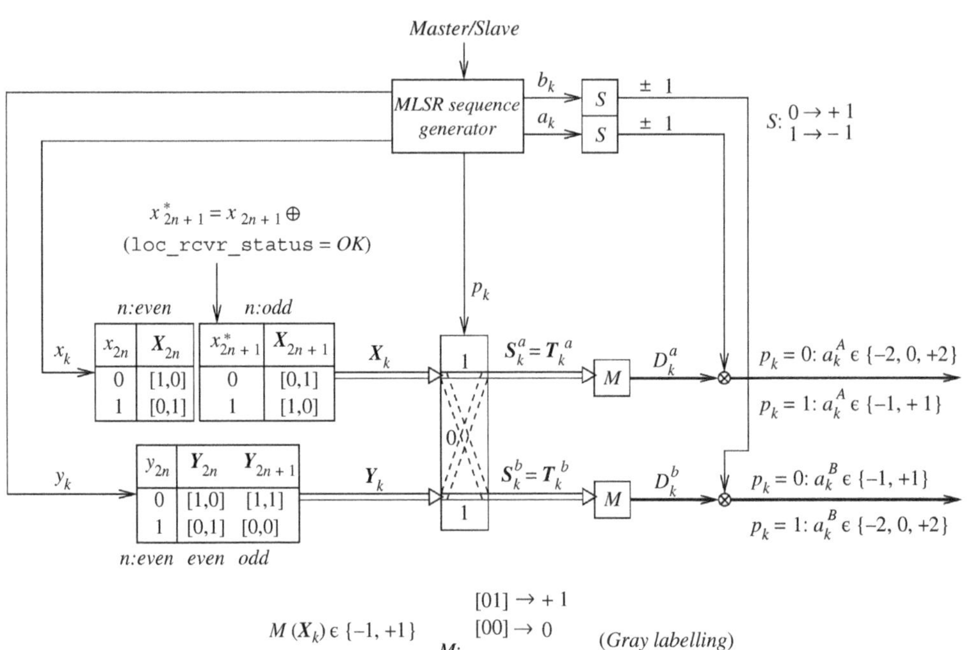

Figure 19.14 Signal encoding during idle mode.

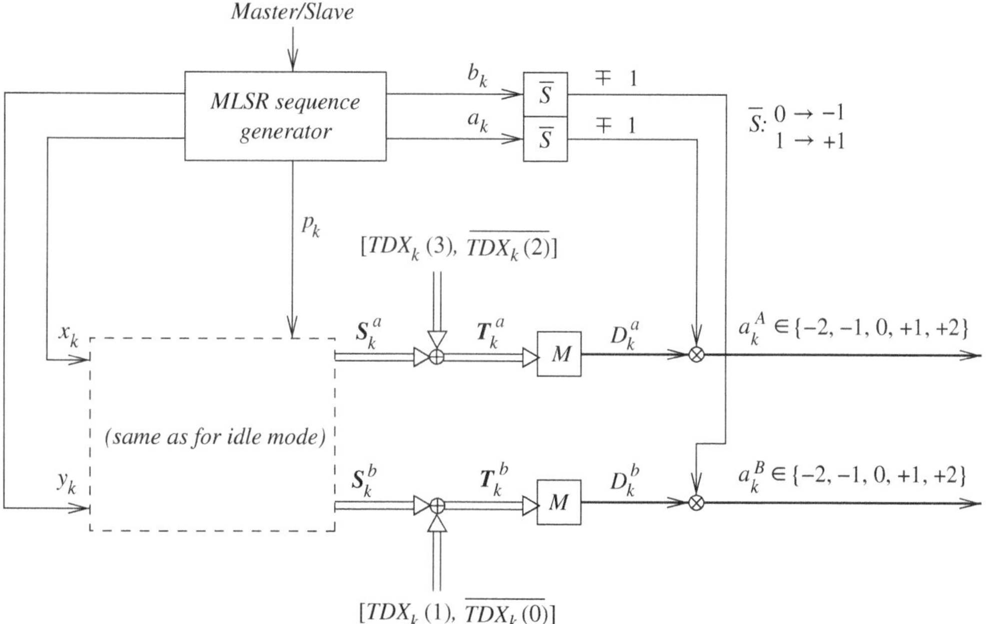

Figure 19.15 Signal encoding during data mode.

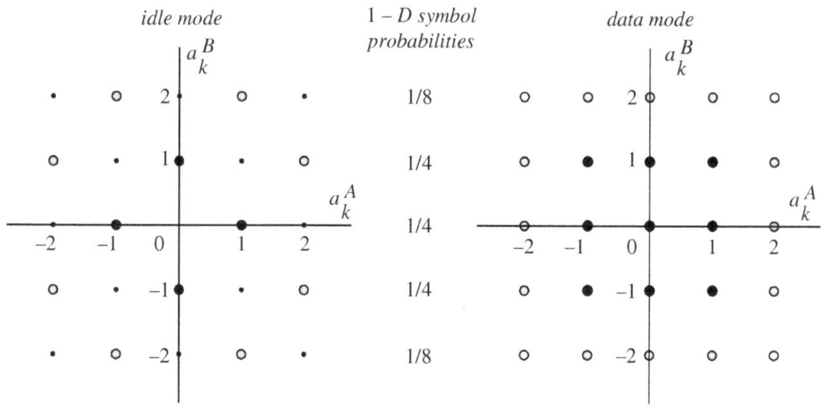

Figure 19.16 Two-dimensional symbols sent during idle and data transmission.

of the sequences $\{x_k\}$, $\{y_k\}$, $\{a_k\}$, and $\{b_k\}$. Once a receiver has recovered $\{p_k\}$, these sequences are known, and correct signal polarities, the even/odd state of the time index k, and remote receiver status can be determined.

In data mode, the two-bit vectors $\boldsymbol{S}_k^a$ and $\boldsymbol{S}_k^b$ are employed to side-stream scramble the data nibble bits. Compared to the idle mode, the signs of the transmitted symbols are scrambled with opposite polarity. In the event that detection of the delimiter marking transitions between idle mode and data mode fails due to noise, a receiver can nevertheless rapidly distinguish an idle sequence from a data sequence by inspecting the signs of the two received ±2 symbols. As mentioned above, during the transmission of idle sequences one symbol ±2, i.e. with absolute value equal to 2, occurs in every two-symbol interval.

The previous description does not yet explain the generation of delimiters. A *start-of-stream delimiter* (SSD) indicates a transition from idle-sequence transmission to sending packet data. Similarly, an *end-of-stream delimiter* (ESD) marks a transition from sending packet data to idle sequence transmission. These delimiters consist of two consecutive symbol pairs ($a_k^A = \pm 2, a_k^B = \pm 2$) and ($a_{k+1}^A = \pm 2, a_{k+1}^B = 0$). The signs of symbols ±2 in a SSD and in an ESD are selected opposite to the signs normally used in the idle mode and data mode, respectively. The choice of these delimiters allows detection of mode transitions with increased robustness against noise.

19.2.1 Signal processing functions

The principal signal processing functions performed in a 100BASE-T2 transmitter and receiver are illustrated in Figure 19.17. The digital-to-analog and analog-to-digital converters operate synchronously, although possibly at different multiples of 25 MBaud symbol rate. Timing recovery from the received signals, as required in slave transceivers, is not shown. This function can be achieved, for example, by exploiting the strongly cyclostationary nature of the received signals (see Chapter 14).

The 100BASE-T2 transmitter

At the transmitter, pairs of quinary symbols ($a_k^{A_T}, a_k^{B_T}$) are generated at the modulation rate of $1/T = 25$ MBaud. The power spectral density (PSD) of transmit signals s_A and s_B has to conform to the spectral template given in Figure 19.18. An excess bandwidth of about 100%, beyond the Nyquist frequency of 12.5 MHz is specified to allow for alien NEXT suppression in the receiver by adaptive fractionally

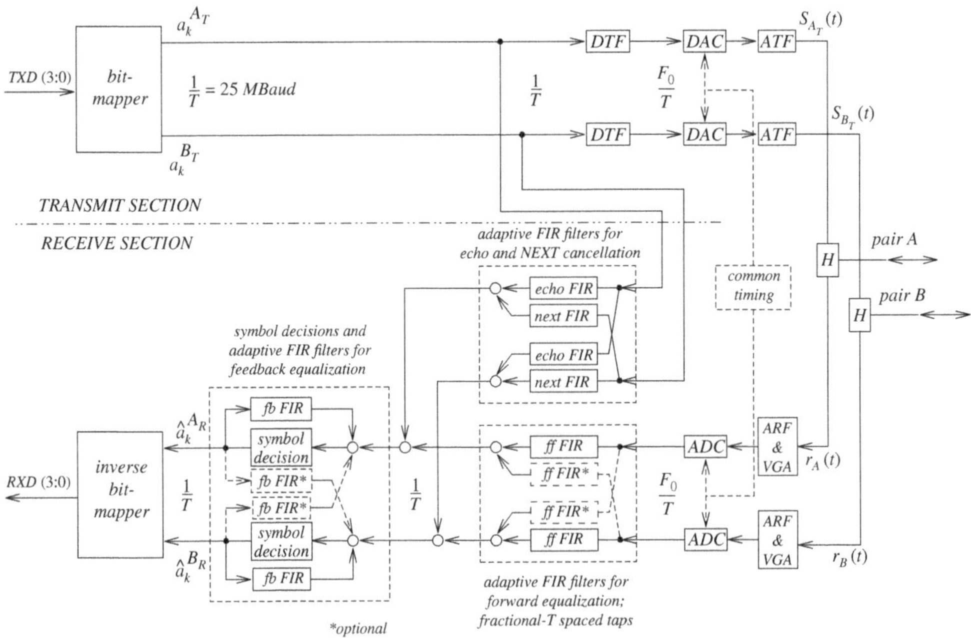

Figure 19.17 Principal signal processing functions performed in a 100BASE-T2 transceiver. Source: Reproduced with permission from Cherubini et al. [2]. ©1997, IEEE.

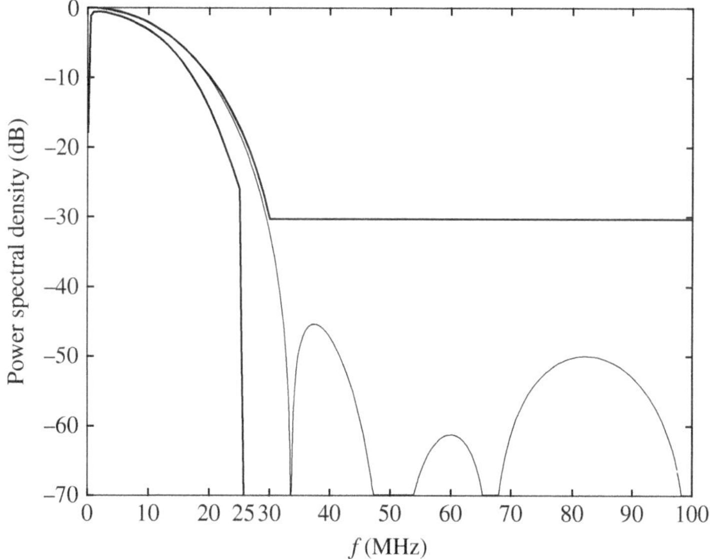

Figure 19.18 Spectral template specified by the 100BASE-T2 standard for the power spectral density of transmit signals and achieved power spectral density for a particular transmitter implementation comprising a 5-tap digital transmit filter, 100 MHz DAC, and a 3° order Butterworth ATF. Source: Reproduced with permission from Cherubini et al. [2]. ©1997, IEEE.

spaced equalization, as explained below. The transmit signals are produced by *digital pulse-shaping and interpolation filters* or digital transmit filters (DTFs), conversion from digital to analog signals at a multiple rate of 25 Msamples/s, and *ATFs*. The PSD obtained with a particular implementation in which the digital-to-analog conversion occurs at a sampling rate of 100 Msamples/s (oversampling factor $F_0 = 4$) is included in Figure 19.18.

The 100BASE-T2 receiver

The receive signals r_A and r_B are bandlimited by *ARFs* to approximately 25 MHz, adjusted in amplitude by *variable gain amplifiers* (VGAs) and converted from analog-to-digital at a multiple rate of 25 Msamples/s. For the following discussion, a sampling rate of 50 Msamples/s will be assumed (oversampling factor $F_0 = 2$, or equivalently sampling at the rate $2/T$).

The remaining receiver operations are performed digitally. Before detection of the pairs of quinary symbols $(a_k^{A_R}, a_k^{B_R})$ transmitted by the remote transceiver, an adaptive decision-feedback equalizer (DFE) structure together with adaptive echo and self NEXT cancellers is employed, as shown in Figure 19.17. The forward equalizer sections of the DFE operate on $T/2$-spaced input signals. The estimated echo and self NEXT signals are subtracted from the T-spaced equalizer output signals. When signal attenuation and disturbances increase with frequency, as is the case for 100BASE-T2 transmission, a DFE receiver provides noticeably higher noise immunity compared to that achieved by a receiver with linear forward equalization only. In Figure 19.17, additional feedback filters for direct-current (DC) restoration are not shown. In a complete receiver implementation, these filters are needed to compensate for a spectral null at DC introduced by linear transform coupling. This spectral notch may be broadened in a well-defined manner by the ARFs and compensated for by non-adaptive IIR filters.

The use of forward equalizers with $T/2$-spaced coefficients serves two purposes. First, as illustrated in Section 7.4.3, equalization becomes essentially independent of the sampling phase. Second, when the received signals exhibit excess bandwidth, the superposition of spectral input-signal components at frequencies f and $f - 1/T$, for $0 < f < 1/(2T)$, in the T-sampled equalizer output signals, can mitigate the effects of synchronous interference and asynchronous disturbances, as shown in the Appendix 19.4. Interference suppression achieved in this manner can be interpreted as a frequency diversity technique [5]. Inclusion of the optional cross-coupling feedforward and backward filters shown in Figure 19.17 significantly enhances the capability of suppressing alien NEXT. This corresponds to adding space diversity at the expense of higher implementation complexity. Mathematical explanations for the ability to suppress synchronous and asynchronous interference with the cross-coupled forward equalizer structure are given in the Appendix 19.4. This structure permits the complete suppression of the alien NEXT interferences stemming from another 100BASE-T2 transceiver operating in the same multi-pair cable at identical clock rate. Alternatively, the interference from a single asynchronous source, e.g. alien NEXT from 10BASE-T2 transmission over an adjacent pair, can also be eliminated.

The 100BASE-T2 standard does not provide the transmission of specific training sequences. Hence, for initial receiver-filter adjustments, blind adaptation algorithms must be employed. When the MSEs at the symbol-decision points reach sufficiently low values, filter adaptation is continued in decision directed mode based on quinary symbol decisions. The filter coefficients can henceforth be continuously updated by the LMS algorithm to track slow variations of channel and interference characteristics.

The 100BASE-T2 Task Force adopted a symbol-error probability target value of 10^{-10} that must not be exceeded under the worst-case channel attenuation and NEXT coupling conditions when two 100BASE-T2 links operate in a four-pair UTP-3 cable, as illustrated in Figure 19.11. During the development of the standard, the performance of candidate 100BASE-T2 systems has been extensively investigated by computer simulation. For the scheme ultimately adopted, it was shown that by adopting time spans of $32T$ for the echo and self NEXT cancellers, $12T$ for the forward filters, and $10T$ for the feedback filters, the MSEs at the symbol-decision points remain consistently below a value corresponding to a symbol-error probability of 10^{-12}.

Computational complexity of digital receive filters

The digital receive filters account for most of the transceiver implementation cost. It is worthwhile to compare the filter complexities for a dual-duplex and a mono-duplex scheme. Intuitively, the dual-duplex scheme may appear to be more complex, because it requires two transceivers. We define the complexity of a finite-impulse response finite impulse response (FIR) as

$$\begin{aligned} \text{Filter complexity} &= \text{time span } \times \text{input sampling rate } \times \text{output sampling rate} \\ &= \text{number of coefficients } \times \text{output sampling rate} \\ &= \text{number of multiply-and-adds per second.} \end{aligned} \tag{19.19}$$

Note that the time span of an FIR filter is given in seconds by the product of the number of filter coefficients times the sampling period of the input signal. Transmission in a four-pair cable environment with suppression of alien NEXT from a similar transceiver is considered. Only the echo and self NEXT cancellers and forward equalizers will be compared. Updating of filter coefficients will be ignored.

For dual-duplex transmission, the modulation rate is 25 MBaud and signals are transmitted with about 100% excess bandwidth. Echo and self NEXT cancellation requires four FIR filters with time spans T_C and input/output rates of 25 Msamples/s. For equalization and alien NEXT suppression, four forward FIR filters with time spans T_E, an input rate of 50 Msamples/s and an output rate of 25 Msamples/s are needed.

The modulation rate for mono duplex transmission is $1/T = 50$ MBaud and, signals are transmitted with no significant excess bandwidth. Hence, both schemes transmit within a comparable bandwidth

Table 19.1: Complexities of filtering for two transmission schemes.

	Dual duplex	Mono duplex
Echo and self NEXT cancellers	$4 \times T_C \times 25 \times 10^{12}$	$1 \times T_C \times 50 \times 50 \times 10^{12}$
Forward equalizers	$4 \times T_E \times 50 \times 25 \times 10^{12}$	$2 \times T_E \times 50 \times 50 \times 10^{12}$

($\leq$ 25 MHz). For a receiver structure that does not allow alien NEXT suppression, one self NEXT canceller with time span T_C and input/output rates of 50 Msamples/s, and one equalizer with time span T_E and input/output rates of 50 Msamples/s will be needed. However, for a fair comparison, a mono duplex receiver must have the capability to suppress alien NEXT from another mono duplex transmission. This can be achieved by receiving signals not only from the receive pair but also in the reverse direction of the transmit pair, and combining this signal via a second equalizer with the output of the first equalizer. The additionally required equalizer exhibits the same complexity as the first equalizer.

The filter complexities for the two schemes are summarized in Table 19.1. As the required time spans of the echo and self NEXT cancellers and the forward equalizers are similar for the two schemes, it can be concluded that the two schemes have the same implementation complexity. The same arguments can be extended to the feedback filters. Finally, we note that with the filter time spans considered in the preceding section ($T_C = 32T$, $T_E = 12T$ and $T_{Fb} = 10T$), in a 100BASE-T2 receiver on the order of 10^{10} multiply-and-add operations/s need to be executed.

Bibliography

[1] Cherubini, G., Ölçer, S., and Ungerboeck, G. (1995). A quaternary partial response class-IV transceiver for 125 Mbit/s data transmission over unshielded twisted-pair cables: principles of operation and VLSI realization. *IEEE Journal on Selected Areas in Communications* 13: 1656–1669.

[2] Cherubini, G., Ölçer, S., Ungerboeck, G. et al. (1997). 100BASE-T2: a new standard for 100Mb/s ethernet transmission over voice-grade cables. *IEEE Communications Magazine* 35: 115–122.

[3] Cherubini, G., Ölçer, S., and Ungerboeck, G. (1996). Adaptive analog equalization and receiver front-end control for multilevel partial-response transmission over metallic cables. *IEEE Transactions on Communications* 44: 675–685.

[4] IEEE Standard 802.3y (1997). Supplement to carrier sense multiple access with collision detection (CSMA/CD) access method and physical layer specifications: physical layer specification for 100 Mb/s operation on two pairs of Category 3 or better balanced twisted pair cable (100BASE-T2, Clause 32).

[5] Petersen, B.R. and Falconer, D.D. (1991). Minimum mean-square equalization in cyclostationary and stationary interference - analysis and subscriber line calculations. *IEEE Journal on Selected Areas in Communications* 9: 931–940.

Appendix 19.A Interference suppression

Figure 19.19 illustrates the interference situations considered here. Equalization by linear forward filters only is assumed. Reception of 100BASE-T2 signals is disturbed either by alien NEXT from another synchronous 100BASE-T2 transmitter or by crosstalk from a single asynchronous source. Only one of these disturbances may be present. The symbol sequences $\{a_k^{A_R}\}$ and $\{a_k^{B_R}\}$ denote the sequences transmitted by the remote 100BASE-T2 transceiver, whereas $\{a_k^{A'_T}\}$ and $\{a_k^{B'_T}\}$ denote the sequences transmitted

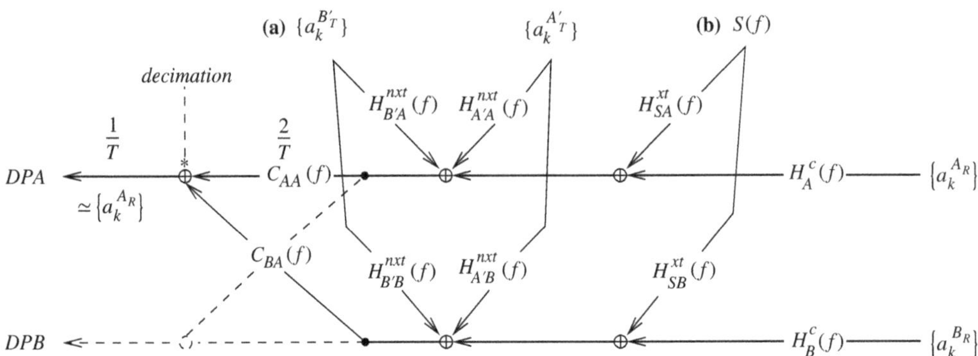

Figure 19.19 Crosstalk disturbance by: (a) alien NEXT from another synchronous 100BASE-T2 transmitter, (b) an asynchronous single source, for example a 10BASE-T transmitter. Source: Reproduced with permission from Cherubini et al. [2]. ©1997, IEEE.

by an adjacent synchronous 100BASE-T2 transmitter. The spectrum $S(f)$ of the asynchronous source may be aperiodic or exhibit a period different from $1/T$. The functions $H(f)$ represent the frequency responses of the signal or crosstalk paths from the respective sources to the inputs of the forward equalizer filters with transfer functions $C_{AA}(f)$ and $C_{BA}(f)$. Because of $2/T$ sampling rate, these functions exhibit $2/T$-periodicity. All signals and filter coefficients are real-valued. It is therefore sufficient to consider only frequencies f and $f - 1/T$, for $0 < f < 1/(2T)$. We will concentrate on the signals arriving at decision point A (DPA); the analysis for signals at decision point B (DPB) is similar.

Intersymbol-interference free reception of the symbol sequence $\{a_k^{A_R}\}$ and the suppression of signal components stemming from $\{a_k^{B_R}\}$ at DPA requires

$$\begin{aligned} H_A^c(f)\, C_{AA}^c(f) + H_A^c\left(f - \frac{1}{T}\right) C_{AA}\left(f - \frac{1}{T}\right) &= 1 \\ H_B^c(f)\, C_{BA}^c(f) + H_B^c\left(f - \frac{1}{T}\right) C_{BA}\left(f - \frac{1}{T}\right) &= 0 \end{aligned} \tag{19.20}$$

To suppress alien NEXT from a 100BASE-T2 transmitter, two additional conditions must be met:

$$\begin{aligned} \sum_{\ell=0,1} \left[H_{A'A}^{nxt}\left(f - \frac{\ell}{T}\right) C_{AA}\left(f - \frac{\ell}{T}\right) + H_{A'B}^{nxt}\left(f - \frac{\ell}{T}\right) C_{BA}\left(f - \frac{\ell}{T}\right)\right] &= 0 \\ \sum_{\ell=0,1} \left[H_{B'A}^{nxt}\left(f - \frac{\ell}{T}\right) C_{AA}\left(f - \frac{\ell}{T}\right) + H_{B'B}^{nxt}\left(f - \frac{\ell}{T}\right) C_{BA}\left(f - \frac{\ell}{T}\right)\right] &= 0 \end{aligned} \tag{19.21}$$

Alternatively, the additional conditions for the suppression of crosstalk caused by a single asynchronous source become

$$\begin{aligned} H_{SA}^{xt}(f)\, C_{AA}(f) + H_{SB}^{xt}(f)\, C_{BA}(f) &= 0 \\ H_{SA}^{xt}\left(f - \frac{1}{T}\right) C_{AA}\left(f - \frac{1}{T}\right) + H_{SB}^{xt}\left(f - \frac{1}{T}\right) C_{BA}\left(f - \frac{1}{T}\right) &= 0 \end{aligned} \tag{19.22}$$

Therefore, in each case the interference is completely suppressed if for every frequency in the interval $0 < f < 1/(2T)$ the transfer function values $C_{AA}(f)$, $C_{AA}(f - (1/T))$, $C_{BA}(f)$ and $C_{BA}(f - (1/T))$ satisfy four linear equations. It will be highly unlikely that the crosstalk responses are such that the coefficient matrix of these equations becomes singular. Hence a solution will exist with high probability. In the absence of filter-length constraints, the $T/2$-spaced coefficients of these filters can be adjusted to achieve these transfer functions. For a practical implementation, a trade-off between filter lengths and achieved interference suppression has to be made.

Index

Algorithms for Communications Systems and their Applications, Second Edition.
Nevio Benvenuto, Giovanni Cherubini, and Stefano Tomasin.